To get extra value from this book for no additional cost, go to:

http://www.thomson.com/wadsworth.html

thomson.com is the World Wide Web site for Wadsworth/ITP and is your direct source to dozens of on-line resources. *thomson.com* helps you find out about supplements, experiment with demonstration software, search for a job, and send e-mail to many of our authors. You can even preview new publications and exciting new technologies.

thomson.com: *It's where you'll find us in the future.*

The Marriage and Family Experience

The Marriage and Family Experience

Intimate Relationships in a Changing Society

7TH EDITION

Bryan Strong
University of California, Santa Cruz

Christine DeVault
Cabrillo College

Barbara W. Sayad
California State University, Monterey Bay

Wadsworth Publishing Company
I(T)P™ An International Thomson Publishing Company

Belmont, CA • Albany, NY • Bonn • Boston • Cincinnati • Detroit • Johannesburg • London • Madrid
Melbourne • Mexico City • New York • Paris • Singapore • Tokyo • Toronto • Washington

Sociology Editor: Denise Simon
Editorial Assistant: Angela Nava
Development Editor: Robert Jucha
Marketing Manager: Chaun Hightower
Project Editor: Jerilyn Emori
Print Buyer: Karen Hunt
Permissions Editor: Veronica Oliva
Production Coordinator: Patty O'Connell, Electronic Publishing Services Inc., NYC
Interior and Cover Designer: Gladys Rosa-Mendoza/Rosa+Wesley Design Associates
Photo Researcher: Francis Hogan, Electronic Publishing Services Inc., NYC
Copy Editor: Gerry Madigan, Electronic Publishing Services Inc., NYC
Cover: Original quilt design, "Address Unknown," by artist Freddy Moran
Compositor: Electronic Publishing Services Inc., NYC
Printer: World Color Book Services/Taunton

Printed in the United States of America
2 3 4 5 6 7 8 9 10

For more information, contact Wadsworth Publishing Company, 10 Davis Drive, Belmont, CA 94002, or electronically at http://www.thomson.com/wadsworth.html

International Thomson Publishing Europe
Berkshire House 168-173
High Holborn
London, WC1V 7AA, England

Thomas Nelson Australia
102 Dodds Street
South Melbourne 3205
Victoria, Australia

Nelson Canada
1120 Birchmount Road
Scarborough, Ontario
Canada M1K 5G4

International Thomson Publishing GmbH
Königswinterer Strasse 418
53227 Bonn, Germany

International Thomson Editores
Campos Eliseos 385, Piso 7
Col. Polanco
11560 México D.F., México

International Thomson Publishing Asia
221 Henderson Road
#05-10 Henderson Building
Singapore 0315

International Thomson Publishing Japan
Hirakawacho Kyowa Building, 3F
2-2-1 Hirakawacho
Chiyoda-ku, Tokyo 102, Japan

International Thomson Publishing
Southern Africa
Building 18, Constantia Park
240 Old Pretoria Road
Halfway House, 1685 South Africa

Library of Congress Cataloging-in-Publication Data

Strong, Bryan.
The marriage and family experience : intimate relationships in a changing society / Bryan Strong, Christine DeVault, Barbara W. Sayad. — 7th ed.
p. cm.
Includes bibliographical references and index.
ISBN 0-534-53757-X
1. Marriage. 2. Family. I. DeVault, Christine. II. Sayad, Barbara Werner. III. Title.
HQ734.S9738 1997
306.8—dc21 97-35925

DEDICATION

To my mother Norma Kathryn Rannie and to the memory of her mother Hilma Calkins (Mimi), strong and compassionate women, my teachers.

CHRISTINE DEVAULT

To family … my own: Bob, Sarah, Elizabeth, and Sam; and to my parents Robert and Elise, and to Bob and Flo—who lovingly gave us wings and a will to fly.

BARBARA WERNER SAYAD

All that a person does or thinks is of consequence.

WALT WHITMAN,
PREFACE TO *Leaves of Grass*

Brief Contents

Table of Contents

UNIT II Intimate Relationships

UNIT III Family Life

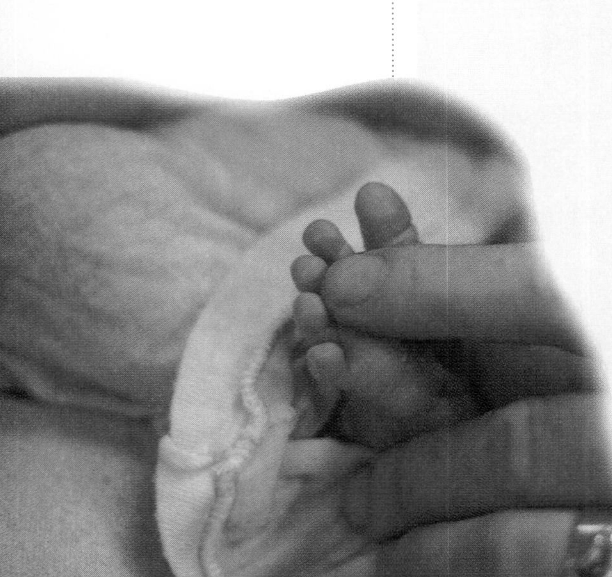

UNIT IV Family Challenges and Strengths

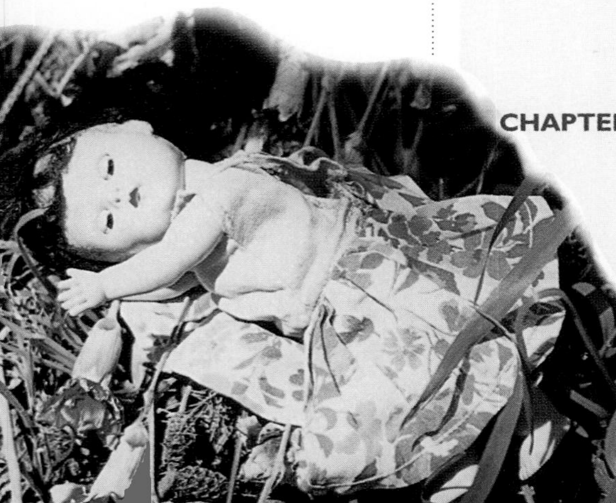

Preface

This seventh edition is an updated, expanded, and reorganized revision of *The Marriage and Family Experience* and comes with a new subtitle: *Intimate Relationships in a Changing Society*. We hope that instructors who have used earlier editions of this text will find this edition both comfortably familiar and provocatively new. For instructors who are just now adopting this text, we welcome you and your students to an experience which we hope will support and encourage their reflections and decisions about what constitutes well-being among families. The underlying theme of our textbook remains unchanged: our enduring belief that our families, whatever their form, are the crucibles in which our humanity is born, nurtured, and fulfilled. They are what makes us human, and they need to be cherished, honored, and supported.

As in earlier editions, we have attempted to present the study of marriage and the family in a manner that enlarges both personal and intellectual understanding. A functional approach need not exclude an academic approach; nor does an academic approach necessarily exclude a functional one. We have tried to combine the virtues of both. Ideas allow us to see beyond our own limited experiences, and our personal experience breathes life into ideas. We believe that good scholarship encourages individual understanding and that personal exploration encourages intellectual growth. *The Marriage and Family Experience* reflects a unique attempt to unify functionally oriented and academically oriented approaches into a single textbook.

We have continued our interdisciplinary approach, since the perspective of a single discipline, such as sociology or psychology, creates only a partial picture of marriage and the family. We have incorporated work from not only sociology and psychology but also history, economics, health, communication, folklore, literature, and ethnic studies. To limit ourselves to but one branch of knowledge would leave us much like the blind man who tried to describe the nature of the elephant: each mistook a part for the whole and, while they argued among themselves, the elephant walked away.

We have meticulously revised our textbook to reflect the most current research available. We have also reorganized it, making it easier to teach from.

New to the Seventh Edition

You and Your Well-Being

The concept of wellness, defined as optimal health and vitality, encompasses physical, emotional, intellectual, spiritual, social, and environmental well-being. When marriage and the family is examined from this additional perspective, it becomes apparent that the wellness model of health has a far-reaching effect on how we view ourselves, our relationships, and our community. Issues such as the effects of stress on family dynamics, dealing with parent burnout, and the role of self-esteem in communication and conflict resolution are spotlighted from a wellness perspective. This material appears as boxes titled "You and Your Well-Being" and is integrated throughout the text.

Marriage, Family, and Popular Culture

As instructors, most of us have found that our students rely on the media for much of their knowledge of the world, including marriage and family. Though not new to edition, the topic of popular culture has been updated and discussed in terms of how it affects not only adolescents and adults, but children as well.

We believe that it is important to make students aware of the role of popular culture—television, magazines, movies, and so on—in shaping their views of marriage and the family. Consequently, we have developed the theme of marriage, family, and popular culture, weaving it throughout the textbook. In the first chapter, "The Meaning of Marriage and Family," we introduce the various norms and family structures presented in the archetypal TV family, the sitcom family. In the second chapter, "Studying Marriage and the Family," we introduce students to pop culture's advice/information genre (such as talk shows, advice columns, and tabloid TV), which transforms information into entertainment. In the same chapter, we introduce students to critical thinking skills which may be applied to the study of marriage and the family. Throughout the remainder of the textbook, where appropriate, we discuss various media and their role in socialization, the establishment of sexual norms, and so on.

New or Expanded Topics Include:

- Further integration and expansion of America's diverse family systems—both in terms of structure and ethnicity—as an organizing principle throughout the textbook.
- Updated and expanded coverage of African-American, Latino, Asian-American, and Native American families.
- Expansion of the historical perspective of marriage and the family. (Chapter 1)
- Inclusion of the feminist and family development perspective of theories of marriage and family. (Chapter 2)
- The relationship between friendship and love. (Chapter 4)
- Getting psychological help. (Chapter 5)
- Family problem solving loop and communication loop. (Chapter 5)
- Expanded discussion of cohabitation. (Chapter 6)
- Singlehood as a lifestyle. (Chapter 6)
- Choices in unwanted pregnancy. (Chapter 7)
- Preconception care. (Chapter 8)
- Selecting and evaluating day care. (Chapter 10)
- Women's health. (Chapter 12)
- Abuse in gay and lesbian relationships. (Chapter 13)

- Intervention for child abuse and neglect. (Chapter 13)
- Divorce-related stressors. (Chapter 14)
- Stepfamily development process. (Chapter 15)
- Ethnic family strengths. (Chapter 16)

Reorganization

Based on our own teaching experience and that of our colleagues, we reorganized the chapters into teaching units which for us seem to present a more logical flow of material. Though the same number of chapters exist, they have been rearranged and divided into four units: *Meanings of Marriage and Family*, which includes definitions, studying marriage and the family, and gender roles; *Intimate Relationships*, including love, communication, pairing and singlehood, and sexuality; *Family Life*, covering pregnancy, family life cycles, parents and children, and work and economics; and finally, *Family Challenges and Strengths*, which encompasses family health, violence and sexual abuse, separation and divorce, single-parent families and stepfamilies, and marriage and family strengths.

Based on feedback from reviewers, we have also expanded Chapter 1 to include a broadened historical perspective of marriage and the family. While some of this material came from the 6th edition, considerably more information has been added. We hope instructors find this useful in expanding students' understanding of their roots.

Pedagogy

Over the years we have developed a number of pedagogical features to engage students in exploring the chapters from both an intellectual and personal perspective. In addition to the new "You and Your Well-Being" boxes, each chapter includes the following learning tools:

Previews

Previews open each chapter with self-quizzes that challenge students' preconceptions about marriage and the family.

Chapter Outline

Chapter outlines at the beginning of each chapter help students organize their learning.

Other Places/Other Times

Cross-cultural and historical perspectives provide depth and breadth to the textbook, showing students the cultural and historical diversity of marriage and the family. The cross-cultural perspectives were written by anthropologist Janice Stockard, Connecticut College, who specializes in the cross-cultural study of family systems. The historical perspectives demonstrate how marriages and families have changed over time in our society.

Understanding Yourself

Understanding Yourself sections use research topics and instruments as starting points for students to examine their own lives. We have found that integrating research, methodology, and self-examination brings to life what could be burdensome abstractions devoid of personal meaning.

Did You Know?

This new feature appears in the side-margin of various pages and is used to highlight or encapsulate facts and figures related to the topic. Most of the information appearing in this format is current and intended to call out a particular statistic or fact that might otherwise go unnoticed.

Perspectives

Perspectives focus on high-interest topics, such as ethnicity and communication, examining martial satisfaction, and the relationship between love and sexuality.

Reflections

Reflections are found within the margins of the text. They ask students to reflect on how the ideas discussed in the previous section may provide insights into their own lives, families, and relationships. Students are asked, for example, to look at the pluses and minuses of a relationship to illustrate exchange theory, to examine how their own families meet their intimacy needs, and to think about the kinds of traditions and rituals their families practice.

Key Terms

Key terms are boldfaced within the textbook as they appear. At the end of each chapter they are listed alphabetically along with the page number where each term first occurred. A complete list of key terms used in the textbook is found in the glossary.

Chapter Summary

At the end of each chapter, the main ideas are summarized to assist students in the reviewing the chapter material. Key terms appearing in the summary are italicized.

Suggested Readings

An annotated suggested readings list provides material for personal interest or further research.

Margin Quotes

Quotes in the margin offer unusual, thoughtful, humorous, or provocative insights. A single quotation may spark an intense class discussion or lead a person to reexamine long-held beliefs. Instructors and students alike have remarked that the margin quotes are almost like having a second book.

Glossary

A comprehensive glossary of key terms is included at the back of the textbook.

Appendixes

Three appendixes on sexual anatomy, sexual physiology, and fetal development appear at the back of the textbook. These can be used to supplement Chapter 7, "Understanding Sexuality," and Chapter 8, "Pregnancy and Childbirth."

Resource Center

The newly updated Resource Center at the end of the textbook contains a self-help directory and practical information on finances and budgeting, personal health, birth control, sexually transmitted diseases, infertility, and other topics relating to individual and family well-being. In addition, it contains study guides for marriage and family studies, women's studies, and African-American, Latino, Asian-American, Native American, and ethnic studies to assist students in their research. A new feature of the 7th edition is a listing of web sites by topic.

Readings in Marriage and the Family

The readings that formerly appeared at the end of each chapter have been taken out of the textbook and now appear in an expanded reader: *Readings in Marriage and Family Experience: Intimate Relationships in a Changing Society.* They include essays, articles, and excerpts from books, journals, magazines, and newspapers. We have retained a number of popular readings, but many are new. All were carefully chosen to present ideas, information, and points of view that both professors and students will find diverse and stimulating. In order to encourage discussion and introspection, reflections (critical thinking questions) are included for each reading.

Study Guide

The *Study Guide* by Carol Mertens of the University of Iowa, has been revised and updated with detailed outlines of each chapter, review questions, self-discoveries, and other helpful devices to reinforce chapter material. The guide provides mini-assignments, which sometimes involve activities outside the classroom, and "just for fun sections," which include self-tests of ideas, values, or personality, and humorous inserts.

Acknowledgments

Many hands assisted us in the production of this new edition. We were fortunate to have the expertise of Professor Gregory Kennedy, Department of Family Studies, Central Missouri State University, in researching and revising Chapter 2, "Studying Marriage and the Family," and Chapter 11, "Family, Work, and Economics." Professor Janice Stockard, Department of Anthropology, Connecticut College, contributed the cross-cultural material for the "Other Places/Other Times" features.

In doing our research we have been kindly assisted by the reference staff of the Dean McHenry Library at the University of California, Santa Cruz, and the library staff at California State University, Monterey Bay. Terence Crowley, Professor of Library Science at California State University, San Jose, continues to assist us with difficult research questions. Thanks are also due to his students Thom Ball, Evelyn Kobayashi, and Ruth Ann Moore for tracking down reference sources for the Resource Center. Eddy Goldberg was most diligent in the pursuit of elusive statistics. Professor Arthur Aron of State University of New York at Stonybrook reviewed Chapter 4, "Friendship, Love, and Commitment." Fran Bussard, LCSW, remains a rich source of insight into personal relationships and communication. Lynne DeSpelder of Cabrillo College helped us redesign the order of the chapters into more coherent teaching units. Grateful acknowledgment is also due to the staff at Networking and Computing at California State University, Monterey Bay, for on-the-spot help, follow-up, and trouble shooting.

Very special appreciation is due to our former editor at West Publishing, Carole Grumney, who shepherded this book through five editions. Her belief in *The Marriage and Family Experience* and her steady support and encouragement are largely responsible for the success the book enjoys today. At Wadsworth Publishing we thank Eve Howard, Publisher, and Susan Badger, President, for their enthusiasm and confidence in us. Many thanks are also due to editors Denise Simon and Bob Jucha, project editor Jerilyn Emori, editorial assistant Angela Nava, permissions editor Veronica Oliva, and marketing manager Chaun Hightower. At Electronic Publishing Services, Patty O'Connell and Jason Jones were diligent and helpful. Gerry Madigan was our intrepid copyeditor.

James Honeycutt of Louisiana State University and Marsesa Murray and Kathleen Gilbert of Indiana University helped us by sharing their unpublished work on TV families and by reading relevant material from the textbook. Daniel Friedman of Antioch College was especially helpful in the early stages by pointing us to many fine sources. He suggested the idea of "Television and the World We Live In" found in Chapter 1. Members of FAMLYSCI, an Internet group for family scholars founded by Greg Brock, Ph.D., University of Kentucky, gave valuable critiques of material we posted on the family and popular culture.

Each edition has benefited from the insightful comments and thoughtful suggestions of our many reviewers. They have been exceptionally aware of the dual requirements of a good textbook—academic integrity and student interest—and have helped us maintain our commitment to both. The professors and consultants who assisted us in this edition are listed below in alphabetical order. We are greatly indebted to them all.

Ginna Babcock
University of Idaho

Carol Campbell
California State University, Long Beach

Sandra Caron
University of Maine at Orono

John M. Deaton
Northland Pioneer College

Craig Forsyth
University of Southwestern Louisiana

David Gay
University of Central Florida

Linda Green
Normandale Community College

Lingzhi Huang
Wichita State University

Gary E. Jepson
Monterey Peninsula College

M. Cathey Maze
Franklin University

Sheila Nelson
College of St. Johns

Janette K. Newhouse
Radford University

Loretta P. Prater
The University of Tennessee at Chattanooga

Shulamit N. Ritblatt
San Diego State University

HernanVera
University of Florida

Frank R. Williams
University of Arizona

John Worobey
Rutgers University

From the Authors

It has been my privilege to co-author this book with Bryan Strong, my husband, since its second edition in 1983. In May of 1993, Bryan was diagnosed with malignant melanoma. Our next few years were filled with love, work, tears, laughter, and the deep appreciation of life and humanity that we seem to find when we are face to face with our own mortality. Bryan died at home on August 10, 1996. He was valiant, steadfast, optimistic, and generous always. Through his teaching and writing, he inspired and encouraged countless students, both known and unknown to him. Part of his legacy is the seventh edition of this textbook. My hope is that through these pages Bryan will continue to touch the lives of those who encounter him.

Throughout Bryan's illness we were supported by the loving hands and hearts of family and friends. I want to especially acknowledge our children—Gabe, Will, and Maria, and Bryan's daughter Kristin—for their love, patience, understanding, and help during difficult times. I am also immeasurably grateful to Barbara Sayad, a friend indeed, who with great good will shouldered a major portion of the work of this revision.

Christine DeVault

I take special pride in thanking Bryan and Christine for the opportunity to become a part of this book. Their scholarly research, poetry, inspiration and hard work have made my involvement in this book a joy and a testament to the goodness in individuals and family.

At home, a shared office phone and computer line sometimes created competition for time and space in order to get work done on time. I thank each member of my family for their patience and support but more importantly for the love that flows so freely though our family.

Barbara Werner Sayad

About the Authors

Bryan Strong received his doctorate from Stanford University and taught at the University of California, Santa Cruz. His fields of expertise included marriage and the family, human sexuality, and American social history.

Christine DeVault, a Certified Family Life Educator, is an educational writer and consultant and an instructor at Cabrillo College. She received her degree in sociology from the University of California, Berkeley. She lives with her children in Felton, California.

Barbara Werner Sayad is a wife, mother, teacher, and writer. She holds a master's degree in Public Health and currently teaches wellness, human sexuality, and women's health at California State University, Monterey Bay. Other areas of research and expertise include marriage and the family and adolescent health. Three young children and a husband share her days in their hillside home overlooking Carmel Valley.

CHAPTER 1

The Meaning of Marriage and the Family

P R E V I E W

To gain a sense of what you already know about the material covered in this chapter, answer "True" or "False" to the statements below.

❶ The majority of American families are traditional nuclear families in which the husband works and the wife stays at home caring for the children. True or false?

❷ TV sitcoms depict working-class and middle-class husband-wife relationships differently. True or false?

❸ *Wellness* refers exclusively to physical health. True or false?

❹ All cultures traditionally divide at least some work into male and female work. True or false?

❺ For immigrant families, family goals generally took precedence over individual ones. True or false?

❻ If women were paid for their household and homemaking responsibilities, they would make over $50,000 a year. True or false?

❼ The majority of cultures throughout the world prefer monogamy, the practice of having only one husband or wife. True or false?

❽ Married men tend to live longer than single men. True or false?

❾ African Americans, Latinos, and women are overrepresented on television. True or false?

❿ Nuclear families, single-parent families, and stepfamilies are equally valid family forms. True or false?

Answers

❶ False, ❷ True, ❸ False, ❹ True, ❺ True, ❻ True, ❼ False, ❽ True, ❾ False, ❿ True

OUTLINE

For many of us, watching TV reruns of *Leave It to Beaver, The Brady Bunch*, or *The Cosby Show* is something like watching old home movies or videos of our own families. In a certain sense, we grew up with the Beav, laughed with Alice, or listened attentively to Dr. Huxtable. In some ways we seem to know their families as well as we do our own. We know their foibles. We follow their weekly adventures. Sometimes we may have wished that our families were more like the Cleavers or the Huxtables. Years later we might be channel grazing and come across a scene from *The Brady Bunch* or *The Wonder Years* that we remember as well as anything that may have happened in our own families. Many of our own ideas about families are not formed by real-world experiences but by TV experiences. Television has had a significant influence in shaping many of our ideas, values, and beliefs about the world, including marriage, family, and other relationships (Gerbner, Gross, Morgan, and Signorielli, 1986).

In this chapter, we examine the role television plays in forming our views of marriage, family, and intimate relationships. Then we study how marriage and family are defined by individuals and society, paying particular attention to the discrepancy between images of the traditional family and real families today. We provide an historical perspective of marriage and the family. Then we look at the functions that marriages and families fulfill. Finally, we examine extended families and kinship.

Marriage, Family, and Television

Television is such a force that the only comparable analogy from the past is . . . the medieval church and carnival.

JAMES B. TWITCHELL

For most people there are only two places in the world. Where they live and their TV set.

DON DELILLO

Popular culture—including pop music, magazines, movies, and television programs—helps shape our attitudes and beliefs about the world in which we live. It is one of our key sources of information and misinformation. Popular culture conveys images, ideas, beliefs, values, myths, and stereotypes about every aspect of life and society. It is an important source of information and knowledge for us about marriage, the family, and other intimate relationships. In part, the mass media's influence is so significant because we don't often see how actual families (other than our own) interact. All we usually see is their "correct" or conventional behavior. Because arguing or being overtly sexual in public is socially unacceptable, couples usually engage in such activities in private. But we are privy to those behaviors on television and in movies and magazines.

Television and the Perception of Reality

Perhaps the most pervasive medium in our culture is television. In the average household, the television set is on seven hours a day. Watching television is the most popular leisure activity for American families (Honeycutt, Wellman, and Larson, 1994). The way people relate to one another on television provides models for us to imitate (DeFleur and Ball-Rokeach, 1989; Honeycutt, 1994). Television transmits and reinforces social values (Chesebro, 1987). The TV models serve as guides that tell us how to interact with others and how to expect others to interact with us. Many of us use TV families as guides to how we should act in our own families (Greenberg, 1994). Television provides children and adolescents with models of how to make friends, act in a relationship, and achieve romantic success (Gunter and McAleen, 1990).

One of the most important theories media researchers use in studying television is cultivation theory. **Cultivation theory** asserts that there are consistent images, themes, and stereotypes that cut across programming genres. Together, these images *cultivate*, or form, a more or less consistent world view. According to cultivation theory, for example, images of men and women found in sitcoms will also be found in television dramas, game shows, soap operas, talk shows, news programs, and so on. These images include males as more ambitious and successful than women and women as more nurturing and emotional than men (Signorielli and Morgan, 1990; Vande Berg and Streckfuss, 1992, Greenberg, 1994).

TV sitcoms have influenced our beliefs and attitudes about marriage and the family. What messages and expectations do these programs convey?

Ozzie and Harriet
The Brady Bunch
The Cosby Show
The Simpsons

The more we rely on popular culture for information and descriptions about life, the more likely we are to mistake the world depicted by the entertainment media for the real world. (See "Understanding Yourself: Television and the World We Live In" on pages 8–9.) The influence of the media is so pervasive, however, that it is invisible, like the air we breathe. Yet it affects us. Heavy TV viewers tend to have different beliefs and attitudes about the world than light TV viewers. Heavy TV viewers, for example, are more likely to hold a "mean-world" view of society. Their viewing of crime programs, TV news, special reports, and so on—which focus on crime—create the perception that the world is a dangerous place. Viewing also affects perceptions of sexuality. A variety of programs, from soap operas to crime dramas and from talk shows to game shows—such as *Singled Out*, use sexuality to attract viewers. They may portray sexuality either blatantly or through not-so-subtle innuendoes. The more regular consumers of media sex are more likely to believe that sex acts, including extramarital, premarital, rape, and prostitution, happen more frequently than they do (Greenberg, 1994). Women are more likely to be shown preoccupied with romance and personal appearance than they are having jobs or going to school ("Female Role Models," 1992). It is important to note that TV viewing may not necessarily "cause" unrealistic beliefs about the world. It may be that people who are more fascinated by crime or more concerned about it are more likely to watch programs that confirm their views. Many researchers, however, argue that these beliefs and attitudes result from TV viewing (Gerbner, Gross, Morgan, and Signorielli, 1986).

TV is environmental and imperceptible, like all environments.

MARSHALL MCLUHAN (1911–1980)

Sitcom Families

During the 1970s and 1980s, when the number of people living in intact nuclear families in real life was declining, their number on television was increasing (Cantor, 1991). Until the mid-1980s, the ideal family depicted on television was the nuclear family consisting of husband, wife, and their dependent children. But in the mid-1980s, the primacy of the nuclear family was challenged by alternative families: single-parent families living together, such as *Kate and Allie*, all-female families, such as *The Golden Girls* and *Designing Women;* all-male families, such as *Dads* and *You Again* (as well as *My Two Dads*, in which two men were the "fathers" of one girl); and multiracial families, such as *Diff'rent Strokes.*

In an effort to avoid our actual lives, we spent our formative years watching television. Those rosy memories we all share are actually memories from our favorite TV shows. We've confused our own childhoods with episodes of "Ozzie and Harriet," "Father Knows Best," and "The Brady Bunch."

CYNTHIA HEIMEL

But whatever their family forms, certain themes run through sitcoms. In prime-time marriages, love overcomes all adversity: "If you just love each other enough, you'll overcome any problem." Divorces are relatively rare in situation comedies. If the parents are divorced, the divorce usually took place in the past, prior to the series' beginning. Conflict is easily resolved through manipulation, and problems are solved through humor. Physical appearance and beauty are especially important (Moore, 1992; Yerby, Buerkel-Rothfus, and Bochner, 1990).

Working-Class/Middle-Class Families

Television portrays working-class and middle-class families differently. Although working-class families are relatively uncommon on television, they nevertheless display recurring themes. Working-class fathers and husbands, from Ralph Kramden in *The Honeymooners* to Archie Bunker in *All in the Family* to Al Bundy on *Married with Children*, for example, have typically been portrayed as clumsy, awkward, or inept, especially in contrast to their wives. Working-class mothers and wives are generally the central characters in sitcoms and are usually portrayed as stronger than and superior to their husbands (Cantor, 1991; Glennon and Butsch, 1982).

The family dynamics of middle-class TV families are different from those of working-class TV families. In these families, husbands and fathers are rarely ridiculed. Middle-class men are generally depicted as competent and caring. The relationships between husband and wife are more egalitarian than in working-class families. Furthermore, middle-class TV children are generally more respectful of their fathers.

Single-Parent Families and Stepfamilies

Single-parent families and stepfamilies on television are usually formed as a result of a spouse's death rather than a divorce or birth to an unmarried woman (which is much more likely in the real world). In fact, although 32 percent of all births are to unmarried women, Murphy Brown is the only never-married mother on television in a major role.

There are almost seven times as many male-headed single-parent families on television as there are in real life. They represent over 20 percent of single-parent

TV families but make up about 3 percent of real-life single-parent families. On television, single mothers, much more than single fathers, are looking for partners. They marry or remarry, whereas single fathers tend to be satisfied in their unmarried state. (This replicates the stereotype that single women want to marry, whereas single men enjoy their freedom.)

Family dynamics are significantly different among single-parent families, stepfamilies, and intact families in real life (as we will see in Chapters 14 and 15), but on television they are remarkably similar. In stepfamilies, developing a sense of family harmony takes considerable time, but in prime-time stepfamilies, harmony is a given. Because differences are disregarded, the sitcom "formula" remains similar for all family forms: A problem occurs, and the family struggles to find a solution for it in a single episode.

Put yourself behind a Pepsi . . . You've got a lot to live and Pepsi's got a lot to give. The Pepsi-Cola Company hopes you have enjoyed our ad with tightly edited images of people having far more fun than you ever do. Now back to our program.

DOONESBURY

Marital and Family Interactions

Most family interactions in sitcoms reflect marital stereotypes (Greenberg, 1994). The overwhelming majority of dialogue and interaction takes place between husbands and wives. Husbands offer advice to their wives, whereas wives look to their husbands for support. Mothers are usually responsible for children, who are well behaved and who ask their parents for permission to do activities. Children are usually cute—the source of humor. They often remain in the background (except to provide humor); other times they are more central (as in *Family Matters* or *The Simpsons*), or they are the main characters, as in *The Wonder Years* or *Moesha*.

Relatively little marital sex is depicted on television. Most TV sex is "talking" about sex rather than showing it (Greenberg, 1994). Sex is usually premarital or extramarital. One study of types of TV programs found, for example, that unmarried men and women are shown in sexual situations six times as often as married couples. And extramarital sex is depicted four times as often as marital sex (Hanson and Knopes, 1993). Such portrayals, however, are misleading, as most sexual interactions actually take place within marriage (Michael, Gagnon, Laumann, and Kolata, 1994).

Family Diversity

Whites are overrepresented on television, whereas Latinos, Asian Americans, and Native Americans are underrepresented (Moore, 1992). In recent years, the presence of African Americans has increased substantially in prime time programming. The characters are disproportionately portrayed, however, as stereotypical "homeboys," "gangstas," or not-so-bright buffoons. The numbers of Latinos, Asian Americans, and Native Americans playing significant characters are so low that such groups are virtually invisible. Until recently, gay and lesbian characters have been rare.

African-American Families

Today, there is a sizable number of African-American sitcoms, including such widely popular ones as *Living Single*, *Moesha*, and *Family Matters*.

Until the mid-1980s, African-American TV families continued to be stereotyped negatively. However, since the mid-1980s, TV portrayals of African-American families have improved. The Huxtables (*The Cosby Show*), the Jenkinses (*227*), and the Richmonds (*Charlie Company*) offered a positive view of African-American family life (Merritt and Stroman, 1993).

The family portrayed on Moesha *is a stepfamily, a common family form in the real world but a relatively new one on television.*

UNDERSTANDING YOURSELF

Television and the World We Live In

One way we can get a sense of television's impact is to compare our knowledge of the world with the images conveyed by television. Take a moment to answer the following questions about life in the United States to get a sense of your "TV reality:"

1 What percentage of children live with both biological parents? (a) 21 percent, (b) 35 percent, (c) 60 percent, (d) 83 percent
2 What percentage of single-parent families are headed by fathers? (a) 3 percent, (b) 14 percent, (c) 17 percent, (d) 27 percent
3 What percent of American children are poor? (a) 5 percent, (b) 10 percent, (c) 21 percent, (d) 50 percent
4 What percentage of the population dies each year? (a) 1 percent, (b) 5 percent, (c) 10 percent, (d) 16 percent
5 What is the homicide rate in the U.S. per 100,000 population? (a) 11, (b) 42, (c) 61, (d) 83
6 In what year was the highest percentage of women pregnant at the time of marriage? (a) 1760, (b) 1954, (c) 1968, (d) 1994
7 What percentage of the total population is female? (a) 61 percent, (b) 52 percent, (c) 45 percent, (d) 40 percent
8 What percentage of women are employed in the workforce? (a) 22 percent, (b) 41 percent, (c) 51 percent, (d) 60 percent
9 What percentage of Americans never marry? (a) 3 percent, (b) 10 percent, (c) 19 percent, (d) 28 percent
10 What percentage of single, college-educated women over age thirty-five are likely to marry? (a) 5 percent, (b) 11 percent, (c) 28 percent, (d) 40 percent
11 How many children are kidnapped by strangers each year? (a) 300, (b) 3,000, (c) 30,000, (d) 300,000
12 What percentage of families are intact two-parent families with children in which the mother does not have outside employment? (a) 18 percent, (b) 36 percent, (c) 54 percent, (d) 70 percent
13 What percentage of the American population is Hispanic? (The term *Hispanic*, rather than *Latino*,is used here to correlate with the terminology used by the U.S. Census Bureau.) (a) 2 percent, (b) 8 percent, (c) 16 percent, (d) 24 percent
14 What percentage of births are to unmarried women? (a) 3 percent, (b) 12 percent, (c) 18 percent, (d) 32 percent
15 What percentage of American women have had abortions? (a) 5 percent, (b) 11 percent, (c) 21 percent, (d) 32 percent

The answers to the questions are as follows:

1 The correct answer is **b.** Thirty-five percent of all children live with both biological parents, according to 1995 census data. In TV families, 44 percent live in such families (Dorr, Kovaric, and Doubleday, 1990).
2 The correct answer is **a.** Three percent of single-parent families are headed by fathers. In TV families, 20 percent are headed by fathers (Dorr, Kovaric, and Doubleday, 1990).
3 The correct answer is **c.** According to the 1991 Luxembourg Income Study (Smeeding, O'Higgins, and Rainwater, 1990) more than 21 percent of American children were poor, thereby giving the United States the highest child

Old stereotypes never die; they just move to the late show.

Ron Miller

A notable feature of several African-American TV families is that they are extended families. For example, in *Fresh Prince of Bel Air* (now in reruns), a married couple lives with their three children and teenage nephew. And in *Family Matters*, a married couple and their children live with the children's grandmother, an aunt, a cousin, and—often—a neighbor's child. These programs build on one of the notable strengths of African-American families: Family members can count on one another for emotional and material support. In time of need, other members are often taken into the family (Murray and Gilbert, 1993).

Despite the advances in presentations of African-American families, there is still a solid argument to be made that blacks are all too often portrayed stereotypically. Bill Cosby (1997) believes the media continue to perpetuate racism: "Get people from anywhere in the world and they all have this negative view of the black man, because of what they've seen, not what they know."

Many offensive stereotypes endure in late-night reruns and syndication. Several years ago, TV critic Ron Miller reviewed old films he had watched in 1976, when he was analyzing racial and ethnic stereotypes. He found that of those poverty rate of any industrialized country. The media, however, rarely report poverty among children.

4 The correct answer is **a.** One percent of our population dies each year. The death rate on TV programs is about 5 percent.

5 The correct answer is **a.** According to 1990 census data, there were 10.8 homicides a year per 100,000 population (U.S. Bureau of the Census, 1993). But the average TV viewer witnesses over six killings a day, or over 2,200 a year. The viewer also sees 38 violent acts a day, or over 14,000 a year (Linz, Wilson, and Donnerstein, 1992).

6 The correct answer is **a.** In 1760, almost 40 percent of women were pregnant at the time of marriage (Mintz and Kellogg, 1988).

7 The correct answer is **b.** Women make up the majority of the population but on prime-time television, 55 percent of the characters are male; in movies the figure is 63 percent; and in music videos, 78 percent of performers are men ("Female Role Models," 1997).

8 The correct answer is **d.** The overwhelming majority of women are employed, but only 35 percent of prime-time women are employed (Vande Berg and Streckfuss, 1992).

9 The correct answer is **b.** About 10 percent of Americans never marry. On prime time, however, as many as 40 percent of the characters are single.

10 The correct answer is **d.** Forty percent of single, college-educated women over thirty-five are likely to marry. In the 1980s, however, the media promoted the idea that such women rarely married (Faludi, 1991).

11 The correct answer is **a.** The media and advocacy groups promote the idea that everyone's child is at risk for kidnapping, citing Federal Bureau of Investigation (FBI) statistics that 1.3 million children are missing each year. Missing children are a staple of local TV news. Of missing children, most are runaways or temporarily lost; 350,000 are abducted by family members; and about 300 are kidnapped by strangers each year.

12 The correct answer is **a.** Only 18 percent of families are intact two-parent families in which the mother does not work outside the home. Traditional families have represented the majority of TV families until recently, however.

13 The correct answer is **b.** Eight percent of the U.S. population is Hispanic, but Latinos account for less than 2 percent of prime-time characters.

14 The correct answer is **d.** Nearly 33 out of every 100 births are to unmarried women. However, unmarried women who give birth represent a miniscule portion of TV characters in ongoing series. In the sitcoms, only Murphy Brown is a central character who is an unmarried mother (Saluter, 1994).

15 The correct answer is **c.** Twenty-one percent of American women of reproductive age have had abortions. Abortion rarely occurs on television, though, because of its controversial nature.

What you have just taken is a "TV Reality" test, an idea suggested by Daniel Friedman of Antioch College. Media researchers have analyzed prime-time television programs to determine their demographic content. They found that television reality often did not reflect the real world.

The closer your answers are to TV reality, the more likely you were influenced by television viewing. What other influences might account for an incorrect answer: radio, newspapers, books, magazines, friends, family? What are your sources of information?

still shown on television, none had been edited to eliminate racist jokes or scenes. (Movies are frequently "edited for TV," but only to eliminate obscenity, sexual references, or explicit sex scenes.)

Latino Families

Latino families are virtually absent in prime-time network television. But Latinos are major consumers of Spanish-speaking television, which broadcasts sitcoms, dramas, movies, and soap operas, known as *telenovelas.* Most of these programs, however, are imported from Latin America. They generally depict Mexican or South American families rather than Spanish-speaking U.S. families.

Gays and Lesbians

In the late 1970s, Billy Crystal portrayed a gay father on the popular comedy series *Soap.* After that, portrayals of gay men on regular programs were few and far between—until very recently. Portrayals of lesbians were virtually nonexistent (except for *L.A. Law* in the early '90s) until the principal character in *Ellen* came out in 1997. Today a number of popular programs (22 in February 1997)—from *Melrose Place* to *The Simpsons*—feature gay characters (Handy, 1997).

Sitcom Family Values

The Simpsons are a misunderstood and underrated family.

HOMER SIMPSON

Stereotypes abound in TV families, but it is important to remember that many sitcoms, such as *Roseanne*, *Moesha*, and even *The Simpsons*—as well as reruns of *The Cosby Show* and *Family Ties*—value love and family relationships. They are what Father Andrew Greeley, a sociologist and novelist (1987), calls "paradigms of love." Among the virtues he finds in such programs are "patience, trust, sensitivity, honesty, generosity, flexibility, and forgiveness."

The world that sitcom families inhabit is much less complicated than the real world. Problems are always manageable. Stress, divorce, unemployment, poverty, chronic illness, and death rarely disrupt TV families, as they may actual families.

Reflections

Think of your three favorite sitcoms. What types of families do they depict? How are husbands and wives portrayed? What norms and values are being conveyed?

The message conveyed is that love and family solidarity triumph against adversity. As we shall see, this is often the case. But we shall also see that the realities of daily life do impinge on our intimate relationships. Our lives are far more complex and richer than life depicted in the media. To help us understand the family, it is important to define *marriage* and *family* and discover how diverse family forms reflect the diversity of American families.

What Is Marriage? What Is Family?

The impact of norms (cultural rules or standards) and stereotypes on our perceptions can be seen if we answer two simple, but basic, questions: What is marriage? What is family?

Defining Marriage

Marriage is a great institution, but I'm not ready for an institution yet.

MAE WEST (1893–1980)

A **marriage** is a legally recognized union between a man and a woman in which they are united sexually; cooperate economically; and may give birth to, adopt, or rear children. The union is assumed to be permanent (although in reality it may be dissolved by separation or divorce).

As simple as marriage may seem, it differs among cultures and has changed historically in our society. Among non-Western cultures, who may marry whom and at what age varies considerably from our society. In some areas of India, Africa, and Asia, for example, children as young as six years may marry other children (and sometimes adults), although they may not live together until they are older. In many cultures, marriages are arranged by families who choose their children's partners. And in one region of China, marriages are sometimes arranged between unmarried young men and women who are dead. (See "Other Places/Other Times: The Cross-Cultural Perspective: The Spirit Marriage in Chinese Society," pages 12–13.)

Many Americans believe that marriage is divinely instituted; others, that it is a civil institution involving the state. The belief in the divine institution of marriage is common to many religions, such as Christianity, Judaism, and Islam, as well as many tribal religions throughout the world. But the Christian church only slowly became involved in weddings. In the early Middle Ages, for example, the priest's blessing was not important. As the church increased its power, however, it extended control over marriage. Traditionally, marriages had been arranged between families (the father "gave away" his daughter in exchange for goods or services); by the tenth century, marriages were valid only if they were performed by priests. By the thirteenth century, the ceremony was required to take place in a church (Gies and Gies, 1987). As states competed with organized religion for power, governments began to regulate marriage. In the United States today, for example, in order for marriages to be legal—whether they are performed by ministers, priests, rabbis, or imams—they must be validated through government-issued marriage licenses. This is a right for which many gay men and lesbians are fighting.

Who may marry whom has changed over the last 150 years in the United States. Laws once prohibited enslaved African Americans from marrying because they were regarded as property. Marriages between members of different races were illegal in over half the states until 1966, when the Supreme Court declared such prohibitions unconstitutional. Each state enacts its own laws regulating marriage. In some states, first cousins may marry; other states prohibit such marriages as incestuous. In 1993, in a case now under appeal, the Hawaii Supreme Court ruled that denying gay men and lesbians the right to marry violated the equal protection clause of its state constitution.

In Western cultures, such as the United States, the only legal form of marriage is **monogamy,** in which there are only two spouses, the husband and wife. But monogamy is a minority preference among world cultures, practiced by only 24 percent of the known cultures (Murdock, 1967). The preferred marital arrangement worldwide is **polygamy,** the practice of having more than one wife or husband. One study of 850 non-Western societies found that 84 percent of the cultures studied (representing, nevertheless, a minority of the world's population) practiced or accepted **polygyny,** the practice of having two or more wives (Gould and Gould, 1989). **Polyandry,** the practice of having two or more husbands, is quite rare: Where it does occur, it coexists with polgyny. Even within polygynous societies, however, plural marriages are in the minority, primarily for simple economic reasons: They are a sign of status that relatively few people can afford. Although problems of jealousy may arise in plural marriages—the Fula in Africa, for example, call the second wife "the jealous one"—there are usually built-in control mechanisms to ease the problem. The wives may be related, especially as sisters; if they are unrelated, they usually have separate dwellings. Women in these societies often prefer that there be several wives; plural wives are a sign of status and, more important, ease the workload of individual wives. This last fact is apparent even in our culture; it is not uncommon to hear overworked American homemakers exclaim, "I want a wife."

Because of our culture's traditional roots in Christianity, polygamy is illegal in the United States. As a result, polygamy sounds strange or exotic. However, it may not really seem so strange if we look at actual American marital practices. Considering the high divorce rate in this country, monogamy may no longer be the best way of describing our marriage forms. Our marriage system might be called *serial monogamy* or *modified polygamy* because one person may have several spouses over his or her lifetime. In our nation's past, enslaved Africans tried unsuccessfully to continue their traditional polygamous practices when they first arrived in America; these attempts, however, were rigorously suppressed by their masters (Guttman, 1976). Today some members of the Nation of Islam practice polygamy on religious grounds despite its legal prohibition. Mormons practiced polygamy until the late nineteenth century, when they abandoned the practice as a condition of Utah's becoming a state. Even today, an estimated three thousand to ten thousand fundamentalist Mormons continue to practice polygamy despite its prohibition by society and their church. (Only the first wife has legal status as a wife, however.)

Did you know?

There has been a sevenfold increase in unmarried-couple households since 1970, thereby growing from 523,000 to 3.7 million. About one-third of these have children under fifteen years old present in the home (Saluter, 1994).

Figure 1.1 Household Composition: 1994

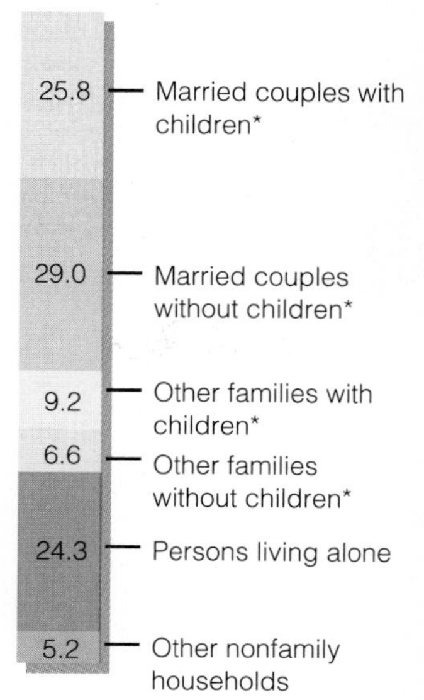

*Own children under 18

SOURCE: Rawlings, Steve, W., *Households and Families: March 1994,* U.S. Bureau of the Census, Current Population Reports, P20-438. U.S. Government Printing Office, Washington, D.C. 1994.

OTHER PLACES OTHER TIMES

The Cross-Cultural Perspective: The Spirit Marriage in Chinese Society

From *Time* magazine comes this report ("A Day in the Life of China," 1989):

> A beautiful day for a wedding—crisp, clear and, for China in midsummer, relatively cool. The latest typhoon's high winds have swept away the air pollution, and under a brilliant blue sky the guests are chatting in the hollow of a terraced field beside a single spindly tree—symbolic decoration in a country whose scant arable land continues to disappear. Arranged neatly alongside the makeshift altar, the gifts intended for the bride's parents include a new refrigerator, a 24-in. color television set and a jet black Yamaha motorcycle. The presents are ogled, but atop the TV a photograph of Margaret Thatcher creates the greatest buzz, a reaction the bride, and perhaps the groom too, would undoubtedly have enjoyed. Were they still alive . . . I am transfixed by the marriage of the two coffins in front of me. The groom died in an automobile accident five days earlier at the age of 23. The body of his bride, dead of cancer for five months, cost $3 to exhume. They had never met.

After overcoming our surprise at the very notion of *spirit marriage,* we wonder why anyone would arrange a marriage between two dead persons. The case of spirit marriage, which was widely practiced in traditional Chinese society, underscores a fundamental point about the family. Anthropologists and historians tell us that families are organized differently in different cultures. Different cultural understandings about kinship, marriage, and residence give the family a decidedly different look across cultures.

Consider, for example, the case of kinship. Who counts as a relative? The answer to this question may seem self-evident to us, but we grew up learning who a relative is from within our own culture. Everyone everywhere learns this way, and cultural understandings of who counts as a relative vary. For example, among the traditional Iroquois of the northeastern United States, the important relatives were those who could be traced through the mother. This was a matrilineal society, and children took their clan membership, their names, and their chiefly titles (if any) from their mothers' families. Sons and daughters inherited from their mothers. A mother's relatives counted more than a father's. Thus, an aunt on your mother's side would be considered a very different kind of relative than one on your father's side.

Consider also the way in which marriage shapes the family. Who marries whom, and why? Who decides? These questions seem like nonquestions to us, because in our society, the individual generally makes these decisions. Yet cultural practices surrounding marriage vary tremendously, contributing further to the distinctive look and character of the family across cultures. In Bedouin society in Egypt, marriages are arranged by the elder males of the bride's and groom's families. The engaged couple themselves may never meet prior to marriage. Love as a "natural" basis for marriage turns out to be another of our cultural assumptions. In Bedouin society, parents may very well decide to arrange a marriage between a bride and groom who are first cousins on the male side—and definitely for reasons other than love! (A marriage between paternal cross-cousins was considered to be a particularly advantageous match.) Thus, who marries whom, why they marry, and who decides are cultural issues that greatly influence the shape of the family across cultures.

Finally, there is the question of residence. Where will the married

Defining Family

Defining *family* is even more complex than defining *marriage.* As contemporary Americans, we live in a society composed of married couples, two-parent families, stepfamilies, single-parent families, multigenerational families, cohabiting adults, child-free families, families headed by gay men and lesbians, and so on. With such variety, how does one define *family?* What are our criteria for identifying some (or all) of these groups as families?

When we asked our students who they included as family members, their lists included (alphabetically) the following:

aunt
best friend
boyfriend
cousin
daughter
father
father-in-law
foster child
foster parent
girlfriend
godchild
godparent
grandfather
grandmother
great-grandparent

couple live? And with whom? The answer to these questions—for us, a matter of individual choice (as well as economics)—gives the family its many different faces across cultures. In traditional Chinese society, it was expected that upon marriage, the bride would live with her husband in his family home, which typically included his father and mother, his father's father (if alive), and his father's brothers and their wives and children. This was the family in traditional Chinese society. It was built upon the male descent line, as anthropologists call it. Unlike Iroquois families, Chinese families were patrilineal, the most important kinship ties being those traced on the father's side of the family. The father's relatives counted more.

In traditional Chinese society, male elders employed matchmakers to arrange marriages for their sons and daughters on the basis of economic and social criteria, not love. Marriages were arranged for sons in order that they themselves might have sons to continue the male descent line. (Marriages were arranged for daughters so that they would, as wives, produce sons for their husbands' descent lines.) Even the male ancestors—as deceased male elders—were thought to be very concerned about the continuation of their descent lines. It was a young man's duty—an almost religious obligation to his male ancestors and father—to continue the descent line unbroken into the future: *a man must have sons.*

Life did not always make it easy for men to continue their male descent lines. Many men in traditional Chinese society were too poor to marry, and thus poverty annihilated their descent lines. Death, too, could interfere with human plans for descent lines. Sons might die before they could marry and produce sons of their own. To remedy this situation, the practice of spirit marriage developed, guaranteeing that the descent line would continue even in the face of death. A family whose young son died would wait until the son's ghost (or spirit) reached proper marriageable age. Then, as in marriages arranged among the living, they engaged the services of a matchmaker to find an appropriate spouse—but in this case, a deceased bride for their dead son to marry.

For the groom's family, this spirit marriage would settle their son's restless spirit, for he would now have a wife. His parents would then adopt a son for him, one that they themselves would raise. This son would one day marry and have a son of his own—fate willing—and thus continue the line of his deceased father. The male descent line would remain intact.

For the bride's family, marrying off their deceased daughter also brought distinct advantages—or rather diverted distinct disadvantages and even disaster. The death of an unmarried daughter not only brought sorrow to her family; her unmarried ghost threatened the fertility of her brothers' wives, and hence their descent lines. The ghost of an unmarried daughter was troublesome to her family and brought misfortune, including bad harvests. But if she could be married off to a husband in a spirit marriage, she could then take her proper place as a married woman at his family home, diverting disaster at her own.

Thus, the practice and meaning of spirit marriage in traditional Chinese society was shaped by cultural notions of kinship (Who counts as a relative?), marriage (Who marries whom, why, and who decides?), and residence (Where does a couple live in marriage, and with whom?). The Chinese answers to all of these questions help to explain a marriage practice that to us may seem unfathomable. Of course, families change with time as new laws, technology, and other cultural influences reshape kinship, marriage, and residence practices. Yet with all the profound changes in Chinese society in recent decades, the continuing practice of spirit marriage speaks to the continuing belief in the male descent line as the foundation of Chinese families (Stockard, 1989).

half-sibling
lover
minister
mother
mother-in-law
neighbor
nephew
niece
pet
priest
rabbi
second cousin
sibling
son
stepfather
stepmother
stepsibling
teacher
uncle

Most family members are related by descent, marriage, remarriage, or adoption, but some are **affiliated kin,** unrelated individuals who *feel* and are *treated* as if they were relatives. (In a couple of instances, the family dog, cat, or bunny was included as a family member.)

Being biologically related or related through marriage is not always sufficient to be counted as a family member or **kin.** One researcher (Furstenberg, 1987) found that 19 percent of the children with biological siblings living with

Reflections

Make a list of whom you consider family. What criteria, such as biological, legal, or affectional, did you use? What biological or legal family members did you exclude? Why?

Did you know?

In 1994, 62 percent of the adult population in America (115 million over eighteen years of age) were married. (Saluter, 1994).

Did you know?

The number of men as well as the number of women maintaining their families with no spouse present has more than doubled since 1970 (Bryson, 1996). Today, the child is just as likely to be living with a divorced parent as with a never-married parent.

them did not identify their brothers or sisters as family members. Sometimes an absent or divorced parent was not counted as a relative. Stepparents or stepchildren were the most likely not to be viewed as family members (Furstenberg, 1987; Ihinger-Tallman and Pasley, 1987). Emotional closeness may be more important than biology or law in defining *family.*

There are also ethnic differences as to what constitutes family. Among Latinos, for example, *compadres* (godparents) are considered family members. Among some Japanese Americans the ***ie*** (pronounced "ee-eh") is the traditional family. The *ie* consists of living members of the extended family (such as grandparents, aunts, uncles, and cousins), as well as deceased and yet-to-be-born family members (Kikumura and Kitano, 1988). Among many traditional Native-American tribes, the **clan,** a group of related families, is regarded as the fundamental family unit (Yellowbird and Snipp, 1994).

A major reason we have such difficulty in defining *family* is that we tend to think that the "real" family is the nuclear family or the traditional family. The **nuclear family** is the family type consisting of mother, father, and children. But the nuclear family is merely an idea or model we have about families. The term itself is less than fifty years old, coined by the anthropologist Robert Murdock in 1949 (Levin, 1993). The **traditional family** is the middle-class nuclear family in which women's primary roles are wife and mother, and men's primary roles are husband and breadwinner. The traditional family is the nuclear family wrapped in nostalgia and inequality. The traditional family exists more in the imagination than it ever did in reality.

Because we believe that the nuclear or traditional family is the "real" family, we compare all other family forms against these models. To include the diverse forms, the definition of **family** needs to be revised. A more contemporary definition describes family as one or more adults related by blood, marriage, or affiliation who cooperate economically, who may share a common dwelling place, and who may rear children. Such a definition more accurately reflects the diversity of contemporary American families.

Functions of Marriages and Families

The word "ordinary" in the dictionary . . . says "familiar; unexceptional; common." Given that set of definitions, it's one of the biggest words in the English language, since life, death, childbirth, love, hate, age, and sex are all familiar, unexceptional and common.

MARY CANTWELL

Whether it is the mother/father/child nuclear family, a married couple with no children, a single-parent family, a stepfamily, a dual-worker family, or a cohabiting family, the family generally performs four important functions. First, it provides a source of intimate relationships. Second, it acts as a unit of economic cooperation and consumption. Third, it may produce and socialize children. Fourth, it assigns social roles and status to individuals. Although these are the basic functions that families are "supposed" to fulfill, families do not necessarily have to fulfill them all (as in families without children), nor do they always fulfill them well (as in abusive families).

Intimate Relationships

Only an animal or a god can live alone.

ARISTOTLE (384–322 B.C.)

Intimacy is a primary human need. Human companionship strongly influences rates of cancer, tuberculosis, suicide, accidents, and mental illness. Studies consistently show that married couples and adults living with others are generally healthier and have a lower mortality rate than divorced, separated, and never-married individuals (Ross, Mirowsky, and Goldsteen, 1991). This holds true for

The "traditional" American family of employed father, homemaker mother, and two children, accounts for less than a quarter of today's families.

both whites and African Americans (Broman, 1988). (See "You and Your Well-Being: Defining Wellness," page 16.)

Family Ties

Marriage and the family usually furnish emotional security and support. This has probably been true from earliest times. Thousands of years ago, in the Judeo-Christian Bible, the book of Ecclesiastes (4:9–12) emphasized the importance of companionship:

> Two are better than one, because they have a good reward for their toil. For if they fall, one will lift up his fellow; but woe to him who is alone when he falls and has not another to lift him up. Again if two lie together, they are warm; but how can one be warm alone? And though a man might prevail against one, two will withstand him. A three-fold cord is not quickly broken.

In our families we generally find our strongest bonds. These bonds can be forged from love, attachment, loyalty, or guilt. The need for intimate relationships, whether they are satisfactory or not, may hold unhappy marriages together indefinitely. Loneliness may be a terrible specter. Among the newly divorced, it may be one of the worst aspects of the marital breakup.

Since the nineteenth century, marriage and the family have become even more important as the source of companionship and intimacy. They have become a "haven in a heartless world" (Lasch, 1978). As society has become more industrialized, bureaucratic, and impersonal, it is within the family that we increasingly expect to find intimacy and companionship. In the larger world around us, we are

Did you know?

A public opinion poll found that only 22 percent of the respondents defined a family solely in terms of blood, marriage, or adoption. Seventy-four percent defined a family as a group whose members loved and cared for one another (Footlick, 1990).

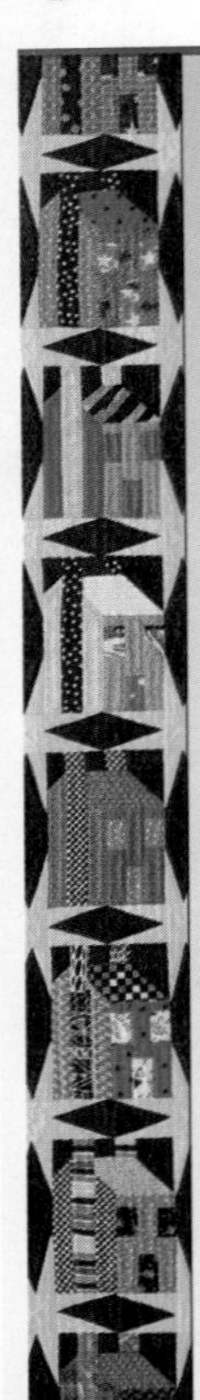

YOU AND YOUR WELL-BEING

Defining Wellness

Can a partnership between two individuals be strong if the two people involved do not come from a place of physical, mental, and social well-being? Probably not. The connection between marriage and **wellness**, or optimal health and well-being, is strong and will be articulated throughout the book. An awareness of the components of well-being may begin to help you understand how far-reaching and important this concept is to an individual, to a family, and to society:

- *Physical:* Maintaining the body's health by making responsible decisions, being aware of changes and symptoms, nourishing it, exercising, and avoiding unhealthy habits.
- *Emotional:* Maintaining positive self-esteem, dealing constructively with feelings, and developing qualities that contribute to both.
- *Intellectual:* Keeping an active, curious, open mind with the ability to think critically about issues, pose questions, identify problems, and seek and find solutions.
- *Spiritual:* Developing faith in something greater than yourself as well as maintaining the capacity for altruism, joy, compassion, forgiveness; finding meaning and purpose in life.
- *Social:* Developing meaningful relationships, cultivating friendships, and contributing to the community.
- *Environmental:* Protecting yourself and taking action to help protect others against environmental hazards.

All the above are interrelated and interconnected, thus a change in one will often affect and impact others. For example, an emotional crisis, such as the loss of a spouse or health problem of a child, will often affect our physical health (how we eat, whether or not we sleep), emotions (how we deal with our feelings), social well-being (whether or not we choose to seek the company of a supportive network of friends), and spiritual health (whether we seek comfort from a higher being).

A combination of behaviors, heredity, environment, and access to adequate health care are important influences on wellness. When combined, they have the capacity to raise or lower the quality of a person's life and the risk of developing certain diseases or problems. Wellness is positive and proactive; it focuses on being healthy and whole and requires attention, preparation, and monitoring. It is a process, not a destination, and requires a conscious decision and commitment.

generally seen in terms of our roles. A professor may see us primarily as students; a used-car salesperson relates to us as potential buyers; a politician views us as voters. Only among our intimates are we seen on a personal level, as Maria or Will. Before marriage, our friends are our intimates. After marriage, our partners are expected to be the ones with whom we are most intimate. With our partners we disclose ourselves most completely, share our hopes, rear our children, grow old.

Reflections

How would you assess your overall wellness? Which components of your wellness do your support and which do you neglect? Why? How does your personal sense of wellness affect your relationships with others?

Pets and Intimacy

The need for intimacy is so powerful that we may even rely upon pets if our intimacy needs are not met by humans. Animals have been important human companions since prehistoric times (Siegel, 1993). They have been important emotional figures in our lives, especially if our other relationships are not fulfilling. Unmarried adults, for example, are more attached to their pets than are married men and women (Stallones, Johnson, Garrity, and Marx, 1990). This does not mean, however, that you reject Fido or Fluffy when you become romantically involved or get married. What happens is that your pet becomes less important—he or she becomes more an "animal" and less "someone" to whom you are emotionally attached. As an object of attachment, your pet is replaced by your partner or children. You do not forget your pet, even in marriage or parenthood; your dog, cat, gerbil, parakeet, or turtle simply becomes less important.

Animals are such agreeable friends—they ask no questions, they pass no criticisms.

George Eliot (1819–1880)

Pets are often considered to be members of the family. They often provide their owners with comfort and a sense of intimacy.

Studies on the role of pets in human relationships suggest that the most prized aspects of pets, especially dogs and cats, are their attentiveness to their owners, their welcoming and greeting behaviors, and their role as confidants—qualities valued in our intimate relationships with humans as well. Pets give children an opportunity to nurture, and they provide a best friend, someone to love.

Reflections

Do you have pets in your family? What role have they played? Are they "like" family? Why?

Economic Cooperation

The family is also a unit of economic cooperation that traditionally divides its labor along gender lines; that is, between males and females (Ferree, 1991; Thompson and Walker, 1989; Voydanoff, 1987). Although the division of labor by gender is characteristic of virtually all cultures, the work that males and females perform (apart from childbearing and breastfeeding) varies from culture to culture. Among the Namibikwara in Africa, for example, the fathers take care of the babies and clean them when they soil themselves; the chief's **concubines,** secondary wives in polygamous societies, prefer hunting over domestic activities. In American society, from the last century until recently, men were expected to work away from home, whereas women were to remain at home caring for the children and house. There is no reason, however, why these roles cannot be reversed. Such tasks are assigned by culture, not biology. Only a woman's ability to give birth and produce milk is biologically determined. And some cultures practice *couvade*, ritualized childbirth in which a male gives birth to the child's spirit while his partner gives physical birth (see Chapter 8).

We commonly think of the family as a consuming unit, but it also continues to be an important producing unit. The husband does not get paid for building a shelf or attending to the children; the wife is not paid for fixing the leaky faucet or cooking. Although children contribute to the household economy by helping around the house, they generally are not paid for such things as cooking,

Did you know?

Unpaid household work by women in the 1990s was worth well over $1 trillion; such work by men was worth $610 billion (Voydanoff, 1991).

cleaning their rooms, or watching their younger brothers or sisters (Coggle and Tasker, 1982; Gecas and Seff, 1991). Yet they are all engaged in productive labor.

Economists have begun to reexamine the family as a productive unit (Ferree, 1991). If men and women were compensated monetarily for the work done in their households, the total would be equal to the entire amount paid out in wages by every corporation in the United States. Household work and assets along with the production of goods (including food) all contribute to the family's productive activities.

As always, the most automated appliance in the household is the mother.

BEVERLY JONES

As a service unit, the family is dominated by women. Because women's work at home is unpaid, the productive contributions of homemakers have been overlooked (Ciancanelli and Berch, 1987; Walker, 1991). Yet women's household work is equal to about 44 percent of the gross domestic product (GDP), and the value of such work is double the reported earnings of women. If a woman were paid wages for her labor as mother and homemaker according to the wage scale for chauffeurs, physicians, baby-sitters, cooks, therapists, and so on, her services today would be worth more than $50,000 a year. Many women would make more for their work in the home than men do for their jobs outside the home. Because family power is partly a function of who earns the money, paying women for their household work might have a significant impact on husband-wife relations.

Reproduction and Socialization

There are no individuals in the world—only fragments of families.

CARL WHITAKER

The family makes society possible by producing (or adopting) and rearing children to replace the older members of society as they die off. Traditionally, reproduction has been a unique function of the married family. But single-parent and cohabiting families also perform reproductive and socialization functions. Technological change has also affected reproduction. Developments in artificial insemination and in vitro fertilization have separated reproduction from sexual intercourse. In addition to permitting infertile couples to give birth, such techniques have also made it possible for lesbian couples to become parents.

The family traditionally has been responsible for socialization. Children are helpless and dependent for years following birth. They must learn how to walk and talk, how to take care of themselves, how to act, how to love, how to touch and be touched. Teaching the child how to fit into his or her particular culture is one of the family's most important tasks.

This socialization function, however, is dramatically shifting away from the family. One researcher (Guidubaldi, 1980) believes that the increasing lack of parental commitment to child rearing may be one of the most significant societal changes in our lifetimes. Since the rise of compulsory education in the nineteenth century, the state has become responsible for a large part of the socialization of children older than age five. The increase in the number of working mothers has placed many infants, toddlers, and small children in day care, further reducing the family's role in socialization. Even while children are at home, television rather than family members often "rears" them (Dorr, Kovaric, and Doubleday, 1989). (Child socialization is discussed more thoroughly in Chapter 10.)

Assignment of Social Roles and Status

We fulfill various social roles as family members, and these roles provide us with much of our identity. During our lifetimes, most of us will belong to two families: the family of orientation and the family of cohabitation. The **family of ori-**

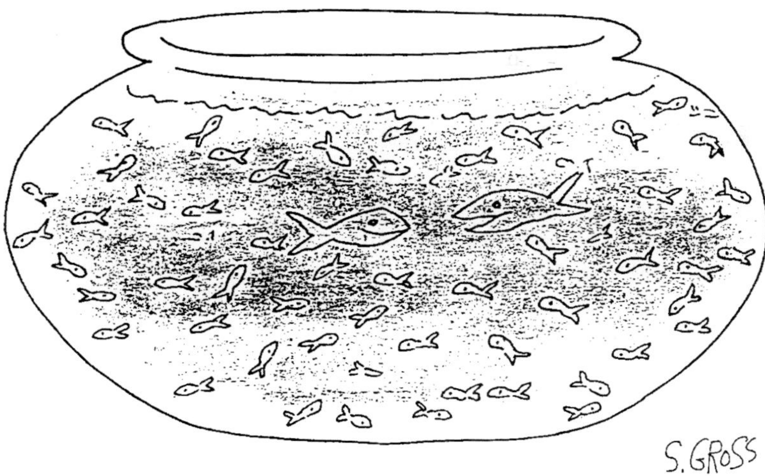

"I guess we'd be considered a family. We're living together, we love each other, and we haven't eaten the children yet."

SOURCE: Drawing by S. Gross; ©1993 The New Yorker Magazine, Inc.

entation (sometimes called the **family of origin**) is the family in which we grow up, the family that orients us to the world. The family of orientation may change over time if the marital status of our parents changes. The family of orientation originally may be an intact nuclear family or a single-parent family; later it may become a stepfamily. The **family of cohabitation** is the family we form through living or cohabiting with another person, whether we are married or unmarried. (This family had traditionally been referred to as the **family of procreation,** but because many families have stepchildren, adopted children, or no children, this older term is no longer an adequate description.) Most Americans will form families of cohabitation sometime in their lives. Much of our identity is formed in the crucibles of these two families.

In our family of orientation, we are given the roles of son or daughter, brother or sister, stepson or stepdaughter. We internalize these roles until they become a part of our being. In each of these roles, we are expected to act in certain ways. For example, children obey their parents, sons roughhouse with their fathers, daughters imitate their mothers, siblings help one another. Sometimes our feelings fit the expectations of our roles; other times they do not.

Our family roles as offspring and siblings are most important when we are living with our parents, brothers, and sisters in our family of orientation. After we leave home, these roles gradually diminish in everyday significance, although they continue throughout our lives. In relation to our parents, we never cease being children; in relation to our siblings, we never cease being brothers and sisters. The roles change as we grow older. They begin the moment we are born and end only when we die.

As we leave our family of orientation, we usually are also leaving adolescence and entering adulthood. Being an adult in our society is defined in part by entering new family roles—those of husband, wife, or partner, or father or mother. These roles are given to us by our family of cohabitation, and they take priority over the roles we had in our family of orientation. When we marry, we transfer our primary loyalties from our parents and siblings to our partners. Later,

Did you know?

The composition of households has changed significantly. Though families have traditionally accounted for a large majority of all households, their proportion of the total has slipped from 40 percent in 1970 to 25 percent in 1995 (Bryson, 1996; Rawlings, 1994).

Children . . . have no choice about being born into a system; nor do parents have a choice, once children are born, as to the existence of the responsibilities of parenthood. . . . In fact, no family relationships except marriage are entered into by choice.

MONICA MCGOLDRICK AND ELIZABETH CARTER

if we have children, we form additional bonds with them. When we assume the role of husband or wife or bonded partner, we assume an entire new social identity linked with responsibility, work, and parenting. In earlier times such roles were considered to be lifelong in duration. Because of divorce or separation, however, these roles today may last for considerably less time.

The status or place we are given in society is acquired in large part through our families. Our families place us in a certain socioeconomic class, such as blue collar (working class), middle class, or upper class. We learn the ways of our class through identifying with our families. Different classes see the world through different eyes. These differences affect their perceptions of the role of women, how they value education, and how they rear children (Rubin, 1976, 1994).

Our families also give us our ethnic identities as African American, Latino, Jewish, Irish American, Asian American, Italian American, and so forth. Families also provide us with a religious tradition as Protestant, Catholic, Jewish, Greek Orthodox, Islamic, Hindu, or Buddhist—as well as agnostic, atheist, or "New Age." These identities help form our cultural values and expectations.

Why Live in Families?

As we look at the different functions of the family, we can see that theoretically, most of them can be fulfilled outside the family. In terms of reproduction, for example, artificial insemination permits a woman to be impregnated by a sperm donor, and embryonic transplants allow one woman to carry another's embryo. Children can be raised communally, cared for by foster families or child-care workers, or sent to boarding schools. Most of our domestic needs can be satisfied by eating frozen or prepared foods or going to restaurants, sending our clothes to the laundry, and hiring help to clean the bathroom and wash the mountains of dishes accumulating (or growing new life-forms) in the kitchen. Friends can provide us with emotional intimacy, therapists can listen to our problems, and sexual partners can be found outside of marriage. With the limitations and stresses of family life, why bother living in families?

Fidelity is part of every human relationship. It is the strain toward permanence and toward public commitment to permanence that is involved in any relationship beyond the most superficial. Fidelity is a longing for love that does not end.

ANDREW GREELEY

Sociologist William Goode (1982) suggests that there are several advantages to living in families.

- First, families offer continuity as a result of emotional attachments, rights, and obligations. Once we choose a partner or have children, we do not have to search continually for new partners or family members who can perform a family task or function better, such as cook, paint the kitchen, provide companionship, or bring home a paycheck. We expect our family members—whether partner, child, parent, or sibling—to participate in family tasks over their lifetimes. If at one time we need to give more emotional support or attention to a partner or child than we receive, we expect the other person to reciprocate at another time. Or if we ourselves are down, we expect our family to help. We further expect that we can enjoy the fruits of our labors together. We count on our family to be there for us in multiple ways. We rarely have the same extensive expectations of friends.
- Second, families offer close proximity. We do not need to travel across town or cross-country for conversation or help. With families, we do not even need to go out of the house; a husband or wife, parent or child, or brother or sister is often right at hand (or underfoot, in the case of children). This close proximity facilitates cooperation and communication.
- Third, families offer us an abiding familiarity with others. Few people know us as well as our family members, for they have seen us in the most intimate

circumstances throughout much of our lives. They have seen us at our best and our worst, when we are kind or selfish, understanding or intolerant. This familiarity and close contact teach us to make adjustments in living with others. As we do so, we also expand our own knowledge of ourselves and others.

- Fourth, families provide us with many economic benefits. They offer us *economies of scale*. Various activities, such as laundry, cooking, shopping, and cleaning, can be done almost as easily for several people as for one. It is almost as easy to prepare a meal for three people as it is for one, and the average cost per person in both time and money is usually less. As an economic unit, a family can cooperate to achieve what a single individual could not. It is easier for a working couple to purchase a house than a single individual, for example, because the couple can pool their resources.

Because most domestic tasks do not take great skill (a corporate lawyer can mop the floor as easily as anyone else), most family members can learn to do them. As a result, members do not need to go outside the family to hire experts. In fact, for many family tasks, ranging from embracing a partner to bandaging a child's small cut or playing peekaboo with a baby, there are no experts to compete with family members—families tend to do the best job.

These are only some of the theoretical advantages families offer to their members. Of course, not all families perform all these tasks or perform them well. But families, based on mutual ties of feeling and obligation, offer us greater potential for fulfilling our needs than do organizations based on profit (such as corporations) or compulsion (such as governments).

Reflections

As you review the functions of the family—intimacy, economic cooperation, reproduction and socialization, and assignment of social roles and status—examine your own family. How does it fulfill these functions? How does your family structure, such as intact two-parent family, single-parent family, or stepfamily, affect how your family fulfills these functions?

Extended Families and Kinship

Society "created" the family to undertake the task of making us human. According to some anthropologists, the nuclear family of man, woman, and child is universal, either in its basic form or as the building block for other family forms (Murdock, 1967). Other anthropologists disagree that the father is necessary, arguing that the basic family unit is the mother and child dyad, or pair (Collier, Rosaldo, and Yanagisako, 1982). The use of artificial insemination and new reproductive technologies, as well as the rise of female-headed single-parent families, are cited in support of the mother/child model.

The biological family isn't the only important unit in society; we have needs and longings that our families cannot meet. Indeed, in some cultures, community is more important than the family.

GLORIA STEINEM

Extended Families

The extended family, as already described, consists not only of the cohabiting couple and their children but also of other relatives, especially in-laws, grandparents, aunts and uncles, and cousins. In the majority of non-European countries, the extended family is often regarded as the basic family unit.

For many Americans, especially those with strong ethnic identification and those in certain groups (discussed in Chapter 2), the extended family takes on great importance. Sometimes, however, we fail to recognize the existence of extended families because we uncritically accept the nuclear family model as our definition of family. We may even be blind to the reality of our own family structure. When someone asks us to name the members of our families, if we are unmarried, most of us will probably name our parents, brothers, and sisters. If we

PERSPECTIVE

Mapping Family Relationships: The Genogram

We can easily identify family structure in terms of parents, children, grandparents, aunts and uncles, and so on, but it is more difficult to discern the emotional structure of families. These emotional structures are especially important, as our families are among the most important influences in our lives. Many of our patterns of interactions, deepest feelings, relationship expectations, and personal standards reflect our families of orientation.

In recent years, family therapists have been using a device called a *genogram* to look at family relationships. A genogram is a diagram of the emotional relationships of a family through several generations. It is something like a family tree of family interactions. If we can understand the ingrained patterns in our family relationships, we can often understand the nature of our present relationships (Bahr, 1990).

If the marital relationships of our parents and grandparents are marked by conflict, for example, we may unconsciously model our own relationships after theirs. We may experience connection with others through conflict. If their relationships are marked by emotional distance, we may find ourselves distant in our "close" relationships. If their relationships, however, are characterized by warmth and sharing, we may choose similar relationships.

Using the genogram in Figure 1.A as a model, draw a genogram of your own family relationships on a large sheet of paper. Use the symbols in the example to indicate relationships. Your genogram should include the following information about you, your parents, your grandparents, your siblings and your partner and children if you have them.

- Name and date of birth. If deceased, year and cause of death.
- Education and occupation.
- Relevant personal information such as health, personality, and style of relating to others.
- Siblings in each generation.
- Marital history, including years of marriage, divorces, years of singlehood (including single parenting), and dates of remarriage.

FIGURE 1.A
A Model Genogram

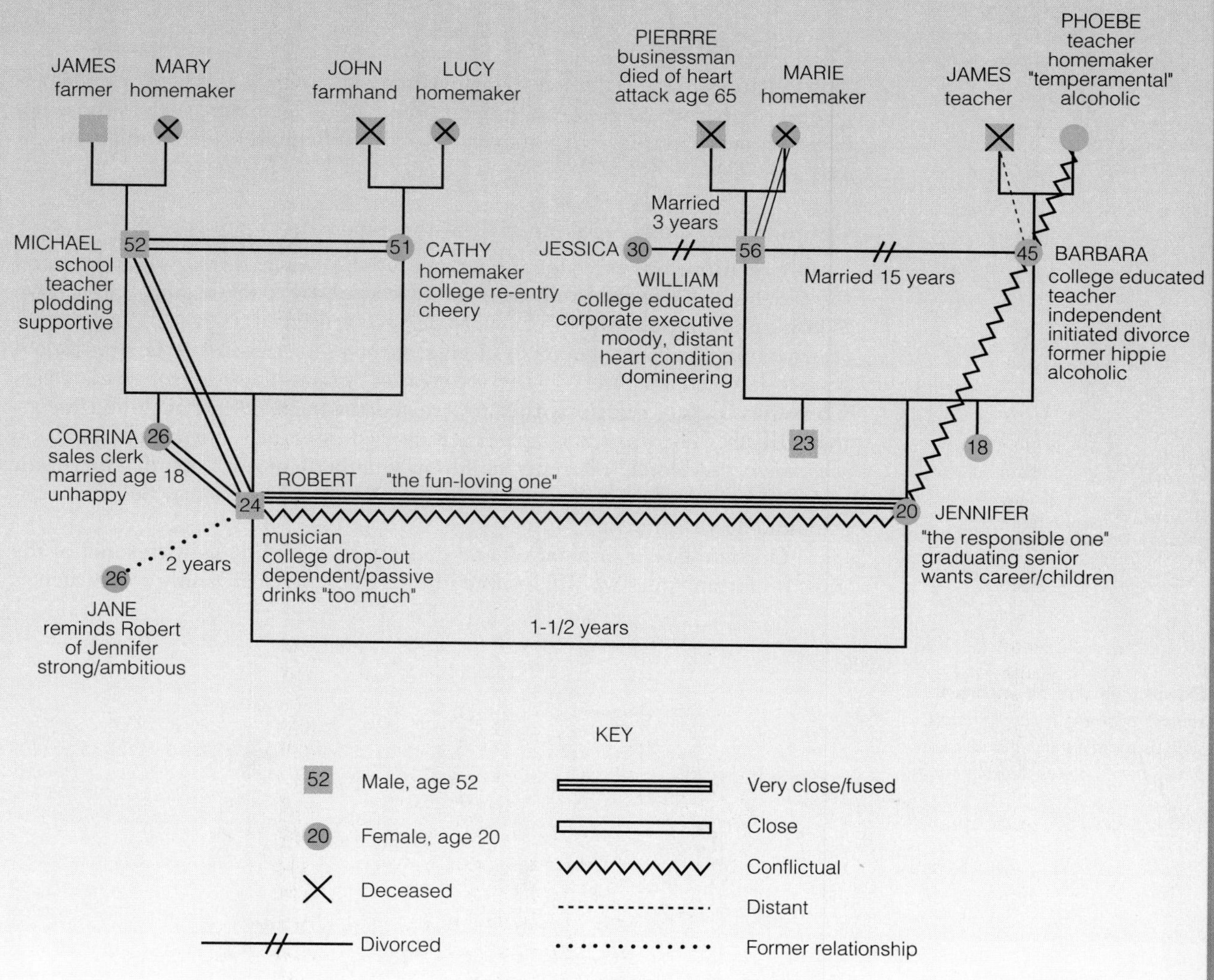
JAMES
farmer
MARY
homemaker
JOHN
farmhand
LUCY
homemaker
PIERRRE
businessman
died of heart
attack age 65
MARIE
homemaker
JAMES
teacher
PHOEBE
teacher
homemaker
"temperamental"
alcoholic
MICHAEL
52
school
teacher
plodding
supportive
51
CATHY
homemaker
college re-entry
cheery
Married
3 years
JESSICA
30
56
WILLIAM
college educated
corporate executive
moody, distant
heart condition
domineering
Married 15 years
45
BARBARA
college educated
teacher
independent
initiated divorce
former hippie
alcoholic
CORRINA
26
sales clerk
married age 18
unhappy
23
18
ROBERT
"the fun-loving one"
24
20
JENNIFER
"the responsible one"
graduating senior
wants career/children
musician
college drop-out
dependent/passive
drinks "too much"
2 years
26
JANE
reminds Robert
of Jennifer
strong/ambitious
1-1/2 years
KEY
52
Male, age 52
20
Female, age 20
Deceased
Divorced
Very close/fused
Close
Conflictual
Distant
Former relationship

are married, we will probably name our husbands or wives and children. Only if questioned further may we include our grandparents, aunts or uncles, cousins, or even friends or neighbors who are "like family." We may not name all our blood relatives, but we will probably name the ones with whom we feel emotionally close, as we saw earlier in the chapter. In 1990, 12 percent of Americans lived in extended-family households, up from 10 percent in 1980. This increase is accounted for by the family structures of immigrants as well as economic necessity (Glick, Bean, and Van Hook, 1997).

Kinship Systems

The **kinship system** is the social organization of the family. It is based on the reciprocal rights and obligations of the different family members, such as those between parents and children, grandparents and grandchildren, and mothers-in-law and sons-in-law.

Conjugal and Consanguineous Relationships

Family relationships are generally created in two ways: through marriage and through birth. Extended family relationships created through marriage are known as **conjugal relationships.** (The word *conjugal* is derived from the Latin *conjungere*, meaning "to join together.") In-laws, such as mothers-in-law, fathers-in-law, sons-in-law, and daughters-in-law, are created by law—that is, through marriage. **Consanguineous relationships** are created through biological (blood) ties—that is, birth. (The word *consanguineous* is derived from the Latin *com-*, "joint," and *sanguineus*, "of blood.") Parents, children, grandparents, and grandchildren, for example, have consanguineous relationships. Aunts and uncles may be either consanguineous or conjugal.

Our families of orientation and cohabitation provide us with some of the most important roles we will assume in life. These nuclear family roles (such as

Extended families are important sources of strength. Family rituals, such as tamale-making in this Latino family, help maintain kinship bonds.

parent/child, husband/wife, and sibling) combine with extended family roles (such as grandparent, aunt/uncle, cousin, and in-law) to form the kinship system.

Kin Rights and Obligations

In some societies, mostly non-Western or nonindustrialized cultures, kinship obligations may be very extensive. In cultures that emphasize kin groups, close emotional ties between a husband and wife are viewed as a threat to the extended family. A remarkable form of marriage that illustrates the precedence of the kin group over the married couple is the institution of spirit marriage, which continues in Canton, China, today. In a **spirit marriage,** according to anthropologist Janice Stockard (1989), a marriage is arranged by two families whose son and daughter died unmarried. After the dead couple is "married," the two families adopt an orphaned boy and raise him as the deceased couple's son to provide family continuity. (See "Other Places/Other Times" in this chapter.) In another Cantonese marriage form, women do not live with their husbands until at least three years after marriage, as their primary obligation remains with their own extended families. Among the Nayar of India, men have a number of clearly defined obligations toward the children of their sisters and female cousins, although they have few obligations toward their own children (Gough, 1968).

The family is a society limited in numbers, but nevertheless a true society, anterior to every state or nation, with rights and duties of its own, wholly independent of the commonwealth.

Pope Leo XIII (1810–1903)

In American society, the basic kinship system consists of parents and children, but it may include other relatives as well, especially grandparents. Each person in this system has certain rights and obligations as a result of his or her position in the family structure. Furthermore, a person may occupy several positions at the same time. For example, an eighteen-year-old woman may simultaneously be a daughter, a sister, a cousin, an aunt, and a granddaughter. Each role entails different rights and obligations. As a daughter, the young woman may have to defer to certain decisions of her parents; as a sister, to share her bedroom; as a cousin, to attend a wedding; and as a granddaughter, to visit her grandparents during the holidays.

Reflections

What are the different kinship roles—such as nephew or niece, aunt or uncle, in-law, grandson or granddaughter—that you play as a result of being a member of your extended family? What rights and obligations does each of these roles entail in your family? Do you have any affiliated kin? How did they become "like family"?

In our own culture, the nuclear family has many norms regulating behavior, such as parental support of children and sexual fidelity between spouses, but the rights and obligations of relatives outside the basic kinship system are less strong and less clearly articulated. Because there are neither culturally binding nor legally enforceable norms regarding the extended family, some researchers suggest that such kinship ties have become more or less voluntary. We are free to define our kinship relations much as we wish. Like friendship, these relations may be allowed to wane (Goetting, 1990).

Despite the increasingly voluntary nature of kin relations, our kin create a rich social network for us. Studies suggest that most people have a large number of kin living in their areas (Mancini and Blieszner, 1989). Adult children and their parents often live close to each other; make regular visits; and help each other with child care, housework, maintenance, repairs, loans, and gifts. The relations between siblings also are often strong throughout the life cycle (Lee, Mancini, and Maxwell, 1990).

The tree of love gives shade to all.

African-American proverb

We generally assume kinship to be lifelong. In the past, if a marriage was disrupted by death, in-laws generally continued to be thought of as kin. But today, divorce is as much a part of the American family system as marriage. Although shunning the former spouse may be no longer be appropriate (or polite), no new guidelines on how to behave have been developed. The ex-kin role is a *role-less role;* that is, it is a role with no clearly defined rules.

An Historical Perspective of Marriage and the Family

American marriages and families have undergone fundamental changes within the last generation (Mintz and Kellogg, 1988). An overview of families in America from the colonial period to the twentieth century may help to provide an historical perspective of the function of tradition, culture, and values and the relationship that exists among them. From this brief history, we hope that connections can be made, genealogy better understood, and most important, an awareness and understanding of our own and other families' histories can be achieved.

The Colonial Era

The colonial era is marked by differences among cultures, family roles, customs, and traditions. These families were the original crucible from which our contemporary families were formed.

Native American Families

The greatest diversity in American family life probably existed during our country's earliest years, when two million Native Americans inhabited what is now the United States and Canada. There were over 240 groups with their own distinct family and kinship patterns. Many groups were **patrilineal:** Rights and property flowed from the father. Others, like the Zuni and Hopi in the southwest and the Iroquois in the northeast, were **matrilineal:** Rights and property descended from the mother.

Native-American families tended to share certain characteristics, although it is easy to overgeneralize. Most families were small. There was a high

In Native-American families, children participated in work from an early age. In this drawing, Iroquois children help with the corn harvest.

child mortality rate, and mothers breastfed their infants; during breastfeeding, mothers abstained from sexual intercourse. Children were often born in special birth huts. As they grew older, the young were rarely physically disciplined. Instead, they were taught by example. Their families praised them when they were good and publicly shamed them when they were bad. Children began working at an early stage. Their play, such as hunting or playing with dolls, was modeled on adult activities. Ceremonies and rituals marked transitions into adulthood. Girls underwent puberty ceremonies at first menstruation. For boys, events such as getting the first tooth and killing the first large animal when hunting signified stages of growing up. A vision quest often marked the transition to manhood.

Man is a history-making creature who can neither repeat his past nor leave it behind.

W.H. AUDEN (1907–1973), *THE DYER'S HAND*

Marriage took place early for girls, usually between the age of twelve and fifteen years; for boys, it took place between the age of fifteen and twenty years. Some tribes arranged marriages; others permitted young men and women to choose their own partners. Most groups were monogamous, although some allowed two wives. Some tribes permitted men to have sexual relations outside of marriage when their wives were pregnant or breastfeeding.

Colonial Families

From earliest colonial times, America has been an ethnically diverse country. In the houses of Boston, the mansions and slave quarters of Charleston, the *maisons* of New Orleans, the *haciendas* of Santa Fe, and the Hopi dwellings of Oraibi (the oldest continuously inhabited place in the United States, dating back to 1150 A.D.), American families have provided emotional and economic support for their members.

We are tomorrow's past.

MARY WEBB (1881–1927)

The Family. Colonial America was initially settled by waves of explorers, soldiers, traders, pilgrims, servants, prisoners, farmers, and slaves. In 1565, at St. Augustine, Florida, the Spanish established the first permanent European settlement in what is now the United States. But the members of these first groups came as single men—as explorers, soldiers, and exploiters.

Colonists who came from Europe preserved their original family systems. They were characterized by Christianity, patriarchy, subordination of women, and family-centered production.

In 1620, the leaders of the Jamestown colony in Virginia, hoping to promote greater stability, began importing English women to be sold in marriage.The European colonists who came to America attempted to replicate their familiar family system. This system, strongly influenced by Christianity, emphasized **patriarchy** (rule by father or eldest male), the subordination of women, sexual restraint, and family-centered production.

The family was basically an economic and social institution, the primary unit for producing most goods and caring for the needs of its members. The family planted and harvested food, made clothes, provided shelter, and cared for the necessities of life. Each member was expected to contribute economically to the welfare of the family. Husbands plowed, planted, and harvested crops. Wives supervised apprentices and servants, kept records, cultivated the family garden, assisted in the farming, and marketed surplus crops or goods, such as grain, chickens, candles, and soap. Older children helped their parents and, in doing so, learned the skills necessary for later life.

Reflections

How far back in history can you trace your family? What would you like to know about your own family's history? What values, traits, or memories do you wish to pass on to your descendants?

As a social unit, the family reared children and cared for the sick, infirm, and aged. Its responsibilities included teaching reading, writing, and arithmetic because there were few schools. The family was also responsible for religious instruction: It was to join in prayer, read scripture, and teach the principles of religion.

Unlike New Englanders, the planter aristocracy that came to dominate the Southern colonies did not give high priority to family life; hunting, entertaining, and politics provided the greatest pleasure. The planter aristocracy continued to idealize gentry ways until the Civil War destroyed the slave system upon which the planters based their wealth.

Marital Choice. Romantic love was not a factor in choosing a partner; one practical seventeenth-century marriage manual advised women that "this boiling affection is seldom worth anything . . ." (Fraser, 1984). Because marriage had profound economic and social consequences, parents often selected their children's mates. Such choices, however, were not as arbitrary as it may seem. Parents tried to choose partners whom their children already knew and with whom they seemed compatible. Children were expected to accept the parents' choices.

Love came after marriage. In fact, it was a person's duty to love his or her spouse. The inability to desire and love a marriage partner was considered a defect of character.

Although the Puritans prohibited premarital intercourse, they were not entirely successful. **Bundling,** the New England custom in which a young man and woman spent the night in bed together, separated by a wooden bundling board, provided a courting couple with privacy; it did not, however, encourage restraint. An estimated one-third of all marriages in the eighteenth century took place with the bride pregnant (Smith and Hindus, 1975).

Family Life. The colonial family was strictly patriarchal. The authority of the husband/father rested in his control of land and property because in an agrarian society like colonial America, land was the most precious resource. The manner in which the father decided to dispose of his land affected his relationships with his children. In many cases, children were given land adjacent to the father's farm, but the title did not pass into their hands until the father died. This power gave father's control over their children's marital choices as well as keeping them geographically close.

This strongly rooted patriarchy called for wives to submit to their husbands. The wife was not an equal, but a "helpmate." This subordination was reinforced by traditional religious doctrine. Like her children, the colonial wife was economically dependent on her husband. Upon marriage, she transferred many of the rights she had held as a single woman to her husband, such as the right to inherit or sell property, to conduct business, and to attend court.

For women, marriage marked the beginning of a constant cycle of childbearing and child rearing. On the average, colonial women had six children and were consistently bearing children until around age forty.

Childhood. The colonial conception of childhood was radically different from ours. First, children, were believed to be evil by nature. The community accepted the traditional Christian doctrine that children were conceived and born in sin.

Second, childhood did not represent a period of life radically different from adulthood. Such a conception is distinctly modern (Aries, 1962; Meckel, 1984; Vann, 1982). In colonial times, a child was regarded as a small adult. When children were six or seven, childhood ended for them. From that time on, they began to be part of the adult world, participating in adult work and play.

Reminiscences make one feel so deliciously aged and sad.

George Bernard Shaw (1856–1950)

Third, when children reached the age of ten, they were often "bound out" for several years as apprentices or domestic servants. They lived in the home of a relative or stranger, where they learned a trade or skill, were educated, and were properly disciplined.

African-American Families

In 1619, a Dutch man-of-war docked at Jamestown in need of supplies. As part of its cargo were twenty Africans who had been captured from a Portuguese slaver. The captain quickly sold his captives as indentured servants. Among these Africans was a woman known by the English as Isabella and a man known as Antony; their African names are lost. In Jamestown, Antony and Isabella married. After several years, Isabella gave birth to William Tucker, the first African-American child born in what is today the United States. William's birth marked the beginning of the African-American family, a unique family system that grew out of the African adjustment to slavery in America.

Strong family ties endured in enslaved African-American families. The extended family, important in West African cultures, continued to be a source of support and stability.

During the eighteenth century and later, West African family systems were severely repressed throughout the New World (Guttman, 1976). At first, some slaves tried unsuccessfully to continue polygamy, which was strongly rooted in many African cultures. Enslaved African Americans were more successful in continuing the traditional African emphasis on the extended family, in which aunts, uncles, cousins, and grandparents played important roles. Although slaves were legally prohibited from marrying, they created their own marriages. Despite the hardships placed on them, they developed strong emotional bonds and family ties. Slave culture discouraged casual sexual relationships and placed a high value on marital stability. On the large plantations, most enslaved people lived in two-parent families with their children. To maintain family identity, parents named their children after themselves or other relatives or gave them African names. In the harsh slave system, the family provided strong support against the daily indignities of servitude. As time went on, the developing African-American family blended West African and English family traditions (see McAdoo, 1996).

Nineteenth-Century Marriages and Families

In the nineteenth century, the traditional colonial family form gradually vanished and was replaced by the modern family.

Industrialization Shatters the Old Family

In the nineteenth century, industrialization transformed the face of America. It also transformed American families from self-sufficient farm families to wage-earning urban families. As factories began producing gigantic harvesters, combines, and tractors, significantly fewer farm workers were needed. Looking for employment, workers migrated to the cities, where they found employment in the ever-expanding factories and businesses. Because goods were now bought rather than made in the home, the family began its shift from being primarily a production unit to being a consumer and service-oriented unit. With this shift, a radically new division of labor arose in the family. Men began working outside the home in factories or offices for wages to purchase the family's necessities and other goods. Men became identified as the family's sole provider or "breadwinner." Their work was given higher status than women's work because it was paid in wages. Men's work began to be identified as "real" work.

To study the history of the American family is to conduct a rescue mission into the dreamland of our national self-concept. No subject is more closely bound up with our sense of a difficult present—and our nostalgia for a happier past.

John Demos

At the same time industrialization made husbands the breadwinner in the family, it also undercut much of their power over their children as fathers. Children were no longer dependent on their fathers.

Industrialization also created the housewife, the woman who remained at home attending to household duties and caring for children. Because much of what the family needed had to be purchased with the husband's earnings, the wife's contribution in terms of *unpaid* work and services went unrecognized, much as it continues today.

Many a family tree needs trimming.

Kin Hubbard (1868–1930)

Marriage and Families Transformed

Without its central importance as a work unit, the family became the focus and abode of feelings.

The Power of Love. This new affectionate foundation of marriage brought love to the foreground. Love as the basis of marriage represented the triumph of individual preference over family, social, or group considerations. Parents had little power in selecting their children's partners, and their children were no longer as economically dependent on them. Women now had a new degree of power; they were able to choose whom they would marry. Women could rule out undesirable partners during courtship; they could choose mates with whom they believed they would be compatible. Mutual esteem, friendship, and confidence became guiding ideals. Without love, marriages were considered empty shells.

The richest love is that which submits to the arbitration of time.

LAWRENCE DURRELL (1912–1990)

Changing Roles for Women. The two most important family roles for middle-class women in the nineteenth century were the housewife and mother roles. As there was a growing emphasis on domesticity in family life, the role of the housewife increased in significance and status. Home was the center of life, and the housewife was responsible for making family life a source of fulfillment for everyone.

Women also increasingly focused their identities on motherhood. The nineteenth century witnessed the most dramatic decline in fertility in American history. Between 1800 and 1900, fertility dropped by 50 percent, falling from an average of seven to about 3.5 children per women. Women reduced their childbearing by insisting that they, not men, control the frequency of intercourse. Child rearing rather than childbearing became one of the most important aspects of a woman's life. Having fewer children allowed more time to concentrate on mothering and opened the door to greater participation in the world outside the family. This outside participation manifested itself in women's heavy involvement in the abolition, prohibition, and women's emancipation movements.

Women have a way of treating people more softly. We treat souls with kid gloves.

SHIRLEY CAESAR

Childhood and Adolescence. A strong emphasis was placed on children as part of the new family. The belief in childhood innocence replaced the idea of childhood corruption. A new sentimentality surrounded the child, who was now viewed as born in total innocence. Protecting children from experiencing or even knowing about the evils of the world became a major part of child rearing.

The nineteenth century also witnessed the beginning of adolescence. In contrast to colonial youths, who participated in the adult world of work and other activities, nineteenth-century adolescents were kept economically dependent and separate from adult activities and often felt apprehensive when they entered the adult world. This apprehension sometimes led to the emotional conflicts associated with adolescent identity crises.

Education also changed as schools, rather than families, became responsible for teaching reading, writing, and arithmetic as well as educating about ideas and values. Conflicts between the traditional beliefs of the family and those of the impersonal school were inevitable. At school, the child's peer group increased in importance.

Ode: Intimations of Immortality from Recollections of Early Childhood

Our birth is but a sleep and a forgetting;
The soul that rises with us, our life's star,
Hath had elsewhere its setting,
And cometh from afar;
Not in entire forgetfulness,
And not in utter nakedness,
But trailing clouds of glory do we come
From God, who is our home.
Heaven lies about us in our infancy;
Shades of the prison-house begin to close
Upon the growing boy,
But he beholds the light, and whence
it flows.
He sees it in his joy;
The youth, who daily farther from the east
Must travel, still is Nature's priest,
And by the vision splendid
Is on his way attended;
At length the man perceives it die away,
And fade into the light of common day.

WILLIAM WORDSWORTH (1770–1850)

The African-American Family: Slavery and Freedom

Although there were large numbers of free African Americans—100,000 in the North and Midwest and 150,000 in the South—most of what we know about the African-American family prior to the Civil War is limited to the slave family.

The Slave Family. By the nineteenth century, the slave family had already lost much of its African heritage. Under slavery, the African-American family lacked two key factors that helped give free African-American and white families stability: autonomy and economic importance. Slave marriages were not recognized as legal. Final authority rested with the owner in all decisions about the lives of slaves. The separation of families was a common occurrence, spreading grief and

Bye baby buntin'
Daddy's gone a-huntin'
Ter fetcha little rabbit skin
Ter wrap de baby buntin' in.

SLAVE LULLABY

despair among thousands of slaves. Furthermore, slave families worked for their masters, not themselves. It was impossible for the slave husband/father to become the provider for his family. The slave women worked in the fields beside the men. When an enslaved woman was pregnant, her owner determined her care during pregnancy and her relation to her infant after birth. The age at which children recognized themselves as slaves depended on whether they were children of field or house slaves. If their parents were field slaves, they would be subject to slave discipline from the beginning. But if the children were offspring of house slaves (or the slave owner), they were often playmates of their master's children. But the day would come when such a child would know that he or she was a slave, and that time would be filled with grieving, anger, and humiliation. The knowledge created a deep crisis in the child's concept of self.

The struggle of man against power is the struggle of memory against forgetting.

MILAN KUNDERA

After Freedom. When freedom came, the formerly enslaved African-American family had strong emotional ties and traditions forged from slavery and from their West African heritage (Guttman, 1976; Lantz, 1980). Because they were now legally able to marry, thousands of former slaves now formally renewed their vows. The first year or so after freedom was marked by what was called "the traveling time," in which African Americans traveled up and down the South looking for lost family members who had been sold. Relatively few families were reunited, although many continued the search well into the 1880s.

African-American families remained poor, tied to the land, and segregated. Despite poverty and continued exploitation, the Southern African-American family usually consisted of both parents and their children. Extended kin continued to be important.

Reflections

As you read through these historical perspectives, what are your feelings about these struggles and triumphs? How does knowledge of your family history affect you, your values, and your behavior?

Immigration: The Great Transformation

The Old and New Immigrants. In the nineteenth and early twentieth centuries, great waves of immigration swept over America. Between 1820 and 1920, 38 million immigrants came to the United States. Historians commonly divided them into "old" immigrants and "new" immigrants. The old immigrants, who came between 1830 and 1890, were mostly from western and northern Europe. During this period, Chinese also immigrated in large numbers to the West Coast. The new immigrants, who came from eastern and southern Europe, began to arrive in great numbers between 1890 and 1914 (when World War I virtually stopped all immigration). Japanese also immigrated to the West Coast and Hawaii during this time. Today, Americans can trace their roots to numerous ethnic groups.

As the United States expanded its frontiers, surviving Native Americans were incorporated. The United States acquired its first Latino population when it annexed Texas, California, New Mexico, and part of Arizona after its victory over Mexico in 1848.

Except for Native Americans, most of us have ancestors who came to American voluntarily or involuntarily. Between 1820 and 1920, more than 38 million immigrants came to the United States.

The Immigrant Experience. Most immigrants were uprooted; they left only when life in the old country became intolerable. The decision to leave their homeland was never easy. It was a choice between life or death and meant leaving behind ancient ties.

Most immigrants arrived in America without skills. Although most came from small villages, they soon found themselves in the concrete cities of America.

But because families and friends kept in close contact even when separated by vast oceans, immigrants seldom left their native countries without knowing where they were going—to the ethnic neighborhoods of New York, Chicago, Boston, San Francisco, Vancouver, and other cities. There they spoke their own tongues, practiced their own religions, and ate their customary foods. In these cities, immigrants created great economic wealth for America by providing cheap labor to fuel growing industries.

In America, kinship groups were central to the immigrants' experience and survival. Passage money was sent to their relatives at home, information was exchanged about where to live and find work, families sought solace by clustering close together in ethnic neighborhoods, and informal networks exchanged information about employment locally and in other areas.

The family economy, critical to immigrant survival, was based on cooperation among family members. For most immigrant families, as for African-American families, the middle-class idealization of motherhood and childhood was a far cry from reality. Because of low industrial wages, many immigrant families could survive only by pooling their resources and sending mothers and children to work in the mills and factories.

Most groups experienced hostility. Crime, vice, and immorality were attributed to the newly arrived ethnic groups; ethnic slurs became part of everyday parlance. Strong activist groups arose to prohibit immigration and promote "Americanism." Literacy tests required immigrants to be able to read at least thirty words in English. In the early 1920s, severe quotas were enacted that slowed immigration to a trickle.

Probe the earth and see where your main roots run.

Henry David Thoreau (1817–1862)

The New Colossus

Not like the brazen giant of Greek fame.
With conquering limbs astride from land to land,
Here at our sea-washed, sunset gates shall stand
A mighty woman with a torch, whose flame
Is the imprisoned lightning, and her name
Mother of Exiles. From her beacon-hand
Glows world-wide welcome; her mild eyes command
The air-bridged harbor that twin cities frame.
"Keep, ancient lands, your storied pomp!" cries she
With silent lips. "Give me your tired, your poor,
Your huddled masses yearning to breathe free,
The wretched refuse of your teeming shore.
Send these, the homeless, tempest-tossed to me,
I lift my lamp beside the golden door."

Emma Lazarus (1849–1887)

Twentieth-Century Marriages and Families

The Rise of Companionate Marriages: 1900–1960

By the beginning of the twentieth century, the functions of American middle-class families had been dramatically altered from earlier times. Families had lost many of their traditional economic, educational, and welfare functions. Food and goods were produced outside the family, children were educated in public schools, and the poor, aged, and infirm were increasingly cared for by public agencies and hospitals. The primary focus of the family was becoming even more centered on meeting the emotional needs of its members.

The New Companionate Family. Beginning in the 1920s, a new ideal family form was beginning to emerge that rejected the "old" family based on male authority and sexual repression. This new family form was based on the **companionate marriage.** There were four major features of this companionate family (Mintz and Kellogg, 1988). First, men and women were to share household decision making and tasks. Second, marriages were expected to provide romance, sexual fulfillment, and emotional growth. Third, wives were no longer expected to be guardians of virtue and sexual restraint. Fourth, children were no longer to be protected from the world but were to be given greater freedom to explore and experience the world; they were to be treated more democratically and encouraged to express their feelings. The heyday of this family was the 1950s.

***Leave It to Beaver*: The Golden Age of the Fifties.** Many of us look back on the fifties as a kind of golden age of the family when the father was the

The past is never dead; it is not even past.

William Faulkner (1897–1962)

breadwinner and the mother was the housewife, as was true in over 70 percent of all families. Marriage and family seemed to be central to American lives. It was a time of youthful marriages, increased birthrates, and a stable divorce rate.

General and uncritical acceptance of traditional gender and marital roles prevailed; man's place was in the world and woman's place was in the home. Women were expected to place motherhood first; they were to be "properly" flattering of men; they were not to excel.

The fifties were also a time of unprecedented prosperity. Real income had risen 20 percent in the fifteen years following World War II. (By contrast, it has fallen almost that same amount since the early 1970s.) This prosperity fueled the movement to the suburbs, where families readily purchased look-alike homes at affordable prices. By 1960, half the population lived in suburbs.

The world has grown suspicious of anything that looks like a happy married life.

Oscar Wilde (1854–1900), *Lady Windermere's Fan*

Suburbanization profoundly affected family life. Most residents were young couples and their children. Few suburban housewives were employed outside the home. Indeed, children seemed to dominate the suburban landscape. To the degree that mothers subordinated themselves to their children, they found themselves isolated from other interests and people. This isolation, coupled with a transient lifestyle occasioned by company moves, made loneliness one of suburbia's compelling problems.

As you begin studying marriage and the family, you can see that such study is both abstract and personal. It is abstract insofar as you learn about the general structure, processes, and meanings associated with marriage and the family. But the study of marriage and the family is also personal: It is *your* present, *your* past, and *your* future that you are studying. It is the family from which you came, the family in which you are now living, and the family that you will create. What R. D. Laing (1972) wrote some years ago may ring true today: "The first family to interest me was my own. I still know less about it than I know about many other families. This is typical. Children are the last to be told what was really going on."

If you don't know where you're going, you will wind up somewhere else.

Yogi Berra

As you continue your study of marriage and the family, much of what was unknown in your own family may become known. You may discover new understanding, strength, complexity, and love.

Summary

- Television cultivates certain images of the family. Until the mid-1980s, the ideal family depicted on television was the nuclear family. Divorces are rare in situation comedies; conflict is easily resolved through manipulation, and problems are solved through humor. Television portrays working-class and middle-class families differently.
- Most family interactions in sitcoms reflect marital stereotypes. Husbands offer advice to their wives, whereas wives look to their husbands for support. Mothers are usually responsible for children, who are well behaved and who ask their parents for permission to do activities. There is relatively little marital sex depicted on television. Sex is usually premarital or extramarital.
- Whites are overrepresented on television, whereas Latinos, Asian Americans, and Native Americans are underrepresented. Until the mid-1980s, African-American TV families were stereotyped negatively; since the mid-1980s, however, such portrayals have improved. A notable feature of several contemporary African-American TV families is that they are extended. Latino families are virtually absent in prime-time network television (but Latinos are major consumers of Spanish-speaking television). Gay and lesbian characters have only recently begun to appear regularly on a broad spectrum of programs.
- *Marriage* is a legally recognized union between a man and a woman in which they are united sexually; cooperate economically; and may give birth to, adopt, or rear children. The union is assumed to be permanent (although in reality it may be dissolved by separation or divorce). Marriage differs among cultures and has changed historically in our own society. Who may marry whom and at what age varies considerably from one society to another. In Western cultures, the preferred form of marriage is *monogamy*, in which there are only two spouses, the husband and wife. *Polygyny*, the practice of having more than one wife, is commonplace throughout many cultures in the world.
- Defining *family* is complex because of the variety of family forms in contemporary America. Most definitions of *family* include individuals who are related by descent, marriage, remarriage, or adoption; some also include affiliative kin. *Family* may be defined as one or more adults related by blood, marriage, or affiliation who cooperate economically, who may share a common dwelling, and who may rear children. There are also ethnic differences as to what constitutes family. Among Latinos, for example, *compadres* (godparents) are considered family members. Among some Japanese Americans, the *ie* is the traditional family. Among many Native American tribes, the *clan* is regarded as the fundamental family unit.
- Four important family functions are (1) the provision of intimacy; (2) the formation of a cooperative economic unit; (3) reproduction and socialization; and (4) the assignment of social roles and status, which are acquired both in our *families of orientation* (in which we grow up) and our *families of cohabitation* (which we form by marrying or living together).
- The connection between marriage and wellness, or optimal health and well-being, is strong. The six components of well-being include physical, emotional, intellectual, spiritual, social, and environmental. All are interrelated and interconnected.
- Advantages to living in families include (1) continuity of emotional attachments; (2) close proximity; (3) familiarity with family members; and (4) economic benefits.
- The *extended family* consists of grandparents, aunts, uncles, cousins, and in-laws. It may be formed *conjugally* (through marriage), creating in-laws or stepkin, or *consanguineously* (by birth), through blood relationships. Extended families are especially important for African-American, Latino, and Asian-American families.
- The *kinship system* is the social organization of the family. In the *nuclear family*, it generally consists of parents and children, but it may also include members of the extended family, especially grandparents, aunts, uncles, and cousins. Kin can be *affiliated*, as when a nonrelated person is considered "as kin," or a relative may fulfill a different kin role, such as a grandmother's taking the role of a child's mother.
- In the early years of colonization, there were two million Native Americans in what is now called the United States. Many of the families were *patrilineal;* rights and property flowed from the father; some other tribal groups were *matrilineal.* Most families were small.
- Diverse groups settled America, including English, Germans, and Africans. In colonial America, marriages were arranged. Marriage was

an economic institution and the marriage relationship was *patriarchal* and companionate. The family was self-sufficient. Women's economic contributions were recognized. Children were also economically important.

- African-American families began in the United States in the early seventeenth century. They continued the African tradition that emphasized kin relations. Most slaves lived in two-parent families that valued marital stability.
- In the nineteenth century, industrialization revolutionized the family's structure; men became wage earners, and women, once they married, became housewives. Childhood was sentimentalized, and adolescence was invented. Marriage was increasingly based on emotional bonds.
- The stability of the African-American enslaved family suffered because it lacked autonomy and had little economic importance. Enslaved families were broken up by slaveholders, and marriage between slaves was not legally recognized. African-American families formed solid bonds nevertheless.
- Thirty-eight million people immigrated to the United States between 1820 and 1920. Most immigrants were uprooted and experienced hostility. Kinship groups were important for survival. The family economy focused on family survival rather than individual success.
- Beginning in the twentieth century, *companionate marriage* became an ideal. Men and women shared household decision making and tasks, marriages were expected to be romantic, wives were expected to be sexually active, and children were to be treated more democratically.
- The 1950s, the golden age of the companionate marriage, was an aberration. It was an exception to the general trend of rising divorce and nontraditional gender roles. Prosperity was unusually high; suburbanization led to increased isolation.

Key Terms

affiliated kin 13
bundling 28
clan 14
companionate marriage 33
concubine 17
conjugal relationship 24
consanguineous relationship 24
cultivation theory 5
family 12
family of cohabitation 19
family of orientation 19
family of origin 19
family of procreation 19
ie 14
kin 13
kinship system 24
marriage 10
matrilineal 26
monogamy 11
nuclear family 14
patriarchy 28
patrilineal 26
polyandry 11
polygamy 11
polygyny 11
spirit marriage 25
traditional family 14
wellness 16

Suggested Readings

Coontz, Stephanie. *The Way We Never Were: American Families and the Nostalgia Trap.* New York: Basic Books, 1992. A historian's view that contrasts the pop images conveyed through the media with the realities of American families since the 1950s.

Dines, Gail, and Jean Humez, eds. *Gender, Race and Class in the Media.* Thousand Oaks, CA: Sage Publications, 1995. A good introduction to popular culture and the media.

Dinnerstein, Leonard, and David Reimers. *Ethnic Americans: A History of Immigration,* 4th ed. Reading, MA: Addison-Wesley, 1990. An excellent history of immigration, assimilation, and family life from colonial times to the mid-1980s.

Hutter, Mark. *The Changing Family: Comparative Perspectives.* New York: Macmillan, 1988. A cross-cultural comparison of developing nations and American family systems (including African-American, Latino, and Japanese-American families).

Lull, James. *Inside Family Viewing: Ethnographic Research on Television's Audiences.* New York: Routledge, 1990. A scholarly, eye-opening study of how families watch television, including family interaction and television, the social uses of television, family power and dominance, and choosing TV programs.

Marciano, Teresa, and Marvin Sussman. *Wider Families: New Traditional Family Forms.* New York: Haworth Press, 1991. A collection of scholarly essays that challenges the traditional concept of family. It includes essays on close relationships, gay and lesbian relationships, communes, and extended family networks.

Real, Michael R. *Explaining Media Culture: A Guide.* Thousand Oaks, CA: Sage, 1996. A scholarly, insightful, and thought-provoking examination of the role of media in modern life.

Skolnik, Arlene. *Embattled Paradise: The American Family in an Age of Uncertainty.* New York: Basic Books, 1992. A thoughtful discussion of contemporary family diversity, pluralistic values, and the political debate about the family.

Tichi, Cecelia. *Electronic Hearth: Creating an American Television Culture.* New York: Oxford University Press, 1991. How television has replaced the fireplace hearth as the centerpiece of family life. The book discusses TV symbols and images in defining American values.

Tremblay, Helène. *Families of the World: Family Life at the End of the 20th Century.* 2 vols. New York: Farrar, Straus, and Giroux. Vol. 1, *The Americas and the Caribbean* (1988); Vol 2, *Asia and the Pacific Islands* (1990). Beautiful color photographs and stories of families around the world.

Twitchell, James, B. *Carnival Culture: The Trashing of Taste in America.* New York: Columbia University Press, 1992. An amusing, informative, and curmudgeonly exploration of popular culture in America by an insightful critic and self-described former teenager. This is an excellent introduction to the study of popular culture.

Zillman, D., J. Bryant, and A.C. Huston, eds. *Media, Children, and the Family: Social, Scientific, Psychodynamic, and Clinical Perspectives.* Hillsdale, NJ: Erlbaum, 1994. One of the key books on TV reality, with essays on gender, social behaviors, and family portrayals.

CHAPTER 2

Studying Marriage and the Family

P R E V I E W

To gain a sense of what you already know about the material covered in this chapter, answer "True" or "False" to the statements below.

1. Much of what you read or view is meant to entertain you. True or false?
2. The statement "Everyone should get married" is an example of an objective statement. True or false?
3. Many researchers believe that love and conflict can coexist. True or false?
4. Stereotypes about families, ethnic groups, and gays and lesbians are easy to change. True or false?
5. Over the past twenty-five years, the sense of ethnicity among Americans of European descent has been increasing. True or false?
6. In day-to-day living, women in Latino families tend to have substantial power. True or false?
7. Native Americans more often have stronger tribal identities, such as Navajo or Sioux, than ethnic identities, such as Native American or American Indian. True or false?
8. Most knowledge that we have about families comes from clinical sources in which therapists intervene in family problems. True or false?
9. A belief that one's own ethnic group, nation, or culture is innately superior to another is an example of an ethnocentric fallacy. True or false?
10. According to some scholars, in marital relationships we tend to weigh the costs against the benefits of the relationship. True or false?

Answers

1 True, 2 False, 3 True, 4 False, 5 True, 6 True, 7 True, 8 False, 9 True, 10 True

OUTLINE

As you read this chapter, it will be a somewhat different experience than other reading you might do in a scientific textbook. The difference is that you are already an expert on the subject as you begin to read. "How did I become an expert?" you may ask. You became an expert through the many years of growing up and living in your own family. You have gained authority in thinking about family because of the continuing series of REAL experiences you have had as a family member. The study of family is not about an abstract concept; it is about your present day-by-day life situation.

The knowledge you have about family is vividly real. However, it is also unique. No other family is exactly like your family. It is good to know about our own experiences, but effective planning and decision making require that we also have a larger background of information. That is the value of the study of family: It enables us to learn about the family experiences of other people. The knowledge of what other families experience and the results of different kinds of responses to family situations enables us to have a more informed understanding of families in general. Such understanding is helpful for persons in terms of their individual lives; it is also helpful for social service, medical, and legal personnel as they deal with family-related issues.

This chapter will help you develop an understanding of the way information about different kinds of families, and families in different circumstances, is obtained. We explore some of the tools and methods of research. We look briefly at two instruments, the "Marriage and Family Life Attitude Survey" (Understanding Yourself, pages 48–49) and the "Family Circumplex Model," (You And Your Well-Being, Wellness page 60) that illustrate ways in which information is obtained and put to use. We also review current theories that help us understand marriage and the family from various perspectives. We look at research issues that are relevant to the study of America's diverse ethnic groups. Finally, we take a brief overview of contemporary families, setting the stage for the in-depth exploration of the chapters that follow.

Thinking Critically about Marriage and the Family

Before we examine the research tools, it is important to emphasize that the attitudes of the researcher (or you, as you read research) are very important. In order to obtain valid research information we need to keep in mind the rules of critical thinking. The term *critical thinking* is another way of saying "clear and unbiased thinking."

As a result of our experiences, we each have our own perspectives, values, and beliefs regarding marriage, family, and relationships. These can create blinders that keep us from accurately understanding the research information. We need to develop a sense of **objectivity** in our approach to information (Kitson, Clark, Rushforth, Brinich, Sudak, and Zyzanski, 1996). This means we need to suspend the beliefs, biases, or prejudices we have about a subject until we really understand what is being said. Then we can take that information and relate it to the information and attitudes we already have. Out of this process a new and enlarged perspective may emerge.

test

One area in which we may need to be alert to maintaining an objective approach is that of family lifestyle. The **values** we have about what makes a successful family can cause us to decide ahead of time that certain family lifestyles are "abnormal" because they differ from our own experience or preference. We may refer to single-parent families as "broken" or say that adoptive parents are "not the *real* parents."

A clue that can sometimes help us "hear" ourselves and detect whether we are making **value judgments** or **objective statements** is as follows: A value judgment usually includes words that mean "should" and imply that *our* way is the correct way. An example is, "Everyone *should* get married." An objective statement presents information based on scientifically measured findings, such as, "About 90 percent of Americans marry."

Opinions, biases, and stereotypes are ways of thinking that lack objectivity. **Opinions** are based on our own experiences or ways of thinking. **Biases** are strong opinions that may create barriers to hearing anything that is contrary to our opinion. **Stereotypes** are sets of simplistic, rigidly held, and overgeneralized beliefs about the personal characteristics of a group of people. They form the "glasses" with which we "see" people and groups. Stereotypes are resistant to change. Furthermore, stereotypes are often negative. Common stereotypes related to marriages and families include the following:

We learn about families in many ways, including our own experience, the media, and the efforts of family researchers. The most common research method is the survey—using questionnaires or interviews.

- Nuclear families are best.
- Stepfamilies are unhappy.
- Lesbians and gay men cannot be good parents.
- Latino families are poor.
- Husbands are henpecked.

Instead of being presented with stereotypes by age, sex, color, class, or religion, children must have the opportunity to learn that within each range, some people are loathsome and some are delightful.

MARGARET MEAD (1901-1978)

We all have opinions and biases; most of us, to varying degrees, think stereotypically. But the commitment to objectivity requires us to become aware of these opinions, biases, and stereotypes and to put them aside in the pursuit of knowledge.

Fallacies are errors in reasoning. These mistakes come as the result of errors in our basis presuppositions. Two common types of fallacies are egocentric fallacies and ethnocentric fallacies. The **egocentric fallacy** is the mistaken belief that everyone has the same experiences and values that we have and therefore should think as we do. The **ethnocentric fallacy** is the belief that one's own ethnic group, nation, or culture is innately superior to others. We will discuss this fallacy in some detail later in this chapter as we consider the differences and strengths of families from different ethnic backgrounds.

Norms and Popular Culture

We all know we are unique individuals, but we tend to see others as representatives of a group.

DEBORAH TANNEN

Apart from our own families, popular culture may be the most important vehicle by which **norms**—cultural rules and standards—and values about marriage and family are transmitted. But media-transmitted norms and values are often disguised as information and entertainment.

Advice/Information Genre as Entertainment

Everybody gets so much information all day long that they lose their common sense.

GERTRUDE STEIN (1874–1946)

We saw in Chapter 1 that the mass media convey numerous images of marriages, families, and relationships. There are a variety of forms of media communication. In addition to television, film, popular music and advertising, there is another form of media communication about families that we might call the **advice/information genre,** which transmits norms and information. A veritable industry exists to support the advice/information genre. It produces self-help and child-rearing books, advice columns, radio and television shows, and numerous articles in magazines and newspapers.

The advice/information genre ostensibly is concerned with transmitting information that is factual and accurate. In newspapers the genre is represented by such popular columnists as Abigail Van Buren (whose column "Dear Abby" is now written by a man), Ann Landers, Beth Winship, and Pat Califia. Self-help and pop psychology books written by "experts" frequent the best-seller lists, ranging from *The Power Principle* to *The Pleasure Prescription* to *The Rules.*

The purpose of television is to keep you watching television long enough to see the advertising.

JAMES B. TWITCHELL

These books, columns, articles, and programs share several features. First, their primary purpose is to sell books or periodicals or to raise program ratings. This is in marked contrast to scholarly research, whose primary purpose is the pursuit of knowledge. Even the inclusion in magazines of survey questionnaires asking readers about their relationships or behaviors is ultimately designed to promote sales. We fill out the questionnaires for fun, much as we would crossword puzzles. Then we buy the subsequent issue or watch a later program to see how we compare with others.

Second, the media must entertain while disseminating information about marriage, family, and relationships. Because the genre seeks to entertain, the information and advice must be simplified. Complex explanations and analyses must be avoided because they would interfere with the entertainment purpose. Furthermore, the genre relies on high-interest or shocking material to attract readers or viewers. Consequently, we are more likely to read or view stories about finding the perfect mate or protecting our children from strangers than stories about new research methods or the process of gender stereotyping.

TV seeps into your brain when you're not looking. Those fantasy values you laugh at? If you're not careful they take over. A daily dose of sexism and stupidity has a surprising effect on the brain.

ALICE HOFFMAN

Third, the advice/information genre focuses on how-to-do-it information or morality. The how-to-do-it material advises us on how to improve our relationships, sex lives, child-rearing abilities, and so on. Advice and normative judgments (evaluations based on norms) are often mixed together. Advice columnists often give advice on issues of sexual morality: "Is it all right to have sex without commitment?" ("Yes, if you love him/her," "No, casual sex is empty," and so on.) Advice columnists act as moral arbiters, much as do ministers, priests, rabbis, and other religious leaders.

Fourth, the genre uses the trappings of social science without its substance. Writers and columnists interview social scientists and therapists to give an aura of scientific authority to their material. They rely especially heavily on therapists with clinical rather than academic backgrounds. Because clinicians tend to deal with people with problems, they often see relationships as problematical.

He who knows nothing doubts nothing.

FRENCH PROVERB

To reinforce their authority, the media also incorporate statistics, which are key features of social science research. But as Susan Faludi (1991) notes,

> The statistics that the popular culture chooses to promote most heavily are the very statistics we should view with the most caution. They may well be in wide circulation not because they are true but because they support widely held media preconceptions.

Evaluating the Advice/Information Genre

After you have read a certain number of advice books on marriage and the family or watched so many experts on television, you discover that they tend to be repetitive. There are several reasons for this. First, the media repeatedly report more or less the same stories because there is relatively little new in the world of family research. Scientific research is a painstakingly slow and tedious process. Research results rarely dramatically change the way we view a topic; usually they flesh out what we already know. Although research is seldom revolutionary, the media must nevertheless continually produce new stories to fill their pages and programs. Consequently, they report similar material in different guises: as interviews, survey results, and first-person stories, for example.

The American people don't believe anything's real until they see it on television.

RICHARD M. NIXON (1913–1994)

Second, the media are repetitive because their scope is narrow. There are only so many ways how-to-do-it books can tell you how to do it. Similarly, most of us are remarkably similar in the personal and moral dilemmas we face: Am I normal? How can I improve my relationship? Is sex without love moral?

With the media awash in advice and information about relationships, marriage, and family, how can we evaluate what is presented to us? Here are some guidelines:

Sometimes I've believed as many as six impossible things before breakfast.

LEWIS CARROLL (1832–1898)

- *Be skeptical.* Remember: much of what you read or see is meant to entertain you. Are the sources scholarly or popular? Do they rely on self-described

Knowledge is of two kinds. We know a subject ourselves, or we know where we can find information upon it.

JAMES BOSWELL (1740–1795) *LIFE OF JOHNSON*

"experts" or "victims"? How representative are the people interviewed? If the story seems superficial, it probably is.

- *Search for biases, stereotypes, and lack of objectivity.* Information is often distorted by points of view. What conflicting information may have been omitted? How are women and members of ethnic groups portrayed? Does the media's idea of family include diverse family forms? Are nontraditional families stigmatized?
- *Look for moralizing.* Many times what passes as fact is really disguised moral judgment. What are the underlying values of the article or program?
- *Go to the original source or sources.* The media simplifies. Find out for yourself what the studies really said. How valid were their methodologies? What were their strengths and limitations?
- *Seek additional information.* The whole story is probably not told. Look for additional information in scholarly books and journals, reference books, or college textbooks.

In all matters of opinion, our adversaries are insane.

MARK TWAIN [SAMUEL CLEMENS] (1835–1910)

Research Methods

Scholarly research about the family brings together information and formulates generalizations about certain areas of experience. These generalizations help us to predict what happens when certain conditions or actions occur. As we have mentioned, from the day of your birth you have been forming impressions about human relationships and developing ways of behaving based on these impressions. Our personal ongoing research process is vital to our effective coping with life and to understanding our life's meaning. There are some limitations, however, to our individualized research process. These limitations may be described as having too small a "database" and insufficient "comparison data."

Students often feel a sense of "been there, done that" as they read about an aspect of personal development or family life. They may find it difficult to become interested in obtaining further background and insights. However, we would like to challenge you to remember that you are in a continuing "research mode" throughout life. You are constantly forming new insights and remodeling and reforming your thinking and behaviors. Your study of the information in this text will provide you the opportunity to consider your present attitudes and past experiences and relate them to the experiences of others. As you do this you will be able to use the logic and problem-solving skills of critical thinking so that you can effectively apply that which is relevant to your life.

Rather than love, than money, than fame, give me truth.

HENRY DAVID THOREAU (1817–1862)

Family science researchers use the **scientific method**—well-established procedures used to collect information about family experiences. Then, with scientifically accepted techniques, they analyze this information in a way that allows other people to know the source of the information and to be confident of the accuracy of the findings.

Some researchers ask the same questions of great numbers of people. They collect information from people of different age, sex, living situation, and ethnic backgrounds. This is known as "representative sampling." In this way researchers can discover whether age or other background characteristics influence peoples' responses. This approach to research is called **quantitative** research because it deals with large quantities of information. Survey research and experimental research (discussed in the following sections) are examples of quantitative research.

Other researchers study smaller groups or sometimes individuals in a more in-depth fashion. They may place observers in family situations, conduct intensive interviews, do case studies involving information provided by several people, or analyze letters or diaries or other records of people whose experiences represent special aspects of family life. This form of research is known as **qualitative** research because it is concerned with a detailed understanding of the object of study. The sections on case methods and observational research are examples of qualitative research (Ambert, Adler, and Detzner, 1995).

What is morality at any given time or place? It is what the majority then and there happens to like and immorality is what they dislike.

Alfred North Whitehead (1861–1947)

In addition to using information provided specifically by people participating in a research project, researchers also utilize information from public sources. This research is called **secondary data analysis.** It involves the reanalyzing of data originally collected for another purpose. Examples are analyzing U.S. Census data and official statistics, such as state marriage, birth, and divorce records. Secondary data analysis also includes content analysis of various communication media such as newspapers, magazines, letters, and television programs.

Family science researchers conduct their investigations using ethical guidelines agreed upon by professional researchers. These guidelines protect the privacy and safety of people who provide information in the research. For example, any research conducted with college students requires the investigator to present the plan and method of the research to a "human subjects review committee." This assures that students are participating voluntarily and their privacy is protected. Research ethics also require researchers to conduct their studies and report their findings in ways that assure readers of the trustworthiness of their reports.

Nothing is so firmly believed as that which is least known.

Michel de Montaigne (1533–1592)

What researchers know about marriage and the family comes from four basic research methods: (1) survey research, (2) clinical research, (3) observational research, and (4) experimental research. There is a continual debate as to which method is best for studying marriage and the family. But such arguments may miss an importrant point: Each method may provide important and unique information that another method may not (Cowan and Cowan, 1990).

Survey Research

The **survey research** method, using questionnaires or interviews, is the most popular data-gathering technique in marriage and family studies. Surveys may be conducted in person, over the telephone, or by written questionnaires. The purpose of survey research is to gather information from a small, representative group of people and to infer conclusions that are valid for a larger population. Questionnaires offer anonymity, may be completed fairly quickly, and are relatively inexpensive to administer. Questionnaires usually do not allow for an in-depth response, however; a person must respond with a short answer, a yes, a no, or a choice on a scale of 1 to 10. Unfortunately, marriage and family issues are generally too complicated for questionnaires to explore in depth.

Lies, damned lies and statistics.

Benjamin Disraeli (1804-1881)

Interview techniques avoid some of the shortcomings of questionnaires because interviewers are able to probe in greater depth and follow paths suggested by the interviewee. Interviewers, however, may allow their own preconceptions to bias their interpretation of responses, as well as the way in which they frame questions.

Whether done by questionnaires or interviews, surveys have certain inherent problems. First, how representative is the sample (the chosen group) that volunteered to take the survey? Self-selection (volunteering to participate) tends to bias a sample. Second, how well do people understand their own behavior? Third, are people underreporting undesirable or unacceptable behavior? They may be reluctant to admit that they have extramarital affairs or that they are alcoholics, for example.

Surveys are well suited for determining the incidence of certain behaviors or for discovering traits and trends. Surveys are more commonly used by sociologists than by psychologists, because they tend to deal on a general or societal level rather than on a personal or small-group level. But surveys are not able to measure very well how people interact with each other. For researchers and therapists interested in studying the dynamic flow of relationships, surveys are not as useful as clinical, experimental, and observational studies.

Better to doubt what is obscure than argue about uncertainties.

AUGUSTINE, BISHOP OF HIPPO (354–430)

Clinical Research

Clinical research involves in-depth examination of a person or a small group of people who come to a psychiatrist, psychologist, or social worker with psychological or relationship problems. The **case-study method,** consisting of a series of individual interviews, is the most traditional approach of all clinical research; with few exceptions, it was the sole method of clinical investigation through the first half of the twentieth century (Runyan, 1982). The advantage of clinical approaches is that they offer long-term, in-depth study of various aspects of marriage and family life. The primary disadvantage is that we cannot necessarily make inferences from them about the general population. People who enter psychotherapy are not a representative sample. They may be more motivated to solve their problems or have more intense problems than the general population (Kitson, et al., 1996).

Clinical studies, however, have been very fruitful in developing insight into family processes. Such studies have been instrumental in the development of family systems theory, discussed later in this chapter. By analyzing individuals and families in therapy, psychiatrists, psychologists, and therapists such as R. D. Laing, Salvador Minuchin, and Virginia Satir have been able to understand how families create roles, patterns, and rules that family members follow without being aware of them.

Observational Research

Observational and experimental studies (discussed in the next section) account for less than 5 percent of recent research articles (Nye, 1988). In **observational research,** scholars attempt to study behavior systematically through direct observation while remaining as unobtrusive as possible. To measure power in a relationship, for example, an observer-researcher may sit in a home and videotape exchanges between a husband and wife. The obvious disadvantage of this method is that the couple may hide unacceptable ways of dealing with decisions, such as threats of violence, while the observer is present.

You can observe a lot just by watching.

YOGI BERRA

Another problem with observational studies is that a low correlation often exists between what observers see and what the people observed report about themselves (Bray, 1995). Researchers (Jacob, Tennenbaum, Seilhamer, Bargiel, and Sharon, 1994) have suggested that self-reports and observations really measure two different views of the same thing: A self-report is an insider's view, whereas an observer's report is an outsider's view. Some observational research involves family members being given structured activities to carry out. These activities will involve interaction that can be observed between family members (Milner and Murphy, 1995). They may include problem-solving tasks, putting together puzzles or games, or responding to a contrived family dilemma. Different tasks are intended to elicit different types

Seeing is believing. I wouldn't have seen it if I hadn't believed it.

ASHLEIGH BRILLIANT

What really teaches a man is not experience but observation.

H. L. MENCKEN (1880–1956)

of family interaction, which will provide the researchers opportunities to observe behaviors of interest.

Discovery consists of seeing what everybody has seen and thinking what nobody has thought.

ALBERT SZENT-GYORGYI (1893–1986)

Experimental Research

In **experimental research,** researchers isolate a single factor under controlled circumstances to determine its influence. Researchers are able to control their experiments by using **variables,** aspects or factors that can be manipulated in experiments. There are two types of variables: independent and dependent variables. **Independent variables** are factors that can be manipulated or changed by the experimenter; **dependent variables** are factors that are affected by changes in the independent variable.

Because it controls variables, experimental research differs from the previous methods we have examined. Clinical studies, surveys, and observational research are correlational in nature. **Correlational studies** measure two or more naturally occurring variables to determine their relationship to each other. Because correlational studies do not manipulate the variables, they cannot tell us which variable causes the other to change. But because experimental studies manipulate the independent variables, researchers can reasonably determine which variables affect the other variables.

Every truth is true only up to a point. Beyond that, by way of counterpoint, it becomes untruth.

SOREN KIERKEGAARD (1813–1855)

Experimental findings can be very powerful because such research gives investigators control over many factors and enables them to isolate variables. Researchers believing that stepmothers and stepfathers are stigmatized, for example, tested their hypothesis experimentally (Ganong, Coleman, and Kennedy, 1990). They devised a simple experiment in which subjects were asked to evaluate twenty traits of a person in a family who was described in a short paragraph. The person was variously identified as a father or mother in a nuclear family, a biological father or mother in a stepfamily, or a stepfather or stepmother in a stepfamily. When identified as a biological parent in either a nuclear family or a stepfamily, the individual was rated more favorably than when identified as a stepfather or stepmother. This paper-and-pencil experiment confirmed the researchers' hypothesis that stepparents are stigmatized.

The obvious problem with such studies is that we respond differently to people in real life than we do in controlled situations, especially in paper-and-pencil situations. We may not stigmatize a stepparent at all in real life. Experimental situations are usually faint shadows of the complex and varied situations we experience in the real world.

Discovery comes from dialogue that starts with the sharing of ignorance.

MARSHALL MCLUHAN (1911–1980)

Differences in sampling and methodological techniques help explain why studies of the same phenomenon may arrive at different conclusions. They also help explain a common misperception many of us hold regarding scientific studies. Many of us believe that because studies arrive at different conclusions, none are valid. What conflicting studies may show us, however, is that researchers are constantly exploring issues from different perpectives as they attempt to arrive at a consensus.

The fact will one day flower into a truth.

HENRY DAVID THOREAU (1817–1862)

Researchers may discover errors or problems in sampling or methodology that lead to new and different conclusions. They seek to improve sampling and methodologies in order to elaborate on or disprove earlier studies. In fact, the very word *research* is derived from the prefix *re-*, meaning "over again," and *search*, meaning "to examine closely." And that is the scientific endeavor: searching and re-searching for knowledge.

UNDERSTANDING YOURSELF

What Do Surveys Tell You about Yourself?

Survey questionnaires are the leading source of information about marriage and the family. The questionnaire below, called "The Marriage and Family Life Attitude Survey," was developed by Don Martin to gain information about attitudes toward marriage and the family. On a scale of 1 to 5, indicate for each statement below whether you strongly agree (1), slightly agree (2), neither agree nor disagree (3), slightly disagree (4), or strongly disagree (5).

1	2	3	4	5
Strongly Agree	Slightly Agree	Neither Agree Nor Disagree	Slightly Disagree	Strongly Disagree

I *Cohabitation and Premarital Sexual Relations*

1 I have or would engage in sexual intercourse before marriage.
2 I believe it is acceptable to experience sexual intercourse without loving one's partner.
3 I want to live with someone before I marry him/her.
4 If I lived intimately with a member of the opposite sex, I would tell my parents.

II *Marriage and Divorce*

5 I believe marriage is a lifelong commitment.
6 I believe divorce is acceptable except when children are involved.
7 I view my parents' marriage as happy.
8 I believe I have the necessary skills to make a good marriage.

III *Childhood and Child Rearing*

9 I view my childhood as a happy experience.
10 If both my spouse and I work, I would leave my child in a day-care center while at work.
11 If I have a child, I feel only one parent should work so the other can take care of the child.
12 The responsibility for raising a child is divided between both spouses.
13 I believe I have the knowledge necessary to raise a child properly.
14 I believe children are not necessary in a marriage.
15 I believe two or more children are desirable for a married couple.

IV *Division of Household Labor and Professional Employment*

16 I believe household chores and tasks should be equally shared between marital partners.
17 I believe there are household chores that are specifically suited for men and others for women.
18 I believe women are entitled to careers equal to those of men.
19 If my spouse is offered a job in a different locality, I will move with my spouse.

Theories of Marriage and the Family

Facts do not cease to exist because they are ignored.

ALDOUS HUXLEY (1894–1963)

We use surveys, clinical studies, observations, and experiments to gather data on marriages and families. A **theory** is a set of general principles or concepts used to explain data and to make predictions that may be empirically (experimentally) tested (Miller, 1986). Theories are also important because they can suggest directions for research according to the questions they raise.

The theories we discuss in this section are symbolic interaction, social exchange, structural functionalism, conflict, and family systems. We also look at the influence of the feminist perspective on family studies. These theories are currently among the most influential ones used by sociologists and psychologists.

It is a capital mistake to theorize before one has data.

SHERLOCK HOLMES

As you study these different theories, notice how the choice of a theoretical perspective influences the way data are interpreted. As you read this textbook, ask yourself how a different theoretical perspective would lead to different conclusions about the same material. (For deeper exploration of family theories, see Klein and White, 1996).

Symbolic Interaction Theory

Symbolic interaction is a theory that looks at how people interact with each other. Symbolic interactionists, like the rest of us, are concerned with relation-

V *Marital and Extramarital Sexual Relations*

20 I believe sexual relations are an important component of a marriage.

21 I believe the male should be the one to initiate sexual advances in a marriage.

22 I do not believe extramarital sex is wrong for me.

VI *Privacy Rights and Social Needs*

23 I believe friendships outside of marriage with the opposite sex are important in a marriage.

24 I believe the major social functioning in a marriage should be with other couples.

25 I believe married couples should not argue in front of other people.

26 I want to marry someone who has the same social needs as I have.

VII *Religious Needs*

27 I believe religious practices are important in a marriage.

28 I believe children should be made to attend church.

29 I would not marry a person of a different religious background.

VIII *Communication Expectations*

30 When I have a disagreement in an intimate relationship, I talk to the other person about it.

31 I have trouble expressing what I feel toward the other person in an intimate relationship.

32 When I argue with a person in an intimate relationship, I withdraw from that person.

33 I would like to learn better ways to express myself in a relationship.

IX *Parental Relationships*

34 I would not marry if I did not get along with the other person's parents.

35 If I do not like my spouse's parents, I should not be obligated to visit them.

36 I believe each spouse's parents should be seen an equal amount of time.

37 I feel parents should not intervene in any matters pertaining to my marriage.

38 If my parents did not like my choice of a marriage partner, I would not marry this person.

X *Professional Counseling Services*

39 I would seek premarital counseling before I got married.

40 I would like to attend marriage enrichment workshops.

41 I will seek education and/or counseling in order to learn about parenting.

42 I feel I need more education of what to expect from marriage.

43 I believe counseling is only for those couples in trouble.

After you have completed this questionnaire, ask yourself the questions below:

- Were the questions correctly posed so that your responses adequately portrayed your attitudes?
- Were questions omitted that are important for you regarding marriage and the family? If so, what were they?
- Do your attitudes reflect your actual behavior?

ships. When we feel that our partner *really* understands us (or doesn't), that we communicate well (or don't), that we live in harmony with each other (or don't), we are expressing feelings that are at the heart of symbolic interaction research. Symbolic interactionists study the interactions that make up a relationship.

Knowledge rests not upon truth alone, but on error also.

Carl Jung (1875–1961)

An **interaction** is a reciprocal act. Interactions are the everyday words and actions that take place *between* people. For an interaction to occur, there must be at least two people who both act and respond to each other. When you ask someone to pass the potatoes *and* he or she does it, an interaction takes place. Even if the person intentionally ignores you or tells you to get the potatoes yourself, an interaction occurs (even if it is not a positive one). Such interactions are conducted through **symbols,** words or gestures that stand for something else. When we interact with people, we do more than simply react to them. We interpret or define their symbols. If someone didn't respond to your request for the potatoes, what did the nonresponse mean or symbolize? Hostility? Rudeness? Indifference? A hearing problem? We interpret the meaning and act accordingly. If we interpret the nonresponse as not hearing, we may repeat the request. If we believe it symbolizes hostility, rudeness, or indifference, we may become angry.

Family as the Unity of Interacting Personalities

In the 1920s, Ernest Burgess (1926) defined the family as a "unity of interacting personalities." This definition has been critical to symbolic interaction theory and in the development of marriage and family studies. Marriages and families consist of individuals who interact with one another over a period of time. Over time, our interactions and relationships define the nature of our family: a loving family, a dysfunctional family, a conflict-ridden family, an emotionally distant family, a high-achieving family, and so on.

In marital and family relationships, our interactions are partly structured by our social roles. (A **social role** is an established pattern of behavior that exists independently of a person, such as the role of wife or husband existing independently of any particular husband or wife.) Each member in a marriage or family has one or more roles—such as husband, wife, mother, father, child, or sibling. These social roles help give us cues as to how we are supposed to act. They help create a "marriage," "family," or other intimate relationship. When we marry, for example, these roles help us "become" wives and husbands; when we have children, they help us "become" mothers and fathers.

Symbolic interactionists study how the sense of self is maintained in the process of acquiring these roles. We are, after all, more than simply the roles we fulfill. There is a core self that is independent of our being a husband or wife, father or mother, son or daughter. Symbolic interactionists ask how we fulfill our roles and continue to be ourselves. And at the same time, how do our roles contribute to our sense of self? Our identities as human beings emerge from the interplay between our unique selves and our social roles.

Only in the most rudimentary sense are families created by society. Families are "created" by their members, according to symbolic interactionists. Each family has its own unique personality and dynamics created by its members' interactions. To classify families by structure, such as nuclear family, stepfamily, and single-parent family, misses the point of families. Structures are significant only insofar as they affect family dynamics. It is what goes on inside families that is important.

The family is a "unity of interacting personalities," according to symbolic interaction theory. How family members interact with each other is partly determined by their social roles as husband/wife, mother/father, son/daughter, and sibling. Parents and older siblings often fulfill the role of teacher for younger family members.

Critique

Although symbolic interaction theory focuses on the daily workings of the family, it suffers from several drawbacks. First, the theory tends to minimize the role of power in relationships. If a conflict exists, it may take more than simply communicating to resolve it. If one partner strongly wants to pursue his career in Los Angeles and the other just as strongly wants to pursue hers in Boston, no amount of communication and role adjustment may be sufficient to resolve the conflict. Ultimately, the partner with the greater power in the relationship may prevail.

Second, symbolic interaction doesn't fully account for the psychological aspects of human life, especially personality and temperament. It sees us only as the sum of our roles. It neglects the self that exists independently of our roles. Limiting us to the sum of our parts limits our uniqueness as human beings.

Third, the theory emphasizes individualism. It encourages competence in interpersonal relationships, and it values individual happiness and fulfillment over stability, duty, responsibility, and other familial values. As Jay Schvaneveldt (1981) observes, "The welfare and happiness of marital partners are held above the belief that the marital union or family union should stay intact. The happiness of the individual family members appears to be the dominant value."

Fourth, the theory does not place marriage or family within a larger social context. It thereby disregards or minimizes the forces working on families from the outside, such as economic discrimination against minorities and women.

Reflections

Using symbolic interaction theory, examine the different roles you play in your relationship, marriage, or family. How do you feel about these roles? How do you go about changing the contents of these roles? What encourages or discourages such changes?

Social Exchange Theory

According to **social exchange theory,** we measure our actions and relationships on a cost-benefit basis. People maximize their rewards and minimize their costs by employing their resources to gain the most favorable outcome. An outcome is basically figured by the equation Reward - Cost = Outcome.

How Exchange Works

At first glance, exchange theory may be the least attractive theory we use to study marriage and the family. It seems more appropriate for accountants than for lovers. But all of us use a cost-benefit analysis to some degree to measure our actions and relationships.

The reason why many of us don't recognize our use of this interpersonal accounting is that we do much of it unconsciously. If a friend is unhappy with a partner, you may ask, "What are you getting out of the relationship?" Your friend will start listing pluses and minuses: "On the plus side, I get company and a certain amount of security; on the minus side, I don't get someone who really understands me." When the emotional costs outweigh the benefits of the relationship, your friend will probably end it. This weighing of costs and benefits is social exchange theory at work.

One problem many of us have in recognizing our own exchange activities is that we think of rewards and costs as tangible objects, like money. In personal relationships, however, resources, rewards, and costs are more likely to be things such as love, companionship, status, power, fear, loneliness, and so on. As people enter into relationships, they have certain resources—either tangible or intangible—that others consider valuable, such as intelligence, warmth, good looks, or high social status. People consciously or unconsciously use their various resources to obtain what they want, as when they "turn on" the charm. Most of us have had

We should be careful to get out of an experience only the wisdom that is in it and stop there; lest we be like that cat that sits down on a hot stove-lid. She will never sit down on a hot stove-lid again and that is well; but also she will never sit down on a cold one anymore.

MARK TWAIN [SAMUEL CLEMENS] (1835–1910)

friends, for example, whose relationships are a mystery to us. We may not understand what our friend sees in his or her partner; our friend is so much better looking and more intelligent than the partner. (Attractiveness and intelligence are typical resources in our society.) But it turns out that the partner has a good sense of humor, is considerate, and is an accomplished musician, all of which our friend values highly.

Equity

A corollary to exchange is *equity:* exchanges that occur between people have to be fair, to balance out. In the everyday world, we are always exchanging favors: You do the dishes tonight and I'll take care of the kids. Often we don't even articulate these exchanges; we have a general sense that ultimately they will be reciprocated. If in the end we feel that the exchange wasn't fair, we are likely to be resentful and angry.

Some researchers suggest that people are most happy when they get what they feel they deserve in a relationship (Hatfield and Walster, 1981). Oddly enough, both partners feel uneasy in an inequitable relationship:

> While it's not surprising that deprived partners (who are, after all, getting less than they deserve) should feel resentful and angry about their inequitable treatment, it's perhaps not so obvious why their overbenefited mates (who are getting more than they deserve) feel uneasy too. But they do. They feel guilty and fearful of losing their favored position.

When partners recognize that they are in an inequitable relationship, they generally feel uncomfortable, angry, or distressed. They try to restore equity in one of three ways:

- They attempt to restore actual equity in the relationship.
- They attempt to restore psychological equity by trying to convince themselves and others that an obviously inequitable relationship is actually equitable.
- They decide to end the relationship.

Society regards marriage as a permanent commitment. Because marriages are expected to endure, exchanges take on a long-term character. Instead of being calculated on a day-to-day basis, outcomes are judged over time.

An important ingredient in these exchanges is whether the relationship is fundamentally cooperative or competitive. In cooperative exchanges, both husbands and wives try to maximize their "joint profit" (Scanzoni, 1979). These exchanges are characterized by mutual trust and commitment. Thus a husband might choose to work part-time and also care for the couple's infant so that his wife may pursue her education. In a competitive relationship, however, each is trying to maximize his or her own individual profit. If both spouses want the freedom to go out whenever or with whomever each wishes, despite opposition from the other, the relationship is likely to be unstable.

Reflections

Think about the benefits you are receiving from your current (or past) intimate, marital, or family relationships. What are the costs? Make a list to compare the benefits and the costs. Assign a value from 1 to 10 for the various items on your list, with 10 being the highest value and 1 being the lowest. Based on the equation Reward - Cost = Outcome, how would you predict the ultimate outcome? Think about the last time you made a trade-off in a relationship. Was it fair? If it wasn't, how did you feel? How did the other person feel?

Critique

Social exchange theory assumes that individuals are rational, calculating animals, weighing the costs and rewards of their relationships. In reality, sometimes we are rational, and sometimes we are not. Sometimes we act altruistically without expecting any reward. This is often true of love relationships and parent-child interaction.

Social exchange theory also has difficulty ascertaining the value of costs, rewards, and resources. If you want to buy eggs, you know they are $1.15 a dozen, and you can compare buying a dozen eggs with spending the same amount on a notebook. But how does the value of an outgoing personality compare with the value of a compassionate personality? Is a pound of compassion equal to ten pounds of enthusiasm? Compassion may be the trait most valued by one person but may not be important to another. The values that we assign to costs, rewards, and resources are highly individualistic.

Structural Functionalism

Structural functionalism is a theory used to explain how society works, how families work, and how families relate to the larger society and to their own members. The theory is used largely in sociology and anthropology, disciplines that focus on the study of society rather than individuals. When structural functionalists study the family, they look at three aspects: (1) what functions the family serves for society (discussed in Chapter 1), (2) what functional requirements are performed by family members for the family, and (3) what needs the family meets for its individual members.

Many of us are structural functionalists without knowing it. Those who believe social stability is in the best interest of society share this assumption with structural functionalists. Those who believe families must be intact to fulfill their functions share a common view with structural functionalists. Those who feel that changes in gender roles and the increase in single-parent and dual-earner families threaten the family fear social instability.

Have no respect whatsoever for authority; forget who said it and instead look at what he starts with, where he ends up, and ask yourself, "Is it reasonable?"

RICHARD FEYNMAN (1918–1988)

Society as a System

Structural functionalism is deeply influenced by biology. It treats society as if it were a living organism, like a person, animal, or tree. In fact, the theory sometimes uses the analogy of a tree in describing society. In a tree, there are many substructures or parts, such as the trunk, branches, roots, and leaves. Each structure has a function. The roots gather nutrients and water from the soil, the leaves absorb sunlight, and so on. Society is like a tree insofar as it has different structures that perform functions for its survival. These structures are called **subsystems.** The subsystems are the major institutions, such as the family, religion, government, and the economy. Each of these structures has a function in maintaining society, just as the different parts of a tree serve a function in maintaining the tree. Religion, for example, gives spiritual support, the government ensures order, and the economy produces goods. The family provides new members for society through procreation and socializes its members so that they fit into society. In theory, all institutions work in harmony for the good of society and one another.

Men do not seek the truth. It is the truth that pursues men who run away and will not look around.

LINCOLN STEFFENS (1866–1936)

The Family as a System

Families themselves may also be regarded as systems. In looking at families, structural functionalists examine (1) how the family organizes itself for survival, and (2) what functions the family performs for its members. For the family to survive, its members must perform certain functions, which are traditionally divided along gender lines. Men and women have different tasks: Men work outside the home to provide an income, whereas women perform household tasks and child rearing.

According to structural functionalists, the family molds the kind of personalities it needs to carry out its functions. It encourages different personality traits for men and women to ensure its survival. Men develop instrumental traits, and women develop expressive traits. *Instrumental traits* encourage competitiveness, coolness, self-confidence, and rationality—qualities that will help a person succeed in the outside world. *Expressive traits* encourage warmth, emotionality, nurturing, and sensitivity—qualities appropriate for someone caring for a family and a home.

Critique

Although structural functionalism has been an important theoretical approach to the family, it has declined in significance in recent decades for several reasons. First, because the theory cannot be empirically tested, we'll never know if it is "right" or "wrong." We can only discuss it theoretically, arguing whether it accounts for what we know about the family.

Second, it is not always clear what function a particular structure serves. "The function of the nose is to hold the *pince-nez* [eyeglasses] on the face," remarked the eighteenth-century philosopher François Voltaire. What is the function of the traditional division of labor along gender lines? Efficiency, survival, or the subordination of women? Also, how do we know which family functions are vital? The family, for example, is supposed to socialize children, but much socialization has been taken over by the schools, peer groups, and the media.

Third, structural functionalism has a conservative bias against change. Aspects that reflect stability are called functional, and those that encourage instability (or change) are called dysfunctional. Traditional roles are functional, but nontraditional ones are dysfunctional. Employed mothers are viewed as undermining family stability because they should be home caring for the children, cleaning house, and providing emotional support for their husbands. But in reality, employed mothers may be contributing to family stability by earning money; their income often pushes their families above the poverty line.

Finally, structural functionalism looks at the family abstractly. It looks at it formally, from a distance far removed from the daily lives and struggles of men, women, and children. It views the family in terms of functions and roles. Family interactions, the very life blood of family life, are absent. Because of its formalism, structural functionalism often has little relevance to real families in the real world.

Conflict Theory

Whereas structural functionalists tend to believe that "what is, is good," conflict theorists seem to believe that "what is, is wrong"—for at least someone. **Conflict theory** holds that life involves discord. Conflict theorists see society not as basically cooperative but as divided, with individuals and groups in conflict with each other. They try to identify the competing forces.

Sources of Conflict

How can we analyze marriages and families in terms of conflict and power? Marriage and family relationships are based on love and affection, aren't they? Conflict theorists would agree that love and affection are important elements in marriages and families, but they believe that conflict and power are also fundamental. Marriages and families are composed of individuals with different personalities, ideas, values, tastes, and goals. Each person is not always in harmony with every other person in the family. Imagine that you are living at home and want to do something your parents don't want you to do, such as spend the weekend with a friend they don't like. They forbid you to carry out your plan. "As long as you live in this house, you'll have to do what we say." You argue with them, but in the end you stay home. Why did your parents win the disagreement? They did so because they had greater power, according to conflict theorists.

Conflict theorists do not believe that conflict is bad; instead, they think it is a natural part of family life. Families always have disagreements, from small ones, such as what movie to see, to major ones, such as how to rear children. Families differ in the number of underlying conflicts of interest, the degree of underlying hostility, and the nature and extent of the expression of conflict. Conflict can take the form of competing goals, such as a husband's wanting to buy a new CD player and a wife's wanting to pay off credit cards. Conflict can also occur because of different role expectations: An employed mother wants to divide housework fifty-fifty, but her husband insists that household chores are "women's work."

Sources of Power

When conflict occurs, who wins? Family members have different resources and amounts of power. Four important sources of power are legitimacy, money, physical coercion, and love. When arguments arise in a family, a man may want his way "because I'm the head of the house" or a parent "because I'm your mother." These appeals are based on *legitimacy*—that is, the belief that the person is entitled to prevail by right. Money is a powerful resource in marriages and families. "As long as you live in this house . . ." is a directive based on the power of the purse. Because men tend to earn more than women, they have greater economic power; this economic power translates into marital power. Physical coercion is another important source of power. "If you don't do as I tell you, you'll get a spanking" is one of the most common forms of coercion of children. But physical abuse of a spouse is also common, as we will see in Chapter 13. Finally, there is the power of love. Love can be used to coerce someone emotionally, as in "If you really loved me, you'd do what I ask." Or love can be a freely given gift, as in the case of a person's giving up something important, such as a plan, desire, or career, to enhance a relationship.

Everyone in the family has power, although the power may be different and unequal. Adolescent children, for example, have few economic resources, so they must depend on their parents. This dependency gives the parents power. But adolescents also have power through the exercise of personal charm, ingratiating habits, temper tantrums, wheedling, and so on.

Families cannot live comfortably with much open conflict. The problem for families, as for any group, is how to encourage cooperation while allowing for differences. Because conflict theory sees conflict as normal, the theory seeks to channel it and to seek solutions through communication, bargaining, and negotiations. We return to these items in Chapter 5 in our discussion of conflict resolution.

Critique

A number of difficulties arise in conflict theory. First, conflict theory derives from politics, in which self-interest, egotism, and competition are dominant elements. Yet is such a harsh judgment of human nature justified? People's behavior is also characterized by self-sacrifice and cooperation. Love is an important quality in relationships. Conflict theorists don't often talk about the power of love or bonding; yet the presence of love and bonding may distinguish the family from all other groups in society. We often will make sacrifices for the sake of those we love. We will defer our own wishes to another's desires; we may even sacrifice our lives for a loved one.

Second, conflict theorists assume that differences lead to conflict. Differences can also be accepted, tolerated, or appreciated. Differences do not necessarily imply conflict. Third, conflict in families is not easily measured or evaluated. Families live much of their lives privately, and outsiders are not always aware of whatever conflict exists or how pervasive it is. Also, much overt conflict is avoided because it is regulated through family and societal rules. Most children obey their parents, and most spouses, although they may argue heatedly, do not employ violence.

Reflections

Using conflict theory, examine the recurring conflicts in your relationship, marriage, or family. Who wins these various conflicts? What resources do the winners have? The losers? Do they differ according to the type of conflict? What are your resources in relationships? How do you use them?

Family Systems Theory

Family systems theory combines two sociological theories, structural functionalism and symbolic interaction, to form a psychotherapeutic theory. Mark Kassop (1987) notes that family systems theory creates a bridge between sociology and family therapy.

Structure and Patterns of Interaction

Family systems theory views the family as a structure of related parts or subsystems. Each part carries out certain functions. These parts include the spousal subsystem, the parent/child subsystem, the parental subsystem (husband and wife relating to each other as parents), and the personal subsystem (the individual and his or her relationships). One of the important tasks of these subsystems is maintaining their boundaries. For the family to function well, the subsystems must be kept separate (Minuchin, 1981). Husbands and wives, for example, should prevent their conflicts from spilling over into the parent/child subsystem. Sometimes a parent will turn to the child for the affection that he or she ordinarily receives from a spouse. When the boundaries of the separate subsystems blur, as in incest, the family becomes dysfunctional.

As in symbolic interaction, interaction is important in systems theory. A family system consists of more than simply its members. It also consists of the pattern of interactions of family members: their communication, roles, beliefs, and rules. Marriage is more than a husband and wife; it is also their pattern of interactions. The structure of marriage is determined by how the spouses act in relation to each other over time (Lederer and Jackson, 1968). Each partner influences, and in turn is influenced by, the partner. And each interaction is determined in part by the previous interactions. This emphasis on the pattern of interactions within the family is a distinctive feature of the systems approach.

Three blind men once stumbled upon an elephant that was blocking their path. Since they had never encountered an elephant before, each tried to describe it to the other. The first blind man, who had felt the elephant's tail, said an elephant was like a rope; the second blind man felt only the elephant's leg. He said an elephant was like a tree. The third blind man had felt only the trunk, and said an elephant was like a snake. As they quarreled among themselves about the nature of the elephant, the elephant walked away.

HINDU FOLKTALE

Virginia Satir (1988) compared the family system to a hanging mobile. In a mobile, all the pieces, regardless of size and shape, can be grouped together and balanced by changing the relative distance between the parts. The family members, like parts of a mobile, require certain distances between one another to maintain their balance. Any change in the family mobile—such as a child leaving the family, family members forming new alliances, hostility distancing the mother from the father—affects the stability of the mobile. This disequilibrium often manifests itself in emotional turmoil and stress. The family may try to restore the old equilibrium by forcing its "errant" member to return to his or her former position, or it may adapt and create a new equilibrium with its members in changed relations to one another.

Analyzing Family Dynamics

In looking at the family as a system, researchers and therapists believe the following:

- *Interactions must be studied in the context of the family system.* Each action affects every other person in the family. The family exerts a powerful influence on our behaviors and feelings, just as we influence the behaviors and feelings of other family members. On the simplest level, an angry outburst by a family member can put everyone in a bad mood. If the anger is constant, it will have long-term effects on each member of the family, who will cope with it by avoidance, hostility, depression, and so on.
- *The family has a structure that can only be seen in its interactions.* Each family has certain preferred patterns of acting that ordinarily work in response to day-to-day demands. These patterns become strongly ingrained "habits" of interactions that make change difficult. A warring couple, for example, may decide to change their ways and resolve their conflicts peacefully. They may succeed for a while, but soon they fall back into their old ways. Lasting change requires more than changing a single behavior; it requires changing a pattern of relating.
- *The family is a purposeful system; it has a goal.* In most instances, the family's goal is to remain intact as a family. It seeks **homeostasis,** or stability. This goal of homeostasis makes change difficult, for change threatens the old patterns and habits to which the family has become accustomed.
- *Despite resistance to change, each family system is transformed over time.* A well-functioning family constantly changes and adapts to maintain itself in response to its members and the environment. The family changes through the family life cycle, for example, as partners age and as children are born, grow older, and leave home. The parent must allow the parent/child relationship to change. A parent must adapt to an adolescent's increasing independence by relinquishing some parental control. The family system adapts to stresses to maintain family continuity while making restructuring possible. If the primary wage earner loses his or her job, the family tries to adapt to the loss in income; the children may seek work, recreation may be cut, or the family may be forced to move.

Reflections

How does your family deal with change? Does it accept or resist change? How does its reaction to change affect you? To get a visual sense of your family's response to change, draw a picture of your family as a mobile. Imagine a current or impending change, and redraw the mobile to reflect how your family deals with the change. Does your family try to maintain the old equilibrium, or does it adjust to form a new one?

Critique

It is difficult for researchers to agree on exactly what family systems theory is. Many of the basic concepts are still in dispute, even among the theory's adherents (Melito, 1985).

Family systems theory originated in clinical settings in which psychiatrists, clinical psychologists, and therapists tried to explain the dynamics of dysfunctional families. Although its use has spread beyond clinicians, its greatest success is still in the analysis and treatment of dysfunctional families. As with clinical research, however, the basic question is whether its insights apply to healthy families as well as to dysfunctional ones. Do healthy families, for example, seek homeostasis as their goal, or do they seek individual and family well-being?

You think my fallacy is all wrong?

MARSHALL MCLUHAN (1911–1980) IN *ANNIE HALL*

Feminist Perspective

As a result of the feminist movement of the past two decades, new questions and ways of thinking about the meaning and characteristics of family have arisen. Although there is not a unified "feminist family theory," the **feminist perspective** expresses a central concern regarding family life: They are concerned about

There is a hidden fear that somehow, if they are given a chance, women will suddenly do as they have been done by.

EVA FIGES

traditional gender roles. **Gender**—the social aspects of being female or male—is the orienting focus in most feminist writing, research, and advocacy. Feminists maintain that family and gender roles have been constructed by society and do not derive from biological or absolute conditions. They believe that family and gender roles have been created by men in order to maintain power over women. Basically, the goals of the feminist perspective are to work to accomplish changes and conditions in society that remove barriers to opportunity and oppressive conditions and are "good for women" (Thompson and Walker, 1995).

Gender and Family Are Concepts Created by Society

Who or what constitutes family cannot be taken for granted. The "traditional family" is no longer the predominant family lifestyle. Today's families have great diversity. What we think family should be is influenced by our own values and family experiences. Research demonstrates that couples actually may construct gender roles in their marriages (Zvonkovic, Greaves, Schmiege, and Hall, 1996).

People call me a feminist whenever I express sentiments that differentiate me from a doormat or a prostitute.

REBECCA WEST (1892-1983)

Are there any basic biological or social conditions that require the existence of a particular form of family? Some feminists would emphatically say no. Some object to the fact that we even try to study the family because to do so accepts as "natural" the inequalities built into the traditional concept of family life. Feminists urge a more extended view of family to include all kinds of sexually interdependent adult relationships regardless of the legal, residential, or parental status of the partnership. For example, families may be formed of committed relationships between lesbian or gay individuals, with children obtained through adoption, from previous marriages, or through artificial insemination.

Feminist Agenda

Feminists strive to raise society's level of awareness regarding the oppression of women. Furthermore, they make the point that all groups defined on the basis of age, class, race, ethnicity, disability, or sexual orientation are oppressed; they extend their concern for greater sensitivity to all disadvantaged groups (Allen and Baber, 1992). Feminists assume that the experiences of an individual are influenced by the social system in which the individual lives. Therefore, the experiences of the individual must be analyzed in order to form the basis for political action and social change. The feminist agenda is to attend to the social context as it impacts on personal experience and to work to translate personal experience into community action and social critique.

Feminists believe it is imperative to speak out and challenge the system that exploits and devalues women. They are aware of the dangers of speaking out, but feel their own integrity will be threatened if they fail to speak out. Some feminists have described themselves as having "double vision." The double vision is the ability to be successful in the existing social system while simultaneously working to change oppressive practices and institutions.

Critique

The feminist perspective is not a unified theory; rather, it represents thinking across the feminist movement. It includes a variety of viewpoints that have, however, an integrating focus relating to the inequity of power between men and women in society and especially in family life (MacDermid, Jurich, Myers-Walls, and Pelo, 1992).

Some family scholars who conceptualize family life and work as a "calling" have taken issue with feminists' focus on power and economics as a description of family. This has created a moral dialogue concerning the place of family work in "the good society" (Ahlander and Bahr, 1995; Sanchez, 1996). Feminists today

Did you know?

While some feminists question the necessity of family, other feminist researchers report findings that a feminist ideology upgrades marriage for women by promoting a vigilance to equality (Blaisure and Allen, 1995).

recognize considerable diversity within their ranks, and the ideas of feminist theorists and other family theorists often overlap.

Family Development Theory

The application of family development theory to the family over its lifetime is discussed in detail in Chapter 9, so we will discuss it only briefly here. **Family development theory** looks at the changes in the family beginning with marriage and proceeding through seven more sequential stages. The stages are identified by the primary or orienting event characterizing a period of the family history. The eight stages are: (1) beginning family, (2) childbearing families, (3) families with preschool children, (4) families with schoolchildren, (5) families with adolescents, (6) families as launching centers, (7) families in middle years, and (8) aging families.

As we grow, each of us responds to certain universal developmental challenges (Person, 1993). For example, all people encounter normative age-graded influences, such as the biological processes of puberty and menopause or sociocultural markers such as the beginning of school and the advent of retirement. Normative history-graded influences come from historical facts that are common to a particular generation, such as the political and economic influences of wars and economic depressions that are similar for individuals in a particular age group (Santrock, 1995).

The life-cycle model give us insights into the complexities of family life and of the different tasks that families perform. This model describes the interacting influences of changing roles and circumstances through time. It provides awareness of ways in which these changes produce corresponding changes in family responsibilities and needs. Planning that utilizes the developmental model alerts the family to seek resources appropriate to the upcoming needs and to be aware of vulnerabilities associated with each family stage (Higgins, Duxbury, and Lee, 1994).

Critique

A criticism sometimes made of the family development theory is that it assumes the sequential processes of a traditional family. For example, lesbian-headed families are likely to experience a life-cycle pattern that is quite different from the traditional one (Slater, 1995). Similarly, stepfamiles experience different stages and tasks (Ahrons and Rogers, 1987). Nevertheless, the universality of the family life cycle may transcend the individuality of the family form. Single-parent and two-parent families go through many of the same development tasks and transitions. They may differ, however, in the timing and length of those transitions.

According to the Family Circumplex Model (discussed and viewed on pages 60–61), families function best when cohesion and adaptability are balanced rather than extreme. Depending on the family stage, between 50 and 60 percent of families are balanced, 25 to 40 percent are mid-range, and about 12 to 18 percent are extreme (Olson et al., 1983).

Ethnicity and Family Research

To study the history of the American family is to conduct a rescue mission into the dreamland of our national self-concept. No subject is more closely bound up with our sense of a difficult present and our nostalgia for a happier past.

John Demos

According to recent census data (U.S. Bureau of the Census, 1996), 26 percent of the U.S. population are people of color: 12 percent African American, 10 percent Hispanic, 3 percent Asian/Pacific Islander, and 1 percent Native American.

YOU AND YOUR WELL-BEING

Measuring Family Functioning

The **Family Circumplex Model** is an important tool for identifying and mapping family relationships (Olson et al., 1983; Olson and DeFrain, 1997). It is a typological model based on systems theory. According to the Family Circumplex Model, there are three important components of family dynamics: cohesion, adaptability, and communication. Cohesion and adaptability are central dimensions to family functioning (see Chapter 16 for discussion of cohesion and adaptability). **Cohesion** refers to the emotional bonding that family members have with one another—the amount of connectedness among family members (Olson et al., 1993). **Adaptability** is the ability of the marital or family system to change in response to situational and developmental stress. Communication facilitates or discourages cohesion and adaptability. Positive communication enables family members to understand, empathize with, and support each other. Negative communication, such as criticism, blaming, and yelling, detracts from family members' ability to be cohesive and adaptable.

Cohesion and adaptability are not "all or nothing" measures. Instead, their levels vary from high to low. Levels of cohesion, for example, may be *disengaged* (very low), *connected* (low to moderate), *cohesive* (moderate to high), or *enmeshed* (very high). Levels of adaptability may be *rigid* (very low), *structured* (low to moderate), *flexible* (moderate to high), or *chaotic* (very high).

Cohesion is determined by many factors. A family may be cohesive in one aspect and not so cohesive in another. It is possible, however, to assess the relative cohesiveness of any family by measuring the elements discussed so far. **Enmeshment** occurs when individuals overidentify with the family and cannot move independently of the others or of the whole. Individuals become so entangled that movement is impossible. Moreover, subsystems in enmeshed families have difficulty functioning because of interference from other parts of the family. At the other extreme, disengagement is equally dysfunctional. **Disengagement** occurs when family members do not feel close and do not communicate. Individuals are disconnected, and the family cannot get together enough to carry out minimum family tasks. Strong families communicate and seek a balance between enmeshment and disengagement.

Putting together the four levels of cohesion and four levels of adaptability creates sixteen possible combinations, as you can see in Fig. 2.1. These combinations can be reduced to three main types of families: *balanced, midrange,* and *extreme.* Balanced families fall into the central cells of cohesion and adaptability. Midrange families fall into an extreme cell on one dimension and into a central cell on another dimension. Extreme families fall into extreme cells on both cohesion and adaptability dimensions.

Projections of recent increases suggest that by the year 2050 whites will constitute just over 50 percent of the American population, Hispanics 24 percent, African Americans 16 percent, Asian/Pacific Islanders 9 percent, and Native Americans 1 percent.

As discussed earlier in this chapter, it is important to be aware of the danger of thinking in terms of ethnocentric fallacies, beliefs that one's own ethnic group, nation, or culture is innately superior to others. In this section we will consider briefly some of the distinctive characteristics and strengths of families from various ethnic and cultural groups.

We begin by noting several important terms. An **ethnic group** is a group of people distinct from other groups because of cultural characteristics, such as language, religion, and customs that are transmitted from one generation to another. A **racial group** is a group of people, such as whites, blacks, and Asians, classified according to their **phenotype** (anatomical and physical characteristics). Racial groups

FIGURE 2.1
The Family Circumplex Model

Balanced families, those with moderate levels of cohesion and adaptability, are seen as having the greatest marital and family strengths across the family life cycle.

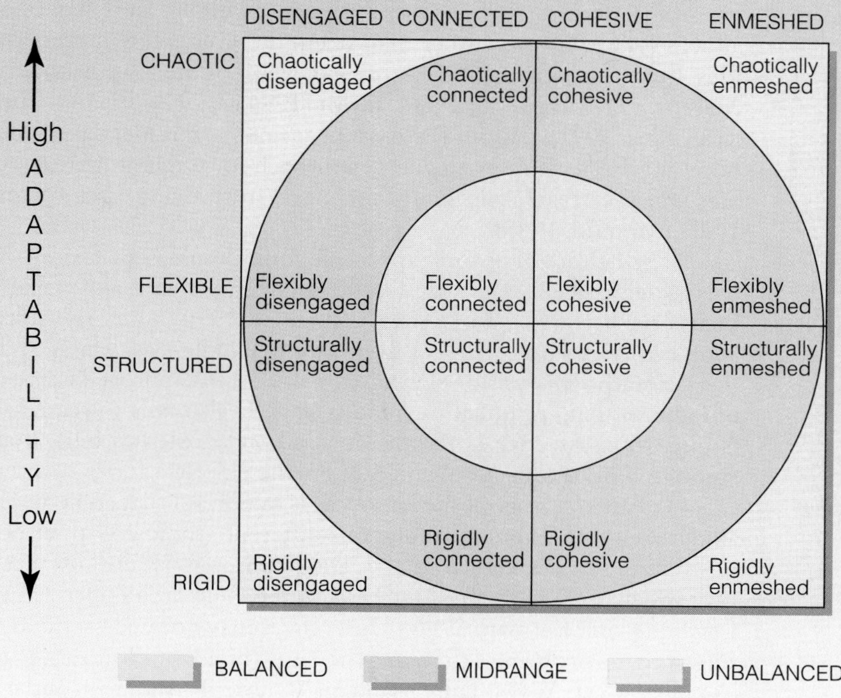

SOURCE: David H. Olson and John DeFrain, *Marriage and the Family,* 2nd ed. Copyright ©1997 by Mayfield Publishing Company. Used by permission of the publisher.

share common phenotypical characteristics, such as skin color and facial structure. A **minority group** is a group of people whose **status** (position in the social hierarchy) places them at an economic, social, and political disadvantage (Taylor, 1994b). These terms often overlap. For example, African Americans are simultaneously an ethnic, racial, and minority group. The term *African American,* used increasingly in place of *black,* reflects the growing awareness of the importance of ethnicity (culture) in contrast to race (skin color) (Smith, 1992; but see Taylor, 1994b).

Changing Perspectives on Ethnicity and Family

Until the last twenty-five years, most research about American marriages and families was limited to the white, middle-class family. The nuclear family was the norm against which all other families were evaluated. As we have seen, such a

Figure 2.2 Top Ten Ancestry Groups, 1990

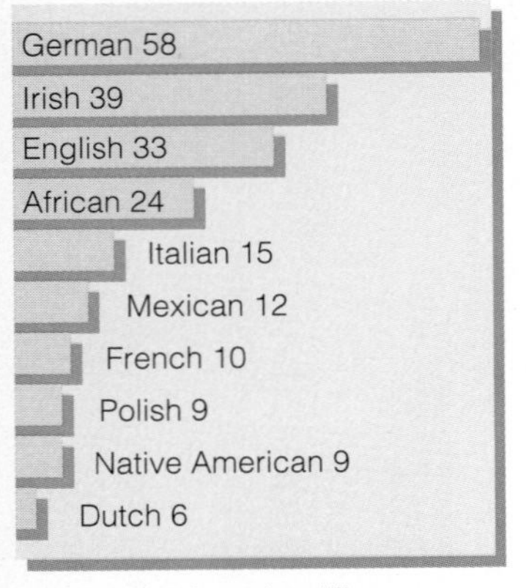

Numbers in millions

SOURCE: U.S. Bureau of the Census. *Statistical Abstract of the United States.* Washington, D.C.: U.S. Government Printing Office, 1996.

It's not a "melting pot" anymore. . . . (It's) a fusion chamber. We are creating American culture daily.

BHARATI MUKHERJEE

I am the dream and hope of the slave.

MAYA ANGELOU

Did you know?

Although stereotypes suggest that *all* African Americans are low income, at least 30 percent are middle-class in their lifestyle and income (U.S. Bureau of the Census, 1994).

perspective distorts our understanding of other family forms, such as single-parent families and stepfamilies. These family forms were often viewed as pathological because they differed from the traditional norm. A similar distortion also has influenced our understanding of African-American, Latino, Asian-American, and Native-American families. Instead of recognizing the strengths of diverse ethnic family systems, we viewed these families as "tangles of pathology" for failing to meet the model of the traditional nuclear family (Moynihan, 1965).

Part of this distortion is a result of the scarcity of studies on families from African-American, Latino, Asian-American, Native-American, and other ethnic groups. Furthermore, many of the earlier studies were flawed or distorted by a focus on weaknesses rather than strengths, giving the impression that all families from a particular ethnic group were riddled by problems (Dilworth-Anderson and McAdoo, 1988; Taylor, 1994a, 1994b; Taylor, Chatters, Tucker, and Lewis, 1991). Two of the most prominent examples of ethnocentric distortions are the "culture of poverty" approach to studying African-American families and the "machismo syndrome" used in studying Latino families (Demos, 1990; Mirandé, 1985).

The culture of poverty approach, for example, sees African-American families as being deeply enmeshed in illegitimacy, poverty, and welfare as a result of their slave heritage. As one scholar (Demos, 1990) notes, the culture of poverty approach "views black families from a white middle-class vantage point and results in a pejorative analysis of black family life." This approach ignores the majority of families that are intact or middle-class. It also fails to see African-American family strengths, such as strong kinship bonds, role flexibility, love of children, commitment to education, and care for the elderly.

In America's pluralistic society, it is important that students and researchers alike reexamine diversity among our different ethnic groups as possible sources of strength rather than pathology (DeGenova, 1997). For instance, cultures may vary widely in how the best interests of the child are defined (Murphy-Berman, Levesque, and Berman, 1996). Differences may not necessarily be problems but solutions to problems; they may be signs of adaptation rather than weakness (Adams, 1985). As two family scholars (Dilworth-Anderson and McAdoo, 1988) point out, "Whether a phenomenon is viewed as a problem or a solution may not be objective reality at all but may be determined by the observer's values."

African-American Families

The largest ethnic group in the United States is African Americans. In 1995, there were over 33 million African Americans in the United States (U.S. Bureau of the Census, 1996), accounting for nearly 13 percent of the population.

Socioeconomic status, rank in society based on a combination of occupational, educational, and income levels, is an important element in African-American life (Staples and Johnson, 1993).

The high rates of divorce and births to unmarried women seen in African-American families are associated with poverty rather than being inherent problems in African-American families. Although there seems to be greater marital instability in African-American families than in white families, when divorce rates are adjusted according to socioeconomic status, the differences are minimal. Poor African Americans have divorce rates similar to poor whites, and middle-class African Americans have divorce rates similar to middle-class whites (Raschke, 1987). Thus, understanding socioeconomic status, especially poverty, is critical in examining African-American life (Bryant and Coleman, 1988; Julian, McKenry, and McKelvey, 1994; Wilkinson, 1997).

Figure 2.3

Geographic Distribution of Ethnic Groups by Population Density According to 1990 Census

These maps indicate the highest concentrations by state of specific ethnic groups and the proportion they form of a state's population.

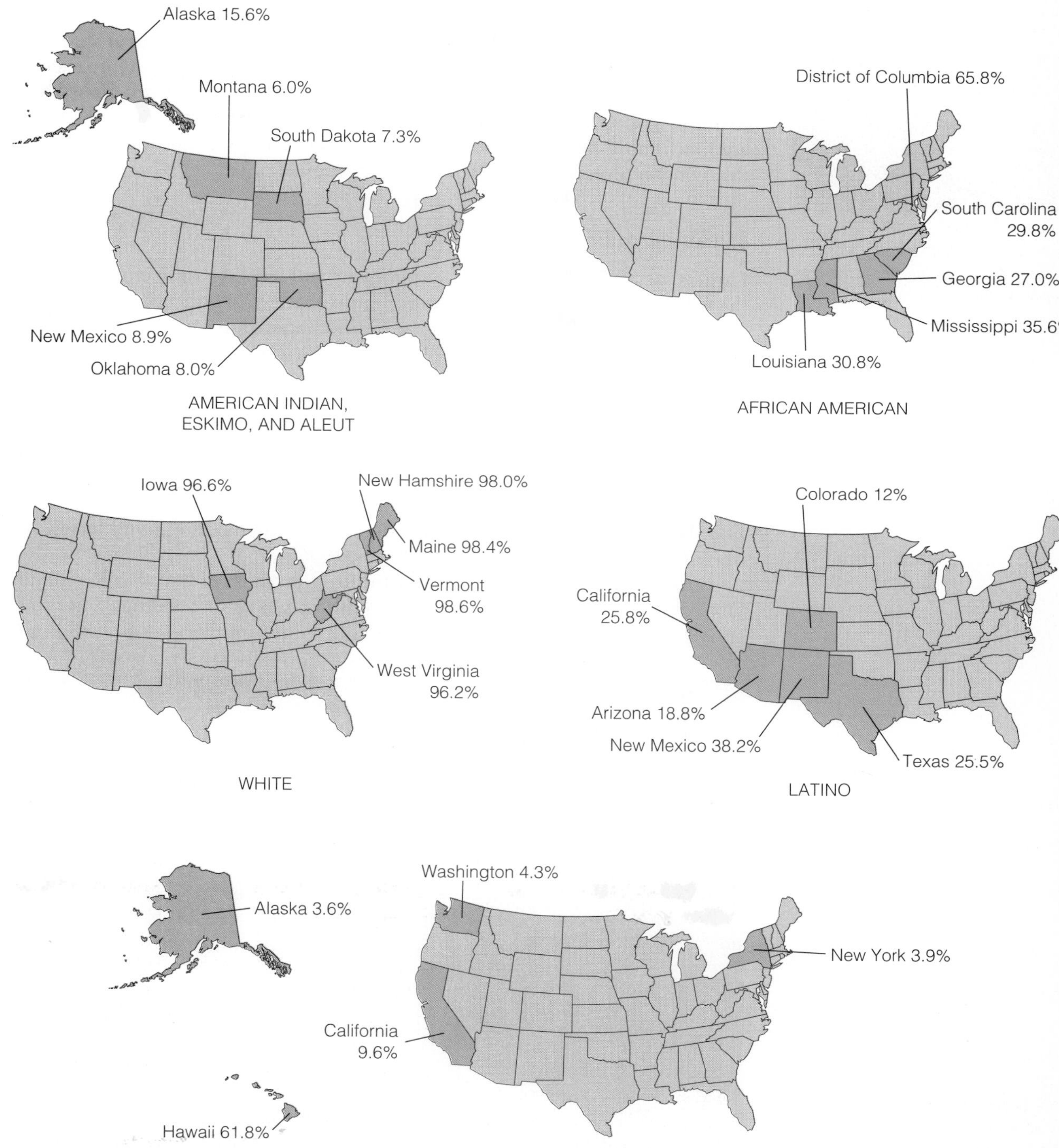

SOURCE: U.S. Bureau of the Census. *Statistical Abstract of the United States*. Washington, D.C.: U.S. Government Printing Office, 1994.

One of the notable features of African-American families is the high value they place on children.

There are several striking features of African-American families. First, African-American families, in contrast to white families, have a long history of being dual-earner families as a result of economic need. As a consequence, employed women have played important roles in the African-American family. They also have more egalitarian family roles. Research finds that responsibilities and tasks are more equally divided between males and females (McAdoo, 1996). Second, kinship bonds are especially important, for they provide economic assistance and emotional support in times of need (Taylor, 1994c; Taylor et al., 1991). Third, African Americans have a strong tradition of **familialism** (emphasis on family and family loyality), with an emphasis on intergenerational ties. Fourth, the African-American community values children highly. Fifth, African Americans are much more likely than whites to live in **extended households,** households that contain several different families (Taylor, 1994c).

Latino Families

Latinos (or Hispanics) are the fastest growing and second largest ethnic group in the United States. In 1995, there were approximately 26.9 million Latinos in the country. Between 1980 and 1995, the Latino population increased by approximately 85 percent, compared to 16 percent for the general population. This increase was a result of both immigration and a higher birthrate among Latinos. (U.S. Bureau of the Census, 1996; Vega, 1991).

Currently, 65 percent of Latinos are of Mexican descent. Latinos, mostly of Mexican and Central American descent, are concentrated in California and the Southwest. Latinos of Puerto Rican descent are concentrated in the Northeast, especially New York. The greatest numbers of Cuban Americans are found in Florida. There are significant Latino populations also in Illinois, New Jersey, and Massachusetts (U.S. Bureau of the Census, 1996).

Continued immigration has transformed the nature of Latino culture in the United States. First, immigration makes both Latino culture and the larger society a "permanently unfinished" society. The newer immigrants are urban and

For Latinos, their ancient heritage, such as Aztec, Mayan, or Taino (Puerto Rican), is often a source of great pride.

Figure 2.4
Families by Type for Selected Hispanic Groups: 1995

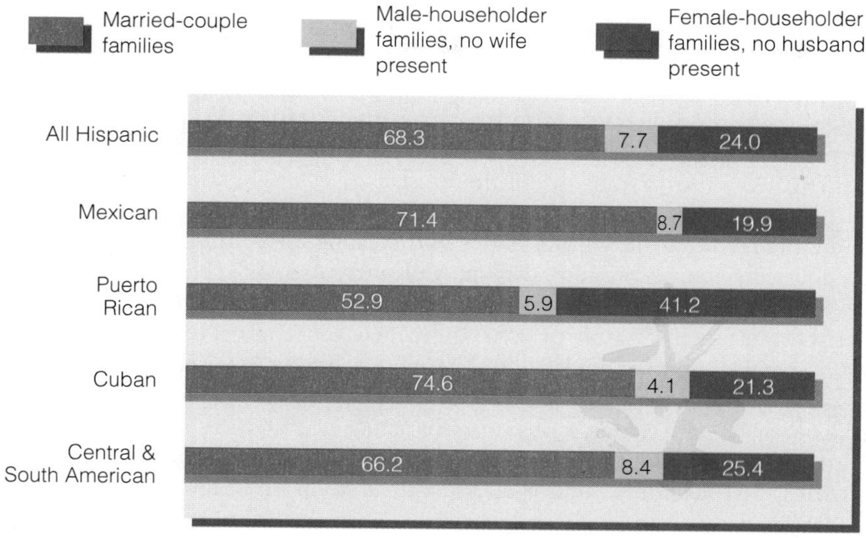

SOURCE: U.S. Bureau of the Census. *Statistical Abstracts of the United States*. Washington, D.C: U.S. Government Printing Office, 1996.

overwhelmingly workers and laborers rather than professionals. Second, in some areas, immigration is changing the proportion of U.S.-born and foreign-born Latinos. In 1960 in California, for example, four out of five Mexicans were born in the United States; today, because of the massive influx of immigrants, only about half are U.S. born (Zinn, 1994).

It is important to remember that there is considerable diversity among Latinos, both in terms of ethnic heritage (such as Mexican, Cuban, or Puerto Rican) and socioeconomic status (Sanchez, 1997; Walker, 1993). The middle class is strongest among Cuban Americans, followed by Puerto Ricans and then Mexican Americans. For example, 40 percent of Cuban Americans, 23 percent of Mexican Americans, and 15 percent of Puerto Ricans earn over $35,000 annually. Furthermore, there are two-and-a-half times as many single-parent families among Puerto Ricans as there are among Mexican Americans. (For an overview of Mexican-American families, see Sanchez, 1997; for Cuban-American families, see Suarez, 1997; and for Puerto Rican families, see Carrasquillo, 1997). This diversity is further accentuated by the varying proportions of U.S.-born and foreign-born Latinos in each group.

La familia is based on the nuclear family, but it also includes the extended family of grandparents, aunts, uncles, and cousins, who tend to live close by, often in the same block or neighborhood. There is close kin cooperation and mutual assistance, especially in times of need, when the family bands together. Children are especially important; over one-third of Latino families have five or more children. Because Spanish is important in maintaining ethnic identity, many Latinos, as well as educators, support bilingualism in schools and government. Catholicism is also an important factor in Latino family life. Although there is a tradition of male dominance, day-to-day living patterns suggest that women have considerable power and influence in the family (Griswold del Castillo, 1984).

Frijoles, tortillas, y chile are more American than the hamburger.
LUIS VALDEZ

Out of poverty, poetry
Out of suffering, song.
MEXICAN PROVERB

Did you know?
Of the 32 million Americans who speak languages other than English at home, 17.4 million speak Spanish (U.S. Bureau of the Census, 1996.)

Asian Americans are an especially diverse group. Chinese Americans and Japanese Americans have been in the United States since the nineteenth century. Southeast Asians, dislocated by the Vietnam War, are a more recent group. (For further discussion of Chinese-American families, see Ishii-Kuntz, 1997a; for Japanese Americans, see Ishii-Kuntz, 1997b; for Filipino Americans, see Root, 1997; and for Asian Indians, see Pais, 1997.)

Asian-American Families

Asian Americans are an especially diverse group, comprised of Chinese, Filipino, Japanese, Vietnamese, Cambodian, Hmong, and other groups. The largest Asian-American groups are Chinese Americans, Filipino Americans, and Japanese Americans. Other groups, such as Vietnamese, Cambodians, Laotians, and Hmong, are more recent arrivals, first coming to this country in the 1970s as refugees from the upheavals resulting from the Vietnam War. In the 1980s, Koreans, Filipinos, and Asian Indians began immigrating in larger numbers. The majority of Asian Americans live in the West.

Much diversity can be observed within the Asian-American families, based on time of arrival in the United States and reasons for coming to this country (e.g., political versus economic) (Julian, McKenry, and McKelvey, 1994). Values that continue to be important to Asian Americans in general include a strong sense of importance of family over the individual, self-control to achieve societal goals, and appreciation of one's cultural heritage. Chinese Americans tend to exercise strong parental control while encouraging their children to develop a sense of independence and strong motivation for achievement (Ishii-Kuntz, 1997; Lin and Fu, 1990). The more recent immigrants retain more culturally distinct characteristics, such as family structure and values, than do older groups, such as Chinese Americans and Japanese Americans. Asian-American families tend to be slightly larger than the average U.S. family (U.S. Bureau of the Census, 1996).

The most dramatic change affecting Chinese Americans has been their sheer increase in numbers over the last twenty-five years. The Chinese-American population increased fourfold between 1970 and 1990, from 431,000 to 1,645,000. More recent immigrants tend to be from Taiwan or Hong Kong rather than mainland China (Glenn and Yap, 1994). Because of the large numbers of new immigrants, it is important to distinguish between American-born and foreign-born Chinese Americans; little research is available concerning the latter.

Did you know?

Over half of Asian Americans do not speak English well; about one-third are linguistically isolated (U.S. Bureau of the Census, 1994).

Figure 2.5
Asian Population for Selected Groups: 1990

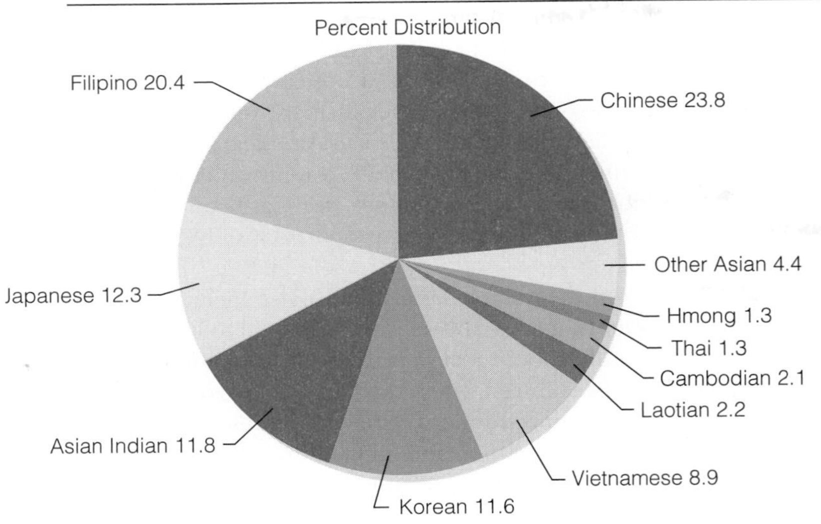

SOURCE: U.S. Bureau of the Census. "We the American Asians." Economics and Statistics Administration. Washington, D.C.: U.S. Government Printing Office, 1993.

We are increasingly becoming a world of migrants, made up of bits and fragments from here, there. We are here. And we have not really left anywhere we have been.

SALMAN RUSHDIE

Contemporary American-born Chinese families continue to emphasize familialism, although filial piety and strict obedience to parental authority have become less strong. Chinese Americans tend to be better educated, have higher incomes, and have lower rates of unemployment than the general population. Their sexual values and attitudes toward gender roles tend to be more conservative. Chinese-American women are expected to be employed and to contribute to the household income. Over 1.2 million speak Chinese at home.

Native-American Families

Approximately 2 million Americans identify themselves as being of native descent. Those who continue to be deeply involved with their own traditional culture give themselves a tribal identity, such as Dine (Navajo), Lakota, or Cherokee (Kawamoto and Cheshire, 1997). Those who are more acculturated, such as urban dwellers, tend to give themselves an ethnic identity as Native Americans or Indians. Most Americans of native descent consider themselves members of a tribal group rather than an ethnic group. And, observes John Price (1981), "Specific tribal identities are almost universally stronger and more important than identity as a Native American."

A people without history is like wind on the buffalo grass.

TETON SIOUX PROVERB

There has been a considerable migration of Native Americans to urban areas since World War II because of poverty on reservations and pressures toward acculturation. Today, 1.2 million Americans of native descent live outside tribal lands; most live in cities. In the cities, they are separated from their traditional tribal cultures and may experience great cultural conflict as they attempt to maintain traditional values. Not surprisingly, those in the cities are more acculturated than those remaining on the reservations. Urban Native Americans may attend powwows, intertribal social gatherings centering around drumming, singing, and traditional dances. Powwows are important mechanisms in the development of the Native-American ethnic identity in contrast to the tribal identity. Urban Native Americans, however, may visit their home reservations regularly.

Those who do not remember the past are condemned to repeat it.

GEORGE SANTAYANA (1863–1952)

Because of the importance of tribal identities and practices, there is no single type of Native-American family. Although there is considerable variation among different tribal groups, two aspects of Native-American families are important. First, extended families are significant. These extended families may be different from what the larger society regards as an extended family (Wall, 1993). They often revolve around complex kinship networks based on clan membership rather than birth, marriage, or adoption. Concepts of kin relationships may also differ. A child's "grandmother" may be an aunt or great-aunt in European-based conceptualization of kin (Yellowbird and Snipp, 1994).

Second, increasingly large numbers of Native Americans are marrying non-Indians. Among married Native Americans, 53 percent have non-Indian spouses. With such high rates of intermarriage, a key question is whether Native Americans can sustain their ethnic identity. Michael Yellowbird and Matthew Snipp (1994) wonder if "Indians through their spousal choices, may accomplish what disease, western civilization, and decades of Federal Indian policy failed to achieve."

European Ethnic Families

The sense of ethnicity among Americans of European descent has been growing in recent years. This is especially true among working-class Germans, Italians, Greeks, Poles, Irish, Croats, and Hungarians. This increasing awareness seems to

OTHER PLACES OTHER TIMES

The Ohlone: Native American Marriage and Family Ways

In many ways, the Ohlone of the San Francisco and Monterey Bay region in present-day California reflected many basic characteristics of Native American groups (Margolin, 1978). When an Ohlone boy sought marriage, for example, he presented his request to his parents. He and his prospective bride knew and liked each other but custom called for the marriage to be arranged by both sets of parents. His parents conferred with other important relatives; the family as a whole decided whether the match was good. A relative informally approached the girl's parents to sound out their feelings. Her family called in other members to discuss the proposal. If the proposal was accepted, the boy's family made a formal visit. At this visit, baskets, beads, feathers, furs, and other valuable gifts were brought. Valuable gifts conferred status on the girl and her family because they indicated the family's high standing in the village.

If the marriage proposal was accepted, the girlís family prepared a wedding feast that lasted for several days. Following the feast, the newly married couple wandered outside the village to a thicket of willows, where they consummated their marriage. After a while they returned, with the groom's face scratched and clawed. The marks were a sign that the marriage had been consummated; they showed that the bride was modest. After the marriage, the youth stayed with the wife's family for a few months. Then, if the girl's parents approved of their son-in-law, she left to live with her husband.

Sometimes men had two wives, often in separate villages. Within the village the wives were often sisters, as they preferred sharing husband and housework with a family member. In marriage, fidelity was expected and divorce was discouraged. If divorce was necessary, however, the man or woman simply moved out of the family dwelling. They could then remarry.

The Ohlone restricted their sexual activities because they believed sexuality interfered with their relationship with the spirit world; it weakened their powers. Men were sometimes overcome with desire, however, and ignored the various taboos surrounding sexuality, such as sexual restrictions during menstruation and breast-feeding (which lasted about two years). If a man or woman violated these taboos and later fell ill or had bad luck, it was said that he or she was being punished for weakness.

Although the Ohlone restricted their sexuality, they accepted same-sex relationships. If a boy began imitating women and began wearing women's clothes as he grew older, he was allowed to marry another man. Sometimes he became the second "wife" of a wealthy man whose other wives were women. Women attracted to other women were also accepted, but they were

be part of a general rise in ethnic identification over the last two decades (Rubin, 1994). Earlier, members of European ethnic groups sought to assimilate—to adopt the attitudes, beliefs, and values of the dominant culture. Most white ethnic groups have assimilated to a considerable degree—they have learned English, moved away from their ethnic neighborhoods, and married outside their group, but many continue to be bound emotionally to their ethnic roots. These roots are psychologically important, giving them a sense of community and a shared history. This common culture is manifested in shared rituals, feast days, and saint's days, such as St. Patrick's Day.

Except for some West Coast enclaves, such as Little Italy in San Francisco, white ethnicity is strongest in the East and Midwest. The Irish neighborhoods of Boston, the Polish areas of Chicago, and the Jewish sections of Brooklyn, for example, have strong ethnic identities. Common languages and dialects are spoken in the homes, stores, and parks. (At home, French is spoken by 1.7 million persons, German by 1.5 million, Italian by 1.3 million, and Polish by 723,000.) Traditional holidays are celebrated; the foods are prepared from recipes passed down through generations. Elders speak of the old country and their villages—even if it was their parents or grandparents who immigrated.

not allowed to adopt male roles; their same-sex orientation was restricted to sexual activities.

When a woman became pregnant, both she and her husband followed prescribed rituals. They abstained from sex and were both careful to avoid meat, fish, and salt, fearing that these would cause a difficult birth. The husband curbed his anger for fear of injuring the baby by disturbing the harmony of nature. He hunted little because it was bad to hurt living things during pregnancy. When the woman went into labor, she went to her hut and was attended by the old women, who caressed, massaged, and encouraged her. Following birth, the mother was led to a stream, where she splashed cold water on herself and the baby. A few days later, she began breastfeeding. From then on, until she weaned her child two years later, both she and her husband refrained from physical contact.

As the Ohlone raised their children, they sought to strengthen the bonds that linked the child to his or her family, clan, and tribe. The child's identity, strength, and fulfillment were found in belonging to family and clan. Selfishness and extreme individualism were discouraged because they weakened the bonds on which the family and community depended.

The child was watched over by the immediate and extended family. Children were not physically punished; instead, good behavior was taught by example. By age five, children were expected to engage in useful work, such as gathering berries and carrying wood. At age eight, boys and girls entered separate worlds. Boys began hunting, working rope and nets, and attending rituals in the sacred sweathouse, where saunalike heat purified their bodies and spirits. During puberty, the boys passed into manhood through a series of ceremonies.

As girls grew older, they helped to grind acorns and gather roots and herbs; they also learned to weave intricate baskets. A girl's passage into womanhood was marked by *menarche*, her first menstrual bleeding. Menarche was one of the most important events for the Ohlone girl, marking the beginning of her spiritual power. At menarche, she retired to a menstrual hut and began fasting in order to gather her spiritual power. The village women visited her and shared their secrets of female power. In the night, both male and female members of her family performed sacred menstrual dances. From then on, whenever she menstruated, she withdrew to her menstrual hut, where she communed with the spirit world through dreams and fasting.

The Ohlone flourished until the late eighteenth century and the arrival of the Catholic missionaries led by Junipero Serra (who was beatified in 1988 in preparation for sainthood). The padres, seeking to convert and "civilize" the Ohlones, uprooted and herded them into missions and destroyed their culture. After reducing the Ohlones to servitude, the padres became their "defenders" against the demands of Spanish and Mexican settlers and soldiers. Under the mission system, an estimated 100,000 native Californians—half the population—died (Fogel, 1988). Another quarter perished within ten years of the arrival of the Forty-Niners during the California gold rush.

As children grow up and move away from their neighborhoods, their ethnic identity becomes weaker in terms of language and marriage to others within their group—but they often retain some of the elements of ethnic pride. Their ethnicity is what Herbert Gans (1979) calls "symbolic ethnicity"—an ethnic identity that's used only when the individual chooses. Symbolic ethnicity has little impact on one's day-to-day life. It is not linked to neighborhoods, accents, the use of a foreign language, or one's working life. Others cannot easily identify the person's ethnicity; he or she "looks" American. Nevertheless, ethnicity has emotional significance. A person *is* Irish, Jewish, Italian, or German, for example—not only an American.

We boast of being immigrants . . . when we are no longer subject to the immigrants' ordeal, after we have become certifiable natives and, often, at a time when we and our kind are busily closing the door to others.

MEG GREENFIELD

European ethnic groups differ from one another in many ways. However, a major study of contemporary American ethnic groups (Lieberson and Waters, 1988) found that European ethnic groups are much more similar to one another than they are to African Americans, Latinos, Asian Americans, and Native Americans. The researchers concluded that a European–non-European distinction remains a central division in our society. There are several reasons for this. First, most European ethnic groups no longer have **minority status**, that is, unequal access to economic and political power. Second, because most European

A great many people think they are thinking when they are merely rearranging their prejudices.

WILLIAM JAMES (1842–1910)

Figure 2.6
U.S. Ethnic Composition, 1950–2050 (projected)

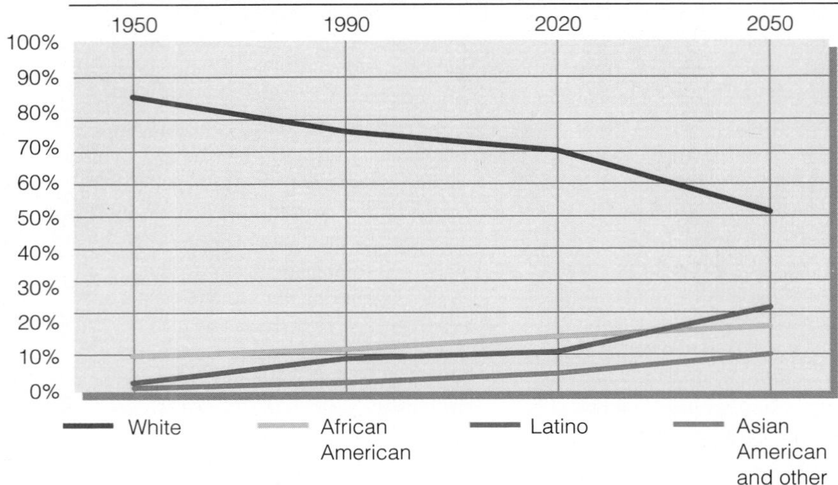

SOURCE: U.S. Bureau of the Census. *Statistical Abstract of the United States.* Washington, D.C.: U.S. Government Printing Office, 1996.

ethnic groups are not physically distinguishable from other white Americans, they are not discriminated against racially.

If anyone means to deliberate successfully about anything, there is one thing he must do from the outset. He must know what it is he is deliberating about; otherwise he is bound to go utterly astray. Now most people fail to realize that they don't know what this or that really is; consequently when they start discussing something, they dispense with any agreed definition, assuming that they know the thing; then later on they naturally find, to their cost, that they agree neither with each other or with themselves.

PLATO, (CA. 428–348 B.C) *PHAEDRUS*

Contemporary American Marriages and Families

Before we begin examining marriages and families in detail, we will look at some areas in which dramatic changes have occurred.

- Cohabitation. In its technical sense, *cohabitation* refers to individuals sharing living arrangements in an intimate relationship, whether these individuals are married or unmarried. In common usage, **cohabitation** refers to relationships in which unmarried individuals share living quarters and are sexually involved. (*Cohabitation* and *living together* are often used interchangeably.) A cohabitating relationship may be similar to marriage in many of its functions and roles, but it does not have equivalent legal sanctions or rights.
- Marriage. A combination of factors including the women's movement, shifting demographics, family policy, and changing values, particularly as they relate to sexuality, have altered the meaning of marriage and the role it plays in people's lives.
- Separation and Divorce. **Separation** occurs when two married people no longer live together. It may or may not lead to divorce. Many more people separate than divorce. **Divorce** is the legal dissolution of a marriage. Over the last thirty years, divorce has changed the face of marriage and the family in America. The divorce rate is two to three times what it was for our parents and grandparents. About half of all those who currently marry will divorce within seven years.
- Divorce has become so widespread that many scholars are beginning to view it as one variation of the normal life course of American marriages (Raschke, 1987). The high divorce rate does not indicate that Americans devalue

As to marriage or celibacy, let a man take which course he will, he will be sure to repent it.

SOCRATES (4TH CENTURY B.C.)

Figure 2.7
Marital Status By Ethnicity: 1970 and 1993

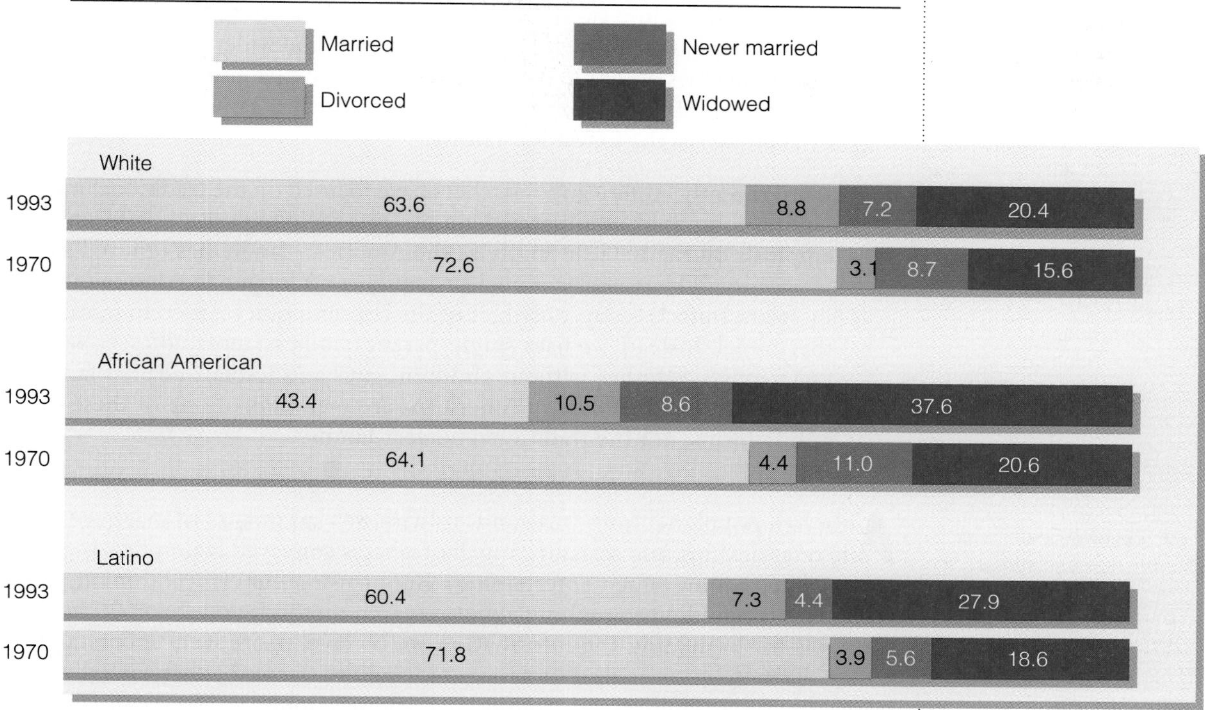

SOURCE: Saluter, Arlene F. *Marital Status and Living Arrangement*: March, 1993, U.S. Bureau of the Census, Current Population Reports, Series P29-478. U.S. Government Printing Office, Washington, D.C., 1994.

marriage, however. Paradoxically, Americans may divorce because they value marriage so highly: If a marriage does not meet their standards, they divorce to marry again. They hope that their second marriages will fulfull the expectations that their first marriages failed to meet (Furstenberg and Spanier, 1987).

- Contemporary divorce patterns create three common experiences for American marriages and families: single-parent families, remarriage, and stepfamilies. Because of their widespread incidence, these variations are becoming part of our normal marriage and family patterns. They no longer represent part of the normal life cource. The majority of young Amerians will probably have some experience of single-parent families, remarriage, or stepfamilies either as children or adults.
- Remarriage. Half of all recent marriages are remarriages for at least one partner (Ganong and Coleman, 1994). Although remarriages are increasing at a faster rate than first marriages, remarriage rates are dramatically different for different ethnic groups. Remarriage rates for African Americans have been dropping disproportionately compared with rates for whites during the last twenty years. African Americans remarry at about one-quarter the rate of whites. Latinos remarry at about half the rate of non-Latino whites.

    Those who remarry are usually older, have more experience in both life and work, and have different expectations than those who marry for the first time. Remarriages also may create stepfamilies. When remarriages include children, a person may become not only a husband or a wife but also a stepfather or stepmother. Ironically, despite the hopes and experience of those who remarry, their divorce rate is about the same as for those who marry for the first time.

- Family. During the twentieth century, Americans have tended to identify the nuclear family as the family. But the older (and wider) definition of *family* also includes other kin, especially in-laws, grandparents, aunts and uncles, and cousins. To distinguish it from the the nuclear family, this wider family is known as the **extended family**.

Until recently, American researchers have focused on the traditional nuclear family, composed of a married couple and their biological children. The emphisis on the nuclear family as the American family has ignored a simple reality: The traditional nuclear family is no longer the dominant family form. In fact, today, only half of the contemporary Amerian families fit this model. Instead, we have single-parent families, stepfamilies, cohabitating families, families without children, gay- and lesbian-headed families, and so on. The majority of Americans are members of one of these other family forms, not the traditional nuclear family.

Everybody gets so much information all day long that they lose their common sense.

GERTRUDE STEIN (1874-1946)

Although popular culture surrounds us with myriad images of marriage, family relationships, and sexuality, much of what is conveyed is simplified, stereotypical, and shallow (albeit entertaining). But by using our critical thinking skills and by understanding something about research methods and theories, we have the tools for evaluating the information we receive. Moreover, understanding ethnic diversity gives us the basis for exploring the strengths that such diversity brings to us, our families, our community, and our country. The methodologies and theories we use allow us to examine objectively both marriage and the family. They allow us to step outside our own individual experiences to view marriage and family from a broader perspective.

Summary

- We need to be alert to maintain *objectivity* in our consideration of different forms of family lifestyle. The *values* we have about what makes a successful family can cause us to decide ahead of time that certain family lifestyles are "abnormal," because they differ from our own experience or preferences. *Opinions*, *biases*, and *stereotypes* are ways of thinking that lack objectivity.
- *Fallacies* are errors in reasoning. Two common types of fallacies are *egocentric fallacies* and *ethnocentric fallacies.* These reflect errors in our basic presuppositions that come either from thinking all people are, or should be, the same as we are, or that our way of living is superior to all others.
- The *advice/information genre* transmits information both to entertain and to inform. The information is generally oversimplified so that it does not interfere with the genre's primary purpose, entertainment. Much of the information or advice conveys *norms*—cultural rules or standards.
- Family researchers use the **scientific method**—well-established procedures used to collect information. They may ask the same questions of large numbers of persons, using the *quantitative* method of research, or they may study smaller groups or individuals in more depth, using *qualitative* methods of research. Analyzing information from public sources (newspapers, court records, and media) is termed *secondary data analysis.*
- Research data come from surveys, clinical studies, and direct observation, which are all *correlational* types of studeies, in which naturally occurring variables are measured against each other. Data are also obtained from experimental research.
- *Surveys* use questionnaires and interviews. They are most useful for dealing with societal or general

issues rather than personal or small-group issues. Inherent problems with the survey method include (1) volunteer bias or an unrepresentative sample, (2) individuals' lack of self-knowledge, and (3) underreporting of undesirable or unconventional behavior.

- *Clinical research* involves in-depth examinations of individuals or small groups who have entered a clinical setting for the treatment of psychological or relationship problems. The primary advantage of clinical studies is that they allow in-depth *case studies;* their primary disadvantage is that the people coming into a clinic are not representative of the general population.
- *Observational research* consists of studies in which interpersonal behavior is examined in a natural setting, such as the home, by an unobtrusive observer.
- In *experimental research*, the researcher manipulates *variables.* Such studies are of limited use in marriage and family research because of the difficulty of controlling behavior and duplicating real-life conditions.
- Theories attempt to provide frames of reference for the interpretation of data. Theories of marriage and families include the symbolic interaction theory, social exchange theory, structural functionalism, conflict theory, and family systems theory. The *symbolic interaction theory* examines how people interact and how we interpret or define others' actions through the symbols they communicate (their words, gestures, and actions). Social interactionists study how *social roles* and personality interact. Drawbacks to the approach include (1) the tendency to minimize power in relationships, (2) the failure to account fully for the psychological aspects of human life, (3) an emphasis on individualism and personal fulfillment at the expense of the marital or family unit, and (4) inadequate attention to the social context.
- *Social exchange theory* suggests that we measure our actions and relationships on a cost-benefit basis. People seek to maximize their rewards and minimize their costs to gain the most favorable outcome. A corollary to exchange is *equity*: Exchanges must balance out, or hard feelings are likely to ensue. Exchanges in marriage can be either cooperative or competitive. Exchange theory has been criticized because (1) it assumes that individuals are rational and calculating in relationships, and (2) it says that the value of costs, rewards, and resources can be gauged.
- *Structural functionalism* looks at society and families as an organism containing different structures. Each structure has a function. Structural functionalists study three aspects of the family: (1) the functions the family serves for society, (2) the functional requirements performed by the family for its survival, and (3) the needs of individual members that are met by the family. Family functions are usually divided along gender lines. Criticisms include (1) the inability to test the theory empirically, (2) the difficulty in ascertaining what function a particular structure serves, and (3) the theory's conservative bias against viewing change as functional.
- *Conflict theory* assumes that individuals in marriages and families are in conflict with each other. Power is often used to resolve the conflict. Four important sources of power are legitimacy, money, physical coercion, and love. Criticism of conflict theory includes (1) its politically based view of human nature, (2) its assumption that differences lead to conflict, and (3) the difficulty in measuring and evaluating conflict.
- *Family systems theory* approaches the family in terms of its structure and pattern of interactions. Systems analysts believe that (1) interactions must be studied in the context of the family, (2) family structure can be seen only in the family's interactions, (3) the family is a purposely system seeking *homeostasis* (stability) and (4) family systems are transformed over time. Family systems theory is not a coherent, systematic theory, and it has been criticized because it is based on clinical studies of nonrepresentative families.
- The *feminist perspective* provides an orienting focus for considering *gender* differences relating to family and social issues. In the writing, research, and advocacy of the feminist movement, the goals are to help clarify and remove oppressive conditions and barriers to opportunities for women. The postmodern feminist position has been expanded to include constraints affecting black/white and gay/straight dichotomies.
- *Family development theory* looks at the changes in the family, beginning with marriage and proceeding through seven more sequential stages. This life-cycle model gives us insights into the complexities of family life and the interacting influences of changing roles and circumstances through time. Family development theory has been criticized for assuming that all families follow the sequence of traditional two-parent families. However, family life-cycle models

have been developed to include nontraditional forms such as single-parent families, stepfamilies, and families headed by lesbian or gay parents.

- The *Family Circumplex Model* examines family fucntions based on levels of *cohesion* and *adaptability*. The extremes of cohesion are *disengagement* and *enmeshment*. Levels of cohesion and adaptability vary throughout the family life cycle. Familes function best when cohesion and adaptability are balanced.
- Contemporary families are culturally diverse. The terms *ethnic group*, *racial group*, and *minority group* are conceptually distinct. An *ethnic group* is a group of people distinct from other groups because of cultural characteristics, such as language, religion, and customs that are transmitted from one generation to another. A *racial group* is a group of people, such as whites, blacks, and Asians, classified according to *phenotype*, anatomical and physical characteristics. A *minority group* is a group whose *status* (position in the social hierarchy) places its members at an economic, social, and political disadvantage.
- African Americans are the largest ethnic group in the United States. *Socioeconomic status* is an important element in understanding African-American families. Many of the problems African-American families experience are the result of low socioeconomic status rather than family structure.
- Several features of African-American families are notable. Because of economic necessity, women traditionally have been employed, which has given them important economic roles in the family and more egalitarian relationships. Kinship bonds are especially important; they provide emotional and economic assistance in times of need. African Americans have a strong tradition of *familialism*, with the emphasis on intergenerational ties. The African-American community values children highly. African Americans are much more likely than whites to live in *extended households*.
- Latinos are the fastest growing and second largest ethnic group, as a result of immigration and a higher birthrate than the general population. Important factors in understanding Latino culture are (1) ethnic diversity within the culture, and (2) the role of socioeconomic status. Latinos emphasize extended kin relationships, cooperation, and mutual assistance. *La familia* includes not only the nuclear family but also the extended family. Bilingualism helps maintain ethnic identity.
- Asian Americans are the third largest ethnic group in the United States. Immigration has contributed heavily to the dramatic recent increase in the Asian-American population. Sixty-six percent of Asian Americans are foreign born. The largest Asian-American groups are Chinese Americans, Filipino Americans, and Japanese Americans. The more recent immigrants retain more culturally distinct characteristics, such as family structure and values, than do older groups.
- Native Americans include nearly 2 million Americans. Tribal identity is their key identity. Powwows are social gatherings of diverse tribes that center around drumming, singing, and dancing. Over half of Native Americans live in cities, although many remain in contact with their home reservation. Extended families are important and are often based on clan membership. About 53 percent of Native Americans are married to non-Indians.
- Ethnic identity among Americans of European descent has been growing, especially among working-class families. For many, their ethnicity is symbolic and has little impact on day-to-day life. Most members of European ethnic groups are physically indistinguishable from other white Americans and no longer have minority status.
- Unmarried *cohabitation* is when a couple lives together and is sexually involved. *Divorce* is the legal dissolution of marriage following separation. High divorce rates have made single-parent families and stepfamilies important family forms in the contemporary United States. Divorce, remarriage, single-parent families, stepfamilies, and extended families are normal aspects of the contemporary American marriage system.

Key Terms

adaptability 60
advice/information genre 42
bias 41
case-study method 46
clinical research 46
cohabitation 70
cohesion 60
conflict theory 54
correlational study 47
dependent variable 47
disengagement 60
divorce 70
egocentric fallacy 42
enmeshment 60
ethnic group 60
ethnocentric fallacy 42
experimental research 47
extended family 72
extended household 64
fallacy 42
familialism 64
Family Circumplex Model 60
family development theory 59
family systems theory 56
feminist perspective 57
gender 58
homeostasis 57
independent variable 47
interaction 49
minority group 61
minority status 69
norm 42
objective statements 41
objectivity 41
observational research 46
opinion 41
phenotype 60
qualitative 45
quantitative 44
racial group 60
scientific method 44
secondary data analysis 45
separation 70
social role 50
social exchange theory 51
socioeconomic status 62
status 61
stereotype 41
structural functionalism 53
subsystem 53
survey research 45
symbol 49
symbolic interaction 48
theory 48
value judgment 41
values 41
variable 47

Suggested Readings

Billingsley, Andrew. *Climbing Jacob's Ladder.* New York: Simon & Schuster, 1992. A leading family scholar examines the African-American family in historical context.

Chow, Esther Ngan-Ling, Doris Wilkinson, and Maxine Baca Zinn, eds. *Race, Class, and Gender: Common Bonds, Different Voices.* Newbury Park, CA: Sage, 1996. An interdisciplinary look at the intersecting patterns that occur among peoples from different backgrounds and experiences.

Copeland, Ann, and Kathleen White. *Studying Families.* Newbury Park, CA: Sage Publications, 1991. Research issues in studying diverse families, such as cohabiting parents, gay and lesbian families, and childless marriages.

DeGenova, Mary Kay. *Families in Cultural Context: Strengths and Challenges in Diversity.* Mountain View, CA: Mayfield, 1997. A look at the experiences of eleven different American ethnic groups—their cultures, family systems, strengths, challenges, and the myths and stereotypes they confront.

Klein, David M., and James M. White. *Family Theories: An Introduction.* Thousand Oaks, CA: Sage, 1996. An in-depth survey of six major theoretical frameworks.

Kohis, R., and J. Knight. *Developing Intercultural Awareness.* Yarmouth, ME: Intercultural Press, 1995. An excellent guide for those who wish to broaden their knowledge and understanding about other cultures.

Maynard, Mary, and Jan Purvis, eds. *Researching Women's Lives from a Feminist Perspective.* Bristol, PA: Taylor & Francis, 1994. The authors examine and discuss research techniques from a feminist perspective.

McAdoo, Harriette, ed. *Black Families*, 3rd ed. Thousand Oaks, CA: Sage Publications, 1996. Examines conceptual, historical, demographic, and economic aspects of African-American family systems.

——.*Family Ethnicity: Strength in Diversity.* Newbury Park, CA: Sage Publications, 1993. Explores the role of ethnicity in family life among African Americans, Latinos, Native Americans, Muslims, and Asian Americans in a collection of essays edited by a leading family scholar.

Root, Maria P., ed. *Racially Mixed People in America.* Newbury Park, CA: Sage Publications, 1992. A collection of essays on racially mixed marriages and children—whose births are increasing 26 times as fast as births in any other group—that argues that the multiracial experience is healthy and normal.

Staples, Robert, ed. *The Black Family: Essays and Studies*, 4th ed. Belmont, CA: Wadsworth Publishing, 1994. A valuable collection of essays.

Sussman, Marvin, and Suzanne Steinmetz, eds. *Handbook of Marriage and the Family.* New York: Plenum Press, 1987. The definitive reference book on marriage and the family.

Taylor, Ronald, ed. *Minority Families in the United States.* Englewood Cliffs, NJ: Prentice-Hall, 1994. An outstanding collection of essays on African-American, Latino, Asian-American, and Native American families.

CHAPTER 3

Contemporary Gender Roles

PREVIEW

To gain a sense of what you already know about the material covered in this chapter, answer "True" or "False" to the statements below.

1. Gender roles generally reflect the instinctive nature of males and females. True or false?
2. Gender roles are influenced by ethnicity. True or false?
3. Psychologists believe that the increasing similarity of male and female gender roles represents a decline in mental health. True or false?
4. Among immigrants, men generally experience greater gender-role stress than women. True or false?
5. It is often possible to tell the gender of an infant based on his or her level of activity. True or false?
6. Peers are the most important influence on gender-role development from adolescence through old age. True or false?
7. For African Americans, the traditional female gender role includes both employment and motherhood. True or false?
8. Women are significantly underrepresented on television. True or false?
9. It is relatively easy to change one's gender-role behaviors if one wants to do so. True or false?
10. As they age, middle-aged and older men gain status, whereas women lose it. True or false?

Answers

1 False, 2 True, 3 False, 4 False, 5 False, 6 False, 7 True, 8 True, 9 False, 10 True

OUTLINE

Are men and women basically the same or different? The traditional view of masculinity and femininity sees men and women as polar opposites. Our popular terminology, in fact, reflects this view. Men and women refer to each other as the "opposite sex." Our gender stereotypes fit this pattern of polar differences: Men are aggressive, and women are passive; men are **instrumental** (task oriented), and women are **expressive** (emotion oriented); men are rational, and women are irrational; men want sex, and women want love; men are from Mars, and women are from Venus (Duncombe and Marsden, 1993; Lips, 1997).

As you will see in this chapter, men are not from Mars, and women are not from Venus. We both inhabit planet Earth. The *perception* of male-female differences is far greater than the actual differences themselves (Hare-Mustin and Murecek, 1990b).

In this chapter we examine several gender and socialization theories and discover how we learn to act masculine and feminine. Next, we explore changing gender roles and the various limitations that gender roles place on us. Finally, we discuss androgynous gender roles.

Understanding Gender and Gender Roles

Studying Gender

Before we continue, we need to define several key terms to avoid confusion. These terms are *gender, role, gender role, gender-role stereotype, gender-role attitude,* and *gender-role behavior.* **Gender** refers to male or female, often in a social sense; *sex* refers to male or female in a biological sense. **Role** refers to the culturally defined expectations that an individual is expected to fulfill in a given situation in a particular culture. **Gender roles** are the roles that a person is expected to perform as a result of being male or female in a particular culture. (The term *gender role* is gradually replacing the traditional term *sex role.*) A **gender-role stereotype** is a rigidly held and oversimplified belief that all males and females, as a result of their sex, possess distinct psychological and behavioral traits. Stereotypes tend to be false not only for the group as a whole but also for any individual member of the group. The "men are aggressive" stereotype not only is untrue for men

A map is not the territory.

Alfred Korzybski (1879–1950)

Roles come with costumes and speeches and stage directions. In a role, we don't have to think.

Ellen Goodman

as a group, but it also cannot be applied randomly to any individual man: Michael or Danny, in fact, may not be aggressive. Even if the generalization is statistically valid in describing a group average (males are taller than females), such generalizations do not necessarily predict whether Jason will be taller than Tanya. **Gender-role attitudes** refer to the beliefs we have of ourselves and others regarding appropriate male and female personality traits and activities. **Gender-role behaviors** refer to the actual activities or behaviors we engage in as males and females. When we discuss gender roles, it is important not to confuse stereotypes with reality or confuse attitudes with behavior.

Most gender-role studies have focused on the white middle class. Very little is known about gender roles among African Americans, Latinos, Asian Americans, and other ethnic groups (Binion, 1990; Reid and Comas-Diaz, 1990; True, 1990). Students and researchers must be careful not to project onto other groups the gender-role concepts or aspirations based on their own groups. Too often such projections can lead to distortions or moral judgments.

Throughout history the more complex activities have been defined and re-defined, now as male, now as female—sometimes as drawing equally on the gifts of both sexes. When an activity to which each sex could have contributed is limited to one sex, a rich, differentiated quality is lost from the activity itself.

Margaret Mead (1901–1978)

Gender and Gender Roles

At birth, we are identified as either male or female. This identity, based on genitalia, is called **gender identity,** and we learn it at a very young age. It is perhaps the deepest concept we hold of ourselves. The psychology of insults suggests this depth, for few things offend a person as much as to be tauntingly characterized as a member of the "opposite" sex. Gender identity determines many of the directions our lives will take—for example, whether we will fulfill the role of husband or wife, father or mother. When the scripts are handed out in life, the one you receive depends largely on your gender.

Each culture determines the content of gender roles in its own way. Among the Arapesh of New Guinea, both males and females possess what we consider feminine traits. Men and women alike tend to be passive, cooperative, peaceful, and nurturing. The father is said to "bear a child" as well as the mother; only the father's continual care can make a child grow healthily, both in the womb and in childhood. Eighty miles away, the Mundugumor live in remarkable contrast to the peaceful Arapesh. "Both men and women," Margaret Mead (1975) observes, "are expected to be violent, competitive, aggressively sexed, jealous, and ready to see and avenge insult, delighting in display, in action, in fighting." Mead concludes, "Many, if not all of the personality traits which we have called masculine or feminine are as lightly linked to sex as are the clothing, the manners, and the form of head-dress that a society at a given period assigned to either sex. . . ." Biology creates males and females, but culture creates masculinity and femininity.

Generally the only limit on women's performing traditional male work is social custom rather than individual ability. Most jobs can be done equally well by women or men.

Masculinity and Femininity: Opposites or Similar?

Until the last generation, the **bipolar gender role** was the dominant model used to explain male-female differences. In this model, males and females are seen as polar opposites, with males possessing exclusively instrumental traits and females possessing exclusively expressive ones. Although sociologists no longer use this model, a review of gender-role stereotypes between 1972 and 1988 found little change in American beliefs (Bergen and Williams, 1991).

Traditional views of masculinity and femininity as opposites have several implications. First, if one differs from the male or female stereotype, he or she is seen to become more like the other gender. If a woman is sexually assertive, for example, not only is she less feminine but she is believed to be more masculine. Similarly, if a man is nurturing, not only is he less masculine but he is also more feminine. Second, because males and females are opposites, they cannot share the same traits or qualities. A "real man" possesses exclusively masculine traits and behaviors, and a "real woman" possesses exclusively feminine traits and behaviors. A man is assertive, and a woman is receptive; in reality, men and women are both assertive and receptive. Third, because males and females are viewed as opposites, they are believed to have little in common with each other; a "war of the sexes" is the norm. Men and women can't understand each other, nor can they expect to do so. Difficulties in their relationships are attributed to their "oppositeness."

The fundamental problem with the view of men and women as opposites is that it is erroneous. Men and women are significantly more alike than they are different. Our culture, however, has encouraged us to look for differences and, when we find them, to exaggerate their degree and significance. It has taught us to ignore the single most important fact about males and females: that we are both human. As human beings, we are significantly more alike biologically and psychologically than we are different. As men and women, we share similar respiratory, circulatory, neurological, skeletal, and muscular systems. (Even the penis and the clitoris evolved from the same undifferentiated embryonic structure.) Hormonally, both men and women produce androgens and estrogen (but in different amounts). Where men and women biologically differ most significantly is in terms of their reproductive functions: Men impregnate, whereas women gestate and nurse. Beyond these reproductive differences, biological differences are not great.

A recent review of gender differences in social behavior and experiences (Lips, 1997) found very little inherent difference between males and females. In terms of social behavior, studies suggest that men are more aggressive both physically and verbally than women; the gender difference, however, is not large. Most differences can be traced to gender-role expectations, male-female status, and gender stereotyping.

Although we possess traits of both genders, most of us feel either masculine or feminine; we usually do not doubt our gender (Heilbrun, 1982). Unfortunately, when people believe that individuals should *not* have the attributes of the other gender, males suppress their expressive traits, and females suppress their instrumental traits. As a result, the range of human behaviors is limited by a person's gender role. As psychologist Sandra Bem (1975) points out, "Our current system of sex role differentiation has long since outlived its usefulness, and . . . now serves only to prevent both men and women from developing as full and complete human beings."

Reflections

To get a sense of the significance of gender roles, would you be a similar person if you were the other gender? Would your plans be different? Your activities, feelings, and goals?

One half of the world cannot understand the pleasures of the other.

JANE AUSTEN (1775–1817)

The fact that we are all human beings is infinitely more important than all the peculiarities that distinguish humans from one another.

SIMONE DE BEAUVOIR (1908–1986)

If you want anything said, ask a man. If you want anything done, ask a woman.

MARGARET THATCHER

Gender Schema: Exaggerating Male-Female Differences

Although actual differences between males and females are minimal, culture exaggerates them or creates differences where none otherwise existed (Carter, 1987c). One way that culture does this is by creating a gender schema. (A schema, you will recall from Chapter 1, is a set of interrelated ideas that help us process information by categorizing it in useful ways.) We often categorize people by age, ethnicity, nationality, physical characteristics, and so on. Gender is one such way of categorizing. Bem (1983) observes that although gender is not inherent in inanimate objects or in behaviors, we treat many objects and behaviors as if they were masculine or feminine. These gender divisions form a complex structure of associations that affect our perceptions of reality. Bem refers to this cognitive organization of the world as **gender schema** (Bem, 1983; Spence, 1993).

One's only real life is the life one never leads.

Oscar Wilde (1854–1900)

Knowledge about different aspects of gender is acquired early in childhood. It is usually completed between ages two and four. One study found that the earlier they learn about it, the more important gender is for children (Fagot, Leinbach, and O'Boyle, 1992). Similarly, such children are more likely to view the world stereotypically (Levy, 1994).

Adults who have a strong gender schema quickly divide people's behavior, personality characteristics, objects, and so on into masculine versus feminine categories. They disregard information that does not fit their gender schema. One study (Hudak, 1993) found, for example, that men who schematized gender tended to view women in stereotypically feminine terms. Another study found that men, more than women, tended to view certain types of occupations—such as physician, nurse, and secretary—as more appropriate for one gender than for the other (Beggs and Doolitle, 1993).

Did you know?

In a recent review of literature concerning gender, neither men nor women were found to be more likely to dominate, more susceptible to influence, or more nurturing, altruistic, or empathetic (Lips, 1997).

When we initially meet a person, we unconsciously note whether the individual is male or female and respond accordingly (Skita and Maslach, 1990). But what happens if we can't immediately classify a person as male or female? Many of us feel uncomfortable because we don't know how to act if we don't know the gender. This is true even if gender is irrelevant, as in a bank transaction, walking past someone on the street, or answering a query about the time. ("Was that a man or woman?" a person may ask in exasperation, although it really makes no difference.) An inability to tell a person's gender may provoke a hostile response. As Hilary Lips (1997) writes:

> It is unnerving to be unsure of the sex of the person on the other end of the conversation. The labels *female* and *male* carry powerful associations about what to expect from the person to whom they are applied. We use the information the labels provide to guide our behavior toward other people and to interpret their behavior toward us.

He is playing masculine. She is playing feminine. He is playing masculine because she is playing feminine. She is playing feminine because he is playing masculine. He is playing the kind of man that she thinks the kind of woman she is playing ought to admire. She is playing the kind of woman that he thinks the kind of man he is playing ought to desire.

Betty Roszak and Theodore Roszak

Our need to classify people as male or female and its significance is demonstrated in the well-known Baby X experiment (Condry and Condry, 1976). In this experiment, three groups played with an infant known as Baby X. The first group was told that the baby was a girl, the second group was told that the baby was a boy, and the third group was not told what gender the baby was. The group that did not know what gender Baby X was felt extremely uncomfortable, but the group participants then made a decision based on whether the baby was "strong" or "soft." When the baby was labeled a boy, its fussing behavior was called "angry"; when the baby was labeled a girl, the same behavior was called "frustrated." Once the baby's gender was determined (whether correctly or not), a train of responses followed that could have profound consequences in his or her socialization. The study was replicated numerous times with the same general results. (Even birth congratulations cards reflect gender stereotyping of newborns [Bridges, 1993]). A review of studies on infant labeling found that gender stereotyping is strongest among children, adolescents, and college students (Stern and Karraker, 1989). Stereotyping diminishes among adults, especially among infants' mothers (Vogel et al., 1991).

He who knows the masculine but keeps to the feminine will be in the whole world's channel.

Taoist aphorism

Processing information by gender is important in cultures such as ours. First, gender-schema cultures make multiple associations between gender and

The war between the sexes is the only one in which both sides regularly sleep with the enemy.

QUENTIN CRISP

other non-gender-linked qualities, such as affection and strength. Our culture regards affection as a feminine trait and strength as a masculine one. Second, such cultures make gender distinctions important, using them as a basis for norms, status, taboos, and privileges. Men are assigned leadership positions, for example, whereas women are placed in the rank and file (if not kept in the home); men are expected to be sexually assertive, and women sexually passive.

Gender and Socialization Theories

There are several prominent theories used to explain the significance of gender in our culture and how we learn our gender roles. These are gender theory, social learning theory, and cognitive development theory.

Gender Theory

If sex is a war, I am a conscientious objector; I will not play.

MARGE PIERCY

In studying gender, feminist scholars begin with two assumptions: First, that male-female relationships are characterized by power issues, and second, that society is constructed in such a way that males dominate females. They argue that on every level, male-female relationships—whether personal, familial, or societal—reflect and encourage male dominance, putting females at a disadvantage. Male dominance is neither natural nor inevitable, however. Instead, it is created by social institutions, such as religious groups, government, and the family (Acker, 1993; Ferree, 1991). The question is, how is male-female inequality created?

Social Construction of Gender

The great question . . . which I have not been able to answer, despite my thirty years of research into the feminine soul, is "What does a woman want?"

SIGMUND FREUD (1856–1939)

In the 1980s, gender theory emerged as the most important model explaining inequality. According to gender theory, gender is a **social construct,** an idea or concept created by society through the use of social power. The theory asserts that society may be best understood by how it is organized according to gender, and that social relationships are based on the *socially perceived* differences between females and males that are used to justify unequal power relationships (Scott, 1986; White 1993). Imagine, for example, an infant crying in the night. In the mother/father parenting relationship, which parent gets up to take care of the baby? In most cases, the mother does because (1) women are socially perceived to be nurturing and (2) it's the woman's "responsibility" as mother (even if she hasn't slept in two nights and is employed full-time).

Gender theory focuses on (1) how specific behaviors (such as nurturing or aggression) or roles (such as child rearer, truck driver, or secretary) are defined as male or female; (2) how labor is divided into man's work and woman's work, both at home and in the workplace; and (3) how different institutions bestow advantages on men (such as male-only clergy in many religious denominations or women receiving less pay than men for the same work).

The key to the creation of gender inequality is the belief that men and women are "opposite" sexes—that they are opposite each other in personalities, abilities, skills, and traits. Furthermore, the differences between the genders are unequally valued: Reason and aggression (defined as male traits) are considered more valuable than sensitivity and compliance (defined as female traits). Making men and women appear to be opposite and of unequal value requires the suppression of natural similarities by the use of social power. The exercise of social power might take the form of greater societal value being placed on looks than on achievement for women, sexual harassment of women in the workplace or university, patronizing attitudes toward women, and so on.

Reflections

Using gender theory, examine the division of labor in your family. How is housework divided? How is unpaid household work valued in comparison with employment in the workplace?

Social Learning Theory

Social learning theory is derived from behaviorist psychology. In explaining our actions, behaviorists emphasize observable events and their consequences rather than internal feelings and drives. According to behaviorists, we learn attitudes and behaviors as a result of social interactions with others (hence, the term *social learning*).

The cornerstone of social learning theory is the belief that consequences control behavior. Acts that are regularly followed by a reward are likely to occur again; acts that are regularly followed by a punishment are less likely to recur. Girls are rewarded for playing with dolls ("What a nice mommy!"), but boys are not ("What a sissy!").

This behaviorist approach has been modified recently to include cognition—that is, mental processes (such as evaluation and reflection) that intervene between stimulus and response. The cognitive processes involved in social learning include our ability to (1) use language, (2) anticipate consequences, and (3) make observations. These cognitive processes are important in learning gender roles. By using language, we can tell our daughter that we like it when she does well in school and that we don't like it when she hits someone. A person's ability to anticipate consequences affects behavior. A boy doesn't need to wear lace stockings in public to know that such dressing will lead to negative consequences. Finally, children observe what others do. A girl may learn that she "shouldn't" play video games by seeing that the players in video arcades are mostly boys.

Nobody was born a man; you earned your manhood, provided you were good enough.

NORMAN MAILER

We also learn gender roles by imitation, according to social learning theory. Learning through imitation is called **modeling.** Most of us are not even aware of the many subtle behaviors that make up gender roles—the ways in which men and women use different mannerisms and gestures, speak differently, use different body language, and so on. We don't "teach" these behaviors by reinforcement. Children tend to model friendly, warm, and nurturing adults; they also tend to imitate adults who are powerful in their eyes—that is, adults who control access to food, toys, or privileges. Initially, the most powerful models that children have are their parents. As children grow older and their social world expands, so do the number of people who may act as their role models: siblings, friends, teachers, media figures, and so on. Children sift through the various demands and expectations associated with the different models to create their own unique selves.

Women are made, not born.

SIMONE DE BEAUVOIR (1908–1986)

Cognitive Development Theory

In contrast to social learning theory, cognitive development theory focuses on the child's active interpretation of the messages he or she receives from the environment. Whereas social learning theory assumes that children and adults learn in fundamentally the same way, cognitive development theory stresses that we learn differently, depending on our age. Swiss psychologist Jean Piaget (1896–1980) showed that children's abilities to reason and understand change as they grow older.

The moment of change is the only poem.

Adrienne Rich

Lawrence Kohlberg (1969) took Piaget's findings and applied them to how children assimilate gender-role information at different ages. At age two, children can correctly identify themselves and others as boys or girls, but they tend to base this identification on superficial features, such as hair and clothing. Girls have long hair and wear dresses; boys have short hair and never wear dresses. Some children even believe they can change their sex by changing their clothes or hair length. They don't identify sex in terms of genitalia, as older children and adults do. No amount of reinforcement will alter their views because their ideas are limited by their developmental stage.

When children are six or seven, they begin to understand that gender is permanent; it is not something they can change as they can their clothes. They acquire this understanding because they are capable of grasping the idea that basic characteristics do not change. A woman can be a woman even if she has short hair and wears pants. Oddly enough, although children can understand the permanence of sex, they tend to insist on rigid adherence to gender-role stereotypes. Even though boys can play with dolls, children of both sexes believe they shouldn't because "dolls are for girls." Researchers speculate that children exaggerate gender roles to make the roles "cognitively clear."

According to social learning theory, boys and girls learn appropriate gender-role behavior through reinforcement and modeling. But according to cognitive development theory, once children learn that gender is permanent, they independently strive to act like "proper" girls or boys. They do this on their own because of an internal need for congruence, the agreement between what they know and how they act. Also, children find performing the appropriate gender-role activities rewarding in itself. Models and reinforcement help show them how well they are doing, but the primary motivation is internal.

Learning Gender Roles

Although biological factors, such as hormones, clearly are involved in the development of male and female differences, the extent of biological influences is not well understood. Moreover, it is difficult to analyze the relationship between biology and behavior, for learning begins at birth. In this section, we'll explore gender-role learning from infancy through adulthood.

Gender-Role Learning in Childhood and Adolescence

We unwittingly set the stage for feelings of inadequacy and inferiority. The more rigidly we define ourselves the less likely we are able to cope with the infinite variety of life.

Ernest Rossi

In our culture, infant girls are usually held more gently and treated more tenderly than boys, who are ordinarily subjected to rougher forms of play. The first day after birth, parents tend to describe their daughters as soft, fine featured, and small, and their sons as hard, large featured, big, and attentive. Fathers tend to stereotype their sons more extremely than mothers do (Fagot and Leinbach, 1987). Although it is impossible for strangers to know the gender of a diapered baby, once they learn the baby's gender, they respond accordingly. Such gender-role socialization occurs throughout our lives. By middle childhood, although conforming to gender-role behavior and attitudes becomes increasingly important, there is still considerable flexibility (Absi-Semaan, Crombie, and Freeman,

1993). It is not until late childhood and adolescence that conformity becomes most characteristic. The primary agents forming our gender roles are parents, teachers, peers, and the media, which we describe below.

Parents as Socializing Agents

What are little girls made of?
Sugar and spice
And everything nice.
That's what little girls are made of.

What are little boys made of?
Snips and snails
And puppy dogs' tails.
That's what little boys are made of.
NURSERY RHYME

During infancy and early childhood, a child's most important source of learning is the primary caretaker—often both parents, but also often just the mother, father, grandmother, or someone else. Most parents are not aware that their words and actions contribute to their children's gender-role socialization (Culp et al., 1983). Nor are they aware that they treat their sons and daughters differently because of their gender. Although parents may recognize that they respond differently to sons than to daughters, they usually have a ready explanation—the "natural" differences in the temperament and behavior of girls and boys. Parents may also believe that they adjust their responses to each particular child's personality. In an everyday living situation that involves changing diapers, feeding babies, stopping fights, and providing entertainment, it may be difficult for harassed parents to recognize that their own actions may be largely responsible for the differences they attribute to nature. The role of nature cannot be ignored completely, however. Many parents who have conscientiously tried to raise their children in a nonsexist way have been frustrated to find their toddler sons shooting each other with carrots or their daughters primping in front of the mirror.

Children are socialized in gender roles in many ways. Children's literature, for example, typically depicts girls as passive and dependent, whereas boys are instrumental and assertive (Kortenhaus and Demarest, 1993). In general, children are socialized by their parents through four very subtle processes: manipulation, channeling, verbal appellation, and activity exposure (Oakley, 1985):

- *Manipulation.* Parents manipulate their children from infancy onward. They treat a daughter gently, tell her she is pretty, and advise her that nice girls do not fight. They treat a son roughly, tell him he is strong, and advise him that big boys do not cry. Eventually, children incorporate their parents' views in such matters as integral parts of their personalities.
- *Channeling.* Children are channeled by having their attention directed to specific objects. Toys, for example, are differentiated by gender. Dolls are considered appropriate for girls and cars, for boys.
- *Verbal appellation.* Parents use different words with boys and girls to describe the same behavior. A boy who pushes others may be described as "active," whereas a girl who does the same may be called "aggressive."
- *Activity exposure.* The activity exposure of boys and girls differs markedly. Although both are usually exposed to feminine activities early in life, boys are discouraged from imitating their mothers, whereas girls are encouraged to be "mother's little helpers." Even the chores children do are categorized by gender. Girls may wash dishes, make beds, and set the table; boys are assigned to carry out trash, rake the yard, and sweep the walk (Blair, 1992). The boy's domestic chores take him outside the house, whereas the girl's keep her in it—another rehearsal for traditional adult life. A study of fourth and fifth graders found that television reinforced these stereotypic views Signorielli and Lears, 1992).

Fathers, provoke not your children to anger, lest they be discouraged.
COLOSSIANS 3:21

It is generally accepted that parents socialize their children differently according to gender (Fagot and Leinbach, 1987). Fathers more than mothers pressure their children to behave in gender-appropriate ways. Fathers set higher

One way children learn gender roles is through activity exposure. Girls may be encouraged to help their mothers inside the house, whereas boys may participate with their fathers in outdoor work.

standards of achievement for their sons than for their daughters; fathers emphasize the interpersonal aspects of their relationships with their daughters. Mothers also reinforce the interpersonal aspect of their parent-daughter relationships (Block, 1983).

Both parents of teenagers and the teenagers themselves believe that parents treat boys and girls differently. It is not clear, however, whether parents are reacting to differences or creating them (Fagot and Leinbach, 1987). It is probably both, although by that age, gender differences are fairly well established in the minds of adolescents.

Various studies have indicated that ethnicity and social class are important in socialization (Zinn, 1990; see Wilkinson, Chow, and Zinn, 1992, for scholarship on the intersection of ethnicity, class, and gender). Among whites, working-class families tend to differentiate more sharply than middle-class families between boys and girls in terms of appropriate behavior; they tend to place more restrictions on girls. African-American families tend to socialize their children toward more egalitarian gender roles (Taylor, 1994c). There is evidence that African-American families socialize their daughters to be more independent than white families do. Indeed, among African Americans, the "traditional" female role model may never have existed. The African-American female role model in which the woman is both wage earner and homemaker is more typical and more accurately reflects the African-American experience (Lips, 1997).

Reflections

How did your parents influence the development of your gender role? In what ways did you model yourself after your same-gender parent? In what ways are your conceptions of appropriate gender roles similar to or different from those of your parents?

Teachers as Socializing Agents

As children grow older, their social world expands and so do their sources of learning. Around the time children enter day-care centers or kindergarten, teachers (and peers, discussed next) become important influences. Day-care centers, nursery schools, and kindergartens are often a child's first experience in the wider world outside the family. Teachers become important role models for their students. Because most day-care, nursery school, kindergarten, and elementary school teachers are women, children tend to think of child-adult interactions as primarily taking place with women.

Girls generally excel over boys in all areas during grade school, but by middle school boys are better in math, science, history, geography, reading, and spelling. By high school, boys excel in almost all areas (Sadker and Sadker, 1994). Classroom observations find that boys are louder, are more demanding and receive a disproportionate amount of the teacher's attention. Teachers call on boys more often and are more patient with boys in their explanations and more generous in their praise to boys. Girls are praised for their appearance and neatness of work. As a result, girls become more tentative and hesitant as they enter middle school. By high school, they preface their answers with disclaimers: "I'm probably wrong, but . . ." or "I'm not sure, but. . . ." Intelligent girls are often devalued by boys. Only in all-girl schools, write Myra Sadker and David Sadker (1994), do female students tend to assert themselves vigorously in class. They argue that in gender-segregated schools and classes, girls do not have to compete with boys for the teacher's attention. In such schools, girls are not overly concerned with their appearance, and they do not fear that their intelligence will make them undesirable as dates.

Treat people as if they were what they ought to be and you help them become what they are capable of being.

JOHANN GOETHE (1749–1832)

Schools that have developed nonsexist curricula show that ordinary males and females can act and work in nonstereotypical ways (Bigler and Liben, 1992). Research suggests that if schools consciously structured activities involving both girls and boys, there would be considerably more interaction between the two. (Mead and Ignicio, 1992) Furthermore, both girls and boys would benefit by acquiring the skills ordinarily restricted to one sex. Boys could learn homemaking skills, such as cooking, whereas girls could learn more assertiveness skills, such as political debate.

Peers as Socializing Agents

A child's age-mates, or **peers,** become especially important when the child enters school. By granting or withholding approval, friends and playmates influence what games children play, what they wear, what music they listen to, what television programs they watch, and even what cereal they eat. Peer influence is so pervasive that it is hardly an exaggeration to say that in some cases, children's peers tell them what to think, feel, and do.

Peers also provide standards for gender-role behavior in several ways (Carter, 1987b):

- Peers reinforce gender-role norms through play activities and toys. With their friends, girls play with dolls that cry and wet or with glamorous dolls with well-developed figures and expensive tastes. Boys play together with dolls known as "action figures," such as G. I. Joe and Power Rangers, equipped with automatic weapons and bigger-than-life biceps.
- Peers react with approval or disapproval to others' behavior. Smiles encourage a girl to play with makeup or a boy to play with a football.
- Peers influence the adoption of gender-role norms through verbal approval or disapproval. "That's for boys!" or "Only girls do that!" discourages girls from playing with footballs or boys from playing house.
- Children's perceptions of their friends' gender-role attitudes, behaviors, and beliefs encourage them to adopt similar ones in order to be accepted. If a

We forfeit three-fourths of ourselves to be like other people.

ARTHUR SCHOPENHAUER (1788–1860)

girl's female friends play soccer, she is more likely to play soccer. If a boy's male friends display feelings, he is more likely to display feelings.

During adolescence, peers continue to have a strong influence, but research indicates that parents can be more influential than peers (Gecas and Seff, 1991). Parents influence their adolescent's behavior primarily by establishing norms, whereas peers influence others through modeling behavior.

Even though parents tend to fear the worst from their children's peers, peers provide important positive influences. It is within their peer groups, for example, that adolescents learn to develop intimate relationships (Gecas and Seff, 1991). Also, adolescents tend to be more egalitarian in gender roles than do parents, especially fathers (Thornton, 1989).

Did you know?

Men on television give directions to women by a 7-to-3 ratio (Klassen, Jasper, and Schwartz, 1993).

The Media as Socializing Agents

Much of television programming promotes or condones negative stereotypes about gender, ethnicity, age, and gay men and lesbians. Women are significantly underrepresented on television (Signorielli, 1993). Through the 1970s, men outnumbered women on prime-time television 3 to 1; today it is about 2 to 1. (Even on *Sesame Street*, 84 percent of the characters were male in 1992, compared with 76 percent five years earlier ["Muppet Gender Gap," 1993]). Blonde women are disproportionately represented; they are also more provocatively dressed ("Depiction of Women," 1993). Women in traditional roles as wives, mothers, and homemakers continue to dominate women's magazines. Over the last thirty years, however, career themes also have increased (Demarest and Garner, 1992).

Did you know?

Ninety percent of narrators in commercials are men, especially in commercials directed toward women in family roles (Klassen, Jasper, and Schwartz, 1993).

TV women are portrayed as emotional and needing emotional support; they are also sympathetic and nurturing. Not surprisingly, women are usually portrayed as wives, mothers, or sex objects (Vande Berg and Strekfuss, 1992). Earlier depictions of women as professionals characterized them as "tough broads" with aggressive attitudes; today, their portrayal is more realistic ("Depiction of Women," 1993). Women on television typically are under age forty, well groomed, and attractive.

It is amazing how complete is the delusion that beauty is goodness.

LEO TOLSTOY (1828–1910)

On television male characters are shown as more aggressive and constructive than female characters. They solve problems and rescue others from danger. Only in the last few years in prime-time series have males been shown in emotional, nurturing roles.

Ethnic stereotypes continue to be standard fare in television—the "Native American in full headdress, the black man as villain and Hispanics with lots of children" (Wardle, 1989/1990). Gay men and lesbians are increasing in visibility in TV movies and a few sitcoms, such as *Ellen*, but often they are portrayed stereotypically—as sinister, comic, or the "victims" of AIDS—or solely in terms of their homosexuality (Kalin, 1992; Tharp, 1991; Weir, 1992).

Reflections

When you were a child, who were your favorite TV characters? What male-female stereotypes did they depict? What ethnic stereotypes?

Gender-Role Learning in Adulthood

Researchers have generally neglected gender-role learning in adulthood (Losh-Hesselbart, 1987). Several scholars, however, have formulated a life-span perspective to gender-role development known as *role transcendence* (Hefner, Rebecca, and Oleshansky, 1975).

The role transcendence approach argues that there are three stages an individual goes through in developing his or her gender-role identity: (1) undifferentiated stage, (2) polarized stage, and (3) transcendent stage. Young children

Everybody wants to be somebody; nobody wants to grow.

JOHANN GOETHE (1749–1842)

have not clearly differentiated their activities into those considered appropriate for males or females. As children enter school, however, they begin to identify behaviors as masculine or feminine. They tend to polarize masculinity and femininity as they test the appropriate roles for themselves. As they enter young adulthood, they begin slowly to shed the rigid male-female polarization as they are confronted with the realities of relationships. As they mature and grow older, men and women transcend traditional masculinity and femininity. They combine masculinity and femininity into a more complex transcendent role, a role similar to androgyny (see "Androgynous Gender Roles" later in the chapter).

For adults, gender-role development takes place in contexts outside the family of orientation. In adulthood, new or different sources of gender-role learning include college, marriage, parenthood, and the workplace.

College

The college and university environment, which encourages critical thinking and independent behavior, contrasts markedly with high school. In the college setting, many young adults learn to think critically, to exchange ideas, and to discover the bases for their actions. In colleges, many young adults first encounter alternatives to traditional gender roles, either in their personal relationships or in their courses. A longitudinal study of gender roles found that traditional and egalitarian gender-role attitudes affected dating relationships in college but had little impact on later life (Peplau, Hill, and Rubin, 1993).

Marriage

Chains do not hold a marriage together. It is threads, hundreds of tiny threads, which sew people together through the years.

SIMONE SIGNORET (1921–1985)

Marriage is an important source of gender-role learning, for it creates the roles of husband and wife. For many individuals, no one is more important than a partner in shaping gender-role behaviors through interaction. Our partners have expectations of how we should act as a husband or wife, and these expectations are important in shaping behavior.

Husbands tend to believe in innate gender roles more than wives do. This should not be especially surprising, because men tend to be more traditional and less egalitarian about gender roles (Thornton et al., 1983). Husbands stand to gain more in marriage by believing that women are "naturally" better at cooking, cleaning, shopping, and caring for children.

Parenthood

For most men and women, motherhood alters life more significantly than fatherhood does. For men, much of fatherhood means providing for their children. As a consequence, fatherhood does not create the same work-family conflict that motherhood does. A man's work role allows him to fulfill much of his perceived parental obligation.

Yet contemporary fatherhood has become more complex. Traditional fatherhood was tied to marriage. Today, with almost 30 percent of all current births occurring outside of marriage and half of all current marriages ending in divorce, what is the father's role for a man who is not married to his child's mother or who is divorced and does not have custody? What are his role obligations as a single father as distinguished from those of married fathers? For many men, the answers are painfully unclear, as evidenced by the low rates of contact between unmarried or divorced fathers and their children.

PERSPECTIVE

Gender Roles and Music Videos

Although listening to popular music continues to be important in our lives—adolescents, for example, listen to about 10,500 hours of it between seventh and twelfth grade—television, where everything must be made visual, has remade music with the "invention" of the music video (Twitchell, 1992). Now music is "seen" as well as heard. Visual images are as important in conveying meaning and feeling as is the music itself—indeed, the images may be even more important than the music.

Music videos present popular images of masculinity and femininity: The people in them are youthful, attractive, fashionably attired, hip, and sexual. According to a study by Steven Seidman (1992), men are more adventuresome, generally aggressive, violent, and domineering than women. Women are more affectionate, dependent, and fearful. Men are often construction workers, mechanics, firefighters, or in the military. Women are often secretaries, librarians, or cheerleaders. Women are systematically excluded from most white-collar professions. In Seidman's study, three times as many women were in blue-collar jobs, such as waitress and hairstylist, as in white-collar jobs. Not a single woman was portrayed as a politician, business executive, or manager. The occupational stereotyping is similar to that found in prime-time television and in commercials. Such stereotyping serves to reinforce the status quo (Signorielli, 1989).

There is considerable verbal or physical aggression against both men and women in music videos (Kalis and Neuendorf, 1989). Female aggression is often provoked by jealousy. Male aggression is frequently unprovoked. Aggression is often a part of male swagger—the assertion of power and status—especially in heavy metal and rap videos. Critic James Twitchell (1992) writes of these two forms: "Both are rife with adolescent misogyny, homophobia, and threats of violence. They are rude, bawdy, boastful, with a kind of 'in your face' aggression . . . characteristic of insecure masculinity."

Many female vocalists appear in music videos. However, the majority of music videos are dominated by male singers or male groups, and women provide erotic backdrop or vocal backup (Seidman, 1992; Sommers-Flanagan, Sommers-Flanagan, and Davis, 1993). Most women are depicted as sex objects (Kaloff, 1993). Women are typically pictured condescendingly, are provocatively dressed, or both.

Because television prohibits the explicit depiction of sexual acts, music videos use sexual innuendos and suggestiveness to impart their sexual meanings (Baxter et al., 1985; Sherman and Dominick, 1987). One

Although women today have greater latitude as wives (it is acceptable to work outside the home), they are still under considerable pressure to become mothers. Once a woman becomes a mother, especially if she is white, she may leave the workforce for at least a few months. When children are born, roles tend to become more traditional, even in nontraditional marriages. The wife remains at home, at least for a time, and the husband continues full-time work outside the home. The woman must then balance her roles as wife and mother against her own needs and those of her family.

While most contemporary popular music and music videos depict women stereotypically, some female groups, such as Salt-N-Pepa, present women as strong and independent.

The Workplace

It is well established that men and women are psychologically affected by their occupations (Menaghan and Parcel, 1991; Schooler, 1987). Work that encourages self-direction, for example, makes people more active, flexible, open, and democratic; restrictive jobs tend to lower self-esteem and make people more rigid and less tolerant. If we accept that female occupations are usually low status with little room for self-direction, we can understand why women are not as achievement oriented as men. Because men and women have different opportunities for promotion, they have different attitudes toward achievement. Women typically downplay any desire for promotion, suggesting that promotions would interfere with their family responsibilities. But this really may be related to a need to protect study found that adolescent or male viewers generally rated music videos, especially sexually provocative ones, more positively than did older or female viewers (Greeson, 1991). Another study found that both male and female undergraduates responded with positive emotions to music videos with sexual content; they responded negatively to those with violence. The music videos declined in appeal when sex and violence were combined (Hansen and Hansen, 1991). Females have stronger positive responses to soft rock, while males seem to enjoy hard rock (Toney and Weaver, 1994).

In the past decade, Janet Jackson, Toni Braxton, and Madonna have been among the most popular contemporary female pop performers. Both Jackson and Braxton continue to portray the more traditional female concerns of love and romance. But Madonna represents a different kind of performer. Beginning in the early 1980s, Madonna broke away from the traditional mold, a major reason for her being a controversial performer. Her blending of musical and visual components, especially dance, brought an erotic female presence to music videos. Madonna was neither a sex object nor a woman living for love. Instead, she affirmed herself as a powerful figure in her own right. Men did not control her; rather, they became "boy toys." Her love songs did not follow the traditional romantic formula. She wanted love, sex, or commitment on her own terms. Her videos represented powerful affirmations of female sexuality and strength (Lewis, 1992; McClary, 1990; see also Brown and Schulze, 1990). In "Justify Your Love", originally banned from MTV, Madonna challenged traditional gender stereotypes and sexuality with erotic scenes involving male cross-dressing. One experimental study which showed Madonna and Amy Grant videos found that those who watched Madonna videos were subsequently more liberal in premarital attitudes (Calfin, Carroll, and Schmidt, 1993).

In the late 1980s, MTV's ratings began to decline. Its chairman admitted that its audience was "bored with clips that featured heavy-metal music, smoke-filled sets, and pretty girls in revealing lingerie—but not much imagination" (quoted in Seidman, 1992). As part of an attempt to reinvigorate MTV, its executives introduced rap and hip-hop into its programming. In doing so, they introduced into mainstream America a form of African-American music that had originated in the inner city. Rap and hip-hop, with their emphasis on words, restored lyrics to popular music.

Some of the strongest female musical voices are found in rap and hip-hop. Queen Latifah, for example, lashed back at men for calling women "hos" and "bitches." She demanded respect, fighting rap's *gangsta* attitude that demeaned women and glorified violence. Salt-N-Pepa validated women's erotic power with such songs as *Shoop, What a Man,* and *Champagne*. Such songs and videos portray women as sexual beings who are also competent, strong, and independent.

themselves from frustration because many women are in dead-end jobs in which promotion to management positions is unlikely.

Household work affects women psychologically in many of the same ways that paid work does in female-dominated occupations such as clerical and service jobs (Schooler, 1987). Women in both situations feel greater levels of frustration owing to the repetitive nature of the work, time pressures, and being held responsible for things outside their control. Such circumstances do not encourage self-esteem, creativity, or a desire to achieve.

Reflections

During high school, what influences did your peers have on your gender-role development? How important were your boyfriends or girlfriends in developing your sense of yourself as a woman or man? Is college different? If so, how? Who are the people who most influence your gender-role concepts today?

Changing Gender Roles

Within the past generation, there has been a significant shift from traditional toward more egalitarian gender roles. Women have changed more than men, but men also are changing. These changes seem to affect all classes. Those from conservative religious groups, such as Mormons, Catholics, and fundamentalist and evangelical Protestants, adhere most strongly to traditional roles (Jensen and Jensen, 1993; Spence et al., 1985). The greatest change in gender-role attitudes appears to have occurred in the 1970s. These roles are continuing to change and become more egalitarian but at a slower pace.

Traditional Gender Roles

According to traditional gender-role stereotypes, many of the traits ascribed to one gender are not ascribed to the other. Men show instrumental traits, women display expressive ones (Hort, Fagot, and Leinbach, 1990). These traits theoretically complemented women's and men's traditional roles in the family. Because of assumed basic gender differences, men were expected to participate in the world of work and politics. Their central male role was worker; in the family, this role translated to breadwinner. Because women were thought to be primarily expressive, they were expected to remain in the home as wives and mothers.

The things we admire in men, kindness and generosity, openness, honesty, understanding and feeling are the concomitants of failure in our system. And those traits we detest, sharpness, greed, acquisitiveness, meanness, egotism and self-interest are the traits of success.

JOHN STEINBECK, (1902–1968), *CANNERY ROW*

Traditional Male Gender Role

What is it to be a "real" man in America? Bruce Feirstein parodied him in *Real Men Don't Eat Quiche* (1982):

> QUESTION: How many Real Men does it take to change a lightbulb?
> ANSWER: None. Real Men aren't afraid of the dark.
>
> QUESTION: Why did the Real Man cross the road?
> ANSWER: It's none of your damn business.

This humor contains something both familiar and chilling. Being a male in America carries a certain uneasiness. Men often feel they need advice about how to act out their male roles; they are unsure of what it means to be masculine. The resurgence of bodybuilding—and the rise of Sylvester Stallone and Arnold Schwarzenegger as film stars in the 1980s—reflected an attempt to define masculinity in muscular terms (Jeffords, 1994; Klein, 1993).

Once, power was considered a masculine attribute. In fact, power has no sex.

KATHERINE GRAHAM

Central features of the traditional male role, whether white, African American, Latino, or Asian American, include dominance, work, and family. Males are generally regarded as being more power oriented than females. Statistically, men demonstrate higher degrees of aggression, especially violent aggression (such as assault, homicide, and rape), seek to dominate and lead, and show greater competitiveness. Although aggressive traits are thought to be useful

As contemporary male gender roles allow increasing expressiveness, men are encouraged to nurture their children.

in the corporate world, politics, and the military, such characteristics are rarely helpful to a man in fulfilling marital and family roles requiring understanding, cooperation, communication, and nurturing.

The centrality of men's work identity affects their family roles as husbands and fathers. Traditional men see their primary family function as that of provider, which takes precedence over all other family functions, such as nurturing and caring for children, doing housework, preparing meals, and being intimate. Because of this focus, traditional men are often confused by their spouses' expectations of intimacy; they believe that they are good husbands simply because they are good providers (Rubin, 1983).

Traditional Female Gender Role

Although the main features of traditional male gender roles do not appear to have significant ethnic variation, there are striking ethnic differences in female roles.

Rabbi Zusya said that on the Day of Judgment, God would ask him, not why he had not been Moses, but why he had not been Zusya.

WALTER KAUFMANN

Whites. Traditional white female gender roles center around women's roles as wives and mothers. When a woman leaves adolescence, she is expected to either go to college or to get married and have children. Although a traditional woman may work prior to marriage, she is not expected to defer marriage for work goals, and soon after marriage, she is expected to be "expecting." Once married, she is expected to devote her energies to her husband and family and work but to find her meaning as a woman by fulfilling her roles as wife and mother. Within the household, she is expected to subordinate herself to her husband. Often this subordination is sanctioned by religious teachings.

Wives, submit yourselves unto your own husbands, as unto the Lord.

PAUL OF TARSUS (EPHESIANS 5:22)

Did you know?

Only 3 percent of men in prime-time shows are depicted performing household tasks, whereas 20 percent of the women are. ("Depiction of Women," 1993).

African Americans. We know relatively little about the lives of African-American women, as most research focuses primarily on unmarried mothers and the poor (Wyche, 1993). Yet we do know that the traditional white female gender role does not extend to African-American women. This may be attributed to a combination of the African heritage, slavery (which subjugated women to the same labor and hardships as men), and economic discrimination that pushed women into the labor force. Karen Drugger (1988) notes:

> A primary cleavage in the life experiences of Black and White women is their past and present relationship to the labor process. In consequence, Black women's conceptions of womanhood emphasize self-reliance, strength, resourcefulness, autonomy, and the responsibility of providing for the material as well as emotional needs of family members. Black women do not see labor-force participation and being a wife and mother as mutually exclusive; rather, in Black culture, employment is an integral, normative, and traditional component of the roles of wife and mother.

One study (Leon, 1993) found that African-American women appeared more instrumental than either white or Latina women; they also have more flexible gender and family roles. African-American men are generally more supportive than white or Latino men of egalitarian gender roles.

The man over there says women need to be helped into carriages and lifted over ditches, and to have the best place everywhere. Nobody ever helps me into carriages and over puddles, or gives me the best place—and ain't I a woman? Look at my arm! I have ploughed and planted and gathered into barns, and no man could head me—and ain't I a woman? I could work as much and eat as much as a man . . . and bear the lash as well! And ain't I a woman? I have born thirteen children, and seen most of 'em sold into slavery, and when I cried out with my mother's grief, none but Jesus heard me—and ain't I a woman?

SOJOURNER TRUTH (1797–1883)

Latinas. In traditional Latina gender roles, women subordinate themselves to males (Vasquez-Nuthall, Romero-Garcia, and De Leon, 1987). But this

OTHER PLACES OTHER TIMES

Becoming "American:" Immigrants and Gender Roles in Transition

The enormous rise in immigration beginning in the 1980s brought great numbers of Latinos and Asians to the United States. Today, more than 750,000 immigrants arrive annually; a large percentage are children and adolescents. The process of adaptation to a new environment, culture, and language places individuals under considerable stress. In adjusting to life in the United States, women tend to experience greater stress than men. Mental health workers attribute women's greater stress to conflicting gender-role expectations and family roles (Aneshensel and Pearling, 1987).

The acquisition of "American" gender roles may be difficult and complicated. New immigrants must deal with several problems, including those resulting from (1) their immigrant status, including issues involved in being a newcomer and a member of an ethnic group (which is often a minority group as well); (2) sorting out what it means to be female or male both in American culture and in their original cultures; and (3) integrating their ethnic identity and their gender role (Goodenow and Espín, 1993).

For young immigrants, acquiring a meaningful gender role is complicated by the conflict between the norms of their new and old cultures. Their old social environment has disappeared and a new, alien one taken its place. Most adolescents are able to try out their different roles with peers whose language they understand, but most new immigrants do not have this advantage. English leaves them bewildered. In addition, different meanings are given to everyday things and events. What does a certain gesture or action mean? Whom should I trust? What are appropriate choices? What kinds of clothes should I wear?

Immigrants must adjust to two gender-role cultures: their original culture and the American culture. One notable problem is that many cultures are far less egalitarian than American culture. This creates strains between men and women as they become acculturated. Furthermore, immigrant males and females face different problems in acculturation. Families expect females to maintain traditional roles and values, whereas males are often encouraged to become Americanized quickly. For females, conflict is most likely to arise over appropriate gender-role behavior and sexuality. Conflict is most acute for females who were reared in traditional cultures. Young women often feel pulled in different directions at the same time. Adolescents must become American without alienating their families or losing their cultural heritage.

subordination is based more on respect for the male's role as provider than on subservience (Becerra, 1988). Wives have greater equality if they are employed; they also have more rights in the family if they are educated (Baca Zinn, 1994).

Latino gender roles, unlike those of Anglos, are strongly affected by age roles in which the young subordinate themselves to the old. In this dual arrangement, notes Becerra (1988), "females are viewed as submissive, naive, and somewhat childlike. Elders are viewed as wise, knowledgeable, and deserving of respect." As a result of this intersection of gender and age roles, older women are treated with greater deference than younger women.

Contemporary Gender Roles

Women have served all these centuries, as looking glasses possessing the magic and delicious power of reflecting the figure of man at twice its natural size.

VIRGINIA WOOLF, (1882–1941), *A ROOM OF ONE'S OWN*

Contemporary gender roles are evolving from traditional hierarchical gender roles (in which one gender is subordinate to the other) to more egalitarian roles (in which both genders are treated equally) and to androgynous gender roles (in which both genders display the instrumental and expressive traits previously associated with one gender). Thus, contemporary gender roles often display traditional elements as well as egalitarian and androgynous ones.

Carol Goodenow and Olivia Espín (1993) interviewed adolescent Latina immigrants to see how they adjusted to American gender roles. They found that although the girls did not challenge their original culture's traditional sexual norms, they did challenge its gender roles. The girls said that if they had remained in their native countries, they would have felt pressured to marry and have children early. Each wanted to marry someday, but first each wanted to "make something of myself" or "have a career, not just to get married." They wanted adventure, "to be somebody," to "do things." For the girls, "the personal freedom offered by life in the United States, especially the relative freedom for women, was exhilarating. . . . Their vision of the future focused on careers or jobs—very different from the limited domestic roles that might easily have been theirs."

As Latinas become more acculturated, they adopt more egalitarian gender roles (Vasquez-Nuthall, Romero-Garcia, and De Leon, 1987). First-generation Latinas have held significantly more traditional gender-role attitudes than second- and third-generation Latinas. This traditionalism is reflected in part by the reluctance of first-generation Latinas to work outside the home (Ortiz and Cooney, 1984). Researchers found that the shift from a culture emphasizing definite male-female gender roles to one with greater freedom has affected the self-identities of immigrant Latinas (Salgado de Snyder, Cervantes, and Padilla, 1990). This shift has been a potential source of both personal and family conflict. As a Latina demands greater input in making decisions or asks her partner's assistance in child care—behavior more typical among Anglos, she is likely both to encounter resistance from her husband and to feel guilty for violating traditional Latino norms. Similarly, Asian-American women, notes Reiko True (1990), struggling to establish themselves in American society, "often find themselves trapped within restrictive roles and identities as defined by Asian cultures or by general American stereotypes."

Although immigrant males also experience stress resulting from changed male-female relationships, their greatest stress tends to be occupational. This difference results from the priority of work roles over family roles for males. Latino males have experienced occupational stress, for example, reflecting their difficulty in finding good jobs or job advancement as a result of language differences, education, or discrimination (Salgado de Snyder, Cervantes, and Padilla, 1990). Although many Asian male immigrants tend to be from white-collar backgrounds, they also experience occupational stress resulting from discrimination and language difficulties.

Gender-role stress increases for both men and women as their children adopt more typically American gender roles. Parents and children may come into conflict about appropriate gender-related behavior because parents tend to be more culturally traditional and their children tend to be more Americanized.

Gender roles affect marriage (Huston and Geis, 1993). Within the family, attitudes toward gender roles have become more liberal; in practice, however, gender roles continue to place women at a disadvantage, especially by making them responsible for housekeeping and child-care activities (Atkinson, 1987). Some of the most important changes affecting contemporary gender roles in the family are briefly described in the following sections.

Women as Workers and Professionals

Life shrinks or expands in proportion to one's courage.

Anaïs Nin (1903–1977)

Even though the traditional roles for women have typically been those of wife and mother, in recent years an additional role has been added: employed worker or professional. (This change has most notably affected white women, as African-American women have traditionally combined wife, mother, and worker roles.) It is now generally expected that most women will be employed at various times in their lives. For most women, entrance into work or a career does not exclude their more traditional roles as wives and mothers. Such work, however, may conflict with their family roles. Women generally attempt to reduce the conflict between work and family roles by giving family roles precedence. As a result, they tend to work outside the home in greatest numbers before motherhood and after

Why is it men are permitted to be obsessed about their work, but women are only permitted to be obsessed about men?

BARBRA STREISAND

divorce, when single mothers generally become responsible for supporting their families. After marriage, most women are employed even after the arrival of the first child. Regardless of whether a woman is working full-time, she almost always continues to remain responsible for housework and child care. It spite of the fact that employment is often a necessity for mothers, most feel great emotional conflict as they want to be or believe they should be at home with their children. (Women's employment is discussed in more detail in Chapter 11.)

Questioning Motherhood

Record numbers of women are choosing not to have children because of the conflicts child rearing creates. They are reconceptualizing what it means to be a woman without being a mother (Ireland, 1993).

Women from ethnic and minority groups, however, are less likely than whites to view motherhood as an impediment. African-American women and Latinas tend to place greater value on motherhood than the white or Anglo majority. For African Americans, tradition has generally combined work and motherhood; the two are not viewed as necessarily antithetical (Basow, 1993). For Latinas, the cultural and religious emphasis on family, the higher status conferred on motherhood, and their own familial attitudes have contributed to high birthrates (Jorgensen and Adams, 1988).

Greater Equality in Marital Power

Although husbands were once the final authority, wives have greatly increased their power in decision making. They are no longer expected to be submissive but to have significant, if not equal, input in marital decision making. This trend toward equality is limited in practice by an unspoken rule of marital equality: "Husbands and wives are equal, but husbands are more equal." In actual practice, husbands continue to have greater power than wives. Husbands have become what sociologist John Scanzoni (1982) describes as the "senior partner" of the marriage.

Breakdown of Instrumental/Expressive Dichotomy

The identification of masculinity with instrumentality and femininity with expressiveness appears to be breaking down. As a group, men perceive themselves to be more instrumental than do women, and women perceive themselves as being more expressive than do men. A substantial minority of both genders is relatively high in both instrumentality and expressiveness or is low in both. It is interesting that the instrumental/expressiveness ratings men and women give each other have very little to do with how they rate themselves as masculine or feminine (Spence and Sawin, 1985).

Expansion of Men's Family Roles

The key assumption about male gender roles has been the centrality of work and success to a man's sense of self. This assumption has led many researchers to believe that men's primary family role has been that of provider. As a result, until recently, there has been little research into men's participation in family life. Unlike research into female role conflicts, research in how men's work may conflict with family roles is rare. Although both the public and researchers have expressed concern about the impact of working mothers on the family, few have expressed similar concern about working fathers. In fact, the term *working father* seems redundant because work is what is expected of fathers (Cohen, 1987).

Research indicates that men consider their family role to be much broader than that of the family breadwinner. As researchers have long pointed out, economic responsibilities are only one small part of men's family lives (Goetting, 1982). Other dimensions include emotional, psychological, community, and legal dimensions; they also include housework and child-care activities. Many of those in the evolving Men's Movement share the beliefs of feminism: Equal pay for equal work, more parental leave, and better child-care facilities (Wood, 1994). They resent the social expectations of masculinity and seek the support of other men in defining themselves in their own terms (see Chapter 10 for discussion of men's and women's parenting roles and Chapter 11 for discussion of the division of household labor).

I am in the hand of the unknown God, he is breaking me down to his own oblivion to send me forth on a new morning, a new man.

D. H. LAWRENCE (1885–1930)

Constraints of Contemporary Gender Roles

Even though substantially more flexibility is offered to men and women today, contemporary gender roles and expectations continue to limit our potential. Indeed, there is considerable evidence that stereotypes about gender traits have not changed very much over the last twenty-five years; men are perceived as having more undesirable self-oriented traits (such as being arrogant, self-centered, and domineering) than women. Women are viewed as having more traits reflecting a lack of a healthy sense of self (such as being servile and spineless) than men. Only "intellectual" seems to be applied to women more than in the past (Spence et al., 1985).

No one is a greater slave than he who imagines himself free when he is not free.

JOHANN GOETHE (1749–1842)

Limitations on Men

Men are required to work and to support their families. The male as provider is one of men's central roles in marriage; the provider role is not central for women. As a result, men do not have the same role freedom to choose to work as women have. Therefore, when the man's roles of worker and father come into conflict, usually it is the father role that suffers (Cohen, 1987). A factory worker may want to spend time with his children, but his job does not allow flexibility. Because he must provide income for his family, he will not be able to be more involved in parenting. In a familiar scene, a child comes into the father's home office to play, and the father says, "Not now. I'm busy working. I'll play with you later." When the child returns, the "not now, I'm busy" phrase is repeated. The scene recurs as the child grows up, and one day, as his child leaves home, the father realizes that he never got to know him or her.

In every real man a child is hidden that wants to play.

FRIEDRICH NIETZSCHE (1844–1900)

Men continue to have greater difficulty expressing their feelings than do women. Men tend to cry less and show love, happiness, and sadness less. When men do express their feelings, they are more forceful, domineering, and boastful; women, in contrast, tend to express their feelings more gently and quietly. When a woman asks a man what he feels, a common response is "I don't know," or "Nothing." Such men have lost touch with their inner lives because they have repressed feelings that they have learned are inappropriate. This male inexpressiveness often makes men strangers to both themselves and their partners.

If we do not redefine manhood, war is inevitable.

PAUL FUSSELL

Men continue to expect and, in many cases, are expected to be the dominant member in a relationship. Unfortunately, the male sense of power and command often does not facilitate personal relationships. Without mutual respect and equality genuine intimacy is difficult to achieve. One cannot control another person and at the same time be intimate with that person.

Limitations on Women

Man's love is of man's life a part; it is woman's whole existence.

LORD BYRON (1788–1824)

Research suggests that the traditional female gender role does not facilitate self-confidence or mental health. Both men and women tend to see women as being less competent than men. A study by Lyn Brown and Carol Gilligan (1992), discussed at greater length in Chapter 10, revealed that the self-esteem of adolescent girls plummeted between the age of nine and the time they started high school. Traditional women married to traditional men experience the most symptoms of stress (feeling tired, depressed, or worthless) and express the most dissatisfaction about life as a whole (Spence et al., 1985). The combination of gender-role stereotypes and racial/ethnic discrimination tends to encourage feelings of both inadequacy and lack of physical attractiveness among African-American women, Latinas, and Asian-American women (Basow, 1993).

Because of differences in gender roles, Jessie Bernard (1982) has suggested that each gender experiences marriage differently. There is, she argues, a "his" and a "her" marriage. Men appear to be more satisfied in marriage than women (Rettig and Bubolz, 1983). More wives than husbands report frustration, dissatisfaction, marital problems, and desire for divorce. Unmarried women tend to be happier and better adjusted than married women. (See Chapter 12 for a discussion of marital status and health.)

Finally, there is a "double standard of aging" that treats men and women differently. As women get older, they tend to be regarded as more masculine and as unattractive. Also, our culture treats aging in men and women differently: As men age, they become distinguished; as women age, they simply get older. Masculinity is associated with independence, assertiveness, self-control, and physical ability; with the exception of physical ability, none of these traits necessarily decreases with age. Because they are considered to have lost their attractiveness and because they have fewer potential partners as women get older, they are less likely to marry.

Resistance to Change

The perpetual obstacle to human advancement is custom.

JOHN STUART MILL (1806–1873)

We may think that we want change, but both men and women reinforce traditional gender-role stereotypes among themselves and each other (Hort, Fagot, and Leinbach, 1990). Women's self-help books often reflect a conservative bias by encouraging women to discover their "inner" feminine selves, which mirror traditional gender roles (Schrager, 1993). Both genders react more negatively to men displaying so-called female traits (such as crying easily or needing security) than to women displaying male traits (such as assertiveness or worldliness), and both define male gender-role stereotypes more rigidly than they do female stereotypes. Men, however, do not define women as rigidly as women do men. And both men and women describe the ideal female in very androgynous terms (Hort, Fagot, and Leinbach, 1990).

In large measure we have equality with men; now we need to convince men to have equality with us.

ANNA QUINDLEN

Despite the limitations that traditional gender roles place on us, changing them is not easy. Gender roles are closely linked to self-evaluation. Our sense of adequacy often depends on gender-role performance as defined by parents and peers in childhood ("You're a good boy/girl"). Because gender roles often seem to be an intrinsic part of our personality and temperament, we may defend these roles as being natural, even if they are destructive to a relationship or to ourselves. To

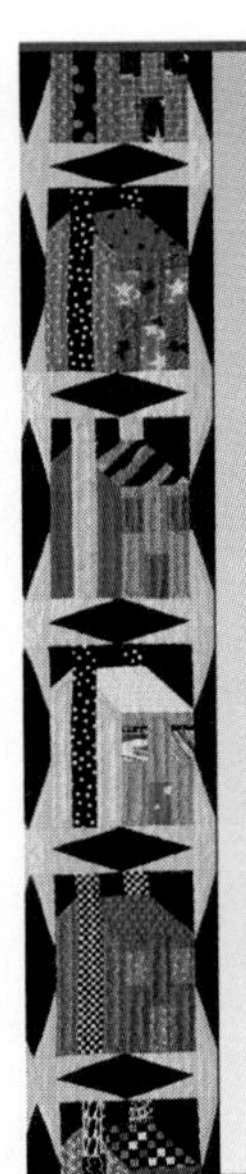

YOU AND YOUR WELL-BEING

Gender and Stress

Not only do contemporary gender roles and expectations limit our potential, they also can result in characteristic stresses. Traditional women's roles, which are often related to family matters and conflicting work pressures, have been demonstrated to result in lowered self-esteem and feelings of helplessness and worthlessness. The result can be seen in an overall decline in well-being: Women experience higher rates of depression, eating disorders, and mental illness (Baslow, 1992). It is probably not a surprise that the attempted suicide rate for women is higher than men's (though men are more likely to succeed in their suicide efforts and show higher rates of suicide). Though traditional sex roles have not been pinpointed as a cause, we cannot eliminate them as a significant factor.

The stressors associated with traditional male gender roles can also contribute to increased health risks. Expectations for achievement, productivity, and competition can create chronic stress which contributes to heart disease, hypertension, and a lower life expectancy. Add to this a world in which employment opportunities and promotion are still not equal for all, the stressors of married minority men are felt particularly hard (Barnett and Baruch, 1987).

Though gender-role differentiation serves a variety of functions, we might choose to question those that impair or infringe upon our options, opportunities, or expression of individualism. (For a further discussion of stress and health, see Chapter 12.)

threaten an individual's gender role is to threaten his or her gender identity as male or female because people do not generally make the distinction between gender role and gender identity. Such threats are an important psychological mechanism that keeps people in traditional but dysfunctional gender roles.

Finally, the social structure itself works to reinforce traditional gender norms and behaviors. Some religious groups, for example, strongly support traditional gender roles. The Catholic church, conservative Protestantism, orthodox Judaism, and fundamentalist Islam, for example, view traditional roles as being divinely ordained. Accordingly, to violate these norms is to violate God's will. The marketplace also helps enforce traditional gender roles. The wage disparity between men and women (women earn about 70 percent of what men earn) is a case in point. Such a significant difference in income makes it "rational" that the man's work role take precedence over the woman's work role. If someone needs to remain at home to care for the children or an elderly relative, it makes "economic sense" for the woman to quit her job because her male partner probably earns more money.

Reflections

Do gender roles, either traditional or contemporary, help us fulfill our potential as human beings? Do they help us love? To rear children well? If you were to decide, what would gender roles be like?

Androgynous Gender Roles

Some scholars have challenged the traditional masculine/feminine gender-role dichotomy, arguing that such models are unhealthy and fail to reflect the real world. Instead of looking at gender roles in terms of the bipolar model, they suggest examining them in terms of androgyny (Roopnarine and Mounts, 1987). **Androgyny** refers to the state of combining male and female characteristics. The term is derived from the Greek *andros*, meaning "man," and *gyne*, meaning "woman." Androgynous gender roles are characterized by flexibility and a unique combination of instrumental and expressive traits (as influenced by individual differences, situations, and stages in the life cycle) (Bem, 1976; Kaplan, 1979; Kaplan and Bean, 1976). An androgynous person combines both the instrumental traits previously associated

Did you know?

As people get older, many men tend to become more androgynous whereas women become more feminine (Jelski, Falk, and Foglesong, 1992).

UNDERSTANDING YOURSELF

Masculinity, Femininity, and Androgyny

Increased interest in androgynous gender roles has led to the development of a number of instruments to measure masculinity, femininity, and androgyny in individuals. The one below is patterned after the Bem Sex Role Inventory (BSRI), which is one of the most widely used tests (Bem, 1974, 1981).

To get a rough idea of how androgynous you are, examine the twenty-one personality traits below. Use a scale of 1 to 5 to indicate how well a personality trait describes yourself.

1	2	3	4	5
NOT AT ALL	SLIGHTLY	SOME-WHAT	QUITE A BIT	VERY MUCH

1 Aggressive
2 Understanding
3 Helpful
4 Decisive
5 Nurturing
6 Happy
7 Risk taker
8 Shy
9 Unsystematic
10 Strong
11 Affectionate
12 Cordial
13 Assertive
14 Tender
15 Moody
16 Dominating
17 Warm
18 Unpredictable
19 Independent minded
20 Compassionate
21 Reliable

Scoring the Sex Role Inventory

Your masculinity, femininity, and androgyny scores may be determined as follows:

1 To determine your masculinity score, add up your answers for numbers 1, 4, 7, 10, 13, 16, and 19, and divide the sum by 7.
2 To determine your femininity score, add up your answers for numbers 2, 5, 8, 11, 14, 17, and 20, and divide the sum by 7.
3 To determine your androgyny score, subtract the femininity score from the masculinity score.

The closer your score is to 0, the more androgynous you are. A high positive score indicates masculinity; a high negative score indicates femininity. A high masculine score not only indicates masculine attributes but also a rejection of feminine attributes. Similarly, a high feminine score indicates not only feminine characteristics but also a rejection of masculine attributes.

with masculinity and the expressive traits associated with traditional femininity. An androgynous lifestyle allows men and women to choose from the full range of emotions and behaviors, according to their temperaments, situations, and common humanity, rather than their gender. Males may be expressive. They are permitted to cry and display tenderness; they can touch, feel, and nurture without being called "effeminate." Women can express the instrumental aspects of their personalities without fear of disapproval. They can be aggressive or career oriented; they can seek leadership, be mechanical, or enjoy physical activities.

Flexibility and integration are important aspects of androgyny. Individuals who are rigidly both instrumental and expressive, despite the situation, are not considered androgynous. A woman who is always aggressive at work and passive at home, for example, would not be considered androgynous, as work may call for compassion and home life, for assertion. Androgynous individuals are flexible. They invoke their instrumental and expressive qualities in appropriate situations (Vonk and Ashmore, 1993).

I don't know why people are afraid of new ideas. I am terrified of the old ones.
John Cage (1912–1992)

There is considerable evidence that androgynous individuals and couples have a greater ability to form and sustain intimate relationships and adopt a wider range of behaviors and values (Ickes, 1993). One study found that androgynous individuals felt more comfortable talking with their parents and felt their parents understood them better. The more negatively men and women felt about their parents, the more likely their offspring were to hold traditional gender roles (Lombardo and Kemper, 1992). Androgynous individuals have shown greater resilience when faced with stress (Roos and Cohen, 1987). They are more aware of feelings of love and more expressive of them (Ganong and Coleman, 1987). Also, androgynous couples may have greater satisfaction in their relationships than gender-typed couples. Androgyny even seems to affect health behaviors. Androgynous individuals in one study, for example, were more likely to exercise and not smoke (Shifren, Bauserman, and Carter, 1993).

A study of African-American and white women found that more African Americans than whites were androgynous (Harris, 1993). Another study (Binion, 1990) of African-American women indicated that 37 percent identified themselves as androgynous, 18 percent as feminine, and 24 percent as masculine. The remaining were undifferentiated. That such a large percentage were androgynous or masculine, the study argues, is not surprising, given the demanding family responsibilities and cultural expectations that require instrumental and active traits.

Not all researchers are convinced, however, that androgyny necessarily leads to greater psychological health and happiness. Moreover, the concept of androgyny itself has come under attack on two fronts: measurement issues concerning personality (Finlay and Scheltema, 1991; Lips, 1997) and charges that androgyny perpetuates stereotypes (Lott, 1994). Bem herself, who was one of the leading proponents of androgyny, acknowledges that expectations to be both feminine and masculine may impose stresses equal to those formerly demanded to be either one of these. "The individual now has not one but two potential sources of inadequacy to contend with" (Bem, 1983). It will probably be some time before we understand the full ramifications of this issue.

When man lives he is soft and tender; when he is dead he is hard and tough. All living plants and animals are tender and fragile; when dead they become withered and dry. Thus it is said: The hard and tough are parts of death; the soft and tender are parts of life. This is the reason why soldiers when they are too tough cannot carry the day; when the tree is too tough it will break. The position of the strong and great is low, but the position of the weak and tender is high.

Lao-Tse (7th century B.C.)

Reflections

Think about your gender role. Would you describe yourself as masculine, feminine, or androgynous? Why? Do you have traits associated with the other gender? What are they? How do you feel about them?

Contemporary gender roles are still in flux. Few men or women are entirely egalitarian or traditional. Even those who are androgynous or who have egalitarian attitudes, especially males, may be more traditional in their behaviors than they realize. Few with egalitarian or androgynous attitudes, for example, divide all labor along lines of ability, interest, or necessity rather than gender. Also, marriages that claim to be traditional rarely have wives who submit to their husbands in all things. Among contemporary men and women, women find that their increasing access to employment puts them at odds with their traditional (and personally valued) role as mother. Women continue to feel conflict between their emerging equality in the workplace and their continued responsibilities at home. Within marriages and families, the greatest areas of gender inequality continue to be the division of housework and child care. But change continues to occur in the direction of greater gender equality, and this equality promises greater intimacy and satisfaction for both men and women in their relationships.

Whatever you can do or dream you can, begin it. Boldness has genius, power, and magic in it. Begin it now.

Johann Goethe (1749–1842)

Summary

- A *gender role* is the role a person is expected to perform as a result of being male or female in a particular culture. *Gender-role stereotypes* are rigidly held and oversimplified beliefs that males and females, as a result of their sex, possess distinct psychological and behavioral traits. *Gender-role attitudes* are beliefs that we have about ourselves and others regarding appropriate male and female personality traits and activities. *Gender-role behaviors* are the actual activities or behaviors that we or others engage in as males and females. *Gender identity* refers to being male or female.
- Although our culture encourages us to think that men and women are "opposites," they are actually more similar than different. Innate gender differences are generally minimal; differences are encouraged by socialization. *Gender schemas* reflect our tendency to divide objects, activities, and behaviors into masculine and feminine categories.
- According to *gender theory*, social relationships are based on the socially perceived differences between males and females that justify unequal power relationships. The key to creating gender inequality is the belief that men and women are opposite in personalities, abilities, skills, and traits. Furthermore, the differences between the sexes are unequally valued in society.
- The two most important socialization theories are social learning theory and cognitive development theory. *Social learning theory* emphasizes learning behaviors from others through rewards and punishments and *modeling*. This behaviorist approach has been modified to include cognitive processes, such as the use of language, the anticipation of consequences, and observation. *Cognitive development theory* asserts that once children learn that gender is permanent, they independently strive to act like "proper" boys or girls because of an internal need for congruence.
- Children learn their gender roles through manipulation, channeling, verbal appellation, and activity exposure. Parents, teachers, and *peers* (age-mates) are important agents of socialization during childhood and adolescence. Ethnicity and social class also influence gender roles. Among African Americans, strong women are important female role models.
- During adolescence, parents can be more important influences than peers. Peers, however, have more egalitarian gender-role attitudes than do adults. The media tend to portray traditional stereotypes of men and women as well as ethnic groups. For students, colleges and universities are important sources of gender-role learning, especially for nontraditional roles. Marriage, parenthood, and the workplace also influence the development of adult gender roles.
- Traditional male roles, whether for whites, African Americans, Latinos, or Asian Americans, emphasize dominance and work. A man's central family role has been viewed as being the provider. For women, there is greater role diversity according to ethnicity. Traditional female roles among middle-class whites emphasize being a wife and mother. Among African Americans, women are expected to be instrumental; there is no conflict between work and motherhood. Among Latinos, women are deferential to men generally from respect rather than subservience; elders, regardless of gender, are afforded respect.
- Contemporary gender roles are more egalitarian than the traditional ones of the past. Important changes affecting contemporary gender roles include (1) the acceptance of women as workers and professionals; (2) increased questioning, especially among white women, of motherhood as a core female identity; (3) greater equality in marital power; (4) the breakdown of the instrumental/expressive dichotomy; and (5) the expansion of male family roles.
- Limitations of contemporary gender roles for men include the primacy of the provider role, which limits men's father and husband roles; difficulty in expressing feelings; and a sense of dominance that precludes intimacy. Limitations for women include diminished self-confidence and mental health; another limitation is the association of femininity with youth and beauty, which creates a disadvantage as women age. Ethnic women may suffer both racial discrimination and gender-role stereotyping, which compound each other. Expectations and limitations reduce our potential as humans and create characteristic stresses.

- Changing gender-role behavior is often difficult because (1) each sex reinforces the traditional roles of its own and the other sex; (2) we evaluate ourselves in terms of fulfilling gender-role concepts; (3) gender roles have become an intrinsic part of ourselves and our roles; and (4) the social structure reinforces traditional roles.
- *Androgyny* is the combination of traditional male and female characteristics into a more flexible pattern of behavior. Evidence suggests that androgyny contributes to psychological and emotional health. Some researchers, however, believe that striving for androgyny may impose a double burden of expectation on people.

Key Terms

androgyny 99
bipolar gender role 79
expressive trait 78
gender 78
gender identity 79
gender role 78
gender-role attitude 79
gender-role behavior 79
gender-role stereotype 78
gender schema 81
instrumental trait 78
modeling 83
peer 87
role 78
social construct 82

Suggested Readings

Bem, Sandra. *The Lenses of Gender: Transforming the Debate on Sexual Inequality.* New Haven, CT: Yale University Press, 1993. A leading gender-role scholar examines gender concepts.

Craig, Steve, ed. *Men, Masculinity, and the Media.* Newbury Park, CA: Sage Publications, 1992. A collection of essays on how the media influences and depicts men and their relationships, as well as how men respond to these media images.

Douglas, Susan J. *Where the Girls Are: Growing Up Female with the Mass Media.* New York: Times Books, 1994. A fast-moving study of the ambivalent and contradictory messages about women in the movies and on television.

Estés, Clarissa Pinkola. *Women Who Run with the Wolves: Myths and Stories of the Wild Woman Archetype.* New York: Ballantine, 1992. A book by a Jungian psychologist and well-known *cantadora* (storyteller) who uses myths, fairy tales, and stories from around the world to explore women's inner lives.

Hess, Beth, and Myra Marx Ferree, eds. *Analyzing Gender.* Newbury Park, CA: Sage Publications, 1987. A valuable collection of feminist essays on gender theory.

Keen, Sam. *Fire in the Belly; On Being a Man.* New York: Bantam Books, Inc., 1992. A modern classic of the men's movement, the book is as much about being human as being a man.

Kimmel, Michael S., and Michael A. Messner. *Men's Lives,* 2d ed. New York: Macmillan, 1992. A collection of reflections, stories, and articles about growing up, sports, intimacy, war, families, masculinity, sexuality, and power.

Lips, Hilary. *Sex and Gender,* 3rd ed. Mountain View, CA: Mayfield, 1997. An excellent introduction to the study of gender roles.

Majors, Richard G., and Janet Billson. *Cool Pose: The Dilemmas of Black Manhood in America.* Lexington, MA: Health, 1992. An examination of male roles among African Americans.

Pittman, Frank. *Man Enough; Fathers, Sons, and the Search for Masculinity.* New York: Putnam Publication Group, 1994. An insightful and provocative analysis of the men's movement, specifically as it relates to the relationship of fathers and sons.

Sadker, Myra, and David Sadker. *Failing at Fairness: How America's Schools Cheat Girls.* New York: Scribner, 1994. An engaging but disturbing study of how, beginning in middle school, girls begin to academically fall behind boys. In coed (but not single-gender) high schools, the gender gap accelerates as girls conceal their intelligence and downplay their achievements.

Sollie, Donna, and Leigh Leslie, eds. *Gender, Families, and Close Relationships.* Newbury Park, CA: Sage Publications, 1994. A collection of essays by feminist scholars that examines the role of gender in intimate relationships.

Tavris, Carol. *The Mismeasure of Woman.* New York: Simon & Schuster, 1992. An examination of various misconceptions and biases that affect our understanding of women.

CHAPTER 4

Friendship, Love, and Commitment

P R E V I E W

To gain a sense of what you already know about the material covered in this chapter, answer "True" or "False" to the statements below.

1. The development of mutual dependence is an important factor in love. True or false?
2. Love and commitment are inseparable from each other. True or false?
3. Friendship and love share many of the same characteristics. True or false?
4. In romantic relationships, caring is less important than physical attraction. True or false?
5. Males and females, whites, African Americans, Latinos, and Asians, heterosexuals, gay men and lesbians, old and young are equally as likely to fall in love. True or false?
6. In many ways, love is like the attachment an infant experiences for his or her parent or primary caregiver. True or false?
7. You always hurt the one you love. True or false?
8. Infatuation is usually equally intense for both parties. True or false?
9. A high degree of jealousy is a sign of true love. True or false?
10. Partners with different styles of loving are likely to have more satisfying relationships because their styles are complementary. True or false?

Answers

1 True, 2 False, 3 True, 4 False, 5 True, 6 True, 7 False, 8 False, 9 False, 10 False

OUTLINE

Only love with its science makes us so innocent.

VIOLETA PARRA

Love is essential to our lives (O'Sullivan and O'Leary, 1992). Love binds us together as partners, husbands and wives, parents and children, and friends and relatives. The importance of romantic love cannot be overrated, for we make major life decisions, such as marrying, on the basis of love (Simpson et al., 1986). Love creates bonds that enable us to endure the greatest hardships, suffer the severest cruelty, overcome any distance. Because of its significance, we may torment ourselves with the question: Is it really love? Many of us have gone through frustrating scenes such as the following (Greenberg and Jacobs, 1966):

YOU: "Do you love me?"
MATE: "Yes, of course I love you."
YOU: "Do you really love me?"
MATE: "Yes, I really love you."
YOU: "You are sure you love me—you are absolutely sure?"
MATE: "Yes, I'm absolutely sure."
YOU: "Do you know the meaning of love?"
MATE: "I don't know."
YOU: "Then how can you be sure you love me?"
MATE: "I don't know. Perhaps I can't."
YOU: "You can't, eh? I see. Well, since you can't even be sure you love me, I can't really see much point in our remaining together. Can you?"
MATE: "I don't know. Perhaps not."
YOU: "You've been leading up to this for a pretty long time, haven't you?"

Love is both a feeling and an activity. We feel love for someone and act in a loving manner. But we can also be angry with the person we love as well as frustrated, bored, or indifferent. This is the paradox of love; it encompasses opposites. Love includes affection and anger, excitement and boredom, stability and change, bonds and freedom. Its paradoxical quality makes some ask whether they are really in love when they are not feeling "perfectly" in love or when their relationship is not going smoothly. Love does not give us perfection; however, it gives us meaning. In fact, as sociologist Ira Reiss (1980a) suggests, an

Was it not by loving that I learned to love?
Was it not by living that I learned to live?
JORGE AMADO

important question to ask is: Is the love I feel the kind of love on which I can build a lasting relationship or marriage?

We can look at love in many ways besides through the eyes of lovers, although other ways may not be as entertaining. Whereas love was once the province of lovers, madmen, poets, and philosophers, social scientists have begun to appear on the scene. Although there is something to be said for the mystery of love, understanding how love works in the day-to-day world may help us keep our love vital and growing. (For an overview of recent theory and research in love and jealousy, see Berscheid, 1994).

Friendship, Love, and Commitment

Friendship, love, and commitment are closely linked in our intimate relationships; they help to create and sustain each other.

Friendship is the foundation for love and commitment. Love reflects the positive factors, such as caring and attraction, that draw two people together and sustain them in a relationship. Commitment reflects the stable factors, including not only love but also obligations and social pressure, that help maintain the relationship for better or for worse. Although love and commitment are related, they are not necessarily connected. One can exist without the other. It is possible to love someone without being committed. It is also possible to be committed to someone without loving that person. Yet, when all is said and done, most of us long for a love that includes commitment and a commitment that encompasses love.

Prototypes of Love and Commitment

Despite centuries of discussion, debate, and complaint by philosophers and lovers, no one has succeeded in finding a definition of love on which all can agree. Ironically, such discussions seem to engender conflict and disagreement rather than love and harmony.

Because of the unending confusion surrounding definitions of love, some researchers wonder whether such definitions are even possible (Fehr, 1988; Kelley, 1983). In the everyday world, however, we do seem to have something in mind when we tell someone we love him or her. We may not have formal definitions of love, but we do have **prototypes** (that is, models) of what we mean by love stored in the backs of our minds. Researchers suggest that instead of looking for formal definitions of love and commitment, we examine people's prototypes; in some ways these prototypes may be more important than formal definitions. For

While it's often difficult to come up with a formal definition of love, we usually know what we mean when we tell someone that we love them, or, as scholars would (unromantically) say, we have a prototype of love.

Love looks not with the eyes,

But with the mind.

William Shakespeare (1564–1616), *A Midsummer Night's Dream*

example, when we say "I love you," we are referring to our prototype of love rather than its definition. By thinking in terms of prototypes, we can study how people actually use the words *love* and *commitment* in real life and how their meanings of love and commitment help define the progress of their intimate relationships.

To discover people's prototypes, researcher Beverly Fehr (1988) asked 172 respondents to rate the central features of love and commitment. The twelve central attributes of love they listed are given in order:

- Trust
- Caring
- Honesty
- Friendship
- Respect
- Concern for the other's well-being
- Loyalty
- Commitment
- Acceptance of the other, the way he or she is
- Supportiveness
- Wanting to be with the other
- Interest in the other

It lies not in our power to love or hate,

For will in us is over-ruled by fate.

Christopher Marlowe (1564–1593), *Hero and Leander*

The twelve central attributes of commitment, listed in order, were as follows:

- Loyalty
- Responsibility
- Living up to your word
- Faithfulness
- Trust
- Being there for the other in good and bad times
- Devotion
- Reliability
- Giving your best effort
- Supportiveness
- Perseverance
- Concern about the other's well-being

Bliss is unnatural.

Emily Dickinson (1830–1886)

There are many other characteristics identified as features of love (euphoria, thinking about the other all the time, butterflies in the stomach) or commitment (putting the other first, contentment). These, however, tend to be peripheral. As relationships progress, the central aspects of love and commitment become more characteristic of the relationship than the peripheral ones. The central features, observes Fehr (1988), "act as true barometers of a move toward increased love or commitment in a relationship." Similarly, violations of central features of love and commitment are considered to be more serious than violations of peripheral ones. A loss of caring, trust, honesty, or respect threatens love, while the disappearance of butterflies in the stomach does not. Similarly, lack of responsibility or faithfulness endangers commitment, whereas discontent is not perceived as threatening. Researchers have found that love *and* commitment are correlated to satisfaction in romantic relationships (Hendrick, Hendrick, and Adler, 1988).

Attitudes and Behaviors Associated with Love

A review of the research on love finds a number of attitudes, feelings, and behaviors associated with love (Kelley, 1983).

Positive Attitudes and Feelings

Positive attitudes and feelings toward the other bring people together. Zick Rubin (1973) found that there were four feelings identifying love:

- Caring for the other—wanting to help him or her.
- Needing the other—having a strong desire to be in the other's presence and to have the other care for one.
- Trusting the other—mutually exchanging confidences.
- Tolerating the other—accepting his or her faults.

Of these, caring appears to be the most important, followed by needing, trusting, and tolerating (Steck, Levitan, McLane, and Kelley, 1982). J. R. Davitz (1969) identified similar feelings associated with love but noted in addition that respondents reported feeling an inner glow, optimism, and cheerfulness. They felt harmony and unity with the person they loved. They were intensely aware of the other person, feeling that they were fully concentrated on him or her.

Loving Behaviors

Love is also expressed in certain behaviors. One study (Swensen, 1972) found that romantic love was expressed in several ways. (Notice that the expression of love often overlaps thoughts of love.) Some of these ways are:

- Verbally expressing affection, such as saying "I love you."
- Self-disclosing, such as revealing intimate facts about oneself.
- Giving nonmaterial evidence, such as offering emotional and moral support in times of need and showing respect for the other's opinion.
- Expressing nonverbal feelings such as happiness, contentment, and security when the other is present.
- Giving material evidence, such as providing gifts or small favors or doing more than one's share of something.
- Physically expressing love, such as by hugging, kissing, and making love.
- Tolerating the other, such as in accepting the other's idiosyncrasies, peculiar routines, or forgetfulness about putting the cap on the toothpaste.

The opposite of loneliness is not togetherness. It is intimacy.

RICHARD BACH, *THE BRIDGE ACROSS FOREVER*

These behavioral expressions of love are consistent with the prototypical characteristics of love. In addition, research supports the belief that people "walk on air" when they are in love. Researchers have found that those in love view the world more positively than those who are not in love (Hendrick and Hendrick, 1988a).

Though little research exists on ethnicity and attitudes and behaviors associated with love, one study of Mexican American college students suggests that they share many of the same attitudes and behaviors described above (Castaneda, 1993). Both females and males valued communication/sharing, trust, mutual respect, shared values and attitudes, and honesty. Data from white, middle-class adults indicate that men and women are quite similar in their love attitudes across adulthood (Montgomery and Sorell, 1997).

I am not one of those who do not believe in love at first sight, but I do believe in taking a second look.

HENRY VINCENT (1813–1878)

Factors Affecting Commitment

Although we generally make commitments to a relationship because we love someone, love alone is not sufficient to make a commitment last. Our commitments seem to be affected by several factors that can strengthen or weaken the relationship. Ira Reiss (1980a) believes that there are three important factors in commitment to a relationship: (1) the balance of costs to benefits, (2) normative inputs, and (3) structural constraints.

Did you know?

Those in committed romantic or marital relationships have higher levels of satisfaction than those in noncommitted relationships. (Hecht, Marston, and Larken, 1994).

PERSPECTIVE

Love and Sexuality

Although love and sex are often thought of as being separate phenomena, recent research suggests that for both men and women, sex includes intimacy and caring, key aspects of love (Aron and Strong, n.d.; Strong and DeVault, 1997). Nevertheless, gender differences do exist, especially in terms of casual relationships. (See Chapter 7 for a further discussion of sexuality.)

Men, Sex, and Love

Men and women who are not in an established relationship have different expectations. Men are more likely than women to separate sex from affection. Studies consistently demonstrate that for the majority of men, sex and love can be easily separated (Blumstein and Schwartz, 1983; Carroll et al., 1985; Laumann, Gagnon, Michael, and Michaels, 1994).

Although men are more likely than women to separate sex and love, Linda Levine and Lonnie Barbach (1985) found in their interviews that men indicated that their most erotic sexual experiences took place in a relational context. The authors quote one man as follows:

> Emotions are everything when it comes to sex. There's no greater feeling than having an emotional attachment with the person you're making love to. If those emotions are there, it's going to be fabulous. . . . They don't call it "making love" for nothing.

Most men in the study responded that it was primarily the emotional quality of the relationship that made their sexual experiences special.

Women, Sex, and Love

Women generally view sex from a relational perspective. In the decision to have sexual intercourse, the quality and degree of intimacy of a relationship were more important for women than men (Christopher and Cate, 1984). Women were more likely to report feelings of love if they were sexually involved with their partners than if they were not sexually involved (Peplau et al., 1977). Love is also more closely related to feelings of self-esteem (Walsh, 1991).

Women generally seek emotional relationships, but men may initially seek physical relationships. This difference in intentions can place women in a bind. Carole Cassell (1984) suggests that today's women face a "damned if you do, damned if you don't" dilemma in their sexual relationships. If a woman has sexual intercourse with a man, he says good-bye; if she doesn't, he says he respects her and still says good-bye. A young woman Cassell interviewed related the following:

> I really hate the idea that, because I'm having sex with a man whom I haven't known for a long, long time, he'll think I don't value myself. But it's hard to know what to do. If you meet a man and date him two or three times and don't have sex, he begins to feel you are either rejecting him or you have serious sex problems. . . . But I really dread feeling that I could turn into, in his eyes, an easy lay, a good-time girl. I want men to see me as a grown-up woman who has the same right as they do to make sexual choices.

Traditionally, women were labeled "good" or "bad" on the basis of their sexual experience and values. "Good" women are virgins, sexually naive, or passive; "bad" women were sexually experienced, independent, and passionate. According to Lillian Rubin (1990), this attitude has not entirely changed. Rather, we are ambivalent about sexually experienced women. One exasperated woman leaped out of her chair and began to pace the floor, exclaiming to Rubin: "I sometimes think what men really want is a sexually experienced virgin. They want you to know the tricks, but they don't like to think you did those things with someone else."

1 *The balance of costs to benefits.* Whether we like it or not, human beings have a tendency to look at romantic and marital relationships from a cost-benefit perspective. Most of the time, when we are satisfied, we are unaware that we judge our relationships in this manner. But as we saw in our discussion of social exchange theory in Chapter 2, when there is stress or conflict, we often ask ourselves, "What am I getting out of this relationship?" Then we add up the pluses and minuses. If the result is on the plus side, we are encouraged to continue the relationship; if the result is negative, we are more likely to discontinue it.

2 *Normative inputs.* Normative inputs for relationships are the values that you and your partner hold about love, relationships, marriage, and family. These values can either sustain or detract from a commitment.

The meeting of two personalities is like the contact of two chemical substances; if there is any reaction, both are transformed.

CARL JUNG (1875–1971)

Gay Men, Lesbians, and Love

Love is equally important for heterosexuals, gay men, lesbians, and bisexuals (Aron and Aron, 1991; Keller and Rosen, 1988; Kurdek, 1988; Peplau and Cochran, 1988). Many heterosexuals, however, perceive lesbian and gay love relationships as less satisfying and less loving than heterosexual ones. In a study of 360 heterosexual undergraduates (Testa et al., 1987), students were presented with identical information about a couple that was variously described as heterosexual, gay, and lesbian. When the couple was identified as heterosexual, it ranked high on love and satisfaction. When the couple was identified as lesbian or gay, it was ranked significantly lower on love and satisfaction. Because the couples were identical except for sexual orientation, the researchers concluded that there was a heterosexual bias in the perception of gay and lesbian love relationships. It is well documented that love is important for lesbians and gay men; like heterosexual relationships, theirs have multiple emotional dimensions (Adler, Hendrick, and Hendrick, 1989).

Men in general are more likely than women to separate love and sex; gay men are especially likely to make this separation. Although gay men value love, they also tend to value sex as an end in itself. Furthermore, they place less emphasis on sexual exclusiveness in their relationships. Researchers suggest, however, that heterosexual males are not very different from gay males in terms of their acceptance of casual sex. Heterosexual males, they maintain, would be as likely as gay males to engage in casual sex if women were equally interested. Women, however, are not as interested in casual sex; as a result, heterosexual men do not have as many willing partners available as do gay men (Foa et al., 1987; Symons, 1979).

For lesbians, gay men, and bisexuals, love has special significance in the formation and acceptance of their identities. Although significant numbers of women and men have had sexual experiences with members of the same sex or both sexes, relatively few identify themselves as lesbian or gay. An important element in solidifying such an identity is loving someone of the same sex. Love signifies a commitment to being gay or lesbian by unifying the emotional and physical dimensions of a person's sexuality (Troiden, 1988). For the gay man or lesbian, it marks the beginning of sexual wholeness and acceptance. In fact, some researchers believe that the ability to love someone of the same sex, rather than having sex with him or her, is the critical element that distinguishes being gay or lesbian from being heterosexual (Money, 1980).

Sex Without Love

Is love necessary for sex? Most of us assume it is, but our assumption is based on our values. The question cannot be answered by reference to empirical or statistical data (Crosby, 1985). A more fundamental question is this: Is sexual activity legitimate in itself, or does it require justification? John Crosby (1985) observes:

> Sexual pleasure is a value in itself and hence capable of being inherently meaningful and rewarding. The search for extrinsic justification and rationalization simply belies our reluctance to believe that sex is a pleasurable activity in and of itself.

To believe that sex does not require love as a justification, argues Crosby, does not deny the significance of love and affection in sexual relations. In fact, love and affection are important and desirable for enduring relationships. They are simply not necessary, Crosby believes, for affairs in which erotic pleasure is the central feature.

Ironically, although sex without love may violate social norms, it is the least threatening form of extrarelational sex. Even those who accept their partners' having sex outside the relationship find it especially difficult to accept their partners' having a meaningful affair. "They believe," note Philip Blumstein and Pepper Schwartz (1983) "that two intense romantic relationships cannot coexist and that one would have to go."

How do you feel about a love commitment? A marital commitment? Do you believe that marriage is for life? Does the presence of children affect your beliefs about commitment? What are the values that your friends, family, and religion hold regarding your type of relationship?

3. *Structural constraints.* The structure of a relationship will add to or detract from commitment. Depending on the type of relationship—whether it is dating, living together, or marriage—different roles and expectations are structured in. In marital relationships, there are partner roles (husband/wife) and economic roles (employed worker/homemaker). There may also be parental roles (mother/father).

These different factors interact to increase or decrease the commitment. Commitments are more likely to endure in marriage than in cohabiting or dating relationships, which tend to be relatively short lived. They are more likely to last in heterosexual relationships than in gay or lesbian relationships (Testa et al., 1987). Ethnicity may also be the greatest predictor of satisfaction and commitment to a friendship (deVries, Jacoby, and Davis, 1996). The reason commitments tend to endure in marriage may or may not have anything to do with a couple being happy. Marital commitments may last because norms and structural constraints compensate for the lack of personal satisfaction.

Reflections

What are the ideas you associate with love? With friendship? With commitment? How do they overlap? Have you ever mistaken love for commitment? Friendship for love?

For most people, love seems to include commitment and commitment seems to include love. Beverly Fehr (1988) found that if a person violated a central aspect of love, such as caring, that person was also seen as violating the couple's commitment, and if a person violated a central aspect of commitment, such as loyalty, it called love into question.

Because of the overlap between love and commitment, we can mistakenly assume that someone who loves us is also committed to us. As one researcher (Kelley, 1983) points out: "Expressions of love can easily be confused with expressions of commitment. . . . Misunderstandings about a person's love versus commitment can be based on honest errors of communication, on failures of self-understanding." Or a person can intentionally mislead the partner into believing that there is a greater commitment than there actually is. Even if a person is committed, it is not always clear what the commitment means: Is it a commitment to the person or to the relationship? Is it for a short time or a long time? Is it for better and for worse?

Friendship and Love

Friendships with members of the other sex are increasingly common, even after marriage.

Friendship and love breathe life into humanity. They bind us together, provide emotional sustenance, buffer us against stress, and help to preserve our physical and mental well-being. The loss of a friend and especially a loved one can lead to illness and even suicide.

What distinguishes love from friendship? Should marriage satisfy all of our needs (including friendship)? What is this thing called friendship?

Two researchers (Todd and Davis, 1985) set out to distinguish the differences between characteristics of love and friendship. In their study of 250 college students and community members, they found that though love and friendship were alike in many ways, some crucial differences make love relationships both more rewarding and more vulnerable. Best friends were similar to spouse/lover relationships in several ways: level of acceptance, trust, respect, levels of confiding, understanding, spontaneity, and mutual acceptance. Levels of satisfaction and happiness with the relationship were also found to be similar for both groups. What separated friends from lovers was that lovers had much more fascination and a sense of exclusiveness with their partners than did friends. Though love had a greater potential for distress, conflict, and mutual criticism, it ran deeper and stronger than friendship.

Did you know?

The majority of men and women do not consider their spouse to be a best friend, though women are somewhat more likely than men to do so (Bell, 1991).

Friendship appears to be the foundation for a strong love relationship. Shared interests and values, acceptance, trust, understanding, and enjoyment are at the root of friendship and form a basis for love. Adding passion and emotional intimacy alters the nature of the friendship.

A friend may well be reckoned the masterpiece of nature.

RALPH WALDO EMERSON (1803-1882), *FRIENDSHIP*

While some believe that marriage should satisfy all their needs, it is important to remember that when people marry they do not cease to be separate individuals. Friendships and patterns of behavior continue, so the effect that maintaining friendship has on marital satisfaction must be understood.

With men and women marrying later than ever before and women being an integral part of the workforce, close friendships, including other-sex friends, are more likely to be a part of the tapestry of relationships in peoples' lives. Partners need to communicate and to seek understanding regarding the nature of activities and degree of emotional closeness they find acceptable in their spouse's friendships. Boundaries should be clarified and opinions shared. Many couples have found cross-sex friendships to be acceptable and even desirable. Like other significant issues that interface with marriage, keys to their success are the ability to communicate concerns and the maturity of all individuals involved.

Reflections

Examine the wheel theory of love diagram and ask yourself whether your love relationships follow the course Reiss suggests. What creates rapport for you? What factors increase or decrease self-revelation? When self-revelation increases, does mutual dependency also increase? If mutual dependency decreases, do self-revelation and rapport decrease? What impact have social background and role conceptions had on the development of your relationships?

The Development of Love: The Wheel Theory

Sociologist Ira Reiss (1980a) suggests that love develops and is maintained through four processes: (1) rapport, (2) self-revelation, (3) mutual dependency, and (4) fulfillment of the need for intimacy. Reiss calls the processes the **wheel theory of love** to emphasize their interdependence. A reduction in any one affects the development or maintenance of a love relationship (see Figure 4.1). If a couple habitually argue, for example, the arguments will affect their mutual dependency and their need for intimacy; this in turn will weaken their rapport.

Rapport

When two people meet, they quickly sense if rapport exists between them. This rapport is a sense of ease, the feeling that they understand each other in some special way. This feeling is dependent on our social environment in two ways. First, rapport tends to depend on sharing the same social and cultural background.

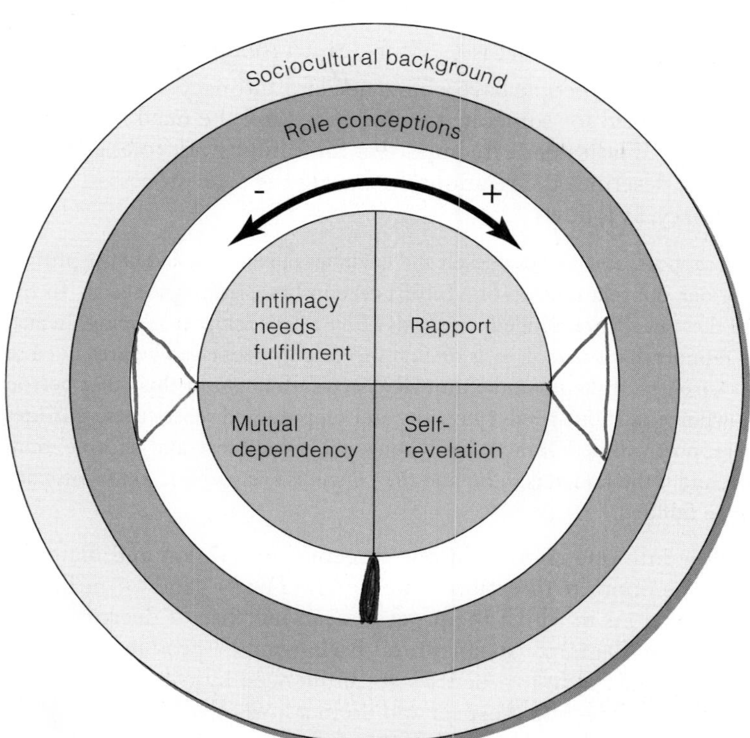

FIGURE 4.1 Graphic Representation of the Wheel Theory of Love.

Reiss's Wheel Theory of Love. In the wheel theory of love, the development of intimacy is most likely to take place between those who share the same sociocultural background and role conceptions. Intimacy develops from a feeling of rapport, which leads to self-revelation; self-revelation leads to mutual dependency, which in turn may lead to intimacy need fulfillment.

If one person has only a grade school education and the other a college education, it is not likely that they will share many of the same values. If one person is upper class and the other is working class, their life experiences have probably been quite different. Second, we tend to feel rapport with those who share our role conceptions. Two people who believe in egalitarian gender roles, for example, are more likely to feel rapport than are a radical feminist and a male chauvinist.

Self-Revelation

If you feel rapport with someone, you are likely to feel relaxed and confident. As a result, self-revelation—the disclosure of intimate feelings—is likely to occur. But self-revelation also depends on what is considered proper by your socialization group. For example, upper-class people have a tendency to be reserved about themselves. Middle-class people feel more comfortable in revealing intimate aspects of their lives and feelings.

Mutual Dependency

We are shaped and fashioned by what we love.

JOHANN GOETHE (1746–1842)

After two people feel rapport and begin revealing themselves to each other, they may become mutually dependent. Each needs the other to share pleasures, fears, and jokes, as well as sexual intimacies; each becomes the other's confidant. Each person develops ways of acting and being that cannot be fulfilled alone. Going for a walk is no longer something you do by yourself; you do it with your partner. Sleeping no longer takes place in a single bed but in a double one with your partner. You are a couple.

We two boys together clinging
One the other never leaving.

WALT WHITMAN (1819–1892), *CALAMUS*

Here, too, in the development of mutual dependency, the social and cultural background is important. The forms of dependent behavior that develop are influenced by your conception of the role of courtship. Interdependency may develop through dating, getting together, or living together. Premarital intercourse may or may not be acceptable.

Intimacy Needs for Fulfillment

According to Reiss (1980a), people have a basic need for intimacy—"the need for someone to love, the need for someone to confide in, and the need for sympathetic understanding." These needs are important for fulfilling our roles as a partner or parent. Reiss describes the relationship among the four processes, which culminate in intimacy, as follows:

> By virtue of rapport, one reveals oneself and becomes dependent, and in the process of carrying out the relationship one fulfills certain basic intimacy needs. To the extent that these needs are fulfilled, one finds a love relationship developing. In fact, the initial rapport that a person feels on first meeting someone can be presumed to be a dim awareness of the potential intimacy need fulfillment of this other person for one's own needs. If one needs sympathy and support, and senses these qualities in a date, rapport will be felt more easily; one will reveal more and become more dependent, and if the hunch is right, and the person is sympathetic, one's intimacy needs will be fulfilled.

The processes flow into one another in one direction to develop and maintain love; they flow in the opposite direction to weaken it. The "+" and "-" in Figure 4.1 indicate the directions in which the processes can increase or decrease love. The outer ring on the diagram, "sociocultural background," produces the next ring, "role conceptions." All four processes are influenced by role conceptions, which define what a person should expect and do in a love relationship.

How Do I Love Thee? Approaches to the Study of Love

Researchers have developed a number of ways to study love (Hendrick and Hendrick, 1987). Some of these ways are discussed in this section.

Love reckons hours for months, and days for years; and every little absence is an age.

JOHN DRYDEN (1631–1700)

Styles of Love

Sociologist John Lee (1973, 1988; Borrello and Thompson, 1990) describes six basic styles of love. These styles, he cautions, are relationship styles, not individual styles. The style of love may change as the relationship changes or when individuals enter different relationships. The styles are:

- **Eros:** love of beauty
- **Ludus:** playful love
- **Storge:** companionate love
- **Mania:** obsessive love
- **Agape:** altruistic love
- **Pragma:** practical love

In addition to these pure forms there are mixtures of the basic types: storgic-eros, ludic-eros, and storgic-ludus. The six basic types can be described as follows:

Eros. Erotic lovers delight in the tactile, the sensual, the immediate; they are attracted to beauty (though beauty may be in the eye of the beholder). They love the lines of the body, its feel and touch. They are fascinated by every detail of their beloved. Their love burns brightly but soon flickers and dies.

Beauty stands
In the admiration of weak minds
Led captive

JOHN MILTON (1608–1674)

Ludus. For ludic lovers, love is a game, something to play at rather than to become deeply involved in. Love is ultimately ludicrous. Love is for fun; encounters are casual, carefree, and often careless. "Nothing serious" is the motto of ludic lovers.

Love gives itself; it is not bought.

HENRY LONGFELLOW (1807–1882)

Storge. Storge (pronounced STOR-gay) is the love between companions. It is, writes Lee, "love without fever, tumult, or folly, a peaceful and enchanting affection." It begins usually as friendship and then gradually deepens into love. If the love ends, it also occurs gradually, and the couple often become friends once again. Of such love Theophile Gautier wrote, "To love is to admire with the heart; to admire is to love with the mind."

Mania. The word *mania* comes from the Greek word for madness. The Russian poet Mikhail Lermontov aptly described a manic lover:

> He in his madness prays for storms
> And dreams that storms will bring him peace.

For manic lovers, nights are marked by sleeplessness and days by pain and anxiety. The slightest sign of affection brings ecstasy for a short while, only to have it disappear. Satisfactions last but a moment before they must be renewed. Manic love is roller-coaster love.

UNDERSTANDING YOURSELF

Your Style of Love

John Lee, who developed the idea of styles of love, also developed a questionnaire that allows men and women to identify their styles of love. Complete the questionnaire to identify your style of love. Then ask yourself to which style of love you find yourself drawn in others. Is it the same or different?

To diagnose your style of love, look for patterns across characteristics. If you consider your childhood less happy than that of your friends, were discontent with life when you fell in love, and very much want to be in love, you have "symptoms" that are rarely typical of eros and almost never true of storge but that do suggest mania. Where a trait does not especially apply to a type of love, the space in that column is blank. Storge, for instance, is not the *presence* of many symptoms of love but precisely their absence; it is a cool, abiding affection.

Graph Your Own Style of Loving

Consider each characteristic as it applies to a current relationship that you define as love, or to a previous one if that is more applicable. For each, note whether the trait is almost always *true (AA),* usually *true (U),* rarely *true (R), or* almost never *true (AN).*

	Eros	Ludus	Storge	Mania	Ludic Eros	Storgic Eros	Storgic Ludus	Pragma
1 You consider your childhood less happy than the average of peers.	R		AN	U				
2 You were discontented with life (work, etc.) at the time your encounter began.	R		AN	U	R			
3 You have never been in love before this relationship.				U	R	AN	R	
4 You want to be in love or have love as security.	R	AN		AA		AN	AN	U
5 You have a clearly defined ideal image of your desired partner.	AA	AN	AN	AN	U	AN	R	AA
6 You felt a strong gut attraction to your beloved on the first encounter.	AA	R	AN	R		AN		
7 You are preoccupied with thoughts about the beloved.	AA	AN	AN	AA			R	
8 You believe your partner's interest is at least as great as yours.		U	R	AN			R	U
9 You are eager to see your beloved almost every day; this was true from the beginning.	AA	AN	R	AA		R	AN	R

Love is patient and kind; love is not jealous or boastful; it is not arrogant or rude. Love does not insist on its own way; it is not irritable or resentful; it does not rejoice at wrong, but rejoices in the right. Love bears all things, believes all things, hopes all things, endures all things.

I CORINTHIANS 13:4–7

Agape. Agape (pronounced AH-ga-pay) is love that is chaste, patient, selfless, and undemanding; it does not expect to be reciprocated. Agape emphasizes nurturing and caring as their own rewards. It is the love of monastics, missionaries, and saints more than that of worldly couples.

Pragma. Pragmatic lovers are, first and foremost, logical in their approach toward looking for someone who meets their needs. They look for a partner who has the background, education, personality, religion, and interests that are compatible with their own. If they meet a person who meets their criteria, erotic or manic feelings may develop. But, as Samuel Butler warned, "Logic is like the sword—those who appeal to it shall perish by it."

Lee believes that to have a mutually satisfying love affair, a person has to find a partner who shares the same style and definition of love. The more different two people are in their styles of love, the less likely it is that they will

Item	Eros	Ludus	Storge	Mania	Ludic Eros	Storgic Eros	Storgic Ludus	Pragma
10 You soon believed this cou								
11 You see "warning signs" of tro								
12 You deliberately restrain frequency								
13 You restrict discussion of your feeling								
14 You restrict display of your feelings with								
15 You discuss future plans with beloved.								
16 You discuss wide range of topics, experiences w								
17 You try to control relationship, but feel you've lost								
18 You lose ability to be first to terminate relationship.								
19 You try to force beloved to show more feeling, commitme								
20 You analyze the relationship, weight it in your mind.								
21 You believe in the sincerity of your partner.								
22 You blame partner for difficulties of your relationship.								
23 You are jealous and possessive but not to the point of angry conflict.								
24 You are jealous to the point of conflict, scenes, threats, etc.								
25 Tactile, sensual contact is very important to you.								
27 Sexual intimacy was achieved early, rapidly in the relationship.								
28 You are willing to work out sex problems, improve technique.	U							
29 You have a continued high rate of sex, tactile contact throughout the relationship.	U							
30 You declare your love first, well ahead of partner.		AN	R					
31 You consider love life your most important activity, even essential.		AN	R	A				
32 You are prepared to "give all" for love once under way.	U	AN	U	AA				
33 You are willing to suffer abuse, even ridicule, from partner.		AN	R	AA				
34 Your relationship is marked by frequent differences of opinion, anxiety.	R	AA	R	AA	R			
35 The relationship ends with lasting bitterness, trauma for you.	AN	R	R	AA	R	AN		

SOURCE: From J. A. Lee, "The Styles of Love," *Psychology Today,* 1974. Reprinted by permission.

understand each other's love. Research confirms that love styles are also linked to gender and ethnicity (Hendrick and Hendrick, 1986).

Research indicates that heterosexual and gay men have similar attitudes toward eros, mania, ludus, and storge. The research also confirms that gay male relationships have multiple emotional dimensions (Adler, Hendrick, and Hendrick, 1989).

Reflections

What is your style of love? Have your partners had the same or different styles? How has your style affected your relationships?

The Triangular Theory of Love

The **triangular theory of love,** developed by Robert Sternberg (1986; Chojnacki and Walsh, 1990), seeks to emphasize the dynamic quality of love relationships. This theory sees love as composed of three elements, as the points of a triangle. These elements are intimacy, passion, and decision/commitment

They can
nt type of
The com-
love rela-
ory (Acker

feelings of
and Grajek

things.

ne elements of
y be fueled by
to affiliate with

mitment com-
The short-term
may not make
to make a com-
term aspect; it is
not necessarily
entail a commitment to maintaining that love.

Passion

Decision/
Commitment

Kinds of Love: The Different Combinations

The intimacy, passion, and decision/commitment components can be combined in eight basic ways, according to Sternberg. These combinations form the basis for classifying love:

- Liking (intimacy only).
- Romantic love (intimacy and passion).
- Infatuation (passion only).
- Fatuous love (passion and commitment).
- Empty love (decision/commitment only).
- Companionate love (intimacy and commitment).
- Consummate love (intimacy, passion, and commitment).
- Nonlove (absence of intimacy, passion, and commitment).

These types represent extremes that probably few of us experience. Not many of us, for example, experience infatuation in its purest form, in which there is

Did you know?

For men and women of all ages, commitment is a more powerful predictor of relationship satisfaction than passion or intimacy, according to a study on the three components of love (Acker and Davis, 1992).

absolutely no intimacy. The categories are nevertheless useful for examining love. We will discuss the various categories below (except for empty love which is not really love at all).

Liking: Intimacy Only. Liking represents the intimacy component alone. It forms the basis for close friendships but is neither passionate nor committed. As such, liking is often an enduring kind of love. Boyfriends and girlfriends may come and go, but good friends remain.

Romantic Love: Intimacy and Passion. Romantic love combines intimacy and passion. It is similar to liking, but it is more intense as a result of physical or emotional attraction. It may begin with an immediate union of the two components—with friendship that intensifies with passion or with passion that also develops intimacy. Although commitment is not an essential element of romantic love, it may develop.

Infatuation: Passion Only. Infatuation is love at first sight. It is the kind of love that idealizes its object; it rarely sees the other as a "real" person who sometimes has bad breath—but is worthy of being loved nonetheless. Infatuation is marked by sudden passion and a high degree of physical and emotional arousal. It tends to be obsessive and all consuming. One has no time, energy, or desire for anything or anyone but the beloved (or thoughts of him or her). To the dismay of the infatuated individual, infatuations are usually asymmetrical: One's passion (or obsession) is rarely returned equally, and the greater the asymmetry, the greater the distress in the relationship.

Fatuous Love: Passion and Commitment. Fatuous, or deceptive, love is whirlwind love; it begins the day a couple meet and quickly results in cohabitation or engagement, and then marriage. It goes so fast one hardly knows what has happened. Often enough, nothing much really did happen that will permit the relationship to endure. As Sternberg (1988) observes, "It is fatuous in the sense that a commitment is made on the basis of passion without the stabilizing element of intimate involvement—which takes time to develop."

Passion fades soon enough, and all that remains is commitment. But commitment that has had relatively little time to deepen is a poor foundation on which to build an enduring relationship. With neither passion nor intimacy, the commitment wanes.

Companionate Love: Intimacy and Commitment. Companionate love is essential to a committed relationship. It often begins as romantic love, but as the passion diminishes and the intimacy increases, it is transformed. Some couples are satisfied with such love; others are not. Those who are dissatisfied in companionate love relationships may seek extrarelational affairs to maintain passion in their lives. They may also end the relationship to seek a new romantic relationship in the hope that it will remain romantic.

Consummate Love: Intimacy, Passion, and Commitment. Consummate love is born when intimacy, passion, and commitment combine to form their unique constellation. It is the kind of love we dream about but do not expect in all our love relationships. Many of us can achieve it, but it is difficult to sustain over time. To sustain it, we must nourish its different components, for each is subject to the stress of time.

Figure 4.3
The Triangles of Love

The greater the intensity of love one experiences, the greater will be one's love triangle in area. The greater a given component of love, the further the point from the center of the triangle. Thus A, B, C, and D experience different kinds of love; as a result, their triangles are differently shaped.

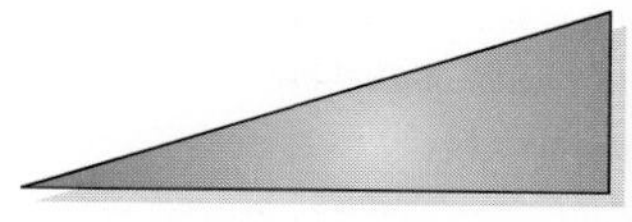

A. Infatuation (passion only)

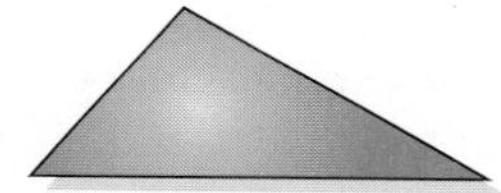

B. Empty love (decision/commitment only)

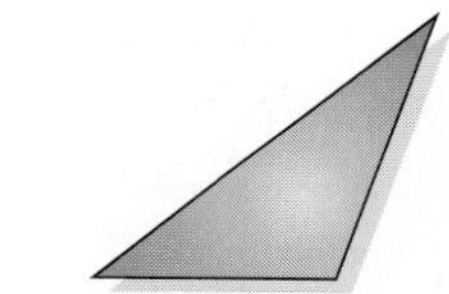

C. Romantic love (intimacy and passion)

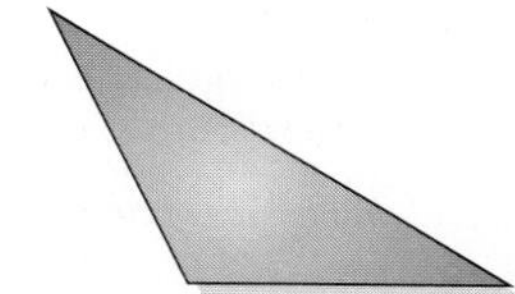

D. Companionate love (intimacy and commitment)

Upon my bed by night
I sought him whom my soul loves;
I sought him, but found him not;
I called him, but he gave no answer.
"I will rise now and go about the city,
in the streets and in the squares;
I will seek him whom my soul loves."
I sought him, but found him not.
The watchmen found me,
as they went about in the city.
"Have you seen him whom my soul loves?"
Scarcely had I passed them,
when I found him whom my soul loves.
I held him, and would not let him go
until I had brought him into my mother's house,
and into the chamber of her that conceived me.
I adjure you, O daughters of Jerusalem,
by the gazelles or the hinds of the field,
that you stir not up nor awaken love until it please.
SONG OF SOLOMON 3:1–5

Speak low, if you speak love.
WILLIAM SHAKESPEARE, (1564–1616)
MUCH ADO ABOUT NOTHING

My love is like to ice, and I to fire:
How comes it then that this her cold so great
Is not dissolved through my so hot desire,
But harder grows the more I her entreat?
Or how comes it that my exceeding heat
Is not allayed by her heart-frozen cold,
But that I burn much more in boiling sweat,
And feel my flames augmented manifold?
What more miraculous thing may be told
That fire, which all things melts, should harden ice,
And ice, which is congeal'd with senseless cold,
Should kindle fire by wonderful device?
Such is the power of love in gentle mind,
That it can alter all the course of kind.
EDMUND SPENSER (1552?–1599)

Nonlove: Nonexistent Love. Nonlove can take many forms, such as attachment for financial reasons, fear, or the fulfillment of neurotic needs.

The Geometry of Love

The shape of the love triangle depends on the intensity of the love and the balance of the parts. Intense love relationships create triangles with greater area; such triangles occupy more of one's life. Just as love relationships can be balanced or unbalanced, so can love triangles. The balance determines the shape of the triangle (see Figure 4.3). A relationship in which the intimacy, passion, and commitment components are equal forms an equilateral triangle. But if the components are not equal, unbalanced triangles form. The size and shape of a person's triangle give a good pictorial sense of how that person feels about another. The greater the match between each person's triangle in a relationship, the more likely each is to experience satisfaction in the relationship.

Love as Attachment

Attachment theory maintains that the degree and quality of attachments we experience in early life influence our later relationships. It has been increasingly used in recent years in studying personal relationships including love. It examines love as a form of attachment that finds its roots in infancy (Hazan and Shaver, 1987; Shaver, Hazan, and Bradshaw, 1988). Research suggests that romantic love and infant/caregiver attachment have similar emotional dynamics. Phillip Shaver and his associates (1988) suggest that "all important love relationships—especially the first ones with parents and later ones with lovers and spouses—are attachments." On the basis of infant/caregiver work by John Bowlby (1969, 1973, 1980), some researchers suggest numerous similarities between attachment and romantic love (Bringle and Bagby, 1992; Shaver et al., 1988). These include the following:

ATTACHMENT

- Attachment bond's formation and quality depend on attachment object's responsiveness and sensitivity.
- When attachment object is present, infant is happier.
- Infant shares toys, discoveries, objects with attachment object.
- Infant coos, talks baby talk, "sings."
- There are feelings of oneness with attachment object.

ROMANTIC LOVE

- Feelings of love are related to lover's interest and reciprocation.
- When lover is present, person feels happier.
- Lovers share experience and goods, give gifts.
- Lovers coo, talk baby talk, and sing.
- There are feelings of oneness with lover.

Studies conducted by Mary Ainsworth and colleagues (1978, cited in Shaver et al., 1988) indicate that there are three styles of infant attachment: secure, anxious/ambivalent, and avoidant. In *secure attachment*, the infant feels secure when the mother is out of sight. He or she is confident that the mother will offer protection and care. In *anxious/ambivalent attachment*, the infant shows separation anxiety when the mother leaves. He or she feels insecure when the mother is not present. In *avoidant attachment*, the infant senses the mother's detachment and rejection when he or she desires close bodily contact. The infant shows avoidance behaviors with the mother as a means of defense. In Ainsworth's study,

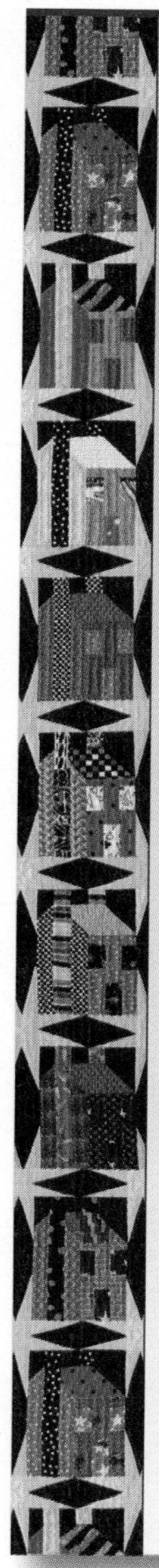

YOU AND YOUR WELL-BEING

SOCIAL SUPPORT AND WELLNESS

Imagine a life alone. For most of us, the thought of this is unfathomable. It may also be unhealthy. Research has proven that social ties help to maintain both physical and mental health. Social support in partnerships consists of both instrumental and emotional elements. Instrumental support includes such things as cooking for the other, doing favors, and providing economic security. But emotional support—the sense of being loved and cared about, esteemed and valued—appears to be even more important. Although spouses or committed partners are the most likely candidates for providing this type of support, we may also receive it from close friends. Numerous studies have demonstrated that those who have close, confiding relationships are more likely to survive health challenges such as heart attacks and surgery and less likely to develop diseases such as cancer and respiratory ailments (Brody, 1992; Justice, 1988). However, when partners or parents are not supportive, are authoritarian, divide labor unfairly, or are degrading or demeaning, their partners and children are likely to be depressed and may become more prone to illness (Gottman and Katz, 1989; Keilcott-Glaser, 1987).

Social support improves psychological and physical health in several ways (Ross, Mirowsky, and Goldsteen, 1991). First of all, a partner or friend who listens and cares helps reduce depression, anxiety, and other psychological problems. With the improvement of a person's emotional well-being, his or her physical health and survival ability increase dramatically. Second, a live-in partner encourages and reinforces protective health behaviors. Compared with their unmarried peers, husbands and wives have better diets and are less likely to smoke, drink heavily, drive too fast, or take risks that increase the likelihood of accidents or injuries. Because women generally have healthier lifestyles than men, marriage improves men's health behaviors and survival rates more significantly than women's (Ross, Mirowsky, and Goldsteen, 1991; Umberson, 1987). Third, a partner provides "secondary prevention" by helping to identify or treat an illness or disease early. Secondary prevention occurs when, for example, a man notices a discoloration on his partner's back, which turns out to be a skin cancer that is curable if treated early. It occurs when you convince your sneezing, wheezing partner to rest instead of going to work or school, thereby helping him or her to speed recovery and build resistance to other infections.

Finally, social support aids recovery. The love and care of a partner or friend, gestures of intimacy, small acts of thoughtfulness, a familiar hand soothing a feverish forehead—all assist in the psychological recovery following an illness. It is well documented, for example, that intimacy encourages emotional recovery following a heart attack (Brody, 1992). And a partner's high level of support and caring can significantly reduce the depression and anxiety that a woman experiences as she copes with breast cancer.

Historically, Americans have valued and passed on from one generation to the next the concept of self-reliance. Holding problems within, time constraints as families try to earn a living, and the lack of an extended family have contributed to creating a generation which suffers from the malady of isolation. Well-being involves a balance between supporting others and seeking social support, otherwise called interdependence.

66 percent of the infants were secure, 19 percent were anxious/ambivalent, and 21 percent were avoidant.

Some researchers (Feeney and Noller, 1990; Shaver et al., 1988) believe that the styles of attachment developed during infancy continue through adulthood. Others, however, question the validity of applying infant research to adults as well as the stability of attachment styles throughout life (Hendrick and Hendrick, 1994). Still others found a significant association between attachment styles and relationship satisfaction (Brennan and Shaver, 1995).

Love cannot save life from death, but it can fulfill life's purpose.

ARNOLD TOYNBEE (1889–1975)

Secure Adults

Secure adults found it relatively easy to get close to others. They felt comfortable depending on others and having others depend on them. They generally didn't worry about being abandoned or having someone get too close to them. More than avoidant and anxious/ambivalent adults, they felt that others generally liked them; they believed that people were generally well intentioned and good hearted. In contrast to others, secure adults were less likely to believe in media images of love and more likely to believe that romantic love can last. Their love experiences tended to be happy, friendly, and trusting. They accepted and supported their partners. On the average, their relationships lasted ten years. About 56 percent of the adults in the study were secure. A recent study reported that as compared to others, secure adults find greater satisfaction and commitment in their relationships (Pistol, Clark, and Tubbs, 1995).

Reflections

What is your attachment style now? From what you know about your infancy, is it the same or different? Do you find yourself attracted to those with similar or different attachment styles? Why?

Anxious/Ambivalent Adults

Anxious/ambivalent adults felt that others did not get as close as they themselves wanted. They worried that their partners didn't really love them or that they would leave them. They also wanted to merge completely with the other person, which sometimes scared that person away. More than others, anxious/ambivalent adults believed that it is easy to fall in love. Their experiences in love were often obsessive and marked by a desire for union, high degrees of sexual attraction and jealousy, and emotional highs and lows. Their love relationships lasted an average of five years. Between 19 and 20 percent of the adults in the study were identified as anxious/ambivalent.

Avoidant Adults

Avoidant adults felt discomfort in being close to others; they were distrustful and fearful of becoming dependent (see also Bartholomew, 1990). More than others, they believed that romance seldom lasted but that at times it could be as intense as it was at the beginning. Their partners wanted more closeness than they did. Avoidant lovers feared intimacy and experienced emotional highs and lows and jealousy. Their relationships lasted six years on the average. Between 23 and 25 percent of the adults in the study were avoidant.

In adulthood, the attachment styles developed in infancy combine with sexual desire and caring behaviors to give rise to romantic love.

Unrequited Love

Love is unjust; justice loveless.

DELMORE SCHWARTZ

As most of us know from painful experience, love is not always returned. We may reassure ourselves that, as Tennyson wrote 150 years ago, "'Tis better to have loved and lost/Than never to have loved at all." Too often, however, such words sound like a rationalization. Who among us does not sometimes think, "'Tis better never to have loved at all"? **Unrequited love**—love that is not returned—is a common experience.

Several researchers (Baumeister, Wotman, and Stillwell, 1993) accurately captured some of the feelings associated with unrequited love in the title of their research article: "Unrequited Love: On Heartbreak, Anger, Guilt, Scriptlessness, and Humiliation." They found that unrequited love was distressing for both the would-be lover and the rejecting partner. Would-be lovers felt both positive and intensely negative feelings about their unlucky attempt at a relationship. The rejectors, however, felt uniformly negative about the experience. Unlike the rejectors, the would-be lovers felt that the attraction was mutual, that they had been led on, and that the rejection had never been clearly communicated. Rejectors, by contrast, felt that they had not led the other person on; moreover,

Did you know?

About 75 percent of respondents surveyed experienced unrequited love and about 20 percent were currently experiencing it (Aron, Aron, and Allen, 1989b).

they felt guilty about hurting him or her. Nevertheless, many found the other person's persistence intrusive and annoying; they wished the other would have simply gotten the hint and gone away. Rejectors saw would-be lovers as self-deceptive and unreasonable; would-be lovers saw their rejectors as inconsistent and mysterious.

Unrequited love presents a paradox: If the goal of loving someone is an intimate relationship, why should we continue to love a person with whom we could not have such a relationship? Arthur Aron and his colleagues addressed this question in a study of almost 500 college students (Aron et al., 1989b). The researchers found three different attachment styles underlying the experience of unrequited love:

- *The Cyrano style*—the desire to have a romantic relationship with a specific person regardless of how hopeless the love is. In this style, the benefits of loving someone are so great that it does not matter how likely the love is to be returned. Being in the same room with the beloved—because he or she is so wonderful—may be sufficient. This style is named after Cyrano de Bergerac, a seventeenth-century musketeer, whose love for Roxanne was so great that it was irrelevant that she loved someone else.
- *The Giselle style*—the misperception that a relationship is more likely to develop than it actually is. This might occur if one misreads the other's cues, such as in mistakenly believing that friendliness is a sign of love. This style is named after Giselle, the tragic ballet heroine who was misled by Count Albrecht to believe that her love was reciprocated.
- *The Don Quixote style*—the general desire to be in love, regardless of whom one loves. In this style, the benefits of being in love—such as being viewed as a romantic or the excitement of extreme emotions—are more important than actually being in a relationship. This style is named after Cervantes's Don Quixote, whose love for the common Dulcinea was motivated by his need to dedicate knightly deeds to a lady love. "It is as right and proper for a knight errant to be in love as for the sky to have stars," Don Quixote explained.

Using attachment theory, the researchers found that some people were predisposed to be Cyranos, others Giselles, and still others Don Quixotes. Anxious/ambivalent adults tended to be Cyranos, avoidant adults often were Don Quixotes, and secure adults were likely to be Giselles. Those who were anxious/ambivalent were most likely to experience unrequited love; those who were secure were least likely to experience such love. Avoidant adults experienced the greatest desire to be in love in general; yet they had the least probability of being in a specific relationship. Anxious/ambivalent adults showed the greatest desire for a specific relationship; they also had the least desire to be in love in general.

No disguise can long conceal love where it exists, nor long feign it where it is lacking.

François de la Rochefoucauld (1613–1680)

I will reveal to you a love potion, without medicine, without herbs, without any witch's magic. If you want to be loved, then you must love.

Hecaton of Rhodes

Abstinence makes the heart grow fonder.

Anonymous

Absence makes the heart grow fonder—then you forget.

Floyd Zimmerman (1942–1996)

Reflections

Have you experienced unrequited love? How did it differ from requited love? Do you have a "style" of unrequited love? Have you been the object of someone's unrequited love? How did you handle it?

Jealousy: The Green-Eyed Monster

Does jealousy prove love? Many of us think it does. We may try to test someone's interest or affection by attempting to make him or her jealous by flirting with another person. If our date or partner becomes jealous, the jealousy is taken as a sign of love (Salovey and Rodin, 1991; White, 1980a). But provoking jealousy proves nothing except that the other person can be made jealous. Making jealousy a litmus test of love is dangerous, for jealousy and love are not necessarily related.

Beware, my Lord, of jealousy.
It is the green-eyed monster
that mocks the meat it feeds on.

William Shakespeare, (1564–1616)
Othello

Jealousy is not necessarily a sign of love.

Jealousy may be a more accurate yardstick for measuring insecurity or possessiveness than love (see Mullen, 1993, for a discussion of changing cultural attitudes toward jealousy).

It's important to understand jealousy for several reasons. First of all, jealousy is a painful emotion filled with anger and hurt. Its churning can turn us inside out and make us feel out of control. If we can understand jealousy, especially when it is irrational, then we can eliminate some of its pain. Second, jealousy can help cement or destroy a relationship. It helps maintain a relationship by guarding its exclusiveness, but in its irrational or extreme forms, it can destroy a relationship by its insistent demands and attempts at control. We need to understand when and how jealousy is functional and when and how it is not. Third, jealousy is often linked to violence (Follingstad et al., 1990; Laner, 1990; Riggs, 1993). It is a factor in precipitating violence or emotional abuse in dating relationships among both high school and college students (Burcky, Reuterman, and Kopsky, 1988; Stets and Pirog-Good, 1987). Marital violence and rape are often provoked by jealousy (Russell, 1990). It is often used by abusive partners to justify their violence (Adams, 1990). Rather than being directed at a rival, jealous aggression is often used against one's partner (Paul and Galloway, 1994).

What Is Jealousy?

He who is not jealous is not in love.

AUGUSTINE OF HIPPO (354–430)

Jealousy is an aversive response that occurs because of a partner's real, imagined, or likely involvement with a third person (Bringle and Buunk, 1985; Sharpsteen, 1993). It functions as a boundary-making mechanism. Jealousy sets the boundaries for what an individual or group feels are important relationships; others cannot trespass these limits without evoking it (Reiss, 1986).

Jealousy: The Psychological Dimension

Jealousy is a painful experience. It is an agonizing compound of hurt, anger, depression, fear, and doubt. We feel less attractive and acceptable to our partner when we are jealous (Bush, Bush, and Jennings, 1988). Jealous responses are most intense in committed or marital relationships because both partners assume "specialness." This specialness occurs because one's intimate partner is different from everyone else: It is with him or her that you are most confiding, revealing, vulnerable, caring, and trusting. There is a sense of exclusiveness. To have sex outside the relationship violates that sense of exclusiveness because sex symbolizes "specialness." Words such as *unfaithfulness*, *cheating*, and *infidelity* reflect the sense that an unspoken pledge has been broken. This pledge is the normative expectation that serious relationships, whether dating or marital, will be sexually exclusive (Lieberman, 1988).

A man was jealous of his wife, who, despite his jealousies, was always faithful. She gave him no reason to be jealous, but it did not matter; he tormented her with his constant tirades. Finally, one evening she left and was never seen again. But outside the camp, her sons discovered a great rock which resembled their mother. Their father's jealousy, they said, had turned her to stone.

PLAINS INDIAN TALE

As our lives become more and more intertwined, we become less and less independent. For some, this loss of independence increases the fear of losing the partner. But it takes more than simple dependency to make a person jealous. The relationship must be threatened or we must suspect, rightly or wrongly, that it is being threatened. Those who are most jealous lack self-esteem and feel insecure, either about themselves or about their relationships (Berscheid and Frei, 1977; McIntosh and Tate, 1990). This is true for both Whites and African Americans (McIntosh, 1989).

Social psychologists suggest that there are two types of jealousy: suspicious and reactive (Bringle and Buunk, 1991). **Suspicious jealousy** is jealousy that occurs where there is either no reason to be suspicious or only ambiguous evidence to suspect that a partner is involved with another. **Reactive jealousy** is jealousy that occurs when a partner reveals a current, past, or anticipated relationship with another person.

Love is as strong as the grave; jealousy is as cruel as the grave.

SONG OF SOLOMON 8:6

Suspicious jealousy generally occurs when a relationship is in its early stages. The relationship is not firmly established, and the couple is unsure about its future. The smallest distraction, imagined slight, or inattention can be taken as evidence of interest in another person. Even without any evidence, a jealous person may worry ("Is she seeing someone else but not telling me?"). This person may engage in vigilance, watching the partner's every move ("I'd like to audit your marriage and family class"). He may snoop, unexpectedly appearing in the middle of the night to see if someone else is there ("I was just passing by and thought I'd say hello"). The partner may try to control the other's behavior ("If you go to your friend's party without me, we're through"). Suspicious jealousy may have both legitimate and negative functions in a relationship. While it may be a reasonable response to circumstantial evidence and warn the partner what will happen if there are serious transgressions, if unfounded, it can be self-defeating.

Reactive jealousy occurs when one partner learns of the other's present, past, or anticipated sexual involvement with another. This usually provokes the most intense jealousy. If the affair occurred in the early part of the present relationship, the unknowing partner may feel that the primary relationship has been based on a lie. Trust is questioned. Every word and event must be reevaluated in light of this new knowledge: "If you slept with him when you said you were going to the library, did you also sleep with him when you said you were going to the laundromat?" Or "How could you say you loved me when you were seeing her?" The damage can be irreparable.

Boundary Markers

As we noted earlier, jealousy represents a boundary marker. It points out what the boundaries are in a particular relationship. It determines how, to what extent, and in what manner others can interact with members of the relationship. It also shows the limits within which the members of the relationship can interact with those outside the relationship. Culture prescribes the general boundaries of what evokes jealousy, but individuals adjust them to the dynamics of their own relationships.

There is more self-love than love in jealousy.

FRANÇOIS DE LA ROCHEFOUCAULD (1613–1680)

OTHER PLACES OTHER TIMES

Jealousy in Polygamous Cultures: Tibetan Society

In traditional Tibetan society, polyandry (the less common variety of polygamy in which a woman has more than one husband) was the cultural ideal. This, however, was a special sort of polyandry—fraternal polyandry. For Tibetans in Tibet and Nepal, the best form of marriage was that of one woman married to several men who were brothers. Fraternal polyandry was what every household sought to accomplish, but of course a family had to have more than one son to arrange this kind of marriage. With only one son, the best a family could do was arrange a monogamous marriage, which was considered a less fortunate match.

For Westerners, polyandry has always needed more explaining than polygyny. Having several wives seems, even to Westerners, more natural. Certainly co-wives may be jealous of one another—novels and histories have described the hostility felt by one wife toward another in traditional Chinese society, for example—but they must carry on. To conceive of men overcoming their jealousy at sharing one wife takes more imagination, because co-husbands would "naturally" be jealous of one another, and jealousy would inevitably lead to fighting or killing. How, therefore, could polyandry ever work?

Anthropologists and historians have explored several hypotheses to explain the practice of polyandry among Tibetans. The leading hypothesis is that polyandry is particularly adaptive to certain kinds of ecological and economic conditions. Traditionally, Tibetans, using plow and oxen, farmed very difficult terrain. Their primary crop was buckwheat. The most economically successful households were those that maximized the number of adult males, who not only did the plowing but also herded sheep and engaged in long-distance salt trading. According to this hypothesis, polyandry created households composed of more than one adult male to handle these various economic tasks.

But what about "natural" jealousy? Tibetan boys were raised to place supreme value on loyalty and cooperation among brothers. Households built on the marriage of several brothers to one wife not only were more successful economically but were also prestigious. These households were living proof of the solidarity of brothers. Of course, there were tensions of various kinds with Tibetan families, but the brothers sought to resolve

There is no greater glory than love nor greater punishment than jealousy.

LOPE DE VEGA (1562–1635)

Boundaries may vary, depending on the type of relationship, gender, sexual orientation, and ethnicity. Sexual exclusiveness is generally important in serious dating relationships and cohabitation; it is virtually mandatory in marriage (Blumstein and Schwartz, 1983; Buunk and van Driel, 1989; Hansen, 1985; Lieberman, 1988). Men are generally more restrictive toward their partners than women; heterosexuals are more restrictive than gay men and lesbians. Although we know very little about jealousy and ethnicity, traditional Latinos and new Latino and Asian immigrants appear to be more restrictive than Anglos and African Americans (Mindel, Habenstein, and Wright, 1988). Despite variations on where the boundary lines are drawn, jealousy functions to guard those lines.

Although our culture sets down general marital boundaries, each couple evolves its own boundaries. For some, it is permissible to carve out an area of individual privacy. In some relationships, partners may have few or many friends of their own (of the same or other sex), activities, and interests apart from the couple. In others there are no separate spheres because of jealousy or a lack of interest. But wherever a married couple draws its boundaries, each member understands where the line is drawn. The partners implicitly or explicitly know what behavior will evoke a jealous response (Bringle and Buunk, 1991). For some, it is having lunch with a member of the other sex (or same sex, if they are gay or lesbian); for others, it is having dinner; for still others, it is having dinner and seeing a movie. It is often disingenuous for a married partner to say that he or she didn't know that a particular action (a them in order to preserve the common residence and marriage.

The eldest brother chose the woman whom he and his brothers would marry. He looked for a woman who was attractive to him, who came from a good family, and who was known to be a hard worker. Upon marriage, the woman left her parents' home and settled with her husbands in theirs. Initially, she slept mostly with the eldest brother. Some of the younger brothers might be awkward adolescents, or even younger. However, over time, all brothers took turns sleeping with their wife. The guiding spirit in polyandrous marriage was that the eldest brother was to defer to his younger brothers, so that they too might form an attachment to their wife. According to anthropologist Nancy Levine (1988), who studied Tibetan polyandry for many years, the younger husbands, who at first might be too young to be of any interest to the wife, would with time become increasingly attractive to her. However, the wife, too, was concerned about promoting equal time and good feeling among the brothers. Protocol required that during breakfast the wife would signal the husband with whom she would sleep that night. Thus it would be clear to all whose turn it was.

The wife in a polyandrous marriage kept track of her menstrual cycles and believed she knew which of the brothers fathered which children. Generally, of course, the eldest brother fathered the first children. But after that, it was the wife who assigned paternity at the birth of a child. Although the brothers believed she knew who the father was, the wife also knew that it was important that each brother have a son.

Levine found in her study that statistically, households experiencing the most domestic difficulty—for example, households whose problems of equal time and equal attraction were chronic—were those with three or more brothers. In some of these households, problems of time and attraction could be resolved only by marriage to an additional wife. This situation was considered to be a great failure by both the family and the community. Ideally, one wife was best, as all the brothers understood. Families struggled to resolve the troublesome issues and avoid the second marriage. However, in some cases, a second marriage was inescapable. In those instances, the ideal second wife was a sister of the first. Anthropologists refer to this kind of plural marriage as sororal polygynous fraternal polyandry.

Fraternal polyandry as practiced in Tibetan society (and elsewhere, among the Todaz of India and the Sinhalese of Sri Lanka) raises several questions: What is a family? What is marriage? Who is a parent? Among the most interesting issues, however, is the "natural fact" of sexual jealousy. What role does culture play in its creation?

flirtatious suggestion, a lingering touch, or dinner with someone else) would provoke a jealous response.

But jealous souls will not be answer'd so;
They are not ever jealous for the cause;
But jealous for they are jealous; 'tis a monster
Begot upon itself, born on itself.

William Shakespeare, (1564–1616)
Othello.

Gender Differences

Both men and women fear that their partner might be attracted to someone else because of dissatisfaction with the relationship and the desire for sexual variety. Women, however, feel especially vulnerable to losing their partner to an attractive rival (Nader and Dotan, 1992; White, 1981). Men and women become jealous about different matters. Men tend to experience jealousy when they feel their partner is sexually involved with another man. Women, by contrast, tend to experience jealousy over intimacy issues (Buss, Larsen, Westen, and Semmelroth, 1992). Women feel the most jealousy when they believe their partner is both emotionally and physically involved with another woman (White, 1981).

Both men and women react to jealousy with anger. But men are more likely to express anger, and women are more likely to suppress it and feel depressed. Ira Reiss (1986) suggests that this difference in expressing jealousy is consistent with cultural restraints prohibiting women from displaying anger. At the same time, it reflects their greater powerlessness vis-à-vis men. As a result, women may turn their anger inward, transforming it into depression.

There may be some truth in the belief that women are more jealous than men, but such jealousy is not inherent in being female. It is more likely related to the greater sexual freedom that men are permitted. (Even in marriage, it is more acceptable for men to "roam" because they are believed to be "naturally" more sexual than women.) If women appear to be more jealous than men, it may be because men, granted greater autonomy than women, have more opportunities to evoke jealous responses from women (Reiss, 1986).

Jealousy is not a barometer by which the depth of love can be read. It merely records the depth of the lover's insecurity.

MARGARET MEAD (1902–1978)

Managing Jealousy

Jealousy can be unreasonable or a realistic reaction to genuine threats. Unreasonable jealousy can become a problem when it interferes with an individual's well-being or that of the relationship. Dealing with irrational suspicions can often be difficult, for such feelings touch deep recesses in ourselves. As noted earlier, jealousy is often related to personal feelings of insecurity and inadequacy. The source of such jealousy lies within ourselves, not within the relationship.

Did you know?

Men and women who have among other things high self-esteem, a sense of control, and a feeling of power within the relationship are best able to cope with jealousy (McIntosh and Matthews, 1992).

If we can work on the underlying causes of our insecurity, then we can deal effectively with our irrational jealousy. Excessively jealous persons may need considerable reassurance, but at some point they must confront their own irrationality and insecurity. If they do not, they emotionally imprison their partner. Their jealousy may destroy the very relationship they were desperately trying to preserve.

But jealousy is not always irrational. Sometimes there are real reasons for it, such as the violation of relationship boundaries. In this case, the cause lies not within ourselves but within the relationship. Gordon Clanton and Lynn Smith (1977) write:

> Jealousy cannot be treated in isolation. *Your* jealousy is not *your* problem alone. It is also a problem for your partner and for the person whose interest in your partner sparks your jealousy. Similarly, when your partner feels jealous, you ought not to dismiss the matter by pointing a finger and saying "That's your problem." Typically, three or more persons are involved in the production of jealous feelings and behaviors. Ideally, all three should take a part of the responsibility for minimizing the negative consequences.

Here lies the body of Mannie
They put him here to stay;
He lived the life of Riley
While Riley was away.

BOOT HILL EPITAPH

Managing jealousy requires the ability to communicate, the recognition by each partner of the feelings and motivations of the other, and a willingness to reciprocate and compromise (Ridley and Crowe, 1992). If the jealousy is well founded, the partner may need to modify or end the relationship with the "third party" whose presence initiated the jealousy. Modifying the third-party relationship reduces the jealous response and, more important, symbolizes the partner's commitment to the primary relationship. If the partner is unwilling to do this—because of lack of commitment, unsatisfied personal needs, or other problems in the primary relationship—the relationship is likely to reach a crisis. In such cases, jealousy may be the agent for profound change.

The Transformation of Love: From Passion to Intimacy

Intense, passionate love does not last forever at the same high level. Instead, it fades or transforms itself into a more enduring love based on intimacy.

Love is a flickering flame between two darknesses. . . . Whence does it come? . . . From sparks incredibly small. . . . How does it end? . . . In nothingness equally incredible . . . The more raging the flame, the sooner it is burnt out.

HEINRICH HEINE (1797–1856)

The Instability of Passionate Love

Ultimately, romantic love may be transformed or replaced by a quieter, more lasting love. In fact, those in secure companionate love relationships, according to one study, experience the highest levels of satisfaction; they are much more satisfied than those in traditional romantic relationships (Hecht et al., 1994).

The Passage of Time: Changes in Intimacy, Passion, and Commitment

According to researcher Robert Sternberg (1988), time affects our levels of intimacy, passion, and commitment.

Intimacy over Time. When we first meet someone, intimacy increases rapidly as we make critical discoveries about each other, ranging from our innermost thoughts of life and death to our preference for strawberry or chocolate ice cream. As the relationship continues, the rate of growth decreases and then levels off. After the growth levels off, the partners may no longer consciously feel as close to each other. This may be because they are beginning to drift apart, or it may be because they are becoming intimate at a different, less conscious, deeper level. This kind of intimacy is not easily observed. It is a latent intimacy that nevertheless is forging stronger, more enduring bonds between the partners.

Jump out the window if you are the object of passion. Flee it if you feel it. . . . Passion goes, boredom remains.

Coco Chanel (1883–1971)

Passion over Time. Passion is subject to habituation. What was once thrilling—whether love, sex, or roller coasters—becomes less so the more we get used to it. Once we become habituated, more time with a person (or more sex or more roller coaster rides) does not increase our arousal or satisfaction.

If the person leaves, however, we experience withdrawal symptoms (fatigue, depression, anxiety), just as if we were addicted. In becoming habituated, we have also become dependent. We fall beneath the emotional baseline we were at when we met our partner. Over time, however, we begin to return to that original level.

Commitment over Time. Unlike intimacy and passion, time does not necessarily diminish, erode, or alter commitments. Our commitment is most affected by how successful our relationship is. Even initially, commitment grows more slowly than intimacy or passion. As the relationship becomes long term, the growth of commitment levels off. Our commitment will remain high as long as we judge the relationship to be successful. If the relationship begins to deteriorate, after a time the commitment will probably decrease. Eventually, it may disappear and an alternative relationship may be sought.

Disappearance of Romance as Crisis

The disappearance (or transformation) of passionate love is often experienced as a crisis in a relationship. A study of college students (Berscheid, 1983) found that half would seek divorce if passion disappeared from their marriage. But intensity of feeling does not necessarily measure depth of love. Intensity, like the excitement of toboggan runs, diminishes over time. It is then that we begin to discover if the love we experience for each other is one that will endure.

Our search for enduring love is complicated by our contradictory needs. Elaine Hatfield and William Walster (1981) write:

> What we really want is the impossible—a perfect mixture of security and danger. We want someone who understands and cares for us, someone who will be around through thick and thin, until we are old. At the same time, we long for sexual excitement, novelty, and danger. The individual who offers just the right combination of both ultimately wins our love. The problem, of course, is that, in time, we get more and more security—and less and less excitement—than we bargained for.

There is no love apart from the deeds of love; no potentiality of love than that which is manifested in loving. . . .

Jean-Paul Sartre (1905–1980), *Existentialism as Humanism*

The disappearance of passionate love, however, enables individuals to refocus their relationship. They are given the opportunity to move from an intense one-on-one togetherness that excludes others to a togetherness that includes family, friends, and external goals and projects. They can look outward on the world together.

It is only with the heart that one can see rightly; what is essential is invisible to the eye.

ANTOINE DE SAINT-EXUPÉRY (1900–1944), *THE LITTLE PRINCE*

The Reemergence of Romantic Love

Contrary to what pessimists believe, many people find that they can have both love and romance and that the rewards of intimacy include romance.

Romantic love may be highest during the early part of marriage and decline as stresses from child rearing and work intrude on the relationship. Most studies suggest that marital satisfaction proceeds along a U-shape curve, with highest satisfaction in the early and late periods (see Chapter 9). Romantic love may be affected by the same stresses as general marital satisfaction. In fact, romantic love begins to increase as children leave home. In later life, romantic love may play an important role in alleviating the stresses of retirement and illness.

Generally, by the time you are Real, most of your hair has been loved off, and your eyes drop out and you get loose in the joints and very shabby.

But these things don't matter at all, because once you are Real you can't be ugly, except to people who don't understand.

MARGERY WILLIAMS, *THE VELVETEEN RABBIT*

New research on the differences in love attitudes across family life stages reveals some unexpected and perhaps encouraging news for older romantics. Montgomery and Sorell (1997) write:

> The love attitudes endorsed by the broad age-range sample contradicts notions that romantic, passionate love is the privilege of youth and young relationships, functioning to bring partners together. Instead, individuals throughout the life-stages of marriage consistently endorse the love attitudes involving passion, romance, friendship, and self-giving love, and these results indicate that any popularization of young single adulthood as the enviable passionate idea is erroneous.

So it is that, among those whose marriages survive, passion and romance which do not necessarily decline over time.

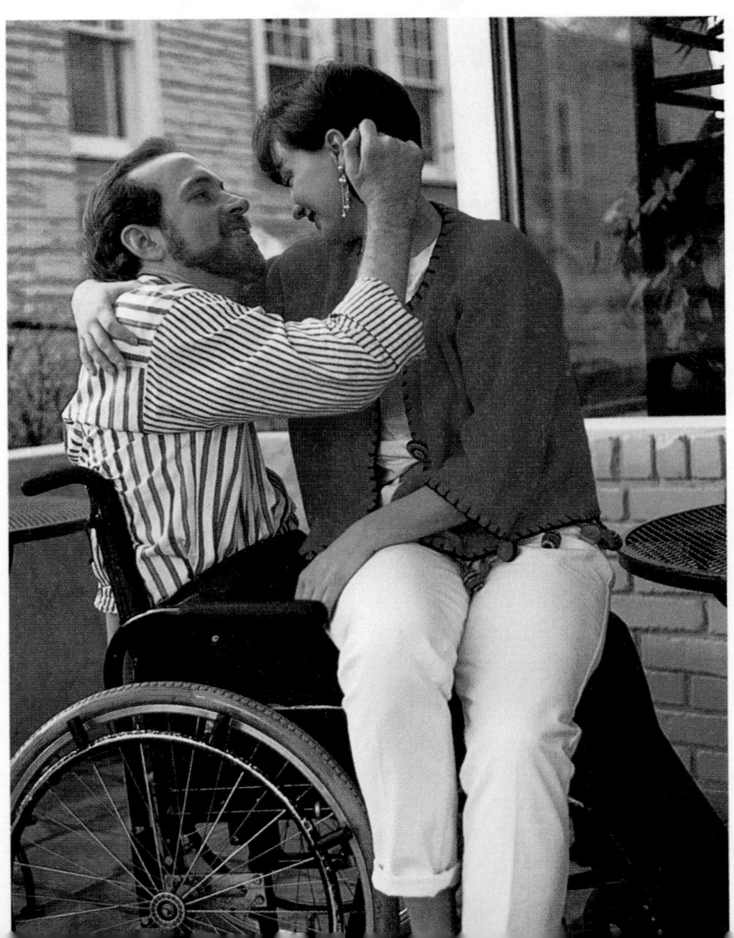

Being physically limited does not necessarily inhibit love and sexuality any more than being able-bodied guarantees them.

Intimate Love: Commitment, Caring, and Self-Disclosure

Perhaps one of the most profound questions we can ask about love is how to make it stay. The key to making love stay seems to be not in love's passionate intensity but in the transformation of that intensity into intimate love. Intimate love is based on commitment, caring, and self-disclosure.

It is easier to stay out than get out.

MARK TWAIN (1835–1910)

Commitment

Commitment is an important component of intimate love because it is a "determination to continue" a relationship or marriage in the face of bad times as well as good (Reiss, 1980a). It is based on conscious choice rather than on feelings, which, by their very nature, are transitory. Commitment is a promise of a shared future, a promise to be together come what may.

Commitment has become an important concept in recent years. We seem to be as much in search of commitment as we are in search of love or marriage. We speak of "making a commitment" to someone or to a relationship. (Among singles, commitment is sometimes referred to as "the C-word.") A committed relationship has become almost a stage of courtship, somewhere between dating and being engaged or living together.

Love doesn't sit there like a stone, it has to be made, like bread; remade all the time, made new.

URSULA LEGUIN

Caring

Caring is placing another's needs before your own. As such, caring requires treating your partner as valued for simply being himself or herself. It requires what the philosopher Martin Buber called an I-Thou relationship. Buber described two fundamental ways of relating to people: I-Thou and I-It. In an I-Thou relationship, each person is treated as a Thou—that is, as a person whose life is valued as an end in itself. In an I-It relationship, each person is treated as an It; a person has worth only as someone who can be used. When a person is treated as a Thou, his or her humanity and uniqueness are paramount.

Love and you shall be loved. All love is mathematically just, as much as two sides of an algebraic equation.

RALPH WALDO EMERSON (1803–1882)

Self-Disclosure

When we self-disclose, we reveal ourselves—our hopes, our fears, our everyday thoughts—to others. Self-disclosure deepens others' understanding of us. It also deepens our own understanding, for we discover unknown aspects as we open ourselves to others. (Self-disclosure is discussed in detail in Chapter 5.)

Without self-disclosure, we remain opaque and hidden. If others love us, such love leaves us with anxiety. Are we loved for ourselves or for the image we present to the world?

Together, commitment, caring, and self-disclosure help transform love. But in the final analysis, perhaps the most important means of sustaining love is our words and actions. Caring words and deeds provide the setting for maintaining and expanding love (Byrne and Murnen, 1988).

While we increasingly understand the dynamics and varied components of love, the experience of love itself remains ineffable, the subject of poetry rather than scholarship. A journal article is not a love poem, and romantics should not forget that love exists in the everyday world. Researchers have helped us increasingly understand love in the light of day—its nature, its development, its varied aspects—so that we may better be able to enjoy it in the moonlight.

Summary

- *Prototypes* of love and commitment are models of how people define these two ideas in everyday life. The central aspects of the love prototype include trust, caring, honesty, friendship, respect, and concern for the other; central aspects of the commitment prototype include loyalty, responsibility, living up to one's word, faithfulness, and trust.
- Attitudes and feelings associated with love include caring, needing, trusting, and tolerating. Behaviors associated with love include verbal, nonverbal, and physical expressions of affection; self-disclosure; giving of nonmaterial and material evidence; and tolerance.
- Commitment is affected by the balance of costs to benefits, normative inputs, and structural constraints.
- Though friendship and love share many of the same characteristics, love contains passion, fascination, and instability. Friendship is the foundation for loving relationships.
- The *wheel theory of love* emphasizes the interdependence of four processes: (1) rapport, (2) self-revelation, (3) mutual dependency, and (4) fulfillment of intimacy needs.
- According to John Lee, there are six basic styles of love: *eros*, *ludus*, *storge*, *mania*, *agape*, and *pragma*.
- The *triangular theory of love* views love as consisting of three components: intimacy, passion, and decision/commitment.
- The *attachment theory of love* views love as being similar in nature to the attachments we form as infants. The attachment (love) styles of both infants and adults are secure, anxious/ambivalent, and avoidant.
- *Unrequited love* is a common experience. There are three styles of unrequited love: (1) the Cyrano style—the desire to have a relationship with another, regardless of how hopeless it is; (2) the Giselle style—the misperception that a relationship is more likely to develop than it actually is; and (3) the Don Quixote style—the general desire to be in love. Anxious/ambivalent adults are most likely to be Cyranos, avoidant adults to be Don Quixotes, and secure adults to be Giselles.
- *Jealousy* is an aversive response that occurs because of a partner's real, imagined, or likely involvement with a third person. Jealousy acts as a boundary marker for relationships. Jealous responses are most likely in committed or marital relationships because of the presumed "specialness" of the relationship. Specialness is symbolized by sexual exclusiveness. As individuals become more interdependent, there is a greater fear of loss. Fear of loss, coupled with insecurity, increases the likelihood of jealousy.
- Time affects romantic relationships. The rapid growth of intimacy tends to level off, and we become habituated to passion. Commitment tends to increase, provided that the relationship is judged to be rewarding.
- Romantic love tends to diminish. It may either end or be replaced by intimate love. Many individuals experience the disappearance of romantic love as a crisis. Romantic love seems to be most prominent in adolescence and in early and later stages of marriage. Intimate love is based on commitment, caring, and self-disclosure.

Key Terms

Suggested Readings

Ackerman, Diane. *A Natural History of Love*, New York: Random House, 1994. An historical and cultural perspective on love.

Bradshaw, John. *Creating Love: The Next Great Stage of Growth*. New York: Bantam, 1992. The popular lecturer's views on and insights into intimate relationships.

Buscaglia, Leo. *Born for Love: Reflections on Loving*. Thorofare, NJ: Slack/Random House, 1992. A series of thoughtful reflections about love by America's most popular proponent of love.

Cancian, Francesca. *Love in America: Gender and Self-Development*. New York: Cambridge University Press, 1987. A look at love as it relates to gender: feminine, masculine, and androgynous.

Clark, Don. *The New Loving Someone Gay*. Rev. and exp. ed. Berkeley, CA: Celestial Arts, 1987. An exploration of the many aspects of gay and lesbian love (including love in gay and lesbian relationships, love between gay men—or lesbians—and their families and friends by a gay therapist.

Feeney, Judith, and Patricia Noller. *Adult Attachment*. Thousand Oaks, CA: Sage Publications, Inc., 1996. A coherent account of diverse strands of attachment research.

Osherson, Samuel. *Wrestling with Love: How Men Struggle with Intimacy with Children, Parents, and Each Other*. New York: Fawcett Columbine, 1992. A book focusing on men's issues, but recommended for women as well.

Peck, M. Scott. *The Road Less Traveled: A New Psychology of Love, Traditional Values, and Spiritual Growth*. New York: Simon & Schuster, 1978. A psychological/spiritual approach to love that sees love's goal as spiritual growth.

_____. *Further Along the Road Less Traveled*. New York: Simon & Schuster, 1994. More on love and spiritual growth from the best-selling author.

Solomon, Robert C. *Love: Emotion, Myth, & Metaphor*. Buffalo, NY: Prometheus Books, 1990. A book that seeks to separate love as an emotion from the many myths and illusions surrounding it.

Sternberg, Robert, and Michael Barnes, eds. *The Psychology of Love*. New Haven, CT: Yale University Press, 1988. An excellent collection of essays on love by some leading researchers.

Weber, Ann L., and John Harvey, eds. *Perspectives on Close Relationships*. Boston: Allyn & Bacon, 1994. A scholarly collection of essays by scholars on various aspects of intimate relationships.

White, Greg, and Paul Mullen. *Jealousy: A Clinical and Multidisciplinary Approach*. New York: Guilford Press, 1989. A comprehensive examination of what we know about jealousy (and its treatment).

Woods, Paula, and Felix Liddell, eds. *I Hear a Symphony: African Americans Celebrate Love*. New York: Anchor, 1994. A collection of letters, poetry, essays, stories, and art.

CHAPTER 5

Communication and Conflict Resolution

P R E V I E W

To gain a sense of what you already know about the material covered in this chapter, answer "True" or "False" to the statements below.

1. Touching is one of the most significant means of communication. True or false?
2. Always being pleasant and cheerful is the best way to avoid conflict and sustain intimacy. True or false?
3. Studies suggest that those couples with the highest marital satisfaction tend to disclose more than those who are unsatisfied. True or false?
4. Negative communication patterns before marriage are a poor predictor of marital communication because people change once they are married. True or false?
5. Conflict and intimacy go hand in hand in intimate relationships. True or false?
6. Good communication is primarily the ability to offer excellent advice to your partner to help him or her change. True or false?
7. Physical coercion is the method men use most frequently when disagreement arises between them and their partners. True or false?
8. The party with the least interest in continuing a relationship generally has the power in it. True or false?
9. Latinos and Asian Americans tend to rely on the nonverbal expression of intense feelings in contrast to direct verbal expressions. True or false?
10. Wives tend to give more negative messages than husbands. True or false?

Answers

1 True, 2 False, 3 True, 4 False, 5 True, 6 False, 7 False, 8 True, 9 True, 10 True

OUTLINE

Intimacy and communication are inextricably connected. When we speak of communication, we mean more than just the ability to discuss problems and resolve conflicts. We mean communication for its own sake: the pleasure of being in each other's company, the excitement of conversation, the exchange of touches and smiles, the loving silences. Through communication we disclose who we are, and from this self-disclosure, intimacy grows.

One of the most common complaints of married partners, especially unhappy partners, is "We don't communicate." But it is impossible not to communicate—a cold look may communicate anger as effectively as a fierce outburst of words. What these unhappy partners mean by "not communicating" is that their communication drives them apart rather than bringing them together, feeds conflict rather than resolving it. Communication patterns are strongly associated with marital satisfaction (Noller and Fitzpatrick, 1991).

In this chapter we explore how communication brings people together: how to develop communication skills, how to self-disclose, how to give feedback and affirm your partner. We also discuss the relationship between conflict and intimacy, exploring the types of conflict and the role of power in marital relationships. We look at common conflicts about sex and money. Finally, we explore some ways of resolving conflicts.

Communication Patterns and Marriage

There has been an explosion of research on premarital and marital communication in the last decade. Researchers are finding significant correlations between the nature of communication and satisfaction, as well as finding differences in male versus female communication patterns in marriage.

Premarital Communication Patterns and Marital Satisfaction

"Drop dead, you creep!" is hardly the way to resolve a disagreement among dating couples. But it may be an important clue as to whether you and your partner should marry. Many couples who communicate poorly before marriage are likely to continue the same way after marriage, and the result can be disastrous for their marriages. Researchers have found that how well a couple communicates before marriage can be an important predictor of later marital satisfaction (Cate and Lloyd, 1992; Filsinger and Thoma, 1988). If communication is poor before marriage, it is not likely to get significantly better after marriage—at least not without a good deal of effort and help.

Whenever a feeling is voiced with truth and frankness . . . a mysterious and far-reaching influence is exerted. At first it acts on those who are inwardly receptive. But the circle grows larger and larger. . . . The effect is but the reflection of something that emanates from one's own heart.

I CHING

Self-disclosure—the revelation of deeply personal information about one's self—prior to or soon after marriage is related to relationship satisfaction later. In one study (Surra, Arizzi, and Asmussen, 1988), men and women were interviewed shortly after marriage and four years later. The researchers found that self-disclosure was an important factor for increasing each other's commitment later. Talking about your deepest feelings, revealing yourself to your partner, builds bonds of trust that help cement a marriage.

Whether a couple's interactions are basically negative or positive can also predict later marital satisfaction. In a notable experiment by John Markham (1979), fourteen premarital couples were evaluated using "table talk," sitting around a table and simply engaging in conversation. Each couple talked about various topics. Using an electronic device, each partner electronically recorded whether the message was positive or negative. Markham found that the negativity or positivity of the couple's communication pattern had little impact on their marital satisfaction during their first year. This protective quality of the first year is known as the **honeymoon effect.** (You can say almost anything during the first year, and it will not have a serious impact on marriage [Huston, McHale, and Crouter, 1986].) But after the first year, couples with negative premarital communication patterns were less satisfied than those with positive communication patterns. A more recent study (Julien, Markman, and Lindahl, 1989) found that those premarital couples who responded more to each other's positive communication than to each other's negative communication were more satisfied in marriage four years later.

Marital Communication Patterns and Satisfaction

Researchers have found a number of patterns that distinguish the communication patterns in satisfied marriages in contrast to dissatisfied marriages (Gottman, 1994; Hendrick, 1981; Noller and Fitzpatrick, 1991; Schaap, Buunk, and Kerkstra, 1988). Couples in satisfied marriages tend to have the following characteristics:

- Willingness to accept conflict but to engage in conflict in nondestructive ways.
- Less frequent conflict and less time spent in conflict. Both satisfied and unsatisfied couples, however, experience conflicts about the same topics, especially about communication, sex, and personality characteristics.
- The ability to disclose or reveal private thoughts and feelings, especially positive ones, to one's partner. Dissatisfied spouses tend to disclose mostly negative thoughts to their partners.
- Expression by both partners of more or less equal levels of affection, such as tenderness, words of love, and touch.
- More time spent talking, discussing personal topics, and expressing feelings in positive ways.

Touch is one of our primary means of communication. It conveys intimacy, immediacy, and emotional closeness.

PERSPECTIVE

Ethnicity and Communication

Different ethnic groups within our culture have different language patterns that affect the way they communicate. African Americans, for instance, have distinct communication patterns (Hecht, Collier, Ribeau, 1993). Language and expressive patterns are characterized by emotional vitality, realness, and valuing direct experience, among other things (White and Parham, 1990). Emotional vitality is expressed in the animated use of words. Realness refers to "telling it like it is" using concrete nonabstract words. Direct experience is valued because "there is no substitute in the Black ethos for actual experience gained in the course of living" (White and Parham, 1990). "Mother wit"—practical or experiential knowledge—may be valued over knowledge gained from books or lectures.

Latinos, especially traditional Latinos, assume that intimate feelings will not be discussed openly (Guerrero Pavich, 1986). One researcher (Falicov, 1982) writes of Mexican Americans: "Ideally, there should be a certain formality in the relationship between spouses. No deep intimacy or intense conflict is expected. Respect, consideration, and curtailment of anger or hostility are highly valued." Confrontations are to be avoided; negative feelings are not to be expressed. As a consequence, nonverbal communication is especially important. Women are expected to read men's behavior for clues to their feelings and for discovering what is acceptable. Because confrontations are unacceptable, secrets are important. Secrets are shared between friends but not between partners.

Asian-American ethnic groups are less individualistic than the dominant American culture. Whereas the dominant culture views the ideal individual as self-reliant and self-sufficient, Asian-American subcultures are more relationally oriented. Researchers Steve Shon and Davis Ja (1982) note of Asian Americans:

> They emphasize that individuals are the products of their relationship to nature and other people. Thus, heavy emphasis is placed on their relationship with other people, generally with the aim of maintaining harmony through proper conduct and attitudes.

Asian Americans are less verbal and expressive in their interactions than are both African Americans and whites; instead, they rely to a greater degree on indirect and nonverbal communication, such as silence and the avoidance of eye contact as signs of respect (Del Carmen, 1990). Because harmonious relationships are highly valued, Asian Americans tend to avoid direct confrontation if possible. Japanese Americans, for example, "value implicit, nonverbal intuitive communication over explicit, verbal, and rational exchange of information" (Del Carmen, 1990). To avoid conflict, verbal communication is often indirect or ambiguous; it skirts around issues instead of confronting them. As a consequence, in interactions Asian Americans rely on the other person to interpret the meaning of a conversation or nonverbal clues.

- The ability to encode (send) verbal and nonverbal messages accurately and to decode (understand) such messages accurately from their spouses. This is especially important for husbands. Unhappy partners may actually decode the messages of strangers more accurately than those from their partners.

Talking is the major way we establish, maintain, monitor, and adjust our relationships.

DEBORAH TANNEN

Gender Differences in Partnership Communication

For some time, researchers have been aware of gender differences in general communication patterns. More recently, they have discovered specific gender differences in marital communication (Klinetob and Smith, 1996; Noller and Fitzpatrick, 1991; Thompson and Walker, 1989).

First, wives tend to send clearer messages to their husbands than their husbands send to them. Wives are often more sensitive and responsive to their husbands' messages, both during conversation and during conflict. They are more likely to reply to either positive messages ("You look great") or negative messages ("You look awful") than are their husbands, who may not reply at all.

Second, husbands more than wives tend to give either neutral messages or to withdraw. In this case, as the discontented spouse (the wife) demands change, the one who stands the most to gain (the husband) will withdraw (Klinetob and Smith, 1996). Wives tend to give more positive or negative messages, however because they tend to smile or laugh when they send messages, they send fewer clearly neutral messages. Husbands' neutral responses make it more difficult for wives to decode what their partners really are trying to say. If a wife asks her husband if they should go to dinner or see a movie and he gives a neutral response, such as, "Whatever," does he really not care, or is he pretending he doesn't to avoid possible conflict?

A little sincerity is a dangerous thing, and a great deal of it is absolutely fatal.

OSCAR WILDE (1854–1900)

Reflections

Do you find that husband-wife differences in communication tend to hold true in your intimate relationships? If so, how? Are there other differences? What kinds of differences or problems might exist in same-sex relationships?

Third, although communication differences in arguments between husbands and wives are usually small, they nevertheless follow a typical pattern. Wives tend to set the emotional tone of an argument. They escalate conflict with negative verbal and nonverbal messages ("Don't give me that!") or de-escalate arguments by setting an atmosphere of agreement ("I understand your feelings"). Husbands' inputs are less important in setting the climate for resolving or escalating conflicts. Wives tend to use emotional appeals and threats more than husbands, who tend to reason, seek conciliation, and try to postpone or end an argument. A wife is more likely to ask, "Don't you love me?" whereas a husband is more likely to say, "Be reasonable."

Studies suggest that poor communication skills precede the onset of marital problems (Gottman, 1994; Markman, 1981; Markman et al., 1987). The material that follows will assist you in understanding and developing good communication skills.

Silence is the one great art of conversation.

WILLIAM HAZLITT (1778–1830)

Nonverbal Communication

There is no such thing as not communicating. Even when you are not talking, you are communicating by your silence (for example, an awkward silence, a hostile silence, or a tender silence). You are communicating by the way you position your body and tilt your head and through your facial expressions, your physical distance from the other person, and so on. Look around you. How are the people in your presence communicating nonverbally?

One of the problems with nonverbal communication, however, is the imprecision of its messages. Is a person frowning or squinting? Does the smile indicate friendliness or nervousness? A person may be in reflective silence, but we may interpret the silence as disapproval or distance.

The cruelest lies are often told in silence.

ROBERT LOUIS STEVENSON (1850–1890)

Functions of Nonverbal Communication

An important study of nonverbal communication and marital interaction found that nonverbal communication has three important functions in marriage (Noller, 1984): (1) conveying interpersonal attitudes, (2) expressing emotions, and (3) handling the ongoing interaction.

It was fortunate that love did not need words, or else it would be full of misunderstandings and foolishness.

HERMANN HESSE (1877–1962), *NARCISSUS AND GOLDMUND*

Conveying Interpersonal Attitudes

Nonverbal messages are used to convey attitudes. Gregory Bateson describes nonverbal communication as revealing "the nuances and intricacies of how two people are getting along" (quoted in Noller, 1984). Holding hands can suggest intimacy; sitting on opposite sides of the couch can suggest distance. Not looking at each other in conversation can suggest awkwardness or a lack of intimacy.

PERSPECTIVE

Family Rules and Communication

Family Rules

According to family systems theory (discussed in Chapter 2), **rules** are patterned or characteristic responses. Family rules are generally unwritten; in fact, most of the time we probably don't even know we have them. They are formed from habit; like any habit, they are difficult to change.

All families have a **hierarchy of rules,** the ranking of rules in order of significance. Individuals have their rules; so does each family member in his or her roles as mother, father, son, daughter, and so on. Above individual and member rules are family rules. **Family rules** are the combined members' rules. They are arrived at either consensually or through power struggles—both of which may be unconscious. These family rules are "policies" that evolve over time, such as "No one will discuss Daddy's drinking." The rules may be overt (openly recognized) or they may be covert (hidden and unrecognized).

Superior to family rules in the hierarchy are **meta-rules.** Meta-rules are abstract, general, unarticulated rules at the apex of the hierarchy of rules. Meta-rules are different from individual and family rules primarily because they are more abstract. If not discussing Daddy's drinking is a family rule, the meta-rule is "Don't discuss anything that will cause a problem." When we talk of conspiracies of silence, we are usually talking about meta-rules. Meta-rules are much more difficult to change.

Feedback

The way each person responds influences every other person through feedback. Let us say the family rule is that no one confronts the father about his yelling at other family members. The wife responds by withdrawing affection, the son goes into his room and shuts the door, the daughter tries to placate her father. These responses become new input. The father feels that his daughter is supportive of him; he is angry at his wife for withdrawing. The mother is angry at her daughter for responding to her father's yelling and ignoring her. The son is an absent member of the family.

Feedback is the basic principle for processing information. Feedback permits the system to alter its activities, structure, and direction to further its own goals. Information is processed through a feedback loop. The input becomes the system's output; the output in turn becomes new input. There are two types of feedback: negative feedback and positive feedback. *Negative feedback* tells the system that change is unnecessary, that things are normal or returning to normal. In the example above, the feedback the father received was basically negative feedback because no one told him that his behavior was unacceptable. The feedback did not tell him that he had to change.

Positive feedback tells the system that things are changing, that the system is deviating from its normal state. Positive feedback *amplifies* the original input. If the mother had responded to her husband's yelling behavior by telling him to stop, he would have received positive feedback. But because the family rule is to avoid confrontation, the father received negative feedback, that is, none. Imagine, however that the family rule was to meet outburst with outburst. If the family rule called for anger, then the output would be quarreling and discussion in the family. This angry output then would become new input into the family system. If the family rule allows the

Expressing Emotions

When a thing is funny, search it for a hidden truth.

GEORGE BERNARD SHAW (1856–1950)

Our emotional states are expressed through our bodies. A depressed person walks slowly; a happy person walks with a spring. Smiles, frowns, furrowed brows, tight jaws, tapping fingers—all express emotion. Expressing emotion is important because it lets our partner know how we are feeling so that he or she can respond appropriately. It also allows our partner to share our feelings—to laugh or weep with us.

Handling the Ongoing Interaction

Nonverbal communication helps us handle the ongoing interaction by indicating interest and attention. An intent look indicates our interest in the conversation; a yawn indicates boredom. Posture and eye contact are especially important. Are we leaning toward the person with interest or slumping back, thinking about unbridled expression of anger, then the anger will be amplified until it goes out of control. This process causes polarization and escalation. It is known as the *positive feedback spiral.*

Family rules and meta-rules are also known as *rules of transformation* because they transform or interpret input as it becomes output. If the husband loses his job, a family rule might call for blaming. As a result of the family rule, the job loss might generate hostility, family disintegration, and a major crisis. A different family could have different rules dealing with similar input. If the husband loses his job, the family rule might call for unity. In this case, the output would be family solidarity, which might reduce the crisis (Broderick and Smith, 1979).

Changing Family Rules

To show how the family processes information, let's take an extended example used by Mary Hicks and her colleagues (Hicks et al., 1983). As you read this example, remember that we are discussing an interactive process. The family reacts to input, processes the input, and then sends it back to the environment as output, which becomes additional input. Here is an illustration concerning a woman working outside the home:

- *Environmental input:* "Men and women are equal."
- *Wife's rule:* "I should be able to work outside the home and get help from my husband in household work."
- *Husband's rule:* "I should give my full energy to my career; my wife should support me in this."
- *Family rule:* "Husband should work and wife should be responsible for child care and housework."
- *Meta-rule:* "Husband is primary and wife is subordinate."
- *System output:* "Husband is breadwinner; wife supports husband through child care, housekeeping, and additional income."

Look at these rules carefully. This family system is experiencing stress because the wife is out of step with the rest of the system. The wife is providing positive feedback. Through a process of negotiation between husband and wife, the rules in the hierarchy will change, although it will probably take considerable time and negotiation.

New rules evolve as society continues to change and move toward more egalitarian norms. "After many cycles and over an extended period of time," note Hicks and her colleagues, "the combination of inputs, rule transformations, and outputs may result in a situation such as the one that exists in contemporary Western societies, in which an egalitarian norm is evolving, but with husbands' and families' rules still slow to change." If the family evolves these norms, the rules for the preceding example may be somewhat like the following:

- *Environmental input:* "Men and women are equal."
- *Wife's rule:* "I should be able to work and receive help from my husband."
- *Husband's rule:* "My wife should be able to work, but my career is primary."
- *Family rule:* "Wife can work, husband will share in household work and child care as long as it does not interfere with his career."
- *Meta-rule:* "Husband is senior partner; wife is junior partner."
- *Family system output:* "Wife will increase financial contribution and husband will increase child care and household work."

As Hicks and her co-workers point out, some changes have been made in all components, but the husband's, family's, and meta-rules are still resistant to an egalitarian ethic. Because disparity still exists, stress will continue.

something else? Do we look at the person who is talking, or are we distracted, glancing at other people as they walk by or watching the clock?

Relationship between Verbal and Nonverbal Communication

The messages that we send and receive contain both a verbal and a nonverbal component. The verbal part expresses the *basic content* of the message, whereas the nonverbal part expresses the *relationship* part of the message. The relationship part of the message tells the attitude of the speaker (friendly, neutral, or hostile) and indicates how the words are to be interpreted (as a joke, request, or command). The full content of any message has to be understood according to both the verbal and nonverbal parts.

For a message to be most effective, both the verbal and nonverbal components must be in agreement. If you are angry and say "I am angry," and your facial

Reflections

What rules does your family have regarding family members' ability to speak their minds freely on any subject? What can you say about communication patterns in your family? To whom can you talk about what is happening?

Words must be supported by one's entire conduct. If words and conduct are not in accord and not consistent, they will have no effect.

I CHING

expression and voice both show anger, the message is clear. But if you say "I am angry" in a neutral tone of voice and smile, your message is ambiguous. If you say "I'm not angry" but clench your teeth and use a controlled voice, your message is also unclear.

Proximity, Eye Contact, and Touch

Three of the most important forms of nonverbal communication are proximity, eye contact, and touch.

Proximity

Nearness in terms of physical space, time, and so on, is referred to as **proximity.** Where we sit or stand in relationship to another person signifies levels of intimacy or relationship. Many of our words conveying emotion relate to proximity, such as feeling "distant" or "close," or being "moved" by someone. We also "make the first move," "move in" on someone else's partner, or "move in together."

Did you know?

Satisfied couples spend more time together (seven hours per day, in contrast to five hours per day for distressed couples) (Kirchler, 1989).

In a social situation, the face-to-face distances between people when starting a conversation are clues to how the individuals wish to define the relationship. All cultures have an intermediate distance in face-to-face interactions that are neutral. In most cultures, decreasing the distance signifies an invitation to greater intimacy or a threat. Moving away denotes the desire to terminate the interaction. When you stand at an intermediate distance from someone at a party, you send the message that intimacy is not encouraged. If you want to move closer, however, you risk the chance of rejection. Therefore, you must exchange cues, such as laughter or small talk, before moving closer in order to avoid facing direct rejection. If the person moves farther away during this exchange ("Excuse me, I think I see a friend"), he or she is signaling disinterest. But if the person moves closer, then there is the "proposal" for greater intimacy. As relationships develop, there is close gazing into each other's eyes, holding hands, and walking with arms around each other—all of which require close proximity.

But because of cultural differences, there can be misunderstandings. The neutral intermediate distance for Latinos, for example, is much closer than for Anglos, who may misinterpret the distance as close (too close for comfort). In social settings, this can lead to problems. As Carlos Sluzki (1982) points out, "A person raised in a non-Latino culture will define as seductive behavior the same behavior that a person raised in a Latin culture defines as socially neutral." Because of the miscue, the Anglo may withdraw or flirt, depending on his or her feelings. If the Anglo flirts, the Latino may respond to what he or she believes is the other's initiation. Additionally, the neutral responses of people in cultures that have greater intermediate distances and less overt touching, such as Asian-American culture, may be misinterpreted negatively by people with other cultural backgrounds.

Eye Contact

Drink to me only with thine eyes,

And I will pledge with mine.

BEN JONSON (1573–1637)

Much can be discovered about a relationship by watching how people look at each other. Making eye contact with another person, if only for a split second longer than usual, is a signal of interest. Brief and extended glances, in fact, play a significant role in women's expression of initial interest (Moore, 1985). (The word *flirting* is derived from the old English word *fliting*, which means "darting back and forth," as so often occurs when one flirts with his or her eyes.) When you can't take your eyes off another person, you probably have a strong attraction to him or her. In fact, you can often distinguish people in love by their prolonged looking into each other's eyes. In addition to eye contact, dilated pupils may be an indication of sexual interest (or poor lighting).

Research suggests that the amount of eye contact between a couple in conversation can distinguish between those who have high levels of conflict and those who don't. Those with the greatest degree of agreement have the greatest eye contact with each other (Beier and Sternberg, 1977). Those in conflict tend to avoid eye contact (unless it is a daggerlike stare). As with proximity, however, the level of eye contact may differ by culture.

Keep thy eyes wide open before marriage, and half shut afterwards.

BENJAMIN FRANKLIN (1706–1790), *POOR RICHARD'S ALMANAC*

Reflections

As you examine your own nonverbal communication, think about instances in which you and another person have had significant eye contact. What did the eye contact mean? Think about how you and your partner touch. What are the different kinds of touching you do? The kinds your partner does? What are the different meanings you ascribe to touch given and touch received?

Touch

A review of the research on touch finds it to be extremely important in human development, health, and sexuality (Hatfield, 1994). It is the most basic of all senses; it contains receptors for pleasure and pain, hot and cold, rough and smooth. "Skin is the mother sense and out of it, all the other senses have been derived," writes anthropologist Ashley Montagu (1986). Touch is a life-giving force for infants. If babies are not touched, they may fail to thrive and may even die. We hold hands with small children and those we love. Many of our words for emotion are derived from words referring to physical contact: *attraction*, *attachment*, and *feeling*. When we are emotionally moved by someone or something, we speak of being "touched."

But touch can also be a violation. A stranger or acquaintance may touch us as if they were more familiar than they are. Your date or partner may touch you in a manner you don't like or want. And sexual harassment includes unwelcome touching.

Touch often signals intimacy, immediacy, and emotional closeness. In fact, touch may very well be the most initimate form of nonverbal communication. One researcher (Thayer, 1986) writes, "If intimacy is proximity, then nothing comes closer than touch, the most intimate knowledge of another." Touching seems to go "hand in hand" with self-disclosure. Those who touch seem to self-disclose more; in fact, touch seems to be an important factor in prompting others to talk more about themselves (Heslin and Alper, 1983; Norton, 1983).

The amount of contact, from almost imperceptible touches to "hanging all over" each other, helps differentiate lovers from strangers. How and where a person is touched can suggest friendship, intimacy, love, or sexual interest. Levels of touching differ between cultures and ethnic groups. Members of Latin cultures (both European and American) and Jews touch more than do Anglo-Americans, whereas Asian Americans touch less (Henley, 1977). Touching is important among African Americans. Writes June Dobbs Butt (1981):

> Perhaps it is in the touching and the enjoyment of contact with human bodies that black culture is most alive, and is introduced into the life of the growing child. The fondness for touch permeates black culture from cradle to grave.

Despite stereotypes of women's touching and men's avoiding touch, studies suggest that there are no consistent differences between males and females in the amount of overall touching (Andersen, Lustig, and Andersen, 1987). Men do not seem to initiate touch with women any more than women do with men. Women are markedly unenthusiastic, however, about receiving touches from strangers and express greater concern about being touched.

Popcorn enjoys a metaphysical bond with humanness. Popcorn is also the world's most social food. Consider that popcorn is the only food more often shared than eaten alone. Since that is so, the act of sharing brings people together, even if that togetherness is no more poignant than two greasy fingers touching and sliding off each other way down in the bottom of the bucket.

JOHN V. CHERVOKAS

Sexual behavior relies above all else on touch: the touching of self and others, and the touching of hands, faces, chests, arms, necks, legs, and genitals. Sexual behavior is skin contact. In sexual interactions, touch takes precedence over sight, as we close our eyes to caress, kiss, and make love. In fact, we shut our eyes to focus better on the sensations aroused by touch; we shut out visual distractions to intensify the tactile experience of sexuality.

Married couples who love each other, tell each other a thousand things without talking.

CHINESE PROVERB

The ability to interpret nonverbal communication correctly appears to be an important ingredient in successful relationships. The statement "I can tell when something is bothering him or her" reveals the ability to read nonverbal clues, such as body language or facial expressions. This ability is especially important in ethnic groups and cultures that rely heavily on nonverbal expression of feelings, such as Latino and Asian-American cultures, and although the value placed on nonverbal expression may vary between groups and cultures, the ability to communicate and understand nonverbally remains important in all cultures. A comparative study of Chinese and American romantic relationships, for example, found that shared nonverbal meanings were important for the success of relationships in both cultures (Gao, 1991).

Developing Communication Skills

While we cannot *not* communicate, we can enhance the quality of our communication so that we can understand each other and enhance our relationships. We can learn to communicate constructively rather than destructively. What follows, we hope, will help you develop good communication skills so that your relationships will be mutually rewarding.

Styles of Miscommunication

Virginia Satir noted in *Peoplemaking* (1988), her classic work on family communication, that people use four styles of miscommunication: placating, blaming, computing, and distracting.

Placaters. Placaters are always agreeable. They are passive, speak in an ingratiating manner, and act helpless. If a partner wants to make love when a placater does not, the placater will not refuse, because that might cause a scene. No one knows what placaters really want or feel—and they themselves often do not know, either.

Blamers. Blamers act superior. Their bodies are tense, they are often angry, and they gesture by pointing. Inside, they feel weak and want to hide this from everyone (including themselves). If a blamer runs short of money, the partner is the one who spent it; if a child is conceived by accident, the partner should have used contraception. The blamer does not listen and always tries to escape responsibility.

Reflections

Do you find that any of these styles of miscommunication characterize your own communication patterns? Your partner's style? Your parents', siblings', or children's? What happens if you and your partner have similar styles? Different styles?

Computers. Computers are very correct and reasonable. They show only printouts, not feelings (which they consider dangerous). "If one takes careful note of my increasing heartbeat," a computer may tonelessly say, "one must be forced to come to the conclusion that I'm angry." The partner who is interfacing, also a computer, does not change expression and replies, "That's interesting."

Distractors. Distractors act frenetic and seldom say anything relevant. They flit about in word and deed. Inside, they feel lonely and out of place. In difficult situations, distractors light cigarettes and talk about school, politics, business—

PERSPECTIVE

Family Types and Communication

Family systems, discussed in Chapter 2, can be categorized as closed, open, and random systems (Kantor and Lehr, 1975). The type of family affects each member's ability to communicate.

Closed-Type Families

In closed-type systems, families emphasize obligations, conformity to tradition, and stability. David Kantor and William Lehr (1975) describe the affective qualities of the closed-type system:

> The closed-type family strives after an intimacy and nurturance which is stable. Affections are characterized by earnestness and sincerity rather than passion. . . . Loyalties based on blood ties are usually honored above those to friends. Affections are deeply rooted in each member's strong and enduring sense of belonging. Feelings of tenderness predominate. . . . Members' relations with one another are fastidious and sensible. The emotional, or affect, mandate is to care deeply but be composed. . . . Durability, fidelity, and sincerity are the closed system's ideals in the affect dimensions.

Family rules are covert, rigid, and out-of-date; family members are required to change their needs to conform to the family rules. Communication is indirect, unclear; members have a tendency to blame, placate, distract, or be excessively rational. Family members are restricted in their ability to comment about what is going on in their family. Such families often have low self-esteem (Satir, 1972).

Open-Type Families

Open-type families emphasize consent in opinions and feelings in running the family. They seek intimacy and nurturance but not at the expense of the individual's identity. Kantor and Lehr write:

> Members are encouraged to reveal their honest feelings and thoughts to each other. Feelings of all kinds are permissible, as long as they are true ones. Members may communicate a greater intensity as well as a larger range of emotions than can those in a closed system. In addition, emotions may be more readily tapped. If a member is not showing his or her feelings, others are free to ask him to do so. . . . Its emotional mandate is to share and not withhold whatever is being felt. . . . In sum, responsiveness, authenticity, and the legitimacy of emotional latitude are the open system's major affect ideals.

Families in such a system have high self-esteem. Communication is clear, direct, specific, and honest. The family rules are overt, up-to-date, human; they change as the need arises. Family members are free to comment on anything going on in the family (Satir, 1972).

Random-Type Families

The random-type family emphasizes individual expression; its highest value is for each member to exercise his or her freedom unfettered by either tradition or consent. Free exploration by the individual is the basic purpose of the random-type family. Kantor and Lehr (1975) describe the random-type family:

> Emotions wander and are characterized by passion. They can be trenchant, profound, caustic, electric, tender, romantic, or hysterical. In short, members' emotions are unlimited. Random family affections are penetrating and penetrable, rapt and quick. Intense emotional moments, which spring unplanned from nowhere, are preferred to planned or habitual experiences. . . . The random affect mandate is to raise experience to levels of originality and inspiration.

The random-type family is really a roller-coaster family. Its rules calls for no rules, for doing what each member wants regardless of the consequences; "I do my thing and you do your thing" is its motto. Little family cohesion exists; self-esteem is basically low, despite appearances of megalomania. Communication is erratic, high decibel, hysterical.

anything to avoid discussing relevant feelings. If a partner wants to discuss something serious, a distractor changes the subject.

Why People Don't Communicate

We can learn to communicate, but it is not always easy. Traditional male gender roles, for example, work against the idea of expressing feelings. This role calls for men to be strong and silent, to ride off into the sunset alone. If men talk, they talk about things—cars, politics, sports, work, money—but not about feelings. Also, both men and women may have personal reasons for not expressing their feelings. They may have strong feelings of inadequacy: "If you really knew what I was like, you wouldn't like me." They may feel ashamed of, or guilty about, their feelings: "Sometimes I feel attracted to other people, and it makes me feel guilty because

Reflections

As you look at these family types, how would you identify your family? (Remember, however, that families often combine aspects of these different models.) What are the consequences of being in such a family? If your family of orientation is a certain type of family, will the family you create be similar? Why? Can a closed family be transformed into an open family?

It is almost as important to know what is not serious as to know what is.

JOHN KENNETH GALBRAITH

I should only be attracted to you." They may feel vulnerable: "If I told you my real feelings, you might hurt me." They may be frightened of their feelings: "If I expressed my anger, it would destroy you." Finally, people may not communicate because they are fearful that their feelings and desires will create conflict: "If I told you how I felt, you would get angry."

Obstacles to Self-Awareness

No man, for any considerable period, can wear one face to himself, and another to the multitude, without finally getting bewildered as to which may be true.

NATHANIEL HAWTHORNE (1804–1864)

Before we can communicate with others, we must first know how we ourselves feel. Though feelings are valuable guides for actions, we often place obstacles in the way of expressing them. First, we suppress "unacceptable" feelings, especially feelings such as anger, hurt, and jealousy. After a while, we may not even consciously experience them. Second, we deny our feelings. If we are feeling hurt and our partner looks at our pained expression and asks us what we're feeling, we may reply, "Nothing." We may actually feel nothing because we have anesthetized our feelings. Third, we project our feelings. Instead of recognizing that we are jealous, we may accuse our partner of being jealous; instead of feeling hurt, we may say our partner is hurt.

It happens often enough that the lie begun in self-defense slips into self-deception.

JEAN-PAUL SARTRE (1905–1980)

Becoming aware of ourselves requires us to become aware of our feelings. Perhaps the first step toward this self-awareness is realizing that feelings are simply emotional states—they are neither good nor bad in themselves. As feelings, however, they need to be felt, whether they are warm or cold, pleasurable or painful. They do not necessarily need to be acted on or expressed. It is the acting out that holds the potential for problems or hurt.

Self-Disclosure

Self-disclosure creates the environment for mutual understanding (Derlega, Metts, Potronio, and Margulis, 1993). We live much of our lives playing roles—as student, worker, husband, wife, son, or daughter. We live and act these roles conventionally. They do not necessarily reflect our deepest selves. If we pretend that we are only these roles and ignore our deepest selves, we have taken the path toward loneliness and isolation. We may reach a point at which we no longer know who we are. In the process of revealing ourselves to others, we discover who we are. In the process of our sharing, others share themselves with us. Self-disclosure is reciprocal.

Speech is civilization itself. . . . It is silence which isolates.

THOMAS MANN (1875–1955)

Keeping Closed

Having been taught to be strong, men may be more reluctant to express feelings of weakness or tenderness than women. Many women find it easier to disclose their feelings, perhaps because from earliest childhood they are more often encouraged to express them (see Notarius and Johnson, 1982).

A half truth is a whole lie.

YIDDISH PROVERB

If distinct differences exist, they can drive wedges between men and women. One sex does not understand the other. The differences may plague a marriage until neither partner knows what the other wants; sometimes partners don't even know what they want for themselves. In one woman's words (quoted in Rubin, 1976):

> I'm not sure what I want. I keep talking to him about communication, and he says, "Okay, so we're talking; now what do you want?" And I don't know what to say then, but I know it's not what I mean. I sometimes get worried because I think maybe I

> want too much. He's a good husband; he works hard; he takes care of me and the kids. He could go out and find another woman who would be very happy to have a man like that, and who wouldn't be all the time complaining at him because he doesn't feel things and get close.

What is missing is the intimacy that comes from self-disclosure. People live together, or are married, but they feel lonely. There is no contact, and the loneliest loneliness is to feel alone with someone with whom we want to feel close.

If you can't say something nice, don't say anything.

THUMPER'S MOTHER

How Much Openness?

Can too much openness and honesty be harmful to a relationship? How much should intimates reveal to each other? Some studies suggest that less marital satisfaction results if partners have too little or too much disclosure; a happy medium offers security, stability, and safety. But a review of studies on the relationship between communication and marital satisfaction finds that a linear model of communication is more closely related to marital satisfaction than the too-little/too-much curvilinear model (Boland and Follingstad, 1987). In the linear model of communication, the greater the self-disclosure, the greater the marital satisfaction, provided that the couple are highly committed to the relationship and willing to take the risks of high levels of intimacy. High self-disclosure can be a highly charged undertaking. Studies suggest that high levels of negativity are related to marital distress (Noller and Fitzpatrick, 1991). It is not clear whether the negativity reflects the marital distress or causes it. Most likely, the two interact and compound each other's effects.

If you haven't got anything nice to say about anybody, come sit next to me.

ALICE ROOSEVELT LONGWORTH (1884–1980)

When in doubt, tell the truth.

MARK TWAIN (1835–1910)

"If we want this relationship to work, we'll have to start communicating. I'll go first—get your feet off the table."

Self-disclosure is reciprocal.

Trust

When we talk about intimate relationships, the two words that most frequently pop up are *love* and *trust.* As we saw in our discussion of love in Chapter 4, trust is an important part of love. But what is trust? **Trust** is the belief in the reliability and integrity of a person.

Well done is better than well said.

BENJAMIN FRANKLIN (1706–1790)

When a person says, "Trust me," he or she is asking for something that does not easily occur. For trust to develop, three conditions must exist (Book et al., 1980). First, a relationship has to exist and have the likelihood of continuing. We generally do not trust strangers or people we have just met with information that makes us vulnerable, such as our sexual anxieties. We trust people with whom we have a significant relationship.

Ninety-nine lies may save you, but the hundredth will give you away.

WEST AFRICAN PROVERB

Second, we must be able to predict how a person will likely behave. If we are married or in a committed relationship, we trust that our partner will not do something that will hurt us, such as having an affair. In fact, if we discover that our partner is involved in an affair, we often speak of our trust's being violated or destroyed. If trust is destroyed in this case, it is because the predictability of sexual exclusiveness is no longer there.

Did you know?

The happiest couples are those who balance autonomy with intimacy and negotiate personal and couple boundaries thorough supportive communication (Scarf, 1995).

Third, the person must also have other acceptable options available to him or her. If we were marooned on a desert island alone with our partner, he or she would have no choice but to be sexually monogamous. But if a third person, who was sexually attractive to our partner, swam ashore a year later, then our partner would have an alternative. Our partner would then have a choice of being sexually exclusive or nonexclusive; his or her behavior would then be evidence of trustworthiness—or the lack of it.

Trust is critical in close relationships for two reasons (Book et al., 1980). First, self-disclosure—which is vital to closeness—requires trust because self-disclosure makes you vulnerable. A person will not self-disclose if he or she believes that the information may be misused, such as a partner mocking you or revealing a secret. Second, the degree to which you trust a person influences the way you are likely to interpret ambiguous or unexpected messages from him or her. If your partner says he or she wants to study alone tonight, you are likely to take the statement at face value if you have a high trust level. But if you have a low

For neither man nor angel can discern
Hypocrisy, the only evil that walks
Invisible. . . .

JOHN MILTON (1608–1674)

trust level, you may believe he or she is going to meet someone else while you are studying in the library.

Trust in personal relationships has both a behavioral and a motivational component (Book et al., 1980). The behavioral component refers to the probability that a person will act in a trustworthy manner. The motivational component refers to the reasons a person engages in trustworthy actions. Whereas the behavioral element is important in all types of relationships, the motivational element is important in close relationships. One has to be trustworthy for the "right" reasons. As long as you trust your mechanic to charge you fairly for rebuilding your car's engine, you don't care why he or she is trustworthy. But you do care why your partner is trustworthy. For example, you want your partner to be sexually exclusive to you because he or she loves you or is attracted to you. Being faithful because of duty or because your partner can't find anyone better is the wrong motivation. Disagreements about the motivational bases for trust are often a source of conflict. "I want you because you love me, not because you need me" or "You don't really love me; you're just saying that because you want sex" are typical examples of conflict about motivation.

Consider what life would be like if everyone could lie perfectly or if no one could lie at all. . . . If we could never know how someone really felt, and if we knew that we couldn't know, life would be more tenuous. Certain in the knowledge that every show of emotion might be a mere display put on to please, manipulate, or mislead, individuals would be more adrift, attachments less firm. And if we could never lie, if a smile was reliable, never absent when pleasure was felt, and never present without pleasure, life would be rougher than it is, many relationships harder to maintain. Politeness, attempts to smooth matters over, to conceal feelings one wished one didn't feel—all that would be gone. There would be no way not to be known, no opportunity to sulk or lick one's wounds except alone.

Paul Eckman, *Telling Lies*

Giving Feedback

Self-disclosure is reciprocal. If we self-disclose, we expect our partner to self-disclose as well. As we self-disclose, we build trust; as we withhold self-disclosure, we erode trust. To withhold ourselves is to imply that we don't trust the other person, and if we don't, he or she will not trust us.

A critical element in communication is **feedback,** the ongoing process in which participants and their messages create a given result and are subsequently modified by the result. If someone self-discloses to us, we need to respond to his or her self-disclosure. The purpose of feedback is to provide constructive information to increase another's self-awareness of the consequences of his or her behavior toward you. (Note that this type of feedback is different from that discussed on pages 140–141.)

If your partner discloses to you his or her doubts about your relationship, for example, you can respond in a number of ways. Among these are remaining silent, venting anger, expressing indifference, or giving constructive feedback. First, you can remain silent. Silence, however, is generally a negative response, perhaps as powerful as saying outright that you do not want your partner to self-disclose this type of information. Second, you can respond angrily, which may convey the message to your partner that self-disclosing will lead to arguments rather than understanding and possible change. Third, you can remain indifferent, responding neither negatively nor positively to your partner's self-disclosure. Fourth, you can acknowledge your partner's feelings as being valid (rather than right or wrong) and disclose how you feel in response to his or her statement. This acknowledgment and response is constructive feedback. It may or may not remove your partner's doubts, but it is at least constructive in that it opens up the possibility for change, whereas silence, anger, and indifference do not.

Marriage is one long conversation, chequered by disputes.

Robert Louis Stevenson (1850–1894)

Some guidelines (developed by David Johnston for the Minnesota Peer Program) may help you engage in dialogue and feedback with your partner:

Don't use a hatchet to remove a fly from your friend's forehead.

Chinese proverb

1 Focus on "I" statements. An "I" statement is a statement about your feelings: "I feel annoyed when you leave your dirty dishes on the living room floor." "You" statements tell another person how he or she is, feels, or thinks: "You are so irresponsible. You're always leaving your dirty dishes on the living room floor." "You" statements are often

blaming or accusatory. Because "I" messages don't carry blame, the recipient is less likely to be defensive or resentful.

2 Focus on behavior rather than on the person. If you focus on a person's behavior rather than on the person, you are more likely to secure change. A person can change behaviors but not himself or herself. If you want your partner to wash his or her dirty dishes, say, "I would like you to wash your dirty dishes; it bothers me when I see them gathering mold on the living room floor." This statement focuses on behavior that can be changed. If you say, "You are such a slob; you never clean up after yourself," then you are attacking the person. He or she is likely to respond defensively: "I am not a slob. Talk about slobs, how about when you left your clothes lying in the bathroom for a week?"

There is perhaps no phenomenon which contains so much destructive feeling as moral indignation, which permits envy or hate to be acted out under the guise of virtue.

Erich Fromm (1900–1980)

3 Focus feedback on observations rather than on inferences or judgments. Focus your feedback on what you actually observe rather than on what you think the behavior means. "There is a towering pile of your dishes in the living room" is an observation. "You don't really care about how I feel because you are always leaving your dirty dishes around the house" is an inference that a partner's dirty dishes indicate a lack of regard. The inference moves the discussion from the dishes to the partner's caring. The question "What kind of person would leave dirty dishes for me to clean up?" implies a judgment: only a morally depraved person would leave dirty dishes around.

To say what we think to our superiors would be inexpedient; to say what we think to our equals would be ill-mannered; to say what we think to our inferiors is unkind. Good manners occupy the terrain between fear and pity.

Quentin Crisp

4 Focus feedback on observations based on a more-or-less continuum. Behaviors fall on a continuum. Your partner doesn't *always* do a particular thing. When you say that he or she does something sometimes or even most of the time, you are actually measuring behavior. If you say that your partner always does something, you are distorting reality. For example, there were probably times (however rare) when your partner picked up the dirty dishes. "Last week I picked up your dirty dishes three times" is a measured statement. "I always pick up your dirty dishes" is an exaggeration that will probably provoke a hostile response.

5 Focus feedback on sharing ideas or offering alternatives rather than on giving advice. No one likes being told what to do. Unsolicited advice often produces anger or resentment because advice implies that you know more about what a person needs to do than the other person does. Advice implies a lack of freedom or respect. By sharing ideas and offering alternatives, however, you give the other person the freedom to decide based on his or her own perceptions and goals. "You need to put away your dishes immediately after you are done with them" is advice. To offer alternatives, you might say, "Having to walk around your dirty dishes bothers me. What are the alternatives other than my watching my step? Maybe you could put them away after you finish eating, clean them up before I get home, or eat in the kitchen. What do you think?"

6 Focus feedback according to its value to the recipient. If your partner says something that upsets you, your initial response may be to

PERSPECTIVE

Ten Rules for Avoiding Intimacy

If you want to avoid intimacy, here are ten rules that have proved effective in nationwide testing with lovers, husbands and wives, parents and children. Follow these guidelines, and you'll never have an intimate relationship.

1. *Don't talk.* This is the basic rule for avoiding intimacy. If you follow this one rule, you will never have to worry about being intimate again. Sometimes, however, you may be forced to talk. But don't talk about anything meaningful. Talk about the weather, baseball, class, the stock market—anything but feelings.
2. *Never show your feelings.* Showing your feelings is almost as bad as talking because feelings are ways of communicating. If you cry or show anger, sadness, or joy, you are giving yourself away. You might as well talk, and if you talk, you could become intimate. So the best thing to do is remain expressionless (which, it must be admitted, is a form of communication, but at least it is giving the message that you don't want to be intimate).
3. *Always be pleasant.* Always smile, and always be friendly, especially if something's bothering you. You'll be surprised at how effective hiding negative feelings from your partner is in preventing intimacy. It may even fool your partner into believing that everything's okay in your relationship.
4. *Always win.* If you can't be pleasant, try this one. Never compromise, and never admit that your partner's point of view may as good as yours. If you start compromising, that's an admission that you care about your partner's feelings, which is a dangerous step toward intimacy.
5. *Always keep busy.* Keeping busy at school or work will take you away from your partner, and you won't have to be intimate. Your partner may never figure out that you're using work to avoid intimacy. Because our culture values hard work, he or she will feel unjustified in complaining. Incidentally, devoting yourself to your work will nevertheless give your partner the message that he or she is not as important as your work. You can make your partner feel unimportant in your life without saying a word!
6. *Always be right.* There is nothing worse than being wrong because it is an indication that you are human. If you admit you're wrong, you might have to admit that your partner is right, and that will make him or her as good as you. If he or she is as good as you, then you might have to consider your partner, and before you know it, you will be intimate!
7. *Never argue.* If you can't always be right, don't argue at all. If you argue, you might discover that you and your partner are different. If you're different, you may have to talk about the differences so that you can make adjustments, and if you begin making adjustments, you may have to tell your partner who you really are and what you really feel. Naturally, these revelations may lead to intimacy.
8. *Make your partner guess what you want.* Never tell your partner what you want. That way, when your partner tries to guess and is wrong (as he or she often will be), you can tell your partner that he or she doesn't really understand or love you. If your partner did love you, he or she would know what you wanted without asking. Not only will this prevent intimacy, but it also will drive your partner crazy.
9. *Always look out for number one.* Remember; you are number one. All relationships exist to fulfill your needs, no one else's. Whatever you feel like doing is just fine. You're okay—your partner's not okay. If your partner can't satisfy your needs, he or she is narcissistic; after all, you are the one making all the sacrifices in the relationship.
10. *Keep the television on.* Keep the television turned on at all times: during dinner, while you're reading, when you're in bed, and while you're talking (especially if you're talking about something important). This rule may seem petty compared with the others, but it is good preventive action. Watching television keeps you and your partner from talking to each other. Best of all, it will keep you both from even noticing that you don't communicate. If you're cornered and have to talk, you can both be distracted by a commercial, a seduction scene, or the sound of gunfire. When you actually think about it, wouldn't you rather be watching *King of the Hill* anyway?

We want to caution individuals that this list is not complete. Everyone knows additional ways to avoid intimacy. You have your own inventions or techniques that you have learned from your boyfriend or girlfriend, friends, or parents. To round out this compilation, list additional rules for avoiding intimacy on a separate sheet of paper. The person with the best list wins—and never has to be intimate again.

Excuses are always mixed with lies.

ARABIC PROVERB

lash back. A cathartic response may make you feel better for the time being, but it may not be useful for your partner. If, for example, your partner admits lying to you, you can respond with rage and accusations, or you can express hurt and try to find out why he or she didn't tell you the truth.

7 Focus feedback on the amount the recipient can process. Don't overload your partner with your response. Your partner's disclosure may touch deep, pent-up feelings in you, but he or she may not be able to comprehend all that you say. If you respond to your partner's revelation of doubts with a listing of all the doubts you have ever experienced about yourself, your relationship, and relationships in general, you may overwhelm your partner.

8 Focus feedback at an appropriate time and place. Choose a time when you are not likely to be interrupted. Turn the television off and the phone-answering machine on. Also, choose a time that is relatively stress free. Talking about something of great importance just before an exam or a business meeting is likely to sabotage any attempt at communication. Finally, choose a place that will provide privacy; don't start an important conversation if you are worried about people's overhearing or interrupting you. A dormitory lounge during the soaps, Grand Central Station, a kitchen teeming with kids, or a car full of friends is an inappropriate place.

Did you know?

A repeated cycle of negative verbal expression by a wife and withdrawal by the husband is a pattern that commonly manifests itself in distressed marriages (Kurdeck, 1995).

We are all inclined to judge ourselves by our ideals; others by their acts.

HAROLD NICHOLSON

The only way to speak the truth is to speak lovingly.

HENRY DAVID THOREAU (1817–1862)

Mutual Affirmation

Good communication in an intimate relationship involves mutual affirmation, which includes three elements: (1) mutual acceptance, (2) liking each other, and

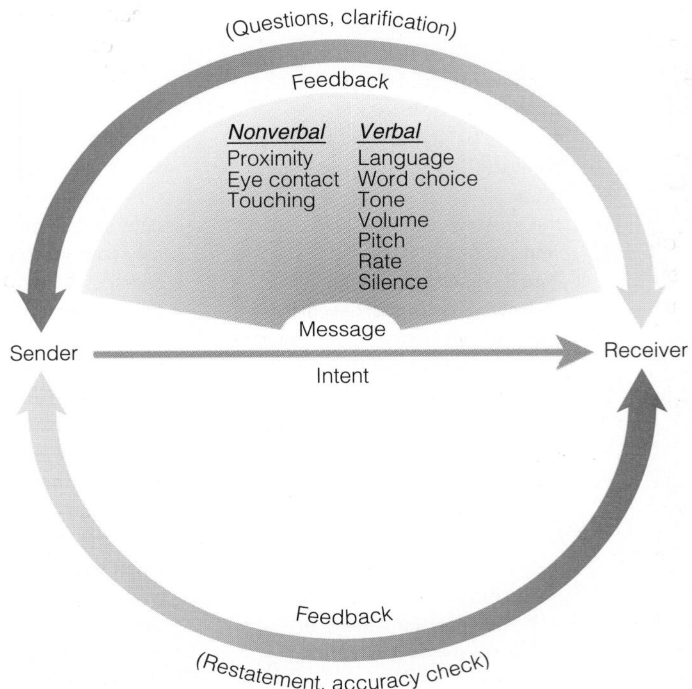

Figure 5.1 Communication Loop

In successful communication, feedback between the sender and receiver ensures that both understand (or are trying to understand) what is being communicated. For communication to be clear, the message and the intent behind the message must be congruent. Nonverbal and verbal components must also support the intended message. Verbal aspects of communication include not only language and word choice but also characteristics such as tone, volume, pitch, rate, and periods of silence.

(3) expressing liking in both words and actions. Mutual acceptance consists of people accepting each other as they are, not as they would like each other to be. People are who they are, and they are not likely to change in fundamental ways without a tremendous amount of personal effort, as well as a considerable passage of time. The belief that an insensitive partner will somehow magically become sensitive after marriage, for example, is an invitation to disappointment and divorce.

If you accept people as they are, you can like them for their unique qualities. Liking someone is somewhat different from being romantically involved. It is not rare for people to dislike those with whom they are romantically linked.

We also need to express our feelings of warmth, affection, and love. To one's partner, unexpressed words, actions, thoughts, kindnesses, deeds, touches, caresses, and kisses can be the same as nonexistent or unfelt ones. "You know that I love you" without the expressions of love is a meaningless statement. A simple rule of thumb for communicating love is: If you love, show love.

Mutual affirmation entails our telling others that we like them for who they are, that we appreciate the little things as well as the big things that they do. Think about how often you say to your partner, your parents, or your children, "I like you," "I love you," "I appreciate your doing the dishes," or "I like your smile." Affirmations are often most frequent during dating or the early stages of marriage or living together. As you get to know a person better, you may begin noting things that annoy you or are different from you. Acceptance turns into negation and criticism: "You're selfish," "Stop bugging me," "You talk too much," or "Why don't you clean up after yourself?"

If you have a lot of negatives in your interactions, don't feel too bad. Many of our negations are habitual. When we were children, our parents may have been negating: "Don't leave the door open," "Why can't you get better grades?" "Stand straight and pull in your stomach." How often did they affirm? Once you become aware that negations are often automatic, you can change them. Because negative communication is a learned behavior, you can unlearn it. One way is to make the decision consciously to affirm what you like; too often we take the good for granted and feel compelled to point out only the bad.

Reflections

To get a sense of how much you affirm or negate someone, keep track of your affirmations and negations. On a sheet of paper, label one column affirmations and the other column negations. Each time you make an affirmation of that person, give yourself a plus; each time you make a negation, give yourself a minus. At the end of the day, compare the numbers of pluses and minuses. As a result of your comments, how do you think the other person feels?

The facts are always friendly. Every bit of evidence one can acquire, in any area, leads one that much closer to what is true.

CARL RODGERS, *ON BECOMING PARTNERS*

Conflict and Intimacy

Conflict between people who love each other seems to be a mystery. The coexistence of conflict and love has puzzled human beings for centuries. An ancient Sanskrit poem reflected this dichotomy:

In the old days we both agreed
That I was you and you were me.
But now what has happened
That makes you, you
And me, me?

We expect love to unify us, but sometimes it doesn't. Two people do not become one when they love each other, although at first they may have this feeling. Their love may not be an illusion, but their sense of ultimate oneness is. In reality, they retain their individual identities, needs, wants, and pasts while loving each other—and it is a paradox that the more intimate two people become, the more likely they may be to experience conflict. But it is not conflict

The awareness of sameness is friendship; the awareness of difference is love.

W. H. AUDEN (1907–1973)

itself that is dangerous to intimate relationships; it is the manner in which the conflict is handled.

Conflict is natural in intimate relationships. If this is understood, the meaning of conflict changes, and it will not necessarily represent a crisis in the relationship. David and Vera Mace (1979), prominent marriage counselors, observe that on the day of marriage, people have three kinds of raw material with which to work. First, there are things you have in common—the things you both like. Second are the ways in which you are different, but the differences are complementary. Third, unfortunately, are the differences between us that are not at all complementary and that cause us to meet head-on with a big bang. In every relationship between two people, there are a great many of those kinds of differences. So when we move closer together to each other, those differences become disagreements. The presence of conflict within a marriage or family does not necessarily indicate that love is going or gone. It may mean just the opposite.

APOLOGIZE, v.i. To lay the foundation for a future offense.

AMBROSE BIERCE (1842–1914), *THE DEVIL'S DICTIONARY*

Argument is the worst sort of conversation.

JONATHAN SWIFT (1667–1745)

Basic versus Nonbasic Conflicts

Two types of conflict—basic and nonbasic—affect the stability of a relationship. Basic conflicts challenge the fundamental assumptions or rules of a relationship, whereas nonbasic conflicts do not.

It is a luxury to be understood.

RALPH WALDO EMERSON (1803–1892)

Basic Conflicts

Basic conflicts revolve around carrying out marital roles and the functions of marriage and the family, such as providing companionship, working, and rearing children. It is assumed, for example, that a husband and wife will have sexual relations with each other. But if one partner converts to a religious sect that forbids sexual interaction, a basic conflict is likely to occur because the other spouse considers sexual interaction part of the marital premise. No room for compromise exists in such a matter. If one partner cannot convince the other to change his or her belief, the conflict is likely to destroy the relationship. Similarly, despite recent changes in family roles, it is still expected that the husband will work to provide for the family. If he decides to quit work altogether and not function as a provider in any way, he is challenging a basic assumption of marriage. His partner is likely to feel that his behavior is unfair. Conflict ensues. If he does not return to work, his wife is likely to leave him.

Did you know?
When the communication patterns of newly married African Americans and Whites were examined, couples who believed in avoiding marital conflict were less happy two years later than those who confronted their problems (Crohan, 1992).

Nonbasic Conflicts

Nonbasic conflicts do not strike at the heart of a relationship. The husband wants to change jobs and move to a different city, but the wife may not want to move. This may be a major conflict, but it is not a basic one. The husband is not unilaterally rejecting his role as a provider. If a couple disagree about the frequency of sex, the conflict is serious but not basic because both agree on the desirability of sex in the relationship. In both of these cases, resolution is possible.

The mind is its own place, and in itself can make a heaven of Hell, a hell of Heaven.

JOHN MILTON (1608–1674)

Situational versus Personality Conflicts

Some conflicts occur because of a situation, and others occur because of the personality of one (or both) of the partners.

Situational Conflicts

Situational conflicts occur when at least one partner needs to make changes in a relationship. They are based on specific demands, like putting the cap on the toothpaste, dividing housework equitably, sharing child-care responsibilities, and so on. Conflict arises when one person tries to change the situation about the toothpaste cap, housework, or child care.

For a marriage to be peaceful the husband should be deaf and the wife blind.

SPANISH PROVERB

Personality Conflicts

Personality conflicts arise not because of situations that need to be changed but because of personality, such as needs to vent aggression or to dominate. Such conflicts are essentially unrealistic. They are not directed toward making changes in the relationship but simply toward releasing pent-up tensions. Whereas situational conflicts can be resolved through compromise, bargaining, or mediation, personality conflicts often require a therapeutic approach. Such personality conflicts may pit a compulsive-type individual against a free spirit or a fastidious personality against a sloppy one.

Power Conflicts

The politics of family life—who has the power, who makes the decisions, who does what—can be every bit as complex and explosive as politics at the national level. **Power** is the ability or potential ability to influence another person or group. Most of the time we are not aware of the power aspects of our relationships. One reason for this is that we tend to believe that intimate relationships are based on love alone. Another reason is that the exercise of power is often subtle. When we think of power, we tend to think of coercion or force; as we shall see, however, marital power takes many forms. A final reason why we are not always aware of power is that power is not constantly exercised. It comes into play only when an issue is important to both people and they have conflicting goals.

Power corrupts and absolute power corrupts absolutely.

LORD ACTON (1834–1902)

Changing Sources of Marital Power

Traditionally, husbands have held authority over their wives. In Christianity, the subordination of wives to their husbands has its basis in the New Testament. Paul (Colossians 3:18–19) states: "Wives, submit yourselves unto your husbands, as unto the Lord." Such teachings reflected the dominant themes of ancient Greece and Rome. Western society continued to support wifely subordination to husbands. English common law stated, "The husband and wife are as one and that one is the husband." A woman assumed her husband's identity, taking his last name on marriage and living in his house.

The U.S. courts have institutionalized these power relationships. The law, for example, supports the traditional division of labor in many states, making the husband legally responsible for supporting the family and the wife legally responsible for maintaining the house and rearing the children. She is legally required to follow her husband if he moves; if she does not, she is considered to have deserted him. But if she moves and her husband refuses to move with her, she is also considered to have deserted him (Leonard and Elias, 1990).

Legal and social support for the husband's control of the family has declined since the 1920s and especially since the 1960s. An egalitarian standard for sharing power in families has taken much of its place (Sennett, 1980). The wife who works has especially gained more power in the family. She has greater influence in deciding family size and how money is to be spent.

Be to her virtues very kind,

Be to her faults a little blind.

MATTHEW PRIOR (1664–1721)

The formal and legal structure of marriage makes the male dominant, but the reality of marriage may be quite different. Sociologist Jessie Bernard (1982) makes an important distinction between authority and power in marriage. Authority is based in law, but power is based in personality. A strong, dominant woman is likely to exercise power over a more passive man simply by the force of her personality and temperament.

If we want to see how power really works in marriage, we must look beneath the stereotypes. Women have considerable power in marriage, although they often feel that they have less than they actually do. They may fail to recognize the extent of their power; because cultural norms theoretically put power in the hands of their husbands, women may look at norms rather than at their own behavior. A woman may decide to work, even against her husband's wishes, and she may determine how to discipline the children. Yet she may feel that her husband holds the power in the relationship because he is *supposed* to be dominant. Similarly, husbands often believe that they have more power in a relationship than they actually do because they see only traditional norms and expectations.

Bases of Marital Power

Power is not a simple phenomenon. Researchers generally agree that family power is a dynamic, multidimensional process (Szinovacz, 1987). Generally speaking, no single individual is always the most powerful person in every aspect of the family. Nor is power necessarily always based on gender, age, or relationship. Power often shifts from person to person, depending on the issue.

According to J.P. French and Betram Raven (1959), there are six bases of marital power:

Those who warp the truth must inevitably be warped and corrupted themselves.

Arthur Miller

1. *Coercive power* is based on the fear that one partner will punish the other. Coercion can be emotional or physical. A pattern of belittling, threatening, or being physical can intimidate and threaten another. This is the least common form of power but is used in partner rape or abuse.

2. *Reward power* is based on the belief that the other person will do something in return for agreement. If, for example, your partner attempts to understand your feelings about a specific issue, he or she may expect you to do the same.

3. *Expert power* is based on the belief that the other has greater knowledge. If you believe that your partner has more wisdom about child rearing, for instance, you may defer the rewards, incentives, and discipline to him or her.

4. *Legitimate power* is based on acceptance of roles giving the other person the right to demand compliance. Gender roles are an important part of legitimacy as they give an aura to rights based on gender. Traditional gender roles legitimize male initiation in dating and female acceptance or refusal rights.

5. *Referent power* is based on identifying with the partner and receiving satisfaction by acting similarly. If you have great respect in your partner's

communication skills, his or her ability to actively listen, provide feedback, and disclose in an honest manner, you are more likely to model yourself after him or her.

6 *Informational power* is based on the partner's persuasive explanation. If, for example, your partner refuses to use a condom, you can provide information about the prevalence and danger of STDs and AIDS.

Relative Love and Need Theory

Another way of looking at the sources of marital power is through the **relative love and need theory,** which explains power in terms of the individual's involvement and needs in the relationship. Each partner brings certain resources, feelings, and needs to a relationship. Each may be seen as exchanging love, companionship, money, help, and status with the other. What each gives and receives, however, may not be equal. One partner may be gaining more from the relationship than the other. The person gaining the most from the relationship is the one who is most dependent. Constantina Safilios-Rothschild (1970) observes:

> The relative degree to which the one spouse loves and needs the other may be the most crucial variable in explaining the total power structure. The spouse who has relatively less feeling for the other may be the one in the best position to control and manipulate all the "resources" that he has in his command in order to effectively influence the outcome of decisions.

If you are afraid of loneliness, don't marry.

ANTON CHEKHOV (1860–1904)

Love is a major power resource in a relationship. Those who love equally are likely to share power equally (Safilios-Rothschild, 1976). Such couples are likely to make decisions according to referent, expert, and legitimate power.

Principle of Least Interest

Akin to relative love and need as a way of looking at power is the **principle of least interest.** Sociologist Willard Waller (Waller and Hill, 1951) coined this term to describe the curious (and often unpleasant) situation in which the partner with the least interest in continuing a relationship has the most power in it. At its most extreme form, it is the stuff of melodrama. "I will do anything you want, Charles," Laura says pleadingly, throwing herself at his feet. "Just don't leave me." "Anything, Laura?" he replies with a leer. "Then give me the deed to your mother's house." Quarreling couples may unconsciously use the principle of least interest to their advantage. The less involved partner may threaten to leave as leverage in an argument: "All right, if you don't do it my way, I'm going." The threat may be extremely powerful in coercing a dependent partner. It may have little effect, however, if it comes from the dependent partner because he or she has too much to lose to be persuasive. The less involved partner can easily call the bluff.

If I speak in the tongues of men and of angels, but have not love, I am as a noisy gong or a clanging cymbal.

I CORINTHIANS 13:1

Rethinking Family Power

Even though women have considerable power in marriages and families, it would be a serious mistake to overlook the inequalities between husbands and wives. As feminist scholars have pointed out, major aspects of contemporary marriage point to important areas where women are clearly subordinate to men: the continued female responsibility for housework and child rearing, inequities in sexual gratification (sex is often over when the male has his orgasm), the extent of violence against women, and the sexual exploitation of children are examples.

Feminist scholars suggest several areas that require further consideration (Szinovacz, 1987). First, they believe that too much emphasis has been placed on the marital relationship as the unit of analysis. Instead, they believe that researchers should explore the influence of the larger society on power in marriage—specifically, the relationship between the social structure and women's position in marriage. Researchers could examine, for example, the relationship of women's socioeconomic disadvantages, such as lower pay and fewer economic opportunities than men, to female power in marriage.

In love it is enough to please each other by loveable qualities and attractions; but to be happy in marriage it is necessary to love each other's faults, or at least to adjust to them.

Nicholas Chamfort (1740–1794)

Second, these scholars argue that many of the decisions that researchers study are trivial or insignificant in measuring "real" family power. Researchers cannot conclude that marriages are becoming more egalitarian on the basis of joint decision making about such things as where a couple goes for vacation, whether to buy a new car or appliance, or which movie to see. The critical decisions that measure power are such issues as how housework is to be divided, who stays home with the children, and whose job or career takes precedence.

Some scholars suggest that we shift the focus from marital power to family power. Researcher Marion Kranichfeld (1987) calls for a rethinking of power in a family context. Even if women's marital power may not be equal to men's, a different picture of women in families may emerge if we examine power within the entire family structure, including power in relation to children. The family power literature has traditionally focused on marriage and marital decision making. Kranichfeld, however, feels that such a focus narrows our perception of women's power. Marriage is not family, she argues, and it is in the larger family matrix that women exert considerable power. Their power may not be the same as male power, which tends to be primarily economic, political, or religious. But if *power* is defined as the ability to change the behavior of others intentionally, "women in fact have a great deal of power, of a very fundamental and pervasive nature, so pervasive, in fact, that it is easily overlooked," according to Kranichfeld (1987). She further observes:

> Women's power is rooted in their role as nurturers and kinkeepers, and flows out of their capacity to support and direct the growth of others around them through their life course. Women's power may have low visibility from a nonfamily perspective, but women are the lynchpins of family cohesion and socialization.

We thought, because we had power, we had wisdom.

Stephen Vincent Bénet (1898–1943)

Power versus Intimacy

The problem with power imbalances or the blatant use of power is the negative effect on intimacy. As Ronald Sampson (1966) observes in his study of the psychology of power, "To the extent that power is the prevailing force in a relationship—whether between husband and wife or parent and child, between friends or between colleagues—to that extent love is diminished." If partners are not equal, self-disclosure may be inhibited, especially if the powerful person believes his or her power will be lessened by sharing feelings (Glazer-Malbin, 1975). Genuine intimacy appears to require equality in power relationships. Decision making in the happiest marriages seems to be based not on coercion or tit for tat but on caring, mutuality, and respect for the other person. Women or men who feel vulnerable to their mates may withhold feelings or pretend to feel what they do not. Unequal power in marriage may encourage power politics. Each partner may struggle with the other to keep or gain power.

It is not easy to change unequal power relationships after they become embedded in the overall structure of a relationship; yet they can be changed. Talking, understanding, and negotiating are the best approaches. Still, in attempting changes, a person may risk estrangement or the breakup of a relationship. He or she must weigh the possible gains against the possible losses in deciding whether change is worth the risk.

The return of understanding after estrangement: Everything must be treated with tenderness at the beginning so that the return may lead to understanding.

I CHING

Conflict Resolution

Differences and conflicts are part of any healthy relationship. If we handle conflicts in a healthy way, they can help solidify our relationships. But conflicts can go on and on, consuming the heart of a relationship, turning love and affection into bitterness and hatred. In the following section, we will look at ways of resolving conflict in constructive rather than destructive ways. In this manner, we can use conflict as a way of building and deepening our relationships.

A number of porcupines huddled together for warmth on a cold day in winter; but because they began to prick each other with their quills, they were obliged to disperse. However, the cold drove them together again, when just the same thing happened. At last, after many turns of huddling and dispersing, they discovered that they would be best off by remaining at a little distance from each other.

ARTHUR SCHOPENHAUER (1788–1860)

Dealing with Anger

Differences can lead to anger, and anger transforms differences into fights, creating tension, division, distrust, and fear. Most people have learned to handle anger by either venting or suppressing it. David and Vera Mace (1980) suggest that many couples go through a love-anger cycle. When a couple comes close to each other, they may experience conflict, and they recoil in horror, angry at each other because just at the moment they were feeling close, their intimacy was destroyed. Each backs off; gradually they move closer again until another fight erupts, driving them away from each other. After a while, each learns to make a compromise between closeness and distance to avoid conflict. They learn what they can reveal about themselves and what they cannot.

Another way of dealing with anger is suppressing it. Suppressed anger is dangerous because it is always there, simmering beneath the surface. Ultimately, it leads to resentment, that brooding, low-level hostility that poisons both the individual and the relationship.

It is human nature to think wisely and act foolishly.

ANATOLE FRANCE (1844–1924)

We can learn to use conflict as a way to build and deepen our relationships.

Anger can be dealt with in a third way: recognizing it as a symptom of something that needs to be changed. If we see anger as a symptom, we realize that what is important is not venting or suppressing the anger but finding its source and eliminating it. David and Vera Mace (1980) write:

Anger causes a man to be far from the truth.

HASIDIC SAYING

> When your disagreements become conflict, the only thing to do is to take anger out of it, because when you are angry you cannot resolve a conflict. You cannot really hear the other person because you are just waiting to fire your shot. You cannot be understanding; you cannot be empathetic when you are angry. So you have to take the anger out, and then when you have taken the anger out, you are back again with a disagreement. The disagreement is still there, and it can cause another disagreement and more anger unless you clear it up. The way to take the anger out of disagreements is through negotiation.

Conflict Resolution and Marital Satisfaction

Happy couples tend to act in positive ways to resolve conflicts, such as changing behaviors (putting the cap on the toothpaste rather than denying responsibility) and presenting reasonable alternatives (purchasing toothpaste in a dispenser). Unhappy or distressed couples, in contrast, use more negative strategies in attempting to resolve conflicts (if the cap off the toothpaste bothers you, then *you* put it on). A study of happily and unhappily married couples found distinctive communication traits as these couples tried to resolve their conflicts (Ting-Toomey, 1983). The communication behaviors of happily married couples displayed the following traits:

The greatest thing in family life is to take a hint when a hint is intended—and not to take a hint when a hint isn't intended.

ROBERT FROST (1874–1963)

- *Summarizing.* Each person summarized what the other said: "Let me see if I can repeat the different points you were making."
- *Paraphrasing.* Each put what the other said into his or her own words: "What you are saying is that you feel bad when I don't acknowledge your feelings."
- *Validation.* Each affirmed the other's feelings: "I can understand how you feel."
- *Clarification.* Each asked for further information to make sure that he or she understood what the other was saying: "Can you explain what you mean a little bit more to make sure that I understand you?"

Love can be angry . . . with a kind of anger in which there is no gall, like the dove's and not the raven's.

AUGUSTINE OF HIPPO (354–430)

In contrast, unhappily married couples displayed the following reciprocal patterns:

- *Confrontation.* Each member of the couple confronted the other: "You're wrong!" "Not me, buddy. It's you who's wrong."
- *Confrontation and defensiveness.* One confronted while the other defended himself or herself: "You're wrong!" "I only did what I was supposed to do."
- *Complaining and defensiveness.* One complained while the other was defensive: "I work so hard each day to come home to this!" "This is the best I can do with no help."

Reflections

How do you handle conflict? Whom do you pattern your communication patterns after? How does your sense of power in a relationship influence the way you communicate and the way you handle conflict?

Attachment style (discussed in Chapters 4 and 10) seems to influence the way conflict is expressed in relationships (Pistole, 1989). In contrast to anxious/ambivalent and avoidant adults, secure adults are more satisfied in their relationships and use conflict strategies that focus on maintaining the relationship. Helping the relationship stay cohesive is more important than "winning" the battle. Secure adults are more likely to compromise than are anxious/ambivalent adults, and anxious ambivalent adults are more likely than avoidant adults to give in to their partners' wishes, whether they agree with them or not.

Fighting about Sex and Money

Even if, as the Russian writer Leo Tolstoy suggested, every unhappy family is unhappy in its own way, marital conflicts still tend to center around certain issues, especially communication, children, sex, money, personality differences, how to spend leisure time, in-laws, infidelity, and housekeeping. In this section, we focus on two areas: sex and money. Then we discuss general ways of resolving conflicts.

All happy families are happy in the same way. Each unhappy family is unhappy in its own way.

LEO TOLSTOY (1828–1910), *ANNA KARENINA*

Fighting about Sex

Fighting and sex can be intertwined in several different ways (Strong and DeVault, 1997). A couple can have a specific disagreement about sex that leads to a fight. One person wants to have sexual intercourse and the other does not, so they fight. A couple can have an indirect fight about sex. The woman does not have an orgasm, and after intercourse, her partner rolls over and starts to snore. She lies in bed feeling angry and frustrated. In the morning she begins to fight with her partner over his not doing his share of the housework. The housework issue obscures why she is really angry. Sex can also be used as a scapegoat for nonsexual problems. A husband is angry that his wife calls him a lousy provider. He takes it out on her sexually by calling her a lousy lover. They fight about their lovemaking rather than about the issue of his provider role. A couple can fight about the wrong sexual issue. A woman may berate her partner for being too quick during sex, but what she is really frustrated about is that he is not interested in oral sex with her. She, however, feels ambivalent about oral sex ("Maybe I smell bad"), so she cannot confront her partner with the real issue. Finally, a fight can be a cover-up. If a man feels sexually inadequate and does not want to have sex as often as his partner, he may pick a fight and make his partner so angry that the last thing she would want to do is to have sex with him.

Did you know?

The way in which a couple deals with conflict resolution both reflects and perhaps contributes to their marital happiness (Boland and Follingstad, 1987).

A word is not a sparrow. Once it flies, you can't catch it.

RUSSIAN PROVERB

In power struggles, sexuality can be used as a weapon, but this is generally a destructive tactic (Szinovacz, 1987). A classic strategy for the weaker person in a relationship is to withhold something that the more powerful one wants. In male-female struggles, this is often sex. By withholding sex, a woman gains a certain degree of power. A small minority of men also use sex in its most violent form: They rape (including date rape and marital rape) to overpower and subordinate women. In rape, aggressive motivations displace sexual ones.

We are always willing to fancy ourselves within a little of happiness and when, with repeated efforts we cannot reach it, persuade ourselves that it is intercepted by an ill-paired mate since, if we could find any other obstacle, it would be our own fault that it was not removed.

SAMUEL JOHNSON (1709–1784)

It is hard to tell during a fight if there are deeper causes than the one about which a couple is currently fighting. Are you and your partner fighting because you want sex now and your partner doesn't? Or are there deeper reasons involving power, control, fear, or inadequacy? If you repeatedly fight about sexual issues without getting anywhere, the ostensible cause may not be the real one. If fighting does not clear the air and make intimacy possible again, you should look for other reasons for the fights. It may be useful to talk with your partner about why the fights do not seem to accomplish anything. Step back and look at the circumstances of the fight; what patterns occur; and how each of you feels before, during, and after a fight.

Against all the evidence of his senses the man in love declares that he and his beloved are one, and is prepared to behave as if it were a fact.

SIGMUND FREUD (1856–1939)

Money Conflicts

An old Yiddish proverb addresses the problem of managing money quite well: "Husband and wife are the same flesh, but they have different purses." Money is a major source of marital conflict. Intimates differ about spending money probably as much as, or more than, any other single issue.

Why People Fight about Money. Couples disagree or fight over money for a number of reasons. One of the most important reasons has to do with power. Earning wages has traditionally given men power in families. A woman's work in the home has not been rewarded by wages. As a result, full-time homemakers have been placed in the position of having to depend on their husbands for money. In such an arrangement, if there are disagreements, the women is at a disadvantage. If she is deferred to, the old cliché "I make the money but she spends it" has a bitter ring to it. As women increasingly participate in the workforce, however, power relations within families are shifting. Studies indicate that women's influence in financial and other decisions increases if they are employed outside the home.

Some people are proud of their handful of justice and commit outrages against all things for its sake, till the world is drowned in their injustice.

FRIEDRICH NIETZSCHE (1844–1900)

Another major source of conflict is allocation of the family's income. Not only does this involve deciding who makes the decisions but it also includes setting priorities. Is it more important to pay a past-due bill or to buy a new television set to replace the broken one? Is a dishwasher a necessity or a luxury? Should money be put aside for long-range goals, or should immediate needs (perhaps those your partner calls "whims") be satisfied? Setting financial priorities plays on each person's values and temperament; it is affected by basic aspects of an individual's personality. A miser probably cannot be happily married to a spendthrift. Yet we know so little of our partner's attitudes toward money before marriage that a miser might very well marry a spendthrift and not know it until too late.

Dating relationships are a poor indicator of how a couple will deal with money matters in marriage. Dating has clearly defined rules about money: Either the man pays, both pay separately, or each pays alternatively. In dating situations, each partner is financially independent of the other. Money is not pooled, as it usually is in a committed partnership or marriage. Power issues do not necessarily enter spending decisions because each person has his or her own money. Differences can be smoothed out fairly easily. Both individuals are financially independent before marriage but financially interdependent after marriage. Even cohabitation may not be an accurate guide to how a couple would deal with money in marriage, as cohabitors generally do not pool all (or even part) of their income. It is the working out of financial interdependence in marriage that is often so difficult.

Epitaph on His Wife

Here lies my wife; here let her lie!

Now she's at rest.

And so am I.

JOHN DRYDEN (1631–1700)

Talking about Money. Talking about money matters is difficult. People are very secretive about money. It is considered poor taste to ask people how much money they make. Children often do not know how much money is earned in their families; sometimes spouses don't know either. One woman remarked that it is easier to talk with a partner about sexual issues than about money matters: "Money is the last taboo," she said. But, as with sex, our society is obsessed with money.

We find it difficult to talk about money for several reasons. First of all, we don't want to appear to be unromantic or selfish. If a couple is about to marry, a discussion of attitudes toward money may lead to disagreements, shattering the illusion of unity or selflessness. Second, gender roles make it difficult for women to express their feelings about money because women are traditionally supposed to defer to men in financial matters. Third, because men tend to make more money than women, women feel that their right to disagree about financial matters is limited. These feelings are especially prevalent if the woman is a homemaker and does not make a financial contribution, but they devalue her child-care and housework contributions.

Resolving Conflicts

There are a number of ways to end conflicts. You can give in, but unless you believe that the conflict ended fairly, you are likely to feel resentful. You can try to impose your will through the use of power, force, or the threat of force, but using power to end conflict leaves your partner with the bitter taste of injustice. Finally, you can end the conflict through negotiation. In negotiation, both partners sit down and work out their differences until they come to a mutually acceptable agreement. Conflicts can be solved through negotiation in three major ways: (1) agreement as a gift, (2) bargaining, and (3) coexistence.

An eye for an eye makes the whole world blind.

Mohandas K. Gandhi (1869–1948)

Agreement as a Gift

If you and your partner disagree on an issue, you can freely agree with your partner as a gift. If you want to go to the Caribbean for a vacation and your partner wants to go backpacking in Alaska, you can freely agree to go to Alaska. An agreement as a gift is different from giving in. When you give in, you do something you

Figure 5.2
Family Problem-Solving Loop

Most family problem-solving occurs in the ebb and flow of daily family events. Though family dynamics and transition take various forms, it is interesting to note which types might have relevance for family issues.

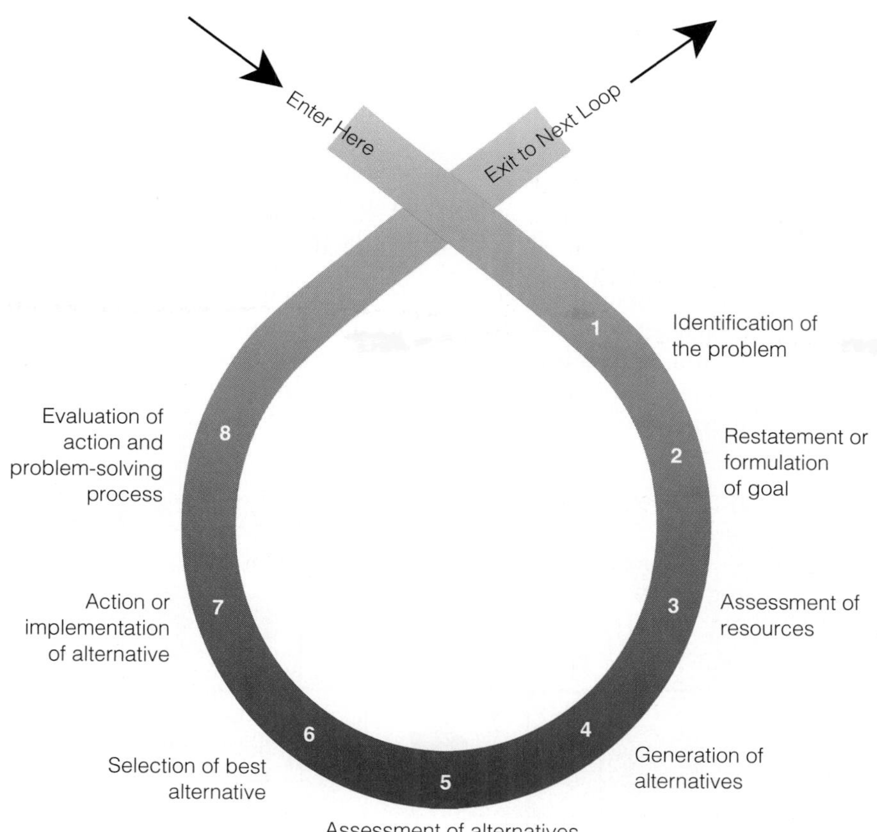

SOURCE: Kieren, D., T. O. Maguire, and N. Hurlbut. "A Marker Method to Test a Phasing Hypothesis in Family Problem-Solving Interaction." *Journal of Marriage and the Family* 58, 2 (May 1996): 442–455. Copyright © 1996 by the National Council on Family Relations. Used by permission.

Reflections

How did (does) your family solve problems? How realistic would this model be if you introduced it to your family?

don't want to do. When you agree without coercion or threats, the agreement is a gift of love, given freely without resentment. As in all exchanges of gifts, there will be reciprocation. Your partner will be more likely to give you a gift of agreement. This gift of agreement is based on referent power, discussed earlier.

Hatred does not cease by hatred at any time. Hatred ceases by love. This is an unalterable law.

SIDDARTHA GAUTAMA, THE BUDDHA (C. 563–483 B.C.)

Bargaining

Bargaining means making compromises, but bargaining in relationships is different from bargaining in the marketplace or in politics. In relationships, you don't want to get the best deal for yourself but rather the most equitable deal for both of you. At all points during the bargaining process, you need to keep in mind what is best for the relationship as well as for yourself, and you need to trust your partner to do the same. In a marriage, both partners need to win. The result of conflict in a marriage should be to solidify the relationship, not to make one partner the winner and the other the loser. To achieve your end by exercising coercive power or withholding love, affection, or sex is a destructive form of bargaining. If you get what you want, how will that affect your partner and the relationship? Will your partner feel you are being unfair and become resentful? A solution has to be fair to both, or it won't enhance the relationship.

Without forgiveness life is governed by . . . an endless cycle of resentment and retaliation.

ROBERT ASSAGIOLI

Coexistence

Sometimes differences can't be resolved, but they can be lived with. If a relationship is sound, differences can be absorbed without undermining the basic ties. All too often we regard a difference as a threat rather than as the unique expression of two personalities. Rather than being driven mad by the cap left off the toothpaste, perhaps we can learn to live with it.

Always forgive your enemies—nothing annoys them so much.

OSCAR WILDE (1854–1900)

If you can't talk about what you like and what you want, there is a good chance that you won't get either one. Communication is the basis for good relationships. Communication and intimacy are reciprocal: Communication creates intimacy, and intimacy, in turn, helps create good communication.

If we fail to communicate, we are likely to turn our relationships into empty facades, with each person acting a role rather than revealing his or her deepest self. But communication is learned behavior. If we have learned *how not to* communicate, we can learn *how to* communicate. Communication will allow us to maintain and expand ourselves and our relationships.

It doesn't cost anything to have loving speech.

VIETNAMESE SAYING

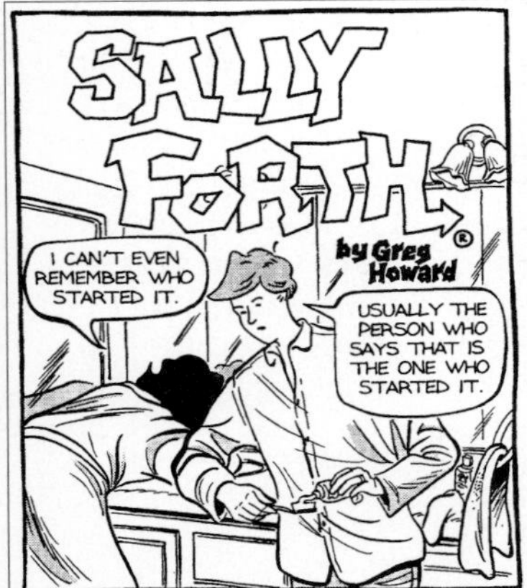

SOURCE: Reprinted with special permission of King Features Syndicate.

YOU AND YOUR WELL-BEING

Getting Psychological Help

In spite of good intentions and communication skills, not all couples are able to resolve their problems on their own. Accepting the need for professional assistance may be a significant first step toward reconciliation and change. Experts advise counseling when communication is hostile, conflict goes unresolved, individuals cannot resolve their differences, and/or a partner is thinking about leaving.

Marriage and partners counseling are professional services whose purpose is to assist individuals, couples, and families gain insight into their motivations and actions within the context of a relationship while providing tools and support to make positive changes. A skilled counselor offers objective, expert, and discreet help. Much of what counselors do is crisis or intervention oriented.

A valuable and perhaps more effective approach is preventive—undertaken to explore dynamics and behaviors before they cause more significant problems. This may occur at any point in relationships; during the engagement, prior to an anticipated pregnancy, or at the departure of a last child.

Each state has its own degree and qualifications for marriage counselors. The American Association for Marital and Family Therapy (AAMFT) is one association that provides proof of education and special training in marriage and family therapy. Graduate education from an accredited program in either social work, psychology, psychiatry, or human development coupled with a license in that field assures both education and training as well as offering the consumer recourse if questionable or unethical practices occur. This recourse is, however, only available if the practitioner holds a valid license issued by the state in which he or she practices. Mental health workers belong to any one of several professions:

- *Psychiatrists* are licensed medical doctors who, in addition to completing at least six years of post-baccalaureate medical and psychological training, can prescribe medication.
- *Clinical psychologists* have usually completed a Ph.D. degree requiring at least six years of postbaccalaureate course work. A license requires additional training and the passing of state boards.
- *Marriage and Family Counselors* typically have a masters degree and additional training to be eligible for state board exams.
- *Social Workers* have masters degrees requiring at least two years of graduate study plus additional training to be eligible for state board exams.
- *Pastoral Counselors* are clergy who have special training in addition to their religious studies.

Financial considerations may be one consideration when selecting which one of the above to see. Typically, the more training a professional has, the more he or she will charge for services. Be sure to check to see what kind of mental health benefits your personal health insurance carrier offers. Because there is variation in costs for services based on the professional's training, location, affiliation, and willingness to accept a sliding-fee scale, it is important to discuss fees and payment schedule in advance.

Finding a therapist can be done via a referral from a physician, school counselor, family, friend, clergy, or by the state department of mental health. In any case, it is important to meet personally with the counselor in order to decide if he or she is right for you. Besides inquiring about his or her basic professional qualifications, it is important to feel comfortable with this person, to decide whether your value and belief systems are compatible, and to assess his or her psychological orientation. Shopping for the right counselor may be as important a decision as deciding to enter counseling in the first place.

Marriage or partnership counseling has a wide variety of approaches: Individual counseling focuses on one partner at a time; joint marital counseling involves both people in the relationship; and family systems therapy includes as many family members as possible. Regardless of the approach, all share the premise that in order to be effective, those involved should be willing to cooperate. Additional logistical questions such as the number and frequency of sessions depends on the type of therapy.

At any time during the therapeutic process, you have the right to stop or change therapists. Before you do, however, ask yourself whether your discomfort is personal or if it has to do with the techniques or personality of the therapist. Discuss this issue with the therapist prior to making a change. Finally, if you believe that your therapy is not benefiting you, change therapists.

Summary

- Researchers are finding that how well a couple communicates before marriage can be an important predictor of later marital satisfaction. *Self-disclosure* prior to marriage is related to relationship satisfaction later. Whether a couple's premarital interactions are basically negative or positive can also predict later marital satisfaction.
- Research indicates that happily married couples (1) are willing to engage in conflict in nondestructive ways, (2) have less frequent conflict and spend less time in conflict, (3) disclose private thoughts and feelings to partners, (4) express equal levels of affection, (5) spend more time together, and (6) accurately encode and decode messages.
- There are gender differences in partnership communication. Wives tend to send clearer messages. Husbands may give neutral messages or withdraw whereas wives tend to give more positive or negative messages. Also, wives tend to set the emotional tone and escalate arguments more than husbands do.
- Communication includes both verbal and nonverbal communication. The functions of nonverbal communication are to convey interpersonal attitudes, express emotions, and handle the ongoing interaction. For communication to be clear, verbal and nonverbal messages must agree. *Proximity*, eye contact, and touch are important forms of nonverbal communication. Levels of touching differ between cultures and ethnic groups.
- Virginia Satir identified four styles of miscommunication. Placaters are passive, helpless, and always agreeable; blamers act superior, are often angry, do not listen, and try to escape responsibility; computers are correct, reasonable, and expressionless; and distractors are frenetic and tend to change the subject.
- Barriers to communication include the traditional male gender role (because it discourages the expression of emotion); personal reasons, such as feelings of inadequacy; and the fear of conflict. To express yourself, you need to be aware of your own feelings. We prevent self-awareness through suppressing, denying, and projecting feelings. A first step toward self-awareness is realizing that our feelings are neither good nor bad but simply emotional states.
- According to some researchers, both low and high levels of self-disclosure are related to low marital satisfaction (the curvilinear model). Other researchers maintain that a high level of self-disclosure is related to a high level of marital satisfaction, although it entails greater risks (linear model).
- *Trust* is the belief in the reliability and integrity of a person. For trust to develop, (1) a relationship has to exist and have the likelihood of continuing, (2) we must be able to predict how a person will likely behave, and (3) the person must also have other acceptable options available to him or her. Trust is critical in close relationships because self-disclosure requires trust; how much you trust a person influences the way you are likely to interpret ambiguous or unexpected messages from him or her.
- *Feedback* is the ongoing process in which participants and their messages create a given result and are subsequently modified by the result. Constructive feedback includes (1) focusing on "I" statements, (2) focusing on behavior rather than on the person, (3) focusing feedback on observations rather than on inferences or judgments, (4) focusing feedback on the observed incidence of behavior, (5) focusing feedback on sharing ideas or offering alternatives rather than on giving advice, (6) focusing feedback according to its value to the recipient, (7) focusing feedback on the amount the recipient can process, and (8) focusing feedback at an appropriate time and place.
- The basis of good communication in a relationship is mutual affirmation. Mutual affirmation includes mutual acceptance, mutual liking, and expressing liking in words and actions.
- Conflict is natural in intimate relationships. Types of conflict include basic versus nonbasic conflicts and situational versus personality conflicts. Basic conflicts may threaten the foundation of a marriage because they challenge fundamental rules; nonbasic conflicts do not threaten basic assumptions and may be negotiable. Situational conflicts are based on specific issues. Personality conflicts are unrealistic conflicts based on the need of the partner or partners to release pent-up feelings or on their fundamental personality differences.

- *Power* is the ability or potential ability to influence another person or group. Traditionally, legal as well as de facto power rested in the hands of the husband. Recently, wives have been gaining more actual power in relationships, although the power distribution still remains unequal. The six bases of marital power are coercive, reward, expert, legitimate, referent, and informational power. Other theories of power include the *relative love and need theory* and the *principle of least interest.*
- People usually handle anger in relationships by suppressing or venting it. Anger, however, makes negotiation difficult. When anger arises, it is useful to think of it as a signal that change is necessary.
- Happily married couples use certain techniques to resolve conflict: summarizing, paraphrasing, validation, and clarification. Unhappy couples use confrontation, confrontation and defensiveness, and complaining and defensiveness.
- Major sources of conflict include sex and money. Conflicts about sex can be specific disagreements about sex, indirect disagreements in which a partner feels frustrated or angry and takes it out in sexual ways, disagreements about the wrong sexual issue, or arguments that are ostensibly about sex but that are really about nonsexual issues. Money conflicts occur because of power issues, disagreements over the allocation of resources, or differences in values.
- Conflict resolution may be achieved through negotiation in three ways: agreement as a freely given gift, bargaining, or coexistence.

Key Terms

family rules 140
feedback 149
hierarchy of rules 140
honeymoon effect 137
meta-rules 140
power 155
principle of least interest 157
proximity 142
relative love and need theory 157
rules 140
self-disclosure 137
trust 148

Suggested Readings

Cupach, W. R., and B. H. Spitzberg, eds. *The Darkside of Interpersonal Communication.* Hillsdale, NJ: Lawrence Erlbaum, 1994. A discussion of conversational dilemmas, distressed marital relationships, and other issues that stress families.

Hecht, Michelle, Mary Jane Collier, and Sidney Ribeau. *African American Communication.* Newbury Park, CA: Sage Publications, 1993. A synthesis of research on African-American communication and culture, including effective and ineffective communication patterns.

Montagu, Ashley. *Touching: The Human Significance of the Skin.* 3d ed. New York: Harper & Row, 1986. The classic study by a distinguished anthropologist on the significance of touch in human relationships.

Notarius, Clifford, and Howard Markman. *We Can Work It Out: Making Sense of Marital Conflict.* New York: Putnam, 1993. Tips on improving communication and resolving conflicts from leading researchers on marital communication.

Satir, Virginia. *The New Peoplemaking.* Rev. ed. Palo Alto, CA: Science and Behavior Books, 1988. One of the most influential (and easy-to-read) books of the last twenty-five years on communication and family relationships.

Scarf, Maggie. *Intimate World: Life Inside the Family.* New York: Random House, 1995. Issues, such as conflict, power, and intimacy, explored by a counseling psychologist.

Tannen, Deborah. *You Just Don't Understand: Women and Men in Conversation.* New York: Morrow, 1990. A best-selling, intelligent, and lively discussion of how females use communication to achieve intimacy and males use communication to achieve independence.

Ting-Toomey, Stella, and Felipe Korzenny, eds. *Cross-Cultural Interpersonal Communication.* Newbury Park, CA: Sage Publications, 1991. A groundbreaking collection of scholarly essays on communication and relationships among different ethnic and cultural groups, including African-American, Latino, Korean, and Chinese ethnic groups and cultures.

CHAPTER 6

Pairing and Singlehood

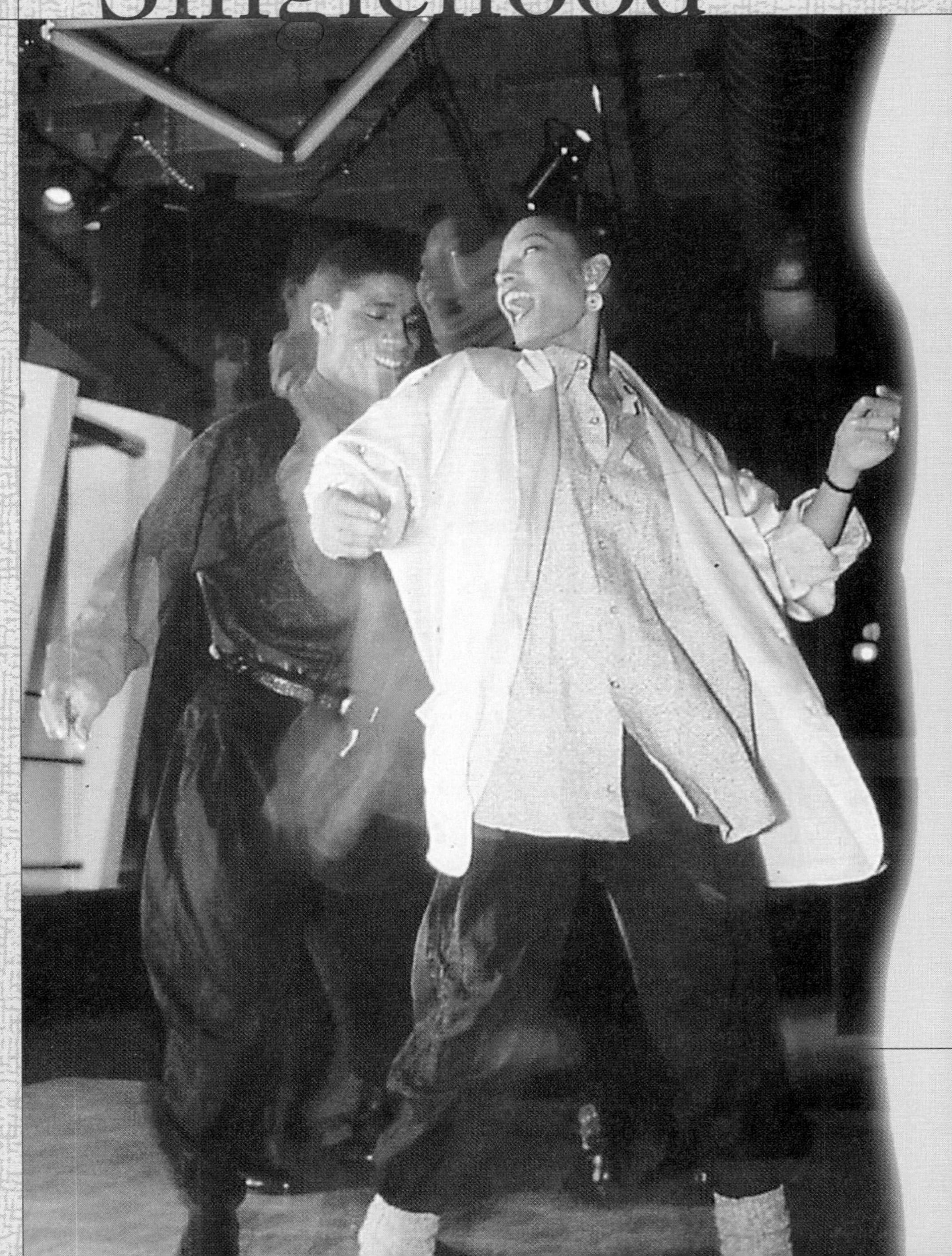

PREVIEW

To gain a sense of what you already know about the material covered in this chapter, answer "True" or "False" to the statements below.

1. Looking for a mate can be compared to shopping for goods in a market. True or false?
2. Generally, the most important factor in judging someone at the first meeting is how he or she looks. True or false?
3. There is a significant shortage of single eligible African-American men that makes marriage less likely for African-American women. True or false?
4. If a woman asks a man out on a first date, it is generally a sign that she wants to have sex with him. True or false?
5. The lesbian subculture values being single and unattached more than being involved in a stable relationship. True or false?
6. Singles, compared to their married peers, tend to be more dependent on their parents. True or false?
7. An important dating problem that men cite is their own shyness. True or false?
8. Cohabitation has become part of the courtship process among young adults. True or false?
9. Pooling money in cohabiting relationships indicates a high degree of commitment. True or false?
10. Previously married cohabitants are more likely than never-married cohabitors to view living together as a test of marital compatibility. True or false?

Answers

1 True, 2 True, 3 True, 4 False, 5 False, 6 False, 7 True, 8 True, 9 True, 10 True

OUTLINE

In theory, we have free choice in selecting our partners. In reality, however, our choices are somewhat limited by rules of mate selection. If we know what the rules are, we can make deductions about whom we are likely to choose as dates or partners.

There is a game you can play if you understand some of the principles of mate selection in our culture. Without ever having met a friend's new boyfriend or girlfriend, you can deduce many things about him or her, using basically the same method of deductive reasoning that Sherlock Holmes used to astound Dr. Watson. For example, if a female friend at college has a new boyfriend, it is safe to guess that he is about the same age or a little older, taller, and a college student; he is probably about as physically attractive as your friend (if not, they will probably break up within six months); his parents probably are of the same ethnic group and social class as hers; and most likely, he is about as intelligent as your friend. If a male friend has a new girlfriend, many of the same things apply, except that she is probably the same age or younger and shorter than he is. After you have described your friend's new romantic interest, don't be surprised if he or she exclaims, "Good grief, Holmes, how did you know that?" Of course, not every characteristic may apply, but you will probably be correct in most instances. These are not so much guesses as deductions based on the principle of homogamy discussed in this chapter.

In this chapter we not only look at the general rules by which we choose partners, but we also examine romantic relationships, the singles world, and living together. Over the last several decades, many aspects of pairing, such as the legitimacy of premarital intercourse and cohabitation, have changed considerably, radically affecting marriage. Today large portions of American society accept and approve of both premarital sex and cohabitation. Marriage has lost its exclusiveness as the only legitimate institution in which people can have sex and share their everyday lives. As a result, it has lost some of its power as an institution.

Choosing a Partner

How do we choose the people we date, live with, or marry? At first glance, it seems that we choose them on the basis of love, but other factors also influence us. These factors have to do with bargaining and exchange. We select our partners in what might be called a marriage marketplace.

The Marriage Marketplace

In the marriage marketplace, as in the commercial marketplace, people trade or exchange goods. Unlike a real marketplace, however, the marriage marketplace is not a place but a process. In the marriage marketplace, the goods that are exchanged are not tomatoes, waffle irons, or bales of cotton; we are the goods. Each of us has certain resources—such as socioeconomic status, looks, and personality—that make up our marketability. We bargain with these resources. We size ourselves up and rank ourselves as a good deal, an average package, or something to be remaindered; we do the same with potential dates or mates.

If you would marry suitably, marry your equal.

OVID (43 B.C.–A.D. 17)

The idea of exchange as a basis for choosing marital partners may not be romantic, but it is deeply rooted in marriage customs. In some cultures, for example, arranged marriages take place after extended bargaining between families. The woman is expected to bring a dowry in the form of property (such as pigs, goats, clothing, utensils, or land) or money, or a woman's family may demand a bride-price if the culture places a premium on women's productivity. Traces of the exchange basis of marriage still exist in our culture in the traditional marriage ceremony when the bride's father pays the wedding costs and "gives away" his daughter.

Gender Roles

The traditional marital exchange is related to gender roles. Traditionally, men have used their status, economic power, and role as protector in a trade-off for women's physical attractiveness and nurturing, childbearing, and housekeeping abilities; women, in return, have gained status and economic security in the exchange.

As women enter careers and become economically independent, the terms of bargaining change. When women achieve their own occupational status and economic independence, what do they ask from men in the marriage exchange? Constantina Safilios-Rothschild (1976b) suggests that many women expect men to bring more expressive, affective, and companionable resources into marriage. An independent woman does not have to settle for a man who brings little more to the relationship than a paycheck; she wants a man who is a companion, not simply a provider.

But even today, a woman's bargaining position is not as strong as a man's. In 1996, women earned only 71 percent of what men made (Smith, 1996). Women are still significantly underrepresented in the professions. Furthermore, many of the things that women traditionally used to bargain with in the marital exchange—children, housekeeping services, or sexuality—are today devalued or available elsewhere. Children are not the economic assets they once were. A man does not have to rely on a woman to cook for him, sex is often accessible in the singles world, and someone can be paid to do the laundry and clean the apartment.

Although we feel we choose our partners on the basis of "love," many other factors influence our choices.

Who ever heard of an ugly or pretty man? The beauty of a man is not in his face, but in his character, in his social standing, in his money. Who ever heard of a rich man that was ugly?

JORGE AMADO

Miss, *n. A title with which we brand unmarried women to indicate that they are in the market.*

Miss, Missis (Mrs.) and Mister (Mr.) are the three most distinctly disagreeable words in the language, in sound and sense. Two are corruptions of Mistress, the other of Master. In the general abolition of social titles in this country they miraculously escaped to plague us. If we must have them, let us be consistent and give one to the unmarried man. I venture to suggest Mush, abbreviated to Mh.

AMBROSE BIERCE (1842–1914?)

Women are at a further disadvantage because of the double standard of aging. Physical attractiveness is a key bargaining element in the marital marketplace, but the older a woman gets, the less attractive she is considered. For women, youth and beauty are linked in most cultures. As women get older, their field of eligible partners declines because men tend to choose younger women as mates.

Marriage Squeeze and Gradient

An important factor affecting the marriage market is the ratio of men to women. Researchers Guttentag and Secord (1983) argue that whenever there is a shortage of women in society, marriage and monogamy are valued; when there is an excess of women, marriage and monogamy are devalued. The scarcer sex is able to weight the rules in its favor. It gains bargaining power in the marriage marketplace.

The **marriage squeeze** refers to the gender imbalance reflected in the ratio of available unmarried women and men. Because of this imbalance, members of one gender tend to be "squeezed" out of the marriage market. The marriage squeeze is distorted, however, if we look at overall figures of men and women without distinguishing between age and ethnicity. Overall, there are significantly more unmarried women than men: 89 single men for every 100 single women (Saluter, 1995). This figure, however, is somewhat deceptive. From ages fifteen to thirty-nine years, the prime years for marriage, there are significantly more unmarried men than women, reversing the overall marriage squeeze. Women in this age group have greater bargaining power and are able to demand marriage and monogamy. But once ethnicity is taken into consideration, the marriage squeeze "squeezes" many African-American women of all ages out of the marriage market. With eligible males scarce, African-American men have greater bargaining power and are less likely to marry because of more attractive alternatives (Guttentag and Secord, 1983). (See Figure 6.1 and Perspective: "The African-American Male Shortage.")

"All the good ones are taken" is a common complaint of women in their mid-thirties and beyond, even if there are still more men than women in that age bracket. The reason for this is the **marriage gradient,** the tendency for women to marry men of higher status. Sociologist Jessie Bernard (1982) writes:

> In our society, the husband is assigned a superior status. It helps if he actually is superior in ways—in height, for example, or age or education or occupation—for such superiority, however slight, makes it easier for both partners to conform to the structural imperatives. The [woman] wants to be able to "look up" to her husband, and he, of course, wants her to. The result is a situation known sociologically as the marriage gradient.

Although we tend to marry those with the same socioeconomic status and cultural background, men tend to marry women slightly below them in age, education, and so on. Bernard continues:

> The result is that there is no one for the men at the bottom to marry, no one to look up to them. Conversely, there is no one for the women at the top to look up to; there are no men superior to them. . . . [T]he never-married men . . tend to be "bottom-of-the-barrel" and the women . . . "cream-of-the-crop."

The marriage gradient puts high-status women at a disadvantage in the marriage marketplace.

We tend to overlook the bargaining aspect of relationships because, as we said before, it doesn't seem very romantic. We bargain nevertheless, but usually

Figure 6.1
Ratio of Unmarried Men to Unmarried Women by Age and Ethnicity, 1991

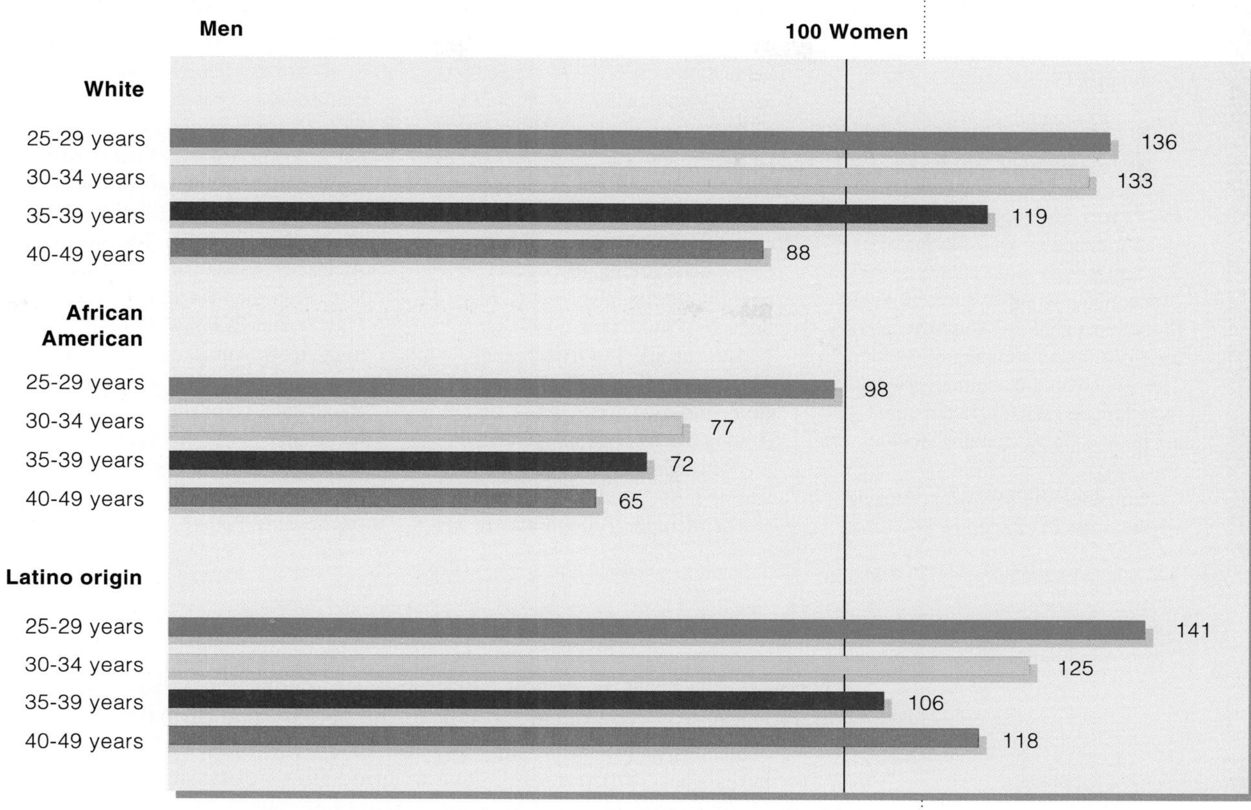

SOURCE: Saluter, Arlene F., "Marital Status and Living Arrangements: March 1991." U.S. Bureau of the Census, *Current Population Reports*, Series P20, U.S. Government Printing Office, Washington, D.C., 1992.

not openly or consciously. We exchange our social and personality characteristics. To get a sense of the complexity of bargaining in our culture, let's examine the role of physical attractiveness.

Physical Attractiveness

What is this thing called beauty? How does each gender define it and what role does it play in our relationship with others?

The Halo Effect

Imagine yourself unattached at a party. You notice that someone is standing next to you as you reach for some chips. He or she says hello. In that moment, you have to decide whether to engage him or her in conversation. On what basis do you make that decision? Is it looks, personality, style, sensitivity, intelligence, or something else?

The beautiful bird gets caged.

CHINESE PROVERB

PERSPECTIVE

The African-American Male Shortage

The ratio of males to females is an important factor affecting relationships and marriage rates. For most groups, there is an abundance of available males in the age groups most likely to marry. Among African Americans, however, there is a significant shortage of eligible males and an excess of women. This abundance of women may affect male/female relationships as well as the likelihood of many women marrying.

Consider the following: Among whites aged 25–29, there are 136 single men per 100 women; among Latinos, there are 141 men per 100 women. Among whites aged 30–34, there are 133 single men per 100 single women; among Latinos, the ratio is 125 single men for 100 women. By the simple law of supply and demand, white and Latina women can be more selective about their partners because the pool of single men is large. The percentage of married individuals is 63 percent among whites and 59 percent among Hispanics (U.S. Bureau of the Census, 1996).

But the story is different for African Americans: Single African-American men are in short supply. Among blacks aged 25–29, there are 98 single men per 100 women; among those aged 30–34, there are 77 men per 100 women. By age 40–44, there are only 65 single men per 100 women. There are an estimated 1.5 million more African-American women than men (Staples and Johnson, 1993).

There are several important consequences resulting from this. First, the percentage of married African Americans has declined significantly, from 64percent in 1971 to 43.2 percent in 1995 (U.S. Bureau of the Census, 1996). Second, single African-American women have sex less often than single white women (Tanfer and Cubbins, 1992). Third, many black women give birth while single and raise their children with the assistance of their extended family and the biological father (Staples, 1988). Births to single mothers are associated with enduring poverty. Fourth, interracial mar-

If you're like most of us, you consciously or unconsciously base this decision on appearance. If you decide to talk to the person, you probably formed a positive opinion about his or her appearance. In other words, he or she looked "cute," looked like a "fun person," gave a "good first impression," or seemed "interesting." As Elaine Hatfield and Susan Sprecher (1986) point out:

> Appearance is the sole characteristic apparent in every social interaction. Other information may be more meaningful but far harder to ferret out. People do not have their IQs tattooed on their foreheads, nor do they display their diplomas prominently about their persons. Their financial status is a private matter between themselves, their bankers, and the Internal Revenue Service.

Physical attractiveness is particularly important during the initial meeting and early stages of a relationship. If you don't know anything else about a person, you tend to judge him or her on appearance.

Most people would deny that they are attracted to others just because of their looks. We do so unconsciously, however, by inferring qualities based on looks. This inference is based on a **halo effect:** the assumption that good-looking people possess more desirable social characteristics than unattractive people. In one well-known experiment (Dion et al., 1972), students were shown pictures of attractive people and asked to describe what they thought these people were like. Attractive men and women were assumed to be more sensitive, kind, warm, sexually responsive, strong, poised, and outgoing than others; they were assumed to be more exciting and to have better characters than "ordinary" people. Research indicates that overall, the differences between perceptions of very attractive and average people are minimal. It is when attractive and average people are compared to those considered to be unattractive that there are pronounced differences, with those perceived as unattractive being rated more negatively (Hatfield and Sprecher, 1986).

Taught from infancy that beauty is woman's sceptre, the mind shapes itself to the body, and roaming around its gilt cage, only seeks to adorn its prison.

MARY WOLLSTONECRAFT (1759–1797), *VINDICATION OF THE RIGHTS OF WOMEN* riages have risen dramatically since 1980, especially between blacks and whites. Currently, 12.1 percent of all new marriages involving an African American are interracial with black men more likely to marry white women than the reverse (Besharov and Sullivan, 1996).

Inner-city African Americans have higher rates of singlehood than middle-class blacks. Among college-educated African Americans, the gender ratio is more extreme as people tend to marry those with similar educational backgrounds. Overall, the ratio of single college-educated black women is 2 to 1. For divorced black women over 35 with more than 5 years of college, the ratio of comparable men is 38 to 1 (Staples, 1991). Furthermore, significantly fewer African Americans marry because of the lack of eligible men. Not only are there fewer males because of the gender ratio, but because of lack of jobs or skills, they are often unemployed. Marriage among blacks is often a function of the male's being employed (Tucker and Taylor, 1989). More African Americans than whites are single because of social problems, such as the gender ratio and high unemployment, rather than a rejection of marriage (Tucker and Taylor, 1989).

The proportion of single males to females decreased sharply between 1970 and 1995 among African Americans, while it rose among both whites and Latinos (U.S. Bureau of the Census, 1996). Why the decline in African-American males? Sociologist Robert Staples (1988) points to the effects of institutional racism: high infant mortality, premature death, devastating homicide rates, poor health-care access, HIV/AIDS, and illegal drugs. High unemployment, disproportionate incarceration rates, increasing school dropout rates, and drug abuse further make affected young African-American men less desirable as mates. As Staples (1988) writes: "Due to the operational effects of institutional racism, large numbers of black males are incarcerated, unemployed, narcotized, or fall prey to early death. . . ."

It is important to establish social policies to reverse the devastation visited upon African Americans by discrimination and poverty. Such policies are important in reversing the gender imbalance and the consequences that have developed over the past 25 to 30 years.

The Rating and Dating Game

In casual relationships, the physical attractiveness of a romantic partner is especially important. Hatfield and Sprecher (1986) suggest three reasons why people prefer attractive people over unattractive ones. First, there is an "aesthetic appeal," a simple preference for beauty. Second, there is the "glow of beauty," in which we assume that good-looking people are more sensitive, kind, warm, modest, self-confident, sexual, and so on. Third, we achieve status by dating attractive people.

Several studies (cited in Hatfield and Sprecher, 1986) have demonstrated that good-looking companions increase our status. In one study, men were asked their first impressions of a man seen alone, arm-in-arm with a beautiful woman, and arm-in-arm with an unattractive woman. The man made the best impression with the beautiful woman. He ranked higher alone than with an unattractive woman. In contrast to men, women do not necessarily rank as high when seen with a handsome man. A study in which married couples were evaluated found that it made no difference to a woman's ranking if she was unattractive but had a strikingly handsome husband. If an unattractive man had a strikingly beautiful wife, it was assumed that he had something to offer other than looks, such as fame or fortune.

A proposal of marriage in our society tends to be a way in which a man sums up his social attributes and suggests to a woman that hers are not so much better as to preclude a merger or partnership in these matters.

Erving Goffman

Trade-Offs

People don't necessarily gravitate to the most attractive person in the room. Instead, they tend to gravitate to those who are about as attractive as themselves. Sizing up someone at a party or dance, a man may say, "I'd never have a chance with her; she's too good-looking for me." Even if people are allowed to specify the qualities they want in a date, they are hesitant to select anyone notably different from themselves in social desirability. We tend to choose people who are our equals in terms of looks, intelligence, education, and so on (Hatfield and Walster,

People tend to choose partners who are about as attractive as themselves.

1981). However, if two people are different in looks or intelligence, usually the individuals make a trade-off in which a lower-ranked trait is exchanged for a higher-ranked trait. A woman who values status, for example, may accept a lower level of physical attractiveness in a man if he is wealthy or powerful.

Are Looks Important to Everyone?

For all of us ordinary-looking people, it's a relief to know that looks aren't everything. Looks are most important to certain types or groups of people and in certain situations or locations (for example, in classes, at parties, and in bars, where people do not interact with one another extensively on a day-to-day basis). Looks are less important to those in ongoing relationships and to those older than young adults. Looks are also less important if there are regular interactions between individuals—for example, those who work together or commute in the same automobile (Hatfield and Sprecher, 1986). Looks tend to be especially important in adolescence and youth because of our need to conform. It is at this time that we are most vulnerable to pressure from our peers to go out with handsome men and beautiful women.

Men tend to care more about how their partners look than do women. This may be attributed to the disparity of economic and social power. Because men tend to have more assets (such as income and status) than women, they do not have to be concerned with their potential partner's assets. Therefore, they can choose partners in terms of their attractiveness. Because women lack the earning power and assets of men, they have to be more practical. They have to choose a partner who can offer security and status.

Love built on beauty, soon as beauty dies.

JOHN DONNE (1572–1631)

Most research on attractiveness has been done on first impressions or early dating. Researchers are finding, however, that attractiveness is important in established relationships as well as in beginning or casual ones. Most people expect looks to become less important as a relationship matures, but Philip Blumstein and Pepper Schwartz (1983) found that the happiest people in cohabiting and married relationships thought of their partners as attractive. People who found their partners attractive had the best sex lives. Physical attractiveness continues to be important throughout marriage.

I'm tired of all this nonsense about beauty being only skin-deep. That's deep enough. What do you want—an adorable pancreas?

JEAN KERR

The Field of Eligibles

The men and women we date, live with, or marry usually come from the **field of eligibles**—that is, those whom our culture approves of as potential partners. The field of eligibles is defined by two principles: **endogamy** (marriage within a particular group) and **exogamy** (marriage outside a particular group).

Endogamy

People usually marry others from a large group—such as the nationality, ethnic group, or socioeconomic status with which they identify—because they share common assumptions, experiences, and understandings. Endogamy strengthens group structure. If people already have ties as friends, neighbors, work associates, or fellow church members, a marriage between such acquaintances solidifies group ties. To take an extreme example, it is easier for two Americans to understand each other than it is for an American and a Fula tribesperson from Africa. Americans are monogamous and urban, whereas the Fula are polygamous wandering herders. But another, darker force may lie beneath endogamy: the fear

Reflections

How important are looks to you? Have you ever mistakenly judged someone by his or her looks? How did you discover your error? Have you ever made trade-offs in a relationship? What did you and your partner trade?

and distrust of outsiders, those who are different from ourselves. Both the need for commonality and the distrust of outsiders urge people to marry individuals like themselves.

Exogamy

The principle of exogamy requires us to marry outside certain groups—specifically, outside our own family (however defined) and outside our same sex. Exogamy is enforced by taboos that are deeply embedded within our psychological makeup. The violation of these taboos may cause a deep sense of guilt. A marriage between a man and his mother, sister, daughter, aunt, niece, grandmother, or granddaughter is considered incestuous; women are forbidden to marry their corresponding male relatives. Beyond these blood relations, however, the definition of incestuous relations changes. One society defines marriages between cousins as incestuous, whereas another may encourage such marriages. Some states prohibit marriages between stepbrothers and stepsisters, as well as cousins; others do not. The cultural presumption and legal precedent that we must marry someone of the other sex is currently being challenged by some lesbian and gay couples and gay rights organizations.

Heterogamy and Homogamy

Endogamy and exogamy interact to limit the field of eligibles. (See this chapter's "Understanding Yourself.") The field is further limited by society's encouragement of **homogamy,** the tendency to choose a mate whose personal or group characteristics are similar to ours. **Heterogamy** refers to the tendency to choose a mate whose personal or group characteristics differ from our own. The strongest pressures are toward homogamy. As a result, our choices of partners tend to follow certain patterns. These homogamous considerations generally apply to heterosexuals, gay men, and lesbians alike in their choice of partners.

There has been a growing tendency toward allowing individuals choice of partners without state interference. In 1966, the U.S. Supreme Court, for example, declared unconstitutional laws prohibiting marriage between individuals of different races (*Loving v. Virginia*). (About 20 percent of whites and 8 percent of African Americans continue to believe intermarriage should be illegal [Wilkerson, 1991].) Although state laws continue specifically to prohibit gay or lesbian marriages, the Hawaii Supreme Court has upheld a lower court decision declaring that such laws violate Hawaii's constitution, which guarantees equal protection not based on gender. As of this writing, the ruling is under appeal (*Daily Record,* 1997).

The most important elements of homogamy are race and ethnicity, religion, socioeconomic status, age, and personality characteristics. These elements are strongest in first marriages and weaker in second and subsequent marriages (Glick, 1988). They also strongly influence our choice of sexual partners, as our sexual partners are often potential marriage partners (Michael, Gagnon, Laumann, and Kolata, 1994).

Race and Ethnicity. Most marriages among whites and African Americans are between members of the same group. Of all new marriages in 1993 involving African Americans, 12.1 percent were interracial, up from 6.6 percent in 1980 and 2.6 percent in 1970 (Besharov and Sullivan, 1996). It is suggested that the reasons why both black groom/white bride and black bride/white groom are increasing is the rise of a black middle class, making both African-American men and women more attractive to middle-class whites. Black women still face obstacles to marriage of any kind; they are more than twice as likely to have children born

Did you know?

As husbands and wives age, changes in the wife's physical appearance seem to have greater impact on the quality of marital sexuality than do changes in the husband's appearance. Researchers found that changes in appearance led to greater sexual disinterest, a decline in sexual satisfaction, and, to a lesser extent, unfaithfulness, especially on the part of the husband (Margolin and White, 1987).

Fashionable ladies fall in love with acrobats and have to marry colonels. Shop assistants fall in love with countesses and are obliged to marry shop girls.

George Bernard Shaw (1856–1950)

After all there is but one race—humanity.

George Moore (1852–1933)

Figure 6.2
Percentages of Adults Living Alone by Age: 1970 and 1994

As more adults are choosing to live alone or in unmarried households, definitions of family are being expanded.

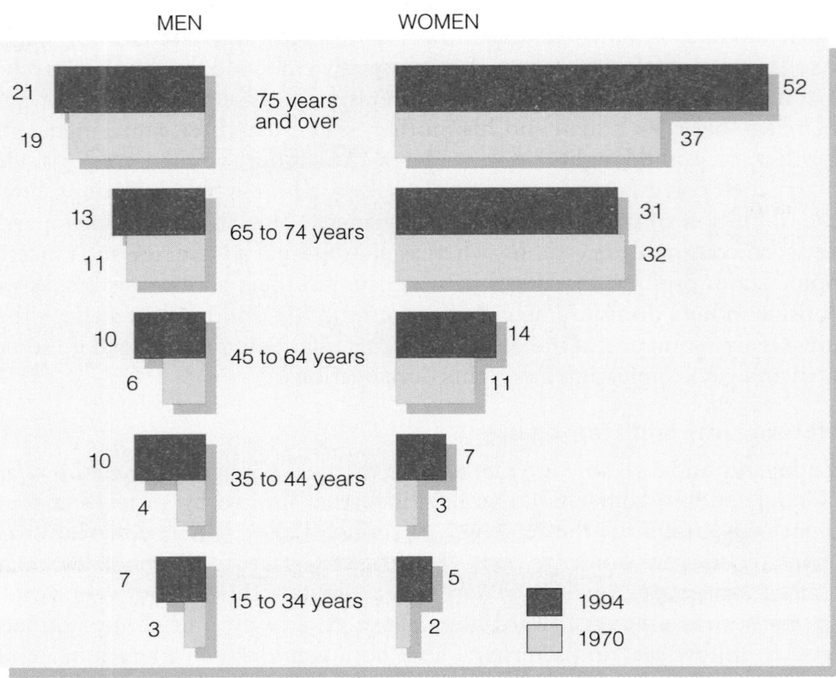

SOURCE: Glick and Norton, 1979; U.S. Population Reference Bureau, 1995.

out of wedlock. About 1.4 percent of marriages consist of one partner who is white and one from an Asian, Native American, or other nonwhite group.

The degree of intermarriage between ethnic groups is of concern to some members of these groups because it affects the rate of assimilation and continued ethnic identity (Stevens and Schoen, 1988). Almost half of all Japanese Americans marry outside their ethnic group (Takagi, 1994). Over half of all Native Americans are married to non-Native Americans (Yellowbird and Snipp, 1994). For both Japanese Americans and Native Americans, intermarriage leads to profound questions about their continued existence as distinct ethnic groups in the twenty-first century. Among European ethnic groups in this country, such as Italians, Poles, Germans, and Irish, only one in four marries within his or her ethnic group. The ethnic identity of these groups has decreased considerably since the beginning of this century.

Did you know?

Those without religious affiliations tend to accept cohabitation more than those who identify with religious groups (Thorton, Axinn, and Hill, 1992; Wu and Balakrishnan, 1992).

Religion. Until the late 1960s, religion was a significant factor in marital choice. Today, most religions still oppose interreligious marriage because they believe it weakens individual commitment to the faith. Nonetheless, interreligious dating and marriage have been increasing. Almost half of all Catholics marry outside their faith (Maloney, 1986). Almost 40 percent of Jews choose a non-Jewish partner, up from 6 percent in the early 1960s (Mindel, Haberstein, and Wright, 1988).

Those who marry from different religious backgrounds are at slightly greater risk of divorce than those from similar backgrounds (Bumpass, Martin, and Sweet, 1991; Lehrer and Chiswick, 1993; Sander, 1993).

Religious groups tend to discourage interfaith marriages, believing that such marriages, in addition to weakening individual beliefs, lead to children's being reared in a different faith or secularize the family. Such fears, however, may be overstated. Among Catholics who marry Protestants, for example, there seems to be little secularization by those who feel themselves to be religious (Petersen, 1986). Some who are from different religious backgrounds, however, do convert to their spouses' religions.

God enters by a private door into every individual.

RALPH WALDO EMERSON (1803–1882)

Socioeconomic Status. Most people marry those of their own socioeconomic status because of shared values, tastes, goals, occupations, and expectations. People tend to marry those who have the same educational background as themselves, reflecting the fact that educational attainment is related to socioeconomic status. Even if a person marries someone from a different ethnic group, religion, or age group, the couple will probably be from the same socioeconomic group. Women, more than men, sometimes marry below their group, possibly as a result of women's slightly lower status.

The man who marries for money earns it.

YIDDISH PROVERB

Age. People tend to marry others within their age group, although the man is usually slightly older than the woman. Age is important because we view ourselves as members of a generation, and each generation's experience of life leads to different values and expectations. Furthermore, different developmental and life tasks confront us at different ages. A twenty-year-old woman wants something different from marriage and from life than a sixty-year-old man does. By marrying people of similar ages, we often ensure congruence for developmental tasks.

Personality Characteristics. Social scientists have put to rest the old saying about the attraction of opposites. Instead, they have found that people tend to choose partners who share similar characteristics. Individuals who were impulsive, accurate reasoners, quarrelsome, or extroverted chose similar partners. The researchers hypothesized that choosing partners who share similar personality characteristics provides greater communication, empathy, and understanding.

These factors in the choice of partner interact with one another. Ethnicity and socioeconomic status, for example, are often closely related because of discrimination. Many African Americans and Latinos are working-class and are not as well educated as whites. Whites generally tend to be better off economically and are usually better educated. Thus, a marriage that is endogamous in terms of ethnicity is also likely to be endogamous in terms of education and socioeconomic status.

Marriages that are homogamous tend to be more stable than heterogamous ones. There are three possible explanations as to why heterogamous relationships tend to be unstable (Udry, 1974):

1 Heterogamous couples may have considerably different values, attitudes, and behaviors, which may create a lack of understanding and promote conflict.

2 Heterogamous marriages may lack approval from parents, relatives, and friends. Couples are then cut off from important sources of support during crises.

3 Heterogamous couples are probably less conventional and therefore less likely to continue an unhappy marriage for the sake of appearances.

Reflections

Think, in terms of heterogamy and homogamy, about those who are or have been your romantic or marital partners. In what respects have your partners shared the same racial, ethnic, religious, socioeconomic, age, and personal traits with you? In what respects have they not? Have shared or differing characteristics affected your relationships? How?

The Stages of Mate Selection: Stimulus-Value-Role Theory

Let's say that you meet someone who fits all the criteria of homogamy: same ethnic group, religion, socioeconomic background, age, and personality traits. Homogamously speaking, he is "Mr. Right," she is "Ms. Right," and your children would be "Little Rights." He or she is the person your parents dream of your marrying. And, of course, you can't stand each other. Homogamy by itself doesn't work. You need to start with "magic," or, more precisely, according to Bernard Murstein (1976, 1987), you need to start with stimulus, the chemistry of attraction that stimulates you to want to have a relationship with another.

Murstein developed a three-stage theory—the **stimulus-value-role theory**—to explain the development of romantic relationships. This theory is based on exchange. Each partner evaluates the other; if the exchange seems more or less equitable, the two will progress to the next stage. According to the theory, couples go through the following stages in a defined sequence.

In the *stimulus* stage, each person is drawn or attracted to the other before actual interaction. This attraction can be physical, mental, or social; it can be based on a person's reputation or status. The stimulus stage, according to Murstein, is most prominent on the first encounter, when one has little information with which to evaluate the other person.

In the next stage, the *value* stage, partners weigh each other's basic values, trying to determine if they are compatible. Each person tries to figure out the other's philosophy of life, politics, sexual values, religious beliefs, and so on. If both highly value rap music, it is a plus for the relationship. If they disagree on religion—one is a dedicated fundamentalist and the other, a goddess follower—it is a minus for the relationship. Each person adds or subtracts the pluses and minuses along value lines. Based on the outcome, the couple will either disengage or go on to the next stage. Values are usually determined between the second and seventh meetings.

In the *role* stage, each person analyzes the other's behaviors, or how the person fulfills his or her roles as lover, companion, friend, worker—and potential husband or wife, mother or father. Are the person's behaviors consistent with marital roles? Is he or she emotionally stable? This aspect is evaluated in the eighth and subsequent encounters.

Although the stimulus-value-role theory has been one of the most prominent theories explaining relationship development, it has been criticized by some scholars. Most notably, scholars ask if men and women really test their degree of fit (Huston et al., 1981). For example, religious fundamentalists and goddess worshipers may sometimes believe that they are compatible. They may not discuss religion; instead they might focus on the "incredible" physical attraction in their relationship. Or they may make errors in arithmetic; they may mistakenly believe that religion is not that important, only to discover after they are married that it is very important.

Romantic Relationships

Dreams are the touchstones of our character.

Henry David Thoreau (1817–1862)

As more and more people delay marriage, never marry, or seek to remarry after divorce or widowhood, romantic relationships, observes one scholar (Surra, 1991), "will take different shapes at different points in time, as they move in and out of marriage, friendship, romance, cohabitation, and so on." As a result, researchers are shifting from the traditional emphasis on mate selection toward the study of the formation and development of romantic relationships, such as the dynamics of heterosexual dating, cohabitation, postdivorce relationships, and gay and lesbian relationships. The field of personal relationships is developing a broad focus that explores relationship dynamics (Duck, 1993; Kelley et al., 1983; Perlman and Duck, 1987).

Beginning a Relationship: Seeing, Meeting, and Dating

Although the general rules of mate selection are important in the abstract, they do not tell us how relationships begin. The actual process of beginning a relationship is discussed below.

Seeing

On a typical day, we may see dozens, hundreds, or thousands of men and women. But seeing isn't enough; masses of people blur into one another. We must become aware of someone for a relationship to begin. It may only take a second from the moment of noticing to meeting, or sometimes it may take days, weeks, or months until you meet. Sometimes "noticing" each other is simultaneous; other times it may take considerable time; sometimes it never happens.

The setting in which you see someone can facilitate or discourage meeting each other (Murstein, 1976, 1987). **Closed fields,** such as small classes or seminars, dormitories, parties, and small workplaces, are characterized by a small number of people who are likely to interact whether they are attracted or not. In such settings, you are likely to "see" and interact more or less simultaneously. **Open fields,** in contrast, are characterized by large numbers of people who do not ordinarily interact with one another, such as beaches, shopping malls, bars, raves, amusement parks, and large university campuses.

Did you know?

Among those who are willing to try nontraditional means of meeting a partner, such as video dating services, computer bulletin boards, and so on, there may be less emphasis on homogamy (LaBeff et al., 1989).

Meeting

How is a meeting initiated? Among heterosexuals, does the man initiate it? On the surface, the answer appears to be yes, but in reality, the woman often "covertly initiates . . . by sending nonverbal signals of availability and interest" (Metts and Cupach, 1989). A woman will glance at a man once or twice and catch his eye; she may smile or flip her hair. If the man moves into her physical space, the woman then relies on nodding, leaning close, smiling, or laughing (Moore, 1985).

If the man believes the woman is interested, he then initiates a conversation using an opening line. The opening line tests the woman's interest and availability. Men use an array of opening lines. According to women, the most effective are innocuous, such as "I feel a little embarrassed, but I'd like to meet you" or "Are you a student here?" The least effective are sexual come-ons, such as "You really turn me on. Do you want to have sex?" Women, more than men, prefer direct but innocuous opening lines over cute, flippant ones, such as "What's a good-looking babe like you doing in a college like this?"

Men are the most likely to initiate a meeting directly, whereas women are more likely to wait for the other to introduce himself or herself or to be introduced by a friend (Berger, 1987). About a third or half of all relationships rely on introductions (Sprecher and McKinney, 1993). An introduction has the advantage of a kind of prescreening, as the mutual acquaintance may believe that both may hit it off. Parties are the most common settings in which young adults meet, followed by classes, work, bars, clubs, sports settings, or events centered around hobbies, such as hiking (Marwell et al., 1982; Shostak, 1987; Simenauer and Carroll, 1982). Technology via the Internet is now gaining popularity as people introduce themselves in fantasylike images.

UNDERSTANDING YOURSELF

Computer Dating and Homogamy

Compatibility Plus, a computer dating service, distributes this questionnaire to its clients. Complete the questionnaire and then examine it in terms of the social and personality characteristics you and your "ideal match" share. To what extent do you find homogamy operating? What do you think would be your chances of liking a person who was your "ideal match?" What other characteristics would be important to you?

Questionnaire Instructions

Each question has two headings: "YOURSELF" and "IDEAL MATCH."

- For "YOURSELF" mark the box on the left that best describes you.
- Then, mark all the boxes on the right that best describe your "IDEAL MATCH."
- Please answer all the questions. If none of the choices is the exact answer you wish to give, then mark the answer that comes closest.

1 Age _____yrs.

2 Minimum age acceptable _____yrs.
Maximum age acceptable _____yrs.

3 Height _____ft. _____in.

4 Minimum height acceptable _____ft. _____in.

5 Sex
- ☐ 1) Male ....☐
- ☐ 2) Female ....☐

6 Race
- ☐ 1) White ....☐
- ☐ 2) Hispanic ....☐
- ☐ 3) Asian ....☐
- ☐ 4) African American ......☐
- ☐ 5) Other _____ ....☐ (write in)

7 Body Build
- ☐ 1) Heavy ....☐
- ☐ 2) Moderately heavy ......☐
- ☐ 3) Average ....☐
- ☐ 4) Moderately thin ....☐
- ☐ 5) Other _____ ....☐ (write in)

8 Religion
- ☐ 1) Catholic ....☐
- ☐ 2) Protestant ....☐
- ☐ 3) Jewish ....☐
- ☐ 4) Eastern Mysticism .....☐
- ☐ 5) Atheist or Agnostic ...☐
- ☐ 6) Other _____ ....☐ (write in)

9 Economic Bracket
- ☐ 1) Low income ....☐
- ☐ 2) Average income ....☐
- ☐ 3) Above average ....☐
- ☐ 4) Much above average ....☐

10 Furthest Education
- ☐ 1) Grade School ....☐
- ☐ 2) High School ....☐
- ☐ 3) Some College ....☐
- ☐ 4) Graduated College .....☐
- ☐ 5) Post-graduate study ...☐

11 Diet
- ☐ 1) No restrictions ....☐
- ☐ 2) No red meat ....☐
- ☐ 3) Kosher ....☐
- ☐ 4) Macrobiotic ....☐
- ☐ 5) Vegetarian ....☐

12 Tobacco use
- ☐ 1) Nonsmoker (only) ......☐
- ☐ 2) Light smoker ....☐
- ☐ 3) Regular smoker ....☐

Single men and women increasingly rely on personal classified ads, where men advertise themselves as "success objects" and women advertise themselves as "sex objects" (Davis, 1990). Their ads tend to reflect stereotypical gender roles. Men advertise for women who are attractive and deemphasize intellectual, work, and financial aspects. Women advertise for men who are employed, financially secure, intelligent, emotionally expressive, and interested in commitment. Men are twice as likely as women to place ads.

Additional forms of meeting others include video dating services, introduction services, computer bulletin boards, and 900 party-line phone services. (One video dating service claims to have over 150,000 members, who pay an average of $2,000 for its services; over 9,000 of its members have married each other.)

Single men and women often rely on their churches and church activities to meet other singles. Black churches are especially important for middle-class African Americans, as they have less chance of meeting other African Americans in integrated work and neighborhood settings. They also attend concerts, plays, film festivals, and other social gatherings oriented toward African Americans (Staples, 1991).

For lesbians and gay men, the problem of meeting is exacerbated by the fact that they cannot necessarily assume that the person in whom they are interested shares their orientation. Instead, they must rely on identifying cues, such as

13 Children Living with You
- ☐ 1) None .... ☐
- ☐ 2) One .... ☐
- ☐ 3) Two .... ☐
- ☐ 4) Three or more .... ☐

14 Political Outlook
- ☐ 1) Liberal .... ☐
- ☐ 2) Moderate .... ☐
- ☐ 3) Conservative .... ☐
- ☐ 4) Progressive-Radical ..... ☐
- ☐ 5) Other _____ .... ☐ (write-in)

15 Affection
- ☐ 1) Very affectionate .... ☐
- ☐ 2) Moderately affectionate .... ☐
- ☐ 3) Mildly affectionate ...... ☐
- ☐ 4) Nondemonstrative ...... ☐
- ☐ 5) Other _____ .... ☐ (write in)

16 Sense of Humor
- ☐ 1) Very funny or witty ..... ☐
- ☐ 2) Average .... ☐
- ☐ 3) Mild .... ☐

17 Assertiveness
- ☐ 1) Very assertive .... ☐
- ☐ 2) Moderately assertive ..... ☐
- ☐ 3) Mildly assertive .... ☐
- ☐ 4) Nonassertive .... ☐

18 General Disposition
- ☐ 1) Easygoing-flexible .... ☐
- ☐ 2) Moderate .... ☐
- ☐ 3) Firm .... ☐

19 Favorite Parties
- ☐ 1) Loud and lively .... ☐
- ☐ 2) Quiet and formal .... ☐
- ☐ 3) Small and intimate .... ☐
- ☐ 4) None—prefer one to one .... ☐

20 Dancing
- ☐ 1) Love it .... ☐
- ☐ 2) Like it .... ☐
- ☐ 3) Mild interest .... ☐
- ☐ 4) Dislike it .... ☐

21 Sexual Activity
- ☐ 1) Very important .... ☐
- ☐ 2) Moderately important .... ☐
- ☐ 3) Not essential .... ☐

22 Relationship Desired
- ☐ 1) Committed one to one ... ☐
- ☐ 2) Friends and lovers .... ☐
- ☐ 3) Nonsexual friendship .... ☐
- ☐ 4) Just sexual .... ☐

23 Favorite Activities and Interests
(Check all you enjoy)
☐ swimming ☐ dancing ☐ yoga
☐ surfing ☐ walking
☐ meditation ☐ sailing
☐ partying ☐ astrology
☐ bicycling ☐ playing music
☐ photography ☐ jogging
☐ listening to music ☐ art
☐ hiking ☐ concerts ☐ cooking
☐ skiing ☐ plays ☐ politics
☐ tennis ☐ movies ☐ shopping
☐ racquetball ☐ beach ☐ reading
☐ volleyball ☐ traveling ☐ bars
☐ aerobics ☐ golf ☐ television

24 Personality Traits
(Check all that describe you)
☐ sexy ☐ intelligent ☐ emotional
☐ spontaneous ☐ considerate
☐ possessive ☐ romantic
☐ imaginative ☐ frugal ☐ playful
☐ conventional ☐ polite
☐ positive ☐ patient ☐ decisive
☐ energetic ☐ talkative
☐ reserved ☐ inquisitive
☐ generous ☐ moody ☐ sociable
☐ sincere ☐ serious ☐ athletic
☐ tolerant ☐ anxious ☐ relaxed
☐ loyal ☐ lazy ☐ open
☐ witty ☐ demanding

25 Additional Remarks:

____
____
____
____

meeting at a gay or lesbian bar, wearing a gay/lesbian pride button, participating in lesbian/gay events, or being introduced by friends to others identified as being gay or lesbian (Tessina, 1989). Once a like orientation is established, gay men and lesbians usually engage in nonverbal processes to express interest. Lesbians and gay men both tend to prefer innocuous opening lines. To prevent awkwardness, the opening line usually does not make an overt reference to orientation unless the other person is clearly lesbian or gay.

Dating

For many of us, asking someone out for the first time is not easy. Shyness, fear of rejection, and traditional gender roles that expect women to wait to be asked may fill us with anxiety and nervousness. (Sweaty palms and heart palpitations are not uncommon when asking someone out the first time.) Both men and women contribute, although sometimes differently, to initiating a first date. Men are more likely to ask directly for a date: "Want to go see a movie?" Women are often more indirect. They hint or "accidentally on purpose" run into the other person: "Oh, what a surprise to see you *here* studying for your marriage and family midterm!"

Although women may initiate dates, they do so less frequently than do men (Berger, 1987). Many women believe that the male prerogative for initiating the

Reflections

What are some of the settings in which you "see" people? How do the settings affect the strategies you use to meet others? How do you move from meeting someone to going out with him or her? What are your feelings at each stage of seeing, meeting, and dating?

YOU AND YOUR WELL-BEING

The Role of Self-Esteem in Partner Selection

Given the choice between a partner who has a strong sense of worth, uniqueness, connectedness, and power, and one who lacks these qualities, the most obvious would be the one with high self-esteem. Are the above listed qualities necessary for a successful partnership? Most psychologists seem to think so (Dion and Dion, 1988).

Ideally, positive self-concept begins in childhood where a sense of being loved and loving are developed and nurtured by parents. Personality characteristics and mannerisms combine with stability to form the groundwork for self-image and self-esteem. Terms like "good," "capable," and "lovable"—the ideal self—hopefully take root over "bad," "incapable," and "unworthy of love." Each of the qualities of self-concept profoundly influence interpersonal relations. However, since one cannot undo the past, it is important to start from where you are and concentrate on those challenges that can lead to long-term mental wellness.

People who feel good about themselves are likely to live up to their positive self-image and reap the benefits of successes. A realistic view of personal self-worth, however, should not be confused with conceit or egocentrism. These latter are founded on low self-esteem and involve the concern with oneself, without regard for the well-being of others.

Individuals with positive self-esteem are capable of physical and emotional intimacy. They can expose their feelings and thoughts and accept others'. In contrast, people with low self-esteem often experience an insatiable need for affection, are more sensitive to criticism, and appear to be more vulnerable to rejection.

People who believe in themselves and in the people around them are the same ones who develop successful intimate relationships. They are willing to give of themselves, have the capacity to change, and are more trusting, accepting, and appreciative of others. Only those who can accept themselves can truly love others.

---

first date is outdated in an age of gender equality, yet they are often reluctant to violate it. In her study of college women, Mirra Komarovsky (1985) writes:

> The strongest sanction against violating the male prerogative of the first move was the male interpretation of such initiative as a sexual come-on. Men described such aggressive women as "sluts." Indignant as the women were to this inference, they hesitated to expose themselves to the risks, unless they were among the very few who did so with full knowledge of the implications.

Power in Dating Relationships

Power doesn't seem to be a concern for most people in dating relationships. In fact, one study of dating couples found that both men and women thought that they had about equal power in their relationships (Sprecher, 1985). If power does become a source of conflict, it may not become as intense an issue as in a marriage. Strong power conflicts can be avoided by dissolving the relationship in question.

An honest man can feel no pleasure in the exercise of power over his fellow citizens.

THOMAS JEFFERSON (1743–1826)

In marriage, the person with the most economic resources usually has the most power. Generally, this means the man because he is more likely to be employed and to earn more money than the woman. In dating relationships, however, economic resources do not appear to be as important a power resource for men as their ability to date other women (Sprecher, 1985). The easier men think it will be for them to go out with other women, the more power they have in a dating relationship.

The fact that men have more power, prestige, and status than do women cannot help but to affect heterosexual interactions. (Henley and Freeman, 1995). The question of who initiates, who touches, and who terminates sexual advances prescribes male leadership and dominance. Even though many people do not wish to have unequal sexual relationships, modes of expression and resistance and difficulty in changing communication patterns help to maintain an edge of inequality and imbalance among women. For equality to occur, women need to determine what they wish to express and how they wish to keep those behaviors that give them strength.

Problems in Dating

Dating is often a source of both fun and intimacy, but a number of problems may be associated with it.

Female/Male Differences

Divergent gender-role conceptions may complicate dating relationships. Often when two people lack complementary gender-role conceptions, the woman is more egalitarian and the man, more traditional.

As we saw earlier, it is difficult for women to initiate dates directly. They usually wait to be asked; and when they are asked, it may be by the wrong person. Another problem is who pays when going out on a date. Some women in Mirra Komarovsky's (1985) study feared that male acquaintances would be put off if they offered to pay their share. Other women who offered to pay, whether traditional or egalitarian, found themselves mocked by their dates. Some men who allowed their dates to pay still insisted on choosing where they went, whether the women wanted to go there or not. Still other men allowed their dates to pay but not publicly; instead, for example, the women secretly slipped money to them under the dinner table.

In a random study of 227 women and 107 men in college, nearly 25 percent of the women said that they received unwanted pressure to engage in sex, usually before the establishment of an emotional bond between the couple. Women rated places to go (23 percent) and communication (22 percent) almost as high on their list of dating problems as sexual pressure (Knox and Wilson, cited in Knox, 1991). Komarovsky (1985) cites male sexual pressure as a major barrier between men and women. A woman who wants to see a man again faces a dilemma: how to encourage him to ask her out again without engaging in more sexual activity than she wants. In Komarovsky's study, men whose sexual advances were rejected by their dates often salved their hurt egos by accusing the women of having sexual hang-ups or being lesbians.

For men, the number one problem, cited by 35 percent, was communicating with their dates (Knox and Wilson, cited in Knox, 1991). Men often felt that they didn't know what to say, or they felt anxious about the conversation's dragging. Communication may be a particularly critical problem for men because traditional gender roles do not encourage the development of intimacy and communication skills among males. A second problem, shared by almost identical numbers of men and women, was where to go. A third problem, named by 20 percent of the men but not mentioned by women, was shyness. Although men can take the initiative to ask for a date, they also face the possibility of rejection. For shy men, the fear of rejection is especially acute. A final problem—and again one not shared by women—was money, cited by 17 percent of the men. Men apparently accept the idea that they are the ones responsible for paying for a date.

Extrarelational Sex in Dating and Cohabiting Relationships

You don't have to be married to be unfaithful (Blumstein and Schwartz, 1983; Hansen, 1987; Laumann, Gagnon, Michael, and Michaels, 1994). Both cohabiting couples and couples in committed relationships usually have expectations of

Reflections

Look at power in your romantic relationships. When a disagreement occurs, who generally wins? Does it depend on the issue? When one person wants to go to the movies and the other wants to go to the beach, where do you end up going? If one wants to engage in sexual activities and the other doesn't, what happens?

OTHER PLACES OTHER TIMES

Love vs. Arranged Marriage in Cross-Cultural Perspective

Is love the natural or ideal basis for marriage? Certainly in American society, marriage without love is considered somewhat shocking, even scandalous. It is the subject of soap operas and whispered gossip: "He married her for her money." "She married his family name." Although we might consider marriage without love an exceptional case, anthropologists tell us that in traditional cultures most people do not consider love a particularly sound basis for marriage.

Marriage customs vary dramatically across cultures, and marriage means very different things in different cultures. If we consider how marriages come about—how they are "arranged"—we find that it is usually not the bride and groom themselves who have decided they would marry, as is the case in our own society today. Typically, the elders have done the matchmaking. In most cultures, the parents of the bride and groom were charged with arranging the marriage of their children. In some cultures, mothers were the primary matchmakers, as in traditional Iroquois culture. In others, fathers had a dominant voice in arranging marriage, as in traditional Chinese society. In still other cultures, the pool of elders involved in matchmaking was more extensive, including grandparents, aunts, uncles, and even local political and religious authorities, such as tribal chiefs and clan leaders.

In traditional societies, marriage was important family business. In fact, marriage was a major event in the life of two families—both the bride's and the groom's—as well as for the clan, tribe, and community to which each family belonged. Because marriage united two families, there were important matters to be taken into account before agreeing to any particular match. Typically, the economic or class standing of both families, as well as their social reputation or honor, was a paramount concern. Families needed to know how a particular marriage would affect the family as a whole. Will this marriage create ties to the chief's family that will bring political advantage to our family? How many cattle will the groom's family send us? Will the bride bring a large dowry of household goods and furniture? Will this marriage bring trade with our tribe? Can we count on his or her clan for additional workers at harvest time and additional strength in times of war?

With issues of this magnitude at stake, marriage could not be left up to the young people themselves. Sentimental feelings of love would certainly cloud their judgment. Marriage was not primarily a personal or intimate event focused on a young couple alone. The feelings and love between an individual bride and groom were subordinate to the greater interests and welfare of the family, clan, and community.

Anthropologists report that in most traditional cultures, newly married couples did not live separately from their parents. Part of the experience of getting married in our society is striking out on our own and setting up house together, independent of our parents. However, in traditional societies the new couple was expected to reside with the parents of either the bride or the groom, depending on the culture. In some cultures, aunts and uncles also lived in the extended family households into which the newly married couple settled. For instance, in many matrilineal cultures—where related females formed the backbone of families and clans—it was customary for the bride and groom to reside with the bride's family. In traditional Iroquois society, it was the groom who moved into the bride's longhouse, which included the bride's mother and father, her maternal grandparents, her mother's sisters,

There are two tragedies in life. One is to lose your heart's desire. The other is to gain it.

GEORGE BERNARD SHAW (1856–1950)

sexual exclusiveness. But, like some married men and women who take vows of fidelity, they do not always remain sexually exclusive. Blumstein and Schwartz (1983) found that those involved in cohabiting relationships had similar rates of extrarelational involvement as did married couples, except that cohabiting males had somewhat fewer partners than husbands did. Gay men had more partners than did cohabiting and married men, and lesbians had fewer partners than any other group.

Large numbers of both men and women have sexual involvements outside dating relationships that are considered exclusive. One study of college students (Hansen, 1987) indicated that over 60 percent of the men and 40 percent of the women had been involved in erotic kissing outside a relationship; 35 percent of the men and 11 percent of the women had had sexual intercourse with someone and their husbands and children. By contrast, in patrilineal cultures—where related males formed the basis of families and clans—the newly married couple settled in the home of the groom's family. In traditional Chinese society, the bride moved in with her husband, his parents and paternal grandparents, his father's brothers, and their wives and children. In traditional societies like these, the newly married couple was not independent but rather dependent on family elders, remaining under their authority and protection.

Let us look more closely at the meaning of marriage and the role of love in one culture: Bedouin society in northern Egypt along the Libyan border. Traditionally, the Bedouin were nomadic pastoralists whose livelihood depended on herding sheep and goats and on trade with neighboring peoples for agricultural products. Bedouin culture is patrilineal. Bedouin camps are composed of extended families formed around related males—typically a grandfather, his adult sons and their wives, and his grandsons and unmarried granddaughters. On marriage, the bride leaves her family and goes to live with her husband and his relatives in their camp.

If marriage in American culture is based on cultural notions about individualism and love, marriage in Bedouin society is based on notions of family honor and duty. The elder males in the Bedouin family arrange a marriage to maximize family honor. They try to establish a link through marriage with an honorable family—one that is strong in numbers of men; has large herds of sheep and established business connections; is independent but able to support an impressive number of wives, children, and clients; and is of good blood and overall reputation. Moreover, the bride herself must have personal qualities that will enhance her husband's family. She is first under the authority of her husband's parents and his older male relatives, and she must know her place as a daughter-in-law. She must work hard and obey and serve her in-laws. She must show proper respect for all of her husband's relatives, conducting herself modestly through proper veiling and deferential behavior. In so doing, she will reflect honor on her husband and his family.

As in other patrilineal societies, an important concern of the male elders in Bedouin society is the protection of their authority and control over junior family members. They are also concerned with the promotion of a sense of loyalty and duty among the next generation of brothers, who will one day form the core of the family. This demands that the relationship between husband and wife be subordinated to family interests. A wife must not cause trouble between her husband and his father or his brothers. She should not win her husband's heart to the extent that he will forget his primary loyalties to his father, elders, and brothers. Thus, romantic attachment or love is viewed by Bedouin elders as potentially threatening to their own control over junior family members. Excessive love and passion can make a man disobedient, disrespectful, and even dependent, lessening both his and his family's honor.

Bedouin marriages are usually arranged between a young man and woman who belong to different camps, thus creating blind marriages. That is, bride and groom typically have not met prior to their engagement and marriage. The practice of arranging blind marriages further enhances the control and authority of the older generation over the young couple. Without ever having met, two people can hardly be in love at the time of their marriage.

In the less frequent case where bride and groom have perhaps made slight acquaintance—on the occasion of other weddings or celebrations—it is still very unlikely that they will be in love at the time of their own marriage. Indeed, if family elders were to learn that affection or attraction existed between a young man and woman, they would not agree to the marriage. That would be asking for trouble. Of course, the emotion, attraction, and commitment that we mean by the word *love* may in time develop between husband and wife. However, in Bedouin society, as in most traditional societies, love is neither a necessary nor an advisable condition in arranging a good marriage.

else. Of those who knew of their partner's affair, a large majority felt that it had hurt their own relationship. When both partners had engaged in affairs, each believed that their partner's affair had harmed the relationship more than their own had. Both men and women seem to be unable to acknowledge the negative impact of their own outside relationships. It is not known whether those who tend to have outside involvement in dating relationships are also more likely to have extramarital relationships after they marry.

Breaking Up

"Most passionate affairs end simply," Hatfield and Walster (1981) note. "The lovers find someone they love more." Love cools; it changes to indifference or

Did you know?

In 1995, 20.3 million persons lived alone, or 12 percent of all adults. (U.S. Bureau of the Census, 1996).

hostility. This may be a simple ending, but it is also painful because few relationships end by mutual consent. For college students, breakups are more likely to occur during vacations or at the beginning or end of the school year. Such timing is related to changes in the person's daily living schedule and the greater likelihood of quickly meeting another potential partner.

Breaking up is rarely easy, whether you are initiating or "receiving" it. If you initiate a breakup, thinking about the following may help:

- *Be sure that you want to break up.* If the relationship is unsatisfactory, it may be because conflicts or problems have been avoided or have been confronted in the wrong way. Conflicts or problems, instead of being a reason to break up, may be a rich source of personal development if they are worked out. Sometimes people erroneously use the threat of breaking up as a way of saying, "I want the relationship to change."
- *Acknowledge that your partner will be hurt.* There is nothing you can do to erase the pain your partner will feel; it is only natural. Not breaking up because you don't want to hurt your partner may actually be an excuse for not wanting to be honest with him or her or with yourself.
- *Once you end the relationship, do not continue seeing your former partner as "friends" until considerable time has passed.* Being friends may be a subterfuge for continuing the relationship on terms wholly advantageous to yourself. It will only be painful for your former partner because he or she may be more involved in the relationship than you. It may be best to wait to become friends until your partner is involved with someone else (and by then, he or she may not care if you are friends or not).
- *Don't change your mind.* Ambivalence after ending a relationship is not a sign that you made a wrong decision; neither is loneliness. Both indicate that the relationship was valuable for you.

The bonds that unite another person to ourself exist only in our mind.

MARCEL PROUST (1871–1922)

If your partner breaks up with you, keep the following in mind:

- *The pain and loneliness you feel are natural.* Despite their intensity, they will eventually pass. They are part of the grieving process that attends the loss of an important relationship, but they are not necessarily signs of love.
- *You are a worthwhile person, whether you are with a partner or not.* Spend time with your friends; share your feelings with them. They care. Do things that you like; be kind to yourself.
- *Keep a sense of humor.* It may help ease the pain. Repeat these clichés: No one ever died of love. (Except me.) There are other fish in the ocean. (Who wants a fish?)

Reflections

When you have broken up with someone, what coping strategies have you used? Which were effective? Which were ineffective? Why?

Singlehood

If you are lonely while you are alone, you are in bad company.

JEAN-PAUL SARTRE (1905–1980)

Each year more and more adult Americans are single. Rates of singlehood vary by ethnicity: In 1995 among whites, 20.6 percent had never married; among African Americans; 38.4 percent were single; and among Latinos; 28.6 percent had not married (U.S. Bureau of the Census, 1996).

The happiest of all lives is a busy solitude.

VOLTAIRE (1694–1778), LETTER TO FREDERICK THE GREAT, 1751.

Almost 75 million adult Americans are unmarried (U.S. Bureau of the Census, 1996). They represent a diverse group: never married, divorced, young, old, single parents, gay men, lesbians, widows, widowers, and so on, and they live in diverse situations that affect how they experience their singleness. The varieties of unmarried lifestyles in the United States are too complex to examine under any single category. In research on singles, however, those who are generally regarded as "single" are young or middle-aged, heterosexual, not living with someone, and working rather than attending school or college. Although there are, of course, numerous single lesbians and gay men, they have not traditionally been included in such research.

Singles: An Increasing Minority

A quick observation of demographics in this country point to a new and increasing way of life: singlehood. The trend, which has taken root and grown substantially since 1960, includes divorced, widowed, and never-married individuals.

The percentage of never-married adults has grown from 21.4 percent of the total population in 1970 to 43.9 percent in 1995 (U.S. Bureau of the Census, 1996). This increase cuts across all population groups and represents a change in the way in which society now views this way of life. Further review of demographics reveals an increasing number of never-married young adults and of formerly married singles. With nearly 75 million singles over the age of seventeen, singles appear to be postponing marriage to an age which, for many, makes better economic and social sense (U.S. Bureau of the Census, 1996).

Did you know?

The majority of unmarried adults under twenty-four years of age live with someone else, usually their parents (U.S. Bureau of the Census, 1996).

The growing divorce rate is also contributing to the numbers of singles. In 1995, 8 percent of men and 10.3 percent of women 18 and over were divorced (U.S. Bureau of the Census, 1996). The proportion of widowed men and women has declined somewhat but still remains similar to past numbers. Among older people, singlehood is most often obtained by the death of a spouse rather than by choice. Nevertheless, as society moves toward valuing individualism and choice, the numbers composing singlehood will most likely continue to grow.

The number of single adults is rising as a result of several factors (Buunk and van Driel, 1989; Macklin, 1987):

- Delayed marriage, with a median age at first marriage of 26.7 years for men and 24.5 years for women in 1994 (Saluter, 1996). The longer one postpones marriage, the greater the likelihood of never marrying. It is estimated that between 8 and 9 percent of men and women now in their twenties will never marry.
- Expanded lifestyle and employment options currently open to women.
- Increased rates of divorce and decreased likelihood of remarriage, especially among African Americans.
- Increased number of women enrolled in colleges and universities.
- More liberal social and sexual standards.
- Uneven ratio of unmarried men to unmarried women.

Relationships in the Singles World

When people form relationships within the singles world, both the man and woman tend to remain highly independent. Singles work, and as a result, the individuals tend to be economically independent of each other. They may also be more emotionally independent because much of their energy may already be heavily invested in their work or careers. The relationship that forms consequently tends to emphasize autonomy and egalitarian roles. The fact that single women work is especially important. Single women tend to be more involved in their work, either from choice or necessity, but the result is the same: They are accustomed to living on their own without being supported by a man.

The emphasis on independence and autonomy blends with an increasing emphasis on self-fulfillment, which, some critics argue, makes it difficult for some

Did you know?

People who favor cohabitation also tend to favor marriage (Hobart, 1993).

Table 6.1 Pushes and Pulls Toward Marriage and Singlehood

Pushes	Pulls
Toward Marriage	
Cultural norms	Love and emotional security
Loneliness	Physical attraction and sex
Parental pressure	Desire for children
Economic pressure	Desire for extended family
Social stigma of singlehood	Economic security
Fear of independence	Peer example
Media images	Social status as "grown-up"
Guilt over singlehood	Parental approval
Toward Singlehood	
Fundamental problems in marriage	Freedom to grow
Stagnant relationship with spouse	Self-sufficiency
Feelings of isolation with spouse	Expanded friendships
Poor communication with spouse	Mobility
Unrealistic expectations of marriage	Career opportunities
Sexual problems	Sexual exploration
Media images	

SOURCE: Adapted from Peter J. Stein. "Singlehood: An Alternative to Marriage." *The Family Coordinator* 24:4 (1975).

Reflections

For you, what are the pushes and pulls between singlehood and marriage? If you are unmarried, what type of single would you classify yourself? Why?

to make commitments. Commitment requires sacrifice and obligation, which may conflict with ideas of "being oneself." A person under obligation can't necessarily do what he or she "wants" to do; instead, a person may have to do what "ought" to be done (Bellah et al., 1985).

According to Barbara Ehrenreich (1984), men are more likely to flee commitment, because they need women less than women need men. They feel oppressed by their obligation to be the family breadwinner. In the marital exchange, argues Ehrenreich, men need women less than women need men, because men make more money than women and can obtain many of the "services" provided by wives—such as cooking, cleaning, intimacy, and sex—outside marriage without being tied down by family demands and obligations. Thus, men may not have a strong incentive to commit, marry, or stay married.

Nevertheless, a new study (Marker, 1996) revealed that flying solo at midlife for men appears to be more problematic than for women. Single women appeared to have better psychological well-being than did single men. For those who were socialized during an era of traditional gender roles and family values with marriage as the norm, there seemed to be a degree of mental health risk associated with singlehood, especially for men.

Culture and the Individual versus Marriage

Despite the importance of intimate relationships for our development as human beings, our culture is ambivalent about marriage. This is nothing new. Paul of Tarsus (1 Cor. 7:7–9) declared that it was best for people to remain chaste, as he himself had done. "But if they cannot," he wrote, "let them marry: For it is better to marry than to burn."

The tension between the alternatives of singlehood and marriage is diminishing as society begins to view singlehood as an option rather than a deviant lifestyle. The singles subculture is glorified in the mass media; the marriages portrayed on television are situation comedies or soap operas abounding in extramarital affairs. Yet many are rarely fully satisfied with being single and yearn for marriage. They are pulled toward the idea of marriage by their desires for intimacy, love, children, and sexual availability. They are also pushed toward marriage by parental pressure, loneliness, and fears of independence. Married persons, at the same time, are pushed toward singlehood by the limitations they feel in married life. They are attracted to singlehood by the possibility of creating a new self, having new experiences, and achieving independence.

The land of marriage has this peculiarity, that strangers are desirous of inhabiting it, whilst its natural inhabitants would willingly be banished from thence.

MICHEL DE MONTAIGNE (1533–1592)

Types of Never-Married Singles

Much depends on whether a person is single by choice and whether he or she considers being single a temporary or permanent condition (Shostak, 1987). If one is voluntarily single, his or her sense of well-being is likely to be better than that of a person who is involuntarily single. Arthur Shostak (1981, 1987) has divided singles into four types. Singles may shift from one type to another at different times. The four types are as follows:

- *Ambivalents.* Ambivalents are voluntarily single and consider their singleness temporary. They are not seeking marital partners, but they are open to the idea of marriage. These are usually younger men and women who are actively pursuing education, career goals, or "having a good time." Ambivalents may be included among those who are cohabiting.
- *Wishfuls.* Wishfuls are involuntarily and temporarily single. They are actively seeking marital partners but have been unsuccessful so far. They consciously want to be married.
- *Resolveds.* Resolved individuals regard themselves as permanently single. A small percentage are priests, nuns, or single parents who prefer rearing their children alone. The largest number, however, are "hard-core" singles who simply prefer the state of singlehood.
- *Regretfuls.* Regretful singles would prefer to marry but are resigned to their "fate." A large number of these are well-educated, high-earning women over forty who find a shortage of similar men as a result of the marriage gradient.

A number of sitcoms, such as Seinfeld, Living Single, Ellen, Martin, *and* Friends *(shown here) portray singles as young professionals in search of romance.*

All but the resolveds share an important characteristic: They want to move from a single status to a romantic couple status. "The vast majority of never-married adults," writes Shostak (1987), "work at securing and enjoying romance." Never-married singles share with married Americans "the high value they place on achieving intimacy and sharing love with a special one."

Singles: Myths and Realities

Cargan and Melko (1982), in a study of 400 households in Dayton, Ohio, examined various myths and realities about singlehood. They found that the following statements are myths about singlehood:

- *Singles are dependent on their parents.* Few differences exist between singles and marrieds in their perceptions of their parents and relatives. They do not differ in perceptions of parental warmth or openness and differ only slightly in the amount and nature of parental conflicts.

- *Singles are self-centered.* Singles value friends more than do married people. Singles are also more involved in community service projects.
- *Singles have more money.* Fewer than half the singles interviewed made more than $20,000 a year (adjusted to 1995). Married couples were better off economically than singles, in part because both partners often worked.
- *Singles are happier.* Singles tend to believe that they are happier than marrieds, whereas marrieds believe that they are happier than singles. Single men, however, exhibited more signs of stress than did single women.
- *Singles view singlehood as a lifetime alternative.* The majority of singles expected to be married within five years. They did not view singlehood as an alternative to marriage but as a transitional time in their lives.

Did you know?

In 1995, 20.3 million persons lived alone, or 12 percent of all adults (U.S. Bureau of the Census, 1996).

Cargan and Melko also determined that the following statements do characterize singlehood accurately:

- *Singles don't easily fit into married society.* Singles tend to socialize with other singles. Married people think that if they invite singles to their home, they must match them up with an appropriate single member of the other sex. Married people tend to think in terms of couples.
- *Singles have more time.* Singles are more likely to go out twice a week and much more likely to go out three times a week compared with their married peers. Singles have more choices and more opportunities for leisure-time activities.
- *Singles have more fun.* Although singles tend to be less happy than marrieds, they have more "fun." Singles go out more often, engage more in sports and physical activities, and have more sexual partners than do marrieds. Apparently, fun and happiness are not equated.
- *Singles are lonely.* Singles tend to be more lonely than married people; the feeling of loneliness is more pervasive for the divorced than for the never married.

Gay and Lesbian Singlehood

In the late nineteenth century, groups of gay men and lesbians began congregating in their own clubs and bars. There, in relative safety, they could find acceptance and support, meet others, and socialize. By the 1960s, some neighborhoods in the largest cities (such as Christopher Street in New York and the Castro district in San Francisco) became identified with gay men and lesbians. These neighborhoods feature not only openly lesbian or gay bookstores, restaurants, coffee houses, and bars, but also clothing stores, physicians, lawyers, hair salons—even driver's schools. They have gay churches, such as the Metropolitan Community Church, where gay men and lesbians worship freely; they have their own political organizations, newspapers, and magazines (such as *The Advocate*). They have family and child-care services oriented toward the needs of the gay and lesbian communities; they have gay and lesbian youth counseling programs.

In these neighborhoods men and women are free to express their affection as openly as heterosexuals; they experience little discrimination or intolerance; they are more involved in lesbian or gay social and political organizations. More recently, with increasing acceptance in some areas, many middle-class lesbians and gay men are moving to suburban areas. In the suburbs, however, they remain more discreet than in the larger cities (Lynch, 1992).

The urban gay male subculture that emerged in the 1970s emphasized sexuality. Although relationships were important, sexual experiences and variety were more important (Weinberg and Williams, 1974). This changed with the rise of the HIV/AIDS epidemic. Beginning in the 1980s, the gay subculture placed an increased emphasis on the relationship context of sex (Carl, 1986; Isensee, 1990). Relational sex has become normative among large segments of the gay population (Levine, 1992). Most gay men have sex within dating or love relationships. (In fact, some AIDS organizations are giving classes on gay dating in order to encourage safe sex.) One researcher (Levine, 1992) says of the men in his study: "The relational ethos fostered new erotic attitudes. Most men now perceived coupling, monogamy, and celibacy as healthy and socially acceptable."

Beginning in the 1950s and 1960s, young and working-class lesbians developed their own institutions, especially women's softball teams and exclusively female gay bars as places to socialize (Faderman, 1991). During the late 1960s and 1970s, **lesbian separatists,** lesbians who wanted to create a separate "womyn's" culture distinct from heterosexuals *and* gay men, rose to prominence. They developed their own music, literature, and erotica; they had their own clubs and bars. But by the middle of the 1980s, according to Lilian Faderman (1991), the lesbian community underwent a "shift to moderation." The community became more diverse, including Latina, African-American, Asian-American, and older women. It has developed closer ties with the gay community. They now view gay men as sharing much in common with them because of the common prejudice directed against both groups.

In contrast to the gay male subculture, the lesbian community centers its activities around couples. Lesbian therapist JoAnn Loulan (1984) writes: "Being single is suspect. A single woman may be seen as a loser no one wants. Or there's the 'swinging single' no one trusts. The lesbian community is as guilty of these prejudices as the world at large."

Lesbians tend to value the emotional quality of relationships more than the sexual components. Lesbians usually form longer-lasting relationships than gay men (Tuller, 1988). Lesbians' emphasis on emotions over sex and the more enduring quality of their relationships reflects their socialization as women. Being female influences a lesbian more than being gay.

Cohabitation

Cohabitation appears to be here to stay. Its increase across all socioeconomic, racial, and age groups makes it no longer a moral issue but rather a family lifestyle.

The Rise of Cohabitation

In 1994, more than 3.6 million, or 5 percent of the population, of unmarried heterosexual couples were living together in the United States (U.S. Population Reference Bureau, 1995; Saluter, 1994). In contrast, only 400,000 heterosexual couples were cohabiting in 1960 (See Figure 6.3). About half of people married for the first time have cohabited some time before marriage. This percentage is even higher among separated and divorced people (Bumpass and Sweet, 1995).

Figure 6.3
Cohabitation: 1960 to 1994

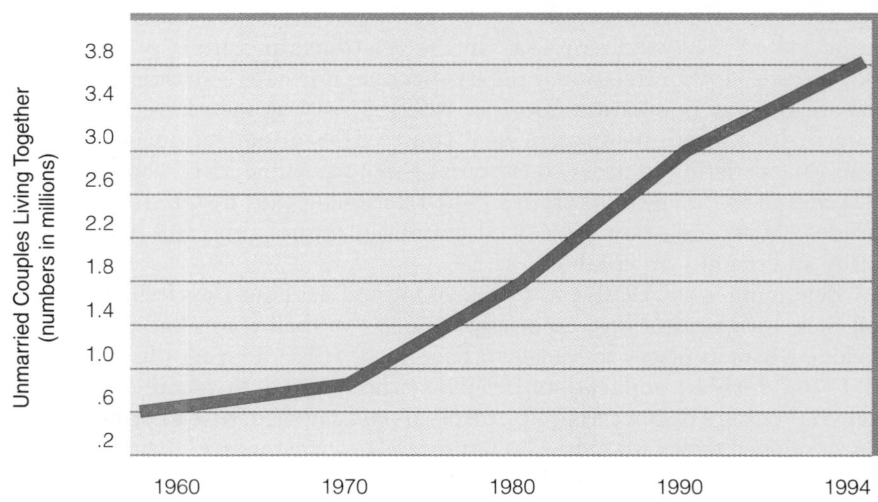

SOURCE: Glick and Norton, 1979; U.S. Population Reference Bureau, 1995.

Types of Cohabitation

There is no single reason to cohabit, just as there is no single type of person who cohabits. At least seven different reasons to cohabit are described below:

- *Temporary casual convenience.* Two people share the same living quarters because it is expedient and convenient to do so.
- *Affectionate dating or going together.* If two people live together because they enjoy being with each other, they will continue living together as long as it is mutually satisfying.
- *Economic advantage or necessity.* Even though approximately 80 percent of cohabitants are under age forty-five, older people find that they can retain their financial benefits by cohabiting if they are not married.
- *Trial marriage.* This includes persons who are "engaged to be engaged," as well as couples who are trying to discover if they want to marry each other.
- *Respite from being single.* For many, cohabiting offers a comfortable and oftentimes more familiar domestic lifestyle.
- *Temporary alternative to marriage.* Cohabiting may offer an alternative when the timing of a marriage must be postponed.
- *Permanent alternative to marriage.* Living together is an acceptable alternative for those who reject traditional marriage.

Living together has become more widespread and accepted in recent years for several reasons:

- *The general climate regarding sexuality is more liberal than it was a generation ago.* Sexuality is more widely considered to be an important part of a person's life, whether or not he or she is married. The moral criterion for judging sexual intercourse has shifted; love rather than marriage is now widely regarded as making a sexual act moral.
- *The meaning of marriage is changing.* Because of the dramatic increase in divorce over the last twenty-five years, marriage is no longer thought of as a necessarily permanent commitment. Permanence is increasingly replaced

by serial monogamy—a succession of marriages. Because the average marriage now lasts only seven years, the difference between marriage and living together is losing its sharpness.

- *Men and women are delaying marriage longer.* As long as children are not desired, living together offers advantages for many couples. When the couple want children, however, they will usually marry so that the child will be "legitimate."

Cohabitation does not seem to threaten marriage. Eleanor Macklin (1987), one of the major researchers in the field, notes:

> Nonmarital cohabitation in the United States serves primarily as a part of the courtship process and not as an alternative to marriage. The great majority of young persons plan to marry at some point in their lives . . . and most cohabiting relationships either terminate or move into legal marriage within a year or two.

The most notable social impact of cohabitation is that it delays the age of marriage for those who live together. As a consequence, cohabitation may actually encourage more stable marriages because the older a person is at the time of marriage, the less likely he or she is to divorce. Concomitantly, as the age of marriage continues to increase, cohabitation may become the preferred form for premarital bonding among those who are beyond the early adult years (Cate and Lloyd, 1992).

The course of true love never did run smooth.

SHAKESPEARE (1564–1616), *A MIDSUMMER NIGHT'S DREAM*

Although there are a number of advantages to cohabitation, there may also be disadvantages. Parents may refuse to provide support for school as long as their child is living with someone, or they may not welcome their child's partner into their home. Cohabiting couples may also find that they cannot easily buy houses together, as banks may not count their income as joint; they also usually don't qualify for insurance benefits. If one partner has children, the other partner is usually not as involved with the children as he or she would be if they were married. Cohabiting couples who live together may find themselves socially stigmatized if they have a child. Finally, cohabiting relationships generally don't last more than two years; couples either break up or get married.

Living together takes on a different quality among those who have been previously married. About 40 percent of cohabiting relationships have at least one previously married partner, and about 30 percent of all cohabiting couples have children from their earlier relationships. As a result, the motivation in these relationships is often colored by painful marital memories and the presence of children (Bumpass and Sweet, 1990). In these cases, men and women tend to be more cautious about making their commitments. The majority of remarriages are preceded by cohabitation (Ganong and Coleman, 1994). Even though cohabiting couples are less likely to stay together compared to married couples, having children in the household somewhat stabilizes the couples (Wu, 1995).

Domestic Partnerships

Domestic partners, cohabiting heterosexual, lesbian, and gay couples in committed relationships are gaining some legal rights. Domestic partnership laws, which grant some of the protection of marriage to cohabiting partners, are increasing the legitimacy of cohabitation. In 1997, San Francisco extended health insurance and other benefits to their employees' domestic (which includes same-sex) partners. It is the first such city ordinance in the nation. Individual employers, such as the Gap, Levi Strauss & Co., and the Walt Disney Company already have started domestic partner policies. In Europe, Sweden, Iceland, and Denmark legally recognize domestic partnerships. In Denmark and Iceland lesbians and gay men have the same rights and responsibilities in domestic partnerships as do people in marriage, except in adoption and child custody.

Domestic partners, whether heterosexual, gay, or lesbian, are denied many legal rights that come automatically with marriage. According to Curry, Clifford, and Leonard (1994), these include the right to:

- File joint tax returns.
- Automatically make medical decisions if your partner is injured or incapacitated.
- Automatically inherit your partner's property if he or she dies without a will.
- Enter hospitals, jails, and other places restricted to "immediate family."
- Create a marital life estate trust.
- Claim the unlimited marital deduction from estate taxes.
- Receive survivor's benefits.
- Obtain health and dental insurance, bereavement leave, and other employment benefits.
- Collect unemployment benefits if you quit your job to move with a partner who has obtained a new job.
- Claim family partnership income.
- Recover damages based on an injury to your partner.
- Live in neighborhoods zoned "family only."
- Get residency status for a noncitizen partner to avoid deportation.

Gay and Lesbian Cohabitation

In 1995, there were over 1.5 million gay or lesbian couples living together. The relationships of gay men and lesbians have been stereotyped as less committed than heterosexual couples because (1) lesbians and gay men cannot legally marry, (2) they may not emphasize sexual exclusiveness, and (3) heterosexuals misperceive love between gay and lesbian couples as being somehow less "real" than love between heterosexual couples.

There is enormous pressure to keep gay people defined solely by our sexuality, which prevents us from presenting our existence in political terms.

VITO RUSSO

Numerous similarities exist between gay and heterosexual couples, according to Ann Peplau (1981, 1988). Regardless of their sexual orientation, most people want a close, loving relationship with another person. For lesbians, gay men, and heterosexuals, intimate relationships provide love, romance, satisfaction, and security. There is one important difference, however. Heterosexual couples tend to adopt a traditional marriage model, whereas gay couples tend to have a "best friend" model. Peplau (1988) observes:

> A friendship model promotes equality in love relationships. As children, we learn that the husband should be the "boss" at home, but friends "share and share alike." Same-sex friends often have similar interests, skills, and resources—in part because they are exposed to the same gender-role socialization in growing up. It is easier to share responsibilities in a relationship when both partners are equally skilled or inept at cooking, making money, and disclosing feelings.

There is this illusion that homosexuals have sex and heterosexuals fall in love. That's completely untrue. Everybody wants to be loved.

BOY GEORGE

With this model, tasks and chores are often shared, alternated, or done by the person who has more time. Usually, both members of the couple support themselves; rarely does one financially support the other (Peplau and Gordon, 1982).

Few lesbian and gay relationships are divided into the traditional heterosexual provider/homemaker roles. Among heterosexuals, these divisions are gender linked as male or female. But in cases in which the couple consists of two men

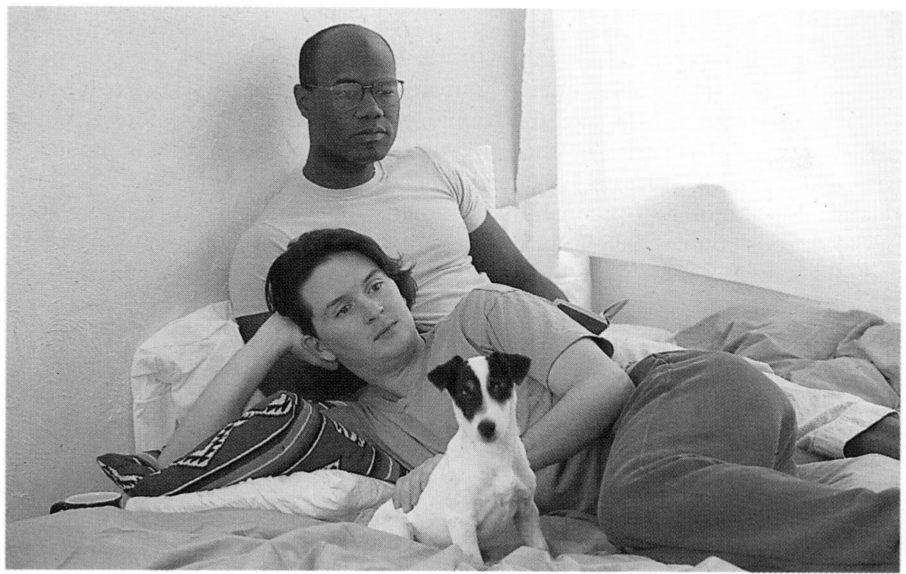

Heterosexuals, gay men, and lesbians cohabit. A significant difference between heterosexual and gay cohabitation is that many gay men and lesbians who would like to marry are prohibited by law from doing so.

or two women, these traditional gender divisions make no sense. As one gay male remarked, "Whenever I am asked who is the husband and who is the wife, I say, 'We're just a couple of happily married husbands.'" Tasks are often divided pragmatically, according to considerations such as who likes cooking more (or dislikes it less) and work schedules (Marecek, Finn, and Cardell, 1988). Most gay couples are dual-worker couples; neither partner supports or depends on the other economically. Furthermore, because gay and lesbian couples are the same gender, the economic discrepancies based on greater male earning power are absent. One partner does not necessarily have greater power than the other based on income. Although gay couples emphasize egalitarianism, if there are differences in power they are attributed to personality; if there is a significant age difference, the older partner is usually more powerful (Harry, 1988).

Cohabitation and Marriage Compared

Different Commitments

Living together tends to be more transitory than marriage (Teachman and Polonko, 1990). When a couple live together, their primary commitment is to each other. As long as they feel they love each other, they will stay together. In marriage, the couple make a commitment not only to each other but to their marriage. Marriage often seems to become a third party that enters the relationship between a man and a woman. Each partner will do things to save a marriage; they may give up dreams, work, ambitions, and extramarital relationships to make a marriage work. A man and woman who are living together may not work as hard to save their relationship. Although society encourages married couples to make sacrifices to save their marriage, unmarried couples rarely receive the same support. Parents may even urge their "living together" children to split up rather than give up plans for school or a career. If the couple is beginning to encounter sexual difficulties, it is more likely that they will split up if they are cohabiting than if they are married. It may be easier to abandon a problematic relationship than to change it.

Finances

A striking difference between cohabiting and married couples is the pooling of money as a symbol of commitment (Blumstein and Schwartz, 1983). People generally assume that in marriage, the couple will pool their money. This arrangement suggests a basic trust or commitment to the relationship: the individual is willing to sacrifice his or her particular economic interests to the interests of the relationship. Among most cohabiting couples, money is not pooled. In fact, one of the reasons couples cohabit rather than marry is to maintain a sense of financial independence. One man said, "A strong factor in the success of the relationship is the fact that we're economically independent of each other. We make no decisions that involve joint finances and that simplifies life a great deal" (Blumstein and Schwartz, 1983).

Only if the couple expect to be living together for a long time or to marry do they pool their income. As Blumstein and Schwartz (1983) point out:

> Since the majority of cohabitors do not favor pooling, these facts say important things about pooling and commitment. When couples begin to pool their finances, it usually means they see a future for themselves. The more a couple pools, the greater the incentive to organize future financial dealings in the same way. As a "corporate" sense of the couple emerges, it becomes more difficult for the partners to think of themselves as unattached individuals.

Work

Traditional marital roles call for the husband to work; it is left to the discretion of the couple whether the woman works. The husband is basically responsible for supporting his wife and family. In cohabiting relationships, the man is not expected to support his partner (Blumstein and Schwartz, 1983). If the woman is not in school, she is expected to work. If she is in school, she is nevertheless expected to support herself. Married couples may fight about the wife's going to work; such fights do not generally occur among cohabiting couples. Cohabiting women spend less time on housework than their married peers, as the homemaker role is not as significant for them (Shelton and John, 1993).

Societal Support

Compared with marriage, cohabitation receives less social support, except from peers. It may be considered an inferior or immoral relationship—inferior to marriage because it does not symbolize lifetime commitment, and immoral because it involves a sexual relationship without the sanction of marriage.

This lack of social reinforcement is an important factor in the greater instability of living-together relationships. Parents usually do not support cohabitation with the same enthusiasm as they would marriage. (If they do not like their son's or daughter's partner, however, they may console themselves with the thought that "at least they are not married.") Research indicates that parents' attitudes toward cohabitation in general become more positive if their own children cohabit (Axinn and Thornton, 1993).

Unmarried couples often find the greatest amount of social support from their friends, especially other couples who are living together. They are able to share similar problems with fellow cohabitants, such as whether to tell parents, how to handle visiting home together, difficulties in obtaining housing, and so forth. Because unmarried couples tend to have similar values, commitments, and uncertainties, they are able to give one another support in the larger non-cohabiting world.

Impact of Cohabitation on Marital Success

There is no consensus on whether cohabitation significantly increases or decreases later marital stability (DeMaris and Rao, 1992; Teachman and Polonko, 1990). Although couples who are living together often argue that cohabitation helps prepare them for marriage, such couples are statistically as likely to divorce as those who do not live together before marriage. People who live together before marriage tend to be more liberal, more sexually experienced, and more independent than people who do not live together before marriage. No conclusive research exists on the link between cohabitation and marital satisfaction.

More and more individuals experience a longer period of singlehood as the age of marriage increases. For many people, being single is a transitional state before cohabitation or marriage. For others singlehood also occurs following divorce or at the end of a cohabiting relationship. Still others choose to remain single throughout their lives. Many of those who are single are lesbian or gay, but they often form cohabiting relationships.

Reflections

Have you ever cohabited? If you have, what were the advantages and disadvantages compared with marriage? If you have not cohabited, would you consider it? For yourself, what would be the pros and cons?

As we have seen, whom we choose as a partner is a complex matter. Our choices are governed by rules of homogamy and exogamy as much as by the heart. But the process of dating or cohabiting helps us determine how well we fit with each other. While these relationships may sometimes be viewed as a prelude to marriage, they are important in their own right. Whatever their outcome, these relationships provide a context for love and personal development.

Summary

- The marriage marketplace refers to the selection activities of men and women when sizing up someone as a potential date or mate. In this marketplace, each person has resources, such as social class, status, age, and physical attractiveness. The marital exchange is based on gender roles. Traditionally, men offer status, economic resources, and protection; women offer nurturing, childbearing, homemaking skills, and physical attractiveness. Recent changes in women's economic status have given women more bargaining power; the decline in the value of housekeeping and children and the increased availability of sexual relations in the singles world have given men more bargaining power.
- The ratio of unmarried men to women also may affect bargaining power in the marriage marketplace. The *marriage squeeze* refers to the gender imbalance reflected in the ratio of available unmarried women to men. Overall, there are significantly more unmarried women than men, but in the age group of fifteen to thirty-nine years, there are significantly more unmarried men than women (except among African Americans, where women significantly outnumber men). Marital choice is also affected by the *marriage gradient*, the tendency for women to marry men of higher status.
- Initial impressions are heavily influenced by physical attractiveness. A *halo effect* surrounds attractive people, from which we infer that they have certain traits, such as warmth, kindness, sexiness, and strength. The "rating and dating" game is the evaluation of men and women by their appearance. Most people, however, choose equals in terms of looks, intelligence, and education. If there is an appreciable difference in looks, there is some kind of trade-off in which a lower-ranked trait is exchanged for a higher-ranked one.

- The *field of eligibles* consists of those of whom our culture approves as potential partners. It is limited by the principles of *endogamy* (marriage within a particular group) and *exogamy* (marriage outside a particular group). The field of eligibles is further limited by *homogamy* (the tendency to choose a mate whose individual or group characteristics are similar to ours) and by *heterogamy* (the tendency to choose a mate whose individual or group characteristics are different from ours). Homogamy is especially powerful in our culture, particularly in terms of race and ethnicity, religion, socioeconomic status, and age.
- According to Bernard Murstein, romantic relationships go through a three-stage development process: the *stimulus*, *value*, and *role* stages. In the stimulus stage, each person is attracted to the other before the actual interaction. In the value stage, each weighs the other's basic values for compatibility. In the role stage, each person analyzes the other's behaviors in roles as lover, companion, and so on.
- Beginning a relationship typically depends on seeing, meeting, and dating another person. The setting in which you see someone can facilitate or discourage a meeting. A *closed field* allows you to see and interact more or less simultaneously. An *open field*, characterized by large numbers of people who do not ordinarily interact, makes meeting more difficult. When meetings do occur, women often covertly initiate them by sending nonverbal signals of availability and interest. Men then initiate conversation with an opening line. African-American churches are especially important for middle-class African Americans as meeting places. For lesbians and gay men, the problem of meeting is accentuated because they cannot assume that the person in whom they are interested shares their orientation. Instead, they must rely on identifying cues.
- Men tend to initiate dates directly, whereas women tend to initiate dates indirectly. Power tends to be more equal in dating relationships than in marriage. Having dating alternatives is an important element in male power; women must determine what they wish to express and how they wish to keep those behaviors that give them strength. Strong self-esteem is essential for healthy relationships. Divergent gender-role conceptions are problems for both genders in dating. For women, problems in dating include sexual pressure, communication, and where to go on the date; for men, problems include communication, where to go, shyness, and money.
- Relationships in the singles world tend to stress independence and autonomy. The emphasis on individual self-fulfillment works against the making of commitments. The marital exchange favors men and may also work against their making a commitment. Singles may be classified into ambivalents, who are voluntarily and temporarily single; wishfuls, who are involuntarily and temporarily single; resolveds, who are permanently single by choice; and regretfuls, who are permanently and involuntarily single.
- Women and men in the gay world often gather in gay neighborhoods where they may openly socialize. Single gay men increasingly focus on relationships rather than sex alone. Lesbians emphasize relationships and are less supportive of women outside relationships.
- *Domestic partnership* laws grant some legal rights to cohabiting couples, including gay and lesbian couples. Cohabitation has become increasingly accepted because of a more liberal sexual climate, the changed meaning of marriage, and delayed marriage. Reasons for cohabitation include: temporary casual convenience, affectionate dating or going together, trial marriage, temporary alternative to marriage, and permanent alternative to marriage.
- Over 1.5 million gay men and lesbians cohabit. Whereas heterosexual cohabiting couples tend to adopt a traditional marriage model, lesbians and gay men utilize a "best friend" model that promotes equality in roles and power.
- Compared with marriage, cohabitation is more transitory, has different commitments, lacks economic pooling, and has less social support. It does not seem to have much impact on marital success.

Key Terms

closed field 181
domestic partner 195
endogamy 176
exogamy 176
field of eligibles 176
halo effect 174
heterogamy 177
homogamy 177
lesbian separatist 193
marriage squeeze 172
marriage gradient 172
open field 181
stimulus-value-role theory 180

Suggested Readings

Cargan, Leonard, and Matthew Melko. *Singles: Myths and Realities.* Ann Arbor, MI: Books on Demand, 1982. Though the data is over seventeen years old, one of the best studies of the singles lifestyle.

Cate, Rodney M., and Sally Lloyd. *Courtship.* Thousand Oaks, CA: Sage, 1992. A concise overview of contemporary courtship, including dating, sexuality, mate selection, and aggression in dating relationships.

Crohn, John. *Mixed Matches: How to Create a Successful Interracial, Interethnic, and Interfaith Relationship.* New York: Fawcett Columbine, 1995. Advice by a psychologist on how to make these kinds of matches successful.

Duck, Steve, ed. *Dynamics of Relationships.* Thousand Oaks, CA: Sage, 1994. A collection of scholarly essays exploring the interpersonal skills necessary to build and maintain successful relationships.

———. *Meaningful Relationships: Talking, Sense, and Relating.* Thousand Oaks, CA: Sage, 1994. An examination of relationship dynamics from a symbolic interactionist perspective: Relationships are based on shared meanings conveyed through our everyday conversations and symbols.

Goss, Robert. *Our Families, Our Values: Snapshots of Queer Kinship.* Binghampton, NY: Haworth Press, Inc., 1997. An exploration of the various ongoing efforts to give religious pride to the various configurations of gay relationships, families, and values.

Hatfield, Elaine, and Susan Sprecher. *Mirror, Mirror: The Importance of Looks in Everyday Life.* Albany, NY: State University of New York Press, 1985. An important survey concerning the significance of looks in relationships and our daily lives.

Hendrick, Susan, and Clyde Hendrick. *Liking, Loving, and Relating,* 2d ed. Pacific Grove, CA: Brooks/Cole, 1992. An excellent academic introduction to the field of close relationships, including the close-relationships perspective, an overview of current research, research methodology, and ethical issues in research.

Miell, Dorothy and Rudi Dallos, eds. *Social Interactions and Personal Relationships.* Newbury Park, CA: Sage, 1996. A clearly written book which explores the interactions between people.

Staples, Robert. *The World of Black Singles: Changing Patterns of Male/Female Relations.* Westport, CT: Greenwood Publishing Group, 1982. A study of contemporary African-American relationships.

CHAPTER 7

Understanding Sexuality

P R E V I E W

To gain a sense of what you already know about the material covered in this chapter, answer "True" or "False" to the statements below.

1. Unlike most human behavior, sexual behavior is instinctive. True or false?
2. A significant number of women require manual or oral stimulation of the clitoris to experience orgasm. True or false?
3. It is normal for children to engage in sexual experimentation with other children of both sexes. True or false?
4. Both men and women report feelings of obligation or pressure to engage in sexual intercourse. True or false?
5. A decline in the frequency of intercourse almost always indicates problems in the marital relationship. True or false?
6. About a fifth of American women of reproductive age have had an abortion. True or false?
7. The American Psychiatric Association has rejected the idea of homosexuality as a form of mental disorder. True or false?
8. Latinos are generally less permissive about sex than African Americans or Anglos. True or false?
9. Because of their knowledge, college students rarely put themselves at risk for HIV/AIDS. True or false?
10. Condoms are not very effective as contraceptive devices. True or false?

Answers

1 False, 2 True, 3 True, 4 True, 5 False, 6 True, 7 True, 8 True, 9 False, 10 False

OUTLINE

Learning about sex in our society is learning about guilt.

JOHN GAGNON AND WILLIAM SIMON

Being sexual is an essential part of being human. Through our sexuality we are able to connect with others on the most intimate levels, revealing ourselves and creating deep bonds and relationships. Sexuality is a source of great pleasure and profound satisfaction. It is the means by which we reproduce, bringing new life into the world and transforming ourselves into mothers and fathers. Paradoxically, sexuality also can be a source of guilt and confusion, a pathway to infection, and a means of exploitation and aggression. Examining the multiple aspects of sexuality helps us understand our own sexuality and that of others. It provides the basis for enriching our relationships.

In this chapter, we will examine the sources of sexual learning; psychosexual development in young, middle, and later adulthood, including sexual scripts and the gay/lesbian/bisexual identity process; sexual behavior; sexual enhancement; sexual relationships; sexual problems and dysfunctions; birth control; sexually transmissible diseases and HIV/AIDS; and sexual responsibility. We hope that this chapter will help you make sexuality a positive element in your life and relationships.

Psychosexual Development in Young Adulthood

At each period in our psychosexual development, we are presented with different challenges. Adolescents are sexually mature (or close to it) in a physical sense, but they are still learning their gender and social roles; they may also be struggling to understand the meaning of their sexual feelings for others and their sexual orientation. During young adulthood—from the late teens through mid-thirties—many of the same tasks continue and new ones are added.

Sources of Sexual Learning

Before we examine the developmental tasks of young adulthood, let's look at some of the sources of our sexual learning.

Parental Influence

Children learn a great deal about sexuality from their parents. For the most part, however, they learn not because their parents set out to teach them but because they are avid observers of their parents' behavior. Much of what they learn concerns the hidden nature of sexuality (Roberts, 1983): "The silence that surrounds sexuality in most families and in most communities carries its own important messages. It communicates that some of the most important dimensions of life are secretive, off limits, bad to talk about or think about."

Most mothers think that to keep young people from love making it is enough not to speak of it in their presence.

Marie Madeline de La Fayette (1678)

As young people enter adolescence, they are especially concerned about their own sexuality, but they are often too embarrassed or distrustful to ask their parents directly about these "secret" matters, and most parents are ambivalent about their children's developing sexual nature. They are often fearful that their children (daughters especially) will become sexually active if they have too much information. They tend to indulge in wishful thinking: "I'm sure Jenny's not really interested in boys yet;" "I know Joey would never do anything like that." Parents may put off talking seriously with their children about sex, waiting for the "right time," or they may bring up the subject once, say their piece, breathe a sigh of relief, and never mention it again. Sociologist John Gagnon calls this the "inoculation" theory of sex education: "Once is enough" (cited in Roberts, 1983). But children need frequent "boosters" where sexual knowledge is concerned. When a parent does undertake to educate a child about sex, it is usually the mother. Thus, most children grow up believing that sexuality is an issue that men don't deal with unless they have a specific problem.

Although parental norms and beliefs are generally influential, they do not appear to have a strong effect on an adolescent's decision to become sexually active. (Peers seem to be the more important factor.) But a lack of rules and structure also seems to be related to more permissive sexual attitudes and premarital sex among adolescents (Forste and Heaton, 1988; Hovell, Sipan, Blumberg, Atkins, Hofstetter, and Kreitner, 1994). Parental communication does have considerable impact on whether an adolescent will use contraception (Baker, Thalberg, and Morrison, 1988). A study of African-American adolescents (Scott-Jones and Turner, 1988) found that the majority of parents not only gave their daughters information about sex but also instructed them about contraception. Mothers and grandmothers are especially important sources of sex information for African-American adolescents (Tucker, 1989). Although parents may seem to have little impact on their children's decision making, the strategy of advising "Don't have sex; but if you do, use a condom" may actually be effective in helping prevent adolescent pregnancies and sexually transmissible diseases. Sex educator Sol Gordon (1984) recommends this "double standard of sex education."

Research indicates that parental concern and involvement with sons and daughters is a key factor in preventing teenage pregnancy (Hanson et al., 1987). Instilling values such as respect for others and responsibility for one's actions provides a context in which young people can use their knowledge about sex. A strong bond with parents appears to lessen teens' dependence on the approval of their peers and to lessen the need for interpersonal bonding that may lead to sexual relationships (DiBlasio and Benda, 1992; Miller and Fox, 1987). But parents also need to be aware of the importance of friends in their teenagers' lives. Families that stress familial loyalty and togetherness over interactions with others outside the family may set the stage for rebellion.

Women's magazines, such as Redbook, Cosmopolitan, *and* Mademoiselle, *use sex to sell their publications and products. How do women's magazines differ from men's magazines, such as* Playboy *and* Penthouse, *in their treatment of sexuality?*

Oh, what a tangled web do parents weave
When they think their children are naive.
OGDEN NASH (1902–1971)

One reason sex education is not being taught at home today is that parents have very good and very bad memories. They forget what they were like when they were young except when kids come in late from a date. And then parents have memories like elephants, and in Technicolor.
JAMES MERRILL

Adam was but human—this explains it all. He did not want the apple for the apple's sake, he wanted it only because it was forbidden.
MARK TWAIN [SAMUEL CLEMENS] (1835–1910)

Peer Influence

Adolescents garner a wealth of misinformation from one another about sex. They also put pressure on one another to carry out traditional gender roles. Boys encourage other boys to be sexually active even if the others are unprepared or uninterested. Those who are pressured must camouflage their inexperience with bravado, which increases misinformation; they cannot reveal sexual ignorance. Bill Cosby (1968) recalled the pressure to have sexual intercourse as an adolescent: "But how do you find out how to do it without blowin' the fact that you don't know how to do it?" On his way to his first sexual encounter he realized that he didn't have the faintest idea of how to proceed:

> So now I'm walkin', and I'm trying to figure out what to do. And when I get there, the most embarrassing thing is gonna be when I have to take my pants down. See, right away, then, I'm buck naked . . . buck naked in front of this girl. Now, what happens then? Do . . . do you just . . . I don't even know what to do. . . . I'm gonna just stand there and she's gonna say, "You don't know how to do it." And I'm gonna say, "Yes, I do, but I forgot." I never thought of her showing me, because I'm a man and I don't want her to show me. I don't want nobody to show me, but I wish somebody would kinda slip me a note.

Even though many teenagers find their early sexual experiences less than satisfying, they still seem to feel a great deal of pressure to conform, which means continuing to be sexually active (DiBlasio and Benda, 1992). The following are typical statements from a group of teenagers in one study (De Armand, 1983): "I had to do it. I was the only virgin" (from a fifteen-year-old girl); "If you want a boyfriend, you have to put out" (from a thirteen-year-old girl); "You do what your friends do or you will be bugged about it" (from a twelve-year-old boy). The students interviewed also said that "everyone" has to make the decision about having sexual intercourse by age fourteen. Researcher Charlotte De Armand comments that although, of course, "not 'everyone' is involved, certainly a large number of young people are making this decision at 14 or younger."

Reflections

Think of your three favorite TV programs. What images do they convey about sexuality? What sexual norms do they affirm? Do you agree or disagree with their underlying sexual messages?

Media Influence

The media have a profound impact on our sexual attitudes (Wolf and Kielwasser, 1991; McMahon, 1990). Anthropologist Michael Moffatt (1989) noted that about a third of the students he studied at Rutgers University mentioned the impact of college and college friends on their sexual development and another third mentioned their parents and religious values. But he found that the major influence on their sexuality was contemporary American pop culture. Moffatt observes:

> The direct sources of the students' sexual ideas were located almost entirely in mass consumer culture: the late-adolescent/young-adult exemplars displayed in movies, popular music, advertising, and on TV; Dr. Ruth and sex manuals; *Playboy*, *Penthouse*, *Cosmopolitan*, *Playgirl*, etc; Harlequins and other pulp romances (females only); the occasional piece of real literature; sex education and popular psychology as it had filtered through these sources, as well as through public schools, and as it continued to filter through the student-life infrastructure of the college; classic soft-core and hard-core pornographic movies, books, and (recently) home videocassettes.

Our collective fantasies center on mayhem, cruelty, and violent death. Loving images of the human body—especially of bodies seeking pleasure or expressing love—inspire us with the urge to censor.
BARBARA EHRENREICH

Partner Influence

Parents, peers, and the media become less important in our sexual learning as we get older, being replaced by our sexual partners. The experience of interpersonal sexuality is ultimately the most important source of modifying traditional sexual scripts. Describing the sources of men's sexual learning, Linda Levine and Lonnie Barbach (1985) note:

> Before their first sexual encounter, men could only rely on secondary sources for information about sex. But once they lost their virginity, women became their primary source of information. It was their continued sexual experience that ultimately expanded and enriched men's sexual repertoire. Their skill at the game of love evolved over time, through trial and error. Each experience left them with a clearer sense of themselves as sexual men. But until they acquired this self-confidence, many men were reluctant to drop their he-man façades and reveal to their partners that they were less than skilled lovers.

In relationships, men and women learn that the sexual scripts and models they learned from parents, peers, and the media do not necessarily work in the real world. They adjust their attitudes and behaviors in everyday interactions. If they are married, sexual expectations and interactions become important factors in their sexuality.

Reflections

Think about your different sources of sexual learning—parents, peers, the media, and partners. How did each influence your sexual development? Did their messages compliment or compete with each other? Which influences were strongest? Why?

Developmental Tasks in Young Adulthood

Several tasks challenge young adults as they develop their sexuality:

- *Integrating love and sex*. Traditional gender roles call for men to be sex oriented and women to be love oriented. In adulthood, we need to develop ways of uniting sex and love instead of polarizing them as opposites.
- *Forging intimacy and commitment*. Young adulthood is characterized by increasing sexual experience. Through dating, cohabitation, and courtship, we gain knowledge of ourselves and others as potential partners. As our relationships become more meaningful, the degree of intimacy and interdependence increases. Sexuality can be a means of enhancing intimacy and self-disclosure as well as a means of obtaining physical pleasure. As we become more intimate, we need to develop our ability to make commitments.
- *Making fertility/childbearing decisions*. Childbearing is socially discouraged during adolescence but becomes increasingly legitimate for young adults in their twenties, especially if they are married. Fertility issues are often critical but unacknowledged, especially for single young adults. If these adults are sexually active, how important is it for them to prevent or defer pregnancy? What will they do if the woman unintentionally gets pregnant?
- *Establishing a sexual orientation*. As children and adolescents, we may engage in sexual experimentation such as playing doctor, kissing, and fondling members of both sexes. Such activities are not necessarily associated with sexual orientation. But by young adulthood a heterosexual, gay, or lesbian orientation emerges. Most young adults develop a heterosexual orientation. Others find themselves attracted to members of the same sex and begin to develop a gay, lesbian, or bisexual identity.
- *Developing a sexual philosophy*. As we move from adolescence to adulthood, we reevaluate our moral standards, moving from moral decision making based on authority to standards based on our own personal principles of right and wrong and caring and responsibility (Gilligan, 1982; Kohlberg, 1969). We become responsible for developing our own moral code, which includes sexual issues. In doing so, we need to develop a philosophical perspective to give coherence to our sexual attitudes, behaviors, beliefs, and values. We need to place sexuality within the larger framework of our lives

In the beginner's mind there are many possibilities, in the expert's mind there are few.

SHUNRYO SUZUKI, *ZEN MIND, BEGINNER'S MIND*

Conscience is the inner voice which warns us that someone may be looking.

H. L. MENCKEN (1880–1956)

Reflections

As you look at the developmental tasks of young adults, which ones are you currently undertaking? Which ones have you completed? If you are middle aged or older, how did resolving (or not resolving) these tasks affect your later development?

and relationships. We need to integrate our personal, religious, spiritual, or humanistic values with our sexuality.

Sexual Scripts

Our gender roles are critical in learning sexuality. Gender roles tell us what behavior (including sexual behavior) is appropriate for each gender. Our sexual impulses are organized and directed through sexual scripts, which we learn and act out. Scripts are the acts, rules, and expectations associated with a particular role. A script is like a road map or blueprint in that it gives general directions; it is more a sketch than a detailed picture of how our culture expects us to act. But even though a script is generalized, it is often more important than our own experiences in guiding our actions. Over time, we may modify or change our scripts, but we will not throw them away. A **sexual script** is a set of expectations of how one is to behave sexually as a female or male and as a heterosexual, lesbian, or gay male.

The scripts we are given for sexual behavior tend to be traditional. These scripts are most powerful during adolescence, when we are first learning to be sexual. Gradually, as we gain experience, we modify and change our sexual scripts. As children and adolescents, we learn our sexual scripts primarily from our parents, peers, and the media. As we get older, interactions with our partners become increasingly important. In adolescence, both middle-class whites and middle-class African Americans appear to share similar values and attitudes about sex and male-female relationships (Howard, 1988).

Female Sexual Scripts

Whereas traditional male sexual scripts focus on sex more than feelings, traditional female sexual scripts focus on feelings more than sex, on love more than passion. The traditional female sexual scripts include the following ideas (Barbach, 1982):

Nothing is good or bad, but thinking makes it so.

William Shakespeare (1564–1616)

- *Sex is both good and bad.* Women are taught that sex is both good and bad. What makes sex good? Marriage or a committed relationship. What makes sex bad? A casual or uncommitted relationship. Sex is so good that you need to save it for your husband (or for someone with whom you are deeply in love). If it is not sanctioned by love or marriage, you'll get a bad reputation.
- *Don't touch me "down there."* Girls are taught not to look at their genitals, not to touch them, especially not to explore them. As a result, women know very little about their genitals. They are often concerned about vaginal odors, which makes them uncomfortable about cunnilingus (oral sex).
- *Sex is for men.* Men are supposed to want sex; women are supposed to want love. Women are supposed to be sexually passive, waiting to be aroused. Sex is not supposed to be a pleasurable activity as an end in itself; it is something performed by women for men.
- *Men should know what women want.* Men are supposed to know what women want, even if women don't tell them. Women are supposed to remain pure and sexually innocent. It is up to the man to arouse the woman, even if he doesn't know what a particular woman finds arousing. To keep her image of sexual innocence, she does not tell him what she wants.
- *Women shouldn't talk about sex.* Many women cannot talk about sex easily because they are not expected to have strong sexual feelings. Some women may know their partners well enough to have sex with them but not well enough to communicate their needs to them.

- *Women should look like beautiful models.* The media present ideal women as having slender hips, firm and full breasts, and no fat; these women are always young, with never a pimple, wrinkle, or gray hair. As a result of these media images, many women are self-conscious about their physical appearance. They worry that they are too fat, too plain, too old. They often feel awkward without clothes on to hide their imagined flaws.
- *Women are nurturers.* Women are supposed to give and men are supposed to receive. Women give themselves, their bodies, their pleasures to men. His needs come first—his desire over hers, his orgasm over hers. If a woman always puts her partner's enjoyment first, she may be depriving herself of her own enjoyment.
- *There is only one right way to experience orgasm.* Women often "learn" that there is only one right way to experience orgasm: during sexual intercourse as a result of penile stimulation. But there are many ways to reach orgasm: through oral sex; manual stimulation before, during, or after intercourse; masturbation; and so on. Women who rarely or never have an orgasm during heterosexual intercourse but believe intercourse is the only legitimate route to orgasm are deprived of expressing themselves sexually in other ways.

Male Sexual Scripts

Therapist Bernie Zilbergeld (1993) suggests that the male sexual script includes the following:

Sex, depersonalized, allows us to avoid the challenge of using our whole self, our total energies and feelings, to present and communicate ourselves to another. Sex is the victim of the fear of love.

Rosemary Reuther

- *Men should not have (or at least should not express) certain feelings.* Men should not express doubts; they should be assertive, confident, and aggressive. Tenderness and compassion are not masculine feelings.
- *Performance is the thing that counts.* Sex is something to be achieved, at which to win. Feelings only get in the way of the job to be done. Sex is not for intimacy but for orgasm.
- *The man is in charge.* As in other things, the man is the leader, the person who knows what is best. The man initiates sex and gives the woman her orgasm. A real man doesn't need a woman to tell him what women like; he already knows.
- *A man always wants sex and is ready for it.* It doesn't matter what else is going on; a man wants sex. He is always able to become erect. He is a machine.
- *All physical contact leads to sex.* Because men are basically sexual machines, any physical contact is a sign for sex. Touching is the first step toward sexual intercourse, not an end in itself. There is no physical pleasure except sexual pleasure.
- *Sex equals intercourse.* All erotic contact leads to sexual intercourse. Foreplay is just that: warming up, getting your partner excited for penetration. Kissing, hugging, erotic touching, oral sex are only preliminaries to intercourse.
- *Sexual intercourse always leads to orgasm.* The orgasm is the proof of the pudding. The more orgasms, the better the sex. If a woman does not have an orgasm, she is not sexual. The male feels that he is a failure because he was not good enough to give her an orgasm. If she requires clitoral stimulation to have an orgasm, she is considered to have a problem.

Common to all these myths is a separation of sex from love and attachment. Sex is seen as a performance.

UNDERSTANDING YOURSELF

Your Sexual Scripts

Sexual scripts tell us the whos, whats, whens, wheres, and whys of sexuality. These scripts change over time, depending on our age, sexual experience, and interaction with intimate partners and others. Let's examine them in relation to ourselves.

Who

The factors of homogamy and heterogamy are almost as strong in selecting sexual partners as they are in choosing marital partners, in part because sexual and marital partners are often the same. Society tells you to have sex with people who are unrelated, around your age, and of the other sex (heterosexual). Less acceptable is having sex with yourself (masturbation), with members of the same sex (gay or lesbian sexuality), and with members of different ethnic groups.

- *Examine the whos in your sexual script.*
- *With whom do you engage in sexual behaviors?*
- *How do your choices reflect homogamy and heterogamy?*
- *What social factors influence your choice?*
- *Does your autoerotic behavior change if you are in a relationship? How? Why?*

What

The whats are the types of sexual behaviors in which you engage. Society classifies sexual acts as good and bad, moral and immoral, appropriate and inappropriate. Although these designations may seem to be absolute, they are culturally relative.

- *What sexual acts are part of your sexual script?*
- *How are they regarded by society?*
- *How important is the level of commitment in a relationship in determining your sexual behaviors?*
- *What level of commitment do you need for kissing? Petting? Sexual intercourse?*
- *What occurs if you and your partner have different sexual scripts for engaging in various sexual behaviors?*

When

When refers to timing. You might make love when your parents are out of the house or, if a parent yourself, when your children are asleep.

A study of sexual stereotypes found the following eight traits to be associated with the traditional male role: sexual competence, the ability to give partners orgasms, sexual desire, prolonged erection, being a good lover, fertility, reliable erection, and heterosexuality (Riseden and Hort, 1992). The researchers observe that their results "offer the rather sad suggestion that men's gender [role] identity may be heavily dependent on the vagaries of a capricious physiological event (getting and maintaining an erection)."

Contemporary Sexual Scripts

As gender roles change, so do sexual scripts. Traditional sexual scripts have been challenged by more liberal and egalitarian ones. Sexual attitudes and behaviors have become increasingly liberal for both white and African-American males and females, but African-American attitudes and behaviors have been and continue to be somewhat more liberal than those of whites (Belcastro, 1985; Weinberg and Wilson, 1988; Wyatt et al., 1988). We do not know how Latino sexuality and Asian-American sexuality have changed, as there is almost no research on the sexual scripts, values, and behaviors in those cultures.

The degree and kind of a person's sexuality reaches up into the ultimate pinnacle of his spirit.

FRIEDRICH NIETZSCHE (1844–1900)

Many college-age women have made an explicit break with the more traditional scripts, especially the good girl/bad girl dichotomy and the belief that "nice" girls don't enjoy sex (Moffatt, 1989). Older professional women who are single also appear to reject the old images (Davidson and Darling, 1988b).

Contemporary sexual scripts include the following elements for both sexes (Gagnon and Simon, 1987; Reed and Weinberg, 1984; Rubin, 1990; Seidman, 1989):

Usually, this type of when is related to privacy. But whens are also related to age.

- *When do you engage in sexual activities?*
- *Are the times related to privacy?*
- *When did you experience your first erotic kiss?*
- *At what age did you first have sexual intercourse? If you have not had intercourse, at what age do you think it would be appropriate?*
- *How was the timing for your first intercourse determined, or how will it be determined?*
- *What influences (friends, parents, religion) are brought to bear on the age timing of sexual activities?*

Where

Where do sexual activities occur with society's approval? In our society, usually in the bedroom, where a closed door signifies privacy. For adolescents, they may also occur in an automobile. Fields, beaches, motels, and drive-in theaters may be identified as locations for sex. Churches, classrooms, and front yards usually are not.

- *For yourself, where are the acceptable places to be sexual?*
- *What makes them acceptable for you?*
- *Have you ever had conflicts with partners about the wheres of sex? Why?*

Why

The whys are the explanations you give yourself and others about your sexual activities. There are many reasons for having sex: procreation, love, passion, revenge, intimacy, exploitation, fun, pleasure, relaxation, boredom, achievement, relief from loneliness, exertion of power, and on and on. Some of these reasons are approved by society; others are not. Some we conceal; others we do not.

- *What are your reasons for sexual activities?*
- *Do you have different reasons for different activities, such as masturbation, oral sex, and sexual intercourse?*
- *Do the reasons change with different partners? With the same partner?*
- *Which reasons are approved by society and which are disapproved?*
- *Which reasons do you make known, and which do you conceal? Why?*

- Sexual expression is positive.
- Sexual activities are a mutual exchange of erotic pleasure.
- Sexuality is equally involving of both partners, and the partners are equally responsible.
- Legitimate sexual activities are not limited to sexual intercourse but also include masturbation and oral-genital sex.
- Sexual activities may be initiated by either partner.
- Both partners have a right to experience orgasm, whether through intercourse, oral-genital sex, or manual stimulation.
- Nonmarital sex is acceptable within a relationship context.
- Gay, lesbian, and bisexual orientations and relationships are increasingly open and accepted or tolerated, especially on college campuses and in large cities.

These scripts give recognition to female sexuality. They are also relationship centered rather than male centered. Society, however, still does not grant women full equality with males. Women who have several concurrent sexual partners or casual sexual relationships, for example, are much more likely to be regarded as promiscuous than are men in similar circumstances (Williams and Jacoby, 1989).

The sexual act without intimacy retains a separateness that cannot be forgotten by orgasm.

Erich Fromm, (1900–1980) *The Art of Loving*

Gay, Lesbian, and Bisexual Identities

In contemporary America, people are generally classified as **heterosexual** (sexually attracted to members of the other gender) or **homosexual** (sexually attracted to members of the same gender). Some identify themselves as **bisexual** (attracted to both genders). In discussing gay men and lesbians, we (like many researchers)

Sin is geographical.

BERTRAND RUSSELL (1872–1970)

generally do not use the term *homosexual* because it often conveys negative or pathological connotations. Instead, we use the neutral terms **gay male** to refer to men and **lesbian** to refer to women. Replacing the term *homosexual* is critical in expanding our understanding of lesbians and gay men. It helps us see individuals as whole persons by emphasizing the fact that sexuality is not the only aspect of their lives. Their lives also include love, commitment, desire, caring, work, children, religious devotion, passion, politics, loss, and hope. Sex is important, obviously, but it is not the only significant aspect of their lives, just as it is not the only significant aspect of the lives of heterosexuals.

Morality is the custom of one's country and the current feeling of one's peers. Cannibalism is moral in a cannibal country.

SAMUEL BUTLER (1612–1680)

Those with lesbian or gay orientations have been called sinful, sick, or perverse, reflecting traditional religious, medical, and psychoanalytic approaches. Contemporary thinking in sociology and psychology has rejected these older approaches as biased and unscientific. Instead, sociologists and psychologists have focused their work on how women and men come to identify themselves as lesbian or gay, how they interact among themselves, and what impact society has on them (Heyl, 1989). Researchers reject the idea that lesbians and gay men are inherently deviant or pathological. As noted sociologist Howard Becker (1963) has pointed out, "Deviant behavior is behavior that people so label." Deviance is created by social groups that make rules whose violation results in violators being labeled deviant and treated as outsiders. Lesbian and gay behavior, then, is deviant only insofar as it is called deviant.

Two significant factors in identifying sexual orientation are (1) the gender of one's partner, and (2) the label one gives oneself (lesbian, gay, bisexual, or heterosexual).

How does one "become" gay, lesbian, bisexual—or even heterosexual, for that matter? A person's **sexual orientation**—sexual identity as heterosexual, gay, lesbian, or bisexual—is complex. It depends on the interaction of numerous factors. These factors, which may be a combination of social, biological, and personal ones, lead to the unconscious formation of a person's sexual orientation. Two of the most important factors are the gender of one's sexual partner and whether one labels oneself as heterosexual, gay, lesbian, or bisexual.

The actual percentage of the population that is lesbian, gay, or bisexual is not known. Among women, about 13 percent have had orgasms with other women, but only 1–3 percent identify themselves as lesbian (Fay, Turner, Klassen, and Gagnon, 1989; Kinsey, Pomeroy, and Martin, 1948, 1953; Marmor, 1980c). Among males, including adolescents, as many as 20–37 percent have had orgasms with other males, according to Kinsey's studies. Ten percent were predominantly gay for at least three years; 4 percent were exclusively gay throughout their entire lives (Kinsey et al., 1948). A review of studies on male same-sex behavior between 1970 and 1990 estimated that a minimum of 5–7 percent of adult men had had sexual contact with other men in adulthood. Based on their review, the researchers suggested that about 4.5 percent of men are exclusively gay (Rogers and Turner, 1991). A more recent large-scale study of 3,300 men age 20–39 reported that 2 percent had engaged in same-sex sexual activities and 1 percent considered themselves gay (Billy, Tanfer, Grady, and Klepinger, 1993). And in 1994, a survey found that of the participants, 2.8 percent men and 1.4 percent women described themselves as homosexual or bisexual, although 9 percent of men and over 4 percent of women said they had had a sexual experience with someone of the same sex (Laumann, Gagnon, Michael, and Michaels, 1994).

What are we to make of these differences between studies? In part, the variances may be explained by different methodologies, interviewing techniques, sampling, or definitions of homosexuality. Furthermore, sexuality is more than simply sexual behaviors; it also includes attraction and desire. One can be a virgin or celibate and still be gay or heterosexual. Finally, sexuality is varied and changes over time; its expression at any one time is not necessarily its expression at another.

Identifying Oneself as Gay or Lesbian

Many researchers believe that a person's sexual interest or direction as heterosexual, gay, or lesbian is established by age four or five (Marmor, 1980a, 1980b). But identifying oneself as lesbian or gay takes considerable time and includes several phases, usually beginning in late childhood or early adolescence (Blumenfeld and Raymond, 1989; Troiden, 1988). **Homoeroticism**—erotic attraction to members of the same gender—almost always precedes gay or lesbian activity by several years.

Stages in Acquiring a Lesbian or Gay Identity. The first stage in acquiring a lesbian or gay identity is marked by fear and suspicion that somehow one's desires are different from others'. At first, the person finds it difficult to label the emotional and physical desires for the same sex. The initial reactions often include fear, confusion, and denial. Adolescents especially fear their family's discovery of their homoerotic feelings. In the second stage, the person labels these feelings of attraction, love, and desire as homoerotic if they recur often enough. The third stage includes the person's self-definition as lesbian or gay. This may take a considerable struggle, for it entails accepting a label that society generally calls deviant. Questions then arise about whether to tell parents or friends, whether to hide one's identity ("to be in the closet") or make the identity known ("come out of the closet").

Some gay men and lesbians may go through two additional stages. One stage is to enter the gay subculture. A gay person may begin acquiring exclusively gay friends, going to gay bars and clubs, or joining gay activist groups. In the gay world, gay and lesbian identities incorporate a way of being in which sexual orientation is a major part of the identity as a person. As Michael Denneny (quoted in Altman, 1982) says: "I find my identity as a gay man as basic as any other identity I can lay claim to. Being gay is a more elemental aspect of who I am than my profession, my class, or my race." Similarly, Pat Califia (quoted in Weeks, 1985) explains: "Knowing I was a lesbian transformed the way I saw, heard, perceived the whole world. I became aware of a network of sensations and reactions that I had ignored all my life."

The final stage begins with a person's first lesbian or gay love affair. This marks the commitment to unifying sexuality and affection. Sex and love are no longer separated. Most lesbians and gay men have had such love affairs, despite the stereotypes of anonymous gay sex.

Coming Out. Being lesbian or gay is increasingly associated with a total lifestyle and way of thinking. In making the gay or lesbian orientation a lifestyle, **coming out** (publicly acknowledging one's gayness) has become especially important as an affirmation of one's sexuality. Coming out is a major decision because it may jeopardize many relationships, but it is also an important means of self-validation. By publicly acknowledging one's gay or lesbian orientation, a person begins to reject the stigma and condemnation associated with it. Generally, coming out occurs in stages, first involving family members, especially the mother and siblings and later the father. Coming out to the family often creates a crisis, but generally the family accepts the situation and gradually adjusts (Holtzen and Agresti, 1990). Religious beliefs, prejudice,

Did you know?

Studies from the 1970s indicated that the average gay male did not identify himself as gay until age nineteen to twenty-one. More recently, with increasing societal awareness of homosexuality, lesbian and gay adolescents are recognizing their orientation at a younger age (Savin-Williams and Rodriguez, 1993; Troiden and Goode, 1980).

Nowadays the polite form of homophobia is expressed in the safeguarding of families and family ties. . . . The estrangement between individual homosexuals and their families is a direct result of homophobia.

Dennis Altman

It is better to be hated for what one is than loved for what one is not.

André Gide (1869–1951)

and misinformation about gay and lesbian sexuality, however, often interfere with a positive parental response, initially making adjustment difficult (Borhek, 1988; Cramer and Roach, 1987). After the family, friends may be told and, in fewer cases, employers and co-workers.

Something is happening to gay men, and we are suddenly no longer affiliated with the family. Where do they think we came from? The cabbage patch?

LARRY KRAMER

Gay men and women are often "out" to varying degrees. Some may be out to no one, not even themselves. Some are out only to their lovers, others to close friends and lovers but not to their families, employers, associates, or fellow students. Still others may be out to everyone. Because of fear of reprisal, dismissal, or public reaction, lesbian and gay schoolteachers, police officers, members of the military, politicians, and members of other such professions are rarely out to their employers, co-workers, or the public.

Anti-Gay/Lesbian Prejudice and Discrimination

Anti-gay prejudice is a strong dislike, fear, or hatred of lesbians and gay men because of their homosexuality. **Homophobia** is an irrational or phobic fear of gay men and lesbians. Not all anti-gay feelings are phobic in the clinical sense of being excessive and irrational. They may be unreasonable or biased. (Nevertheless, they may be within the norms of a biased culture.) Because prejudice may not be clinically phobic, the term *homophobia* is being increasingly replaced by the less clinical *anti-gay prejudice* (Haaga, 1991).

Did you know?

A large-scale survey conducted by *The New York Times* found that 55 percent of the respondents believed behavior between adult gay men or lesbians was morally wrong. At the same time, 78 percent believed gay men and lesbians should have equal job opportunities; 43 percent supported gay men and lesbians in the military (and an equal percentage opposed it); and 42 percent believed laws should be passed to guarantee equal rights for gay men and lesbians (Schmalz, 1993).

As a belief system, anti-gay prejudice justifies discrimination based on sexual orientation. In his classic work on prejudice, Gordon Allport (1958) states that social prejudice is acted out in three stages: (1) offensive language, (2) discrimination, and (3) violence. Gay men and lesbians experience each stage. They are called *faggot*, *dyke*, *queer*, and *homo*. They are discriminated against in terms of housing, equal employment opportunities, insurance, adoption, parental rights, family acceptance, and so on, and they are the victims of violence known as gay bashing or queer bashing. Among college students, anti-gay prejudice extends to heterosexuals who voluntarily choose to room with a lesbian or gay man. They are assumed to have "homosexual tendencies" and to have many of the negative stereotypical traits of gay men and lesbians, such as poor mental health (Sigelman et al., 1991).

Anti-gay prejudice is derived from several factors (Marmor, 1980): (1) a deeply rooted insecurity concerning the person's own sexuality and gender identity, (2) a strong fundamentalist religious orientation, and (3) simple ignorance concerning homosexuality.

In half the states, gay men and lesbians may face criminal prosecution for engaging in oral and anal sex, for which heterosexuals are seldom charged. Medical and public health efforts against HIV/AIDS were inhibited initially because HIV/AIDS was perceived as "the gay plague" and was considered punishment against gay men for their "unnatural" sexual practices (Altman, 1985). The fear of HIV/AIDS has contributed to increased anti-gay prejudice among some heterosexuals (Lewes, 1992). Anti-gay prejudice influences parents' reactions to their lesbian and gay children, often leading to estrangement (Holtzen and Agresti, 1990).

Anti-gay prejudice adversely affects heterosexuals too. First, it creates fear and hatred—aversive emotions that cause distress and anxiety. Second, it alienates heterosexuals from gay family members, friends, neighbors, and co-workers (Holtzen and Agresti, 1990). Third, it limits their range of behaviors and feelings, such as hugging or being emotionally intimate with same-sex friends, for fear that such intimacy may be "homosexual" (Britton, 1990). Heterosexuals may restrict displays of affection with their same-sex friends for fear that such displays could be misinterpreted (Garnets et al., 1990). Fourth, anti-gay prejudice may lead to exaggerated displays of masculinity by heterosexual men trying to prove they are not gay (Mosher and Tomkins, 1988).

What can be done to reduce prejudice against lesbians and gay men? Education and positive social interactions appear to be important vehicles for change. Education can affect negative attitudes. Two researchers (Serdahely and Ziemba, 1984) studied the impact of including a unit on homosexuality in their human sexuality course. They found that students who, at the beginning of the course, scored above the class mean on homophobic attitudes, by the end had a significant decrease in their scores. Other researchers also have reported increased tolerance following human sexuality courses (Stevenson, 1990). Negative attitudes about homosexuality may be reduced by the arranging of positive interactions between heterosexuals and gay men and lesbians. These interactions should be in settings of equal status, common goals, cooperation, and a moderate degree of intimacy. Such interactions may occur when family members or close friends come out. Other interactions may emphasize common group membership (religious, social, ethnic, or political, for example) on a one-to-one basis. Religious volunteers working with people with HIV/AIDS often decrease their homophobia through their caring for and comforting of those with the infection (Kayal, 1992).

Morality is simply the attitude we adopt toward people we don't like.

Oscar Wilde (1854–1900)

Be not too hasty to trust or admire the teachers of morality: they discourse like angels but live like men.

Oscar Wilde (1854–1900)

Bisexuality

As we noted earlier, bisexuals are attracted to members of both genders. Becoming bisexual requires the rejection of two recognized categories of sexual identity: heterosexual and homosexual. In a nationwide study conducted by Samuel Janus and Cynthia Janus (1993), about 5 percent of the men and 3 percent of the women identified themselves as bisexual.

Because they reject both heterosexuality and homosexuality, bisexuals often find themselves stigmatized by gay men and lesbians as well as by heterosexuals. Heterosexuals view bisexuals as really homosexual. Gay men and lesbians view bisexuals as "fence-sitters" not willing to admit their homosexuality or as people simply "playing" with their orientation. Thus, bisexuality may not be taken seriously by either group. Loraine Hutchins and Lani Kaahumanu (1991b) believe that bisexuality arouses hostility because it "challenges current assumptions about the immutability of people's orientations and society's supposed divisions into discrete groups."

Research on homosexuality is extensive, but there is little research on bisexual identity. (There is considerable HIV/AIDS research, however, on bisexual behavior among men who identify themselves as heterosexual.) It was not until 1994 that the first model of bisexual identity formation was developed (Weinberg, Williams, and Pryor, 1994). According to this model, bisexual women and men go through several stages in developing their identity. The first stage, often lasting years, is *initial confusion.* Many are distressed by being sexually attracted to both sexes; others believe that their attraction to the same sex means an end to their heterosexuality; still others are disturbed by their inability to categorize their feelings as either heterosexual or homosexual. The second stage is *finding and applying the bisexual label.* For many, discovering there is such a thing as bisexuality is a turning point. Some find that their first heterosexual or same-sex experience permits them to view sex with both sexes as pleasurable; others learn of the term *bisexuality* from friends and are able to apply it to themselves. The third stage is *settling into the identity*. This stage is characterized as feeling at home with the label *bisexual*. For many, self-acceptance is critical. The fourth stage is *continued uncertainty*. Bisexuals don't have a community or social environment that reaffirms their identity. Despite being settled in, many feel persistent pressure from gay men and lesbians to relabel themselves as homosexual and to engage exclusively in same-sex activities.

Psychosexual Development in Middle Adulthood

Psychosexual development and change does not end with young adulthood. It continues throughout our lives. In middle age and old age, our lives, bodies, sexuality, relationships, and environment continue to change. New tasks and new satisfactions arise to replace or supplement older ones.

Developmental Tasks in Middle Adulthood

In the middle adult years, some of the tasks in psychosexual development begun in young adulthood may continue. These tasks, including issues of intimacy and childbearing, may have been deferred or only partly completed in young adulthood. Because of separation or divorce, we may find ourselves facing the same intimacy and commitment tasks at age forty that we thought we completed fifteen years earlier (Cate and Lloyd, 1992). But life does not stand still; it moves steadily forward, whether we're ready or not. Other developmental issues appear, including the following:

For middle-aged couples, intimacy, communication, and shared interests may become increasingly important.

- *Redefining sex in marital or other long-term relationships*. In new relationships, sex is often passionate, intense; it may be the central focus. But in long-term marital or cohabiting relationships, the passionate intensity associated with sex is often eroded by habituation, competing parental and work obligations, fatigue, and unresolved conflicts. Sex may need to be redefined as a form of intimacy and caring. Individuals may also need to decide how to deal with the possibility, reality, and meaning of extramarital or extrarelational affairs.
- *Reevaluating one's sexuality*. Single men and women may need to weigh the costs and benefits of sex in casual or lightly committed relationships. In long-term relationships, sexuality often becomes less central to relationship satisfaction. Nonsexual elements, such as communication, intimacy, and shared interests and activities, become increasingly important to relationships. Women who have deferred their childbearing begin to reappraise their decision: Should they remain childfree, "race" against their biological clocks, or adopt a child? Some individuals may redefine their sexual orientation. The sexual philosophy of those in middle adulthood continues to be reexamined and to evolve.
- *Accepting the biological aging process*. As we age, our skin wrinkles, our flesh sags, our hair grays (or falls out), our vision blurs—and we become in the eyes of society less attractive and less sexual. By our forties, our physiological responses have begun to slow noticeably. By our fifties, society begins to "neuter" us, especially if we are women who have gone through

menopause. The challenges of aging are to accept its biological mandate and to reject the stereotypes associated with it.

Sexuality and Middle Age

Men and women view aging differently. As men approach their fifties, they fear the loss of their sexual capacity but not their attractiveness; in contrast, women fear the loss of their attractiveness but generally not their sexuality. As both age, purely psychological stimuli, such as fantasies, become less effective for arousal. Physical stimulation remains effective, however.

Among American women, sexual responsiveness continues to grow from adolescence until it reaches its peak in the late thirties or early forties; it is usually maintained at more or less the same level into the sixties and beyond. Men's physical responsiveness is greatest in late adolescence or early adulthood; beginning in men's twenties, responsiveness begins to slow imperceptibly. Changes in male sexual responsiveness become apparent only when men are in their forties and fifties. As a man ages, achieving erection requires more stimulation and time and the erection may not be as firm.

Around the age of fifty, the average American woman begins menopause, which is marked by a cessation of the menstrual cycle. Menopause is not a sudden event. Usually, for several years preceding menopause, the menstrual cycle becomes increasingly irregular. Although menopause ends fertility, it does not end interest in sexual activities. The decrease in estrogen, however, may cause thinning and dryness of the vaginal walls, which makes intercourse painful. The use of vaginal lubricants will remedy much of the problem. A review of the literature about the effects of hormone replacement therapy (HRT) on sexual functioning indicates that such therapy leads to gynecological improvement and thus improves the context for unimpaired sexual activity (Walling, Andersen, and Johnson, 1990). (Women contemplating HRT should inform themselves about it. Recent studies suggest that progesterone should be taken in addition to estrogen to reduce the possibility of uterine cancer.) There is no male equivalent to menopause. Male fertility slowly declines, but men in their eighties are often fertile.

Because of physical changes, notes Herant Katchadourian (1987), "middle-aged couples may be misled into thinking that this change heralds a sexual decline as an accompaniment to aging." Katchadourian continues:

> Sexual partners who have been together for a long time have the benefits of trust and affection. In the younger years of marriage, sex tends to be a battleground where scores are settled and peace is made, but if a couple has stuck together until middle age, sex should become a demilitarized zone. . . . They continue to enjoy the physical pleasures of sex but do not stop there. . . . [T]he sensual quality of the person, rather than the body as such, becomes the main course.

Psychosexual Development in Later Adulthood

As we leave middle age, new tasks confront us, especially dealing with the process of aging itself. Our health and the presence or absence of a partner are key aspects of this time in our lives.

Developmental Tasks in Later Adulthood

Many of the psychosexual tasks older Americans must undertake are directly related to the aging process:

For human beings, the more powerful need is not for sex per se, but for relationship, intimacy and affirmation.

ROLLO MAY

- *Changing sexuality.* As older men's and women's physical abilities change with age, their sexual responses change as well. A seventy-year-old person, though still sexual, is not sexual in the same manner as an eighteen-year-old. As men and women continue to age, their sexuality tends to be more diffuse, less genital, and less insistent. Chronic illness and increasing frailty understandably result in diminished sexual activity. These considerations contribute to the ongoing evolution of the individual's sexual philosophy.
- *Loss of partner.* One of the most critical life events is the loss of a partner. After age sixty, there is a significant increase in spousal deaths. As having a partner is the single most important factor determining an older person's sexual interactions, the death of a partner signals a dramatic change in the survivor's sexual interactions.

The developmental tasks of later adulthood are accomplished within the context of continuing aging. Their resolution helps prepare us for acceptance of our own eventual mortality.

Sexuality and the Aged

The sexuality of older Americans tends to be invisible; that is, society tends to discount their sexuality (Libman, 1989). In fact, one review of the literature on aging concludes that the decline in sexual activity among aging men and women is more cultural than biological in origin (Kellett, 1991). Several reasons for this exist in our culture (Barrow and Smith, 1992). First, we associate sexuality with the young, assuming that sexual attraction exists only between those with youthful bodies. Interest in sex is considered normal and virile in twenty-five-year-old men, but in seventy-five-year-old men it is considered lecherous. Second, we associate the idea of romance and love with the young; many of us find it difficult to believe that the aged can fall in love or love intensely. Third, we associate sex with procreation, measuring a woman's femininity by her childbearing and motherhood and a man's masculinity by the children he sires. Finally, the aged do not have sexual desires as strong as those of the young and they do not express their desires as openly. Intimacy is especially valued and is important for an older person's well-being (Mancini and Blieszner, 1992).

Sexuality is one of the least understood aspects of life in old age. Many older people continue to adhere to the standards of activity or physical attraction they held when they were young (Creti and Libman, 1989). They need to become aware of the taboos and stereotypes about aging that they held when they were younger so they can enjoy their sexuality in their later years (Kellett, 1991).

Aging lesbians and gay men face a double bias: They are old *and* gay. But like other stereotypes of aging Americans, theirs reflects myths rather than realities. A study of gay men over age sixty found that over 80 percent accepted their gayness and about half worried about growing old (Berger, 1982). A study of aging lesbians found that 71 percent were satisfied with being lesbian and about half were concerned about aging (Kehoe, 1988).

Aged men and women face different sexual problems. Physiologically, men are less responsive than they used to be. The decreasing frequency of intercourse and the increasing time required to attain an erection produce anxieties in many older men about erectile dysfunction (impotence)—anxieties that may very well lead to such dysfunction. When the natural slowing down of sexual responses is interpreted as the beginning of erectile dysfunction, this self-diagnosis triggers a vicious spiral of fears and even greater difficulty in attaining or maintaining an erection. One study (Weitzman and Hart, 1987) found that about 31 percent of elderly male respondents were unable to have an erection.

Women have different concerns. They face greater social constraints than men. Women are confronted with an unfavorable gender ratio (twenty-nine unmarried men per hundred unmarried women over age sixty-five), a greater likelihood of widowhood, and norms against marrying younger men. Grieving over the death of a partner, isolation, and depression also affect their sexuality (Rice, 1989). Finally, there is a double standard of aging. In our culture, as men age, they become distinguished; as women age, they simply get older. Femininity is connected with youth and beauty. But as women age, they tend to be regarded as more masculine. A young woman, for example, is "beautiful," but an older woman is "handsome"—a term ordinarily used for men of any age.

The greatest determinants of an aged individual's sexual activity are health and the availability of a partner. Researchers studied more than eight hundred married whites and African Americans over age sixty (Marisiglio and Donnelly, 1991). They found that over half the sample (and 24 percent of those older than seventy-six) had sexual intercourse within the previous month. Those who had sex during the month had it an average of four times. Among the sexually active older people, there were no differences by gender or ethnicity. Those who do not have partners may turn to masturbation as an alternative to sexual intercourse (Pratt and Schmall, 1989).

After age seventy-five, a significant decrease in sexual activity takes place. This seems to be related to health problems, such as heart disease, arthritis, and diabetes. Often older people indicate that they continue to feel sexual desires; they simply lack the ability to express them because of their health. In a study

Sexuality among the aged tends to be sensual and affectionate. Older couples may experience an intimacy forged by years of shared joys and sorrows that is as intense as the passion of young love.

(White, 1982) of men and women in nursing homes whose ages averaged eighty-two years, 91 percent reported no sexual activity immediately prior to their interviews; 17 percent of these men and women, however, expressed a desire for sexual activity. Unfortunately, most nursing homes make no provision for the sexuality of the aged. Instead, they actively discourage sexual expression—not only sexual intercourse but also masturbation—or try to sublimate their clients' sexual interests into crafts or television.

Sexual Behavior

In this section we examine various sexual behaviors. For a discussion of sexual structure and the sexual response cycle, see Appendixes A and B.

Autoeroticism

Two monks, Tanzan and Ekido, were traveling down the road in the heavy rain. As they turned a bend, they came upon a beautiful young woman in a silk kimono. She was unable to pass because the rain had turned the road to mud.

Tanzan said to her, "Come on," and lifted her in his arms and carried her across the mud. Then he put her down and the monks continued their journey.

The two monks did not speak again until they reached a temple in which to spend the night. Finally, Ekido could no longer hold back his thoughts and he reprimanded Tanzan. "It is not proper for monks to go near women," he said. "Especially young and beautiful ones. It is unwise. Why did you do it?"

"I left the woman behind," replied Tanzan. "Are you still carrying her?"

ZEN TALE

Autoeroticism consists of sexual activities that involve only the self. It includes sexual fantasies, masturbation, and erotic dreams. A universal phenomenon in one form or another, autoeroticism is one of our earliest expressions of sexual stirrings. It is also one that traditionally has been condemned in our society. By condemning it, however, our culture sets the stage for the development of deeply negative inhibitory attitudes toward sexuality.

Sexual Fantasies

Erotic fantasizing is probably the most universal of all sexual behaviors. Nearly everyone has experienced erotic fantasies, but because they may touch on feelings or desires considered personally or socially unacceptable, they are not widely discussed. They may also interfere with an individual's self-image, causing a loss of self-esteem as well as confusion.

Sexual fantasies serve a number of important functions in maintaining our psychic equilibrium. First of all, sexual fantasies help direct and define our erotic goals. They take our generalized sexual drives and give them concrete images and specific content. Second, sexual fantasies allow us to plan or anticipate erotic situations that may arise. They provide a form of rehearsal, allowing us to practice in our minds how to act in various situations. Third, sexual fantasies provide escape from a dull or oppressive environment. Routine or repetitive labor often gives rise to sexual fantasies as a way of coping with the boredom in which we are trapped. Fourth, even if our sexual lives are satisfactory, we may indulge in sexual fantasies to bring novelty and excitement into a relationship. Many people fantasize things they would not actually do in real life. Fantasy offers a safe outlet for sexual curiosity. Fifth, sexual fantasies also have an expressive function in somewhat the same manner as dreams do. Our sexual fantasies may offer a clue to our current interests, pleasures, anxieties, fears, or problems.

A fantasy is a map of desire, mastery, escape, and obscuration, the navigational path we invent to steer ourselves between the reefs and shoals of anxiety, guilt, and inhibition.

NANCY FRIDAY

Various studies report that between 60 and 90 percent of respondents fantasize during sex—the percentage depending on gender and ethnicity (Knafo and Jaffe, 1984; Price and Miller, 1984). A recent large-scale study (Michael, Gagnon, Laumann, and Kolata, 1994) found that 54 percent of the men and 19 percent of the women thought about sex daily. Twenty-three percent of the men and 11 percent of the women bought X-rated videos. Despite their prevalence, fantasies during intercourse provoke guilty feelings in a number of people (Cado and Leitenberg, 1990).

Erotic Dreams

Almost all of the men and two-thirds of the women in Kinsey and his colleagues' studies (1948, 1953) reported having had overtly sexual dreams. Sexual images in dreams are frequently very intense. Although people tend to feel responsible for fantasies, which occur when they are awake, they are usually less troubled by sexual dreams.

Dreams almost always accompany nocturnal orgasm. The dreamer may awaken, and men usually ejaculate. Of a woman's orgasms, 2 to 3 percent may be nocturnal, while for men the number may be around 8 percent of the total (Kinsey et al., 1948, 1953). Although the dream content may not be overtly sexual, it is always accompanied by sensual sensations. Women seem to feel less guilt or fear about nocturnal orgasms than do men, accepting them more easily as pleasurable experiences. Men tend to worry about them, perhaps because they emit semen.

Masturbation

Masturbation is the manual stimulation of one's genitals. Individuals masturbate by rubbing, caressing, or otherwise stimulating their genitals to bring themselves sexual pleasure. Masturbation is an important means of learning about our bodies. Girls, boys, women, and men may masturbate during particular periods or throughout their entire lives. An analysis of research articles on gender roles and sexual behavior found that the greatest male-female difference was in masturbation (Oliver and Hyde, 1993). Males had significantly more masturbatory experience than females.

By the end of adolescence, virtually all males and about two-thirds of females have masturbated to orgasm (Knox and Schact, 1992; Lopresto, Sherman, and Sherman, 1985). Masturbation continues after adolescence. In recent years, there seems to have been a slight increase in the incidence and frequency of masturbation. Gender differences, however, continue to be significant (Atwood and Gagnon, 1987; Leitenberg, Detzer, and Srebnik, 1993).

Most people continue to masturbate after they marry, although the rate is significantly smaller than for unmarried people. There are many reasons for continuing the activity during marriage: Masturbation is a pleasurable form of sexual excitement, a spouse may be away or unwilling to engage in sex, sexual intercourse may not be satisfying, the partners may fear sexual inadequacy, or the individual may act out fantasies. In marital conflict, masturbation may act as a distancing device, with the masturbating spouse choosing masturbation over sexual intercourse as a means of emotional protection (Betchen, 1991).

Attitudes toward masturbation vary along ethnic lines. Whites are the most accepting of masturbation, for example; African Americans are less accepting. The differences can be explained culturally. Whites tend to begin their coital activities later than African Americans; whites therefore regard masturbation as an acceptable alternative to sexual intercourse. African Americans, by contrast, tend to accept interpersonal sexual activity at an earlier age. They therefore may view masturbation as a sign of personal and sexual inadequacy. As a result, many African Americans see sexual intercourse as normal and masturbation as deviant (Cortese, 1989; Kinsey et al., 1948; Wilson, 1986). Recently, masturbation has become more accepted as a legitimate sexual activity within the African-American community (Wilson, 1986; Wyatt et al., 1988).

Latinos, like African Americans, hold less permissive attitudes than Anglos about masturbation (Cortese, 1989; Padilla and O'Grady, 1987). In Mexican culture, masturbation is not considered an acceptable sexual option for either males

Did you know?

A study of college students in a human sexuality class found that 87 percent of the men and 58 percent of the women had masturbated (Knox and Schact, 1992). In a larger study, among adults of all ages, 63 percent of the men and 42 percent of the women had masturbated in the previous year (Michael et al., 1994).

Sin is whatever obscures the soul.

André Gide (1869–1951)

or females (Guerrero Pavich, 1986). In part, this is because of the cultural emphasis on sexual intercourse and the influence of Catholicism, which regards masturbation as sinful. As with other forms of sexual behavior, acceptance becomes more likely as Latinos become more assimilated. A study of Mexican-American college students found their attitudes toward masturbation more liberal than those of the Mexican-American community as a whole (Padilla and O'Grady, 1987).

Interpersonal Sexuality

The sexual act is in time what the tiger is in space.

Georges Bataille

We often think that sex is sexual intercourse and that sexual interactions end with orgasm (usually the male's). But sex is not limited to sexual intercourse. Heterosexuals engage in a wide variety of sexual activities, which may include erotic touching, kissing, and oral and anal sex. Except for sexual intercourse, gay and lesbian couples engage in more-or-less the same sexual activities as heterosexuals.

Touching

Because touching, like desire, does not in itself lead to orgasm, it has largely been ignored as a sexual behavior. Sex researchers William Masters and Virginia Johnson (1970) suggest a form of touching they call "pleasuring." **Pleasuring** is nongenital touching and caressing. Neither partner tries to stimulate the other sexually; the partners simply explore each other. Such pleasuring gives each a sense of his or her own responses; it also allows each to discover what the other likes or dislikes. We can't assume we know what any particular individual likes, for there is too much variation among people. Pleasuring opens the door to communication; couples discover that the entire body is erogenous, rather than just the genitals.

As we enter old age, touching becomes increasingly significant as a primary form of erotic expression. Touching in all its myriad forms, ranging from holding hands to caressing, massaging to hugging, walking with arms around each other to fondling, becomes the touchstone of eroticism for the elderly. One study found touching to be the primary form of erotic expression for married couples over eighty years old (Bretschneider and McCoy, 1988).

Kissing

Praise be to Eros! who loves only beauty and finds it everywhere.

Lenore Kandel, "Eros/Poem"

Kissing as a sexual activity is probably the most acceptable of all premarital sexual activities (Jurich and Polson, 1985). The tender lover's kiss symbolizes love, and the erotic lover's kiss, of course, simultaneously represents and is passion. Both men and women in one study regarded kissing as a romantic act, a symbol of affection as well as attraction (Tucker, Marvin, and Vivian, 1991). A cross-cultural study of jealousy found that kissing is also associated with a couple's boundary maintenance: In each culture studied, kissing a person other than the partner evoked jealousy (Buunk and Hupka, 1987).

The lips and mouth are highly sensitive to touch and are exquisitely erotic parts of our bodies. Kisses discover, explore, and excite the body. They also involve the senses of taste and smell, which are especially important because they activate unconscious memories and associations. Often we are aroused by familiar smells associated with particular sexual memories: a person's body smells, perhaps, or perfumes associated with erotic experiences. In some cultures—among the Borneans, for example—the word *kiss* literally translates as "smell." In fact, among traditional Eskimos and Maoris there is no mouth kissing, only the nuzzling that facilitates smelling.

Kissing is probably the most acceptable premarital sexual activity.

Although kissing may appear innocent, it is in many ways the height of intimacy. The adolescent's first kiss is often regarded as a milestone, a rite of passage, the beginning of adult sexuality (Alapack, 1991). Philip Blumstein and Pepper Schwartz (1983) report that many of their respondents found it unimaginable to engage in sexual intercourse without kissing. In fact, they found that those who have a minimal (or nonexistent) amount of kissing feel distant from their partners but engage in coitus nevertheless as a physical release.

The amount of kissing differs according to orientation. Lesbian couples tend to engage in more kissing than heterosexual couples, and gay male couples kiss less than heterosexual couples (Blumstein and Schwartz, 1983).

Oral-Genital Sex

In recent years, oral sex has become part of our sexual scripts. It is engaged in by heterosexuals, gay men, and lesbians. The two types of oral-genital sex are cunnilingus and fellatio. **Cunnilingus** is the erotic stimulation of a woman's vulva by her partner's mouth and tongue. **Fellatio** is the oral stimulation of a man's penis by his partner's sucking and licking. Cunnilingus and fellatio may be performed singly or simultaneously. Oral sex is an increasingly important and healthy aspect of adults' sexual selves (Wilson and Medora, 1990).

Although oral-genital sex is increasingly accepted by white middle-class Americans, it remains less permissible among certain ethnic groups. African Americans, for example, have lower rates of oral-genital sex than do whites; many African Americans consider it immoral (Wilson, 1986). Oral sex is becoming increasingly accepted, however, by African-American women (Wyatt, Peters, and Guthrie, 1988). This is especially true if they have a good relationship and communicate well with their partners (Wyatt and Lyons-Rowe, 1990). Among married Latinos, oral sex is relatively uncommon. When it occurs, it is usually at the instigation of men, as women are not expected to be interested in erotic

Did you know?

Between 60 and 95 percent of the men and women in various studies report that they have engaged in oral sex. Among adults of all ages, 27 percent of the men and 19 percent of the women had oral sex in the previous year (Delamater and MacCorquodale, 1979; Michael et al, 1994; Peterson et al., 1983).

variety (Guerrero Pavich, 1986). Although little is known about older Asian Americans and Asian immigrants, college-age Asian Americans appear to accept oral-genital sex to the same degree as middle-class whites (Cochran, Mays, and Leung, 1991).

For both sexes, fellatio is less common than either sexual intercourse or cunnilingus (Newcomer and Udry, 1985). A study (Moffatt, 1989) of university students of both sexes found that oral sex was regarded as an egalitarian, mutual practice. Students felt less guilty about it than about sexual intercourse because oral sex was not "going all the way."

Sexual Intercourse

The conventional position makes me claustrophobic. And the others give me either a stiff neck or lockjaw.

TALLULAH BANKHEAD (1903–1968)

Sexual intercourse or **coitus**—the insertion of the penis into the vagina and subsequent stimulation—is a complex interaction. As with many other types of activities, the anticipation of reward triggers a pattern of behavior. The reward may not necessarily be orgasm, however, because the meaning of sexual intercourse varies considerably at different times for different people. There are many motivations for sexual intercourse; sexual pleasure is only one. Other motivations include showing love, having children, giving and receiving pleasure, gaining power, ending an argument, demonstrating commitment, seeking revenge, proving masculinity or femininity, or degrading someone (including oneself).

Although sexual intercourse is important for most sexually involved couples, its significance is different for men and women. For men, sexual intercourse appears to be only one of several activities, such as fellatio and cunnilingus, that they enjoy. For women, however, intercourse is often central to their sexual satisfaction. More than any other heterosexual sexual activity, sexual intercourse involves equal participation by both partners. Ideally, both partners equally and simultaneously give and receive. Many women report that the sense of sharing during intercourse is important to them.

Men tend to be more consistently orgasmic than women in sexual intercourse. Part of the reason may be that the clitoris frequently does not receive sufficient stimulation from penile thrusting alone to permit orgasm. Many women need manual stimulation during intercourse to be orgasmic. They may also need to be more assertive. A woman can manually stimulate herself or be stimulated by her partner before, during, or after intercourse. But to do so, she has to assert her own sexual needs and move away from the idea that sex is centered around male orgasm. The sexual script has to be redefined.

Did you know?

According to a scientific, nationwide study of adults of all ages, about one-third of Americans have sexual intercourse twice a week, one-third a few times a month, and one-third a few times a year or not at all. Married couples are more likely to engage in coitus than singles; married women are more likely to be orgasmic. About 40 percent of married couples and 25 percent of singles report having coitus twice a week (Michael et al., 1994).

Anal Eroticism

Sexual activities involving the anus are known as **anal eroticism. Anal intercourse** is the male's insertion of his erect penis into his partner's anus. Both heterosexuals and gay men may participate in this activity. For heterosexual couples who engage in it, anal intercourse is generally an experiment or occasional activity rather than a common mode of sexual expression. About 10 percent of men and 9 percent of women report engaging in anal sex in the previous year (Michael et al., 1994). Among gay men, anal intercourse is less common than oral sex. It is, nevertheless, an important ingredient in the sexual satisfaction of many gay men (Blumstein and Schwartz, 1983). From a health perspective, anal intercourse is the riskiest form of sexual interaction. It is the most prevalent sexual means of transmitting the human immunodeficiency virus (HIV) among both gay men and heterosexuals. Because the delicate rectal tissues are easily torn, HIV (carried within semen) can enter the bloodstream. (HIV will be discussed in detail later in the chapter.)

Sexual Enhancement

Sex is emotion in motion.

Mae West (1893–1980)

Sexual behavior cannot be isolated from our personal feelings and relationships. If our sexuality is not a source of personal growth, meaning, and pleasure, we need to think about its role in our lives and our relationships. Sometimes dissatisfaction arises because the relationship itself is unsatisfactory. At other times the relationship itself is good but the erotic fire needs to be lit or rekindled. Such relationships may grow through sexual enhancement. **Sexual enhancement** is improving the quality of a sexual relationship, especially by providing accurate information about sexuality, developing communication skills, fostering positive attitudes, and increasing self-awareness.

What is it men in women do require?
The lineaments of Gratified Desire.
What is it women do in men require?
The lineaments of Gratified Desire.

William Blake (1757–1827)

Conditions for Good Sex

According to noted sex therapist Bernie Zilbergeld, enhancing our sexual relationships requires the following:

1. Accurate information about sexuality, especially your own and your partner's.
2. An orientation toward sex based on pleasure (including arousal, fun, love, and lust) rather than on performance and orgasm.
3. Being involved in a relationship that allows each person's sexuality to flourish.
4. An ability to communicate verbally and nonverbally about sex, feelings, and relationships.
5. Being equally assertive and sensitive about your own sexual needs and those of your partner.
6. Accepting, understanding, and appreciating differences between partners.

Being aware of your own sexual needs is often critical to enhancing your sexuality. Gender-role stereotypes and negative learning about sexuality often cause us to lose sight of our own sexual needs. Following these sexual stereotypes may impede our ability to have what therapist Carol Ellison calls "good sex." Ellison (1985) writes that you will know you are having good sex if you feel good about yourself, your partner, your relationship, and what you're doing. It's good sex if, after a while, you still feel good about yourself, your partner, your relationship, and what you did. Good sex does not necessarily include orgasm or intercourse. It can be kissing, holding, masturbating, oral sex, anal sex, and so on. It can be heterosexual, gay, lesbian, or bisexual.

Let's talk about sex.
Let's talk about you and me.
Let's talk about all the good
And bad things that can be.
Let's talk about sex.

Salt-N-Pepa

Zilbergeld (1993) suggests that to fully enjoy our sexuality, we need to explore our "conditions for good sex." There is nothing unusual about requiring conditions for any activity. For a good night's sleep, for example, each of us has certain conditions. We may need absolute quiet, no light, a feather pillow, an open window. Others, however, can sleep during a loud dormitory party, curled up in the corner of a stuffy room. Of conditions for good sex, Zilbergeld writes:

I want a lover with a slow hand.

The Pointer Sisters

In a sexual situation, a condition is anything that makes you more relaxed, more comfortable, more confident, more excited, more open to your experience. Put differently, a condition is something that clears your nervous system of unnecessary clutter, leaving it open to receive and transmit sexual messages in ways that will result in a good time for you.

Different individuals report different conditions for good sex. Some common conditions include the following:

- *Feeling intimate with your partner*. This is often important for both men and women, despite stereotypes of men wanting only sex. If the partners are feeling distant from each other, they may need to talk about their feelings before becoming sexual. Emotional distance can take the heart out of sex.
- *Feeling sexually capable*. Generally this relates to an absence of anxieties about sexual performance. For men, this includes anxiety about erections or about ejaculating too soon. For women, it includes worry about painful intercourse or lack of orgasm. For both men and women, it includes worry about whether one is a good lover.
- *Feeling trust*. Both men and women may need to know they are emotionally safe with the other. They need to feel confident that they will not be judged or ridiculed or talked about.
- *Feeling aroused*. A person does not need to be sexual unless he or she is sexually aroused or excited. Simply because one's partner wants to be sexual does not mean that you have to be.
- *Feeling physically and mentally alert*. Alertness requires a person not to feel particularly tired, ill, stressed, or preoccupied. It requires the person not to be under the influence of excessive alcohol or drugs.
- *Feeling positive about the environment and situation*. A person may need privacy, to be in a place where he or she feels protected from intrusion. Each needs to feel that the other is sexually interested and wants to be sexually involved.

Reflections

What are your conditions for good sex? What do you do to meet them? What happens if they are not met?

Intensifying Erotic Pleasure

One of the most significant elements in enhancing one's physical experience of sex is intensifying arousal. Intensifying arousal requires us to focus on increasing erotic pleasure rather than on sexual performance. This can be done in many ways, some of which are described here.

Sexual arousal is the physiological responses, fantasies, and desires associated with sexual anticipation and sexual activity. We have different levels of arousal. These levels are not necessarily associated with particular types of sexual activities. Sometimes we feel more sexually aroused when we kiss than when we have sexual intercourse or oral sex. Masturbation may sometimes be more exciting than oral sex or coitus.

It's not the men in my life, it's the life in my men that counts.

MAE WEST (1893–1980)

The first element in increasing your sexual arousal is having your conditions for good sex met. If you need a romantic setting, go for a walk at the beach by moonlight or listen to music by candlelight. If you want limits on your sexual activities, tell your partner. If you need a certain kind of physical stimulation, show or tell your partner what you like.

A second element in increasing arousal is focusing on the sensations you are experiencing. Once you begin an erotic activity, such as massaging or kissing,

Common conditions for a satisfying sexual relationship include feelings of intimacy, capability, trust, arousal, alertness, and positiveness about the environment and situation.

do not let yourself be distracted. When you're kissing, don't think about what you're going to do next or about an upcoming test in your marriage and family class. Instead, focus on the sensual experience. Zilbergeld (1993) explains:

> Focusing on sensations means exactly that. You put your attention in your body where the action is. When you're kissing, keep your mind on your lips. This is *not* the same as thinking about your lips or the kiss; just put your attention in your lips. As you focus on your sensations, you may want to convey your pleasure to your partner. Let him or her know through your sounds and movements that you are excited.

Sexual Relationships

Sexuality exists in various relationship contexts that may influence our feelings and activities. These include nonmarital, marital, and extramarital contexts.

Nonmarital Sexuality

Nonmarital sex is sexual activities, especially sexual intercourse, that take place outside of marriage. We use the term *nonmarital sex* rather than *premarital sex* to describe sexual behavior among unmarried adults in general. We will use the term **premarital sex** when we are referring to never-married adults under the age of thirty. There are several reasons to make premarital sex a subcategory of nonmarital sex. First, because increasing numbers of never-married adults are over thirty years of age, "premarital sex" does not adequately describe the nature of their sexual activities. Second, at least 10 percent of adult Americans will never marry; it is misleading to describe their sexual activities as "premarital." Third, many adults are divorced, separated, or widowed; 30 percent of divorced women and men will never remarry. Fourth, between 3 and 10 percent of the population is lesbian or gay; gay and lesbian sexual relationships cannot be categorized as "premarital."

Sentiments are for the most part traditional; we feel them because they were felt by those who preceded us.
WILLIAM HAZLITT (1778–1830)

Sexuality in Dating Relationships

Over the last several decades, there has been a remarkable increase in the acceptance of premarital sexual intercourse. For adolescents and young adults, the advent of effective birth control methods, changing gender roles that permit females to be sexual, and delayed marriages have played a major part in the rise of premarital sex. For middle-aged and older adults, increasing divorce rates and longer life expectancy have created an enormous pool of once-married men and women who engage in nonmarital sex. Only **extramarital sex**—sexual interactions that take place outside the marital relationship—continues to be consistently frowned upon.

The increased legitimacy of sex outside of marriage has transformed both dating and marriage. Sexual intercourse has become an acceptable part of the dating process for many couples, whereas only petting was acceptable before. Furthermore, marriage has lost some of its power as the only legitimate setting for sexual intercourse (Sprecher and McKinney, 1993). One important result is that many people no longer feel that they need to get married to express their sexuality in a relationship (Scanzoni, Polonko, Teachman, and Thompson, 1989).

The value of premarital chastity [is] . . . almost as dead as the dodo.

MICHAEL MOFFATT

There appears to be a general expectation among students that they will engage in sexual intercourse sometime during their college careers. Although college students expect sexual involvement to occur within an emotional or loving relationship (Robinson,, Ziss, Ganza, and Katz, 1991), this emotional connection may be relatively transitory.

Factors Leading to Premarital Sexual Involvement. What factors lead individual men and women to have premarital sexual intercourse? One study indicated that among men and women who had premarital sex, the most important factors were their love (or liking) for each other, physical arousal and willingness of both partners, and planning and arousal prior to the encounter (Christopher and Cate, 1985). Among nonvirgins, as with virgins, love or liking between the partners was extremely important (Christopher and Cate, 1984), but feelings of obligation or pressure were about as important as actual physical arousal. Women reported affection as being slightly more important than did men. An interesting finding is that men perceived slightly more pressure or obligation to engage in intercourse than women.

Examining the sexual decision process more closely, researcher Susan Sprecher (1989) identifies individual, relationship, and environmental factors affecting the decision to have premarital intercourse:

- *Individual factors.* A number of individual factors influence the decision to have premarital intercourse. They include previous sexual experience, sexual attitudes, personality characteristics, and gender. The more premarital sexual experience a man or woman has had, the more likely he or she is to engage in sexual activities. Once the psychological barrier against premarital sex is broken, sex appears to become less taboo. This seems to be especially true if the earlier sexual experiences were rewarding in terms of pleasure and intimacy. Those with liberal sexual attitudes are more likely to engage in sexual activity than those with restrictive attitudes. In terms of personality characteristics, men and women who do not feel high levels of guilt about sexuality are more likely to engage in sex, as are those who value erotic pleasure.
- *Relationship factors.* Two of the most important factors determining sexual activity in a relationship are the level of intimacy and the length of time the couple has been together. Even those who are less permissive in their sexual

attitudes accept sexual involvement if the relationship is emotionally intimate and long standing. Individuals who are less committed (or not committed) to a relationship are less likely to be sexually involved. Finally, persons in relationships in which power is shared equally are more likely to be sexually involved than those in inequitable relationships.

- *Environmental factors*. In the most basic sense, the physical environment affects the opportunity for sex. Because sex is a private activity, the opportunity for it may be precluded by the presence of parents, friends, roommates, or children (Tanfer and Cubbins, 1992). The cultural environment, too, affects premarital sex. The values of one's parents or peers may encourage or discourage sexual involvement. A person's ethnic group also affects premarital involvement. Generally, African Americans are more permissive than whites and Latinos are less permissive than non-Latino (Baldwin, Whitely, and Baldwin, 1992). Furthermore, a person's subculture—such as the university or church environment, the singles world, the gay and lesbian community—exerts an important influence on sexual decision making.

Initiating a Sexual Relationship. As we saw in Chapter 6, after we meet someone, we weigh each other's attitudes, values, and philosophy to see if we are compatible. If the relationship continues in a romantic vein, we may include physical intimacy. To signal the transition from nonphysical to physical intimacy, one of us must make the first move. Making the first move marks the transition from a potentially sexual relationship to an actual one.

If the relationship develops according to traditional gender-role patterns, the male will make the first move to initiate sexual intimacy, whether it is kissing, petting, or engaging in sexual intercourse (O'Sullivan and Byers, 1992). At what point this occurs generally depends on two factors: the level of intimacy and the length of the relationship (Sprecher, 1989). The more emotionally involved the couple are, the more likely it is that they will be sexually involved as well. The duration of the relationship also affects the likelihood of sexual involvement.

Initial sexual involvement can occur as early as the first meeting or much later, as part of a well-established relationship. Some people become sexually involved immediately ("lust at first sight"), but the majority begin their sexual involvement in the context of an ongoing relationship. Strategies for making the first move vary, depending on the motives of each individual and the nature of the relationship. If the people do not know each other well, both are likely to rely on traditional sexual scripts, with the man making the first overt move. If the motive for both partners is sexual pleasure, both may acknowledge their lack of interest in commitment. But if one desires pleasure and the other commitment, different strategies may be used. The pleasure-oriented partner, for example, may cease making overtures, feign commitment, or utilize sexual pressure. The other partner may withhold sex unless a commitment is made.

In new or developing relationships, communication about sexuality is generally indirect and ambiguous. Direct strategies are sometimes used to initiate sexual involvement, but they usually are used when the person is confident in the other's interest or is not concerned about being rejected. A study of sexual initiation among college students found that males and females used similar strategies to initiate sex (O'Sullivan and Byers, 1992).

In new relationships, we communicate indirectly about sex because we want to become sexually involved with the other person but we also want to avoid rejection. By using indirect strategies—such as turning down the lights, moving closer, touching the other's face or hair—we may test the other's interest in sexual involvement. If the other person responds positively to our cues, we can initiate a sexual encounter. At the same time, if the other turns the lights up, moves away, or does not respond to our touching, he or she gives a message of disinterest. Because we have not made direct overtures, we can save face. The sexual cues, innuendoes, and signals can pass unacknowledged. Consequently a direct refusal does not occur. We can breathe a sigh of relief because we have avoided rejection.

Because so much of our sexual communication is indirect, ambiguous, or nonverbal, we run a high risk of being misinterpreted. There are four reasons for our communications being misinterpreted (Cupach and Metts, 1991). First, men and women tend to disagree about when sexual activities should take place in a relationship. Men tend to want sexual involvement earlier and with lower levels of intimacy than do women. Second, men may be skeptical about women's refusals. Men often misinterpret women's cues. Men also believe that women often say "no" when they actually mean "coax me," so they interpret it as token resistance. Third, because women communicate indirectly, they may be unclear in signaling their disinterest. They may turn their face aside, move a man's hand back to its proper place, say it's getting late, or try to change the subject. Research indicates that women are most effective when they make strong, direct verbal refusals; men become more compliant if women are persistent in their direct verbal refusals (Christopher and Frandsen, 1990; Murnen, Perot, and Byrne, 1989). Fourth, men are more likely than women to interpret nonsexual behavior or cues as sexual. Although both men and women flirt for fun, men are more likely to flirt with a sexual purpose and to interpret a woman's flirtation as sexual or "teasing."

Men may wear sex-colored glasses.

WILLIAM CUPACH AND SANDRA METTS

Directing Sexual Activity. As we begin a sexual involvement, we have several tasks to accomplish. First, we need to practice safe sex. We need to gather information about our partners' sexual history and determine whether he or she practices safe sex, including the use of condoms. Unlike much of our sexual communication, which is nonverbal or ambiguous, we need to use direct verbal discussion in practicing safe sex. Second, unless we are intending a pregnancy, we need to discuss birth control. Although the use of condoms will help prevent the spread of sexually transmissible diseases, condoms alone are only moderately effective as contraception. To be highly effective, they must be used in conjunction with contraceptive foam or jellies or other devices. Responsibility for contraception, like safe sex, generally requires verbal communication.

In addition to communicating about safe sex and contraception, we also need to communicate about what we like and need sexually. What kind of foreplay do we like? What kind of afterplay? Do we like to be orally or manually stimulated during intercourse? If so, how? What does each partner need to be orgasmic? Many of our needs and desires can be communicated nonverbally by our movements or by other physical cues. But if our partner does not pick up our nonverbal signals, we need to discuss them directly and clearly to avoid ambiguity.

Sexuality in Cohabiting Relationships

As we saw in Chapter 6, cohabitation has become a widespread phenomenon in American culture. Despite its increasing importance for men and women of all ages, little research exists on sexuality in such relationships. In contrast to married men and women, cohabitants have sexual intercourse more frequently, are more egalitarian in initiating sexual activities, and are more likely to be involved in sexual activities outside their relationship (Blumstein and Schwartz, 1983). The higher frequency of intercourse, however, may be due to the "honeymoon" effect: Cohabitants may be in the early stages of their relationship, the stages when sexual frequency is highest. Blumstein and Schwartz found that 22 percent of female cohabitants but only 9 percent of married women had been involved in extrarelational sex in the previous year; 25 percent of male cohabitants and 11 percent of married men were similarly involved. The differences in frequency of extrarelational sex may result from a combination of two factors: Norms of sexual fidelity may be weaker in cohabiting relationships, and men and women who cohabit tend to conform less to conventional norms.

Sexuality in Gay and Lesbian Relationships

Because of their socialization as males, gay men are likely to initiate sexual activity earlier in the relationship than are lesbians. In large part, this is because both partners are free to initiate sex and because men are not expected to refuse sex, as women are (Isensee, 1990). Lesbians do not initiate sex as frequently as do gay or heterosexual men. They often feel uncomfortable because women have not been socialized to initiate sex.

In both gay and lesbian relationships, the more emotionally expressive partner is likely to initiate sexual interaction. The gay or lesbian partner who talks more about feelings and who spontaneously gives the partner hugs or kisses is the one who more often begins sexual activity.

One of the major differences between heterosexuals and gay men and lesbians is in how they handle extrarelational sex. In the gay and lesbian culture, sexual exclusivity is negotiable. Sexual exclusiveness is not necessarily equated with commitment or fidelity.

As a result of these differing norms, gay men and lesbians must decide early in the relationship whether they will be sexually exclusive (Isensee, 1990). If they choose to have a nonexclusive relationship, they need to discuss how outside sexual interests will be handled. They need to decide whether to tell each other, whether to have affairs with friends, what degree of emotional involvement will be acceptable, and how to deal with jealousy.

Marital Sexuality

Marriage has many pains but celibacy has no pleasures.

SAMUEL JOHNSON (1709–1784)

When people marry, they may discover that their sexual life is very different than it was before marriage. Sex is now morally and socially sanctioned. It is in marriage that the great majority of heterosexual interactions take place. Yet as a culture we feel ambivalent about marital sex. On the one hand, marriage is the only relationship in which sexuality is fully legitimized. On the other hand, marital sex is an endless source of humor and ridicule: "Marital sex? What's that?" Two journalists watched prime-time television on the major networks for one week (Hanson and Knopes, 1993) and found that of the forty-five sexual scenes depicted, only four were between married couples. There were almost six times as many depictions of sexual activities between unmarried men and women as between married couples. There was four times as much extramarital sex as marital sex. On television, men have sex more often with prostitutes than with their wives. Erotic activity is often linked with violence. Sex research is not much different from popular culture. The empirical research devoted to healthy marital sexuality is virtually nonexistent.

Sexual Interactions

Sexual intercourse tends to diminish in frequency the longer a couple is married. For newly married couples, the average rate of sexual intercourse is about three times a week. As the couples get older, the frequency drops. In early middle age, married couples make love an average of one and a half to two times a week. After age fifty, the rate is about once a week or less. This decreased frequency, however, does not necessarily mean that sex is no longer important or that the marriage is unsatisfactory. It often means simply that one or both members are too tired. For dual-worker families and families with children, fatigue and lack of private time may be the most significant factors in the decline of frequency (Olds, 1985). Blumstein and Schwartz (1983) found that most people attributed their decline in sexual intercourse to lack of time or physical energy or to "being accustomed" to each other. Also, activities and interests other than sex engage them.

It is better to marry than to burn.

CORINTHIANS 7:9

Most married couples don't seem to believe that declining frequency is a major problem if they rate their overall relationship as good (Cupach and Comstock, 1990). Sexual intercourse is only one erotic bond among many in marriage. There are also kissing, caressing, nibbling, stroking, massaging, dining by candlelight, walking hand in hand, looking into each other's eyes, and talking intimately.

Married men continue to initiate sexual encounters overtly more frequently than do women, but women signal their interest or willingness. They pace the frequency of intercourse by showing their interest through nonverbal cues, such as a "certain look" or lighting candles by the bed; they may also overtly suggest "doing the wild thing." Their partners pick up on the cues and "initiate" sexual interactions. In marital relationships, many women feel comfortable about initiating sex. In part this may be related to the decreasing significance of the double standard as relationships continue. In marital relationships, the woman's initiation may be viewed positively, as an expression of love; it may also be the result of couples becoming more egalitarian in their gender-role attitudes.

Positive responses to initiation are usually nonverbal, such as beginning or continuing the sexual interaction by kissing or touching erotically. In most cases, when a partner refuses the sexual initiation, the couple "agree" not to have sex. They may decide to have sex at a different time, or they may "agree to disagree"—that is, they may find disagreement acceptable and nonthreatening. Partners were most satisfied in the way the disagreement was resolved if the initiation was made verbally, such as with the question: "Do you want to make love?" than if the initiation was made physically, such as by erotic touching or kissing. Partners find it easier to say or accept no to a verbal request than to a physical one. Contrary to the common stereotype, it appears that women do not restrict sexual activities any more than men do (Byers and Heinlein, 1989; O'Sullivan and Byers, 1992).

New Meanings to Sex

Sex within the marriage is significantly different from premarital sex in at least three ways. First, sex in marriage is expected to be monogamous. Second, procreation is a legitimate goal. Third, such sex takes place in the everyday world. These differences present each person with important tasks.

The vow of fidelity is an absurd commitment, but it is the heart of marriage.

ROBERT CAPON

Monogamy. One of the most significant factors shaping marital sexuality is the expectation of monogamy. Before marriage or following divorce a person may have various sexual partners, but within marriage all sexual interactions are expected to take place between the spouses. This expectation of monogamy lasts a lifetime; a person marrying at twenty commits himself or herself to forty to sixty years of sex with the same person. Within a monogamous relationship, each partner must decide how to handle fantasies, desires, and opportunities for extramarital sexuality. Do you tell your spouse that you have fantasies about other people? That you masturbate? Do you flirt with others? Do you have an extramarital relationship? If you do, do you tell your spouse? How do you handle sexual conflicts or difficulties with your partner? How do you deal with sexual boredom or monotony?

Socially Sanctioned Reproduction. Sex also takes on a procreative meaning within marriage. Although it is obviously possible to get pregnant before marriage, in most segments of society, marriage remains the only socially sanctioned setting for having children. At marriage, partners are confronted with the task of deciding whether and when to have children. It is one of the most crucial decisions they will make, for having children profoundly alters a relationship. If the couple decide to have a child, their lovemaking may change from simply an erotic activity to an intentionally reproductive act as well.

Changed Sexual Context. The sexual context changes with marriage. Because married life takes place in a day-to-day living situation, sex must also be expressed in the day-to-day world. Sexual intercourse must be arranged around working hours and at times when the children are at school or asleep. One or the other partner may be tired, frustrated, or angry. The emotions associated with premarital sex may disappear. Some of the passion of romantic love eventually disappears as well, to be replaced with a love based on intimacy, caring, and commitment. Although we may tend to believe that good sex depends on good techniques, it really depends more on the quality of the marriage. As humorist Garrison Keillor (1994) reminds us:

> Despite jobs and careers that eat away at their evenings and weekends and nasty whiny children who dog their footsteps and despite the need to fix meals and vacuum the carpet and pay bills, [married] couples still manage to encounter each other regularly in a lustful, inquisitive way and throw their clothes in the corner and do thrilling things in the dark and cry out and breathe hard and afterward lie sweaty together feeling *extreme pleasure.*

A Frontier Guard

The bamboo leaves rustle
On this cold, frosty night.
I am wearing seven layers of clothing
But they are not as warm, not as warm
As the body of my wife.

ANONYMOUS, EIGHTH-CENTURY JAPANESE

Extramarital Sexuality

As we saw earlier, extramarital sex consists of sexual interactions that take place outside the marital relationship. A fundamental assumption in our culture is that marriages are monogamous. Each person remains the other's exclusive intimate partner in terms of both emotional and sexual intimacy. Extramarital relationships violate that assumption. Figure 7.1 shows the percentage of Americans who have had extramarital affairs according to a 1994 survey.

Personal characteristics and the quality of the marriage appear to be the most important factors associated with extramarital relationships. The personal characteristics—feelings of alienation, need for intimacy, emotional dependence, and egalitarian gender roles—are stronger correlates of extramarital sex than is the quality of the marriage. Generally, the lower the marital satisfaction and the lower the frequency and quality of marital intercourse, the greater the likelihood of extramarital sexual relationships. Most people become involved in extramarital sex because they feel something is missing in their marriage. They have judged

To be faithful to one is to be cruel to all others.

WOLFGANG AMADEUS MOZART (1756-1791), *DON GIOVANNI*

Did you know?

Over 21 percent of American men and over 11 percent of American women reported having extramarital affairs (National Opinion Research Center, 1994).

Adultery in your heart is committed not only when you look with concupiscence [strong or excessive sexual desire] at a woman who is not your wife. The husband must not use his wife, her femininity, to fulfill his instinctive desire. Concupiscence . . . diminishes the richness of the perennial attraction of persons for interpersonal communication. Through such a reduction, the other person becomes the mere object for satisfying a sexual need and it touches the dignity of the person.

POPE JOHN PAUL II

Figure 7.1
Lifetime Incidence of Infidelity by Gender and Age

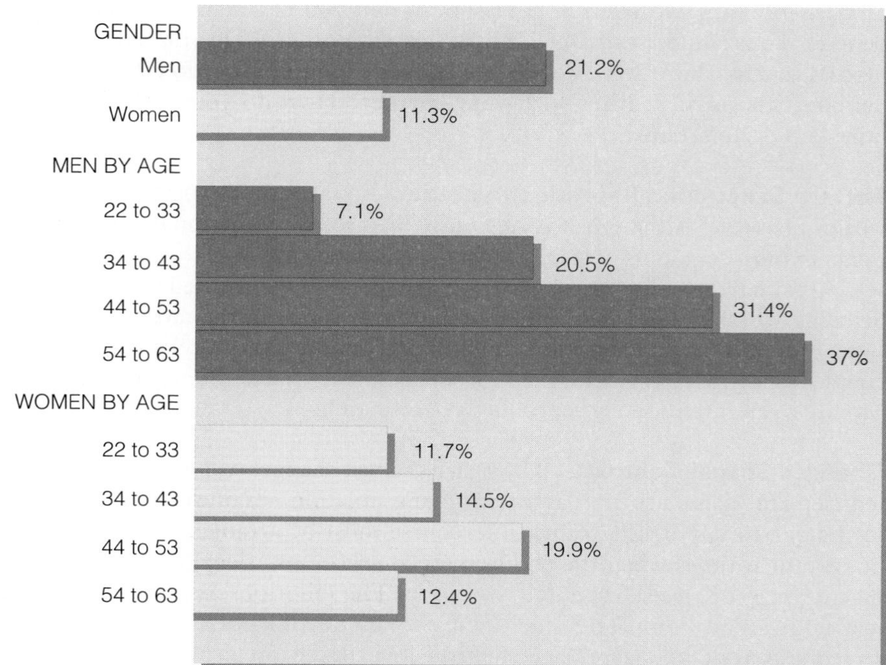

SOURCE: National Opinion Research Center, 1994 Survey.

it defective, although not defective enough to consider divorce. Extramarital relationships are a compensation for these deficiencies.

We tend to think of extramarital involvements as being sexual, but they may actually assume several forms (Moore-Hirschl, Parra, Weis, and Laflin, 1995; Thompson, 1993). They may be (1) sexual but not emotional, (2) sexual and emotional, or (3) emotional but not sexual (Thompson, 1984). Very little research has been done on extramarital relationships in which the couple are emotionally but not sexually involved. Anthony Thompson's study, however, found that among married and cohabiting individuals, the three types of extrarelational involvement were about equally represented.

As a result of marital assumptions, both sexual and nonsexual extramarital relationships take place without the knowledge or permission of the other partner. If the marital assumptions are violated, David Weiss (1983) points out, we have "guidelines" on how to handle the violation: "These guidelines encourage the 'adulterer' to be secretive and discreet, suggest that guilt will be a consequence, and maintain that the spouse will react with feelings of jealousy and rejection if the [extramarital sex] is discovered."

Why should we take advice on sex from the Pope? If he knows anything about it, he shouldn't.

GEORGE BERNARD SHAW (1856–1950)

If the extramarital relationship is discovered, a marital crisis ensues. Many married people believe that the spouse who is unfaithful has broken a basic trust. Sexual accessibility implies emotional accessibility. When one spouse learns that the other is having an affair, the emotional commitment of the spouse having the affair is brought into question. How can the partner prove that he or she is still trustworthy? It cannot be done. Trust is assumed; it can never be proved. Furthermore, the extramarital relationship of one partner may imply to the other (rightly or wrongly) that he or she is sexually inadequate or uninteresting.

People who engage in extramarital affairs have a number of different motivations, and these affairs satisfy a number of different needs (Adler, 1996; Moultrup, 1990). Research by Ira Reiss and his colleagues (1980) suggests that extramarital affairs are related to two variables: unhappiness in the marriage and premarital sexual permissiveness. Generally speaking, in happy marriages, a partner is less likely to seek outside sexual relationships. A person who had premarital sex is more likely to have extramarital sex; once the first prohibition is broken, the second holds less power.

Reflections

Is extramarital sexuality ever justifiable? What about nonsexual affairs? Should partners be honest with each other about close involvements outside of the relationship?

Characteristics of Extramarital Sex

The majority of extramarital sexual involvements are sporadic. Most extramarital sex is not a love affair; it is generally more sexual than emotional. Affairs that are both emotional and sexual appear to detract more from the marital relationship than do affairs that are only sexual or only emotional (Thompson, 1984). More women than men consider their affairs emotional; almost twice as many men as women consider their affairs only sexual. About equal percentages of men and women are involved in affairs that they view as both sexual and emotional.

Immorality may be fun, but it isn't fun enough to take the place of one-hundred percent virtue and three square meals a day.

Noel Coward, (1899–1973), *Design for Living*

An emotionally significant extramarital affair creates a complex system of relationships among the three individuals (Moultrup, 1990). Long-lasting affairs can form a second but secret "marriage." In some ways, these relationships resemble polygamy, in which the outside person is a "junior" partner with limited access to the other. Such relationships form a triangular system. The two involved in the affair continually negotiate their relationship with each other and with the uninvolved partner (whose needs, demands, or possible presence or suspiciousness must always be considered). Meanwhile, the uninvolved partner mistakenly believes he or she is involved in a dyadic (two-person) system. As a result, he or she misinterprets situations. The partner's absence is believed to be the result of working late rather than an affair. The involved partners, who know their system is triadic, must try to meet each other's needs for time, affection, intimacy, and sex while taking the uninvolved partner into consideration. Such extramarital systems are stressful and demanding. Most people find great difficulty in sustaining them. If both people involved in the affair are married, the dynamics become even more complex.

A cuckold is the husband of an unfaithful wife—a far nastier and more humiliating state . . . than being the wife of a philanderer, for which in fact no word exists.

Anne Fausto-Sterling, *Myths of Gender*

Sexually Open Marriages

Open marriage is marriage in which partners agree to allow each other to have openly acknowledged and independent sexual relationships with others. Blumstein and Schwartz (1983) found that 15 to 26 percent of the couples in their sample had "an understanding" that permitted extramarital relations under certain conditions, such as having affairs only out of town, never seeing the same person twice, and never having sex with a mutual friend. Two researchers (Knapp and Whitehurst, 1977) found that successful open marriages required (1) a commitment to the primacy of the marriage, (2) a high degree of affection and trust between the spouses, (3) good interpersonal skills to manage complex relationships, and (4) nonmarital partners who did not compete with the married partner.

When you prevent me from doing anything I want to do, that is persecution; but when I prevent you from doing anything you want to do, that is law, order, and morals.

George Bernard Shaw (1856-1950)

A study (Rubin and Adams, 1986) attempted to measure the impact of sexually open marriages on marital stability. It matched eighty-two couples in 1978 and followed up seventy-four of them five years later. It found no significant difference in marital stability related to whether the couples were sexually open or monogamous in their marriages. Among the marriages that broke up, the reasons given were not related to extramarital sex. No appreciable differences were found in terms of marital happiness and jealousy.

Sexual Problems and Dysfunctions

Sex problems are normal and typical. I know, I know, all of your buddies are functioning perfectly and never have a problem. If you really believe that, I have a nice piece of oceanfront property in Kansas I'd like to talk to you about.

BERNIE ZILBERGELD

Many of us who are sexually active may sometimes experience sexual difficulties or problems. Some problems are recurring, causing distress to the individual or the partner. Such persistent sexual problems are known as **sexual dysfunctions.** Although some sexual dysfunctions are physical in origin, many are psychological. Some dysfunctions have immediate causes, others originate in conflict within the self, and still others are rooted in a particular sexual relationship.

Both men and women may suffer from hypoactive (low or inhibited) sexual desire (Hawton, Catalan, and Fagg, 1991). Other dysfunctions experienced by women are orgasmic dysfunction (the inability to attain orgasm), arousal difficulties (the inability to become erotically stimulated), and dyspareunia (painful intercourse). The most common dysfunctions among men include erectile dysfunction (the inability to achieve or maintain an erection), premature ejaculation (the inability to delay ejaculation after penetration), and delayed orgasm (difficulty in ejaculating) (Spector and Carey, 1990). Figure 7.2 shows the percentage of heterosexual adults in the general U.S. population who reported experiencing sexual problems during the previous year, in response to a recent survey (Laumann et al., 1994).

Origins of Sexual Problems

Physical Causes

It is generally believed that between 10 and 20 percent of sexual dysfunctions are structural in nature. Physical problems may be *partial* causes in another 10 or 15 percent; (Kaplan, 1983; LoPiccolo, 1991). Various illnesses may have an adverse effect on a person's sexuality (Wise, Epstein, and Ross, 1992). Alcohol and some prescription drugs, such as medication for hypertension, may affect sexual responsiveness (Buffum, 1992; "Drugs," 1992).

Among women, diabetes, hormone deficiencies, and neurological disorders, as well as alcohol and alcoholism, can cause orgasmic difficulties. Painful intercourse may be caused by an obstructed or thick hymen, clitoral adhesions, a constrictive clitoral hood, or a weak pubococcygeus muscle. Coital pain caused by inadequate lubrication and thinning vaginal walls often occurs as a result of decreased estrogen associated with menopause. Lubricants or hormone replacement therapy often resolve the difficulties.

Among males, diabetes and alcoholism are the two leading physical causes of erectile dysfunctions; atherosclerosis is another important factor (LoPiccolo, 1991; Roenrich and Kinder, 1991). Smoking may also contribute to sexual difficulties (Rosen et al., 1991).

Psychological/Relationship Causes

Two of the most prominent causes of sexual dysfunctions are performance anxiety and conflicts within the self. *Performance anxiety*—the fear of failure—is probably the most important immediate cause of erectile dysfunctions and, to a lesser extent, of orgasmic dysfunctions in women (H. Kaplan, 1979). If a man

Figure 7.2
Heterosexual Sexual Dysfunctions in a Nonclinical Sample

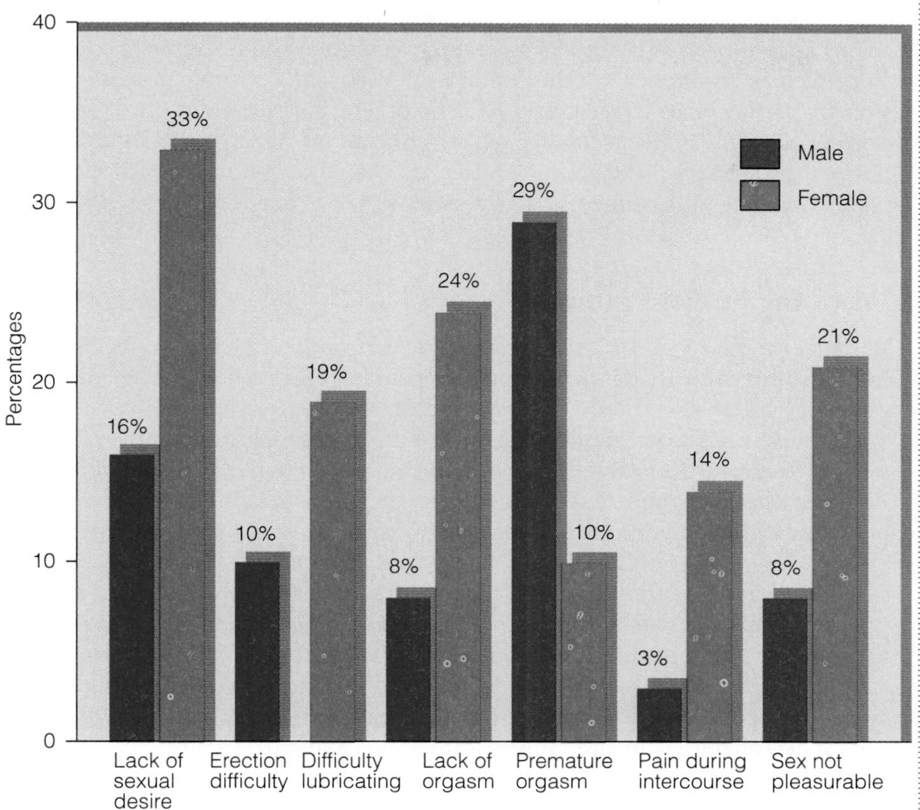

SOURCE: Adapted from Laumann et al., 1994, pp. 370-371.

does not become erect, anxiety is a fairly common response. Some men experience their first erectile problem when a partner initiates or demands sexual intercourse. Women are permitted to say no, but many men have not learned that they too may say no to sex. Women suffer similar anxieties, but they tend to center around orgasmic abilities rather than the ability to have intercourse. If a woman is unable to experience orgasm, a cycle of fear may arise, preventing future orgasms. A related source of anxiety is an excessive need to please one's partner.

Conflicts within the self are guilt feelings about one's sexuality or sexual relationships. But guilt and emotional conflict do not usually eliminate a person's sexual drive; rather, they inhibit the drive and alienate the person from his or her sexuality. These psychic conflicts often are deeply rooted; they may be unconscious. Among gay men and lesbians, concerns about sexual orientation may be an important cause of such conflicts (George and Behrendt, 1987).

The relationship itself, rather than either individual, sometimes can be the source of sexual problems. Disappointment, anger, or hostility may become integral parts of a deteriorating or unhappy relationship. Such factors ultimately affect sexual interactions, for sex can become a barometer for the whole relationship. Helen Kaplan (1979) suggests that relationship discord affects our sexuality in several ways. First, we may *transfer* or redirect feelings we have about someone else (usually parents, former partners, or other important persons) to our current partner. Second, *power struggles* may be a central theme in a relationship. In such cases, sexuality becomes a tool in struggles for control. Third, we may have *unwritten rules* in our relationships. These unwritten rules are the usually unconscious assumptions and expectations about how each should act in the relationship, such as reading each other's minds, putting one's partner first, and so on. Fourth, partners may engage in *sexual sabotage* by asking for sex at the wrong time, putting pressure on each other, or frustrating or criticizing each other's sexual desires and fantasies. People most often do this unconsciously. Fifth, *poor communication* may undermine our ability to express our needs and desires.

Resolving Sexual Problems

Perhaps the first step in dealing with a sexual problem is to turn to your own immediate resources. Begin by discussing the problem with your partner; find out what he or she thinks. Discuss specific strategies that might be useful. Sometimes simply communicating your feelings and thoughts will resolve the difficulty. Seek out friends with whom you can share your feelings and anxieties. Find out what they think. Ask them whether they have had similar experiences and how they handled them. Try to keep your perspective—and your sense of humor.

Partners, friends, and books may provide permission for you to engage in sexual exploration and discovery. From these sources we may learn that many of our sexual fantasies and behaviors are very common. Such methods are most effective when the dysfunctions arise from a lack of knowledge or mild sexual anxieties.

If you are unable to resolve your sexual difficulties yourself, seek professional assistance. It is important to realize that seeking such assistance is not a sign of personal weakness or failure. Rather, it is a sign of strength, for it demonstrates an ability to reach out and a willingness to change. It is a sign that you care for your partner, your relationship, and yourself.

Therapists can help deal with sexual problems on several levels. Some focus directly on the problem, such as lack of orgasm, and suggest behavioral exercises, such as pleasuring and masturbation, to develop an orgasmic response. Others focus on the couple relationship as the source of difficulty. If the relationship improves, they believe, then sexual responsiveness will also improve. Still others work with the individual to help develop insight into the origins of the problem in order to overcome it. Therapy can also take place in a group setting. Group therapy may be particularly valuable for providing us with an open, safe forum in which we can discuss our sexual feelings and experience and discover our commonalities with others.

Birth Control

The command to be fruitful and multiply was promulgated according to our authorities when the population of the world consisted of two persons.

W. R. INGE, DEAN OF ST. PAUL'S (1860–1954)

Most of us think of sexuality in terms of love, passionate embraces, and entwined bodies. Sex involves all of these, but what we so often forget (unless we are worried) is that sex is also a means of reproduction. Whether we like to think about it or not, many of us (or our partners) are vulnerable to unintended pregnancies. Not thinking about pregnancy does not prevent it; indeed, not thinking about it may even contribute to the likelihood of its occurring. Unless we practice **abstinence,** refraining from sexual intercourse, we need to think about unintended pregnancies and then take the necessary steps to prevent them. (See the Resource Center at the back of this book for a discussion of contraceptive and abortion methods.)

Contraception

A woman has about a 2 to 4 percent chance of becoming pregnant during intercourse without contraception. **Contraception** is the prevention of pregnancy using any of a number of devices, techniques, or drugs. If intercourse occurs the day before ovulation, the chance of conception (becoming pregnant) is 30 percent; if it occurs on the day of ovulation, there is a 50 percent chance of conception. Over the period of a year, a woman in a sexual relationship in which contraception is not used has an 85 to 90 percent change of becoming pregnant.

It is now vitally important that we find a way of making the condom a cult object of youth.

Germaine Greer

The key to contraceptive effectiveness is diligent and consistent use. A diaphragm in the bathroom, a condom in the wallet, or pills in the dispenser are useless for preventing pregnancies. Good intentions are not good contraceptives. Numerous studies indicate that the most consistent users of contraception are men and women who explicitly communicate about contraception. Those at greatest risk are those in casual dating relationships and those who infrequently discuss contraception with their partners or others. A review of the literature on the interpersonal factors in contraceptive use concluded, "Individuals in stable, serious relationships of long duration who had frequent, predictable patterns of sexual activity were most likely to use contraception" (Milan and Kilmann, 1987).

Did you know?

An estimated four in ten American women will become pregnant by the time they turn twenty (Alan Guttmacher Institute, 1997).

Our society seems to obstruct rather than encourage the use of contraception. Contraceptives are strikingly absent in sexual depictions in the mass media, for example. (How often do you see passionate love scenes on television or in movies in which a condom is used?) Right-wing groups preach about the "dangers" of condoms and oppose the inclusion of birth control information in sex education courses.

Abstinence sows sand all over the ruddy limbs and flaming hair.

William Blake (1757–1827)

Here are some of the issues each of us must deal with to be contraceptively responsible:

- *Acknowledging sexuality*. On the surface, it may seem fairly simple to acknowledge that we are sexual beings, especially if we have conscious sexual desires and engage in sexual intercourse. Yet acknowledging our sexuality is not necessarily easy, for sexuality may be surrounded by feelings of guilt, conflict, and shame. The younger or less experienced we are, the more difficult it is to acknowledge our sexuality.
- *Planning contraception*. Planning contraception requires us to admit not only that we are sexual but also that we plan to be sexually active. Without such planning, we can pretend that sexual intercourse "just happens" in a moment of passion, when we have been drinking, or when the moon is full—even though it happens frequently.
- *Obtaining contraception*. Difficulty in obtaining contraception may be a deterrent to using it. It is often embarrassing for sexually inexperienced persons to obtain contraceptives. Buying condoms or contraceptive foam at the local drugstore is a public announcement that you are sexual. Who knows if your mother, teacher, or minister might be down the aisle buying toothpaste (or contraceptives, for that matter) and might see you?
- *Continuing contraception*. Many people, especially women using the birth control pill, practice contraception consistently and effectively within a steady relationship but give up their contraceptive practices if the

How did people decide to get pregnant, I wondered. It was such an awesome decision. To undertake responsibility for a new life when you had no way of knowing what it would be like. I assumed that women got pregnant without thinking about it, because if they ever once considered what it really meant, they would surely be overwhelmed with doubt.

Erica Jong, *Fear of Flying*

If toothpaste tasted as disgusting as spermicide, the teeth of the nation would have fallen out years ago.

Germaine Greer

relationship breaks up. They define themselves as sexual only within the context of a relationship. When men or women begin a new relationship, they may not use contraception because the relationship has not yet become long term; they do not expect to have sexual intercourse or to have it often. They are willing to take chances.

- *Dealing with the lack of spontaneity*. Using contraceptive devices such as condoms or diaphragms may destroy the feeling of spontaneity in sex. For those who justify their sexual behavior with romantic impulsiveness, using these devices seems cold and mechanical. Others do not use them because they would have to untangle bodies and limbs, interrupting the passion of the moment.

Because women bear children and have most of the responsibility for raising them, they may have a greater interest then their partners in controlling their fertility. Nevertheless, it is unfair to assume, as people generally do in our society, that the total responsibility for contraception should be the woman's. Men may participate in contraception by using condoms, which are quite effective when used properly, especially in combination with a spermicide (Albert et al., 1995). (See the Resource Center for hints on how to use condoms properly.) Condoms have the additional advantage of being effective in helping prevent the spread of sexually transmissible diseases. (Condoms are also known

He who wears his morality as his best garment were better naked.

KAHLIL GIBRAN, (1983–1931), *THE PROPHET*

Table 7.1 Annual Failure Rates and Outcomes of Unintended Pregnancy for 14 Methods of Contraception

This table shows method failure rates calculated in terms of the number of pregnancies avoided (defined as the difference between the number of pregnancies expected to occur if no method was used and the number expected to occur with that method). Unintended pregnancy outcomes are estimates based on data collected by the Alan Guttmacher Institute.

		Unintended Pregnancy Outcomes, %			
Method	**Failure Rate, %**	**Induced Abortion**	**Spontaneous Abortion**	**Ectopic Pregnancy**	**Term Pregnancy**
Tubal ligation	0.17	23.75	6.20	50.00	20.05
Vasectomy	0.04	47.03	12.28	1.00	39.70
Oral contraceptives	3.00	47.03	12.28	1.00	39.70
Implant	0.32	40.85	10.66	14.00	34.49
Injectable contraceptive (DMPA)	0.30	41.33	10.79	13.00	34.89
Progesterone–T IUD	2.00	39.90	10.42	16.00	33.68
Copper–T IUD	0.42	46.08	12.03	3.00	38.90
Diaphragm	18.00	47.03	12.28	1.00	39.70
Male condom	12.00	47.03	12.28	1.00	39.70
Female condom	21.00	47.03	12.28	1.00	39.70
Spermicides	21.00	47.03	12.28	1.00	39.70
Cervical cap	30.00	47.03	12.28	1.00	39.70
Withdrawal	19.00	47.03	12.28	1.00	39.70
Periodic abstinence	20.00	47.03	12.28	1.00	39.70
No method	85.00	47.03	12.28	1.00	39.70

SOURCE: Trussell, James, et al. (April, 1995) "The Economic Value of Contraception: A Comparison of 15 Methods." *American Journal of Public Health* 85 (4):495.

as *prophylactics* because they protect against disease.) Both men and women report that using a condom gives them "peace of mind" during sexual intercourse (Juran, 1995). In addition to using a condom, a man can take contraceptive responsibility in other ways. These include (1) exploring ways of making love without intercourse; (2) helping pay the woman's medical bills related to contraception and sharing the cost of birth control pills or other contraceptive supplies; (3) checking on contraceptive supplies, helping keep track of the partner's menstrual cycle, and helping the partner acquire supplies; (4) in a long-term relationship, if no (or no more) children are wanted, having a vasectomy.

Reflections

Is it important to you (or your partner) for the male to be an active participant in the practice of contraception? Why?

Abortion

Since 1973, women have been guaranteed the constitutional right to **abortion,** the termination of a pregnancy. The constitutional decision, ***Roe v. Wade***, set off a firestorm of opposition by those who wanted to keep it illegal. The abortion debate has been a major feature of American politics and women's lives for over two decades.

Under safe, clean, and legal conditions, abortion is a very safe medical procedure. Self-administered or under illegal, clandestine conditions, abortion can be very dangerous. The continued availability of legal abortion is considered by most physicians, psychologists, and public health professionals as critical to women's physical and mental well-being (Susser, 1992; Stephenson, Wagner, Baden, and Serbaneson, 1992).

The detriment that the State would impose upon the pregnant woman by denying this choice altogether is apparent. . . . Maternity, or additional offspring, may force upon a woman a distressful life and future. Psychological harm may be imminent. Mental and physical health may be taxed by child care. There is also the distress, for all concerned, associated with the unwanted child, and there is the problem of bringing a child into a family already unable, psychologically or otherwise, to care for it. . . . The additional difficulties and continuing stigma of unwed motherhood may be involved.

ROE V. WADE, 1973

Characteristics of Women Having Abortions

A review of various studies found the following characteristics of women having abortions in the United States (Russo, Horn, and Schwartz, 1992):

- *Race and ethnicity*. The majority of women (69 percent) having abortions are white. Although nonwhite women (aged fifteen through forty-four) represent 17 percent of the population, they have 31 percent of the abortions. Latinas represent 8 percent of all women, but 13 percent of women seeking abortions. The lower abortion rates of whites represent fewer unintended pregnancies rather than less acceptance of abortion.
- *Socioeconomic circumstances*. One out of every three women having an abortion is poor. The abortion rate for poor women is three times that for women with incomes over $25,000 a year. Almost a third are attending school or college; a third are unemployed.
- *Minors*. Nearly 12 percent of abortions are obtained by minors, over 98 percent of whom are unmarried. Most are white, in school, have no children, and have had no previous abortions. Typically they are avoiding having children in order to remain in school and to become more mature before becoming mothers. Thirteen percent of adolescents who have abortions have had prior abortions; 9 percent are already mothers.
- *Adults*. Eighty percent of adult women having abortions are separated, divorced, or never married; 20 percent are married. Among adult women seeking abortions, almost half are already mothers. Women who are already mothers have significant family responsibilities. Of mothers having abortions, nearly half of the single mothers and two-thirds of the married mothers already have at least two children.

Did you know?

The latest data on abortions indicate that in 1992 there were about 6 million pregnancies and 1.5 million abortions. Twenty-one percent of American women of reproductive age have had abortions (Alan Guttmacher Institute, 1994).

If men could get pregnant, then abortion would be a sacrament.

GLORIA STEINEM

YOU AND YOUR WELL-BEING

Choices in Unwanted Pregnancy

When a woman unintentionally becomes pregnant, she is usually faced with a number of decisions. The first is whether or not she wants a child (see "Understanding Yourself: Am I Parent Material?" in Chapter 10). If she determines that she does not want to raise a child at this point in her life, she must choose between terminating the pregnancy or continuing the pregnancy and then having the child adopted. Religious, moral, health, and practical considerations all come into play. Her partner's feelings may also influence her decision, depending upon his level of committment. (Abortion techniques are discussed in the Resource Center; adoption is discussed in Chapter 10 and in the Resource Center.)

A woman considering an abortion may ask herself the following questions:

1. Whom can I talk to about this? Whom can I trust to help me through the process? Who will put my welfare first in this situation?
2. What facilities are available for me? How far will I have to travel? How many visits will I need to make? Is there a waiting period between my first visit and the abortion?
3. Is there a possibility of getting a "medical abortion" using RU-486 or another drug, rather than a vacuum aspiration or a surgical technique?
4. How much will this cost? Will my partner help pay for it?
5. Is this a choice I can live with for the rest of my life? Will I be able to be at peace with myself?

From a medical standpoint, the safest abortions are those performed earliest in the pregnancy. Therefore, it is important for the woman's health and well-being that she get a pregnancy test as soon as she suspects she might be pregnant and that she not put off the decisions that she will ultimately have to make.

Reasons for Having Abortions

There are many stereotypes about women who have abortions: for example, they are selfish, promiscuous, single, unwilling to accept family responsibilities, childless, nonmaternal, depressed, sinners, or immoral (Gordon, 1990; Petchesky, 1990). Furthermore, national public opinion surveys depict a "simplistic image" of women's reasons for abortion (Russo, Horn, and Schwartz, 1992). Because few women openly discuss their abortion experiences, stereotypical views of abortion continue unchallenged.

Women generally have multiple reasons for wanting an abortion. An important study by Aida Torres and Jacqueline Forrest (1988) came to the following conclusions:

In the abortion debate, both sides tend to claim moral rightness. The pro-life movement equates the abortion of an embryo with murder. Pro-choice activists defend abortion as a moral choice that the pregnant woman herself must make.

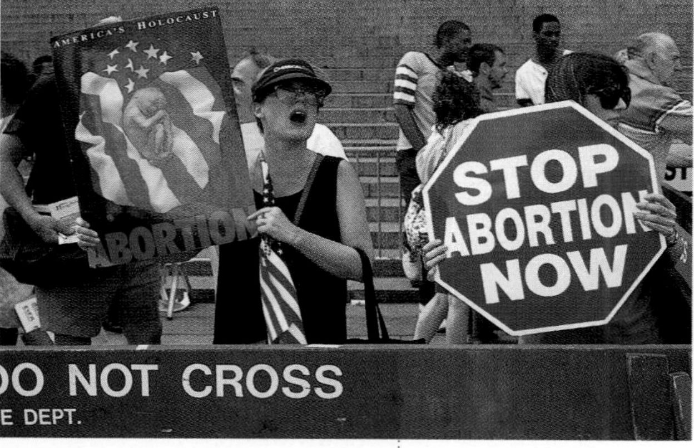

If a woman is considering having her baby adopted, she must likewise begin to make plans at once. She must take care of her health—get prenatal care, eat nutritiously, and avoid cigarettes, alcohol, and other substances that may harm the fetus.

She may ask herself the following questions:

1 Whom can I talk to about this? Whom can I trust to help me through my pregnancy and to support my decision to have my baby adopted?
2 Where will I live during the pregnancy? Will I continue my school or work?
3 How will I arrange the adoption? Can I find a pregnancy or adoption counselor to guide me? Will I go through an adoption agency, a doctor, or a lawyer?
4 Do I want an "open" or "closed" adoption? Do I want to be able to have some continuing contact with the child? Can I help chose the adoptive parents?
5 Can I get financial help during my pregnancy? How can I get the best health care?
6 Do I understand my legal rights and obligations in this situation? Where can I get sound legal advice?
7 What kind of birth plan do I want to make (see Resource Center)? Do I want the adoptive parents to be present at the birth?
8 Is this a decision I can live with in peace for the rest of my life?

An unwanted pregnancy can be a frightening and difficult event for a woman and for her partner. If you should find yourself in this situation, it is important to seek help at once and not allow yourself to be "paralyzed" by fear. You probably have close friends or relatives who want to help you and will not judge you. Seek them out. There are also free counseling services available through many local health agencies and through Planned Parenthood. These agencies can help you deal with your options and with the pregnancy in the way that you feel is best for you.

1 *The abortion decision is complex.* Abortion is not undertaken lightly. The reasons for abortion most commonly cited by women include concern about how a child would change their life, not being able to afford a child, problems in the relationship, and wanting to avoid being a single parent.
2 *A woman's developmental life stage is important.* Eleven percent of the women in the study stated they were "too immature" or "too young" to have a child; 21 percent said they were "unready for responsibility."
3 *Relationships with other people and educational and economic circumstances are important.* Decisions do not reflect only the woman's personal qualities. Over two-thirds said they couldn't afford a child. Twenty-three percent said their husband or partner didn't want a child. Many felt they wouldn't be able to continue their education if they had a child. These reasons point to the significance of outside circumstances and to women's ongoing sense of responsibility to others.

I have noticed that all the people who favor abortion have already been born.

RONALD REAGAN

Reflections

Is it important to you (or your partner) for the male to be an active participant in the practice of contraception? Why?

Men and Abortion

Men are often forgotten people in an abortion; attention is usually focused on the woman who is undergoing the agony of decision. If men are thought of, they are often regarded with hostility and blame, and yet they, like the women, may be undergoing their own private travail, experiencing guilt and anxiety, feeling ambiguous about the possibility of parenthood.

A common feeling men experience is powerlessness. They may try to remain cool and rational, believing that if they reveal their confused feelings they will be unable to give their partners emotional support. Because the drama is within the woman and her body, a man may feel he must not influence her decision.

There is the lure of fatherhood, all the same. A pregnancy forces the man to confront his own feelings about parenting. Parenthood for males, as for females, is a profound passage into adulthood. There is a mixture of pride and fear about being a potential father and a potential adult.

It is not uncommon for some men to temporarily experience erectile or ejaculatory difficulties following an abortion. It is fairly common for couples to split up after an abortion: The stress, conflict, and guilt can be overwhelming. Many abortion clinics now provide counseling for men, as well as women, involved in abortion.

Reflections

If you or your partner became unintentionally pregnant at this time, what impact would this have on your life? What would you do, given your values, beliefs, and situation?

Sexually Transmissible Diseases and HIV/AIDS

O Rose thou art sick.
The invisible worm,
That flies in the night
In the howling storm:
Has found out thy bed
Of crimson joy:
And his dark secret love
Does thy life destroy.
WILLIAM BLAKE (1757–1827)

"Do you have chlamydia, gonorrhea, herpes, syphilis, HIV, or any other sexually transmissible disease?" is hardly a question you want to ask someone on a first date. But it is a question to which you need to know the answers before you become sexually involved. Just because a person is nice is no guarantee that he or she does not have a **sexually transmissible disease (STD),** a disease spread through sexual contact, such as sexual intercourse or oral or anal sex. No one can tell by a person's looks, intelligence, or moral fervor whether he or she has contracted a sexually transmissible disease, and the costs are too great for anyone to become sexually involved with a person without knowing about the presence of any of these diseases.

Americans are in the middle of the worst STD epidemic in our history. The Centers for Disease Control estimate that 12 to 13 million people acquire STDs each year in the United States. College students are as vulnerable as anyone else. Untreated chlamydia and gonorrhea can lead to pelvic inflammatory disease (PID) in women, a major cause of infertility. (The effects of PID on fertility are discussed in the Resource Center.)

Did you know?

Annually, PID affects 1.7 million women, many of whom become sterile (Hilts, 1990a).

Overall, HIV (which will be described in the next section) has infected as many as one million Americans (one in every 250 people). By June 1996, over 548,400 Americans had developed or died from AIDS (which results from HIV infection) (CDC, 1997). HIV and AIDS cases are increasing at a disproportionate rate for African Americans and Latinos; sexually transmitted cases among heterosexuals are increasing at a greater rate than among gay men. These figures indicate that virtually all adults in the United States are or will soon be related to, personally know, work with, or go to school with people who are infected with HIV or will know others whose friends, relatives, or associates test HIV-positive. (See Chapter 12 for a discussion of HIV/AIDS and the family and the Resource Center for a discussion of the different types of sexually transmissible diseases.)

Principal STDs

I had the honor
To receive, worse luck!
From a certain empress
A boiling hot piss.
FREDERICK THE GREAT OF PRUSSIA (1712–1786)

The most prevalent STDs in the United States are chlamydia, gonorrhea, genital warts, genital herpes, syphilis, hepatitis, and HIV/AIDS. Conditions that may be sexually transmitted include urethritis (in both women and men) and vaginitis and PID (in women). Table 7.2 (on pages 248–249) briefly describes the symptoms, exposure intervals, treatments, and other information regarding the principal STDs.

PERSPECTIVE

The Abortion Debate

Those supporting the prohibition of abortion generally identify themselves as "pro-life." Those supporting a woman's right to choose for herself to have an abortion generally identify themselves as "pro-choice." Figure 7.3 indicates the opinions of Americans on this sensitive issue.

The Pro-Life Argument

For those who oppose abortion, there is a basic principle from which their arguments follow: The moment an egg is fertilized, it becomes a human being with the full rights and dignity afforded other humans. A zygote is no less human than a fetus, and a fetus is no less human than a baby. Morally, aborting a zygote or embryo is the equivalent of killing a person.

They generally oppose sex education in the public schools that teaches about contraception. They believe contraceptive knowledge will encourage adolescents to become sexually active. For them, abstinence is the only legitimate alternative for unmarried adolescents and adults.

Even though the majority of those opposing abortion would allow rape and incest (and sometimes a defective embryo or fetus) as exceptions, the pro-life leadership generally opposes any exception except to save the life of the pregnant woman. To abort the zygote of a rape or incest survivor, they reason, is still taking an innocent human life.

Finally, pro-life advocates argue that there are thousands of couples who want to adopt children but are unable to do so because pregnant women abort rather than give birth.

The Pro-Choice Argument

Those who believe that abortion should continue to be legal present a number of arguments. First, for pro-choice men and women, the fundamental issue is who decides whether a woman will bear children: the woman or the state. Because women continue to bear the primary responsibility for rearing children, pro-choice advocates believe that women should not be forced to give birth to unwanted children. Becoming a mother alters a woman's role more profoundly than almost any other event in her life; it is more significant than marrying. When women have the choice of becoming mothers, they are able to decide the timing and direction of their lives.

Second, while pro-choice advocates support sex education and contraception to eliminate much of the need for abortion, they believe that abortion should continue to be available as birth control back-up. Because no contraceptive is 100 percent effective, unintended pregnancies occur even among the most conscientious contraceptive users.

Third, if abortion is made illegal, large numbers of women nevertheless will have illegal abortions, substantially increasing the likelihood of dangerous complications, infections, and death. Those who are unable to have an abortion may be forced to give birth to and rear a child they did not want.

Figure 7.3
American Opinions About Abortion

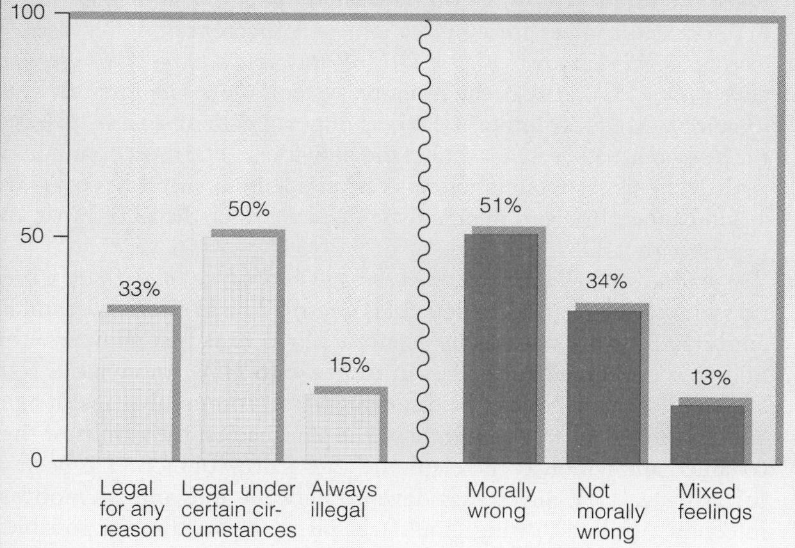

SOURCE: "Legality, Morality of Abortion," (March, 1995) *Gallup Poll Monthly*, 354: pp. 30-31.

The poor homosexuals—they have declared war upon nature and now nature is exacting an awful retribution.

PAT BUCHANAN

HIV/AIDS

The **human immunodeficiency virus (HIV)** is the virus that causes AIDS. **Acquired immune deficiency syndrome (AIDS)** is so termed because of its characteristics.

Acquired—because people are not born with it.

Immune—because the disease relates to the immune system.

Deficiency—because of the body's lack of immunity.

Syndrome—because the symptoms occur together as a group.

Although there is no vaccine to prevent or cure HIV or to prevent the subsequent AIDS symptoms, we have considerable knowledge about the nature of the virus and how to prevent its spread:

- *HIV attacks the body's immune system.* HIV is carried in the blood, semen, and vaginal secretions of infected persons. A person may be HIV-positive (infected with HIV) for many years before developing AIDS symptoms.
- *HIV is transmitted only in certain clearly defined circumstances.* It is transmitted through the exchange of blood (as by shared needles), through sexual contact involving semen or vaginal secretions, and, prenatally, from an infected woman to her fetus through the placenta.
- *Heterosexuals, bisexuals, gay men, and lesbians are susceptible to the sexual transmission of HIV.* Currently 6.5 percent of AIDS cases are attributed to heterosexual transmission of HIV. The rate of heterosexual HIV transmission, however, is rising at three times the rate of gay transmission and almost twice the rate of transmission by contaminated needles. Among women, heterosexual contact accounts for almost 35 percent of AIDS cases.
- *There is a definable progression of HIV infection and a range of illnesses associated with AIDS.* HIV attacks the immune system. Once the immune system is impaired, AIDS symptoms occur, as opportunistic diseases—diseases that the body normally resists—infect the individual. The most common opportunistic diseases are pneumocystis carinii pneumonia and Kaposi's sarcoma, a skin cancer. It is an opportunistic disease rather than HIV that kills the person with AIDS.
- *The presence of HIV can be detected through antibody testing.* To date there are no widely available tests to detect HIV itself. The Western Blot and ELISA antibody tests are reasonably accurate blood tests that show whether the body has developed antibodies in response to HIV. Anonymous testing is available at many college health centers and community health agencies. Self-test kits may also be purchased at pharmacies; the results of these are obtained anonymously by telephone after a drop of blood has been sent in for analysis. HIV antibodies develop between one and six months after infection. Antibody testing should take place one month after possible exposure to the virus and, if the results are negative, again six months later. If the antibody is present, the test will be positive. That means that the person has been infected with HIV and an active virus is present. The presence of HIV does not mean, however, that the person necessarily will develop AIDS symptoms in the near future; symptoms generally occur seven to ten years after the initial infection.
- *All those with HIV (whether or not they have AIDS symptoms) are HIV carriers.* They may infect others through unsafe sexual activity or by sharing needles; if they are pregnant, they may infect the fetus.

I think God did send AIDS for a reason. It was to show how mean and sinful a healthy man can be toward a sick man.

JOE BOB BRIGGS

The AIDS crisis pits two basic human goals and needs against one another: seeking love and avoiding death.

LYNN MILLER

What we learn in times of pestilence [is] that there are more things to admire in men than to despise.

ALBERT CAMUS, (1913–1960), *THE PLAGUE*

Volo comparare nonnulla tegumembra [Latin]. "I'd like to buy some condoms."

Protecting Yourself and Others

Although many people have changed their behaviors to reduce the risk of STDs and HIV infection, many continue to jeopardize their health and lives—as well as the health and lives of their partners and loved ones—by failing to take adequate precautions. They worry about STDs and HIV but do not take the necessary steps—such as always using condoms—to prevent infections. One study of sexually active young adults found that 44 percent had not changed their behavior in any way to reduce the risk of HIV infection (Cochran, Keidan, and Kalechsteir, 1989). A study of female college students found that except for causing an increase in regular condom use—from 21 percent in 1986 to 41 percent in 1989—public health campaigns have not had a substantial influence on the habits and behavior of these well-educated young adults (DeBuono, Zinner, Daamen, and McCormack, 1990).

Said one male university student from Illinois who knew that heterosexuals are at risk for contracting HIV, "I just don't see AIDS as being much of a threat to heterosexuals, and I don't find a lot of pleasure in using a condom" (Johnson, 1990). Another study of college students found that they believe they can "identify" infected men and women (Maticka-Tyndale, 1991). An Ohio female student explained why she does not insist that her partner use a condom: "I have an attitude—it may be wrong—that any guy I would sleep with would not have AIDS" (Johnson, 1990).

Abstinence is the best protection from STDs and HIV. If you are sexually active, however, the key to protecting yourself and others is to talk with your partner about STDs in an open, nonjudgmental way and to use condoms. The best way of finding out whether your partner has an STD is by asking. If you feel nervous about broaching the subject, you can rehearse talking about it. It may be sufficient to ask in a lighthearted manner, "Are you as healthy as you look?" or because many people are uncomfortable asking about STDs, you can open up the topic by revealing your anxiety: "This is a little difficult for me to talk about because I like you and I'm embarrassed, but I'd like to know whether you have herpes, or HIV, or whatever." If you have an STD, you can say, "Look, I like you, but we can't make love right now because I have a chlamydial infection and I don't want you to get it."

Remember, however, that every person who believes he or she doesn't have an STD may honestly not know. Women with chlamydia and gonorrhea, for example, generally don't exhibit symptoms. Both men and women infected with HIV may not show any symptoms for years, although they are capable of spreading the infection through sexual contact.

If you don't know whether your partner has an STD, use a condom. Even if you don't discuss STDs, condoms are simple and easy to use without much discussion. Both men and women can carry them. A woman can take a condom from her purse and give it to her partner. If he doesn't want to use it, she can say, "No condom, no sex." (Condoms and safer sex practices are discussed in greater detail in the Resource Center.)

Sexual Responsibility

Because we have so many sexual choices today, we need to understand what responsibilities our sexuality entails. Sexual responsibility includes the following:

- *Disclosure of intentions.* Each person needs to reveal to the other whether a sexual involvement indicates love, commitment, recreation, and so on.

Did you know?

According to a Centers for Disease Control study, regular condom use almost tripled among women after they were counseled at clinics—from 13 percent to 36 percent ("Urban Women," 1997).

Figure 7.4
AIDS and Ethnicity, 1996

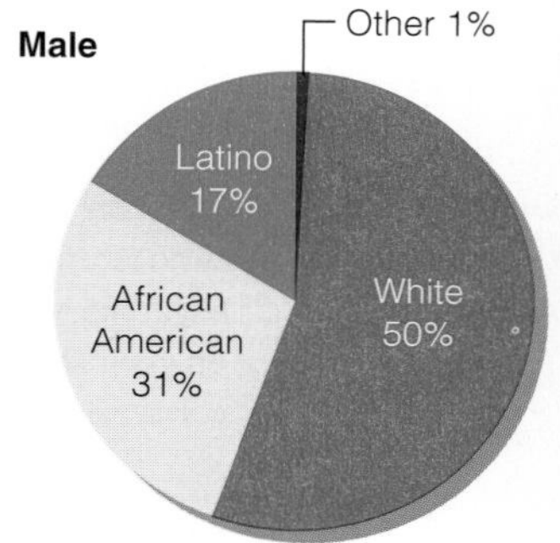

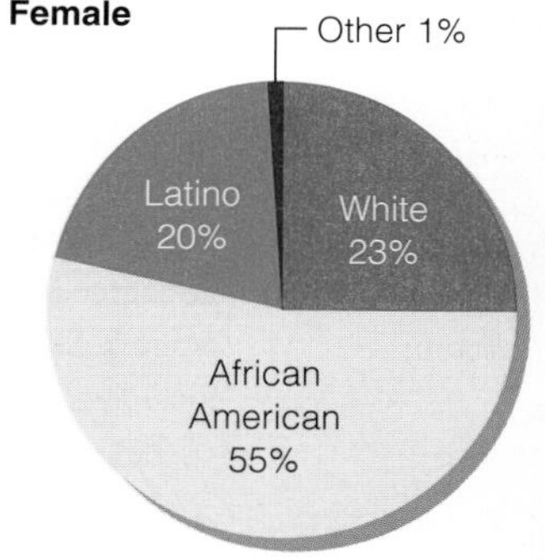

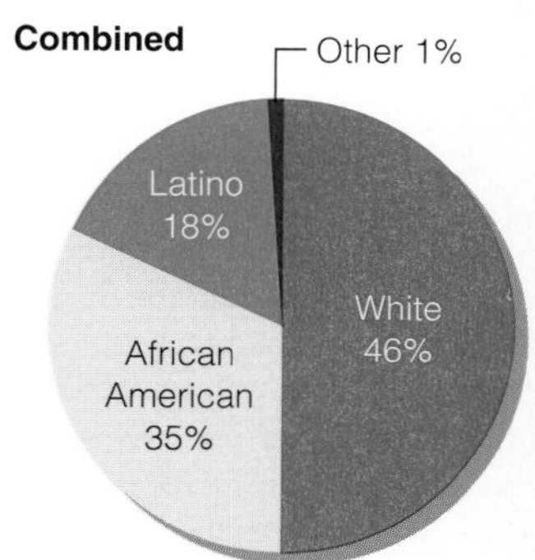

SOURCE: HIV/AIDS Surveillance Report, June, 1997.

Table 7.2 Principal Sexually Transmitted Diseases

STD and Infecting Organism	Symptoms	Time from Exposure to Occurrence	Medical Treatment	Comments
Chlamydia (*Chlamydia trachomatis*)	Women: 80% asymptomatic; others may have vaginal discharge or pain with urination. Men: 30–50% asymptomatic; others may have discharge from penis, burning urination, pain and swelling in testicles, or persistent low fever.	7–21 days	Doxycycline, tetracycline, erythromycin	If untreated, may lead to pelvic inflammatory disease (PID) and subsequent infertility in women.
Gonorrhea (*Neisseria gonorrhoeae*)	Women: 50–80% asymptomatic; others may have symptoms similar to chlamydia. Men: itching, burning or pain with urination, discharge from penis ("drip").	2–21 days	Penicillin, tetracycline, or other antibiotics	If untreated, may lead to pelvic inflammatory disease (PID) and subsequent infertility in women.
Genital warts (Human papilloma virus)	Variously appearing bumps (smooth, flat, round, clustered, fingerlike, white, pink, brown, etc.) on genitals, usually penis, anus, vulva, vagina, or cervix.	1–6 months (usually within 3 months)	Surgical removal by freezing, cutting, or laser therapy. Chemical treatment with podophyllin (80% of warts eventually reappear).	Virus remains in the body after warts are removed.
Genital herpes (Herpes simplex virus)	Small, itchy bumps on genitals, becoming blisters that may rupture, forming painful sores; possibly swollen lymph nodes; flulike symptoms with first outbreak.	3–20 days	No cure although acyclovir may relieve symptoms. Nonmedical treatments may help relieve symptoms.	Virus remains in the body, and outbreaks of contagious sores may recur. Many people have no symptoms after the first outbreak
Syphilis (*Treponema pallidum*)	Stage 1: Red, painless sore (chancre) at bacteria's point of entry. Stage 2: Skin rash over body, including palms of hands and soles of feet.	Stage 1: 1–12 weeks Stage 2: 6 weeks to 6 months after chancre appears	Penicillin or other antibiotics	Easily cured, but untreated syphilis can lead to ulcers of internal organs and eyes, heart disease, neurological disorders, and insanity.

Reflections

How do you deal with the issue of STDs in your relationships? Do you discuss the topic with a partner prior to becoming sexually involved? How do you or would you bring up the topic of STDs? Have you ever contracted an STD? If so, how did you feel about your partner? Yourself?

- *Freely and mutually agreed-upon sexual activities.* Each individual has the right to refuse any or all sexual activities without the need to justify his or her feelings. There can be no physical or emotional coercion.
- *Use of mutually agreed-upon contraception in sexual intercourse if pregnancy is not intended.* The persons in a sexual relationship are equally responsible for preventing an unintended pregnancy in a mutually agreed-upon manner.
- *Use of "safer sex" practices.* Each person is responsible for practicing safer sex unless both have been monogamous with each other for at least five years or have recently tested negative for HIV. Safer sex practices guard against sexually transmissible diseases, especially HIV/AIDS. Such practices do not transmit semen, vaginal secretions, or blood during sexual activities.

Table 7.2 (continued)

STD and Infecting Organism	Symptoms	Time from Exposure to Occurrence	Medical Treatment	Comments
Hepatitis (Hepatitis A or B virus)	Fatigue, diarrhea, nausea, abdominal pain, jaundice, darkened urine due to impaired liver function.	1–4 months	No medical treatment available; rest and fluids are prescribed until the disease runs its course.	Hepatitis B more commonly spread through sexual contact; can be prevented by vaccination.
Urethritis (various organisms)	Painful and/or frequent urination; discharge from penis; women may be asymptomatic.	1–3 weeks	Penicillin, tetracycline, or erythromycin, depending on organism.	Laboratory testing is important to determine appropriate treatment.
Vaginitis (*Gardnerella vaginalis, Trichomonas vaginalis,* or *Candida albicans*)	Intense itching of vagina and/or vulva, unusual discharge with foul or fishy odor, painful intercourse. Men who carry organisms may be asymptomatic.	2–21 days	Depends on organism; oral medications include metronidazole and clindamycin. Vaginal medications include clotrimazole and miconazole.	Not always acquired sexually. Other causes include stress, oral contraceptives, pregnancy, tight pants or underwear, antibiotics, douching, and dietary imbalance.
HIV infection and AIDS (Human immunodeficiency virus)	Possible flulike symptoms but often no symptoms during early phase. Variety of later symptoms including weight loss, persistent fever, night sweats, diarrhea, swollen lymph nodes, bruiselike rash, persistent cough.	Several months to several years	No cure available, although many symptoms can be treated with medications. Antiviral drugs may strengthen immune system. Good health practices can delay or reduce the severity of symptoms.	Cannot be self-diagnosed; a blood test must be performed to determine the presence of the virus.
Pelvic inflammatory disease (PID) (women only)	Low abdominal pain, bleeding between menstrual periods, persistent low fever.	Several weeks or months after exposure to chlamydia or gonorrhea (if untreated)	Penicillin or other antibiotics; surgery.	Caused by untreated chlamydia or gonorrhea; may lead to chronic problems such as arthritis and infertility.

SOURCE: Bryan Strong and Christine DeVault. *Human Sexuality.* Mountain View, CA: Mayfield Publishing Company, 1997.

- *Disclosure of infection from or exposure to STDs*. Each person must inform his or her partner about personal exposure to an STD because of the serious health consequences, such as infertility or AIDS, that may follow untreated infections. If you are infected, you must refrain from behaviors—such as sexual intercourse, oral-genital sex, and anal intercourse—that may infect your partner. To help ensure that STDs are not transmitted, use a condom.
- *Acceptance of the consequences of sexual behavior*. Each person needs to be aware of and accept the possible consequences of his or her sexual activities. These consequences can include emotional changes, pregnancy, abortion, and sexually transmissible diseases.

To approach sex carelessly, shallowly, with detachment and without warmth is to dine night after night in erotic greasy spoons. In time, one's palate will become insensitive, one will suffer (without knowing it) emotional malnutrition, the skin of the soul will fester with scurvy, the teeth of the heart will decay.

TOM ROBBINS, *STILL LIFE WITH WOODPECKER*

A man, on entering the waiting room of a veterinarian's office with his sick dog, sat next to a lady with a beautiful wolfhound. The wolfhound was extremely high spirited and happily gamboled around the waiting room as the man's own dog lay limply on the floor. Finally, curious as to why such an apparently healthy dog should be in a veterinarian's office, he turned to the lady and said:

"You certainly have a beautiful dog."

"Oh, thank you," she replied.

"He looks so healthy," said he, "that I am surprised to see him in a veterinarian's office. What is wrong with him?"

"Oh," she said with some embarrassment, "he has syphilis."

"Syphilis!" he said. "How did he get syphilis?"

"Well," she said, "he claims he got it from a tree."

Dorothy Parker (1893–1967)

Sex contains all, bodies, souls,
Meanings, proofs, purities, delicacies, results, promulgations,
Songs, commands, health, pride, the seminal milk
All hopes, benefactions, bestowals, all loves, beauties, delights of the earth,
These are contain'd in sex as parts of itself and justifications of itself.

Walt Whitman (1819–1892)

Responsibility in many of these areas is facilitated when sex takes place within the context of an ongoing relationship. In that sense, sexual responsibility is a matter of values. Is responsible sex possible outside an established relationship? Are you able to act in a sexually responsible way? Sexual responsibility also leads to the question of the purpose of sex in your life. Is it for intimacy, erotic pleasure, reproduction, or other purposes?

As we consider the human life cycle from birth to death, we cannot help but be struck by how profoundly sexuality weaves its way through our lives. From the moment we are born, we are rich in sexual and erotic potential, which begins to take shape in our sexual experimentations of childhood. As children, we were still unformed, but the world around us haphazardly helped give shape to our sexuality. In adolescence, our education continued as a mixture of learning and yearning. But as we enter adulthood, with greater experience and understanding, we undertake to develop a mature sexuality: we establish our sexual orientation as heterosexual, gay, lesbian, or bisexual; we integrate love and sexuality; we forge intimate connections and make commitments; we make decisions regarding our fertility and sexual health; we develop a coherent sexual philosophy. Then, in our middle years, we redefine sex in our intimate relationships, accept our aging, and reevaluate our sexual philosophy. Finally, as we become elderly, we reinterpret the meaning of sexuality in accordance with the erotic capabilities of our bodies. We come to terms with the possible loss of our partner and our own end. In all these stages, sexuality weaves its bright and dark threads through our lives.

Summary

- Our primary sources of sexual learning are parents, peers, and the media. As we get older, interactions with our partners become increasingly important.
- There are several tasks that we must undertake in developing our sexuality as young adults, including (1) integrating love and sex, (2) forging intimacy and commitment, (3) making fertility/childbearing decisions, (4) establishing a sexual orientation, and (5) developing a sexual philosophy.
- The traditional female *sexual scripts* include the following ideas: Sex is both good and bad (depending on the context), don't touch me "down there," sex is for men, men should know what women want, women shouldn't talk about sex, women should look like beautiful models, women are nurturers, and there is only one right way to experience an orgasm. Traditional male sexual scripts include the following: Men should not have (or at least should not express) certain feelings, performance is the thing that counts, the man is in charge, a man always wants sex and is ready for it, all physical contact leads to sex, sex equals intercourse, and sexual intercourse always leads to orgasm.
- Contemporary sexual scripts are more egalitarian, consisting of the following beliefs: Sexual expression is positive, sexual activities are a mutual exchange of erotic pleasure, sexuality involves both partners equally and the partners are equally responsible, legitimate sexual activities include masturbation and oral-genital sex, sexual activities may be initiated by either partner, both have a right to experience orgasm, and nonmarital sex is acceptable within a relationship context.
- Between 1 and 10 percent of American men are *gay* and between 1 and 3 percent of American women are *lesbian* at one time or another in their lives. Identifying oneself as gay or lesbian occurs over considerable time, usually beginning in late childhood or early adolescence. The first stage is marked by fear and suspicion that one's sexual desires are different from others'. In the second stage the person labels these feelings as gay or lesbian feelings. The third stage includes the person's self-definition as gay or lesbian. Two additional stages include entering the gay subculture and having a gay or lesbian love affair.
- Lesbian, gay, and bisexual individuals may be subject to *antigay prejudice* or *homophobia*, leading to verbal abuse, discrimination, or violence. Education and positive social interactions can reduce such prejudice.
- *Bisexuals* are attracted to members of both genders. In developing a bisexual identity men and women go through several stages: (1) initial confusion, (2) finding and applying the bisexual label, (3) settling into the identity, and (4) continued uncertainty. Bisexuals don't have a community or social environment that reaffirms their identity.
- Developmental tasks in middle adulthood include (1) redefining sex in marital or other long-term relationships, (2) reevaluating one's sexuality, and (3) accepting the biological aging process. In middle age, women tend to reach their sexual peak, which is often maintained into their sixties and beyond; they also experience menopause. The sexual responsiveness of men declines somewhat, causing men to require greater stimulation and time to become aroused. There is no male equivalent to menopause.
- Many of the psychosexual tasks older Americans must undertake are directly related to the aging process. They include changing sexuality and loss of the partner. The sexuality of older Americans tends to be invisible because (1) we associate sexuality with youth, (2) we associate romance and love with youth, (3) we associate sex with procreation, and (4) the elderly themselves do not have desires as strong as those of the young. The main determinants of sexual activity in old age are health and the availability of a partner.
- *Autoeroticism* consists of sexual activities that involve only the self. It includes sexual fantasies, masturbation, and erotic dreams. Erotic fantasizing is the most universal of all sexual behaviors. Sexual fantasies serve several functions: (1) they help direct and define our erotic goals, (2) they allow us to plan or anticipate erotic situations, (3) they provide escape from a dull or oppressive environment, and (4) they bring novelty and excitement into a relationship. Erotic dreams are widely experienced.
- *Masturbation* is an important means of learning about our bodies. By the end of adolescence, most men and the majority of women have masturbated to orgasm. Most people continue to masturbate during marriage, although married men tend to masturbate to supplement their sexual activities whereas women tend to masturbate as a substitute for such activities.

- Oral-genital sex, which includes *cunnilingus* and *fellatio*, is practiced by heterosexuals, gay men, and lesbians.
- *Sexual intercourse (coitus)* is the insertion of the penis into the vagina and the stimulation that follows. It is a complex interaction, involving more than erotic pleasure or reproduction. It is a form of communication that may have many motivations and express a host of feelings.
- *Anal eroticism* is practiced by both heterosexuals and gay men. From a health perspective, *anal intercourse* is dangerous because it is the most common means of sexually transmitting HIV.
- Sexual enhancement means improving the quality of a sexual relationship. It is based on accurate information about sexuality, developing communication skills, fostering positive attitudes, and increasing self-awareness. Awareness of your own sexual needs is often critical to enhancing your sexuality. Enhancing sex includes the intensification of arousal, which requires focusing on erotic pleasure.
- *Nonmarital* sex is sexual activities, especially sexual intercourse, that take place outside of marriage. *Premarital sex* is sexual activities between younger, never-married adults under the age of thirty. Premarital sexual intercourse is widely accepted. In sexual decision making, love or liking is often the most important factor leading men and women to have premarital intercourse. In contrast to married couples, cohabitants have a higher frequency of sex, greater equality in initiating sexual activities, and more extrarelational sex.
- In gay and lesbian relationships, gay men are likely to initiate sexual activity earlier in the relationship than are lesbians. Lesbians do not initiate sex as often as do gay or heterosexual men. In the gay and lesbian culture, sexual exclusivity is negotiable.
- Marital sex tends to decline in frequency over time, but this does not necessarily signify marital deterioration. Sex is only one bond among many in marriage. Sex within marriage is different from premarital sex in the following ways: (1) Sex in marriage is expected to be monogamous, (2) procreation is a legitimate goal, and (3) the sex takes place in the everyday world.
- *Extramarital* sexual involvements assume three basic forms: sexual but not emotional, sexual and emotional, and emotional but not sexual. Women are increasingly engaging in extramarital affairs; male involvement has remained relatively high. Extramarital affairs appear to be related to two variables: unhappiness in the marriage and premarital sexual permissiveness.
- *Sexual dysfunctions* are recurring persistent problems in giving and receiving erotic satisfaction. Sexual dysfunctions may be physiological or psychological in origin. Both men and women are subject to hypoactive (low or inhibited) sexual desire. The most common female problems are orgasmic dysfunction, arousal difficulties, and dyspareunia (painful intercourse). The most common male problems are erectile dysfunction, premature ejaculation, and delayed orgasm. Two of the most prominent causes of sexual dysfunction are performance anxiety and conflicts within the self. Relationship discord affects sexuality through transference, lack of trust, power struggles, contractual disappointments, sexual sabotage, and lack of communication.
- The most consistent contraceptive users are men and women who explicitly communicate about contraception and those who are in serious relationships of long duration with frequent, predictable patterns of sexual activity. Issues in contraceptive use that each of us must deal with include acknowledging sexuality; planning, obtaining, and continuing contraception; and dealing with the lack of spontaneity.
- Women are guaranteed the constitutional right to abortion in *Roe v. Wade* (1973). About 21 percent of all women of reproductive age have had abortions. The majority of women having abortions are white; one-third are poor. Eighty percent of adult women having abortions are unmarried; 20 percent are married. Almost half of adult women seeking abortions already have at least one child.
- The reasons for having abortion are complex, including a woman's developmental/life stage, relationships with others, educational goals, and economic circumstances. Most women suffer the greatest distress prior to the abortion; the most common feeling afterwards is relief. There is no scientific evidence supporting a so-called post-abortion trauma syndrome.
- *Sexually transmissible diseases (STDs)*, especially chlamydia and gonorrhea, are epidemic. *Acquired immune deficiency syndrome (AIDS)* is caused by the *human immunodeficiency virus (HIV)*, which attacks the body's immune system. HIV is carried in the blood, semen, and vaginal fluid of infected persons. Heterosexuals, bisexuals, and gay men and lesbians are susceptible to the sexual transmission of HIV. If one is sexually active, the keys to protection against STDs, including HIV/AIDS, are communication and condom use.

Key Terms

Suggested Readings

Boston Women's Health Collective. *The New Our Bodies, Ourselves.* New York: Simon & Schuster, 1992. A landmark self-help book on women's sexuality and health. Still as good as ever, still the object of censorship.

Chapple, Steve, and David Talbot. *Burning Desires: Sex in America—A Report from the Field.* New York: Doubleday, 1989. An entertaining and informative work on contemporary sexuality and popular culture by two journalists.

Ellenberg, Daniel, and Judith Bell. *Lovers for Life.* Santa Rosa, CA: Aslan Publishing, 1995. Practical and creative advice and "homework" for couples in long-term relationships.

Hooper, Anne. *Anne Hooper's Kama Sutra.* New York: Dorling Kindersley, 1994. Classic lovemaking techniques reinterpreted for modern lovers; illustrated with explicit, but tasteful, photographs and drawings.

Isensee, Rik. *Love between Men: Enhancing Intimacy and Keeping Your Relationship Alive.* New York: Prentice-Hall Press, 1990. A best-selling self-help book on enhancing gay relationships through communication and problem solving.

Klein, Marty. *Ask Me Anything: A Sex Therapist Answers the Most Important Questions for the 90's.* New York: Simon & Schuster, 1992. An informative and compassionate guide to sexuality and relationships.

Kroll, Ken, and Erica Levy Klein. *Enabling Romance: A Guide to Love, Sex, and Relationships for the Disabled.* New York: Harmony Books, 1992. Interviews, with hundreds of men and women with disabilities, about stereotypes, different ways of being sexual, use of sex toys, and strengthening.

Loulan, JoAnn. *Lesbian Sex.* San Francisco: Spinsters, 1984. One of the most popular books on lesbian sexuality and enhancement.

Michael, Robert, John Gagnon, Edward Laumann, and Gina Kolata. *Sex in America: The Definitive Survey.* Boston: Little, Brown, 1994. A random, cross-sectional survey, conducted under the supervision of noted sociologists, in which over 3,200 men and women were personally interviewed. Emphasizes the social context of sexuality.

Pittman, Frank. *Private Lives: Infidelity and the Betrayal of Intimacy.* New York: Norton, 1989. An intelligent, thoughtful look at extramarital affairs.

Reiss, Ira. *The End of Shame: Shaping Our Next Sexual Revolution.* Buffalo, NY: Prometheus, 1990. An impassioned plea for transforming sexuality into a positive force in contemporary society—made by a leading sociological researcher in human sexuality.

Rubin, Lillian. *Erotic Wars.* New York: Farrar, Strauss, & Giroux, 1990. A fine portrayal by a thoughtful therapist of the yearnings of and misunderstandings between men and women.

Shilts, Randy. *And the Band Played On: People, Politics, and the AIDS Epidemic.* New York: St. Martin's Press, 1987. A compelling story of the medical, political, and human responses to the AIDS crisis.

Steinberg, David. *The Erotic Impulse: Honoring the Sensual Self.* New York: Jeremy Tarcher, 1992. An outstanding collection of essays and poems by writers, poets, teachers, and psychologists.

Strong, Bryan, and Christine DeVault. *Human Sexuality.* 2nd ed. Mountain View, CA: Mayfield, 1997. A comprehensive introduction to human sexuality.

Weinberg, Martin, Colin Williams, and Douglas Pryor. *Dual Attraction: Understanding Bisexuality.* New York: Oxford University Press, 1994. An important empirical work.

CHAPTER 8

Pregnancy and Childbirth

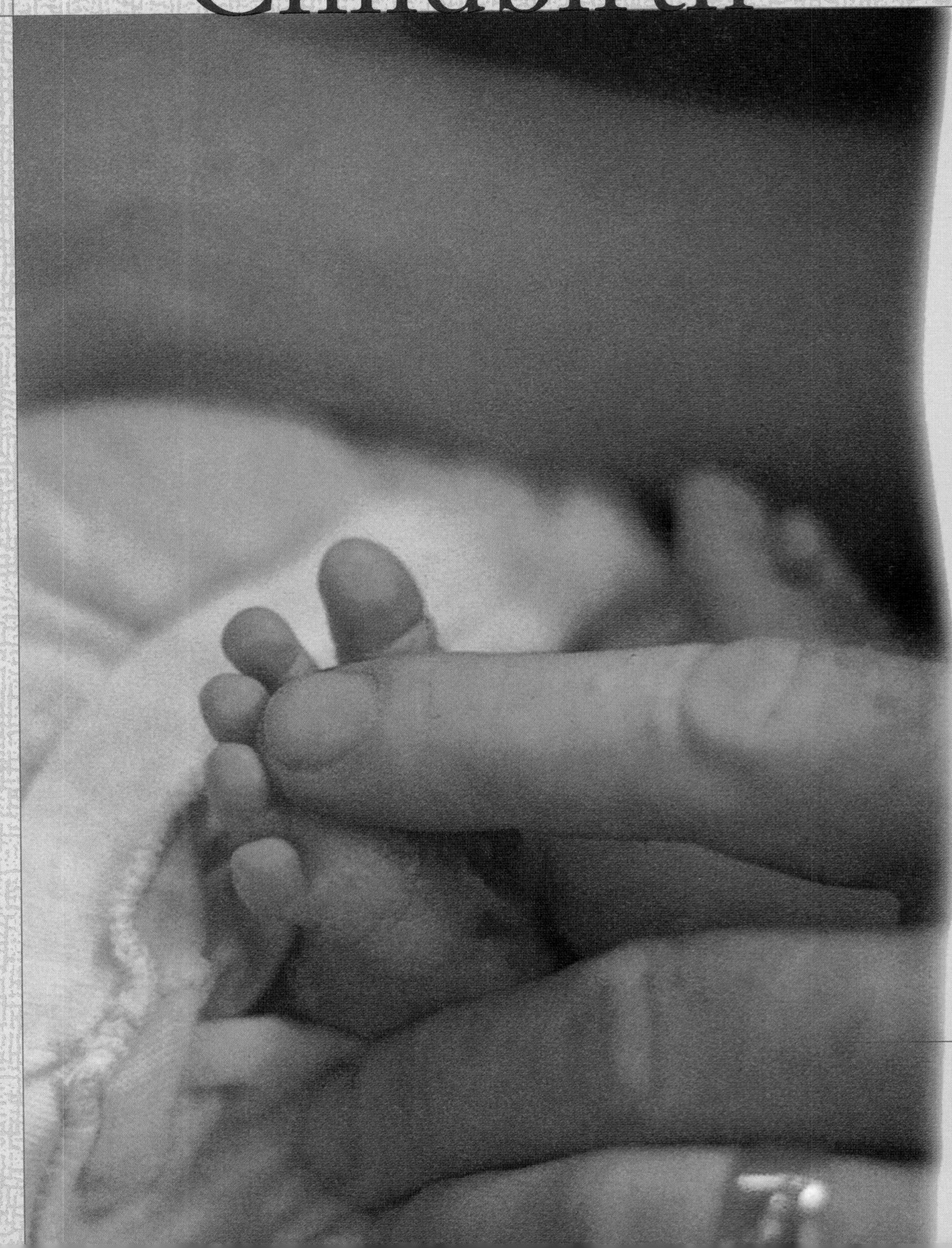

P R E V I E W

To gain a sense of what you already know about the material covered in this chapter, answer "True" or "False" to the statements below.

1. Home pregnancy tests are about 95 percent accurate. True or false?
2. It is usually unsafe for a woman to have sexual intercourse during the last two months of pregnancy. True or false?
3. Even a moderate amount of alcohol consumption affects the fetus. True or false?
4. If a pregnant woman is HIV-positive, there is a 90 percent chance her baby will be positive as well. True or false?
5. Miscarriage and stillbirth are major life events for parents. True or false?
6. The United States ranks twentieth for low infant mortality. True or false?
7. About one-fourth of all U.S. births involve cesarean sections. True or false?
8. In "prepared childbirth," a woman learns how to control the birth process through breathing and exercises. True or false?
9. Circumcision, the surgical removal of the foreskin of the penis, is performed mainly as a medical necessity for health reasons. True or false?
10. Breastfeeding is recommended by the American Academy of Pediatricians because it provides better nutrition and protection from diseases than formula. True or false?

Answers:

1 True, 2 False, 3 True, 4 False, 5 True, 6 True, 7 True, 8 False, 9 False, 10 True

O U T L I N E

The birth of a wanted child is considered by many parents to be the happiest event of their lives. Today, however, pain and controversy surround many aspects of this altogether natural process. As we struggle to balance the rights of the mother, the father, the fetus, and society itself, we find ourselves considering the quality of life as well as life's mere existence.

For most American women, pregnancy will be relatively comfortable and the outcome predictably joyful. For others, especially among the poor, the prospect of having children raises the specters of drugs, disease, malnutrition, and familial chaos.

In this chapter, we will view pregnancy and childbirth from biological, social, and psychological perspectives. We will examine the physical and emotional aspects of pregnancy, giving special attention to maternal and fetal health. We will look at childbirth practices today, including several areas of controversy. We will also consider pregnancy loss, and we will look at the challenges of the transition to parenthood. Detailed information on fetal development and on infertility and its treatment may be found in the Resource Center.

Did you know?

Every year about 2.3 million American couples seek help for infertility (Begley, 1995).

Being Pregnant

What was your original face before you were born?

ZEN KOAN (RIDDLE)

Pregnancy is an important life event for both women and their partners. From the moment it is discovered, a pregnancy affects people's feelings about themselves, their relationship with their partner, and the interrelationships of other family members as well.

Pregnancy Tests

Chemical tests designed to detect the presence of human chorionic gonadatropin (HCG), the hormone secreted by the implanted blastocyst, usually determine pregnancy approximately two weeks following a missed (or spotty) menstrual period. In the *agglutination test*, a drop of the woman's urine causes a test solution to coagulate if she is not pregnant; if she is pregnant, the solution will become smooth and milky in consistency. Home pregnancy tests to detect HCG may be purchased in most drugstores. The directions must be followed closely. Blood analysis can also be used to determine if a pregnancy exists. Although such

Yes—the history of man for the nine months preceding his birth would, probably, be far more interesting, and contain events of greater moment, than all the three score and ten that follow.

SAMUEL COLERIDGE (1772–1834)

tests diagnose pregnancy with better than 95 percent accuracy, no absolute certainty exists until a fetal heartbeat and movements can be detected or ultrasonography is performed.

The first reliable physical sign of pregnancy can be distinguished about four weeks after a woman misses her period. By this time, changes in her cervix and pelvis are apparent during a pelvic examination and the woman is considered to be eight weeks pregnant according to medical terminology. As noted earlier, physicians calculate pregnancy as beginning at the time of the woman's last menstrual period rather than at the time of actual fertilization (because the date of fertilization is often difficult to determine). Another signal of pregnancy, called **Hegar's sign,** is a softening of the uterus just above the cervix; it can be felt during a vaginal examination. Additionally, a slight purple hue colors the labia minora, and the vagina and cervix take on a purplish color rather than showing the usual pink.

Emotional, Psychosocial, and Physical Changes During Pregnancy

Since mother is not all there is to any woman, once she becomes a mother, how does a woman weave the mother into her adult self?

Andrea Eagan

A woman's feelings during pregnancy will vary dramatically according to who she is, how she feels about pregnancy and motherhood, whether the pregnancy was planned, whether she has a secure home situation, and many other factors. Her feelings may be ambivalent; they will probably change over the course of the pregnancy.

A woman's first pregnancy is especially important because it has traditionally symbolized the transition to maturity. Even as social norms change and it becomes more common and acceptable for women to defer childbirth until they've established a career or to choose not to have children, the significance of first pregnancy should not be underestimated. It is a major developmental milestone in the lives of mothers—and fathers as well (Marsiglio, 1991; Notman and Lester, 1988; Snarey et al., 1987).

A couple's relationship is likely to undergo changes during pregnancy. It can be a stressful time, especially if the pregnancy was unanticipated. Communication is particularly important at this time because each partner may have preconceived ideas about what the other is feeling. Both partners may have fears about the baby's well-being, the approaching birth, their ability to parent, and the ways in which the baby will affect their own relationship. All of these concerns are normal. Sharing them, perhaps in the setting of a prenatal group, can deepen and strengthen the relationship (Kitzinger, 1989). If the pregnant woman's partner is not supportive or if she does not have a partner, it is important that she find other sources of support—family, friends, women's groups—and that she not be reluctant to ask for help.

A pregnant woman's relationship with her own mother may also undergo changes. In a certain sense, becoming a mother makes a woman the equal of her own mother. She can now lay claim to treatment as an adult. Women who have depended on their mothers tend to become more independent and assertive as their pregnancy progresses. Women who have been distant, hostile, or alienated from their own mothers may begin to identify with their mothers' experiences of pregnancy. Even women who have delayed childbearing until their thirties may be surprised to find their relationships with

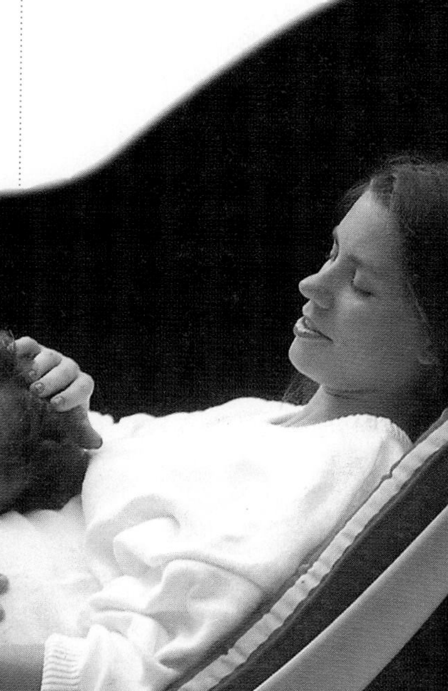

Both expectant parents may feel that the fetus is already a member of the family. They begin the attachment process well before birth.

Figure 8.1
Embryonic and Fetal Development

Fetal development, or gestation, takes approximately 266 days from fertilization of the ovum to birth. These photographs chronicle various stages of the process.

a) After ejaculation, several million sperm move through the cervical mucus toward the fallopian tubes; an ovum has descended into one of the tubes. En route to the ovum, millions of sperm are destroyed in the vagina, uterus, or fallopian tubes. Some go the wrong direction in the vagina and others swim into the wrong tube.

b) The ovum has divided for the first time following fertilization; the mother's and father's chromosomes have united. In subsequent cell divisions the genes will be identified. After about a week, the blastocyst will implant itself into the uterine lining.

c) The embryo is five weeks old and is two-fifths of an inch long. It floats in the embryonic sac. The major divisions of the brain can be seen as well an an eye, hands, arms, and a long tail.

d) The embryo is now seven weeks old and is almost an inch long. Its outer and inner organs are developing. It has eyes, a nose, a mouth, lips, and a tongue.

e) At twelve weeks, the fetus is over three inches long and weighs almost an ounce.

f) At sixteen weeks, the fetus is more than six inches long and weighs about seven ounces. All organs have been formed. The time that follows is now one of simple growth.

(More information on fetal development may be found in the Resource Center.)

Did you know?

It is fairly simple to figure out the date on which a baby's going to be born: add seven days to the first day of the last menstrual period. Then subtract three months and add one year. For example, if a woman's last menstrual period began on July 17, 1998, add seven days (July 24). Next subtract three months (April 24). Then add one year. This gives the expected date of birth as April 24, 1999. Few births actually occur on the date predicted, but 60 percent of babies are born within five days of the predicted time.

their mothers becoming more "adult." Working through the changing relationships is a kind of "psychological gestation" that accompanies the physiological gestation of the fetus (Silver and Campbell, 1988).

The first trimester (three months) of pregnancy may be difficult physically for the expectant mother. She may experience nausea, fatigue, and painful swelling of the breasts. She may also fear that she will miscarry or that the child will not be normal. Her sexuality may undergo changes, resulting in unfamiliar needs (for more, less, or differently expressed sexual love), which may in turn cause anxiety. (Sexuality during pregnancy is discussed later in this section.) Education about the birth process and her own body's functioning and support from partner, friends, relatives, and health-care professionals are the best antidotes to her fear.

During the second trimester, most of the nausea and fatigue disappear and the pregnant woman can feel the fetus move within her. Worries about miscarriage will probably begin to diminish, for the riskiest part of fetal development has passed. The pregnant woman may look and feel radiantly happy. She will very likely feel proud of her accomplishment and be delighted as her pregnancy begins to show. She may feel in harmony with life's natural rhythms. One mother writes (in C. Jones, 1988):

> I love my body when I'm pregnant. It seems round, full, complete somehow. I find that I am emotionally on an even keel throughout; no more premenstrual depression and upsets. I love the feeling that I am never alone, yet at the same time I am my own person. If I could always be five months pregnant, life would be bliss.

Some women, however, may be concerned about their increasing size; they may fear that they are becoming unattractive. A partner's attention and reassurance will ease this fear.

The third trimester may be the time of the greatest difficulties in daily living. The uterus, originally about the size of the woman's fist, has now enlarged to fill the pelvic cavity and is pushing up into the abdominal cavity, exerting increasing pressure on the other internal organs. Water retention (edema) is a fairly common problem during late pregnancy; it may cause swelling in the face, hands, ankles, and feet. It can often be controlled by cutting

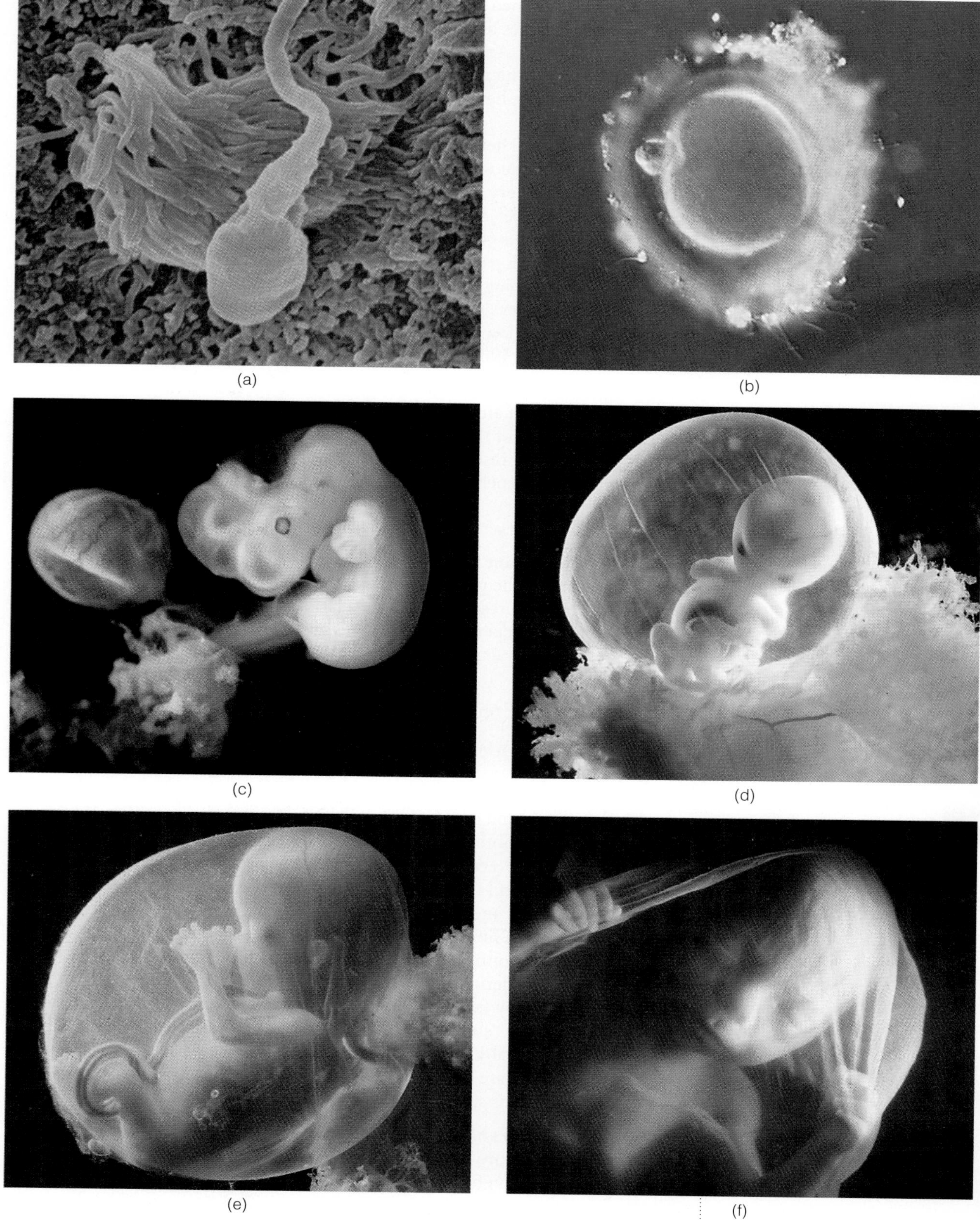
(a)
(b)
(c)
(d)
(e)
(f)

down on salt and refined carbohydrates (such as bleached flour and sugar) in the diet. If dietary changes do not help this condition, however, the woman should consult her physician. Another problem is that the woman's physical abilities are limited by her size. She may also be required by her employer to stop working at some point during her pregnancy. A family dependent on her income may suffer hardship.

The woman and her partner may become increasingly concerned about the upcoming birth. Some women experience periods of depression in the month preceding their delivery; they may feel physically awkward and sexually unattractive. Many, however, feel an exhilarating sense of excitement and anticipation marked by energetic bursts of industriousness. They feel that the fetus is a member of the family. Both parents may begin talking to the fetus and "playing" with it by patting and rubbing the expectant mother's belly.

The principal developmental tasks for the expectant mother and father may be summarized as follows (Valentine, 1982; also see Notman and Lester, 1988; Silver and Campbell, 1988; Snarey et al., 1987).

Tasks of Expectant Mother

- Development of an emotional attachment to the fetus
- Differentiation of the self from the fetus
- Acceptance and resolution of the relationship with own mother
- Resolution of dependency issues (generally involving parents or husband/partner)
- Evaluation of practical/financial responsibilities

Tasks of Expectant Father

- Acceptance of the pregnancy and attachment to the fetus
- Acceptance and resolution of the relationship with own father
- Resolution of dependency issues (involving wife/partner)
- Evaluation of practical/financial responsibilities

Sexuality during Pregnancy

It is not unusual for a woman's sexual feelings and actions to change during pregnancy, although there is great variation among women in these expressions of sexuality. Some women feel beautiful, energetic, sensual, and interested in sex; others feel awkward and decidedly "unsexy." A woman's feelings may also fluctuate during this time. Some studies indicate a lessening of women's sexual interest during pregnancy and a corresponding decline in coital frequency. A study of 219 pregnant women found that although libido, intercourse, and orgasm declined, the frequency of oral and anal sex and masturbation remained at prepregnancy levels (Hart, Cohen, Gingold, and Homburg, 1991).

Men may feel confusion or conflicts about sexual activity during this time. They, like many women, may have been conditioned to find the pregnant body unerotic. Or they may feel deep sexual attraction to their pregnant partner, yet fear their feelings are "strange" or unusual. They may also worry about hurting their partner or the baby.

Although there are no "rules" governing sexual behavior during pregnancy, a few basic precautions should be observed:

- If the woman has had a prior miscarriage, she should check with her health practitioner before having intercourse, masturbating, or engaging in other activities that might lead to orgasm. Powerful uterine contractions could possibly induce a spontaneous abortion in some women, especially during the first trimester.

- If there is bleeding from the vagina, the woman should refrain from sexual activity and consult her physician or midwife at once.
- If the insertion of the penis into the vagina causes pain that is not easily remedied by a change of position, the couple should refrain from intercourse.
- Pressure on the woman's abdomen should be avoided, especially in the final months of pregnancy.
- During oral sex, care should be taken not to blow air into the vagina, as there is a possibility of causing an embolism (an air bubble in the bloodstream).
- Late in pregnancy, an orgasm is likely to induce uterine contractions. Generally this is not considered harmful, but the pregnant woman may want to discuss it with her practitioner. (Occasionally, labor is begun when the waters break as the result of orgasmic contractions.)

A couple, especially during their first pregnancy, may be uncertain as to how to express their sexual feelings. The following guidelines may be helpful (Strong and DeVault, 1997):

- Even during a normal pregnancy, sexual intercourse may be uncomfortable. The couple may want to try positions such as side by side or rear entry to avoid pressure on the woman's abdomen and to facilitate more shallow penetration.
- Even if intercourse is not comfortable for the woman, orgasm may still be intensely pleasurable. She may wish to consider masturbation (alone or with her partner) or cunnilingus.
- Both partners should remember that there are no rules about sexuality during pregnancy. This is a time for relaxing, enjoying the woman's changing body, talking a lot, touching each other, and experimenting with new ways—both sexual and nonsexual—of expressing affection.

Complications of Pregnancy and Dangers to the Fetus

Usually, pregnancy proceeds without major complications. Good nutrition is one of the most important factors in having a complication-free pregnancy (See the Resource Center for nutritional guidelines during pregnancy.) However,

A woman's pregnancy can change and even enhance her and her partners sexuality.

some women experience minor to serious complications, which we will now examine.

Effects of Teratogens

Substances other than nutrients may reach the developing embryo or fetus through the placenta. Although few extensive studies have been done on the subject, toxic substances in the environment can also affect the health of the fetus. Whatever a woman breathes, eats, or drinks is eventually received by the embryo or fetus in some proportion. A fetus's blood-alcohol level, for example, is equal to that of the mother (Rosenthal, 1990). **Teratogens**—substances that cause defects (such as brain damage and physical deformities) in developing embryos or fetuses—are directly traceable in only about 2 or 3 percent of the cases of birth defects. They are thought to be linked to 25 to 30 percent of such cases, however; the causes of the remaining cases remain unknown (Healy, 1988).

Studies have linked chronic ingestion of alcohol during pregnancy to fetal alcohol syndrome (FAS), which can include unusual facial characteristics, small head and body size, congenital heart defects, defective joints, poor mental capabilities, and abnormal behavior patterns. Lesser amounts of alcohol may result in fetal alcohol effect (FAE), the most common problem of which is growth retardation (Waterson and Murray-Lyon, 1990).

About one out of every 1,000 babies born in the United States is affected by FAS. The rate is higher among those with low socioeconomic status (Abel, 1995). The Centers for Disease Control report that FAS is six times more common in African Americans than in whites and thirty times more common in Native Americans than in whites (CDC, 1995; Rosenthal, 1990). Studies have also found that African-American and Latina women are more likely than white women to abstain from alcohol during pregnancy and that a woman's level of alcohol consumption rises with her education and income (Rosenthal, 1990). Most experts counsel pregnant women to abstain entirely from alcohol during pregnancy because there is no safe dosage known at this time.

Pregnant women who regularly use opiates (heroin, morphine, codeine, and opium) are likely to have infants who are addicted at birth. A 1989 study in Rhode Island found that 35 (7.5 percent) out of the 465 women in labor who were studied had traces of drugs in their urine (Centers for Disease Control, 1990). (The urine samples were provided to the Department of Health without names to protect the women's anonymity, although various demographic data were supplied.) Many drug-exposed infants have been subjected to alcohol exposure as well.

Cigarette smoking affects the unborn child (Ellard et al., 1996). Babies born to women who smoke during pregnancy are an average of one-fourth to one-half pound lighter at birth than babies born to nonsmokers. Secondhand smoke is also considered dangerous to the developing fetus (Healy, 1988). Smoking has been implicated in sudden infant death syndrome, respiratory disorders in children, and various adverse pregnancy outcomes.

Caffeine, a powerful stimulant, should be used conservatively by pregnant women. It puts both mother and fetus under stress by raising the level of the hormone epinephrine. Caffeine also reduces the blood supply to the uterus. Coffee, colas, strong black tea, and chocolate are high in caffeine, as are over-the-counter medications such as Excedrin and NoDoz. A pregnant woman should not drink more than two cups of coffee (or the equivalent in other caffeinated foods and beverages) per day.

Prescription drugs should be used only under careful medical supervision, as some may cause serious harm to the fetus. Isotretinoin (Accutane), a popular antiacne drug, has been implicated in more than a thousand cases of severe birth defects over the past several years (Kolata, 1988). Certain vitamins and over-the-counter drugs, such as aspirin, should be avoided or used only under medical supervision. Vitamin A in large doses can cause serious birth defects.

Chemicals and environmental pollutants are also potentially threatening. Continuous exposure to lead, most commonly in paint products or water from lead pipes, has been implicated in a variety of learning disorders. Mercury, from fish contaminated by industrial wastes, is a known cause of physical deformities. Solvents, pesticides, and certain chemical fertilizers should be avoided or used with extreme caution both at home and in the workplace. A 1992 study of pregnant women who worked with computers suggested a strong correlation between exposure to certain types of video display terminals and miscarriage. Researchers in Finland linked the tripled rate of miscarriage to high levels of a particular electromagnetic field (Lindbohm, Hietanan, Kyronen, and Sallmen, 1992).

Infectious Diseases

Infectious diseases can also damage the fetus. If a woman contracts German measles (rubella) during the first three months of her pregnancy, her child may be born with physical or mental disabilities. Immunization against rubella is available, but it must be done before the woman is pregnant; otherwise the injection is as harmful to the fetus as the disease itself. Group B streptococcus, a bacterium carried by 15 to 40 percent of pregnant women, is harmless to adults but can be fatal to newborns. Each year, about twelve thousand infants are infected; sixteen hundred to two thousand of them die. The American Academy of Pediatrics has recommended that all pregnant women be screened for strep B. Antibiotics administered to the newborn during labor can greatly reduce the danger ("Prenatal Tests," 1992).

Sexually transmissible diseases may also damage the fetus. In 1988, the Centers for Disease Control recommended that all pregnant women be screened for hepatitis B, a virus that can be passed to the infant at birth. If the mother tests positive, her child can be immunized immediately following birth (Moore, 1988). A woman with gonorrhea may expose her child to blindness from contact with the infected vagina; the baby will need immediate antibiotic treatment. A woman with HIV/AIDS has a 30 to 40 percent chance of passing the virus to the fetus via the placenta (Cowley, 1990; Fischl, Dickinson, Segal, Flannagan, and Rodriguez, 1987). The number of infants with HIV/AIDS is increasing as more women become infected with the virus either before or during pregnancy. Over six thousand children under age five have been diagnosed with AIDS. Many more were HIV-positive at birth (Centers for Disease Control, 1997). Not all infants who are born with a positive HIV status remain that way, however; some "reconvert" to negative status once their mother's blood is out of their system. Moreover, researchers have found that the drug AZT significantly reduces the transmission of HIV from a pregnant woman to her fetus ("Routine AZT Use," 1996).

The increasingly widespread incidence of genital herpes may present some hazards for newborns. The herpes simplex virus may cause brain damage and is potentially life threatening for these infants. Careful monitoring by a physician can determine whether or not a vaginal delivery should take place. Charles Prober (1987) and colleagues at the Stanford Medical School have extensively studied births to mothers with herpes. They found that a number of infants born to infected mothers had herpes antibodies in their blood; these infants did not contract herpes during vaginal delivery. Although the harmful potential of neonatal herpes should not be underestimated, if the appropriate procedures are followed by the infected mother and her physician or midwife, the chance of infection from genital lesions is minimal ("Neonatal Herpes," 1984).

The current incidence of death caused by neonatal herpes is about two thousand cases annually (Petit, 1992). An initial outbreak of herpes can be especially dangerous if the woman is pregnant; the virus may be passed through the placenta to the fetus. Fathers may also inadvertently infect their newborns (Kulhanjian, Soroush, Au, Bronzan, and Yasukawa, 1992). Infected men who have been symptomless for years and have not previously transmitted the virus to their wives may do so during pregnancy, placing their infants at great risk. In rare cases, herpes-infected infants may be born to mothers who show no symptoms themselves. This indicates that undiagnosed intrauterine herpes infections may be passed to the fetus (Stone, Brooks, Guinan, and Alexander, 1989). Testing for genital herpes is therefore recommended for both expectant parents.

Once the baby is born, a mother who is experiencing an outbreak of either oral or genital herpes should wash her hands often and carefully and not permit contact between her hands, contaminated objects, and the baby's mucous membranes (inside of eyes, mouth, nose, penis, vagina, vulva, and rectum). If the father is infected, he should do likewise until the lesions have subsided.

In recent years, congenital syphilis has increased alarmingly. At the beginning of the 1980s, syphilis was rarely seen in infants. In 1995, however, 7,200 cases were reported (Centers for Disease Control, 1996). Lack of prenatal care, which normally includes testing for syphilis (and treatment, if needed), is cited as the principal reason for the increase (Lewin, 1992). Most infants with syphilis have no obvious symptoms and suffer no long-term effects if they are promptly treated with antibiotics. But some are severely affected with skin rashes, deafness, blindness, bone deformities, or damage to the liver, spleen, or brain.

Ectopic Pregnancy

In **ectopic pregnancy** (tubal pregnancy), the incidence of which has more than quadrupled in the last twenty years, the fertilized egg implants itself in the fallopian tube. Generally this occurs because the tube is obstructed, most often as a result of pelvic inflammatory disease (Hilts, 1991). The pregnancy will never come to term. The embryo may spontaneously abort, or the embryo and placenta will continue to expand until they rupture the fallopian tube. Salpingectomy (removal of the tube) and abortion of the embryo may be necessary to save the mother's life.

Toxemia and Preeclampsia

Toxemia, which may appear in the twentieth to twenty-fourth week of pregnancy, is characterized by high blood pressure and edema (fluid retention and swelling). It can generally be treated through nutritional means, such as reducing or eliminating salt, sugar, and refined carbohydrates. If untreated, toxemia can develop into preeclampsia after the twenty-fourth week.

Preeclampsia is characterized by increasingly high blood pressure. Toxemia and preeclampsia are also known as gestational edema-proteinuria-hypertension complex. If untreated, they can lead to *eclampsia*—maternal convulsions that pose a serious threat to mother and child. Eclampsia is not common; it is prevented by keeping the blood pressure down through diet, rest, and sometimes medication. It is important for a pregnant woman to have her blood pressure checked regularly.

Low Birth Weight

Prematurity or **low birth weight (LBW)** is a major complication in the third trimester of pregnancy, affecting about 7.5 percent of newborns yearly in the United States (Pear, 1992). The most fundamental problem of LBW is that many of the infant's vital organs are insufficiently developed. An LBW baby usually weighs less than 5.5 pounds at birth. Most premature infants will grow normally, but many will experience disabilities. Such infants are subject to various respiratory problems as well as infections. Feeding, too, is a problem because LBW infants may be too small to suck a breast or bottle and their swallowing mechanisms may be too underdeveloped to permit them to drink. As the infants get older, problems such as low intelligence, learning difficulties, poor hearing and vision, and physical awkwardness may become apparent.

Premature delivery is one of the greatest problems confronting obstetrics today. About half the cases are related to teenage pregnancy, smoking, poor nutrition, and poor health in the mother (Pear, 1992; Schneck, Sideras, Fox, and Dupuis, 1990); the causes of the other half are unknown. One study of LBW babies found a sixfold increase in the risk of low birth weight if the mother had financial problems during pregnancy (Binsacca et al., 1987). A 1987 study of 1,382 women indicated that delaying childbearing until after age thirty-five does not increase the risk of delivering an LBW infant (Barkan and Bracken, 1987).

Prenatal care is extremely important as a means of preventing prematurity. A third of all LBW births could be averted with adequate prenatal care (Scott, 1990b). We need to understand that if children's needs are not met today, we will all face the consequences of their deprivation tomorrow. The social and economic costs are bound to be very high.

Reflections

What concerns do (or will) you and your partner have about the prenatal health of a child you have conceived? (Think in terms of your own health, medical history, genetic background, drinking and smoking habits, exposure to disease, chemicals, radiation, and so on.) If you have concerns, where can you get more information? In what practical ways can you deal with your concerns?

Diagnosing Abnormalities of the Fetus

In addition to the desire to bear children, the desire to bear healthy (even perfect) children has encouraged the development of new diagnostic technologies. In cases where serious problems are suspected, these technologies may be quite helpful, but often it seems that they are used simply because they're there.

Ultrasonography is the use of high-frequency sound waves to create a picture of the fetus in utero (in the uterus). The sound waves are transmitted through a quartz crystal; when they detect a change in the density, the sound waves bounce back to the crystal. The results are interpreted on a televisionlike screen; the picture is called a **sonogram.** Sonograms are used to determine fetal age and the location of the placenta; when used with amniocentesis (which will be described shortly), they help determine the fetus's position so the needle can be inserted safely. Often, it is possible to determine the fetus's sex. More extensive ultrasound techniques can be used to gain further information if there is the possibility of a problem with the fetus's development. Although no problems in humans have been noted to result from ultrasonography, high levels of ultrasound have created problems in animal fetuses.

In **amniocentesis,** amniotic fluid is withdrawn from the uterus with a long, thin needle inserted through the abdominal wall. (The position of the fetus is determined by ultrasonography.) The fluid is then examined for evidence of birth defects such as Down syndrome, cystic fibrosis, Tay-Sachs disease, spina bifida, and other conditions caused by chromosomal abnormalities. The sex of the fetus can also be determined. Most amniocentesis tests (80 to 90 percent) are performed in cases of "advanced" maternal age (usually when the mother is over thirty-five) because the possibility of certain birth defects increases significantly with the mother's age after the mid-thirties. Amniocentesis is performed at about sixteen weeks of gestation. It carries a risk—a 0.5 to 2 percent chance of fetal death (Brandenburg, Jahoda, Pijpers, Reuss, Kleyer, and Waldmiroff, 1990; Hanson, Happ, Tennant, Hune, and Peterson, 1990; McCormack et al., 1990).

An alternative to amniocentesis recently introduced in the United States is **chorionic villus sampling (CVS).** This procedure involves removal through the abdomen (by needle) or through the cervix (by catheter) of tiny pieces of the membrane that encases the embryo; it can be performed between nine and eleven weeks of pregnancy. There may be a somewhat greater than normal chance of miscarriage with CVS because of the possible introduction of infectious microorganisms at the time of the procedure (Baumann, Jovanovick, Gellert, and Rauskolb, 1991). A worldwide 1992 study of eighty thousand women who had been examined by CVS found no greater evidence of malformations in their children than in the general population ("Survey," 1992). An advantage of CVS is that it can be used earlier in the pregnancy than amniocentesis.

Alpha-feto protein (AFP) screening is a test (or series of tests) done on the mother's blood after sixteen weeks of gestation. It reveals neural tube defects such as anencephaly and spina bifida. It is much simpler than amniocentesis but sometimes yields false positive results (Samuels and Samuels, 1996). If the results are positive, other tests, such as amniocentesis and ultrasound, will be performed to confirm or negate the findings of the AFP screening.

For approximately 95 percent of women who undergo prenatal testing, the results are negative. The results of amniocentesis and AFP screening can't be determined, however, until approximately twenty weeks of pregnancy; consequently, if the pregnancy is terminated through abortion at this stage, the process is likely to be physically and emotionally difficult. If a fetus is found to be defective, it may be carried to term, aborted, or, in rare but increasing instances, surgically treated while still in the womb (Kolata, 1990a). Gene therapy, in which defective enzymes in the genes of embryos are replaced by normal ones, is considered a challenging frontier of reproductive medicine.

The 1988 National Commission to Prevent Infant Mortality stated that although "sophisticated technologies and great expense can save babies born at risk . . . , the lack of access to health care, the inability to pay for health care, poor nutrition, unsanitary living conditions, and unhealthy habits such as smoking, drinking, and drug use all threaten unborn children" (cited in Armstrong and Feldman, 1990).

Pregnancy Loss

The loss of a child through miscarriage, stillbirth, or death during early infancy is a devastating experience that has been largely ignored in our society. The statement "You can always have another one" may be meant as consolation, but it is particularly chilling to the ears of a grieving mother. In the past few years, however, the medical community has begun to respond to the emotional needs of parents who have lost a pregnancy or an infant.

Spontaneous Abortion

Spontaneous abortion (miscarriage) is a powerful natural selective force toward bringing healthy babies into the world. About one out of four women is aware she has miscarried at least once (Beck, 1988). Studies indicate that at least 60 percent of all miscarriages are due to chromosomal abnormalities in the fetus (Adler, 1986). Furthermore, as many as three-fourths of all fertilized eggs do not mature into viable fetuses (Beck, 1988). One study found that 32 percent of implanting embryos miscarried (Wilcox et al., 1988). The first sign that a pregnant woman may miscarry is vaginal bleeding ("spotting"). If a woman's symptoms of pregnancy disappear and she develops pelvic cramps, she may be miscarrying; the fetus is usually expelled by uterine contractions. Most miscarriages occur between the sixth and eighth weeks of pregnancy. Evidence is increasing that certain occupations involving exposure to chemicals or high levels of electromagnetism increase the likelihood of spontaneous abortions. Miscarriages may also occur because of uterine abnormalities or hormonal levels that are insufficient for maintaining the uterine lining.

Infant Mortality

The U.S. infant mortality rate, while at its lowest point ever, remains far higher than in most of the developed world. The U.S. Public Health Service reported only 6.68 deaths for every 1,000 live births in 1995 (National Center for Health Statistics, 1996; U.S. Bureau of the Census, 1996). Nevertheless, among developed nations, the United States ranks twentieth in low infant mortality. (This means that nineteen countries have *lower* infant mortality rates than the United States.) In some inner-city areas the infant mortality rate approaches that of nonindustrialized countries, with more than 20 deaths per 1,000 births (Petit, 1990).

Of the more than 35,000 American babies less than one year old who die each year, most are victims of the poverty that often results from racial or ethnic discrimination. Up to a third of these deaths could be prevented if mothers were given adequate health care (Scott, 1990a). The infant mortality rate for African Americans is more than twice that for whites. Native Americans are also at high risk; for example, about 1 out of every 67 Navajo infants dies each year (Wilkerson, 1987). A study by the U.S. Centers for Disease Control found that mortality rates for many ethnic minorities have been severely underestimated because the infants were mistakenly classified as white ("Death Rates," 1992; Hahn, Mulinare, and Teutsch, 1992).

In 1981, the federal government began to cut pregnancy and infant-care programs such as WIC (Special Supplemental Food Program for Women, Infants and Children) dramatically to fund its record-breaking military budget. Medicare and Aid to Families with Dependent Children were also cut, leaving many families without pre- or postnatal care. In recent years, the government has begun to restore funding to many of these programs, but many of them are currently threatened. The United States is far behind many other countries in terms of providing health care for children and pregnant women. In France, Sweden, and Japan, for example, all pregnant women are entitled to free prenatal care. Free

Did you know?

About 32,000 American babies under one year old die every year, mostly of causes related to poverty (National Center for Health Statistics, 1996).

health care and immunizations are also provided for infants and young children. Working Swedish mothers are guaranteed one year of paid maternal leave, and French families in need are paid regular government allowances (Scott, 1990a). A recent analysis of seventeen studies done between 1971 and 1988 found that WIC recipients were 25 percent less likely than comparable nonrecipients to have a low-birth-weight baby and 44 percent less likely to have a very low-birth-weight baby. The researchers calculated that one year of WIC expenditures may save society as much as $800 million in federal and state Medicaid expenditures (Avruch and Cackley, 1995).

Although many infants die of poverty-related conditions, others die from congenital problems (conditions appearing at birth) or from infectious diseases, accidents, or other causes. Sometimes, the causes of death are not apparent. Provisional figures from the CDC and the National Center for Health Statistics for 1994 attribute 4,073 infant deaths to **sudden infant death syndrome (SIDS)**—a perplexing phenomenon wherein an apparently healthy infant dies suddenly while sleeping (SIDS Resource Center, 1997). A study from Australia identified four factors that appear to increase the chances of SIDS (Ponsonby, Disyer, Gibbons, Cochrane, and Wang, 1993): a soft, fluffy mattress, the baby being wrapped in a blanket, the baby having a cold or other minor illness, and allowing the baby to become too warm. Exposure to secondhand smoke also has been implicated (Klonoff-Cohen, et al., 1995). It is also very important that an infant not be placed to sleep on its stomach until it is strong enough to turn over ("Sleeping on Back," 1996).

Dear Auntie will come with presents and will ask, "Where is our baby, sister?" And, mother, you will tell her softly, "He is in the pupils of my eyes. He is in my bones and in my soul."

RABINDRANATH TAGORE (1861–1941)

Coping with Loss

The depth of shock and grief felt by many who lose a child before or during birth is sometimes difficult to understand for those who have not had a similar experience. What they may not realize is that most women form a deep attachment to their children even before birth. At first, the attachment may be a "fantasy image of [the] future child" (Friedman and Gradstein, 1982). During the course of the pregnancy, the mother forms an actual acquaintance with her child through the physical sensations she feels within her. Thus, the death of the fetus can also represent the death of a dream and of a hope for the future. This loss must be acknowledged and felt before psychological healing can take place.

Women (and sometimes their partners) who lose a pregnancy or a young infant generally experience similar stages in their grieving process. Their feelings are influenced by many factors: the supportiveness of the partner and other family members, the reactions of social networks, life circumstances at the time of the loss, circumstances of the loss itself, whether other losses have been experienced, the prognosis for future childbearing, and the woman's unique personality. Physical exhaustion and, in the case of miscarriage, hormone imbalance often compound the emotional stress of the grieving mother.

In traditional sexual thought, sex was justified primarily, sometimes exclusively because it was necessary to the propagation of the species. Since one of the promises of the modern tradition in sexual thought has been that sex is a valuable end in itself, modern theorists, by way of reaction, have generally inclined to belittle the reproductive perspective on sexuality.

PAUL ROBINSON, *THE MODERNIZATION OF SEX*

The initial stage of grief is often one of shocked disbelief and numbness. This stage gives way to sadness, spells of crying, preoccupation with the loss, and perhaps loss of interest in the rest of the world. Emotional pain may be accompanied by physical sensations and symptoms, such as tightness in the chest or stomach, sleeplessness, and loss of appetite. It is not unusual for parents to feel guilty, as if they had somehow caused the loss, although this is rarely the case. Anger (toward the physician, perhaps, or God) is also a common emotion.

Experiencing the pain of loss is part of the healing process (Vredevelt, 1994). This process takes time—months, a year, perhaps more for some. Support groups and counseling are often helpful, especially if healing does not seem to be progressing—if, for example, depression and physical symptoms don't appear to be diminishing. Keeping active is another way to deal with the pain of loss, as long as it isn't a way to avoid facing feelings. Projects, temporary or part-time work, or travel (for those who can afford it) can be ways of renewing energy and interest in life. Planning the next pregnancy may be curative, too, though we must keep in mind that the body and spirit need some time to heal. It is important to have a physician's input before proceeding with another pregnancy; specific considerations may need to be discussed, such as a genetic condition that may be passed to the child or a physiological problem of the mother. If future pregnancies are ruled out, the parents need to take stock of their priorities and consider other options that may be open to them, such as adoption. Counselors and support groups can be invaluable at this stage.

Giving Birth

Throughout pregnancy, numerous physiological changes occur in order to prepare the woman's body for childbirth, or **parturition.** Hormones secreted by the placenta regulate the growth of the fetus, stimulate maturation of the breasts for lactation, and ready the uterus and other parts of the body for labor. During the later months of pregnancy, the placenta produces the hormone relaxin, which increases flexibility in the ligaments and joints of the pelvic area. In the last trimester, most women occasionally feel uterine contractions that are strong but generally not painful. These are called **Braxton Hicks contractions.** They exercise the uterus, preparing it for labor.

Union between man and woman is a creative act and has something divine about it. . . . The object of love is a creative union with beauty on both the spiritual and physical levels.

PLATO, *THE SYMPOSIUM* (c. 429–348 B.C.)

Labor and Delivery

During labor, contractions begin the **effacement** (thinning) and **dilation** (opening up) of the cervix. It is difficult to say exactly when labor starts, which helps explain the great differences reported in lengths of labor for different women. When the uterine contractions become regular, true labor begins. During these contractions, the lengthwise muscles of the uterus involuntarily pull open the circular muscles around the cervix. This process generally takes from two to thirty-six hours. Its duration depends on the size of the baby, the baby's position in the uterus, the size of the mother's pelvis, and the condition of the uterus. The length of labor tends to shorten after the first birth experience.

Labor can generally be divided into three stages (see Figure 8.2). The first stage is usually the longest, lasting from four to sixteen hours. An early sign of first-stage labor is the expulsion of a plug of slightly bloody mucus that has blocked the opening of the cervix during pregnancy. At the same time or later on, there is a second fluid discharge from the vagina. This discharge, referred to as the "breaking of the waters," is the amniotic fluid, which comes from the ruptured amnion. (Because the baby is subject to infection after the protective membrane breaks, the woman should receive medical attention soon thereafter, if she has not already.)

[Power] sounds in our bodies. Contractions creep up, seizing ever stronger until they make a mockery of all the work we have done on our own. Birth can silence our ego and, for the moment, we feel ourselves overcome by a larger life pounding through our own.

PENNY ARMSTRONG AND SHERYL FELDMAN, *A WISE BIRTH*

The hormone oxytocin, produced by the fetus, along with prostaglandins from the placenta stimulate strong, regularly spaced, uterine contractions. At the end of the first stage of labor, which is called **transition,** the contractions come more quickly and are much more intense than in the early stages of labor. Many women report this stage as the most difficult stage of labor. At the end of first-stage labor, the baby's head is poised to enter the birth canal. This marks the shift

OTHER PLACES OTHER TIMES

Couvade: How Men Give Birth

Throughout the world, men envy and imitate both pregnancy and childbirth. In our own culture, there are sympathetic pregnancies in which a man develops physical characteristics similar to those of his pregnant partner. If she has morning sickness, so does he; if her belly begins to swell, so does his. As someone wrote, "Man is a rational animal, but only women can have babies." Thus, men often use images of pregnancy and childbirth to describe their creative work. A man "conceives" an idea. He "gives birth" to a new theory. He is in a "fertile" period in his artistic development. His book is "pregnant" with ideas. Less fortunate men find their careers "aborted," while others have "stillborn" ideas. It has even been suggested that men often discriminate against women in the arts and sciences precisely because, in these areas of highly creative work, they feel most jealous of women's biological creativity and thus attempt to keep them from entering this domain.

Other cultures have the ritual of **couvade**. The word comes from the French *couver*, "meaning to hatch or brood." The couvade involves a number of different activities in different cultures. The father must often follow certain rest patterns, work restrictions, and dietary practices. Among the Hopi, for example, a man is required to be careful not to hurt animals. If he does, his child may be born deformed. Among the Ifage in the Philippines, the husbands of pregnant women are prohibited from cutting or killing anything.

Perhaps the most startling to the Westerner are male imitations of childbirth. For instance, in many tribes throughout the world, a man wraps his arms around his belly and imitates his wife having labor contractions. After the birth, he may pretend he is entering a postpartum period. Among the Chaorti in South America, the man takes to his hammock for several days during labor and after the delivery.

The Huichol of Mexico traditionally practiced a ritual of couvade in which the husband squatted in the rafters of the house or the branches of a tree above his wife during labor. When the woman experienced a contraction, she pulled on ropes attached to his scrotum. In this way, the man shared the experience of childbirth.

The couvade is a dramatic symbol of the man's paternity and his "magical" relation to the child. By pretending he is pregnant, he distracts evil spirits from harming his baby. Describing the magical impact of the couvade, Arthur and Libby Colman (1971) write:

> The couvade phenomena have the important side effects of helping a husband play an important part in pregnancy and childbirth. . . . They help a man cope with the envy and competitiveness which he may feel at his wife's ability to perform such a fundamental and creative act. . . . In his activities to deceive the evil spirits, a man may also find a reasonable outlet for his own desire to take on something of the female role in life.

Some American men may experience what medical researchers call the "couvade syndrome." A study of the partners of 267 postpartum women found that 22.5 percent of the men experienced nausea, vomiting, anorexia, abdominal pain and bloating, and other symptoms of pregnancy that could not be objectively explained (Lipkin and Lamb, 1982). These men had four times more symptoms than they had had prior to their partners' pregnancies. Another study suggests that 11–65 percent of men with pregnant partners experience sympathetic pregnancies (White and Bulloch, 1980).

The Huichol people of Mexico traditionally practiced a form of couvade shown in this yarn painting by artist Guadalupe Medina.

from dilation of the cervical opening to expulsion of the infant. The cervix is now almost fully open, but the baby is not yet completely in position to be pushed out.

Some women may feel despair, isolation, and anger at this point. Many appear to lose faith in those assisting in the birth. A woman may feel that management of the contractions is beyond her control; she may be afraid that something is wrong. At this time, she needs the full support and understanding of her helpers. Transition is usually, though not always, brief (a half hour to one hour).

Figure 8.2
Childbirth

During the last phase of pregnancy, the fetus usually settles in a head-down position. In the first stage of labor (a), Braxton Hicks contractions begin effacement and dilation of the cervix. When the cervix is fully dilated (b), the baby's head can enter the birth canal; usually the amniotic membrane has ruptured by this time. During second-stage labor (c), the baby moves through the birth canal and into the outside world. The soft bones of the head may be "molded" by pressure in the birth canal, resulting in a temporarily unusual appearance. The placenta (d) is expelled during third-stage labor.

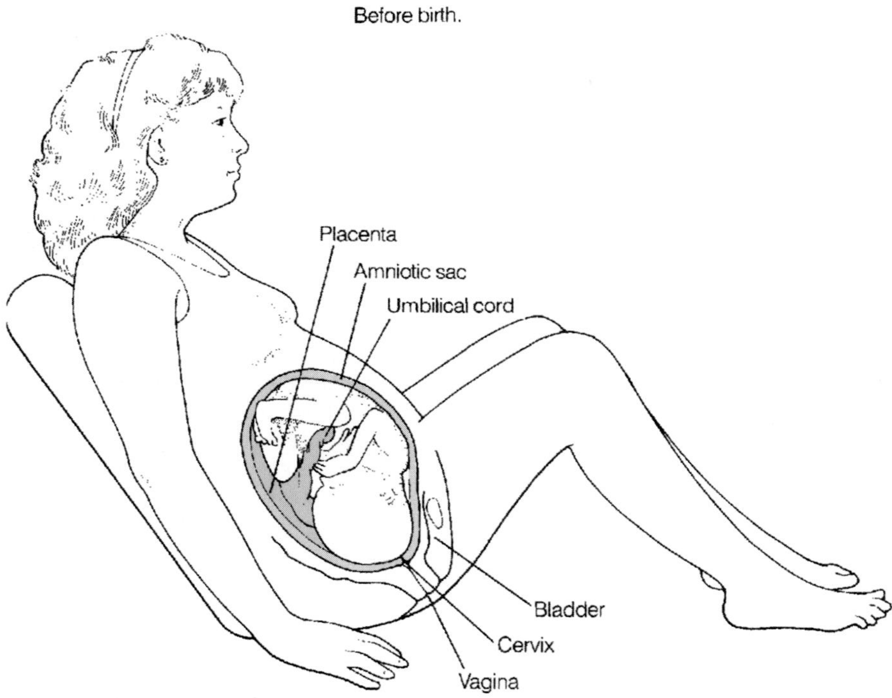

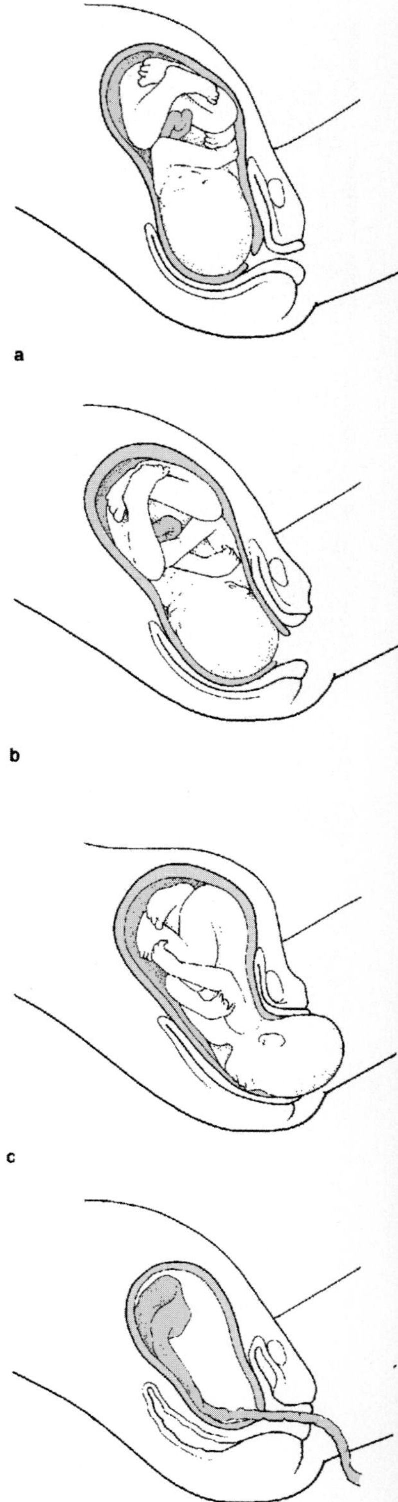

Second-stage labor begins when the baby's head moves into the birth canal and ends when the baby is born. During this time, many women experience a great force in their bodies. Some women find this the most difficult part of labor. Others find that the contractions and bearing down bring a sense of euphoria. One father describes his wife's second-stage labor (Jones, 1986): "When the urge to bear down swept through her, Jan felt as if she were one with all existence and connected with the vast primordial power that summons life forth from its watery depths."

The baby is usually born gradually. With each of the final few contractions, a new part of the infant emerges. The baby may even cry before he or she is completely born, especially if the mother did not have medication. Sheila Kitzinger (1985) describes the moment of birth as a sexual event:

> The whole body slips out in a rush of warm flesh, a fountain of water, a peak of overwhelming surprise and the little body is against her skin, kicking against her thighs or swimming up over her belly. She reaches out to hold her baby, firm, solid, with bright, bright eyes. A peak sexual experience, the birth passion becomes the welcoming of a new person into life.

YOU AND YOUR WELL-BEING

Planning for Pregnancy: Preconception Care

To ensure the healthiest pregnancies with the best outcomes, the U.S. Public Health Service recommends that women prepare for pregnancy by considering the following health issues and taking the appropriate actions before conceiving (American College of Obstetricians and Gynecologists, 1995; National Center for Health Statistics, 1995).

General health

How is your general health? Do you have any conditions or infections, including sexually transmissible diseases, that may affect a pregnancy or jeopardize your own health should you become pregnant? Have you had a recent medical examination? Are you up to date on your vaccinations, including those for rubella (German measles) and hepatitis B? Are you currently taking any prescribed medications that may affect the fetus? Are you exposed to radiation or toxic substances on the job?

Age

Does your age place you at risk for any problems during pregnancy or for infertility? Teenagers may need nutritional supplements to provide for their own growing bodies' needs. Women over 35 have a higher than average risk of having a child with Down syndrome; they are also more likely to have difficulty becoming pregnant.

Nutrition

Is your diet healthy? Is your weight within a range that is normally considered healthy? Do you or have you had problems with an eating disorder? Your health practitioner can give you advice on a healthy, balanced diet that will meet your needs and those of the developing fetus. You will require nutritional supplements—vitamins and minerals, including folic acid which reduces the risk of birth defects.

Alcohol, tobacco, and other drugs

Do you smoke, drink alcohol, or use other nonprescribed drugs? As we discuss in this chapter, these substances can all be harmful to a developing fetus. Do you need to join a program to help you stop using any of these substances? (See the Resource Center for program listings.)

Family health history

Do you or your partner have any condition that your child might inherit? Is there a history in either family of genetically transmitted diseases or conditions, such as diabetes, deafness, cystic fibrosis, Tay-Sachs disease, or sickle-cell disease? Talk to your parents to see if there are such conditions and to your health practitioner for information about genetic testing and possible treatments. (For information on genetic counseling, see the Resource Center.)

The baby will still be attached to the umbilical cord connected to the mother, which is not cut until it stops pulsating. He or she will appear wet, often covered by a milky substance called **vernix.** The head may look oddly shaped at first, from the molding of the soft plates of bone during birth. This shape is temporary, and the baby's head usually achieves a normal appearance within twenty-four hours.

After the baby has been delivered, the uterus will continue to contract, expelling the **placenta** (afterbirth) and completing the third and final stage of labor. The doctor or midwife will examine the placenta to make sure it is whole. The practitioner may also examine the uterus to ensure that no parts of the placenta remain to cause adhesions or hemorrhage. Immediately following birth, the attendants assess the physical condition of the **neonate** (newborn). Heart rate, respiration, color, reflexes, and muscle tone are individually rated from 0 to 2. The total of the individual ratings, called the **Apgar score,** will be at least 8 if the child is healthy. For a few days following labor (especially if it is a second or subsequent birth), the mother will probably feel strong contractions as the uterus begins to return to its prebirth size and shape. This process, called **involution,** takes about six weeks. She will also have a bloody discharge, called **lochia,** which continues for several weeks.

Following birth, if the baby has not been drugged, he or she will probably be alert and ready to nurse. Breastfeeding, discussed further on, provides benefits for both mother and child. If the infant is a boy, the parents will need to decide about circumcision—the surgical removal of the foreskin (the sleeve of skin that covers the glans of the penis).

Choices in Childbirth

Women and couples planning the birth of a child have decisions to make in a variety of areas—birth place, birth attendants, medications, preparedness classes, circumcision, and breastfeeding, to name but a few. The "childbirth market" is beginning to respond to consumer concerns, so it's important for prospective parents to fully understand their options.

Hospital Birth

The impersonal, routine quality of hospital birth (and hospital care in general) is increasingly being questioned. One woman described her initial feelings in the hospital (in Leifer, 1990):

> When they put that tag around my wrist and put me into that hospital gown, I felt as if I had suddenly just become a number, a medical case. All of the excitement that I was feeling on the way over to the hospital began to fade away. It felt like I was waiting for an operation, not about to have my baby. I felt alone, totally alone, as if I had just become a body to be examined and not a real person.

Another woman was shocked at the impersonal treatment (in Leifer, 1990):

> And then this resident gave me an internal [examination], and it was quite painful then. And I said: "Could you wait till the contraction is over?" And he said he had to do it now, and I was really upset because he didn't even say it nicely, he just said: "You'll have to get used to this, you'll have a lot of this before the baby comes."

Some hospitals are responding to the need for family-centered childbirth. Fathers and other relatives or close friends often participate today. Some hospitals permit rooming-in (the baby stays with the mother rather than in the nursery) or a modified form of rooming-in. Regulations vary as to when the father and other family members and friends are allowed to visit.

But the norm is still all too often the impersonal birth. During one of the most profound experiences of her life, a woman may have her baby among strangers to whom birth is merely business as usual. There are likely to be bright lights, loud noises, and people coming and going, moving her, poking at her, and asking questions when she's in the middle of a contraction. She may be given a routine enema, even though surgery is not anticipated. She and her unborn child are likely to be attached to various types of monitoring machines. Studies have shown that although fetal monitoring is helpful in high-risk cases, it is generally not helpful in normal (low-risk) situations. Generally, the baby's heart rate can be detected by stethoscope against the mother's abdomen.

Some form of anesthetic is administered during most hospital deliveries, as well as various hormones (to intensify the contractions and to shrink the uterus after delivery). The mother isn't the only recipient of the drugs, however; they go directly through the placenta to the baby, in whom they may reduce heart and respiration rates.

During delivery, the mother will probably be given an **episiotomy**—a surgical procedure to enlarge the vaginal opening by cutting through the perineum toward the anus. Although an episiotomy may be helpful if the infant is in distress, it is usually performed to reduce the risk of tearing, give the obstetrician more control over the birth, and speed up the delivery. Studies show that episiotomies are performed in about 80 percent of first vaginal births in hospitals (Hetherington, 1990; Klein et al., 1992); yet one midwife who has assisted at over twelve hundred births reports a rate of less than 1 percent (Armstrong and Feldman, 1990). A Canadian study of 703 uncomplicated births found no advantage to routine episiotomies; the disadvantages were pain and bleeding (Klein et al., 1992). The authors recommended that "liberal or routine use of episiotomy be abandoned."

The baby is usually delivered on a table, against the force of gravity. He or she may be pulled from the womb with a vacuum extractor (which has a small suction cup that fits onto the baby's head) or forceps. (In some cases of acute fetal distress, these instruments may be crucial in order to save the infant's life, but too often they are used by physicians as a substitute for patience and skill.) In most cultures, a woman gives birth while sitting in a birthing chair, kneeling, or squatting. Until the present century, most American women used birthing chairs; the delivery table was instituted for the convenience of the physician. A few hospitals use a motorized birthing chair that can be raised, lowered, or tilted according to the physician's and the woman's needs. In about one-fourth of births in the United States, the baby is not delivered vaginally but is surgically removed from the uterus.

Cesarean Section

Cesarean section (C-section) is the removal of the fetus by an incision in the mother's abdominal and uterine walls. The first reported cesarean section performed on a living woman occurred in the seventeenth century, when a butcher cut open his wife's uterus to save her and their child. In 1970, 5.5 percent of American births were done by cesarean section. Today cesarean births account for 22.8 percent of all births ("Rates of Cesarean Delivery," 1995).

Did you know?

The rate of cesarean-section births has increased by 400 percent in the past 25 years ("Rates of Cesarean Delivery," 1995).

Although there is a decreased mortality rate for infants born by C-section, the mother's mortality rate is higher. As with all major surgeries, complications are possible and recovery can be slow and difficult.

Hoping to reduce the alarming increase in C-sections, the National Institutes of Health has issued the following guidelines:

- Just because a woman has had a previous cesarean delivery does not mean that subsequent deliveries must be C-sections; whenever possible, women should be given the option of vaginal birth.
- Abnormal labor does not mean that a C-section is necessary. Sleep or medication may resolve the problems. Only after other measures have been tried should a physician perform a C-section, unless the infant is clearly in danger.
- Breech babies (those who enter the birth canal buttocks or feet first) do not necessarily require C-sections. A physician's experience in using his or her hands to deliver such babies vaginally is crucial.

If a woman does not want a C-section unless it is absolutely necessary, she should learn about her physician's attitude and record on cesareans. It is noteworthy that the greatest number of cesareans are performed in the socioeconomic group of women with the lowest medical risk (Hurst and Summey, 1984). In a study of 245,854 births, the C-section rate for middle- and upper-income women was 22.9 percent, whereas for lower-income women it was 3.2 percent (Gould et al., 1989). Therefore, it can be assumed that cesareans are often performed for reasons other than medical risk. The U.S. Public Health Service has set a goal for reducing the overall C-section rate to no more than 15 percent and the repeat C-section rate to 65 percent by the year 2000 (National Center for Health Statistics, 1995). To achieve these goals, the following strategies are suggested: (1) address concerns about physician malpractice; (2) eliminate physicians' financial incentives for cesareans, which are currently more lucrative than vaginal deliveries; (3) publish C-section delivery rates of physicians and hospitals; and (4) increase physician training in normal labor and vaginal delivery.

Prepared Childbirth

Increasingly, Americans are choosing from alternatives including prepared childbirth, rooming-in, birthing centers, and nurse-midwives. The chief of obstetrics at a major hospital observed: "It's a major upheaval in medicine, and the conflict has tended to polarize the consumer and the caregiver. The basic problem is that we have changed obstetrics with the latest medical advances and not incorporated essentially humanistic considerations" (quoted in Trafford, 1980).

Women can give birth by the action of their own bodies, as animals do. Women can enjoy the process of birth and add to their dignity by being educated to follow the example set by instinctive animals.

ROBERT A. BRADLEY, M.D.

Prepared childbirth (or natural childbirth) was popularized by Grantly Dick-Read (1972) in the first edition of his book *Childbirth without Fear* in the 1930s. Dick-Read observed that fear causes muscles to tense, which in turn increases pain and stress in childbirth. He taught both partners about childbirth and gave them physical exercises to ease muscle tension. Encouraged by Dick-Read's ideas, women began to reject anesthetics during labor and delivery and were consequently able to take a more active role in childbirth as well as being more aware of the process.

In the 1950s, Fernand Lamaze (1956, 1970) developed a method of prepared childbirth based on knowledge of conditioned reflexes. Women learn to mentally separate the physical stimulus of uterine contractions from the conditioned response of pain. With the help of a partner, women use breathing and other exercises throughout labor and delivery. Although Lamaze did much to advance the cause of prepared childbirth, he has been criticized by other childbirth educators as too controlling, or even "repressive," according to Armstrong and Feldman (1990). A woman does not give birth, they write, "by direction, as if it were a flight plan." Prepared childbirth, then, is not so much a matter of controlling the birth process as of understanding it and having confidence in nature's plan.

It is important for both the mother and the father to learn about pregnancy, labor, delivery, and infant care. Prepared childbirth classes are offered by many hospitals.

Clinical studies consistently show better birth outcomes for mothers who have attended prepared childbirth classes (Conway, 1980). Prepared mothers (who usually attend classes with the father or other partner) handle pain more successfully, use less medication and anesthesia, express greater satisfaction with the childbirth process, and experience less postpartum depression than women undergoing routine hospital births (Hetherington, 1990). For guidelines on making a birth plan, see the Resource Center.

Birth Centers and Birthing Rooms

Birth (or maternity) centers, long-standing institutions in England and other European countries (see Odent, 1984), now are being developed in the United States. In 1991, 0.3 percent of U.S. births occurred in freestanding birth centers (National Center for Health Statistics, 1993). Although they vary in size, organization, and orientation, birth centers share the view that childbirth is a normal, healthy process that can be assisted by skilled practitioners (midwives or physicians)

We encourage women in labor to give in to the experience, to lose control, to forget all they have learned—all the cultural images, all the behavioral patterns. The less a woman has learned about the "right" way to have a child, the easier it will be for her.

MICHEL ODENT

in a homelike setting. The mother (or couple) has considerable autonomy in deciding the conditions of birth—lighting, sounds, visitors, delivery position, and so on. Some of these centers can provide some kinds of emergency care; all have procedures for transfer to a hospital if necessary.

An extensive survey of 11,814 births in birth centers concluded that "birth centers offer a safe and acceptable alternative to hospital confinement for selected pregnant women, particularly those who have previously had children, and that such care leads to relatively few cesarean sections" (Rooks, Weatherby, Ernst, Rosen, and Rosenfield, 1989). Another large study showed that freestanding birth centers are associated with "a low cesarean section rate, low neonatal mortality [or] no neonatal mortality" (Eakins, 1989). Some hospitals now have their own birth centers or birthing rooms that provide for labor and birth in a comfortable setting and allow the mother or couple considerable autonomy. Hospital practices vary widely, however; prospective parents should carefully determine their needs and thoroughly investigate their options.

Home Birth

Home births constitute a small fraction of total births, amounting to 0.7 percent, according to available data (National Center for Health Statistics, 1993). Home births tend to be safer than hospital births if they are supervised by midwives or physicians. This is, in part, the result of careful medical screening and planning that eliminate all but the lowest-risk pregnancies. A couple can create their own birth environment at home, and home births usually cost at least one-third less than hospital delivery. With the supervision of an experienced practitioner, the couple have little to worry about. But if a woman is at risk, she is wiser to give birth in a hospital, where medical equipment is readily available.

Midwifery

The United States has an increasing number of certified nurse-midwives who are trained not only as registered nurses but also in obstetrical techniques. They are well qualified for routine deliveries and minor medical emergencies. They also often operate as part of a medical team that includes a backup physician, if needed. Their fees are generally considerably less than a doctor's. Some women prefer the attendance of midwives because they are almost always female, in contrast to physicians who are more likely to be male. A nurse-midwife may participate in hospital births, home births, or both, depending on hospital policy, state law, and the midwife's preference. Midwife-attended births, whether in home or hospital, have "distinctly better than average outcomes" in terms of birth weight and Apgar score (Declerq, 1992).

Lay midwifery, in which the midwife is trained by experienced midwives rather than being formally trained by the medical establishment, has also increased in popularity in the past two decades. Many satisfactory births with lay midwives in attendance have been reported. Extensive, reliable information is not available, however, owing to the "underground" nature of lay midwifery, which is often practiced outside (and without the support of) the medical establishment. Many midwives belong to organized groups; a number of these groups use some form of self-certification to ensure that high professional standards are maintained (Butter and Kay, 1990).

If a woman decides she wants to give birth with the aid of a midwife outside a hospital setting, she should have a thorough medical screening to make sure neither she nor her infant will be at risk during delivery. She should learn about the midwife's training and experience, what type of backup services the midwife has in the event of complications or emergencies, and how the midwife will handle a transfer to a hospital if it becomes necessary. (See the Resource Center for more information on childbirth.)

The Medicalization of Childbirth

Today, although only 5 to 10 percent of births actually require medical procedures, we seem to assume that childbirth is an inherently dangerous process. Of course, in high-risk cases, advances in technology can and do save lives. But the following questions remain: Why do women accept unnecessary, uncomfortable, demeaning, and even dangerous interventions in the birth process? Why do they tolerate a 24 percent cesarean rate, a 61 percent episiotomy rate, the almost universal administration of drugs, and the use of intrusive fetal monitoring—not to mention routine enemas? Why do they allow their infant boys to have their penises circumcised? Why do they accept, often without question, the physician's opinion over their own gut feelings?

Rarely is moral queasiness a match for the onslaught of science.

SHARON BEGLEY

Penny Armstrong and Sheryl Feldman (1990) suggest that society's increasing dependence on technology has hampered women's ability to view birth as a natural process for which they are naturally equipped. Women have allowed themselves to be persuaded that technology can do the job better than they can on their own.

Society expects birth technology to deliver a "product"—a "perfect" baby—without understanding that nature has already equipped women to deliver that product without much outside interference in the physical process (although encouragement and emotional support are paramount). Sheila Kitzinger (1989) writes:

> It is not advances in medicine but improved conditions, better food, and general health which have made childbirth much safer for mothers and babies today than it was 100 years ago. The rate of stillbirths and deaths in the first week of life is directly related to a country's gross national product and to the position of the mother in the social class.

. . . in the 19th century, the possibility of eliminating "pain and travail" created a new kind of prison for women—the prison of unconsciousness, of numbed sensations, or amnesia, and complete passivity.

ADRIENNE RICH. *OF WOMAN BORN*

The idea that the pain of childbirth is to be avoided at all costs is a relatively new one. When England's Queen Victoria accepted chloroform in 1853, she undoubtedly had little idea of the precedent she was setting. Today's advocates of prepared childbirth do not deny that there is pain involved; they argue, however, that it is a different pain from that of injury and that normally it is worth experiencing. This "pain with a purpose" is an intrinsic part of the birth process (Kitzinger, 1989). To obliterate it with drugs is to obliterate the mother's awareness and the baby's as well, depressing the child's breathing, heart rate, and general responsiveness in the process.

Another aspect of dependence on technology is that we get the feeling we are omnipotent and should be able to solve any problem. Thus, if something goes wrong with a birth—if a child is stillborn or has a disability, for example—we look around for something or someone to blame. We have become unwilling to accept that some aspects of life and death are beyond human control.

Prospective parents face a daunting array of decisions. The more informed they are, however, the better able they will be to decide what is right for them. According to Armstrong and Feldman (1990):

Reflections

In having a baby, what factors are (or will be) important to you in choosing a birth method, setting, practitioner, and so on? If you are a man, how have you chosen (or will you choose) to be involved in the birth of your child? If you are a woman, what extent of participation in childbirth do (or will) you expect from your partner?

UNDERSTANDING YOURSELF

The Ethics of Reproductive Technology

The desire to have healthy, happy, beautiful children is understandable, basic, and human. But in the face of today's reproductive technology, many of us may have to search our souls for the answers to a number of ethical questions. When dealing with infertility, the possibility of birth defects, or the likelihood of a difficult labor, we need to weigh the possible benefits of a given technique against the costs and risks involved. What are your feelings about the following reproductive issues? (For more information on intrauterine insemination, in vitro fertilization, and surrogate motherhood, see the Resource Center.)

- *Fetal diagnosis.* If you (or your partner) were pregnant, would you choose to have ultrasound? Should ultrasound be used routinely in all pregnancies or only in selected cases, such as when the mother is over thirty-five or there is a suspected problem? When would you consider amniocentesis appropriate? Should all fetuses be electronically monitored during birth (entailing the attachment of electrodes to the infant's scalp in the uterus), or should this technique be reserved for problem births? Would you want your child to be routinely monitored in this way?
- *Cesarean section.* How would you determine that a C-section was necessary for yourself or your partner? Should vaginal delivery be attempted first? What if it was a breech birth? What if you (or your partner) had had a previous C-section?
- *Intrauterine insemination* and *in vitro fertilization (IVF).* Should these techniques be available to anyone who wants them? Under what circumstances, if any, would you use these techniques? What are the rights of all the parties involved: parents, donors, physicians, child? What are the embryo's or fetus's rights? For example, does a frozen embryo have the "right to life" if its parents die before it is implanted in a surrogate? Can you envision a situation in which you would want to donate ova or sperm? What if your partner wanted to?
- *Older motherhood.* Modern reproductive technology has made it possible for women who have gone through menopause to give birth following IVF. What issues does this raise for these older mothers? For fathers? For the children? Are these decisions solely up to the mother? Both parents? Or should society be involved?
- *Surrogate motherhood.* Should a woman be allowed to "carry" a

Teaching women that they have a say—whether they consciously exercise it or not—is a major educational undertaking, one that requires breaking the hold obstetrical medicine has on the American imagination and helping women to rediscover their natural power at birth.

Minor surgery is one that is performed on someone else.

EUGENE ROBIN, M.D.

The Question of Circumcision

In 1975, when about 93 percent of newborn boys were circumcised, the American Academy of Pediatrics and the American College of Obstetricians and Gynecologists issued a statement declaring that there is "no absolute medical indication" for routine **circumcision.** This surgical operation, which involves slicing and removing the sleeve of skin (foreskin) that normally covers the glans penis, has been performed routinely on newborn boys in hospitals in the United States since at least the 1930s. The increasing popularity of circumcision, which began in the late 1800s, was apparently due to the belief that any boy who washed his penis would discover masturbation.

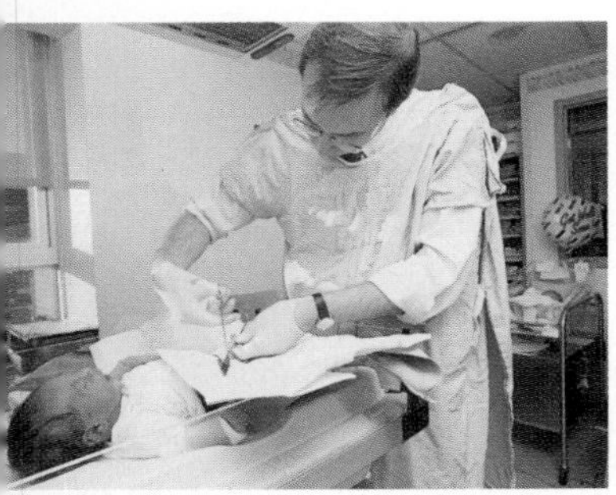

Circumcision: The routine surgical removal of the foreskin is increasingly being questioned.

Although circumcision is obviously painful, it is often done without anesthesia. Recent studies recommend using an anesthetic cream to reduce pain during the procedure (Benini, Johnston, Faucher, and Aranda, 1993; Weatherstone et al., 1993). The principal medical risks of circumcision are excessive bleeding, infection, and surgical error; in rare cases (fortunately), it can also be life threatening (Wallerstein, 1990). The emotional scars—of both infants and parents—are pregnancy for a couple using father-donated sperm? Does it matter if profit is involved? Should a contract involving surrogate motherhood be legally binding? Whose rights should take precedence, those of the surrogate (who is the biological mother) or of the couple (which includes the biological father)? What are the child's rights? Who decides? Can you think of a circumstance when you might wish to have the services of a surrogate mother? Could you be a surrogate or accept your partner's being one?

For all the preceding techniques, consider the following questions: Who is profiting (scientists, physicians, businesspeople, donors, parents, children)? Who is bearing the greatest risks? How great are the costs—financial and psychological—and who is paying them? What are the long-range goals of this technology? Are we "playing God?" How might this technology be abused? Are there certain techniques you think should be outlawed?

- *Abortion.* When do you think human life begins? Do we ever have the right to take human life? Under what conditions, if any, should abortion be permitted? Whenever the pregnant woman requests it? If there is a serious birth defect? If there is a minor birth defect (such as a shortened limb)? If the fetus is the "wrong" sex? If rape or incest led to the pregnancy? Under what conditions, if any, would you have an abortion or want your partner to have one? If you had an unmarried pregnant teenage daughter, would you encourage her to have one?
- *Tissue donation.* Is it appropriate to use the tissues or organs of human corpses for medical purposes? Is it appropriate to use aborted fetuses for tissue donation? For research?
- *Life and death.* Is prolonging the life of an infant always the most humane choice? What if it also prolongs suffering? Should life be prolonged whenever possible, at all costs? Who decides?
- *Human cloning.* Scientists expect that cloning—manipulating a cell from an organism so that it develops into an exact duplicate of the original organism—will be technologically possible for human replication within ten years. Should the cloning of humans be allowed? Can you think of any circumstances when it might be acceptable? Should the use of human cells or human embryos in cloning research be allowed?
- *Fertility and fulfillment.* Do you think your reproductive values and feelings about your own fertility (or lack of it) are congruent with reality, given the world's population problems? For you, are there viable alternatives to conceiving or bearing a child? What are they? Would adoption be one alternative? Why?

not visible or measurable but may be deep, according to some authorities (Milos, 1989; Milos and Macris, 1992).

In 1989, the American Academy of Pediatrics modified its stance on circumcision, stating that "newborn circumcision has potential medical benefits and advantages as well as disadvantages and risks." The organization recommends that "the benefits and risks . . . be explained to the parents and informed consent obtained." This change occurred at least in part in response to several studies indicating a *possible* connection between lack of circumcision and urinary tract infections, penile cancer, and sexually transmissible diseases (Schoen, 1990; Wiswell, 1990). These studies have been contradicted by others, however (Altschul, 1989; Canadian Paediatric Society, 1996). Circumcision clearly does not *guarantee* protection from STDs or infections. Physician George Denniston (1992) puts the issue in perspective:

> Performing 100 mutilative surgeries to possibly prevent one treatable urinary tract infection is not valid preventive medicine. . . . Penile cancer occurs in older men at the rate of approximately 1 per 100,000. The idea of performing 100,000 mutilating procedures on newborns to possibly prevent cancer in one elderly man is absurd. Applying this type of reasoning to women would seem to lead to the conclusion that breast cancer should be prevented by removing breasts at puberty.

There is a clear need for long-range comparative studies of circumcised and intact males (Maden et al., 1993; Poland, 1990). Factors such as hygiene and

And you shall circumcise the flesh of your foreskin: and it shall be a token of the covenant between Me and you.

GENESIS (17:9–14)

number of sexual partners need to be taken into consideration. In addition, as the foreskin contains numerous nerve endings, the effects of its removal on sexual functioning need to be assessed (Taylor, Lockwood, and Taylor, 1996).

According to the National Center for Health Statistics, 59 percent of newborn boys were circumcised in 1990. Although this represents a substantial drop from the 93 percent circumcised in 1975, it still places the United States far ahead of other developed countries, which circumcise less than 1 percent of their newborn boys. The exception is Israel; in Judaism, ritual circumcision, called the *brit milah* (or *bris*), is an important religious event. Circumcision has religious significance for Muslims as well (Bullough, 1976). Within the Jewish community, some parents have developed alternative *brit milah* ceremonies (Bivas, 1988; Karsenty, 1988; Rothenberg, 1991).

Wash, don't amputate.

ALEX COMFORT, M.D.

Aside from religious reasons, the reasons given by parents for circumcising their infants are "cleanliness" and "so he'll look like his dad." A circumcised penis, however, is not necessarily any cleaner than an intact one. Infants do not require cleaning under their foreskins. Adults do, but as sex therapist Louanne Cole (1993) writes, "Isn't it insulting to the average male's intelligence to think that surgery is preferable because he can't be entrusted with washing his genitals when somehow he manages to brush his teeth, clean his ears, and blow his nose?" If reasonable cleanliness is observed, an intact penis poses no more threat of disease to a man's sexual partner than a circumcised one would. As for "looking like dad," most parents can probably find ways to keep their son's self-esteem intact along with his foreskin. There is no evidence we know of to suggest that little boys are seriously traumatized if dad's penis doesn't look exactly like theirs. We know one dad (circumcised) who says to his sons (uncircumcised): "Boy, you guys are lucky. You should've seen what they did to me!"

An analysis of the potential financial costs and health benefits of circumcision concludes that "there is no medical indication for or against circumcision" and that decisions "may most reasonably be made on [the basis of] nonmedical factors such as parent preference or religious convictions" (Lawler, Bisonni, and Holtgrave, 1991). (See the Resource Center for organizations to contact regarding circumcision.)

Breastfeeding

Sometimes a breast is a sexual object, and sometimes it's a food delivery system, and one need not preclude nor color the other.

ANNA QUINDLEN

In the case of hospital birth, the infant will generally be brought to the mother at scheduled intervals for feeding and visiting. In the last few years, some hospitals have tried to comply with mothers' wishes for "demand" feeding (when the baby feels hungry, as opposed to when the hospital wants to feed the child). But because most hospital maternity wards are busy places, they often find it difficult to meet the individual needs of many infants and mothers. If the mother is breastfeeding, a rigid hospital schedule can make it difficult for her to establish a feeding routine with her child.

About three days after childbirth, **lactation**—the production of milk—begins. Before lactation, sometimes as early as the second trimester, a yellowish liquid called **colostrum** is secreted by the nipples. It is what nourishes the newborn infant before the mother's milk comes in. Colostrum is high in protein and contains antibodies that help protect a baby from infectious diseases. Hormonal changes during labor begin the changeover from colostrum to milk, but unless a mother nurses her child, her breasts will soon stop producing milk. If she chooses not to breastfeed, she is usually given an injection of estrogen soon after delivery

Did you know?

To the alarm of public health officials, the rate of breastfeeding has declined in the last 15 years from 63 to 50 percent (Brody, 1994).

Breastfeeding offers the best possible nutrition for an infant as well as providing protection from many infectious diseases. It is a source of emotional well-being for babies and mothers alike.

to stop lactation. It is not certain, however, whether estrogen is actually effective; furthermore, it may cause an increased risk of blood clotting.

Nutritionists and public health officials are alarmed at the decline in breastfeeding in recent years (Brody, 1994; Weiss, 1992). A mother's milk—if she is healthy and has a good diet—offers the best nutrition for the baby. In addition, her milk contains antibodies that will protect her child from infectious diseases such as respiratory infections and meningitis. Finally, a breastfed baby is less likely to become constipated or contract skin diseases. Low-birth-weight babies, similarly, do best with their own mother's milk rather than formula, cow's milk, or mature milk from a donor with an older infant, because mother's milk is naturally adapted to meet infant's needs. A benefit to the mother is that hormonal changes stimulated by breastfeeding cause the uterus to contract and help ensure its return to a normal state. The American Academy of Pediatricians endorses total breastfeeding for a baby's first six months. Breastfeeding has psychological as well as physical benefits. Nursing provides a sense of emotional well-being for both mother and child through close physical contact. A woman may feel that breastfeeding gives her assurance that she is capable of nourishing—able to sustain the life of another through her milk.

Many mothers who work outside the home find that they can breastfeed by using a breast pump (either manual or electric) to express milk at home, which can then be refrigerated and fed to the baby later on. Mothers can also express milk at work, either to refrigerate and take home or to discard; expressing milk on a regular basis relieves the pressure of overfull breasts and keeps up the rate of milk production.

American mothers may worry about whether they will be able to breastfeed "properly." But what the distinguished physician Niles Newton wrote in 1955 still applies today:

> Successful breast-feeding is the type of feeding that is practiced by the vast majority of mothers all over the world. It is a simple, easy process. When the baby is hungry,

All is beautiful
All is beautiful
All is beautiful, yes!
Now Mother Earth
And Father Sky
Join one another and meet
forever helpmates
All is beautiful
All is beautiful
All is beautiful, yes!
Now the night of darkness
And the dawn of light
Join one another and meet
forever helpmates
All is beautiful
All is beautiful
All is beautiful, yes!
Now the white corn
And the yellow corn
Join one another and meet
forever helpmates
All is beautiful
All is beautiful
All is beautiful, yes!
Life that never ends
Happiness of all things
Join one another and meet
Forever helpmates
All is beautiful
All is beautiful
All is beautiful, yes!
NAVAJO NIGHT CHANT

it is simply given a breast to suck. There is an abundance of milk, and the milk supply naturally adjusts itself to the child's growth and intake of other foods.

(See the Resource Center for hints on successful breastfeeding.)

Many American women choose not to breastfeed. Their reasons include the inconvenience of not being able to leave the baby for more than a few hours at a time, tenderness of nipples (which generally passes within several days to two weeks), and inhibitions about nursing a baby, especially in public. If a woman works, bottle feeding may be her only practical alternative, as American companies rarely provide leaves, part-time employment, or nursing (or breast-pump) breaks for their female employees. Some women may have a physical condition that precludes breastfeeding.Some women may choose not to breastfeed because their husbands feel jealous of the intimate relationship between baby and mother. Some husbands may feel incompetent because they cannot contribute to nourishing the child.

Bottle feeding an infant does make it possible for the father and other caregivers to share in the nurturing process. Some mothers (and fathers) have discovered that maximum contact and closeness can be enjoyed when the infant is held against the naked breast while nursing from a bottle.

Reflections

Do you feel you have enough information at this point to make decisions regarding circumcision and breastfeeding for a child you might have in the future? What other information would you like to have? If you made these decisions now, what would you choose and why?

Becoming a Parent

Men and women who become parents enter a new phase of their lives. Even more than marriage, parenthood signifies adulthood—the final, irreversible end of youthful roles. A person can become an ex-spouse but never an ex-parent. The irrevocable nature of parenthood may make the first-time parent doubtful and apprehensive, especially during the pregnancy. Yet, for the most part, parenthood has to be learned experientially, although ideas can modify practices. A person may receive assistance from more experienced parents, but ultimately all new parents have to learn on their own.

The time immediately following birth is a critical period for family adjustment. No amount of reading, classes, and expert advice can prepare expectant parents for the real thing. The three months or so following childbirth (the "fourth trimester") constitute the **postpartum period.** This time is one of physical stabilization and emotional adjustment. The abrupt transition from a nonparent to a parent role may create considerable stress. Parents take on parental roles literally overnight, and the job goes on without relief around the clock. Many parents express concern about their ability to meet all the responsibilities of child rearing.

New mothers, who may well have lost most of their interest in sexual activity during the last weeks of pregnancy, will probably find themselves returning to prepregnancy levels of desire and coital frequency. Some women, however, may have difficulty reestablishing their sexual life because of fatigue, physiological problems such as continued vaginal bleeding, and worries about the infant (Reamy and White, 1987).

The postpartum period also may be a time of significant emotional upheaval. Even women who had easy and uneventful births may experience a period of "postpartum blues" characterized by alternating periods of crying, unpredictable mood changes, fatigue, irritability, and occasional mild confusion or lapses of memory. A woman may have irregular sleep patterns because of the needs of her newborn, the discomfort of childbirth, or the strangeness of the hospital environment. Some mothers may feel lonely, isolated from their familiar world. Many women blame themselves for their fluctuating moods. They may feel that they have lost control over their lives because of the dependency of their newborns.

Biological, psychological, and social factors are all involved in postpartum depression. Biologically, during the first several days following delivery, there is an abrupt fall in certain hormone levels. The physiological stress accompanying labor, dehydration, blood loss, and other physical factors contribute to lowering the woman's stamina. Psychologically, conflicts about her ability to mother, ambiguous feelings toward or rejection of her own mother, and communication problems with the infant or partner may contribute to the new mother's feelings of depression and helplessness. Finally, the social setting into which the child is born is important, especially if the infant represents a financial or emotional burden for the family. Postpartum counseling prior to discharge from the hospital can help couples gain perspective on their situation so they will know what to expect and can evaluate their resources.

Although the postpartum blues are felt by many women, they usually don't last more than a couple of weeks. Interestingly, men seem to get a form of postpartum blues as well. When infants arrive, many fathers do not feel prepared for their new parenting and financial responsibilities. Some men are overwhelmed by the changes that take place in their marital relationship. Fatherhood is a major transition for them, but their feelings are overlooked because most people turn their attention to the new mother. (Parental roles are explored more thoroughly in Chapter 10.)

It looks increasingly as if a number of the things we assumed about mothers and motherhood are really typical of parenthood.

Martha Zaslow, M.D.

The transition to parenthood can be made easier if the new parents understand in advance that a certain amount of tiredness and stress is inevitable. They need to ascertain what sources of support will be helpful to them, such as friends or family members who can help out with preparing meals or running errands. They also need to keep their lines of communication open—to let each other know when they are feeling overwhelmed or left out, and it's very important that they plan time to be together, alone or with the baby—even if it means telling a well-meaning relative or friend they need time to themselves.

For many women and men, the arrival of a child is one of life's most important events. It fills mothers and fathers alike with a deep sense of accomplishment. The experience itself is profound and totally involving. A father describes his wife (Kate) giving birth to their daughter (Colleen) (in Armstrong and Feldman, 1990):

> Toward the end, Kate had her arms around my neck. I was soothing her, stroking her, and holding her. I felt so close. I even whispered to her that I wanted to make

Although becoming a parent is stressful, the role of mother or father is deeply fulfilling for many people.

love to her—It wasn't that I would have or meant to—it's just that I felt that bound up with her.

Colleen was born while Kate was hanging from my neck. . . . I looked down and saw Mimi's [the midwife's] hands appearing and then, it seemed like all at once, the baby was in them. I had tears streaming down my face. I was laughing and crying at the same time. . . . Mimi handed her to me with all the goop on her and I never even thought about it. She was so pink. She opened her eyes for the first time in her life right there in my arms. I thought she was the most beautiful thing I had ever seen. There was something about that, holding her just the way she was. . . . I never felt anything like that in my life.

Summary

- The first chemical pregnancy test—the *agglutination test*—can be made about two weeks after a woman misses her menstrual period. Pregnancy is confirmed by the detection of fetal heartbeats and movements or examination by ultrasound. *Hegar's sign,* another signal of pregnancy, can be detected during a vaginal examination.
- A woman's feelings vary greatly during pregnancy. It is important for the woman to share her fears and to have support from her partner, relatives, friends, and possibly, women's groups. Her feelings about sexuality are likely to change during pregnancy. Men may also experience conflicting feelings. Sexual activity is generally safe during pregnancy unless there is a prior history of miscarriage, bleeding, or pain.
- Harmful substances may be passed to the embryo or fetus through the placenta. Substances that cause birth defects are called *teratogens.* Alcohol, opiates, and cigarettes can damage the unborn child; so can chemicals and environmental pollutants. Infectious diseases, such as rubella, may damage the fetus. Sexually transmissible diseases may be passed to the infant through the placenta or through the birth canal during childbirth.
- *Ectopic pregnancy, spontaneous abortion, toxemia* and *preeclampsia,* and *low birth weight* (prematurity) are the most important complications of pregnancy.
- Abnormalities of the fetus may be diagnosed using *ultrasonography, amniocentesis, chorionic villus sampling (CVS),* or *alphafeto protein (AFP) screening.*
- About one out of four women is aware of having had a spontaneous abortion (miscarriage). Infant mortality rates in the United States are extremely high compared to those in other industrialized nations. Loss of pregnancy or death of a young infant is recognized as a serious life event.
- Throughout pregnancy, a woman feels *Braxton Hicks contractions.* Contractions also begin the *effacement* and *dilation* of the cervix to permit delivery. Labor can be divided into three stages. First-stage labor begins when uterine contractions become regular. When the cervix has dilated approximately 10 centimeters, the baby's head enters the birth canal; this is called *transition.* In second-stage labor, the baby emerges from the birth canal. In third-stage labor, the placenta (afterbirth) is expelled.
- *Cesarean section* is the removal of the fetus by an incision in the mother's abdominal and uterine walls. A dramatic increase in C-sections in recent years has led to criticism that the procedure is used more often than necessary.
- *Prepared childbirth* encompasses a variety of methods that stress the importance of understanding the birth process and of relaxation of the mother during childbirth. *Birth centers* and special birthing rooms in hospitals are providing attractive alternatives to impersonal hospital birth settings for normal births. Carefully planned home births attended by a physician or midwife are another alternative. Nurse-midwives and lay midwives are trained in obstetric techniques. Many women are now choosing *midwives* to deliver their babies because they want home births, cannot afford physicians, or prefer the attendance of a woman.
- The medicalization of childbirth—making this natural process into a medical "problem"—has caused an overdependence on technology and an alienation of women from their bodies and feelings. Women can empower themselves by becoming informed about childbirth and their options.
- *Circumcision* has been performed routinely in the United States for many years. There are usually no medical reasons for this procedure, and the practice is being increasingly debated. Circumcision holds religious meaning for Jews and Muslims.

- About 50 percent of American women breastfeed their children. Mother's milk is more nutritious than formula or cow's milk and provides immunities to many diseases. Nursing offers emotional rewards to mother and infant.
- A critical adjustment period follows the birth of a child. The mother may experience feelings of depression (sometimes called postpartum blues) that are a result of biological, psychological, and social factors. Participation of the father in nurturing the infant and performing household duties may help alleviate both the mother's and the father's feelings of confusion and inadequacy.

Key Terms

alpha-feto protein (AFP) screening 266
amniocentesis 266
Apgar score 272
Braxton Hicks contractions 269
cesarean section (C-section) 274
chorionic villus sampling (CVS) 266
circumcision 278
colostrum 280
couvade 270
dilation 269
ectopic pregnancy 264
effacement 269
episiotomy 273
Hegar's sign 257
involution 272
lactation 280
lochia 272
low birth weight (LBW) 265
neonate 272
parturition 269
placenta 272
postpartum period 282
preeclampsia 265
prepared childbirth 275
sonogram 265
sudden infant death syndrome (SIDS) 268
teratogen 262
toxemia 264
transition 269
ultrasonography 265
vernix 272

Suggested Readings

For the most current research findings in obstetrics, see *Obstetrics and Gynecology*, the *New England Journal of Medicine*, and *JAMA: Journal of the American Medical Association*.

Armstrong, Penny, and Sheryl Feldman: *A Wise Birth.* New York: Morrow, 1990. Written with intelligence and warmth, a thought-provoking book exploring the effects of medical technology and technological thinking on modern childbirth.

Baruch, Elaine Hoffman, et al., eds. *Embryos, Ethics and Women's Rights: Exploring the New Reproductive Technologies.* New York: Harrington Park Press, 1988. A collection of essays from a feminist viewpoint.

Boone, Margaret S. *Capital Crime: Black Infant Mortality in America.* Newbury Park, CA: Sage Publications, 1989. A sobering look at the devastating effects of discrimination and poverty on African-American children.

Dorris, Michael. *The Broken Cord.* New York: Harper Perennial, 1990. A moving account of the author's experience with his adopted son, who was affected by fetal alcohol syndrome.

Eisenberg, Arlene, et al. *What to Expect When You're Expecting.* Rev. ed. New York: Workman Publishing, 1994. A thorough and thoroughly readable "encyclopedia" for expectant and new parents.

Harper, Barbara. *Gentle Birth Choices.* Rochester, VT: Healing Arts Press, 1994. A thoughtful exploration of birth alternatives, with photographs.

Leach, Penelope. *Your Baby and Child: From Birth to Age Five.* New York: Knopf, 1994. Complete, concise, down-to-earth advice for parents and parents-to-be.

Nilsson, Lennart, and Lars Hamberger. *A Child Is Born.* New York: DPT/Seymour Lawrence, 1993. The story of birth, beginning with fertilization, told in stunning photographs and text.

Profet, Margie. *Protecting Your Baby to Be: Preventing Birth Defects in the First Trimester.* Reading, MASS: Addison-Wesley, 1995. A thorough guide to prenatal care.

Pryor, Karen, and Gale Pryor. *Nursing Your Baby.* New York: Pocket Books, 1991. A comprehensive and compassionate guide to breastfeeding.

Samuels, M., and N. Samuels. *The New Well Pregnancy Book.* New York: Fireside, 1996. A comprehensive, user-friendly guide to pregnancy and childbirth.

Vredevelt, P. *Empty Arms: Emotional Support for Those Who Have Suffered Miscarriage or Stillbirth.* Sisters, OR: Questar, 1994. A sensitive guide to dealing with grief after a pregnancy loss.

Weschler, T. *Taking Charge of Your Fertility.* New York: Harper Perennial, 1995. An up-to-date, sensitive, and holistic approach to understanding fertility.

CHAPTER 9

Marriage as Process: Family Life Cycles

P R E V I E W

To gain a sense of what you already know about the material covered in this chapter, answer "True" or "False" to the statements below.

1. More women than men tend to live with their parents. True or false?
2. It is a myth that marriage changes a person for the better. True or false?
3. Couples who are unhappy before marriage significantly increase their happiness after marriage. True or false?
4. Marriage, more than parenthood, radically affects a woman's life. True or false?
5. The advent of children generally increases a couple's marital satisfaction. True or false?
6. Grandparents play a significant role in the caretaking of their grandchildren. True or false?
7. In-law relationships tend to be characterized by low emotional intensity. True or false?
8. The empty nest syndrome, characterized by maternal depression after the last child leaves home, is more a myth than a problem for American women. True or false?
9. The vast majority of long-term marriages involve couples who are blissful and happily in love. True or false?
10. The key to marital satisfaction in the later years is continued good health. True or false?

Answers

1 False, 2 True, 3 False, 4 False, 5 False, 6 True, 7 True, 8 True, 9 False, 10 True

OUTLINE

Marriage is a process in which people interact with each other, create families, and give each other companionship and love. Marriage is not static; it is always changing to meet new situations, new emotions, new commitments, and new responsibilities. The marriage process may begin informally with cohabitation or formally with engagement. Marriage itself ends with divorce or with the death of a partner.

We may begin marriage thinking we know how to act within it, but we find that the reality of marriage requires us to be more flexible than we had anticipated. We need flexibility to meet our needs, our partners' needs, and the needs of the marriage. We may have periods of great happiness and great sorrow within marriage. We may find boredom, intensity, frustration, and fulfillment. Some of these may occur because of our marriage; others may occur in spite of it. But, as we shall see, marriage encompasses many possibilities.

In this chapter we examine the developmental perspective from both an individual and family life cycle perspective. Next we look at beginning marriages, including factors predicting marital success, cohabitation, engagement, and weddings, as well as the establishment of marital roles and boundaries. Then we look at youthful marriages, especially the impact of children and the individual changes we may experience. We turn next to middle-aged marriages, examining families with young children and adolescents, families as launching centers of the young, and the process of reevaluation. Then we review later-life marriages, especially the extended family, grandparenting, retirement, and widowhood. Finally, we survey the different patterns of lasting marriages.

The Developmental Perspective

The developmental perspective is an interdisciplinary approach that unites sociology and related disciplines with child and human development (Duvall, 1988). It sees individual development and family development as interacting with each other.

The Individual Life Cycle

Our identity, our sense of who we are, is not fixed or frozen. It changes as we mature. At different points in our lives, we are confronted with different developmental tasks, such as acquiring trust and becoming intimate. Our growth as human beings depends on the way we perform these tasks. Erik Erikson (1963) describes the human life cycle as containing eight developmental stages. At each stage, we have an important developmental task to accomplish, and each stage intimately involves the family. As we enter young adulthood, these stages may also involve marriage or other intimate relationships (Nichols and Pace-Nichols, 1993).

. . . if one advances confidently in the direction of his dreams, and endeavors to live the life which he has imagined, he will meet with a success unexpected in his common hours.

HENRY DAVID THOREAU (1817–1862)

The way we deal with these stages, which are summarized here, is strongly influenced by our families, marriages, or other intimate relationships. We cannot separate our identity from our relationships.

- *Infancy: Trust versus Mistrust*. In the first year of life, children are wholly dependent on their parenting figures for survival. It is in this stage that they learn to trust by having their needs satisfied and by being loved, held, and caressed. Without loving care, an infant may develop a mistrusting attitude toward others and toward life in general.
- *Toddler: Autonomy versus Shame and Doubt*. Between ages one and three, children learn to walk and talk; they also begin toilet training. At this stage they need to develop a sense of independence and mastery over their environment and themselves.
- *Early Childhood: Initiative versus Guilt*. Ages four to five are years of increasing independence. The family must allow the child to develop initiative while at the same time directing the child's energy. The child must not be made to feel guilty about his or her desire to explore the world.

The young are slaves to dreams; the old servants of regrets. Only the middle aged have all their senses.

HERVY ALLEN

- *School Age: Industry versus Inferiority*. Between ages six and eleven, children begin to learn that their activities pay off and that they can be creative. The family needs to encourage the child's sense of accomplishment. Failing to do so may lead to feelings of inferiority in the child.
- *Adolescence: Identity versus Role Confusion*. The years of puberty, between ages twelve and eighteen, may be a time of turmoil as well as discovery and growth. Adolescents try new roles as they make the transition to adulthood. To make a successful transition, they need to develop goals, a philosophy of life, and a sense of self. The family needs to be supportive as the adolescent tentatively explores adulthood. If the adolescent fails to establish a firm identity, he or she is likely to drift without purpose.
- *Young Adulthood: Intimacy versus Isolation*. In young adulthood, the adolescent leaves home and begins to establish intimate ties with other people through cohabitation, marriage, or other important intimate relationships. A young adult who does not make other intimate connections may be condemned to isolation and loneliness.
- *Adulthood: Generativity versus Self-Absorption*. Generativity is the bearing of offspring, productiveness, or creativity. In adulthood, the individual establishes his or her own family and finds satisfaction in family relationships. It is a time of creativity. Work becomes important as a creative act, perhaps as important as family or an alternative to family. The failure to be generative may lead to self-centeredness and a "what's-in-it-for-me" attitude toward life.
- *Maturity: Integrity versus Despair*. In old age, the individual looks back on life to understand its meaning—to

Families as well as individuals experience different stages of growth. The presence of children profoundly influences a couple's relationship.

Reflections

As you look at these stages, where are you in the life cycle? What is the psychological task that Erikson says you need to accomplish? Is it an important issue for you? How are you resolving it?

assess what has been accomplished and to gauge the meaning of his or her relationships. Those who can make a positive judgment have a feeling of wholeness about their lives. The alternative is despair.

Throughout our life cycle, our goals and concerns change. Among young adults, goals include education and family-related goals, such as marriage and having children. Among middle-aged adults, goals shift to concern about children's lives and about property, such as buying or maintaining homes. Among the elderly, health, retirement, leisure activities, and interest in the world predominate (Nurmi, 1992).

The Family Life Cycle

Just as individuals have life cycles with specific stages, so do marriages and families. Within the stages of the **family life cycle,** each marriage and family has its own unique history (Aldous, 1978). The concept of the family life cycle uses a developmental framework to explain people's behavior in families. According to the developmental framework, families change over time both in terms of the people who are members of the family and in terms of the roles they play. At various stages in the family life cycle, the family has different developmental tasks to perform. Much of the behavior of family members can be explained in terms of the family's developmental stage. The key factor in such developmental stages is the presence of children. The family organizes itself around its child-rearing responsibilities. A woman's role as wife is different when she is childless than when she has children. A man's role is different when he is the father of a one-year-old than when he is the father of a fifteen-year-old.

We arrive at the various stages of life quite as novices.

François de La Rochefoucauld (1613–1680)

The family life-cycle approach gives us important insights into the complexities of family life. Not only is the family performing various tasks during its life cycle, but also each family member takes on various developmental tasks during each stage of the life cycle. The task for families with adolescents is to give their children greater autonomy and independence. While the family is coping with this new developmental task, an adolescent daughter has her own individual task of trying to develop a satisfactory identity. Meanwhile, her older brother is struggling with intimacy issues, her younger sister is developing industry, her parents are dealing with issues of generativity, and her grandparents are confronting issues of integrity.

The whole world is a comedy to those who think, a tragedy to those who feel.

Horace Walpole (1717–1797)

Stages of the Family Life Cycle

One of the most widely used approaches divides the family life cycle into eight stages (Duvall and Miller, 1985). The eight-stage approach reflects the model life cycle. It does not encompass the family life cycle of single-parent or remarried families, as we discuss shortly.

- *Stage I: Beginning Families.* During the first stage, the married couple has no children. In the past, this stage was relatively short because children soon followed marriage. Today, however, this stage may last until a couple is in their late twenties or thirties. On the average, this stage is about two or three years. In 1994, the average age for women to have their first child was twenty-seven (U.S. Bureau of the Census, 1996). Most studies on marital satisfaction agree that couples experience their greatest satisfaction in this stage (Glenn, 1991).
- *Stage II: Childbearing Families.* Because families tend to space their children about thirty months apart, the family in Stage II is still considered to be forming, as new births are likely during this period. By the second stage, the

average family has two children. Although mothers are deeply involved in childbearing and child rearing, 60 percent of married women are in the labor force (U.S. Bureau of the Census, 1995, Table 1). Stage II lasts around two and a half years. Marital satisfaction begins to lessen and continues to decline through the stage of families with schoolchildren (Stage IV) or those with adolescents (Stage V).

- *Stage III: Families with Preschool Children.* The family's oldest child is thirty months to six years in Stage III. The parents, especially the mother, are still deeply involved in child rearing. This stage lasts about three and a half years.
- *Stage IV: Families with Schoolchildren.* In Stage IV, the family's oldest child is between six and thirteen years old. With the children in school and more free time, the mother has more options available to her. By now, most women have reentered the job market. This stage lasts about seven years.
- *Stage V: Families with Adolescents.* In a family with adolescents, the oldest child is between thirteen and twenty years old. Marital satisfaction reaches its nadir. This stage lasts about seven years.
- *Stage VI: Families as Launching Centers.* By Stage VI, the first child has been launched into the adult world. This stage lasts until the last child leaves home, a period averaging about eight years. Virtually all studies show that marital satisfaction begins to rise for most couples during this stage.
- *Stage VII: Families in the Middle Years.* Stage VII lasts from the time the last child has left home to retirement. It is commonly referred to as the "empty nest" stage. It is a distinct and relatively new phase in the family life cycle. Until this century, most parents continued to have children until middle age, and the child-rearing and launching periods were extended into old age. More recently, however, because of deferred marriage and the high cost of housing, many adult children continue to live at home or return home after being away for a number of years. This "not-so-empty nest" phase, argue some scholars, constitutes a more recent variation in Stage VII (Glick, 1989a; Mattessich and Hill, 1987). At this time, many families begin caretaking activities for elderly relatives, especially parents and parents-in-law.
- *Stage VIII: Aging Families.* The working members of the aging family have retired. Usually, the husband retires before the wife because he tends to be older. Ill health begins to take its toll. Eventually, one of the spouses dies, usually the husband. The surviving spouse may live alone or with other family members or be cared for by them.

For everything there is a season, and a time to every purpose under heaven:
a time to be born and a time to die;
a time to plant, and a time to pluck up what is planted;
a time to kill, and a time to heal;
a time to break down, and a time to build up;
a time to weep, and a time to laugh;
a time to mourn, and a time to dance;
a time to cast away stones and a time to gather stones together;
a time to embrace, and a time to refrain from embracing;
a time to seek, and a time to lose;
a time to keep, and a time to cast away;
a time to rend, and a time to sew;
a time to keep silence, and a time to speak:
a time to love, and a time to hate;
a time for war, and a time for peace.
ECCLESIASTES 3: 1–8

Transitions from one stage to another in the family life cycle seem to affect relationships. Change is marked by stress as we adjust to new situations and roles. A more positive outcome is likely when couples have other meaningful roles in their lives, such as work, hobbies, or school.

Family Life Cycle Variations

The critical period in matrimony is breakfast time.
A. P. HERBERT

A major limitation of the family life cycle approach is its tendency to focus on the intact nuclear family as *the* family. Paul Glick, who coined the term *family life cycle* in 1947, has noted that a number of social changes have affected the traditional family life cycle (Glick, 1989a). Deferred marriages, cohabitation, divorce, remarriage, single parenthood, and gay and lesbian families introduce notable variations in the life course of the family.

As marriages are increasingly deferred, cohabitation rates have skyrocketed. It may be useful, Glick suggests, to include a cohabitational stage that recognizes the significance of living together in family development. In addition, the family

Table 9.1 Individual and Marital Stages of Development

Item	Stage 1 (18–21 years)	Stage 2 (22–28 years)	Stage 3 (29–31 years)
Individual stage	Developing roots	Provisional adulthood	Transition at age 30
Individual task	Developing autonomy	Developing intimacy and occupational identification; getting into the adult world	Deciding about commitment to work and marriage
Marital task	Shift from family of origin to new commitment	Provisional marital commitment	Commitment crisis; restlessness
Marital conflict	Original family ties conflict with adaptation	Uncertainty about choice of marital partner, stress over parenthood	Doubts about choice come into sharp conflict, rates of growth may diverge if spouse has not successfully negotiated stage 2 because of parental obligations
Intimacy	Fragile intimacy	Deepening but ambivalent intimacy	Increasing distance while partners make up their minds about each other
Power	Testing of power	Establishment of patterns of conflict resolution	Sharp vying for power and dominance
Marital boundaries	Conflicts over in-laws	Friends and potential lovers; work versus family	Temporary disruptions including extramarital sex or reactive "fortress building"

life cycle of single adults is considerably different from that of their married counterparts. For childless adults who have never married, the families of orientation may become the central family focus. Single women, for example, may be intimately tied to their parents as caregivers; these women also may be connected to children as aunts (Allen and Pickett, 1987; Allen, 1989).

Families that include children with disabilities must negotiate the family life cycle differently, depending on the child's unique abilities and limitations (see Chapter 12). For example, independence may come later, or the child may never be able to live outside the family (Hanline, 1991). Adoptive families have their own unique life cycles (Hajal and Rosenberg, 1991). They may simultaneously become beginning families and families with schoolchildren with the adoption of a five-year-old child. Lesbian and gay families also have special family life cycle issues as a result of having two mothers or two fathers, sperm donor fathers, or parents who are the heterosexual partners from earlier marriages or relationships (Slater and Mencher, 1991). Such families often encompass members of their communities as part of their extended families.

Scholars are also recognizing the importance of ethnicity to family life cycles. The timing of the various stages, for example, is affected by ethnicity. African Americans are less likely to marry than are whites, and they are likely to marry at a later age than whites. Single parenthood is more prevalent among African-American

Table 9.1 (continued)

Stage 4 (32–39 years)	Stage 5 (40–42 years)	Stage 6 (43–59 years)	Stage 7 (60 years and over)
Settling down	Midlife transition	Middle adulthood	Older age
Deepening commitments; pursuing more long-range goals	Searching for "fit" between aspirations and environment	Restabilizing and reordering priorities	Dealing effectively with aging, illness, and death while retaining zest for life
Productivity: children, work, friends, and marriage	Summing up; success and failure are evaluated and future goals sought	Resolving conflicts and stabilizing the marriage for the long haul	Supporting and enhancing each other's struggle for productivity and fulfillment in face of the threats of aging
Husband and wife have different and conflicting ways of achieving productivity	Husband and wife perceive "success" differently; conflict between individual success and remaining in the marriage	Conflicting rates and directions of emotional growth; concerns about losing youthfulness may lead to depression and/or acting out	Conflicts are generated by rekindled fears of desertion, loneliness, and sexual failure
Marked increase in intimacy in "good" marriages; gradual distancing in "bad" marriages	Tenuous intimacy as fantasies about others increase	Intimacy is threatened by aging and by boredom vis-à-vis a secure and stable relationship; departure of children may increase or decrease intimacy	Struggle to maintain intimacy in the face of eventual separation; in most marriages this dimension achieves a stable plateau
Establishment of definite patterns of decision making and dominance	Power in outside world is tested vis-à-vis power in the marriage	Conflicts often increase when children leave, and security appears threatened	Survival fears stir up needs for control and dominance
Nuclear family closes boundaries	Disruption due to reevaluation; drive versus restabilization	Boundaries are usually fixed except in crises such as illness, death, job change, and sudden shift in role relationships	Loss of family and friends leads to closing in of boundaries, physical environment is crucial in maintaining ties with the outside world

SOURCE: Levinson et al., 1974. Reprinted with permission from Berman, F. and Lief, H. "Marital Therapy from a Psychiatric Perspective: An Overview." *American Journal of Psychiatry* 132, 6:586, June 1975.

families than among white ones. In 1995, for example, 64 percent of African-American families, 36 percent of Hispanic families, and 25 percent of white families were headed by single parents (U.S. Bureau of the Census, 1996). African Americans are also more likely than whites or Latinos to begin their family life cycles with unmarried single-parent families (U.S. Bureau of the Census, 1994).

Immigrant families find themselves experiencing unique issues in the family life cycle as old values clash with new ones, such as whether unmarried adult children (especially women) should leave home and whether aged parents should live with their children. Furthermore, if the family is split—with some members living in the country of origin and some in the United States—family members may differ in their expectations of what should occur during various stages of the family life cycle (Hong and Ham, 1992). Religious commitments of parents can also clash with the predominant society, especially with such issues as sex education and discipline.

Of all these variations, researchers are increasingly focusing on two widely prevalent alternatives to the traditional intact family life cycle: the single-parent and the stepfamily life cycles. Today, nearly one out of three families is a single-parent family and one out of six is a stepfamily.

Reflections

As you examine the different family life cycle stages, find the stage in which your family of orientation or cohabitation is. What are the tasks your family confronts? In what individual developmental tasks are the different members of your family engaged?

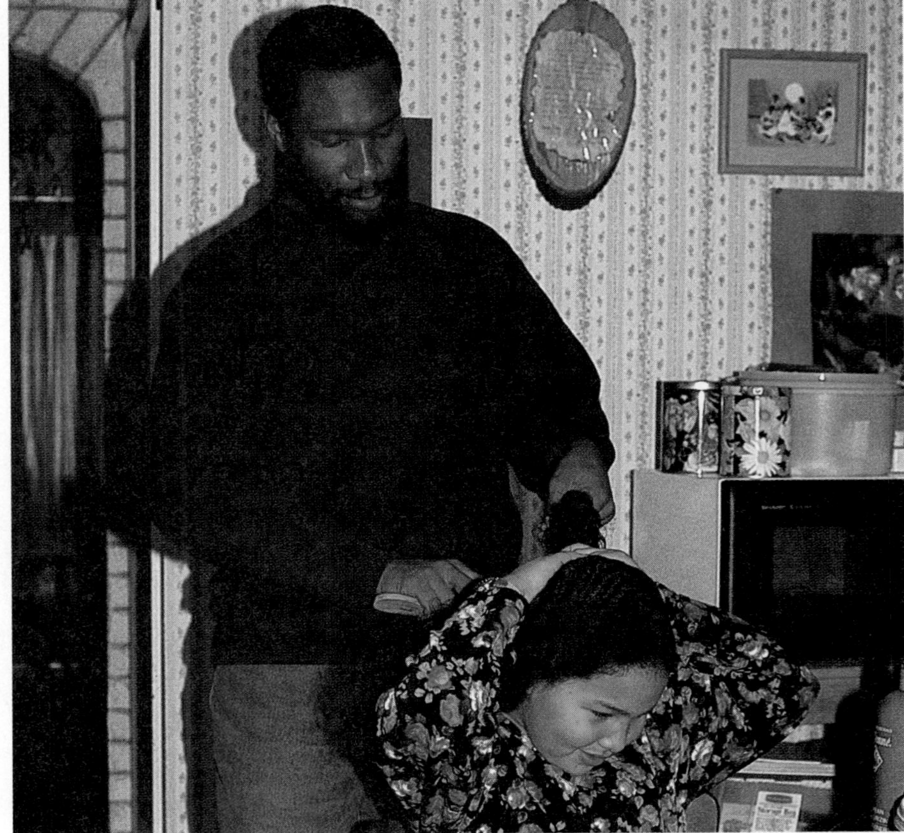

The single-parent family experiences a different life-cycle pattern than the two-parent family.

Did you know?

In the 1990s, nearly one in three births in the United States have been to unmarried mothers (Bumpass and Raley, 1995).

The number of marriages is greater in proportion to the ease and convenience of supporting a family. When families can be easily supported, more persons marry and earlier in life.

BENJAMIN FRANKLIN (1706–1790)

Let there be spaces in your togetherness.

KAHIL GIBRAN (1883–1931), *THE PROPHET*

About half of single-parent families are headed by divorced women, who usually divorce between Stages II and IV, during the childbearing through school-age stages. About 72 percent of recently divorced women will remarry. If they remarry, the single-parent stage generally lasts between three and six years. If these divorced women do not remarry, they continue their single parenting until their adolescent children are launched (Mattessich and Hill, 1987).

The second type of single-parent family originates with unmarried mothers. Birth to unmarried parents now rivals divorce as a pathway in which children enter family structures (Acquilino, 1996). Fifty-seven percent of African-American children, 32 percent of Hispanic children, and 21 percent of white children were born to an unmarried parent (Saluter, 1994). About a third of the mothers are adolescents, and another third are between ages twenty and twenty-four years. They usually have become pregnant unintentionally. Their family life cycle begins with a single-parent family as its first stage. In this stage, the young mother (especially if she is an adolescent) and her child often live with the child's grandmother or cohabit with the child's father. The second stage may be marriage. After marriage, the family life cycle may follow more traditional patterns or diverge back into the single-parent pattern. Most studies indicate multiple disadvantages to both mothers and children, including poverty, lack of education, and less productive lives (Polakow, 1993).

Stepfamilies are formed when the husband or the wife (or both) has children from a previous marriage or relationship. After single parenthood, the stepfamily generally forms during the school-age and adolescent family stages. The parents and stepparents, who are usually aged thirty years or older, have a double developmental task: they must simultaneously enter both the beginning family and child-rearing stages.

Beginning Marriages

Americans are waiting longer to marry today than in previous generations. Whatever the reason, increasing age at time of marriage probably results in young adults beginning marriage with more maturity, independence, work experience, and education. These are important assets to bring into marriage.

A man who marries a woman to educate her falls a victim to the same fallacy as the woman who marries a man to reform him.

ELBERT HUBBARD (1859–1915)

Predicting Marital Success

The period before marriage is especially important because couples learn about each other—and themselves. Courtship sets the stage for marriage. Many of the elements important for successful marriages, such as the ability to communicate in a positive manner and to compromise and resolve conflicts, develop during courtship. They are often apparent long before a decision to marry has been made (Cate and Lloyd, 1992).

The time and patterns that precede marriage can often predict how happy a couple will be in marriage. Happy couples are more likely than unhappy couples to be satisfied in their marriages, and couples who are unhappy before marriage are more likely to be unhappy after marriage as well (Olson and DeFrain, 1994). Each individual brings into marriage the same strengths and weaknesses that he or she brought into the earlier relationship. Marriage permits relationships to grow—or to deteriorate.

Whether marriage is an arena for growth or disenchantment depends on the individuals and the nature of their relationship. It is a dangerous myth that marriage will change a person for the better: An insensitive single person simply becomes an insensitive husband or wife. In fact, undesirable traits tend to become magnified in marriage because one has to live with them in close, unrelenting, and everyday proximity.

Family researchers have found numerous premarital factors to be important in predicting later marital happiness and satisfaction. Although they may not necessarily apply in all cases—and when we are in love, we believe we are the exceptions—they are worth thinking about. These premarital factors include background, personality, and relationship factors, according to Rodney Cate and Sally Lloyd (1992).

Did you know?

When newlyweds were first interviewed and then again four years later, those with higher levels of self-disclosure and interdependence were happier than those with lower levels (Surra, Arizzi, and Asmussen, 1988).

Background Factors

Age at marriage is important. Adolescent marriages are especially likely to end in divorce. Those marrying after age thirty are also slightly more likely to divorce than those marrying in their twenties. Most research shows that up to age twenty-five, the older you are when you marry, the greater the likelihood of marital happiness. After you reach twenty-five years or so, however, the relationship levels off. The correlation is largely a function of youthful marriages.

Length of courtship is also related to marital happiness. The longer you date and are engaged to someone, the more likely you are to discover whether you are compatible with each other. But you can also date "too long." Those who

Married in haste, we repent at leisure.

WILLIAM CONGREVE (1670–1729)

OTHER PLACES OTHER TIMES

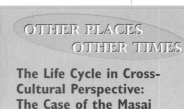

The Life Cycle in Cross-Cultural Perspective: The Case of the Masai

In American society, the concept of the *life cycle* is tied to our understanding of individual physiological and psychological development, as well as related cultural notions of life stages: childhood, adolescence, marriage and adulthood, and old age. Underlying all these measures of the life cycle is a focus on the individual. In U.S. society, each person progresses through the stages of life—through the life cycle—individually. Our cultural emphasis on the individual is, from a cross-cultural perspective, extreme. To us, of course, the focus on the individual seems natural, precisely because our society is shaped and permeated by notions of the individual: individual rights, responsibilities, opinions, votes, and identities.

Anthropologists and historians have found that persons in other cultures are not seen as "individuals" as we define the term. Because we take the individual as a fundamental and universal identity, our cultural bias leads us to project individual identities onto persons in other cultures. But in most traditional societies, a person's primary identity is a clan identity. Which clan you belong to—whether birth established membership in your mother's clan or your father's—was determined by custom. Thus, you as a person would be known and identified by others first as a son or daughter of a particular clan. Your relationships with other people, both your behavior toward them and their behavior toward you, would be determined by your primary identity, your clan identity.

In addition to kinship, age sets in some traditional societies were also important in shaping a person's identity, as well as his or her position in the life cycle. (An age set consists of members of society who are approximately the same age.) Most societies that have age set systems are found in Africa, but they are also found among the Cheyenne of the Great Plains in North America and the Chavante of Brazil. Let's look at age sets by focusing our cross-cultural lens on the Masai peoples of Kenya and Tanzania (Paul, 1988; Saitoti, 1986).

Traditionally, the Masai based their livelihood almost totally on pastoralism; specifically, the herding of shorthorned zebu cattle. The Masai themselves did no farming but instead traded with their neighbors for needed agricultural products. As with many other pastoral peoples, the Masai do not depend on their cattle primarily for meat but in fact derive most of their caloric intake from the blood of their cattle and dairy products. The Masai refer to themselves as "the people of the cattle." All of life revolves in some way around the management and accumulation of cattle. Boys grow up helping their fathers herd cattle; girls grow up assisting their mothers with milking and caring for calves. Anthropologists have found that pastoral societies like the Masai are in general more androcentric, or male focused, than other kinds of societies. They are usually patrilineal: You inherit your father's clan identity, not your mother's.

Among the Masai, marriages are made of cattle. A man needs cattle to marry, for the bride's father demands many cattle for his daughter. Thus, a young man, without a substantial herd of his own, must borrow heavily from his own father. Herd size both confers status on a man and enables him to marry several wives. The ability to marry wives and accumulate cattle is tied to the age set system and a man's place in it.

In essence, Masai society is organized into fraternities. When a boy is physically mature and around fifteen years of age, he will undergo a public circumcision ceremony along with other local boys of about the same age. (Circumcision

have long, slow-to-commit, up-and-down relationships are likely to be less satisfied in marriage. They are also more likely to divorce. Such couples may torture themselves (and their friends) with the familiar dilemma of whether to split up or get married—and then get married, to their later regret.

Level of education seems to affect both marital adjustment and divorce. Education may give us additional resources, such as income, insight, or status, that contribute to our ability to carry out our marital roles.

Childhood environment, such as attachment to family members, parents' marital happiness, and low parent-child conflict, is associated with one's own marital happiness. This is especially true for women: Some studies indicate that it is the woman's relationship with her family of orientation that is crucial to later marital happiness. In fact, it may spell trouble if the man is too close to his family of orientation. Most studies on childhood environment, however, are ceremonies are held only every few years; schedules are staggered in neighboring Masai areas.) Those boys who are initiated in the same "season" will henceforth and for the duration of their lives be members of the same fraternity or age set. They share a basic lifelong identity and loyalty to one another as brothers. Together, they will progress through a fixed series of graduations from one life stage, or grade, to the next, each fifteen years in duration. The first and most important age grade is the warrior grade, into which boys are initiated at their circumcision ceremony.

Traditionally, it was men of the warrior grade who were the first line of defense against hostile incursions and cattle raids by neighboring peoples. Warriors lived separately from the camps of their fathers, together in their own camp in the bush. The warrior identity was marked by ritual—warriors drank milk communally in the forest and were not permitted to drink it alone—as well as by special dress and ornamentation, weapons, dances, songs, and calls. Only warriors could wear the lion headdress, carry a warrior's spear, and give the lion's call at night. Young boys looked forward to the day they would become warriors, and old men fondly recalled their own warrior days. The warrior was the cultural ideal of manhood. Warriors were considered beautiful, virile, and fierce—and were much admired by women.

Generally, men did not marry until around thirty years of age, when they entered the next stage of life, junior elderhood. As warriors, young men owned only a few cattle apiece. Although they might try to add additional head to their herds—even raiding their own fathers' herds in a show of defiance—warriors were occupied with their defense duties and ritual obligations. With the initiation of a new class of warriors, however, and the reluctant graduation of the "old" class into junior elderhood, there came new responsibilities. Junior elders worked hard to build up their cattle herds, married, and began to father children. Elderhood followed junior elderhood at around age forty-five; senior elderhood commenced at around age sixty. Each graduation was marked with ritual and, of course, the initiation of a new group of boys as warriors. For all elders, herd management and marriage were important business. As an elder or senior elder, a man might—if sufficiently wealthy in cattle—marry a third or fourth wife. An extremely wealthy man might have as many as seven wives. However, cattle that could be used to acquire perhaps a fourth or fifth wife for a senior elder could also, of course, be used to acquire a first wife for his son.

Women did not have their own separate age system. Their progress through life was in a sense more individual but was tied to the male age grades in certain important respects. A young girl was initially free to admire and be courted by warrior boyfriends. With her first period, however, a young girl's life changed radically. She immediately underwent a clitoridectomy, making her marriageable, which was celebrated by her female relatives with gifts and feasting. Soon afterward, she was married in an arranged marriage to a much older man. You will recall that most men did not marry until after graduation into junior elderhood, at age thirty. Thus, one general effect of the age system was to concentrate wives at the upper or older end of the male age hierarchy, by denying men of the youngest age grade the right to marry. In addition, men were forbidden to take the daughters of men of their own age set as wives. This was considered incestuous because these women were the daughters of "brothers." However, men of an age set could, as brothers, share wives with one another if given permission.

In societies with age set systems like the Masai, the life cycle was divided into fixed stages through which all males who were initiated together progressed as a group. A young man did not move through the life cycle by meeting individual psychological or physiological criteria, as in our society. Instead, the age set system created fixed group identities and life stages, through which men of the age set moved together, as had their fathers in their own age sets before them.

based on men and women who came of age prior to the 1960s. The social context of marriage has changed dramatically since then, with the rise of divorce, smaller families, and changing gender roles. It is not clear, for example, how today's young adults may be affected by parental divorce. Parental divorce may cause one either to shy away from marriage or to marry with the determination not to repeat the parents' mistakes.

Personality Factors

How does having a flexible personality affect marital success? A contentious one? A giving one? An obnoxious one? As can be imagined, our partner's personality affects our lives considerably. But there is little research on personality characteristics and marital success. We do know, however, that opposites do not usually

Honor—a moral cousin of manners—requires your telling the truth about yourself before the wedding.

Miss Manners

Marriage often unites for life two people who scarcely know each other.

HONORÉ DE BALZAC (1799–1850)

attract; instead, they repel. We choose partners who share similar personality characteristics because similarity allows greater communication, empathy, and understanding (Antill, 1983; Buss, 1984; Kurdek and Smith, 1987; Lesnick-Oberstein and Cohen, 1984). It may be that personality characteristics are most significant during courtship. It is then that those with undesirable or incompatible personalities are weeded out—at least in theory.

Because researchers tend to focus on relationship process and change, they are reluctant to examine personality. Personality seems fixed and unchanging. Nevertheless, it does affect marital processes. A rigid personality may prevent negotiation and conflict resolution. A dominating personality may disrupt the give-and-take necessary to making a relationship work.

To some extent, marital success can be predicted based on the partners' backgrounds, personalities, and relationship factors.

Relationship Factors

Most recent research has focused on aspects of premarital relationships that might predict marital success. Loving each other did not seem to have much impact on whether couples fought. Couples who had other partners simultaneously prior to marriage or who compared their partners with others had lower levels of satisfaction. Another study on communication and marital satisfaction examined the same couples after one, two and a half, and five and a half years of marriage (Markman, 1981, 1984). During the first year, there was no relationship between communication and marital satisfaction, but after two and a half and five and a half years, the more negative the communication, the less satisfactory the marriage. Researchers suggest that negative interactions did not significantly affect the first year of marriage because of the **honeymoon effect,** the tendency of newlyweds to overlook problems (Huston, McHalle, and Crouter, 1986). Failure to fulfill one's partner's expectations about marital roles, such as intimacy and trust, predicted marital dissatisfaction (Kelley and Burgoon, 1991).

Researchers (Olson and DeFrain, 1994) assert that they can more or less predict an engaged couple's eventual marital satisfaction based on their current relationship. The factors they find significant in reviewing the research literature include the ability to do the following:

- Communicate well with each other.
- Resolve conflicts in a constructive way.
- Develop realistic expectations about marriage.
- Like each other as people.
- Agree on religious and ethical issues.
- Balance individual and couple leisure activities with each other.

In addition, how each person's parents related to each other and to their daughter or son is also an important predictor. It is in our families of orientation that we learn our earliest (and sometimes most powerful) lessons about intimacy and relationships (Larsen and Olson, 1989).

Engagement, Cohabitation, and Weddings

The first stage of the family life cycle may begin with engagement or cohabitation followed by a wedding, the ceremony that represents the beginning of a marriage.

Engagement

Engagement is the culmination of the formal dating process. Today, in contrast to the past, engagement has greater significance as a ritual than as a binding commitment to be married. Engagement is losing even its ritualistic meaning, however, as more couples start out in the less formal patterns of "getting together" or living together. These couples are less likely to become formally engaged. Instead, they announce that they "plan to get married." Because it lacks the formality of engagement, "planning to get married" is also less socially binding.

Engagements currently average between twelve and sixteen months (Carmody, 1992). They perform several functions:

- Engagement signifies a commitment to marriage and helps define the goal of the relationship as marriage.
- Engagement prepares couples for marriage by requiring them to think about the realities of everyday married life: money, friendships, religion, in-laws, and so forth. They are expected to begin making serious plans about how they will live together as a married couple.
- Engagement is the beginning of kinship. The future marriage partner begins to be treated as a member of the family. He or she begins to become integrated into the family system.
- Engagement allows the prospective partners to strengthen themselves as a couple. The engaged pair begin to experience themselves as a social unit. They leave the youth or singles culture and prepare for the world of the married, a remarkably different world.

Men and women typically need to deal with several key psychological issues during engagement (Wright, 1990). The issues include (1) anxiety, a general uneasiness that comes to the surface when you decide to marry; (2) maturation and dependency needs, questions about whether you are mature enough to marry and to be interdependent; (3) losses, or regret over what you give up by marrying, such as the freedom to date and responsibility for only yourself; (4) partner choice, worry about whether you're marrying the right person; (5) gender-role conflict, disagreement over appropriate male/female roles; (6) idealization and disillusionment, the tendency to believe that your partner is "perfect" and to become disenchanted when she or he is discovered to be "merely" human; (7) marital expectations, or beliefs that the marriage will be blissful and conflict free and that your partner will be entirely understanding of your needs; and (8) self-knowledge, an understanding of yourself, including your weaknesses as well as your strengths.

Cohabitation

The rise of cohabitation has led to its becoming an alternative beginning to the contemporary family life cycle (Glick, 1989a; Surra, 1991). More than half (52.8 percent) of first cohabitating unions result in marriage (London, 1991).

Although cohabiting couples may be living together before marriage, their relationship is not legally recognized until the wedding, nor is the relationship afforded the same social legitimacy. Most relatives do not consider cohabitants, for example, as kin, such as sons-in-law or daughters-in-law. At the same time there is some evidence that marriages that follow cohabitation have a higher divorce rate than do marriages that begin without cohabitation (De Maris and Rao, 1992; Hall and Zhao, 1995). Cohabitation does, however, perform some of the same functions as engagement, such as preparing the couple for some of the realities of marriage and helping them think of themselves in terms of being a couple as well as individuals.

Weddings

Weddings are ancient rituals that symbolize a couple's commitment to each other. The word *wedding* is derived from the Anglo-Saxon *wedd*, meaning "pledge." But it also included a pledge to the bride's father to pay him in money,

Reflections

As you look at the factors predicting marital success, look at your own past relationships. Retrospectively, what factors, such as background, personality characteristics, and relationship characteristics, might have predicted the quality of your relationship? Were any particular characteristics especially important for you? Why?

Nothing in the world is single
All things by a law devine
In one spirit meet and mingle
Why not I with thine?

Percy Bysshe Shelley (1792–1822); *Love's Philosophy*

To speak frankly, I am not in favour of long engagements. They give people the opportunity of finding out each other's character before marriage, which I think is never advisable.

Oscar Wilde (1854–1900), *The Importance of Being Earnest*

Ceremonies are the outward expression of inner feeling.

Lao-Tzu (d. 531 b.c.)

cattle, or horses for his daughter (Ackerman, 1994; Chesser, 1980). When the father received his pledge, he "gave the bride away." The exchanging of rings dates back to ancient Egypt and symbolizes trust, unity, and timelessness because a ring has no beginning and no end. It is a powerful symbol. To return a ring or take it off in anger is a symbolic act. Not wearing a wedding ring may be a symbolic statement about a marriage. Another custom, carrying the bride over the threshold, was practiced in ancient Greece and Rome. It was a symbolic abduction growing out of the belief that a daughter would not willingly leave her father's house. The eating of cake is similarly ancient, representing the offerings made to household gods; the cake made the union sacred (Coulanges, 1960). The African tradition of jumping the broomstick, carried to America by enslaved tribespeople, has been incorporated by many contemporary African Americans into their wedding ceremonies (Cole, 1993).

Did you know?

In 1994, the average formal church wedding cost close to $18,000 (John, 1994). The average cost for a honeymoon was $3,000. Travel outside the United States can increase this amount to $5,000 or more (Mank, 1994).

The honeymoon tradition can be traced to a pagan custom for ensuring fertility: Each night after the marriage ceremony, until the moon completed a full cycle, the couple drank mead, honey wine. The honeymoon was literally a time of intoxication for the newly married man and woman. Flower girls originated in the Middle Ages; they carried wheat to symbolize fertility. Throughout the world, gifts are exchanged, special clothing is worn, and symbolically important objects are used or displayed in weddings (Werner, Brown, Altman, and Staples, 1992).

Weddings are important rituals acknowledging the transition from singlehood to couplehood. Tradition plays an important role, as seen in this Korean ceremony.

Weddings are big business. Not all couples, however, have formal church weddings. Civil weddings now account for almost one-third of all marriage ceremonies (Ravo, 1991). Because of the expense, many couples are opting for civil ceremonies. Civil ceremonies sometimes cost no more than $30, in addition to the marriage license. Ceremonies for second marriages are more likely to be civil than religious. Fewer religious ceremonies may also be a result of declining religious homogamy; either the bride and groom are from different religious backgrounds or are not religious.

Whether a first or second (or third) marriage, the central meaning of a wedding is that it symbolizes a profound life transition. Most significantly, the partners take on marital roles. For young men and women entering marriage for the first time, marriage signifies a major step into adulthood. Some of the apprehension felt by those planning to marry may be related to their taking on these important new roles and responsibilities. Many will have a child in the first year of marriage. (In about 24 percent of all couples, the woman is pregnant at marriage.) Therefore, the wedding must be considered a major rite of passage. Before two people exchange their marriage vows, they are single, and their primary responsibilities are to themselves. Their parents may have greater claims on them than the couple have on each other. But with the exchange of vows, they are transformed. When they leave the wedding scene, they leave behind singlehood. They are now responsible to each other as fully as they are to themselves and more than they are to their parents.

Reflections

Do you plan to have (or did you have) a religious or civil wedding ceremony? Why? What will the wedding symbolize for you? What will it symbolize for your families of orientation? How will it assist the transition to marital roles?

Establishing Marital Roles

The expectations that two people have about their own and their spouse's marital roles are based on gender roles and their own experience. There are four traditional assumptions about husband/wife responsibilities: (1) the husband is the head of the household, (2) the husband is responsible for supporting the family, (3) the wife is responsible for domestic work, and (4) the wife is responsible for child rearing.

The traditional assumptions about marital responsibilities do not necessarily reflect marital reality, however. For example, the husband traditionally may be regarded as head of the family, but power tends to be shared. In dual-earner families, both men and women contribute to the financial support of the family; in fact, women earn more than men in about 20 percent of marriages (U.S. Bureau of the Census, 1996). Although responsibility for domestic work still tends to reside with women, men are beginning to share housework. The mother is generally still responsible for child rearing, but fathers are beginning to participate more.

Our gender-role attitudes and behaviors contribute to our marital roles. They create marital roles that reflect both traditional and nontraditional beliefs about men and women (Huston and Geis, 1993; Thoits, 1992). Although there have been significant changes over the last generation concerning gender and marital roles, many expectations have not significantly changed. According to one study (Ganong and Coleman, 1992), although both men and women expected their future partners to be successful, both expected the husband to be more successful than the wife. In addition, women expected their husbands to make significantly more money than the women did, to be better educated and more intelligent, and to be more competent in general. Even among dual-earner couples, husbands tend to continue holding traditional role expectations for themselves and their wives, such as the idea that wives are responsible for household tasks (Blair, 1993).

Although lesbian and gay marriages are not legally recognized, some couples nevertheleess celebrate their relationships with a marriage ceremony. As couples in such relationships cannot divide tasks or allocate roles on the basis of gender, they must use other criteria such as expertise or preference.

Marital Tasks

Newly married couples need to begin a number of marital tasks in order to build and strengthen their marriages. The failure to complete these tasks successfully may contribute to what researchers identify as the "duration-of-marriage effect." The **duration-of-marriage effect** refers to the accumulation over time of various factors—such as unresolved conflicts, poor communication, grievances, role overload, heavy work schedules, and child-rearing responsibilities—that cause marital disenchantment (see the Perspective "Examining Marital Satisfaction"). These tasks are primarily adjustment tasks and include the following:

- *Establishing marital and family roles.* Discuss marital-role expectations for self and partner, make appropriate adjustments to fit each other's needs and the needs of the marriage, discuss childbearing issues, and negotiate parental roles and responsibilities.
- *Providing emotional support for the partner.* Learn how to give and receive love and affection, support the other emotionally, and fulfill one's own identity as both an individual and a partner.
- *Adjusting personal habits.* Adjust to each other's personal ways by enjoying, accepting, tolerating, or changing personal habits, tastes, and preferences, such as differing sleep patterns, levels of personal and household cleanliness, musical tastes, and spending habits.
- *Negotiating gender roles.* Adjust gender roles and tasks to reflect individual personalities, skills, needs, interests, and equity.
- *Making sexual adjustments with each other.* Learn how to physically show affection and love, discover mutual pleasures and satisfactions, negotiate timing and activities, and decide on the use of birth control.
- *Establishing family and employment priorities.* Balance employment and family goals; recognize the importance of unpaid household labor as work; negotiate child-care responsibilities; decide on whose employment, if either, receives priority; and divide household responsibilities equitably.

Matrimony is a process by which a grocer acquires an account the florist had.

Francis Rodman

Venus, a beautiful good-natured lady, was the goddess of love; Juno, a terrible shrew, the goddess of marriage; and they were always mortal enemies.

Jonathan Swift (1667–1745)

Marriage is that relation between men and women in which the independence is equal, the dependence mutual and the obligation reciprocal.

L. Anspacher

PERSPECTIVE

Examining Marital Satisfaction

Because marriage and the family have moved to the very center of people's lives as a source of personal satisfaction, we generally evaluate them according to how well they fulfill emotional needs (although such fulfillment is not the only measurement of satisfaction). Marital satisfaction influences not only how we feel about our marriages and our partners but also how we feel about ourselves. If we have a good marriage, we tend to feel happy and fulfilled (Glenn, 1991).

Considering the various elements that make up or affect a marriage—from identity bargaining to economic status—it should not be surprising that marital satisfaction ebbs and flows. The ebb, however, begins relatively early for many couples. Researchers have found significant declines in the average level of marital satisfaction beginning in the first year of marriage (McHale and Huston, 1985). In some cases the decline in marital satisfaction during the first stage of marriage may mean that we have chosen the wrong partner. In fact, those with low marital satisfaction during the beginning marriage stage are four to five times as likely to divorce as those with high satisfaction (Booth et al., 1986). Satisfaction generally continues to decline during the first ten years of marriage, or perhaps longer (Glenn, 1989). Studies consistently indicate that marital satisfaction changes over the family life cycle, following a U-shape or curvilinear curve (Finkel and Hansen, 1992; Glenn, 1991; Suitor, 1991; but see Vaillant and Vaillant, 1993). Satisfaction is highest during the initial stages and then begins to decline but rises again in the later years. There seems to be little difference in marital satisfaction between first marriages and remarriages (Vemer et al., 1989).

It was once thought that couples with average or higher marital satisfaction would have stable marriages, whereas those with low satisfaction would have divorce-prone marriages. Although low marital satisfaction may make a marriage more likely to end in divorce, marital dissatisfaction alone cannot predict eventual divorce (Kitson and Morgan, 1991). Many unhappy marriages continue to endure in the face of misery and discord; sometimes they outlast much happier marriages. Unhappy marriages continue if there are too many barriers to divorce (such as a potential decline in the standard of living) and if the available alternatives seem less attractive than the current marriage. Happier marriages sometimes end in divorce if there are few barriers and better alternatives (Glenn, 1991; see Chapter 16 for a discussion of marital cohesiveness).

Decline in Marital Satisfaction

Why does marital satisfaction tend to decline soon after marriage? Two explanations for changes in marital satisfaction have been given by researchers. The first explanation ascribes the changing patterns of satisfaction to the presence of children. The second points to the effects of time on marital satisfaction.

Children and Marital Satisfaction

Traditionally, researchers have attributed decline in marital satisfaction to the arrival of the first child: Children take away from time a couple spends together, are a source of stress, and cost money. The decline reaches its lowest point when the oldest child enters adolescence (or school, according to some studies). When children begin leaving home, marital satisfaction begins to rise again.

It seems paradoxical that children cause marital satisfaction to decline. For many people, children are among the things they value most in their marriages. First, attributing the decline to children creates a single-cause fallacy—that is, attributes a complex phenomenon to one factor when there are probably multiple causes. Second, the arrival of children at the same time that marital satisfaction declines may be coincidental, not causal. Other undetected factors may be at work.

Although many societal factors make child rearing a difficult and sometimes painful experience for some families, it is also important to note that children create parental roles and the family in its most traditional sense. For some, the marital relationship may be less than fulfilling with children present, but many

- *Developing communication skills*. Share intimate feelings and ideas with each other; learn how to talk to each other about difficulties; share moments of joy and pain; establish communication rules; and learn how to negotiate differences to enhance the marriage.
- *Managing budgetary and financial matters*. Establish a mutually agreed-upon budget; make short-term and long-term financial goals, such as saving for vacations or home purchase; and establish rules for resolving money conflicts.

couples may make a trade-off for fulfillment in their parental roles. In times of marital crisis, parental roles may be the glue that holds the relationship together until the crisis passes. Many couples will endure intense situational conflict, not for the sake of the marriage but for the sake of the children. If the crisis can be resolved, the marriage may be even more solid than before. We need to balance marital satisfaction with family satisfaction.

The Duration of Marriage and Marital Satisfaction

More recently, researchers have looked for factors besides children that might explain decline in marital satisfaction. The most persuasive alternative is the duration-of-marriage effect.

The duration-of-marriage effect is most notable during the first stage of marriage rather than during the transition to parenthood that follows (White and Booth, 1985). This early decline may reflect the replacement of unrealistic expectations about marriage by more realistic ones—a challenge to be intimate and loving in the everyday world. Because beginning marriage requires us to undertake numerous relational tasks, the transition from singlehood to marriage is a time filled with challenges. How we handle these challenges may set the tenor for our marriages for years to come. Those who handle the challenges successfully have the tools with which to enhance their marriages; those who fail may see their marriages decline in satisfaction.

Social and Psychological Factors in Marital Satisfaction

Social factors are important ingredients in marital satisfaction. Income level, for example, is a significant factor. Blue-collar workers have less marital satisfaction than white-collar, managerial, and professional workers because their lower income creates financial distress. Unemployment and economic uncertainty as sources of stress and tension also directly affect marital satisfaction. If a couple have an insufficient income or are deeply in debt, how to allocate their resources—for rent, repairing the car, or paying dental bills—becomes critical, sometimes involving conflict-filled decisions.

Psychological factors also affect marital satisfaction (London, Wakefield, and Lewak, 1990). Although it was once believed that marital satisfaction was dependent on a partner's fulfilling complementary needs and qualities (an introvert's marrying an extrovert, for example), research has failed to substantiate this assertion. Instead, marital success seems to depend on partners' being similar in their psychological makeup and personalities. Outgoing people are happier with outgoing partners; tidy people like tidy mates. Furthermore, a high self-concept (how a person perceives himself or herself), as well as how the spouse perceives the person, also contributes to marital satisfaction. Finally, similarity in perception, such as "seeing" events, relationships, and values through the same lenses, may be critical in marital satisfaction (Deal, Wampler, and Halverson, 1992).

Attitudes toward gender and marital roles may have an impact on marital satisfaction. There are numerous studies that approach this relationship from different angles. Together they indicate the significance of social roles in marriage. One study found that the discrepancy between how you expect your partner to behave and his or her actual behavior could predict marital satisfaction. Discrepancies in expectations were particularly significant in terms of intimacy, equality, trust, and dominance. Interestingly, discrepancies were more important in predicting dissatisfaction than was the fulfillment of expectations (Kelley and Burgoon, 1991). This finding is not entirely surprising. We seem to take for granted that our partner will fulfill our expectations, so it may be an unpleasant surprise to discover that our spouse is not interested in (or lacks the ability for) intimacy or that he or she is untrustworthy.

Expressiveness seems to be an important quality in marital satisfaction (L. King, 1993). One study found that expressive traits were more closely related to higher marital satisfaction than were instrumental ones (Juni and Grimm, 1993). Wives whose husbands discussed their relationships tended to be more satisfied with their marriages than other wives (Acitelli, 1992).

A psychological perspective may help explain why many people in middle-years marriages experience low levels of satisfaction. The middle stages of the family life cycle are also when adults enter midlife, a time characterized by psychological crises and reevaluation. The causes of decreased marital satisfaction may be the result of the adults' own psychological concerns and distress. Steinberg and Silverberg (1987) suggest, "It is reasonable to assume that these feelings may provoke disenchantment in the marital relationship, regardless

(Continued on following page)

- *Establishing kin relationships*. Participate in extended family and manage boundaries between family of marriage and family of orientation.
- *Participating in the larger community*. Make friends, nurture friendships, meet neighbors, and become involved in community, school, church, or political activities.

As you can see, a newly married couple must undertake numerous tasks as their marriage takes form. Marriages take different shapes according to how different

PERSPECTIVE

Examining Marital Satisfaction (continued)

of changes in the adolescent or parent-adolescent relationship."

Even though much of the literature points to declines in marital satisfaction over time, we must remember that not all marriages suffer a significant decline. For many marriages, the decline is small or satisfaction rises after a relatively short period. Even for those whose marriages decrease significantly in satisfaction, the decline may be offset by other satisfactions, such as pleasure in parental roles or a sense of security. Studies consistently demonstrate, for example, that the psychological and physical health of married men and women tends to be better than that of individuals who are unmarried (Gove, Styles, and Hughes, 1990; Waite, 1995).

It is important to understand that marital satisfaction is not static. It fluctuates over time, battered by stress, enlarged by love. Husband and wife continuously maneuver through myriad tasks, roles, and activities—from sweeping floors to kissing each other—to give their marriages form. Children, who bring us both delight and frustration, constrain our lives as couples but challenge us as mothers and fathers and enrich our lives as a family. Trials and triumphs, laughter and tears punctuate the daily life of marriage. If we are committed to each other and to our marriage, work together in a spirit of flexibility and cooperation, find time to be alone together, and communicate with each other, we lay the groundwork for a rich and meaningful marriage.

tasks are shared, divided, or resolved. It is no wonder that about 40 percent of newlyweds find marriage harder than they expected (Arond and Pauker, 1987). But if the tasks are undertaken in a spirit of love and cooperation, they offer the potential for marital growth, richness, and connection (Whitbourne and Ebmeyer, 1990). If the tasks are avoided or undertaken in a selfish or rigid manner, however, the result may be conflict and marital dissatisfaction.

Identity Bargaining

It is an easier thing to be a lover than a husband, for the same reason that it is more difficult to be witty everyday than now and then.

HONORÉ DE BALZAC (1799–1850)

People carry around idealized pictures of marriage long before they meet their marriage partners. They have to adjust these preconceptions to the reality of the partner's personality and the circumstances of the marriage. The interactional process of role adjustment is called **identity bargaining** (Blumstein, 1975). The process is critical to marriage. A study of African-American and white newlyweds, for example, found that marital interactions that affirmed a person's identity predicted marital well-being (Oggins, Veroff, and Leber, 1993). Mirra Komarovsky (1987) points out that a spouse has a "vital stake" in getting his or her partner to fulfill certain obligations. "Hardly any aspect of marriage is exempt from mutual instruction and pressures to change," she writes.

Identity bargaining is a three-step process. First, a person has to identify with the role he or she is performing. A man must feel that he is a husband, and a woman must feel that she is a wife. The wedding ceremony acts as a catalyst for role change from the single state to the married state.

Second, a person must be treated by the other as if he or she fulfills the role. The husband must treat his wife as a wife; the wife must treat her husband as a husband. The problem is that a couple rarely agree on what actually constitutes the roles of husband and wife. This is especially true now as the traditional content of marital roles is changing.

Did you know?

Those who hold more traditional attitudes toward family life are more satisfied in their marriages than those with nontraditional attitudes (Llye and Biblarz, 1993).

Third, the two people must negotiate changes in each other's roles. A woman may have learned that she is supposed to defer to her husband, but if he makes an unfair demand, how can she do this? A man may believe that his wife is supposed to be receptive to him whenever he wishes to make love, but if she is not, how should he interpret her sexual needs? A woman may not like housework (who does?), but she may be expected to do it as part of her marital role. Does she then do all the housework, or does she ask her husband to share responsibility with her? A man believes he is supposed to be strong, but sometimes he feels weak. Does he reveal this to his wife?

Eventually, these adjustments must be made. At first, however, there may be confusion; both partners may feel inadequate because they are not fulfilling their role expectations.

Although some may fear losing their identity in the give and take of identity bargaining, the opposite may be true: One's sense of identity may actually grow in the process of establishing a relationship. A major study by Susan Whitbourne and Joyce Ebmeyer (1990) on marriage and identity concludes:

> Couples can grow as individuals and as a pair through the operation of the identity processes. . . . The attempt that couples make to accommodate each other has its benefits as the alliance between them is strengthened and solidified. . . . The relationship between identity and intimacy is reciprocal. Perhaps the greatest impetus for growth of identity is through the unique and intense bond that the intimate relationship can offer.

In the process of forming a relationship, we discover ourselves. An intimate relationship requires us to define who we are.

Reflections

Using the identity bargaining process as a model, think about how you and an intimate partner went about defining and redefining your roles in your relationship. What part did you and your partner play in the process?

Establishing Boundaries

When young people marry, they often have strong ties to their parents. Until the wedding, their family of orientation has greater claim to their loyalties than their spouse-to-be. Once the marriage ceremony is completed, however, the newlyweds can establish their own family independent of their families of orientation. The couple must negotiate a different relationship with their parents, siblings, and in-laws. Loyalties shift from their families of orientation to their newly formed family. The families of orientation must accept and support these breaks. Indeed, opening themselves to outsiders who have become in-laws places no small stress on families (Carter and McGoldrick, 1989). At the same time, however, many so-called in-law problems may actually be problems between the couple. It's easier

When a young man marries, he divorces his mother.

YIDDISH PROVERB

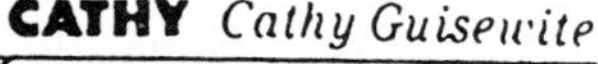

SOURCE: *Cathy* © 1994 Cathy Guisewite. Reprinted with permission of Universal Press Syndicate. All rights reserved.

UNDERSTANDING YOURSELF

Marital Satisfaction

An important question in studying marital satisfaction is how to measure it (Fincham and Bradbury, 1987). One measure widely used is Graham Spanier's Dyadic Adjustment Scale, which we have reprinted here.

The Dyadic Adjustment Scale is an example of the type of questionnaire scholars use as they examine marital adjustment. What are the advantages of a questionnaire such as this? The disadvantages?

Answer the questions and then ask yourself if you think these questions can measure marital satisfaction. (Hint: You must first define what marital satisfaction is.) If you are currently involved in a relationship or marriage, you and your partner might be interested in answering the questions separately and comparing your answers. Do you have similar perceptions of your relationship? At the end of this course, answer the questions again without referring to your first set of answers. Then compare your responses. What do you infer from this comparison?

	Always Agree	Almost Always Agree	Occasionally Disagree	Frequently Disagree	Almost Always Disagree	Always Disagree
1 Handling family finances	5	4	3	2	1	0
2 Matters of recreation	5	4	3	2	1	0
3 Religious matters	5	4	3	2	1	0
4 Demonstrations of affection	5	4	3	2	1	0
5 Friends	5	4	3	2	1	0
6 Sex relations	5	4	3	2	1	0
7 Conventionality (correct or proper behavior)	5	4	3	2	1	0
8 Philosophy of life	5	4	3	2	1	0
9 Ways of dealing with parents or in-laws	5	4	3	2	1	0
10 Aims, goals, and things believed important	5	4	3	2	1	0
11 Amount of time spent together	5	4	3	2	1	0
12 Making major decisions	5	4	3	2	1	0
13 Household tasks	5	4	3	2	1	0
14 Leisure time interests and activities	5	4	3	2	1	0
15 Career decisions	5	4	3	2	1	0

to complain about a mother-in-law, for example, than it is to deal with troubling issues in one's own relationship (Silverstein, 1992).

The new family must establish its own boundaries. The couple should decide how much interaction with their families of orientation is desirable and how much influence these families may have. The addition of extended family can bring into contact people who are very different from one another in culture, life experiences, and values. There are often important ties to the parents that may prevent new families from achieving their needed independence. First is the tie of habit. Parents are used to being superordinate; children are used to being subordinate. The tie between mothers and daughters is especially strong; daughters often experience greater difficulty separating themselves from their mothers than do sons. These continuing ties may cause an adult child to feel conflicting loyalties toward parents and spouse (Cohler and Geyer, 1982). Much conflict occurs when a spouse feels that an in-law is exerting too much influence on his or her partner (for example, a mother-in-law's insisting that her son visit each

	All the Time	Most of the Time	More Often Than Not	Occasionally	Rarely	Never
16 How often do you discuss or have you considered divorce, separation, or terminating your relationship?	0	1	2	3	4	5
17 How often do you or your mate leave after a fight?	0	1	2	3	4	5
18 In general, how often do you think that things between you and your partner are going well?	5	4	3	2	1	0
19 Do you confide in your mate?	5	4	3	2	1	0
20 Do you ever regret that you married (or live together)?	0	1	2	3	4	5
21 How often do you and your partner quarrel?	0	1	2	3	4	5
22 How often do you and your mate "get on each other's nerves?"	0	1	2	3	4	5

	Every Day	Almost Every Day	Occasionally	Rarely	Never
23 Do you kiss your mate?	4	3	2	1	0

	All of Them	Most of Them	Some of Them	Very Few of Them	None of Them
24 Do you and your mate engage in outside interests together?	4	3	2	1	0

How often would you say the following events occur between you and your mate?

	Never	Less Than Once a Month	Once or Twice a Month	Once or Twice a Week	Once a Day	More Often
25 Have a stimulating exchange of ideas	0	1	2	3	4	5
26 Laugh together	0	1	2	3	4	5
27 Calmly discuss something	0	1	2	3	4	5
28 Work together on a project	0	1	2	3	4	5

(Continued on following page)

Sunday and the son's accepting despite the protests of his wife; or a father-in-law's warning his son-in-law to establish himself in a career or risk losing his wife). If conflict occurs, husbands and wives often need to put the needs of their spouses ahead of those of their parents.

Another tie to the family of orientation may be money. Newly married couples often have little money or credit with which to begin their families. They may turn to parents to borrow money, co-sign loans, or obtain credit. But financial dependence keeps the new family tied to the family of orientation. The parents may try to exert undue influence on their children because it is their money, not their children's money, that is being spent. They may try to influence their children's purchases, or they may refuse to loan money to buy something of which they disapprove.

A review of research on in-laws found that in-law relationships generally had little emotional intensity (Goetting, 1989). The relationship between married women and their mothers-in-law and mothers seems to change with the birth of a

All tragedies are finished by a death. All comedies are ended by a marriage.

Lord Byron (1788–1824)

UNDERSTANDING YOURSELF

Marital Satisfaction (continued)

These are some things about which couples sometimes agree and sometimes disagree. Indicate if either item below caused differences of opinions or were problems in your relationship during the past few weeks (check yes or no).

	Yes	No	
29	0	1	Being too tired for sex.
30	0	1	Not showing love.

31 The dots on the following line represent different degrees of happiness in your relationship. The middle point, "happy," represents the degree of happiness of most relationships. Please circle the dot that best describes the degree of happiness, all things considered, of your relationship.

0	1	2	3	4	5	6
•	•	•	•	•	•	•
Extremely Unhappy	Fairly Unhappy	A Little Unhappy	Happy	Very Happy	Extremely Happy	Perfect

32 Which of the following statements best describes how you feel about the future of your relationship?

5 I want desperately for my relationship to succeed, and would go to almost any length to see that it does.
4 I want very much for my relationship to succeed, and will do all I can to see that it does.
3 I want very much for my relationship to succeed, and will do my fair share to see that it does.
2 It would be nice if my relationship succeeded, but I can't do much more than I am doing now to help it succeed.
1 It would be nice if it succeeded, but I refuse to do any more than I am doing now to keep the relationship going.
0 My relationship can never succeed, and there is no more that I can do to keep the relationship going.

first child (Fischer, 1983). Mother-daughter relationships seem to improve as the mother shifts some of her maternal role onto the grandchild. In-laws gave minimal direct support. Bonding between in-laws tends to be between women, and if there is a divorce, divorced women are more likely than their ex-husbands to maintain supportive ties with former in-laws (Serovich, Price, and Chapman, 1991).

The critical task is to form a family that is interdependent rather than totally independent or dependent. It is a delicate balancing act as parents and their adult children begin to make adjustments to the new marriage. We need to maintain bonds with our families of orientation and to participate in the extended family network, but we cannot let those bonds turn into chains.

Any marriage, happy or unhappy, is infinitely more interesting and significant than any romance, however passionate.

W. H. Auden (1907–1973

Youthful Marriages

Youthful marriages represent Stages II through IV in the family life cycle: childbearing families (Stage II), families with preschool children (Stage III), and families with schoolchildren (Stage IV).

Impact of Children

Husband and wife both usually work until their first child is born; about half of all working women leave the workplace for at least a short period of time to attend

SOURCE: *Jumpstart*. Reprinted by permission of UFS, Inc.

to child-rearing responsibilities after the birth of the first child. The husband continues his job or career. Although the first child makes the husband a father, fatherhood generally does not radically alter his relationship with his work. The woman's life, however, changes dramatically with motherhood. If she continues her outside employment, she is usually responsible for arranging child care and juggling her employment responsibilities when her children are sick, and, if her story is like that of most employed mothers, she continues to have primary responsibility for the household and children. If she withdraws from the workplace, her contacts during most of the day are with her children and possibly other mothers. This relative isolation requires her to make a considerable psychological adaptation in her transition to motherhood, leading in some cases to unhappiness or depression. (The transition to parenthood and the roles of mother and father are discussed at greater length in Chapter 10.)

Typical struggles in families with young children concern child-care responsibilities and parental roles. The woman's partner may not understand her frustration or unhappiness because he sees her fulfilling her roles as wife and mother. She herself may not fully understand the reasons for her feelings. The partners may increasingly grow apart during this period. During the day they move in different worlds, the world of the workplace and the world of the home; during the night they cannot relate easily because they do not understand each other's experiences. Research suggests that men are often overwhelmed by the emotional intensity of this and other types of conflict (Gottman, 1994).

For adoptive families, the transition to parenthood may differ somewhat from that of biological families (Levy-Shiff, Goldschmidt, and Har-Even, 1991). Adoptive parents report more positive expectations about having a child, as well as more positive experiences in their transition to parenthood. In part this may be explained by adoptive parents' being able to fulfill parental roles that they vigorously sought. For them, parenting is a much more conscious decision than for many biological parents, for whom a pregnancy sometimes just "happens." For adoptive parents to become parents, considerable effort and expense must be undertaken; they are less likely to question their decision to become parents.

Individual Changes

Around the time people are in their thirties, the marital situation changes substantially. The children have probably started school and the mother, who usually has the lioness's share of duties, now begins to have more freedom from child-rearing responsibilities. She evaluates her past and decides on her future.

Seldom, or perhaps never, does a marriage develop into an individual relationship smoothly and without crises; there is no coming to consciousness without pain.

CARL JUNG (1875–1961)

The majority of women who left jobs to rear children return to the workplace by the time their children reach adolescence. By working, women generally increase their marital power.

Husbands in this period may find that their jobs have already peaked; they can no longer look forward to promotions. They may feel stalled and become depressed as they look into the future, which they see as nothing more than the past repeated for thirty more years. Their families may provide emotional satisfaction and fulfillment, however, as a counterbalance to workplace disappointments.

Middle-Aged Marriages

Middle-aged marriages generally represent Stages V and VI in the family life cycle: families with adolescents (Stage V) and families as launching centers (Stage VI). Some parents may continue to raise young children while others, especially if one partner is considerably younger than the other, may choose to start a new family. Couples in these stages are usually in their forties and fifties.

Families with Young Children

Increasing dramatically since 1970 are the over-thirty-five women who have chosen to postpone childbearing until they are emotionally or financially ready. In 1995, 375,000 babies were born to women over age thirty-five (U.S. Bureau of the Census, 1996). While there have always been older women having children, in the past these mothers were having their last child, not their first. With the majority of these over-thirty-five women having a higher education, job status, and income, they also experience a lower divorce rate, are more stable, and are frequently more attentive to their young. (For a further discussion about this topic, see Chapter 10.)

Did you know?

Family values, such as support, communication, and respect, face their greatest challenge in families with adolescent children (Larson and Richards, 1994).

Families with Adolescents

Adolescents require considerable family reorganization on the part of the parents: They stay up late; play loud music; infringe upon their parents' privacy; and leave a trail of empty pizza cartons, popcorn, dirty socks, and Big Gulp cups in their wake. As Carter and McGoldrick (1989) point out:

> Families with adolescents must establish qualitatively different boundaries than families with younger children. . . . Parents can no longer maintain complete authority. Adolescents can and do open the family to a whole array of new values as they bring friends and new ideas into the family arena. Families that become derailed at this stage are frequently stuck at an earlier view of their children. They may try to control every aspect of their lives at a time when, developmentally, this is impossible to do successfully. Either the adolescent withdraws from the appropriate involvements for this developmental stage or the parents become increasingly frustrated with what they perceive as their own impotence.

Did you know?

Women who are ambivalent about employment or feel their jobs are secondary to their husbands tend to have low levels of marital satisfaction (Perry-Jenkins, Seery, and Crouter, 1992).

While the majority of teenagers do not cause "storm and stress" (Larson and Ham, 1993), increased family conflict may occur as adolescents begin to assert their autonomy and independence. Conflicts over tidiness, study habits, communication, and lack of responsibility may emerge. Adolescents want rights and privileges but have difficulty accepting responsibility. Conflicts are often contained, however, if both parents and adolescents tacitly agree to avoid "flammable" topics, such as how the teenager spends his or her time or money. Such tactics may be useful in maintaining family peace, but in the extreme they can backfire by decreasing family closeness and intimacy. Despite the growing pains accompanying adolescence, parental bonds generally remain strong (Gecas and Seff, 1991).

After the kids leave home, some parents suffer from the empty-nest syndrome. Others change the locks!

ANONYMOUS

Families as Launching Centers

Some couples may be happy or even grateful to see their children leave home, some experience difficulties with this exodus, and some continue to accommodate their adult children under the parental roof.

The Empty Nest

As children are "launched" from the family (or ejected, as some parents wryly put it), the parental role becomes increasingly less important in daily life. The period following the child's exit is commonly known as the **empty nest** period. Most parents make the transition reasonably well (Anderson, 1988). In fact, marital satisfaction generally begins to rise for the first time since the first stage of marriage (Glenn, 1991). For some parents, however, the empty nest is seen as the end of the family. Children have been the focal point of much family happiness and pain, and now they are gone.

The best way to keep children at home is to make the home atmosphere pleasant—and let the air out of the tires.

DOROTHY PARKER (1893–1967)

Traditionally, it has been asserted that the departure of the last child from home leads to an "empty nest syndrome" among women, characterized by depression and identity crisis. However, there is little evidence that the syndrome is widespread. Rather, it is a myth that reinforces the traditional view that women's primary identity is found in motherhood. Once deprived of their all-encompassing identity as mothers, the myth goes, women lose all sense of purpose. (In reality, however, mothers may be more likely to complain that their adult children have not left home.)

The couple must now re-create their family minus their children. Their parental roles become less important and less stressful on a day-to-day basis (Anderson, 1988). The husband and wife must rediscover themselves as man and woman. Some couples may divorce at this point if the children were the only reason the pair remained together. The outcome is more positive when parents have other more meaningful roles, such as school, work, or other activities to turn to (Lamanna and Riedmann, 1997).

The proverbial minister, priest and rabbi were debating the point at which life begins. The priest said, "Life begins at the moment of conception, when the sperm invades the egg." The minister insisted that "Life begins at birth, with that first breath of God's air." But the rabbi, older and wiser, said "You're both too young to know this, but life doesn't really begin until the kids leave home and the dog dies."

FRANK PITTMAN

The Not-So-Empty Nest: Adult Children at Home

Just how empty homes actually are after children reach age eighteen is open to question. In fact, the 1990 census data revealed that 21 percent of twenty-five-year-olds were living with one or both parents, as compared to 15 percent in 1970 (American Demographics, 1996). Some aren't moving out at all before their mid-twenties and many are doing an extra rotation through their family home after a temporary or lengthy absence. This later group is sometimes referred to as the **boomerang generation.**

Just 19 percent of full-time, first-time college freshmen surveyed in 1995 said wanting to get away from home was a very important reason to go to school. A larger share (25 percent) were living at home while they attended school, according to University of California at Los Angeles' Annual American Freshman Study.

When I was young, I was told: "You'll see when you're fifty." I am fifty and I haven't seen a thing.

ERIK SATIE (1866–1925)

The young are slaves to dreams; the old servants of regrets. Only the middle aged have all their senses.

HERVY ALLEN

Hispanics are more likely than other young adults to take a traditional route of staying home until they marry. Blacks are less likely than whites or Hispanics to leave home before marriage. Though family income may influence nest-leaving, ethnic or racial tradition seem to be more important in determining whether a young adults will leave home (American Demographics, 1996). Most, however, move away from home when they marry.

Researchers note that there are important financial and emotional reasons for this trend (Mancini and Bliezner, 1991). High unemployment, housing costs, and poor wages lead to adult children's returning home. High divorce rates, as well as personal problems, push adult children back to the parental home for social support and child care, as well as cooking and laundry service.

Young adults at home are such a common phenomenon that one of the leading family life cycle scholars suggests a new family stage: *adult children at home* (Aldous, 1990). This new stage generally is not one that parents have anticipated. Almost half reported serious conflict with their children. For parents, the most frequently mentioned problems were the hours of their children's coming and going and their failure to clean and maintain the house. Most wanted their children to be "up, gone, and on their own."

Reevaluation

Middle-aged people find that they must reevaluate relations with their children, who have become independent adults, and must incorporate new family members as in-laws. Some must also begin considering how to assist their own parents, who are becoming more dependent as they age (discussed in greater detail in Chapter 12).

Couples in middle age tend to reexamine their aims and goals (Steinberg and Silverberg, 1987). On the average, husbands and wives have thirteen more years of marriage without children than they used to, and during this time their partnership may become more harmonious or more strained. The man may decide to stay at home or not work as hard as before. The woman may commit herself more fully to her job or career, or she may remain at home, enjoying her new child-free leisure. Because the woman has probably returned to the workplace, wages and salary earned during this period may represent the highest amount the couple will earn.

As people enter their fifties, they probably have advanced as far as they will ever advance in their work. They have accepted their own limits, but they also have an increased sense of their own mortality. Not only do they feel their bodies aging but they also begin to see people their own age dying. Some continue to live as if they were ageless—exercising, working hard, keeping up or even increasing the pace of their activities. Others become more reflective, retreating from the world. Some may turn outward, renewing their contacts with friends, relatives, and especially their children and grandchildren.

Reflections

Recall your family of orientation when you were an adolescent. How did you and your parents deal with establishing new family boundaries and with issues of autonomy and independence? What was the process of "launching" like? Has it been completed? If you continue to live at home, what difficulties has it caused your parents? Yourself?

Later-Life Marriages

You can only hold your stomach in for so many years.

BURT REYNOLDS

Later-life marriages represent the last two stages (Stages VII and VIII) of the family life cycle. A later-life marriage is one in which the children have been launched and the partners are middle aged or older. Later-life families tend to be significantly more satisfied than families at earlier stages in the family life cycle (Mathis and Tanner, 1991). Compared with middle-aged couples, older couples showed less potential for conflict and greater potential for engaging in pleasurable activities together and separately, such as dancing, travel, or reading (Levenson, Cartensen, and Gottman, 1993).

During this period, the three most important factors affecting middle-aged and older couples are health, retirement, and widowhood (Brubaker, 1991). In addition, these women and men must often assume roles as caretakers of their own aging parents or adjust to adult children who have returned home. Later-middle-aged men and women tend to enjoy good health, are firmly established in their work, and have their highest discretionary spending power because their children are gone (Voydanoff, 1987). As they age, however, they tend to cut back on their work commitments for both personal and health reasons.

As they enter old age, men and women are better off, on the average, than young Americans (Peterson, 1991). Beliefs that the elderly are neglected and isolated tend to reflect myth more than reality (Woodward, 1988). Over half of all people aged sixty-five and over live in either the same house or in the same neighborhood as one of their adult children (Troll, 1994). In addition, a national study of people over sixty-five found that 41 percent of those with children see or talk with them daily; 21 percent, twice a week; and 20 percent, weekly. Over half have children within thirty minutes' driving time (U.S. Bureau of the Census, 1988).

Beliefs that the elderly are a particularly poverty-stricken group are also misleading. Because of government programs initiated during the 1970s and 1980s benefiting the elderly, the poverty rate for the elderly declined from 25 percent in 1970 to 11.7 percent in 1994. In the meantime, the poverty level of young children soared to 21.2 percent; among young African-American and Latino children it skyrocketed to 43.4 percent and 41.1 percent respectively (Proctor, 1994). We must not think, however, that all aged people benefited from the decline in poverty. The poverty rate in 1994 for aged African Americans was 27.4 percent, and 22.6 percent for Latinos (U.S. Bureau of the Census, 1996). Elderly African Americans tend to have lower levels of life satisfaction than do elderly whites (Krause, 1993). In part, this is because of greater financial strain and economic dependence on relatives among African Americans.

The health of the elderly also appears to be improving as they increase their longevity. Half of those between seventy-five and eighty-four years are free of health problems that require special care or limit their activities. Bernice Neugarten notes, "Even in the very oldest group, those above eighty-five, more than one-third report no limitation due to health" (Toufexis, 1988). More married couples are living into old age, and there are fewer widows at younger ages.

Did you know?

Average life expectancy for men is 72.0; for women, 78.8. When individuals reach forty-five, their life expectancy increases to seventy-eight. When they reach age sixty-five, their life expectancy rises to eighty-two. If they reach eighty-five, they can anticipate living another six years (Metropolitan Life Insurance Company, 1993).

The Intermittent Extended Family: Sharing and Caring

Although many later-life families contract in size as children are launched, pushed, or cajoled out of the nest, other families may expand as they come to the assistance of family members in need. Families are most likely to become an intermittent extended family during their later-life stage (Beck and Beck, 1989). **Intermittent extended families** are families that take in other relatives during a time of need. These families "share and care" when younger or older relatives are in need or crisis: they help daughters who are single mothers; a sick parent, aunt, or uncle; or an unemployed cousin. (See Chapter 12 for a discussion of family caregiving.) When the crisis passes, the dependent adult leaves, and the family resumes its usual structure.

Did you know?

The size of the 50–59 age group will increase about 50% by the year 2006 (U.S. Bureau of the Census, 1995).

The incidence of intermittent extended families tends to be linked to ethnicity. Using national population studies, researchers estimate that the families of almost two-thirds of African-American women and a third of white women were extended for at least some part of the time during their middle age (Beck and Beck, 1989; Minkler and Roe, 1993). Latina women are more likely than non-Latina women to form extended households (Tienda and Angel, 1982). Asian-American families are also more likely to live at some time in extended families. There are two reasons for the prevalence of extended families among certain ethnic groups. First, extended families are by cultural tradition more significant to African Americans, Latinos, and Asian Americans than to whites. Second, ethnic families are more likely to be economically disadvantaged. They share households and pool resources as a practical way to overcome short-term difficulties. In addition, there is a higher rate of single parenthood among African Americans, which makes mothers and their children economically vulnerable. These women often turn to their families of orientation for emotional and economic support until they are able to get on their own feet.

The Sandwich Generation

A relatively new phenomena, now referred to as the **sandwich generation,** are those middle-aged (or older) individuals who are sandwiched between the simultaneous responsibilities of raising both their dependent children and their aging parents. Given the number of baby boomers now in their middle years, coupled with the increased longevity among their parents, we can anticipate that this type of dual care will become increasingly common. An estimated 20 to 30 percent of workers over age thirty are currently involved in caregiving to their parents, and this percentage is expected to grow (Field and Minkler, 1993). Daughters outnumber sons as caretakers by more than three to one (Cox, 1993), though among Asian Americans, the eldest son may be expected to be responsible for his elders (Kamo and Zhou, 1994).

One's prime is elusive.

Muriel Spark, The Prime of Miss Jean Brodie

As people live longer, their disabilities, dependency, and the number of their long-term chronic illnesses increase. Complicating this is the shrinking number of young workers, facilities, and resources to care for the old and frail. All of this puts additional pressure on families to provide support for their elders. Care that was traditionally handled by health-care professionals—injections, monitoring of medications, bathing, and physical therapy—is now often in the hands of family members.

The trend today, whenever possible, is for the dependent aged to be cared for in the home (Freedman, 1993). Placing added demands on family members' time, energy, and emotional commitment often results in exhaustion, anger, and in some cases, violence. Most people, however, are amazingly adept at meeting the needs of both their parents and their children. It is going to be an increasing challenge for society to acknowledge this phenomenon and provide services and support to both the elderly and those who care for them.

Parenting Adult Children

A few years ago, a Miami Beach couple reported their son missing (Treas and Bengtson, 1987):

> Joseph Horowitz still doesn't understand why his mother got so upset. He wasn't "missing" from their home in Miami Beach: he had just decided to go north for the

winter. Etta Horowitz, however, called authorities. Social worker Mike Weston finally located Joseph in Monticello, N.Y., where he was visiting friends. Etta, 102, and her husband, Solomon, 96, had feared harm had befallen their son Joseph, 75.

Parenting does not end when children grow up. Most elderly parents still feel themselves to be parents, but they are parents in different ways. Their parental role is considerably less important in their daily lives. They generally have some kind of regular contact with their adult children, usually by letters or phone calls; parents and adult children also visit each other fairly frequently and often celebrate holidays and birthdays together. Researcher Joan Aldous (1987b) found that middle-class parents typically assisted their children financially or provided services. Parents made loans, gave gifts, or paid bills for their children around six times a year; they also provided child care about the same number of times. They assisted in shopping, house care, and transportation and also helped in times of illness.

Parents tend to assist those whom they perceive to be in need, especially children who are single or divorced. Parents perceive their single children as being "needy" when they have not yet established themselves in occupational and family roles. These children may need financial assistance and may lack intimate ties; their parents may provide both until their children are more firmly established. Parents often assist divorced children, especially if grandchildren are involved, by providing financial and emotional support. They may also provide child-care and housekeeping services. Parents generally provide the greatest assistance to their children who are single mothers.

In providing assistance to their children, the parents in Aldous's study exercised considerable control over whom they would help and how involved they would become. Despite their continued parental concern, they tended to be maritally rather than parentally oriented. They valued their independence from their children. When one woman was asked how she would feel about an adult child's coming back home to live, she exclaimed, "Oh, God, I wouldn't like it. I'm tired of waiting on people" (Aldous, 1987b). In Aldous's study, adult children generally reciprocated in terms of physical energy. They helped thir parents in household chores and yard work about six times a year; they also assisted to some extent with physical care during illness.

Some elderly parents never cease being parents because they provide home care for children who are severely limited either physically or mentally. Many elderly parents, like middle-aged parents, are taking on parental roles again as children return home for financial or emotional reasons. Although we don't know how elderly parents "parent," presumably they are less involved in traditional parenting roles. The presence of adult children in aging families does not seem to detract from marital quality.

Grandparenting

The image of the lonely, frail grandmother in a rocking chair needs to be discarded. Grandparents are often not very old, nor are they very lonely, and they are certainly not absent in contemporary American family life. Grandparents are "a very present aspect of family life, not only for young children but young adults as well," writes Gregory Kennedy (1990).

Grandparenting is expanding tremendously these days, creating new roles that relatively few Americans played a few generations back. Three-quarters of people aged sixty-five and over are grandparents (Aldous, 1995). Grandparents

All the world's a stage
And all the men and women merely players:
They have their exits and their entrances:
And one man in his time plays many parts,
His acts being seven ages. At first the infant,
Mewling and puking in the nurse's arms.
Then the whining school-boy, with his satchel
And shining morning face, creeping like snail
Unwillingly to school. And then the lover,
Sighing like furnace, with a woeful ballad
Made to his mistress's eyebrow. Then a soldier,
Full of strange oaths, and bearded like the pard,
Jealous in honor, sudden and quick in quarrel,
Seeking the bubble reputation,
Even in the cannon's mouth. And then the justice,
In fair round belly with good capon lined,
With eyes severe and beard of formal cut,
Full of wise saws and modern instances;
And so he plays his part. The sixth age shifts
Into lean and slippered pantaloon,
With spectacles on nose and pouch on side,
His youthful hose, well saved, a world too wide
For his shrunk shank; and his big manly voice,
Turning again toward childish treble, pipes
And whistles in his sound. Last scene of all,
That ends this strange eventful history,
Is second childishness and mere oblivion,
Sans teeth, sans eyes, sans taste, sans everything.

WILLIAM SHAKESPEARE (1564–1616), *AS YOU LIKE IT*

Grandparents are important to their grandchildren as caregivers, playmates, and mentors.

play important emotional roles in American families; the majority appear to establish strong bonds with their grandchildren (Kennedy, 1990; Strom, Collingsworth, Strom, and Griswold, 1992–1993). They help to achieve family cohesiveness by conveying family history, stories, and customs. Grandparents influence grandchildren directly when they act as caretakers, playmates, and mentors. They influence indirectly when they provide psychological and material support to parents who may consequently have more resources for parenting (Brooks, 1996).

Grandparents seem to take on even greater importance in single-parent and stepparent families. They frequently act as a stabilizing force for their children and grandchildren when the families are divorcing and reforming as single-parent families or stepfamilies. Kennedy and Kennedy (1993) found that the significance of grandparents varies by family form. When compared with children from intact families, children in single-parent families report greater closeness and active involvement with their grandparents; children in stepfamilies are even closer.

Grandparents, especially grandmothers, are often involved in the daily care of their grandchildren. A recent study found that one in ten grandparents provides primary care for grandchildren for at least six months—if not longer. ("When Parenthood Extends," 1997). The study also found that African Americans had twice the odds of becoming caregiving grandparents, partly reflecting the long tradition of caregiving that goes back to West African cultures. In the crack cocaine epidemic, grandmothers and great-grandmothers play critical roles in rearing the children of addicted parents (Minkler and Roe, 1993).

Grandparenting tends to fall into three distinct styles, according to a national study of 510 grandparents by Andrew Cherlin and Frank Furstenberg (1986):

- *Companionate*. Most grandparents perceive their relationships with their grandchildren as companionate. The relationships are marked by affection, companionship, and play. Because these grandparents tend to live relatively close to their grandchildren, they can have regular interaction with them. Companionate grandparents do not perceive themselves as rule makers or enforcers; they rarely assume parentlike authority. Companionate grandparents accounted for 55 percent of the grandparents in the survey.
- *Remote*. Remote grandparents are not intimately involved in their grandchildren's lives. Their remoteness, however, is due to geographic remoteness rather than emotional remoteness. Geographic distance prevents the regular visits or interaction with their grandchildren that would bind the generations together more closely. About 29 percent of the grandparents were classified as remote.
- *Involved*. Involved grandparents are actively involved in what have come to be regarded as parenting activities: making and enforcing rules and disciplining children. Involved grandparents (most often grandmothers) tend to emerge in times of crisis, such as when the mother is an unmarried adolescent or enters the workforce following divorce. Some involved grandparents may become overinvolved, however. They may cause confusion as the family tries to determine who is the real head of the family. Involved grandparents made up 16 percent of the grandparents in the study.

What determines the amount of interaction between grandparents and grandchildren? Cherlin and Furstenberg (1986) succinctly sum up the three most important factors: "distance, distance, and distance." Other factors include age, health, employment, personality, and other responsibilities (Troll, 1985a). The mid-

Reflections

Think about your grandparents. How many are alive? What kind of relationship do (or did) you have with them? What role do (or did) they play in your life and your family's life?

dle generation is usually responsible for determining how much interaction takes place between grandparents and grandchildren. If rivalry rather than cooperation ensues, the middle generation may restrict grandparent-grandchildren interaction. But most families gain from such interactions (Barranti and Ramirez, 1985).

Single parenting and remarriage have made grandparenthood more painful and problematic for many grandparents. Stepfamilies have created stepgrandparents, who are often confused about their grandparenting role. Are they really grandparents? The grandparents whose sons or daughters do not have custody often express concern about their future grandparenting role (Goetting, 1990). Although research indicates that children in stepfamilies tend to do better if they continue to have contact with both sets of grandparents, it is not uncommon for the parents of the noncustodial parent to lose contact with their grandchildren (Bray and Berger, 1990).

Figure 9.1
Percent of Children under Age 18 Years Living with Grandparent (with or without parent present), 1993

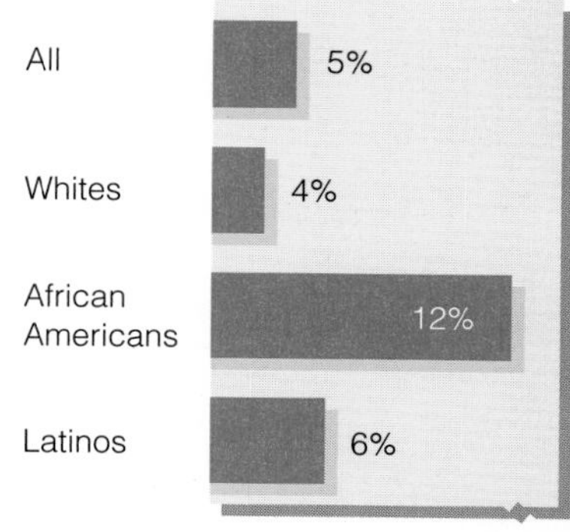

SOURCE: Saluter, Arlene F., "Marital Status and Living Arrangements: March 1993." U. S. Bureau of the Census, *Current Population Reports*, Series P-20, No. 461, U. S. Government Printing Office, Washington, D. C., 1994.

Retirement and Widowhood

Retirement, like other life changes, has the potential for satisfactions and problems. In a time of relative prosperity for the elderly, retirement is an event to which older couples generally look forward. The key to marital satisfaction in these later years is continued good health (Brubaker, 1991).

Retirement

Retirement needs to be viewed as a process. People's experiences of retirement are often different and multilayered (Jensen-Scott, 1993). Usually as retirement approaches, men and women look forward to it; they slowly disengage themselves from their work roles. They experience less interest and greater dissatisfaction in their work. This disengagement prepares them for their actual retirement (Ekerdt and DeViney, 1993).

There is a growing trend toward early retirement among financially secure men and women. (Health limitations, however, increase both the likelihood of retirement as well as the chance that people will be unhappy during this time [Gradman, 1994; Solomon and Szwabo, 1994].) Changing social mores have led many to value leisure more than increased buying power and other job benefits. In fact, today, three-fourths of men and more than four-fifths of women choose to collect their Social Security checks before they reach the age of sixty-five years (Dentzer, 1990). Two scholars (Treas and Bengtson, 1987) note, "Together with greater prosperity, early retirement has probably reordered the preoccupations of later life toward greater concern with leisure activities—a development most compatible with the historic shift to companionate marriages." The bumper sticker "We're spending our children's inheritance" seen on campers and RVs captures this shift in sentiment.

This increased financial status is not true, however, for all older Americans (Gibson, 1993). Fewer African Americans than whites are financially able to retire. Within each racial group, older women are poorer than their male counterparts, with African American older women the most likely to be poor.

When men reach their sixties, they usually retire from their jobs, losing a major activity through which they defined themselves. In spite of this, most men look favorably on retirement. Their role as husband becomes more important as they focus on leisure activities with their wives. In addition, many retirees are not really retired; many continue working part-time (Dentzer, 1990). Volunteer activities become more important. At home, the division of labor becomes somewhat

Advice on Growing Old

1. *Avoid fried meats which angry up the blood.*

2. *If your stomach disputes you, lie down and pacify it with cool thoughts.*

3. *Keep the juices flowing by jangling around gently as you move.*

4. *Go very lightly on the vices, such as carrying on in society. The social ramble ain't restful.*

5. *Avoid running at all times.*

6. *Don't look back. Something might be gaining on you.*

SATCHEL PAIGE (1906–1982)

There is a certain part of all of us that lives outside of time. Perhaps we become aware of our age at exceptional moments and most of the time we are timeless.

Milan Kundera

more egalitarian; men participate in more household activities than they did when they were working (Rexroat and Shehan, 1987).

A study of older white and African-American women found that the psychological well-being of those who identified themselves as homemakers versus retirees did not differ significantly (Adelmann, 1993). A woman who retires may not find herself facing the same identity issues as a man, however, because women often have had (or continue to have) important family roles as wife, homemaker, grandmother, and possibly mother. In fact, women who identified themselves with dual roles as both homemakers and retirees had greater self-esteem and less depression than women who identified themselves in a single role. But retiring women also face greater financial insecurity because their traditionally lower wages result in lower retirement benefits (Perkins, 1992). Retired women who have adequate sources of income, high marital satisfaction, and overall good health have higher levels of satisfaction during retirement than do men (Slevin and Wingrove, 1995).

The marital relationship generally continues along the same track following retirement: Those who had vital, rewarding marriages will probably continue to have happy marriages, whereas those whose marriages were difficult will continue to have unsatisfying relationships (Brubaker, 1991). Various factors affect marital satisfaction during retirement (Higginbottom, Barling, and Kelloway, 1993). Good health is important, and financial security, contact with others, a sense of purpose, and the ability to structure time meaningfully contribute to marital satisfaction. Retired couples experience the highest degree of marital satisfaction since the first family stage, when they had no children (Johnson, White, Edwards, and Booth, 1986).

Tho' much is taken, much abides;
and tho'
We are not now that strength which
in old days
Moved earth and heaven; that which
we are, we are—
One equal temper of heroic hearts,
Made weak by time and fate, but
strong in will
To strive, to seek, to find, and not
to yield.

Alfred, Lord Tennyson (1809–1892)

Although retirement affects marital interaction, changes in health are far more important over the long run. As long as both partners are healthy, the couple can continue their marital relationship unfettered. If one becomes ill or disabled, the other generally comes to his or her aid, providing care and nurturance. Ill spouses may also receive help from other family members, including their adult children, siblings, and other relatives.

Widowhood

Marriages are finite; they do not last forever. Eventually, every marriage is broken by divorce or death. Despite high divorce rates, most marriages end with death, not divorce. "Until death do us part" is a fact for most married people.

In 1995, 72 percent of those between sixty-five and seventy-four years old were married. Among those seventy-five years old and older, however, only 49 percent were married; 43 percent were widowed. Because women live about seven years longer on average than men, most widowed persons are women. Among women from sixty-five to seventy-four, 55 percent lived with their husbands, but only 27 percent over age seventy-five lived with a spouse. In contrast, among men sixty-five to seventy-four years old, 81 percent lived with their wives; among those over seventy-five, 71 percent lived with a spouse (U.S. Bureau of the Census, 1996). Three out of four wives will become widows.

When we look at the image of our own future provided by the old we do not believe it: an absurd inner voice whispers that that will never happen to us—when that happens it will no longer be ourselves that it happens to. Until the moment it is upon us, old age is something that only affects other people. So it is understandable that society prevents us from seeing our own kind, or fellowmen, when we look at the old.

Simone de Beauvoir (1908–1986)

Widowhood is often associated with a significant decline in income, plunging the grieving spouse into financial crisis and hardship in the year or so following death. This is especially true for poorer families (Smith and Zick, 1986). Feelings of well-being among both elderly men and elderly women are related to their financial situations. If the surviving spouse is financially secure, she or he does not have the added distress of a dramatic loss of income or wealth.

Recovering from the loss of a spouse is often difficult and prolonged. (See Chapter 12 for a discussion of death and the grieving process.) A woman may experience considerable disorientation and confusion from the loss of her role as a wife and companion. Having spent much of her life as part of a couple—having mutual friends, common interests, and shared goals—a widow suddenly finds herself alone. Whatever the nature of her marriage, she experiences grief, anger, distress, and loneliness. Physical health appears to be tied closely to the emotional stress of widowhood. Widowed men and women experience more health problems over the fourteen months following their spouses' deaths than do those with spouses. Over time, however, widows appear to regain much of their physical and emotional health (Brubaker, 1991).

Love seems the swiftest, but it is the slowest of all growths. No man or woman really knows what perfect love is until they have been married a quarter of a century.

Mark Twain (1835–1910)

Eventually widows adjust to the loss. Some enjoy their new freedom. Others believe that they are too old to date or remarry; still others cannot imagine living with someone other than their former husband. (Those who had good marriages think of remarrying more often than those who had poor marriages.) A large number of elderly men and women live together without remarrying. For many widows, widowhood lasts the rest of their lives.

Enduring Marriages

The pursuit of happiness is a most ridiculous phrase: if you pursue happiness you'll never find it.

C. P. Snow (1905–1980)

Examining marriages and families in terms of the family life cycle is an important way of exploring the different tasks we must undertake at different times in our relationships. A number of those who have studied long-term marriages lasting fifty years or more have discovered several common patterns. Two researchers (Rowe and Lasswell, cited in Sweeney, 1982) have divided relationships into three categories: (1) couples who are happily in love, (2) unhappy couples who continue marriage out of habit and fear, and (3) couples in between who are neither happy nor unhappy and accept the situation. Lasswell and Row found that approximately 20 percent of long-term marriages to be very happy, while 20 percent were very unhappy.

About 20 percent of long-term marriages can be classified as "very happy."

The greatest happiness you can have is knowing that you do not necessarily require happiness.

WILLIAM SAROYAN (1908–1981)

Another way to look at marriage is according to stability rather than satisfaction. What researchers find is what many of us already know: Little correlation exists between happy marriages and stable ones. Many unhappily married couples stay together while some happily married couples undergo a crisis in breakup. In general, however, the quality of the marital relationship appears to show continuity over the years.

The influence of retirement on marital satisfaction acts as a double-edged sword. While marital satisfaction may be enhanced by a reduction in work commitments, an adjustment of roles, and an increase in the amount of time available for companionship and other activities, this reshuffling may also create marital dissatisfaction (Ward, 1993).

Marriage is an empty box. It remains empty unless you put in more than you take out.

H. JACKSON BROWN

Over time, causes of conflict don't appear to change, proving once again that issues facing couples prior to retirement are often the same ones that face couples in their later years. (See Figure 9.2 for sources of marital conflicts.)

At each stage in our individual and family life cycle we are presented the opportunity for growth and change as we enter our roles as husband/wife/partner, parent, stepparent, or grandparent. In all these stages, marriage requires a deep commitment. As David Mace and Vera Mace (1979) observe:

Figure 9.2
Sources of Marital Conflict for Middle-Aged and Older Couples

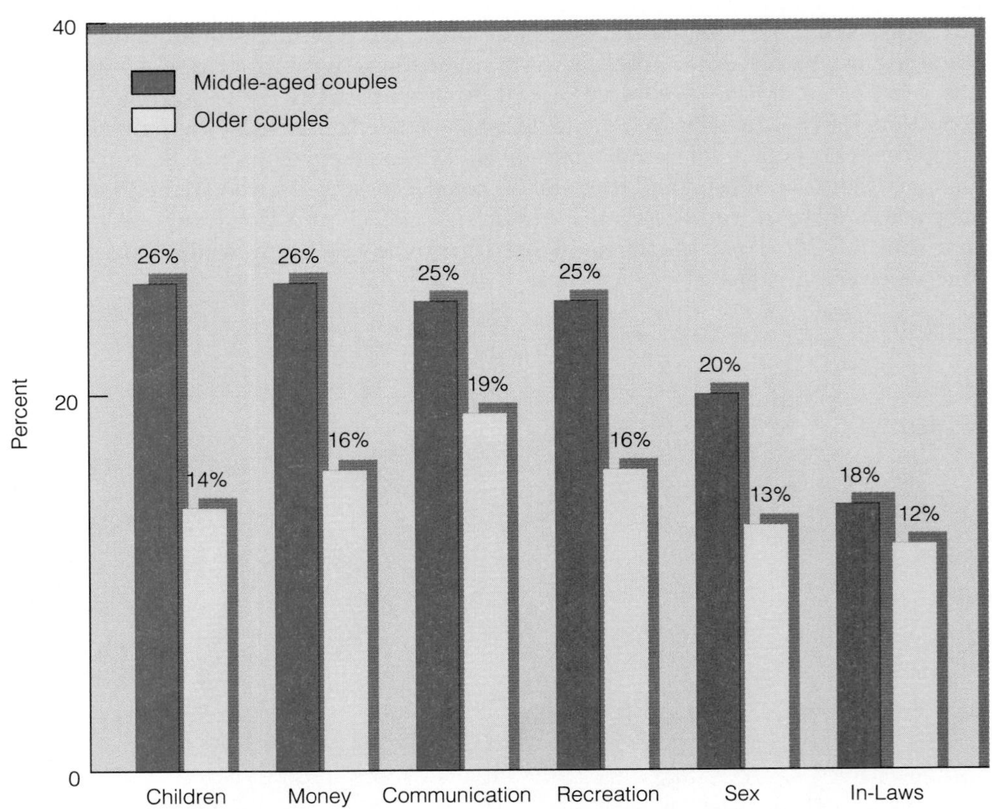

SOURCE: Levenson, R.W., L.L. Carstensen, and J.M. Gottman. "Long-Term Marriage: Age, Gender, and Satisfaction." *Psychology and Aging* 8 (1993), p. 307.

> Until two people, who are married, look into each other's eyes and make a solemn commitment to each other—that they will stop at nothing, that they will face any cost, any pain, any struggle, go out of their way so that they may learn and seek in order that they may make their marriage a continuously growing experience—until two people have done that they are not in my judgment married.

We have not finished thinking, imagining, acting. It is still possible to know the world; we are unfinished men and women.

Carlos Fuentes

As we have seen, marriages and families never remain the same. They change as we change; as we learn to give and take; as children enter and exit our lives; as we create new goals and visions for ourselves and our relationships. In our intimate relationships, we are offered the opportunity to discover ourselves.

Summary

- The eight developmental stages of the human life cycle described by Erik Erikson include (1) infancy: trust versus mistrust; (2) toddler: autonomy versus shame and doubt; (3) early childhood: initiative versus guilt; (4) school age: industry versus inferiority; (5) adolescence: identity versus role confusion; (6) young adulthood: intimacy versus isolation; (7) adulthood: generativity versus self-absorption; and (8) maturity: integrity versus despair. Each stage is intimately interconnected with family.
- The *family life cycle* perspective uses a developmental framework to look at how families change over time both in terms of the people who are members of the family and in terms of the roles they play. The family life cycle consists of eight stages: beginning families, childbearing families, families with preschool children, families with schoolchildren, families with adolescents, families as launching centers, families in the middle years, and aging families. Variations on the traditional life cycle include disabled, adoptive, gay and lesbian, ethnic, immigrant, and single-parent families, as well as stepfamilies.
- The time and patterns that precede marriage often predict marital success because marital patterns emerge during these times. Premarital factors correlated with marital success include (1) background factors (age at marriage, length of courtship, level of education, and childhood environment), (2) personality factors, and (3) relationship factors (communication, self-disclosure, and interdependence).
- Engagement is the culmination of the formal dating pattern. It prepares the couple for marriage by involving them in discussions about the realities of everyday life, it involves family members with the couple, and it strengthens the couple as a social unit. Individuals must deal with key psychological issues, such as anxiety, maturation and dependency needs, losses, partner choice, gender-role conflict, idealization and disillusionment, marital expectations, and self-knowledge. Cohabitation serves many of the same functions as engagement.
- A wedding is an ancient ritual that symbolizes a couple's commitment to each other. About two-thirds are formal church weddings. The wedding marks a major transition in life as the man and woman take on marital roles. Marriage involves many powerful traditional role expectations, including assumptions that the husband is head of the household and is expected to support the family and assumptions that the wife is responsible for housework and child rearing.
- Gender-role attitudes and behaviors contribute to marital roles. Women are more egalitarian than men in marital-role expectations, but both genders expect men to earn more money. Marital tasks include establishing marital and family roles, providing emotional support for the partner, adjusting personal habits, negotiating gender roles, making sexual adjustments, establishing family and employment priorities, developing communication skills, managing budgetary and financial matters, establishing kin relationships, and participating in the larger community.
- Couples undergo *identity bargaining* in adjusting to marital roles. This is a three-step process: a person must identify with the role, the person must be treated by the other as if he or she fulfills that role, and both people must negotiate changes in each other's roles.

- A critical task in early marriage is to establish boundaries separating the newly formed family from the couple's families of orientation. Ties to the families of orientation may include habits of subordination and economic dependency. In-law relationships tend to have little emotional intensity.
- In youthful marriages, about half of all working women leave the workforce to attend to child-rearing responsibilities. Motherhood more radically alters a woman's life than fatherhood changes a man's. Parental roles and child-care responsibilities need to be worked out.
- Middle-aged families must deal with issues of independence in regard to their adolescent children. Most women do not suffer from the *empty nest* syndrome. In fact, for many families, there is no empty nest because of the increasing presence of adult children in the home. As children leave home, parents reevaluate their relationship with each other and their life goals.
- In later-life marriages, usually no children are present. Marital satisfaction tends to be highest during this time. The most important factors affecting this life cycle stage are health, retirement, and widowhood. As a group, the aged have regular contact with their children, the lowest poverty level of any group, and good health through the early years of old age. Many families, especially among African Americans, Latinos, and Asian Americans, are *intermittent extended families* in which aging parents, adult children, or other relatives periodically live with them during times of need. This differs from the *sandwich generation* which finds itself caring for both their children and their aging parents at the same time.
- Parenting roles continue through old age. Elderly parents provide financial and emotional support to their children; they often take active roles in child care and housekeeping for their daughters who are single parents. Divorced children and those with physical or mental limitations may continue living at home.
- Grandparenting is an important role for the middle-aged and aged; it provides them and their grandchildren with a sense of continuity. Grandchildren report feeling close to grandparents; grandparents are important role models. Grandparents often provide extensive child care for grandchildren. Grandparents take on greater importance in single-parent and stepparent families. Grandparenting can be divided into three styles: companionate, remote, and involved.
- Long-term marriages may be divided into three categories: (1) couples who are happily in love, (2) unhappy couples who stay together out of habit or fear, and (3) couples who are neither happy nor unhappy. The percentage of couples who are happily in love is approximately 20 percent, the same as that of those who are unhappy.

Key Terms

boomerang generation 311
duration-of-marriage effect 301
empty nest 311
family life cycle 290
honeymoon effect 298
identity bargaining 304
intermittent extended family 313
sandwich generation 314

Suggested Readings

Aldrous, John. *Family Careers; Rethinking the Developmental Perspective*, Newbury Park, CA: Sage, 1996. An examination of the expectable changes in today's families from the time it is first formed until it is dissolved.

Arond, Marian, and Samuel L. Parker. *The First Year of Marriage.* New York: Warner Books, 1987. An authoritative and insightful book which illustrates the opportunities and challenges facing first-year marriages.

Bellah, Robert. *Habits of the Heart: Individualism and Commitment in American Life.* Berkeley, CA: University of California Press, 1985. A reflective book examining the traditional American conflict between individuality and community; in marriage and the family, this conflict leads to a tension between independence and commitment.

Baber, Kristine, and Katherine Allen. *Women and Families: Feminist Reconstructions.* New York: Guilford Publications, 1992. A multidisciplinary examination of the diversity of women's experiences in family relationships, including intimacy, sexuality, reproduction, caregiving, and work.

Carter, Betty, and Monica McGoldrick, eds. *The Changing Family Life Cycle, 2d ed.* Boston: Allyn and Bacon, 1989. A fine collection of essays on the family life cycle from a family-therapy perspective.

Cate, Rodney, and Sally Lloyd. *Courtship.* Newbury Park, CA: Sage Publications, 1992. A concise scholarly overview of current knowledge on courtship.

DeGenova, Mary Kay, ed. *Families in Cultural Context.* Mountain View, CA: Mayfield, 1997. This collection of personal accounts explores cultural variation in family structure, life cycles, functions and controls.

Eskridge, William. *The Case For the Same-Sex Marriage: From Sexual Liberty to Civilized Commitment.* New York: Free Press, 1996. A consideration of legalizing same-sex marriages as a means toward encouraging same-gender couples to model their relationships on heterosexual marriage.

Hansson, Robert O., and B. N. Carpenter. *Relationships in Old Age: Coping with the Challenge of Transitions.* New York: Guilford, 1994. Discusses the common transitions people face as they age and the social network that is necessary to support aging.

Markides, Kyiados, and Charles Mindel. *Aging and Ethnicity; Perspectives on Gender, Race, Ethnicity and Class.* Newbury Park, CA: Sage Publications, 1989. A good introduction to aging among African Americans, Latinos, Asian Americans, and other ethnic groups.

Mathabane, Mark. *Love in Black and White: The Triumph of Love over Prejudice and Taboo.* New York: HarperCollins, 1992. The acclaimed black South African author of *Kafir Boy* describes his marriage to an American white woman.

Preston, John, ed. *Member of the Family: Gay Men Write about Their Closest Relations.* New York: Dutton, 1992. A moving collection of stories and memoirs about family love, alienation, and loyalty.

Whitbourne, Susan, and Joyce Ebmeyer. *Identity and Intimacy in Marriage: A Study of Couples.* New York: Springer-Verlag, 1990. An outstanding book describing how the individual identities of two people interact to create a relationship. Scholarly but readable.

CHAPTER 10

Parents and Children

P R E V I E W

To gain a sense of what you already know about the material covered in this chapter, answer "True" or "False" to the statements below.

1. Egalitarian marriages usually remain so after the birth of the first child. True or false?
2. Specific strategies may be developed to cope with parental stress. True or false?
3. A maternal instinct has been proved to exist in humans. True or false?
4. Playing with dolls can help both girls and boys to become good parents. True or false?
5. The great majority of married men say that they get greater satisfaction from the husband-father role than from their work. True or false?
6. All studies show that regular day care by nonfamily members is detrimental to intellectual and social development. True or false?
7. A link between television violence and aggressive behavior in children has not been scientifically proven. True or false?
8. Many parents follow the advice of "experts" even though it conflicts with their own opinions, ideas, or beliefs. True or false?
9. Children who are raised with authoritarian parents tend to be less cheerful, more moody, and more vulnerable to stress than others. True or false?
10. Research shows that the children of gay parents are just as well adjusted as those raised by heterosexual parents. True or false?

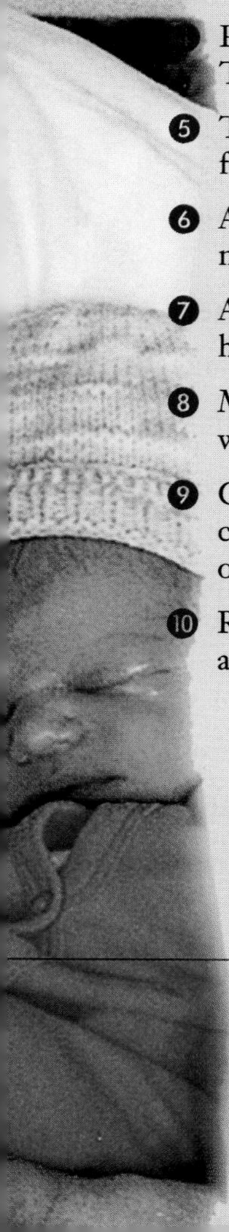

Answers:

1 False, 2 True, 3 False, 4 True, 5 True, 6 False, 7 False, 8 True, 9 True, 10 True

OUTLINE

Toward the end of the eighteenth century, three sportsmen hunting in the woods of Caune, France, were startled by the sight of a young boy gathering roots and acorns. The boy hunched over the ground like a wild animal, naked but for a tattered shirt. As soon as the boy sighted the hunters, he fled and attempted to climb a tree. The hunters quickly captured him and brought him to a neighboring village. Eventually he was taken to Paris. The boy, who apparently had been abandoned in the woods at an early age, became known as "the Young Savage of Aveyron." He had lived alone, sleeping in the fields and foraging for food. Without human contact, his mental development was little more than that of a one-year-old child. A contemporary (quoted in Itard, [1801] 1972) described the boy as

> incapable of attention . . . and consequently of all the operations of the mind which depended upon it; destitute of memory, of judgment, even of a disposition to imitation; and so bounded were his ideas, even those which related to his immediate wants, that he could not even open a door, nor get on a chair to obtain the food which was put out of reach of his hand; in short, destitute of every means of communication . . . his pleasure, and agreeable sensation of the organs of taste, his intelligence, a susceptibility of producing incoherent ideas, connected with his physical wants; in a word, his whole existence was a life purely animal.

The boy, now named Victor, was put under the tutelage of a young physician, Jean Itard (Malson, 1972). Itard painstakingly undertook to teach the boy human speech. He also taught him to walk and sit, groom himself, open doors, and use instruments and tools. Over the years, Itard endeavored to teach Victor to think conceptually; eventually Victor was able to express his simplest needs in writing.

The story of Victor and Itard is a reminder of what we tend to take for granted: that the family is the nursery of our humanity. The traits that make us human are not given to us full blown; rather, they exist potentially in us and are first developed within the family or its surrogate. When Itard took in Victor, he began to undertake the task of socialization, of making him human—one of the basic tasks of the family. If it were not for families that socialize and educate us, we, too, would be like the wild boy of Aveyron. We would not know how to speak, how to interact with others, or how to live.

Although today's families come in a variety of forms, they all seek to fulfill the needs of their members. In families with children (which includes most families at one time or another), these needs are complex and ever changing. In this chapter we explore how the needs of both parents and children may be addressed. We begin with discussions of the process of deciding to have children, of child-free relationships, and of deferred parenthood. We next discuss theories of socialization and the developmental needs of children. This is followed by a look at the meaning and special challenges of parenthood and at the components of contemporary motherhood and fatherhood. We examine the roles of child care, the schools, and television in child socialization. The unique issues that are dealt with in ethnic families, adoptive families, and families headed by gay men and lesbians are also addressed. Finally, we discuss various styles and strategies of child rearing.

A child is like a precious stone, but also a heavy burden.

SWAHILI PROVERB

Should We or Shouldn't We? Choosing Whether to Have Children

Parenthood may now be considered a matter of choice because of the widespread use of birth control. If men and women want to have children, they can decide when to have them. As a result, the U.S. birthrate has fallen to an average of two children per marriage. Among Latino couples, 2.9 is the average number; among African-American couples, it is 2.47, and among white couples, 1.97 (U.S. Bureau of the Census, 1995).

The United States has experienced a decrease in the fertility rate, the number of births per year per 1,000 women, from 118 in 1960 to 64 in 1994. As seen below, fertility rates vary considerably by race and ethnicity. Within the Hispanic American population, rates vary from a high of 111 among Mexican Americans to fifty among Cuban Americans. Cultural, social, and economic factors play a significant part in influencing the number of children a family has. Statisticians predict that with higher fertility rates and continuing immigration patterns Hispanics could, within fifteen years, become our nations' largest minority group, thereby surpassing African Americas (U.S. Bureau of the Census, 1996).

Approximately 9.3 percent of American women between the ages of eighteen and thirty-four do not plan to have children (U.S. Bureau of the Census, 1995). About two-thirds of women currently without children, however, expect to have at least one child ("Advance Report," 1993). Even if fewer couples choose to have children, the average number of

Many couples today (especially those in middle- and upper-income brackets) defer having children until they have established their own relationships and built their careers. These parents are usually quite satisfied with their choice.

Figure 10.1
Fertility Rates by Race and Ethnicity: 1994

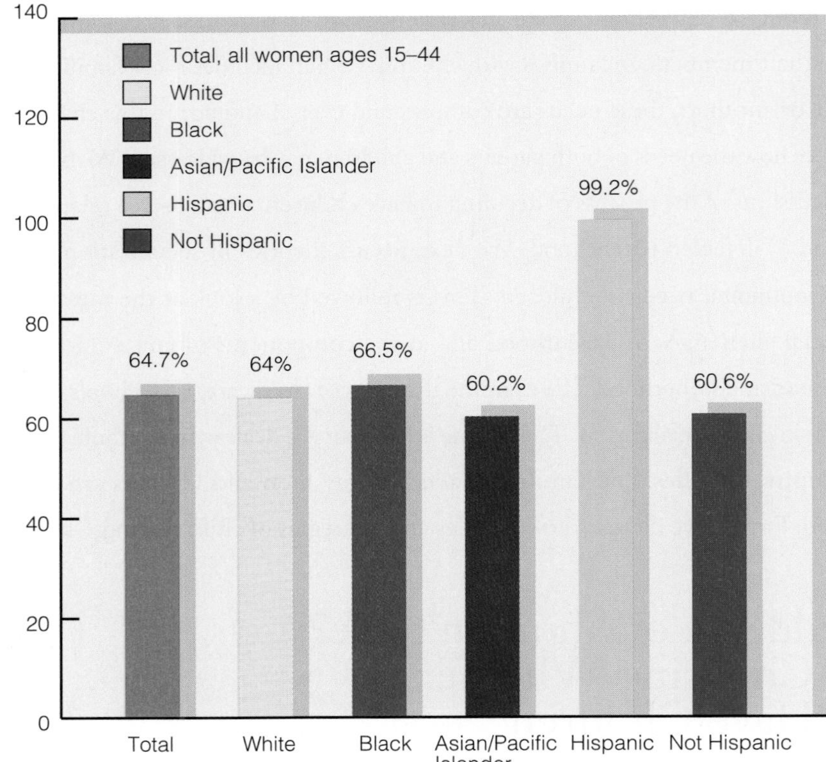

SOURCE: Bachu, A. "Current Populations Report," Series pp. 20–482, U.S. Government Printing Office, U.S. Bureau of the Census, Washington, D.C. 1995.

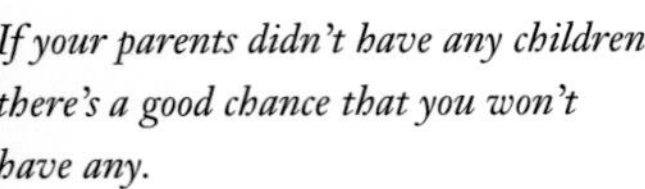

If your parents didn't have any children, there's a good chance that you won't have any.

CLARENCE DAY (1874–1935)

children per couple is expected to remain fairly stable in the coming years. Decreases will be offset by births to immigrant families (who tend to prefer more children) and to older women who have deferred childbearing (Vobejda, 1993).

Child-free Marriage

Aunt Tillie: Do you like children?

Uncle Gus: I do if they're properly cooked.

W.C. FIELDS (1880–1946)

In recent articles and discussions of marriages in which there are no children, the term *childless* is often replaced by **child-free.** This change in terminology reflects a shift of values in our culture. Couples who do not choose to have children need no longer be viewed as lacking something hitherto considered essential for personal fulfillment. Indeed, the use of the suffix *free* suggests liberation from the bonds of a potentially oppressive condition (see Callan, 1985). Women who choose to be child-free are generally well educated and career oriented (Ambry, 1992).

Even today, however, with less familial and societal pressure to reproduce, the decision to remain without children is not always easily made. Although the partners in some child-free marriages have never felt that they wanted to have children, for most the decision seems to have been gradual. Jean Veevers (1980) identifies four stages of this decision process:

1. The couple decides to postpone having children for a definite time period (until he gets his degree, until she gets her promotion, and so on).

2 When the time period expires, they decide to postpone having children indefinitely (until they "feel like it").

3 They increasingly appreciate the positive advantages of being child-free (as opposed to the disadvantages of being childless).

4 The decision is made final, generally by the sterilization of one or both partners.

Studies of child-free marriages show nine basic categories of reasons given by those who have chosen this alternative (Houseknecht, 1987). The categories, in descending order of importance, are as follows:

- Freedom from child-care responsibility and greater opportunity for self-fulfillment.
- More satisfactory marital relationship.
- Wife's career considerations.
- Monetary advantages.
- Concern about population growth.
- General dislike of children.
- Early socialization experiences and doubts about parenting ability.
- Concern about physical aspects of childbirth.

Couples usually have some idea that they will or will not have children before they marry. If the intent isn't clear from the start or if one partner's mind changes, the couple may have serious problems ahead.

Many studies of child-free marriages indicate a higher degree of marital adjustment or satisfaction than is found among couples with children. These findings are not particularly surprising if we consider the great amount of time and energy that child rearing entails. It has also been observed that divorce is more probable in child-free marriages perhaps because, unlike some other unhappily married couples, the child-free couples do not stay together "for the sake of the children."

Did you know?

A family with a child born in 1995 will spend more than $145,000 ($239,000 when adjusted for inflation) for food, shelter, and other necessities over the next seventeen years (U.S. Department of Agriculture, 1996).

Deferred Parenthood

Although most women still begin their families while in their twenties, demographers predict that the trend toward later parenthood will continue to grow, especially in middle- and upper-income groups (Edmondson, Waldrop, Crispell, and Jacobssen, 1993; Price, 1982; Whitehead, 1990). A number of factors contribute to this. More career and lifestyle options are available to single women today than in the past. Marriage and reproduction are no longer economic or social necessities. People may take longer to search out the "right" mate (even if it takes more than one marriage to do it), and they may wait for the "right" time to have children. Increasingly effective birth control (including safe, legal abortion) has also been a significant factor in the planned deferral of parenthood.

Besides giving parents a chance to complete education, build careers, and firmly establish their own relationship, delaying parenthood can also be advantageous for other reasons. Maternity and medical expenses, food, furniture and equipment, clothes, toys, baby-sitters, lessons, and summer camp are costly. In the early 1990s, the cost of raising a child from birth to seventeen years of age ranged from $86,000 to $168,480 depending on the family's income level, age of the child, and region of residence (Lino, 1991). (These costs are based on two-parent, middle-income families; with three or more children, the cost per child declines.) Obviously, parents who have had a chance to establish themselves financially will be better able to bear the economic burdens of child rearing. Ann

Reflections

If you don't have children, do you want children? When? How many? What factors do you need to take into consideration when contemplating a family for yourself? Does your partner (if you have one) agree with you about having children?

UNDERSTANDING YOURSELF

Am I Parent Material?

The following questions are designed to stimulate ideas that you may not have thought about. There are no right answers and no grades—your answers are right for you and may help you decide for yourself whether or not you want to be a parent. Because we all change, your answers to some of these questions may change two, five, or even ten years from now. You do have a choice. Investigate what you know, and give it some thought. Then do what seems right for you.

Does Having and Raising a Child Fit the Lifestyle I Want?

1 What do I want out of life for myself? What do I think is important?
2 Could I handle a child and a job at the same time? Would I have time and energy for both?
3 Would I be ready to give up the freedom to do what I wanted to do, when I wanted to do it?
4 Would I be willing to cut back my social life and spend more time at home? Would I miss my free time and privacy?
5 Can I afford to support a child? Do I know how much it takes to raise a child?
6 Do I want to raise a child in the neighborhood where I live now? Would I be willing and able to move?
7 How would a child affect my growth and development.
8 Would a child change my educational plans? Would I have the energy to go to school and raise a child at the same time?
9 Am I willing to give a great part of my life—at least eighteen years—to being responsible for a child? And am I prepared to spend a large portion of my life being concerned about my child's well-being?

What's in It for Me?

1 Do I like doing things with children? Do I enjoy activities that children can do?
2 Would I want a child to be "like me"?
3 Would I try to pass on to my child my ideas and values? What if my child's ideas turned out to be different from mine?
4 Would I want my child to achieve things that I wish I had achieved but didn't?
5 Would I expect my child to keep me from being lonely in my old age? Do I do that for my parents? Do my parents do that for my grandparents?
6 Do I want a boy or a girl child? What if I didn't get what I wanted?

Goetting's (1986a) review of parental satisfaction research revealed that "those who postpone parenthood until other components of their lives—especially their careers—are solidified" express enhanced degrees of satisfaction. Older parents may also be more emotionally mature and thus more capable of dealing with parenting stresses (although age isn't necessarily indicative of emotional maturity). In addition, as Jane Price (1982) writes, "Combating the aging process is something of a national preoccupation. . . . In our society, the power of children to revitalize and refresh is part of the host of forces encouraging men and women to become parents much later than they did in the past."

If I pay the basic price of $80,000 to raise a child for the next eighteen years, plus all the extras that you've mentioned, is there any guarantee it will turn out all right?

There is no warranty with the price whatsoever. You pay your money and you take your chances. We no longer guarantee that when it gets to be eighteen, it will be able to read and write. Frankly, you'll be lucky if it volunteers to cut your lawn or do the dishes once a week. And if you think for eighty grand it's going to clean up its own room, you're living in a dream world.

ART BUCHWALD

The Transition to Parenthood

Women or men who become parents enter a new phase of their lives. More than marriage, parenthood signifies adulthood—the final, irreversible end of youthful roles. A person can become an ex-spouse but never an ex-parent. The irrevocable nature of parenthood may make the first-time parent doubtful and apprehensive, especially during the pregnancy. Yet people have few ways of preparing for parenting. Parenthood has to be learned experientially (although ideas can modify practices). A person may receive assistance from more experienced parents, but each new parent has to learn on his or her own.

Sources of Stress

The abrupt transition from a nonparent to a parent role may create considerable stress. Parents take on parental roles literally overnight, and the job goes on with-

7 Would having a child show others how mature I am?
8 Will I prove I am a man or a woman by having a child?
9 Do I expect my child to make my life happy?

Raising a Child: What's There to Know?

1 Do I like children? When I'm around children for a while, what do I think or feel about having one around all the time?
2 Do I enjoy teaching others?
3 Is it easy for me to tell other people what I want or need, or what I expect of them?
4 Do I want to give a child the love he or she needs? Is loving easy for me?
5 Am I patient enough to deal with the noise, the confusion, and the twenty-four-hour-a-day responsibility? What kind of time and space do I need for myself?
6 What do I do when I get angry or upset? Would I take things out on a child if I lost my temper?
7 What does discipline mean to me? What does freedom, setting limits, or giving space mean? What is being too strict or not strict enough? Would I want a perfect child?
8 How do I get along with my parents? What will I do to avoid the mistakes my parents made?
9 How would I take care of my child's health and safety? How do I take care of my own?
10 What if I have a child and find out I made a wrong decision?

Have My Partner and I Really Talked about Becoming Parents?

1 Does my partner want to have a child? Have we talked about our reasons?
2 Could we give a child a good home? Is our relationship a happy and strong one?
3 Are we both ready to give our time and energy to raising a child?
4 Could we share our love with a child without jealousy?
5 What would happen if we separated after having a child, or if one of us died?
6 Do my partner and I understand each other's feelings about religion, work, family, child raising, and future goals? Do we feel pretty much the same way? Will children fit into these feelings, hopes, and plans?
7 Suppose one of us wanted a child and the other didn't. Who would decide?
8 Which of these questions do we need to discuss in depth before making a decision?

SOURCE: Adapted with permission from "Am I Parent Material?" ETR Associates, P.O. Box 1830, Santa Cruz, CA 95061.

out relief around the clock. Many parents express concern about their ability to meet all the responsibilities of child rearing (see Klinman and Vukelich, 1985).

Many of the stresses felt by new parents closely reflect gender roles. Overall, mothers seem to experience greater stress than fathers (Harriman, 1983). Although a couple may have an egalitarian marriage before the birth of the first child, the marriage usually becomes more traditional once a child is born. If the mother, in addition to the father, must continue to work, or if the woman is single, she will have a dual role as both homemaker and provider. She will also probably have the responsibility for finding adequate child care, and it will most likely be she who stays home with the child when he or she is sick. Multiple role demands are the greatest source of stress for mothers. In a *New York Times* survey, 83 percent of women acknowledged they (sometimes) felt torn between the demands of their job and wanting to spend more time with their family (Belkin, 1989).

There are various other sources of parental stress. Sixty-four percent of fathers in one study (Ventura, 1987) described severe stress associated with their work. A number of mothers and fathers were concerned about not having enough money. Other sources of stress involve infant health and care, infant crying, interactions with the spouse (including sexual relations), interactions with other family members and friends, and general anxiety and depression (Harriman, 1983; McKim, 1987; Ventura, 1987; Wilkie and Ames, 1986).

Changes in marital quality and marital conflict were studied among a sample of white and African-American spouses as they transitioned to parenthood (Crehan, 1996). The results of this study showed a decline in marital happiness and more frequent conflicts among both white and African-American spouses. White parents also reported higher marital tension and a greater likelihood to

Having a family is like having a bowling alley installed in your head.

MARTIN MULL

Did you know?

Most, though not all, mothers would choose parenthood again. Interviews with two thousand mothers found that eight out of ten would have children if they had the chance to do it again (Genevie and Margolies, 1987).

Warning! The Surgeon General has determined that trying to be a good parent can be hazardous to your health.

ART DWORKEN

become quiet and withdrawn after the birth of their child. This increase in avoidance behaviors may be due to the limited time and energy that new parents have to devote to conflict resolution.

Coping with Stress

Cleaning and scrubbing can wait till tomorrow.
For babies grow up we've learned to our sorrow.
So quiet down cobwebs, dust go to sleep.
I'm rocking my baby and babies don't keep.
ANONYMOUS

Although the first year of child rearing is bound to be stressful, the couple experience less stress if they (1) have already developed a strong relationship, (2) are open in their communication, (3) have agreed on family planning, and (4) originally had a strong desire for the child. In spite of planning, the reality for most is that this is a very stressful time. Accepting this fact while developing time management skills, patience with oneself, and a sense of humor can be most beneficial. Psychiatrist and researcher Jerry M. Lewis and colleagues (Lewis, Owen, and Cox, 1989) stress the importance of "marital competence" in the "successful incorporation of the child into the family." Another important factor in maintaining marital quality during this time is the father's emotional and physical support (Tietjen and Bradley, 1985). The more involved the father is in the care of the baby, the easier it is on the recovering mother.

Cleaning your house with the kids still around is like shoveling the walk before it stops snowing.
PHYLLIS DILLER

Ventura (1987) suggests that families can be assisted by improved health care and support in three areas. First, during the prenatal period, parents need help locating community support resources, such as La Leche League and child-care assistance, so that they can "restore their energy" during the postpartum period. Fathers need to be encouraged in their nurturing role. Employers need to restructure schedules, develop child-care programs, and institute parental leave policies.

Second, coordination of care is important. Parents need good communication and support from practitioners following the birth. They need "concrete explanations and demonstrations." Health practitioners, child-care providers, and community professionals need to coordinate their communication with each other.

Third, families need anticipatory care and help in problem solving. Teaching about development, childhood illnesses, and safety should be an ongoing part of clinic and child-care services. New parents should reach out to others with similar experiences and concerns for support and understanding.

Reflections

If you have children, did you plan to have them? What considerations led you to have them? What adjustments have you had to make? How did your relationship with your partner change?

Theories of Child Socialization

Before I got married, I had six theories about bringing up children. Now I have six children and no theories.
JOHN WILMOT, EARL OF ROCHESTER

Twentieth-century attitudes about children have been influenced by a number of theories concerning learning and socialization. The issue of "nature versus nurture" (whether the developing child is influenced most by inborn traits or by the environment) continues to be vigorously debated.

Psychological Theories

Psychological theories of human development give prime importance to the role of the mind, particularly the subconscious mind, which, according to psychoanalytic theory, motivates much of our behavior without our being consciously aware of the process. According to these theories, many aspects of our psychological makeup are inborn; the mind is programmed to grow and develop just as the body is.

Psychoanalytic Theory

Sigmund Freud's (1856–1939) contribution to the understanding of the human psyche is profound. His emphasis on the importance of unconscious mental processes and on the stages of psychosexual development has greatly influenced modern psychology. Freud's **psychoanalytic theory** of personality development holds that we are driven by instinct to seek pleasure, especially sexual pleasure. This part of the personality, called the **id,** is kept in check by the **superego**—what we might call the conscience. The third component of personality, the rational **ego,** mediates between the demands of the id and the constraints of society. Freudian theory views the uninhibited id of the infant as gradually becoming controlled as the individual internalizes societal restraints. Too much restraint, however, leads to repression and the development of **neuroses**—psychological disorders characterized by anxiety, phobia, and so on.

Freud viewed the parents as mainly responsible for the child's psychological development. He posited that between the ages of four and six, the child identifies with the parent who is of the same sex. Not becoming like that parent was seen as a failure to reach maturity. Freud divided psychosexual development into five stages spanning the time from birth through adolescence. These stages are called oral, anal, phallic, latency, and genital. (See Table 10.1.) The scientific thought of Freud's time was greatly influenced by Charles Darwin's theories on evolution. As Jerome Kagan (1984) has pointed out, Freud viewed evolution as an apt metaphor for human behavior and constructed his theories in accordance with the scientific thought of his day.

Psychosocial Theory

Erik Erikson (1902–1994) based much of his work on psychoanalytic theory, but he emphasized the effects of society on the developing ego (Erikson, 1963). His theory is thus called a **psychosocial theory** of development. It stresses parental and societal responsibilities in children's development. Each of Erikson's life cycle stages (see Table 10.1) is centered around a specific emotional concern based on individual biological influences and external sociocultural expectations and actions. (See Chapter 9 for an in-depth discussion of Erikson's theory of development.)

Learning Theory

Learning theorists emphasize the aspects of behavior that are acquired rather than inborn or instinctual.

Behaviorism

Pioneering American psychologists such as John B. Watson (1878–1958) and B. F. Skinner (1904–1990) sought to explain human behaviors entirely on the basis of what could be observed; this model of human development is known as **behaviorism.** Behaviorists reject the concept of hidden "drives" that cannot be seen. Skinner developed the concept of **reinforcement** to explain how behaviors may be increased or decreased. He labeled the process of increasing the frequency of a behavior by adding a reinforcing stimulus **operant conditioning.** When we praise a child for picking up her blocks, we are positively reinforcing her behavior, increasing the likelihood that she will pick them up in the future.

We are born princes and the civilizing process makes us frogs.

Eric Berne

The chief cause of human error is to be found in the prejudices picked up in childhood.

RENÉ DESCARTES (1596–1650)

Social Learning Theory

Developed by psychologists such as Julian Rotter and Albert Bandura, **social learning theory** emphasizes the role of cognition, or thinking, in learning (Bandura, 1977; Rotter, Liverant, and Crowne, 1961). Human nature is formed by the interactions of culture, society, and the family with the inner qualities of the individual (Bandura, 1986, 1989). Children are first socialized through parental direction of their behavior. Parents teach children what is good, what is bad, what to eat, what not to eat, what to keep, what to share, how to talk, what to feel, and what to think. Parents influence their children by modeling (serving as examples for their children) and by defining (establishing expectations for them) (Cohen, 1987). Social learning theory accepts many of the tenets of behavioral psychology but adds to it the individual's innate ability to think and make choices to change his or her environment (Schickedanz, Schickedanz, Hansen, and Forsyth, 1993).

Cognitive Developmental Theory

Beginning in the 1930s, Swiss psychologist Jean Piaget (1896–1980) began intensively observing and interviewing children. Piaget observed that cognitive development occurs in discrete stages through which all infants and children pass. Based on the development of the brain and the nervous system, these stages occur at about the same time in the development of all children (unless they are mentally impaired), according to Piaget's **cognitive developmental theory.** The stages can be seen as building blocks, each of which must be completed before the next one can be put into place. In Piaget's view, children develop their cognitive abilities through interaction with the world and adaptation to their surroundings. Children adapt by **assimilation** (making new information compatible with their world understanding) and **accommodation** (adjusting their cognitive framework to incorporate new experiences) (Dworetsky, 1990). Piaget labeled the stages of cognitive development sensorimotor, preoperational, concrete operational, and formal operational. (See Table 10.1.)

Table 10.1 Stages of Development: Freud, Piaget, and Erikson Compared

	Freud	Piaget	Erikson
Infancy	Oral	Sensorimotor	Trust vs. mistrust
	Anal		Autonomy vs. shame and doubt
Early childhood	Phallic	Preoperational	Initiative vs. guilt
Late-middle childhood	Latency	Concrete operational	Industry vs. inferiority
Adolescence	Genital	Formal operational	Identity vs. confusion
Early adulthood			Intimacy vs. isolation
Middle adulthood			Generativity vs. stagnation
Late adulthood			Ego integrity vs. despair

The Developmental Systems Approach

Parents do not simply give birth to children and then "bring them up." According to the **developmental systems approach,** the growth and development of children takes place within a complex and changing family system that both influences and is influenced by the child. The family system is part of a number of larger systems (extended family, friends, health care, education, and local and national government, to name a few), all of which mutually interact with one another. Models or theories that use a developmental systems approach include Bronfenbrenner's (1979) ecological model, Lerner's (1986) developmental contextual theory, Dewey and Bentley's (1949) transactional model, and Magnuson's (1988) interactive approach. (Family development theory was discussed in Chapter 2.)

Parent-Child Interactions

Children begin by loving their parents; after a time, they judge them; rarely, if ever, do they forgive them.

Oscar Wilde (1854–1900)

Not only are children socialized by their parents but they are also socializers in their own right. When an infant cries to be picked up and held, to have a diaper changed, or to be burped, or when he or she smiles when being played with, fed, or cuddled, the parents are being socialized. The child is creating strong bonds with the parents (see the discussion of attachment later in the chapter). Although the infant's actions are not at first consciously directed toward reinforcing parental behavior, they nevertheless have that effect. In this sense, even very young children can be viewed as participants in creating their own environment and in contributing to their further development (see Peterson and Rollins, 1987).

In the developmental systems model of family growth, social and psychological development are seen as lifelong processes, with each family member having a role in the development of the others. In terms of the eight developmental stages of the human life cycle described by Erikson, parents are generally at the seventh stage (generativity) during their children's growing years, and the children are probably anywhere from the first stage (trust) to the fifth (identity) or sixth (intimacy). The parents' need to establish their generativity is at least partly met by the child's need to be cared for and taught. The parents' approach to child rearing will inevitably be modified by the child's inherent nature or temperament.

Sibling Interactions

Over 80 percent of American children have one or more siblings. Siblings influence one another according to their particular needs and personalities. They are also significant agents for socialization. While rivalry and aggression may appear to be the foundation of such interactions, young siblings at home spend a large percentage of their time actually playing together. Sibling influence (or the lack of it, in the case of only children) is important in subtle yet powerful ways as the result of birth order and spacing (the number of years between sibling births). A study at Colorado State University, for example, found that a firstborn's self-esteem suffers if a sibling is born two or more years later but is not affected if the sibling is born less than two years later (Goleman, 1985). Furthermore, if the firstborn child is already five or six years old, the birth of a second sibling does not have the same impact. A study of African-American and white families indicated that race has no impact on the dynamics of birth order (Steelman and Powell, 1985). (See Sulloway, 1996, for a study of the effects of birth order on personality and achievement.)

The quality of sibling interaction may have consequences for the child's later behavior (Newcombe, 1996). Close, affectionate sibling relationships contribute to the development of desirable characteristics such as social sensitivity, communication skills, cooperation, and understanding of social roles.

The Sense of Self

There is, in short, something within you—that deep inward sense of I-ness—that is not memory, thoughts, mind, body, experience, surroundings, feelings, conflicts, sensations or moods. . . . It was not born with your body, nor will it perish upon death. It does not recognize time nor cater toward its distresses. It is without color, without shape, without form, without size, and yet it beholds the entire majesty before your own eyes. It sees the sun, clouds, stars and moon, but cannot itself be seen. It hears the birds, crickets, the singing waterfall, but cannot itself be heard. It grasps the fallen leaf, the crusted rock, the knotted branch, but cannot itself be grasped.

KEN WILBER

The models we have described leave out something very crucial about human personality. Each person knows that his or her basic motives are not simply the result of biological drives, nor is everything he or she does just the result of socialization. Something more human about human beings exists than these models acknowledge. This is the "I" within each person; it may be what philosophers have called the soul, or it may be something else, but it has an existence of its own that resists biological and environmental pressures. Erik Erikson (1968) writes:

> What the "I" reflects on when it sees or contemplates the body, the personality, and the roles to which it is attached for life—not knowing where it was before or will be after—are the various selves which make up our composite Self. One should really be decisive and say that the "I" is all-conscious, and that we are truly conscious only insofar as we can say "I" and mean it. . . . to ignore the conscious "I" . . . means to delete the core of human self-awareness.

Children's Developmental Needs

Although the relative effects of physiology and environment on human development are much debated by today's experts, it is clear that both nature and nurture play important roles. In addition to biological factors, important factors affecting early development include the formation of attachments (especially maternal) and individual temperamental differences.

Biological Factors

We can expect a conscience of every child. We don't have to build it in. All we have to do is arrange the environment so they don't lose it.

JEROME KAGAN

A biological determinist believes that much of human behavior is guided by genetic makeup, physiological maturation, and neurological functioning. Jerome Kagan (1984) has presented a strong case for the role of biology in early development. He holds that the growth of the central nervous system in infants and young children ensures that such motor and cognitive abilities as walking, talking, using symbols, and becoming self-aware will occur "as long as children are growing in any reasonably varied environment where minimal nutritional needs are met and [they] can exercise emerging abilities." Furthermore, according to Kagan, children are biologically equipped for understanding the meaning of right and wrong by the age of two, but although biology may be responsible for the development of conscience, social factors can encourage its decline.

Attachment

The strong bond forged between an infant and his or her primary caregiver or caregivers is called **attachment.** Babies appear to be equipped to build relationships with those who care for them. They signal their needs by methods such as gazing, crying, and smiling, and they bond with the people who are most responsive to them (in most cases, the mother). Based on her observations of babies in "the strange situation"—that is, in the presence of a stranger both with and without the mother—researcher Mary Ainsworth discovered three patterns of infants' attachment to their mothers. These were labeled *secure*, *anxious/ambivalent*, and *anxious/avoidant* attachment (Ainsworth, Blehar, Waters, and Wall, 1978). (See Table 10.2.) A person's style of attachment in infancy is thought to affect his or her style of relating in adulthood, as was discussed in Chapter 4.

Individual Temperament

A child's unique temperament, such as "inhibited/restrained/watchful" or "uninhibited/energetic/spontaneous," also influences the way in which he or she develops (Kagan, 1984). Temperamental differences may be rooted in the biology of the brain (Kagan and Snidman, 1991), but temperament is also developed by interaction with the environment. For example, a baby who is vigorous, strong, and outgoing will probably encourage her parents to play with her vigorously. She and her parents will reinforce the lively, extroverted, and spontaneous aspects of her personality. An infant who is shy, fearful, and cries easily, however, will not encourage his parents to play energetically with him and may, in fact, inhibit them from interacting with him, thus causing him to become more shy and fearful. It is important for parents to understand "how they create the meaning of the child's individuality by their own temperaments, and their demands, attitudes, and evaluations," according to psychologists Richard Lerner and Jacqueline Lerner (Brooks, 1994). Lerner and Lerner stress the importance of what they call "goodness of fit" between the child's characteristics and those of the parents and other environmental influences. If parents are sensitive to a child's unique temperament, they are better able to understand the child and to seek appropriate ways to influence the child's behavior.

Table 10.2 Attachment Patterns in 12- to 18-Month-Olds in the "Strange Situation"

Attachment Pattern	Behavior before Separation	Behavior during Separation	Reunion Behavior	Behavior with Stranger
Secure	Separates from mother to explore toys; shares play with mother; is friendly toward stranger when mother is there	May cry; play is subdued for a while, usually recovers, plays	If distressed during separation, contact with mother ends distress; if not distressed, greets mother with affection	Somewhat friendly; may play with stranger
Anxious/ ambivalent	Has difficulty separating to explore toys; wary of new situations and people; stays close to mother	Is very distressed; cries hysterically	Seeks comfort but then rejects it; may continue to cry; may be passive	Wary of stranger; rejects offers to play
Anxious/ avoidant	Separates to explore toys, but does not share with parent; shows little preference for parent over stranger	Shows no distress; continues to play; interacts with stranger	Ignores or moves away from mother	Does not avoid stranger

SOURCE: Compiled from Ainsworth, M. D. S., & Wittig, B. A. (1969). "Attachment and Exploratory Behavior of One-Year-Olds in a Strange Situation." In B. M. Foss (ed.), *Determinants of Infant Behavior* (Vol. 4). London: Methuen.

OTHER PLACES OTHER TIMES

Families of Mexican Origin

The Socialization of Children

The Mexican-origin home is usually child-centered when children are young; yet the role of children is based on the belief that they should "be seen and not heard." Although both parents tend to be permissive, boys and girls are raised very differently in Mexican-American families. Boys are granted far more liberty, and loud, aggressive behavior is generally tolerated. The male child is often overindulged and accorded greater status than the female. Young girls are expected to be demure and feminine, and girls are usually taught feminine roles, just as boys are taught masculine roles. Playmates are often segregated by gender, rather than by age, particularly as they grow older.

Adolescence marks differences in behavior patterns between boys and girls (Locke, 1992). The adolescent male is given much more freedom, and his decisions and actions are seldom questioned. His activities are seldom restricted and are not closely monitored. He is free to date and pursue intimate relationships, and he is expected to take on the role of the protector for his sisters.

Activities for girls are often restricted and closely monitored. They are expected to remain much closer to home. Regardless of age, but particularly during adolescence, girls are expected to be subservient to their older brothers. At this stage, a strong relationship with her mother is encouraged and is viewed as a means of preparing a girl for the role of wife and mother (Mirandé, 1985). In families of Mexican origin, the *quinceanera,* or fifteenth birthday, marks the coming of age of a young girl. The *quinceanera* is celebrated by a mass or prayer service with a sermon, reminding the young girl of her future responsibilities.

Assimilation and acculturation have had an impact on socialization patterns. Whereas adolescent males still maintain greater degrees of independence, many adolescent females have challenged traditional gender roles and are exerting greater autonomy. However, the servile attitude of females remains present in many Mexican-origin families and is still encouraged and expected, regardless of the level of acculturation and assimilation.

As Mexican-American children make the transition to adolescence, just like all teenagers, they may encounter difficulties with issues related to autonomy. Adolescence is a period during which individuals are attempting to assert their independence, and for many Mexican-origin adolescents, this is compounded by conflicts that arise between first and second generations.

When demands and desires accumulate and are in conflict with traditional expectations, teenagers have difficulty dealing with the competing influences in their lives. Adolescence is particularly problematic for children who are raised in homes with traditional cultural values. They often encounter a different set of values at school than at home. External expectations may be in direct opposition to traditional values, which reinforce that assimilation should be avoided altogether and that *la cultura* (the culture) should be maintained (Klor de Alva, 1988). The extended family system has been a mainstay

Basic Needs

As we grow up, we need to get different strokes from different folks in different settings to become sentient, capable, competent, and compassionate human beings.

URIE BRONFENBRENNER

Parents often want to know what they can do to raise healthy children. Are there specific parental behaviors or amounts of behaviors (say, twelve hugs a day?) that all children need to grow up healthy? Apart from saying that basic physical needs must be met (adequate food, shelter, clothing, and so on), along with some basic psychological ones, experts cannot give us detailed instructions.

Noted physician Melvin Konner lists the needs for optimal child development—which, he writes, "parents, teachers, doctors, and child development experts with many different perspectives can fairly well agree on"—as follows (Konner, 1991):

- Adequate prenatal nutrition and care.
- Appropriate stimulation and care of newborns.
- The formation of at least one close attachment during the first five years.
- Support for the family "under pressure from an uncaring world," including child care when a parent or parents must work.
- Protection from illness.

in many Mexican-origin families, and supportive institutions of *la familia*—including *parentesco,* the concept of family, *compadrazgo,* godparents, and *confianza,* trust—have helped set the standard for families. These ideologies reinforce the view that family is the central and most important institution in life. Although these concepts influence actual behavior, their ideals can never be fully realized; however, they provide a basis that helps establish norms and expectations (Kane, 1993).

Mexican-origin families have an extended family structure that serves as a vital link between family and community in Mexican-American society. While it is a very strong institution, it varies from one generation to the next. *Compadrazgo,* or godparents, who have a moral obligation to act as guardian, provide financial assistance in times of need, and substitute as parents in the event of death. The *compadrazgo* relationship is formed usually through baptism and confirmation ceremonies in the church. The parental relationship is maintained throughout life and extends beyond the child. Godparents often refer to each other as *comadre* (co-mother) or *compadre* (co-father) and are expected to maintain a reciprocal relationship of support and mutual assistance.

Parentesco is a kinship concept that extends family sentiment to kin and nonkin, ensuring that there is an automatic family network. This concept often helps build networks of support and reciprocity and also helps establish a sense of community support among individuals who share regional or geographic origins.

Confianza, commonly referred to as trust, is essential to the relationship between *compadrazgo* and *parentesco.* Yet it means more than trust and includes the notions of respect and intimacy. It builds relationships and provides the foundation for reciprocity. *Confianza* is seen as an institution that has facilitated adaptation after immigration.

Much of what the culture condemns is related to kinship relationships. The family is perceived as more important than the individual. Selfishness is condemned, and its absence is considered a virtue. The strength of the family in providing security to its members is sometimes expressed through the sharing of material things with other relatives, even when there is precious little to meet one's own immediate needs (Locke, 1992). The concept that one should sacrifice everything for family has its costs, however.

In traditional Mexican life, a set of family, religious, and community obligations was significant. Women had certain legal and property rights that acknowledged the importance of their work, their families of origin, and their children. However, the imposition of American law and custom ignored and ultimately undermined many aspects of the extended family in Mexican culture. Because of economic changes that occurred in the family as a result of these laws, many Mexican-American women were forced to participate in the economic support of their families by working outside the home. The preservation of traditional customs, such as language, celebrations, and healing practices, became an important element in maintaining and supporting familial ties (Thornton-Dill, 1994).

The culture and identity of Mexican Americans will continue to change as they are affected by inevitable generational fusion with Anglo society and the influence of immigrants.

SOURCE: Yolanda M. Sanchez. "Families of Mexican Origin." In Mary Kay DeGenova, ed, *Families in Cultural Context.*

- Freedom from physical and sexual abuse.
- Supportive friends, both adults and children.
- Respect for the child's individuality and the presentation of appropriate challenges leading to competence.
- Safe, nurturing, and challenging schooling.
- An adolescence "free of pressure to grow up too fast, yet respectful of natural biological transformations."
- Protection from premature parenthood.

In today's society, especially in the absence of adequate health care and schools in so many communities, it is difficult to see how even these minimal needs can all be met. Even when the necessary social supports are present, parents may find themselves confused, discouraged, or guilty because they don't live up to their own expectations of perfection. Yet children have more strength, resiliency, and resourcefulness than we may ordinarily think. They can adapt to and overcome many difficult situations. A mother can lose her temper and scream at her child, and the child will most likely survive, especially if the mother later apologizes and shares her feelings with the child. A father can turn his child away with a grunt because he is too tired to listen, and the child will not necessarily grow up neurotic, especially if the father spends some "special time" with the child later on.

Self-Esteem

Respect yourself, if you would have others respect you.

BALTASAR GRACIÁN (1601–1658)

High self-esteem—what Erik Erikson called "an optimal sense of identity"—is essential for growth in relationships, creativity, and productivity in the world at large. Low self-esteem is a disability that afflicts children (and the adults they grow up to be) with feelings of powerlessness, poor ability to cope, low tolerance for differences and difficulties, inability to accept responsibility, and impaired emotional responsiveness. Self-esteem has been shown to be more significant than intelligence in predicting scholastic performance.

Don't compromise yourself. You're all you got.

JANIS JOPLIN (1943–1970)

A study of three thousand children found that adolescent girls had lower self-images, lower expectations from life, and less self-confidence than boys (Brown and Gilligan, 1992). At age nine, most of the girls felt positive and confident, but by the time they entered high school, only 29 percent said they felt "happy" the way they were. The boys also lost some sense of self-worth, but not nearly as much as the girls. Ethnicity was an important factor in this study. African-American girls reported a much higher rate of self-confidence in high school than did white or Latina girls. Two reasons were suggested for this discrepancy. First, African-American girls often have strong female role models at home and in their communities; African-American women are more likely than others to have a full-time job and run a household. Second, many African-American parents specifically teach their children that "there is nothing wrong with them, only with the way the world treats them" (Daley, 1991). According to researcher Carole Gilligan, this survey "makes it impossible to say that what happens to girls is simply a matter of hormones. . . . [It] raises all kinds of issues about cultural contributions, and it raises questions about the role of the schools, both in the drop of self-esteem and in the potential for intervention" (quoted in Daley, 1991).

Harris Clemes and Reynold Bean (1983) describe four conditions necessary for developing and maintaining high self-esteem:

1. A sense of connectedness, of being an important part of a family, class, team, or other group, and of being connected ("in touch") with our bodies.
2. A sense of uniqueness, a feeling that our specialness and differentness are supported and approved by others.
3. A sense of power, the belief that we have the capability to influence others, solve problems, complete tasks, make our own decisions, and satisfy our needs. Children develop a sense of power through sharing duties and responsibilities in the home and having clear limits and rules set for them.
4. A sense of models. Human, philosophical, and operational models (mental constructs and images derived from experience) help us establish meaningful values and goals and clarify our own standards.

A study of 655 adolescents suggested that family support is "crucial for the development and maintenance of self-esteem among high-school-aged adolescents" (Hoelter and Harper, 1987; see also Gecas and Seff, 1991). Parents can foster high self-esteem in their children by (1) having high self-esteem themselves, (2) accepting their children as they are, (3) enforcing clearly defined limits, (4) respecting individuality within the limits that have been set, and (5) responding to their child with sincere thoughts and feelings.

It's also important to single out the child's behavior—not the whole child—for comment (Kutner, 1988). Children (and adults as well) can benefit from specific information about how well they've performed a task. "You did a lousy job" not only makes us feel bad, but it also gives us no useful information about what would constitute a good job.

Misusing the concept of self-esteem with superficial praise is probably the most common way parents have it backfire. Children notice when praise is insincere. If, for instance, Martha refuses to comb her hair, yet we continually tell her how shiny it looks, Martha quickly realizes that we either have very low expectations or do not have a clue about hair care. Instead, parents can accomplish more by giving kids timely, honest, specific feedback. For example, "That's a terrific essay you wrote" is more effective than "You're wonderful!" Each time you treat your child like an intelligent, capable person, you increase your child's self-esteem.

Reflections

How would you rate your self-esteem? What have been the most positive and the most negative factors which have influenced it? How might families, schools, and other institutions increase children's self-esteem?

Psychosexual Development in the Family Context

Conscience is the inner voice which warns us that someone may be looking.

H.L. Mencken (1880–1956)

Within the context of our overall growth, and perhaps central to it, our sexual selves develop. Within the family we learn how we "should" feel about our bodies—whether we should be ashamed, embarrassed, proud, or indifferent. Some families are comfortable with nudity in a variety of situations: swimming, bathing, sunbathing, dressing, or undressing. Others are comfortable with partial nudity from time to time: when sharing the bathroom, changing clothes, and so on. Still others are more modest and carefully guard their privacy. Most researchers and therapists would allow that all these styles can be compatible with the development of sexually well-adjusted children as long as some basic needs are met:

1 The child's body (and nudity) is accepted and respected.
2 The child is not punished or humiliated for seeing the parent naked, going to the toilet, or making love.
3 The child's needs for privacy are respected.

Families also vary in the amount and type of physical contact in which they participate. Some families hug and kiss, give back rubs, sit and lean on each other, and generally maintain a high degree of physical closeness. Some parents extend this closeness into their sleeping habits, allowing their infants and small children in their beds each night. (In many cultures, this is the rule rather than the exception.) Other families limit their contact to hugs and tickles. Variations of this kind are normal. Concerning children's needs for physical contact, we can make the following generalization. First, all children (and adults) need a certain amount of freely given physical affection from those they love. Although there is no prescription for the right amount or form of such expression, its quantity and quality both affect children's emotional well-being and the emotional and sexual health of the adults they will become.

Second, children should be told, in a nonthreatening way, what kind of touching by adults is "good" and what kind is "bad." They need to feel that they are in charge of their own bodies, that parts of their bodies are private property and that no adult has the right to touch them with sexual intent. It is not necessary to frighten a child by going into great detail about the kinds of things that might happen. A better strategy is to instill a sense of self-worth and confidence in children so that they will not allow themselves to be victimized (Pogrebin, 1983). We also should learn to listen to children and to trust them. They need to know that if they are sexually abused, it is not their fault. They need to feel that they can tell about it and still be worthy of love.

Parenthood

Over the last three decades or so, major changes in society have profoundly influenced parental roles. Parents today cannot necessarily look to their own parents as models. Of course, the mothers and fathers of today's children have some things, such as the desire for their children's well-being, in common with mothers and fathers throughout history. But in some areas they must chart a new course.

Women become mothers for many reasons. Most mothers report overall satisfaction with their role.

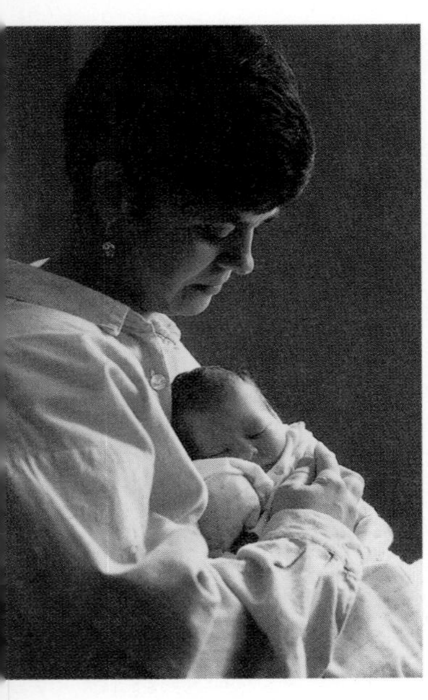

Motherhood

Many women see their destiny as motherhood. Given the choice of becoming mothers or not (made possible through birth control), most women would probably choose to become mothers at some point in their lives, and they would make this choice for very positive reasons (see Cook et al., 1982; Gallup and Newport, 1990; Genevie and Margolies, 1987). But many women make no conscious choice; they become mothers without weighing their decision or considering its effect on their own lives and the lives of their children and partners. The consequences of a nonreflective decision—bitterness, frustration, anger, or depression—may be great. Yet it is also possible that a woman's nonreflective decision will turn out to be "right" and that she will experience unique personal fulfillment as a result.

Although researchers are unable to find any instinctual motivation for having children among humans (which does not necessarily mean that such motivation does not exist), they recognize many social motives impelling women to become mothers. When a woman becomes a mother, she may feel that her identity as an adult is confirmed. Having a child of her own proves her womanliness because from her earliest years, she has been trained to assume the role of mother. She has changed dolls' diapers and pretended to feed them, practicing infant care. She played house while her brother built forts. The stories a girl has heard, the games she has played, the textbooks she has read, the religion she has been taught, the television she has watched—all have socialized her for the mother role. Jessie Bernard, a pioneer in family studies, writes, "An inbred desire is no less potent than an instinctive one. The pain and anguish resulting from deprivation of an acquired desire for children are as real as the pain and anguish resulting from an instinctive one" (Bernard, 1982). Whatever the reason, most women choose motherhood.

It is so still in the house.
There is a calm in the house;
The snowstorm wails out there,
And the dogs are rolled up with snouts under the tail
My little boy is sleeping on the ledge,
On his back he lies, breathing through his open mouth.
His little stomach is bulging round—
Is it strange if I start to cry with joy?
ESKIMO MOTHER'S SONG

Still, many mothers feel ambivalence. Liz Koch (1987) writes:

> We fear we will lose ourselves if we stay with our infants. We resist surrendering even to our newborns for fear of being swallowed up. We hear and accept both the conflicting advice that bonding with our babies is vital, and the opposite undermining message that to be a good mother, we must get away as soon and as often as possible. We hear that if we mother our own babies full time, we will have nothing to offer society, our husbands, ourselves, even our children. We fear isolation, lack

of self-esteem, feelings of entrapment, of emotional and financial dependency. We fear that we will be left behind—empty arms, empty home, empty women, when our children grow away. . . . The reality is that in many ways contemporary America does not honor Mothering.

Koch observes that the "job" of mother is not valued because it is associated with "menial tasks of housekeeper, cook, laundrymaid," and so on. Whether a mother is employed outside the home or works at home "full-time," her role deserves to be valued by society. Koch says of the "special state" of motherhood:

> Being mothers is truly immersing ourselves in a special state, a moment to moment state of being. It is difficult to look at our day and measure success quantitatively. The day is successful when we have shared moments, built special threads of communication, looked deeply into our children's eyes and felt our hearts open. . . . It is important that we see our job as vitally important to our own growth, to our community, to society, and to world peace. Building family ties, helping healthy, loved children grow to maturity is a worthwhile pursuit. . . . The transmission of values is a significant reason to raise our own children. We are there to answer their questions and to show children, through our example, what is truly important to us.

The expectation that mothering comes naturally can be frustrating and guilt-producing among many who are struggling with the new roles and responsibilities that motherhood brings. Add to this socialization, the lack of confidence by both parties in a father's ability to parent, and the inherent ability of women to breastfeed and one can quickly see the enormous pressures that can face new mothers. Because the strains of parenthood seem to fall more heavily upon women, guilt, depression, and conflict with other children are often the result. Patience with oneself, support from family and friends, and a partner who takes equal responsibility in child rearing can go a long way toward alleviating the stresses and bringing greater joy to motherhood (Levy-Schiff, 1994).

Fatherhood

When we speak of *mothering* a child, everyone knows what we mean: nurturing, caring for, feeding, diapering, soothing, loving. Mothers generally "mother" their children almost every day of the year for at least eighteen consecutive years. The meaning of *fathering* is quite different. Fathering a child need take no more than a few minutes if we understand the term in its traditional sense—that is, impregnating the child's mother. Nurturant behavior by a father toward his child has not typically been referred to as *fathering*. (*Mothering* doesn't seem appropriate, either, in this context.) *Parenting*, a relatively new word, more adequately describes the child-tending behaviors of both mothers and fathers (Atkinson and Blackwelder, 1993).

Fathers are increasingly involved in parenting roles—not just playing with their children, but changing their diapers, bathing, dressing, feeding, and comforting them.

As we have seen, the father's traditional roles of provider and protector are instrumental; they satisfy the family's economic and physical needs. The mother's role in the traditional model is expressive; she gives emotional and psychological support to her family. However, the lines between these roles are becoming increasingly blurred because of economic pressures and new societal expectations and desires. From a developmental viewpoint, the father's importance to the family derives not only from his role as a representative of society, connecting his family and his culture, but also from his role as a developer of self-control and

OTHER PLACES OTHER TIMES

The Rights of Biological Parents versus Social Parents: The Nuer of the Sudan

Courtroom custody battles, replayed on the front pages of our newspapers and featured on the evening news, have brought the issue of parental rights into our living rooms. The very definition of *parent* has been brought into question. *Which* parents have a greater claim to a child? Is it the biological parents, who have contributed to the very genetic makeup of the child? Do they have superior claim to the child because they are its "natural" parents? What about the "social" parents, who through adoption, stepparenting, or fosterage have created a loving and enduring family for the child? Are they "unnatural" parents with only secondary rights because they do not share genes with the child? These are not only complex legal and social issues but also agonizing ones that pit "parents" by different definitions against one another.

A different light can be brought to these issues by looking at the solutions to these problems in other societies. For anthropologists, an especially illuminating case is provided by the Nuer of the Sudan in East Africa, a Nilotic (Nile basin) people whose livelihood has traditionally depended on pastoralism (Evans-Pritchard, 1951). The Nuer herd cattle, and the most important relationships in their society—as in other pastoralist societies—are created by the exchange of animals. (See the discussions of the pastoralist Bedouin and Masai societies for comparison, on pages 000 and 000, respectively.) For the Nuer, even marriages are created by cattle. This has, as we will see, important implications for the rights of particular parents. In the arrangement of Nuer marriages, the elder males of the two different families engage in lengthy negotiations. They must decide how many cattle the bride's family will receive from the groom's family, as well as the timing of the cattle "installments." At marriage, the bride will leave her parents' home and move to the groom's family home. The marriage cattle move in precisely the other direction, from the groom's home to the bride's.

The transfer of cattle at marriage establishes a husband's inalienable rights to the children his wife will bear, regardless of who is the biological father. Take, for example, the case of adultery in Nuer society. Adultery is not socially condoned. It causes conflict between spouses and makes a wife's lover liable for expensive compensation payments. If it is established that a man has committed adultery with another man's wife, he becomes obligated to pay the wronged husband cattle as compensation. But there will be no contest over children born of an adulterous union. They be-long unquestionably to the wife's husband. In essence, the Nuer distinguish *genitor,* or biological father, from *pater,* or legal and social father. The cattle paid at marriage establish a husband's rights in the children of his wife, regardless of who fathers them. Because she is a married woman, a woman's children are her husband's, inheriting membership in his clan and rights in his cattle at death. The contribution of the biological father is not unacknowledged, however. For example, upon the marriage of his biological daughter, the biological father receives one

autonomy in his children. Research indicates that although mothers are inclined to view both sons and daughters as "simply children" and to apply similar standards to both sexes, fathers tend to be involved differently with their male and female children. Fathers tend to be more closely involved with their sons than their daughters (Morgan, Lye, and Condran, 1988; Smith and Morgan, 1994). This involvement generally involves sharing activities rather than sharing feelings or confidences (Cancian, 1989; Starrels, 1994). This may place a daughter at a disadvantage because she has less opportunity to develop instrumental attitudes and behaviors. It may also be disadvantageous to a son, as it can limit the development of his own expressive patterns and interests (Gilbert et al., 1982; Starrels, 1994).

Nevertheless, it appears that the family today emphasizes the expressive qualities of all its members, including the father, much more than in the past (Lamb, 1986, 1993). Feminist ideology is credited with being largely responsible for this change (Griswold, 1993). The "new nurturant father," as Michael Lamb refers to him, is able to participate in virtually all parenting practices (except gestation and lactation). Further, such participation is viewed as beneficial to the

Don't be the man you think you should be, be the father you wish you'd had.

LETTY COTTIN POGREBIN

or two head of the many marriage cattle that are received by the bride's legal father. The legal father receives most of the cattle because he is pater.

Several interesting cultural practices further underscore the distinction the Nuer make between genitor and pater, as well as the superior claim of the pater over children in this society. Divorce does occur in Nuer society, and when it does take place, the marriage cattle must be returned to the husband's family. (This can be difficult if the couple have been married several years and many of the cattle have been parceled out to various of the bride's kin.) Once the cattle are returned, there is no particular stigma attached to remarriage, and a divorced woman will usually marry again. If, however, two or more children have been born to a married couple, divorce will probably not occur. Instead, the unhappy couple will separate and the wife will set up house with a lover in a different village. The children born to a wife and her lover will belong not to the genitor-lover but to the separated husband, who is their pater. Although these children may be raised by their mother's lover—who then becomes their foster father—at adolescence they will return to their pater's home to herd his cattle and marry.

Because this is a patrilineal society, it is important that a man have a son to carry on his name and descent line. (See the discussion of patrilineal societies in "Other Places, Other Times" on pages 186–187.) If a man dies before marriage and therefore dies without an heir, a spirit marriage will be arranged to remedy the situation. In this case, a brother will use some of the cattle belonging to his deceased brother in order to marry a wife for him. The brother will father "his brother's" children. The genitor will become foster father to his own biological children. These offspring are considered the legal children of the deceased brother by whose cattle the wife was acquired. In other words, the children of the spirit marriage belong to their deceased pater. If male, they inherit cattle from him; if female, they are married with his cattle. The foster father, who is actually the biological father, has no rights to his biological children by that wife. His own legal children and heirs will be children born to his own wife, who is acquired with his own cattle.

The unquestioned claim of a husband to all children born to his wife is further illustrated in the case where a man dies, leaving behind a widow. She may choose to live with one of her deceased husband's brothers or leave her husband's family completely and set up housekeeping with a lover of her choosing. In either event, the payment of cattle by the deceased husband created rights in her children—even those unborn at the time of his death. All children born to this wife will belong not to their genitor or biological father but to their legal father, who, it is believed in Nuer society, is the true and rightful father.

Family Education Network, now supported solely by advertising, will start charging fees sometime in 1997, Carson says, probably from $10 to $35 a year—about the same price as a magazine subscription. The company also prints a profit-making newsletter on the same subjects, Education Today, which has 150,000 readers.

But no matter what the success rate of this newest generation of sites, they will undoubtedly help lure more new users to the Web. According to one September study, 24 percent of all U.S. households have computers with modems, up from 20 percent just six months earlier. Millions more explore the Web at work, where fast lines eliminate the irritating download waiting time. Now, if only someone would create a site that does the grocery shopping and folds the laundry.

development and well-being of both children and adults. The implicit contradiction between the terms *real man* and *good father* needs to be resolved if boys are to develop into fathers who feel their "manhood enlarged and not depleted by active, caring fatherhood" (Pogrebin, 1982a). Most men today compare themselves favorably with their own fathers in both the quality and quantity of involvement they have with their children.

But even though fathers are clearly more involved than in the past, many are unsure about what is expected of them, and women often can't understand why men don't automatically know what to do. Such stresses between mothers and fathers are common, according to a study by the Families and Work Institute (Levine, 1997; Martin, 1993). Although men are often willing to "help out" their wives, this can pose a problem (as we will discuss at greater length in Chapter 11). Although women generally appreciate any help they can get, they often wish their partners would take on an equal share of the work, rather than simply "helping." Fathers still see their role as breadwinner an important contribution because doing so provides for the family (Cohen, 1993). Though not as

Did you know?

Fathering is not necessarily a responsibility that all undertake with zest. In a study of fathers, 74 percent thought they should share child-rearing equally with mothers, however, in practice only 13 percent actually did (Thomas and Wilcox, 1987).

involved as their wives, most are also still very emotionally involved with their children.

Most mothers have acquired some of their parenting skills by modeling their own mothers, but many fathers have not had the advantage of such a parenting model. Because their own fathers were not highly involved in nurturing roles, today's fathers tend "to focus on being a model to their children to create for them a new set of standards for *who the father is*" (Daly, 1993). The creation of a new role is understandably accompanied by doubt and anxiety. Recently, a number of fathers have written books to help guide their peers through the joys and perils of involved fatherhood.

Did you know?

In a random telephone survey of 420 fathers it was found that almost 70 percent said they would like the opportunity to stay home with their children while their wives worked (Lunsberg, 1993).

Based on his intensive study, Kyle Pruett (1987) has determined that "fatherhood is changing, with fastball speed, especially compared with the languid pace of social evolution." He concludes that fathers must be encouraged to develop the nurturing quality in themselves in order to experience the "unimagined rewards" of parenthood. Psychologist Jerrold Shapiro, who writes about fatherhood, says, "Whether men have been enticed or cajoled, the fact is that we're around our kids a lot more. And when you're around your kids, you get to like it" (Landsberg, 1993; Shapiro, 1993).

Parents' Needs

We learn from experience. A man never wakes up his second baby just to see it smile.

GRACE WILLIAMS

Although some needs of parents are met by their children, parents have other needs as well. Important needs of parents during the child-rearing years are personal developmental needs (such as social contacts, privacy, and outside interests) and the need to maintain marital satisfaction. Yet so much is expected of parents that they often neglect these needs. Parents may feel a deep sense of guilt if their child is not happy or has some defect, an unpleasant personality, or even a runny nose. The burden is especially heavy for mothers because their success is often measured by how perfect their children are. Children have their own independent personalities, however, and many forces affect a child's development and behavior.

Reflections

How should child-rearing tasks be delegated between spouses (or partners)? Are there any particular tasks that you believe either men or women should not do? How are tasks delegated in your household? What was the role of your father in the care and nurturing of you and your siblings?

Accepting our limitations as parents (and as human beings) and accepting our lives as they are (even if they haven't turned out exactly as planned) can help us cope with the many stresses of child rearing in an already stressful world. Contemporary parents need to guard against the "burnout syndrome" of emotional and physical overload. Parents' careers and children's school activities, organized sports, Scouts, and music, art, or dance lessons compete for the parents' energy and rob them of the unstructured (and energizing) time that should be spent with others, with their children, or simply alone.

Other Important Child Socializers

. . . we have attributed the influence to [the mother] as if she was a sorceress. She's a good witch if the child turns out fine and a bad witch if he doesn't.

JEROME KAGAN

With the rise of single-parent families and maternal employment, children are increasingly socialized by influences outside the immediate family. Child-care providers and schools play a major role in children's development, and the television is a veritable living presence in most American homes. In this section we take a closer look at the influences of child care, schools, and television on child socialization.

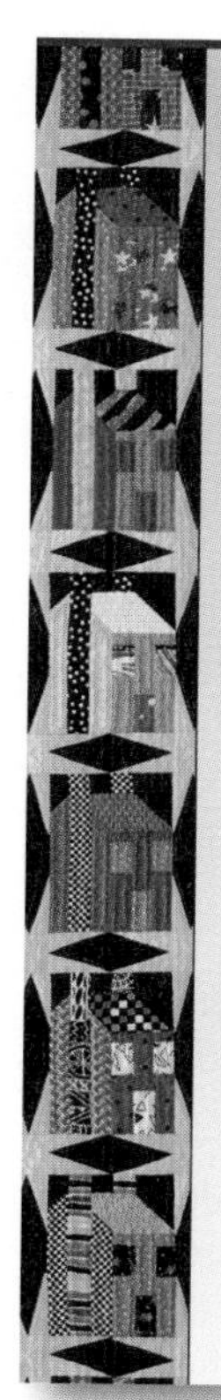

YOU AND YOUR WELL-BEING

Dealing with Parent Burnout

Living up to your own expectations as a parent, trying to maintain a marriage or partnership, observing your children's difficulties, and trying to earn a living can provide a strong basis for parental burnout. In most families, this translates into stress, guilt, and frustration—emotions which all parents feel from time to time. When these feelings are intense, prolonged, or interfere with the ability to effectively parent, it is important to deal with them.

Involved, devoted, and enthusiastic parents are those most likely to burn out because they invest so much of themselves into the process of parenting. Jane Brooks (1996) suggests eight ingredients for reducing burnout:

- Get information about children and parenting skills from books.
- Connect with a significant other for support.
- Become part of a small social group.
- Engage in some goal-oriented activity.
- Gain knowledge of self.
- Have access to money or credit.
- Develop spiritual or intellectual beliefs that provide meaning to life, and
- Maintain self-nourishing activity.

Brooks notes that only one ingredient, the first, bears directly on rearing children. All the rest focus on helping the parent to become more competent and well-rounded.

Research supports the importance of training in parenting techniques and reducing parental distress when managing children's behavior (Forehand, Walley, and Furey 1984). It is important for parents to learn how to develop confidence and self-esteem, reward themselves for doing a good job, define and meet their personal goals, and forgive themselves when they are less than perfect. It is also important for them to recognize when they need outside help—whether from friends, relatives, self-help groups, counselors, or other sources. The ability to ask for assistance is a major component of healthy families, as we will discuss in Chapter 16.

Child Care

I take my children everywhere, but they always find their way back home.

Robert Orhen

Supplementary child care is a crucial issue for today's parents of young children. By the end of the century, an estimated 80 percent of families will consist of two working parents. Many parents must look outside their homes for assistance in child rearing. Thirty-one percent of parents use day care (U.S. Bureau of the Census, 1996). An additional 21 percent use grandparents and other relatives. The percentage for this later group is expected to rise as mothers continue to return to the workplace. (See Chapter 11 for a discussion of the role of day care in maternal employment.) Day-care homes and centers, nursery schools, and preschools can relieve parents of some of their child-rearing tasks and also furnish them with some valuable time of their own.

Most experts agree that the ideal environment for raising a child is in the home with the parents and family. Intimate daily parental care of infants for the first several months to a year is particularly important. Because this ideal is often not possible, the role of day care needs to be considered.

What is the effect of child care on children? The results of research are mixed. In evaluating such data, it is important to keep in mind the family's education, personality, and interests as these factors play a part in which parents choose to or must return to work once a child is born (Crouter and McHale, 1993). Furthermore, a child's personality, age at which the custodial parent reentered the workforce, involvement of the other parent in the home, quantity of time, nature of work, along with the quality of care all contribute to how child care effects the child.

Most adoptive families feel greatly enriched by the experience of adoption, as witnessed by this proud big sister and her new brother.

As more women return to the workforce, a critical issue is the quality of the day care for their children. High-quality day care can facilitate the development of positive social qualities.

You can learn things from children—how much patience you have, for instance.

FRANKLIN P. JONES

When mothers of infants enter the workforce, there is some evidence that these infants are at risk for insecure attachments between the ages of twelve to eighteen months (Brooks, 1996). They are also at risk of being considered noncompliant and aggressive at age three to eight years (Howes, 1990). Other consequences, such as behavior problems, lowered cognitive performance, distractibility, and inability to focus attention have also been noted. These negative effects are not necessarily the consequence of maternal employment. Rather, they may be the result of poor-quality child care. It has been noted that high-quality care, that given by sensitive, responsive, and stimulating caregivers in a safe and low teacher-to-student ratio, can actually facilitate the development of positive social qualities, consideration, and independence (Field, 1991). In school-age and adolescent children, maternal employment is associated with self-confidence and independence, especially for girls whose mothers become role models of competence (Hoffman, 1979).

National concern periodically is focused on day care by revelations of sexual abuse of children by their caregivers. Although these revelations have brought providers of child care under close public scrutiny and have alerted parents to potential dangers, they have also produced a backlash within the child-care profession. Some caregivers are now reluctant to have physical contact with the children; male child-care workers feel especially constrained and may find their jobs at risk (Chaze, 1984). A national study (Finkelhor, 1988), however, found that children have a far greater likelihood of being sexually abused by a father, stepfather, or other relative than by a day-care worker. In 1985, the U.S. Department of Health and Human Services announced day-care guidelines, calling for training of staff in the prevention and detection of child abuse, thorough checks on prospective employees, and allowances of parental visits at any time. Critics believe that the government should go further in establishing standards for day care itself.

What can parents do to ensure quality care for their children? In addition to the obvious requirements of cleanliness, comfort, good food, and a safe environment, parents should be familiar with the state licensure regulations for child care. They should also check references and observe the caregivers with the child.

Although the needs of young children differ from those of older ones, the American Academy of Child and Adolescent Psychiatry (1992) suggests that parents seek day-care services with:

- More adults per child than older children require.
- A lot of individual attention.
- Trained, experienced teachers who enjoy, understand, praise, and enjoy children.
- The same day-care staff for a long period of time.
- Opportunity for creative work, imaginative play, and physical activity.
- Space to move indoors and out.
- Enough teachers and assistants—ideally, at least one for every five (or fewer) children.
- Lots of drawing and coloring materials and toys, as well as equipment such as swings, wagons, jungle gyms, etc.
- Small rather than large groups if possible. (Studies have shown that five children with one caregiver is better than twenty children with four caregivers.)

Although parents may worry about how the child will do, they should show pleasure in helping them succeed. If the child shows persistent terror about leaving home, parents should discuss the problem with the child-care provider and their pediatrician.

It is a wise father that knows his own child.

WILLIAM SHAKESPEARE (1564–1616)

As with a number of critical services in our society, those who most need supplementary child care are those who can least afford it. Child care done properly is a costly business (even though child-care work remains a relatively low-paying, low-status job). The United States is one of the few industrialized nations that does not have a comprehensive national day-care policy. In fact, beginning in 1981, the federal government dramatically cut federal contributions to day care; many state governments followed suit.

Schools

As discussed in Chapter 3, schools play a tremendously important role in children's socialization; yet in most areas there is less and less money allocated for education. Many of today's schools face myriad challenges beyond their basic goals of education and preparation of students for success in adulthood. They must deal with issues of sexuality, violence, drug and alcohol abuse, and racial and ethnic prejudice. Poverty is a serious problem in many school districts. Not only are schools in these areas often run down, poorly supplied, and inadequately staffed, but the students are at a distinct disadvantage as well. For a variety of reasons, success in schools is very highly correlated with socioeconomic status, especially during the first year or so of school (Hamburg, 1992). In addition to these problems, many schools are beleaguered by onslaughts from the religious right; curricula involving fairy tales or mythology becomes "witchcraft," teaching critical thinking skills becomes "disrespect for authority," and family life education is seen as "promoting promiscuity."

Only the educated are free.

EPICTETUS (484–406 B.C.)

Entering kindergarten is an important transition for children. If they have been in day care or preschool, however, the environment may not seem so alien. Nevertheless, it often marks the beginning of profound changes in a child's life. As a child progresses through school, he or she will spend less and less time with parents and come to rely more on peers for companionship and emotional support. Interactions with teachers and performance of the tasks of learning will become the bases for the child's evaluation of his or her self-worth. In elementary school, children generally have some sense of security because they develop a relationship with a single teacher over the course of each year, and they remain with more or less the same peer group throughout their time in their particular school. Moving on to middle school or junior high is often more problematic. The pressures of adolescence are often intensified rather than alleviated by school experiences. Physician and educator David Hamburg (1992) says of adolescents:

> These young people are the victims of stereotypes—they are often described as cranky, rebellious, turbulent, self-indulgent, incapable of learning anything serious or civilized. . . . In fact, adolescents are full of curiosity, energy, imagination, and emerging idealism. They also tend to be uncertain, groping, impressionable.

Yet, Hamburg writes, "as currently constituted, [American] middle-grade schools constitute an arena of casualties—damaging to both students and teachers." Middle school reform is a critical challenge for our educational system.

It is vitally important to the future of our children, our country, and our world that the role of schools as a positive socializing force in children's lives be strengthened. It will take a great deal of dedication, involvement, and political will on the part of today's and tomorrow's parents to meet this challenge.

Television

To be able to fill leisure intelligently is the last product of civilization.

BERTRAND RUSSELL (1872–1970)

In most American households, the television is on at least seven hours each day. Starting around age four, most children watch an average of two to three hours of television each day. No other activity except sleep occupies more time for children and adolescents (Brown, 1990; Singer, 1983). An average viewing week of twenty-one hours includes about four hundred commercials and provides a kaleidoscopic mix of Barbie, blood, bloopers, Big Bad Beetle Borgs, fast cars, fast food, Fred Flintstone, Freddy Krueger, Peg Bundy, and Pepsi. Twenty-one hours of watching television is twenty-one hours *not* spent interacting with family members, playing outside, playing creatively, reading, fantasizing, doing homework, exploring, or even napping. Besides limiting children's time for such pursuits, television has also been implicated in a number of individual, familial, and societal disorders.

. . . TV causes . . . serious and pervasive violence in normal . . . youngsters who watch TV attentively and regularly. What results from this psychic abuse is the impoverishment of personality and the trivialization of life.

HERB KOHL

Because of the constant pressure to buy and the necessity of regulating their children's viewing habits, some parents choose not to have a television. Once the television is gone, it is often not missed. It may be missed more by adults who use it for their own entertainment and as a baby-sitter for their children. Children are peer oriented and have their friends to play with. The use of television as a baby-sitter underscores a dilemma for many parents: They disapprove of what their children see, but they also want to have the time to themselves that may be obtained by letting their children watch television.

We are drowning our youngsters in violence, cynicism and sadism piped into the living room and even the nursery. The grandchildren of kids who used to weep because the Little Match Girl froze to death now feel cheated if she isn't slugged, raped, and thrown into a Bessemer converter.

JENKIN LLOYD JONES

A long-awaited age-based programming rating was put into place in January 1997, not however without much fanfare and controversy. While many were thankful for some guidance about program viewing, others rejected the vagueness and commercial interests upon which the program ratings were based (Miller, 1996). Programs for television are divided into the following six categories:

Ratings for Children's Programs:

TV-K: Appropriate for all children.

TV-K7: Directed to children seven and older. Some material—like mild comedic violence—might upset younger kids.

YOU AND YOUR WELL-BEING

Children and Television: What's the Verdict?

The impact of television depends on the nature of person watching it, the content of program, and the family context in which it is viewed (Andreasen, 1994). A five-year study published by the American Psychological Association (1992) concludes that television is not inherently harmful. If it is used wisely, it can promote learning and social skills. For example, if a family views a program about nature together and discusses what they have learned, the process can not only increase social interaction among family members but also can facilitate language and thinking skills among the children while providing all with a source of relaxation and pleasure.

On the other hand, television can profoundly influence (usually in a negative way) children's views about how the sexes should behave (Richmond-Abbott, 1992). Furthermore, the more television children watch, the more they subscribe to the male-female stereotypes (Lips, 1997). Given the fact that positive portrayals of women are uncommon while women of color, poor working women, and elderly women are rare and often stereotypic, it becomes apparent that the negative impact of unmonitored television can override the potential benefits. Music television videos (MTV) are probably most guilty of presenting negative attitudes toward women (Sommers-Flanagan and Davis, 1993). The American Academy of Child and Adolescent Psychiatry (1992) points out the following troublesome yet prominent themes in some rock music and music videos:

- Advocating and glamorizing abuse of drugs and alcohol.
- Pictures and explicit lyrics presenting suicide as an "alternative" or "solution."
- Graphic violence.
- Preoccupation with the occult; songs about Satanism and human sacrifice, and the apparent enactment of these rituals in concerts.
- Sex which focuses on controlling sadism, masochism, incest, devaluing women, and violence toward women.

By perpetuating social stereotypes such as violence and aggression as a norm, children may become hardened. Children's viewing of violent programs has been shown to cause difficulties with impulse control, delay of gratification, and ability to stay focused on a task (Kobey, 1994).

There are measures that parents can take to minimize the negative effects of television.

- First, it is important that parents regulate the amount of time and content of programs a child watches. This may not be entirely feasible for parents whose demands often exceed the minute-by-minute monitoring this kind of task demands, but having some idea about what the child has observed as well as discussion about the program's content can be helpful in dispelling some of the fears and myths a child may be obtaining from television.
- Second, by modeling the quality and quantity of time parents themselves spend watching television, children can learn by example the values that are placed on television.
- Third, watching programs with children and discussing the content of what they see can turn television viewing into not only a social occasion but one in which stereotypes, violence, and fears are openly discussed and dealt with.

Psychologist Diana Zuckerman suggests (quoted in Oldenberg, 1992):

> TV can be made a shared experience where you learn important lessons. If parents talk to their kids, ask them, "What do you think happened there? . . ." You can teach kids how to be skeptical about the TV commercials they watch, how to tell the difference between reality and fantasy, how to think of alternative solutions to the ones they're seeing on TV.

Neither music nor television is a danger for a child or teenager whose life is happy and healthy. If, however, a child is persistently preoccupied with either and there are changes in behavior such as isolation, night terror, depression, or substance use, then parents should consider psychological intervention.

Ratings for General-Audience Programs:

TV-G: Appropriate for audiences of all ages, containing little or no violence, strong language, or sexual material.

A body at rest tends to remain at rest.

NEWTON'S FIRST LAW OF MOTION

TV-PG: Parental guidance suggested. Shows may contain some violence, sexual material, or coarse language. Most sitcoms would earn this rating.

TV-14: Parents might deem material—including more intense violence and sexual content—inappropriate for kids under fourteen.

TV-M: Not for children under seventeen. Shows may contain explicit sexual material, profane language, or graphic violence.

The verdict on the utility and effectiveness of this system is still out.

Issues of Diverse Families

The diversity of family forms in our country creates a variety of experiences, needs, and possibilities. The concerns of families with disabled members are discussed in Chapter 12. The problems and strengths of single-parent and stepfamilies are discussed in Chapter 15. Here we look at the influences of ethnicity, adoption, and lesbian and gay parenthood on today's families.

Ethnicity and Child Socialization

A person's ethnicity is not necessarily fixed and unchanging. Researchers generally agree that ethnicity has both objective and subjective components. The objective component refers to one's ancestry, cultural heritage, and to varying degrees physical appearance. The subjective component refers to whether one feels he or she is a member of a certain ethnic group, such as African American, Latino, Asian American, Native American, and so on. If both parents are from the same ethnic group, the child will probably identify as a member of that group. But if a child has parents from different ethnic groups, ethnic identification becomes more complex. In such cases, one may identify with both groups, only one, or according to the situation—Latino when with Latino relatives and friends or Anglo when with Anglo friends and relatives, for example. However we choose to identify ourselves, our families are the key to the transmission of ethnic identification.

A child's ethnic background affects how he or she is socialized. Latinos and Asian Americans, for example, stress the authority of the father in the family. In both groups, parents command considerable respect from their children, even when the children become adults. Older siblings, especially brothers, have authority over younger siblings and are expected to set a good example (Becerra, 1988; Tran, 1988; Wong, 1988). Asian Americans tend to discourage aggression in children and for discipline rely on compliance based on the desire for love and respect.

Groups with minority status in the United States may be different from one another in some ways, but they also have much in common. For one thing, they tend to emphasize education as the means for the children to achieve success. Studies show that immigrant children tend to excel as students until they become acculturated and discover that it's not "cool." For another, they are often dual-worker families, so the children may have considerable exposure to television while the parents are away from home. This may be viewed as a mixed blessing: On the one hand, television may help children who need to acquire English language skills; on the other, it can promote fear, violence, and negative stereotypes

If we are to achieve a richer culture, rich in contrasting values, we must recognize the whole gamut of human potentialities, and so weave a less arbitrary social fabric, one in which each diverse human gift will find a fitting place.

MARGARET MEAD (1901–1978)

of women and minority-status groups. Television can discourage creativity and encourage passivity.

Some American children are raised with a strong sense of ethnic identification, whereas others are not. Identification with a particular group gives a sense of pride, security, and belonging. At the same time, it can also give a sense of separateness from the mainstream society. Often, however, that sense of separateness is imposed by the greater society. Discrimination and prejudice shape the lives of many American children. For minority status children, there may be a lack of congruence between society's assumptions and values and their own healthy psychological development (Spencer, 1985). Parents of ethnic minority children may try to prepare their children for the harsh realities of life beyond the family and immediate community (Peterson, 1985). According to Mary Kay Genova (1997), one of the first steps in changing racist attitudes is to admit to racist thinking. People must examine their racist attitudes and the reasons for them. DeGenova (1997) writes about the search for similarity:

> No matter how many differences there may be, beneath the surface there are even more similarities. It is important to try to identify the similarities among various cultures. Stripping away surface differences will uncover a multiplicity of similarities: people's hopes, aspirations, desire to survive, search for love, and need for family—to name just a few. While superficially we may be dissimilar, the essence of being human is very much the same for all of us.

There are no elements so diverse that they cannot be joined in the heart of a man.

Jean Giraudoux (1882–1944)

Suggestions for raising children to be free of prejudice are given in this chapter's Perspective. (The relationship between ethnicity and family strength is discussed in Chapter 16.)

Adoptive Families

It is estimated that a little over 2 percent of the American population is adopted (Samuels, 1990). Adoption is the traditionally acceptable alternative to pregnancy for infertile couples. In recent years, many Americans—married and single, child-free or with children—have been choosing to adopt for other reasons as well. They may have concerns about overpopulation and the number of homeless children in the world. They may wish to provide families for older or disabled children. Although tens of thousands of parents and potential parents are currently waiting to adopt, there is a shortage of available healthy babies (especially healthy white babies) in this country. In 1970, the number of adoptions per year peaked at 175,000. Today the number has declined, owing to more effective birth control and an increase in the number of single mothers who choose to keep their children.

The mother that carries you in her heart is your true mother.

Heather Guffee

The costs of open adoption tend to run between $6,000 and $20,000 with some paying up to $100,000 (Waldman and Caplan, 1994), depending on lawyer and agency fees and the birth mother's expenses (if the state allows these to be covered by the adoptive parents). Adoption laws vary widely from state to state; six states prohibit private adoption, whereas California and Texas have laws that are considered quite supportive of it.

With confidentiality no longer the norm, the trend is toward open adoption in which there is contact between the adoptive family and the birth parents (McRoy, Grotevant, and Ayers-Lopez, 1994). This involvement can be either mediated (i.e., through an adoption agency) or direct, where the birth mother and adoptive family have contact with each other. Many adoption experts agree that some form of open adoption is usually in the best interests of both the child and the birth parents.

PERSPECTIVE

Raising Children to Be Prejudice Free

The world in which our grandparents grew up was simpler than our own in many ways. Towns and neighborhoods stayed the same from year to year, and customs didn't change much either. People passed their prejudices on to their children, who often accepted them without question because they had little or no experience to teach them otherwise. George Simpson and Milton Yinger (1985) explain, "If asked why he dislikes a member of a minority group, [the child] will recite the reasons taught him by his culture, by the adults around him—always assuming, of course, that they are strictly his own reasons and that they are the true reasons."

In today's fast-changing world, we cannot afford to cling complacently to the narrow-mindedness and prejudices of the past. Technological advances in communication and transportation are quickly shrinking the distances between us and other peoples of the world. At home we are becoming more diverse ethnically and linguistically. By the year 2020, it is expected that one out of every three people in the United States will be from what we now call a "minority" group (Sobol, 1990). Timothy Bergen comments, "The future will be in a culturally pluralistic nation and a rapidly shrinking world" (quoted in Fuller, 1989/1990). We must learn to live together with respect and tolerance for our ethnic, cultural, and personal differences. The following guidelines are suggested to help parents foster an appreciation of cultural differences in their children.

One way to instill cultural appreciation is to explore your family's heritage with your children. Look at old photos, and talk to grandparents and other older family members. Find out about the hopes and dreams they had and the obstacles they overcame. Learn more about your cultural background through its art, music, and literature.

Another technique might be to examine your own attitudes and knowledge about other ethnic groups. Expand your understanding of other cultures, and encourage your child's interest in new cultural experiences. Possible resources include libraries, museums, language classes, travel agents, festivals, dances, concerts, music tapes, cookbooks, ethnic markets, and restaurants. Be sure that you have an up-to-date world map or globe for reference. Consider finding pen pals in other countries or helping to support a foreign child in need through an agency (such as Child-Reach or Save the Children) that encourages correspondence

One study of 720 adoptive families and birth mothers found that those participating in open adoptions reported more awareness of the adoption, an increased empathy toward the birth parents and child, a stronger feeling of permanence in the relationship with their child, and fewer fears that the birth mother might try to reclaim the child (Grotevant, McRoy, Elde, and Fravel, 1994).

Several states have now enacted laws allowing adoptees to get copies of their original birth certificates. Other states have set up voluntary registries to assist adoptees and biological parents who wish to meet each other. Although many adoptees do not feel the need to search for their biological parents, others do. Those who succeed report a variety of outcomes. Many find "long-buried family problems that led to their being put up for adoption in the first place"; others need to figure out "what kind of relationship to have with a stranger who is also a parent" (Chira, 1993). Adoptees must also often deal with the bruised feelings of adoptive parents who wonder where they've "gone wrong."

Did you know?

Whereas 80 percent of U.S. babies born out of wedlock were once adopted, rates today have dropped to 2 or 3 percent (*U.S. News and World Report,* 1994).

Some professionals in the adoption field believe that it is not in the child's best interest to tell him or her of the adoption. Psychiatrist Dennis Donovan thinks that children should be given only the information they ask for (only if they ask) and no more. He believes that telling a child he or she was loved by the birth mother but was still given up blocks the child's ability to attach (cited in "At a Glance," 1990). Most adoption authorities, however, disagree with this view.

In addition to open adoptions, foreign adoptions are also increasingly favored. Approximately 15 percent of U.S. adoptions (about 8,000 per year) are of children born outside this country (Bogert, 1994). International adoption between participants. Think about hosting a visiting high school or university student—for a school term or just for dinner.

Whenever possible, take the opportunity to celebrate diversity and contradict stereotypes. Help children develop critical thinking so that they can recognize stereotyping—for example, in TV shows that always portray "the Native American in full headdress, the black man as the villain and Hispanics with lots of children" (Wardle, 1989/1990). Get your children books, posters, videos, puzzles, and dolls that honor cultural and ethnic diversity.

It's important to be open to questions about race and ethnicity so that children don't get the idea that these are taboo subjects. If you don't know an answer, do some research, or talk to your child's teacher or a librarian. If your son or daughter has a conflict with someone from another ethnic group, guide the child to deal with personal issues and not to blame the other person's culture or ethnic group. If you hear children using stereotyped or derogatory terms, intervene. Let them know it's not right to put others down or tease them about their differences. Watch your own language! Call ethnic groups by the names they prefer rather than using terms from common usage or slang that may be offensive.

Select child-care groups and schools with ethnically diverse staffs and children. Support your children's friendships with others of diverse backgrounds. Make sure that schools teach positively about diversity and that events such as holiday programs don't enforce stereotypes. For example, Native Americans should be portrayed as a diverse people with unique ways of life, legends, philosophies, and contributions and not as a group represented merely as "guests" of the Pilgrims on Thanksgiving. In addition, schools need to be sensitive to religious differences when planning holiday programs. With imagination, these events can be conducted so that they include presentations from diverse cultural groups and can be both entertaining and educational for all.

Take action. Don't ignore bigotry when it happens. If someone makes a racist remark, say, "Please don't talk that way around me or my children." When your children see you standing up for your beliefs, they will learn to do the same. Encourage your child to respond to put-downs or aggressive teasing with a stock response such as, "Don't call me that. It's not fair" ("Talking to Children," 1989/1990).

Remember: it all starts with self-esteem. If we let children know that they are valued, they are more likely to find value in others. If they are sensitive to others' feelings, they are less likely to form prejudices.

Although these guidelines have been formulated with ethnic and cultural differences in mind, many of them can be applied to, or adapted for, learning to appreciate and accept people with other kinds of differences—such as those who have disabilities or who are gay or lesbian.

agencies are working to expand their programs in India, the Philippines, Romania, Colombia, and a number of other Asian and Latin American countries. For foreign adoptions, the waiting period is usually around a year, although it may be longer, depending on the political climate of the country in question. The costs vary widely, depending on the number and kind of agencies involved and whether or not the parents travel to pick up their child.

Families with children from other cultures face challenges in addition to those faced by other adoptive families. There is usually little information about the birth parents and no opportunity for continued contact. Older children from foreign countries must deal with the loss of their birth parents and other significant people. They must also adjust to different customs, strange food, and a baffling new language. To combat a sense of rootlessness, many parents of foreign-born children endeavor to give them an understanding of their birth country and its culture. They often participate in supportive networks with other adoptive families.

Adoptive families face unique problems and stresses. They may struggle with the physical and emotional strains of infertility; they endure uncertainty and disappointment as they wait for their child; and they may have spent all their savings and then some in the process. They often face insensitivity or prejudice. For example, an adopted child may be asked, "Who is your *real* mother?" or "Are you their *real* daughter?" Adoptive parents may be congratulated by well-meaning folks: "Oh, you're doing such a good thing!" as though they had made a sacrifice of some kind in choosing to build a family in this way. Even grandparents may reject adopted grandchildren (at least initially), especially if the

Life is full of unfamiliar, even unprecedented relationships, and adopted or not, part of life is adjusting, seeing that diversity is enriching, even when it's not all good.

PHYLLIS THEROUX

adoption is interracial. The idea that adoption is not quite "natural" is all too common in our society.

Adopted children may feel uniquely loved. Suzanne Arms (1990) recounts, "When Joss was six, he was overheard explaining to a friend how special it was to be adopted. Apparently," she adds, "he made a good case for it, because when his friend got home, he told his mother he wanted to be adopted so he could be special too." (See the Resource Center for organizations to contact about adoption.)

Gay and Lesbian Parenting

Reflections

Is the ability to create a child important to your sense of self-fulfillment? If you discovered that you were infertile, what do you think your responses would be? Would adoption be an option for you? Why or why not?

Although numbers are difficult to obtain, researchers believe the number of gay families range from 6 million to 14 million children with at least one gay parent (Kantrowitz, 1996). Most of these parents are, or have been, married. Heterosexual concerns about gay and lesbian parents center around parenting abilities, fear of sexual abuse, and worry that the children will become gay or lesbian.

Lesbian Mothers

Although their sexual orientation is different from that of heterosexual single mothers, lesbian mothers share many similarities with them. One study of lesbian and heterosexual single mothers (Kirkpatrick et al., 1981) found that their lifestyles, child-rearing practices, and general demographic data were strikingly similar. Furthermore, in a summary of studies of gay parenting, psychologist Charlotte Patterson (1992) concluded that the children are just as well adjusted as the offspring of heterosexual parents. As adults, those children who were raised by gay parents are no more likely to be gay than are children of straight parents.

Many lesbians fear losing their children in custody battles. The courts, however, have increasingly taken the position that the sexual orientation of mothers is not an issue if the children are well cared for and well-adjusted (Lamanna and Riedman, 1997). On the other hand, gay men have seldom been awarded custody unless the quality of care by the mother was contested.

When children learn that their mothers are lesbians, according to one study of families with formerly married lesbian mothers, they are initially shocked (Lewis, 1980). At first, they tend to deny any pain or anger. Generally, they have more difficulty in accepting their mother's lesbian identity than their parents' divorce or separation because there is no community of support. It is easier to talk with others about their parents' divorcing than about the mother's being a lesbian. In Karen Lewis's study (1980), children between nine and twelve years of age felt a deep need to keep the mother's lesbian identity a secret. As a result, these children felt a sense of isolation; they were unable to talk over their feelings with their friends. Both children and older adolescents worried that they might become gay or lesbian or that others might think they were gay or lesbian. The boys tended to become angry at the mother's partner, blaming her for the mother's sexual orientation. But whatever the children's response, all expressed a desire to accept their mothers. "Problems between the mother and children," observed Lewis, "seemed secondary to the children's respect for the difficult step she had taken." Therapist Saralie Bisnovitch Pennington (1987) writes, "I cannot emphasize enough that children fare best in homes where the mothers are secure in both their lesbian identity and their parental role, and where they have a strong support system that includes other lesbian mothers."

Some lesbians, especially those in committed relationships, are choosing to create families through artificial insemination. Nearly a third of lesbians have

Families headed by lesbians or gay men generally experience the same joys and pains as those headed by heterosexuals—with one exception: They are likely to face insensitivity or discrimination from society.

become mothers through some form of assisted reproductive technology (Salholz, 1990). The nonbiological (or nonadoptive) parent usually has no legal tie to the child, although recently "second-parent adoptions" by lesbians and gays have been approved in California, Oregon, Alaska, Washington, New York, New Jersey, Massachusetts, and Washington, D.C. (Schenden, 1993). Even so, society may not recognize the "second parent" as a "real" parent because children are expected to have only one real mother and one real father.

Gay Fathers

Gay men and heterosexual men often marry for similar reasons (Bozett, 1987). These may include genuine love or sense of companionship or pressure from family, friends, or a girlfriend. Additionally, a gay man may marry in an attempt to "cure homosexuality" or out of a lack of awareness that he is gay. Most marriages of gay men appear to start off satisfactorily. Over time, however, the sexual relationship declines, and the man begins to participate in clandestine same-sex liaisons. For those whose wives accept their sexual orientation, the marriage may be fairly stable. But for most, it is likely to be characterized by deceit, shame, and increasing anger. If such marriages do end in divorce, however, it is not necessarily a direct result of the homosexuality; it seems that gay men's marriages, "like most other marriages, deteriorate for a variety of reasons" (Bozett, 1987).

Studies of gay fathers indicate that "being gay is compatible with effective parenting" (Bozett, 1987; Harris and Turner, 1985/1986). Furthermore, it appears that gay men who disclose their orientation to their children and who have a stable gay relationship tend to provide better-quality parenting than those who remain married and keep their orientation hidden. One study of gay fathers (Turner, Scadden, and Harris, 1985) concluded that (1) most gay fathers have positive relationships with their children, (2) the father's sexual orientation is relatively unimportant to the relationship, and (3) gay fathers endeavor to create stable home environments for their children.

Although the research on children of gay fathers is fairly limited at this time, it seems to indicate that the children's responses are similar to those of children of lesbian parents. Their primary concerns, especially at first, are about what others may think. They are embarrassed if their fathers are obviously gay, and they worry that others will think they are gay, too. Nevertheless, gay fathers can be effective role models. Research shows that good fathers can help children focus on the quality, rather than the sex, of a partner, encourage children to make choices independent of societal norms, and help decrease prejudice, especially among gays and lesbians (Bozett and Sussman, 1990).

Did you know?

In a recent *Newsweek* poll, 57 percent of the adults surveyed said they think gay people can be as good at parenting as straight people; only 31 percent said they didn't think so (*Newsweek*, November 4, 1996).

Fears about Gay and Lesbian Parenting

Heterosexual fears about the parenting abilities of lesbians and gay men are unwarranted. Fears of the sexual abuse of children by gay parents or their partners are completely unsubstantiated. A review of the literature on the children of gay men and lesbians found that there were virtually no documented cases of sexual abuse by gay parents or their lovers; such exploitation appears to be committed disproportionately by heterosexuals (Cramer, 1986).

Fears about gay parents' rejecting children of the other sex also seem unfounded. Such fears reflect the popular misconception that being gay or lesbian is a rejection of members of the other sex. Many gay and lesbian parents go out of their way to make sure that their children have role models of both sexes (Kantrowitz, 1996). Many also say that they hope their children will develop heterosexual identities in order to be spared the pain of growing up gay in a homophobic society. Research finds children of gay males and lesbians to be well adjusted, and no more likely to be gay as adults (Golemann, 1992; Flaks, Ficher, Masterpasqua, and Joseph, 1995; Kantrowitz, 1996). Ultimately, it is the quality of parenting—not the lifestyle of the parents—that matters most to kids.

Styles and Strategies of Child Rearing

A parent's approach to training, teaching, nurturing, and helping a child will vary according to cultural influences, the parent's personality, the parent's basic attitude toward children and child rearing, and the role model that the parent presents to the child.

Authoritarian, Permissive, and Authoritative Parents

The three basic styles of child rearing may be termed authoritarian, permissive, and authoritative (Baumrind, 1971, 1983).

Parents who practice **authoritarian child rearing** typically require absolute obedience. The parents' maintaining control is of first importance. "Because I said so" is a typical response to a child's questioning of parental authority, and physical force may be used to ensure obedience. Working-class families tend to be more authoritarian than middle-class families. Diana Baumrind found that children of authoritarian parents tend to be less cheerful than other children and correspondingly more moody, passively hostile, and vulnerable to stress.

Permissive child rearing is a more popular style in middle-class families than in working-class families. The child's freedom of expression and autonomy are valued. Permissive parents rely on reasoning and explanations. Yet permissive parents may find themselves resorting to manipulation and justification. The child is free from external restraints but not from internal ones. The child is supposedly free because he or she conforms "willingly," but such freedom is not authentic. This form of socialization creates a bind: "Do what we tell you to do because you want to do it."

Baumrind (1983) found that although children of permissive parents are generally cheerful, they exhibit low levels of self-reliance and self-control.

Parents who favor **authoritative child rearing** rely on positive reinforcement and infrequent use of punishment. They direct the child in a manner that shows awareness of his or her feelings and capabilities. Parents encourage the development of the child's autonomy within reasonable limits and foster an atmosphere of give-and-take in parent-child communication. Parental support is a crucial ingredient in child socialization. It is positively related to cognitive development, self-control, self-esteem, moral behavior, conformity to adult standards, and academic achievement (Gecas and Seff, 1991). Control is exercised in conjunction with support by authoritative parents. Children raised by authoritative parents tend to approach novel or stressful situations with curiosity and show high levels of self-reliance, self-control, cheerfulness, and friendliness (Baumrind, 1983). The child-rearing strategies discussed in the following sections may be used by parents who take an authoritative approach to child rearing.

Because it's good for you. Reason given to make child eat food it does not want.

MISS MANNERS

Do you want to be sure that your children are telling the truth? Then limit your questions to the names of their schools.

BILL COSBY

Reflections

In your family, what child-rearing attitudes (authoritarian, permissive, or authoritative) predominated? Do you think these attitudes influenced your own development? If so, how? Which might (or do) you find useful in raising your own child?

How to Raise Your Child: A Few Words on Experts

About 150 years ago, Americans began turning to books to learn how to act and live rather than turning to one another. They began to lose confidence in their own abilities to make appropriate judgments. The vacuum that formed when traditional ways broke down under the impact of industrialization was filled by the so-called expert. The old values and ways had been handed down from parents to child in an unending cycle; men and women had learned how to be mothers and fathers from their own parents. But with increasing mobility, the continuity of generations ceased. A woman's mother was often not physically present to help her with her first child. New mothers were not able to turn to their more experienced kin for help. Instead, they enlisted the aid of new authorities—the experts who, through education and training, supposedly knew what to do. If your baby was colicky, for example, the experts recommended a drop of laudanum in his or her bottle. (The laudanum would put the baby to sleep, but it would also make him or her a heroin addict by the end of a year.)

Contemporary parents may still follow experts' advice even if it conflicts with their own beliefs. Yet if an expert's advice counters their own understanding, parents should carefully examine that advice, as well as their own beliefs. All parents should take an expert's advice with at least a grain of salt. It is the parents' responsibility to raise their children, not the expert's.

The perniciousness of so much of the advice from experts that pervades the media is that it undermines the confidence of parents in their own abilities and values, overemphasizes the significance of specific child-rearing techniques, and grossly misrepresents the contribution the expert in psychiatry or education can make to the conduct of ordinary family life.

RITA KRAMER

Contemporary Child-Rearing Strategies

Parents are the bones on which children sharpen their teeth.

Peter Ustinov

One of the most challenging aspects of child rearing is knowing how to change, stop, encourage, or otherwise influence children's behavior. We can request, reason, explain, command, cajole, compromise, yell and scream, or threaten with physical punishment or the suspension of privileges; or we can just get down on our knees and beg. Some of these approaches may be appropriate at certain times; others clearly are never appropriate. Some may prove effective some of the time, some may never work very well, and no technique will work every time.

The techniques of child rearing currently taught or endorsed by educators, psychologists, and others involved with child development are included in programs such as Parent Effectiveness, Assertive Discipline, Positive Discipline, and numerous others. Although these approaches differ somewhat in their emphasis, they share most of the tenets that follow. (Also see Chapter 16 for a discussion of the traits of psychologically healthy families.)

Respect

The secret of dealing successfully with a child is not to be its parent.

Mell Lazarus

Mutual respect between children and parents must be fostered in order for growth and change to occur. One important way to teach respect is by modeling—treating the child and others respectfully. Child psychologist Rudolph Dreikurs stresses the importance of treating children with kindness and firmness simultaneously. Counselor Jane Nelsen (1987) writes, "Kindness is important in order to show respect for the child. Firmness is important in order to show respect for ourselves and the situation."

Consistency and Clarity

Consistency is crucial in child rearing. Without it, children become hopelessly confused and parents become hopelessly frustrated. Patience and teamwork (a united front) on the parents' part help ensure consistency. Because consistency means following through with what we say, parents should beware of making promises or threats they won't be able to keep. Clarity is important for the same reason. A child needs to know the rules and the consequences for breaking them. This eliminates the possibility of the child's being unjustly disciplined or wiggling out through loopholes. ("But, Mom, I didn't know you meant not to walk on *this* clean carpet with my muddy shoes!")

Logical Consequences

One of the most effective ways to learn is by experiencing the logical consequences of our actions. Some of these consequences occur naturally—we forget our umbrella, and then we get wet. Sometimes parents need to devise consequences that are appropriate to their child's misbehavior. Dreikers and Soltz (1964) distinguish between logical consequences and punishment. The "Three R's" of logical consequences dictate that the solution must be Related to the problem behavior, Respectful (no humiliation), and Reasonable (designed to teach, not to induce suffering).

Open Communication

The lines of communication between parents and children must be kept open. Numerous techniques exist for fostering communication. Among these are active listening and the use of "I" messages, both important components of Thomas Gordon's (1978) Parent Effectiveness program. In *active listening*, the parent ver-

bally feeds back the child's communications in order to understand the child and
338 339
tant because they impart facts without placing blame and are less likely to promote rebellion in children than are "you" messages. (Communication techniques are discussed in greater detail in Chapter 5.)

The chief tools of proper child rearing are example and nagging.

MISS MANNERS

Family meetings are another important way in which families can communicate. Regular weekly meetings provide an opportunity for being together and a forum for airing gripes, solving problems, and planning activities. Decisions are best reached by consensus rather than majority vote, as majority rule can lead to a "tyranny of the majority" in which the minority is consistently oppressed. If consensus can't be reached, the problem can be put on the next meeting's agenda, allowing time for family members to come up with alternative solutions (Nelsen, 1987).

No Physical Punishment

The American Psychological Association notes that "physical violence imprinted at an early age, and the modeling of violent behavior by punishing adults, induces habitual violence in children" (cited in Haferd, 1986). The American Medical Association also opposes physical punishment of children. Although such punishment may "work" in the short run by stopping undesirable behavior, its long-range results—anger, resentment, fear, and hatred—are appalling (Dodson, 1987). Besides, it often makes parents feel confused, miserable, and degraded right along with their kids.

Behavior Modification

More effective types of discipline use some form of behavior modification. Rewards (hugs, stickers, or special activities) are given for good behavior, and privileges are taken away when misbehavior is involved. Good behavior can be kept track of on a simple chart listing one or several of the desired behaviors. Undesirable behavior may be met with the revocation of TV privileges or the curtailment of other activities. *Time-outs*—sending the child to his or her room or to a "boring" place for a short time or until the misbehavior stops—are useful for particularly disruptive behaviors. They also give the parent an opportunity to cool off (Dodson, 1987; see also Canter and Canter, 1985).

Although no child-rearing technique is guaranteed to be 100 percent successful all the time, it is important for families to keep seeking ways to improve their communication and satisfaction. It is also important for parents to develop and maintain confidence in their own parenting skills, their common sense, and especially their love for their children.

Child rearing is the only task in the world where your goal is to make your own job obsolete.

MISS MANNERS

Summary

- Parenthood may now be considered a matter of choice because of effective methods of birth control. *Child-free marriage* and deferred parenthood are alternatives that are growing in popularity.
- Parental roles are acquired virtually overnight and can create considerable stress. The main sources of stress for women involve traditional gender roles and multiple role demands (parent, spouse, and provider). Other sources of stress for mothers and fathers are associated with work; not having enough money; worries about infant care and health; and interactions with spouse, family, and friends.
- Theories of child socialization include Sigmund Freud's *psychoanalytic theory*, in which the personality is seen to be composed of the pleasure-seeking *id*, the controlling *superego*, and the rational *ego;* Erik Erikson's *psychosocial theory*, in which society and family are seen as influencing the individual during specific life cycle stages; *behaviorism*, propounded by psychologists such as B. F. Skinner, who developed the concept of *reinforcement* and stressed the importance of observing actual behaviors; and *social learning theory*, which emphasizes the role of thinking in human development. Jean Piaget's *cognitive developmental theory* emphasizes the importance of specific stages of mental development. Children's abilities are developed through the processes of *assimilation* and *accommodation*. In the *developmental systems approach*, family members are viewed as being interdependent in their growth, which continues throughout their lives. Birth order, spacing between births, and sibling interactions are also important. Each of us also has a sense of self that resists biological and environmental pressures.
- Important factors in children's development include *attachment* (the bond between the infant and the primary caregiver or caregivers) and temperament (the child's basic personality and way of relating to the world).
- Children have a number of basic physical and psychological needs, including adequate prenatal care; formation of close attachments; protection from illness and abuse; and respect, education, and support from family, friends, and community. High self-esteem is essential for growth in relationships, creativity, and productivity. Adolescent girls, especially nonblacks, are likely to be low in self-esteem. Parents can foster high self-esteem in their children by encouraging the development of a sense of connectedness, uniqueness, power, and by providing models.
- Psychosexual development begins in infancy. Infants and children learn from their parents how they should feel about themselves as sexual beings.
- Many women find considerable satisfaction and fulfillment in motherhood. Although there is no concrete evidence of a biological maternal drive, it is clear that socialization for motherhood does exist. The role of the father in his children's development is currently being reexamined. The traditional instrumental roles are being supplemented, and perhaps supplanted, by expressive ones.
- Supplementary child care outside the home is a necessity for many families. The development and maintenance of quality day-care programs should be a national priority. Schools are also important agents of child socialization. Today's schools face challenges beyond those of education; they must deal with issues of sexuality, drugs, violence, and racial and ethnic prejudice.
- Television has been implicated in a number of individual, familial, and societal disorders. Violence on television has been demonstrated to cause aggressive behavior in children. Families are often portrayed in a stereotypical or romanticized manner.
- Ethnicity profoundly influences the way children are socialized. Identification with a group gives a sense of pride and belonging. Minority-status parents may try to give their children special skills for dealing with prejudice and discrimination.
- Adoptive families face unique problems and stresses; nevertheless, most report feeling greatly enriched. Issues faced by adoptive families include choosing open or closed adoption, dealing with feelings about the biological parents, and dealing with insensitivity and prejudice from society.
- Most gay and lesbian parents are, or have been, married. Lesbian mothers share many similarities with heterosexual mothers. Studies indicate that children of both lesbians and gay men fare best when the parents are secure in their sexual orientation. Although the children of gay and lesbian parents may have difficulty accepting their parents' sexual orientation at first, they generally maintain close relationships with their parents, are well-

adjusted, and develop the same sexual orientations and gender roles as children of heterosexuals.

- Basic attitudes toward child rearing can be classified as *authoritarian*, *permissive*, and *authoritative*. Today's parents often rely on expert advice. It needs to be tempered by parents' confidence in their own parenting abilities and in their children's strength and resourcefulness. Contemporary strategies for child rearing include the elements of mutual respect, consistency and clarity, logical consequences, open communication, and behavior modification in place of physical punishment.

Key Terms

accommodation 334
assimilation 334
attachment 336
authoritarian child rearing 358
authoritative child rearing 359
behaviorism 333
child-free marriage 328
cognitive developmental theory 334
developmental systems approach 335
ego 333
id 333
neurosis 333
operant conditioning 333
permissive child rearing 359
psychoanalytic theory 333
psychosocial theory 333
reinforcement 333
social learning theory 334
superego 333

Suggested Readings

Biller, H.B. *Fathers and Families: Paternal Factors in Child Development.* Westport, CT: Auburn House, 1993. An insightful book from an author who has written and studied extensively about the father's role in child development.

Brazelton, T. Berry. *Touchpoints—The Essential Reference: Your Child's Emotional and Behavioral Development.* Needam Heights, MA: Addison Wesley, 1994. A valuable book by a well-known child specialist on the developmental and behavioral milestones of children.

Coles, Robert. *The Spiritual Life of Children.* Boston: Houghton Mifflin, 1990. Interviews with children throughout the world by a leading educator and child psychologist. The text reveals their concern about ultimate truths and their search for meaning in life.

Daly, Kerry J. *Families and Time.* Thousand Oaks, CA: Sage, 1996. An examination of time as a pervasive influence in the changing experimental world of families.

Dinkmeyer, Don, and Gary D. McKay. *Parenting Teenagers*, 2nd ed. Circle Pines, MN: American Guidance Service, 1990. One of a series of books from two psychologists that takes a useful and encouraging look at parenting teens.

Gottfried, A. E. and Allen W. Gottfried. *Redefining Families: Implications for Children's Development.* New York: Plenum, 1994. A collection of articles on the diversity of families in the United States.

Hawkins, Alan J., and David C. Dollahite, eds. *Generative Fathering; Beyond Deficient Perspectives.* Thousand Oaks, CA: Sage, 1996. A look at the work fathers do for their children in terms of caring for and contributing to the life of the next generation.

Hopson, Darlene, and Derek Hopson. *Different and Wonderful: Raising Black Children in a Race-Conscious Society.* New York: Prentice-Hall, 1990. A child-rearing guide for African-American families that addresses the issues of ethnicity and racism.

Hutter, Mark, ed. *The Family Experience*, 2nd ed. Boston: Allyn and Bacon, 1997. A fascinating and well-rounded look at the family, including women, the poor, working people, and racial and ethnic minorities.

Konner, Melvin. *Childhood.* Boston: Little Brown, 1991. An exploration of several aspects of childhood, written to accompany the PBS series *Childhood*, but it stands well on its own.

Levine, James. *Working Fathers: Strategies for Balancing Work and Family.* Reading, MA: Addison Wesley Longman, 1997. An important book for involved fathers by the director of the Fatherhood Project of the Families and Work Institute.

Marzollo, Jean. *Fathers and Babies.* New York: HarperCollins, 1993. A hands-on guide for dads of infants from birth through the first eighteen months.

McMahon, Martha. *Engineering Motherhood: Identity and Self-Transformation in Women's Lives.* New York: Guiford, 1996. An insightful book which examines the process by which women become mothers and the meaning that motherhood brings to them.

Pollack, Jill. *Lesbian and Gay Families: Redefining Parenting in America.* Danbury, CT: Frank Watts Inc., 1995. A look at parenting in the '90s from a personal, social, and psychological viewpoint.

Shapiro, Jerrold Lee. *The Measure of a Man: Becoming the Father You Wish Your Father Had Been.* New York: Delacorte Press, 1993. Practical, empathetic advice from one dad to others.

CHAPTER 11

Marriage, Work, and Economics

P R E V I E W

To gain a sense of what you already know about the material covered in this chapter, answer "True" or "False" to the statements below.

1. In contrast to single-worker couples, dual-career couples tend to divide household work almost evenly. True or false?
2. About 500,000 American men are full-time homemakers with no outside employment. True or false?
3. It is generally agreed by economists that welfare encourages poverty. True or false?
4. Families make up about 25 percent of the homeless. True or false?
5. Women in the United States currently make ninety cents for every dollar that men earn. True or false?
6. Family economic well-being is a national priority. True or false?
7. The majority of female welfare recipients are on welfare as a result of a change in their marital or family status. True or false?
8. The majority of families are dual-earner families. True or false?
9. Women tend to interrupt their work careers for family reasons over thirty times as often as men. True or false?
10. Married women tend to earn more and have higher-status jobs than single women. True or false?

Answers

1 False, 2 True, 3 False, 4 True, 5 False, 6 False, 7 True, 8 True, 9 True, 10 False

OUTLINE

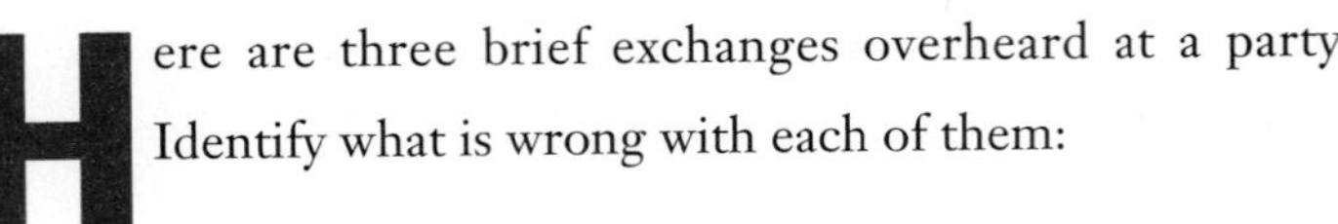

Here are three brief exchanges overheard at a party. Identify what is wrong with each of them:

- *Exchange number 1:* "Do you work?" the man asks. "No, I'm a mother," the woman replies.
- *Exchange number 2:* "What do you do?" the first woman inquires politely as she is introduced to another woman. "Nothing. I'm a housewife," the second responds. "Oh, that's nice," the first replies, losing interest.
- *Exchange number 3:* "What do you do?" the man asks. "I'm a doctor," the woman responds as she picks up her child, who is impatiently tugging on her. "And I'm an architect," her husband says while nursing their second child with a bottle.

In the first exchange, the woman ignores that as a mother, she works. In the second exchange, both women ignore the fact that as a homemaker, the second woman works. They also devalue such unpaid work in comparison with paid work. In the third exchange, the woman identifies herself as a physician without acknowledging that she is also a mother. Her husband makes the same mistake as he identifies himself as an architect without also noting that he is a father participating in the care of his infant. As husband and wife, father and mother, both the architect and the physician are unpaid family workers making important—but generally unrecognized—contributions to the family's economy.

These examples point to a curious fact: When we think of work, we tend to recognize only paid work. Family work, such as caregiving activities or household duties, is not regarded as "real" work. Because it is unpaid, family work is ignored as being somehow inferior to paid work, regardless of how difficult, time consuming, creative, rewarding, and important it is for our lives and future as human beings. This is not surprising, for in the United States, employment takes precedence over family.

The role of work in families requires many of us to rethink the meaning of *family*. We ordinarily think of families in terms of relationships and feelings—the family as an emotional unit. But families are also economic units bound together by emotional ties (Ross, Mirowsky, and Goldsteen, 1991). Paid work and unpaid family work, as well as the economy itself, profoundly affect the way we live as families. Our most intimate relationships vary according to how husbands and wives participate in paid work and family responsibilities (Voydanoff, 1987).

Money is a good thing to have. It frees you from doing things you dislike. Since I dislike doing nearly everything, money is handy.

GROUCHO MARX (1895–1977)

Family work is the work we perform in the home without pay: It includes housekeeping; household maintenance; and caring for children, the ill, and the aged. Women do the overwhelming majority of family work. Paid work is the work we do for salary or wages. Traditionally, men have been responsible for paid work, but currently, the vast majority of women—whether single or married, with children or without—are employed. Women either support themselves or make primary or significant contributions to family income.

Our paid work helps shape the quality of family life: It affects time, roles, incomes, spending, leisure, and even individual identities. Within our families, the time we have for one another, for fun, for our children, and even for sex is the time that is not taken up by paid work. The main characteristic of our paid work in relation to the family is its inflexibility. Work regulates the family, and for most families, as in the past, a woman's work molds itself to her family, whereas a man's family molds itself to his work (Ross, Mirowsky, and Goldsteen, 1991). We must constantly balance work roles and family roles.

Workplace/Family Linkages

Outside of sleeping, probably the single activity to which the majority of employed men and women devote the most time is their jobs, and we are working more and more. Compared to 1970, the average worker today puts in an additional 164 hours of work a year. This is the equivalent of an extra month a year (Schor, 1991). As we work more and more, we have less time for our families and leisure. Most of us know from our own experiences that our work or studies affect our personal relationships. We carry our workplace stress home with us. Work-family conflict, especially relating to work schedules and demands and to children at home, is often a painful source of stress (Menaghan and Parcel, 1991). To date, however, most research has been devoted to descriptions of home life or

Work spillover and role strain affect many employed women.

work life, not to how the two interact with each other. Only recently have scholars begun to look at how our jobs affect our family life, causing moodiness, fatigue, irritability, and so on (Crouter and Manke, 1994; Small and Riley, 1990).

Work Spillover

Work is the refuge of people who have nothing better to do.

OSCAR WILDE (1854–1900)

Common sense (as well as our own fatigue) suggests that **work spillover**—the effect of work on other aspects of our lives—is important in family life. Work spillover affects individuals and families by absorbing their time and energy and impinging on their psychological states. It links our home lives to our workplace (Small and Riley, 1990). Work is as much a part of marriage as love is. What happens at work—frustration or worry, a rude customer, an unreasonable boss—affects our moods, making us irritable or depressed. We take these moods home with us, affecting the emotional quality of our relationships. Workplace stress often causes us to focus on our problems at work rather than on our families. It can lead to fatigue, depression, stomach ailments, and increased drug and alcohol use (Crouter and Manke, 1994).

If you don't want to work you have to work to earn enough money so that you won't have to work.

OGDEN NASH (1902–1971)

Work spillover especially affects women. Whereas for men, excessive work time is the major cause of conflict between work and family, fatigue and irritability are the major causes of conflict for women.

As a result of work spillover, an employed woman's leisure is rarely spontaneous: It is fitted in or planned around her household work and paid work (Kinnunen, Gerris, and Vermulst, 1996). In addition, the stress a husband experiences at his work can diminish his wife's emotional health (Dooley and Catalano, 1991).

The family can help alleviate some workplace stress. The way family members respond helps shape the way in which the stressed worker expresses his or her frustration, anger, or fatigue. If, for example, your partner comes home feeling frustrated from a bad day at work, you can permit him or her to withdraw temporarily from family obligations, listen empathetically, and help him or her think of solutions or strategies for dealing with the work stress. Responding to your partner's stress with anger or a list of your own complaints is likely to add to the stress. If it's your partner's turn to make dinner, make dinner yourself, or order Chinese takeout.

Role Strain: Role Overload and Interrole Conflict

For families in which both partners are employed, work-related problems are more severe for parents than for nonparents. Being a working parent means performing three roles simultaneously: worker, parent, and spouse (Voydanoff and Donnelly, 1989). Because the time, energy, and commitment needed to carry out these roles are interdependent, individuals may experience difficulties such as depression, fatigue, and irritability (Gruelzow et al., 1991; Menaghan and Parcel, 1991; Voydanoff and Donnelly, 1989). Such difficulties are referred to as **role strain.** Role strain is the most pressing problem for working families (Hansen, 1991).

If you make money your god, it will plague you like the devil.

HENRY FIELDING (1707–1754)

Two types of role strain are especially likely to occur among working parents: role overload and interrole conflict (interference). **Role overload** occurs when the total prescribed activities for one or more roles are greater than an individual can comfortably or adequately handle. In role overload—what might be called the "too-many-hats" syndrome—a person can easily feel overwhelmed with the multiple responsibilities of parent/spouse/worker. He or she must care for children, be intimate with a spouse, work twelve hours a day, shop, clean house, take children to school, and so on. **Interrole conflict** exists when responsibilities from the parent/spouse/worker roles conflict. In interrole conflict, the expectations of the parent/spouse/worker roles may be contradictory or may require an individual to do two things at the same time. In this "too-many-balls-to-juggle" syndrome, for example, a parent may be expected to be in two places at once, such as at the job and at home caring for a sick child. Individuals high in self-esteem seem to feel less interrole conflict than those with low self-esteem (Long and Martinez, 1994). Women with high self-esteem, for example, accept lower housekeeping standards as realistic adjustments to their multiple roles rather than as signs of inadequacy.

Men experience role strain when trying to balance their family and work roles. Because the workplace expects men to give priority to their jobs over their families, it is not easy for men to be as involved in their families as they may like. A recent study examined role strain among men (O'Neil and Greenberger, 1994; also see Marks, 1994, and Greenberger, 1994). It found that men with the least role strain fell into two groups. One group consisted of men who were highly committed to both work and family roles. They were determined to succeed at both. The other group consisted of men who put their family commitments above their job commitments. They were willing to work at less demanding or more flexible jobs, spend less time at work, and put their family needs first. In both instances, however, the men received strong encouragement and support from their spouses.

It is not enough to be busy . . . the question is: what are we busy about?

Henry David Thoreau (1817–1862)

Figure 11.1
Work-Family Conflicts Reported by Men and Women

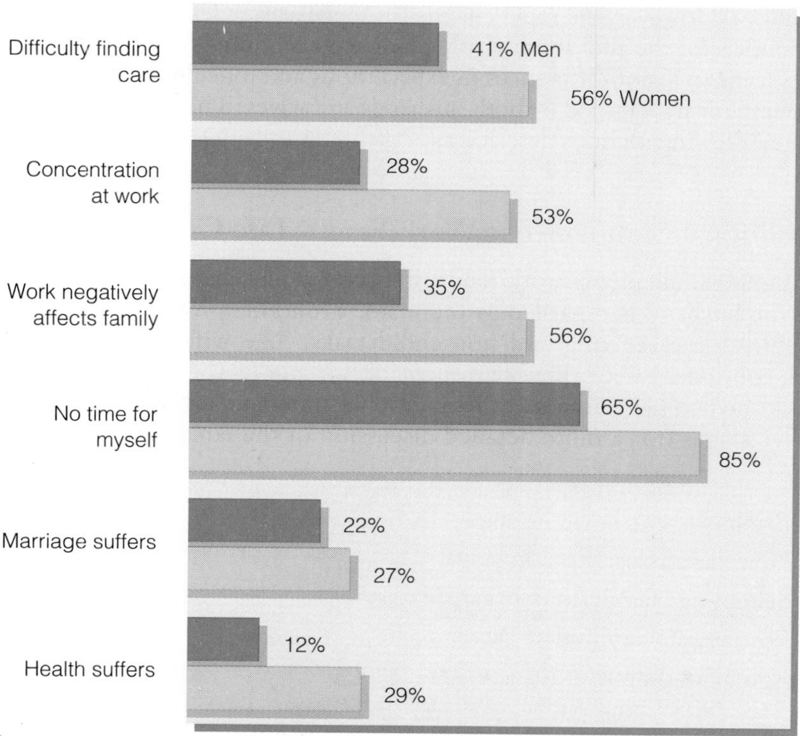

SOURCE: Rodgers & Associates. Surveys conducted in 1985 and 1988.

Reflections

Much of the workplace-family linkage concept can be applied to the college environment. If you think of your student role as a work role and the college as the workplace, what types of work spillover do you experience in your personal or family life? If you are a homemaker or are employed (or both), what kinds of role strain do you experience?

Married women employed full-time often prefer working fewer hours as a means of reducing role strain (Warren and Johnson, 1995). Not surprisingly, because they have less role strain, single women (including those who are divorced) are often more advanced in their careers than married women (Houseknecht, Vaughan, and Statham, 1987). They are more likely to be employed full-time and have higher occupational status and incomes. They are also more highly represented in the professions and hold higher academic positions.

Employment and the Family Life Cycle

Throughout the life cycle, families try to balance work needs and family needs, which change as we go through different stages of the family life cycle. Studies indicate that women spend considerably more total time in work and family roles than do men. This is because, in addition to out-of-home employment, they often spend more time with work in the home. Women have higher levels of role overload in every stage of the life cycle (Higgins, Duxbury, and Lee, 1994).

Much marital conflict in the contemporary family grows out of inequities between male and female work and family role experiences and expectations (Rogers, 1996). What it means to be male and to be female is not only influenced by biology, but even more by the way in which families define those roles in their work and home life. Role taking and role making are negotiated and renegotiated all through the family life cycle and are influenced by changing patterns in society (Zvonkovic, Greaves, Schmiege, and Hall, 1996).

Families may follow several patterns in combining family and employment responsibilities. However, the rapid changes in society often leave couples without clear models for the allocation of roles and responsibilities. Such uncertainty sometimes leads to major differences in expectations and to feelings of exploitation and being misunderstood by both husbands and wives. The three main patterns in the following sections describe approaches in today's families.

The Traditional-Simultaneous Work/Family Life Cycle Model

Work expands so as to fill the time available for its completion.

NORTHCOTE PARKINSON

The traditional-simultaneous work/family life cycle model characterizes families in which the husband is regarded as the major economic provider with little responsibility for caregiving and household tasks. The wife, even if she is employed, continues to be primarily responsible for family work.

Work and family life intersect the traditional-simultaneous work/family life cycle in five stages (for a more detailed discussion of the family life cycle, see Chapter 9):

1. Establishment/novitiate stage.
2. New parents/early career stage.
3. School-age family/middle career stage.
4. Postparental family/late career stage.
5. Aging family/post exit stage.

Stages 1 through 4, in which paid work affects family life most significantly, are briefly described below.

Establishment/Novitiate Stage

In the establishment/novitiate stage, men and women begin to establish their families and begin their paid work. Role sharing may be especially high during this stage for both men and women traditionally. Men are not expected to marry, for example, until they are able to support a family. (Women do not have this work-related constraint on when to marry.) Women may experience role conflict if they believe that they must subordinate their work or career needs to their husbands' advancement needs. During this stage and the next, families are the most likely to be poor, as responsibilities and expenses tend to outrun income (Voydanoff, 1991).

New Parents/Early Career Stage

In the new parents/early career stage, role overload becomes an acute problem. As men and women are developing their career or work skills, they tend to emphasize their work roles: this brings with it long hours, fatigue, the need for additional education, work-related socializing, networking, and work brought home. At the same time, as they have small children, their roles as parents become especially demanding (as well as rewarding): There are babies to diaper, noses to wipe, grandparents to visit, lullabies to sing, doctors to see, baby-sitters to find, and so on. Role strain is greatest at this stage for employed women. Interrole conflict comes to the foreground for women as they try to fulfill both child-rearing and work roles (Higgins, Duxbury, and Lee, 1994).

A number of factors determine whether women leave or remain in the workforce at this stage (Menaghan and Parcel, 1991; Spitze, 1991). These factors include (1) how much their income is needed, (2) the meaning and importance they attach to mother roles versus work roles, (3) whether they accept traditional or nontraditional gender roles, and (4) the degree of work spillover they experience.

In dual-earner families interrole conflict is often high as parents try to balance family and work obligations.

School-Age Family/Middle Career Stage

Interrole conflict is high as work and parent roles conflict with each other in the school-age family/middle career stage: parent/teacher meetings versus business meetings, sick children versus work obligations, Little League versus overtime. Parental roles are more likely than spousal roles to contribute to role overload and interrole conflict. In part, this may be because some harried husbands and wives neglect or "forget" their spousal roles altogether when attempting to cope with their parenting and work obligations. Interrole conflict, however, begins to diminish for women after their children reach thirteen years of age. After that point, male and female levels of interrole conflict are about the same (Higgins, Duxbury, and Lee, 1994).

Postparental Family/Late Career Stage

Many parents experience a paradox as they enter the postparental family/late career stage. Although they have more time to develop their parent-child relationships because they are now firmly established in their employment, they find that their children have grown up and gone. At this time, husbands and wives reestablish their spousal relationships, which may have been languishing in neglect. Too often they have difficulty in renewing marital ties because they have grown apart emotionally and have developed separate interests. In addition, many middle-aged women who also planned to establish careers after their children left home find themselves responsible for the care of aged parents.

Alternative Work/Family Life Cycle Approaches

The burdens and strains of combining paid work and family work have led many women who are both wives and mothers to modify the traditional pattern. The two alternative patterns are sequential work/family role staging and symmetrical work/family role allocation.

Sequential Work/Family Role Staging

The pattern of sequential work/family role staging is probably the most common for dual-earner marriages. This pattern reflects the adjustments women try to make in balancing work and family demands. Many of women's choices about employment and careers are based on their plans for a family and whether and when they will want to work. The key event is first pregnancy. Prior to pregnancy, most married women are employed. When they become pregnant, however, they begin leaving their jobs and careers to prepare for the transition to parenthood. By the last month of pregnancy, 80 percent have left the workforce. Within a year, more than half of these women have returned to employment. Most women who leave their paid work do so because of impending birth. Those who return to employment are strongly motivated by economic considerations or need.

There are four common forms of sequential work/family patterns:

- *Conventional*, in which a woman quits her job after marriage or the birth of her first child and does not return.
- *Early interrupted*, in which a woman stops working early in her career in order to have children and resumes working later.
- *Later interrupted*, in which a woman first establishes her career, quits to have children (usually in her thirties), and then returns.
- *Unstable*, in which a woman goes back and forth between full-time paid employment and homemaking, usually according to economic need.

A major decision for a woman who chooses sequential work/family role staging is at what stage in her life to have children. Should she have them early or defer them until later? As with most things in life, there are pros and cons. Early parenthood allows women to have children with others in their age group; they are able to share feelings and common problems with their peers. It also enables them to defer or formulate career decisions. At the same time, however, if they have children early, they may increase economic pressures on their beginning families. They also have greater difficulty in reestablishing their careers.

Women who defer parenthood until they reach their middle career stage often are able to reduce the role strain and economic pressures that accompany the new parent/early career stage of the traditional pattern. Such women, however, may not easily find other new mothers of the same age with whom to share their experiences. They may find the physical demands to be greater than anticipated. Some may decide that they do not want children at all because motherhood would interfere with their careers.

Symmetrical Work/Family Role Allocation

Families that try to reduce role overload and interrole conflict may reallocate traditional family roles. Because both husband and wife are employed, both share in family work. Equity and fairness are thought to be more important than gender in dividing responsibilities and tasks. Women undertake some traditional male responsibilities, such as outside employment, car maintenance, and hauling trash. Men do some traditional female tasks, such as cooking, shopping, cleaning, and child care. Although such families tend to be more egalitarian, men rarely share family responsibilities fifty-fifty (but they often think they do).

What distinguishes symmetrical families from traditional families is the extent of the responsibilities husbands and wives undertake. In traditional families, wives may help contribute to family income through their employment, and men may help with the caregiving and household duties. But neither goes beyond "helping"; neither is responsible for these tasks. In symmetrical families, husbands are responsible for certain aspects of family work, and wives are responsible for contributing to family income.

Although symmetrical arrangements encourage gender equality, they are often resisted by men, who tend to be more committed to work roles than family roles. Such arrangements also require more commitment to work roles by women. Although married partners may consciously believe that such shifts are desirable, on an unconscious level they may resist such changes because of deeply held beliefs about the "proper" roles of husbands and wives. Both men and women may have difficulty in relinquishing beliefs that the traditional provider role is necessarily the most important male family role. They may be unable to envision gifts of love and time as being equally important as (or more important than) paychecks.

Reflections

Which work/family pattern will you adopt (or have you adopted)? What would its benefit be for you? Its drawbacks? Which pattern did your family of orientation adopt? What were its benefits for your parents? Its drawbacks? Does their experience influence your choice of patterns? How would single parenting affect the work/family pattern?

The Family's Division of Labor

Families divide their labor in numerous ways. Some follow traditional male-female patterns, others more egalitarian ones. How families divide labor has a tremendous impact on how a family functions.

The Traditional Division of Labor: The Complementary Model

In the traditional division of labor in the family, work roles are complementary: The husband is expected to work outside the home for wages, and the wife is expected to remain at home caring for children and maintaining the household. (We discussed the child-rearing aspect of family work in Chapter 10.) A man's family role is secondary to his provider role, whereas a woman's employment role is secondary to her family role (Blair, 1993). This difference in primary roles between men and women profoundly affects such basic family tasks as who cleans the toilet, mops the floors, does the ironing, and washes the baby's diapers. Women—whether or not they are employed outside the home—remain primarily responsible for household tasks (Demo and Acock, 1993). The division of family roles along stereotypical gender lines is more characteristic of white families than African-American families. African-American women, for example, are less likely than white women to be exclusively responsible for household tasks (Taylor, Chatters, Tucker, and Lewis, 1991).

Men's Family Work

Men's work traditionally takes place outside the home, where men fulfill their primary economic role as provider. The husband's role as provider is probably the male's most fundamental role in marriage. The basic equation is that if the male is a good provider, he is a good husband and a good father. This core concept seems to endure despite trends toward more egalitarian and androgynous gender roles. In fact, a woman's marital satisfaction is often related to how well she perceives her husband as fulfilling his provider role (Blair, 1993). It is not uncommon for women to complain of husbands who do not work to their full potential. They feel their husbands do not contribute their fair share to the family income.

Did you know?
In 1990, approximately 26.5 million women and 541,000 men were full-time homemakers with no outside employment (Bird and Ross, 1993).

Men are traditionally expected to contribute to family work by providing household maintenance. Such maintenance consists primarily of repairs, light construction, mowing the lawn, and other activities that are consistent with instrumental male norms. (But, as one woman asked, how often do you have to repair the toaster or paint the porch?) Men often contribute to housework and child care, although their contribution may not be notable in terms of the total amount of work to be done. Men tend to see their role in housekeeping or child care as "helping" their partner, not as assuming equal responsibility for such work. For husbands to become equal partners in the family work, wives, as well as their husbands, must have an egalitarian view of family work (Greenstein, 1996). The more husbands define themselves as the provider, regardless of whether their wives are employed, the fewer household tasks they perform (Mederer, 1993).

Women's Family Work

A woman's work may consist of unpaid family work as homemaker and paid work outside the home as an employee or professional. Her work as homemaker has been traditionally considered to be her primary role. Because this work is performed by women and is unpaid, however, it has been denigrated as "women's work"—inconsequential and unproductive.

I hate to be called a homemaker; I prefer "domestic goddess."

Roseanne

Women's family work is considerably more diverse than men's. Women's work permeates every aspect of the family. It ranges from housekeeping to child

Researchers Linda Thompson and Alexis Walker observe, "Family work is unseen and unacknowledged because it is private, unpaid, commonplace, done by women, and mingled with love and leisure."

care, maintaining kin relationships to organizing recreation, socializing children to caring for aged parents and in-laws, and cooking to managing the family finances, to name but a few of the tasks. Ironically, family work is often invisible to the women who do most of it (Brayfield, 1992).

Although most women do paid work, contributing about 30 percent to the family's income, neither women nor their partners regard employment as a woman's fundamental role. Women are not duty bound to provide; they are duty bound to perform household tasks (Thompson and Walker, 1991). No matter what kind of work the woman does outside the home or how nontraditional she and her husband may consider themselves to be, there is seldom equality when it comes to housework.

The greatest determinant of the amount of time a woman spends on housework is her employment status. Women spend less time on housework if they are employed (Greenstein, 1996). Men may help their partners with washing dishes, vacuuming, or doing the laundry, but as long as they merely "help," household responsibility continued to fall on women.

Sociologist Ann Oakley (1985) describes four aspects of the **homemaker role:**

- Exclusive allocation to women, rather than to adults of both sexes.
- Association with economic dependence.
- Status as nonwork, or its opposition to "real," economically productive work that is paid.
- Primacy to women—that is, its priority over other women's roles.

Did you know?

Women constitute half the world's population, perform nearly two-thirds of its work hours, receive one-tenth of the world's income, and own less than one-hundredth of the world's property (United Nations Report, 1985).

Housework consists primarily of household work and, if children are present, child-rearing tasks. It has the following characteristics (Bird and Ross, 1993; Hochschild, 1989; Oakley, 1985):

1. Housework tends to isolate a woman at home. She cleans alone, cooks alone, launders alone, and cares for children alone. Loneliness is a common complaint.

A man's work is from sun to sun, but woman's work is never done.

FOLK SAYING

OTHER PLACES OTHER TIMES

Industrialization "Creates" the Traditional Family

The American family has radically altered its economic functions since colonial times. During the colonial era, the family was basically an economic and social institution, the primary unit for producing most goods and caring for the needs of its members. The family planted and harvested food, made clothes, provided shelter, and cared for the necessities of life. Each member was expected to contribute economically to the welfare of the family. Husbands plowed, planted, and harvested crops. Wives supervised apprentices and servants, kept records, cultivated the family garden, assisted in the farming, and marketed surplus crops or goods, such as grain, chickens, candles, and soap. Older children helped their parents and, in doing so, learned the skills necessary for later life.

In the nineteenth century, industrialization transformed the face of America. It also transformed American families from self-sufficient farm families to wage-earning urban families. As factories began producing farm machinery such as harvesters, combines, and tractors, significantly fewer farm workers were needed. Looking for employment, workers migrated to the cities, where they found employment in the ever-expanding factories and businesses. Families no longer worked together in the fields or in the home producing food, clothing, and other necessities; instead these were produced by large-scale farms and factories. Food and other goods were purchased with wages earned in factories and stores. Because goods were now bought rather than made in the home, the family began to shift from being primarily a production unit to being a consumer and service-oriented unit.

With this shift, a radically new division of labor arose in the family. Men began working outside the home in factories or offices for wages to purchase the family's necessities and other goods. Men became identified as the family's sole providers or "breadwinners." Their work was given higher status than women's work because it was paid in wages. Men's work began to be identified as "real" work.

Industrialization also created the housewife, the woman who remained at home attending to household duties and caring for children. Previously, men and women were interdependent within the family unit because women produced many of the family's necessities. Because much of what the family needed had to be purchased with the husband's earnings, the wife's contribution in terms of unpaid work and services went unrecognized, much as it continues today.

Industrialization removed many of women's productive tasks from the home and placed them in the factory. Even then, although women made fewer things at home, they nevertheless continued their service and child-rearing roles: cooking, cleaning, sewing, raising children, and nurturing the family. With the rise of a money economy in which only paid work was recognized as real work, women's work in the home went unrecognized as necessary and important labor (Ferree, 1991).

Without its central importance as a work unit, the family became the focus and abode of feelings. In earlier times, the necessities of family-centered work gave marriage and family a strong center based on economic need. The emotional qualities of a marriage mattered little as long as the marriage produced an effective working partnership. Without its productive center, however, the family focused on the relationships between husband/wife and parent/child. Affection, love, and emotion became the defining qualities of a good marriage.

2 Housework is unstructured, monotonous, and repetitive. A homemaker never feels that her work is done. There is always more dust, more dishes, more dirty laundry.

3 The full-time homemaker role is restricted. For a woman who is not also employed outside the home, homemaking (which may also include child rearing) is essentially her only role. By contrast, a man's role is dual—employed worker and husband/father. If the man finds one role to be unsatisfactory, he may find satisfaction in the other. It is much more difficult for a woman to separate the satisfaction she may get from being a mother from the dissatisfaction of household work.

4 Housework is autonomous. This is one of the most well-liked aspects of the homemaker role. Being her own boss allows a woman

to direct a large part of her own life. In contrast to employment, this is a definite plus.

5 Homemakers work long days and nights. This is especially true for employed mothers. For employed women, there is a "second shift" of work that awaits them at home, requiring anywhere from fifty to one hundred or more hours a week.

6 Homemaking can involve child rearing. For many homemakers, this is their most important and rewarding work. They enjoy the interactions with their infants and children. Although they can also feel overwhelmed and exhausted by the demands of child rearing, relatively few would choose not to have children.

7 Homemaking often involves role strain, especially for employed mothers as they try to perform their various obligations as parent, spouse, and worker.

8 Housework is unpaid. The homemaker is "paid" in goods and services, primarily from her husband. Because she is unpaid, she is often dependent. She gets no increase in pay regardless of her skill. Her standard of living depends on her husband, not her own efforts. If the homemaker is a single woman, the husband-wife reciprocity is absent. There is no payment in goods and services from a spouse.

Both women and men who are full-time housekeepers feel the same about housework: It is routine, unpleasant, unpaid, and unstimulating, but it provides a degree of autonomy. Full-time male houseworkers, however, do not often call themselves *housekeepers* or *homemakers*. Instead, they identify themselves as retired, unemployed, laid off, or disabled (Bird and Ross, 1993).

Many women find satisfaction in the homemaker role, even in housework. Young women, for example, may find increasing pleasure as they experience a sense of mastery over cooking, entertaining, or rearing happy children. If homemakers have formed a network among other women—such as friends, neighbors, or relatives—they may share many of their responsibilities. They discuss ideas and feelings and give one another support. They may share tasks as well as problems.

Reflections

List the different tasks that make up family work in your family. What family work is given to women? To men? On what basis is family work divided? Is it equitable?

Women in the Labor Force

Women have always worked outside the home. Since the early nineteenth century, single women have traditionally been members of the workforce, and there have always been large numbers of working mothers, especially among African Americans and many other ethnic groups. It was not until the late 1970s that the employment rate of white women began to converge with that of African-American women (Herring and Wilson-Sadberry, 1993).

What has changed significantly is the emergence since the 1960s of a predominant family form in which both husbands and wives/mothers work outside the home. Until the 1980s, nonemployed wives and mothers were viewed as the norm. Recent research indicates that women's employment tends to have positive rather than negative effects on marriage.

In 1995, 59 percent of adult women and 75 percent of adult men were employed (U.S. Bureau of the Census, 1996). Between 1960 and 1995, the percentage of married women in the labor force almost doubled—from 32 percent to 61 percent. During the same period, the number of employed married women

Did you know?

By 2005, female/male employment rates will be nearly equal: 63 percent of adult women and 69 percent of adult men will be employed (U.S. Bureau of the Census, 1996).

UNDERSTANDING YOURSELF

The Division of Labor: A Marriage Contract

How do you expect to divide household and employment responsibilities in marriage? More often than not, couples live together or marry without ever discussing basic issues about the division of labor in the home. Some think that things will "just work out." Others believe that they have an understanding, although they may discover later that they do not. Still others expect to follow the traditional division of labor. Often, however, one person's expectations conflict with the other's.

The following questions cover important areas of understanding for a marriage contract. These issues should be worked out before marriage. Although marriage contracts dividing responsibilities are not legally binding, they make explicit the assumptions that couples have about their relationships.

Answer these questions for yourself. If you are involved in a relationship, live with someone, or are married, answer them with your partner. Consider putting your answers down in writing.

- Which has the highest priority for you: marriage or your job? What will you do if one comes into conflict with the other? How will you resolve the conflict? What will you do if your job requires you to work sixty hours a week? Would you consider that such hours conflict with your marriage goals and responsibilities? What would your partner think? Do you believe that a man who works sixty hours a week shows care for his family? Why? What about a woman who works sixty hours a week?
- Whose job or career is considered the most important—yours or your partner's? Why? What would happen if both you and your partner were employed and you were offered the "perfect" job five hundred miles apart from your partner? How would the issue be decided? What effect do you think this would have on your marriage or relationship?
- How will household responsibilities be divided? Will one person be entirely, primarily, equally, secondarily, or not at all responsible for housework? How will this be decided? Does it matter whether a person is employed full-time as a salesclerk or a lawyer in deciding the amount of housework he or she should do? Who will take out the trash? Vacuum the floors? Clean the bathroom? How will it be decided who does these tasks?
- If you are both employed and then have a child, how will the birth of a child affect your employment? Will one person quit his or her job or career to care for the child? Who will that be? Why? If both of you are employed and a child is sick, who will remain home to care for the child? How will that be decided?

between twenty-four and thirty-four years of age (the ages during which women are most likely to bear children) rose from 29 percent to 72 percent. (In 39 percent of dual-worker families, however, both husband and wife preferred that the woman not work [Ross, Mirowsky, and Goldsteen, 1991].)

Almost two million employed women in 1995 were single mothers; slightly over half had preschool children (U.S. Bureau of the Census, 1996). More than 60 percent of all mothers with young children are currently in the workforce.

Why Women Enter the Labor Force

Four factors influence a woman's decision to enter the labor force (Herring and Wilson-Sadberry, 1993):

- *Financial factors.* To what extent is income significant? For unmarried women and single mothers, employment may be their only source of income. The income of married women may be primary or secondary to their husbands' incomes.
- *Social norms.* How accepting is the social environment for married women and mothers working at paid jobs? Does the woman's partner support her?

If she has children, do her partner, friends, and family believe that working outside the home is acceptable? After the 1970s, social norms changed to make it more acceptable for white mothers to hold a job.

- *Self-fulfillment*. Does a job meet needs for autonomy, personal growth, and recognition? Is it challenging? Does it provide a change of pace?
- *Attitudes about employment and family*. Does the woman believe she can combine her family responsibilities with her job? Can she meet the demands of both? Does she believe that her partner and children can do well without her as a full-time homemaker?

Economic necessity is a driving force for many women in the labor force. Economic pressures traditionally have been powerful influences on African-American women. Many married women and mothers enter the labor force or increase their working hours to compensate for their husband's loss in earning power due to inflation. Additionally, the social status of the husband's employment often influences the level of employment chosen by the wife (Smits, Utee, and Lammers, 1996).

Among the psychological reasons for employment are an increase in a woman's self-esteem and sense of control. A comparison between African-American and white women found that personal preference was the primary employment motivation for about 42 percent of African-American women and 46 percent of white women (Herring and Wilson-Sadberry, 1993). Employed women are less depressed and anxious than nonemployed homemakers; they are also physically healthier (Gecas and Seff, 1989; Ross, Mirowsky, and Goldsteen, 1991). A thirty-four-year-old Latina, mother of three, told social psychologist Lillian Rubin (1994) the following:

> I started to work because I had to. My husband got hurt on the job and the bills started piling up, so I had to do something. It starts as a necessity and it becomes something else.
>
> I didn't imagine how much I'd enjoy going to work in the morning. I mean, I love my kids and all that, but let's face it, being mom can get pretty stale. . . . Since I went to work I'm more interested in life, and life's more interested in me.
>
> I started as a part-time salesperson and now I'm assistant manager. One day I'll be manager. Sometimes I'm amazed at what I've accomplished; I had no idea I could do all this, be responsible for a whole business.

There are two reasons why employment improves women's emotional and physical well-being (Ross, Mirowsky, and Goldsteen, 1991). First, employment decreases economic hardship, alleviating stress and concern not only for the woman herself but for other family members as well. A single parent's earnings may constitute her entire family's income. Second, an employed woman receives greater domestic support from her partner. The more a woman earns relative to what her partner earns, the more likely he is to share housework and child care.

Women's Employment Patterns

The employment of women has generally followed a pattern that reflects their family and child-care responsibilities. Because of these demands, women must consider the number of hours they can work and what time of day to work. Their decision to work in the labor force is determined, in part, by the availability of timely work hours and adequate child care.

Traditionally, women's employment rates dropped during their prime child-bearing years, from ages twenty to thirty-four years. But this is no longer true. In fact, women's employment during these years has more than doubled. In 1995, about 61 percent of all married women were employed. During that same time 59.0 percent of married women with children under one year were employed; 63.5 percent of married women with children under six years were employed. Fifty-three percent of single mothers with children under six years were employed (U.S. Bureau of the Census, 1996). Women no longer leave the job market when they become mothers (O'Connell and Bachu, 1992). Either they need the income, or they are more committed to work roles than in previous generations.

But because of family responsibilities, many of the employed women—about 40 percent—work part-time. Furthermore, when family demands increase, wives, not husbands, cut back on their job commitments (Folk and Beller, 1993). As a result of family commitments, women tend to interrupt their job and career lives more frequently than do men.

In a study of over a thousand married couples with children at home, women devoted over forty hours a week to household tasks, while men averaged less than eleven hours. These tasks included preparing meals, cleaning house, washing dishes, laundry, shopping, paying bills, and outdoor work (Demo and Acock, 1993). Most of the women in this study were employed outside the home over thirty hours per week. Whether or not she was employed outside the home, the woman contributed nearly the same major proportion of the total family time spent in housework. Employed wives' proportion of total family hours spent on household chores was still about 72 percent, compared with 81 percent for nonemployed wives.

Men tended to perceive that they do more housework than they actually do. While husbands of employed women contributed slightly more hours to household work than did husbands of nonemployed women, these differences were relatively small.

Researchers have found that a woman's decision to remain in the workforce or to withdraw from it during her childbearing and early child-rearing years is critical for her later workforce activities. If a woman chooses to work at home caring for her children, she is less likely to be employed later. If she later returns to the workforce, she will probably earn substantially less than women who have remained in the workforce.

Dual-Earner Marriages

The chief value of money lies in the fact that one lives in a world in which it is overestimated.

H. L. MENCKEN (1880–1956)

Since the 1970s, inflation, a dramatic decline in real wages, the flight of manufacturing, and the rise of a low-paying service economy have altered the economic landscape. In the last twenty years, wages have plummeted for male high school graduates by 25 to 30 percent. Although women's wages declined only between 15 and 18 percent, the smaller decline occurred because women started at lower wages. And although 2 million new jobs are being created annually, the typical wage is $8 an hour, or $16,000 a year (Kilborn, 1994). The income of college-educated men has fallen 3 percent, but that of college-educated women has increased 15 percent, still significantly less than the earnings of college-educated men (Vobejda, 1994). In 1993, among families with a single male wage earner, the median income was $37,500. Families with two employed members earned a median income of nearly $52,000 (U.S. Bureau of the Census, 1996). Even being a dual-earner family doesn't guarantee prosperity; almost a third earned less than $35,000, the equivalent of two poor-paying jobs (U.S. Bureau of the Census, 1996).

These economic changes have led to a significant increase in dual-earner marriages. Even though two-thirds of all married women held jobs in 1995, the great majority of these women were employed in low-paying, low-status jobs—secretaries, clerks, nurses, factory workers, and the like. Rising prices and declining wages pushed most of them into the job market. Employed mothers generally do not seek personal fulfillment in their work as much as they do additional family income. Their families remain their top priorities.

Dual-career families are a subcategory of dual-earner families. They differ from other dual-earner families insofar as both husband and wife have high achievement orientations, greater emphasis on gender equality, and a stronger desire to exercise their capabilities. Unfortunately, these couples may find it difficult to achieve both their professional and their family goals. Often they have to compromise one goal to achieve the other because the work world generally is not structured to meet the family needs of its employees. As one study points out (Berardo et al., 1987),

> The traditional "male" model of career involvement makes it extremely difficult for both spouses to pursue careers to the fullest extent possible, since men's success in careers has generally been made possible by their wives' assuming total responsibility for the family life, thus allowing them to experience the rewards of family life but exempting them from this competing set of responsibilities.

Domestic Labor: Housework and Child Care

We are increasingly seeing that marital satisfaction is tied to fair division of household labor (Blair, 1993; Pina and Bengston, 1993; Suitor, 1991). A husband's wielding a vacuum cleaner or cooking dinner while his partner takes off her shoes to relax a few moments after returning home from work is sometimes better than his presenting her a bouquet of flowers—it may show better than any material gift that he cares. In a world where both spouses are employed, dividing household work fairly may be a key to marital success (Hochschild, 1989; Perry-Jenkins and Folk, 1994; Suitor, 1991).

Money is a terrible master but an excellent servant.

P.T. BARNUM (1810–1891)

Although we traditionally separate housework, such as mopping and cleaning, from child care, in reality the two are inseparable (Thompson, 1991). Although fathers have increased their participation in child care slightly, they have done little in terms of swinging a mop or scrubbing a toilet. If we continue to separate the two domains, men will take the more pleasant child-care tasks of playing with the baby or taking the kids to the playground, while women take on the more unpleasant duties of washing diapers, cleaning ovens, and ironing. Furthermore, someone must do behind-the-scenes dirty work in order for the more pleasant tasks to be performed. As Alan Hawkins and Tomi-Ann Roberts (1992) note:

> Bathing a young child and feeding him/her a bottle before bedtime is preceded by scrubbing the bathroom and sterilizing the bottle. If fathers want to romp with their children on the living room carpet, it is important that they be willing to vacuum regularly. . . . Along with dressing their babies in the morning and putting them to bed at night comes willingness to launder jumper suits and crib sheets.

If we are to develop a more equitable division of domestic labor, we need to see housework and child care as different aspects of the same thing: domestic labor that keeps the family running. (See Hawkins and Roberts, 1992, and Hawkins, 1994, for a description of a program to increase male involvement in household labor.)

"My wife works, and I sit on the eggs. Want to make something of it?"

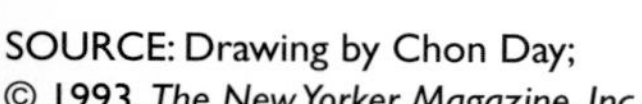
SOURCE: Drawing by Chon Day; © 1993, *The New Yorker Magazine, Inc.*

Housework

Standards of housework have changed over the last few generations. Writes Barbara Ehrenreich (1993):

> Recall that not long ago, in our mother's day, the standards were cruel but clear: Every room should look like a motel room. The floors must be immaculate enough to double as plates, in case the guests prefer to eat doggie-style. The kitchen counters should be clean enough for emergency surgery, should the need at some time arise, and the walls should ideally be sterile. The alternative, we all learned in Home Economics, is the deadly scorn of the neighbors and probably the plague.

Housekeeping ain't no joke.

LOUISA MAY ALCOTT (1832–1888)

The engine of change was not the vacuum cleaner—which, in fact, seemed to increase hours spent in housework, as it promised the possibility of immaculateness if one worked hard enough. What changed was that working women could no longer hold up the standards of their mothers—or of household product advertisers. They now spend less time on housework. But Ehrenreich advises those who miss the good old days: "For any man or child who misses the pristine standards of yesteryear, there is a simple solution. Pitch in!"

Whether or not married women are employed seems to have little impact on the division of housework. Studies suggest that employed women do twice as much housework as men (Spitze, 1991). Cohabiting women, however, do significantly less housework than married women. Whether men are married or cohabiting does not significantly affect their housework—it remains minimal (Shelton and John, 1993). It seems that marriage, rather than living with a man, transforms a woman into a homemaker. Marriage seems to change the house from a space to keep clean to a home with husband/family to care for.

You can say this about ready-mixes—the next generation isn't going to have any trouble making pies exactly like mother used to make.

EARL WILSON

Various factors seem to affect men's participation in housework. Men tend to contribute more to household tasks when they have the fewest time demands from their jobs—that is, early in their employment careers and after retirement (Rexroad and Shehan, 1987). As wives' income rises, they report more participation by husbands in household tasks; increased income and job status motivate women to secure their husbands' sharing of tasks. Men who are relatively expressive help their wives more than men who are relatively aggressive, dominant, and emotionally "tough." Husbands in one study (Benin and Agostinelli, 1988) felt most satisfied in the division of household labor if tasks were divided equally, especially if their total time spent on chores was small. Women appeared to be more satisfied if their husbands shared traditional women's chores (such as laundry) rather than limiting their participation to traditional male tasks (such as mowing the lawn). But even men who contribute many hours to household labor tend to do traditional male tasks (Blair and Lichter, 1991). Children appear to be less gender segregated with their household work than their parents (Simons and Whitbeck, 1991). African Americans are less likely to divide household tasks along gender lines than whites.

Cleaning . . . should be done from the top down—starting with the ceiling, which is ridiculous. Gravity takes care of that. If there were any dirt on the ceiling, it would fall off and land on the floor. The same goes for walls. Dirt falls right off them and lands on the floor. And you shouldn't fool around with the dirt on the floor, because it will get all over the walls and ceiling.

P.J. O'Rourke

If both husband and wife work outside the home, it is generally agreed that household tasks should be divided equitably; that is, fairly. Most studies find that the majority of husbands and wives believe their family's division of labor is fair (Spitze, 1991). For couples who can afford household help, the husband may be excused from many household chores, such as cleaning and mopping. His income allows him to "hire" a substitute to do his share of the housework (Perry-Jenkins and Folk, 1994).

It is important to note that an equitable division is not the same as an equal division. Relatively few couples, in fact, divide housework fifty-fifty. For women, a fair division of household work is more important than both spouses' putting in an equal number of hours. There is no absolute standard of fairness, however (Thompson, 1991). What is fair is determined differently by different couples. Because most women work fewer hours than men in paid work, and wives tend to work more hours in the home, some women believe that the household labor should be divided proportionately to hours worked outside the home. Other women believe that it is equitable for higher-earning husbands to have fewer household responsibilities. Still others believe that the traditional division of labor is equitable, as household work is women's work by definition. Middle-class women are more likely to demand equity, whereas equity is less important for working-class women, who are more traditional in their gender-role expectations (Perry-Jenkins and Folk, 1994; Rubin, 1994).

I hate housework! You make the beds, you do the dishes—and six months later you have to start over again.

Joan Rivers

Child-Rearing Activities

Men increasingly believe that they should be more involved fathers than men have been in the past. Yet the shift of attitudes has not greatly altered men's behavior. One study (Darling-Fisher and Tiedje, 1990) found that the father's time involved in child care is greatest when the mother is employed full-time (fathers responsible for 30 percent of the care compared with mothers' 60 percent; the remaining 10 percent of care is presumably provided by other relatives, baby-sitters, or child-care providers). The father's involvement is less when the mother is employed part-time (fathers' 25 percent versus 75 percent for mothers) and least when she is a full-time homemaker (fathers' 20 percent versus 80 percent for mothers).

A review of studies (Lamb, 1987) on parental involvement in two-parent families concluded the following:

- Mothers spend from three to five hours of active involvement for every hour fathers spend, depending on whether the women are employed or not.
- Mothers' involvement is oriented toward practical daily activities, such as feeding, bathing, and dressing. Fathers' time is generally spent in play.
- Mothers are almost entirely responsible for child care: planning, organizing, scheduling, supervising, and delegating.

- Women are the primary caretakers; men are the secondary.

Although mothers are increasingly employed outside the home, fathers are doing relatively little to pick up the slack at home. As a result, children suffer from the lack of parental time and energy because their fathers do not participate more. If children are to be given the emotional care and support they need to develop fully, their fathers must become significantly more involved.

Marital Power

Money is a source of power that supports male dominance in the family. . . . Money belongs to him who earns it, not to her who spends it, since he who earns it may withhold it.

Reuben Hill and Howard Becker

An important consequence of women's working is a shift in the decision-making patterns in a marriage. Although decision-making power in a family is not based solely on economic resources (personalities, for instance, also play a part), economics is a major factor. A number of studies suggest that employed wives exert greater power in the home than nonemployed wives (Blair, 1991). Marital decision-making power is greater among women who are employed full-time than among those who are employed part-time. Wives have the greatest power when they are employed in prestigious work, are committed to it, and have greater income than their husbands.

Some researchers are puzzled about why many employed wives, if they do have more power, do not demand greater participation in household work on the part of their husbands. Joseph Pleck (1985) suggests several reasons for women's apparent reluctance to insist on their husbands' equal participation in housework. These include (1) cultural norms that housework is the woman's responsibility, (2) fears that demands for increased participation will lead to conflict, and (3) the belief that husbands are not competent.

Money talks.

Folk saying

Marital Satisfaction

How does employment affect marital satisfaction? Traditionally, this question was asked only of wives, not husbands; even then, it was rarely asked of African-American wives, who had a significantly higher employment rate. In the past, married women's employment, especially maternal employment, was viewed as a problem. It was seen as taking away from a woman's time, energy, and commitment for her children and family. In contrast, nonemployment or unemployment was seen as a major problem for men. But it is possible that the husband's work may increase marital and family problems by preventing him from adequately fulfilling his role as a husband or father: He may be too tired, too busy, or never there. It is also possible that a mother's not being employed may affect the family adversely: Her income may be needed to move the family out of poverty, and she may feel depressed from lack of stimulation (Menaghan and Parcel, 1991).

Money doesn't talk,
It swears.

Bob Dylan

How does a woman's employment affect marital satisfaction? There does not seem to be any straightforward answer when comparing dual-earner and single-earner families (Piotrkowski et al., 1987). In part, this may be because there are trade-offs: a woman's income allows a family a higher standard of living, which compensates for the lack of status a man may feel for not being the "sole" provider. Whereas men may adjust (or have already adjusted) to giving up their sole-provider ideal, women find current arrangements less than satisfactory. After all, women are bringing home additional income but are still expected to do the overwhelming majority of household work. Role strain is a constant factor for

What shall it profit a man, if he shall gain the whole world, and lose his own soul?

Mark 8:36

women, and in general, women make greater adjustments than men in dual-earner marriages.

Studies of the effect of women's employment on the likelihood of divorce are not conclusive, but they do suggest a relationship (Spitze, 1991; White, 1991). Many studies suggest that employed women are more likely to divorce. Employed women are less likely to conform to traditional gender roles, which potentially causes tension and conflict in the marriage. They are also more likely to be economically independent and do not have to tolerate unsatisfactory marriages for economic reasons. Other studies suggest that the only significant factor in employment and divorce is the number of hours the wife works. Hours worked may be important because full-time work for both partners makes it more difficult for spouses to share time together. Numerous hours may also contribute to role overload on the part of the wife (Greenstein, 1990).

African-American women, however, are not more likely to divorce if they are employed. This may be because of their historically high employment levels and their husbands' traditional acceptance of such employment (Taylor et al., 1991).

Overall, despite an increased divorce rate, in recent years the overall effect of wives' employment on marital satisfaction has shifted from a negative impact to no impact or even a positive impact. If there are negative effects, they generally result from specific aspects of a woman's job, such as long hours or work stress (Spitze, 1991).

The effect of a wife's full-time employment on a couple's marital satisfaction is affected by such variables as social class, the presence of children, and the husband's and wife's attitudes and commitment to her working. Thus, the more the wife is satisfied with her employment, the higher their marital satisfaction will be. Also, the higher the husband's approval of his wife's employment, the higher the marital satisfaction.

Coping in Dual-Earner Marriages

Dual-earner marriages are here to stay. They are particularly stressful today because society has not pursued ways to alleviate the work-family conflict. The three greatest social needs in dual-earner marriages are (1) redefining gender roles to eliminate role overload for women, (2) providing adequate child-care facilities for working parents, and (3) restructuring the workplace to recognize the special needs of parents and families.

Coping strategies include reorganizing the family system and reevaluating household expectations. Husbands may do more housework. Children may take on more household tasks than before. Household standards—such as a meticulously clean house, elaborate meal preparation, and washing dishes after every meal—may be changed. Careful allocation of time and flexibility assist in coping.

Dual-earner couples often hire outside help, especially for child care, which is usually a major expense for most couples. One of the partners may reduce his or her hours of employment, or both partners may work different shifts to facilitate child care (but this usually reduces marital satisfaction as a result) (White and Keith, 1990).

The goal for most dual-earner families is to manage their family relationships and their paid work to achieve a reasonable balance that allows their families to thrive rather than merely survive. Achieving such balance will continue to be a struggle until society and the workplace adapt to the needs of dual-earner marriages and families.

Reflections

The chances are very good that if you cohabit or marry, you will be in a dual-earner relationship. How will you balance your employment and relationship or family needs?

Family Issues in the Workplace

Many workplace issues, such as economic discrimination against women, occupational stratification, adequate child care, and an inflexible work environment, directly impact families. They are more than economic issues—they are also family issues.

Discrimination against Women

A woman's earnings have a significant impact on family well-being, whether the woman is the primary or secondary contributor to a dual-earner family or the sole provider in a single-parent family. Thus, economic discrimination against women and sexual harassment are also important family issues.

Economic Discrimination

There is a simple way to define a woman's job. Whatever the duties are—and they vary from place to place and from time to time—a woman's job is anything that pays less than a man will accept for comparable work.

CAROLINE BIRD

The effects of economic discrimination can be devastating for women. In 1996, women in the United States made seventy-five cents for every dollar that men earned. (In Sweden, women make ninety cents for every dollar made by men.) For the 46.3 million American men aged twenty-five to sixty-four years, the median income is $31,200. For the 35 million women in the same age bracket, median income is $23,000 (U.S. Bureau of Labor Statistics, 1996).

Because of the great difference in women's and men's wages, many women are condemned to poverty and are forced to accept welfare and its accompanying stigma. Wage differentials are especially important to single women. Twenty-three percent of all families are headed by single mothers who are usually responsible for supporting them. In 1993, 54.1 percent of these families made less than $15,000 a year; 19 percent made less than $5,000 (U.S. Bureau of the Census, 1996).

Women face considerable barriers in their access to well-paying, higher-status jobs (Bergen, 1991). Although employment and pay discrimination is prohibited by Title VII of the 1964 Civil Rights Act, the law did not end the pay discrepancy between men and women. Much of the earnings gap is the result of occupational differences, gender segregation, and women's tendency to interrupt their employment for family reasons and to take jobs that do not interfere extensively with their family lives. Earnings are about 30 to 50 percent higher in traditionally male occupations, such as truck driver or corporate executive, than in predominantly female or sexually integrated occupations, such as secretary or schoolteacher. The more an occupation is dominated by women, the less it pays.

Sexual Harassment

Sexual harassment is a mixture of sex and power; power may often be the dominant element. Such harassment may be a way to keep women in their place. **Sexual harassment** refers to two distinct types of harassment: (1) the abuse of power for sexual ends and (2) the creation of a hostile environment. In abuse of power, sexual harassment consists of unwelcome sexual advances, requests for sexual favors, or other verbal or physical conduct of a sexual nature as a condition of instruction or employment. Only a person with power over another can commit the first kind of harassment. In a **hostile environment,** someone acts in sexual ways so as to interfere with a person's performance by creating a hostile or offensive learning or work environment. Sexual harassment is illegal.

Issues of sexual harassment are complicated in the workplace because work, like college, is one of the most important places where adults meet potential partners. As a consequence, sexual undercurrents or interactions often are present. Flirtations, romances, and affairs are common in the work environment. Drawing the line, especially for men, between flirtation and harassment can be filled with ambiguity.

There are significant gender differences that may contribute to sexual harassment. First, men are generally less likely to perceive activities as being harassing than are women (Jones and Remland, 1992; Popovich et al., 1992). Second, men misperceive women's friendliness as sexual interest (Johnson et al., 1991; Stockdale, 1993). Third, men are more likely to perceive male-female relationships as adversarial (Reilly et al., 1992). In addition to gender differences, power differences also affect perception. Behaviors exhibited by a supervisor, such as personal questions, are more likely to be perceived as sexual harassment than are the same behaviors of a co-worker.

Sexual harassment can have a variety of consequences. One study of the workplace (Gutek, 1985) found that 9.1 percent of the women and 1 percent of the men quit their jobs because of harassment; almost 7 percent of the women and 2 percent of the men were dismissed from their jobs as part of their harassment. Victims often report depression, anxiety, shame, humiliation, and anger (Paludi, 1990).

Occupational Stratification and Ethnicity

Occupational differences exist along ethnic lines, and these differences in turn affect the economic well-being of families (Billingsley, 1992; Malveaux, 1988). **Occupational stratification** is the hierarchical ranking of jobs by income and status. A person's position in the occupational hierarchy is often correlated to gender and ethnicity. It is generally accepted by labor economists that the relative success of native-born whites in contrast to African Americans and Latinos is due to whites' having greater access to quality education and job training and the political leverage to protect their social and economic advantages (Hacker, 1992). In addition, members of some ethnic groups lack proficiency in standard English, a critical skill, and this prohibits them from obtaining employment in high-paying jobs (Bean and Tienda, 1987).

Almost one-third of African Americans work in middle-class or professional occupations; the remainder are typically employed in declining industries or the lowest-paying jobs. Among African-American men, for example, the majority are employed in manufacturing and service jobs. Technological changes and foreign competition, especially in the automobile, steel, and industrial machines industries, have decimated American manufacturing, where African Americans once were employed in well-paying jobs. The economic basis of the male African-American worker has been dramatically undercut, causing African-American unemployment rates to be at 10.4 percent, or more than double that of whites (U.S. Bureau of the Census, 1996).

Affirmative action programs appear to have little effect on the income of African Americans (Hacker, 1992). Although opponents complain that because of these programs African Americans were given jobs, opportunities, and promotions that whites deserved, the earnings of African-American, male college graduates decreased 11 percent between 1979 and 1989, whereas the earnings of white male college graduates increased by 11 percent.

Occupational stratification also has an important effect on the potential earnings of Latinos. About 17 percent of native-born Latinos are in professional or managerial jobs, in contrast to 29 percent of Anglos. This number drops to 6 percent if the Latinos were foreign born. With the notable exception of Cuban Americans, Latinos tend to be concentrated in blue-collar or service work. Among immigrants, farm laborers account for 12 percent of the employed (Bean and Tienda, 1987).

Employed women tend to be concentrated in relatively low-paying clerical and service occupations. At first glance, there do not appear to be major differences in the job distribution between white and African-American women, but African-American women tend to be employed in the lower-paying levels of these low-paying jobs. Writes Julianne Malveaux (1988), "In addition to being employed in jobs that are 'typically female,' Black women are also employed in jobs that are 'typically' or disproportionately Black female." For example, 41 percent of African-American women in the service industry work as chambermaids, welfare service aides, cleaners, and nurse's aides (overrepresented by a factor of three or four). Among the forty-eight types of clerical occupations, African Americans are overrepresented by a factor of four in six categories: file clerks, typists, teacher aides, key-punch operators, calculating machine operators, and social welfare clerical assistants. These are among the lowest-paying clerical jobs (Malveaux, 1988).

The experience of Latinas is similar in many ways. Cuban-American and Puerto Rican women both have traditionally worked in large numbers; only in the last two decades have Mexican-American women significantly increased their rates of employment. Puerto Ricans and Mexican Americans tend to be concentrated in the lower-paying jobs. Mexican immigrant women enter agricultural jobs in large numbers. All are important in the low-paying service industries.

The overall result is that African-American, Mexican-American, and Puerto Rican families have lower incomes than do Anglos. Cuban Americans tend to do well economically because of the influx of upper-class and middle-class refugees who fled the Castro regime in the early 1960s. Native-born Asian Americans usually have higher median incomes than do non-Asians, including whites.

Lack of Adequate Child Care

As mothers enter the workforce in ever-increasing numbers, high-quality, affordable child care has become even more important. For many women, especially those with younger children and single mothers, the availability of child care is critical to their employment. Almost 10 million children under age five were in some form of child care in 1993 (U.S. Bureau of the Census, 1994).

Inadequate Child-Care Options

For about 70 percent of employed mothers with children aged five to fourteen years, school attendance is their primary day-care solution. Women with preschool children, however, do not have that option; in-home care by a relative is their most important resource. As more mothers with preschool children become employed, families are struggling to find suitable child-care arrangements. This may involve constantly switching arrangements, depending on who or what is available and the age of the child or children (Atkinson, 1994). One

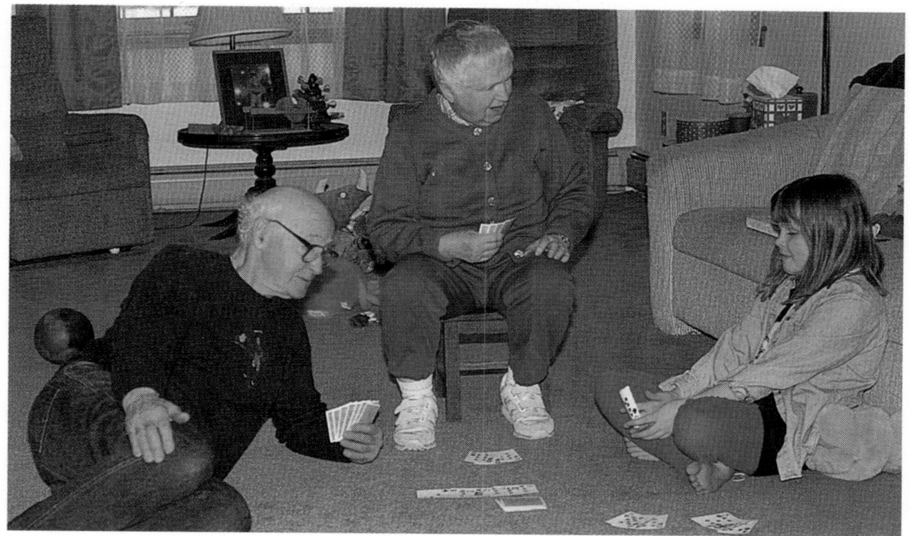

About 10 percent of children are regularly cared for by grandparents.

study found that 23 percent of employed mothers had to change child-care arrangements at least once in the previous year (Casper, Hawkins, and O'Connell, 1994). Relatives, especially fathers and grandparents, are important in caring for children, tending to the needs of almost 40 percent of children (Brayfield, 1995). Married women are more likely to use fathers, and single women to use other relatives (Folk and Beller, 1993). About 14 percent of children are cared for by their fathers; 10 percent, by their grandparents; and 8 percent, by other relatives, such as older brothers and sisters or aunts. The younger the child, the more likely he or she is to be cared for by a grandmother (Presser, 1989; Minkler et al., 1977). In 1991, the primary child-care arrangements for employed women with children under age five were care in the child's home (36 percent), care in another home (31 percent), and organized child-care facilities (23 percent). About 8 percent of employed mothers care for their children at work; most of these mothers work at home (U.S. Bureau of the Census, 1994).

Women often use three or four different arrangements—the child's father, relatives living in or outside the household, day care, or a combination of these—before a child reaches school age. For African-American and Latina single mothers, living in an extended family in which they are likely to have other adults to care for their children is an especially important factor allowing them to find jobs (Rexroat, 1990; Tienda and Glass, 1985).

Frustration is one of the most common experiences in finding or maintaining day care. Changing family situations, such as unemployed fathers' finding work or grandparents' becoming ill or overburdened, may lead to these relatives being unable to care for the children. Family day-care homes and child-care centers often close because of low wages or lack of funding. Furthermore, child care may be quite expensive. In 1991, the average weekly cost was $55 for one child, $66 for two, and $79 for three. This amounts to 7 to 9 percent of the average family's income. For poor families with one child, it amounts to 17 percent of their income (Casper, Hawkins, and O'Connell, 1994). In fact, the high cost of child care is a major force keeping mothers on welfare from working (Joesch, 1991).

In recent years, day care inside the child's home by relatives or baby-sitters has decreased. Over half the children of employed parents are cared for outside their own homes. This shift from parent care at home to day care outside the home may have profound effects on how children are socialized (see Chapter 10). It certainly disturbs many parents, who wonder whether they are, in fact, good parents. Many parents express concern about their children's being cared for outside their own family environment. Concerns about their children's development tends to negatively affect parental well-being and feelings about work (Greenberger and O'Neil, 1990).

Parents who accept the home-as-haven belief—that the home provides love and nurturing—prefer placing their children in family day-care homes. They believe that a homelike atmosphere is more likely to exist in family day care than in preschools or children's centers, where greater emphasis is placed on education (Rapp and Lloyd, 1989). But we know relatively little about family day care: It is the most widely used and least researched form of American child care (Goelman, Shapiro, and Pence, 1990).

Impact on Employment and Educational Opportunities

The lack of child care or inadequate child care has the following consequences:

- It prevents women from taking paid jobs.
- It keeps women in part-time jobs, most often with low pay and little career mobility.
- It keeps women in jobs for which they are overqualified and prevents them from seeking or taking job promotions or the training necessary for advancement.
- It conflicts sometimes with women's ability to perform their work.
- It restricts women from participating in education programs.

For women, lack of child care or inadequate child care is one of the major barriers to equal employment opportunity. Many women who want to work are unable to find adequate child care or to afford it. Child-care issues play a significant role in women's choices concerning work schedules, especially among women who work part-time. Eighteen percent of women with preschool children indicated that the need for better child-care arrangements dictated whether they worked day or night shifts. Among women working part-time, 47 percent said that the availability of child care was the prime consideration in choosing a work shift (Casper, Hawkins, and O'Connell, 1994). Studies of women in welfare-to-work programs report that the mothers' staying in the training program depends not only on the supply of child care but, also, on the quality and convenience of the care (Meyers, 1993).

Child care is an important consideration for students with children. In 1991, child-care costs averaged $45 weekly—about 6 percent of their income—for mothers enrolled in school (Casper, Hawkins, and O'Connell, 1994). At the University of Michigan, 20 to 25 percent of students with children said they would seek more employment or education if child-care services were available. Forty percent of colleges and universities—approximately one thousand—provide day-care facilities for children of students, and waiting lists are long. One student testified before the Select Committee (1985) about her child-care difficulties:

> For a year I had to drive 40 miles a day to take my infant son to a Title XX [child care for low- and moderate-income families] licensed child-care provider. Several times

> I was nearly forced to terminate my schooling because I had no infant care. I have often missed exams and have had to take incomplete grades because of a sick child.

For some, child care determines the pace of their academic careers; for others, it determines whether additional education can be pursued.

Children's Self-Care

Because of the lack of adequate after-school programs or prohibitive costs, **self-care** among five- to eleven-year-olds has quickly become a major form of child care (Koblinsky and Todd, 1989). Self-care increased 40 percent during the 1980s (U.S. Bureau of the Census, 1990). It is a rapidly growing phenomenon, largely the result of inadequate or costly child care and the increasing numbers of married and single mothers who work outside the home (Folk and Yi, 1994). Self-care exists in families of all socioeconomic classes. In fact, there is little difference in the rates of self-care between the poor and the middle class (Casper, Hawkins, and O'Donnell, 1994).

The popular image of self-care arrangements is negative; such children are known pejoratively as "latchkey children." Research on self-care arrangements has been sketchy and often contradictory. Some studies find no differences in levels of achievement, fear, and self-esteem between children in day-care and self-care arrangements. Other studies do find differences. Some children are frightened, others get into mischief, and still others enjoy their independence. These differences may result from the child's own readiness to care for himself or herself, the family's readiness, and the community's suitability (Cole and Rodman, 1987). Rather than rejecting self-care altogether, researchers suggest that it may be appropriate for some children and not for others. Children who are mature physically, emotionally, and mentally, whose families are able to maintain contact, and who live in safe neighborhoods, for example, may have no problem in self-care. But if children are immature or live in unsafe neighborhoods, they may find themselves overwhelmed by anxiety and fear. Parents need to evaluate whether self-care is appropriate for their children (see Cole and Rodman, 1987, for suggested guidelines). Because of the rise in self-care, educators are developing programs to teach children and their parents such self-care skills as basic safety, time management, and other self-reliance skills.

Did you know?

Almost 3.7 percent of five- to eleven-year-olds care for themselves after school for some part of the time while their parents are at their jobs (Casper, Hawkins, and O'Donnell, 1994).

Inflexible Work Environment

In dual-worker families, the effects of the work environment stem from not just one workplace but two. While some companies and unions are developing programs that are responsive to family situations (Crouter and Manke, 1994), in general, the workplace has failed to recognize that the family has been radically altered during the last fifty years. Most businesses are run as if every worker were male with a full-time wife at home to attend to his and his children's needs. But the reality is that women make up a significant part of the workforce, and they don't have wives at home. Allowances are not made in the American workplace for flexibility in work schedules, day care, emergency time off to look after sick children, and so on. Many parents would reduce their work schedules to minimize work-family conflict. Unfortunately, many don't have that option.

Carol Mertensmeyer and Marilyn Coleman (1987) note that our society provides little evidence that it esteems parenting. This seems to be especially true in the workplace, where corporate needs are placed high above family needs. Mertensmeyer and Coleman write:

Reflections

Of the family economic issues discussed previously, have any affected you or your family? How? How were they handled?

Homeless families represent about a quarter of the homeless population; they are the fastest-growing segment of that population. Homeless children are likely to suffer depression, anxiety, and malnutrition.

> Family policymakers should encourage employers to be more responsive in providing parents with alternatives that alleviate forced choices that are incongruent with parents' values. For example, corporate-sponsored child care may offset the conflict a mother feels because she is not at home with her child. Flextime and paid maternal and paternal leaves are additional benefits that employers could provide employees. These benefits would help parents fulfill self and family expectations and would give parents evidence that our nation views parenting as a valuable role.

Employees who feel supported by their employer with respect to their family responsibilities are less likely to experience work-family role strain. It seems that having a family-friendly atmosphere is an integral part of how business organizations can help employees balance work and family obligations (Warren and Johnson, 1995). A model corporation would provide family-oriented policies that would benefit both its employees and itself, such as flexible work schedules, job-sharing alternatives, extended maternity and/or paternity leaves and benefits, and child-care programs or subsidies. Such policies would increase employee satisfaction, morale, and commitment.

Unemployment

Unemployment is a major source of stress for individuals, with its consequences spilling over into their families (Voydanoff, 1991). Even employed workers suffer anxiety about possible job loss due to economic restructuring and downsizing (Larson, Wilson, and Beley, 1994). Job insecurity leads to uncertainty that affects the well-being of both worker and spouse. They feel anxious, depressed, and unappreciated. For some, the uncertainty prior to losing one's job causes more emotional and physical upset than the actual job loss itself. (Refer to the "Money Matters" section of the Resource Center at the back of the text for information on money management, and other issues of economic concern.)

Economic Distress

Those aspects of economic life that are potential sources of stress for individuals and families make up **economic distress** (Voydanoff, 1991). Major economic sources of stress include unemployment, poverty, and economic strain (such as financial concerns and worry, adjustments to changes in income, and feelings of economic insecurity).

In times of hardship, economic strain increases, and the rates of infant mortality, alcoholism, family abuse, homicide, suicide, and admissions to psychiatric institutions and prisons also sharply increase. Patricia Voydanoff (1991), one of the leading researchers in family-economy interactions, notes:

> A minimum level of income and employment stability is necessary for family stability and cohesion. Without it many are unable to form families through marriage and others find themselves subject to separation and divorce. In addition, those experiencing unemployment or income loss make other adjustments in family composition such as postponing childbearing, moving in with relatives, and having relatives or boarders join the household.

Further, economic strain is related to lower levels of marital satisfaction as a result of financial conflict, the husband's psychological instability, and marital tensions.

Did you know?

In 1996, as much as 5.4 percent of the workforce over age twenty was unemployed. By ethnicity, 4.6 percent of whites, 10.6 percent of African Americans, and 8.3 percent of Latinos were unemployed (U.S. Bureau of Labor Statistics, 1996).

The emotional and financial cost of unemployment to workers and their families is high. A common public policy assumption, however, is that unemployment is primarily an economic problem. Joblessness also seriously affects health and the family's well-being.

The families of the unemployed experience considerably more stress than those of the employed. Reports by Ramsey Liem and his colleagues (cited in Gnezda, 1984) indicate that in the first few months of the male primary wage earner's unemployment, mood and behavior changes cause stress and strain in family relations. As families adapt to unemployment, family roles and routines change. The family spends more time together, but wives often complain of their husbands' "getting in the way" and not contributing to household tasks. Wives may assume a greater role in family finances by seeking employment if they are not already employed. After the first few months of their husbands' unemployment, wives of the unemployed begin to feel emotional strain, depression, anxiety, and sensitiveness in marital interactions. The children of the unemployed are more likely to avoid social interactions and tend to be more distrustful; they report more problems at home than do children in families with employed fathers. Families seem to achieve stable but sometimes dysfunctional patterns around new roles and responsibilities after six or seven months. If unemployment persists beyond a year, dysfunctional families become highly vulnerable to marital separation and divorce; family violence may begin or increase at this time (Teachman, Call, and Carver, 1994).

When, in a city of 100,000, only one man is unemployed, that is his personal trouble, and for its relief we properly look to the character of the man, his skills, and his immediate opportunities. But when in a nation of 50 million employees, 15 million men are unemployed, that is an issue, and we may not hope to find its solution within the range of opportunities open to any one individual. The very structure of opportunities has collapsed. Both the correct statement of the problem and the range of possible solutions require us to consider the economic and political institutions of the society, and not merely the personal situation and character of a scatter of individuals.

C. Wright Mills, *The Sociological Imagination*

The types of families hardest hit by unemployment are single-parent families headed by women, African-American and Latino families, and young families. Wage earners in African-American, Latino, and female-headed single-parent families tend to remain unemployed longer than other types of families. Because of discrimination and the resultant poverty, they may not have important education and employment skills. Young families with preschool children often lack the seniority, experience, and skills to regain employment quickly. Therefore, the largest toll in an economic downturn is paid by families in the early years of childbearing and child rearing.

Coping with Unemployment

Economic distress does not necessarily lead to family disruption. In the face of unemployment, some families experience increased closeness (Gnezda, 1984). Families with serious problems, however, may disintegrate. Individuals and families use a number of coping resources and behaviors to deal with economic distress. Coping resources include an individual's psychological disposition, such as optimism; a strong sense of self-esteem; and a feeling of mastery. Family coping resources include a family system that encourages adaptation and cohesion in the face of problems and flexible family roles that encourage problem solving. In addition, social networks of friends and family may provide important support, such as financial assistance, understanding, and a willingness to listen.

Several important coping behaviors assist families in economic distress caused by unemployment. These include the following:

- *Defining the meaning of the problem.* Unemployment means not only joblessness. It can also mean diminished self-esteem if the person feels the job loss was his or her fault. If a worker is unemployed because of layoffs or plant closings, the individual and family need to define the unemployment in terms of market failure, not personal failure.

Therefore I tell you, do not be anxious about your life, what you shall eat or what you shall drink, nor about your body, what you shall put on. Is not life more than food, and the body more than clothing? Look at the birds of the air; they neither sow nor reap nor gather into barns, and yet your Heavenly Father feeds them. Are you not of more value than they? And which of you by being anxious can add one cubit to his span of life? And why are you anxious about clothing? Consider the lilies of the field, how they grow; they neither toil nor spin; yet I tell you, even Solomon in all his glory was not arrayed like one of these.

Matthew 6:25–29

Reflections

Have you or your family experienced unemployment or job insecurity? How did it affect you? Your family? What coping mechanisms did you use?

- *Problem solving*. An unemployed person needs to attack the problem by beginning the search for another job; dealing with the consequences of unemployment, such as by seeking unemployment insurance and cutting expenses; or improving the situation, such as by changing occupations or seeking job training or more schooling. Spouses and adolescents can assist by increasing their paid work efforts. Studies suggest that about a fifth of spouses or other family members find employment after a plant closing.
- *Managing emotions*. Individuals and families need to understand that stress may create roller-coaster emotions, anger, self-pity, and depression. Family members need to talk with one another about their feelings; they need to support and encourage one another. They need to seek out individual or family counseling services to cope with problems before they get out of hand.

Poverty

One man eats very little and is always full. But another man constantly eats and is always hungry. Why is that?

ZEN RIDDLE

Although poverty and unemployment may appear to be only economic issues, they are not. The family and economy are intimately connected to each other, and economic inequality directly affects the well-being of America's disadvantaged families. Poverty drives families into homelessness. The poor have traditionally been isolated from the mainstream of American society (Goetz and Schmiege, 1996). Poverty is consistently associated with marital and family stress, increased divorce rates, low birth weight and infant deaths, poor health, depression, lowered life expectancy, and feelings of hopelessness and despair ("Poverty Helps Break Up Families," 1993). Poverty is a major contributing factor to family dissolution. In 1991, 12 percent of poor two-parent white families dissolved within two years, compared with only 7 percent of white families above the poverty line. Among African Americans, 21 percent of poor families split up, compared with 11 percent of those above poverty levels. Among Latinos, 11 percent of poor families split up, compared with 9 percent above the poverty line.

Catherine Ross and her colleagues (Ross, Mirowsky, and Goldsteen, 1991) suggest the connection between poverty and divorce:

> It is in the household that the larger social and economic order impinges on individuals, exposing them to varying degrees of hardship, frustration, and struggle. The struggle to pay the bills and to feed and clothe the family on an inadequate budget takes its toll in feeling run-down, tired, and having no energy, feeling that everything is an effort, that the future is hopeless, that you can't shake the blues, that nagging worries make for restless sleep, and that there isn't much to enjoy in life.

Poverty has been increasing since 1981, primarily as a result of sharp swings in employment, an increase in single-parent families, and government cutbacks in assistance to low-income families. Despite stereotypes of the poor being African Americans and Latinos, the majority of the poor—and of welfare recipients—are white (Hacker, 1992). Although we tend to think of poverty as primarily an urban phenomenon, over 9 million poor live in America's rural areas. In some Iowa counties, poverty rates approach 30 percent. It is particularly ironic that hunger is a common problem in America's farming heartland (Davidson, 1990).

Inequality is as dear to the American heart as liberty itself.

WILLIAM DEAN HOWELLS (1837–1930)

Poverty levels differ according to certain characteristics, such as ethnicity and family type. See Figure 11.2 for poverty levels by ethnicity. By family type

5.6 percent of two-parent families are poor, compared to 32.4 percent of single-mother families (U.S. Bureau of the Census, 1996).

Spells of Poverty

The majority of welfare recipients tend to be in poverty for spells of time rather than permanently (Rank and Cheng, 1995). About a quarter of the American population, in fact, requires welfare assistance at one time or another during their lives because of changes in families caused by divorce, unemployment, illness, disability, or death. About half of our children are vulnerable to poverty spells at least once during their childhood (Duncan and Rodgers, 1988). Many families receiving welfare are in the early stages of recovery from an economic crisis caused by the death, separation, divorce, or disability of the family's major wage earner. Many who accept government assistance return to self-sufficiency within a year or two. Only about 2 percent of the population depends heavily on welfare for more than seven out of ten years. Most of the children in these families do not receive welfare after they leave home.

Two major factors are related to the beginning and ending of spells of poverty: changes in income and changes in family composition. Thirty-eight percent of poverty spells begin with a decline in earnings of the head of the household, such as a job loss or a cut in work hours. Other causes include a decline in earnings of other family members (11 percent), the transition to single parenting (11 percent), the birth of a child to a single mother (9 percent), and the move of a youth to his or her own household (15 percent). One study of recipients of **Aid to Families with Dependent Children (AFDC)** found that most women required assistance as a result of changes in their family situations—45 percent after separation or divorce, and 30 percent after becoming unmarried mothers. (AFDC is a government program designed to support poor families.) One-third of the women left the program within a year, half left at the end of two years, and two-thirds left within four years. About a third left the program because their income had increased, another third left when they remarried or reconciled with their mates, and 14 percent left when their children moved away from home or grew up.

Poverty spells are shorter if they begin with a decline in income than if they begin with transition to single parenthood or the birth of a child to a single mother. Half of poverty spells end with an increase in the earnings of the head of the household, and 23 percent end with an increase in earnings to other family members. Fifteen percent end when the family receives public assistance, and 10 percent end when a single mother marries.

Income and Wealth

There are vast disparities in income and wealth between white and African-American, Latino, and other ethnic groups. In fact, in 1990, the U.S. Bureau of the Census reported that over the last two decades, there has been a "growing inequality of income distribution" (Pear, 1990). In 1993, the median income for white families was $39,300, for African-American families it was $21,542, and for Latino families it was $23,654 (U.S. Bureau of the Census, 1996).

The disparity in wealth is even greater than the disparity in income. **Wealth** is a person's net worth, which represents decades of differences in income, investment, and the inheritance of property. This disparity grew significantly during the

Did you know?

In 1995, 38 million Americans—13.8 percent of the population—lived in poverty, the lowest figure since 1980, but still higher than any time during the 1970s (U.S. Bureau of the Census, *March Current Population Survey*, 1996).

Figure 11.2
Percent of Persons below Poverty Level By Race and Hispanic Origin, 1995

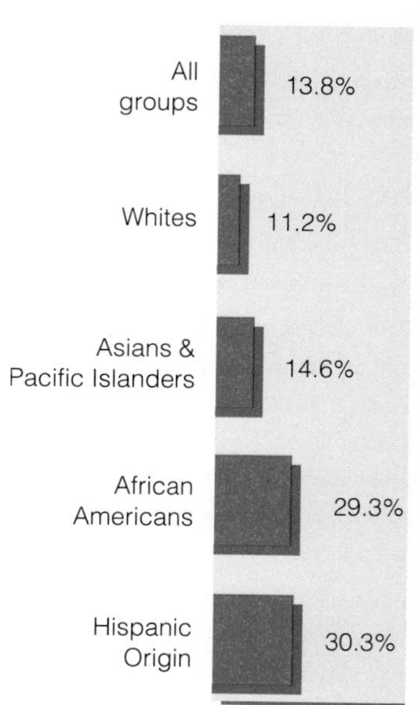

SOURCE: U.S. Bureau of the Census. *March Current Population Survey, 1996.* Washington, DC: U.S. Government Printing Office, 1996.

A Puerto Rican who didn't speak English came to the United States. He went along the streets of New York and came upon an expensive house. He went up to it and asked the butler in Spanish, "Who does this house belong to?"

The butler replied, "What did you say?"

The Puerto Rican misunderstood the butler, thinking he had said "Juan José."

The Puerto Rican went on and saw a beautiful car. He asked the man standing next to it, "Whose car is this?"

"What did you say?" asked the man.

"Dios mio!" exclaimed the Puerto Rican. "It belongs to Juan José."

Then he saw a prosperous clothing store. "Who owns this store?" asked the Puerto Rican.

"What did you say?" was the reply.

Again the Puerto Rican thought the man had said Juan José.

The Puerto Rican saw a group of people surrounding two men fighting. "Who is fighting?" he asked.

"What did you say?" replied the man.

"Juan José is fighting," the Puerto Rican said to himself.

Shortly afterward he came upon a funeral procession. "Whose funeral is this?" he asked.

"What did you say?" a mourner asked.

"Oh, Juan José has died," the Puerto Rican murmured to himself.

Then the Puerto Rican saw a friend and said to him in Spanish, "Juan José has died. He was the richest man in all the United States. He wanted everything and would even fight to get it. And now he is dead. Poor Juan José that he didn't know better."

PUERTO RICAN FOLKTALE. (ORIGINALLY WEST AFRICAN ANANSE [SPIDER] TALE BROUGHT TO THE AMERICAS BY ENSLAVED BLACKS AND MODIFIED OVER TIME TO CHANGED CONDITIONS.)

1980s (Pear, 1991). The net worth of whites is ten times that of African Americans and eight times that of Latinos. The median net worth for whites was $43,280; for African Americans, $4,170; and for Latinos, $5,520. Today, one percent of Americans own 37 percent of the wealth.

The Working Poor

Since 1979, the largest increase in the numbers of poor has been among the working poor because of low wages, occupational segregation, and the dramatic rise in single-parent families (Ellwood, 1988). Over 11.5 million families in 1992 were "working poor." Although their family members were working or were looking for work, these families could not earn enough to raise themselves out of poverty (U.S. Bureau of the Census, 1994).

An individual working full-time at minimum wage earns only 78 percent of the poverty-line income for a family of three. Almost half of two-parent working poor families had at least one adult working full-time. Four out of five poor two-parent families are poor because of problems in the economic structure—low wages, job insecurity, or lack of available jobs (see Chilman, 1991, for a literature review). The young are especially hard hit. Families headed by men and women younger than thirty years of age are experiencing "a frightening cycle of plummeting earnings and family incomes, declining marriage rates, rising out-of-wedlock birth rates, increasing numbers of single-parent families, and skyrocketing poverty rates," writes Marian Wright Edelman (1988).

Women, Children, and Poverty

The **feminization of poverty** is a painful fact. It has resulted primarily from high rates of divorce, increasing numbers of unmarried women with children, and women's lack of economic resources in contrast to men's (Starrels, Bould, and Nicholas, 1994). When women with children divorce, their income falls dramatically.

In 1995, 23.7 percent of children living in families and under age six were poor; their poverty rate is the highest of any group. Like their parents, they move in and out of spells of poverty, depending on major changes in family structure, employment status of family members, or the disability status of the family head (Duncan and Rodgers, 1988). These variables affect ethnic groups differently and account for differences in poverty rates. African Americans, for example, have significantly higher unemployment rates and numbers of never-married single mothers than do other groups. As a result, their childhood poverty rates are markedly higher. Being poor puts the most ordinary needs—from health care to housing—out of reach, jeopardizes the children's schooling, and undermines their sense of self (Edelman, 1989).

The Ghetto Poor

In the last dozen years, the homeless and **ghetto poor,** inner-city residents, primarily African Americans and Latinos, who live in poverty, have become deeply disturbing features of American life, destroying cherished images of wealth and economic mobility. (For information about homelessness, see the "Perspective" on pages 400–401.) One commentator (DeParle, 1991) observes why many find the ghetto poor and homeless so disturbing:

> They suggest a second, separate America, a nation within a nation whose health, welfare, and social mobility evoke the Third World. Indeed visitors startled by rows of homeless lying in Grand Central Terminal in New York strain for analogies and produce the word "Calcutta."

It is not clear exactly who the ghetto poor are. They comprise not simply the poor, who have always existed in great numbers in the United States. They are primarily a phenomenon of the ghettoes and barrios of decaying cities, where poor African Americans and Latinos are overrepresented. The ghetto poor feel excluded from society; indeed, they are often rejected by a society that neither understands nor empathizes with their plight (Appelbome, 1991). Theirs is not a culture of poverty, however; the ghetto poor's behaviors, actions, and problems are often a response to lack of opportunity, urban neglect, and inadequate housing and schooling.

With the flight of manufacturing, few job opportunities exist in the inner cities; the jobs that do exist are usually service jobs that fail to pay their workers sufficient wages to allow them to rise above poverty. Schools are substandard. The infant death rate approaches that of Third World countries, and HIV infection and AIDS are epidemic. The housing projects are infested with crime and drug abuse, turning them into kingdoms of despair. Gunfire punctuates the night. A woman addicted to crack explained, "I feel like I'm a different person when I'm not here. I feel good. I feel I don't need drugs. But being in here, you just feel like you're drowning. It's like being in jail. I hate the projects. I hate this rat hole" (DePerle, 1991).

Within the inner city, residents struggle to maintain their dignity against surging hopelessness. They live day to day, fighting the forces that threaten to engulf them. A mother waiting for her child to return home from school said: "Mostly, you try to keep them away from the drugs and violence, but it's hard. I tell my oldest boy I don't want him hanging out with the boys who are getting in trouble, and he says, 'Aw, mama, ain't nobody else for me to be with'" (DeParle, 1991).

Welfare and the War on the Poor

Since the 1960s, when massive social programs known as the war on poverty cut the poverty rate almost in half, national priorities have shifted. The war on poverty has become the war on welfare—or, as some describe it, the war on the poor. Instead of viewing poverty as a structural feature of our society—caused by low wages, lack of opportunity, and discrimination—we increasingly blame the poor for their poverty (Aldous and Dumon, 1991; Katz, 1990). They are viewed as poor because they are "losers," "cheats," "lazy," "welfare queens," and "drug abusers"—people undeserving of assistance. Poverty is viewed as the result of individual character flaws—or even worse, as something inherently racial (Katz, 1990).

Almost 5 million families (nearly 14 million people) received AFDC (Aid to Families with Dependent Children) benefits in 1994 (U.S. Bureau of the Census, 1996). Additionally, 27 million people received food stamps; their monthly value averaged $71. About 6.3 million children received free school breakfasts and 6.9 million pregnant women, infants, and children under two

Figure 11.3 Money Income of Families, 1994

The distribution of family income is subject to great inequalities in the United States. The wealthiest 20 percent of families receive over ten times the total income of the poorest 20 percent of families. The wealthiest 5 percent receive four times the total income of the poorest 20 percent.

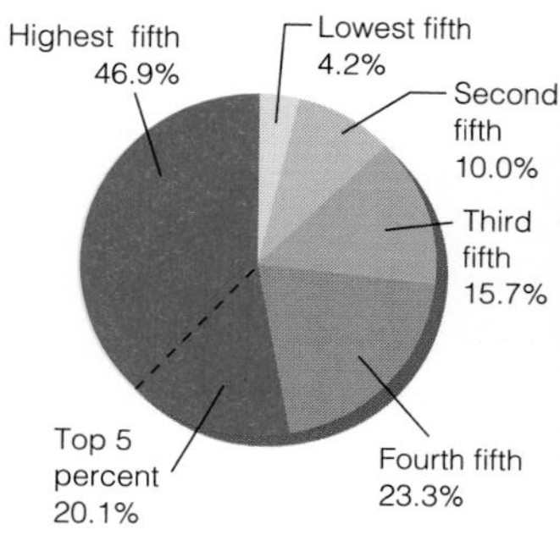

SOURCE: U.S. Bureau of the Census. *Statistical Abstract of the United States, 1996.* Washington, DC: U.S. Government Printing Office, 1996.

The rich get richer and the poor get poorer.

MIGUEL DE CERVANTES (1547–1616)

Did you know?

In 1993, the median income for single working mothers was $12,411 (U.S. Bureau of the Census, 1996).

years of age participated in supplemental food programs known as WIC (Women, Infants, and Children) (U.S. Bureau of the Census, 1996).

Much of the anti-welfare sentiment is based on stereotypes of welfare recipients, especially young unmarried mothers. (While women receiving welfare are often described as "welfare queens," there are no equivalent "welfare kings.")

Joel Handler, a longtime welfare researcher (quoted in Herbert, 1994) describes the stereotype of welfare recipients as "young women, without education, who are long-term dependents and whose dependency is passed on from generation to generation." He further notes: "The subtext is that these women are inner-city substance abusing blacks spawning a criminal class." Furthermore, single mothers receiving welfare are stigmatized as incompetent and uncaring; some suggest that their children be placed in orphanages (Seeyle, 1994). Conservative thinker Charles Murray, for example, believes most adolescent girls "don't know how to be good mothers. A great many of them have no business being mothers and their feelings don't count as much as the welfare of the child" (quoted in Waldman and Shackelford, 1994).

Prosperity doth best discover vice; but adversity doth best discover virtue.

FRANCIS BACON (1561–1626)

Welfare has become a central issue in contemporary politics. It is an emotional "hot-button" issue that generates intense feelings but not necessarily great insight. Welfare is what political commentator Mickey Kaus (1994) calls a "values issue." Many Americans who oppose welfare view it as violating the work ethic and destroying the traditional family. They believe that a person uses welfare as a way to avoid working and that welfare undermines the traditional family by "encouraging" women to become single mothers (Waldman and Shackelford, 1994). Unmarried adolescent mothers are accused of getting pregnant in order to collect welfare benefits. But it is doubtful that adolescents are thinking of welfare benefits as they contemplate premarital sex. In fact, part of the problem is that adolescents often don't make the connection between sex and pregnancy. Furthermore, the birthrate of children to unmarried women between 1979 to 1992 has doubled not only among the poor, but also among those who aren't poor. (Since 1990, the birthrate for unmarried African-American adolescents has declined while that of whites has increased [Bureau of the Census, 1996].) Finally, studies indicate that government welfare policies have had little to do with the rise of divorce, single-parent families, and births to single mothers (Aldous and Dumon, 1991). Indeed, welfare benefits help stabilize families; those states with the most generous welfare benefits also have the lowest divorce rates (Zimmerman, 1991).

Did you know?

Unmarried adolescent mothers account for only 5 percent of AFDC recipients (U.S. Bureau of the Census, 1996).

Numerous approaches to welfare reform have been suggested on both the federal and state level. The general outlines of debate have been set by conservatives who propose that the states or the federal government (1) deny welfare benefits to unmarried adolescent mothers and their children, (2) deny increased benefits if a woman has more children while receiving welfare, (3) require welfare recipients to enroll in work programs, (4) limit the amount of time one can receive welfare benefits, and (5) deny welfare benefits to legal (as well as illegal) immigrants. In addition, they propose to eliminate federal participation in welfare programs by transferring the programs to the states and cutting funding. The states then would receive a limited amount of funds called "block grants," from the federal government to use as they see fit ("Contract with America," 1994; DeParle, 1994).

The true test of civilization is a decent provision for the poor.

SAMUEL JOHNSON (1709–1784)

Moderates and liberals have responded with various proposals that encourage self-sufficiency. In general, their proposals stress education and work training to prepare welfare recipients for employment. They believe that affordable child care should be made available in order for parents to work. Such solutions, however, entail spending public monies at a time many are demanding tax cuts and limits on spending. Moderates and liberals also criticize welfare programs that make children's welfare support dependent on their parents' reproductive or employment behavior (such as not having children if they are unmarried adolescents or finding employment [regardless of its low pay]). They point out that such programs penalize children if their parents "misbehave." Finally, they note that state bureaucracies may be as or more inefficient and unresponsive as the federal government. More important, states may not be equally willing to devote resources to helping welfare recipients out of poverty. In California, welfare recipients receive an average of $556 a month; in Texas the amount is $159 (U.S. Bureau of the Census, 1996). A national system is required to ensure equity. A welfare reform package incorporating both conservative and moderate elements was signed by President Clinton in 1996.

Other progressives argue that the problem is not welfare but poverty. People use welfare for the simple reason that they are poor. The best way to resolve welfare issues is by focusing on the poverty issues underlying it: low wages, unemployment, the high cost of housing, lack of affordable child care, economic discrimination against women and ethnic groups, and a deteriorating education system.

No doubt our welfare system is in trouble. But punitive approaches that blame the poor for their poverty do not resolve the problem. More imaginative

The stoical scheme of supplying our wants by lopping off our desires is like cutting off our feet when we want shoes.

JONATHAN SWIFT (1667–1745)

Did you know?

In 1994, 7.7 percent of the population received public assistance; almost 40 percent of the poor, however, received no such assistance. The average recipient received $382 a month, less than a seventh of the average monthly family income (Hacker, 1992; U.S. Bureau of the Census, 1996).

Figure 11.4
Children under Eighteen Years Old below Poverty Level by Ethnicity, 1994

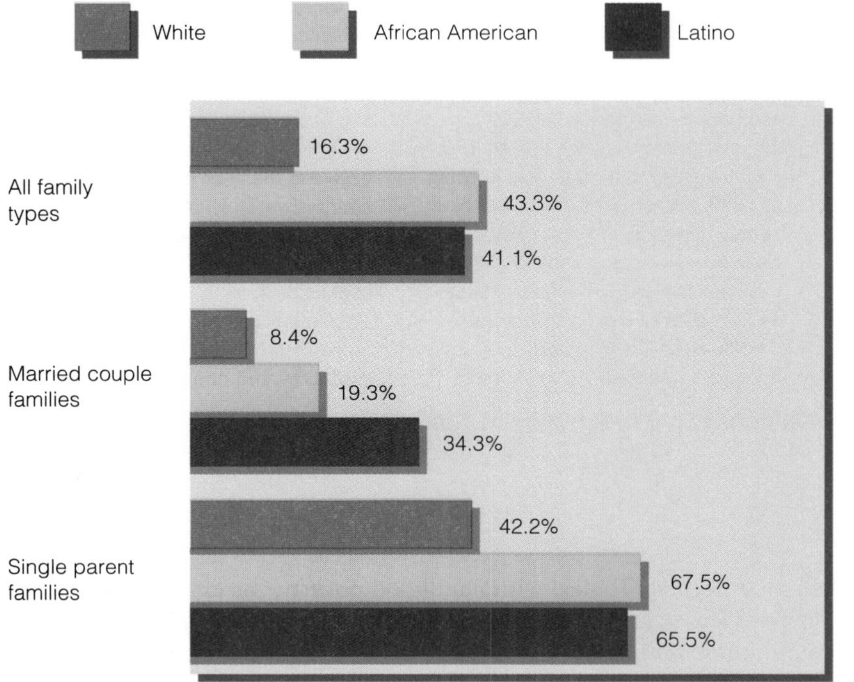

SOURCE: U.S. Bureau of the Census. *Statistical Abstract of the United States, 1996.* Washington, DC: U.S. Government Printing Office, 1996.

The law, in its majestic equality, forbids the rich as well as the poor to sleep under bridges, to beg in the streets, and to steal bread.

ANATOLE FRANCE (1844–1924)

We should not let ourselves be driven into supporting the bad in the hopes of fending off the worse. We stand against policies which deprive poor children and scapegoat poor mothers. A politics of blaming the poor fosters a downward cycle of impoverishment, stigmatization, and despair.

LINDA GORDON, FRANCES FOX PIVEN, AND LOUISE TRUBEK

PERSPECTIVE

Hungry and Homeless in America

I don't know how I could express hunger," the mother said. "It was not to the point where I'd pass out. The children always ate. Maybe I didn't. Maybe not for two or three nights. Maybe some juice during the day, but that's not eating." She had lost her job because of a difficult pregnancy; welfare benefits arrived too late to prevent eviction. For a while she stayed with relatives, but then she was shunted from one welfare hotel to another. Finally, after over a year of living in welfare hotels in New York City, she was moved to a city-renovated apartment. She was lucky, for there are few such apartments available (Kozol, 1988; Roberts, 1988).

Others are not as lucky. They sleep in parks, on steps, in abandoned buildings. They are society's outcasts. A middle-aged, homeless artist, who used to sleep beneath the outstretched arms of Jesus at the front of St. Patrick's Cathedral in New York City, moved to the south side of the cathedral "because the Fifth Avenue Association says I'm bad for the tourists." He now occupies a "cardboard condominium." "It's not bad," he says. "With your body heat you won't freeze to death, even if you cut a couple of holes to breathe" (Hevesi, 1986). One researcher (Rivlin, 1990) describes homelessness:

> Picture a day when you cannot be certain where you will sleep, how you will clean yourself, how you will find food, how you will hang on to your belongings . . . , how you will be safe, how you will dress properly to get to work (for many homeless persons do, indeed, work), and for some, how you will fill up the long hours of a day and do so in a place that will not tolerate your presence. Ordinary activities become major struggles: finding places to wash hair or clothing, locating toilets when needed, surfaces on which to raise swollen feet, or finding food, places to keep warm in cold weather and to escape the rain, snow or summer heat.

The rising numbers of homeless are disturbing. Writes sociologist Christopher Jencks (1994b):

> The faces of the homeless often suggest depths of despair that we would rather not imagine, much less confront in the flesh. Daily contact with the homeless also raises troubling questions about our moral obligations to strangers.

There is considerable debate as to how many Americans are homeless (Jencks, 1994a). The best current estimate is that about 100,000 become homeless each month, and about the same number find housing. In 1990, a national study found that about 3 percent of Americans had been homeless for a time during the previous five years. Between 1985 and 1990, between 6 million and 7 million people spent some time homeless (Jencks, 1994b). About 2 million single-parent families live in someone else's home (Jencks, 1994c). Probably more than 3 million Americans go hungry at least occasionally (Marin, 1987).

The face of hunger and homelessness has changed since the 1970s. Then the homeless were most often single individuals who were drug addicts, alcoholics, or mentally ill persons forced into the streets with the closing of many mental health facilities. Today, however, more than 25 percent of the homeless are families; they represent the fastest-growing sector of the homeless (Kozol, 1988). Most homeless families consist of single mothers with two or three children, most of whom are preschoolers (Bassuk et al., 1986). In some metropolitan areas, as many as three-fourths of the homeless are single mothers (Axelson and Dail, 1988). The families became homeless when the mother lost a job, fled an abusive partner, separated or divorced, became physically disabled or ill, was unable to get welfare, was cut from welfare or had benefits reduced, or was unable to make ends meet on a fixed income.

According to one study (Axelson and Dail, 1988), about half the mothers heading homeless families are between the ages of seventeen and twenty-five. They are equally distributed between whites and African Americans; about 7 percent are Latinas or Asian Americans. Only 10 percent are currently married. More than half are high school graduates, and one-fifth have had

approaches are needed. To deal with childhood poverty, for example, we might use the approach used by all Western industrial nations (except ours): the provision of a minimum children's allowance. A children's allowance is based on the belief that a nation is responsible for the well-being of its children (Meyer, Phillips, and Maritato, 1991). A children's allowance goes to all families. By being universal, no poor child is missed nor is his or her family stigmatized as being "on welfare."

some college. The majority of women became homeless because they fled a relationship rather than because of eviction or job loss. Frequently, a violent incident was the precipitating cause of their homelessness, and they escaped with their children. Homeless mothers report a lack of ongoing family, social, and emotional support. The overwhelming majority say their children are the most important source of support. Twenty-five percent of homeless mothers suffer severe depression or substance abuse, but it is not clear whether these are reactions to homelessness or causes of it.

As a makeshift response to housing needs, some cities provide emergency shelters, usually barrack-style warehouses or single-occupancy hostels. Many shelters permit only women and children, a policy that forces husbands and fathers in intact families to separate from their wives and children. Unless time limits are imposed, emergency shelters tend to become permanent residences. The shelters, however, are generally located in dangerous neighborhoods inhabited by transients, drug abusers, petty criminals, and prostitutes.

Shelters are often frightening environments for families. A twelve-year-old girl paints a desperate picture of life in a residential hotel (quoted in Edelman, 1989):

> I don't like the hotel because there is always a lot of trouble there. I don't go down into the street because there is no place to play. . . . The streets are dangerous, with all kinds of sick people who are on drugs or crazy. My mother is afraid to let me go downstairs. Only this Saturday, my friend, the security guard at the hotel, Mr. Santiago, was killed on my floor. The blood is still on the walls and on the floor.

Under such conditions, children suffer developmental delays, severe depression and anxiety, and learning difficulties. They also suffer from malnutrition. One study (Bassuk and Rubin, 1987) found that as many as half of the school-age children in the sample were in need of psychiatric evaluation; they displayed symptoms of depression, including suicide attempts. Although most of the mothers are intensely concerned about their children's well-being and future, the women feel helpless to prevent the same misfortunes from befalling their own children (Axelson and Dail, 1988).

How did this crisis in hunger and homelessness originate? The major factors were a lack of adequate income, a decline in affordable housing, welfare cuts, and drug and alcohol abuse (Edelman and Mihaly, 1989; Jencks, 1994b). First, poverty increased dramatically beginning in the early 1980s. Second, a housing crisis was slowly growing (Jencks, 1994c). While housing prices skyrocketed, the availability of low-income rentals decreased at the rate of 125,000 apartments a year (Whitman, 1987). Today there are more than twice as many poor households as there is affordable housing. The monthly AFDC housing allowance pays less than half the actual cost of rent. Third, the traditional welfare system underwent a relentless attack; single mothers, demeaned as "welfare queens" and "welfare breeders," were the most direct targets. Although reductions in cash benefits did not reduce the incidence of single motherhood (which actually rose), the reductions in benefits reduced single mothers to homelessness. As a result of draconian welfare cuts, the number of single mothers receiving less than $5,000 a year rose from 600,000 in 1979 to 1.4 million in 1989 (Jencks, 1994b). Without the ability to pay rent, they were unable to find housing or were forced to leave their homes. Fourth, drug and alcohol abuse made marginally employable men and women unemployable, ate up rent money, and made friends and relatives unwilling to shelter them. The arrival of crack in the 1980s decimated the poor (Jencks, 1994b).

We like to find moral fault in the individual; we attribute homelessness to character defects or drugs rather than seek solutions. A commitment to ending homelessness would include the following: (1) the development of low-cost housing and housing subsidy programs, (2) the establishment of child-care programs that would allow homeless single mothers to work, (3) the provision of mental health support to deal with psychological problems, and (4) the initiation of job training and education programs to develop employable skills. As homeless families continue to live and sleep in the streets, in abandoned cars, and in rat-infested shelters, we are confronted with a basic question: Is this the best America can do for its most vulnerable families?

Welfare reform continues to be an issue of acute concern. Evaluation of legislative changes enacted by Congress in 1996 will continue for several years, along with various experimental programs. Each state is in the process of developing its own plans for assisting families and children in poverty. The ongoing challenge is to find ways for people to have adequate food and shelter in an environment that facilitates the development of life skills and assists parents to succeed in the labor force, while at the same time providing for safety, care, and

Reflections

Do you believe that welfare helps or hinders families? Have you, your family, or your friends received welfare assistance? If so, were its effects positive, negative, or both? Why?

guidance of their children. (See Chapter 16 for information on TANF, the new federal welfare program.)

Family Policy

Family policy is a set of objectives concerning family well-being and the specific government measures designed to achieve those objectives. As we examine America's priorities, it is clear that we have an implicit family policy that directs our national goals. Although it has never been articulated, it is very powerful in determining government and corporate policies. The policy is very simple: Families are not a national priority. Its corollary is equally simple: Neither are women and children.

Over the years, a confusing combination of laws affecting families have been passed by state and federal governments. These laws include policies relating to family planning, abortion, sex education, foster care, maternal and child health, child support, AFDC, food stamps, and Medicare (Zimmerman and Owens, 1989). The issues underlying some of these laws, such as abortion, are controversial; others, such as child support, are not. But none of the family support programs that require substantial monies, such as AFDC and WIC (Women, Infants, and Children) is adequately funded.

The question is not whether or not government will intervene. It will. The question is will it intervene for enhancement and prevention or respond to breakdown, problems, and deviance.

ALFRED KAHN AND SHEILA KAMERMAN

Since the late 1970s, interest has been rising in systematic, family-oriented legislation. A systematic approach to family policy would consist of clearly defined objectives concerning family well-being and specific measures initiated by government to achieve them. Interest has been increasing for several reasons (Aldous and Dumon, 1991; Wisensale and Allison, 1989): (1) there has been a major increase in female-headed families because of the rise in divorce rates and births to single mothers; (2) women, especially mothers, have entered the labor force in unprecedented numbers; (3) problems confronting families—ranging from poverty to abuse, homelessness to inadequate child care—are increasingly viewed as social rather than individual problems; and (4) "the family" as symbol has become an ideological battlefield for both liberals and conservatives.

If our families were truly a national priority, the following policies might be instituted by government and business in the areas of health care, social welfare, education, and the workplace (see Edelman, 1989; Macchiarola and Gartner, 1989).

Health-care policies might include the following:

1. Guaranteed adequate medical care for every citizen, with a national health-care policy.
2. Prenatal and infant care for all mothers; adequate nutrition, immunizations, and "well baby" clinics to monitor infant health.
3. Medical and physical care of the aged and the disabled, including support for family caregivers.
4. Education for young people about sexuality and pregnancy prevention; implementation of comprehensive and realistic programs that address the economic and social realities of youth.
5. Education for all Americans about the realities of STDs, HIV, and AIDS and their prevention through abstinence or condom use; guaranteed access to treatment.
6. Drug and alcohol rehabilitation programs available to all who want them.

Social welfare would be enhanced by these policies:

1. Tax credits or income maintenance programs for families.
2. Child allowances to ensure a basic standard of living.
3. Child care for working or disabled parents; temporary respite child care when parents are ill or unable to care for their children; training of neighborhood day-care providers.
4. Advocacy for children, the aged, the disabled, and others who may not be able to speak for themselves.
5. Attention given to problems of the homeless, such as food, shelter, and medical and psychiatric care.
6. Regulation of children's television to promote literacy, good nutrition, humane values, and critical thinking.

In the area of education, the following policies would help everyone:

1. Implementation of preschool programs such as Head Start wherever needed.
2. Stress on the teaching of basic skills; use of innovative programs to reach all students, including outreach programs for adult literacy and English as a Second Language (ESL).
3. Work exposure for students who wish to work through programs such as the Job Corps.
4. Guarantees that all Americans receive the benefits of education through the implementation of bilingual and multicultural programs, special education for the developmentally disabled, and adult education.

Finally, workplace policies might include the following:

1. Paid parental leave for pregnancy and sick children; paid personal days for child and family responsibilities.
2. Flexible work schedules for parents whenever possible; job-sharing alternatives.
3. Increased minimum wage so that workers can support their families.
4. Policies to ensure fair employment for all, regardless of ethnicity, gender, sexual orientation, or disability.
5. Pay equity between men and women for the same or comparable jobs; affirmative action programs for women and ethnic groups.
6. Corporate child-care programs or subsidies for families.
7. Individual and family counseling services; provision of flexible benefit programs.

Our marriages and families are not simply emotional relationships—they are also work relationships in which we divide or share many household and child-rearing tasks, ranging from changing diapers, washing dishes, cooking, and fixing leaking faucets to planning a budget and paying the monthly bills. These household tasks are critical to maintaining the well-being of our families. They are also unpaid and insufficiently honored. In addition to household work and

Reflections

If you were to construct a coherent family policy that meets your needs and reflects your values, what would it be like? How would it compare to the author's suggestions?

A modern nation's honor is not the honor of a warrior; it is the honor of a father providing for his children, it is the honor of a mother providing for her children.

E. L. Doctorow

child rearing, there is our employment, the work we do for pay. Our jobs usually take us out of our homes from twenty to eighty hours a week. They are not only a source of income; they also help our self-esteem and provide status. They may also be a source of work/family conflict.

As we approach the end of the twentieth century, we need to rethink the relationship between our work and our families. Too often, household work, child rearing, and employment are sources of conflict within our relationships. We need to rethink how we divide household and child-rearing tasks so that our relationships reflect greater mutuality. For many, poverty and chronic unemployment lead to distressed and unhappy families. We need to develop policies that help build strong families.

Summary

- Families may be examined as economic units bound together by emotional ties. Families are involved in two types of work: paid work at the workplace and *family work* (unpaid work in the household).
- Employment affects family life. *Work spillover* is the effect that employment has on the time, energy, and psychological functioning of workers and their families at home. *Role strain* refers to the difficulties that individuals have in carrying out multiple roles. Two types of role strain are *role overload*, which occurs when the total prescribed activities for one or more roles are greater than an individual can handle, and *interrole conflict*, which occurs when roles conflict with each other.
- Families must balance family and work needs throughout the family life cycle. The three basic work/family life cycle models are (1) the traditional-simultaneous work/family life cycle, (2) sequential work/family role staging, and (3) symmetrical work/family role allocation. The five stages of the traditional-simultaneous work/family life cycle model are (1) establishment/novitiate, (2) new parents/early career, (3) school-age family/middle career, (4) postparental family/late career, and (5) aging family/post exit. Major problems in this model are related to role strain. In the sequential pattern, women alternate work and mother roles rather than combine them. In the symmetrical pattern, men assume greater household and child-rearing responsibilities.
- The traditional division of labor in the family follows a complementary pattern: The husband works outside the home for wages and the wife works inside the home without wages. Men's participation in household work is traditionally limited to repairs, construction, and yard work. Women's primary responsibility for household work and child rearing is part of the traditional marriage contract.
- Four aspects of the *homemaker role* are (1) its exclusive allocation to women, (2) its association with economic dependence, (3) its status as nonwork, and (4) its priority over other roles for women. Characteristics of housework are that it (1) isolates the person at home; (2) is unstructured, monotonous, and repetitive; (3) is often a restricted, full-time role; (4) is autonomous; (5) is "never done"; (6) may involve child rearing; (7) often involves role strain; and (8) is unpaid.
- Women enter the workforce for economic reasons and to raise their self-esteem. Employed women tend to have better physical and emotional health than do nonemployed women. In 1995, 59 percent of adult women and 75 percent of adult men were employed. Women's employment tends to be influenced by family needs; their labor-force participation is interrupted for family reasons over thirty times as often as is men's participation.
- More than half of all married women are in dual-earner marriages. Husbands generally do not significantly increase their share of household duties when their wives are employed. Employed mothers remain primarily responsible for child rearing. Working wives are more independent than nonemployed women and have increased power in decision making. Women's employment has little or a slightly positive impact on marital satisfaction; there does seem to be a slightly greater likelihood of divorce when the woman is employed.
- Family issues in the workforce include economic discrimination against women; *sexual harassment; occupational stratification* for members of ethnic groups and especially for women, placing them in the lowest levels; lack of adequate child care; and an inflexible work environment.
- Almost 14.0 percent of the population of the United States lives in poverty. There are significant differences between whites and members of ethnic groups in terms of income and wealth. The disparity increased dramatically during the 1980s. As many as 25 percent of Americans go through spells of poverty, during which time they need welfare assistance. Poverty spells generally occur because of divorce; the birth of a child to an unmarried mother; or unemployment, illness, disability, or death of the head of the household. Young families are particularly vulnerable to poverty. The majority of poor people are women and children. The *ghetto poor* are inner-city poor, disproportionately African American and Latino.
- Economic distress refers to aspects of a family's economic life that may cause stress, including unemployment, poverty, and economic strain. Unemployment causes family roles to change; families spend more time together, but wives complain that unemployed husbands don't participate

in housework. Unemployment most often affects female-headed single-parent families, African-American and Latino families, and young families. Coping resources for families in economic distress include individual family members' positive psychological characteristics, an adaptive family system, and flexible family roles. Coping behaviors consist of defining the problem in a positive manner, problem solving, and managing emotions.

- National priorities have shifted from the war on poverty to the war on welfare. Much of antiwelfare sentiment is based on stereotypes of welfare recipients, especially young unmarried mothers. Many Americans who oppose welfare view it as violating the work ethic and destroying the traditional family but research does not support these beliefs.
- Moderates and liberals have responded with various proposals that encourage self-sufficiency, such as education, work training, and affordable child care. Other progressives argue that the problem is not welfare but poverty. The best way to resolve welfare issues is by focusing on the poverty issues underlying them.
- Family policy is a set of objectives concerning family well-being and the specific government measures designed to achieve those objectives. Family policy would contain provisions affecting health care, social welfare, education, and the workplace.

Key Terms

Aid to Families with Dependent Children (AFDC) 395
economic distress 392
family policy 402
family work 367
feminization of poverty 396
ghetto poor 396
homemaker role 375
hostile environment 386
interrole conflict 369
occupational stratification 387
role overload 368
role strain 368
self-care 391
sexual harassment 386
wealth 395
work spillover 368

Suggested Readings

Danziger, Sheldon, Gary Sandefur, and Daniel Weinberg, eds. *Confronting Poverty: Prescriptions for Change.* New York: Harvard University Press, 1994. A well-written book which reviews what antipoverty programs and policies work and where we go from here.

Davidson, Osha Gray. *Broken Heartland: The Rise of America's Rural Ghetto.* New York: Free Press, 1990. A discussion of how America's agricultural heartland has become increasingly impoverished.

Davison, Jane, and Leslie Davison. *To Make a House a Home: Four Generations of American Women and the Houses They Lived In.* New York: Random House, 1994. A rich narrative of what it meant to be a woman, wife, mother, and daughter in one family, chronicled from the turn of the century to the 1980s. Personal experiences, diaries, and poems are interwoven with the larger social, political, and historical context.

Gilbert, Lucia Albino. *Two Careers/One Family.* Newbury Park, CA: Sage Publications, 1993. An excellent description of the two-career family, including research and theory, female and male perspectives, life in dual-career families, workplace policies, and future trends.

Hacker, Andrew. *Two Nations: Black and White, Separate, Hostile, Unequal.* New York: Scribner, 1992. A powerful analysis of the political, economic, and social chasm between whites and African Americans by a noted political scientist.

Hood, Jane C., ed. *Men, Work, and Family.* Newbury Park, CA: Sage Publications, 1993. A collection of essays focusing on the interaction between men's family roles and work roles.

Jenks, Christopher. *The Homeless.* Cambridge, MA: Harvard University Press, 1994. A concise examination of the complex causes of homelessness that sorts out facts from myths.

Kozol, Jonathan. *Amazing Grace: The Lives of Children and the Conscience of A Nation.* New York: Crown, 1995. A powerful and poignant exploration of the social conditions of children in New York's inner city minority, poverty families.

Kurz, Demie. *For Richer, for Poorer: Mothers Confront Divorce.* New York: Routledge, 1995. An examination of the social and economic contexts of divorce and paints a troubling picture for divorced women and their children.

Lerner, Jacqueline. *Working Women and Their Families.* Newbury Park, CA: Sage Publications, 1993. A brief overview of research on the effects of maternal employment on children.

Mahoney, Rhona. *Kidding Ourselves: Breadwinning, Babies, and Bargaining Power.* New York: Basic Books, 1995. A presentation of the thesis that women will not achieve economic equality until men do half the work of raising children.

Rank, Mark. *Living on the Edge: The Realities of Welfare in America.* New York: Columbia University Press, 1994. A work that puts to rest the myth that welfare recipients are only the ghetto poor.

Rubin, Lillian B. *Families on the Fault Line: America's Working Class Speaks about the Family, the Economy, Race, and Ethnicity.* New York: HarperCollins, 1994. The experience of working-class Americans artfully recorded and described, revealing the interaction between work and family life.

Smith, Sheila, ed. *Two Generation Programs for Families in Poverty: A New Intervention Strategy.* Norwood, NJ: Ablex Publishing, 1995. A description of five social service programs that work to integrate the improving of the employability of parents with quality services for children. It is particularly pertinent in light of the recent welfare "reform."

Zavella, Patricia. *Women's Work and Chicano Families.* Ithaca, NY: Cornell University Press, 1987. A well-written anthropological study of Latina cannery workers and their family lives.

CHAPTER 12

Families and Wellness

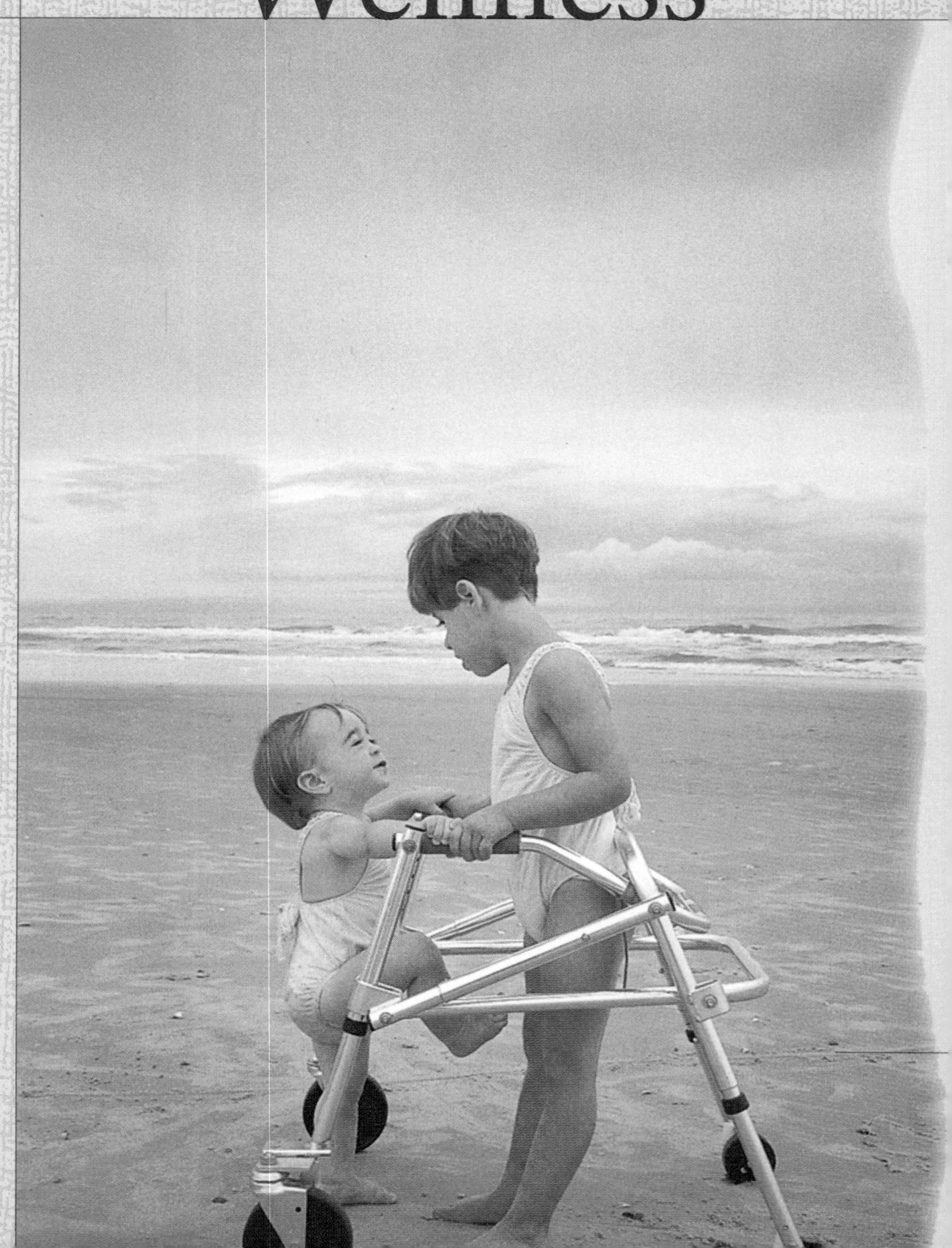

PREVIEW

To gain a sense of what you already know about the material covered in this chapter, answer "True" or "False" to the following statements below.

1. The relationship between our family and our health is casual and independent. True or false?
2. Most of what we learn about health-related behaviors, such as exercise and diet, comes from the media. True or false?
3. Happily married men and women tend to be healthier than unmarried individuals. True or false?
4. About one-fifth of all American children do not receive basic medical care, such as immunizations and checkups. True or false?
5. Getting married is considered a stressful event in people's lives. True or false?
6. Disabilities or chronic illnesses are not necessarily viewed negatively by families experiencing them. True or false?
7. In contrast to women, men provide an insubstantial amount of caregiving for aged parents and in-laws. True or false?
8. Bargaining (with God) is often a normal part of the death and dying process. True or false?
9. A beer is equal in alcohol content to a rum and Coke. True or false?
10. The use of heroin and cocaine is decreasing. True or false?

Answers

1 False, 2 False, 3 True, 4 True, 5 True, 6 True, 7 True, 8 True, 9 True, 10 False

OUTLINE

When a parent brings us soup as we lie sick in bed, when stress at work leaves us too tired to play with our children or enjoy our partner, when our spouse urges us to stop drinking, when a chronically ill child or aging parent requires constant attention, or when a family member dies—we are reminded that families consist of individuals with bodies. Our physical health and emotional health are critical elements of family life. But until fairly recently, family researchers ignored this aspect of families. Families were viewed as having roles, values, and relationships—but not bodies (Doherty, 1993a). During the last twenty or so years, however, research on the interrelationship of families and health has burgeoned.

In this chapter we focus on the social and psychological dimensions of families and health, for, as William Doherty and Thomas Campbell (1988) observe, "The family affects the individual's health and the individual's health affects the family." To understand the interactions between the two, we first examine how marriages and families can encourage mental and physical health among men and women. We also explore the current health-care crisis, and we look at the effects of stress on the health and well-being of families. Then we examine family caregiving for the disabled, the chronically ill, and the elderly. Next we explore death and grieving. We then look at alcohol and drug abuse from a family perspective. Finally, we speculate on the future of health care in the United States.

Families and Wellness

Health and good estate of body are above all good.

ECCLESIASTES 30:15

As we have seen throughout this book, our attitudes toward our bodies, health, and lives are influenced to a great extent by our family. In a healthy family, the six dimensions of wellness (physical, emotional, intellectual, spiritual, social, and environmental) interrelate to create and support optimal health and well-being. Making healthy choices in each of these domains empowers individuals to move away from passivity and illness toward a higher quality of life.

The relationship between the family and health should be viewed as interpenetrating rather than causal. The connections are likely to be subtle and complex, and we need to be wary of oversimplification. Our family does not "make" us ill or well directly, although it does provide the context for our beliefs and behaviors regarding illness and wellness. Our cultural background also affects our relationship to health and illness. How we answer the following questions depends largely on what we've learned in our family and culture: Do we take Sissy to the doctor (or the emergency hospital) with a sore throat and 100-degree temperature, or do we wait and see what develops? Do we go ahead with the leg amputation suggested by Dr. Savage, or do we get a second opinion? Do we trust the medical establishment, or do we believe in alternative or holistic approaches to health care? How much control do we believe we actually have over our health? What we do and feel very likely depends on what we learned in our family, unless we have been independent of our family long enough to form different ideas (which, in turn, we will pass on to our own children).

Life is not merely living but living in health.

MARTIAL (A.D. 40–104)

The Family Context of Health and Illness

"The family has a powerful influence on health beliefs and behaviors because it is the primary social agent in the promotion of health and well-being," writes Thomas Campbell (1993), a professor and physician. It is within our families that we receive our earliest and most powerful messages regarding healthy behaviors and risk reduction, especially as they relate to diet, exercise, stress management, and substance use and abuse. If our family serves healthful foods, encourages exercise, manages their stress, discourages smoking, and models little or no alcohol use, we have a better chance for a long and healthy life than if we grow up eating french fries in front of the television, with Mom smoking a cigarette on one end of the sofa and Dad chugging a six-pack on the other. It is estimated that fully half of all premature deaths are behavior related and could be prevented. Cancer and heart disease are often linked to unhealthful behaviors. (For guidelines to nutrition, see the Resource Center.) Each year in the United States, smoking accounts for over 430,000 deaths plus another 50,000 deaths among nonsmokers exposed to environmental tobacco smoke (APA Position Statement, 1995). This is equivalent to three jumbo jets crashing every day.

The psychological state of the family and its members is also interrelated with the family's health. The family's emotional climate affects the health of its members. For example, chronic distress and conflict can negatively affect health by raising blood pressure and impairing immune response (Campbell, 1993; Sgoutas-Emch, 1994). As we will discuss at greater length later in the chapter, the health or illness of individual family members also affects family psychological well-being. Declining family health appears to affect marital quality negatively in several ways (Booth and Johnson, 1994). Family finances are often strained, the division of labor shifts, spouses may spend less quality time together, and the afflicted person's behavior may be worrisome or require extra attention. It is important for us to acquire skills for coping with long-term illness or disability in the family in order to minimize the deleterious effects and to maximize our enjoyment of life.

The psychological state of the family is interrelated with its health. Families with a positive emotional climate are likely to have better general health than those that experience chronic distress and conflict.

Marriage and Health

The institution of marriage is good for your health, at least if you are happily married. Overall, married men and women (especially men) tend to be healthier and happier than their unmarried peers; they live longer, are less depressed, and have a higher general sense of well-being. An important study found that marriage provides individuals-especially men-with someone who helps to monitor their health, manage their stress, and provide a sense of meaning and obligation, all of which help to decrease risky behaviors and encourage healthy ones (Waite, 1995). Linda Waite, a scholar and researcher from the University of Chicago, goes on to say:

> If we think of marriage as an insurance policy—which it is, in some respects—does it matter if more people are uninsured or are insured with a term rather than a whole-life policy? I argue that it does matter, because marriage typically provides important and substantial benefits.

Cohabitation has some but not all the characteristics of marriage, so it carries some but not all the benefits. Cohabitants are much less likely than married couples to pool financial resources, more likely to assume that each partner is responsible for supporting him or herself financially, and more likely to spend free time separately (Blumstein and Schwartz, 1993). This independence makes investment in the relationship and specialization with this partner much riskier than in marriage and so reduces them (Waite, 1995).

New research has found that, in part, unhealthy behaviors and characteristics influence marriage rates (Fu and Goldman, 1996). Persons with unhealthy behaviors (such as high levels of alcohol consumption or the abuse of drugs) and with poor physical health status (such as obesity) face a greater difficulty in finding an acceptable spouse.

Most researchers suggest that marriage encourages people to be healthy by enabling them to participate in certain behaviors or situations that promote good health (Gove, Style, and Hughes, 1990; Waite 1995). There appear to be three categories of such enabling factors: (1) living with a partner, (2) social support (also discussed in Chapter 4), and (3) economic well-being (Ross, Mirowsky, and Goldsteen, 1991).

Reflections

What kinds of health habits did your family have when you were growing up? How have your personal health habits been influenced by those in your family?

Living with a Partner

Initially, researchers believed that the major health difference between married and single men and women resulted from the presence of another person, whether it was a roommate, friend, or spouse. More recently, it has been suggested that not just anyone will do. It is the person's marital partner who is critical to his or her health—a spouse who will sustain the other in time of need, make chicken soup when the partner is sick, and dance with him or her when happy. Many of these same benefits likely accrue to cohabiting partners as well.

Social Support

At every age, people who feel connected with others experience better physical and psychological health. The fact that individuals with few social contacts face two to four times the mortality rate of others and that married individuals face lower risks of dying at any point than those who have never married or whose previous marriage has ended give credence to the role of social support in a relationship (Glasser and Kiecolt-Glaser, 1994; Waite, 1995. (For a further discussion about the role of social support in a relationship, see the Perspective, "Social Support and Wellness.") To begin to get an idea of the value of social support, just talk to anyone who has suffered from a bad case of the flu and been nursed back to health by a loyal friend or companion or who has received a back rub when the tension of school and work have felt overwhelming.

Economic Well-Being

Married men and women tend to have higher household incomes than unmarried people: an average of $45,041 annually for married couples compared with $24,593 for single men and $14,498 for single women in 1994 (U.S. Bureau of the Census, 1996). With higher incomes and economies of scale, married men and women are more likely to visit physicians and have health insurance. They are also less likely to experience high levels of stress occasioned by the daily grind of poverty. Poverty is associated with decreased life expectancy; higher infant mortality; and higher rates of infectious diseases, disability, and mortality. The poor are more likely to get sick; if they do get sick, they are less likely to survive. Because of the close tie between race or ethnicity and income, African Americans, Native Americans, and Latinos are especially vulnerable to higher disease and mortality rates (Daley, 1993, Waite, 1995). These issues are discussed at greater length later in the chapter.

Health Care in Crisis

America's health-care system is a strange paradox. Health costs account for nearly 14 percent of the U.S. gross national product. Although the latest technological developments are available for those who are well-to-do or well insured (usually the same people), millions of Americans do not have access to such care. About 39 million people (14 percent) in this country have no insurance coverage, and millions more are underinsured. By contrast, other major industrialized countries offer universal health care for all citizens, regardless of their ability to pay; generally health is better and costs less to maintain than in the United States. The overall health of the Japanese population, for example, is significantly better than that of Americans; yet Japan spends, as a percentage of its gross domestic product, half as much as the United States—and provides health care for all its citizens (Tsuda, Aoyama, and Froom, 1994).

Health and Economics

Because the United States does not offer universal health-care coverage, the greatest determinant of health is socioeconomic status, and the poorer one is, the greater the likelihood of debilitating illnesses and early death. Despite our vast outlay of money and our ever-expanding medical technology, our infant death rate is among the highest in the industrialized world. The health status of many of our African-American communities is that of third-world countries, according to the *Western Journal of Medicine* (Nickens, 1991). A young African-American male in Detroit, for example, has a shorter life expectancy than his counterpart in Bangladesh. Almost half the poor are frozen out of the medical system, and people in the uninsured middle class are petrified of losing their homes to medical bills. Many insured Americans don't seek treatment because of high deductibles and copayments (Freeman and Corey, 1993). We are in crisis, and it is getting worse.

The first wealth is health.

RALPH WALDO EMERSON (1803–1882)

UNDERSTANDING YOURSELF

Creating a Family Health Tree

The genetic inheritance that each of us receives from our parents—and that our children receive from us—contains more than just physical characteristics such as eye and hair color. Heredity also contributes directly to our risk of developing certain diseases and disorders.

For certain uncommon diseases such as hemophilia and sickle-cell disease, heredity is the primary cause; if your parents give you the necessary genes, you'll almost always get the disease. But heredity plays a subtler role in many other diseases, which are caused at least in part by "environmental" influences such as infection, cancer-causing chemicals, or an artery-clogging diet. Although your genes alone will not produce those diseases, they can determine how susceptible you are to them. Researchers have found a genetic influence in many common disorders, including coronary heart disease, diabetes, certain forms of cancer, depression, and alcoholism.

Knowing that a specific disease runs in your family can save your life. It allows you to watch for early warning signs and get screening tests more often than you otherwise would. Changing health habits, too, can be valuable for people with a family history of certain diseases. A smoker with a close relative who had lung cancer, for example, is fourteen times more likely to get the disease than other smokers.

In general, the more relatives that had a genetically transmitted disease and the closer they are to you, the greater your risk. However, nongenetic factors, such as health habits, can also play a role. Signs of strong hereditary influence include early onset of the disease, appearance of the disease largely or exclusively on one side of the family, onset of the same disease at the same age in more than one relative, and occurrence of the disease despite good health habits.

You can put together a simple family tree by compiling a few key facts on your primary relatives: siblings, parents, aunts and uncles, and grandparents. Those facts include the date of birth; major diseases; health-related conditions and habits; and, for deceased relatives, the date and cause of death. (For a free family medical history form and guidelines on what to ask, write The March of Dimes Birth Defects Foundation, 1275 Mamaroneck Ave., White Plains, NY 10605.) Once you've collected the information you want, create a tree using the example here as a guide. Then show your tree to your physician to get a full picture of what the information means for your or your children's health.

A Sample Family Health Tree and What It Means

In this sample family tree, the prostate cancer that killed the man's father means that he should be tested for a prostate tumor at a younger age and more frequently than is generally recommended.

His sisters may need to have earlier, more frequent mammograms because of their mother's breast cancer. If they're overweight, they can reduce their risk by losing weight.

One grandmother and one uncle each died of a heart attack. There are several reasons not to worry too much about that: The two relatives were from different sides of the family; both had the attack at a relatively old age; both had two other major risk

Ethnicity, Economics, and Health

Our ethnicity or race per se does not determine our health as much as does our economic status. Studies find that poor and uneducated whites are as unhealthy as poor and uneducated blacks; the two groups are equally at risk for major diseases (Guralnik, Land, Blazer, Fillenbaum, and Branch, 1993). The factors that affect health the most are socioeconomic (Daley, 1993). Poorer people have less access to preventive care and treatment. They are also likely to be less informed about health issues. Because disproportionate numbers of African Americans, Latinos, and other minority-status groups live in discrimination-enforced poverty, health is a particularly important issue in these communities.

Many of the victims of our inadequate health-care system are children. Fewer than half of American two-year-olds are appropriately vaccinated. In some inner-cities the rate is as low as 10 percent. ("Immunization Information," 1995). Because they miss out on immunizations, screening tests, and medical checkups, these children are at risk for a number of preventable diseases—such as whooping cough, mumps, measles, and rubella—which have now reached epidemic levels in some areas. Uninsured children are three times less likely to receive treatment for acute factors for coronary heart disease (smoking and either diabetes or obesity); and neither of the man's parents had any apparent heart trouble. He should check to see whether either relative had highly elevated cholesterol levels, a possible sign of familial hypercholesterolemia.

The colon cancer that struck another grandmother and uncle is a different story. Two factors suggest a possible hereditary link: They were mother and son, and they both developed the disease at nearly the same, comparatively young age. So the man should be screened early and often.

Finally, alcoholism seems to run in the family. The man should be aware that such a history could indicate a hereditary susceptibility to the problem, though the habit might simply have been passed down by example.

Source: "What's Lurking in Your Family Tree?" *Consumer Reports on Health,* September 1992.

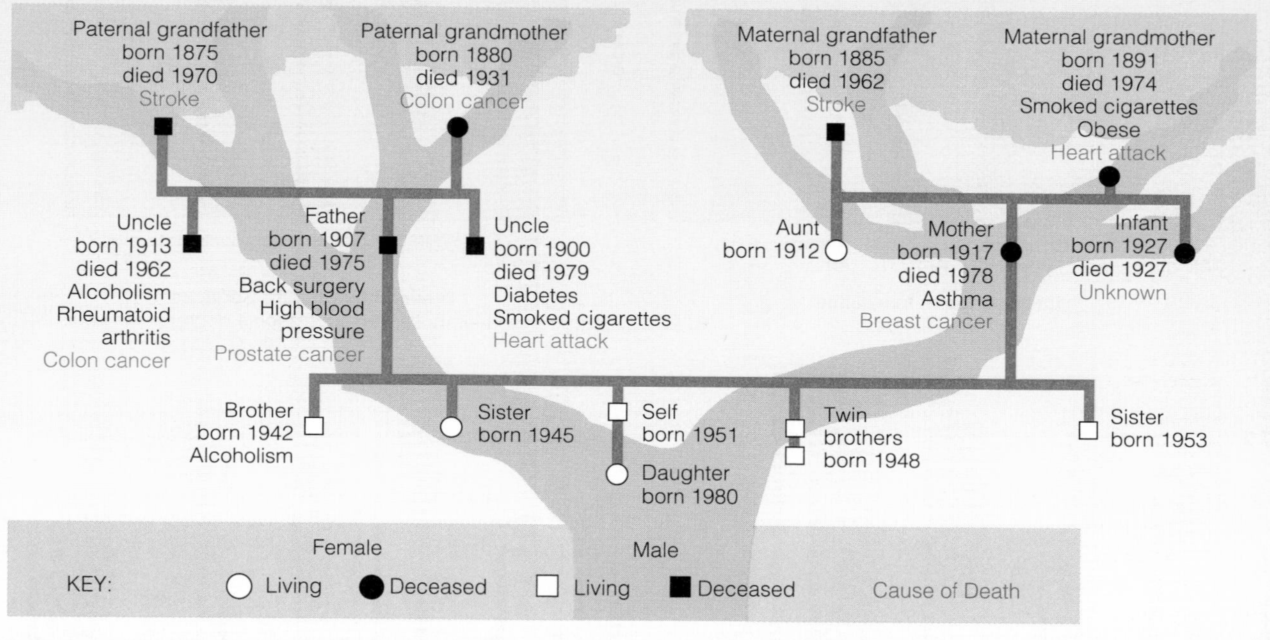

ailments, such as recurrent ear infections and asthma (Stoddard, St. Peter, and Newacheck, 1994). Minority adolescents have greater health problems than do white adolescents; yet they make significantly fewer visits to physicians because of lack of insurance (Lieu, Newacheck, and McManus, 1993).

Poor children are also at increased risk for HIV infection because intravenous (IV) drug users are more prevalent among lower-income groups. As of the end of 1996, 7,629 children under age thirteen had been diagnosed with AIDS and many thousands more are HIV-infected (Centers for Disease Control, 1997). Most HIV-positive children acquire the virus from their mothers before birth (the mother is likely to be the partner of an IV drug user or to be a user herself) (Strong and DeVault, 1997). If a pregnant woman is HIV-infected, there is a 20 to 50 percent chance that her infant will be infected. (See Chapter 8 for more information on HIV and pregnancy.)

Improving children's health appears to have bipartisan congressional support and is one of President Clinton's priorities. A 43-cent-per-pack cigarette tax, in addition to helping reduce the budget deficit, would enable five million additional children to have access to health care ("Children's Health," 1997).

Figure 12.1
A Look at Minority Health

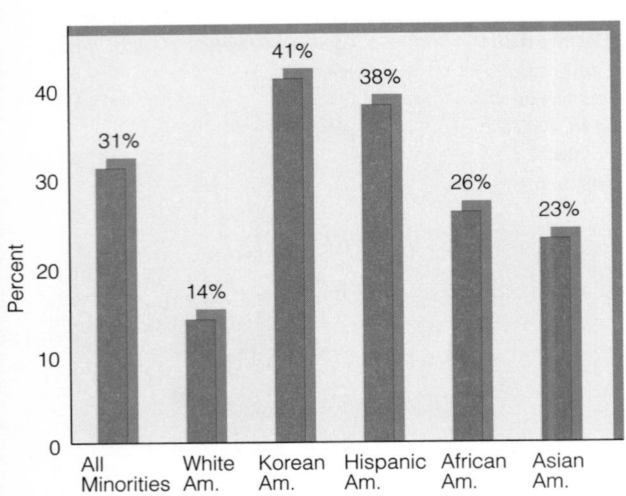

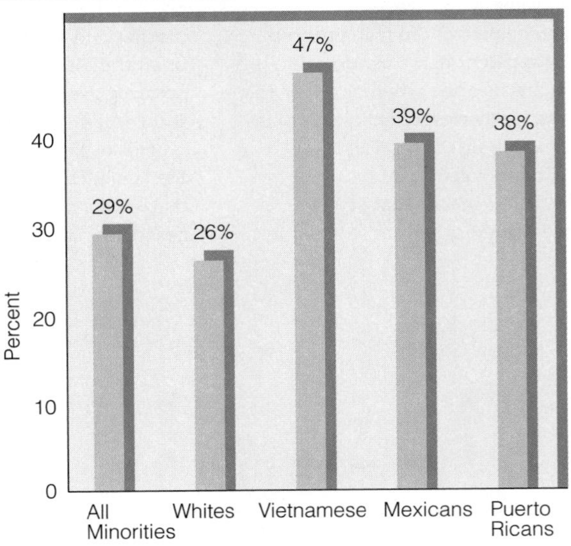

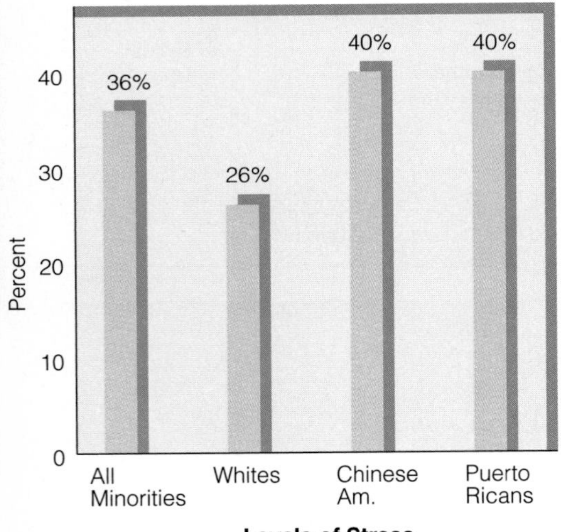

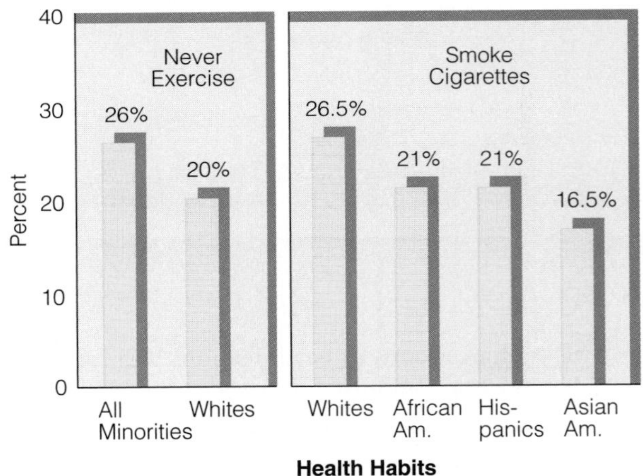

Charts adapted from "A Comparative Survey of Minority Health." New York, NY, The Commonwealth Fund, 1996. [Website: http://www.cmwf.org/minhilit.html]

Environmental Equity

Have you heard about the plans to build a nuclear waste disposal station in Beverly Hills, a wood pulp mill and rayon manufacturing plant in Grosse Pointe, and a chemical waste incinerator in Scarsdale? No, of course you haven't. Factories and facilities that produce toxic and polluting wastes are not constructed in our affluent communities. They are concentrated disproportionately in minority communities (Cushman, 1992; Schneider, 1991). A report by the Federal Environmental Protection Agency presents evidence that racial and ethnic minorities suffer disproportionate exposure to hazardous substances, including lead, carbon monoxide, ozone, and sulfur (Suro, 1993). Workers in our fields and factories are exposed to toxic substances and dangerous working conditions. And poor children of all ethnic backgrounds have higher-than-normal levels of lead in their blood from old paint, old plumbing, and contaminated soil, according to a study by the Federal Agency for Toxic Substances and Disease Registry. African-American children, regardless of socioeconomic status, are more likely to have dangerous levels of lead in their blood (Suro, 1993).

Grassroots organizations around the country are calling for "environmental equity" (Cushman, 1992). "We are the real endangered species in America, people of color," says a community leader in Texas. "We're the ones who are dying with the cancer clusters and the birth defects because of the air we breathe" (quoted in Suro, 1993). Alliances among groups from African-American, Latino, Asian-American, and Native American communities have succeeded in focusing attention on some of the major industrial offenders. Legal teams from such groups as the Natural Resources Defense Council, the American Civil Liberties Union, the National Association for the Advancement of Colored People, and the Sierra Club have filed lawsuits on behalf of threatened communities. These cases, viewing "freedom from pollution" as a basic civil right, demonstrate the connection between environmental protection and social justice.

Individuals, rather than doctors or other medical professionals, are the first line of defense in maintaining good health. Women need to inform themselves about diseases such as breast cancer and take appropriate steps to detect it and treat it if necessary.

Health Insurance

Imagine that you are riding your bicycle to class. You are late, so you are pedaling especially fast, but you hit a rock, causing you to flip off your bicycle. You spin through the air and tumble against the concrete curb. You hear a crunching sound. You get up; you are dizzy and nauseated, and you feel intense pain in your shoulder. You may have suffered a concussion, and you think your shoulder may be broken. Across the street is a private hospital called Mount Profit, which will take care of you immediately; ten miles away is the county hospital, which will examine you in its emergency room, though you will probably have to wait two to six hours while the more serious gunshot and accident victims are treated. Which hospital will you choose? (*Hint:* It probably depends on whether you have medical insurance.)

Because the United States does not provide universal health care, health insurance has become one of the prime determinants of an individual's or family's access to medical care. The spiraling costs of medical care—up to thousands of dollars a day for hospital care—have made visits to the doctor or hospital out of the reach of most Americans without medical insurance. Yet in 1994, 40 million Americans were uninsured at any given time (Davis, 1996). Overall, 31 percent of minority Americans, ages eighteen to sixty-four, and 14 percent of white Americans lack health insurance, with 41 percent of Korean-American, 38 percent of Hispanic-American, 26 percent of African-American, and 23 percent of Asian-American adults uninsured ("Comparative Survey," 1996).

Did you know?
The future looks bleak for the uninsured poor. The Council of Economic Impact of Health System Change estimates that 67 million American individuals will be uninsured in the year 2002 (Davis, 1996).

Most insured people are covered either through their employment, through Medicaid if they are very poor (only one-third of the poor are covered), or through Medicare if they are elderly. Health insurance is closely tied to the workplace, through an employer's benefit program, usually a group insurance plan. An employee pays part or all of the insurance through his or her employer; the amount may be nominal or run into thousands of dollars per year. Because of increasing insurance costs, employers are curtailing their coverage. (The recent rise in the uninsured is largely attributable to cutbacks in employer coverage.) Even among the insured, insurance companies often try to curtail doctor visits and limit access to other types of health care (Bashshur, Homan, and Smith, 1994; Terry, 1994). Furthermore, employee group plans are usually not extended to part-time or temporary workers. Millions of those in low-paying service, retail,

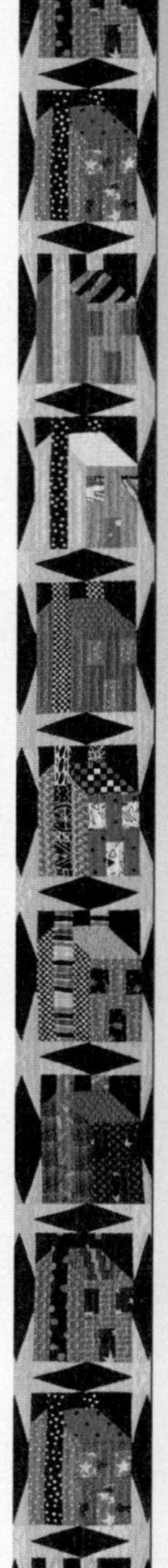

YOU AND YOUR WELL-BEING

Women's Health: A Closer Look

A critical look at women's health, the gender biases that have been built into it, the lack of information about it, and the barriers women face in obtaining it raise important questions for women, policy makers, scientists, and health-care professionals.

Because there appear to be more questions than there are answers, we must be extremely cautious in interpreting the literature and taking at face value the recommendations and biases of those who have been traditionally responsible for women's health. Women must demand complete explanations for any and all procedures that are recommended to them. They should research gender-appropriate diagnoses, procedures, treatments, and methods of prevention, and they need to advocate for themselves and others if they desire change. The health care system that has advised and guided women's decisions is also one that has traditionally been male dominated and discriminatory.

Discrimination is not new to women and comes in many forms. When, for instance, a woman visits her doctor because of lightheadedness and shortness of breath, she is often told that she is under stress, should relax, take tranquilizers, and return in three months. On the other hand, when a man reports these same symptoms, he is more likely than not tested for a heart condition and treated appropriately. Keep in mind that heart disease is the number one killer of women. Because discrimination—based upon sex, age, race, and socioeconomic status—is so pervasive and long-standing, women are often not aware of it until they realize that their health status is threatened.

In spite of their increased life expectancy (79 vs. 72.3 years) women suffer from a higher incidence of disease and debility and spend more money and time on health care than men (Semler, 1995). We must look at women's traditional roles when we examine profiles related to women's health. Economic pressures due to lower salaries, stress from balancing home and work, the likelihood that a women is single and raising children, and the fact that women have traditionally been responsible for the care of children combine to influence and often subsume their own health needs to those of their family members.

The result of putting their health needs behind those of others is that they are at considerable risk of falling prey to detectable and preventable disease. Breast and uterine cancer, heart disease, osteoporosis, and rheumatoid arthritis are not only debilitators and killers of women; they are also diseases for which the male health model is not applicable. With more knowledge and aggressive prevention strategies, the illness and death rates for each of these conditions could be significantly reduced.

Important findings related to women's health came as a result of a 1993 Commonwealth Fund Survey of 2,525 women and 1,000 men. Asked about a variety of subjects, including the access and use of health services, risk factors, and domestic violence, the survey found that the likelihood of women receiving preventive services were influenced by:

- *Insurance coverage.* Insured women with coverage for clinical breast exams have rates of annual screening one-and-a-half times greater than those of uninsured women.
- *Regular sources of care.* Women who do not have a regular source of care are less likely to receive preventive services: 52 percent of women without a regular

or agricultural jobs receive no benefits (Pear, 1993). Employees who lose their jobs often lose their coverage as well. Many employees experience "job lock"—the need to remain with their current employers because they fear loss of health insurance (Cooper and Monheit, 1993).

Universal health coverage has been proposed since the end of World War II, but it has not been until the last few years that it has become a critical social and political issue. Major health insurance reform came to the foreground in the 1992 presidential election and continues to haunt us. Politically, the debate around health-care centers on (1) whether all Americans should be covered by health insurance and (2) how the health insurance is to be paid. It is important to note, however, that without universal coverage, the most vulnerable, such as the poor and members of minority-status ethnic groups, are most likely to suffer. They are the very ones who most require access to health care. Despite the critical need for universal health care, efforts for such coverage have been stalled source of care received a Pap smear in the last year, compared with 68 percent of women with a regular source of care.

- *Financial barriers.* Women with lower incomes have lower preventive screening rates. For example, only 58 percent of women with incomes less than $25,000 had a clinical breast exam in the past year, compared with 75 percent of women with incomes greater than $25,000.
- *Age.* Half of women over age sixty-five failed to get a mammogram within the past year. Failure to obtain timely mammography increases with age (Wyn, Brown, and Yu, 1996).

Of special concern are African-American women who are more likely than white women to be living in poverty (33 percent vs. 12 percent) and to have less than twelve years of education (25 percent vs. 14 percent). African-American women experience more health-care access problems and are more likely than white women to express frustration about finding care (Lillie-Blanton, Bowie, and Ro, 1996). Preventive screening rates among African-American women are discouraging. One in four did not receive a Pap smear in the past year, one-third failed to receive a clinical breast exam, and more than half between the ages of fifty and sixty-four did not receive a mammography screening within the prior year.

Most Latina women in this country are relatively young, live in larger-than-average size households, and have children under age eighteen (Falk and Collins, 1996). Their age alone would suggest that they have a better-than-average health status. However, survey findings indicate otherwise.

Many Latinos have poor health and a compromised sense of well-being. According to Annette Ramirez de Arrellano (1996), one in four Latina women reports her overall health status as fair to poor, compared with about one in seven among all women. They are less likely than other women to report engaging in positive health activities, such as exercise and more than half report having been depressed in the last week. With one in five uninsured and one in four without a source of care, few Latinas get the health care they need. Latinas have the lowest preventive screening rates among the ethnic groups surveyed. "The fact that 42 percent had never been screened for blood cholesterol is particularly problematic, since a comparatively high proportion of Latinas are overweight (77 percent) and do not exercise regularly (38 percent)," writes Ramirez de Arrellano (1996).

Another major health issue for women is violence. The Commonwealth Fund's Commission on Women's Health has identified the pervasiveness of violence in the lives of many American women as a major public health problem (Davis, 1996):

- More than one out of every ten women reported sexual abuse as a child, and nearly 13 percent reported physical abuse.
- Almost 3 percent of women reported that they had been raped in the last five years.
- More than 8 percent of women surveyed who were between the ages of eighteen and sixty-five and were living with a man reported physical abuse by their domestic partners.
- The women who reported abuse were far more likely than other women interviewed to be in poor health, make frequent visits to physicians, abuse drugs, express low satisfaction with life, experience depression, and think about suicide.

As increasing numbers of women enter the fields of politics and health care, women's health issues are becoming more prominent. Too many women suffer from social isolation, frailty, depression, cancer, heart disease, and osteoporosis. Scientists, policy makers, and women *must* recognize the importance and share the task in keeping our nation's women mentally and physically healthy throughout all the years of their lives.

by the corporate health-care establishment, partisan politics, the fear of government and increased taxes, and anxiety over change.

The effects of the insurance crisis are threefold. First, the current health-care situation forces the uninsured to live in constant dread of a major medical expense that can destroy their savings or force them to sell their homes. Second, it leads to deteriorating health because the uninsured do not seek timely medical attention. For the uninsured, health problems require a triage approach (assigning priorities on the basis of urgency). In one family, for example, the oldest child needs a tooth filled; the mother has abdominal pain that requires ultrasonography for diagnosis; and the father has a sprained ankle he keeps on ice, hoping to avoid the cost of an X-ray procedure. The family puts off treatment as long as possible. Finally, they decide to seek treatment for one member. On what basis? The basis of pain. The member with the most pressing pain—the father, in this case—sees a doctor. Yet the mother's abdominal pain may indicate a life-threatening tumor.

UNDERSTANDING YOURSELF

Do You Know Your Medical Rights?

The first step toward protecting your medical rights is to know what they are. Answer "yes" or "no" to the following:

1 Your eight-year-old son is injured in a ball game and rushed to the hospital. At the emergency room, staff members ask you to step outside while they examine the frightened, sobbing boy. Do you have to leave? ____

2 You're going to the hospital for an operation, and you want your partner to be with you through all tests and consultations and to be able to visit you at any time of the day or night. Can the hospital refuse? ____

3 You're moving to another city, and you ask your physician for copies of your medical records. She refuses but agrees to send them to your new doctor. Do you have the right to see them yourself? ____

4 Upon entering a hospital, you're asked to sign a blanket consent form, allowing physicians to administer whatever treatments they deem necessary and relieving the hospital and your doctor of all liability for your care. Do you have to sign it? ____

5 While on a weekend trip out of your home state, you're seriously injured in a car accident. At the nearest emergency room, you're asked for proof of insurance coverage; you have none. Can the hospital refuse to treat you? ____

6 You have a complex medical problem, and your doctor recommends that you enter a large teaching hospital that provides training for doctors, nurses, and other health professionals. You're afraid that you won't have any privacy. Can you refuse to be interviewed or examined by anyone but your own physician? ____

7 You're hesitant about a treatment your physician strongly recommends. You'd like to seek a second opinion and spend more time discussing your options, but you're afraid that your doctor might drop you as a patient if you consulted another doctor or demanded more attention. Can a physician drop a patient for challenging him or her? ____

8 Hospital staff members routinely call you by your first name or a nickname, such as "Sweetie" or "Sonny," which makes you feel childlike and powerless. Can you insist on more respectful treatment? ____

9 As you prepare to leave the hospital, you're informed that you won't be released until you pay all charges not covered by your insurance policy. Can the hospital keep you until you ante up? ____

10 You agreed to participate in a research test of an experimental drug mainly because you didn't want to say no to your doctor. However, the side effects are much worse than you anticipated. Can you drop the study? ____

11 When you get a bill from your doctor or hospital, you're stunned by the total amount and can't make sense of the specific items. Can you refuse to pay it? ____

12 You've had a biopsy, and you think your overly protective family doesn't want your doctor to tell you if you have cancer. You'd rather know the facts, however upsetting they might be. Do you have a right to insist that your doctor tell you the whole truth? ____

Third, the insurance crisis affects the ability of our public hospitals and emergency rooms to function. Because so many Americans are uninsured, they turn to public hospitals for medical treatment. As a result of budget cutbacks and growing numbers of uninsured, waits in public hospitals for nonemergency visits are routinely three to six months. Emergency rooms are crowded, leading some private hospitals to turn away walk-in patients or to "dump" those requiring critical care by sending them to public hospitals, which are often already overcrowded (Olson, 1994).

Stress and the Family

Marriage, no less than life in general, is just one damned thing after another.

FRANK PITTMAN

Life involves constant change, to which both the individual and the family must adjust. Change brings **stress**—psychological or emotional responses or adaptations made to any demands that disturbs homeostasis. It does not matter whether the change is for the better or the worse; stress is produced in either case. Because

Answers

Compare your answers with those below. A high total of correct replies (ten or more) indicates that you're well prepared to deal with the health-care system. The lower your score, the more you may need to learn about protecting your medical rights.

1 *No.* Parents have a right to stay with a child twenty-four hours a day, in either a hospital or a doctor's office. Health professionals can't stop you from remaining with your child during all tests and treatments or from being in the recovery room when your child regains consciousness after surgery. You can be forced to leave your child alone with health-care providers under only two circumstances: if they suspect child abuse by a parent or guardian or if you're interfering with medical treatment (for example, getting in the way during a diagnostic test).

2 *No.* Adult patients have the right to request that a spouse, grown child, relative, or friend stay with them during an exam, diagnostic test, or treatment. As long as your partner doesn't get in the way, he or she can remain with you and act as your advocate, asking for information and making sure you understand what's happening.

3 *That depends on your state's law.* In most states patients are entitled by law to copies of their medical records. All they have to do is sign a release form and pay a copying fee. Other states guarantee patients access to their records only under certain conditions, such as if they show good cause for wanting to see them, or exclude lab reports, X-rays, prescriptions, and technical information. Some states require patients to obtain access through an attorney or physician.

4 *No.* Legally, you have the right not to sign. However, unless you require emergency care, a hospital can refuse to admit you unless you provide some sort of authorization. Remember that the actual content of hospital forms is not set by law, but by the hospital, and you can challenge or change it. Some consumer advocates suggest signing the blanket form to give consent for routine hospital procedures, such as taking blood pressure, and asking for a separate consent form for each invasive procedure, such as certain diagnostic tests or surgery.

5 *No.* A hospital with emergency facilities cannot turn away anyone requiring immediate treatment, regardless of ability to pay. It doesn't matter whether you actually have insurance but no proof or have no insurance at all. You still have a legal right to prompt attention in a medical emergency (any situation that is likely to cause death, disability, or serious illness if not attended to immediately).

6 *Yes.* Patients always have the right to refuse to be examined by anyone not involved directly in their care. However, in a teaching hospital, you will have more people treating you than in a smaller community hospital. Your own doctor (your "attending" physician) will be in charge, but recent medical school graduates (interns and residents) will provide most of your day-to-day care. In addition, as part of their training, medical students also regularly examine patients. By law, they must identify themselves as students.

7 *No.* Doctors can't drop patients because they sought another opinion or didn't follow their advice. According to the American

(Continued on following page)

all families undergo periods of stress, the ability to cope successfully is seen as an important factor in measuring family health.

Family Stress and Coping

Families constantly face stress—tension resulting from real or perceived demands that require the family system to adjust or adapt its behavior. (See Chapter 2 for a discussion of family systems.) Not all stress, however, is bad for you. In fact, there are many who believe that humans need some degree of stress to stay well. Stress be beneficial when it serves as a positive motivator. An example of this type of stress is when a child is trying out for a part in a school play. Sure, she feels frightened, but that stress can often motivate her to be prompt for the audition, practice her lines, and utilize all the gestures necessary in order to make her character believable. Beyond this optimal point, stress does more harm than good (Seaward, 1997).

There are three types of stress: **eustress,** or good stress, that results in a person feeling motivated or inspired; **neustress,** or stimuli that have no consequential

Did you know?

Three quarters of Americans believe the amount of stress in their lives is within their control. How do they handle this stress? Exercise is the preferred way (42%), others slow down (17%), take time off (11%), watch television (7%) or meditate (3%). ("A Nation Out of Balance," 1994.)

UNDERSTANDING YOURSELF

Do You Know Your Medical Rights? (continued)

Medical Association's Principles of Medical Ethics, doctors must comply with a patient's request for a second opinion or a referral to a specialist. Many insurance carriers now require a second opinion for hospitalization or surgery, and most physicians consider them routine.

8 *Yes.* The Joint Commission for the Accreditation of Hospitals (JCAH), which licenses the nation's hospitals, recognized a patient's "right to considerate, respectful care at all times and under all circumstances, with recognition of his personal dignity." Rude, patronizing, or indifferent treatment is not only disrespectful but demeaning and demoralizing.

9 *No.* Anyone can leave the hospital at any time—even against medical advice or without paying the bill. Trying to keep patients against their will for any reason represents false imprisonment.

10 *Yes.* You always have the right to say no or stop any treatment, whether or not it is experimental. You don't have to provide any justification. By law, a physician or hospital can never administer a treatment against your will.

11 *Yes.* Regardless of who is paying for their care, patients have the right to receive and examine an itemized, detailed bill for all services rendered in a doctor's office or hospital. If the charges seem exorbitant, it's recommended that you write a letter to the doctor, clinic, or hospital stating that you are withholding payment (which is your right) until you get an explanation of the charges. If the explanation is unsatisfactory, document the reasons why you believe the charges to be exorbitant. For instance, compile a list of comparable charges for the same services from other health-care providers.

12 *Yes.* By law, physicians must tell patients what their diagnosis and prognosis are—and in language they can understand. The only exceptions are when the patients specifically say that they don't want to know and when the doctor has good reason to believe that the patient could be harmed by the information (for instance, if a patient is extremely depressed and might consider suicide after learning of a fatal disease).

Source: Dianne R. Hales, *An Invitation to Health,* 7th ed. Copyright © 1977, The Benjamin/Cummings Publishing Company. Used by permission of Brooks/Cole Publishing Company.

effect; and **distress,** or bad stress. This type, often used interchangeably with the term *stress*, will be discussed in greater detail throughout this chapter.

When a family is under stress, a psychological and physical state involving conditions such as intense upset, mood changes, headache, and muscle tension may result.

Both individuals and families must adapt to stressful events, but they must also maintain their equilibrium. They must keep their identities intact. As a result, individuals and families are involved in a delicate balancing act to adapt to changing situations while maintaining their integrity. They respond to stress in many different ways. How we cope with stressful demands determines whether or not we experience a crisis.

Don't trouble trouble till trouble troubles you.

TRADITIONAL AFRICAN-AMERICAN PROVERB

Variables Affecting the Response to Stress

The provoking event of stress is known as a **stressor.** Six variables are involved in a family's response to stress:

1. *Stressor.* A life event—such as the birth, adoption, or departure of a family member; illness; or unemployment—that affects that family at a certain point in time and produces change in the family system.
2. *Family hardship.* Difficulties specifically associated with the stressor, such as loss of income in the case of unemployment.
3. *Strains.* The tensions lingering from previous stressors or the tensions inherent in family roles, such as being a parent or spouse. Strains

include the emotional scars from bitter or unresolved fights or the fatigue of parenting.

4 *Resources.* The material, psychological, or social assets that can be used by the family to influence others or to cope with stress, such as money, emotional support, or friends.

5 *Meaning.* How the family defines the event, such as perceiving it as a stressful but manageable problem versus "the end of the world."

6 *Coping.* The process of using the resources the family has at its disposal. The family that copes well experiences much less stress than the family that copes poorly.

Families that are unable to cope find themselves moving from a state of stress into crisis (Boss, 1987). In crisis, the family system becomes immobilized, and the family can no longer perform its functions.

Types of Stressors

Stressors are the provoking events of stress, so we examine them in some detail. Stressors may be (1) normative or nonnormative, (2) external or internal, (3) short term or long term, and (4) with norms or normless.

Normative and Nonnormative Stressors. Certain events are common in all families across the life cycle—birth, marriage, retirement, death of elderly members, and so on. These normative stressors are typical. (See Table 12.1 for stress points in the family life cycle.) Although still stressful, normative stressors can be anticipated and some of their consequences alleviated. Other family stressors are nonnormative, or atypical, such as accidental death, conflict over family roles, sudden loss of income, or caring for a disabled child.

I cannot and should not be cured of my stress, but merely taught to enjoy it.

Hans Selye

External and Internal Stressors. Dealing with stress differs according to whether the stressor originates from inside or outside the family. External stressors can foster family unity if they are not so constant or overwhelming that they destroy the family. Natural disasters, such as floods and hurricanes, encourage family cooperation. Families tend to meet these kinds of events head-on. Wars and persecution have similar effects. But internal stressors can break the family apart. Severe illness, role strain, alcoholism, unemployment, extramarital affairs, or unexpected death may destroy the same family that heroically withstood a natural disaster or persecution. The family may search for a scapegoat, fight among themselves as to whom to blame, or take sides.

Short-Term and Long-Term Stressors. A short-term stressor, such as a broken leg or temporary unemployment, creates stress for a limited time. Such stress can be extremely painful, but the family may return to its normal pattern of interactions once the short-term stressor is eliminated. Long-term stressors, however, are generally more disruptive and require considerable family adjustment. A broken leg soon mends, and the family may return to its usual mode of operation; an amputated limb, however, requires permanent change in the way the family works, such as other family members' performing some of the responsibilities of the disabled person while still allowing him or her to be a fully contributing family member.

Stressors with and without Norms. Society often provides us with norms or guidelines that assist us in coping with stressors. If we give birth to a child under typical circumstances, our society validates us and provides us with

Table 12.1 The Stages of the Family Life Cycle: Stress Points

Family Life Cycle Stage	Emotional Process of Transition: Key Principles	Second Order Changes in Family Status Required to Proceed Developmentally
1 Between families: the unattached young adult	Accepting parent-offspring separation	a. Differentiation of self in relation to family of origin b. Development of intimate peer relationships c. Establishment of self in work
2 The joining of families through marriage: the newly married couple	Commitment to new system	a. Formation of marital system b. Realignment of relationships with extended family and friends to include spouse
3 The family with young children	Accepting new members into the system	a. Adjusting marital system to make space for child(ren) b. Taking on parenting roles c. Realignment of relationships with extended family to include parenting and grandparenting roles
4 The family with adolescents	Increasing flexibility of family boundaries to include children's independence	a. Shifting of parent/child relationships to permit adolescent to move in and out of system b. Refocus on midlife marital and career issues c. Beginning shift toward concerns for older generation
5 Launching children and moving on	Accepting a multitude of exits from and entries into the family system	a. Renegotiation of marital system as a dyad b. Development of adult to adult relationships between grown children and their parents c. Realignment of relationships to include in-laws and grandchildren d. Dealing with disabilities and death of parents (grandparents)
6 The family in later life	Accepting the shifting of generational roles	a. Maintaining own and/or couple functioning and interests in face of physiological decline, exploration of new familial and social role options b. Support for a more central role for middle generation c. Making room in the system for the wisdom and experience of the elderly; supporting the older generation without overfunctioning for them d. Dealing with loss of spouse, siblings, and other peers and preparation for own death; life review and integration

SOURCE: E.A. Carter and M. McGoldrick, eds. *The Family Life Cycle: A Framework for Family Therapy.* New York: Gardner, 1980. Reprinted by permission.

You don't get ulcers from what you eat. You get them from what's eating you.

VICKI BAUM (1888–1960)

guidelines as to how to feel and behave as the parents of a newborn, easing some of the stress associated with becoming a parent. Because we have fewer (or no) norms assisting us in the transition to parenthood if we are adoptive parents or gay or lesbian parents, however, we may experience a more stressful transition. Society is supportive of "normal" parenthood but often negative or hostile toward nonnormative parenthood.

The Double ABC-X Model of Family Stress

When sorrows come, they come not single spies, But in battalions.

WILLIAM SHAKESPEARE (1564–1616), *Hamlet*

The most widely used model to explain family stress is the **double ABC-X model** (AaBC-X model) (McCubbin and Patterson, 1982). A key concept in this model is **stressor pileup.** According to the double ABC-X model, during stress or crisis, the family responds not only to a current stressor but also to family hardships

Figure 12.2
The Double ABC-X Model

Crisis
X
b
Existing
Resources
a
Stressor
b
Perception
of
Stressor
Prior
Strains
Stressor
Hardships
Pile-up
Demands
a A
Pile-up of
Demands
b B
Existing and
New Resources
Coping
c C
Perception of
Pile-up of Demands
Good
Adaptation
Adaptation
Poor
Adaptation

Precrisis **Crisis** **Postcrisis**

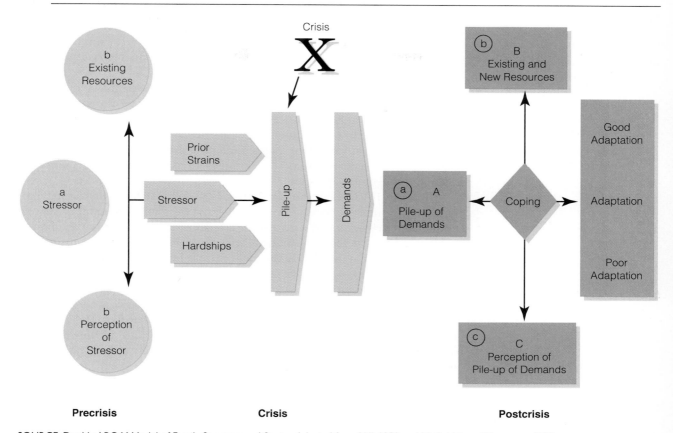

SOURCE: Double ABC-X Model of Family Stressors and Strains. Adapted from Hill, 1958, and McCubbin and Patterson, 1982.

and strains. Stressors, family hardships, and strains combine to create stressor pileup. Stressor pileup magnifies the impact of the stressor event, transforming a relatively minor stressor, such as going on a family vacation, into a catastrophic event. The stress of planning, packing, and leaving on a trip may bring other simmering problems to the surface. These problems may include lack of communi-

cation ("I thought you made the reservations"), inequitable distribution of power ("You always decide where we're going"), financial difficulties ("We can't afford this trip"), and parental role issues ("Why don't you ever help the kids pack?") (McCubbin and Patterson, 1982).

The double ABC-X model has the following components:

Aa represents stressor pileup (the combination of the current stressor, family hardship, and strain).

B represents the family's coping resources.

C represents the family's perception of the stressor pileup.

X represents the outcome of the situation, including the coping strategies that the family brings into play and any disruption that may occur if the family fails to cope adequately.

A family or an individual can draw upon many kinds of resources (the "B factor"). These assets may be material, psychological, emotional, or social.

What's the worst that can happen? The storm picks up our house and plants it in a better neighborhood.

ROSEANNE

They include money, time, health, intelligence, information, job skills, relationships with others, and support from social networks. Lack of economic resources is a significant component in stress and crisis. For the poor, everyday life is stressful as they try to make ends meet. Whereas a crisis such as illness or unemployment may be difficult but manageable for a middle-class family with savings and health insurance, for a poor family, such a crisis can be overwhelming.

The mere possession of resources is no guarantee that they will be used, however. The double ABC-X model of family stress views the actual use of resources (coping) as both a process among the B, C, and X factors and an outcome (part of X). How the family views a particular stressful event (the "C factor") is a significant variable in determining the eventual outcome. Viewing a stressful event as a crisis may very well lead to a crisis, as will utterly ignoring a situation that requires attention.

Research (McCubbin and McCubbin, 1987, 1989) suggests that strong families are much more successful in adapting to stress than weak families. Strong families tend to work together to resolve problems, are more democratic in their problem solving, delegate responsibilities, have members with considerable respect and affection for one another, and share values. These factors encourage family cohesion and support during times of stress. (See Chapter 16 for a discussion of coping mechanisms in strong families.) Weak families, however, tend to blame one another in stressful situations, are unwilling to ask for assistance or help from family members, are hesitant to shift or share burdens, and are reluctant to compromise. As a result, they have less flexibility and fewer family resources to deal effectively with stress and are more likely to have unfavorable outcomes.

Reflections

Using the double ABC-X model, examine a recent family stress or crisis. What were the stressor(s), the strain(s), and the family hardship(s) that combined to form the stressor pileup? What were your family's resources? Its coping behaviors? Was the crisis successfully or unsuccessfully resolved? Why?

For fast-acting relief, try slowing down.

LILY TOMLIN

Family Caregiving

One of the principal tasks of families is to care for their members in times of ill health or incapacity. We often take this aspect of families for granted, not thinking of the impact of the illness or injury on the caregiver or the family system as a whole. But as we have seen from our discussion of family stress, there are apt to be profound and far-reaching effects when a family member can no longer fulfill his or her expected role and the others must adjust to the new situation. Family researchers are giving increased attention to the **caregiver role**—the role of the family member who provides most of the ongoing physical work and decision making that relates to the ill or disabled person. The caregiver, who is usually (no surprise here) Mom, may be at risk for health problems herself, especially if the illness or disability is serious or prolonged. And if the principal caregiver becomes sick, the family may be thrown into turmoil.

Wives are young men's mistresses, companions for middle age, and old men's nurses.

FRANCIS BACON (1561–1626), *Essays*

Caregiving for Family Members with Chronic Illnesses and Disabilities

We have all experienced illnesses and accidents that have laid us low for a few days or longer. Illnesses and accidents affect us not only bodily but also psychologically. When we are sick or incapacitated, we are more prone to depression and mood swings. Illnesses are also stressful because they affect daily living patterns. A bout

Did you know?

Men and women between the ages of thirty and fifty—the peak time for parenting—have less free time than do younger or older individuals. Working mothers face the greatest pressures (Hales, 1997).

of the flu, for example, may put a student out of commission during finals week. A sprained ankle constrains mobility; a cold affects not only a sick child but also the parent who must stay home from work or school. Minor illnesses temporarily disrupt the structure of our lives and families.

Acute illnesses, such as appendicitis, require hospitalization but are of relatively short duration. Chronic illnesses, such as cystic fibrosis, diabetes, heart disease, arthritis, multiple sclerosis, asthma, and certain mental illnesses, are ongoing. A chronically ill person may be better sometimes, and worse other times, but most of the time, he or she continues to have a potentially debilitating or even life-threatening condition. Similarly, physical limitations (such as impaired speech or movement, deafness, and blindness) and developmental impairments (such as Down syndrome or autism) are lifelong. The care of those who are chronically ill, physically limited, or developmentally impaired is usually given at home by the family. Thus, the life courses of those with chronic illnesses or disabilities, their caregivers, and their families are radically affected.

In many cases, the family must shape itself around the needs, limits, and potential of the ill or disabled family member. The magnitude of the problem for families can be imagined once we realize that as many as 49 million Americans have physical or mental impairments, including blindness, paralysis, loss of limbs, disorders of the nerves or muscles, or mental retardation (Hales, 1997).

Many families must also cope with some form of mental illness at some point. A nationwide survey found that nearly half of all American adults have some type of mental illness at least once in their lives (Kessler, 1994). Because many people are reluctant to seek psychological help, these illnesses often go untreated. The most common disorders found by the survey were depression, alcohol dependence, social phobia (a persistent fear of being scrutinized), and simple phobias (such as fear of animals, closed spaces, or heights). Even seriously mentally ill people often live at home. Over 1 million adults with "serious and persistent mental illness" are currently living at home. They may be cared for by family members, or they may be parents who are trying to raise their families without the medical support they desperately need (Cole, 1993).

Sources of Stress

Families with chronically ill or disabled members are subject to various types of stresses, depending on the nature and severity of the illness or disability and whether the chronically ill or disabled person is a child, spouse, or aged parent. Whereas the care of a chronically ill or disabled child or spouse is regarded as an unpredictable stress, care of aging parents is considered a normative or predictable family stress (E. Brody, 1985; Matthews and Rosner, 1988). The general stresses experienced by families caring for ill or disabled members include the following (Darling, 1987; Patterson and McCubbin, 1983; Yura, 1987):

> *It's not easy being green.*
>
> KERMIT THE FROG

- *Strained family relationships*, including resentment by the caregiver or other family members; competition for time between the ill or disabled person, the caregiver, and family members; overt or covert rejection of the ill or disabled family member; and coalitions between the ill or disabled person and the primary caregiver that leave out other family members. Overprotection may be especially prominent in families with children who have disabilities.
- *Modifications in family activities and goals*, such as reduced leisure, travel, or vacation time; change in personal or work goals (especially by the primary caregiver); and concern over having additional children if the illness or disability is genetic.

UNDERSTANDING YOURSELF

Stress: How Much Can Affect Your Health?

Change, both good and bad, can create stress and stress, if sufficiently severe, can lead to illness. Drs. Thomas Holmes and Minoru Masuda, psychiatrists at the University of Washington in Seattle, have developed the Social Readjustment Rating Scale. In their study, they gave a point value to stressful events. The psychiatrists discovered that in 79 percent of the persons studied, major illness followed the accumulation of stress-related changes totaling over 300 points in one year. Examine the scale that follows. Notice how most directly or indirectly relate to marriage and family. What stresses have you experienced in the past six months? What stresses do you expect to experience in the next six months?

Life Event	Past 6 mos.	Value	Future 6 mos.
Death of spouse	☐	100	☐
Divorce	☐	73	☐
Marital separation from mate	☐	65	☐
Detention in jail or other institution	☐	63	☐
Death of a close family member	☐	63	☐
Major personal injury or illness	☐	53	☐
Marriage	☐	50	☐
Being fired at work	☐	47	☐
Marital reconciliation with mate	☐	45	☐
Retirement from work	☐	45	☐
Major change in the health or behavior of a family member	☐	44	☐
Pregnancy	☑	40	☐
Sexual difficulties	☐	39	☐
In-law troubles	☐	29	☐
Outstanding personal achievement	☐	28	☐
Wife beginning or ceasing work outside the home	☐	26	☐
Beginning or ceasing formal schooling	☐	26	☑
Major change in living conditions (e.g., building a new home, remodeling, deterioration of home or neighborhood)	☑	25	☐
Revision of personal habits (dress, manners, association, etc.)	☑	24	☐
Troubles with the boss	☐	23	☐
Major change in working hours or conditions	☑	20	☐
Change in residence	☑	20	☐

- *Increased tasks and time commitments*, such as the provision of special diets, daily therapy, or treatment; appointments and transportation to medical facilities; and possible need for constant attendance.
- *Increased financial costs* resulting from medication, therapy, medical consultation and treatment, special equipment needs, and so on. Baby-sitting is the largest single out-of-pocket expense for families with physically disabled children.
- *Special housing requirements*, such as close proximity to medical facilities, optimal climate conditions, and special housing features (such as wheelchair ramps).
- *Social isolation* resulting from the reactions of friends and relatives, individual or family embarrassment, limited mobility, fear of exposure to infections or conditions that might exacerbate the illness, inability to predict behavior, or lack of available time for social interactions.
- *Medical concerns*, such as the individual's willingness or ability to follow prescribed treatment, obtain competent medical care, or minimize pain and discomfort; and uncertainty of the medical prognosis.
- *Grieving over disabilities, limitations, and restricted life opportunities;* for some, anticipation of early or painful death.

Coping Strategies

When illness or disability unexpectedly strikes our family, we are likely to feel confused, betrayed, angry, or overwhelmed. But we can take action to combat

Event		Value	
Gaining a new family member (e.g., through birth, adoption, oldster moving in, etc.)	☑	39	☐
Major business readjustment (e.g., merger, reorganization, bankruptcy, etc.)	☐	39	☐
Major change in financial state (e.g., a lot worse off or a lot better off than usual)	☑	38	☐
Death of a close friend	☐	37	☐
Changing to a different line of work	☐	36	☐
Major change in the number of arguments with spouse (e.g., either a lot more or a lot less than usual regarding child rearing, personal habits, etc.)	☐	35	☐
Taking out a mortgage or loan for a major purchase (e.g., for a home, business, etc.)	☐	31	☐
Foreclosure on a mortgage or loan	☐	30	☐
Major change in responsibilities at work (e.g., promotion, demotion, lateral transfer)	☐	29	☐
Son or daughter leaving home (e.g., marriage, attending college, etc.)	☐	29	☐
Changing to a new school	☐	20	☐
Major change in usual type and/or amount of recreation	☐	19	☐
Major change in church activities (e.g., a lot more or a lot less than usual)	☐	19	☑
Major change in social activities (e.g., clubs, dancing, movies, visiting, etc.)	☐	18	☐
Taking out a mortgage or loan for a lesser purchase (e.g., for a car, TV, freezer, etc.)	☐	17	☐
Major change in sleeping habits (a lot more or a lot less sleep, or change in part of day when asleep)	☐	16	☐
Major change in number of family get-togethers (e.g., a lot more or less than usual)	☐	15	☐
Major change in eating habits (a lot more or a lot less food intake, or very different meal hours or surroundings)	☐	15	☐
Vacation	☐	13	☐
Christmas	☐	12	☐
Minor violations of the law (e.g., traffic tickets, jaywalking, disturbing the peace, etc.)	☑	11	☐

Thomas Holmes and Minoru Masudu. *Social Readjustment Rating Scale.*

SOURCE: T.H. Holmes, M.D.; The Social Readjustment Rating Scale, *Journal of Psychosomatic Research,* Vol. 11, pp. 213–218. © 1967, Pergamon Press Ltd.

our feelings of helplessness and despair. Some strategies for coping with serious health problems in the family include the following (adapted from Doherty, 1993a).

- Make a place for the illness, and keep it in its place. William Doherty likens illness to an uninvited guest: We may not like it, but we have to let it in. We need to accept the reality of the problem, grieve our losses, and make the necessary adjustments. Even though we had to let the guest in, we get to decide which room it occupies. It doesn't have to take over our entire life. We need to keep balance in the family by focusing on our strengths, maintaining other family relationships, and having a part of our life that is not involved with the illness.
- Keep communication open. Both negative and positive feelings need to be expressed by all family members (see Chapter 5 for communication guidelines). We need to develop our tolerance for differences in feelings among family members.
- Develop good working relationships with health-care professionals. We need to learn how to find out what to expect from the disease or condition and what the treatment options are; for example, we may want to utilize medical libraries, computer databases, or local or national support groups. We need to develop our skills in communicating with health-care professionals; this may require some patience, depending on our particular situation.

By the same token, health-care professionals need to be able to listen and communicate with families. Many doctors were taught in medical school to be "detached." There is, in fact, says Doherty, "an implicit anti-family bias in much medical training, with families frequently viewed as interfering with the efforts of practitioners to help patients" (Doherty, 1993b). But to truly accomplish the task of healing, involvement and genuine expressions of caring by doctors are crucial. There is currently a movement in some medical schools to address this issue. Former Surgeon General C. Everett Koop, now at Dartmouth Medical School, told an interviewer (Ryan, 1994):

> The patient is more than the aches and pains he or she brings to you. You may come in to see the doctor about a pain in your elbow, but you may have problems that are economical, emotional, psychological or spiritual. . . . A smart doc will talk about the elbow but also help settle the spiritual problem.

- Cultivate sources of support. It is important that we acknowledge and support the primary caregiver both emotionally and practically, and while we need to be open to, or to seek out, support among family and friends, at the same time we should be prepared to accept that some people "will never get it" (Doherty, 1993a). If we accept this, then we won't waste our energy trying to get them to understand something they may simply not be capable of grasping. Support groups made up of people with similar situations may likely be found within the community (or even on the Internet). Ask your doctor or other health-care practitioner, your public health department, or the reference desk at the local library for information about support groups. Just knowing that we're not alone can be a comfort. (See the Resource Center for sources of support for the disabled and their families.)

Family Outcomes

Most studies suggest that chronic illnesses have negative effects on the family. Family disruption and decreased marital satisfaction, for example, are common consequences (Hafstrom and Schram, 1984). Other studies find positive or inconsequential effects (Masters et al., 1983; Shapiro, 1983). Mark Peyrot and his colleagues (1988) found that the ultimate outcome of chronic illness is not necessarily bad. They found that after a period of disruption, some families resolved the crisis in a positive manner. Such families perceive the crisis as manageable and find the personal and social resources to create a favorable outcome. Several persons reported increased cohesiveness because they were compelled to spend more time with their families. Says one man with diabetes (Peyrot et al., 1988): "I'm closer to my family than I was. Prior to this [the onset of diabetes] I was working eighty hours a week. I spent more hours at work than I did at home. For the first ten years my daughter never knew me because I was never at home." Other families find hidden strengths and make their priorities more clear. Another respondent says (Peyrot et al., 1988), "The traumatic experiences that we've been through pulled us closer together, made our marriage stronger. It's pointed out the more important things."

Reflections

Does your family, extended family, or the family of friends have a disabled or chronically ill member? What coping strategies are used? Are they effective or ineffective? How does the family perceive the impact of the disability or illness?

Caregiving for the Aged

Although most elderly men and women are healthy and live independently, almost 85 percent of the 25.5 million elderly have at least one chronic illness. About 2 million older people need assistance with such daily living activities as bathing, dressing, eating, shopping, and managing money (Blieszner and Alley, 1990). At least 5 million American men and women care for an aged parent. These numbers can all be expected to increase dramatically over the coming decades.

Americans are living longer but "appear to be trading longer life for worsening health" (Olshansky, 1991). Medical science is increasingly able to prolong life, but it hasn't given much attention to the quality of that life. Some researchers are suggesting that medical research needs to shift its priorities, concentrating less on fighting fatal diseases and more on understanding the biology of aging and the diseases of the aged, such as arthritis and Alzheimer's disease (Goleman, 1991).

The Caregivers

Most elder care is provided by women, generally daughters or daughters-in-law (Mancini and Blieszner, 1991). Psychologist Rita Ghatak estimates that "eighty percent of the time it's the female sibling who is taking most of the responsibility" (quoted in Rubin, 1994). Elder caregiving seems to affect husbands and wives differently. Women report greater distress and depression from caregiving than do men (Fitting, Rabins, Lucas, and Eastham, 1986). In part, this may be because men approach their daily caregiving activities in a more detached, instrumental way. Another factor may be that women frequently are not only mothers but also workers; an infirm parent can sometimes be an overwhelming responsibility to an already burdened woman (Rubin, 1994). Fortunately, most adult children in a given family participate in parental caregiving in some fashion, whether it involves doing routine caregiving, providing backup, or giving limited or occasional care (Mancini and Blieszner, 1991).

A recent study (Silverstein, Chen, and Heller, 1996) of 539 older participants found that although there are psychological benefits associated with intergenerational support, excessive support received from adult children may be harmful by virtue of eroding competence and imposing excessive demands. The authors point out that, "given the norm of intergenerational independence that prevails between adults in many American families, it is reasonable to expect that an overabundance of social support from adult children may sometimes do more harm than good." In the process of balancing personal needs with those of families, it is important to define the level of care that is both appropriate and necessary.

Successful caregiving to elders necessitates a balance between our personal and family needs and those of our elders. For most of us, this is not an easy process.

PERSPECTIVE

HIV, AIDS, and the Family

In contemporary America, those who are terminally ill are often stigmatized because they are dying. People feel uncomfortable in their presence and often avoid them. It is as if being with person with a life-threatening illness forces us to confront our own mortality and triggers irrational fears of somehow "catching" the death (DeSpelder and Strickland, 1994). Those with HIV or AIDS often carry an extra burden: Not only are they presumably dying, but they are also presumably members of one of two highly stigmatized groups—gay men or IV drug users.

Families are affected by HIV and AIDS in numerous ways. The following are real-life examples of how HIV and AIDS affect families and other intimate relationships (Macklin, 1988, 1989; Shilts, 1987):

- Parents reject their gay son because of his sexual orientation and reconcile with him only as he is dying.
- A grandmother whose daughter has died from AIDS now cares for her two orphaned HIV-positive grandchildren.
- A young executive who experimented with IV drugs five years earlier discovers he is HIV-positive after he infects his wife, who unknowingly has infected their two children before birth; in three years, the entire family is dead.
- A mother discovers she is HIV-positive only after the birth of her infected son. She worries about who will care for her children if she dies.
- A woman's husband has AIDS, and she fears that she is infected, although she repeatedly tests HIV-negative.
- A married woman finds a notice in her lover's desk informing him that he is HIV-positive but is afraid of confronting him because she was snooping; she is afraid to talk with her husband because he believes she is monogamous.

The family system is disrupted when the family learns that one of its members has HIV or AIDS (Carter, 1989; Cates, Graham, Boeglin, and Tickler, 1990). The family may unite to support its infected member. Alternatively, old issues may be resurrected, renewing conflict between family members, especially if members focus on how the HIV was transmitted. HIV and AIDS not only transform the life of the infected individual but also the lives of those emotionally involved with him or her: lovers (past and present), partners or spouses, children, parents, brothers and sisters, friends, and caregivers. Furthermore, when people test HIV-positive or have AIDS, they may have to confront partners who have been unaware of their drug use or sexual activities outside of their primary relationships. Both the infected and uninfected partners may have to deal with a variety of issues (Carter, 1989). A wife may be unaware of her husband's bisexual orientation and his extramarital involvement with other men or women. She will have to deal simultaneously with his infection, his sexual orientation, and his extramarital involvement.

Problems Faced by People with AIDS and Their Families

The potential problems experienced by both a person with HIV or AIDS and his or her family and friends include facing social stigma and isolation, fear of contagion, fear

To be a man is precisely to be responsible.

ANTOINE DE SAINT-EXUPÉRY (1900–1944)

Caregiver Conflicts

Even though elder care is often done with love, it can be the source of profound stress.

Caregivers often experience conflicting feelings about caring for an elderly relative. The conflicts experienced by primary caregivers include the following (Springer and Brubaker, 1984):

- Earlier unresolved antagonisms and conflicts.
- The caregiver's inability to accept the relative's increasing dependence.
- Conflicting loyalties between spousal or child-rearing responsibilities and caring for the elderly relative.
- Resentment toward the older relative for disrupting family routines and patterns.
- Resentment by the primary caregiver for lack of involvement by other family members.
- Anger or hostility toward an elderly relative who tries to manipulate others.
- Conflicts over money or inheritance.

of infection, fear of abandonment, guilt, anger, grief, and economic hardship (Macklin, 1988). (For a discussion of children's understanding of AIDS from a developmental perspective, see Schvaneveldt, Lindauer, and Young, 1990).

- *Social stigma and isolation.* Unlike other individuals with life-threatening illnesses, most of those with HIV or AIDS must deal with personal and societal hostility because of their sexual orientation or drug use. (Even those who are infected through blood transfusions, heterosexual intercourse, or prenatal exposure suffer stigmatization.) Some view AIDS as divine retribution for violating religious teachings against homosexuality; others believe that those with AIDS deserve the affliction because of their gay lifestyle or drug activities. Some people with HIV or AIDS are fired from their jobs, evicted from their apartments, rejected by their lovers, or refused admission to school. Some families support their child through the illness but do not tell others that their son or daughter has AIDS (Cleveland et al., 1988).
- *Fear of contagion.* Many people who have a family member with HIV or AIDS experience some fear of becoming infected themselves, even when they know there is no rational basis for the fear. Partners may hesitate to share the same bed, relatives may be reluctant to visit or dine together, or grandparents may be fearful of baby-sitting for an infected child. Relationships are often severely tested. Many relatives and friends feel unable to talk openly about their anxieties.
- *Fear of infection.* Anxieties are high among members of high-risk groups. They may be reluctant to become involved in relationships for fear of transmitting the virus or becoming infected themselves. They may fear being tested lest they discover that they are infected. If they are tested, they fear learning the results, and if they are HIV-positive, they wonder how long it will be before AIDS symptoms occur.
- *Fear of abandonment.* Many with HIV or AIDS face rejection by their families, not only because of their infection but also because of their sexual orientation. The families of many gay men learn that they are gay at the same time they learn that they have AIDS. Some families are unable to cope with the needs of their dying child or sibling and consequently pull away. Others, however, stand by with increased resolve and love (Cleveland et al, 1988).
- *Guilt.* The person with HIV or AIDS may feel guilty for being gay or an IV drug user. He or she may feel somehow "deserving" of the infection because of "immorality," "stupidity," or lack of awareness. Families that once rejected a dying family member because he or she was gay may feel guilt over their earlier rejection, or they may feel guilty because they are unable to accept or care for their ill member.
- *Anger.* Guilt and anger may alternate. At one time a person may feel angry for being infected; at other times he or she may feel guilty about it. He or she may feel angry at family, friends, associates, and society for not being more supportive. Family

(continued on following page)

Coping Strategies

Affection certainly eases the burdens, but it does not necessarily decrease the strains of caregiving that relatives experience (Sheehan and Nuttall, 1988). Elders are especially vulnerable to abuse by their children (see Chapter 13). "Granny dumping," the abandonment of elderly relatives, is not unusual.

Caregiver education and training programs, self-help groups, caregiver services, and family therapy can provide assistance in dealing with the problems encountered by caregivers. Of these, family therapy appears to be the most effective for dealing with the emotional aspects of caregiving (Sheehan and Nuttall, 1988). In addition, elders receiving Medicaid may be eligible for respite care and homemaker/housework assistance. Because elder care involves complex emotions raised by issues of dependency, adult children and their parents often postpone discussions until a crisis occurs. (See the Resource Center for referral services and organizations for seniors.)

The best way for adult children to deal with elder care is to plan ahead with their siblings and with their aging parents, if they are willing. The major tasks include the following (Levin, 1987): (1) planning for legal and financial incapacities;

PERSPECTIVE

HIV, AIDS, and the Family (continued)

members may feel angry at the person with HIV or AIDS for "getting" infected because of sexual orientation or drug behavior.

- *Grief.* There is no cure for HIV or AIDS. Also, researchers increasingly believe that most people infected with HIV will develop AIDS. Those with HIV and AIDS, along with their families and friends, feel grief for the impending loss of health, potential, and life. The grief is compounded for partners because some may be infected themselves. Because those with AIDS tend to be young adults, the feelings of grief and loss tend to be more intense. (In New York City, AIDS is the leading cause of death among men aged twenty-one to twenty-four years.)
- *Economic hardship.* AIDS is a costly disease. Medical and hospital expenses may run between $50,000 and $100,000 or higher. The loss of earning power, the cost of treatment, and the lack of adequate financial assistance spell financial ruin for many individuals and families. Those without health insurance face even greater hardships.

One consequence of these costs and the increasing lack of sufficient numbers of treatment facilities is a movement toward home care of persons with AIDS. More and more programs are experimenting with ways of helping individuals remain in their homes with the combined care of family, partners, and professional caregivers. Hospice care is becoming a major alternative for persons in the last stages of AIDS (Magno, 1990).

Supporting Friends and Relatives with AIDS

If you have a friend or relative with HIV or AIDS, you can provide needed support without fear of contracting the infection yourself. Some things that you can do to show your caring are described here (Martelli et al., 1987; Moffatt et al., 1987):

- Show your caring by visiting and keeping in contact.
- Touch your friend or relative; touch communicates love, warmth, and hope. Don't hesitate or be afraid to touch, despite whatever apprehension you might at first have.
- Call before visiting to make sure that your friend or relative feels like having a visitor at a particular time.
- Be willing to talk about the disease and its prognosis if the affected person wants to express his or her feelings about it. Find HIV or AIDS support groups for your friend or relative.
- Take your friend or relative out to dinner, a movie, the park or beach, or a favorite spot if his or her health permits.
- Offer to shop or do household chores, banking, pet care, or other tasks. Provide transportation when needed.
- Organize your own support group to assist or to be on call for your friend or relative. In this manner, you can create a network for sharing tasks and feelings.
- Be available. Encourage your friend or relative to reach out when he or she needs help.

In coping with AIDS, we must attempt to reduce the stigma and discrimination associated with it; work against homophobia; recognize the impact of the illness on the family; and encourage social, health, and economic programs that support persons with AIDS. We must focus on enhancing the quality of life for those with AIDS, remembering that they are *living* with AIDS. We are all living until we are dead.

(2) managing income and expenses; (3) arranging for long-term care; (4) assessing capabilities of the whole family unit; (5) dividing responsibilities among parents, siblings, adult children, friends, and neighbors; and (6) determining community backup services, such as meals-on-wheels programs, visiting nurses, and housekeeping services.

Reflections

As your parents become dependent, will you become the primary caregiver? Why? Are you, your parents, or other close relatives involved in caregiving activities with an aging relative? If so, what effect has that had on you, your family, or your relative's family?

Death and Dying in America

Even though on a cognitive level we know that death comes to us all, when we actually confront death or dying, we are likely to be surprised, shocked, or at a loss about what to do.

Attitudes toward Death

The nature and extent of our feelings and fears about death have to do with who we are—our age, sex, personality, spiritual beliefs, and so on. Our feelings also have to do with the person who has died—whether he or she was young or old, whether or not the death was expected, and what our relationship was.

Of all the wonders that I yet have heard,
It seems to me most strange that men should fear;
Seeing that death, a necessary end,
Will come when it will come.

William Shakespeare, (1564–1616)
Julius Caesar

Cultural and Religious Context

Our cultural and religious context is important in determining our responses to death. Although Americans today seem somewhat more willing to talk and act realistically about death than they were a decade or two ago, as a society we remain ambivalent about the subject. Our responses to death and dying fall into three categories: denial, exploitation, and romanticization (Rando, 1987).

Denial. Although the existence of death is not specifically denied, there are practices in our culture that make it hard to view death as the real part of life it actually is. The removal of old or dying people from the home to the hospital has encouraged the belief that death is unnatural, frightening, or even disgusting. In a way, we treat death like sex: We try to shield children from it, and we use euphemisms when we speak of it ("Aunt Helen passed away," or "Fluffy was put to sleep").

No man is an island, entire of itself; every man is a piece of the continent, a part of the main; if a clod be washed away by the sea, Europe is the less, as well as if a promontory were, as well as if a manor of thy friends or of thine own were. Any man's death diminishes me, because I am involved in mankind. And therefore never send to know for whom the bells tolls. It tolls for thee.

John Donne (1572–1631)

Exploitation. At the opposite extreme of denial is the exploitation of death. Therese Rando (1987) writes:

> While there has been a decline in the average individual's personal contact with death and dying, due to sociological changes and advances in technology, there has been an increase in exposure to violence and death through television, movies, and the print media. . . . Acts of murder, war, violence, terrorism, abuse, rape, crime, natural disasters, and so forth are exploited and sensationalized in the name of the public's right to know. Financial gains for the media and its sponsors are the obvious results.

The effects of this exploitation of death can be damaging. We may become overwhelmed by images of death and suffer from "annihilation anxiety." We may overcompensate and become aggressive or may withdraw and become desensitized to human pain. We may end up denying the realities of death altogether.

Romanticization. Sometimes our images of death are glamorized and glorified. Although some deaths are in fact peaceful and "beautiful," many are not. Those who have been led to expect a beautiful death (their own or another's) may be shocked and feel betrayed to discover the messy, frustrating, or ugly aspects of dying.

Death is our eternal companion. . . . It has always been watching you. It will always until the day it taps you. . . .

Death is the only wise advisor that we have. Whenever you feel . . . that everything is going wrong and you're about to be annihilated, turn to your death and ask if it is so. Your death will tell you that you're wrong; that nothing really matters outside its touch. Your death will tell you, "I haven't touched you yet."

This, whatever you're doing right now, may be your last act on earth. There is no power which could guarantee that you are going to live one minute more.

Carlos Castañeda

Fear of Death and Dying

Many of us share a number of fears and anxieties regarding the dying process and death itself. When we think about dying, we may worry about being unable to care for ourselves and becoming a burden on those we love. We may worry about pain or physical impairment. We may fear isolation, loneliness, and separation from people who are dear to us. The finality of death is often fearsome; it implies the loss of relationships, the abandonment of goals, and an end to pleasure. For some, contemplating the possibilities of an unknown afterlife or the idea of eternal nothingness may be frightening. Others may be concerned about how their bodies will look after death and what will happen to them. One of the greatest fears aroused by the thought of death is the fear of losing control: The world will go on, but we will no longer have any effect on what happens in it.

Thanatologists—those who study death and dying—tell us that a certain amount of fear of death is a good thing. It certainly helps keep us alive. A certain amount of denial is healthy, too, for it prevents us from dwelling morbidly on the subject of death. What we need to develop is a realistic, honest view of death as part of life. Acknowledging that death exists can enrich our lives greatly. It can show us the importance of getting on with the "business of living," ordering our priorities, and appreciating what is around us.

I'm not afraid to die. I just don't want to be there when it happens.

WOODY ALLEN

The Process of Dying

Unless death comes very quickly and unexpectedly, the person who is dying will undergo a number of emotional changes in addition to physical ones. These changes are often referred to as "stages," although they do not necessarily happen in a particular order. Indeed, some of them may not be experienced at all, or a person may return again and again to a particular part of the process.

Stages of Dying

The common stages of reaction that a person goes through when facing death include (Kubler-Ross, 1969, 1982):

1. *Denial and isolation* ("No, not me"). Coping with a diagnosis of a terminal illness often results in rejection of the news. Denial helps to overcome the shock, allows the person to begin to prepare himself or herself, and acts as a useful coping mechanism to allow the person to begin to gather together his or her resources.
2. *Anger* ("Why me?"). When the truth can no longer be denied, the dying person begins to feel rage and resentment. This anger is often directed toward family, physicians, and God. As caretakers, we can do little but provide comfort and support.
3. *Bargaining* ("Yes, me but . . ."). In this third stage, a person may try to bargain, usually with God, for a way to prolong life. In exchange, the patient often makes promises or amends. At this point, the person may also be his or her most vulnerable—grasping for anything that might help cure the condition.
4. *Depression* ("Yes, it's me"). Once people begin to accept their fate and face their impending death, they often become depressed about the unfinished business they are leaving behind. This anticipatory grieving is the most difficult time in the dying process, yet is also an appropriate step in coming to terms with losses. It is eased when the patient is allowed to express sorrow among supportive and nonjudgmental friends.
5. *Acceptance* ("Yes, me; and I'm ready"). In this final stage, people facing their death come to some resolution and understanding. This time is neither frightening nor painful, neither sad nor happy—only inevitable. At the end, when death is near, patients may choose not to talk much, nor to visit with friends or family. This is part of the process of letting go.

Once again, this is only a general description of a process. While several stages may occur at the same time, others may happen out of sequence or not at all. This is not a schedule to be anticipated or imposed; a dying person should not be expected to behave in a certain manner. Rather, each person facing death should be allowed to do so in his or her own way (DeSpelder and Strickland, 1996).

All goes onward and outward,
Nothing collapses
And to die is different from
What anyone supposes
And luckier.

WALT WHITMAN (1819–1892)

Happy is he who dies before his children.

YORUBA PROVERB

Thus shall you think of all this fleeting world: a star at dawn, a bubble in a stream, a flash of lightning in a summer cloud, a flickering lamp, a phantom, and a dream.

BUDDHA (563–483 B.C.), *Diamond Sutra*

Needs of the Dying

Robert Kavanaugh (1972) writes:

> No matter how we measure [one's] worth, a dying human being deserves more than efficient care from strangers, more than machines and antiseptic hands, more than a mouthful of pills, arms full of tubes and a rump full of needles. . . . More than furtive eyes, reluctant hugs, medical jargon, ritual sacraments or tired Bible quotes, more than all the phony promises for a tomorrow that will never come.

Aside from basic physical care and perhaps relief from pain, what a dying person needs can be summed up very succinctly: to be treated like a human being. Surgeon Sherwin Nuland (1994) writes that although "death with dignity" is the deepest wish of terminally ill patients, the realities of death for most people are far from dignified. Our society's own ambivalence and fear of death is all too often evident in the attitudes and actions of its doctors, nurses, and other medical personnel. Far too often, the dying are isolated from their loved ones during their final hours, until the machines that keep them in this world or measure their progress as they depart it emit their final beeps and blinks.

In response to the impersonality that has generally characterized death and dying in the hospital setting, the hospice movement has gained momentum in the last two decades. There are now more than a thousand hospice programs in the United States (DeSpelder and Strickland, 1996). A **hospice** may be an actual place where terminally ill people can be cared for with respect for their dignity, but it is also much more than that. It is a medical program that emphasizes both patient care (including management of pain and symptoms) and family support. The hospice provides education, grief counseling, financial counseling, and various types of practical assistance for home care of the dying person. It may also provide in-patient care in a noninstitutional setting. (See the Resource Center for information on locating a hospice program.)

We don't know life: how can we know death?

CONFUCIUS (551–479 B.C.)

It is not death that a man should fear, but he should fear never beginning to live.

MARCUS AURELIUS (A.D. 121–180)

Every shut eye ain't sleep and every goodnight ain't gone.

TRADITIONAL AFRICAN-AMERICAN PROVERB

Bereavement

Bereavement is our response to the death of a loved one. It includes the customs and rituals that we practice within our culture or subculture. It also includes the emotional responses and expressions of feeling that we call the grieving process.

Mourning Rituals

Our culture, religion, and personal beliefs all influence the types of rituals we participate in after someone dies. By prescribing a specific set of formalized behaviors, bereavement rituals can give us security and comfort; we don't have to think, "What do I do now?" Social rituals, such as funerals and wakes, give us the opportunity to share our sorrow, to console and be consoled. A funeral also clearly marks the end of a life. Because we must face up to the fact that an important person in our life is gone, we can begin to move ahead to our "new" life. Religious rituals affirm a spiritual relationship for those who believe in them; for those who don't, they may be a source of tension or embarrassment.

For some Americans, rituals having to do with the dead consist basically of a funeral service followed by burial, entombment, or cremation. For others, there are important practices to be observed long after the burial (or cremation). Under Jewish law, for example, there are three successive periods of mourning. The first of these, **shiva,** is a seven-day period during which the immediate family undergoes certain austerities, such as refraining from haircutting, shaving, and using cosmetics; going to work; or engaging in sexual relations. During the second period,

The bridge between the known and the unknown is always love.

STEPHEN LEVINE

The believer, not the belief, brings peace.

ROBERT KAVANAUGH

On the whole I'd rather be in Philadelphia.

EPITAPH SUGGESTED BY W.C. FIELDS (1880–1946)

your hair is falling out, and
you are not so beautiful;
your eyes have dark shadows
your body is bloated; arms covered
with bruises and needlemarks;
legs swollen and useless . . .
your body and spirit
are weakened with toxic chemicals
urine smells like antibiotics,
even the sweat
that bathes your whole body
in the early hours of morning
reeks of dicloxacillin and
methotrexate.

you are nauseous all the time
i am afraid to move on the bed
for fear of waking you to moan
and lean over the edge
vomiting into the bag

i curl up fetally
withdraw into my dreams
with a frightened back to you . . .
and i'm scared
and i'm hiding
but i love you so much;
this truth does not change . . .
years ago,
when i met you, as we were falling in love,
your beauty attracted me:
long golden-brown hair
clear and peaceful green eyes
high cheekbones and long smooth muscles
but you know—and this is true—
i fell in love with your soul
the real essence of you
and this cannot grow less beautiful . . .
sometimes these days
even your soul is cloudy

i still recognize you
we may be frightened
be hiding our sorrow
it may take a little longer
to acknowledge the truth,
but i would not want to be anywhere else
i am here with you
you can grow less beautiful to the world
you are safe
i will always love you.

CHRISTINE LONGAKER

shloshim, the prohibitions becomes less strict, and the final period, *avelut*, applies only if one's mother or father has died. During this eleven-month period, sons are to say *kaddish* (a form of prayer) for the parent daily. When the year of mourning is over, it is forbidden to continue practices that demonstrate grief (Kearl, 1989).

Among Latinos, *el Dia de los Muertos* (the Day of the Dead), an ancient ritual with Indian and Catholic roots that is observed on November 1, is making a comeback (Garcia, 1990). In Mexico and other Latin American countries, the dead are honored with prayers, gifts of food, and a nightlong graveside vigil; in the United States, parades and special exhibits not only commemorate the dead but also celebrate the cultural heritage of the participants. At home, altars may be set up with pictures of those who have died, religious figures or pictures, candles, orange marigolds (the Aztec flower of death), offerings of the honored one's favorite food or drink, and perhaps cartoons or darkly humorous verses that "laugh in the face of death."

Grief as a Healing Process

Like dying, grieving is a process. Thanatologists have variously described the stages of the grieving process (Kavanaugh, 1972; Kubler-Ross, 1982; Tallmer, 1987). There are also certain emotions or psychological states that may commonly be expected. Among these are shock, denial, depression, anger, loneliness, and feelings of relief. Guilt is also often experienced as part of the grieving process. If we felt resentment or anger toward the dead person, we feel guilty when he or she dies, as if we somehow caused the death. If we are spared and another dies—in an accident, for example—we feel guilty for surviving, and if the deceased was a burden to us while alive, we feel compelled to shoulder a load of guilt now that the burden has been lifted.

For some, most of the grieving process will be over in a matter of weeks or months. For others, it will occupy a year or two or maybe more. The first year will undoubtedly be the most difficult as holidays, birthdays, and anniversaries are experienced without the loved person. Grieving may occur sporadically for years to come, touched off by memories evoked by a particular date, a special piece of music, or a beautiful view that can no longer be shared. Healing, which is the goal of grieving, does not appear suddenly as the reward for all our suffering. Rather, it comes little by little as we work through grief (and around it and over it and under it and back through it again), until we look at ourselves one day and find we are whole.

Consoling the Bereaved

When a friend or relative is bereaved, we may feel awkward or embarrassed. We may want to avoid the family of the person who has died because we "just don't know what to say."

That's okay. We really don't have to say much, except "I'm so sorry" and, perhaps, "How can I help?" Here are some suggestions for helping someone who is grieving:

- Listen, listen, listen. This may be the most helpful thing you can do.
- Express your own sadness about the death and your caring for the bereaved person, but don't say "I know how you feel" unless you really do—that is, unless you've experienced a similar loss.
- Talk about the person who has died. Recall special qualities he or she possessed and the good times you may have experienced. If the bereaved person begins to speak of the one who has died, don't change the subject.
- Give practical support: Help with household tasks, do shopping or other errands, cook a meal, or help with child care.

- If there are children, involve them in remembering. Support their grieving process (it will be different from an adult's).
- Don't avoid the bereaved person because you are uncomfortable, worry about mentioning the dead person, or attempt to point out the "bright side."

Being able to share one's grief with others is a crucial part of healing. By "just" being around and "just" listening, we can be a positive part of the process.

For el Dia de los Muertos, an elaborate altar may be set up to honor the dead.

Alcohol, Drug Abuse, and Families

"Alcohol and drug abuse are the most prevalent and costly mental disorders currently facing society," says Cornell psychiatry professor Peter Steinglass, who has done extensive research on alcoholism and the family (Steinglass, 1993). He goes on to state:

> Community and government commitment to alcoholism and substance abuse prevention should be commensurate with its immense negative impact on families and society. . . . Families are tremendous potential resources in combatting substance abuse; social policy needs to be designed to support and preserve this resource.

Alcohol Abuse, Alcohol Dependence, and Families

Alcohol is a drug. Many people, however, do not think of alcohol as a drug; instead, it is regarded as "a drink": a beer, a glass of wine, a rum and Coke. Yet, like certain illegal drugs such as marijuana and cocaine, alcohol alters an individual's mood and perceptions. The altered state produced by alcohol is a "high;" its more extreme manifestation is drunkenness. Because alcohol is a culturally accepted drug, the current antidrug climate largely excludes alcohol from national concern.

Reflections

What experiences have you had with death and dying? How is the subject of death handled in your family? Have you made a will? Why or why not?

The abuse of alcohol can be seen as a symptom of a disorder that is both physical and emotional. Health professionals view alcoholism as a disease rather than a personal shortcoming. Recent definitions create a distinction between alcohol abuse and alcohol dependence, or alcoholism. **Alcohol abuse,** as defined by the American Psychiatric Association's *Diagnostic and Statistical Manual of Mental Disorders*, 4th ed. (*DSM-IV*), involves continued use of alcohol despite awareness of social, occupational, psychological, or physical problems related to drinking, or drinking in dangerous ways or situations.

Alcohol dependence, or **alcoholism,** is a separate disorder involving more extensive problems with alcohol use. Over time, individuals experience changes that lead to tolerance or involve withdrawal. A person does not have to be an alcoholic, however, to have problems with alcohol. If, for instance, a person drinks once a month but blacks out each time he drinks, he is an alcohol abuser.

Numbness, separation, anxiety, and despair are common of grieving—not rigidly fixed, but rather part of the process of healing.

Although physiological and genetic factors may predispose a person to alcoholism (Blakeslee, 1984; Kumpfer and Hopkins, 1993), it appears that problem drinking, like ordinary drinking, is a learned behavior. Most people in the United States learn to drink during adolescence; most adolescents have their first drink at home but do most of their drinking away from home, with their friends. The drinking behavior that children observe at home will greatly influence their own behavior. Many alcoholics have at least one alcoholic parent. Whereas light or moderate drinkers may drink on social occasions, problem drinkers may drink as a means of coping with stress, anxiety, or low self-esteem.

Alcohol Use and Abuse

To understand the scope of alcohol abuse in the United States, we should consider the following (Celis, 1994; Hales, 1997):

- Alcohol is the number one drug problem among our nation's youth. About 30 percent of teenagers experience negative consequences of alcohol abuse, including accidents, arrests, or impaired performance ("Bibulous America," 1995).
- Fifteen and a half percent of adults meet the criteria for alcohol abuse or dependence.
- Every twenty-one minutes, someone in the United States is killed by a drunk driver. This adds up to sixty-eight men, women, and children who die every day as a result of drunk driving. Drunk drivers kill 25,000 people and injure another 530,000 each year.
- Alcoholism is the fourth-ranked cause of death in this country. It is a factor in nearly half of U.S. murders, suicides, and accidental deaths.
- An estimated 7 million children in the United States live with an alcoholic parent. Studies have linked child abuse and neglect to parental drinking.
- Alcohol abuse by prospective mothers (and fathers) can cause serious fetal damage, resulting in fetal alcohol syndrome, low birth weight, and other impairments (see Chapter 8 for a discussion of birth complications).
- Nearly half of all college students are binge drinkers (defined as those who had at least four drinks at one time on at least three occasions in the previous two weeks), according to a study from the Harvard School of Public Health (Wechsler et al., 1994).
- About 35 percent of college women report "drinking to get drunk." This figure represents a threefold increase from twenty years ago (Celis, 1994). (The Columbia University Center on Addiction and Substance Abuse reports that 90 percent of campus rapes occurred when alcohol had been used by the assailant, the victim, or both. Sixty percent of women with STDs reported being drunk at the time of infection.)

Did you know?

Having a drug-abusing parent increases sevenfold one's risk of becoming a drug abuser oneself (Steinglass, 1993).

These are not abstract facts. Alcohol abuse affects millions of American families, undermining their relationships and hopes. Yet the popularly perceived image of the alcoholic as a "bum in the gutter" prevails because families tend to deny their alcoholism. (See Table 12.2 for signs of alcoholism.)

Even though Americans drink substantially less distilled liquor than in the past, we still consume vast quantities of beer and increasing amounts of wine. Although the percentage of alcohol in beer and wine is lower than that in distilled liquor, we often don't take into account the fact that distilled liquor is generally served diluted in mixed drinks. A five-ounce glass of wine contains as much alcohol as a standard mixed drink; a can of beer is equal in alcohol content to a rum and Coke.

The heaviest drinkers are men between eighteen and twenty-five years old, although studies indicate that young college women may be narrowing the gap (Celis, 1994). Ten percent of the nation's drinkers consume over half its alcohol (Lord et al., 1987).

A Family Disease

Alcoholism is sometimes a "family disease" because it involves all members of the family in a complex interactional system. (See Table 12.3 for a "Children of Alcoholics Screening Test.") Regarding the family's role in alcoholism (and drug abuse), Robert Lewis (1989) notes:

Alcoholism isn't a spectator sport.

Eventually the whole family gets to play.

JOYCE REBETA-BURDITT

> Not only do dysfunctional families often produce addictive behavior in their members, but these addictions, in turn, then may affect the quality of life, negatively impacting

> the behavior of family members and devitalizing and fracturing family relationships. The most demoralizing aspect of this reciprocity . . . is that addictions are often passed from one generation to later generations, unless there is successful intervention.

The principle of *homeostasis*—the tendency toward stability in a system—operates in alcoholic families, maintaining established behavior patterns and strengthening resistance to change. Many alcoholic families do not progress through the normal family life cycle stages but remain in an unhealthy (yet stable) cycling between "sober and intoxicated interactional states" (Steinglass, 1983).

Treatment

In one sense alcoholism is not curable. Alcoholics must stay sober to stay well. For most alcoholics, this means no more drinking, ever. Although recovery is possible, a major stumbling block exists to motivating the alcoholic to pursue it. Denial is used by the alcoholic and his or her family and friends. "I can quit any time," says the alcoholic, or "It's not my drinking that's the problem, it's . . . (anything else)." The spouse, parent, or lover colludes in maintaining the alcoholism by denying it: "Jane is not really like this. She's a wonderful person when she's sober." People often deny the fact of alcoholism because of the social stigma attached to the term *alcoholic*.

We are never deceived: we deceive ourselves.

Johann Goethe (1749–1842)

Treatment for alcoholism is generally not considered possible until the alcoholic makes the conscious choice to become well. Many must first "hit bottom" before they are willing to admit that they are alcoholics. But because families are often organized around protecting the alcoholic from the consequences of his or her alcoholism, the alcoholic is able to deny the problem. Although adolescent alcoholics also deny their problems, research indicates that they are likely to perceive their need for help if they experience serious physiological consequences of their drinking, experience serious personal problems, or have another substance abuser in the family (Lorch and Dukes, 1989). It is important for the alcoholic to understand his or her feelings about, and responses to, alcohol and to develop "personal resiliency" for coping with the desire for alcohol (Kumpfer and Hopkins, 1993).

An educated public and an enlightened approach by professionals are essential to achieving progress in freeing millions from the destructive grip of alcoholism. Family support is very important; the best results seem to be obtained when the whole family is treated. Indeed, for many families it is imperative that the family be treated as a unit because its structure and stability may be organized around the alcoholism (Steinglass, 1987). Over the past decade, family systems intervention has become one of the most important means of treating substance abuse. Family "love bonds" are utilized to work for the abuser rather than against the abuser; strong family feelings are focused on breaking the abuse rather than denying it or sustaining it. Self-help groups such as Alcoholics Anonymous and Al-Anon (for the families of alcoholics) have had good success rates. Organizations such as Adult Children of Alcoholics offer self-help for children of alcoholic families. (See the Resource Center for more information.)

Drug Abuse and Families

Although alcohol is the most commonly used and misused mind-altering substance in the United States, a number of other drugs have achieved popularity among Americans. Within the family, parents, adolescents, or both may be drug users. Adolescents (or adults, for that matter) may use drugs to resolve conflicts, escape stressful situations, express defiance of authority, or elicit sympathy. Survivors of sexual abuse are significantly more likely to abuse drugs and alcohol as a form of self-medication and as a way to escape family problems (Harrison, Hoffman, and Edwall, 1989). Drug use may also be a symptom of serious disturbance and a threat to health or life. Teenage drug use is of particular concern to many families.

Table 12.2 The Signs of Alcoholism

Yes	No	Questions
		1 Do you occasionally drink heavily after a disappointment, a quarrel, or when the boss gives you a hard time?
		2 When you have trouble or feel under pressure, do you always drink more heavily than usual?
		3 Have you noticed that you are able to handle more liquor than you did when you were first drinking?
		4 Did you ever wake up on the "morning after" and discover that you could not remember part of the evening before, even though your friends tell you that you did not "pass out"?
		5 When drinking with other people, do you try to have a few extra drinks when others will not know it?
		6 Are there certain occasions when you feel uncomfortable if alcohol is not available?
		7 Have you recently noticed that when you begin drinking you are in more of a hurry to get the first drink than you used to be?
		8 Do you sometimes feel a little guilty about your drinking?
		9 Are you secretly irritated when your family or friends discuss your drinking?
		10 Have you recently noticed an increase in the frequency of your memory "blackouts"?
		11 Do you often find that you wish to continue drinking after your friends say they have had enough?
		12 Do you usually have a reason for the occasions when you drink heavily?
		13 When you are sober, do you often regret things you have done or said while drinking?

Drugs and Drug Use

There are many kinds of psychoactive ("mind-affecting") drugs and many reasons for using them. After alcohol, the most commonly used psychoactive drugs include marijuana, hallucinogens (such as LSD, mescaline, and psilocybin), cocaine (and crack, its smokable form), phencyclidine (PCP), narcotics ("sleep-inducing" drugs—principally heroin, and also other opiates such as morphine and codeine), and inhalants (such as toluene, found in spray paints, and nitrous oxide, or "laughing gas"). Psychotherapeutic ("mind-healing") drugs, such as antidepressants, stimulants, and sedatives, are widely used. Among the most commonly abused drugs in this group are amphetamines ("speed" or "crank") and the sedative-hypnotics such as methaqualone (quaaludes), barbiturates (such as secobarbital [Seconal] or "reds"), and diazepam (Valium). "Designer drugs," such as "Ecstasy" (MDMA), are popular among some groups of affluent or middle-class college students and other young adults. Drug use is prevalent in virtually all socioeconomic, sociocultural, and age groups (beginning with preteens). What varies is the drug of choice and the usage pattern.

Cigarettes kill more people each year than AIDS, heroin, crack, cocaine, alcohol, car accidents, fire, and murder combined.

Iris Shannon

Whereas overall drug use appeared to decline after peaking in the late 1970s to mid-1980s, studies indicate that the use of certain drugs is once again rising (Holder, 1994; Treaster, 1993). Marijuana, LSD, and amphetamines are becoming increasingly popular on high school and college campuses. Although advocates of marijuana and hallucinogens argue that their dangers have been exaggerated, medical experts fear that there may be serious health consequences for the young users

Table 12.2 *(continued)*

Yes	No	Questions
		14 Have you tried switching brands or following different plans for controlling your drinking?
		15 Have you often failed to keep the promises you have made to yourself about controlling or cutting down on your drinking?
		16 Have you ever tried to control your drinking by making a change in jobs, or moving to a new location?
		17 Do you try to avoid family or close friends while you are drinking?
		18 Are you having an increasing number of financial and work problems?
		19 Do more people seem to be treating you unfairly without good reason?
		20 Do you eat very little or irregularly when you are drinking?
		21 Do you sometimes have the "shakes" in the morning and find that it helps to have a little drink?
		22 Have you recently noticed that you cannot drink as much as you once did?
		23 Do you sometimes stay drunk for several days at a time?
		24 Do you sometimes feel very depressed and wonder whether life is worth living?
		25 Sometimes after periods of drinking, do you see or hear things that aren't there?
		26 Do you get terribly frightened after you have been drinking heavily?

If you have answered "yes" to any of these questions, you have some of the symptoms that may indicate alcoholism.

Questions 1–8 relate to the early stages of alcoholism.

Questions 9–21 relate to the middle stage.

Questions 22–26 mark the beginning of the final stage.

SOURCE: *Signs of Alcoholism* is published by the National Council on Alcoholism.

of these drugs who "have no memory of the awful LSD nightmares of the 60's and 70's" (Mitchell Rosenthal, quoted in Treaster, 1993).

The use of heroin and crack continues to maintain a steady level. These drugs continue to ravage Americans, especially poor Americans. There are numerous consequences, including birth defects (discussed in Chapter 8) and the spread of HIV to drug users, their sexual partners, and their children. Users of IV drugs represent the fastest growing segment of HIV infection and AIDS cases. The vast majority of those infected through IV drug use are the inner-city poor. African-American sociologist Robert Staples cites substance abuse as a particular threat to the black family because of its disruptive effect on the integrity of both the nuclear and extended family (Staples, 1990). Mitchell Rosenthal, the chairperson of the New York State Advisory Council on Drug Abuse, warns of the current dangers of "a shrinking political interest in the problem [that means] the most vulnerable and high-risk populations will not get the kind of services they need" (quoted in Kolata, 1991). Drug use is a problem that must be dealt with on multiple levels: societal, familial, and individual.

The Family's Role in Drug Abuse

Researchers are increasingly recognizing the role of the family in beginning, continuing, stopping, and preventing drug use by its members (Costantini, Wermuth, Sorensen, and Lyons, 1992; Freeman, 1993; Joanning, Quinn, Thomas, and

Table 12.3 Children of Alcoholics Screening Test (CAST)

CAST can be used to identify latency age, adolescent, and grown-up children of alcoholics.

Please check (✓) the answer below that best describes your feelings, behavior, and experiences related to a parent's alcohol use. Take your time and be as accurate as possible. Answer all thirty questions by checking either "yes" or "no."

Sex: Male _____ Female _____ Age: _____

Yes	No	Questions
_____	_____	**1** Have you ever thought that one of your parents had a drinking problem?
_____	_____	**2** Have you ever lost sleep because of a parent's drinking?
_____	_____	**3** Did you ever encourage one of your parents to quit drinking?
_____	_____	**4** Did you ever feel alone, scared, nervous, angry, or frustrated because a parent was not able to stop drinking?
_____	_____	**5** Did you ever argue or fight with a parent when he or she was drinking?
_____	_____	**6** Did you ever threaten to run away from home because of a parent's drinking?
_____	_____	**7** Has a parent ever yelled at or hit you or other family members when drinking?
_____	_____	**8** Have you ever heard your parents fight when one of them was drunk?
_____	_____	**9** Did you ever protect another family member from a parent who was drinking?
_____	_____	**10** Did you ever feel like hiding or emptying a parent's bottle of liquor?
_____	_____	**11** Do many of your thoughts revolve around a problem-drinking parent or difficulties that arise because of his or her drinking?
_____	_____	**12** Did you ever wish that a parent would stop drinking?
_____	_____	**13** Did you ever feel responsible for and guilty about a parent's drinking?
_____	_____	**14** Did you ever fear that your parents would get divorced due to alcohol misuse?

Mullen, 1992). Some believe that parents with marital problems use their adolescent's drug abuse as a means of avoiding their own problems. These families may deny the drug abuse and sometimes encourage it by providing money. In order to solicit care and attention from their parents, abusers may engage in antisocial or self-destructive behavior. In such cases, drug abuse is a symptom of family pathology.

In families where there is chronic drug abuse, members are likely to feel anxious, depressed, guilty, or enraged. Communication is probably poor, involving unclear messages, vague information giving, lack of direct talk, and avoidance of eye contact. There may be little direct expression of positive or negative emotions, because family members are intensely concerned about controlling feelings. Such families defend against their emptiness and anger "by heavy alcohol consumption, self-medication, or overeating, all of which serve as anesthetics, tranquilizers, antidepressants" (Textor, 1987). The parents in such families not only rationalize their own misuse of drugs—which may be legal drugs, such as alcohol or tranquilizers—but also often deny their child's substance abuse. If the drug abuse is acknowledged, neither the parents nor the child accepts responsibility for the abuse. Each

Table 12.3 (continued)

Yes	No	Questions
_____	_____	**15** Have you ever withdrawn from and avoided outside activities and friends because of embarrassment and shame over a parent's drinking problem?
_____	_____	**16** Did you ever feel caught in the middle of an argument or fight between a problem-drinking parent and your other parent?
_____	_____	**17** Did you ever feel that you made a parent drink alcohol?
_____	_____	**18** Have you ever felt that a problem-drinking parent did not really love you?
_____	_____	**19** Did you ever resent a parent's drinking?
_____	_____	**20** Have you ever worried about a parent's health because of his or her alcohol use?
_____	_____	**21** Have you ever been blamed for a parent's drinking?
_____	_____	**22** Did you ever think your father was an alcoholic?
_____	_____	**23** Did you ever wish your home could be more like the homes of your friends who did not have a parent with a drinking problem?
_____	_____	**24** Did a parent ever make promises to you that he or she did not keep because of drinking?
_____	_____	**25** Did you ever think your mother was an alcoholic?
_____	_____	**26** Did you ever wish that you could talk to someone who could understand and help the alcohol-related problems in your family?
_____	_____	**27** Did you ever fight with your brothers and sisters about a parent's drinking?
_____	_____	**28** Did you ever stay away from home to avoid the drinking parent or your other parent's reaction to the drinking?
_____	_____	**29** Have you ever felt sick, cried, or had a "knot" in your stomach after worrying about a parent's drinking?
_____	_____	**30** Did you ever take over any chores and duties at home that were usually done by a parent before he or she developed a drinking problem?
_____		TOTAL NUMBER OF "YES" ANSWERS

Score of 6 or more means that more than likely this child is a child of an alcoholic parent.

SOURCE: ©1983 by John W. Jones, Ph.D., *Family Recovery Press.*

blames the other or outside influences. They may attempt to absolve both the user and the family from responsibility by blaming the problem on the drug itself.

Although the world is full of suffering, it is full also of the overcoming of it.

HELEN KELLER (1880–1968)

The Chronic Drug User

Drug abuse often begins as an adolescent problem; it may be tied to normal adolescent experimentation (Textor, 1987). The use of certain drugs by adolescents may be an important component of their peers' culture, making it especially difficult for them to "just say no."

In preventing and treating such drug abuse, it is crucial to understand the social and cultural influences that are involved (Gordon, 1993; Ja and Aoki, 1993; Mayers, Kail, and Watts, 1993). Adolescents must deal with issues of family loyalty, separation and identity, and new ways of relating with their parents. Future drug abusers may fail at most of these tasks. Early in life they may be identified as problem children; they may be viewed as weak, immature, and needing help. They may not have sufficient coping skills; they are likely to be nonassertive and feel that they have little control over their lives. They may fail at school, be unable to find suitable work, and refuse to accept social responsibilities. They develop negative self-images.

Chronic drug abusers often remain intimately involved with their families. They rarely achieve real independence. Because they feel trapped, they may attempt to rebel against their parents or punish them for not allowing them to be free of their emotional bondage.

Treatment

In the treatment of drug abuse, as with alcohol abuse, many factors must be taken into account. The family context is of paramount importance, and it is crucial to understand the ways in which the family contributes to or supports the abuse in addition to the ways in which it can help stop it (Freeman, 1993; Piercy, Volk, Trepper, and Sprankle, 1991).

Parents may have a vital stake in maintaining their child's drug abuse as a means of avoiding their own problems. Their child's drug use keeps them from focusing on their marriage. Many drug abuse programs, like alcohol abuse programs, involve intensive family therapy. If the user begins recovery, however, the rest of the family may suffer a crisis: Members may become depressed, parents may threaten divorce, and siblings may act out. Family members may sabotage the treatment of chronic drug abusers. These crises will dissipate as soon as the drug user returns to his or her habit. These families "need" their drug-abusing member to maintain their cohesiveness. Their task is to discover healthy, positive ways of meeting their needs so that individuals may pursue their own growth and fulfillment.

Therapists, counselors, teachers, physicians, and others who work with drug-affected families also need to be aware of the cultural context in which the drug use is occurring. Norms and pressures in various communities—urban, rural, middle-class, poor, white, Latino, African American, Asian American, Native American, and so on—may differ significantly (Gordon, 1993; Ja and Aoki, 1993; Mayers, Kail, and Watts, 1993). Additionally, individual factors influence a person's drug use patterns and recovery process. A person who understands his or her drug abuse from physiological, psychological, and social perspectives is better equipped to combat it (Kumpfer and Hopkins, 1993). Physiological factors include the physical feelings experienced before, during, and after drug use. Psychological factors include emotions ranging from pain and anger to euphoria to numbness and guilt. Social factors include the community environment, peer pressure, and the availability of drugs. Although many drug treatment programs are forced to turn people away because of budget limitations, many communities have a chapter of Narcotics Anonymous or a similar group that can provide understanding and support for those who wish to become free of the grip of addiction.

The Future of Health Care

There remains much to be done to make adequate health care uniformly available to all Americans. Some of the policy issues that need to be addressed follow (Anderson and Feldman, 1993; Daley, 1993; Elders and Tuteur, 1993; Gilliss, 1993):

- *Access to care.* Inability to pay or preexisting medical conditions should not preclude anyone from receiving basic medical care, including preventive care.
- *Poverty as a health issue.* Social issues cannot be separated from health issues. Many of our health problems are directly or indirectly related to poverty. These include preventable childhood diseases, teenage pregnancy, drug and alcohol abuse, domestic violence, and HIV/AIDS.

- *Preventive care and health maintenance.* Programs that encourage the maintenance of good health and provide preventive care and health education must be supported and expanded if we are to control the spiraling economic, social, and personal costs of health care.
- *Mental health.* Psychiatric and psychological problems, including substance abuse, need to be addressed on a national level and treated in community-based programs.
- *Diversity.* Health practitioners and others who work in the health-care system need to be educated regarding issues of family diversity and cultural differences.
- *Support for families.* Family members need education and support so that they may more effectively carry out their caregiving tasks, whether they are caring for family members with temporary acute conditions, chronic illnesses, disabilities, or terminal illnesses.
- *United effort.* No single group is responsible for reforming our health-care system. Within our communities, individual volunteers, businesses, schools, religious groups, activist organizations, government officials, and health-care providers must work together.

As we have seen, it is not possible to separate health issues from the family. Families have people, and people have bodies. It is inevitable that our families have had, or will have, to deal with many of the issues discussed in this chapter. Although there is much we can do as individuals to understand and support our families during health-related crises, we also need to have support from the larger community. Doctors, nurses, and other health-care workers; insurance companies; government agencies; public clinics and programs—all these individuals and organizations must work with families. And they must work not only to cure disease or heal injuries but also to maximize the quality of life for patients and their families.

Summary

- Our attitudes toward our bodies and health are learned in the family. In a healthy family, the six dimensions of wellness (physical, emotional, intellectual, spiritual, social and environmental) interrelate to create optimal health and well-being.
- The relationship between the family and health is interpenetrating rather than causal. It is within the family that we learn behaviors relating to diet, exercise, stress, and substance use and abuse. The emotional climate of the family affects the health of its members, and the health of family members affects the family's psychological well-being.
- Happily married people tend to have better health than unmarried individuals. Marriage encourages health and well-being through living with a partner, social support, and economic well-being.
- The United States is undergoing a severe health-care crisis. We are the only major industrialized country that does not provide universal health care. The factors that most affect health are socioeconomic. Poorer people have less access to preventive care and medical treatment. Minority-status groups are particularly affected. Children in these communities are at risk for numerous diseases, many of which are preventable with immunizations. Poor communities are disproportionately threatened by polluting industries and toxic wastes.
- Because of the high cost of medical care, the possession of health insurance has become a prime determinant of access to such care. Millions of Americans do not possess health insurance, while many millions more are underinsured. Most health insurance is linked to employment. Unemployed, part-time, and low-paid workers in service industries are usually not covered.
- *Stress*—psychological or emotional responses made to demands—occurs as a result of change. Not all stress is bad. Both families and individuals attempt to maintain equilibrium in the face of stress. Normative stress, such as death of elderly members, is predictable and common to all families. Nonnormative stress, such as drug abuse or unemployment, is neither predictable nor common to all families. *Stressor pileup,* or prior strains

and hardships converging with new stressors, is important in determining stress outcome.

- The *double ABC-X model* is the most widely used model describing family stress. *Aa* represents stressor pileup, the combination of the current stressor, family hardship, and strain. *B* represents family coping resources. *C* represents the family's perception of the stressor pileup. *X* represents the outcome.
- Disabilities and chronic illnesses create many forms of stress, including strained family relationships, modifications in family activities and goals, increased tasks and time commitments, increased financial costs, special housing requirements, social isolation, medical concerns, and grieving. The *caregiver role*, usually performed by women, involves considerable stress; a woman must often juggle the demands of multiple roles and may be at risk for illness herself. Coping strategies include making a place for the illness, keeping communication open, developing good working relationships with professionals, and cultivating sources of support.
- Family caregiving activities often begin when an aged parent becomes infirm or dependent. Conflicts that may arise involve previous unresolved problems, the caregiver's inability to accept the parent's dependence, conflicting loyalties, resentment, anger, and money or inheritance conflicts. Family therapy may facilitate coping; other coping strategies include planning ahead with siblings to manage expenses, arrange long-term care, assess family capabilities, divide responsibilities, and determine what services are available.
- Cultural influences on our perception of death may cause us to respond with denial, exploitation, and romanticization. *Thanatologists*, people who study death and dying, tell us that the stages of dying are likely to include denial and isolation, anger, bargaining, depression, and acceptance. Apart from physical care and relief from pain, the most important need of a dying person is to be treated like a human being. *Hospices* provide care for terminally ill individuals, emphasizing both patient care and family support.
- *Bereavement* is the response to the death of a loved one, including customs and rituals and the grieving process (emotional responses and expressions of feeling). Mourning rituals include the funeral service and burial or cremation. The grieving process varies for different people; experiencing grief is a necessary part of healing.
- The costs of alcohol and drug abuse—in both economic and human terms—are enormous. *Alcohol abuse* involves continued use of alcohol despite consequences. *Alcohol dependence* or *alcoholism* involves more extensive problems along with tolerance and withdrawal. Alcoholism is primarily a learned behavior; it often begins as a means of coping with stress, anxiety, or low self-esteem. It is often referred to as a "family disease," as family members may unconsciously help maintain the alcoholic's drinking. Denial plays an important part in the lives of many alcoholics. An alcoholic can recover only if he or she stops drinking permanently.
- Like alcohol, drug use is a serious threat to individuals' and families' lives. The family may be deeply implicated in sustaining drug abuse and addiction. As with alcohol, treating drug abuse may involve the entire family. Drug use must be understood in terms of the cultural context in which it occurs.
- Critical policy issues relating to the future of health care include access to care, poverty, preventive care and health maintenance, mental health (including substance abuse), diversity of families, and support for family caregivers. A united effort among individual volunteers, health-care workers, schools, religious groups, activists, businesses, and government is necessary to effect positive changes within our health-care system.

Key Terms

Suggested Readings

Useful journals for topics covered in this chapter include *American Journal of Public Health*, *Caregiving* (a newsletter published by the National Council on Aging), *Death Studies*, *JAMA: Journal of the American Medical Association*, *Journal of Health and Social Behavior*, *New England Journal of Medicine*, *Omega: Journal of Death and Dying*, and *Psychology of Addictive Behaviors*.

Barasch, Marc Ian. *The Healing Path*. NY: Putnam, 1994. A thorough and thought-provoking exploration of the path through disease toward wholeness, focusing on the patient's role in the healing process.

Biegel, David, Esther Sales, and Richard Shulz. *Family Caregiving in Chronic Illness*. Newbury Park, CA: Sage Publications, 1991. An examination of caregiving research and strategies related to care of family members living with Alzheimer's disease, mental illness, cancer, or other chronic health problems.

Black, Claudia, ed. *Children of Alcoholics, Selected Readings* Rockville, MD: National Association for Children of Alcoholics, 1996. A collection of recent articles written by a host of distinguished researchers and clinicians who look at the question of whether there are differences in functioning between the population of children of alcoholics and other Americans. A fresh approach to this and other issues is offered.

Candib, Lucy M. *Medicine and the Family: A Feminist Perspective*. New York: Basic Books, 1995. A physician's views on such issues as the family life cycle and doctor-patient relationship.

DeSpelder, Lynne Ann, and Albert Lee Strickland. *The Last Dance: Encountering Death and Dying*, 4th ed. Mountain View, CA: Mayfield, 1996. A thorough and thoroughly readable examination of death and dying in America, including personal, sociocultural, medical, and spiritual aspects.

Doherty, William, and Thomas Campbell. *Families and Health*. Newbury Park, CA: Sage Publications, 1988. An in-depth look at the family's role in promoting health and its response and adaptation to illness.

Larson, David, ed. *The Mayo Clinic Family Health Book*, 2nd ed. New York: Morrow, 1996. A comprehensive "everything-you-ever-wanted-to-know-about-your-health-but-were-afraid-to-ask" type of reference book. It covers the human life cycle, environmental hazards, medical emergencies, health behaviors and disorders, medical problems and treatment options, and organizational aspects of medicine (such as living wills and birth options).

Nuland, Sherwin B. *How We Die: Reflections on Life's Final Chapter*. New York: Knopf, 1994. A description by a surgeon of what he observes as people die from cancer, AIDS, heart failure, lung disease, Alzheimer's disease, and grievous injury, reminding us of the importance of compassion and respect for the quality of life.

Peters, Ray D.V., and Robert J. McMahon, eds. *Preventing Childhood Disorders, Substance Abuse and Delinquency*. Thousand Oaks, CA: Sage Publications, Inc., 1996. The newest research on the effectiveness of prevention and early intervention programs for children and adolescents.

Seaward, Brian Luke. *Managing Stress; Principles and Strategies for Health and Well-Being*, 2nd ed. Boston: Jones and Barlett Publishers, 1997. An excellent summation and resource for material on stress and stress management techniques.

Semler, Tracy Chutorian. *All About Eve: The Complete Guide to Women's Health and Well Being*. New York: HarperCollins, 1995. A readable and definitive how-to, what-to, and when-to guide to women's wellness.

Siegel, Bernie. *Love, Medicine and Miracles*. New York: Harper & Row, 1986. A best-selling book in which a surgeon stresses the importance of an individual's mind and emotions in recovering from illness.

Steinglass, Peter. *The Alcoholic Family*. New York: Basic Books, 1987. An overview of alcoholism and the family by leading authorities.

The Planned Parenthood Women's Health Encyclopedia. New York: Crown Trade Paperbacks, 1996. An excellent resource offering a wide range of health information for women.

Viorst, Judith. *Necessary Losses*. New York: Fawcett Gold Medal, 1987. A wise and witty analysis of the stages of our lives and the "loves, illusions, dependencies and impossible expectations" that we must give up, including those surrounding death.

CHAPTER 13

Family Violence and Sexual Abuse

P R E V I E W

To gain a sense of what you already know about the material covered in this chapter, answer "True" or "False" to the statements below.

1. Intimate relationships of any kind increase the likelihood of violence. True or false?
2. Rape by an acquaintance, date, or partner is less likely than rape by a stranger. True or false?
3. Male aggression is generally considered to be a desirable trait in our society. True or false?
4. Studies of family violence have helped strengthen policies for dealing with domestic offenders. True or false?
5. Physically abused children are often perceived by their parents as "different" from other children. True or false?
6. Sibling violence is the most widespread form of family violence. True or false?
7. Relatively few missing children have been kidnapped by strangers. True or false?
8. Deliberate fabrications of sexual abuse constitute nearly 25 percent of all reports. True or false?
9. Most people who were sexually abused as children at least partially remember the abuse. True or false?
10. Brother/sister incest is generally harmless. True or false?

Answers

1 True, 2 False, 3 True, 4 True, 5 True, 6 True, 7 True, 8 False, 9 True, 10 False

OUTLINE

It is an unhappy fact that intimacy or relatedness in any form can increase the likelihood of violence or sexual abuse. If this seems hard to believe, think about whom our society "permits" us to shove, hit, or kick. It is not a stranger, fellow student, co-worker, or employer; if we assaulted any of them, we would run great risk of being arrested. It is with our intimates that we are allowed to do such things. Dating, loving, or being related seems to give us permission to be violent when we are angry. Those nearest and dearest are also those most likely to be slapped, punched, kicked, bitten, burned, stabbed, or shot (Gelles and Cornell, 1990; Gelles and Straus, 1988). For some partners, husbands, or parents, intimacy seems to confer a right to be physically or sexually abusive. Only war zones and urban riot scenes are more dangerous places than families.

Consider the following points:

- Every thirty seconds, a woman is beaten by her boyfriend or husband.
- In various studies, 30 to 40 percent of college students report violence in dating relationships.
- At least a million American children are physically abused by their parents each year.
- Almost a million parents are physically assaulted by their adolescent or youthful children every year.
- As many as 27 percent of American women and 16 percent of men have been the victims of childhood sexual abuse, much of it in their own families.

Until the 1970s, Americans believed that when they locked their homes at night, they locked out violence; the sad fact is that they also locked in violence. When the image of Nicole Brown Simpson first appeared on our screens, a permanent scar was marked in the minds of most Americans. The abuse and killing of Nicole and thousands like her was a wake-up call for policy makers, prosecutors, and individuals around the country to begin to listen and act on behalf of those who have been ignored too long. Even though domestic violence is beginning to be recognized and understood, there is much work to be done toward reducing and eliminating it.

In this chapter, we look at the models researchers use in studying family violence, and we discuss the dynamics that are present in battering relationships. We look at violence between husbands and wives (including marital rape), between gay and lesbian partners, between dating partners (including acquaintance rape), and between siblings, as well as violence committed against children by parents and against parents by grown children. We also discuss prevention and treatment strategies. The last portion of the chapter is devoted to the discussion of child sexual abuse—its forms, participants, effects, and treatment and prevention strategies.

In violence we forget who we are.

Mary McCarthy (1912–1989)

Family Violence

Researchers have not been in complete agreement about what constitutes **violence.** For the purposes of this book, we use the definition offered by Richard Gelles and Claire Pedrick Cornell (1990): "an act carried out with the intention or perceived intention of causing physical pain or injury to another person." There are other prevalent forms of abuse, of course—such as neglect and emotional abuse, including verbal abuse—but the focus of this chapter is physical violence and sexual abuse. Violence may be seen as a continuum, with "normal" abuse (such as spanking) at one end and abusive violence (acts with high potential for causing injury) at the other extreme. We must look at the continuum as a whole to be concerned with "families who shoot and stab each other as well as those who spank and shove, . . . [as] one cannot be understood without considering the other" (Straus, Gelles, and Steinmetz, 1980). (For a review of the literature on domestic violence and sexual abuse of children, see Gelles and Conte, 1991.)

Our homes may be the most dangerous places for us to be, especially if we are young or female.

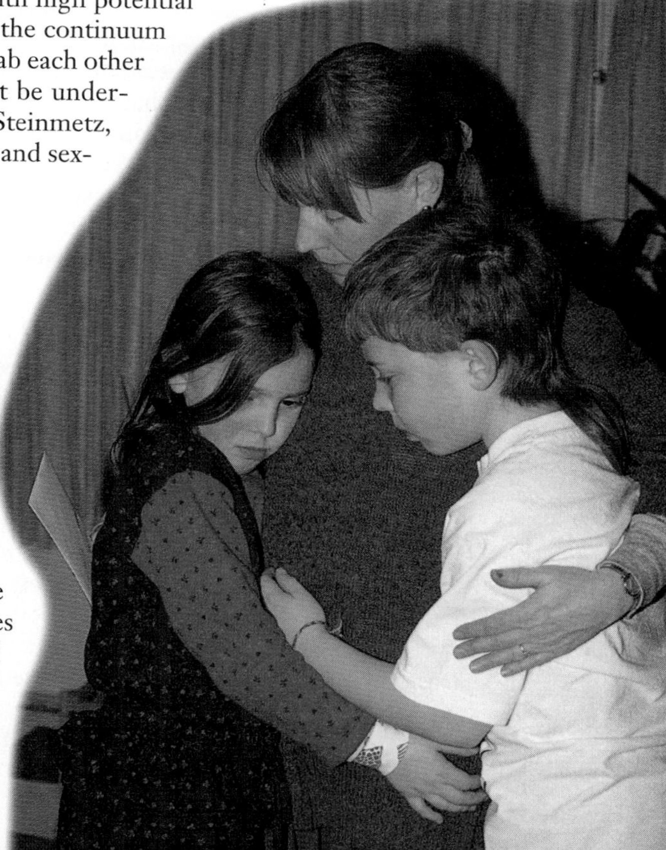

Models of Family Violence

To better understand violence within the family, we must look at its place in the larger sociocultural environment. Aggression is a trait that our society labels as generally desirable, especially for males. Getting ahead at work, asserting ourselves in relationships, and winning at sports are all culturally approved actions. But does aggression necessarily lead to violence? All families have their ups and downs, and all family members at times experience anger toward one another. How do we explain that violence erupts more frequently and with more severe consequences in some families than in others? The principal models used in understanding family violence are discussed in the following sections. Each of these models has valuable insight to offer concerning a very complex problem with no easy or single solution.

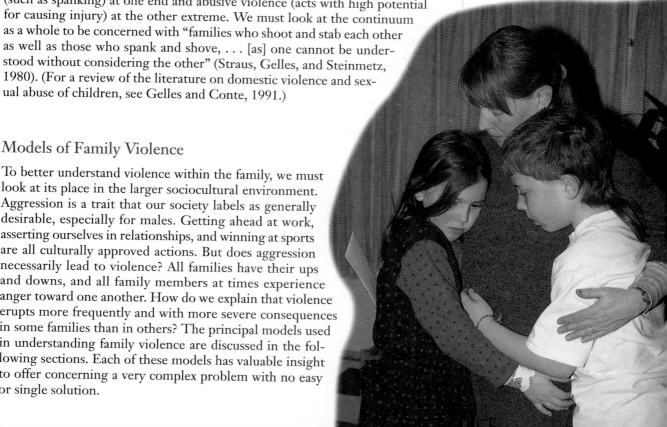

You can always kill a relative with a gun that isn't loaded.

MARK TWAIN (1835–1910)

Psychiatric Model

The psychiatric model finds an important source of family violence to be within the personality of the abuser (O'Leary, 1993). It assumes that he or she is violent as a result of personality disorder, mental or emotional illness, or alcohol or drug misuse. Although research indicates that fewer than 10 percent of family violence cases are attributable to psychiatric causes, the idea that people are violent because they are crazy or drunk is widely held (Gelles and Cornell, 1990). Gelles and Cornell (1990) suggest that this model is compelling because "if we can persist in believing that violence and abuse are the products of aberrations or sickness, and, therefore, believe ourselves to be well, then our acts cannot be hurtful or abusive." But besides looking at the abuser, we must step back and look at the big picture—at the family and society that influence the abuser.

Ecological Model

The ecological model utilizes a systems perspective to look at the child's development within the family environment and the family's development within the community. Psychologist James Garbarino (1982) has suggested that cultural support for physical force against children combines with lack of family support in the community to increase the risk of intrafamily violence. Under this model, a child who doesn't "match" well with the parents (such as a child with emotional or developmental disabilities) and a family that is under stress (from, for example, unemployment or poor health) and that has little community support (such as child care or medical care) is at increased risk for child abuse.

Men are but children of a larger growth.

JOHN DRYDEN (1631–1700)

Patriarchy Model

The patriarchy (male-dominance) model of family violence draws its conclusions from a historical perspective. It holds that most social systems have traditionally placed women in a subordinate position to men, thus condoning or supporting the institution of male violence (Schechter and Gary, 1988; Yllo, 1993). There is no doubt that violence against women and children, and indeed violence in general, has had an integral place in most societies throughout history. Feminist theory must be credited for advancing our understanding of domestic violence by insisting that the patriarchal roots of domestic relations be taken into account. Taken alone, however, the patriarchy model does not adequately explain the variations in degrees of violence among families in the same society (Yllo, 1993).

By our readiness to allow arms to be purchased at will and fired at whim, we have created an atmosphere in which violence and hatred have become popular pastimes.

MARTIN LUTHER KING, JR. (1929–1968)

Social Situational and Social Learning Models

The social models are related to the ecological and patriarchy models in that they view violence as originating in the social structure. The social situational model views family violence as arising from two main factors: (1) structural stress (such as low income or illness) and (2) cultural norms (such as the "spare the rod and spoil the child" ethic) (Gelles and Cornell, 1990). In this model, groups with few resources, such as the poor, are seen to be at greater risk for family violence than those who are well off. The social learning model holds that people learn to be violent from society at large and from their families (Ney, 1992). Although it is true that many perpetrators of family violence were themselves abused as children, it is also true that many victims of childhood violence do not become violent parents. These theories do not account for this discrepancy. (See Egeland, 1993, and Kaufman and Zigler, 1993, for conflicting views on the significance of the intergenerational transmission of abuse.)

Resource Model

William Goode's (1971) resource theory can be applied to family violence. This model assumes that social systems are based on force or the threat of force. A person acquires power by mustering personal, social, and economic resources. Thus, according to Goode, the person with the most resources is the least likely to resort to overt force. Gelles and Cornell (1990) explain, "A husband who wants to be the dominant person in the family but has little education, has a job low in prestige and income, and lacks interpersonal skills may choose to use violence to maintain the dominant position."

Exchange/Social Control Model

Richard Gelles (Gelles, 1993b; Gelles and Cornell, 1990) posits a two-part theory of family violence. The first part, exchange theory, holds that in our interactions, we constantly weigh the perceived rewards against the costs. When Gelles says that "people hit and abuse family members because they can," he is applying exchange theory. The expectation is that "people will only use violence toward family members when the costs of being violent do not outweigh the rewards." (The possible rewards of violence might be getting one's own way, exerting superiority, working off anger or stress, or exacting revenge. Costs could include being hit back, being arrested, being jailed, losing social status, or dissolving the family.)

People hit and abuse family members because they can.

RICHARD GELLES

Social control raises the costs of violent behavior through such means as arrest, imprisonment, loss of status, or loss of income. Three characteristics of families that may reduce social control—and thus make violence more likely—are the following:

1 *Inequality.* Men are stronger than women and often have more economic power and social status. Adults are more powerful than children.
2 *Private nature of the family.* People are reluctant to look outside the family for help, and outsiders (the police or neighbors, for example) may hesitate to intervene in private matters. The likelihood of family violence goes down as the number of nearby friends and relatives increases (Gelles and Cornell, 1990).
3 *"Real man" image.* In some American subcultures, aggressive male behavior brings approval. The violent man in these groups may actually gain status among his peers for asserting his "authority."

One murder makes a villain, millions a hero.

BISHOP BEILBY PORTEUS (1731–1808)

The exchange/social control model is useful for looking at treatment and prevention strategies for family violence, which we discuss later in this chapter.

There is no crime of which one cannot imagine oneself to be the author.

JOHANN GOETHE (1749–1832)

Battered Women and Battering Men

Battering, as used in the literature of family violence, is a catchall term that includes, but is not limited to, slapping, punching, knocking down, choking, kicking, hitting with objects, threatening with weapons, stabbing, and shooting. The use of physical force against women is certainly not a new phenomenon. It may be thought of as a time-honored tradition in our culture. The commonly used term *rule of thumb* derives from the legally sanctioned (until the nineteenth century) practice of "disciplining" one's wife with a switch or rod—provided it was no wider than the disciplinarian's thumb (a good example of the basis for the patriarchy model).

Did you know?

Female victims of violence were more likely to be injured when attacked by someone they knew than female victims of violence who were attacked by a stranger (U.S. Dept. of Justice, 1995).

In recent years the subject of wife battering has gained public notoriety. It remains, however, subordinate in the public mind to the physical and sexual abuse of children. In part, this may be because historically and culturally, women are considered "appropriate" victims of domestic violence (Gelles and Cornell, 1990). Many expect, understand, and accept the idea that women sometimes need to be "put in their place" by men. Such misogynistic ideas provide the cultural basis for the physical and sexual abuse of women.

No one knows for certain how many battered women there are in the United States, but government figures indicate that battering is one of the most common and underreported crimes in the country. Here are some figures concerning violence against women in the United States (Gelles and Cornell, 1990; Marek, 1994; Campbell 1995; U.S. Dept. of Justice, 1995):

- About one woman in twenty-two (at least 2–4 million adult women) are victims of abusive violence each year.
- About half of battered women are beaten at least three times a year.
- Nearly 30 percent of murdered women are killed by their husbands or lovers.
- Women of all races are equally vulnerable to attacks by intimates.
- In a national survey, about one in four wives and one in three husbands said that slapping a partner was "at least somewhat necessary, normal, or good."

Battering occurs at all levels of society. It occurs most frequently among those under thirty years of age (Straus, Gelles, and Steinmetz, 1980), and it is apparently twice as prevalent among African-American couples as among white couples. It should be noted, however, that the rate of violence against African-American women by their partners appears to have dropped considerably in recent years. A 1995 survey showed a 43 percent decrease in such abuse since 1975 (Hampton, 1995). (Possible reasons for this decline will be discussed later on.) Although no social class is immune to it, most studies find that marital violence is more likely to occur in low-income, low-status families (Gelles and Cornell, 1990).

Characteristics of Battered Women

Although early studies of battering relationships seemed to indicate a cluster of personality characteristics constituting a typical battered woman, more recent studies have not borne out this viewpoint. Factors such as self-esteem or childhood experiences of violence do not appear to be necessarily associated with a woman's being in an assaultive relationship (Hotaling and Sugarman, 1990). Two characteristics, however, do appear to be highly correlated with wife assault. First, a number of studies have found that wife abuse is both more common and more severe in families of lower socioeconomic status, a finding partly attributed to the fact that higher income adults have greater privacy and thus are better able to conceal domestic violence (Fineman and Mykitiuk, 1994). Second, marital conflict—and the apparent inability to resolve it through negotiation and compromise—is a factor in many battering relationships. Hotaling and Sugarman (1990) found that conflicts in these marriages often were associated with a difference in expectations about the division of labor in the family, frequent drinking by the husband, and the wife's having attained a higher educational level than the husband. These researchers concluded that it is not useful to focus "primarily on the victim in the assessment of risk to wife assault."

Force cannot give right.

THOMAS JEFFERSON (1743–1826)
The Rights of British America

Characteristics of Batterers

A man who systematically inflicts violence on his wife or lover is likely to have some or all of the following traits (Edelson et al., 1985; Gelles and Cornell, 1990; Goldstein and Rosenbaum, 1985; Margolin, Sibner, and Gleberman, 1988; Vaselle-Augenstein and Erlich, 1992; Walker, 1979, 1984):

- He believes the common myths about battering (see this chapter's "Understanding Yourself").
- He believes in the traditional home, family, and gender-role stereotypes.
- He has low-self esteem and may use violence as a means of demonstrating power or adequacy.
- He may be sadistic, pathologically jealous, or passive-aggressive.
- He may have a "Dr. Jekyll and Mr. Hyde" personality, being capable at times of great charm.
- He may use sex as an act of aggression.
- He believes in the moral rightness of his violent behavior (even though he may "accidentally" go too far).

All they that take the sword shall perish with the sword.

MATHEW ZBISZ

Lenore Walker (1984) believes that a man's battering is not the result of his interactions with his partner or any kind of provocative personality traits of the partner. She writes:

> The best prediction of future violence was a history of past violent behavior. This included witnessing, receiving, and committing violent acts in [the] childhood home; violent acts toward pets, inanimate objects, or other people; previous criminal record; longer time in the military service; and previous expression of violent behavior toward women. If these items are added to a history of temper tantrums, insecurity, need to keep the environment stable, easily threatened by minor upsets, jealousy, possessiveness, and the ability to be charming, manipulative, and seductive to get what he wants, and hostile, nasty and mean when he doesn't succeed—then the risk for battering becomes very high. If alcohol abuse problems are included, the pattern becomes classic.

Battered Husbands

The incidence of battered husbands is unknown. We have, after all, the cartoon image of Blondie chasing Dagwood with a rolling pin. (We don't see Dagwood chasing Blondie with a gun or knife, however, which is a more realistic depiction of family violence.) Although it is undoubtedly true that some men are injured in attacks by wives or lovers, *husband battering* is probably a misleading term. The overwhelming majority of victims of adult family violence are women: Ten times as many women as men are seriously victimized by an intimate partner or ex-partner (Campbell, 1995). In one study (Saunders, 1986) investigating "husband abuse," almost all the women reported that they acted in self-defense; they did not initiate the violence. Their actions did not cause noticeable injury. Indeed, a woman may attempt to inflict damage on a man in self-defense or retaliation, but most women have no hope of prevailing in hand-to-hand combat with a man. A woman may be severely injured simply trying to defend herself. As Gelles and Cornell (1990) observe, although there may be similar rates of hitting, "when injury is considered, marital violence is primarily a problem of victimized women."

Although male violence dramatically overshadows female violence, female violence probably occurs at about the same rate; because women tend to do less damage,

UNDERSTANDING YOURSELF

The Mythology of Family Violence and Sexual Abuse

The understanding of family violence and sexual abuse is often obscured by the different mythologies surrounding these issues. Below are twelve popular myths about family violence and sexual abuse in our society. Some of these myths may apply to some cases, but as generalizations they are not accurate. Many of these myths are accepted by the victims of family violence, as well as by the perpetrators themselves. As you look at the following myths, which ones do you believe (or have you believed)? What was the basis or source of your beliefs? If you no longer believe a particular myth, what changed your mind?

- Family violence is extremely rare.
- Family violence is restricted to families with low levels of education and low socioeconomic status.
- Most family violence is caused by alcohol or drug abuse.
- Violent spouses or parents have psychopathic personalities.
- Violent families are not loving families.
- Battered women cause their own battering because they are masochistic or crazy.
- A battered woman can always leave home.
- Most child sexual abuse is perpetrated by strangers.
- Sexual abuse in families is a fairly rare occurrence.
- Abused children will grow up to abuse their own children.
- The police give adequate protection to battered women.
- Most of society does not condone domestic violence.

These myths hide the extent of physical and sexual abuse that actually takes place inside a painfully large number of American families. Belief in these myths makes it possible to avoid dealing with some of the unhappy realities—at least for a while.

however, we may not consider it as important as that committed by men (Straus, 1993). Suzanne Steinmetz (1987) suggests that some scholars "deemphasize the importance of women's use of violence." As such, there is a "conspiracy of silence [which] fails to recognize that family violence is never inconsequential." Murray Straus (1993) lists four reasons for taking the study of female violence seriously:

1. Assaulting a spouse—either a wife or a husband—is an "intrinsic moral wrong."
2. Not doing so unintentionally validates cultural norms that condone a certain amount of violence between spouses.
3. There is always the danger of escalation. A violent act—whether committed by a man or a woman—may well lead to increased violence.
4. Spousal assault is a model of violent behavior for children. Children are affected as strongly by viewing the violent behavior of their mothers as by viewing that of their fathers.

The Cycle of Violence

Lenore Walker's (1979) research has revealed a three-phase wife-battering cycle. The duration of each phase may vary, but the cycle goes on and on:

- *Phase 1: tension building.* Tension is in the air. The woman tries to do her job well, to be conciliatory. Minor battering incidents may occur. She denies her own rising anger. Tension continues to build.
- *Phase 2: the explosion.* The man loses control. Sometimes the woman will precipitate the incident to "get it over with." He generally sets out to

"teach her a lesson" and goes on from there. This is the shortest phase, usually lasting several hours but sometimes continuing for two or three days or longer.

- *Phase 3: the "honeymoon."* Tension has now been released, and the batterer is contrite, begs forgiveness, and sincerely promises never to do it again. The woman chooses to believe him and forgives him. This "symbiotic bonding" (interdependence) makes intervention, help, or change unlikely during this phase.

Often the battered woman expresses surprise at what has touched off the battering incident. It may have been something outside the home—at the man's job, for example. He may have come home drunk, or he may have been drinking steadily at home. Alcohol is implicated in many battering incidents and possibly in the majority of them. Results of studies of alcohol and battering vary widely; alcohol problems were reported in 35 to 93 percent of cases of assaultive husbands, depending on the study (Leonard and Jacob, 1988; see Edelson et al., 1985; Egeland, 1993; and Gelles, 1993a, for discussions of conflicting studies about battering and alcohol use).

In a battering relationship, the woman may not only suffer physical damage but also be seriously harmed emotionally by a constant sense of danger and the expectation of violence that weaves a "web of terror" about her (Edelson et al., 1985). Walker (1993) suggests that women who are repeatedly abused may develop a set of psychological symptoms similar to those of post-traumatic stress disorder (PTSD). She labels these symptoms *battered woman syndrome.*

Violence in Gay and Lesbian Relationships

Until recently, very little was known about violence in lesbian and gay relationships. One reason is that such relationships have not been given the same social status as those of heterosexuals. (In nine states a partner in a same-sex couple cannot obtain a restraining order against the other partner [King, 1993].) Recent research indicates that the rate of abuse in gay and lesbian relationships is comparable to that in heterosexual relationships; 11 percent to more than 45 percent according to various studies (Renzetti, 1995). Furthermore, Claire Renzetti found that violence in same-sex relationships is rarely a one-time event; once violence occurs it is likely to reoccur. It also appears to be as serious as violence in heterosexual relationships, including physical, psychological, and/or financial abuse. One additional form, that of "outing" (revealing another's gay orientation without consent), may be used as a form of psychological abuse in same-sex relationships.

For battered partners in same-sex relationships, there is often nowhere to go for support. Services for gay men and lesbians are often nonexistent or uniformed about the multifaceted issues that face such victims. Renzetti (1995) points out several policy issues that must be addressed among service providers and domestic violence agencies:

- Consider how homophobia inhibits gay and lesbian victims of abuse from self-identifying as such.
- Recognize that battered gay men and lesbians of color experience a triple jeopardy: as victims of domestic violence, as homosexuals, and as racial/ethnic minorities.
- Address the issue of gay men and lesbians as both batterers and victims who may seek services at the same time from the same agency.

Marital Rape

Rape in marriage is one of the most widespread and overlooked forms of family violence. Marital rape is a form of battering. Most legal definitions of **rape** include "unwanted sexual penetration, perpetrated by force, threat of harm, or when the victim [is] intoxicated" (Koss and Cook, 1993). Rape may be perpetrated by males or females and against males or females; it may involve vaginal, oral, or anal penetration; and it may involve the insertion of objects other than the penis. Approximately 10–14 percent of wives have been forced by their husbands to have sex against their will (Yllo, 1995).

Historically, marriage has been regarded as giving husbands unlimited sexual access to their wives. Beginning in the late 1970s, most states enacted legislation to make at least some forms of marital rape illegal. It was not until 1993, however, that laws were enacted to make marital rape a crime. Throughout the United States, a husband can be prosecuted for raping his wife, although twenty-six states limit the conditions, such as requiring extraordinary violence. Only seventeen states offer full legal protection for wives (Muehlenhard, Powch, Phelps, and Giusti, 1992). The precise definition of **marital rape** differs from state to state, however. In several states, wife rape is illegal only if the couple has separated.

Did you know?

Fifty percent of the men who frequently assaulted their wives also physically abused their children (Tolman, 1995).

Because of the sexual nature of marriage, marital rape has not been regarded as a serious form of assault. According to Kersti Yllo (1995):

> A widely held assumption has been that an act of forced sex in the context of an ongoing relationship in which consensual sex occurs cannot be very significant or traumatic. This assumption is flawed because it overlooks the core violation of rape that is coercion, violence and in the case of wife rape, the violation of trust.

Rape is nothing more or less than a conscious process of intimidation by which all men keep all women in a state of fear.

SUSAN BROWNMILLER

There still remains the problem of enforcing the law. Many people discount rape in marriage as a "marital tiff" that has little to do with "real" rape (Yllo, 1995). Many victims themselves have difficulty acknowledging that their husbands' sexual violence is indeed rape. White females are more likely than African-American females to identify sexual coercion in marriage as rape (Cahoon, Edmonds, Spaulding, and Dickens, 1995, and all too often, judges seem more in sympathy with the perpetrator than with the victim, especially if he is very intelligent, successful, and well educated. There is also the "notion that the male breadwinner should be the beneficiary of some special immunity because of his family's dependence on him" (Russell, 1990). Diana Russell goes on to say:

> On the basis of such an argument, it follows that it would be a violation of the principle of equity to incarcerate men who beat up and/or rape women who are not their wives. Specifically, it would not be fair to the wives and children of employed stranger rapists, acquaintance rapists, date rapists, lover rapists, authority figure rapists, or rapists who rape their friends. Why should these families have to endure the loss of their breadwinners if the families of husbands who are rapists are spared this hardship?

Because these kinds of attitudes are so entrenched in the American psyche, it is estimated that two-thirds of sexual assault victims do not report the crime (U.S. Dept. of Justice, 1997).

Marital rape victims experience feelings of betrayal, anger, humiliation, and guilt. Following their rapes, many wives feel intense anger toward their husbands. One woman recounted, "'So,' he says, 'You're my wife and you're gonna . . .' I just laid there thinking 'I hate him, I hate him so much.'" Another expressed her humiliation and sense of "dirtiness" by taking a shower: "I tried to wash it away, but you can't. I felt like a sexual garbage can" (Finkelhor and Yllo, 1985). Some feel guilt and blame themselves for not being better wives. Some develop negative self-images and view their lack of sexual desire as a reflection of their own inadequacies rather than as a consequence of abuse.

Dating Violence and Rape

In the last decade or so, researchers have become increasingly aware that violence and sexual assault can take place in all forms of intimate relationships. Violence between intimates is not restricted to family members. Even casual or dating relationships can be marred by violence or rape.

Dating Violence

The incidence of physical violence in dating relationships, including those of teenagers, is alarming. Evidence suggests that it approaches or even exceeds the level of marital violence (Lloyd, 1995). A recent survey on physical violence during courtship revealed that almost one-third of young adults (age thirty and under) had experienced or used physical violence in a dating relationship during the past twelve months (Lloyd, 1995). Sexual violence may be even more prevalent. Nearly two in five college women have experienced an actual rape or an attempted rape or have been coerced into intercourse against their will.

Although it may seem logical to assume that dating violence leads to marital violence, little actual research has been done in this area. It does appear, however, that the issues involved in dating violence are different than those generally involved in spousal violence. Whereas marital violence may erupt over domestic issues such as housekeeping and child rearing (Hotaling and Sugarman, 1990), dating violence is far more likely to be precipitated by jealousy or rejection (Lloyd and Emery, 1990; Makepeace, 1989). One young woman recounted the following incident (Lloyd and Emery, 1990):

> I was waiting for him to pick me up in front of school. I was befriended by some guys and we struck up a conversation. When my boyfriend picked me up he didn't say anything. When we got home, physical violence occurred for the first time in our relationship. I had no idea it was coming. He caught me on the jaw, and hit me up against the wall. I couldn't cry or scream or anything—all I could do was look at him. He picked me up and threw me against the wall and then started yelling and screaming at me that he didn't want me talking to other guys.

Lloyd and Emery found that dating violence might also involve the man's use of alcohol or drugs, "unpredictable" reasons, and intense anger.

Although many women leave a dating relationship after one violent incident, others stay through repeated episodes. Women who have "romantic" attitudes about jealousy and possessiveness and who have witnessed physical violence between their own parents may be more likely to stay in such relationships (Follingstad, Rutledge, McNeill-Harkins, and Polek, 1992). Women with "modern" gender-role attitudes are more likely to leave than those with traditional attitudes (Flynn, 1990). Women who leave violent partners cite the following factors in making the decision to break up: a series of broken promises that the man will end the violence, an improved self-image ("I deserve better"), escalation of the violence, and physical and emotional help from family and friends (Lloyd and Emery, 1990). Apparently, counselors, physicians, and law enforcement agencies are not widely used by victims of dating violence (Pirog-Good and Stets, 1989; see Lloyd, 1991, for implications for intervention in courtship violence; see Levy, 1991, for perspectives on violence in adolescent relationships).

Did you know?

Forty-four percent of rape victims are younger than eighteen years old, and two-thirds of violent sex offenders serving time in state prisons said their victims were younger than eighteen years (U.S. Dept. of Justice, 1997).

Date Rape

Sexual intercourse with a dating partner that occurs against his or her will with force or the threat of force—**date rape**—is the most common form of rape. Date rape is also known as **acquaintance rape.** One study found that women were more likely than men to define date rape as a crime. Disturbingly, date rape was considered less serious when the woman was African American (Foley et al., 1995).

Date rapes are usually not planned. Two researchers (Bechhofer and Parrot, 1991) describe a typical date rape:

> He plans the evening with the intent of sex, but if the date does not progress as planned and his date does not comply, he becomes angry and takes what he feels is his right—sex. Afterward, the victim feels raped while the assailant believes that he has done nothing wrong. He may even ask the woman out on another date.

Alcohol or drugs are often involved. When both people are drinking, they are viewed as more sexual. Men who believe in rape myths are more likely to see drinking as a sign that females are sexually available (Abbey and Harnish, 1995). In one study, 79 percent of women who were raped by their date had been drinking or taking drugs prior to the rape. Seventy-one percent said their assailant had been drinking or taking drugs (Copenhaver and Grauerholz, 1991). There are also high levels of alcohol and drug use among middle school and high school students who have unwanted sex (Rapkin and Rapkin, 1991).

Incidence of Date Rape. Lifetime experience of date rape ranges from 15 to 28 percent for women, according to various studies. If the definition is expanded to include attempted intercourse as a result of verbal pressure or the misuse of authority, then women's lifetime incidence increases significantly. When all types of unwanted sexual activity are included, ranging from kissing to sexual intercourse, half to three-quarters of college women report sexual aggression in dating (Cate and Lloyd, 1992). There is also considerable sexual coercion in relationships between gay men. Coercion also exists in lesbian relationships, though less than in gay male and heterosexual ones.

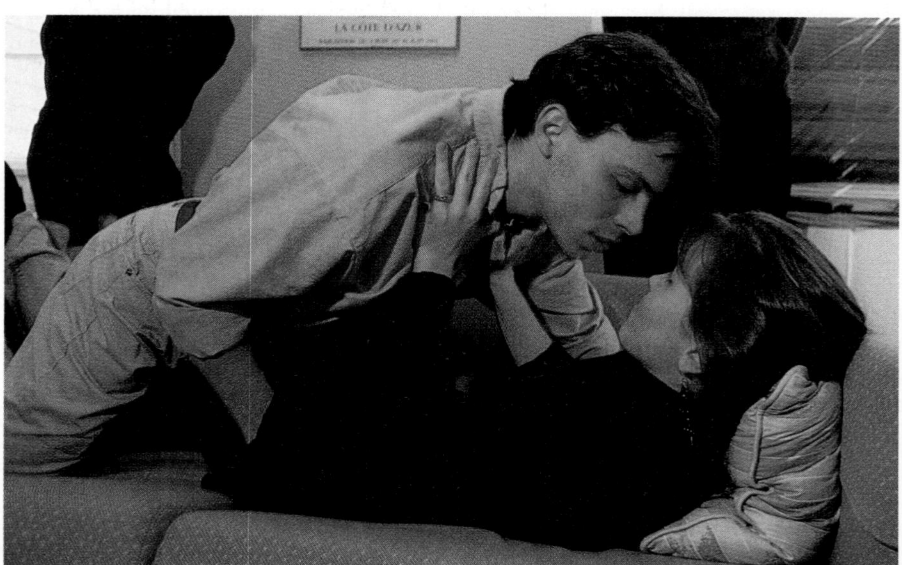

Sexual assault peer educators at Brown University dramatize date rape to make students aware of its dynamics.

In a large-scale study on sexual aggression, Mary Koss (1988) surveyed over 6,100 students in thirty-two colleges. Her findings indicated the following:

- Almost 54 percent of the women surveyed had been sexually victimized in some form. Fifteen percent had been raped.
- A quarter of the women surveyed had been the victims of rape or attempted rape; 84 percent knew their assailants.
- Forty-seven percent of the rapes were by first dates, casual dates, or romantic acquaintances.
- Twenty-five percent of the men had perpetrated sexual aggression. Three percent had attempted rape, and 4 percent had actually raped.
- Almost three-quarters of the raped women did not identify their experiences as rape.

Physical violence often goes hand in hand with sexual aggression. One researcher found, in a study of acquaintance rape victims, that three-fourths of the women sustained bruises, cuts, black eyes, and internal injuries. Some were knocked unconscious (Belknap, 1989).

When No Is No. There is considerable confusion and argument about sexual consent. Much sexual communication is done nonverbally and ambiguously. As Charlene Muehlenhard and her colleagues (1992) note:

> Most sexual scripts do not involve verbal consent. One such script involves two people who are overcome with passion. Another such script involves a male seducing a hesitant female, who, according to the sexual double standard, must not acknowledge her desire for sex lest she be labeled "loose" or "easy." Neither of these scripts involve explicit verbal consent from both persons.

I hate victims who respect their executioners.

JEAN-PAUL SARTRE (1905–1980)

That we don't usually give verbal consent for sex indicates the importance of nonverbal clues. Nonverbal communication is imprecise, however, as we saw in Chapter 5. It can be misinterpreted easily if it is not reinforced verbally. For example, men frequently mistake a woman's friendliness for sexual interest (Johnson, Stockdale, and Saal, 1991; Stockdale, 1993). They often misinterpret a woman's cuddling, kissing, and fondling as wishing to engage in sexual intercourse (Gillen and Muncher, 1995; Muehlenhard, 1988; Muehlenhard and Linton, 1987). A woman needs to make her boundaries clear verbally.

Our sexual scripts often assume yes unless a no is directly stated (Muehlenhard et al., 1992). This makes individuals "fair game" unless a person explicitly says no. The assumption of consent puts women at a disadvantage. First, because men traditionally initiate sex, men may feel it is legitimate to initiate sex whenever they desire without women's explicitly consenting. Second, women's withdrawal can be considered insincere because consent is always assumed. Such thinking reinforces a common sexual script in which men initiate and women refuse so as not to appear promiscuous. In this script, the man continues believing that her refusal is token. One study found that almost 40 percent of the women had offered a token no at least once (Muehlander and Hollabaugh, 1989). Some common reasons for offering token no's include not wanting to appear "loose," unsureness of how the partner feels, inappropriate surroundings, and game playing (Muehlenhard and Hollabaugh, 1989; Muelhenhard and McCoy, 1991). Because some women sometimes say "no" when they mean "coax me," male-female communication may be especially unclear regarding consent (Muehlenhard and Cook, 1991). Furthermore, men are more likely than women to think of male-female relationships as a "battle of the sexes" (Reilly et al., 1992).

Because relationships are conflictual, no's are to be expected as part of the battle. A man, however, "should" persist because it is his role to conquer, even if he is not interested in sex (Muehlenhard and Schrag, 1991; Muehlenhard et al., 1991).

Avoiding Date Rape. To reduce the risk of date rape, women should consider the following points:

- When dating someone for the first time, go to a public place, such as a restaurant, movie, or sports event.
- Share expenses. A common scenario is a date expecting you to exchange sex for his paying for dinner, the movie, drinks, and so on (Muehlenhard and Schrag, 1991; Muehlenhard et al., 1991).
- Avoid using drugs or alcohol if you do not want to be sexual with your date. Their use is associated with date rape (Abbey, 1991).
- Avoid ambiguous verbal or nonverbal behavior. Examine your feelings about sex and decide early if you wish to have sex. Make sure your verbal and nonverbal messages are identical. If you only want to cuddle or kiss, tell your partner that those are your limits. Tell him that if you say no, you mean no. If necessary, reinforce your statement emphatically, both verbally ("No!") and physically (pushing him away) (Muehlenhard and Linton, 1987).
- Be forceful and firm. Don't worry about being polite. Often men interpret passivity as permission and ignore or misunderstand "nice" or "polite" approaches (Hughes and Sandler, 1987).
- If things get out of hand, be loud in protesting, leave, and go for help.

Why Women Stay in Violent Relationships

Violence in relationships generally develops a continuing pattern of abuse over time. We know from systems theory that all relationships have some degree of mutual dependence, and battering relationships are certainly no different. Women stay in or return to violent situations for many reasons. Some common ones are the following:

- *Economic dependence.* Even if a woman is financially secure, she may not perceive herself as being able to cope with economic matters. For low-income or poor families, the threat of losing the man's support—if he is incarcerated, for example—may be a real barrier against change.
- *Religious pressure.* She may feel that the teachings of her religion require her to keep the family together at all costs, to submit to her husband's will, and to try harder.
- *"The children need a father."* She may believe that even a father who beats the mother is better than no father at all. If the abusing husband also assaults the children, the woman may be motivated to seek help (but this is not always the case).
- *Fear of being alone.* She may have no meaningful relationships outside her marriage. Her husband may have systematically cut off her ties to other family members, friends, and potential support sources. She has nowhere to go and no way to get any real perspective on her situation. (See Nielsen, Endo, and Ellington, 1992, for the relationship between social isolation and abuse.)
- *Belief in the American dream.* The woman may have accepted without question the myth of the perfect woman and happy household. Even though her existence belies this, she continues to believe that it is how it should (and can) be.

- *Pity*. She feels sorry for her husband and puts his needs ahead of her own. If she doesn't love him, who will?
- *Guilt and shame*. She feels that it is her own fault if her marriage isn't working. If she leaves, she believes, everyone will know she is a failure, or her husband might kill himself.
- *Duty and responsibility*. She feels she must keep her marriage vows "till death us do part."
- *Fear for her life*. She believes she may be killed if she tries to escape.
- *Love*. She loves him; he loves her. On her husband's death, one elderly woman (a university professor) spoke of her fifty-three years in a battering relationship (Walker, 1979): "We did everything together. . . . I loved him; you know, even when he was brutal and mean. . . . I'm sorry he's dead, although there were days when I wished he would die. . . . He was my best friend. . . . He beat me right up to the end. . . . It was a good life and I really do miss him."
- *Cultural reasons*. Women from nonmainstream cultural backgrounds may face great obstacles to leaving a relationship. They may not speak English; they may not know where to go for help and may fear they will not be understood. They often fear that the husband will lose his job, retaliate against them, or take the children back to their country of origin (Donnelly, 1993). Recent immigrants from Latin America, Asia, and South Asia may be especially fearful that their revelations will reflect badly on the family and community.

When we want to read of the deeds that are done for love, whither do we turn? To the murder column.

George Bernard Shaw (1856–1950)

Learned Helplessness

Lenore Walker (1979, 1993) theorizes that women stay in battering relationships as a result of "learned helplessness." According to Walker, women who are repeatedly battered develop much lower self-concepts than women in nonbattering relationships. They begin to feel that they cannot control the battering or the events that surround them. Through a process of behavioral reinforcement, they "learn" to become helpless. As Walker notes:

> Women are systematically taught that their personal wants, survival, and anatomy do not depend on effective and creative responses to life situations, but rather on their physical beauty to men. They learn that they have no direct control over the circumstances of their lives.

If violence is used against them, women may even become desensitized to the accompanying pain and fear. The more it happens, the more helpless they feel and the less they are able to see alternative possibilities. Walker points out that these women are not totally helpless or passive, but that "they narrow their choice of responses, opting for those that have the highest predictability of creating successful outcomes" (Walker, 1993).

Women's Coping Strategies

Some women think they can stop their partners' violence, and some, in fact, do. Lee Bowker (1983) reported that women used a variety of strategies to stop their husbands' abusiveness. These ranged from passive defense techniques (covering their bodies with arms or hands) to seeking informal help (friends) or formal help (counseling through a social service agency). The particular strategy was not important in stopping the violence, however. What made the crucial difference appeared to be the woman's determination that the violence must cease (Bowker, 1983; Gondolf, 1987, 1988).

Reflections

In your family (including your extended family), has there been spousal violence? Have you experienced violence in a dating relationship? If so, what were the factors involved in causing it? In sustaining it? If you or your family have not been involved in such violence, what factors do you think have protected against it?

Alternatives: Police Intervention, Shelters, and Abuser Programs

Professionals who deal with domestic violence have long debated the relative merits of control versus compassion as intervention strategies (Mederer and Gelles, 1989). Although more understanding of the dynamics of abusive relationships and the deterrence process is clearly needed, we can see that both approaches have their place. Both controlling measures (which raise the "costs" of violent behavior), such as arrest, prosecution, and imprisonment, and compassionate measures, such as shelters, education, counseling, and support groups, have been shown to be successful to varying degrees under varying conditions. Used together, these interventions may be quite effective. Mederer and Gelles (1989) suggest that controlling measures may be used to "motivate violent offenders to participate in treatment programs."

Battered Women and the Law

Family violence studies and feminist pressure have spurred a movement toward the implementation of stricter policies for dealing with domestic offenders. Long ignored, domestic violence has only recently become a top concern for legislators and law enforcement agencies throughout the country (Wilson, 1997). California has introduced measures to crack down on spousal abuse and increase funding for shelters and other related agencies. In 1995 alone, nearly twenty-two laws against domestic violence were passed. Other states have quickly followed. Obviously too late for many like Nicole Brown Simpson, our collective conscience is finally drawn to such issues as policy misconduct, racism, child custody, and spousal abuse.

Still there is resistance by some law enforcement and judicial branches to listen to the victims of abuse. Prevention and law enforcement are necessary measures this society must take in order to reduce the incidence of domestic violence. Today, at least a third of the largest U.S. police forces are instructed to arrest the assailants, although the jury is still out as to whether mandatory arrest is actually effective. (See Berk, 1993, and Buzawa and Buzawa, 1993, for conflicting views on the efficacy of arrest as a deterrent to domestic violence.)

Battered Women's Shelters

At the point where a woman finds that she can leave an abusive relationship, even temporarily, she may have any number of serious needs. If she is fleeing an attack, she may need immediate medical attention and physical protection. She will need accommodation for herself and possibly her children. She will certainly need access to support, counseling, and various types of assistance—money, food stamps or other basic survival items for herself and her children. She will need to deal with informed, compassionate professionals such as police officers, doctors, and social workers.

Battered women's shelters provide safe havens for women in abusive relationships. Shelters provide counseling and emotional support as well as temporary lodging, meals, and other necessities for women and their children.

In the late 1970s, the shelter movement developed to meet the needs of many battered women. The shelter movement has grown slowly, hampered by lack of funding and mixed reaction from the public. There are an estimated one thousand shelters throughout the United States, a vast improvement over the estimated five or six shelters in existence in 1976 (Gelles and Cornell, 1990). Besides offering immediate safe shelter (the locations of safe houses are usually known only to the residents and shelter workers), these refuges let battered women realize that they are not alone in their misery and help them form supportive networks with one another. The shelters also provide many other services for battered women who call, such as information, advice, or referrals.

Women who seem to benefit most from shelter stays are those who have decided to "take charge of their lives," according to one study (Berk, Newton, and Berk, 1986). Hampton, Gelles, and Harrop (1989) suggest that the dramatic drop in wife abuse among African Americans may be the result of African-American women's "increased status" coupled with their apparent willingness to make use of shelters and other programs.

Abuser Programs

"A comprehensive solution to violence against women in intimate relationships demands that perpetrators of abuse be held accountable for their behavior and that direct efforts be made with batterers to change their behavior", says Richard M. Tolman, Associate Professor the University of Michigan (1995). Treatment services for men who batter provide one important component of a coordinated response to domestic violence (see Gondolf, 1993, for program and treatment issues). Psychotherapy, group discussion, stress management, or communication skills classes may be available through mental health agencies, women's crisis programs, or various self-help groups.

The extent to which attending batterers' groups actually changes abusing men's violent behavior is difficult to measure (Gelles and Conte, 1991). What has become apparent is the effectiveness of the "one size fits all" approach and the need to adopt a more sophisticated understanding of individual's violent behaviors (Tolman, 1995). Edward Gondolf's studies of men who have completed voluntary programs (1987, 1988) showed that two-thirds to three-quarters of these men were subsequently nonviolent. Gondolf also confirmed the conclusion of a number of studies of women who are successful in stopping abuse: Battered women play a crucial role in stopping the violence against them. Women's insistence on their partners' getting help (the "woman factor") apparently can influence men to "learn more or try harder" to change.

A coordinated community response which includes proactive police and criminal justice strategies, advocacy and services for battered women and their children, and responses by other community institutions that promote safety for battered women and sanctions for men who batter are necessary interventions (Tolman, 1995).

America has sometimes been described as child-centered; however, any unbiased observer of child life in this nation will find that many millions of children are living and growing up under circumstances of severe social and economic deprivation. . . . Many of these children lack adequate nutrition, medical and dental care, and educational and vocational opportunities. . . . However high the prevalence of physical abuse of individual children within their families and homes may be, the abuse inflicted upon children collectively by society as a whole is far larger in scope and far more serious in its consequences.

David Gil, *Violence against Children*

Did you know?

More than 1 million children are confirmed to be abused and neglected each year in the United States. More realistic perhaps is the estimated number of children who are abused and neglected: nearly 3 million ("Child Abuse and Neglect," 1997).

Child Abuse and Neglect

The history of children is not a particularly happy one. At various times and places, children have been abandoned to die of exposure in deserts, in forests, and on mountainsides or have simply been murdered at birth if they were deemed too sickly, too ugly, of the wrong sex, or just impractical. Male children have been subjected to castration to make them fit for guarding harems or singing soprano in church choirs. Millions of female children in the Middle East and Africa undergo devastating sexual mutilation. These practices (and many others) have all been socially condoned in their time and place. In the societies in which they have existed, they have not been (or are not) recognized as abusive.

Of course, we feel we are more "civilized" today. In our society, some degree of physical force against children, such as spanking, is generally accepted as normal. Most child-rearing experts, however, currently suggest that parents use alternative disciplinary measures.

Child abuse was not recognized as a serious problem in the United States until the 1960s. At that time, C. H. Kempe and his colleagues coined the medical term *battered-baby syndrome* to describe the patterns of injuries commonly observed in physically abused children. The Children's Defense Fund (1996) reports:

In the United States we spend more money on shelters for dogs and cats than for human beings. If we have effective animal rescue shelters for abused dogs, cats, and bunny rabbits we should be able to spare something for people as well.

Murray Straus et al.

Children are the least protected members of our society. Much physical abuse is camouflaged as discipline or as the parent "losing" his or her temper.

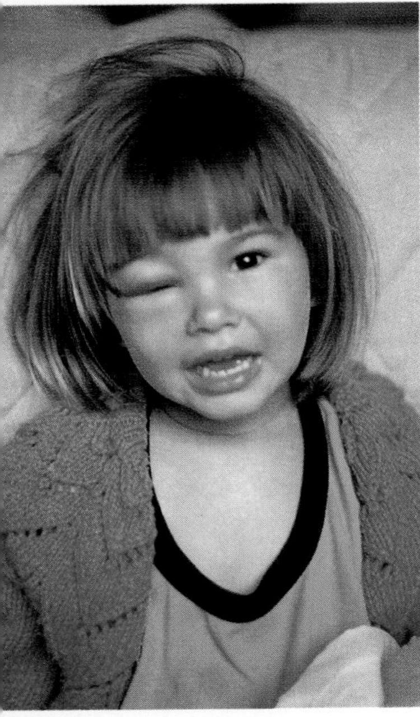

- Every ten seconds, a child is reported abused or neglected.
- Every fourteen seconds, a child is arrested.
- Every two hours a child is killed by firearms.
- Every four hours a child commits suicide.
- Every five hours a child dies from abuse or neglect.

When we look at violence among children from a global perspective, we see a even larger shadow cast over our nation. A recently released study by the Centers for Disease Control and Prevention found that nearly three out of four child slayings in the industrialized world occur in the United States ("Violence Kills," 1997). The statistics show that the epidemic of violence in recent years that has hit younger and younger children is confined almost exclusively to the United States. The suicide rate alone for children fourteen and under is double that of the rest of the industrialized world. No explanation for the huge gap between the rates of violent death for American children and those of other countries were given, though some experts speculate it is due to a growing faction of children who are unsupervised or otherwise at risk. The low level of funding for social programs, sexism, racism, and epidemic rates of poverty among our young are other factors that continue to embarrass our nation. Parental violence is among the five leading causes of death for children between the ages of one and eighteen. About 1,300 children are killed by their parents or other close relatives each year (McCormick, 1994).

Families at Risk

Research suggests that three sets of factors put families at risk for child abuse and neglect: parental characteristics, child characteristics, and the family ecosystem—that is, the family system's interaction with the larger environment (Burgess and Youngblood, 1987; Vasta, 1982). The characteristics described in the next sections are likely to be present in abusive families (Straus, Gelles, and Steinmetz, 1980; Turner and Avison, 1985).

Did you know?

Approximately 80 percent of the perpetrators of child abuse and neglect were parents and other relatives ("Child Abuse and Neglect," 1997).

Parental Characteristics. Some or all of the following characteristics are likely to be present in parents who abuse their children:

- The abusing father was physically punished by his parents, and his father physically abused his mother.
- The parents believe in corporal discipline of children and wives.
- The marital relationship itself may not be valued by the parents. There may be interspousal violence.
- The parents believe that the father should be the dominant authority figure.
- The parents have low self-esteem.
- The parents have unrealistic expectations for the child.
- There is persistent role reversal in which the parents use the child to gratify their own needs, rather than vice versa.
- The parents appear unconcerned about the seriousness of a child's injury, responding, "Oh well, accidents happen."

Do not withhold correction from the child: For if you beat him with the rod, he will not die. Beat him with the rod and deliver him from hell.

PROVERBS 23

Child Characteristics. Who are the battered children? Are they any different from other children? Surprisingly, the answer is often yes; they are different in some way or at least are perceived to be so by their parents. Brandt Steele (1980) notes that children who are abused are often labeled by their parents as "unsatisfactory," a term that may describe any of the following:

- A "normal" child who is the product of a difficult or unplanned pregnancy, is of the "wrong" sex, or is born outside of marriage.
- An "abnormal" child—one who was premature or of low birth weight, possibly with congenital defects or illness.
- A "difficult" child—one who shows such traits as fussiness or hyperactivity.

Steele also notes that all too often, a child's perceived difficulties are a result (rather than a cause) of abuse and neglect.

Family Ecosystem. As discussed earlier in this chapter, the community and the family's relation to it may be relevant to the existence of domestic violence. The following characteristics may be found in families that experience child abuse:

- The family experiences unemployment.
- The family is socially isolated, with few or no close contacts with relatives, friends, or groups.
- The family has a low level of income, which creates economic stress.
- The family lives in an unsafe neighborhood, which is characterized by higher-than-average levels of violence.
- The home is crowded, hazardous, dirty, or unhealthy.
- The family is a single-parent family in which the parent works and is consequently overstressed and overburdened.
- One or more family members have health problems.

There was an old woman who lived in a shoe,
She had so many children she didn't know what to do.
She gave them some broth without any bread,
And whipped them all soundly and sent them to bed.
MOTHER GOOSE RHYME

The likelihood of child abuse increases with family size. Parents of two children have a 50 percent higher abuse rate than do parents of a single child. The rate of abuse peaks at five children and declines thereafter. The overall child abuse rate by mothers has been found to be 75 percent higher than that by fathers (Straus, Gelles, and Steinmetz, 1980). The responsibilities and tensions of mothering and the enforced closeness of mother and child may lead to situations in which women are likely to abuse their children. But, as David Finkelhor (1983) and others have pointed out, if we "calculate [child] vulnerability to abuse as a function of the amount of time spent in contact with a potential abuser, . . . we . . . see that men and fathers are more likely to abuse."

Single parents—both mothers and fathers—are at especially high risk of abusing their children (Gelles, 1989). According to Gelles, "the high rate of abusive violence among single mothers appears to be a function of the poverty that characterizes mother-only families." He states that programs must be developed that are "aimed at ameliorating the devastating consequences of poverty among single parents." Single fathers, who show a higher abuse rate than single mothers, "need more than economic support to avoid using abusive violence toward their children."

Did you know?

American children are twelve times more likely to die by gunfire than their counterparts in the rest of the industrialized world (Meyer, 1997).

Intervention

The goals of intervention in domestic violence are principally to protect the victims and assist and strengthen their families. In dealing with child abuse, professionals and government agencies may be called on to provide medical care, counseling, and services such as day care, child-care education, telephone crisis lines, and temporary foster care. Many of these services are costly, and many of those who require them cannot afford to pay. Our system does not currently

PERSPECTIVE

The Epidemic of Missing Children: Myth or Reality?

In the morning, as they eat their cereal, children in homes throughout America look at milk cartons with photographs of "missing children." The faces stare from the cartons as if warning the children that they, too, may become missing if they are not careful. The photographs on milk cartons, grocery bags, billboards, and television give us the impression, observes pediatrician Benjamin Spock, that "children are being abducted all the time and that this child might be next" (Kilzer, 1985).

To protect their children, parents fingerprint them in massive identification programs. They buy books warning against strangers and have dentists implant computer identification chips in their children's teeth. Schools also often stress the threat posed by strangers.

A 1990 study commissioned by the Justice Department (Finkelhor, Hotaling, and Sedlak, 1990) has helped to put the missing children "problem" into perspective. Based on interviews conducted from July 1988 to January 1989, the researchers estimated that around 1,369,000 children are missing at some time during a given year in the United States.

Of this number, about 350,000 are abducted by a family member as a result of a custody dispute between parents. The researchers found that most of these children were back home within two days. All but 10 percent were home within a month. In 83 percent of the cases, the parent from whom the child was taken knew where the child was. Other major categories of missing children (in terms of numbers of children affected) include runaways (about 450,700 annually) and "thrownaways"—children who are forced by their parents to leave home (about 127,000). Another large category includes children who are lost, injured, or unable to return home for some other reason. This category, consisting of about 438,000 children annually, also includes a large number of "misplaced" children—those who forget or misunderstand where they are supposed to be. Most of these children return home within a day, but the fact that they are included in many estimates of missing children obscures the number that are actually missing.

A very small percentage of missing children are abducted by strangers. The researchers estimated the number to be between 200 and 300 annually. An additional 3,200 to 4,400 are "lured away," often in conjunction with a sexual assault; they are not kidnapped, however. In another study, based on data collected from six different sources, including the National Center for Missing and Exploited Children and various police records, Gerald Hotaling and David Finkelhor (1990) estimated that between 52 and 158 children are murdered yearly in stranger abductions. In contrast, close to 1,300 children die yearly from parental abuse or neglect. (The United States leads the world in homicides for one- to four-year-olds, most of whom are killed by a family member.) About 2,000 children are victims of suicide every year and about 7,000 die in auto accidents.

Contrary to media claims, there is no upsurge in stranger abductions or in murders related to these abductions. Furthermore, there is far greater risk to teenagers than to young children. Hotaling and

provide the human and financial resources necessary to deal with these socially destructive problems.

The first step in treating child abuse is locating the children who are threatened. With heightened public awareness in recent years and mandatory reporting of suspected child abuse required of certain professionals (such as teachers, doctors, and counselors) in all fifty states, identifying these children is much easier now than it was two decades ago. Reported incidents of child abuse have increased greatly during this time, but the actual number of incidents appears to have decreased. This is good news as far as it goes. Still, levels of violence against children are unacceptably high, and not nearly enough resources are available to assist children. Child welfare workers are notoriously overburdened with cases, and adequate foster placement is often difficult to find, (Gelles and Cornell, 1990).

Much of the interventions in child abuse appears to be the equivalent of putting a Band Aid on a huge malignant tumor. We must address this societal cancer from a variety of levels:

A society which is mobilized to keep child molesters, kidnappers, and Satanists away from innocent children is not necessarily prepared to protect children from ignorance, poverty, and ill health.

JOEL BEST

Finkelhor (1990) write, "Runaways or children who are considered to be possible runaways are not regarded with the same solicitude as kidnap victims. Still, the data suggest that if the public is concerned about abducted and murdered children, it will have to broaden its concern to include the adolescent, who appears to be the child at highest risk."

If the epidemic of missing children is a myth, the public reaction to it is not. The myth has touched a very profound fear. In part, this fear may reflect anxieties about changing patterns of parenting and child rearing. With increasing numbers of mothers working, more children are being placed in child care than ever before. One psychologist suggests that "not being home may be tapping into hidden guilt" (Kilzer and Griego, 1985). For some, this fear may reflect anxieties concerning minorities and may touch on xenophobia, as many stories involve specific ethnic groups or foreigners. One man recalled hearing of Mexicans abducting blonde, blue-eyed children to sell to childless couples. "They were also abducted and shipped to Africa or to Arabian harems; there were those who were forced into prostitution by Mafia pimps" (Schneider, 1987).

Whatever the source, the myth of missing children has severe consequences. First of all, it instills fear in children. Psychologist Lee Salk warns, "We are terrifying our children to the point where they are going to be afraid to talk to strangers" (Kantrowitz, 1986). Such fear makes it difficult to instill a basic sense of trust; it makes children anxious toward people they don't know and fearful of new situations. It makes them suspicious of foreigners and people who are different from themselves. Second, it affects our parenting style. It makes us afraid to let our children out of our sight. It creates anxiety in us when we are not at home or they are gone. Finally, it allows us to continue to believe that all families are happy and nurturing. It makes it easier to deny that families can be abusive and violent. (National attention was focused on the issue of parents who kill their children in 1994 when Susan Smith, a mother of two young boys, rolled her car—with the children strapped inside—into a South Carolina lake. And in 1996, the murder of child model JonBenet Ramsey—unsolved at this book's publication—brought speculation of family involvement in the crime.)

Sexual abuse, a far more common problem than kidnapping, is much less often discussed by parents with their children. In one study, Finkelhor (1984) found that 84 percent of parents had discussed kidnapping, but only 29 percent had talked about sexual abuse with their children.

Children need to be taught to keep themselves safe at home and at the homes of friends, neighbors, and relatives as well as on the street. But it is not necessary to terrify them in the process. If they are allowed to develop their own good judgment skills and have high self-esteem, they will feel secure, competent, and in control of their own bodies and have the tools to help them stay out of risky situations. Parents need to look at where the most plausible dangers lie. They need to acknowledge the violence and sexual abuse within families that lead children to become runaways and suicides, or they might become involved in promoting motor-vehicle safety (see the Resource Center for information on Mothers Against Drunk Driving). As Hotaling and Finkelhor (1990) write, "Children being killed by strangers who abduct them are terrible and properly feared tragedies, but they are a small portion of children who die in tragic and preventable circumstances."

- Parents must learn how to deal more positively and effectively with their children.
- Children need to be infused with self-esteem and taught skills in order to recognize and report abuse as soon as it occurs.
- Professionals working with children and families should be required to receive adequate training in child abuse and neglect and to be sensitive to cultural norms.
- Agencies should coordinate their efforts for preventing and investigating child abuse.
- Public awareness of child abuse needs to be created by methods such as posters and public service announcements.
- The workplace should promote educational programs to eliminate sexism, provide adequate child care, and help reduce stress among its workforce.
- Government should support sex education and family life programs in order to help reduce the number of unwanted pregnancies.

Reflections

If you were (or are) a parent, would you consider it violent to spank your child with an open hand on the buttocks if the child was disobedient? To slap your child across the face? Is it acceptable to spank your small child to teach him or her not to run into a busy street? To spank because you are angry?

- Criminal statutes should be developed and enforced to impose felony sentences on those who perpetuate child maltreatment.
- Research efforts concerning family violence and child maltreatment should be supported.

(See the Resource Center for information on Parents Anonymous.)

The Hidden Victims of Family Violence: Siblings, Parents, and the Elderly

Most studies of family violence have focused on violence between spouses and on parental violence toward children. There is, however, considerable violence between siblings, between teenage children and their parents, and between adult children and their aging parents. These are the "hidden victims" of family violence (Gelles and Cornell, 1990).

Sibling Violence

#27 Violence between siblings is by far the most common form of family violence (Straus, Gelles, and Steinmetz, 1980). Most of this type of sibling interaction is simply taken for granted by our culture—"You know how kids are!" Seventy-five percent of siblings experience at least one act of violence per year (Gelles and Cornell, 1985). The National Association of Child Abuse estimated that 29 million siblings physically harm each other annually (Tiede, 1983). Straus and his colleagues report these additional findings:

- The rate of sibling violence goes down with the increasing age of the child.
- Boys of all ages are more violent than girls. The highest rates of sibling violence occur in families with only male children.
- Violence between children often reflects what they see their parents doing to each other and to the children themselves.

The full scope and implications of sibling violence have not been rigorously explored. However, Straus, Gelles, and Steinmetz (1980) conclude:

> Conflicts and disputes between children in a family are an inevitable part of life. . . . But the use of physical force as a tactic for resolving their conflicts is by no means inevitable. . . . Human beings learn to be violent. It is possible to provide children with an environment in which nonviolent methods of solving conflicts can be learned. . . . If violence, like charity, begins at home, so does nonviolence.

Teenage Violence toward Parents

Most of us find it difficult to imagine children attacking their parents because it so profoundly violates our image of parent-child relations. Parents possess the authority and power in the family hierarchy. Furthermore, there is greater social disapproval of a child striking a parent than of a parent striking a child; it is the parent who has the "right" to hit. Finally, parents rarely discuss such incidents because they are ashamed of their own victimization; they fear that others will blame them for the children's violent behavior (Gelles and Cornell, 1985).

Although we know fairly little about adolescent violence against parents, scattered studies indicate that it is almost as prevalent as spousal violence (Gelles and Cornell, 1985; Straus, 1980; Straus, Gelles, and Steinmetz, 1980).

The majority of youthful children who attack parents are between the ages of thirteen and twenty-four. Sons are slightly more likely to be abusive than daughters; the rate of severe male violence tends to increase with age, whereas that of females decreases. Boys apparently take advantage of their increasing size and the cultural expectation of male aggression. Girls, in contrast, may become less violent because society views female aggression more negatively. Most researchers believe that mothers are the primary targets of violence and abuse because they may lack physical strength or social resources and because women are "acceptable" targets for abuse (Gelles and Cornell, 1985).

Abuse of the Elderly

Of all the forms of hidden family violence, only the abuse of elderly parents by their grown children (or, in some cases, by their grandchildren) has received considerable public attention. Elder mistreatment may be an act of commission (abuse) or omission (neglect) (Wolf, 1995). It is estimated that approximately 500,000 elderly people are physically abused annually. An additional 2 million are thought to be emotionally abused or neglected. Though mandatory reporting of suspected cases of elder abuse is the law in forty-two states and the District of Columbia, much abuse of the elderly goes unnoticed, unrecognized, and unreported (Wolf, 1995). Older people generally don't get out much and are often confined to bed or a wheelchair. Many do not report their mistreatment out of fear of institutionalization or other reprisal. Although some research indicates that the abused in many cases were, in fact, abusing parents, more knowledge must be gained before we can draw firm conclusions about the causes of elder abuse (Egeland, 1993; Kaufman and Zigler, 1993; Ney, 1992).

The most likely victims of elder abuse are the very elderly—in the majority of cases, women—who are suffering from physical or mental impairments, especially those with Alzheimer's disease. Their advanced age renders them dependent on their caregivers for many, if not all, of their daily needs. It may be their dependency that increases their likelihood of being abused. Other research indicates that many abusers are financially dependent on their elderly parents; they may resort to violence out of feelings of powerlessness.

While researchers are sorting out the whys and wherefores of elder abuse, battered older people have a number of pressing needs. Karl Pillemer and Jill Suitor (1988) recommend the following services for elders and their caregiving families:

- Housing services, including temporary respite care to give caregivers a break and permanent housing (such as rest homes, group housing, and nursing homes).
- Health services, including home health care; adult day-care centers; and occupational, physical, and speech therapy.
- Housekeeping services, including shopping and meal preparation.
- Support services, such as visitor programs and recreation.
- Guardianship and financial management.

Reducing Family Violence

Based on the foregoing evidence, you may by now have concluded that the American family is well on its way to extinction as we bash, thrash, cut, shoot, and otherwise wipe ourselves out of existence. Statistically, the safest family homes are those with one or no children in which the husband and wife experience little life stress and in which decisions are made democratically (Straus, Gelles, and Steinmetz, 1980). By this definition, most of us probably do not live in homes that are particularly safe. What can we do to protect ourselves (and our posterity) from ourselves?

Prevention strategies usually take one of two paths: eliminating social stress or strengthening families (Swift, 1986). Family violence experts make the following general recommendations (Straus, Gelles, and Steinmetz, 1980):

- Reduce societal sources of stress, such as poverty, racism and inequality, unemployment, and inadequate health care.
- Eliminate sexism. Furnish adequate day care. Promote educational and employment opportunities equally for men and women. Promote sex education and family planning to prevent unplanned and unwanted pregnancies.
- Initiate prevention and early intervention efforts for young males before they become adult batterers.
- End social isolation. Explore means of establishing supportive networks that include relatives, friends, and community.
- Break the family cycle of violence. Eliminate corporal punishment and promote education about disciplinary alternatives. Support parent education classes to deal with inevitable parent-child conflict.
- Eliminate cultural norms that legitimize and glorify violence. Legislate gun control, eliminate capital punishment, and reduce media violence.

(For specific prevention and treatment strategies, see Hampton, Gullota, Adams, Potter, and Weissberg, 1993.)

A . . . society which promotes the ownership of firearms, women and children; which makes homes men's castles; and which sanctions societal and interpersonal violence in the forms of wars, athletic contests, and mass media fiction (and news) should not be surprised to find violence in its homes.

NORMAN DENZIN

Child Sexual Abuse

Child sexual abuse, whether it is committed by relatives or nonrelatives, is widespread. **Child sexual abuse** is any sexual interaction (including fondling, erotic kissing, or oral sex, as well as genital penetration) between an adult or older adolescent and a prepubertal child. It does not matter whether the child is perceived by the adult as freely engaging in the sexual activity. Because of the child's age, he or she cannot legally give consent; the activity can only be considered as self-serving to the adult.

Estimates of the incidence of child sexual abuse vary considerably. A review of small-scale studies found estimates ranging from 6 to 62 percent for females and from 3 to 31 percent for males (Peters et al., 1986). The first national survey found that 27 percent of the women and 16 percent of the men surveyed had experienced sexual abuse as children (Finkelhor, Hotaling, Lewis, and Smith, 1990). Different definitions of abuse, methodologies, samples, and interviewing techniques account for the varied estimates (Gelles and Conte, 1991). Fabricated reports of sexual abuse do occur, but deliberate fabrications constitute only 4–8 percent of all reports (Finkelhor, 1995).

Child sexual abuse is generally categorized in terms of kin relationship. **Extrafamilial abuse** is sexual abuse by nonrelated individuals. **Intrafamilial abuse** is abuse by related individuals, including steprelatives. The abuse may be pedophilic or nonpedophilic. **Pedophilia** is an intense, recurring sexual attraction to prepubescent children. Nonpedophilic sexual interactions with children are not motivated as much by sexual desire as by nonsexual motives, such as power or affection (Groth, 1980). (For sexual abuse from an anthropological perspective, see Konker, 1992.)

The child's victimization may include force or the threat of force, pressure, or the taking advantage of trust or innocence. The most serious forms of sexual abuse include actual or attempted penile-vaginal penetration, fellatio, cunnilingus, and anilingus, with or without the use of force. Other serious forms range from forced digital penetration of the vagina to fondling of the breasts (unclothed) or simulated intercourse without force. The least traumatic sexual abuse ranges from kissing to intentional sexual touching of the clothed genitals, breasts, or other body parts with or without the use of force (Russell, 1984).

General Preconditions for Sexual Abuse

Researchers have found that intrafamilial and extrafamilial sexual abuse share many common elements (Finkelhor, 1984). Because there are so many variables—such as the age and sex of the victims and perpetrators, their relationships, the type of acts involved, and whether there was force—we cannot automatically say that abuse within the family is more harmful than extrafamilial abuse, as we might assume.

David Finkelhor (1984) believes that four preconditions need to be met by the offender for sexual abuse to occur. These preconditions apply to pedophilic, nonpedophilic, intrafamilial, and extrafamilial abuse. According to Finkelhor, all four of these factors must come into play for sexual abuse to occur:

1. *Motivation to sexually abuse a child.* This motivation consists of three components: (a) emotional congruence, in which relating sexually to a child fulfills some important emotional need; (b) sexual arousal toward the child; and (c) blockage, in which alternative sources of sexual gratification are not available or are less satisfying.
2. *Overcoming internal inhibitions against acting on the motivation.* Internal inhibitions may be overcome through the use of alcohol, lack of impulse control, senility, social acceptance of sexual interest in children, and so on.
3. *Overcoming external obstacles to committing sexual abuse.* The most important obstacle appears to be the supervision and protection a child receives from others, such as family members, neighbors, and the child's peers. The mother is especially significant in protecting children. Growing evidence suggests that children are more vulnerable to abuse when the mother is absent, neglectful, or incapacitated in some way through illness, marital abuse, or emotional problems.
4. *Undermining or overcoming the child's potential resistance to the abuse.* The abuser may use outright force or select psychologically vulnerable targets. Certain children may be more vulnerable because they feel insecure, needy, or unsupported and will respond to the abuser's offers of attention, affection, or bribes. Children's ability to resist may be undercut because they are young, naïve, or have a special relationship with the abuser as friend, neighbor, or family member.

Forms of Intrafamilial Child Sexual Abuse

The incest taboo, which is nearly universal in human societies, prohibits sexual activities between closely related individuals. There are only a few exceptions, and these concern brother-sister marriages in the royal families of ancient Egypt, Peru, and Hawaii. **Incest** is generally defined as sexual intercourse between persons too closely related to marry legally (it is usually interpreted to mean father-daughter, mother-son, or brother-sister intercourse). Sexual abuse in families can involve blood relatives (most commonly uncles and grandfathers) and steprelatives (most often stepfathers and stepbrothers). Grandfathers who abuse their granddaughters frequently sexually abused their children as well. Stepgranddaughters

Lot went out of Zo'ar, and dwelled in the hills with his two daughters, for he was afraid to dwell in Zo'ar. So he dwelled in a cave with his two daughters. Then the firstborn said to the other, "Our father is old, and there is not a man on earth to come unto us in the manner of men. Let us make our father drink wine and we will lie with him that we may have offspring through our father." So they made their father drink wine that night and his older daughter went in and lay with her father. He did not know when she lay down or when she arose. And then on the next day, the older daughter said to the younger one, "Behold, I lay last night with our father. Let us make him drink wine again tonight, then you go in and lie with him that we may have children by our father." So they gave their father wine that night, and the younger daughter slept with him. He did not know when she lay down or when she arose. Thus both daughters became great with child by their father.

GENESIS 19:30–36

are at greater risk than are granddaughters (Margolin, 1992). (For a review of assessment and treatment of incest perpetrators, see Cole, 1992.)

It is not clear what type of familial sexual abuse is the most frequent (Peters et al., 1986; Russell, 1986). Some researchers believe that father-daughter (including stepfather-stepdaughter) abuse is the most common; others think that brother-sister abuse is the most common. Still other researchers believe that abuse committed by uncles is the most common (Russell, 1986). Mother-son sexual relations are considered to be rare (or they are underreported).

Father-Daughter Sexual Abuse

There is general agreement that the most traumatic form of sexual victimization is father-daughter abuse, including that committed by stepfathers. Over twice as many daughters abused by fathers reported serious long-term consequences compared to children who were victimized by other family members. Some factors contributing to the severity of reactions to father-daughter sexual relations include the following:

- Fathers were more likely to have engaged in penile-vaginal penetration than other relatives (18 percent versus 6 percent).
- Fathers sexually abused their daughters more frequently than other perpetrators abused their victims (38 percent of the fathers sexually abused their daughters eleven or more times, compared with a 12 percent abuse rate for other abusing relatives).
- Fathers were more likely to use force or violence than others (although the numbers for both fathers and others were extremely low).

In the past, many have discounted the seriousness of sexual abuse by a stepfather because incest is generally defined legally as sexual activity between two biologically related persons. The emotional consequences are just as serious, however. Sexual abuse by a stepfather still represents a violation of the basic parent-child relationship.

Brother-Sister Sexual Abuse

There are contrasting views concerning the consequences of brother-sister incest. Researchers generally have expressed little interest in it. Most have tended to view it as harmless sex play or sexual exploration between mutually involved siblings. The research, however, has generally failed to distinguish between exploitative and nonexploitative brother-sister sexual activity. Sibling incest needs to be taken seriously (Adler and Schultz, 1995). Diana Russell suggests that the idea that brother-sister incest is usually harmless and mutual may be a myth. In her study, the average age difference between the brother (age 17.9 years) and the sister (age 10.7 years) was so great that the siblings could hardly be considered peers (Russell, 1986). The age difference represents a significant power difference. Furthermore, not all brother-sister sexual activity is "consenting;" considerable physical force may be involved. Russell writes:

> So strong is the myth of mutuality that many victims themselves internalize the discounting of their experiences, particularly if their brothers did not use force, if they themselves did not forcefully resist the abuse at the time, if they still continued to care about their brothers, or if they did not consider it abuse when it occurred. And sisters are even more likely than daughters to be seen as responsible for their own abuse.

Two percent of the women in Russell's random sample had at least one sexually abusive experience with a brother.

Uncle-Niece Sexual Abuse

Both Alfred Kinsey (1953) and Diana Russell (1986) found the most common form of intrafamilial sexual abuse to involve uncles and nieces. Russell reported that almost 5 percent of the women in her study had been molested by their uncles, slightly more than the percentage abused by their fathers. The level of severity of the abuse was generally less in terms of the type of sexual acts and the use of force. Although such abuse does not take place within the nuclear family, many victims found it quite distressing. A quarter of the respondents indicated long-term emotional effects (Russell, 1986).

Children at Risk

Not all children are equally at risk for sexual abuse. Although any child can be sexually abused, some groups of children are more likely to be victimized than others. A review (Finkelhor and Baron, 1986) of the literature indicates that children at higher risk for sexual abuse are the following: female children, preadolescent children, children with absent or unavailable parents, children whose relationships with parents are poor, children whose parents are in conflict, and children who live with a stepfather. A variety of studies have found little or no association between sexual abuse and race and socioeconomic status (Finkelhor, 1995).

The majority of sexually abused children are girls, but boys are also victims (Watkins and Bentovim, 1992). The ratio of girls to boys appears to be between 2.5 to 1 and 4 to 1 (Finkelhor and Baron, 1986). We have only recently recognized the sexual abuse of boys (Bera et al., 1991). Finkelhor (1979) speculates that men tend to underreport sexual abuse because they experience greater shame; they feel that their masculinity has been undermined. Boys tend to be blamed more than girls for their victimization, especially if they did not forcibly resist: "A real boy would never let someone do that without fighting back" (Rogers and Terry, 1984).

Most sexually abused children are between eight and twelve years of age when the abuse first takes place. At higher risk appear to be children who have poor relationships with their parents (especially mothers) or whose parents are absent or unavailable and have high levels of marital conflict. A child in such a family may be less well supervised and, as a result, more vulnerable to manipulation and exploitation by an adult. In this type of family, the child may be unhappy, deprived, or emotionally needy; the child may be more responsive to the offers of friendship, time, and material rewards promised by the abuser.

Finally, children with stepfathers are at greater risk for sexual abuse. Russell (1986) found that only 2.3 percent of the daughters studied were sexually abused by their biological fathers. In contrast, 17 percent were abused by their stepfathers. The higher risk may result from the incest taboo's not being as strong in stepfamily relationships and because stepfathers have not built up inhibitions resulting from parent-child bonding beginning from infancy. As a result, stepfathers may be more likely to view the stepdaughter sexually. In addition, stepparents may also bring into the family steprelatives—their own parents, siblings, or children—who may feel no incest-related prohibition about becoming sexually involved with stepchildren.

Effects of Child Sexual Abuse

No one ever keeps a secret so well as a child.

Victor Hugo (1802–1885)

Until recently, much of the literature on child sexual abuse has been anecdotal or based on case studies or small-scale surveys of nonrepresentative groups. Numerous well-documented consequences of child sexual abuse exist for both

This drawing was made by an adolescent who was impregnated by her father. According to psychologists, it expresses her inability to deal with body images, especially genitalia, and her rejection of her body's violation.

intrafamilial and extrafamilial abuse (see Kendall-Tackett et al., 1993, for a review of the literature). These include both initial and long-term consequences. Many abused children experience symptoms of post-traumatic stress disorder (PTSD) (McLeer, Deblinger, Henry, and Ovraschel, 1992).

Initial Effects of Sexual Abuse

The initial consequences of sexual abuse—those occurring within the first two years—include these effects:

- *Emotional disturbances*, including fear, anger, hostility, guilt, and shame.
- *Physical consequences*, including difficulty in sleeping, changes in eating patterns, and pregnancy.
- *Sexual disturbances*, including significantly higher rates of open masturbation, sexual preoccupation, and exposure of the genitals (Hibbard and Hartman, 1992).
- *Social disturbances*, including difficulties at school, truancy, running away from home, and early marriages among abused adolescents.

Ethnicity appears to influence how a child responds to sexual abuse. For example, a study (Rao, Diclemente, and Pouton, 1992) recently compared sexually abused Asian-American children with a random sample of abused white, African-American, and Latino children. The researchers found that Asian-American children suffered less sexually invasive forms of abuse. They tended to be more suicidal and to receive less support from their parents than did non-Asians. They were also less likely to express anger or to act out sexually. These different responses point to the importance of understanding the cultural context when treating ethnic victims of sexual abuse. (For a discussion of child sexual abuse histories among African-American college students, see Priest, 1992.)

Long-Term Effects of Sexual Abuse

Although the initial effects of child sexual abuse can subside to some extent, the abuse may leave lasting scars on the adult survivor (Beitchman et al., 1992). These adults often have significantly higher incidences of psychological, physical, and sexual problems than the general population. Abuse as a child may predispose some women to sexually abusive dating relationships (Cate and Lloyd, 1992).

Long-term problems include the following (Beitchman et al., 1992; Browne and Finkelhor, 1986; Elliott and Briere, 1992; Wyatt, Gutherie, and Notgrass, 1992):

- Depression, the most frequently reported symptom of adults sexually abused as children.
- Self-destructive tendencies, including suicide attempts and thoughts of suicide (Jeffrey and Jeffrey, 1991).
- Somatic disturbances and dissociation, including anxiety and nervousness, eating disorders (anorexia and bulimia), feelings of "spaciness," out-of-body experiences, and feelings that things are "unreal" (DeGroot, Kennedy, Rodin, and McVey, 1992; Walker et al., 1992; Young, 1992).
- Negative self-concept, including feelings of low-self-esteem, isolation, and alienation.
- Interpersonal relationship difficulties, including difficulties in relating to both sexes, parental conflict, problems in responding to their own children, and difficulty in trusting others.
- Revictimization, in which women abused as children are more vulnerable to rape and marital violence (Wyatt, Gutherie, and Notgrass, 1992).

- Sexual problems, in which survivors find it difficult to relax and enjoy sexual activities, or they avoid sexual relations and experience hypoactive (inhibited) sexual desire and lack of orgasm.

In recent years, some adults have been accusing family members or others of abusing them as children. They say that they repressed their childhood memories of abuse and only later, as adults, recalled them. These accusations have given rise to a fierce controversy about the nature of memories of abuse. A review of the research related to this topic was done by the American Psychological Association (1994) and the following conclusions were made:

- Most people who were sexually abused as children at least partially remember the abuse.
- Memories of sexual abuse that have been forgotten may later be remembered.
- False memories of events that never happened may occur.
- The process by which accurate or inaccurate recollections of childhood abuse are made is not well understood.

Because firm scientific conclusions cannot be made at this time, the debate is likely to continue.

Sexual Abuse Trauma

As we have seen, childhood sexual abuse has numerous initial and long-term consequences. Together, these consequences create a traumatic dynamic that affects the child's ability to deal with the world. David Finkelhor and Angela Browne (1986) suggest a model of sexual abuse that contains four components: traumatic sexualization, betrayal, powerlessness, and stigmatization. When these factors converge as a result of sexual abuse, they affect the child's cognitive and emotional orientation to the world. They create trauma by distorting a child's self-concept, world view, and affective abilities. These consequences affect abuse survivors not only as children but also as adults.

In every child who is born, under no matter what circumstances, and of no matter what parents, the potentiality of the human race is born again.

JAMES AGEE (1909–1955)

Traumatic Sexualization. The process in which a sexually abused child's sexuality develops inappropriately and the child becomes interpersonally dysfunctional is referred to as *traumatic sexualization*. Finkelhor and Browne note:

> It occurs through the exchange of affection, attention, privileges, and gifts for sexual behavior, so that the child learns sexual behavior as a strategy for manipulating others to get his or her other developmentally appropriate needs met. It occurs when certain parts of the child's anatomy are fetishized and given distorted importance and meaning. It occurs through the misconceptions and confusions about sexual behavior and morality that are transmitted to the child from the offender. And it occurs when very frightening memories and events become associated in the child's mind with sexual activity.

Sexually traumatized children learn inappropriate sexual behaviors (such as manipulating an adult's genitals for affection), are confused about their sexuality, and inappropriately associate certain emotions—such as loving and caring—with sexual activities.

As adults, sexual issues may become especially important. Survivors may suffer flashbacks, sexual dysfunctions, and negative feelings about their bodies. They may also be confused about sexual norms and standards. A fairly common confusion is the belief that sex may be traded for affection. Some women label themselves as "promiscuous," but this label may be more a result of their negative self-image than of their actual behavior. There seems to be a history of childhood sexual abuse among many prostitutes (Simons and Whitbeck, 1991).

Betrayal. Children feel betrayed when they discover that someone on whom they have been dependent has manipulated, used, or harmed them. Children may also feel betrayed by other family members, especially mothers, for not protecting them from abuse. As adults, survivors may experience depression as a manifestation, in part, of extended grief over the loss of trusted figures. Some may find it difficult to trust others. Other survivors may feel a deep need to regain a sense of trust and become extremely dependent. Distrust may manifest itself in hostility and anger or in social isolation and avoidance of intimate relationships. In adolescents, antisocial or delinquent behavior may be a means of protecting themselves from further betrayal. Anger may express a need for revenge or retaliation.

Powerlessness. Children experience a basic kind of powerlessness when their bodies and personal spaces are invaded against their will. A child's powerlessness is reinforced as the abuse is repeated. In adulthood, powerlessness may be experienced as fear or anxiety; a person feels unable to control events. Adult survivors often believe that they have impaired coping abilities. This feeling of ineffectiveness may be related to the high incidence of depression and despair among survivors. Powerlessness may also be related to increased vulnerability or revictimization through rape or marital violence; survivors may feel unable to prevent subsequent victimization. Other survivors, however, may attempt to cope with their earlier powerlessness by an excessive need to control or dominate others.

Stigmatization. Ideas of badness, guilt, and shame about sexual abuse are transmitted to abused children and then internalized by them. Stigmatization is communicated in numerous ways. The abuser conveys it by blaming the child or, through secrecy, communicating a sense of shame. If the abuser pressures the child for secrecy, the child may also internalize feelings of shame and guilt. Children's prior knowledge that their families or communities consider such activities deviant may contribute to their feelings of stigmatization. As adults, survivors may feel extreme guilt or shame about having been sexually abused. They may have low self-esteem because they feel that the abuse made them "spoiled merchandise." They also feel different from others because they mistakenly believe that they alone have been abused.

Treatment Programs

We are healthy only to the degree that our ideas are humane.

Kurt Vonnegut, Jr.

Child sexual abuse, especially father-daughter incest, is increasingly being treated through therapy programs working in conjunction with the judicial system rather than through breaking up the family by removing the child or the offender (Nadelson and Sauzier, 1986). Because the offender is often also the breadwinner, incarcerating him may greatly increase the family's emotional distress. The district attorney's office may work with clinicians in evaluating the existing threat to the child and deciding whether to prosecute, refer the offender to therapy, or both. The goal is not simply to punish the offender but to try to assist the victim and the family in coming to terms with the abuse.

Many of these clinical programs work on several levels at once: they treat the individual, the father-daughter relationship, the mother-daughter relationship, and the family as a whole. They work on developing self-esteem and improving the family and marital relationships. If appropriate, they refer individuals to alcohol or drug abuse treatment programs.

A crucial ingredient in many treatment programs is individual and family attendance at self-help group meetings. These self-help groups are composed of incest survivors, offenders, mothers, and other family members. Self-help groups such as Parents United and Daughters and Sons United help the offender acknowledge his responsibility and understand the impact of the incest on everyone involved.

Preventing Sexual Abuse

The idea of preventing sexual abuse is relatively new (Berrick and Barth, 1992). Prevention programs began about a decade ago, a few years after programs were started to identify and help child or adult survivors of sexual abuse. (For an evaluation of commercially available materials for preventing child abuse, see Roberts et al., 1990.) Such prevention programs have been hindered, however, by three factors (Finkelhor, 1986a, 1986b):

1. The issue of sexual abuse is complicated by differing concepts of appropriate sexual behavior and partners, which are not easily understood by children.
2. Sexual abuse, especially incest, is a difficult and scary topic for adults to discuss with children. Children who are frightened by what their parents tell them, however, may be less able to resist abuse than those who are given strategies of resistance.
3. Sex education is controversial. Even where it is taught, instruction often does not go beyond physiology and reproduction. The topic of incest is especially opposed.

In confronting these problems, child abuse prevention (CAP) programs have been very creative. These programs typically aim at three audiences: children, parents, and professionals (especially teachers). Children have the right to control their own bodies and genitals and to feel "safe," and they have the right not to be touched in ways that feel confusing or wrong. The CAP programs stress that the child is not at fault when such abuse does occur. They also try to give children possible courses of action if someone tries to sexually abuse them. In particular, children are taught that it's all right to say no, that they should get away from scary situations, and that it's very important to tell someone they trust about what has happened (and to keep telling until they are believed) (Gelles and Conte, 1991).

Other programs focus on educating parents who, it is hoped, will in turn educate their children. These programs aim at helping parents discover abuse or abusers by identifying warning signs. Such programs, however, need to be culturally sensitive, as Latinos and Asian Americans may be reticent about discussing these matters with their children (Ahn and Gilbert, 1992). Parents seem reluctant in general about dealing with sexual abuse issues with their children, according to David Finkelhor (1986a). First, many do not feel that their children are at risk. Second, parents are fearful of unnecessarily frightening their children. Third, parents feel uncomfortable talking with their children about sex in general, much

Reflections

Assume for a moment that a young child disclosed to you the fact that she was hurt by her father. What would you say to her? How would you feel? Whom would you tell?

less about such taboo subjects as incest. In addition, parents may not believe their own children or may feel uncomfortable confronting a suspected abuser, who may be a partner, uncle, friend, or neighbor.

CAP programs have also directed attention to professionals, especially teachers, physicians, mental health professionals, and police officers. Because of their close contact with children and their role in teaching children about the world, teachers are especially important. Professionals are encouraged to watch for signs of sexual abuse and to investigate children's reports of such abuse. A number of schools have instituted programs to educate both students and their parents. (For a research review of child sexual abuse prevention, see Berrick and Barth, 1992.)

We have too many high-sounding words, and too few actions that correspond with them.

Abigail Adams (1744–1818)

In recent years, both the American Medical Association (AMA) and the federal government have become more actively involved in fighting domestic violence. AMA guidelines advise doctors to question female patients routinely as to whether they have been attacked by their partners or forced to have sex. Physicians are also urged to investigate cases of injuries to women that are not well explained. (As we discussed earlier, there are already laws in place regarding the reporting of suspected child abuse.) At the federal level, a system is being set up by the Centers for Disease Control for tracking domestic violence cases and assessing ways to prevent abuse. Additionally, the Family Preservation and Support Act of 1993 provides $930 million for domestic abuse prevention. Donna Shalala, Secretary of Health and Human Services, emphasizes that government alone cannot solve the problem of violence in the home (Marek, 1994). Professionals, such as physicians, law enforcement personnel, and social workers, also have important roles to fulfill, and within local communities, organizations and individuals need to promote awareness of abuse and reach out to affected families. Most important, we all need to look into ourselves to find nonviolent solutions to the problems we face in our own relationships.

Summary

- Any form of intimacy or relatedness increases the likelihood of violence or abuse. *Violence* is defined as an act carried out with the intention or perceived intention of causing physical pain or injury to another person.
- The principal models used to study sources of family violence are the psychiatric model, which finds the source of violence within the personality of the abuser; the ecological model, which looks at both the child's development in the family context and the family's development within the community; the patriarchy model, which finds violence to be inherent in male-dominated societies; the social situational model, which views family violence as arising from a combination of structural stress and cultural norms; the social learning model, in which violence is seen as a behavior learned within the family and larger society; the resource model, which assumes that force is used to compensate for the lack of personal, social, and economic resources; and the exchange/social control model, which holds that people weigh the costs versus the rewards in all their actions and that they will use violence if the social controls (costs) are not strong enough. Three factors that may reduce social control are inequality of power in the family, the private nature of the family, and the "real man" image.
- *Battering* is the use of physical force against another person. It includes slapping, punching, knocking down, choking, kicking, hitting with objects, threatening with weapons, stabbing, and shooting. Wife battering is one of the most common and most underreported crimes in the United States. Although there does not appear to be a "typical" battered woman, two characteristics correlate highly with wife assault: low socioeconomic status and a high degree of marital conflict. A man who batters his wife probably has some or all of the following characteristics: belief in common myths about battering, traditional beliefs about the family, low self-esteem, pathological personality characteristics, use of sex as aggression, and a belief in the moral rightness of his aggression.
- The three-phase cycle of violence in battering relationships proposed by Lenore Walker includes (1) the tension-building phase, (2) the explosive phase, and (3) the resolution ("honeymoon") phase.
- *Marital rape* is a form of battering. Many people, including victims themselves, have difficulty acknowledging that forced sex in marriage is rape, just as it is outside of marriage.
- The incidence of violence and sexual assault in dating relationships is alarming. Violence is often precipitated by jealousy or rejection. *Date rape* or *acquaintance rape* may not be recognized by either the assailant or the victim because they think that rape is something done by strangers.
- Women may stay in, or return to, battering relationships for a number of reasons, including economic dependency, religious pressure or beliefs, the perceived need for a father for the children, pity, guilt, a sense of duty, fear, love, or reasons pertaining to their particular culture. Women may also be paralyzed by "learned helplessness." Women may try to stop their husband's violence. The most important factor in stopping abuse appears to be the woman's own determination that it must stop. Women may leave violent relationships when the level of violence is very high or when their children become threatened.
- Domestic violence intervention can be based on either control or compassion. Arrest, prosecution, and imprisonment are examples of control; shelters and support groups (including abuser programs) are examples of compassionate intervention.
- At least a million children are physically abused and neglected by their parents each year in the United States. The majority of abuse cases are unreported. Parental violence is one of the five leading causes of childhood death. Families at risk for child abuse often have specific parental, child, and family ecosystem characteristics. Parental characteristics include a father who was abused as a child, belief in corporal punishment, a devalued marital relationship and interspousal violence, father dominance, low self-esteem, unrealistic expectations for the child, parent-child role reversal, and lack of parental concern about the child's injury. Child characteristics include a "normal" child who is the product of a difficult or unplanned pregnancy, is the "wrong" sex, or is born outside of marriage; an "abnormal" child with physical or medical problems; or a "difficult" child. The family ecosystem includes the general social and economic environment in which the family lives. Characteristics in families at risk

include such conditions as unemployment, social isolation, poverty, and unsafe neighborhoods.

- Mandatory reporting of suspected child abuse may be helping to decrease the number of abused children in the United States. However, social workers are still overburdened, and services such as foster care are in short supply. Early intervention and education may be successful in reducing abuse, but there is a shortage of government funds for these and other programs to assist the victims of family violence.
- The hidden victims of family violence include siblings (who have the highest rate of violent interaction), parents assaulted by their adolescent or youthful children, and elders assaulted by their middle-aged children.
- Some recommendations for reducing family violence include (1) reducing sources of societal stress, such as poverty and racism; (2) eliminating sexism; (3) establishing supportive networks; (4) breaking the family cycle of violence; and (5) eliminating the legitimization and glorification of violence.
- *Incest* is defined as sexual intercourse between persons too closely related to marry. Sexual victimization of children may include incest, but it can also involve other family members and other sexual activities. The most traumatic form of child abuse is probably father-daughter (or stepfather-stepdaughter) abuse. Stepfathers abuse their stepdaughters at significantly higher rates than biological fathers abuse their daughters. Brother-sister abuse is often traumatic if it is exploitative or violent.
- Children most at risk for sexual abuse include females, preadolescents, children with absent or unavailable parents, children with poor parental relationships, children with parents in conflict, and children living with a stepfather.
- Child sexual abuse has both initial and long-term effects. The initial effects include emotional disturbances, physical consequences, and sexual and social disturbances. The long-term effects include depression, self-destructive tendencies, somatic disturbances and dissociation, negative self-concept, interpersonal relationship difficulties, revictimization, and sexual difficulties. The survivors of sexual abuse frequently suffer from sexual abuse trauma, which is characterized by traumatic sexualization, betrayal, powerlessness, and stigmatization.
- Child sexual abuse offenders are increasingly being sent into treatment programs in an attempt to assist the incest survivor and family in coping with the crisis that incest creates. Self-help groups are important for many survivors of sexual abuse.

Key Terms

acquaintance rape 462
battering 455
child sexual abuse 474
date rape 462
extrafamilial sexual abuse 474
incest 475
intrafamilial sexual abuse 474
marital rape 460
pedophilia 474
rape 460
violence 453

Suggested Readings

Barnett, Ola, Cindy Miller-Penn, and Robin Perrin. *Family Violence Across the Lifespan.* Newbury Park, CA: Sage, 1996. Coverage of all types of abuse and methodology, etiology, prevalence, treatment, and prevention of family violence.

Fontes, Lisa Aronson, ed. *Sexual Abuse in Nine North American Cultures: Treatment and Prevention.* Thousand Oaks, CA: Sage, 1995. Examines the impact of culture on child sexual abuse, including ways in which cultural norms can be used to protect children and help them recover from abuse.

Freeman, Lory. *It's My Body.* Seattle: Parenting Press, 1983. An illustrated booklet for children on the differences between "good" touching and "bad" touching. It is also available in Spanish as *Mi Cuerpo Es Mio.*

Hampton, Robert, ed. *Black Family Violence: Current Research and Theory.* New York: Lexington Books, 1991. A comprehensive, multidisciplinary collection of essays.

Island, David, and Patrick Letellier. *Men Who Beat the Men Who Love Them.* Binghamton, NY: Haworth,

1991. A comprehensive look at violence in gay relationships, including personal narratives and prevention and treatment strategies.

Johann, Sara Lee. *Domestic Abusers: Terrorists in our Homes.* Springfield, IL: Charles C. Thomas, 1994. An examination of judicial policy concerning domestic violence by an attorney.

Kirschner, S., D. A. Kirschner, and R. L. Rappaport. *Working with Adult Incest Survivors: The Healing Journey.* New York: Brunner/Mazel, 1993. A treatment program for adults who have experienced abuse.

Renzetti, Claire. *Violent Betrayal: Partner Abuse in Lesbian Relationships.* Newbury Park, CA: Sage Publications, 1992. An exploration of lesbian relationships and the factors leading to abuse. It includes a section on seeking help.

Russell, Diana E. H. *Rape in Marriage.* Rev. ed. Bloomington, IN: Indiana University Press, 1990. A superbly researched, clear, and sobering view of sexual violence against women by their husbands and lovers.

Sipe, Beth, and Evelyn Hall. *I Am Not Your Victim; Anatomy of Domestic Violence* Thousand Oaks, CA: Sage, 1996. A moving firsthand account by a victim of domestic violence and a system who didn't believe her.

Stark, Evan, and Ann Flitcraft. *Women at Risk.* Thousand Oaks, CA: Sage, 1996. An exploration of the theoretical perspectives as well as health consequences of woman abuse and clinical interventions to reduce the incidence of abuse.

Stith, Sandra, and M. Straus, eds. *Understanding Partner Violence: Prevalence, Causes, Consequences and Solutions.* Minneapolis, MN: National Council on Family Relations, 1995. An informative book which covers a broad range of issues related to domestic violence.

White, Evelyn C. *Chain Chain Change.* Seattle, WA: Seal Press, 1985. A book directed toward the African-American woman who wants to understand the role of violence and emotional abuse in her life. Discusses stereotypes and cultural assumptions and offers practical information about getting help.

Wolfe, David, Christine Wekerle, and Katreena Scott. *Alternatives to Violence: Empowering Youth to Develop Healthy Relationships.* Thousand Oaks, CA: Sage, 1996. A practical and broad-based book addressing the important topic of preventing youth violence. It works well with *The Youth Relationship Manual* (also published by Sage).

CHAPTER 14

Coming Apart: Separation and Divorce

P R E V I E W

To gain a sense of what you already know about the material covered in this chapter, answer "True" or "False" to the following statements.

1. Half of all those currently marrying end up divorcing within seven years. True or false?
2. Divorce occurs as a single event in a person's life. True or false?
3. Americans have one of the highest marriage, divorce, and remarriage rates among industrialized nations. True or false?
4. The critical emotional event in a marital breakdown is the separation rather than the divorce. True or false?
5. Anglos have a higher divorce rate than Latinos. True or false?
6. Divorce is an important element of the contemporary American marriage system because it reinforces the significance of emotional fulfillment in marriage. True or false?
7. The higher an individual's employment status, income, and level of education, the greater the likelihood of divorce. True or false?
8. Many problems assumed to be due to divorce are actually present before marital disruption. True or false?
9. Those whose parents are divorced have a significantly greater likelihood of themselves divorcing. True or false?
10. Marital conflict in an intact two-parent family is generally more harmful to children than living in a tranquil single-parent family or stepfamily. True or false?

Answers

1 True, 2 False, 3 True, 4 True, 5 True, 6 True, 7 False, 8 True, 9 False, 10 True

OUTLINE

Experience is the name everyone gives to their mistakes.

Oscar Wilde (1854–1900)

Americans' feelings about marriage and divorce seem strangely paradoxical. Consider the following (Ganong and Coleman, 1994; White, 1991):

- Americans like marriage: They have one of the highest marriage rates in the industrialized world.
- Americans don't like marriage: They have one of the highest divorce rates in the world.
- Americans like marriage: They have one of the highest remarriage rates in the world.

What sense can we make from the fact that we are one of the most marrying, divorcing, and remarrying nations in the world? What does our high divorce rate actually tell us about how we feel about marriage? In this chapter we hope to explain the paradox of high rates of marriage and divorce as we examine the divorce process, marital separation, divorce consequences, children and divorce, child custody, and divorce mediation. This exploration will help you better understand what parents, children, and families experience and how they cope with what increasingly has become part of our marriage system—divorce.

Scholars suggest that divorce does not represent a devaluation of marriage but, oddly enough, an idealization of it. We would not divorce if we did not have so much hope about marriage fulfilling our various needs (Furstenberg and Spanier, 1987). In fact, divorce may very well be a critical part of our contemporary marriage system, which emphasizes emotional fulfillment and satisfaction. Frank Furstenberg and Graham Spanier (1987) note:

> Divorce can be seen as an intrinsic part of a cultural system that values individual discretion and emotional gratification. Divorce is a social invention for promoting these cultural ideals. Ironically, the more divorce is used, the more exacting the standards become for those who marry. . . . Divorce . . . serves not so much as an escape hatch from married life but as a recycling mechanism permitting individuals a second (and sometimes third and fourth chance) to upgrade their marital situation.

To free oneself is nothing: it's being free that is hard.

André Gide (1869–1951)

Our high divorce rate also tells us that we may no longer believe in the permanence of marriage. Norval Glenn (1991) suggests that there is a "decline in the ideal of marital permanence and . . . in the expectation that marriages will last until one of the spouses dies." Instead, marriages disintegrate when love goes or a potentially better partner comes along. Divorce is a persistent fact of American marital and family life and one of the most important forces affecting and changing American lives today (Furstenberg and Cherlin, 1991).

Before 1974, the view of marriage as lasting "until death do us part" reflected reality. However, a surge by divorce rates that began in the mid-1960s did not level off until the 1990s. In 1974, a watershed in American history was reached when more marriages ended by divorce than by death. Today approximately 50 percent of all new marriages are likely to end in divorce (U.S. Bureau of the Census, 1996).

Not only does divorce end marriages and break up families; it also creates new forms from the old ones. It creates remarriages (which are very different from first marriages). It gives birth to single-parent families and stepfamilies. Today about one out of every five American families is a single-parent family; more than half of all children will become stepchildren by the year 2000, and nearly half of current marriages include at least one spouse who is remarried (U.S. Bureau of the Census, 1996). Within the singles subculture is an immense pool of divorced men and women (most of whom are on their way to remarriage). But divorce does not create these new forms easily. It gives birth to them in pain and travail.

Researchers traditionally looked on divorce from a deviance perspective (Coleman and Ganong, 1991). It was assumed that normal, healthy individuals married and remained married. Those who divorced were considered abnormal, immature, narcissistic, or unhealthy in some manner. Social scientists, however, are increasingly viewing divorce as one path in the normal family life cycle. Those who divorce are not necessarily different from those who remain married. If we begin to regard divorce in this light, social scientists reason, part of the pain accompanying divorce may be diminished, as those involved will no longer regard themselves as "abnormal" (Raschke, 1987).

Social scientists express their greatest concern about divorce being its effect on children (Aldous, 1987; "Study Reveals," 1997; Wallerstein and Blakeslee, 1989). But even

Dating again after divorce announces to the world that one is available to choose a new partner. It can be a way of enhancing self-esteem.

in studies of the children of divorce, the research may be distorted by traditional assumptions about divorce being deviant (Amato, 1991). For example, problems that children experience may be attributed to divorce rather than to other causes, such as personality traits. Although some effects are caused by the disruption of the family itself, others may be linked to the new social environment—most notably poverty and parental stress—into which children are thrust by their parents' divorce (McLanahan and Booth, 1991; Raschke, 1987). Some therapists suggest that we begin looking at those factors that help parents and children successfully adjust to divorce rather than focusing on risks, dysfunctions, and disasters (Abelsohn, 1992).

Factors Affecting the Likelihood of Divorce

A man should not marry a woman with the mental reservation that, after all, he can divorce her.

TALMUD: *Yebamoth*

Almost everyone who marries today knows that he or she has a fifty-fifty chance of divorcing later. The uncertainty of marital success dogs many of us. For some people it creates an underlying sense of fear as they make their commitments; it makes others hesitant to make a commitment for fear of failure. But if we can be aware of some of the factors associated with divorce, we can overcome the disadvantages associated with them. Such knowledge empowers us to have successful marriages.

It may be difficult to discover the underlying reasons for any individual divorce, but researchers have found various factors related to divorce. Some are societal, others are demographic, and still others are related to the nature of marriage or the family itself.

Figure 14.1
Percent Distribution of Divorces by Duration of Marriage: 1990

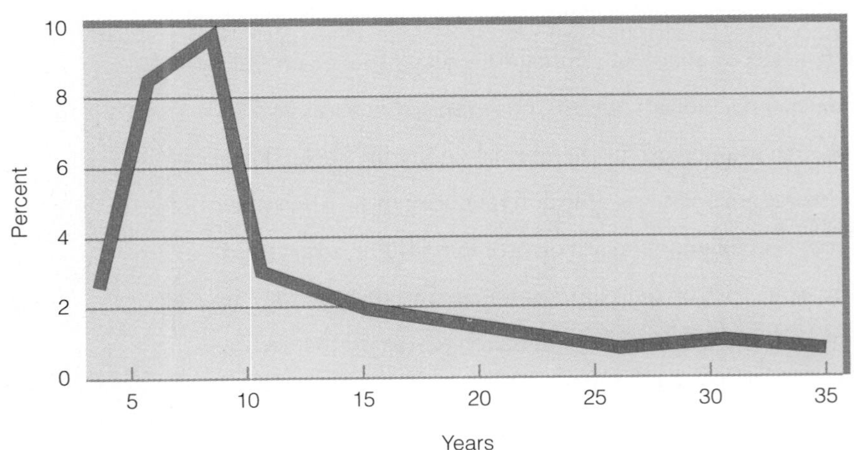

SOURCE: Clarke, Sally. "Advance Report of Final Marriage Statistics, 1989 and 1990." *Monthly Vital Statistics Report* 43, 12 (July 14, 1995). U.S. Department of Health and Human Services, National Center for Health Statistics.

Societal Factors

The divorce rate in our society has shifted from less than 10 percent at the beginning of the twentieth century to approximately 50 percent as we approach the twenty-first century. Such a dramatic change points to the need for societal rather than individual explanations. Although social factors are involved in divorce, it is often difficult for us to view divorce in sociological terms because the pain of divorce seems so uniquely personal. "Social and structural factors," notes Joseph Guttman (1993), "are largely invisible to people who must cope with the personal consequences of these factors on a daily basis."

Changed Nature of the Family

The shift from an agricultural society to an industrial one undermined many of the family's traditional functions. Schools, the media, and peers are now important sources of child socialization and child care. Hospitals and nursing homes manage birth and care for the sick and aged. Because the family pays cash for goods and services rather than producing or providing them itself, its members are no longer interdependent.

As a result of losing many of its social and economic underpinnings, the family is not a necessity. It is now simply one of many choices we have: We may choose singlehood, cohabitation, marriage—or divorce—and if we choose to divorce, we enter the cycle of choices again: singlehood, cohabitation, or marriage and possibly divorce for a second time. A second divorce leads to our entering the cycle for a third time, and so on.

Social Integration

Social integration—the degree of interaction between individuals and the larger community—is emerging as an important factor related to the incidence of divorce. The social integration approach regards such factors as urban residence, church membership, and population change as especially important in explaining divorce rates (Breault and Kposowa, 1987; Glenn and Shelton, 1985; Glenn and Supancic, 1984).

In this country, the rates of divorce increase from east to west. The highest rate is found in California, where two divorces currently occur for every three marriages. The greater likelihood of divorce in the West and Southwest may be caused by the higher rates of residential mobility and lower levels of social integration with extended families, ethnic neighborhoods, and church groups (Glenn and Supancic, 1984; Glenn and Shelton, 1985). Among African Americans, the lowest divorce rate is found among those born and raised in the South; African Americans born and raised in the North and West have the highest divorce rates. One study found that urban residence was the highest correlate of divorce (Breault and Kposowa, 1987). Those who live in urban areas, where the divorce rate is higher than in rural areas, for example, are less likely to be subject to the community's social or moral pressures. They are more independent and have greater freedom of personal choice.

Individualistic Cultural Values

American culture has traditionally been individualistic. We value individual rights; we cherish images of the individual's battling nature; we believe in individual responsibility. It should not be surprising that many view the individual as having priority over the family when the two conflict. Since the 1950s, perhaps as a reaction to the alienation and stifling conformity of the time, we have increasingly

It's one thing marrying the wrong person for the wrong reasons: it's another sticking it out with them.

PHILIP ROTH, *Letting Go*

As for infidelity, which has been blamed for so many divorces, that doesn't seem to be the real issue. It might be closer to the truth to say that people divorce not because one or both partners have been unfaithful, but because they are faithful—not to each other, but to themselves.

ANATOLE BROYARD

valued self-fulfillment and personal growth (Guttman, 1993). As marriage and the family lost many of their earlier social and economic functions, their meaning shifted. Marriage and family are viewed as paths to individual fulfillment. We marry for love and expect marriage to bring us happiness. When individual needs conflict with family demands, however, we no longer automatically submerge our needs to those of the family. We often struggle to balance individual and family needs. But if we are unable to do so, divorce has emerged as an alternative to an unhappy, unfulfilling, mean-spirited, or violent marriage.

Demographic Factors

There are a number of demographic factors that appear to have a correlation with divorce.

Employment Status #3

Among whites, a higher divorce rate is more characteristic of low-status occupations, such as factory worker, than of high-status occupations, such as executive (Greenstein, 1985; Martin and Bumpass, 1989). Unemployment, which contributes to marital stress, is also related to increased divorce rates.

Studies conflict as to whether employed wives are more likely than nonemployed wives to divorce; overall, though, the findings seem to suggest that female employment contributes to the likelihood of divorce because the employed wife is less dependent on her husband's earnings (White, 1991). Wives' employment may lead to conflict about the traditional division of household labor, child-care stress, and other work spillover problems that, in turn, create marital distress.

Income

The higher the family income, the lower the divorce rate for both whites and African Americans. It is interesting, however, that the higher a woman's individual income, the greater her chances of divorce perhaps because with greater incomes women are not economically dependent on their husbands or because conflict over inequitable work and family roles increases marital tension.

Educational Level

For whites, the higher the educational level, the lower the divorce rate. Divorce rates among African Americans are not as strongly affected by educational levels. Men and women with only a high school education are more likely to divorce than those with a college education (Glick, 1984b).

Ethnicity

African Americans are more likely than whites to divorce. The relation between ethnicity and divorce is not surprising because of the strong correlation between socioeconomic status and divorce: The lower the socioeconomic class, the more likely a person is to divorce. As income levels for African Americans increase, in fact, divorce rates decrease; they become similar to those of whites (Raschke, 1987; Garfinkel, McLanahan, and Robins, 1994). In 1995, 9.1 percent of whites, 10.7 percent of African Americans, and 7.9 percent of Latinos were divorced (U.S. Bureau of the Census, 1996).

Among Latino groups, there are different divorce rates. Mexican Americans and Cuban Americans have lower divorce rates than do Anglos; the divorce rate among Puerto Ricans, however, approaches that of African Americans (Frisbie, 1986). If we examine marital disruption rates (which include marital separation,

OTHER PLACES OTHER TIMES

Divorce in the Chinese American Family

Despite long periods of separation in split-household families and the stresses caused by overwork in small-producer families, most Chinese American marriages remained intact. The low divorce rate, however, does not say much about the quality of marriages. It is clear that marriages were looked upon by early Chinese immigrants as unions whose purpose was to produce offspring; the quality of the marital relationship was immaterial. Women who later joined their husbands or who were "hasty brides" suffered from many adjustment problems in their new environment. They often felt isolated because they spoke little English and had no supportive network of friends and relatives. Husbands were often too busy to pay close attention to the problems their wives faced. Long separations made even longtime mates strangers to each other. Brides who came to the United States after arranged marriages hardly knew their husbands, and their relationships were sometimes strained also by a disparity in age. In many cases, men in their forties, who had worked in the United States for years, married women in their teens or early twenties.

The low divorce rate, then, reflects the lack of choices Chinese American women have, rather than a high level of marital quality. In small-producer families, interdependence made it impossible for spouses to survive without each other. At the same time, divorced women were considered to be an embarrassment to the community; a stigma was attached to divorce. Some desperate Chinese American women even took their lives, believing there was no other way out of a miserable marriage and living conditions. In one study, Sung (1967) found that the suicide rate among Chinese Americans in San Francisco was four times that in the city as a whole, and that victims were predominantly women.

The Chinese American divorce rate continues to be low. In 1990, only 2.3 percent and 3.3 percent of Chinese American men and women, respectively, were divorced (U.S. Bureau of the Census, 1990). These figures are considerably lower than those of white men (7.5 percent) and white women (9.4 percent), and slightly lower than those of other Asian American groups (e.g., 4.2 percent and 6.5 percent for Japanese American men and women, respectively, and 3.7 percent and 5.1 percent for Filipino American men and women) (U.S. Bureau of the Census, 1990).

We know little about the prevalence of remarriage among Chinese Americans. However, because most white divorced people eventually remarry, there is no reason to think that this is not the case with Chinese Americans. A high remarriage rate among Chinese Americans can also be predicted on the basis of an increasing number of interracial marriages, the stigma attached to divorce, and the strong belief among Chinese Americans that children fare better with two parents.

There is little information about the frequency of single parenthood in the Chinese American population. According to the U.S. Bureau of the Census (1990), a vast majority of Chinese American children (87.6 percent) reside with two parents. This figure is slightly higher than that for whites (77.2 percent) and considerably higher than that for African Americans (35.6 percent).

as well as divorce), they are more or less the same for Mexican Americans and Cuban Americans as for Anglos (Bean and Tienda, 1987; Vega, 1991). Among Puerto Ricans, the rates of marital disruption are the highest of any ethnic group in the United States. These recent findings stand in contrast to the old belief that Latino families are more stable than Anglo families (Vega, 1991).

Religion

Frequency of attendance at religious services (not necessarily the depth of beliefs) tends to be associated with the divorce rate (Glenn and Supancic, 1984). Among white males, the rate of divorce for those who never attend religious services is three times as high as for those who attend two or three times a month. By religion, the lowest divorce rate is for Jews, followed by Catholics and then

Protestants. Fundamentalist Protestants have a higher divorce rate than those of more moderate Protestant denominations (Guttman, 1993). Because the Roman Catholic church now allows divorce through annulments and no longer excommunicates divorced people by refusing them the sacraments, the annulment rate has increased from 450 in 1968 to over 50,000 in 1994 (Woodward, Quade, and Kantrowitz, 1995).

The greater the involvement in religious activities, the less likelihood there is of divorce. Since the major religions discourage divorce, highly religious men and women are less likely to accept divorce because it violates their values. It may also be that a shared religion and participation in organized religious life affirms the couple relationship (Guttman, 1993; Wineberg, 1994).

Life Course Factors

Different aspects of the life course may affect the divorce probability of some individuals.

Intergenerational Transmission

Both African Americans and whites have a slightly increased likelihood of divorce if their families of origin were disrupted by divorce or desertion (Raschke, 1987). One study (Glenn and Kramer, 1987), however, found a statistically significant relationship for daughters of divorced white parents; these women may be more prone to divorce because they tend to marry at an early age. Another study (Amato, 1988) comparing adults from divorced and intact families found that both groups held similar attitudes toward marriage. All in all, coming from a divorced family appears to have relatively little effect on adult children's divorcing.

Age at Time of Marriage

Adolescent marriages are more likely to end in divorce than are marriages that take place when people are in their twenties or older. (Kurdek, 1993). This is true for both whites and African Americans. Younger partners are less likely to be emotionally mature. After age twenty-six for men and age twenty-three for women, however, age at marriage seems to make little difference (Glenn and Supancic, 1984).

Premarital Pregnancy and Childbirth

Premarital pregnancy by itself does not significantly increase the likelihood of divorce. But if the pregnant woman is an adolescent, drops out of high school, and faces economic problems following marriage, the divorce rate increases dramatically. If a woman gives birth prior to marriage, the likelihood for divorce in a subsequent marriage increases, especially in the early years. This negative effect on marriage is stronger for whites than for African Americans (White, 1991).

Remarriage

The divorce rate among those who remarried in the 1980s is so far about 25 percent higher than it is for those who entered first marriages in that decade (White, 1991). It is not clear why there is a higher divorce rate in remarriages. Some researchers suggest that the cause may lie in a "kinds-of-people" explanation. The probability factors associated with the kinds of people who divorced in first marriages—low levels of education, unwillingness to settle for unsatisfactory marriages, and membership in certain ethnic groups—are present in subsequent marriages, which increases the likelihood of divorce (Martin and Bumpass, 1989).

Others argue that the dynamics of second marriages, especially the presence of stepchildren, increase the chances of divorce (White and Booth, 1985). Stepfamily research, however, does not provide much support for this hypothesis (see Ganong and Coleman, 1994).

Family Processes

The actual day-to-day marital processes of communication—handling conflict, showing affection, and other marital interactions—may be the most important factors holding marriages together or dissolving them (Gottman, 1994).

Marital Happiness

Although it seems reasonable that there would be a strong link between marital happiness (or, rather, the lack of happiness) and divorce, this is true only during the earliest years of marriage. Those who have low marital-happiness scores in the first years of marriage are four or five times more likely to divorce within three years than those with high marital happiness (Booth et al., 1986). The strength of the relationship between low marital happiness and divorce decreases in later stages of marriage, however (White and Booth, 1991). In fact, alternatives to one's marriage and barriers to divorce appear to influence divorce decisions more strongly than does marital happiness.

Nothing proves better the necessity of an indissoluble marriage than the instability of passion.

HONORÉ DE BALZAC (1799–1850)

Children

It is not clear what relation, if any, children have to the likelihood of divorce (Raschke, 1987). Children were once considered a deterrent to divorce—people stayed together for the sake of the children—but 60 percent of all divorces now take place among couples who have children. The birth of the first child reduces the chance of divorce to almost nil in the year following birth; this preventive effect does not hold true, however, for subsequent births (White, 1991). One of the most significant findings indicates that parents of sons are less likely to divorce than parents of daughters. The researchers suggest that fathers participate more in the parenting of sons than daughters, thereby creating greater family involvement for the men (Morgan, Lye, and Condran, 1988; Katzen, Warner, and Acock, 1994).

In some instances, the presence of children may be related to higher divorce rates. Premaritally conceived (during adolescence) children and physically or mentally limited children are associated with divorce. Children in general contribute to marital dissatisfaction and possibly divorce, according to one researcher (Raschke, 1987): "It could be expected that normal children at least contribute to strains in an already troubled marriage, given the consistent findings that children, especially in adolescent years, lower marital satisfaction." At the same time, however, women without children have considerably higher divorce rates than women with children.

Marital Problems

If you ask divorced people to give the reasons for their divorce, they are not likely to say, "I blame the changing nature of the family" or "It was demographics." They are more likely to respond, "She was on my case all the time" or "He just didn't understand me," or if they are charitable, they might say, "It wasn't right for us." Personal characteristics leading to conflicts are obviously very important factors in the dissolution of relationships.

Getting divorced because you don't love a man is almost as silly as getting married just because you do.

ZSA ZSA GABOR

Studies of divorced men and women cite such problems as alcoholism, drug abuse, marital infidelity, sexual incompatibility, and conflicts about gender roles

#5

as leading to their divorces. Kitson and Sussman (1982) found that the four most common reasons given were, in descending order of frequency: personality problems, home life, authoritarianism, and differing values. Extramarital affairs ranked seventh. Complaints associated with gender roles accounted for 35 percent of the men's responses and 41 percent of women's responses, but because the studies included only divorced respondents, it is difficult to tell whether the presence of these factors can predict divorce. We know from studying enduring marriages that marriages often continue in the face of such problems.

The Divorce Process

Divorce is not a single event. You don't wake up one morning and say, "I'm getting a divorce," and then leave. It's a far more complicated process (Kitson and Morgan, 1991). It may start with little things that at first you hardly notice—a rude remark, thoughtlessness, an unreasonable act, a closedness. Whatever these things are, they begin to add up. Other times, however, the sources of unhappiness are more blatant—yelling, threatening, or battering. For whatever reasons, the marriage eventually becomes unsatisfactory; one or both partners become unhappy.

Love, the quest; marriage, the conquest; divorce, the inquest.

HELEN ROWLAND (1876–1950)

Over half of those who marry undergo this distressing process. Yet we know very little about the process of marital breakdown and divorce. We understand more about falling in love and courtship than we do about falling out of love and divorce (Furstenberg and Cherlin, 1991).

Anthropologist Paul Bohannan (1970b) developed one of the most influential psychological models explaining the divorce process. (For a discussion of other models, see Guttman, 1993.) Bohannan views divorce as consisting of six "stations" (processes) or "divorces": emotional, legal, economic, co-parental, community, and psychic. As people divorce, they undergo these divorces more or less simultaneously. The level of intensity of these different divorces varies at different times.

The Emotional Divorce ✓

Leaving is more often an avoidance of change than an instrument of change. The pain of parting misleads people into thinking that they have not chosen the easy way out.

DANIEL GOLDSHUE ET AL., THE DANCE-AWAY LOVER

The emotional divorce, when one spouse (or both) begins to disengage, begins before the legal divorce. But even as divorce papers are filed, the couple may find themselves feeling ambivalent. Because the emotional divorce is not complete, they may try to reconcile.

The partners may undermine each other's self-esteem with indifference or destructive criticism. From the outside, the marriage may appear to be functioning adequately, but its heart is missing. In fact, the marriage may even appear to be functioning better, to be more relaxed, and to have less conflict. But in reality, tension is diminished because one partner has decided to ignore or overlook problems—it is no longer worth the effort. "Why try to get him to do his share of the housework? I'm leaving!" a woman may ask. A husband may quit battling his wife over smoking. "I'm outta here," he thinks. "She can puff herself sick, for all I care." At the same time, ignoring the problems often adds to them, reinforcing the decision to divorce.

The Legal Divorce

The legal divorce is the court-ordered termination of a marriage. It permits divorced spouses to remarry and conduct themselves in a way that is legally independent of each other. The legal divorce also sets the terms for the division of property and child custody, issues that may lead to bitterly contested divorce battles. Many of the unresolved issues of the emotional divorce, such as feelings of hurt and betrayal, may be acted out during the legal divorce.

The Economic Divorce

The economic basis of marriage often becomes most painfully apparent during the economic divorce. Most property acquired during a marriage is considered joint property and is divided between the divorcing spouses. The property settlement is based on the assumption that each spouse contributes to the estate. This contribution may be nonmonetary, as in the case of the homemakers whose "moral assistance and domestic services" permitted their husbands to work outside the home. Alimony and child support may be required. As the partners go their own ways, they often suffer dramatic decreases in their standards of living because they must set up separate households and no longer pool their resources. Women usually experience the greatest decline in their standards of living, as we shall see.

The Co-Parental Divorce

Marriages end, but parenthood does not. Spouses may divorce each other, but they do not divorce their children. (Even those parents who never see their children remain in some sense fathers and mothers.) This may be the most complicated aspect of divorce, for it also gives birth to single-parent families and, in the majority of cases, stepfamilies. Issues of child custody, visitation, and support must be dealt with. The impact of divorce on children must be understood, negative consequences must be minimized as much as possible, and new ways of relating to the children and former spouses must be developed, keeping the children's best interest foremost in mind.

The Community Divorce

When people divorce, their social context changes. In-laws become ex-laws; often they lose (or stop) contact. (This is particularly troublesome when in-laws are also grandparents.) Old friends may choose sides or drop out; they may not be as supportive as one wants. New friends may replace old ones as divorced men and women begin dating again. They may enter the singles subculture, where activities center around dating. Single parents may feel isolated from such activities because child rearing often leaves them no leisure, and diminished income leaves them no money.

One must first have chaos in oneself to give birth to a dancing star.

FRIEDRICH NIETZSCHE (1844–1900)

The Psychic Divorce

The psychic divorce is accomplished when your former spouse becomes irrelevant to your sense of self and emotional well-being. You are psychically divorced when you learn that your ex-spouse has gotten a promotion; married someone smarter, funnier, more sensitive, more understanding, and better looking than you; bought a 4 × 4; received an honorary doctorate; and looks terrific—and you don't care. You have your own life to live. Bohannan regards the psychic divorce as the most important element in the divorce process. It is in this stage that each partner develops her or his own sense of independence, completeness, and stability. Neither misses the other or blames him or her for mistakes or misfortunes. Both are responsible for their own lives and are going forward with them.

The divorce process, as we can see, is complex. It takes place on many different levels. Those who go through divorce experience both pain and liberation. But eventually emerge as new women and men.

Reflections

From what you know about divorce, either from your own experience as a child or partner, or from the experiences of friends or other family members, how well does Bohannon's six-station model describe the experience? Are some stages more difficult than others? Why?

Marital Separation

The crucial event in a marital breakdown is the act of separation. Divorce is a legal consequence that follows the emotional fact of separation (Melichar and Chiriboga, 1988). Although separation generally precedes divorce, not all separations lead to divorce. As many as one couple out of every six that remains married is likely to have separated for at least two days (Kitson, 1985). The majority of separated women are between fifteen and forty-five years old, and a greater proportion of African Americans than whites are separated. Unfortunately, we don't know much about marital separation. We don't know, for example, which separations are likely to lead to reconciliations, divorce, or long-term separation (Morgan, 1988). Those who reconcile may have separated in order to dramatize their complaints, create emotional distance, or dissipate their anger (Kitson, 1985).

Did you know?

A study of separation and reconciliation found that of those who separated, 40 percent reconciled at least once; 18 percent reconciled twice or more. Almost all who reconciled did so within a year; 45 percent reconciled within a month (Bumpass, Martin, and Sweet, 1991).

Uncoupling: The Process of Separation

The trends in divorce are fairly clear, but the causes are not. Sociologists can describe societal, demographic, and family factors that appear to be associated with divorce. Unfortunately, however, such variables tell us about groups rather than individuals. Similarly, divorced men and women can tell us what they believe were the causes of their own divorces. But human beings do not always know the reasons for their own actions. They can deceive themselves, blame others, or remain ignorant of the causes.

Sometimes marital complaints are culled from long-term marital problems as a justification for the splitup. Gay Kitson and Marvin Sussman (1982) observe that the study of marital complaints as causes of divorce is merely the study of people's perceptions: "It is perhaps an impossibility to determine what 'really' broke up the marriage." Robert Weiss (1975) uses the term *account* to describe the individual's personal perception of the breakup. These accounts focus on a few

"That's right, Phil. A separation will mean—among other things—watching your own cholesterol."

SOURCE: Drawing by Maslin; © 1992, *The New Yorker Magazine, Inc.*

dramatic events or factors in the marriage. Because accounts are personal perceptions, each spouse's account is often very different from the other's; what is important to one partner may not be important to the other. Sometimes the accounts of ex-spouses seem to describe entirely different marriages.

People do not suddenly separate or divorce. Instead, they gradually move apart through a set of fairly predictable stages. Sociologist Diane Vaughan (1986) calls this process *uncoupling*. The process appears to be the same for married or unmarried couples and for gay or lesbian relationships. The length of time together does not seem to affect the process. "Uncoupling begins," Vaughan observes, "as a quiet, unilateral process." Usually one person, the initiator, is unhappy or dissatisfied but keeps such feelings to himself or herself. The initiator often ponders fundamental questions about his or her identity: "Who am I? Who am I in this relationship? What do I want out of my life? Can I find it in this relationship?" The dissatisfied partner may attempt to make changes in the relationship, but these are often unsuccessful, as he or she may not really know what the problem is.

Don't leave in a huff. Leave in a minute and a huff. If you can't leave in a minute and a huff, leave in a taxi.

Groucho Marx (1890–1977)

Because the dissatisfied partner is unable to find satisfaction within the relationship, he or she begins turning elsewhere. This is not a malicious or intentional turning away; it is done to find self-validation without leaving the relationship. In doing so, however, the dissatisfied partner "creates a small territory independent of the coupled identity" (Vaughan, 1986). This creates a division within the relationship. Gradually, the dissatisfied partner voices more and more complaints, which make the relationship and partner increasingly undesirable. The initiator begins thinking about alternatives to the relationship and comparing the costs and benefits of these alternatives. Meanwhile, both the initiator and his or her partner try to cover up the seriousness of the dissatisfaction, submerging it in the little problems of everyday living.

Excuses

Are you leaving because you are hungry?
Aha! Is your stomach your master?

Are you leaving me to cover yourself?
Have I not a blanket on my bed?

Are you leaving me because you are thirsty?
Then take my breast, it flows over for you.

Traditional African poem

Eventually, the initiator decides that he or she can no longer go on. There are several strategies for ending the relationship. One way is simply to tell the partner that the relationship is over. Another way is to break—consciously or unconsciously—a fundamental rule in the relationship, such as by having an extramarital affair and letting the partner know or discover it.

Uncoupling does not end when the end of a relationship is announced, or even when the couple physically separate. Acknowledging that the relationship cannot be saved represents the beginning of the last stage of uncoupling. Vaughan (1986) writes:

> Partners begin to put the relationship behind them. They acknowledge that the relationship is unsaveable. Through the process of mourning they, too, eventually arrive at an account that explains this unexpected denouement. "Getting over" a relationship does not mean relinquishing that part of our life that we shared with another; but rather coming to some conclusion that allows us to accept and understand its altered significance. Once we develop such an account, we can incorporate it into our lives and go on.

The New Self: Separation Distress and Postdivorce Identity

Our married self becomes part of our deepest self. Therefore, when people separate or divorce, many feel as if they have "lost an arm or a leg." This analogy, as well as the traditional marriage rite in which a man and a woman are pronounced "one," reveals an important truth of marriage: The constant association of both partners makes each almost a physical part of the other. This dynamic is true even if two people are locked in conflict; they, too, are attached to each other (Masheter, 1991).

After years of advising other people on their personal problems, I was stunned by my own divorce. I only wish I had someone to write to for advice.

Ann Landers

YOU AND YOUR WELL-BEING

Gender and Divorce-Related Stressors

Ask anyone who has experienced a divorce what one word summarizes the process and you probably won't be surprised when you hear: "Stressful!" A short look at the issues faced by divorced couples gives us some idea of the distress they can experience:

- severance of marital bonds
- establishment of a new lifestyle
- economic stressors
- negotiation of custody arrangements
- adjustments in parenting
- changes in social support system

To some degree, gender influences how individuals respond to divorce. Research indicates that divorced men experience greater emotional distress and report more suicidal thoughts than do women (Riesman and Gerstel, 1985; Rosengren, Wedel, and Wilhelmensen, 1989; Wallerstein and Kelly, 1980). Because women are more likely to initiate divorce, research suggests that they experience fewer postdivorce psychological problems. This may be because they have begun the detachment process earlier than men (Lawson and Thompson, 1996). Furthermore, divorced men exhibit higher rates of auto accidents, alcohol abuse, diabetes, heart disease, and mental illness than do divorced women. Higher rates of mortality have been found to exist among divorced men and women, especially if they are remarried or are cohabiting (Hemstrom, 1996).

The immediate impact of divorce on women is economic. This is especially true if they become the primary custodial parent. Approximately 53 percent of women who are granted child support do not receive it (U.S. Bureau of the Census, 1987). A combination of lowered earning power, increased expenses, and lack of financial support results in a decreased standard of living for the divorced mother and her children.

The psychological responses experienced among partners are numerous, ranging from anger to depression to ambivalence. Though some men suffer little distress following divorce (Albrecht, 1980), generally men seem to experience the greater emotional distress, possibly because of their more frequent social isolation (Reismann, 1990). In addition, men report greater attachment to their former spouses and are more likely to desire to rekindle the marriage (Bloom, and Kindle, 1985).

Almost 60 percent of divorces involve children (Kitson and Morgan, 1991), and since 80 percent of these children end up living with their mothers (with an additional 14 percent living with other relatives), fathers must face new emotional territory regarding these issues and their relationships with their children. Single parenting for the mother involves added responsibility to an already overburdened workload. Noncustodial parenting raises new role expectations concerning the quality of the parent-child relationship, normative behaviors, and discipline.

The more resources a person has, the better he or she may handle the separation crisis. Social support is positively correlated with lower distress and positive adjustment. These resources may be emotional as well as social and financial. Parents become an important resource for their divorcing adult children (Johnson, 1988). They may provide economic assistance and emotional support; they are also important for their grandchildren in easing the divorce transition. One study found that reliance on family and friends, involvement in church-related activities, social participation in community activities, and establishment of intimate heterosexual relationships correlated positively to post-divorce adjustment among African-American men (Lawson and Thompson, 1996). Friends are especially important in overcoming the isolation and loneliness that generally accompanies a separation.

As with other stressors in a person's life, it is often the individual's perception of a the event, not the stress itself, that influences how a person adjusts to change. If those experiencing separation and divorce can begin to view and accept their changing circumstances as presenting new challenges and opportunities, there is a greater likelihood that the physiological and psychological symptoms of stress that follow divorce can be reduced.

. . . rather bear those ills we have
Than fly to others we know not of.
WILLIAM SHAKESPEARE (1564–1616)

Separation Distress

Most newly separated people do not know what to expect. There are no divorce ceremonies or rituals to mark this major turning point. Yet people need to understand divorce in order to alleviate some of its pain and burden. Except for the death of a spouse, divorce is the greatest stress-producing event in life (Holmes and Rahe, 1967). The changes that take place during separation are crucial because at this point a person's emotions are rawest and most profound. Men and women react differently during this period. Most people experience **separation distress,** situational anxiety caused by separation from an attachment figure.

Researchers have considerable knowledge about the negative consequences accompanying marital separation, some of which we discuss here. In looking at this negative impact, however, we need to keep in mind Helen Raschke's (1987) caution: "The psychological and emotional consequences of separation and divorce have been more distorted than any of the other consequences as a result of the deviance perspective." The negative aspects of separation are balanced sooner or later, notes Raschke, by positive aspects, such as the possibility of finding a more compatible partner, constructing a better (or different) life, developing new dimensions of the self, enhancing self-esteem, and marrying a better parent for one's children. These positive consequences may follow, or be intertwined with, separation distress. In the pain of separation, we may forget that a new self is being born.

Almost everyone suffers separation distress when a marriage breaks up. The distress is real but, fortunately, does not last forever (although it may seem so). The distress is situational and is modified by numerous external factors. About the only men and women who do not experience distress are those whose marriages were riddled by high levels of conflict. In these cases, one or both partners may view the separation with relief (Raschke, 1987).

During separation distress, almost all attention is centered on the missing partner and is accompanied by apprehensiveness, anxiety, fear, and often panic. "What am I going to do?" "What is he or she doing?" "I need him . . . I need her . . . I hate him . . . I love him . . . I hate her . . . I love her. . . ." Sometimes, however, the immediate effect of separation is not distress but euphoria. This usually results from feeling that the former spouse is not necessary, that one can get along better without him or her, that the old fights and the spouse's criticism are gone forever, and that life will now be full of possibilities and excitement. That euphoria is soon gone. Almost everyone falls back into separation distress.

Whether a person had warning and time to prepare for a separation affects separation distress. An unexpected separation is probably most painful for the partner who is left. Separations that take place during the first two years of marriage, however, are less difficult for the husband and wife to weather. Those couples who separate after two years find separation more difficult because it seems to take about two years for people to become emotionally and socially integrated into marriage and their marital roles (Weiss, 1975). After that point, additional years of marriage seem to make little difference in the spouses' reaction to separation.

As the separation continues, separation distress slowly gives way to loneliness. Eventually, loneliness becomes the most prominent feature of the broken relationship. Old friends can sometimes help provide stability for a person experiencing a marital breakup, but those who give comfort need to be able to tolerate the other person's loneliness.

Establishing a Postdivorce Identity

A person goes through two distinct phases in establishing a new identity following marital separation: *transition* and *recovery* (Weiss, 1975). The transition period begins with the separation and is characterized by separation distress and then loneliness. In this period's later stages, most people begin functioning in an orderly way again, although they still may experience bouts of upset and turmoil. The transition period generally ends within the first year. During this time, individuals have already begun making decisions that provide the framework for new

Reflections

From your own experience, how well does "uncoupling" describe the process of separating from someone you care about? Are there missing elements or elements that should be emphasized? What about separation distress? In your own experience, what was it like? What things were you able to do to alleviate it? What advice would you give others about it?

Things You Told Me

I am a lousy housekeeper
I don't wash the dishes often enough
When I do wash the dishes, I run the water too much
When I clean up, I do it in the wrong order
The mustard is on the wrong shelf in the refrigerator
I arrange our refrigerator wrong
My inconsiderate children step on the papers you have left neatly stacked on the living room floor
I raise my children wrong
My little boy doesn't wash his dish
He washes his dish wrong
We make too much noise
I cut wrapping paper wrong
I massage you wrong
I touch wrong
I kiss wrong
I make love wrong
I am too sexual
My hugs are too sexual
I lay with my arm across the wrong part of your body (which hurts your stomach or makes you itch)
I pull your hair when I embrace you
I take too long to come
Or sometimes (and this is worse) I don't come at all
I sleep too close
I don't sleep close enough

(continued on following page)

I express my feelings wrong
I handle your anger wrong
I communicate wrong
I don't always like the things you like and I say so—this is wrong
I share my enthusiasm wrong
I'm wrong to be upset by your disrespect for me
I buy you the wrong presents
I take the wrong routes when I drive from one place to another
I drive too slow
I am too hesitant when I drive
I shift too late or too early
I drive in the wrong gear
I don't pass the slow cars when I should
I drive in the wrong lane
I ask too much of you
I relate to people wrong
I forget things because I pay attention wrong
I don't always understand what you are trying to say because I listen wrong
I'm illogical
I complain too much
I have bad manners
I'm inconsiderate
My children are untogether
I'm sick too often
I'm too serious
I take too much aspirin
I hurt your feelings
I am insensitive to your vulnerability
S. B. R.

selves. They have entered the role of single parent or noncustodial parent, have found a new place to live, have made important career and financial decisions, and have begun to date. Their new lives are taking shape.

The recovery period usually begins in the second year and lasts between one and three years. By this time the separated or divorced individual has already created a reasonably stable pattern of life. The marriage is becoming more of a distant memory, and the former spouse does not arouse the intense passions she or he once did. Mood swings are not as extreme, and periods of depression are fewer. Yet the individual still has self-doubts that lie just beneath the surface. A sudden reversal, a bad time with the children, or doubts about a romantic involvement can suddenly destroy a divorced person's confidence. By the end of the recovery period, the distress has passed.

It takes some people longer than others to recover because each person experiences the process in his or her own way. But most are surprised by how long the recovery takes—they forget that they are undergoing a major discontinuity in their lives.

Dating Again

A new partner reduces much of the distress caused by separation. A new relationship prevents the loneliness caused by emotional isolation. It also reinforces a person's sense of self-worth. But it does not necessarily eliminate separation distress caused by the disruption of intimate personal relations with the former partner, children, friends, and relatives.

A first date after years of marriage and subsequent months of singlehood evokes some of the same emotions felt by inexperienced adolescents. Separated or divorced men and women who are beginning to date again may be excited and nervous; worry about how they look; and wonder whether or not it is okay to hold hands, kiss, or make love. They may feel that dating is incongruous with their former selves or be annoyed with themselves for feeling excited and awkward. Furthermore, they have little idea of the norms of postmarital dating (Spanier and Thompson, 1987).

For many divorced men and women, the greatest problem is how to meet other unmarried people. They believe that marriage has put them "out of circulation," and many are not sure how to get back in. Because of the marriage squeeze (discussed in Chapter 6), separated and divorced men in their twenties and thirties are at a particular disadvantage: considerably fewer women are available than men. The squeeze reverses itself at age forty when there are significantly fewer single men available. The problem of meeting others is most acute for single mothers who are full-time parents in the home because they lack opportunities to meet potential partners. Divorced men, having fewer child-care responsibilities and more income than divorced women, tend to have more active social lives.

Hope in reality is the worst of all evils, because it prolongs the torments of man.

FRIEDRICH NIETZSCHE (1844–1900)

Dating fulfills several important functions for separated and divorced people. First, it is a statement to both the former spouse and the world at large that the individual is available to become someone else's partner (Vaughan, 1986). Second, dating is an opportunity to enhance one's self-esteem (Spanier and Thompson, 1987). Free from the stress of an unhappy marriage, dating may lead people to discover, for example, that they are more interesting and charming than either they or (especially) their former spouses had imagined. Third, dating initiates individuals into the singles subculture, where they can experiment with the

While there's life there's hope.

CICERO (106–43 B.C.)

freedom about which they may have fantasized when they were married. Interestingly, Spanier and Thompson (1987) found no relationship between dating experience and well-being following separation.

Several features of dating following separation and divorce differ from premarital dating. First, dating does not seem to be a leisurely matter. Divorced people are often too pressed for time to waste time on a first date that might not go well. Second, dating may be less spontaneous if the divorced woman or man has primary responsibility for children. The parent must make arrangements about child care; he or she may wish not to involve the children in dating. Third, finances may be strained; divorced mothers may have income only from low-paying or part-time jobs or AFDC benefits while having many child-care expenses. In some cases a father's finances may be strained by paying alimony or child support. Finally, separated and divorced men and women often have a changed sexual ethic based on the simple fact that there are few (if any) divorced virgins (Spanier and Thompson, 1987).

Sexual relationships are often an important component in the lives of separated and divorced men and women. Engaging in sexual relations for the first time following separation may help people accept their newly acquired single status. Because sexual fidelity is an important element in marriage, becoming sexually active with someone other than one's ex-spouse is a dramatic symbol that the old marriage vows are no longer valid. Men initially tend to enjoy their sexual freedom following divorce, but women generally do not find it as satisfying as do men. For men, sexual experience following separation is linked with their well-being. Sex seems to reassure men and bolster their self-confidence. Sexual activity is not as strongly connected to women's well-being (Spanier and Thompson, 1987).

Did you know?

About one-third of divorced men and women remarry within a year of divorce (Ganong and Coleman, 1994).

Consequences of Divorce

Most divorces are not contested; between 85 and 90 percent are settled out of court through negotiations between spouses or their lawyers. But divorce, whether it is amicable or not, is a complex legal process involving highly charged feelings about custody, property, and children (who are sometimes treated by angry partners as property to be fought over).

Divorce Song

I thought you were good.
I thought you were like silver. But you are lead.
Now look at me on the mountain top
As I walk through the sun.
I am sunlight.

TSIMSHIAN TRADITIONAL SONG

No-Fault Divorce

Since 1970, beginning with California's Family Law Act, all fifty states have adopted no-fault divorce. **No-fault divorce** is the legal dissolution of a marriage in which guilt or fault by one or both spouses does not have to be established. No-fault divorce has changed four basic aspects of divorce (Weiztman, 1985), described below. Although no-fault divorce has had no effect on divorce rates, it has decreased the time involved in the legal process (Kitson and Morgan, 1991).

First, no-fault divorce has eliminated the idea of fault-based grounds. Under no-fault divorce, no one is accused of desertion, cruelty, adultery, impotence, crime, insanity, or a host of other melodramatic acts or omissions. Neither party is found guilty of anything; rather, the marriage is declared unworkable and is dissolved. Husband and wife must agree that they have irreconcilable differences (which they need not describe) and that they believe it is impossible for their marriage to survive the differences.

Sometimes I wonder if men and women really suit each other. Perhaps they should live next door and just visit now and then.

KATHARINE HEPBURN

PERSPECTIVE

Lesbians, Gay Men, and Divorce

Although there are no reliable studies, it is estimated about one-fifth of gay men and one-third of lesbians have been married. Estimates of bisexual men and women who are married run into the millions (Hill, 1987; Gochoros, 1989). Relatively few gay men, lesbians, and bisexuals are consciously aware of their sexual orientation at the time they marry. Those who are aware rarely disclose their feelings to their prospective partners (Gochros, 1989). When married lesbians and gay men acknowledge their gayness to themselves, they often feel that they are "living a lie" in their marriage. While they may deeply love their spouses, the majority eventually divorce.

How is it that lesbians and gay men marry heterosexuals in the first place? As we saw in Chapter 7, the gay/lesbian identity process is difficult and complex. Because of fear and denial, some gay men and lesbians are unable to acknowledge their sexual feelings. They believe or hope they are heterosexual and do their utmost to suppress their same-sex fantasies or behaviors. They often believe that their homosexuality is just a "phase." Typically they hold negative stereotypes about homosexuality and cannot bring themselves to believe or accept that they might be "one of them." Marriage is one way of convincing themselves they are heterosexual. In addition to "curing" or denying one's gayness, their motivations to marry are no different from heterosexuals (Bozett, 1987). Like heterosexuals, gay men and lesbians marry because of pressure from family, friends, and fiancé, genuine love for one's fiancé, the wish for companionship, and the desire to have children.

When husbands or wives discover their partner's homosexuality or bisexuality they may initially experience shock; others experience temporary relief. Mysteries get explained: why one's spouse disappears for periods of time, why mysterious phone calls occur, the spouse's lack of sexual interest. But whether one is shocked or relieved, inevitably the heterosexual spouse feels deceived or stupid. Many feel shame (Hays and Samuels, 1987). One woman, who felt ashamed to tell anyone of her distress, recalled, "His coming out of the closet in some ways put the family in the closet" (Hill, 1987). At the same time, the gay, lesbian, or bisexual spouse often feels deeply grieved (Voeller, 1980):

> Many people date, marry, and become parents, only to realize too late the error they made. They then find themselves deeply pained, fearful of losing their children through court suits, of losing spouses they care for but are ill suited to, of depriving their spouses and themselves of more deeply appropriate and meaningful relationships, and of causing their friends and other relatives deep pain.

When gay men, lesbians, or bisexuals disclose their orientation to their spouses, separation and divorce is the usual outcome. Many gay men and lesbians are also parents at the time they separate from their spouses. It is generally important for them to affirm their identities both as gay or lesbian *and* as a parent (Bozett, 1989c). This is especially important as negative stereotypes portray gay men and lesbians as "anti-family." Men and women begin to fuse their identities as gay or lesbian with their parental role. A study of gay fathers reported that gay men usually do not reveal their orientation to their children unless the parents are separating or the gay father develops a gay love relationship (Bozett, 1989c). As with divorced fathers in general, gay fathers usually do not have custody of the children, but lesbians, like other divorced women, are more likely to have custody (Bozett, 1989b).

Second, no-fault divorce eliminates the legal adversary process. The stress and strain of the courtroom are eliminated under the new procedure. There is little research, however, to document whether no-fault divorce has been successful in lowering the distress level or the conflict among divorcing couples (Kitson and Morgan, 1991).

Third, the bases for no-fault divorce settlements are equity, equality, and need rather than fault or gender. "Virtue" is no longer financially rewarded, nor is it assumed that women need to be supported by men. Community property is to be divided equally, reflecting the belief that marriage is a partnership with each partner's contributing equally, if differently. The criteria for child custody are based on a sex-neutral standard of the "best interests of the child" rather than on a preference for the mother.

Fourth, no-fault divorce laws are intended to promote gender equality by redefining the responsibilities of husbands and wives. The husband is no longer considered head of the household but is an equal partner with his wife. The husband is no longer solely responsible for support, nor is the wife solely responsible for the care of the children. The limitations placed on alimony assume that a woman will work.

Economic Consequences of Divorce

Probably the most damaging consequences of the no-fault divorce laws is that they systematically impoverish divorced women and their children. Following divorce, women are primarily responsible for both child rearing *and* economic support (Maccoby, Buchanan, Mnookin, and Dornbusch, 1993). As a result, women are at a greater risk for poverty than they were during their marriage. Even if a woman is not plunged into poverty, she often experiences a dramatic downward turn in her economic condition (Garrison, 1994; Morgan, 1991). A single mother's income shows about a 27 percent decline, whereas the income of a divorced man results in a 10 percent decline of his predivorce income (Peterson, 1996; Smock, 1993). Because over half of the children born today will live in a single-parent family at some point during their childhood, through no-fault divorce rules "we are sentencing a significant proportion of the next generation of American children to periods of financial hardship" (Weitzman, 1985).

Sociologist Lenore Weitzman (1985) explains how no-fault divorce's gender-neutral rules, designed to treat men and women equally, actually place older homemakers and mothers of young children at a great economic disadvantage:

> Since a woman's ability to support herself is likely to be impaired during marriage, especially if she is a full-time homemaker and mother, she may not be 'equal' to her former husband at the point of divorce. Rules that treat her as if she is equal simply serve to deprive her of the financial support she needs.

One of the most striking differences between two-parent and single-parent families is poverty. More than a third of middle-income women and a quarter of upper-income women find themselves needing welfare following divorce. The majority of single mothers become poor as a result of their marital disruption. Contributing factors are (1) the mother's low earning capacity, (2) lack of child support, and (3) inadequate welfare benefits (discussed in Chapter 11).

Husbands typically enhance their earning capacity during marriage. In contrast, wives generally decrease their earning capacity because they either quit or limit their participation in the workforce to fulfill family roles. This withdrawal from full participation limits their earning capacity when they reenter the workforce. Divorced homemakers have outdated experience, few skills, and no seniority. Furthermore, they continue to have the major responsibility and burden of child rearing.

When marriage ends, many women must face the triple consequences of gender, ethnic, and age discrimination as they seek to support themselves and their children. Because the workplace favors men in terms of opportunity and income, separation and divorce does not affect them as adversely. (See Chapter 11 for a discussion of economic discrimination against women and ethnic groups.)

About a quarter of divorced women enter a spell of poverty sometime during the first five years following divorce. Whereas the disparities in income between white and African-American women are significant during marriage, following divorce white women suffer a relatively greater decline in their standards of living.

Reflections

As you examine the four premises of no-fault divorce, do you agree with each premise? Why? Would you change or delete any of the premises? Would you add new elements to no-fault?

When the legal system treats men and women equally at the point of divorce, it ignores very real economic inequalities between men and women in our society, inequalities that marriage itself creates.

LENORE WEITZMAN, *The Divorce Revolution*

Did you know?

One out of every two single mothers lives below the poverty level, whereas only one out of ten two-parent families is poor (McLanahan and Booth, 1991).

Following divorce, the income levels of white and African-American women converge (Morgan, 1991). Mexican-American women suffer relatively less decline in economic status than do Anglo-American women because Latinas are already more economically disadvantaged. But because their lives have prepared them for greater economic adversity, Latinas' emotional well-being appears to suffer less than does that of Anglo-American women following divorce (Wagner, 1993).

Employment

The economic impact of divorce on women with children is especially difficult, as their employment opportunities are often constrained by the necessity of caring for children (Maccoby, Buchanan, Mnookin, and Dornbusch, 1993). Child-care costs may consume a third or more of a poor single mother's income. Women may work fewer hours because of the need to care for their children.

Separation and divorce dramatically change many mothers' employment patterns (Morgan, 1991). If a mother was not employed prior to separation, she is likely to seek a job following the splitup. The reason is simple: If she and her children relied on alimony and child support alone, they would soon find themselves on the street. Most employed single mothers are still on the verge of financial disaster, however. On the average, they earn only a third as much as married fathers. This is partly because women tend to earn less than men and partly because they work fewer hours, primarily because of child-care responsibilities (Garfinkel and McLanahan, 1986). The general problems of women's lower earnings capacity and lack of adequate child care are particularly severe for single mothers. Gender discrimination in employment and lack of societal support for child care condemn millions of single mothers and their children to poverty.

Alimony and Child Support

You never realize how short a month is until you've paid alimony.

JOHN BARRYMORE (1882–1942)

Alimony is the money payment a former spouse makes to the other to meet his or her economic needs. Alimony is different from **child support,** which is monetary payments made by the noncustodial spouse to the custodial spouse to assist in child-rearing expenses. For many women, their source of income changes upon divorce from primarily joint wages earned during marriage to their own wages, supplemented by child support payments, alimony, help from relatives, and welfare. The Child Support Enforcement Amendments, passed in 1984, together with the Family Support Act of 1988 requires states to deduct delinquent support from fathers' paychecks, authorizes judges to use their discretion when support agreements cannot be met, and mandates periodic reviews of award levels to keep up with the rate of inflation. In addition, all states implemented automatic wage withholding of child support in 1994. Recent research has shown that enforcement has had a beneficial impact on compliance with child support orders (Meyer and Bartfeld, 1996). Nevertheless, child support awards are historically small, usually amounting to 10 percent of the noncustodial father's income and less than half of a child's expenses.

In 1989, about 16 percent of divorcing white women and 11 percent each of divorcing African-American and Latina women received alimony (U.S. Bureau of the Census, 1994). Child support was awarded to 64 percent of white women, 36 percent of African-American women, and 35 percent of Latina women. Altogether about 56 percent of single mothers were awarded child support. Of this group, 76 percent of the mothers received full payment (U.S. Bureau of the

Census, 1996). Even though many men have the financial resources to support their children and ex-wives, they are less inclined to pay it if they disagree with the child-custody arrangements (Finkel and Roberts, 1994).

Even when fathers pay support, however, the amount is generally low, averaging slightly less than $5,500 annually (U.S. Bureau of the Census, 1996). The amount paid generally depends more on the father's circumstances than on the needs of the mother and children (Teachman, 1991). Alimony and child support payments make up only about 10 percent ($1,246) of the income of white single mothers and 3.5 percent ($322) of the income of African-American single mothers (McLanahan and Booth, 1991). Although some argue that divorced and remarried fathers cannot pay additional support without pushing themselves and their new families into poverty, evidence indicates that this is not true in the majority of cases (Duncan and Hoffman, 1985).

People generally approve, at least in principle, of child support, but alimony is more controversial. In the past, alimony represented the continuation of the husband's responsibility to support his wife. Currently, laws suggest that alimony should be awarded on the basis of need to those women who would otherwise be indigent. Although the courts award alimony in about 15 percent of all divorce cases, a much smaller percent actually receive it. At the same time, there is a strong countermovement in which alimony represents the return of a woman's "investment" in marriage (Oster, 1987; Weitzman, 1985). Weitzman argues that a woman's homemaking and child-care activities must be considered important contributions to her husband's present and future earnings. If divorce rules don't give a wife a share of her husband's enhanced earning capacity, then the "investment" she made in her spouse's future earnings is discounted. According to Weitzman, alimony and child support awards should be made to divorced women in recognition of the wife's primary child-care responsibilities and her contribution to her ex-husband's work or career. Such awards will help raise divorced women and children above the level of poverty to which they have been cast as a result of no-fault divorce's specious equality. A landmark court decision, in fact, upheld the "investment" doctrine by ruling that a woman who supported her husband during his medical education was entitled to a portion of his potential lifetime earnings as a physician (Oster, 1987).

Reflections

Why are alimony and child support often such emotional issues in divorce? On what basis should alimony be awarded? Child support? Why do many noncustodial parents fail to pay child support? What could be done to improve their likelihood of supporting their children?

Children and Divorce

Slightly over half of all divorces involve children. Popular images of divorce depict "broken homes," but it is important to remember that an intact nuclear family, merely because it is intact, does not necessarily offer an advantage to children over a single-parent family or a stepfamily. A traditional family racked with spousal violence, sexual or physical abuse of children, alcoholism, neglect, severe conflict, or psychopathology creates a destructive environment that is likely to inhibit children's healthy development. Living in a two-parent family with marital conflict is often more harmful to children than living in a tranquil single-parent family or step-family. Children living in happy two-parent families appear to be the best adjusted, and those from conflict-ridden two-parent families appear to be the worst adjusted. Children from single-parent families are in the middle. The key to children's adjustment following divorce is a lack of conflict between divorced parents (Kline, Johnston, and Tschann, 1991).

When I can no longer bear to think of the victims of broken homes, I begin to think of the victims of intact ones.

Peter DeVries

Telling children that their parents are separating is one of the most difficult and unhappy events in life. Whether or not the parents are relieved about the separation, they often feel extremely guilty about their children. Children are generally aware of parental discord and are upset by the separation, although their distress may not be apparent.

A meta-analysis (a research technique combining statistical data from previous studies and reanalyzing it) used earlier divorce studies on the impact of parental divorce on the well-being of children in adulthood (Amato and Keith, 1991). The study found very little difference in the well-being of children from divorced families and intact families. A recent study by psychologist Judith Wallerstein found that children from divorced families suffered both emotionally and developmentally ("Study Reveals," 1997). Young children and girls fared the worst.

The Three Stages of Divorce for Children

Growing numbers of studies have appeared on the impact of divorce on children, but these studies frequently contradict one another. Part of the problem is a failure to recognize divorce as a process for children as opposed to a single event. Divorce is a series of events and changes in life circumstances. Many studies focus on only one part of the process and identify that part with divorce itself. Yet at different points in the process, children are confronted with different tasks and adopt different coping strategies. Furthermore, the diversity of children's responses to divorce is the result, in part, of differences in temperament, gender, age, and past experiences.

Children experience divorce as a three-stage process, according to Judith Wallerstein and Joan Kelly (1980b). Studying sixty California families during a five-year period, these researchers found that for children, divorce consisted of initial, transition, and restabilization stages:

- *Initial stage.* The initial stage, following the decision to separate, was extremely stressful; conflict escalated, and unhappiness was endemic. The

Children react differently to divorce, depending on their age. Most feel sad, but the eventual outcome for children depends on many factors, including having a competent and caring custodial parent, siblings, and friends, and their own resiliency. The postdivorce relationship between parents and the custodial parent's economic situation are also important factors.

children's aggressive responses were magnified by the parents' inability to cope because of the crisis in their own lives.

- *Transition stage.* The transition stage began about a year after the separation, when the extreme emotional responses of the children had diminished or disappeared. The period was characterized by restructuring of the family and by economic and social changes: living with only one parent and visiting the other, moving, making new friends and losing old ones, financial stress, and so on. The transition period lasted between two and three years for half the families in the study.
- *Restabilization stage.* Finally came the restabilization stage, which the families had reached by the end of five years. Economic and social changes had been incorporated into daily living. The postdivorce family, usually a single-parent family or stepfamily, had been formed.

Children's Responses to Divorce

A decisive element in children's responses to divorce is their developmental stage (Guttman, 1993). A child's age affects how she or he responds to one parent's leaving home, changes (usually downward) in socioeconomic status, moving from one home to another, transferring schools, making new friends, and so on.

Developmental Tasks of Divorce

Children must undertake six developmental tasks when their parents divorce (Wallerstein, 1983). The first two tasks need to be resolved during the first year. The other tasks may be worked on later; often they may need to be reworked because the issues often recur. How children resolve these tasks differs by age and social development. The tasks are as follows:

1. *Acknowledging parental separation.* Children often feel overwhelmed by feelings of rejection, sadness, anger, and abandonment. They may try to cope with them by denying that their parents are "really" separating. They need to accept their parents' separating and to face their fears.
2. *Disengaging from parental conflicts.* Children need to psychologically distance themselves from their parents' conflicts and problems. They require such distance so that they can continue to function in their everyday activities without being overwhelmed by their parents' crisis.
3. *Resolution of loss.* Children lose not only their familiar parental relationship but also their everyday routines and structures. They need to accept these losses and focus on building new relationships, friends, and routines.
4. *Resolution of anger and self-blame.* Children, especially young ones, often blame themselves for the divorce. They are angry with their parents for disturbing their world. Many often "wish" their parents would divorce, and when their parents actually do, they feel responsible and guilty for "causing" it.
5. *Accepting the finality of divorce.* Children need to realize that their parents will probably not get back together. Younger children hold "fairy-tale" wishes that their parents will reunite and "live happily

ever after." The older the child is, the easier it is for him or her to accept the divorce.

6 *Achieving realistic expectations for later relationship success.* Children need to understand that their parents' divorce does not condemn them to unsuccessful relationships as adults. They are not damaged by witnessing their parents' marriage; they can have fulfilling relationships themselves.

Younger Children

Younger children react to the initial news of a parental breakup in many different ways. Feelings range from guilt to anger and from sorrow to relief, often vacillating among all of these. The most significant factor affecting children's responses to the separation is their age. Preadolescent children, who seem to experience a deep sadness and anxiety about the future, are usually the most upset. Some may regress to immature behavior, wetting their beds or becoming excessively possessive. Most children, regardless of their age, are angry because of the separation. Very young children tend to have more temper tantrums. Slightly older children become aggressive in their play, games, and fantasies—for example, pretending to hit one of their parents.

A recent study using longitudinal data collected over a twelve-year period examines parent-child relationships before and after divorce. Researchers found that marital discord may exacerbate children's behavior problems, making them more difficult to manage (Amato and Booth, 1996). Because discord between parents often preoccupies and distracts them from the tasks of parenting, they appear unavailable and unable to deal with their children's needs. This study reinforced a growing body of evidence showing that many problems assumed to be due to divorce are actually present before marital disruption.

Children of school age may blame one parent and direct their anger toward him or her, believing the other one innocent. But even in these cases the reactions are varied. If the father moves out of the house, the children may blame the mother for making him go, or they may be angry at the father for abandoning them, regardless of reality. Younger schoolchildren who blame the mother often mix anger with placating behavior, fearing she will leave them. Preschool children often blame themselves, feeling that they drove their parents apart by being naughty or messy. They beg their parents to stay, promising to be better. It is heartbreaking to hear a child say, "Mommy, tell Daddy I'll be good. Tell him to come back. I'll be good. He won't be mad at me anymore."

A study of 121 six- to twelve-year-old white children found that about a third initially blamed themselves for their parents' divorce. After a year, the figure dropped to 20 percent (Healy, Stewart, and Copeland, 1993). The largest factor in self-blaming was being caught in the middle of parental conflict. Children who blamed themselves displayed more psychological symptoms and behavior problems than those who did not blame themselves.

When parents separate, children want to know with whom they are going to live. If they feel strong bonds with the parent who leaves, they want to know when they can see him or her. If they have brothers or sisters, they want to know if they will remain with their siblings. They especially want to know what will happen to them if the parent they are living with dies. Will they go to their grandparents, their other parent, an aunt or uncle, or a foster home? These are practical questions, and children have a right to answers. They need to know what lies ahead for them amid the turmoil of a family splitup so that they can prepare for the changes. Some parents report that their children seemed to do better psychologically than they themselves did after a splitup. Children often have more strength and inner resources than parents realize.

The outcome of separation for children, Weiss (1975) observes, depends on several factors related to the children's age. Young children need a competent and loving parent to take care of them; they tend to do poorly when a parenting adult becomes enmeshed in constant turmoil, depression, and worry. With older, preadolescent children, the presence of brothers and sisters helps because the children have others to play with and rely on in addition to the single parent. If they have good friends or do well in school, this contributes to their self-esteem. Regardless of the child's age, it is important that the absent parent continue to play a role in the child's life. The children need to know that they have not been abandoned and that the absent parent still cares (Wallerstein and Kelly, 1980b). They need continuity and security, even if the old parental relationship has radically changed.

Adolescents

Many adolescents find parental separation traumatic. Studies indicate that much of what appears to be negative results of divorce for children (personal changes, parental loss, economic hardships, psychological adjustments) are probably the result of parental conflict that precedes and surrounds the divorce (Amato and Keith; 1991; Morrison and Cherlin, 1995; Amato and Booth, 1996).

Adolescents tend to protect themselves from the conflict preceding separation by distancing themselves. Although they usually experience immense turmoil within, they may outwardly appear cool and detached. Unlike younger children, they rarely blame themselves for the conflict. Rather, they are likely to be angry with both parents, blaming them for upsetting their lives. Adolescents may be particularly bothered by their parents' beginning to date again. Some are shocked to realize that their parents are sexual beings, especially when they see a separated parent kiss someone or bring someone home for the night. The situation may add greater confusion to the adolescents' emerging sexual life. Some may take the attitude that if their mother or father sleeps with a date, why can't they? Others may condemn their parents for acting "immorally."

Reflections

As you look at the tasks for children in divorce, are there others you would add? What ones do you believe are the most important? Most difficult? If you were a divorcing parent, what strategies would you use to help your children adjust to divorce? How would your strategies differ according to the age of the child or adolescent? What do you think the experience might be of adult children whose parents divorce?

Helping Children Adjust

Helen Raschke's (1987) review of the literature on children's adjustment after divorce found that the following factors were important:

- Prior to separation, open discussion with the children about the forthcoming separation and divorce and the problems associated with them.
- The child's continued involvement with the noncustodial parent, including frequent visits and unrestricted access.
- Lack of hostility between the divorced parents.
- Good emotional and psychological adjustment to the divorce on the part of the custodial parent.
- Good parenting skills and the maintenance of an orderly and stable living situation for the children.

Continued involvement with the children by both parents is important for the children's adjustment. The greatest danger is that children may be used as pawns by their parents after a divorce. The recently divorced often suffer from a lack of self-esteem and a sense of failure. One means of dealing with the feelings

A man that studieth revenge keeps his own wounds green.

Francis Bacon (1561–1626)

caused by divorce is to blame the other person. To prevent further hurt or to get revenge, divorced parents may try to control each other through their children. A recent study has shown that children are likely to suffer long-term psychological damage—well into adulthood—if the parents do not consider their emotional needs during the divorce process ("Study Reveals," 1997).

Child Custody

Why is happiness such a precious thing? What have we done with our lives so that everywhere we turn, no matter how hard we try not to, we cause other people sorrow?

WILLIAM STYRON

Of all the issues surrounding separation and divorce, custody issues are generating the "greatest attention and controversy" among researchers (Kitson and Morgan, 1991). When the court awards custody to one parent, the decision is generally based on one of two standards: the *best interests of the child* or the *least detrimental of the available alternatives.* In practice, however, custody of the children is awarded to the mother in about 90 percent of the cases. Three reasons can be given for this: (1) women usually prefer custody, and men do not; (2) custody of the mother is traditional; and (3) the law reflects a bias that assumes women are naturally better able to care for children.

Sexual orientation has also been a traditional basis for awarding custody (Baggett, 1992; Beck and Heinzerling, 1993). In the past, a parent's homosexuality per se has been sufficient grounds for denying custody, but increasingly, courts are determining custody on the basis of parenting ability rather than sexual orientation. Interviews with children whose parents are gay or lesbian testify to the children's acceptance of their parents' orientation without negative consequences (Bozett, 1987).

Types of Custody

The major types of custody are sole, joint, and split. In **sole custody,** the child lives with one parent, who has sole responsibility for physically raising the child and making all decisions regarding his or her upbringing. There are two forms of **joint custody:** legal and physical. In **joint legal custody,** the children live primarily with one parent, but both share jointly in decisions about their children's education, religious training, and general upbringing. In **joint physical custody,** the children live with both parents, dividing time more-or-less equally between the two households. Even though joint custody does not necessarily mean that the child's time is evenly divided between parents, it does give children the chance for a more normal and realistic relationship with each parent (Arnetti and Keith, 1993). Under **split custody,** the children are divided between the two parents; the mother usually takes the girls and the father, the boys. Split custody often has harmful effects on sibling bonds and should be entered into only cautiously (Kaplan, Hennon, and Ade-Ridder, 1993).

Parental satisfaction with custody arrangements depends on many factors (Arditti, 1992; Arditti and Allen, 1992). These include how hostile the divorce was, whether the noncustodial parent perceives visitation as lengthy and frequent enough, and how close the noncustodial parent feels to his or her children. In addition, the amount of support payments also affects satisfaction. If parents feel they are paying too much or were "cheated" in the property settlement, they are also likely to feel that the custody arrangements are unfair. Unfortunately, custodial satisfaction is not necessarily related to the best interests of the child.

The anger and conflict surrounding custody arrangements has given rise to a fathers' rights movement (Coltrane and Hickman, 1992). The fathers' rights movement depicts its participants as caring fathers who want equal treatment regarding child custody, visitation, and support (Bertoia and Drakich, 1993). Given the nature of changing gender roles and the reality of economic hardships, more mothers are relinquishing their children to the fathers. This trend of fathers seeking and gaining custody of their children comes in spite of many judges' traditional attitudes about gender and established child-care patterns. Research concerning the effects of a father's custody on the psychological well-being of children reveals no conclusive evidence to preclude or prefer it (Rosenthal and Keshert, 1980). The chances of a father gaining custody are improved when the children are older at the time of the divorce, the oldest is male, and when the father is the plaintiff in the divorce (Fox and Kelly, 1995). Regardless of who gets custody, however, it is important for children, if possible, to maintain close ties with both parents following a divorce (Howell, Brown, and Eichenberger, 1992).

Man is not the enemy here, but the fellow victim.

Betty Friedan

Sole Custody

Sole custody accounts for 85 percent of cases in the United States for several reasons. First, because women have traditionally been responsible for child rearing, sole custody by mothers has seemed the closest approximation to the traditional family, especially if the father is given free access. Second, many men have not had the day-to-day responsibilities of child rearing and do not feel (or are not perceived to be) competent in that role.

Sole custody does not mean that the noncustodial parent is prohibited from seeing his or her children. Wallerstein and Kelly (1980b) believe that if one parent is prohibited from sharing important aspects of the children's lives, he or she will withdraw from the children in frustration and grief. Children experience such withdrawal as a rejection and suffer as a result. Generally, it is considered in the best interests of the child for him or her to have easy access to the noncustodial parent. Changes in the noncustodial parent's relationship with his or her children may be related to the difficulties and psychological conflicts arising from visitation and divorce, the noncustodial parent's ability to deal with the limitations of the visiting relationship, and the age and gender of the child (Wallerstein and Kelly, 1980a).

Joint Custody

Joint custody, in which both parents continue to share legal rights and responsibilities as parents, accounts for about 10 percent of cases.

A number of advantages accrue to joint custody. First, it allows both parents to continue their parenting roles. Second, it avoids a sudden termination of a child's relationship with one of his or her parents. Joint custody fathers tend to be more involved with their children; they spend time with them and share responsibility and decision making (Bowman and Ahrons, 1985). Third, dividing the labor lessens many of the burdens of constant child care experienced by most single parents.

Joint custody, however, requires considerable energy from the parents in working out both the logistics of the arrangement and their feelings about each other. Many parents with joint custody find it difficult, but they nevertheless feel that it is satisfactory. The children do not always like joint custody as much as the parents do. In actual practice, children relatively rarely split their time evenly between parents (Little, 1992).

Reflections

What form of custody do you believe is the most advantageous to a child? What factors would you consider important in deciding which is the best type of custody for a particular child? If two parents constantly battled over their children, what are some of the consequences you might expect for the children? How do children cope in such circumstances?

OTHER PLACES OTHER TIMES

Cross-Cultural Issues in Parenting: Who Gets the Children?

Foster parent, biological parent, adoptive parent, genetic parent, stepparent: Who is the "real" parent? Who has the greatest rights in relation to a child, and who has the most responsibilities toward him or her? Our answers to these questions are deeply conditioned by our own kinship system, which is a product of our own cultural history. We tend to believe that real parents are established by "nature." We believe that if we can learn a child's genetic or biological makeup, we can identify his or her real parents. We also believe that parent-child emotional bonds follow naturally from the biological tie. A child's relationship to his or her biological or genetic mother—"the mother-child bond"—is believed to be the most nurturant, basic, and (some would say) even mystical parental bond. A biological father's bond with his child seems more distant, though protective, and this is somehow also established by nature.

Of course, these cultural beliefs—as anthropologists argue—leave all foster parents, adoptive parents, and stepparents in a secondary category, with lesser claims to their children. Their parental rights are wrestled out in courtrooms across the country.

Kinship Systems

Perhaps the best way to appreciate the cultural character of our assumptions about parents and their relationships with their children is to look at other societies in which these roles and relationships have other meanings. Anthropologists who study kinship systems tell us that families in other societies are shaped by very different understandings of who is related to whom—and which relationships are most important. Learned in families as children grow up, different kinship systems become the natural way of seeing relationships. In each society, the kinship system comes to exist in people's minds as a kind of blueprint or technical manual that instructs people on the different meanings of genealogical relationships. Even words as basic as *mother* and *father* do not have the same meanings the whole world over.

Anthropologists give kinship systems descriptive names. Some are called *patrilineal* because—as was discussed in Chapter 1—the most important relationships are those traced on the father's side of the family, not the mother's. For example, paternal grandparents in these societies play a greater role in the lives of their grandchildren than do maternal grandparents; they have greater say in their grandchildren's futures and a greater claim on their labor and loyalties. In patrilineal societies, property rights and titles are passed down to children from their male relatives on the father's side of the family. Children inherit membership in their father's clan or lineage. They take their family identity and name from their fathers and paternal grandfathers—who took theirs from their own paternal grandfathers and great-grandfathers. Although Bedouin, Chinese, Masai, and Nuer societies have their own distinctive cultures, all these societies are called *patrilineal,* which describes their most important kinship relations.

Patrilineal Kinship

The fact that a particular society has patrilineal kinship also gives strong clues about people's ideas on parenting. Cross-culturally, there is great diversity in who counts as the more significant parent—and which other relatives have important roles in the "parenting" of children. Just knowing that a particular society is patrilineal does not tell us everything about parenting. (It doesn't tell us what work mothers and fathers do in that society, for example.) But it does tell us something basic about a child's identity. At birth, a child is assigned to the father's group—the father's family, descent line, lineage, or clan. This, in turn, tells us which parent and group has greater rights in the child.

In patrilineal societies, one important repercussion of the primary identity with father and his group can be seen on divorce: Children remain with their father and his relatives. They may be raised by another "mother," who is often a female member of the father's family (a paternal aunt or grandmother), or they may be "mothered" by another of the father's wives. Children will not leave their father to go with their mother when she returns to live with her own family after divorce. (In some societies, very young children might accompany their mother until after weaning or

Any custody arrangement has both benefits and drawbacks, and joint custody is no exception. Sometimes, although it may be in the best interests of the parents for each of them to continue parenting roles, it may not necessarily be in the best interests of the child. For parents who choose joint custody, it appears to be a satisfactory arrangement. But when joint custody is mandated by the courts later; then these children are expected to return to their rightful and natural home with their father.) Thus, in patrilineal societies, children's basic identity is established at birth, and it links them to their father and his group, descent line, family, and relatives. Divorce has serious implications for a woman as mother in patrilineal societies: She surrenders her children to their father, who has greater rights in them. A father in these societies is the more fundamental—and even the more "natural"—parent in the lives of children.

Matrilineal Kinship

In this book we have looked at several patrilineal societies. Let's now turn to consider a matrilineal one. Traditional Iroquois society in nineteenth-century New York has been described by anthropologists as *matrilineal.* This means that the female descent line was dominant in the kinship system. The mother's side of the family counted more, and her relatives (both male and female) were more significant in the lives of her children. Important aspects of identity—clan membership and chiefly titles—were inherited through the female descent line. Residence after marriage was the reverse of the situation in most patrilineal societies: In Iroquois society, husbands moved to live with the wife in her family's house, the Iroquois longhouse. That house included the wife's own extended family—her sisters and their husbands and children, the wife's mother and her husband, perhaps the wife's mother's sister and her husband, and the wife's maternal grandmother. Thus, in Iroquois society, the husband was the newcomer and the outsider. In his wife's longhouse, he came under the scrutiny of her family. They were concerned that he be a good worker. A husband helped clear land for his wife's family to farm, and he provided them with meat from his hunting ventures. In addition, a husband was expected to father children to continue their mother's descent line. Thus, in Iroquois society, a man's children continued his wife's family, not his own.

Couples divorced when a husband failed in one of these capacities or if the couple could not get along. A husband's belongings set outside the longhouse signaled that the marriage was over. When this happened, it was the husband who moved; he returned to the longhouse belonging to his own mother and sisters. His children remained behind with his wife, to whose family they belonged.

In matrilineal societies, fathers were important in the parenting of their children. However, maternal uncles (the mother's brothers) were often the more significant male relatives. Maternal uncles were members of the same lineage group as their sisters and their children. Whether a sister divorced or remained married to her husband, her brother assumed an active role in raising and overseeing her children, and he passed on family titles to her sons. Thus, children took their primary identity from the mother and her family. The mere fact that a society is matrilineal thus tells us that relatives on the mother's side of the family are more significant in the lives of children. (Other examples of matrilineal societies include the Pueblo Indians of the U.S. Southwest and Trobriand Islanders of the South Pacific.)

Patrilineal/Matrilineal Kinship

On marriage in traditional Samoan society, the newly married couple could choose to live in either the husband's or the wife's village, with relatives of either side, in a large extended family household. Each person traced kinship on both the mother's and father's side of the family, through blood relationship, marriage, and adoption. Residence was a matter of choice, and people could choose fairly easily to go and live with other relatives on the other side of the family, in another village. (The head of the household had to agree, and all members helped farm the family fields.) There were no exclusive clans or lineages that tied a person to a particular family, household, or locale. In fact, each person could trace kinship to at least several different households in different villages, and thus had several alternative "homes." People could choose to go and live with any of these relatives. According to anthropologist Margaret Mead, Samoans changed homes with relative ease. However, it was not just married couples or single adults who enjoyed this flexibility; children did, too. A child could choose to leave the house where his or her parents lived to go and live with an aunt or uncle, even in another village. There, a child called female relatives by the same term used for "mother;" male relatives were called by the term for "father." These surrogate parents exercised authority over the child, supervising and educating him or her. Hence, children in Samoan society were sometimes raised by multiple sets of parents. This, according to Mead, meant that children could avoid intense conflict with any one set of parents. Diversity in kinship systems in cultures around the world has thus shaped diverse experiences in parenting.

over the opposition of one or both parents, it may be problematic. Joint custody may force two parents to interact (*cooperate* is too benign a word) when they would rather never see each other again, and the resulting conflict and ill will may end up being detrimental to the children. Parental hostility may make joint custody the worst form of custody (Opie, 1993).

Noncustodial Parents

Only recently is research emerging about noncustodial parents. Popular images of noncustodial parents depict them as absent and noncaring, but a more accurate picture depicts varying degrees of involvement (Bray and Depner, 1993; Depner and Bray, 1993). Noncustodial parent involvement exists on a continuum in terms of caregiving, decision making, and parent-child interaction. Involvement also changes depending on whether the custodial family is a single-parent family or a stepfamily (Bray and Berger, 1993).

Researchers know very little about noncustodial and nonresidential fathers (Depner and Bray, 1990). They know even less about noncustodial mothers, who account for about 13 percent of noncustodial parents (Christensen, Dahl, and Rettig, 1990). What we do know about men, however, tells us that they often suffer grievously from the disruption or disappearance of their father roles following divorce. They feel depressed, anxious, and guilt ridden; they feel a lack of self-esteem (Arditti, 1990). The change in status from full-time father to noncustodial parent leaves fathers bewildered about how they are to act; there are no norms for an involved noncustodial parent. This lack of norms makes it especially difficult if the relationship between the former spouses is bitter. Without adequate norms, fathers may become "Disneyland Dads," who interact with their children only during weekends, when they provide treats such as movies and pizza, or they may become "Disappearing Dads," absenting themselves from any contact at all with their children. For many concerned noncustodial fathers, the question is simple but painful: "How can I be a father if I'm not a father anymore?"

Noncustodial fathers often weigh the costs of continued involvement with their children, such as emotional pain and role confusion, against the benefits, such as emotional bonding (Braver et al., 1993a, 1993b). Those fathers who maintain

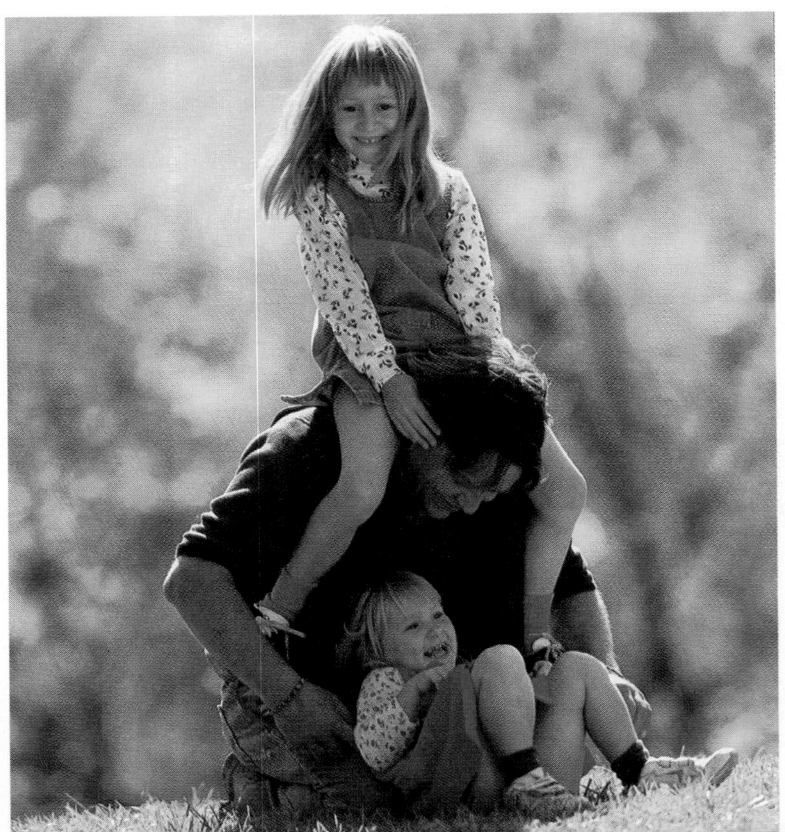

It is usually important for a child's postdivorce adjustment that he or she have continuing contact with the noncustodial parent. Noncustodial parents are involved with their children in varying degrees.

their connections are generally older and remarried; they have little or no conflict with their ex-spouses and no significant problems with their children (Wall, 1992). For others, however, the costs outweigh the benefits. They are not successful in being noncustodial fathers and abandon the role altogether. A study of noncustodial parents in a support group found that common themes included children rejecting parents and parents rejecting children (Greif and Kristall, 1993).

Children tend to have little contact with the nonresidential parent. A national survey revealed that over 60 percent of fathers did not visit their children or have contact with them over a one-year period (Bianchi, 1990).

The reduced contact between nonresidential fathers and children seems to weaken the bonds of affection. A study of eighteen- to twenty-two-year-olds whose parents were divorced found that almost two-thirds had poor relationships with their fathers, and one-third had poor relationships with their mothers—about twice the rate of a comparable group from nondivorced families (Zill, Morrison, and Coiro, 1993).

Divorced fathers are less likely to consider their children as sources of support in times of need (Amato, 1994; Cooney, 1994). Furstenberg and Nord (1982) conclude that "marital dissolution involves either a complete cessation of contact between the nonresidential parent and child or a relationship that is tantamount to a ritual form of parenthood."

Custody Disputes and Child Stealing

As many as one-third of all postdivorce legal cases involve children. Vagueness of the "best interests" and "least detrimental alternative" standards by which parents are awarded custody may encourage custody fights by making the outcome of custody hearings uncertain and increasing hostility. Any derogatory evidence or suspicions, ranging from dirty faces to child abuse, may be considered relevant evidence. As a result, child custody disputes are fairly common in the courts. They are often quite nasty.

As discussed in Chapter 13, about 350,000 children are abducted each year by family members in child custody disputes. Most are returned in two days to a week, and generally the parent from whom the child was taken knows the child's whereabouts (Hotaling, Finkelhor, and Sedlak, 1990; Finkelhor, Hotaling, and Sedlak, 1991). Research on the consequences for children of custody abductions is not reliable because much of it relies on parental impressions and on criminal and clinical populations (Greif and Hegar, 1992). According to researcher David Finkelhor, the number of family abductions could be reduced significantly. Under the present system, courts don't respond to people's needs and fears in bitter custody disputes. "For many people, going into court for custody is too risky, too expensive, and too time-consuming," Finkelhor notes (quoted in Barden, 1990), "so they grab the child." Finkelhor suggests that much child stealing could be prevented by the court's assigning a mediator to whom the distressed parents could turn in times of crisis. Mediation can ease parental anxiety and offer an alternative to legal proceedings, which tend to inflame the situation.

When Jason and the Argonauts returned from their search for the Golden Fleece, he had taken a new wife. His wife Medea was enraged with grief, anger, and jealousy. A sorceress, she wove a golden cloth which she gave to Jason's bride. When she wore it, it turned to flames and burned her to death. Then Medea killed her own two children to take revenge upon the unfaithful Jason and fled the palace in a chariot pulled by dragons.

GREEK MYTH

Divorce Mediation

The courts are supposed to act in the best interests of the child, but they often victimize children by their emphasis on legal criteria rather than on the children's psychological well-being and emotional development (Schwartz, 1994). There is

You never really know a man until you've divorced him.

ZSA ZSA GABOR

increasing support for the idea that children are better served by those with psychological training than by those with legal backgrounds (Miller, 1993). Growing concern about the impact of litigation on children's well-being has led to the development of divorce mediation as an alternative to legal proceedings (Walker, 1993).

Divorce mediation is the process in which a mediator attempts to assist divorcing couples in resolving personal, legal, and parenting issues in a cooperative manner. Over two-thirds of the states offer or require mediation through the courts over such legal issues as custody and visitation. Mediators act as facilitators to help couples arrive at mutually agreed-upon solutions. Mediators can either be private or court ordered. Mediators generally come from marriage counseling, family therapy, and social work backgrounds, though increasing numbers are coming from other backgrounds and are seeking training in a divorce mediation (DeWitt, 1994).

Mediation has many different goals. A primary goal is to encourage divorcing parents to see shared parenting as a viable alternative and to reduce anxiety about shared parenting (Kruk, 1993). Mediators try to help couples develop communication skills to negotiate with each other. They help them clarify their personal, relationship, and parental goals. They suggest ways to minimize conflict. Mediators try to help parents determine whether their demands are based on their anger or on the best interests of their children. They may have parents role-play how their children feel. Mediators can assist parents to develop strategies for helping their children in postdivorce adjustment (Bonney, 1993).

When mediation is court mandated, topics are generally limited to custody and visitation issues. If the mediator is unsuccessful in getting the couple to cooperate, he or she becomes an arbitrator who makes decisions for the couple. The mediator role shifts from facilitator to decision maker. If a couple is unable to negotiate reasonable visitations during the summer, for example, the mediator will decide on a solution for them. As arbitrator, the mediator's decisions are accepted by the court.

Divorcing parents often find mediation helpful for resolving visitation and custody issues. In contrast to court settings, mediation provides an informal setting to work out volatile issues. Men and women both report that mediation is more successful at validating their perceptions and feelings than is litigation. Furthermore, women, the poor, and those from ethnic groups are less likely to experience bias in mediation than in a courtroom setting (Rosenberg, 1992).

Some courts order parents to participate in seminars covering the children's experience of divorce as well as problem solving and building co-parent relationships (Petersen and Steinman, 1994). Parents report that these seminars help them become more aware of their children's reactions and give them more options for resolving child-related disputes.

Divorcing parents also report that mediation helps decrease behavioral problems in their children (Slater, Shaw, and Duquesnel, 1992). If parents can work through their differences apart from their children, their children are less likely to react to the anger and fear they might otherwise observe.

A study of fathers a year after divorce found those who mediated more satisfied with custody than those who litigated. They were also more likely to comply with child support (Emery, Matthews, and Kitzmann, 1994).

It is important, however, not to replace unrealistic images of "conflict-ridden postdivorce parenting" with equally unrealistic pictures of "happy-ever-after-postdivorce-parenting-thanks-to-mediation" (Walker, 1993). The stresses and conflicts of divorce are real and painful. Sometimes co-parenting can't work because of the personalities of the divorcing parents. But mediation is an impor-

Reflections

If you were divorcing, what would be the pros and cons of entering divorce mediation? What would you personally do? Why?

tant step forward in involving parents with therapists rather than with lawyers and courts to resolve difficult family matters.

There is no denying that separation or divorce is filled with pain for everyone involved—husband, wife, and children. But it is not only an end; it is also a beginning of a new life for each person. This new life includes new relationships and possibilities, new families with unique relationships: the single-parent family or the stepfamily. We explore these family forms in the next chapter.

Summary

- Divorce is an integral part of the contemporary American marriage system, which values individualism and emotional gratification. Divorce serves as a recycling mechanism, giving people a chance to improve their marital situations by marrying again. The divorce rate increased significantly in the 1960s but leveled off in the early 1990s. About half of all current marriages end in divorce.
- Researchers are increasingly viewing divorce as a normal part of the family life cycle rather than as a form of deviance. Divorce creates the single-parent family, remarriage, and the stepfamily.
- A variety of factors can affect the likelihood of divorce. Societal factors inlude the changed nature of the family, social integration, and individualistic cultural values. Demographic factors include socioeconomic status, employment status, income, educational level, ethnicity, and religion. Life course factors are intergenerational transmission, age at time of marriage, premarital pregnancy and childbearing, and remarriage. The most important factors may be family processes: marital happiness, presence of children (in some cases), and marital problems.
- Divorce can be viewed as a process involving six "stations" or processes: emotional, legal, economic, co-parental, community, and psychic. As people divorce, they undergo these "divorces" more or less simultaneously. The intensity level of these different stages varies at different times.
- Uncoupling is the process by which couples drift apart in predictable stages. Initially, uncoupling is unilateral; the initiator begins to turn elsewhere for satisfactions, creating an identity independent of the couple. The initiator voices more complaints and begins to think of alternatives. Eventually the initiator ends the relationship. Uncoupling ends when both partners acknowledge that the relationship cannot be saved.
- In establishing a new identity, newly separated people go through transition and recovery. Transition begins with the separation and is characterized by *separation distress.* Separation distress is usually followed by loneliness. Separation distress is affected by (1) whether the person had any forewarning of the separation, (2) the length of time married, (3) who took the initiative in leaving, (4) whether someone new is found, and (5) available resources. The more personal, social, and financial resources a person has at the time of separation, the easier the separation generally will be.
- Dating is important for separated or divorced people. Their greatest social problem is meeting other unmarried people. Dating is a formal statement of the end of a marriage; it also permits individuals to enhance their self-esteem.
- *No-fault divorce* has revolutionized divorce by eliminating fault finding and the adversarial process and by treating husbands and wives as equals. The most damaging unintended consequence of no-fault divorce is the growing poverty of divorced women with children.
- Women generally experience dramatic downward mobility after divorce. The economic consequences of divorce include the impoverishment of women, changed female employment patterns, and very limited *child support* and *alimony.*
- Children in the divorce process go through three stages: (1) the initial stage, lasting about a year, when turmoil is greatest; (2) the transition stage, lasting up to several years, in which adjustments are being made to new family arrangements, living and economic conditions, friends, and social environment; and (3) the restabilization stage, when the changes have been integrated into the children's lives. Children must undertake six developmental tasks when their parents divorce: (1) acknowledging parental separation, (2) disengaging from parental

conflicts, (3) resolving loss, (4) resolving anger and self-blame, (5) accepting the finality of divorce, and (6) achieving realistic expectations for later relationship success.

- A significant factor affecting the responses of children to divorce is their age. Young children tend to act out and blame themselves, whereas adolescents tend to remain aloof and angry at both parents for disrupting their lives. Adolescents may be bothered by their parents' dating again. Many problems assumed to be due to divorce are actually present before marital disruption.
- Factors affecting a child's adjustment to divorce include: (1) open discussion prior to divorce, (2) continued involvement with noncustodial parent, (3) lack of hostility between divorced parents, (4) good psychological adjustment to divorce by custodial parent, and (5) stable living situation and good parenting skills. Continued involvement with the children by both parents is important for the children's adjustment.
- Custody is generally based on one of two standards: the best interests of the child or the least detrimental of the available alternatives. The major types of *custody* are *sole*, *joint*, and *split*. Custody is generally awarded to the mother. Joint custody has become more popular because men are becoming increasingly involved in parenting.
- Noncustodial parent involvement exists on a continuum from absent to intimately and regularly involved. Noncustodial parents often feel deeply grieved about the loss of their normal parenting role. Children tend to have little contact with nonresidential parents.
- As a result of custody disputes, as many as 350,000 children are stolen from custodial parents each year. Most are returned home within a week. *Divorce mediation* is the process in which a mediator attempts to assist divorcing couples in resolving personal, legal, and parenting issues in a cooperative manner. A primary goal of mediation is to encourage divorcing parents to see shared parenting as a viable alternative, to ease parental anxiety, and to reduce custody-related abductions.

Key Terms

alimony 506
child support 506
divorce mediation 518
joint physical custody 512
joint legal custody 512
joint custody 512
no-fault divorce 503
separation distress 501
social integration 491
sole custody 512
split custody 512

Suggested Readings

Ahrons, Constance. *The Good Divorce: Raising Your Family Together When Your Marriage Comes Apart.* New York: HarperCollins, 1994. An easy read based on the author's "Binuclear Family Study" of family functions after divorce.

Ardell, Terry. *Men and Divorce.* Thousand Oaks, CA: Sage, 1995. A look at the legal, economic, and social consequences of divorce, based on interviews with seventy-five men.

———. *Mothers and Divorce: Legal, Economic, and Social Dilemmas.* Berkeley, CA: University of California Press, 1987. A well-written examination of the lives of sixty divorced women that looks at the legal, economic, and social consequences of divorce.

Clapp, Genevieve. *Divorce and New Beginnings.* New York: Wiley, 1992. A useful handbook based on research for coping with divorce, single parenting, and stepparenting.

Depner, Charlene, and James Bray, eds. *Noncustodial Parenting: New Vistas in Family Living.* Newbury Park, CA: Sage Publications, 1993. A collection of essays revealing the diverse patterns of nonresidential parenting following divorce, including multicultural perspectives, mothers as nonresidential parents, and children who refuse visitation.

Furstenberg, Frank, and Andrew Cherlin. *Divided Families.* Cambridge, MA: Harvard University Press, 1991. A concise overview of marital disintegration, economic consequences of divorce, and children's adjustment.

Gottman, John M. *What Predicts Divorce? The Relationship between Marital Processes and Marital Outcomes.* Hillsdale, NJ: Erlbaum, 1994. An examination by a leading marital scholar of why some marriages fail and others thrive.

Guttmann, Joseph. *Divorce in Psychosocial Perspective: Theory and Research.* Hillsdale, NJ: Erlbaum, 1993. An excellent work integrating current knowledge.

Irving, Howard, and Michael Benjamin. *Family Mediation; Contemporary Issues.* Thousand Oaks, CA: Sage, 1995. A broad-based look at mediation that succinctly addresses such issues as diversity of culture, the scope of feminist thought, and gender issues.

Kayser, Karen. *When Love Dies: The Process of Marital Disaffection.* New York: Guilford Press, 1993. A social psychological description of the gradual process of emotional estrangement from one's partner—which may or may not end in divorce. In-depth interviews add poignancy to the work.

Morgan, Leslie. *After Marriage Ends: Economic Consequences for Midlife Women.* Newbury Park, CA: Sage Publications, 1991. A study of the economic impact of separation, divorce, widowhood, and remarriage on a group of women as they undergo these transitions over time.

Simons, Ronald. *Understanding Differences Between Divorced and Intact Families: Stress, Interaction, and Child Outcome.* Thousand Oaks, CA: Sage, 1996. An illustration of the special stresses both divorced and intact families suffer and the impact of divorce on children, based on two large-scale studies of Midwest families

Stevenson, M.R., and K.N. Black. *How Divorce Affects Offspring: A Research Approach.* Madison, WI: Brown and Benchmark, 1995. A review of recent literature regarding the subject of children and divorce.

Vaughan, Diane. *Uncoupling: Turning Points in Intimate Relationships.* New York: Oxford University Press, 1986. A sociological examination of the process by which couples become ex-couples.

Wallerstein, Judith, and Sandra Blakeslee. *Second Chances: Men, Women, and Children a Decade after Divorce.* New York: Ticknor & Fields, 1989. A bestselling book describing the impact of divorce on families written by one of the leading psychologists in the field.

CHAPTER 15

New Beginnings: Single-Parent Families and Stepfamilies

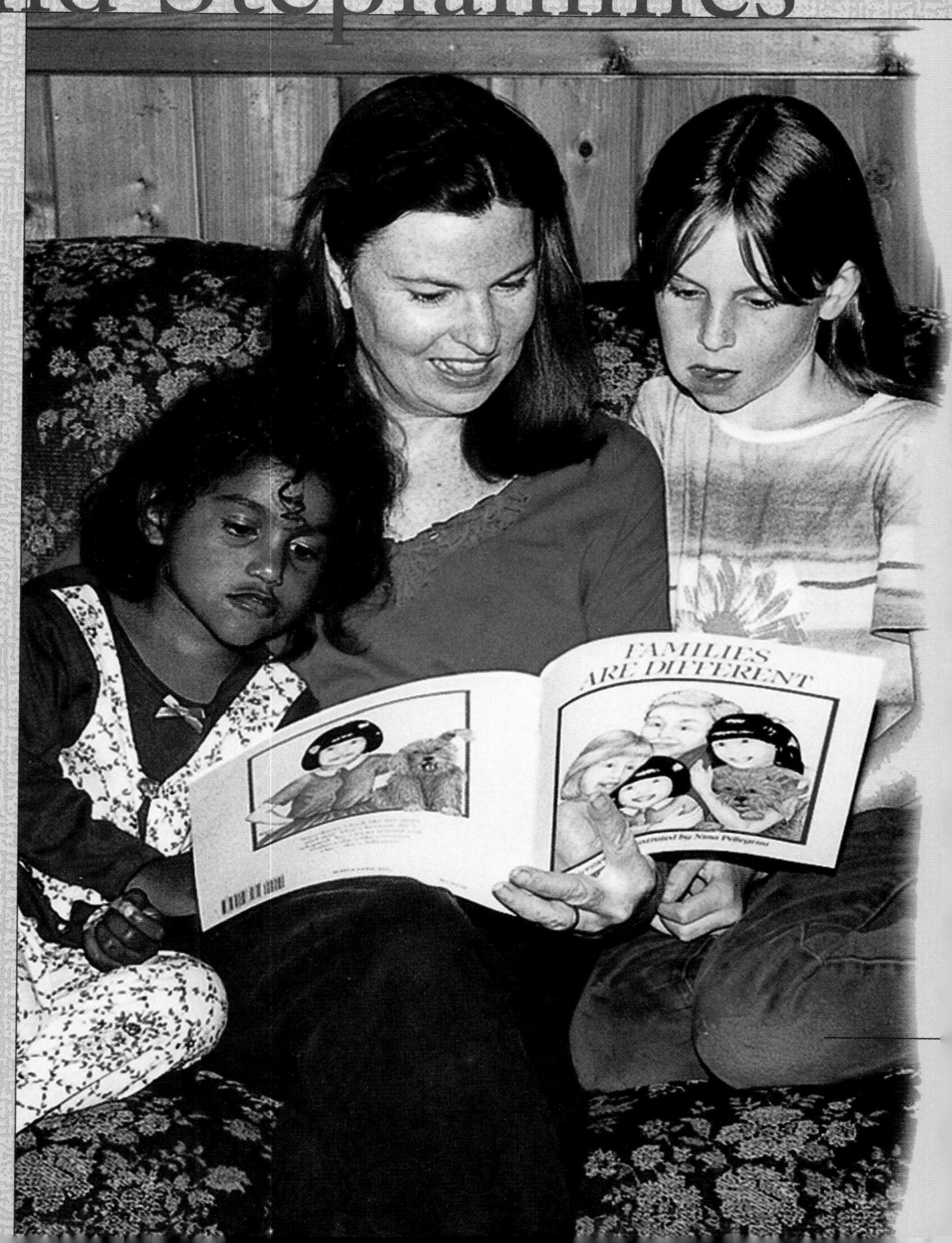

P R E V I E W

To gain a sense of what you already know about the material covered in this chapter, answer "True" or "False" to the statements below.

1. Researchers are increasingly viewing stepfamilies as normal families. True or false?
2. Divorce does not end families. True or false?
3. Shared parenting tends to be the strongest tie holding former spouses together. True or false?
4. Second marriages are significantly happier than first marriages. True or false?
5. Most stepfamilies feel that they have become true families. True or false?
6. Children tend to have greater power in single-parent families than in traditional nuclear families. True or false?
7. Becoming a stepfamily is a process. True or False?
8. Stepmothers generally experience less stress in stepfamilies than stepfathers because stepmothers are able to fulfill themselves by nurturing their stepchildren. True or false?
9. Researchers are increasingly finding that remarried families and intact nuclear families are similar to each other in many important ways. True or false?
10. People who remarry and those who marry for the first time tend to have similar expectations. True or false?

Answers

1 True, 2 True, 3 True, 4 False, 5 True, 6 True, 7 True, 8 False, 9 True, 10 False

OUTLINE

The mid-1990s mark a definitive shift from the traditional family system, based on lifetime marriage and the intact nuclear family, to a pluralistic one, including families created by divorce, remarriage, and births to single women. This new pluralistic family system consists of three major types of families: (1) intact nuclear families, (2) single-parent families (either never married or formerly married), and (3) stepfamilies. **Single-parent families** are families consisting of one parent and one or more children; the parent can be either divorced or never married. **Stepfamilies** are families in which one or both partners have children from a previous marriage or relationship. Stepfamilies are sometimes referred to as **blended families.**

The dominance of the new system is attested to by the following facts (Bumpass, Sweet, and Martin, 1990; Coleman and Ganong, 1991; Demo and Acock, 1991; Dainton, 1993; Ganong and Coleman, 1994; U.S. Bureau of the Census, 1996).

- The chances are more than two out of three that an individual will divorce, remarry, or live in a single-parent family or stepfamily as a child or parent sometime during his or her life.
- **Remarriage,** a marriage in which one or both partners have been previously married, is as common as first marriage. Half of all recent marriages involve at least one previously married partner, and one out of ten marriages is a third marriage for one or both partners.
- Nearly one-fourth of all families are currently single-parent families. Single-parent families are growing faster in number than any other family form. Approximately half of all children born in the 1990s will live in single-parent families sometime during their childhoods.
- Over 2.3 million households have stepchildren living in them; the number of noncustodial stepfamilies without children living with them is significantly higher. One-sixth of

> all children are currently members of stepfamilies. Over a third of all children can expect to live with a biological parent and a stepparent at some time during their childhoods.

To better understand this evolving pluralistic family system, in this chapter we examine single-parent families, binuclear families, remarriage, and stepfamilies. Because of this shift to a pluralistic family system, researchers are beginning to reevaluate single-parent families and stepfamilies and to view them as normal rather than deviant family forms (Coleman and Ganong, 1991; Pasley and Ihinger-Tallman, 1987). It is useful to see these families as different structures pursuing the same goals as traditional nuclear families: the provision of intimacy, economic cooperation, the socialization of children, and the assignment of social roles and status.

If we shift our perspective from structure to function, the important question is no longer whether a particular family form is deviant. (If incidence of a family form determines deviance, logic tells us that the traditional nuclear family may soon become deviant.) The important question becomes whether a specific family—regardless of whether it is a traditional family, a single-parent family, or a stepfamily—succeeds in performing its functions. In a practical sense, as long as a family is fulfilling its functions, it is a normal family.

The dramatic rise in single-parent families and stepfamilies over the last twenty-five years is the result of shifting social values and trends rather than individual shortcomings or pathologies. Single-parent families and stepfamilies have become a natural part of the contemporary American family system. As such, they are not problems in themselves. Instead, to a great extent, many of their problems lie in the stigma attached to them and their lack of support by the larger society (Ahrons and Rodgers, 1987; Gongla and Thompson, 1987). If we are going to strengthen these families, note Constance Ahrons and Roy Rodgers (1987), "We must unambiguously acknowledge and support them as normal, prevalent family types that have resulted from major societal trends and changes."

Reflections

What effect does it have on your views of single-parent families and stepfamilies to think of them as "normal" families? As "abnormal" or "deviant" families? If you were reared in a single-parent family or stepfamily, did your friends, relatives, schools, and religious groups treat your family as normal? Why?

Unmarried adolescent mothers are empowered to build successful families when they have emotional and financial support from their families, educational and employment opportunities, and child care.

Single-Parent Families

Throughout the world, single-parent families are increasing in number (Burns and Scott, 1994). In the United States, they are the fastest-growing family form; no other family type has increased in number as rapidly. Yet single-parent families are treated negatively in the popular imagination (Kissman and Allen, 1993). They are negated as "broken homes" or as headed by "welfare queens" who "breed" children "out of wedlock" only to collect benefits. Neither image is true. The "broken home" image is based on the myth of the "happy" traditional family, and the "welfare queen" mythology is based on a mixture of racism and moralism condemning women for bearing children outside of marriage (as discussed in Chapter 11).

Did you know?

In 1983, a child in a single-parent family was twice as likely to be living with a divorced parent as with a never-married parent. In 1995, the odds were almost equal (U.S. Bureau of the Census, 1996).

Between 1970 and 1995 the percentage of children living in single-parent-families nearly doubled, increasing from 13 percent to 22 percent (U.S. Bureau of the Census, 1996). The single-parent family is a more significant departure from the traditional nuclear family than is the dual-worker family or the stepfamily in three important ways. First, both the dual-worker family and the stepfamily are two-parent families, whereas the single parent family is not. Second, single-parent families are generally headed by women and are thus more vulnerable to poverty. Third, the mother may be unmarried.

In previous generations, the life pattern most women experienced was (1) marriage, (2) motherhood, and (3) widowhood. Single-parent families existed in the past, but they were formed by widowhood rather than divorce or births to unmarried women; significant numbers were headed by men. But a new marriage and family pattern has taken root. Its greatest impact has been on women and their children. Divorce and births to unmarried mothers are key factors creating today's single-parent family.

The life pattern many married women today experience is (1) marriage, (2) motherhood, (3) divorce, (4) single parenting, (5) remarriage, and (6) widowhood. For those who are not married at the time of their child's birth, the pattern may be (1) dating/cohabitation, (2) motherhood, (3) single parenting, (4) marriage, and (5) widowhood.

Characteristics of Single-Parent Families

Single-parent families share a number of characteristics, including the following: (1) creation by divorce or births to unmarried women, (2) usually female headed, (3) significance of ethnicity, (4) poverty, (5) diversity, and (6) transitional character. In addition, some single-parent families are created intentionally through planned pregnancy, artificial insemination, and adoption. Others are headed by lesbians and gay men (Miller, 1992).

Creation by Divorce or Births to Unmarried Women

Single-parent families today are usually created by marital separation, divorce, or births to unmarried women rather than by widowhood. Throughout the world, including the United States, single-parent families created through births to unmarried women are increasing at a higher rate than are single-parent families created through divorce (Burns and Scott, 1994). In 1992, 30 percent of all births were to unmarried women. The number of children living with an unmarried couple more than doubled between 1985 and 1995. Today, 8.1 million children under age 18 years—25 percent of all children—live with one parent (U.S. Bureau of the Census, 1996).

In comparison to single parenting by widows, single parenting by divorced or never-married mothers receives less social support. A divorced mother usually receives less assistance from her own kin and considerably less (or none) from her former partner's relatives. Widowed mothers, however, often receive social support from their husband's relatives. Our culture is still ambivalent about divorce and tends to consider single-parent families deviant (Kissman and Allen, 1993). It is even less supportive of families formed by never-married mothers. Conservatives have recently returned to earlier forms of stigmatization by characterizing children of never-married women as "illegitimate" and their mothers as "unwed mothers." Eighty-seven percent of single-parent families are headed by women. This has important economic ramifications because of gender discrimination in wages and job opportunities, as discussed in Chapter 11.

Significance of Ethnicity

Ethnicity is an important demographic factor in single-parent families. In 1995, among white children, 25 percent lived in single-parent families; among African-American children, 64 percent lived in such families; among Hispanics, 36 percent

Figure 15.1
Single-Parent Families, by Ethnicity: 1970–1995

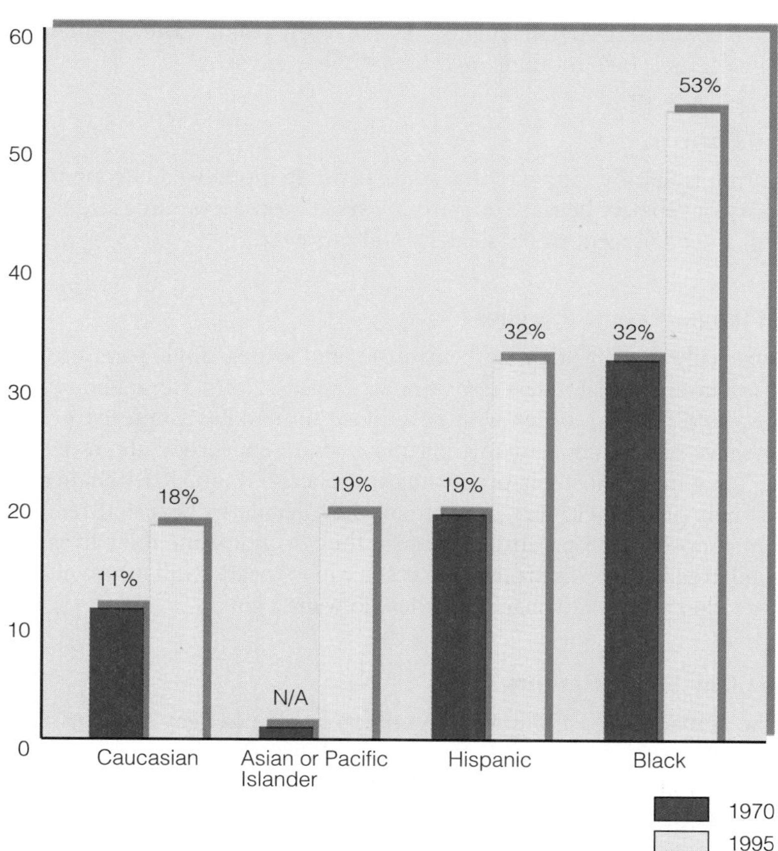

SOURCE: U.S. Bureau of the Census, *Statistical Abstract of the United States*, 116th ed. Washington, DC: U.S. Government Printing Office, 1996.

The poor man's wisdom is despised and his words are not heard.

Ecclesiastes 9:16.

Did you know?

Among children in divorced single-parent families, 32.4 percent live in poverty (U.S. Bureau of Statistics, 1996).

lived in single-parent families (U.S. Bureau of the Census, 1996). White single mothers were more likely to be divorced than their African-American or Latino counterparts, who were more likely to be unmarried at the time of the birth or widowed.

Poverty

Married women usually experience a sharp drop in their income when they separate or divorce (as discussed in Chapter 14). Among unmarried single mothers, poverty and motherhood go hand in hand. Because they are women, because they are often young, and because they are frequently from ethnic minorities, single mothers have few financial resources. They are under constant economic stress in trying to make ends meet (McLanahan and Booth, 1991). They work at low wages, endure welfare, or both. They are unable to plan because of their constant financial uncertainty. They move more frequently than two-parent families as economic and living situations change, uprooting themselves and their children. They accept material support from kin, but often at the price of receiving unsolicited "free advice," especially from their mothers.

Diversity of Living Arrangements

There are many different kinds of single-parent households. Single-parent families show great flexibility in managing child care and housing with limited resources. In doing so, they rely on a variety of household arrangements. In 1991, 20 percent of single-mother families included another male, such as a grandfather, uncle, or boyfriend ("Diverse Living Arrangements," 1994). Many young African-American mothers live with their own mothers in a three-generation setting.

Transitional Form

Single parenting is usually a transitional state. A single mother has strong motivation to marry or remarry because of cultural expectations, economic stress, role overload, and a need for emotional security and intimacy.

Intentional Single-Parent Families

For many, especially single women in their thirties and forties, single parenting has become a more intentional and less transitional lifestyle (Gongla and Thompson, 1987; Miller, 1992). Some older women choose unmarried single parenting because they have not found a suitable partner and are concerned about declining fertility. They may plan their pregnancies or choose donor insemination or adoption. If their pregnancies are unplanned, they decide to bear and rear the child. Others choose single parenting because they do not want their lives and careers encumbered by the compromises necessary in marriage. Still others choose it because they don't want a husband but they do want a child.

Lesbian and Gay Single Parents

There may be as many as 2.5 million to 3.5 million lesbian and gay single parents. The majority were married before they were aware of their sexual orientation or got married with hopes of "curing" it. They became single parents as a result of divorce. Others were always aware of being lesbian or gay; they chose adoption or donor insemination in order to have children. Said one gay adoptive father, "I always knew I wanted to be a father." A lesbian who was artificially inseminated said, "I started to get this baby hunger. I just needed to have a child" (quoted in Miller, 1992).

Children in Divorced Single-Parent Families

More than 60 percent of people who divorce have children living at home. Of children born in 1980 who will live in a single-parent family, white children can expect to spend about 31 percent of their childhood in single-parent families and African-American children, about 59 percent.

We have constructed a family system which depends upon fidelity, lifelong monogamy, and the survival of both parents. But we have never made adequate provision for the security and identity of the children if that marriage is broken.

MARGARET MEAD (1901–1978)

Parental Stability and Loneliness

After a divorce, the single parent is usually glad to have the children with him or her. Everything else seems to have fallen apart, but as long as divorced parents have their children, they retain their parental function. Their children's need for them reassures them of their own importance. A mother's success as a parent becomes even more important to counteract the feelings of low self-esteem that result from divorce.

Feeling depressed, the mother knows she must bounce back for the children. Yet after a short period, she comes to realize that her children do not fill the void left by her divorce. The children are a chore as well as a pleasure, and she may resent being constantly tied down by their needs. Thus, minor incidents with the children—a child's refusal to eat or a temper tantrum—may get blown out of proportion. A major disappointment for many new single parents is the discovery that they are still lonely. It seems almost paradoxical. How can a person be lonely amid all the noise and bustle that accompany children? However, children do not ordinarily function as attachment figures; they can never be potential partners. The attempt to make them so is harmful to both parent and child. Yet children remain the central figures in the lives of single parents. This situation leads to a second paradox: Although children do not completely fulfill a person, they rank higher in most single mothers' priorities than anything else.

Changed Family Structure

A single-parent family is not the same as a two-parent family with one parent temporarily absent. The permanent absence of one parent dramatically changes the way in which the parenting adult relates to the children. Generally, the mother becomes closer and more responsive to her children. Her authority role changes, too. A greater distinction between parents and children exists in two-parent homes. Rules are developed by both mothers and fathers. Parents generally have an implicit understanding to back each other up in child-rearing matters and to enforce mutually agreed-on rules. In the single-parent family, no other partner is available to help maintain such agreements; as a result, the children may find themselves in a much more egalitarian situation. Consequently, they have more power to negotiate rules. They can badger a single parent into getting their way about staying up late, watching television, or going out. They can be more stubborn, cry more often and louder, whine, pout, and throw temper tantrums. Any parent who has tried to get children to do something they do not want to do knows how soon an adult can be worn down. So single parents are more willing to compromise: "Okay, you can have a small box of Cocoa Puffs. Put that large one back, and promise you won't fuss like this anymore." In this way, children acquire considerable decision-making power in single-parent homes. They gain it through default: The single parent finds it too difficult to argue with them all the time.

Children in single-parent homes may also learn more responsibility. They may learn to help with kitchen chores, to clean up their messes, or to be more

OTHER PLACES OTHER TIMES

Single-Parent Families among African-American Adolescents

The Rise of Adolescent Single-Parent Families

In a new phenomenon, single-parent families are increasingly being formed by births to never-married women. Fifteen years ago, most children living in single-parent families lived with a divorced parent. Today, however, a child in a single-parent family is almost as likely to live with a never-married parent (35 percent) as with a divorced parent (37 percent) (Saluter, 1994). Births to unmarried mothers nearly doubled between 1980 and 1993, from about 20 percent to 30 percent of all births; almost one-third of these births are to unmarried adolescents (U.S. Bureau of the Census, 1996).

A global look at adolescent pregnancy reveals that among all industrialized countries, the United States has the dubious distinction of having the highest birthrate. Among girls younger than 15, the birthrate is five times higher (MacFarlane, 1997). The increasing trend toward never-married adolescents becoming mothers is significant because single parenting for them is markedly different from that of divorced adults. Adolescent mothers are more likely to be young, poor, have little or no contact with the child's father, be a high school dropout, receive welfare, and live in a three-generational household.

Adolescent pregnancy in the United States is not unique to any racial or ethnic group. However, African-American teens are twice as likely to be sexually active as whites and birthrates are higher among African-American teens (Yawn and Yawn, 1997). This is partly explained by the way in which the forces of racial discrimination and poverty combine to limit the options of young people of color (Singh, 1986). (Poor whites as well have disproportionately high teen birthrates.) But there are additional factors that contribute to pregnancy among African-American teens, including the following: (1) African Americans tend to use contraception less consistently than whites, (2) African Americans are less likely than whites to abort an unintended pregnancy, (3) the gender ratio of available men to women puts African-American women at a significant disadvantage for marriage, (4) the African-American community is more willing to accept children born to unmarried women, (5) three-generation families are much more common among African Americans, with the result that grandparents often have an active role in child rearing, (6) there is often a great deal of pressure among young African-American men (especially among the ghetto poor) to prove their masculinity by engaging in sexual activity, (7) there is often a lack of role models with whom poor African-American girls can identify, and (8) where a teen lives is very important in determining the age and circumstances of initiation of intercourse (Staples and Johnson, 1993; Strong and DeVault, 1997; Wallis, 1985; Yawn and Yawn, 1997).

Despite the low socioeconomic status of many teenage mothers, images of the adolescent African-American mother as "unemployed, uneducated, and living on welfare with three or more unkempt, poorly motivated and socialized children" reflect a minority of cases (Staples and Johnson, 1993; also see Franklin, 1988; Furstenberg, Morgan, Moore, and Peterson, 1987). While many African-American teenage mothers drop out of school, a decade later nearly 70 percent have received a high school diploma, 30 percent take courses after high school, and at least 5 percent graduate from college. Those most likely to receive

considerate. In the single-parent setting, the children are encouraged to recognize the work their mother does and the importance of cooperation. One single parent related how her husband had always washed the dishes when they were still living together. At that time it had been difficult to get the children to help around the house, particularly with the dishes. Now, she said, the children always do the dishes—and they do the vacuuming and keep their own rooms straightened up, too (also see Greif, 1985).

Although single parents continue to demonstrate love and creativity in the face of adversity, research on their children reveals some negative long-term consequences. (It is not clear, however, if the negative consequences are due to the nature of the family structure or to the debilitating effects of poverty.) A review welfare, live in poverty, and "produce children who fail academically and socially" are the 25 to 30 percent of teen mothers who have not completed high school (Staples and Johnson, 1993).

Some studies show positive outcomes to teen pregnancies if certain criteria are met. A study by Lee SmithBattle (1996) found that adolescent mothers' responsiveness to their children is "ultimately tied to the development of trusted and responsive relationships and the availability of basic resources that support the mothers' emerging skill and responsiveness." Furthermore, teenagers with family support are more likely to have had adequate health care during pregnancy, thereby reducing the risks to their children's health.

Meeting the Needs of Teenage Mothers

The most pressing needs of teen mothers that can be provided within the community are health care and education (Voydanoff and Donnelly, 1990). Regular prenatal care is essential to monitor the fetus's growth and the mother's health, including diet, possible sexually transmitted disease, and possible alcohol or drug use. After the birth, both mother and child need continuing care. The mother may need contraceptive counseling and services, and the child needs regular physical checkups and immunizations. Graduation from high school is an important goal of education programs for teenage mothers, as it impacts directly on their employability and ability to support (or help support) themselves and their children. Other goals of education include developing parenting and life management skills and job training. Programs designed to enhance adolescents' life options may give young women the incentive to delay pregnancy (MacFarlane, 1997). Mentoring programs that connect the mother with a woman from her community who serves as a role model and a special friend have proved beneficial (Stack and Burton, 1994). Many teenage mothers need financial assistance, at least until they complete their education; programs such as AFDC, food stamps, Medicaid, and WIC are often crucial to the survival of young mothers and their children. Conservatives, however, wish to abolish most of these programs because they believe such programs encourage teenage motherhood (as discussed in Chapter 11).

Adolescent Fathers

The incidence of teenage fatherhood is lower than teenage motherhood because almost half the fathers of infants born to teenage mothers are age 20 or older (Sonenstein, 1986). Five to 10 percent of adolescent boys are likely to be responsible for the majority of teen pregnancies (Elster and Panzarine, 1983). It is often assumed that teenage fathers are irresponsible, selfish, and uninterested in their partners and children. Many times this is not the case. Adolescent fathers more typically remain physically or psychologically involved throughout the pregnancy and have "intimate feelings toward both mother and baby" (Robinson, 1988). About 10 percent of pregnant teen couples marry, but the chances of the marriage working out are small (Robinson, 1988). It is usually difficult for teenage fathers to contribute much to the support of their children, although most express the intention of doing so during the pregnancy. Most have lower incomes, less education, and more children than men who postpone having children until age twenty or more. They may feel overwhelmed at the responsibility and may doubt their ability to be good providers. Teenage fathers are often the sons of absent fathers; they may not have a male parent role model and may in fact have numerous models of pregnancy outside of marriage. But most do want to learn to be fathers (Robinson, 1988).

It is critical that single-parent adolescent families be provided the support they need to thrive. Moralistic remedies victimize not only the parents but the children in these families. Instead of moralism, what is needed to empower these young and vulnerable families are educational opportunities, employment, child care—and hope.

(McLanahan and Booth, 1991) of the relevant studies found that young children from mother-only families tended not to do as well academically as those from two-parent families. In adolescence and young adulthood, children from single-parent families had fewer years of education and were more likely to drop out of high school. They had lower earnings and were more likely to be poor. They were more likely to initiate sex earlier, become pregnant in their teens, and cohabitate but not marry earlier (Furstenberg and Teitler, 1994). Furthermore, they were more likely to divorce. These conclusions are consistent for whites, African Americans, Latinos, and Asian Americans. The reviewers note that socioeconomic status accounts for some, but not all, of the effects. Some of the effects are attributed to family structure.

Harriette Pipes McAdoo (1988, 1996) traces the cause to poverty. She notes that African-American families are able to meet their children's needs in a variety of structures. "The major problem arising from female-headed families is poverty," she writes. "The impoverishment of Black families has been more detrimental than the actual structural arrangement" (McAdoo, 1988).

Successful Single Parenting

Single parenting is difficult, but for many single parents, the problems are manageable. Almost two-thirds of divorced single parents found that single parenting grows easier over time (Richards and Schmeige, 1993). As we discuss single parenting it is important to note that many of the characteristics of successful single parents and their families are shared by all successful families, as we shall see in Chapter 16.

Characteristics of Successful Single Parents

In-depth interviews with successful single parents found certain themes running through their lives (Olsen and Haynes, 1993):

- *Acceptance of responsibilities and challenges of single parenthood.* Successful single parents saw themselves as primarily responsible for their families; they were determined to do the best they could under varying circumstances. Following divorce, they were determined to get on with their lives.
- *Parenting as first priority.* In balancing family and work roles, their parenting role ranked highest. Romantic relationships were balanced with family needs.
- *Consistent, nonpunitive discipline.* Single parents realized that their children's development required discipline. They adopted an authoritative style of discipline that respected their children and helped them develop autonomy. They rejected authoritarian discipline as ineffective and damaging to parent-child relationships.
- *Emphasis on open communication.* They valued and encouraged expression of their children's feelings and ideas. Parents similarly expressed their feelings.
- *Fostering individuality that was supported by the family.* Children were encouraged to develop their own interests and goals; differences were valued by the family.
- *Recognition of the need for self-nurturance.* Single parents realized that they needed time for themselves so they would not be submerged by family responsibilities and roles. They needed to maintain an independent self that they achieved through other activities, such as dating, music, dancing, reading, classes, and trips.
- *Dedication to rituals and traditions.* Single parents maintained or developed family rituals and traditions, such as bedtime stories; family prayer or meditation; sit-down family dinners at least once a week; picnics on Sundays, fireworks on the Fourth of July; a special birthday dinner; visits to Grandma's; or watching television or going for walks together.

Single-Parent Family Strengths

Although most studies emphasize the stress of single parenting, some studies view it as building strength and confidence, especially for women. A study of sixty white single mothers and eleven white single fathers (most of whom were divorced) identified five family strengths associated with successful single parenting (Richards and Schmeige, 1993):

1 *Parenting skills.* Successful single parents develop the ability to assume some of the roles and attributes of the absent parent—the ability to take on both expressive and instrumental roles and traits. Single mothers may teach their children household repairs or car maintenance; single fathers may become more expressive and involved in their children's daily lives.
2 *Personal growth.* Developing a positive attitude toward the changes that have taken place in their lives helps single parents, as does feeling success and pride in overcoming obstacles.
3 *Communication.* Through good communication, single parents can develop trust and a sense of honesty with their children, as well as an ability to convey their ideas and feelings clearly to their children and friends.
4 *Family management.* Successful single parents develop the ability to coordinate family, school, and work activities and to schedule meals, appointments, family time, and alone time.
5 *Financial support.* Developing the ability to become financially self-supporting and independent is important to single parents.

Let women be provided with living strength of their own. Let them have the means to attack the world and wrest from it their own subsistence, and their dependence will be abolished—that of man also.

Simone De Beauvoir (1908–1986)

Among the single parents in the study, over 60 percent identified parenting skills as one of their family strengths. Forty percent identified family management as a strength in their families (Richards and Schmeige, 1993). About a quarter identified personal growth and communication among their family strengths.

In another study, Jean Miller (1982) concluded the following about single mothers:

> A significant number of the single women studied have solved many extraordinary problems in the face of formidable obstacles. Their single parenthood has led to personal growth for many. In adulthood they have made major revisions in their roles in life and in their self- and object-representations. Many have become contributors to their community, and their children are often a source of strength rather than difficulty.

Reflections

If you are or have been a member of a single-parent family, what were its strengths and problems? What do you know of the strengths and problems of friends and relatives in single-parent families?

Binuclear Families

One of the most complex and ambiguous relationships in contemporary America is what some researchers call the **binuclear family** (Ahrons and Rodgers, 1987; Ganong and Coleman, 1994). The binuclear family is a postdivorce family system with children. It is the original nuclear family divided in two. The binuclear family consists of two nuclear families—the maternal nuclear family headed by the mother (the ex-wife) and the paternal one headed by the father (the ex-husband). Both single-parent families and stepfamilies are forms of binuclear families.

Divorce ends a marriage, but not a family. It dissolves the husband-wife relationship but not necessarily the father-mother, mother-child, or father-child relationship. The family reorganizes itself into a binuclear family. In this new family, ex-husbands and ex-wives may continue to relate to each other and to their children, although in substantially altered ways. The significance of the maternal and paternal components of the binuclear family varies. In families with joint physical custody, the maternal and paternal families may be equally important to their children. In single-parent families headed by women, the paternal family component may be minimal.

Complexity of Binuclear Families

As an illustration of the complexity of the binuclear family system, consider the family history of two children, whom we have named Paige and Daniel Brickman. (See Figure 15.2 for a diagram of their binuclear family.)

When Paige was six and Daniel eight, their parents separated and divorced. The children continued to live with their mother, Sophia, in a single-parent household while spending weekends and holidays with their father, David. After

Figure 15.2
Diagram of Binuclear Family

The binuclear family consists of five subsystems: former spouse, remarried couple, parent/child, sibling (biological, step, and half), and mother-stepmother/father-stepfather subsystems. Identify these subsystems in the diagram below.

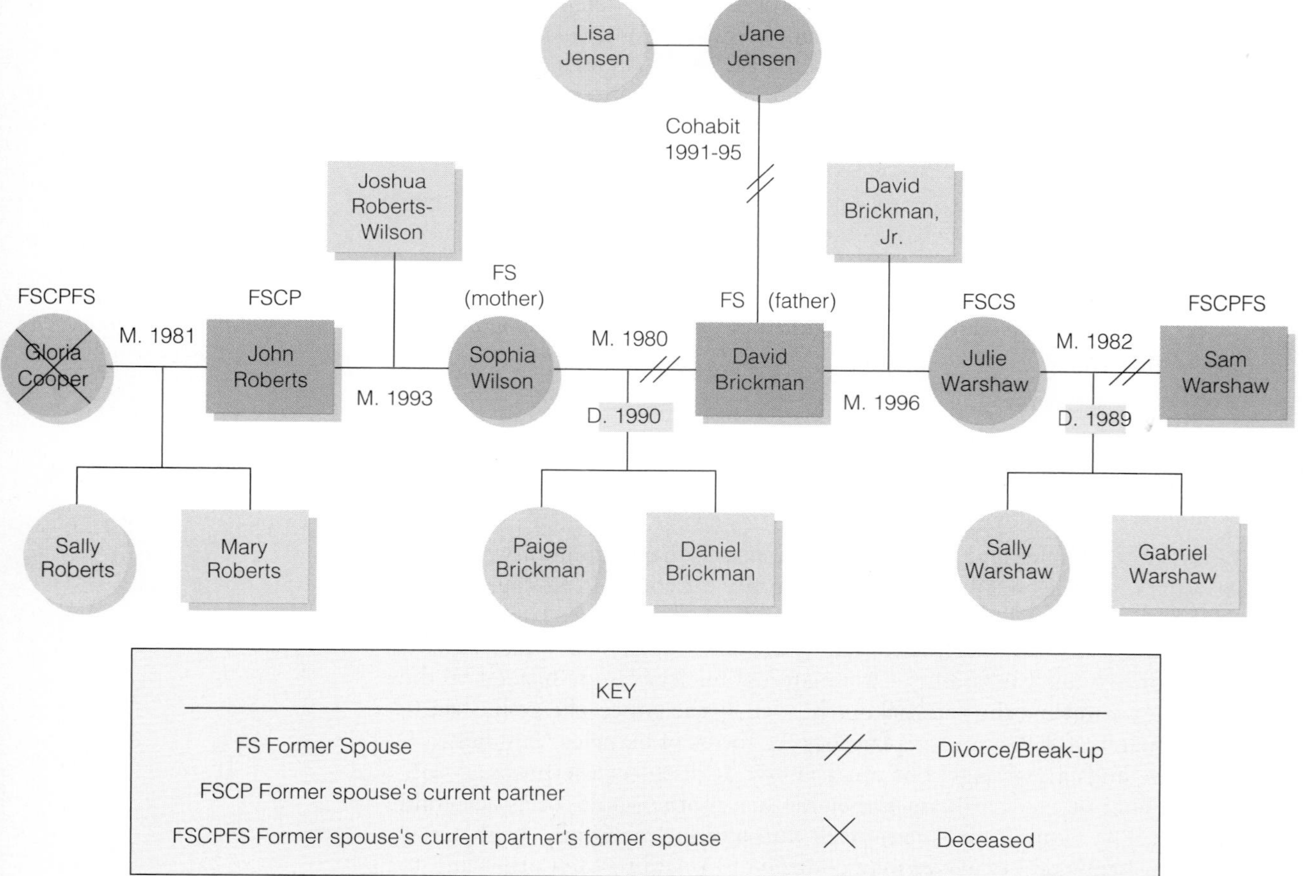

a year, David began living with Jane, a single mother who had a five-year-old daughter, Lisa.

Three years after the divorce, Sophia married John, who had joint physical custody of his two daughters, Sally and Mary, aged seven and nine. Paige and Daniel continued living with their mother, and their stepfather's children lived with them every other week. After two years of marriage, Sophia and John had a son, Joshua. About the same time, David and Jane split up; they continued to maintain close ties because of the bonds formed between David and Lisa. A year later, David married Julie, who had physical custody of two children, Sally and Gabriel; the next year, they had a son, David, Jr. Lisa visits every few weeks.

Although Paige and Daniel's binuclear family is not at all unusual, don't be surprised if it's hard to figure out who's related to whom. As Ahrons and Rodgers (1987) point out, "The variations in family structure that result from remarriage in binuclear families almost defy categorization." Eight years after their parents divorced, Paige and Daniel's family consists of the following family members: two biological parents, two stepparents, three stepsisters, one stepbrother, and two half-brothers. In addition, they had a "cohabiting" stepmother and stepsister with whom they continued to have close ties. Their extended family includes two sets of biological grandparents and stepgrandparents, along with a large array of biological and step aunts, uncles, and cousins. Paige and Daniel continue to have two households to which they belong as children.

Subsystems in the Binuclear Family

To clarify the different relationships, researchers Constance Ahrons and Roy Rodgers (1987) divide the binuclear family into five subsystems: (1) former spouse subsystem; (2) remarried couple subsystems; (3) parent-child subsystems; (4) sibling subsystems: stepsiblings and half-siblings; and (5) mother/stepmother-father/stepfather subsystems.

Former Spouse Subsystem

Although divorce severs the husband-wife relationship, the mother-father relationship endures as former spouses continue their parenting responsibilities. Although the degree of involvement for the noncustodial parent varies, children generally benefit from the continued involvement of both parents. The former spouses, however, must deal with a number of issues. These include the following:

- Anger and hostility toward each other as a result of their previous marriage and separation.
- Conflict regarding child custody, parenting styles, values, and aspirations concerning their children.
- Shifting roles and relationships between former spouses when one or both remarry.
- The need to incorporate others as stepparents, stepsiblings, and stepgrandparents into the family system when one or both former spouses remarry.

As long as former spouses are able to separate parenting from personal issues involving each other, they may form effective co-parenting relationships. Constance Ahrons and Lynn Wallisch (1987) indicated that about half of the former spouses in their study are able to work well with each other; another quarter interact, but with substantial conflict. About a quarter are unable to co-parent because of the high degree of conflict.

Remarried Couple Subsystems

Remarried couples are generally unprepared for the complexities of remarried life. If they have physical custody of children from the first marriage, they must provide access between the children and the noncustodial parents. Both custodial and noncustodial parents must facilitate the exchange of children, money, decision-making power, and time. This may often be difficult. Typical marital issues such as power and intimacy may become magnified because they frequently involve not only the remarried couple but the former spouse as well. Because of custody arrangements, for example, the former spouse may exercise veto power over the remarried couple's plan to take a family vacation because it conflicts with visitation.

Parent-Child Subsystems

Remarriage is probably more difficult for children than for parents. While parents are caught up in the excitement of romance, their children may be reacting with anxiety and distress. Their parents are making choices that affect the children but over which the children themselves have little or no control. Furthermore, remarriage destroys the children's fantasies that their parents will reunite. In addition, children may not want a stepparent to function as a parent for various reasons. The children may feel, for example, that the stepparent is usurping the role of the absent biological parent. Or they may resent the stepparent's intrusion into their restructured single-parent family.

Researchers are still somewhat inconclusive regarding family dynamics among stepfamilies. While some point out that comparing dynamics in nuclear families to those in stepfamilies is like comparing apples to oranges (Olson and DeFrain, 1997), others have found that families with stepchildren do not show more frequent conflict. In fact, contrary to researchers' expectations, less frequent conflict has been found (MacDonald and DeMaris, 1995).

What is certain is that both the biological parent and the stepparent must make adjustments. Former single parents, for example, must adjust to the presence of a second parent in decision-making and child-rearing practices. Stepparents, however, have the greatest adjustment to make. Because remarried families tend to model themselves after traditional nuclear families, stepparents often expect that the stepparent role will be similar to the parent role. Stepfathers who have their own children may feel conflict because they are "more" a father to their stepchildren than to their biological ones. Stepmothers, however, seem to experience greater stress than stepfathers. The stress may result from stepmothers' relatively high degree of involvement in child-care and nurturing activities, which are not adequately acknowledged or appreciated by stepchildren.

Sibling Subsystems: Stepsiblings and Half-Siblings

When parents remarry, their children may acquire "instant" brothers or sisters, who may differ considerably in age and temperament. The sibling relationships may be especially complex in binuclear families. Consider Paige and Daniel's stepsibling and half-sibling relationships. In their mother's family, the two children had to adjust to a half-brother and two stepsisters who lived with them half-time. In their father's family, they adjusted to a "cohabiting" sibling who was "like" a sister. When their father remarried, they gained two stepsiblings and, a little later, a half-brother. To make matters more complex, biological relationships do not guarantee emotional closeness. Paige feels closest, for example, to her "cohabitant" sibling,

The binuclear family is the post-divorce family consisting of two nuclear families, the nuclear family that includes the mother (or ex-wife) and that which includes the father (or ex-husband). The creation of the binuclear family presents both rewards and challenges for its members.

Lisa, and her stepsister Sally, neither of whom live with her. Daniel, however, feels closest to Mary, his stepsister, and Joshua, his half-brother.

All things being equal, stepsiblings are predisposed to bond with each other because family norms require affection between family members (Pasley, 1987). Bonding will occur most rapidly when siblings are of similar age and sex, have similar experiences and values, are interdependent, and perceive greater rewards than costs in their relationships.

Stepsiblings and siblings contend with one another for parental affection, toys, attention, physical space, and dominance. Sharing a parent with a new stepparent is often difficult enough, but to share the parent with a stepsibling can be overwhelming. Visiting biological children compete with stepchildren who are living with the visiting children's biological parent.

Mother/Stepmother-Father/Stepfather Subsystems

The relationship between new spouses and former spouses often influences the remarried family. The former spouse can be an intruder in the new marriage and a source of conflict between the remarried couple. Other times the former spouse is a handy scapegoat for displacing problems. Much of the current spouse-former spouse interaction depends on how the ex-spouses themselves feel about each other.

Reflections

Are you involved in a binuclear family? If so, in what role? Which subsystems are functional or dysfunctional in your binuclear family? How do you imagine that conflict within the former spouse subsystem would affect children in a binuclear family?

Remarriage

The eighteenth-century writer Samuel Johnson described remarriage as "the triumph of hope over experience." Americans are a hopeful people. Almost one-third remarry within a year of their divorces (Ganong and Coleman, 1994). About half of all divorced women remarry within five years of dissolution of their marriage.

Marriage is made in Heaven, but second marriages are arranged by people.

YIDDISH PROVERB

Many newly divorced men and women express great wariness about marrying again; yet at the same time they are actively searching for mates. Women often view their divorced time as important for their development as individuals, whereas men, who often complain that they were pressured into marriage before they were ready, become restless as "born-again bachelors" (Furstenberg, 1980).

Remarriage Rates

Nearly half of all marriages in the United States are marriages in which at least one partner has been previously married (U.S. Bureau of the Census, 1996). Twenty percent remarry other divorced men and women; and approximately 22 percent marry never-married individuals (see Figure 15.3).

Remarriage is more or less standard for divorced persons. The number of remarriages is increasing, but the rate is slightly declining. The decline may be partly the result of the desire on the part of divorced men and women to avoid the legal responsibilities accompanying marriage. Instead of remarrying, many are choosing to cohabit. In addition, African Americans and women under age twenty-five are experiencing declining rates of remarriage. In addition to age and ethnicity, the presence of children may also affect remarriage rates.

Figure 15.3

Percent Distribution of Marriages by Previous Marital Status: 1970 to 1988.

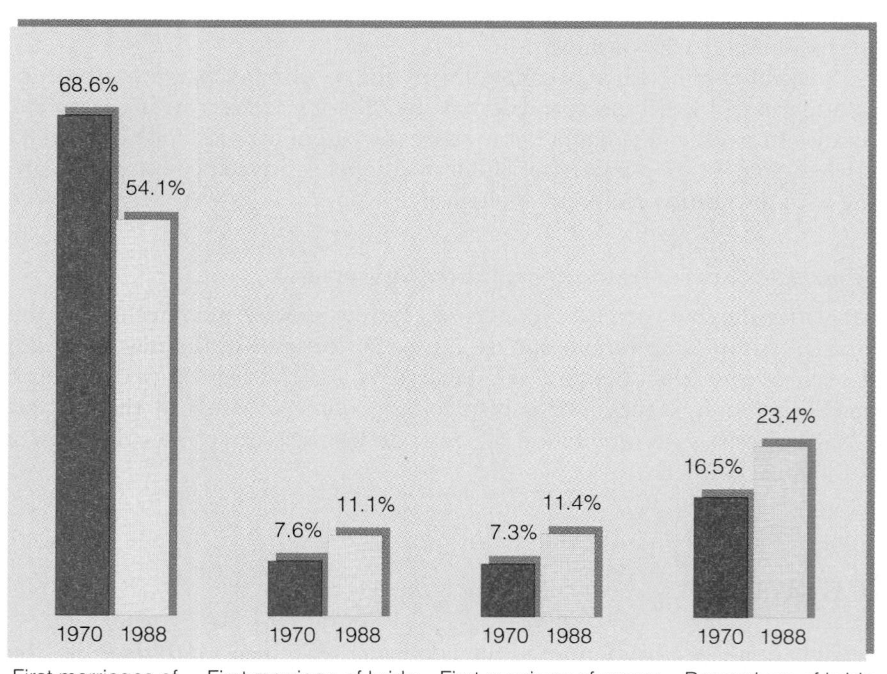

SOURCE: U.S. Bureau of the Census, *Statistical Abstracts of the United States*, 116th ed. Washington, DC: U.S. Government Printing Office, 1996.

Age

Divorced women and men marry when they are about ten to twelve years older than those in first marriages. The average age for women remarrying is thirty-four; for men, it is thirty-seven. A man's or woman's age at the time of separation is the greatest individual factor affecting remarriage. As with the "marriage market" in first marriages, as a divorced person ages, the number of potential partners declines; at the same time, one's own "marketability" decreases with increasing age. For women, the highest remarriage rate takes place in the twenties; it declines by a quarter in the thirties and by two-thirds in the forties. Because the pool of eligible partners is smaller for remarriage, men and women may be willing to "settle for less." They may choose someone they would not have chosen when they were younger (Ganong and Coleman, 1994).

Ethnicity

Ethnicity also affects remarriage rates. About 60 percent of white women, 42 percent of African-American women, and 45 percent of Hispanic women remarry (U.S. Bureau of the Census, 1996). The remarriage rates have been declining for African Americans, although first-marriage rates have also been declining.

Presence of Children

Although children from earlier marriages traditionally have been thought to decrease the likelihood of remarriage, the evidence is mixed. A review of the literature found that the presence of children decreases the rate of remarriage by about one-quarter (Bumpass, Sweet, and Martin, 1990). The effects are most marked when a woman has three or more children. Most of the research, however, is fifteen to twenty years old, and the increased incidence of single-parent families and stepfamilies may have decreased some of the negative impact of children. In fact, whereas researchers generally speculate that children are a "cost" in remarriage, others point out that some men may regard children as a "benefit" in the form of a ready-made family (Ganong and Coleman, 1994). Some research suggests that the stepparent with no biological children experiences the most negative effect (MacDonald and DeMaris, 1995).

Courtship

The norms governing courtship prior to first marriage are fairly well defined. As courtship progresses, individuals spend more time together; at the same time, their family and friends limit time and energy demands because "they're in love." Courtship norms for second marriages, however, are not clear (Ganong and Coleman, 1994; Rodgers and Conrad, 1986). For example, when is it acceptable for formerly married (and presumably sexually experienced) men and women to become sexually involved? What type of commitment validates "premarital" sex among postmarital men and women? How long should courtship last before a commitment to marriage is made? Should the couple cohabit? Without clear norms, courtship following divorce can be plagued by uncertainty about what to expect.

I've learned from my mistakes and I'm sure I could repeat them exactly.

Peter Cook (1937–1994)

Remarriage courtships are short. As noted earlier, almost one-third of divorced individuals marry within a year of their divorces. This may indicate, however, that they knew their future partners before they were divorced. Furthermore, because many cohabit prior to remarriage, the courtship period may be even shorter than marriage dates indicate.

If neither partner has children, courtship for remarriage may resemble courtship before the first marriage, with one major exception: The memory of the earlier marriage exists as a model for the second marriage. Courtship may trigger old fears, regrets, habits of relating, wounds, or doubts. At the same time, having experienced the day-to-day living of marriage, the partners may have more realistic expectations. Their courtship may be complicated if one or both are noncustodial parents. In that event, visiting children present an additional element.

Cohabitation

Larry Ganong and Marilyn Coleman (1994) describe cohabitation as "the primary way people prepare for remarriage." In fact, cohabitation is a major difference between first-time marriages and remarriages. Although 15 to 25 percent of people marrying for the first time may be living together at the time of marriage, the majority of those remarrying are cohabiting (Coleman and Ganong, 1991; Ganong and Coleman, 1994). This larger percentage may reflect the desire to test compatibility in a "trial marriage" to prevent later marital regrets (Buunk and van Driel, 1989). At the same time, however, couples who lived together before remarriage did not discuss stepfamily issues any more than did those who did not cohabit (Ganong and Coleman, 1994).

We know very little about cohabitation prior to remarriage. A study of single mothers who cohabited, however, found that some partially cohabited prior to marriage (Montgomery, Anderson, Hetherington, and Clingempeel, 1992); that is, the potential partner spent several days and nights a week in the mother's home before moving in on a full-time basis. Presumably, this gave both partners the chance to gauge the man's "fit" with his future stepfamily before marriage.

Courtship and Children

Courtship before remarriage differs considerably from that preceding a first marriage if one or both members in the dating relationship are custodial parents. Single parents are not often a part of the singles world because such participation requires leisure and money, which single parents generally lack. Children rapidly consume both of these resources.

Although single parents may wish to find a new partner, their children usually remain the central figures in their lives. This creates a number of new problems. First, the single parent's decision to go out at night may lead to guilt feelings about the children. If a single mother works and her children are in day care, for example, should she go out in the evening or stay at home with them? Second, a single parent must look at a potential partner as a potential parent as well. A person may be a good companion and listener and be fun to be with, but if he or she does not want to assume parental responsibilities, the relationship will often stagnate or be broken off. A single parent's new companion may be interested in assuming parental responsibilities, but the children may regard him or her as an intruder and try to sabotage the new relationship.

A single parent may also have to decide whether to permit a lover to spend the night when children are in the home. This is often an important symbolic act. For one thing, it brings the children into the parent's new relationship. If the couple have no commitment, the parent may fear the consequences of the children's emotional involvement with the lover; if the couple break up, the children may be adversely affected. Single parents are often hesitant to expose their children again to the distress of separation; the memory of the initial parental separation and divorce is often still painful. For another thing, having a lover spend

Did you know?

Recent research has found that having children in the home has a strong positive effect on economic distress and a strong negative effect on income (Shapiro, 1996).

the night reveals to the children that their parent is a sexual being. This may make some single parents feel uncomfortable and may also make the parent vulnerable to moral judgment by his or her children. It may also raise questions about sex outside marriage for younger children. Single parents often fear that their children will lose respect for them under such circumstances. Sometimes children do judge their parents harshly, especially their mothers. Parents are often deeply disturbed at being condemned by children who do not understand their need for love, companionship, and sexual intimacy. Finally, having someone sleep over may trigger the resentment and anger that the children feel toward their parents for splitting up. They may view the lover as a replacement for the absent parent and feel deeply threatened.

Characteristics of Remarriage

Remarriage is different from first marriage in a number of ways. First, the new partners get to know each other during a time of significant changes in life relationships, confusion, guilt, stress, and mixed feelings about the past (Keshet, 1980). They have great hope that they will not repeat past mistakes, but there is usually some fear that the hurts of the previous marriage will recur (McGoldrick and Carter, 1989). The past is still part of the present. A Talmudic scholar once commented, "When a divorced man marries a divorced woman, four go to bed."

Remarriages occur later than first marriages. People are at different stages in their life cycles and may have different goals. A woman who already has had children may enter a second marriage with strong career goals. In her first marriage, raising children may have been more important (see Teachman and Heckert, 1985).

Divorced people have different expectations of their new marriages. Considering divorce has been seen as psychologically detrimental, remarriage has often been regarded as the pathway to well-being. In a study of second marriages in Pennsylvania, Frank Furstenberg (1980) discovered that three-fourths of the couples had a different conception of love than couples in their first marriages. Two-thirds thought they were less likely to stay in an unhappy marriage; they had already survived one divorce and knew they could make it through another. Four out of five believed their ideas of marriage had changed. One woman (quoted in Furstenberg, 1980) said:

> I think second marriages are less idealistic and a little more realistic. You realize that it's going to be tough sometimes but you also know that you have to work them out. You come into a second marriage with a whole new set of responsibilities. It's like coming into a ball game with the bases loaded. You've got to come through with a hit. Likewise, there's too much riding on the relationship; you've got to make it work and you realize it more after you've been divorced before. You just have to keep working out the rules of the game.

Finally, the majority of remarriages create stepfamilies. A single-parent family is generally a transition family leading to a stepfamily, which has its own unique structure, satisfactions, and problems.

Marital Satisfaction and Stability

According to various studies, remarried people are about as satisfied or happy in their second marriages as they were in their first marriages. As in first marriages, marital satisfaction appears to decline with the passage of time (Coleman and

Did you know?

Research has concluded remarriage indeed offers enhanced psychological well-being (Shapiro, 1996).

Reflections

If you were seeking a marital partner, would you consider a previously married person? Why or why not? Would it make a difference if he or she already had children?

A person who has been married many years knows more about marriage than one who has been married many times.

ANONYMOUS

Advice for the Modern Bride:
Photo Session Etiquette
1. Bride 2. Groom 3. Groom's daughter from first marriage 4. Bride's mother 5. Bride's mother's current lover 6. Bride's sperm donor father 7.&8. Sperm donor's parents who sued for visitation rights to bride 9. Bride's mother's lover at time of bride's birth 10. Groom's mother 11. Groom's mother's boyfriend 12. Groom's father 13. Groom's stepmother 14. Groom's father's third wife 15. Groom's grandfather 16. Groom's grandfather's lover 17 Groom's first wife
Signe

SOURCE: © Signe Wilkinson/*San Jose Mercury News/Cartoonists & Writers Syndicate.*

Ganong, 1991). Yet despite the fact that marital happiness and satisfaction are more or less the same in first and second marriages, remarried couples are more likely to divorce, especially if children from a prior relationship are in the home (Booth and Edwards, 1992). How do we account for this paradox? Researchers have suggested several reasons for the higher divorce rate in remarriage. (See Ganong and Coleman [1994] for a discussion of various models explaining the greater fragility of remarriage.)

First, persons who remarry after divorce often have a somewhat different outlook on marital stability and are more likely to use divorce as a way of resolving an unhappy marriage (Booth and Edwards, 1992). Furstenberg and Spanier (1987) note that they were continually struck by the willingness of remarried individuals to dissolve unhappy marriages: "Regardless of how unattractive they thought this eventuality, the great majority indicated that after having endured a first marriage to the breaking point they were unwilling to be miserable again simply for the sake of preserving the union."

Second, remarriage is an "incomplete institution" (Cherlin, 1981). Society has not evolved norms, customs, and traditions to guide couples in their second marriages. There are no rules, for example, defining a stepfather's responsibility to a child: Is he a friend, a father, a sort of uncle, or what? Nor are there rules establishing the relationship between an individual's former spouse and his or her present partner: Are they friends, acquaintances, rivals, or strangers? Remarriages don't receive the same family and kin support as do first marriages (Goldenberg and Goldenberg, 1994).

Third, remarriages are subject to stresses that are not present in first marriages. Perhaps the most important stress is stepparenting. Children can make the formation of the husband-wife relationship more difficult because they compete for their parents' love, energy, and attention. In such families, time together alone becomes a precious and all-too-rare commodity. Furthermore, although children have little influence in selecting their parent's new husband or wife, they have immense power in "deselecting" them. Children have "incredible power" in maintaining or destroying a marriage, observe Marilyn Ihinger-Tallman and Kay Pasley (1987):

> Children can create divisiveness between spouses and siblings by acting in ways that accentuate differences between them. Children have the power to set parent against stepparent, siblings against parents, and stepsiblings against siblings.

Ganong and Coleman (1994) note that the presence of stepchildren is a major contributor to the higher divorce rate in stepfamilies compared with remarried families with no children.

Stepfamilies

Remarriages that include children are very different from those that do not. The families that emerge from remarriage with children are known generally as stepfamilies. They are sometimes called blended, reconstituted, restructured, or remarried families by social scientists—names that emphasize their structural differences from other families. Satirist Art Buchwald, however, calls them "tangled families." He may be close to the truth in some cases. Nevertheless, by the year 2000, there may be more stepfamilies in America than any other family form (Pill, 1990). If we care about families, we need to understand and support stepfamilies.

Stepfamilies: A Different Kind of Family

When we enter a stepfamily, many of us expect to recreate a family identical to an intact family. The intact nuclear family becomes the model against which we judge our successes and failures. But researchers believe that stepfamilies are significantly different from intact families (Ganong and Coleman, 1994; Papernow, 1993; Pill, 1990). If we try to make our feelings and relationships in a stepfamily identical to those of an intact family, we are bound to fail. But if we recognize that the stepfamily works differently and provides different satisfactions and challenges, we can appreciate the richness it brings us and have a successful stepfamily.

Structural Differences

Six structural characteristics make the stepfamily different from the traditional first-marriage family (Visher and Visher, 1979, 1991). Each one is laden with potential difficulties.

1. *Almost all the members in a stepfamily have lost an important primary relationship.* The children may mourn the loss of their parent or parents, and the spouses may mourn the loss of their former mates. Anger and hostility may be displaced onto the new stepparent.
2. *One biological parent lives outside the current family.* He or she may either support or interfere with the new family. Power struggles may

occur between the absent parent and the custodial parent, and there may be jealousy between the absent parent and the stepparent.

3 *The relationship between a parent and his or her children predates the relationship between the new partners.* Children have often spent considerable time in a single-parent family structure. They have formed close and different bonds with the parent. A new husband or wife may seem to be an interloper in the children's special relationship with the parent. A new stepparent may find that he or she must compete with the children for the parent's attention. The stepparent may even be excluded from the parent-child system.

4 *Stepparent roles are ill defined.* No one knows quite what he or she is supposed to do as a stepparent. Most stepparents try role after role until they find one that fits.

5 *Many children in stepfamilies are also members of the noncustodial parent's household.* Each home may have differing rules and expectations. When conflict arises, children may try to play one household against the other. Furthermore, as Visher and Visher (1979) observe:

> The lack of clear role definition, the conflict of loyalties that such children experience, the emotional reaction to the altered family pattern, and the loss of closeness with their parent who is now married to another person create inner turmoil and confused and unpredictable outward behavior in many children.

Numerous researchers have found that children in stepfamilies exhibit about the same number of adjustment problems as children in single-parent families and more problems than children in original, two-parent families (Furstenberg and Cherlin, 1991; McLenahan and Sandefor, 1994).

Figure 15.4
Stepfamily Types as Percent of Families with Children, 1990

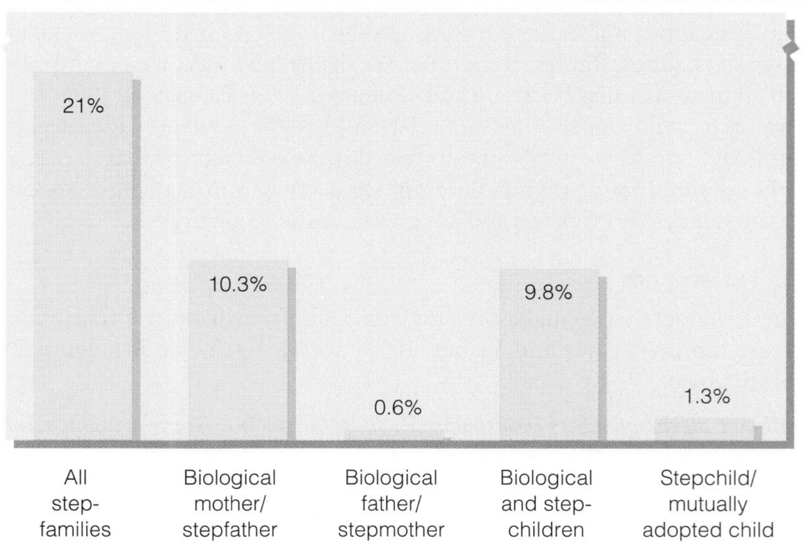

SOURCE: U.S. Bureau of the Census, *Statistical Abstract of the United States.* Washington, DC: U.S. Government Printing Office, 1994.

6 *Children in stepfamilies have at least one extra pair of grandparents.* Children get a new set of stepgrandparents, but the role these new grandparents are to play is usually not clear. A study by Spanier and Furstenberg (1980) found that grandparents were usually quick to accept their "instant" grandchildren.

As we examine the stepfamily further, we should keep in mind that when Ganong and Coleman (1984) reviewed the empirical literature comparing stepfamilies with traditional nuclear families, they found no significant differences in relationships with stepfathers as compared to fathers, in perceptions of parental happiness, in degree of family conflict, and in positive family relationships. Only the clinical literature upholds the image of stepfamilies beset by conflict and traumas (Ganong and Coleman, 1986).

Stepfamilies may well be very similar to traditional nuclear families. We perceive them as being different because we use a "deficit model" in examining them (Ganong and Coleman, 1994). Thus, we look for deficiencies and exaggerate problems or differences.

The Developmental Stages of Stepfamilies

Becoming a stepfamily is a process. Individuals and families become a stepfamily through a series of developmental stages. Each person—the biological parent, the stepparent, and the stepchild (or children)—experiences the process differently. For family members, it involves seven stages, according to a study of stepfamilies by Patricia Papernow (1993). The early stages are fantasy, immersion, and awareness; the middle stages are mobilization and action; and the later stages are contact and resolution.

Early Stages: Fantasy, Immersion, and Awareness

The early stages in becoming a stepfamily include the courtship and early period of remarriage, when each individual has his or her fantasy of their new family. It is a time when the adults (and sometimes the children) hope for an "instant" nuclear family that will fulfill their dreams of how families should be. They have not yet realized that stepfamilies are different from nuclear families.

Fantasy Stage. During the fantasy stage, biological parents hope that the new partner will be a better spouse and parent than the previous partner. They want their children to be loved, adored, and cared for by their new partners. They expect their children to love the new parent as much as they do.

New stepparents fantasize that they will be loving parents who are accepted and loved by their new stepchildren. They believe that they can ease the load of the new spouse, who may have been a single parent for years. One stepmother recalled her fantasy: "I would meet the children and they would gradually get to know me and think I was wonderful. . . . I just knew they would love me to pieces. I mean, how could they not!" (quoted in Papernow, 1993). Of course, they did not.

The children, meanwhile, may have quite different fantasies. They may still feel the loss of their original families. Their fantasies are often that their parents will get back together. Others fear they may "lose" their parent to an interloper, the new stepparent. Some fear that their new family may "fail" again. Still others are concerned about upheavals in their lives, such as moving, going to new schools, and so on.

Immersion Stage. The immersion stage is the "sink-or-swim" stage in a stepfamily. Reality replaces fantasy. "We thought we would just add the kids to this wonderful relationship we'd developed. Instead we spent three years in a sort of Cold War over them," recalled one stepparent (quoted in Papernow, 1993). Once the partners marry, everyday living collides with fantasies as family members begin to interact with one another. As one stepparent said, "The trouble with life is that it's daily" (Visher and Visher, 1991). The fantasy of recreating a traditional family gives way to the reality of creating a stepfamily. For adults, the challenge is to keep swimming through disappointment and doubt, to find out what will work.

For children, a man's transformation from "Mom's date" to stepfather may be the equivalent of the transformation from Dr. Jekyll to Mr. Hyde. Suddenly an outsider becomes an insider—with authority. As a twelve-year-old (whose new stepmother also had children) said: "In the beginning it's fun. Then you realize that your whole life is going to change. Everything changes. We just had fun before, my Dad and me. And now there's all these new people and new rules" (quoted in Papernow, 1993). When children have a new stepparent, they may feel disloyal to their absent biological parent if they show affection. (Biological parents can make a difference: They can let their children know it's okay to love a stepparent.)

Mother's Day or Father's Day can become a stepchild's worst nightmare. A nine-year-old said: "On Mother's Day I didn't know what to do. . . . If I went with my stepmother, my mother would be furious. If I went with my mother, my stepmother would be upset. I couldn't even think about it. It's the worst situation." As the parent and stepparent get more involved with each other, children may feel new losses: The stepparent has replaced them as the parent's number one interest.

Awareness Stage. The awareness stage in stepfamily development is reached when family members "map" the territory. They become aware that they are in new lands; they try to figure out where they are. They gather data about their new family members; they try to understand them—what they like and don't like about them. This stage involves individual and joint family tasks. The individual task is for each member to identify and name the feelings he or she experiences in being in the new stepfamily. A key feeling for stepparents to acknowledge is feeling like an outsider. They need to become aware of feelings of aloneness; they must discover their own needs; and they must set some distance between themselves and their stepchildren. They need to understand why their stepchildren are not warmly welcoming them, as they had expected.

Biological parents need to become aware of unresolved feelings from their earlier marriages and from being single parents. They may feel pulled from the multiple demands of their children and their partners. Biological parents may feel resentment toward their children, their partners, or both. They need to let go of residual feelings of creating the perfect family. They need to understand what it's like for their partners "to create an intimate relationship in the presence of a parent-child relationship that often feels more like ex-lovers" (Papernow, 1993).

Children in the awareness stage often feel "bumped" from their close relationship with the single parent. They miss cuddling in bed in the morning, the bedtime story, the wholehearted attention. When a new stepparent moves in, their feelings of loss over their parents' divorce are often rekindled. They resent the new stepparent for taking the place of their "real" mother or father. Loyalty issues resurface. If they are not pressured into feeling "wonderful" about their new family, however, they can slowly learn to appreciate the benefits of an added parent and friend who will play with them or take them places.

Middle Stages: Mobilization and Action

In the middle stages of stepfamily development, family members are more clear about their feelings and relationships with one another. They have given up many of their fantasies. They understand more of their own needs. They have mapped the new territory. The family, however, remains biologically oriented. Parent-child relationships are central. In this stage, changes involve the emotional structure of the family as a whole.

Mobilization Stage. In the mobilization stage, family members recognize differences. Conflict becomes more open. Members mobilize around their unmet needs. A stepmother described this change: "I started realizing that I'm different than Jim [her husband] is, and I'm going to be a different person than he is. I spent years trying to be just like him and be sweet and always gentle with his daughter. But I'm not always that way. I think I made a decision that what I was seeing was right" (quoted in Papernow, 1993). The challenge in this stage is to resolve differences while building the stepfamily's sense of family.

Let me say to begin with: It is not neurotic to have conflicts. Conflicts within ourselves are an integral part of human life.

KAREN HORNEY (1885–1952)

Stepparents begin to take a stand. They stop trying to be the ideal parent. They no longer are satisfied with being outsiders. Instead, they want their needs met. They begin to make demands on their stepchildren: to pick up their clothes, be polite, do the dishes. Similarly, they make demands on their partners to be consulted; they often take positions regarding their partners' former spouses. Because stepparents make their presence known in this stage, the family begins to change. The family begins to integrate the stepparent into its functioning. In doing so, the stepparent ceases being an outsider and the family increasingly becomes a real stepfamily. Some stepparents, however, become frustrated in their attempts to become an insider. They may metaphorically "move out" and withdraw from the family.

For biological parents, the mobilization stage can be frightening. The stepparents' desire for change leaves biological parents torn. Biological parents feel they must protect their own children and yet satisfy the needs of their partners. One father said of the fighting between his wife and daughter (quoted in Papernow, 1993):

> My daughter is sulking upstairs, crying that nobody loves her. My wife is crying in the bedroom. If I comfort my wife, my daughter will accuse me of abandoning her. If I talk to my daughter, I'll catch hell from Gina for "giving in" to my daughter. What do I do?

At this stage, some parents step out of the middle. They no longer act as buffers between stepparents and children. The stepparent and child are allowed to establish their own relationship. Other parents begin to negotiate family roles with their partners, trying to determine what will work.

Children often attempt to resolve loyalty issues at this stage. They have been tugged and pulled in opposite directions by angry parents too long. Often the adults paid no attention to them. Finally, the children have had enough and can articulate their feelings. After hearing her parents squabble one time too many, one girl reflected: "I thought, this stinks. It's horrible. After the 50 millionth time I said, 'That's your problem. Talk to each other about it,' and they didn't do it again" (quoted in Papernow, 1993).

Action Stage. In the action stage, the family begins to take major steps in reorganizing itself as a stepfamily. It creates new norms and family rituals. Although members have different feelings and needs, they begin to accept each other. The family finds a middle ground. They compromise: Some holidays will be spent with the nonresidential parent; bedrooms do not have to be immaculate; adults can play Barry Manilow, and kids can play rap. Most important, stepfamily members develop shared, realistic expectations and act on them.

Stepcouples begin to develop their own relationship independent from children. They also begin working together as a parental team. Stepparents begin to take on disciplinary and decision-making roles; they are supported by the biological parents. Stepparents begin to develop relationships with their stepchildren independent of the biological parents. Stepparent-stepchild bonds are strengthened.

Later Stages: Contact and Resolution

The later stages in stepfamily development involve solidifying the stepfamily. Much of the hard work has been accomplished in the middle stages.

Contact Stage. In the contact stage, stepfamily members make intimate contact with each other. Their relationships become genuine. They communicate with a sense of ease and intimacy. The couple relationship becomes a sanctuary from everyday family life. The stepparent becomes an "intimate outsider" with whom stepchildren can talk about things "too hot" for their biological parents, such as sex, drugs, their feelings about the divorce, and religion.

For the stepparent, a clear stepparent role finally emerges. The **stepparent role** is an individually created role because, as we saw earlier, it is undefined in our society. The role varies, however, from stepparent to stepparent and from stepfamily to stepfamily. It is mutually suitable to both the individual and the different family members.

Resolution Stage. The stepfamily is solid in its resolution stage. It no longer requires the close attention and work of the middle stages. Family members feel that earlier issues have been resolved. As one stepfather said (quoted in Papernow, 1993):

> I can feel that we've moved. Not easily, because it's been a pain in the ass. But I feel clear that our family works. . . . It's been proved over the years that we could do it, and we're doing it. We're happy for the most part. There's a lot of love.

Stepchildren can feel the benefits of having an "outsider" inside the family. Not all steprelationships in the family are necessarily the same; they may differ according to the personalities of each individual. Some of the relationships develop more closely than others. But in any case, there is a sense of acceptance. The stepfamily has made it and has benefited from the effort.

Reflections

If you are a member of a stepfamily, what were your experiences at the different stages? If you are not, ask friends or relatives who are members what their experiences were at the different stages. If you were to become a stepparent, how would you handle each stage?

It takes most stepfamilies about seven years to complete the developmental process. Some may complete it in four, and others take many, many years. Some only go through a few of the stages and get stuck. Others split up with divorce. But many are successful. Becoming a stepfamily is a slow process that moves in small ways to transform strangers into family members.

Problems of Women and Men in Stepfamilies

Most people go into stepfamily relationships expecting to recreate the traditional nuclear family found on *The Cosby Show* or *7th Heaven;* they are full of love, hope, and energy. Perhaps the hardest adjustment they have to make is realizing that stepfamilies are different from traditional nuclear families—and that being different does not make stepfamilies inferior. A nuclear family is neither morally superior to the stepfamily nor a guarantor of happiness.

Women in Stepfamilies

Stepmothers tend to experience more problematic family relationships than do stepfathers (Santrock and Sitterle, 1987; Kurdeck and Fine, 1993). To various degrees, women enter stepfamilies with certain feelings and hopes. Stepmothers generally expect to do the following (Visher and Visher, 1979, 1991):

- Make up to the children for the divorce.
- Create a happy, close-knit family and a new nuclear family.
- Keep everyone happy.
- Prove that they are not wicked stepmothers.
- Love the stepchild instantly and as much as their own biological children.
- Receive instant love from their stepchildren.

Needless to say, most women are disappointed. Expectations of total love, happiness, and the like would be unrealistic in any kind of family, be it a traditional family or a stepfamily. The warmer a woman is to her stepchildren, the more hostile they may become to her because they feel she is trying to replace their "real" mother. If a stepmother tries to meet everyone's needs—especially her stepchildren's, which are often contradictory, excessive, and distancing—she is likely to exhaust herself emotionally and physically. It takes time for her and her children to become emotionally integrated as a family.

Because she is more often involved in raising children, the stepmother frequently assumes the role of disciplinarian. This is particularly true if she plays a more active domestic role than her husband. Consequently, poorer relationships with her stepchildren may occur.

Stepmothers married to men who have their children full-time often experience greater problems than stepmothers whose children are with them part-time or occasionally (Furstenberg and Nord, 1985). In part, it may be because children whose fathers have full-time custody may be more difficult for a number of reasons. Bitter custody fights may leave children emotionally troubled and hostile to stepmothers, whom they perceive as "forcibly" replacing their mothers. In other instances, children (especially adolescents) may have moved from their mother's home to their father's because their mother could no longer handle them. In either case, the stepmother may be required to parent children who have special needs or problems. Stepmothers may find these relationships especially difficult. Typically, stepmother-stepdaughter relationships are the most problematic (Clingempeel et al., 1984). Relationships become even more difficult when the stepmothers never intended to become full-time stepparents.

Queen: *No, be assur'd you shall not find me, daughter,*
After the slander of most stepmothers,
Evil-ey'd unto you.
Imogen: *Dissembling courtesy! How fine this tyrant*
Can tickle where she wounds.
SHAKESPEARE, *Cymbeline*

Men in Stepfamilies

Different expectations are placed on men in stepfamilies. Because men are generally less involved in child rearing, they usually have few "cruel stepparent" myths to counter. Nevertheless, men entering stepparenting roles may find certain areas particularly difficult at first (Visher and Visher, 1991). A critical factor in a man's stepparenting is whether he has children of his own. If he does, they are more likely to live with his ex-wife. In this case, the stepfather may experience guilt and confusion in his stepparenting because he feels he should be parenting his own children. When his children visit, he may try to be "Superdad," spending all his time with them and taking them to special places. His wife and stepchildren may feel excluded and angry.

A stepfather usually joins an already established single-parent family. He may find himself having to squeeze into it. The longer a single-parent family has been functioning, the more difficult it usually is to reorganize it. The children may resent his "interfering" with their relationship with their mother (Wallerstein and Kelly, 1980). His ways of handling the children may be different from his wife's, resulting in conflict.

Working out rules of family behavior is often the area in which a stepfamily encounters its first real difficulties. Although the mother usually wants help with discipline, she often feels protective if the stepfather's style is different from hers. To allow a stepparent to discipline a child requires trust from the biological parent and a willingness to let go. Disciplining often elicits a child's testing response: "You're not my real father. I don't have to do what you tell me." Homes are more positive when parents include children in decision making and are supportive (Barber and Lyons, 1994). Nevertheless, disciplining establishes legitimacy, because only a parent or parent figure is expected to discipline in our culture. Disciplining, however, may be the first step toward family integration, because it establishes the stepparent's presence and authority in the family.

The new stepfather's expectations are important. Though the motivations to stepparent are often quite different from those of biological parents, research from the 1987–1988 National Survey of Families and Households shows that over half (55 percent) of stepfathers found it somewhat or definitely true that having stepchildren was just as satisfying as having their own children (Sweet, Bumpass, and Call, 1988). In spite of this, stepparents tend to view themselves as less effective than natural fathers view themselves (Beer, 1992). The complex role of that the stepfather brings to his family often create role ambiguity and confusion that takes time to work out.

Conflict in Stepfamilies

Achieving family solidarity in the stepfamily is a complex task. When a new parent enters the former single-parent family, the family system is thrown off balance. Where equilibrium once existed, there is now disequilibrium. A period of tension and conflict usually marks the entry of new people into the family system. Questions arise about them: Who are they? What are their rights and their limits? Rules change. The mother may have relied on television as a baby-sitter, for example, permitting the children unrestricted viewing in the afternoon. The new stepfather, however, may want to limit the children's afternoon viewing, and this creates tension. To the children, everything seemed fine until this stepfather came along. He has disrupted their old pattern. Chaos and confusion will be the norm until a new pattern is established, but it takes time for people to adjust to new roles, demands, limits, and rules.

Conflict takes place in all families: traditional nuclear families, single-parent families, and stepfamilies. If some family members do not like each other, they will bicker, argue, tease, and fight. Sometimes they have no better reason for disruptive behavior than that they are bored or frustrated and want to take it out on someone. These are fundamentally personal conflicts. Other conflicts are about definite issues: dating, use of the car, manners, television, or friends, for example. These conflicts can be between partners, between parents and children,

UNDERSTANDING YOURSELF

Parental Images: Biological Parents versus Stepparents

We seem to hold various images or stereotypes of parenting adults, depending on whether they are biological parents or stepparents. Our images affect how we feel about families and stepparents (Coleman and Ganong, 1987). The following instrument (modeled after one devised by Ganong and Coleman [1983]) will help give you a sense of how you perceive parents and stepparents.

The instrument consists of nine dimensions of feelings presented in a bipolar fashion—that is, as opposites, such as *hateful/affectionate, bad/good,* and so on. You can respond to these feelings on a 7-point scale, with 1 representing the negative pole and 7, the positive pole. For example, say you were using this instrument to determine your perceptions about aardvarks. You might feel that aardvarks are quite affectionate, so you would give them a 7 on the *hateful/affectionate* dimension. But you might also feel that aardvarks are not very fair, so you would rank them 2 on the *unfair/fair* continuum.

To use this instrument, take four separate sheets of paper. On one sheet, write *Stepmother;* on the second, *Stepfather;* on the third, *Biological Mother;* and on the fourth, *Biological Father.* On each sheet, number from 1 to 9 in a column, with each number representing a dimension. Number 1 would represent *hateful/affectionate,* and so on. Then, using the 7-point scale on each sheet, score your general impressions about biological parents and stepparents.

The 7-point scale is as follows:

Negative						Positive
1	**2**	**3**	**4**	**5**	**6**	**7**

The dimensions are as follow:

1 Hateful/affectionate
2 Bad/good
3 Unfair/fair
4 Cruel/kind
5 Unloving/loving
6 Strict/not strict
7 Disagreeable/agreeable
8 Rude/friendly
9 Unlikable/likable

After you've completed these ratings, compare your responses for stepmother, stepfather, biological mother, and biological father. Do you find differences? If so, how do you account for them?

or among the children themselves. Certain types of stepfamily conflicts, however, are of a frequency, intensity, or nature that distinguishes them from conflicts in traditional nuclear families. Recent research on how conflict affects children in stepfather households found that parental conflict does not account for children's lower level of well-being (Hanson, McLanahan, and Thompson, 1996). These conflicts are about favoritism; divided loyalties; discipline; and money, goods, and services.

Favoritism

Favoritism exists in families of first marriages, as well as in stepfamilies. In stepfamilies, however, the favoritism often takes a very different form. Whereas a parent may favor a child in a biological family on the basis of age, sex, or personality, in stepfamilies favoritism tends to run along kinship lines. A child is favored by one or the other parent because he or she is the parent's biological child, or if a new child is born to the remarried couple, they may favor him or her as a child of their joint love. In American culture, where parents are expected to treat children equally, favoritism based on kinship seems particularly unfair.

Divided Loyalties

"How can you stand that lousy, low-down, sneaky, nasty mother (or father) of yours?" asks (or, more accurately, demands) a hostile parent. It is one of the most painful questions children can confront, for it forces them to take sides against someone they love. One study (Lutz, 1983) found that about half of the adolescents studied confronted situations in which one divorced parent talked negatively about the other. Almost half of the adolescents felt themselves "caught in the middle." Three-quarters found such talk stressful.

Divided loyalties put children in no-win situations, forcing them not only to choose between parents but to reject new stepparents. Children feel disloyal to one parent for loving the other parent or stepparent. But divided loyalties, like favoritism, can exist in traditional nuclear families as well. This is especially true of conflict-ridden families in which warring parents seek their children as allies.

Discipline

Researchers generally agree that discipline issues are among the most important causes of conflict among remarried families (Ihinger-Tallman and Pasley, 1987). Discipline is especially difficult to deal with if the child is not the person's biological child. Disciplining a stepchild often gives rise to conflicting feelings within the stepparent. Stepparents may feel that they are overreacting to the child's behavior, that their feelings are out of control, and that they are being censured by the child's biological parent. Compensating for fears of unfairness, the stepparent may become overly tolerant.

The specific discipline problems vary from family to family, but a common problem is interference by the biological parent with the stepparent (Mills, 1984). The biological parent may feel resentful or overreact to the stepparent's disciplining if he or she has been reluctant to give the stepparent authority. As one biological mother who believed she had a good remarriage stated (quoted in Ihinger-Tallman and Pasley, 1987):

> Sometimes I feel he is too harsh in disciplining, or he doesn't have the patience to explain why he is punishing and to carry through in a calm manner, which causes me to have to step into the matter (which I probably shouldn't do). . . . I do realize that it was probably hard for my husband to enter marriage and the responsibility of a family instantly . . . but this has remained a problem.

As a result of interference, the biological parent implies that the stepparent is wrong and undermines his or her status in the family. Over time, the stepparent may decrease his or her involvement in the family as a parent figure.

Money, Goods, and Services

Problems of allocating money, goods, and services exist in all families, but they can be especially difficult in stepfamilies. In first marriages, husbands and wives form an economic unit in which one or both may produce income for the family; husband and wife are interdependent. Following divorce, the binuclear family consists of two economic units: the custodial family and the noncustodial family. Both must provide separate housing, which dramatically increases their basic expenses. Despite their separation, the two households may nevertheless continue to be extremely interdependent. The mother in the custodial single-parent family, for example, probably has reduced income. She may be employed but still dependent on child support payments or AFDC. She may have to rely more extensively on child care, which may drain her resources dramatically. The father in the noncustodial family may make child support payments or contribute to medical or school expenses, which depletes his income. Both households have to deal with financial instability. Custodial parents can't count on always receiving their child support payments, which makes it difficult to undertake financial planning.

When one or both of the former partners remarry, their financial situation may be altered significantly. Upon remarriage, the mother receives less income from her former partner or lower welfare benefits. Instead, her new partner becomes an important contributor to the family income. At this point, a major problem in stepfamilies arises. What responsibility does the stepfather have in supporting his stepchildren? Should he or the biological father provide financial support? Because there are no norms, each family must work out its own solution.

Stepfamilies typically have resolved the problem of distributing their economic resources by using a *one-pot* or *two-pot* pattern (Fishman, 1983). In the one-pot pattern, families pool their resources and distribute them according to need rather than biological relationship. It doesn't matter whether the child is a biological child or a stepchild. One-pot families typically have relatively limited resources and consistently fail to receive child support from the noncustodial biological parent. By sharing their resources, one-pot families increase the likelihood of family cohesion.

In two-pot families, resources are distributed by biological relationship; need is secondary. These families tend to have a higher income, and one or both parents have former spouses who regularly contribute to the support of their biological children. Expenses relating to children are generally handled separately; usually there are no shared checking or savings accounts. Two-pot families maintain strong bonds between members of the first family. For these families, a major problem is achieving cohesion in the stepfamily while maintaining separate checking accounts.

Just as economic resources need to be redistributed following divorce and remarriage, so do goods and services (not to mention affection). Whereas a two-bedroom home or apartment may have provided plenty of space for a single-parent family with two children, a stepfamily with additional residing or visiting stepsiblings can experience instant overcrowding. Rooms, bicycles, and toys, for example, need to be shared; larger quarters may have to be found. Time becomes a precious commodity for harried parents and stepparents in a stepfamily. When visiting stepchildren arrive, duties are doubled. Stepchildren compete with parents and other children for time and affection.

It may appear that remarried families are confronted with many difficulties, but traditional nuclear families also face financial, loyalty, and discipline problems. We need to put these problems in perspective. (After all, half of all current marriages end in divorce, which suggests that first marriages are not problem free.) When all is said and done, the problems that remarried families face may not be any more overwhelming than those faced by traditional nuclear families (Ihinger-Tallman and Pasley, 1987).

Stepfamily Strengths

Because we have traditionally viewed stepfamilies as deviant, we have often ignored their strengths. Instead, we have only seen their problems. Let us end this chapter, then, by focusing on the strengths of stepfamilies.

Reflections

Think about conflicts involving favoritism, loyalty, discipline, and the distribution of resources. Do you experience them in your family of orientation? If so, how are they similar to, or different from, stepfamily conflicts? If you are in a stepfamily, do you experience them in your current family? How are these conflicts similar or different in your original family versus your current family? If you are a parent or stepparent, how are these issues played out in your current family?

A stepfamily may provide more companionship and security than the nuclear family it has replaced.

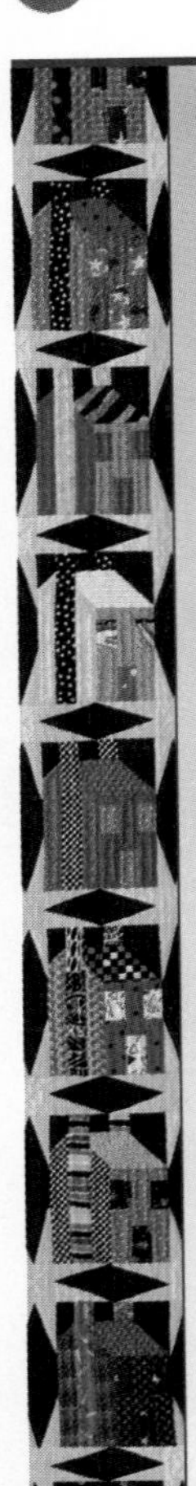

YOU AND YOUR WELL-BEING

Hints for Creating Happy Stepfamilies

A stepfamily often lives in two worlds: that of the biological family and that of the stepfamily. If both worlds coexist peacefully, then children and families can benefit and thrive. On the other hand, when familial relationships are strained and conflict is evident, the world can turn upside for everyone involved. Finding themselves caught between two worlds, children must risk dividing loyalties, often at the expense of their own well-being.

The following steps can help individuals and families anticipate problems, discuss realities, and gain understanding of feelings and emotions:

1. Be honest with yourself and your partner before getting married. Share your motivations for partnering and find out his or hers. Spend lots of time together, both with and without the children, getting to know each other.
2. Acknowledge that changes in lifestyle will occur for everyone involved. Discuss them and how each of you plan to handle the challenges that lie ahead.
3. Discuss similarities and expect differences in child rearing. Decide with one another how you plan to handle differences when they do arise.
4. Invite and encourage the children to share their feelings about the new living arrangement. Expect ambivalence. Discuss with them how their lifestyle is going to affect the relationship they have with their noncustodial parent.
5. Be patient. Love takes time and sometimes it does not occur. Consider your children's feelings, but don't let them dictate the outcome of the relationship between you and your partner. Developing a thick skin can often help during times when children appear to be unfeeling about other family members or are less than enthusiastic about family life.
6. Be persistent. Acknowledging that a stepparent will be compared with a noncustodial parent is important, but just as real are the desires of most stepparents to bond and create a family. Patience and persistence can help to make this real for the child.
7. Acknowledge that as a stepfamily, you need support. Parenting is difficult. Stepparenting is even more difficult. Individuals and couples need family and social support. Don't be afraid to accept it.
8. Make the marriage the primary relationship. The parents as a unit lay the foundation and set the course for a family.
9. Seek help if you need it. Stepfamily associations, resources, groups, and individual family therapy are available for those who are willing to admit that they need help.

Family Functioning

Although traditional nuclear families may be structurally less complicated than stepfamilies, stepfamilies are nevertheless able to fulfill traditional family functions. A binuclear single-parent, custodial, or noncustodial family may provide more companionship, love, and security than the particular traditional nuclear family it replaces. If the nuclear family was ravaged by conflict, violence, sexual abuse, or alcoholism, for example, the single-parent family or stepfamily that replaces it may be considerably better, and because children now see happy parents, they have positive role models of marriage partners (Rutter, 1994). Second families may not have as much emotional closeness as first families, but they generally experience less trauma and crisis (Ihinger-Tallman and Pasley, 1987).

New partners may have greater objectivity regarding old problems or relationships. Opportunity presents itself for flexibility and patience. As family boundaries expand, individuals grow and adapt to new personalities and ways of being. In addition, new partners are sometimes able to intervene between former spouses to resolve long-standing disagreements, such as custody or child-care arrangements.

Impact on Children

Stepfamilies potentially offer children a number of benefits that can compensate for the negative consequences of divorce.

- Children gain multiple role models from which to choose. Instead of having only one mother or father after whom to model themselves, children may have two mothers or fathers: the biological parents and the stepparents.
- Children gain greater flexibility. They may be introduced to new ideas, different values, or alternative politics. For example, biological parents may be unable to encourage certain interests, such as music or model airplanes, because they lack training or interest; a stepparent may play the piano or be a die-hard modeler. In such cases, that stepparent can assist his or her stepchildren in pursuing their development. In addition, children often have alternative living arrangements that enlarge their perspectives.
- Stepparents may act as a sounding board for their children's concerns. They may be a source of support or information in areas in which the biological parents feel unknowledgeable or uncomfortable.
- Children may gain additional siblings, either as stepsiblings or half-siblings, and consequently gain more experience in interacting, cooperating, and learning to settle disputes among peers.
- Children gain an additional extended kin network, which may become at least as important and loving as their original kin network.
- A child's economic situation is often improved, especially if a single mother remarried.
- Finally, children may gain parents who are happily married. Most research indicates that children are significantly better adjusted in happily remarried families than in conflict-ridden nuclear families.

As we near the year 2000 it is clear that the American family is no longer what it was at the beginning of this century. The rise of the single-parent family and stepfamily, however, does not imply an end to the American family. Rather, these forms provide different paths that contemporary families take as they strive to fulfill the hopes, needs, and desires of their members, and they are becoming as American as Beaver Cleaver's family and apple pie.

Summary

- Single parenting is an increasingly significant family form in the United States. Single-parent families tend to be created by divorce or births to unmarried women, are generally headed by women, are predominantly African American or Latino, are usually poor, involve a wide variety of household types, and are usually a transitional stage.
- Relations between the parent and his or her children change after divorce: The single parent generally tends to be emotionally closer but to have less authority. Successful single parents have similar themes running through their lives: (1) acceptance of responsibilities and challenges of single parenthood; (2) parenting as first priority; (3) consistent, nonpunitive discipline; (4) emphasis on open communication; (5) fostering individuality; (6) recognition of the need for self-nurturance; and (7) dedication to rituals and traditions. Family strengths associated with successful single parenting include (1) parenting skills, (2) personal growth, (3) communication, (4) family management, and (5) financial support.
- The *binuclear* family is a postdivorce family system with children. It consists of two nuclear families: the mother-headed family and the

father-headed family. Both single-parent families and stepfamilies are forms of binuclear families. The binuclear family consists of five subsystems: former spouse, remarried couple, parent-child, sibling, and mother/stepmother-father/stepfather subsystems.

- Courtship for second marriage does not have clear norms. If children are not involved from an earlier marriage, the only major difference between first and second marriage courtships is that the first marriage exists as a model in the second courtship. Courtship is complicated by the presence of children because remarriage involves the formation of a stepfamily. In addition, dating poses unique problems for single parents: They may feel guilty for going out, they must look at potential partners as potential parents, and they must deal with their children's judgments or hostility.
- Remarriage differs from first marriage in several ways: Partners get to know each other in the midst of major changes, they remarry later in life, they have different marital expectations, and their marriage often creates a stepfamily. Marital happiness appears to be about the same in first and second marriages. Remarried couples are more likely to divorce than couples in their first marriages. This may be accounted for either by their willingness to use divorce as a means of resolving an unhappy marriage or because remarriage is an "incomplete institution." Stresses accompanying stepfamily formation may also be a contributing factor.
- The stepfamily differs from the original family because (1) almost all members have lost an important primary relationship, (2) one biological parent lives outside the current family, (3) the relationship between a parent and his or her children predates the new marital relationship, (4) *stepparent roles* are ill defined, (5) children often are also members of the noncustodial parent's household, and (6) children have at least one extra pair of grandparents.
- *Stepfamilies* are families in which one or both partners have one or more children from a previous marriage or relationship. Traditionally, scholars have viewed stepfamilies from a "deficit" perspective. As a result, they have assumed that stepfamilies are very different from traditional nuclear families. More recently, scholars have begun to view stepfamilies as normal families; they have found few significant differences in levels of satisfaction and functioning between stepfamilies and traditional nuclear families.
- Becoming a stepfamily is a process. Individuals and families become a stepfamily through a series of developmental stages. Each person—the biological parent, the stepparent, and the stepchild (or children)—experiences the process differently. For family members, it involves seven stages. The early stages are fantasy, immersion, and awareness; the middle stages are mobilization and action; the later stages are contact and resolution.
- Stepmothers tend to experience greater stress in stepfamilies than do stepfathers. In part this may be because families with stepmothers are more likely to have been subject to custody disputes or to include children with a troubled family history. They also may assume the role of the disciplinarian. Stepfathers tend not to be as involved in child rearing as stepmothers. Both often experience difficulty in being integrated into the family.
- A key issue for stepfamilies is family solidarity—the feeling of oneness with the family. Conflict in stepfamilies is often over favoritism; divided loyalties; discipline; and money, goods, and services.
- Stepfamily strengths may include improved family functioning and reduced conflict between former spouses. Children may gain multiple role models, more flexibility, concerned stepparents, additional siblings, additional kin, improved economic situation, and happily married parents.

Key Terms

binuclear family 533
blended family 524
remarriage 524
single-parent family 524
stepfamily 524
stepparent role 548

Suggested Readings

Beer, William R. *American Stepfamilies.* New Brunswick, NJ: Transaction Publishers, 1992. An examination of family complexity and child rearing in stepfamilies.

Bloomfield, Harold, and Robert B. Kory. *Making Peace in Your Stepfamily: Surviving and Thriving as Parents and Stepparents.* New York: Hyperion, 1993. A self-help book offering exercises, case studies, and techniques to help stepfamilies with conflicts and problems.

Burns, Ailsa, and Cath Scott. *Mother-Headed Families and Why They Have Increased.* Hillsdale, NJ: Erlbaum, 1994. An exploration of the worldwide rise of single-parent families and a discussion of the various cultural, economic, and social/psychological theories used to explain it.

Dickerson, Bette J., ed. *African-American Single Mothers: Understanding Their Lives and Families.* Newbury Park, CA: Sage, 1994. A collection of essays presenting an inside view, including the role of children, grandparents, religious values, government, and media stereotypes.

Gangon, Lawrence, and Marilyn Coleman. *Remarried Family Relationships.* Newbury Park, CA: Sage, 1994. A concise description of current scholarship on remarriage and stepfamilies.

Ihinger-Tallman, Marilyn, and Kay Pasley. *Remarriage.* Newbury Park, CA: Sage, 1987. A useful introduction to issues involved in remarriage, including an examination of the strengths of remarried families.

Kelley, Patricia. *Developing Healthy Stepfamilies; Twenty Families Tell Their Stories.* Binghamton, NY: Hayworth Press, 1995. A look at what adults and children in stepfamilies say about such issues as discipline, money, family roles, relationships with ex-spouses, and the development of new traditions and rituals.

Papernow, Patricia. *Becoming a Stepfamily.* San Francisco: Jossey-Bass, 1993. An important book examining the normal development of stepfamilies and stepfamily relationships using a psychodynamic perspective.

Pasley, Kay, and Marilyn Ihinger-Tallman, *Remarriage and Stepparenting: Issues in Theory, Research, and Practice.* 2d ed. Westport, CT: Greenwood Press, 1994. A superb collection of essays by leading researchers in the field.

———. *Stepparenting Issues in Theory, Research, and Practice.* Westport, CT: Greenwood Press, 1994. A collection of articles by various experts, written for an academic audience.

Polakow, Valerie. *Lives on the Edge: Single Mothers and Their Children in the Other America.* Chicago: University of Chicago Press, 1993. A revealing study of the lives and struggles of poor single mothers to rear their children.

Visher, Emily, and John Visher. *How to Win as a Stepfamily.* 2d ed. New York: Brunner/Mazel, 1991. One of the best (and most readable) books available examining the problems confronting parents, stepparents, stepchildren, and stepfamilies in creating a new family.

CHAPTER 16

Marriage and Family Strengths

P R E V I E W

To gain a sense of what you already know about the material covered in this chapter, answer "True" or "False" to the statements below.

1. Researchers generally agree on the basic components of healthy families. True or false?
2. Among psychologically healthy families, family emotional health remains constant over the family life cycle. True or false?
3. A common way of measuring family strength is to see how a family responds to crisis. True or false?
4. The happiness of a couple before marriage is a good indicator of their happiness after they have children. True or false?
5. Although having a perfect family is difficult, it is possible. True or false?
6. Good communication is recognized by the overwhelming majority of researchers as the most important quality for family strength. True or false?
7. Children need to be taught respect and responsibility by their families. True or false?
8. Taking time to play is crucial to family health. True or false?
9. Seeking outside help for problems is a sign of family health. True or false?
10. Blood ties are more important than feelings in determining the strengths of our relationships with kin. True or false?

Answers:

1 True, 2 False, 3 True, 4 False, 5 False, 6 True, 7 True, 8 True, 9 True, 10 Fales

OUTLINE

A good marriage is that in which each appoints the other guardian of his individuality.

RAINER MARIA RILKE (1875–1926)

In this book we have considered the history of family, the diversity of family forms, and the ways in which family life provides the bulwark for the stability and creativity of a society. We have also considered some of the pressures facing families in these late years of the twentieth century and the different approaches that families and communities have used in an attempt to deal with and mitigate those pressures.

In recent years, family life in the United States has taken many forms and endured a variety of changing conditions. However, the vital place of family has never been seriously questioned. Indeed, working with the family is one of the first solutions suggested for virtually all social problems.

Multitudes of committees and agencies studying problems of drugs, violence, teen pregnancy, and other social concerns invariably declare that the first place for prevention and remediation is in the home. To assist families, communities and government have much to do by way of legal, economic, and social reforms to temper social attitudes and treatment of individuals who experience different kinds of family relationships and problems. Also, industry and government can work to ensure that employment schedules complement rather than compete with the work and maintenance of family life.

If we define the family as simply a people-producing factory, we will call all families that produce offspring "successful." Yet we know that a family has other important functions. As you will recall from Chapter 1, the basic tasks of the family are (1) reproduction and socialization, (2) economic cooperation, (3) assignment of status and social roles, and (4) intimacy. How well a family is able to accomplish these tasks appears to depend on a number of characteristics and abilities. In this chapter we discuss the particular characteristics that seem to make marriages and families strong. We explore how culture and ethnicity can contribute to family success, and we examine how relationships with the extended family, with affiliative kin (friends who are like family), and with friends help sustain the family unit. Finally, we look at the ecology of the family—that is, how the family interacts with the community.

Marital Strengths

Marriage may be seen as a forum for negotiating the balance between the desire for intimacy and the need to maintain a separate identity. To negotiate this balance successfully, we need to develop what sociologist Nelson Foote (1955) called *interpersonal competence*, the ability to share and develop an intimate, growing relationship with another. (For practical insights, see Gottman, 1995.)

A Marriage Ring
The ring so worn as you behold
So thin, so pale, is yet of gold;
The passion such it was to prove;
Worn with life's cares, love yet was love.
GEORGE CRABBE (1754–1832)

Marriage Strengths versus Family Strengths

Many of the traits of healthy marriages are also found in healthy families, as discussed later in this chapter. Indeed, marital competence can be viewed as a necessary basis for family success (Epstein, Bishop, and Baldwin, 1982). It might seem ideal if couples could perfect their interpersonal skills before the arrival of children. Child-free couples generally have more time for each other and substantially less psychological, economic, and physical stress. In reality, however, many of our marital skills probably develop alongside our family skills. Thus, as we improve communication with our spouses, for example, our communication with our children also improves.

There have been hundreds of studies of marital happiness and very few (and those in the last decade or so) on family happiness. Still, an argument can be made that certain kinds of strengths accrue only to families with children. The relationships of couples with children generally have greater stability than those of childless couples because the emotional cost of a breakup is much greater when children are present. Also, the relationships of parents to their children are rewarding in and of themselves and fulfill much of the need for intimacy. Sometimes the bonds between the individual parent and the children or between the children themselves can sustain a family during times of marital stress. Furthermore, the growing acceptance of singlehood and child-free marriages notwithstanding, society expects its adults to be parents and rewards the attainment of the parental role with approval and respect.

Vows are an important part of most wedding ceremonies, but commitment to the marriage must be renewed every day.

Essential Marital Strengths

Therapist and researcher David Mace (1980) cites the essential aspects of successful marriage as commitment, communication, and the creative use of conflict. Research indicates that the strongest predictor of marital success may be the effectiveness of communication experienced by the couple before marriage (Goleman, 1985). Numerous studies show that there is a strong correlation between a couple's communication patterns and marital satisfaction (Noller and Fitzpatrick, 1991). Classes and workshops are being developed in many communities to help teach these vital marital skills to couples before they commit themselves to marriage.

Mace defines *commitment* in terms of the relationship's potential for growth. He writes, "There must be a commitment on the part of the couple to ongoing growth in their relationship. . . . There is tremendous potential for loving, for caring, for warmth, for understanding, for support, for affirmation; yet, in so many marriages of today it never gets developed."

Commitment to any endeavor, and marriage is certainly no exception, requires the willingness and ability to work. In their study of strong families, Nick Stinnett and John DeFrain (1985) quote one-half of a married couple:

> You know the stereotypical story of the couple who have the lavish wedding, expensive and exotic honeymoon and then settle down. The work is over. She gets dumpy and nags, he gets sloppy and never again brings flowers. We were like that until one day when we examined our life together and found it lacked something. Then we decided that the wedding, rings, and honeymoon marked the beginning, *not* the end. We had to renew the marriage all along.

Commitment to the sexual relationship within marriage appears to be an important aspect of marital strength. All the families in the Stinnett and DeFrain study recommended sexual fidelity. An extramarital relationship does not mean the end of a strong marriage, however. Sometimes an affair can be a "catalyst for growth" (Stinnett and DeFrain, 1985). A couple may not realize how far they have drifted apart, how much their communication has deteriorated, or how important the marital relationship is until the extramarital relationship brings things into clearer focus. When spouses who love each other have sexual liaisons outside their marriage, it is almost certainly a sign that their relationship is in trouble. Yet they can then take the painful lessons they've learned and use them to forge a stronger and better marriage.

Another aspect of commitment is compromise. Often people give up things—work demands, social activities, or material goods—for those they love without thinking twice about it. Sometimes difficult choices must be made. Within a committed relationship, there is give-and-take so that neither partner ends up in the martyr role (see Stinnett and DeFrain, 1985). When spouses give up something in order to be together, they nurture the marriage relationship, which in turn supports and strengthens the individual.

The great secret of successful marriage is to treat all disasters as incidents and none of the incidents as disasters.

HAROLD NICHOLSON

We can look at marriage as a kind of task to be performed (albeit willingly and joyfully) or as a growing thing to be nurtured. Whatever metaphor we choose, we see that the relationship requires time, energy, patience, thoughtfulness, planning, and perhaps a little more patience in order to grow and prosper. Commitment to this success (and the sacrifices that commitment may entail) is an essential component in the formation of strong marriages, and strong families as well. In studying marriage and family strengths, researchers use models of family functioning such as the Circumplex Model discussed in Chapter 2. By discovering the components of healthy families, researchers hope to show us how to strengthen our own.

Family Strengths

Family strengths are those characteristics that contribute to a family's satisfaction and its perceived success as a family. An important part of the work of being a family and building family strengths is the identification of family goals. Each family is unique. Therefore, the goals by which a family will identify its success and family satisfaction will be described differently by each family.

Our family is not yet so good as to be degenerating.

KURT EWALD

Family scholars integrating feminist principles state how important it is to view family in its full range of forms and compositions (Allen and Farnsworth, 1993; Dilworth-Anderson, Burton, and Turner, 1993; Walker, 1993). In that regard, what is presented in this chapter are the strengths and qualities that are characteristic of successful, healthy families regardless of their structure. Some families may develop these strengths more easily than other families. Yet it is important to underscore that these strengths are available to all families. These are the strengths that are present in families who describe their families as satisfying and successful.

Family as Process

As the saying goes, "nobody's perfect," and neither is any family. Perfection in families, as in most other aspects of life, exists as an ideal, not as a reality. Family quality can be seen as a continuum, with a few very healthy or unhealthy families at either end and the rest somewhere in between. Also, family quality varies over the family life cycle. Generally, families experience increased stress during the childbearing and teenager phases of the cycle. The overall cohesiveness of the family is severely tested at these times, and although families often emerge stronger, they may have experienced periods of distrust, disorder, and unhappiness.

Families also are idiosyncratic: Each is different from all the others. Some may function optimally when the children are very young and not so well when they become teenagers, whereas the reverse may be true for other families. Some families may possess great strengths in certain areas and weaknesses in others. But all families constantly change and grow. The family is a process.

The Work of Families

Family is the irreplaceable means by which most of the social skills, personality characteristics, and values of individual members of society are formed. Hope, purpose, and general attitudes of commitment, perseverance, and well-being are nurtured in the family.

Indeed, even the rudimentary maintenance and survival care provided by families is no small contribution to the well-being of a community. Some of the services provided by families are such a basic part of our existence that we tend to overlook them. These include such essentials as the provision of food and shelter—a place to sleep, rest, and play—as well as caretaking, including supervision of health and hygiene, transportation, and the accountability of family members involving their activities and whereabouts. Without families, communities themselves would need to provide extensive dormitories and many personal-care workers with different levels of training and responsibility.

The "climate" of the family determines the way and the milieu in which caregiving is offered and will thus affect the social, psychological, and spiritual characteristics of the individual members. Personal identity is being formed in even the most elemental acts of family life.

Our society may take healthy families for granted. We also tend to publicize that which is broken down, isn't working, or has problems. News stories inundate us with the weaknesses and tragedies of family lives while ignoring the creative, strong elements of families all around us.

Family life professionals, therefore, have three objectives. One, they attempt to become aware of problems and to devote research and service to assist where help is needed. Two, professionals strive to take the time and the opportunity to recognize and celebrate the tremendous contribution that families make to our society. Three, family life professionals try to educate the public that *strong families do not just happen*. Developing and maintaining family strengths takes work.

The Characteristics of Strong Families

All individuals and families have strengths. All families feel more successful and more satisfied at some times than at other times. Some families go through periods of great stress and come out stronger; other families have great difficulty and nearly disintegrate. Research shows that strong families share some important patterns that enable them to survive and to grow. These patterns are labeled *family strengths.* These relationship patterns, interpersonal skills and competencies, and social and psychological characteristics create a sense of positive family identity. They promote satisfying and fulfilling interaction among family members, encourage the development of the potential of the family group and the individual family members, and contribute to the family's ability to deal effectively with stress and crisis (Stinnett and DeFrain, 1985).

The area of research devoted to the study of the characteristics of successful families has developed descriptions and models that help families examine and strengthen their approaches to family living. As with success in any field, strong families benefit from knowledge and practice. One of the most well known of these research projects is the Family Strengths Research Project directed by Nick Stinnett and John DeFrain. In this project, approximately three thousand families were questioned about their family life. Some families lived in the United States; others were from Latin America, Germany, Switzerland, Austria, South Africa, and Iraq. They were from all socioeconomic levels, both rural and urban. Some were single-parent families. All were families who had been nominated as "successful families" by organizations or people in their communities.

Reflections

What are the strengths of your family? How do these compare with those qualities found by Stinnett and DeFrain? Which of those cited in the literature do you value the most? Why?

When the mass of information from these families was analyzed, six qualities stood out repeatedly, even though the families presented considerable complexity and diversity. Those six qualities were appreciation, spending time together, commitment, good communication patterns, spiritual wellness, and ability to deal with crises. Not all six qualities appeared in every family, nor was the emphasis the same in all families. However, the patterns prevailed. Other researchers have discovered additional areas of family strength (Curran, 1983; McCubbin and McCubbin, 1988).

In surveying the rich collection of research and writing about life in successful families, we have identified ten areas of strength: commitment; affirmation, respect, and trust; communication; responsibility, morality, and spiritual orientation; rituals and traditions; crisis management; ability to seek help; spending time together; a family wellness orientation; and a balance between cohesion and adaptability. These will be discussed in the following sections.

Commitment

The family you come from isn't as important as the family you're going to have.

Ring Lardner (1885–1933)

Interviews and surveys with strong families reveal that the members of those families feel an identification with their families. They are willing to work for the well-being, or defend the unity and continuity, of their families (Bichof, Stith, and Wilson, 1992). (Children and youth appearing in juvenile court for various acts of violence usually express a low sense of commitment to their families.) Most adults view marriage as a commitment for life. With the expectation of an enduring relationship, couples are motivated to develop satisfying styles of relating.

When an individual is committed to his or her family, motivation is high to solve problems and to deal with conditions that threaten the family. When families become busy and fragmented, those with high levels of commitment take the initiative to review their priorities and activities and make changes that will relieve this pressure. The members deliberately work together to set priorities for the use of family time.

Commitment involves the promotion of growth of other family members. Family members are concerned for one another's happiness and well-being. Commitment involves working in behalf of one's family—wanting the best for each person. Committed persons strive to help other family members actualize their potential and achieve in their areas of interest (Schvaneveldt and Young, 1992). Marital satisfaction studies report that the more the partners express a sense of personal responsibility for the success of the relationship, the greater their satisfaction with their marriage. A partner's giving of time or self is most effective when it is in terms of enhancing the quality of the relationship rather than in terms of returns expected to be received individually. From a negative perspective, the less an individual contributes to the well-being of the family, the less that person receives.

Commitment is a prevailing characteristic in strong families of all forms. Single-parent and remarried families show commitment to members of the family even though the makeup of the family constellation has changed due to divorce or remarriage.

A challenging aspect of commitment is finding the right balance between extending ourselves to the priorities and identity of family and maintaining respect for our own individuality. Successful handling of this challenge involves being aware that the family milieu and relationships contribute to our own identity and creativity. It also involves understanding that freedom to choose to be involved in service to the family is vital to the development of both the individual and the family (Carter and McGoldrick, 1989).

As we saw in Chapter 4, commitment and love go together. Commitment to the family involves the participation of family members in a world view that encompasses more than self-centered interest. The members feel they are part of something larger than themselves. Scott Peck (1978) reminds us that although the work of attending to one's loved ones takes concentration and sacrifice, those expressing love do it voluntarily because it is important to their own identity and values, and because it entails voluntary service, commitment differs from obligation and duty. Although vows and promises may be important as symbols of our intent to loyally support those we care about, true commitment is much more than words. Commitment is revealed and renewed in our actions. It is created every day in the choices we make to do things for the benefit of those we love.

Affirmation, Respect, and Trust

For most of us, the phrase that we like best to hear consists of three little words. "I love you" may be a short sentence, but it can go a long way toward soothing a hurt, drying a tear, restoring a crumbling sense of self-worth, and maintaining a feeling of satisfaction and well-being. Supporting others in our family—letting them know we're interested in their projects, problems, feelings, and opinions—and being supported and affirmed in return is essential to family health (Gecas and Seff, 1991).

Another way to show that we care about family members, friends, and fellow human beings is by according them respect for their uniqueness and differences, even if we may not necessarily understand or agree with them. Healthy families encourage the development of individuality in their members, though it may lead to difficulties when the children's views begin to diverge widely from the parents' views on such subjects as religion, premarital sex, consumerism, patriotism, and child rearing. Criticism, ridicule, and rejection undermine self-esteem and severely restrict individual growth. Families that hamper the expression of their children's attitudes and beliefs tend to send into society children who are unable to respect differences in others. In addition to exhibiting respect for others, a healthy family member also insists on being respected in return.

The establishment of trust, you will recall from Chapter 9, is a child's first developmental stage, according to Erik Erikson. Not only must an infant develop trust in his or her parents, but a growing child must also continue to feel that other family members can be relied on absolutely. In turn, children learn to act in ways that make their parents trust them. Children in healthy families are allowed to earn trust as deemed appropriate by their parents. Children who know they are trusted are then able to develop self-confidence and a sense of responsibility for themselves and others.

It is not only children who need to "earn" the right to be called trustworthy; parents do, too. It's important for parents to be realistic in their promises to children and honest about their own mistakes and shortcomings. It's also important for children to see that their parents trust each other.

Many researchers agree that parental role modeling is a crucial factor in the development of qualities that ensure personal psychological health and growth. A loving relationship between parents, observes a family program administrator, "seems to breed security in the children, and, in turn, fosters the ability to take risks, to reach out to others, to search for their own answers, become independent, and develop a good self-image" (quoted in Curran, 1983). The more children observe their elders in situations that demonstrate mutual trust, respect, and care, the more they are encouraged to incorporate these successful and satisfying behaviors into their own lives. In divorced families, a continued positive relationship between the parents (if possible) is crucially important to the developing child (Goldsmith, 1982; Hanson, McLanahan, and Thompson, 1996).

Reflections

How are differences of opinion handled in your family? Do you feel loved and supported by your family? Do you feel trusted? How would you like to be treated differently? Do you let your family know that you love them, that you respect and trust them?

Good Communication

We say "I love you" or "I'm angry with you" in many ways other than just in words. Our tone of voice, body language, eye contact, silences, a touch, or a gift are all forms of communication (Satir, 1988). Many books and manuals are available to help people improve their communication as individuals, families, business associates, or in other groups. Chapter 5 highlights the dimensions of communication that are associated with effective problem solving.

In strong families, communication is direct. Strong families talk a lot. They have much to share, and they enjoy doing it. They trust one another. They are good listeners. Ultimately, when family members have truly been heard—not just the words, but the feelings, also—they know they are respected and appreciated because of the active attention and empathy of the listener.

In times of conflict, strong families seem to be able to keep their communication focused on the issues rather than on the personalities of those involved. In fact, recognizing conflict and actively attempting to resolve it can serve as a catalyst in deepening relationships (Rosenzweig, 1992). Strong families attempt to avoid accusation or labeling of others and to consider together the factors involved in the dispute. Studies of youth who are in trouble with the law or are in juvenile detention report that most youths with conduct disorders come from family situations characterized by poor communication among the members (Bichof, Stith, and Wilson, 1992).

Communication has been described as a huge umbrella that covers all that transpires among human beings. Virginia Satir (1988) estimates that by the time we reach age five, we have had a billion experiences in sharing communications. Communication is both the medium and the message of relationships among family members. Good communication is an art that is well developed in strong families. To some extent, they have developed this art through intentional learning and practice. Communication also facilitates the other family strengths presented in this chapter.

The aspects of healthy communication and the conflict-resolution skills discussed in Chapter 5 apply not only to the parents' relationship with each other but also to parent-child and child-child relationships. Among the most significant things a family can teach its children are the value of expressing their feelings effectively and the importance of truly listening to the expressions of others.

Responsibility, Morality, and Spiritual Orientation

One of the family's principal socializing tasks is to teach the individual responsibility for his or her own actions. Fostering morality—a code of ethics for dealing with our fellow human beings—is also essential. A spiritual orientation, whether defined as adherence to a particular religion or as reverence for life in general, gives family members a sense of being part of a larger whole.

The acquisition of responsibility is rooted in self-respect and an appreciation of the interdependence of people. When we as children develop a sense of our own self-worth, we begin to understand how much difference our own acts can make in the lives of others; this feeling of "making a difference" aids the further growth of self-esteem, as the following true story illustrates.

When Sally was ten years old, she went one very cold Saturday with a neighborhood youth group on an outing to the city shopping mall. There Sally noticed a girl about her own age. The girl had on a short, sleeveless sweater and was shivering in the cold. Impetuously, Sally went to the girl and asked if she were cold. The girl, with teeth chattering, nodded her head yes. Without really thinking of what she was doing, Sally took off her coat, put it on the girl, and ran to catch up with her group. The sponsor of the group told Sally to go back and get her coat. The thought horrified Sally; she said she could not do that. When the group returned to their neighborhood, the sponsor marched Sally home and declared

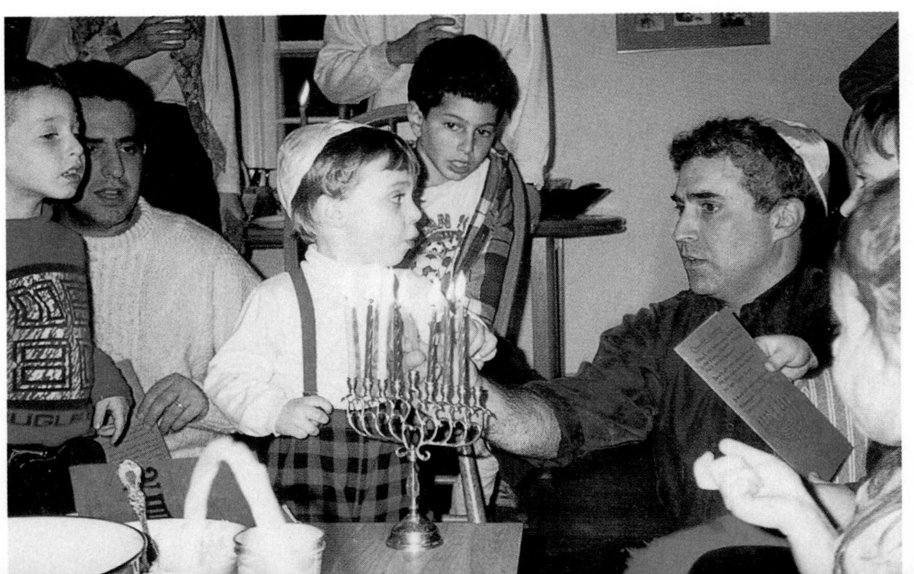

Families can encourage the spiritual development of their members through formal religious education or by passing on their own values, beliefs, and traditions. Here, a family celebrates Hanukkah.

There are only two lasting bequests we can hope to give our children. One is roots; the other, wings.

HODDING CARTER

to her mother that Sally had given her coat to a ragamuffin in the mall, implying that Sally should be severely punished. Sally's mother put her hand on Sally's shoulder and asked her if the little girl had been cold. Sally said yes, that she had been shivering. Sally's mother thanked the sponsor for taking care of the group. Then she took Sally to a clothing store nearby and told her to pick out any coat she wanted. When recalling this event in a college class, Sally recognized that her impulse to share with the shivering girl was an expression of her own personality and the values and attitudes of her family. From that time on, Sally had a new sense of her identity.

Successful families realize the importance of delegating responsibility (participating in household chores, sharing in family decisions, and so on) and of developing responsible behavior (completing homework, remembering appointments, keeping our word, and cleaning up our own messes). Sometimes children may have to accept more responsibility than they want, as when their dual-worker or single parents rely on them for assistance (Kissman and Allen, 1993). Although these situations may be detrimental for children in families already suffering severe economic (or other) hardships, in many families children have come to benefit from the added responsibility; they feel a sense of accomplishment and worthiness. Parental acknowledgment of a job well done goes a long way toward building responsibility in children.

Reflections

Do you consider yourself to be a responsible person? A moral person? How did you get that way? Have you had any noteworthy experiences with hypocrisy (yours or someone else's)? How did these experiences feel?

Healthy families also know the importance of allowing children to make their own mistakes and face the consequences. These lessons can be very painful at times, but they are stepping stones to growth and success. Along with a sense of responsibility, healthy families help develop their children's ability to discriminate between right and wrong. This morality may be grounded in specific religious principles, but it need not be; it is, however, based on a firm conviction that the world and its people must be valued and respected. Harvard psychologists Robert Coles and Jerome Kagan believe that children are naturally equipped with a basic moral sense; parents need to recognize and nurture that sense from early on. Besides helping children resolve their moral dilemmas, healthy parents take care to be responsible for their own behavior. Parental hypocrisy sets the stage for disrespect, disappointment, and rebellion.

Families with a spiritual orientation see a larger purpose for their family than simply their own maintenance and self-satisfaction. They see their families as contributing to the well-being of their neighborhood, city, country, or world, and as being an avenue through which love, caring, and hospitality can be expressed. Many families find support and expression of spiritual strength and purpose in religious associations. These families find in religious activities a transcendent framework on which they formulate family values, behavior patterns, and goals, as well as a source of strength with which they attempt to live out those values.

Enjoying nature's beauty together is one way for families to enrich their spiritual lives.

Spirituality gives meaning, purpose, and hope. In the case of formal religions, it provides a sense of community and support (Abbott, Berry, and Meredith, 1990). In times of adversity, families tap their spiritual resources to gain the strength that they need to sustain themselves. Stinnett and DeFrain (1985) suggest seven ways to help spirituality work in families:

1. Set aside time each day for meditation, prayer, or contemplation. Try to get outdoors to enjoy the beauty of nature.
2. Join (or help form) a discussion group to consider religious or philosophical issues.

3 Examine your own values. Try keeping a journal of your thoughts.
4 Help your children to clarify their values.
5 Identify three of your strengths. Work on developing them more fully. Identify three of your weaknesses. Decide how you can improve in these areas.
6 Have regular family devotional times: "Read . . . inspirational material, pray, sing, count your blessings, reaffirm your love and commitment to each other."
7 Volunteer time, energy, and money to a cause that helps others.

Spiritual wellness, it should be noted, is not necessarily the same as religiosity. Though religion can provide a source of personal, family, and social support as well as an opportunity for a family to engage in religious services together (Robinson and Blanton, 1993), it can also create a wedge between couples, especially when they disagree about the type and place of religion in their lives.

We can only hold to the name of family when we actively celebrate ourselves, tell each other stories, and pass the sense of oneness on to our children—the way we pass on a taste for chestnut stuffing or cranberry relish with orange. It is the traditions that we constantly create that become the special glue to hold us together over time and distance.

Ellen Goodman

Rituals and Traditions

Traditions, especially those rooted in our cultural background, help give us a sense of who we are as a family and as a community. Through tradition and ritual, families find a link to the past and, consequently, a hope for the future.

Speaking to the American Association for Counseling and Development, the renowned African-American writer Alex Haley said, "Americans should study their family histories, hold reunions, interview their elders and commemorate them with pictures" (Haley, 1986). Haley pointed to the far-reaching impact of his book *Roots* to illustrate the deep desire people have for family connectedness. *Roots* captured the interest of the nation. It tapped a process vital to all humankind, a process of telling and nurturing the family story, a process that is continually threatened by rapid social changes. "*Roots* was really born in the evenings of the summer when I was six years old," Haley said. He described how he would sit on his front porch, listening to his grandmother and her sisters talk about their parents, their grandparents, and the lives they had led. "I was greatly imprinted by my grandparents," he added.

Most successful families have someone who maintains and transmits the family story (Gunn, 1980). They have rituals and traditions that help them keep alive their identity as a family. Those traditions and rituals help the family remember their bondedness of commitment—the goals for which the family endured difficulties and the purposes for which the family works. Young people's personal identity, their sense of style and distinctive way of doing things, their sense of who they are and who they might become have a foundation in the sense of continuity and rootedness that comes from their identification with their family history.

Traditions and rituals vary greatly, from elaborate holiday celebrations to daily or weekly routines. Children's bedtime rituals—with blankets, books, bears, or prayers—give security and comfort, as do Sunday morning rituals that find the family together in church or in bed with the funny papers. Large and small rituals, if not passed down from preceding generations, can be freshly created and passed forward. Therapist Mary Whiteside (1989) advises that stepfamilies may benefit particularly from the creation of new rituals, which can "generate a feeling of closeness" and "provide fuel for weathering the more difficult times." She adds, "As these experiments . . . succeed enough to recur and to emerge as customs, the stepfamily unit begins to feel as if it has a life of its own." Rituals,

Prune Cake

Cream: 1 c. brown sugar
1/2 c. shortening
Add: 1 egg
3/4 c. combined orange and prune juice
a little grated orange rind
1 tsp. soda
1 tsp. cinnamon
1/2 tsp. allspice
salt
2 c. flour

Beat and add 1 c. prunes cut in pieces; add nuts if desired. Bake at 350 for 35 min. or until done. [Grandma Mimi's note at bottom follows.] "Found this old recipe. It's the one I used during the Depression—made and sold for profit."

Reflections

Make a list of your family traditions. Where did they come from? Which are your favorites? Your least favorites? Would you like to create any new ones? Make a list of your daily and weekly family rituals. Note your favorites. Are there others you'd like to incorporate into your family routine?

special practices, customs, and techniques unique to the family—whether ways of celebrating birthdays or the "correct" angle for mowing the lawn—imprint their family identity in the hearts and memories of its members.

Crisis Management

Everything comes of itself at the appointed time. This is the meaning of heaven and earth.

I CHING

The growth and development of the family through the life cycle inevitably produces crises. Growth involves change and, with it, a certain amount of stress. Unexpected events and those not prepared for can cause great pressures. Some families are able to cope with these effectively, and some may become immobilized. (Chapter 12 discussed family stress in some detail.)

Research consistently identifies the capacity to deal effectively with family crises as a characteristic of strong families (McCubbin and McCubbin, 1988; Burr and Klein, 1994). Members of strong families unite to face the challenges of a crisis. Each feels a part of the team and says, "What can I do?" No one carries the load alone.

Strong families are more able than other families to face challenges with confidence. The cumulative effect of other family strengths, such as commitment, cohesion, communication, and adaptability, enable strong families to evaluate the obstacle facing the family with resolve and to anticipate the resources and help needed. They are often able to recognize the task before the family as a normal developmental task, whereas less secure families might see the same situation as an unusual and overwhelming disaster.

If you're afraid of the dark, remember the night rainbow. If you lose the key to your house, throw away the house. And if it's the last dance, you better dance backwards.

COOPER EDENS

Perception itself is an important factor in the approach and solution to a problem. Strong families are able to accept changes resulting from crises and to see possibilities for growth in them. Thus, they are able to approach with hope situations that might generate despair and paralysis in other families. Family strengths such as adaptability, communication skills, spiritual resources, and a sense of unity and trust based on family commitment contribute to effective crisis management.

Ability to Seek Help

Experience is not what happens to a man; it is what a man does with what happens to him.

ALDOUS HUXLEY (1894–1963)

Another characteristic of strong families is associated with effective crisis management. This is the family's ability to be open to resources outside itself. Strong families are able to acknowledge their vulnerabilities. They feel themselves to be accepted members of the community and are able to ask for and receive assistance in time of need. Strong families' experiences of interdependence within the family better equip them to recognize, in turn, the interdependence among families and individuals in the neighborhood and community. They feel less compelled to prove their independence and to "go it alone."

A vast range of experiences (borrowing an egg from a neighbor, cooperative work on the school board, serving as a block parent, giving financial assistance when an unexpected illness keeps a neighbor homebound and out of work, and the like) continually involve families in interdependent relationships with their communities. Some needs can be met by normal exchanges among friends. Some can be met by help from extended family, from church, or from other support groups. Some needs require professional assistance. For example, family counseling may provide the context and support a family needs to examine difficult relationships and behavior patterns (Browne, 1997). Many community programs are available, such as parent education, marriage enrichment, and support groups dealing with alcoholism or other specific problems. (A guide to finding family resources is included in the Resource Center at the end of this book.)

Reflections

Does your family seek outside help for problems? Make a list of all the resources available to you that you might want to use sometime. Note the ones you've used in the past. The next time you have a family problem, take a look at your list.

Spending Time Together

When young couples look toward marriage, one of their anticipations is that they will be able to "be together." They anticipate the sharing of interests and activities that are mutually enjoyable. They look forward to companionship and shared psychological support. The expression of affection and meeting one another's emotional needs are sometimes romanticized as the central motivation for marriage. Certainly, these motivations are part of the vital glue that holds couples together.

As children arrive, the range of activities expands and relationships become more rich and complex. Having time to be together becomes increasingly crucial. The very process of making provisions for having time together becomes an important part of the building of family strengths. In some instances, it may be viewed as a "sacrifice" that is made by some members for the well-being of the whole family. Making time for family is an interlocking trait that brings together other characteristics found in strong families. It is an action that expresses commitment to the family.

The maturing of the love relationship that first brought the couple together requires their deliberately setting aside time to spend together. Competition of careers or other friendships may begin to erode the basic substance of their marriage without the couples' awareness unless they stop to listen to each other. New identities are being formed by both members of the marriage. Strong marriages are those in which couples create special occasions to be together. In strong families, "dating" does not stop after the wedding.

Spending time together is necessary to develop adequate communication and to build cohesion. Family traditions and rituals require family time. Family counselors indicate that one of the first signs of family difficulties is lack of family time together. Children and spouses learn they are appreciated, are valued, and have worth when other members spend time with them. The modeling and teaching that parents provide for their children require shared time, as does showing support to family members by attending school events and other occasions of special significance in the lives of family members.

Quality time together need not be spent on lengthy or expensive vacations: A relaxed meal or a game in which everyone participates can serve the same purpose a lot more often. Sometimes parents may have difficulty learning to leave a messy desk at work or a sticky kitchen floor at home in order to relax with each other or with their children. Yet time taken to play and relax with our loved ones pays off in ways that clean desks and shiny floors never can. When we are realistic, we see that the papers will never stop flowing onto the desk and the jam will never stop dripping onto the floor, but our children and our mates will never again be as they are now. If we don't take the time now to enjoy our families, we lose an opportunity forever. Healthy families know this and give play and leisure time high priority.

Reflections

Do you feel that your family has enough time for play and leisure? Can you give up anything to spend more time with your family? Can you reorder your priorities? Can they? Make a list of things you'd like to do with your family if you had the time. Take the time.

A Family Wellness Orientation

Having a *family wellness* orientation means making a conscious decision to live our lives in ways that move us toward optimal health in physical, emotional, intellectual, spiritual, and social dimensions. Wellness is positive and proactive; it focuses on being healthy and whole, as opposed to being fearful of disease and dysfunction.

Families with a wellness orientation identify their strengths. They recognize that strong, healthy family life does not just happen. It requires attention,

PERSPECTIVE

Tradition and Ritual in Modern American Life

If we think about all the rituals and holidays that we take for granted in our lives, we may be forced to conclude that humans are just naturally "party animals." Our lives are measured with life cycle rituals, mainly surrounding birth, the transition to adulthood, marriage, and death. Our years are punctuated with holidays and celebrations, both religious and secular. Weddings and funerals have their own unique characteristics among different ethnic groups, as do "baby-welcoming" rituals such as baptism among Christians and brit milah (circumcision) among Jews. Many ethnic and cultural groups also celebrate the child's transition to adult status with bar mitzvahs or bat mitzvahs, church confirmations, or "coming-out" parties. Other rituals, such as family reunions or tribal gatherings (for example, powwows), may be associated with a particular holiday or convened when the family or community feels the need.

In addition to "traditional" holidays celebrated in the United States, such as Christmas, Passover, and Thanksgiving, holidays introduced by more recent immigrants enrich our culture. Yuan Tan (Chinese New Year), Tet (Vietnamese New Year), Hana Matsuri (the Japanese observance of the Buddha's birthday), Ramadan (the Islamic month of fasting), and *el Dia de los Muertos* (the Latino Day of the Dead) are just a few examples of holidays currently celebrated by American families. Among African Americans, Juneteenth, a celebration of emancipation, and Kwanzaa, a celebration of African heritage, are increasingly observed.

Although many holidays commemorate historic events, their origins may lie in ancient rites having to do with the movements of the sun and moon, the cycle of the seasons, or other natural events, such as the flooding of a river or the ripening of acorns. Our ancestors' celebrations of the beginnings of new life or the giving of thanks for a successful harvest may have since taken on different meanings, but the observances often retain their ancient symbols—fertile bunnies and decorated eggs, for example, or displays of scarecrows along with the harvested grains, fruits, and nuts.

Besides giving people a way of expressing their unity with nature, ritual celebrations also bring them together in a common purpose and strengthen their bonds to one another and to the community. Some ceremonies, such as marriage, formally (we might say "legally") create new relationships (in-laws), with attendant duties and privileges. Rituals, such as coming-of-age ceremonies, also strengthen the ties of the individual to the group. Some elements common to many diverse rituals are the use of food and drink; costumes; music and dance; ritual objects such as candles, flowers, and feathers; representations of people or deities; and special words, such as prayers, chants, incantations, blessings, and vows. A brief sampling of holidays and rituals currently practiced in the United States follows. The holidays are discussed in the order they occur throughout the year.

- **Yuan Tan** Chinese New Year begins on the first day of the lunar year, at the first new moon when the sun is in the constellation known to Westerners as Aquarius (typically between January 10 and February 19). It is celebrated for fourteen days, until the full moon. The holiday culminates in a nighttime parade accompanied by fireworks and a huge, bejeweled dragon with a papier-mâché head; a long, red, fabric body; and the legs of many humans underneath. The dragon represents fertility and prosperity.
- **Tet** The Vietnamese New Year, Tet Nguyen Dan, also coincides with the start of the lunar year, although the celebration begins days earlier. Traditionally, Vietnamese families make new clothes and many new purchases in preparation for the coming year. The father or grandfather lights candles on the ancestral altar, and symbolic objects and many kinds of foods are offered. At midnight on the first day of the year, the ancestral spirits are welcomed. This day is considered portentous, so families carefully monitor their words and actions to ensure a year of harmony and good fortune.
- **Holī** The North Indian festival Holī has ancient roots as a celebration of the arrival of spring. At the first full moon of spring a ritual bonfire is kin-

preparation, and monitoring. Families oriented toward wellness take advantage of educational opportunities that help them gain perspective on family developmental processes. Marriage enrichment, parent education, and family retreats are occasions where families can exercise their family wellness orientation. There they can gain a perspective that helps them recognize, articulate, and prepare for upcoming changes in child development and family stages.

dled at moonrise. Holī is characterized by noise and activity—singing, shouting, dancing, and reveling in the streets. It is referred to as the Festival of Colors because of the brightly hued dyes with which celebrants drench one another. Traditionally, Holī provided an opportunity for those of low social status to be "equal" with those of high status by splashing them with dye or playing practical jokes. Today the celebration serves more as a reminder of the equality of all of humanity.

- **Pesach** The Feast of the Passover is a Jewish festival that begins after the first full moon of spring (Nisan 15 on the Jewish calendar). It commemorates the flight of the Jews from Egypt under the leadership of Moses. Pesach has been celebrated for more than three thousand years. Associated with Passover is the Seder, a family service and meal in remembrance of the ancestors' hardships. Traditional foods include matzoth (unleavened bread) and lamb (a holdover from the ancient sacrificial rite).
- **Hana Matsuri** On April 8, Japanese Buddhists commemorate the birthday of Siddhārtha Gautama (or Shakyamuni), the Buddha. Traditionally, images of the infant Buddha are bathed with hydrangea tea, and flowers and chanting are offered at Buddhist shrines and temples by adults and children alike. Hana Matsuri is known in English as the Flower Festival.
- **Juneteenth** The African-American holiday Juneteenth originated in Texas following the Civil War. Although the Emancipation Proclamation became law in January 1863, the last slaves were not freed until June 19, 1865, when Union troops rode into Galveston, Texas. Traditionally, Juneteenth (along with other Jubilee or Freedom Days) included such events as barbecues, baseball games, speeches, parades, concerts, and dances. As African Americans moved northward, the holidays moved with them. During World War II, however, such celebrations were considered unpatriotic, as they emphasized a separate history from the white majority. Renewed interest in African-American history has sparked the revival of Juneteenth in communities across the United States.
- **Ramadan** The Islamic Fast of Ramadan begins with the new moon of the tenth lunar month and lasts until the next new moon. During Ramadan, the faithful fast daily between daybreak and dusk and spend as much time as possible in the mosque (place of worship) or at prayer. A mosque may be brilliantly lit to commemorate the prophet Mohammed's illumination on the Night of Power. The Koran (the Muslim holy book) is read in its entirety over the course of the month. On the twenty-ninth night, the fast is broken, and the feast of Id al-Fitr begins. There are special foods, confections, and presents for the children.
- **Christmas** Although the exact date of Jesus's birth is not known, it has been celebrated by Christians since the fourth century on December 25, around the time of the winter solstice, the Roman holiday Saturnalia, and the Jewish Chanukah. Over time, Christmas has taken on many quasi-religious and nonreligious aspects, including feasts, the exchange of gifts, decorations (such as evergreen trees and branches, which symbolize eternal life), and rituals. The tradition of visits by Santa Claus (Saint Nicholas) was brought to the New World by early Dutch immigrants. The cartoonist Thomas Nast first gave Santa his sleigh and reindeer to take presents to soldiers in the Civil War. Music and pageantry are important to many people's Christmas celebrations, and a variety of traditions are practiced all over the United States by different ethnic and cultural groups.
- **Kwanzaa** The African-American holiday Kwanzaa, meaning "first fruits of the harvest" in Swahili, is celebrated from the day after Christmas until New Year's Day. Kwanzaa is an amalgam of many African harvest festivals, incorporating symbols such as the kinara, a seven-branched candleholder. Each candle represents an important African-American value. These are *umoja* (unity), *kujichagulia* (self-determination), *ujima* (collective work and responsibility), *ujamma* (cooperative economics), *nia* (purpose), *kuumba* (creativity), and *imani* (faith).

(continued on following page)

Cohesion versus Adaptability

Cohesion and adaptability are central dimensions to family functioning. Families appear to function best when levels of cohesion and adaptability are balanced, but there is often tension between the two as individuals' needs and desires conflict with those of the family as a whole. The Family Circumplex Model, discussed in

PERSPECTIVE

Tradition and Ritual in Modern American Life (continued)

- **La Quinceañera** On their fifteenth birthdays, young women of Latin American descent may be "introduced" into society in a ritual celebration known as the *quinceañera* (fifteenth-year celebration). A great deal of preparation usually goes into the event. It begins with a special Catholic mass involving the girl's family and godparents and proceeds to a dance and feast that is generally as lavish as the parents can afford. There may be both a dance band and a mariachi band; the guest of honor and her attendants, in their matching formal dresses, open the party with a specially choreographed waltz.
- **Powwows** Powwows—large gatherings of various Native American groups—involve demonstrations and contests of singing, dancing, and drumming. Powwows are secular, unlike tribal religious ceremonies, which are usually carried out solemnly in private. Some performers at powwows are professionals who travel the "powwow circuit," keeping the traditions of their particular tribe alive. But anyone can participate—even small children—and learn the steps their people have done for hundreds of years in dances portraying different animals, hunting techniques, or historical events. Traditional songs may be augmented with modern verses—to honor veterans of the wars in Korea and Vietnam, for example (Pareles, 1990). Powwows help keep Native American culture alive. A Mohawk singer expressed his feelings (quoted in Pareles, 1990): "I didn't learn anything about my culture when I was younger. But the more I find out, the more I've been seeing the logic of it and how beautiful and natural it was."

Chapter 2, measures family well-being in terms of the balance of cohesion and adaptability.

Family Cohesion

Cohesion, or closeness, is a common goal for many families. When asked to describe their family, most Americans will typically do so in terms of closeness or close-knittedness. Recipes for increasing closeness are regular features of popular magazines.

Cohesion refers to the emotional bonding that family members have toward one another—the amount of connectedness among family members (Olson, McCubbin, Barnes, Larson, Muxen, and Wilson, 1983). Cohesion is more than just time together; we may spend great amounts of time together and not necessarily be "connected" in any real sense. High degrees of cohesion indicate that family members feel close when they are together, and not lonely and disconnected; feel bonded, but not "stuck together"; have individual autonomy without being isolated; have a sense of unity without a loss of individuality; and have coalitions that perform expected functions without interference. Several elements, then, contribute to a family's cohesion: emotional bonding, boundaries, coalitions, time, space, interests, recreation, friends, and decision making.

- *Emotional bonding.* **Bondedness** is the feeling of closeness family members have for one another; bonding leads to support during difficult times. Unity in families implies pride in membership. Evidence for bondedness can be seen in the intense efforts family members make to preserve family patterns.
- *Boundaries.* **Boundaries** define space and separate family members from one another and from the rest of their environment. Family boundaries are

Time together is one of the family's most precious resources.

physical, emotional, and psychological (for example, doors and private space, role definitions, rules, and expected practices). Each family member requires boundaries that permit individuality and autonomy. Within a family system are many smaller combinations of relationships called subsystems. Subsystems may be formed by generation, by sex, by interest, or by function. Subsystems within the family also require boundaries. If individual boundaries are not permeable (open), the person will be distant, disconnected, and self-centered. If, in contrast, the individual boundaries are too permeable, the person will become fused with another person and have little separate identity. This applies to subsystems within the family as well as to individuals. The boundaries in strong families are open enough for lots of closeness; yet they allow individuals and subsystems to be free to do what they need to do for appropriate functioning and development.

- *Coalitions.* In a family, **coalitions** are alliances between two or more family members. Coalitions can either increase or decrease family cohesiveness. Coalitions may be formed on the basis of generation, gender, age, interest, and so on. Coalitions that promote unity are useful, and coalitions that are divisive are not.
- *Sharing time, space, interests, recreation, and friends.* Time spent together as a family reflects the cohesion of the family. How free time is utilized, how much effort is made to share time each day, and whether eating meals together is a priority are all measures of family cohesion. Similarly, shared space, shared interests and recreation, and knowing each other's friends are also indicators of cohesion. Families with high levels of cohesion tend to have much in common because members' and individual's pursuits are strongly supported by the rest of the family.
- *Decision making.* The way in which families make decisions reflects cohesion as well. When decisions are made with the family in mind, cohesion is promoted.

The level of cohesion in families can be expected to be different at various stages in the family life cycle. An understanding of how the individual development of children and parents influences the nature of cohesion at different stages in the life of the family is important in strong families.

If you cannot get rid of the family skeleton, you may as well make it dance.

GEORGE BERNARD SHAW (1856–1950)

Adaptability

Healthy families continually change and adapt as they grow through the life cycle. This important characteristic, family **adaptability,** is the ability of the marital/family system to change in response to situational and developmental stress. These changes include developmental changes (such as children becoming adolescents), structural changes (the wife becoming employed), or stress (a family member becoming chronically ill). A balance of change and continuity is necessary for a system to be healthy.

Six dimensions appear to be critical to family adaptability: leadership, assertiveness, discipline, negotiation, roles, and rules. These dimensions are all part of the family's organization. Some families can change easily and effectively; others are rigid and have difficulty changing; and yet others are chaotic, changing so much there is little family stability at all. There is always change in families and these are the six elements which are indicative of adaptability to change.

- *Leadership* refers to how guidance and direction are given in family decision making. The leadership continuum extends from families with virtually no recognized leadership to an autocratic family system in which all decisions are arbitrarily made and are adhered to rigidly. Neither extreme is found in strong families. Leadership is clear in strong families but moves from one person to another according to the area of family activity. Input is received from all members.
- *Assertiveness* is a family characteristic that reflects how free the members of the family are to express their opinions. In strong families, members express their requests and feelings without fear or intimidation. The families allow for individual self-confidence and self-expression as legitimate contributions to the decisions and directions to be taken by the family.
- *Discipline* is a definite part of family structure. A combination of nurturance and discipline helps to set limits and create the framework for desired family behavior (Starrels, 1994). Parents and children plan together the guidelines for family activities and individual responsibilities. Consequences from infractions are intended to assist in individual learning and to clarify family goals.
- *Negotiation* refers to patterns of settling family problems. Sometimes negotiations are characterized by clear commitment and a careful establishment of responsibilities. The family here is very dependable and secure, and it matches the developmental needs of its members with good problem-solving techniques. At the extremes of negotiation patterns are, on the one extreme, endless negotiations, and on the other, very limited negotiations. Strong families talk together about new ways of dealing with their problems.
- *Roles* refer to what part each member of the family plays in the family's day-to-day living. Some families may have dramatic role shifts or poorly defined roles; in either case, no one knows what is expected. In other families, roles are arbitrarily fixed with family position or tradition, and change is difficult to achieve. In strong families, roles can change rather easily and appropriately with the interests and abilities of the individuals and the well-being of the family.
- *Rules* can be explicit or implicit. In some families, many of the rules that govern behavior are hidden. These rules may determine individual behavior without the conscious acknowledgment of the members. Other rules may be more explicit. Some rules are open to negotiation, and others seem to be immutable. Strong families discuss rules, along with the reasons for them; rules in these families change as circumstances and family awareness direct.

PERSPECTIVE

Cohesion and Adaptability in Stepfamilies

Do healthy stepfamilies handle issues of cohesion and adaptability differently than members of healthy intact families? This issue is especially important if children are active members of a binuclear family. How do they manage being members of two nuclear families? Do stepfamilies have less cohesion than intact families? Do they have more adaptability?

According to Cynthia Pill (1990), healthy stepfamilies may differ from healthy intact families on both dimensions: Healthy stepfamilies are less cohesive and more adaptable than healthy intact families. Pill found that healthy stepfamilies demonstrated "remarkable capacity to allow for differences among family members and to permit looser, more flexible family relationships." Stepfamilies did not demand family unity at the price of individuality. Pill found that many healthy stepfamilies, for example, allowed their children to decide whether to come on the family vacation together or to spend time with their other birth parent. She also found that the closest emotional relationships generally were formed between biologically related family members. This meant that the stepparent was not necessarily as emotionally close as he or she might have been if the children were biologically related. Yet the stepparents noted this in a matter-of-fact manner, observing that it was understandable. Pill writes:

> While many stepparents wished that they might have been a part of the children's lives at an earlier stage, both because of the influence they would have had in the formation of the children's values and also because of the bonding that would have occurred, most were accepting of the limitations that entering children's lives later in their development brings.

Because stepfamilies have to adapt continually to the coming and going of family members, stepfamilies need to be more adaptable than intact families: They need to adjust expectations, define and redefine roles (such as redefining the father role to fit the stepfather), and create a meaningful definition of "family" for their stepfamilies.

Although stepfamilies are generally less cohesive than intact families, they tend to become more so over time. The overwhelming majority of stepfamilies in Pill's study (1990) felt they had become "family." When asked how they thought they had become a family, the family members replied that they had developed a sense of connectedness and a mutual feeling of caring and support among family members. Sharing major life events, such as birth or death, and living their mundane, everyday lives together were the major factors giving them a sense of family identity and helping to solidify family feelings. Such experiences might include eating spaghetti together when wild noodles, dripping with tomato sauce, flapped against Mom's white dress, or they might include common occurrences such as holding hands in prayer, dancing to favorite music, Dad's stupid but endearing jokes, or stories about dating disasters. Stepfamilies collect trivia, record odd events in the family's collective unconscious, recall revealing incidents, and weave these diverse strands together to create a family story or mythology. In creating family stories, stepfamilies begin to see themselves as uniquely different from all other families, and in doing so, stepfamilies become families in the deepest, most profound sense.

As with all other things, a balance offers the most potential for the adaptability that leads to strong families. At the midpoint there is some flexibility and some structure, allowing for orderly change and accommodation. This balance permits the family to adapt optimally to changing circumstances. (See Chapter 2 for a visual representation of cohesion and adaptability in the Family Circumplex Model.)

Family Form and Ethnic Identity

Each family is unique. However, different forms of families or families of different ethnic backgrounds often have distinctive characteristics. These differences may reflect a special family strength. However, an analysis of family diversity requires first the recognition of the commonality of family processes among families of all types (Walker, 1993).

Whereas an understanding of diversity requires us to appreciate the commonalities of family, we need also to take care to place the different forms of family life within the larger picture of families in general. The commonality that particular families have with other families constitutes their ethnic or cultural identity. It is this commonality that allows us to also appreciate the uniqueness of an individual or family. As we consider this somewhat paradoxical picture of uniqueness and commonality, we need also to remind ourselves of the dangers of stereotyping. Stereotyping results from a failure to appreciate individuality and uniqueness and to exaggerate limited aspects of personal or group characteristics—often for negative purposes.

Studies of different types of families often reveal similarities beneath apparent differences. Bettie Sanderson and Lawrence Kurdek (1993), for example, in a comparison of African-American and white couples, found that despite the existence of ethnic differences, the processes that are linked to relationship satisfaction are similar between these two groups. Another study found relations between a variety of family processes and child well-being to be similar for both traditional (nuclear) families and nontraditional families (Bronstein, Clauson, Stoll, and Abrams, 1993).

Family Strengths and Family Form

In single-parent, never-married families, a special form of loyalty may be found between the parent and child. This involves a conscious consideration of the tasks faced by the single parent and a strong determination to succeed in maintaining an effective family life. A significant contrast has been noted by researchers between single-parent, never-married families with a strong commitment to their families and single-parent, never-married families who do not affirm that commitment. In the committed families, the single parent accepts responsibility for maintaining a supportive family climate and often works with extended family members in developing the kind of psychological and social assistance needed for his or her family's success.

Reflections

What are the various strengths of your family? How do you contribute to them? What role do culture, religion, and extended family play in supporting your family?

Members of single-parent families often develop a sense of confidence and pride in maintaining and providing for their family in very difficult circumstances (Richards and Schmiege, 1993). Strengths of single-parent families include the following: (1) they have a more efficient decision-making system; (2) they have more direct communication, and parent and child share responsibilities in partnership fashion; (3) a greater sense of vitality is present in the work and contribution made by the child; and (4) children develop a more egalitarian view of the roles of men and women.

Resources of the extended family may come into play in unexpected ways as the family moves through life cycle changes or changes in family form. For example, research with college students (Kennedy and Kennedy, 1993) found that college students from stepfamilies tend to feel closer to a particular grandparent than do college students from intact or single-parent families. (Extended family strengths are discussed later in this chapter; stepfamily strengths were discussed in Chapter 15.)

Family Strengths and Ethnic Identity

Ethnicity, writes Monica McGoldrick (1989), "is more than race, religion, and national or geographic origin. . . . It involves conscious and unconscious processes that fulfill a deep psychological need for identity and a sense of historical continuity. It is transmitted by an emotional language within the family and reinforced by the surrounding community."

If you don't know your past
You don't know your future.

ZIGGY MARLEY

In the "melting pot" of American society, ethnicity is a complicated and ever-changing phenomenon. Although some ethnic groups intermarry with, and adapt rapidly to, mainstream society, others, by choice or by pressure from without, have retained many of their traditions and values. In this sense, the United States today can be seen more as a cauldron of very complex stew or a stir-fry dish than as a homogenized melting pot (McAdoo, 1993). Appreciating and respecting diverse types of families may shed light on how families can adapt effectively to adverse circumstances (Fine, 1993).

African-American Families

An ethnic group may possess particular strengths that help its families to survive in a larger society that is often less than welcoming and sometimes overtly hostile. During slavery and its aftermath (and amid the racism and discrimination still present in U.S. society today), African Americans needed to rely on their own families and community strengths. Strong kinship networks characterized the West African families whose members were sent to the Americas as slaves, and such networks have been a major resource of African-American families during times of trouble (Taylor, Chatters, and Jackson, 1993).

African-American families traditionally are centered around the children (creating a "pedi-focal" family system). Thus, the family unit can be defined as including all those involved in the nurturance and support of an identified child, regardless of household membership (Crosbie-Burnett and Lewis, 1993). Because of the high value placed on child rearing, the role of mother for African-American women is often more important than any other role, including that of wife, and because of the strong kinship bonds in the culture, the single mother is not left alone to raise her child (Ellison, 1990; Hampson, Beavers, and Huylgus, 1992). Black fathers also tend to be warm and loving toward their children.

Am I not a man and a brother?

ANONYMOUS, INSCRIPTION ON THE SEAL OF THE ANTISLAVERY SOCIETY OF LONDON, 1770

Family values such as unconditional love for children, respect for self and others, and the "assumed natural goodness of the child" are other strengths of the African-American family (Nobles, 1988). Robert Taylor (1990) and his

For many families, celebrations honoring their cultural or ethnic heritage provide a sense of connection and belonging. This African-American family is celebrating Kwanzaa.

colleagues note that a distinctive task of African-American parents is to attempt to prepare their children for the "realities of being black in America." Family values also include a special consideration for the elderly.

Family strengths associated with African-American families include:

The more hands that are offered, the lighter the burden.

HAITIAN PROVERB

- *An extended kinship network.* The close relationship between family members provides economic and moral support in both day-to-day and crisis situations (Wilkinson, 1993).
- *Flexibility of roles.* Households have the ability to expand and contract in response to external and internal pressures.
- *Resilient children.* Youth are socialized to obtain the best from "both worlds," African American and white, through survival techniques that provide a wide repertoire from which they may choose.
- *Egalitarian parental relationships.* There is a reciprocal exchange of roles, duties, and rights by both parents.
- *Strong motivation to achieve.* African-American parents support the education of their children.

(See also Carter, 1993, and Gary et al., 1986, describing the strong religious orientation that has historically been an integral part of African-American family life.)

Characteristics of African-American extended kin systems include (1) a high degree of geographical closeness; (2) a strong sense of family and familial obligation; (3) fluidity of household boundaries, with a great willingness to absorb relatives—both real and affiliated, adult and minor—if the need arises; (4) frequent interaction with relatives; (5) frequent extended family get-togethers for special occasions and holidays; and (6) a system of mutual aid (Hatchett and Jackson, 1993).

Latino Families

Cuando cuentas cuentos,
cuenta cuantos cuentos cuentas
cuando cuentas cuentos.
(When you tell stories,
count how many stories you tell
when you tell stories.)

SPANISH TONGUE TWISTER

Although there is great diversity among Spanish-speaking cultures, Latino families tend to live in nuclear families near others in the extended family network. Among Latinos, father-child relationships are often playful and companionable. Children's ties with mothers are primary and lifelong no matter what the geographic distance or age. Children are taught to carry family responsibilities, to prize family unity, and to respect their elders (Chilman, 1993). Close relationships with maternal and paternal grandparents are fundamental. Of special importance are the emotional ties with the mother's relatives. Maternal aunts often serve as "brokers," providing a link between parents and other adults in the family (Wilkinson, 1993). Latino culture emphasizes the family as a basic source of emotional support, especially for children. No sharp distinction is made between relatives and friends; in fact, friends are considered virtually kin if a close relationship has been formed. The term *compadrazgo* (co-parentship) is often used for this relationship (Chilman, 1993).

In Puerto Rican families, the role of motherhood is especially central. This is expressed in the term for the culture's ideal feminine role, *marianismo*, so named for the Virgin Mary. In this role, a woman is seen to realize herself and derive her life's greatest satisfaction through motherhood (Zayas and Palleja, 1988). Family interdependence and family unity are central concepts in the Puerto Rican family culture (Sánchez-Ayéndez, 1986).

Families of Mexican heritage integrate family into their daily living. They emphasize strong kinship bonds, and intimate emotional relationships are prominent among family members. Children are at the center of family life (Wilkinson, 1993). Mexican Americans tend to emphasize the needs of the family above those of the individual. A parent may readily stay home from work or a

UNDERSTANDING YOURSELF

Family Strengths Inventory

The Family Strengths Inventory was developed by Nick Stinnett and John DeFrain to identify areas of strengths and weaknesses in families. Complete the inventory, and then compute your score by adding the numbers you have circled. (You might also have other family members complete it independently to get a sense of their perceptions of your family.) The score will fall between 13 and 65.

What does the score mean, according to the researchers? A score below 39 is below average. It indicates that there are areas in your family relationships that need improving. Look at the areas in which you scored low to help you target where you need work. Scores from 39 to 52 are average. Scores above 52 indicate strong families. Consider discussing the inventory results with your family.

Circle on a 5-point scale (with 1 representing the least degree and 5 representing the greatest degree) the degree to which your family possesses each one of the following:

Spending time together and doing things with one another

1 2 3 4 5

Commitment to one another

1 2 3 4 5

Good communication (talking with one another often, listening well, sharing feelings with one another)

1 2 3 4 5

Dealing with crises in a positive manner

1 2 3 4 5

Expressing appreciation to one another

1 2 3 4 5

Spiritual wellness

1 2 3 4 5

Circle the degree of closeness of your relationship with your spouse on a 5-point scale (with 1 representing the least degree and 5 representing the greatest degree):

1 2 3 4 5

Circle the degree of closeness of your relationship with your children on a 5-point scale (with 1 representing the least degree and 5 representing the greatest degree):

1 2 3 4 5

Circle the degree of happiness of your relationship with your spouse on a 5-point scale (with 1 representing the least degree and 5 representing the greatest degree):

1 2 3 4 5

Circle the degree of happiness of your relationship with your children on a 5-point scale (with 1 representing the least degree and 5 representing the greatest degree):

1 2 3 4 5

Some people make us feel good about ourselves; that is, they make us feel self-confident, worthy, competent, and happy about ourselves. What is the degree to which your spouse makes you feel good about yourself? Indicate it on the following 5-point scale (with 1 representing the least degree and 5 representing the greatest degree):

1 2 3 4 5

Indicate on the following 5-point scale the degree to which you think you make your spouse feel good about himself or herself (with 1 representing the least degree and 5 representing the greatest degree):

1 2 3 4 5

Indicate on the following 5-point scale the degree to which you think you make your children feel good about themselves (with 1 representing the least degree and 5 representing the greatest degree):

1 2 3 4 5

child may stay home from school to care for a sick family member. There is strong family pride. In some regions, Mexican-American families live in multifamily houses or nuclear households in close geographic proximity. A child's *padrinos* (godparents) are called *compadres* (co-parents) by his or her parents. The *compadre* (co-father) and *comadre* (co-mother) are an important part of the family support system (Horowitz, 1983). *Compadres* feel "a deep sense of obligation to each other for economic assistance, support, encouragement, and even personal correction" (Fitzpatrick, 1981).

Family strengths associated with Latino families include (Vega, 1995):

- *Family focus.* The family is a major priority and as such receives strong commitment by all members.

- *Strong ethnic identity.* The importance of culture in the Latino family binds and gives families strength and identity.
- *High family flexibility.* In contrast to the family being male dominated, there is flexibility of family roles.
- *Supportive network of kin.* The extended family is both a tradition and a support of family.
- *Equalitarian decision making.* As traditions shift, input by all family members is valued.
- *Family cohesion.* Although cohesion decreases across generations living in the United States, Latino families still remain cohesive.

Asian-American Families

Resiliency marks the lives of many Asian-American families living in the United States. Even though they have faced prejudice and discrimination, Asian Americans have maintained both family values and ties and in many cases, economic stability. Responsibilities to aged parents and to close relatives are fundamental to the family institution among Japanese, Chinese, Filipino, and other Asian families living in the United States (Wilkinson, 1993). Most often, an elderly parent lives in the same household as the adult child.

In Chinese-American families, the family orientation is summarized in the concept of *hsiao*, or filial piety, which involves a series of obligations of child to parent. First, the child is to provide aid, comfort, affection, and contact with the parent in an attitude of loving warmth and reverence. Second, the child is to bring reflected glory to the parent by doing well in education and occupational activity (Lin and Liu, 1993). A child who misbehaves brings shame to the family name (Braun and Chao, 1978). Children are taught that everyone has to work for the welfare of the family. They are given a great deal of responsibility and are assigned specific chores. Adolescents are responsible for supervising young children and for work around the house or in the family business (McLeod, 1986). The concept of *hsiao* provides a stability, purpose, and structure for the activities and commitment of all the family members.

Japanese-American families are often cohesive units that act as agents of both socialization of their children and social control of their members. Typical features that characterize strengths of the Japanese-American family include: close family ties between generations indicated by strong feelings of loyalty to family; low divorce rates, which often are seen as an indicator of family stability; and a complex system of values and techniques of social control including guilt, shame, obligation, and duty, which are transmitted from parents to children. These characteristics reflect the continuities between traditional Japanese culture and contemporary Japanese Americans (Takagi, 1994).

Vietnamese-American families are the most recent Asian population to enter the United States in significant numbers. They bring a strong, traditional Vietnamese extended family structure with them. This includes a group that stretches beyond immediate or nuclear family ties to include a wide range of kin. These extended households mesh in a large and active web of kinship relations in the neighborhood and general vicinity. These relations with kin often function as important sources of economic and social support.

Due to patterns of immigration and disruption of family relations, Vietnamese-American families have developed variations in the traditional extended family household and extended kin system. People are incorporated into the kin network who would not have been part of it in traditional Vietnam. These include more distant relatives and nonrelated friends. These reconstructed kin networks reflect the importance placed on familial relations by Vietnamese Americans. These extended kin networks also provide for mutual aid, exchanging goods, services (like child care and cooking), and information (about government agencies, hospitals, and so on) (Kibria, 1994).

Strengths of Asian-American families include:

- *Filial piety.* A series of obligations and great respect for elders exist among Asian-American families.
- *Family as a cohesive unit.* Both the nuclear family and the extended family play important roles in the lives of Asian-American families.
- *Value of education.* From preschool through college, parents support and encourage their children's education.
- *Feelings of loyalty.* Close family ties, low divorce rate, and maintenance of traditional values all reflect loyalty to family and culture.
- *Extended family support.* Emotional and financial support exists among nuclear and extended family.

Native-American Families

Native Americans are a diverse group. There are hundreds of Indian tribes, and although they share many traditions and beliefs, they also differ from one another in significant ways. Generally, however, Native-American families see human life as being in harmony with nature. They are usually group oriented and emphasize cooperation (Markstrom-Adams, 1990). Relations with kin are often characterized by residential closeness, obligatory mutual aid, active participation in life cycle events, and the presence of central figures around whom family ceremonies revolve. Relatives tend to live near one another and become involved in the daily lives of the members of the kinship unit. Women play fundamental roles in these extended systems. One of the enduring features of the Native-American family has been its dynamic quality. Native-American families have a great capacity to adapt to a changing social environment.

Extended family networks may include several households of significant relatives that assume a village type of character. Transactions within and among these households occur within a community context. Despite a history of severe dislocation and disenfranchisement among Native Americans, the community-family configuration is a mark of an integrated social organization that serves to offset the ravages of perpetual displacement (Wilkinson, 1993). In a study of Native Americans' adjustments to urban life, anthropologist John Price (1981) found that kinship networks remained "strong and supportive." In Los Angeles, he found that Indians from over a hundred tribes shared pride in their common heritage through Native-American sports leagues and dance groups, powwows, and traditional crafts. Many urban Indian families also commute frequently to their home reservations to renew their family and cultural ties.

A special role for the elderly has historically been recognized as a strength in Native-American families. Despite the tenuous economic conditions under which many Native American elderly live, they maintain social and cultural ties to their families. They perform a variety of important and beneficial roles, including instructing the young and helping care for children. For example, one-fourth of elderly Native Americans take charge of caring for at least one grandchild and two-thirds reside within five miles of family, with whom they share socialization, chores, and routine obligations (Edwards, 1983). Elders are seen as a resource for young parents to assist them in understanding traditional roles of discipline

Reflections

Do you consider yourself part of a particular ethnic group? Are there special strengths you see within your group? Would you consider your group to be assimilated into American society? Accepted by American society? If you don't consider yourself part of a particular group, do you nevertheless preserve certain aspects of your ancestors' ethnicity? If so, what are they?

and child rearing. They also maintain responsibility for remembering and relating tribal philosophies, myths, traditions, and stories peculiar to their tribal groups (Yellowbird and Snipp, 1994).

Contemporary research emphasizes the continuing importance of certain "core family values" that enable Native Americans to preserve important aspects of their culture (Yellowbird and Snipp, 1994). The concepts of *time*, *cooperation*, *leadership*, *sharing*, and *harmony with nature* are often viewed quite differently by Native Americans than by those in the dominant culture. For example, Native Americans tend to view the role of a leader as more sacred, more humanistic, more person oriented, more honest, more intuitive, and less ambitious. The leader's role is one of servitude rather than one of assertiveness (Lewis and Gingerich, 1980).

Major family strengths of Native-American families include:

- *Extended family network.* Strong ties exist among relatives as well as with extended kin and tribe.
- *Value placed on cooperation and groups.* Geographic proximity and solid family values give relatives the opportunity to support and become involved in each others' daily lives.
- *Respect for the elderly.* Social and cultural ties help to maintain respect for elders, who are seen as teachers and resources for the young.
- *Tribal support system.* There is reliance and trust in tribal support for advice and resolution of problems.
- *Preservation of culture.* Core family values are supported by means of the maintenance of native language, harmony with nature, respect for leadership, and high family cohesion.

Kin and Community

In addition to our immediate family, other networks of caring and identity provide strength and a sense of belonging to us as individuals and families. These networks offer a larger context of purpose, continuity, posterity, and support. Relationships with extended family members and family units (aunts, uncles,

A variety of cultural traditions unify families and communities. Navajo women pass their weaving skill (and the stories and cultural teachings that accompany it) from generation to generation.

cousins), enduring "familylike" (affiliated) relationships, abiding friendships, and participation in social organizations and their attendant roles and responsibilities in the community create meaning and promote security.

Raising children isn't an individual act. It is a social and communal enterprise, involving kin, neighbors, other parents, friends, and many other unrelated adults.

BARBARA DAFOE WHITEHEAD

Intimacy Needs

Whether we are married or single, we need relationships in which we can be intimate. Neither couples nor individuals can function well in isolation. Robert Weiss (1969) notes that people have needs that can be met only in relationships with other people. These needs can be summarized as follows:

1. *Nurturing others.* This need is filled through caring for a partner, children, or other intimates, both physically and emotionally.
2. *Social integration.* We need to be actively involved in some form of community; if we are not, we feel isolated and bored. We meet this need through knowing others who share our interests and participating in community or school projects.
3. *Assistance.* We need to know that if something happens to us, there are people we can depend on for help. Without such relationships, we feel anxious and vulnerable.
4. *Intimacy.* We need people who will listen to us and care about us; if such people are not available, we feel emotionally isolated and lonely.
5. *Reassurance.* We need people to respect our skills as persons, workers, parents, and partners. Without such reassurance, we lose our self-esteem.

The Extended Family: Helping Kin

Few aspects of family life exist to which relatives (especially parents and siblings) do not make a significant contribution. Among African-American families, for example, grandparents are often a valuable resource for child care and child socialization (Flaherty, Facteau, and Garver, 1990; Kennedy, 1990). Adult siblings are also important sources of help in times of need (Chatters, Taylor, and Neighbors, 1989).

In many families, parents loan money to their adult children at low or no interest. They may give or loan them the down payment for a home, for example. The obligations that the loans and gifts entail differ according to a person's age and marital status. If the children are young and single, parents may still expect to exercise considerable control over their children's behavior in return for their support. But if the children are older or, more important, married, there are fewer obligations.

Even when extended families are separated geographically, they continue to provide emotional support. Contacts with kin are especially important in the lives of the aged (Kivett, 1993).

Families today create their own relatives as needed.

BARBARA SETTLES

Affiliative Kin

Blood relationships do not define the type of feelings that a person will have. Instead, they provide a framework to encourage brotherly feelings toward a brother, motherly feelings toward a child, and so on. As we discussed earlier, the strength of kinship ties ultimately depends more on feeling than on biology. A brother or sister can seem like a stranger; a grandmother can be more of a mother

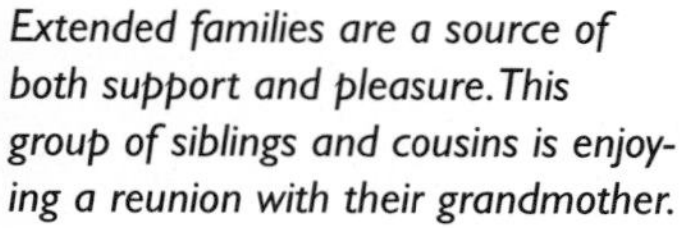

Extended families are a source of both support and pleasure. This group of siblings and cousins is enjoying a reunion with their grandmother.

than one's biological mother; a parent can be like a brother or sister. Feelings of kinship can extend beyond traditional kin. We form affiliative kin by transforming friends and neighbors into kin. "He is like a brother to me;" "We are like cousins."

Among African Americans, affiliative kin are important in child rearing. Especially in single-parent African-American families, one or more people may be responsible for providing money for a child; others may contribute clothing or meals and still others, nurturance and guidance. Unrelated men and women take on family roles and often have the same rights and privileges as related family members (Crosbie-Burnett and Lewis, 1993). In Latino culture, *compadres* (godparents, literally co-parents) are important figures in the family. The *compadre* (godfather) and *comadre* (godmother) are considered to be responsible for the child's spiritual development, and they are available as resources in time of need. Among lesbians and gay men, affiliative kin are important as a means of including one's partner and his or her relatives and friends as family.

Because divorce and geographic mobility are breaking down our abililty to interact with in-laws and biological kin, we are beginning to form new kinds of kin (Lindsey, 1982). Single persons may attempt to create families from friends by sharing time, problems, meals, and housing with one another. Single parents may form networks with other single parents for emotional support and exchange of child care. Family networks may be formed in which three or four families from the same neighborhood share problems, exchange services, and enjoy leisure time together. Sometimes involved neighbors may be called by a family name, such as "Auntie" or "Gramps."

My friends are my estate.

EMILY DICKINSON (1830–1886)

Friendship

Families today find close friends, casual acquaintances, and temporary support within a complex and changing social network. Traditional sources of community support, such as church groups, are augmented by other sources. Parents meet one another and form friendships in groups that center around their children: parent-teacher groups, Scouts and Campfire, athletic groups, and so on. Peer support groups and self-help groups are available to fill a variety of short- or long-term needs; lasting friendships are often formed within such contexts. People may join political groups to work with others for change in society; they also join hobby

PERSPECTIVE

Welfare Reform: Hardship or Hope?

When we examine our most recent attempt to revamp the welfare system, we can't help but wonder what effect the interplay between politics and economics will have on our children. The welfare reform package, signed into law by President Clinton in 1996, is an attempt at solving a vexing problem: how to wean individuals and families off the welfare system and help them to become more productive citizens. New federal guidelines now give each state money to manage its own welfare program, as long as it meets new federal goals. The guidelines stipulate that recipients will only be eligible to receive aid for five years. Aid to Families with Dependent Children, the existing welfare grant program, will be replaced by a program called "Temporary Assistance for Needy Families" or TANF. Under TANF, participants will be expected to get involved in welfare department-operated work training programs. As the state creates jobs for parents, it must also pave the way to providing available and affordable child care. But licensed family care is unlikely to meet the needs of the four million welfare families and working poor who are mandated to work (Kilborn, 1997). Furthermore, in cities such as New York, Chicago, and Boston, the cost of care for even one child may be almost equal to the earnings of a minimum-wage worker. This situation could encourage wider use of unqualified child-care providers. Nevertheless, the welfare law requires that recipients be working within two years and terminates their cash assistance in five years, so by then they must be able to support themselves. Critics of welfare reform say the poor who have legitimate reasons for a parent's unemployment will get caught without a safety net (Livernois, 1997). On the other hand, there is hope that the emphasis of such reform, which is to put more people to work, will be good for the individuals and families who do succeed under the new system. Time will tell.

groups, crafts clubs, or dance classes in order to share their special interests with others. Some networks of support provide specific information or services for dealing with specific needs, whereas others provide more general help (Cooke et al., 1988). Social networks enrich our day-to-day lives and provide vital assistance in times of stress and crisis.

Family in the Community

Urie Bronfenbrenner (1979) has proposed that we look at the family in an "ecological environment." He suggests that we think of this environment in terms of a set of nested Russian dolls. The developing child is the tiny innermost doll contained inside the various systems, such as home or school, that influence him or her. The interplay among the many systems profoundly affects the child in their midst. Furthermore, the influences of external factors, such as the flexibility of the parents' job schedules or the availability of good health care, also play a critical role.

If you have built castles in the air, your work need not be lost; that is where they should be. Now put the foundations under them.

HENRY DAVID THOREAU (1817–1862)

The well-being of the family depends not solely on its own resources, then, but also on the support it receives from the community in which it is embedded (Unger and Sussman, 1990). This community includes extended family, friends, schools, employers, health care providers, and government agencies at local, state, and federal levels. As discussed in Chapter 11, the United States is far behind most industrialized nations in terms of the support it gives families in the areas of health, education, social welfare, and workplace policy. The need for creativity and energy in these areas is great and presents challenges and opportunities for those who wish to work with the families of today and tomorrow.

Despite the complexities of modern life, the families that love, shelter, and teach us remain America's greatest national resource. They deserve to be nurtured, strengthened, and protected.

Here is the test to find whether your mission on earth is finished: If you're alive, it isn't.

RICHARD BACH, *Illusions*

Summary

- Marital success requires the development of interpersonal competence, the ability to share and develop an intimate, growing relationship with another. Commitment (including sexual fidelity and the willingness to sacrifice), communication, and the creative use of conflict are essential aspects of success in marriage.
- The family should be seen as a process. Family health may change over the course of the family cycle.
- The immense work of families includes socializing children and sustaining the care, supervision, and accountability of all family members. The "climate" of the family determines how these services are carried out. Strong families do not just happen. Developing and maintaining family strengths take work.
- *Family strengths* are the patterns that enable successful families to survive and grow. Strengths identified in this chapter include (1) sustaining a commitment to the family; (2) giving affirmation, respect, and trust; (3) developing communication skills; (4) promoting responsibility, morality, and a spiritual orientation; (5) preserving rituals and traditions; (6) managing crises creatively and effectively; (7) utilizing the ability to seek help; (8) spending time together; (9) processing a family wellness orientation; and (10) balancing levels of cohesion and adaptability.
- Families appear to function best when levels of *cohesion* (connectedness among family members) and *adaptability* (the ability to change in response to stress) are balanced. Elements of family cohesion include emotional bonding, boundaries, coalitions, time, space, interests, recreation, friends, and decision making.
- Specific family strengths may be associated with families that correspond to particular family structures. Single-parent families may benefit from efficient decision making, direct communication, a strong sense of contribution to the family by children, and more egalitarian views of roles for women and men. Extended families, stepfamilies, and other types of families may be likely to profit from strengths specific to those forms.
- A sense of ethnicity or cultural heritage is an i=mportant source of strength for many American families. Often an ethnic group possesses particular strengths that help its families survive in an indifferent or unwelcoming larger society.
- We all have needs that can be met only in close relationships with others. These needs are to nurture others, to be socially integrated in some form of community, to have assistance we can rely on, to be intimate with others, and to be reassured about our skills. Kin, friends (including affiliative kin), neighbors, and a variety of social networks all contribute to our well-being. They provide assistance and emotional support that give our families added strength.
- The family exists in the environment of community and is influenced directly and indirectly by many systems existing outside of itself. The well-being of the family depends not only on its own resources but also on the support of the larger community. For this reason, policies supportive of the family are crucial to its survival.

Key Terms

adaptability 576
bondedness 574
boundary 574
coalition 575
cohesion 574
family strengths 562

Suggested Readings

Acock, Alan, and John Demo. *Family Diversity and Well-Being.* Thousand Oaks, CA: Sage, 1994. A comprehensive examination of families of varying structures by two leading family researchers.

Beavers, Robert W., and Robert Hampton. *Successful Families: Assessment and Intervention.* New York: Norton, 1990. A systems approach to building successful families.

Borenstein, M.H., ed. *Cultural Approaches to Parenting.* Hillsdale, NJ: Lawrence Erlbaum Associates, 1991. This book discusses child-rearing practices in a wide variety of cultures in the United States and abroad.

Gottman, John M. *Why Marriages Succeed or Fail, and How You Can Make Yours Work.* New York: Simon and Schuster, 1995. This popular author focuses his research on marriage and provides food for thought about what makes it strong.

Hogan, M.J., ed. *Initiatives for Families: Research, Policy, Practice, Education.* St. Paul, MN: National Council on Family Relations, 1995. An excellent and well-respected contribution to the United States International Year of the Family.

Kagan, Sharon L., and Bernice Weissbourd, eds. *Putting Families First: America's Support Movement and the Challenge of Change.* San Francisco: Jossey-Bass, 1994. Contributing authors discuss how families can be strengthened through policy in education, health care, social services, and religious organizations.

McAdoo, Harriette Pipes, ed. *Family Ethnicity: Strength in Diversity.* Newbury Park, CA: Sage Publications, 1993. A collection of twenty-one essays on major American ethnic groups, stressing their strengths.

McCubbin, Hamilton I., ed. *Family Types and Strengths.* Edina, MN: Burgess International Group, 1988. An examination of family strengths; emerging family types; and recent developments in family stress therapy, coping, and support.

McGoldrick, Monica, John Pearce, and Joseph Giordano, eds. *Ethnicity and Family Therapy.* 2nd ed. New York: Guilford Press, 1996. Separate chapters examining over twenty-five ethnic groups. Each chapter includes a historical overview, cultural traits, and values. The book treats ethnicity as an important aspect of psychological functioning.

Appendix A

Sexual Structure

The Female Reproductive System

External Genitalia

The female external genitalia are known collectively as the *vulva*, which includes the mons veneris, labia, clitoris, urethra, and introitus. The *mons veneris* (literally, "mountain of Venus") is a protuberance formed by the pelvic bone and covered by fatty tissue. The *labia* are the vaginal lips surrounding the entrance to the vagina. The *labia majora* (outer lips) are two large folds of spongy flesh extending from the mons veneris along the midline between the legs. The outer edges of the labia majora are often darkly pigmented and are covered with pubic hair beginning in puberty. Usually the labia majora are close together, giving them a closed appearance. The *labia minora* (inner lips) lie within the fold of the labia majora. The upper portion folds over the clitoris and is called the *clitoral hood*. During sexual excitement, the labia minora become engorged with blood and double or triple in size. The labia minora contain numerous nerve endings that become increasingly sensitive during sexual excitement.

The *clitoris* is the center of erotic arousal in the female. It contains a high concentration of nerve endings and is highly sensitive to erotic stimulation. The clitoris becomes engorged with blood during sexual arousal and may increase greatly in size. Its tip, the *clitoral glans*, is especially responsive to touch.

Between the folds of the labia minora are the urethral opening and the *introitus*. The introitus is the opening to the vagina; it is often partially covered by a thin perforated membrane called the *hymen*, which may be torn accidentally or intentionally before or during first intercourse. On either side of the introitus is a tiny *Bartholin's gland* that secrets a small amount of moisture during sexual arousal.

Internal Genitalia

The *vagina* is an elastic canal extending from the vulva to the cervix. It envelops the penis during sexual intercourse and is the passage through which a baby is normally delivered. The vagina's first reaction to sexual arousal is "sweating," that is, producing lubrication through the vaginal walls.

A few centimeters from the vaginal entrance, on the vagina's anterior (front) wall, there is, according to some researchers, an erotically sensitive area that they have dubbed the "Grafenberg spot" or "G-spot." The spot is associated with female ejaculation, the expulsion of clear fluid from the urethra, which is experienced by a small percentage of women.

A female has two *ovaries*, reproductive glands (gonads) that produce *ova* (eggs) and the female hormones *estrogen* and *progesterone*. At the time a female is born, she already has all the ova she will ever have—more than forty thousand of them. About four hundred will mature during her lifetime and be released during ovulation; ovulation begins in puberty and ends at menopause.

The Path of the Egg

The two *fallopian tubes* extend from the uterus up to, but not touching the ovaries. When an egg is released from an ovary during the monthly *ovulation*, it drifts into a fallopian tube, propelled by waving *fimbriae* (the finger-like projections at the end of each tube). If it is fertilized by sperm, fertilization usually takes place within the fallopian tube. The fertilized egg will then move into the uterus.

The *uterus* is a hollow, muscular organ within the pelvic cavity. The pear-shape uterus is normally about 3 inches long, 3 inches wide at the top, and 1 inch at the bottom. The narrow, lower part of the uterus projects into the vagina and is called the *cervix*. If an egg is fertilized, it will attach itself to the inner lining of the uterus, the *endometrium*. Inside the uterus it will develop into an embryo and then into a fetus. If an egg is not fertilized, the endometrial tissue that developed in anticipation of

Figure A.1
External Female Genitalia

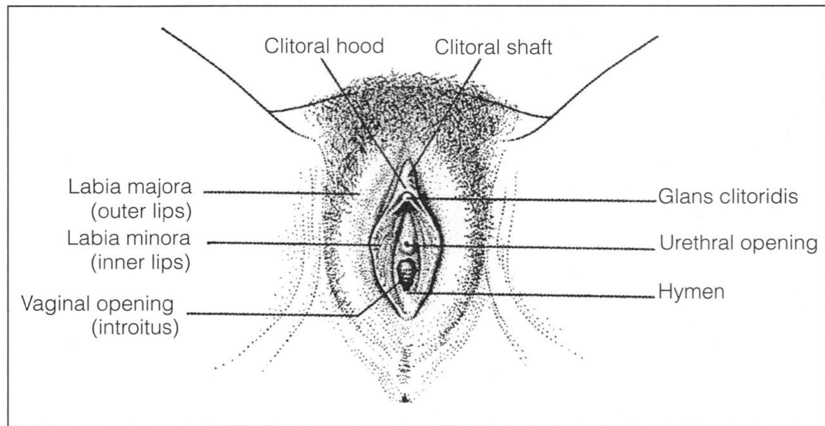

fertilization will be shed during *menstruation.* Both the unfertilized egg and the inner lining of the uterus will be discharged in the menstrual flow.

The Male Reproductive System

The Penis

Both urine and semen pass through the penis. Ordinarily, the penis hangs limp and is used for the elimination of urine because it is connected to the bladder by the urinary duct (urethra). The penis is usually between 2.5 and 4 inches in length. When a man is sexually aroused, it swells to about 5 to 8 inches in length, is hard, and becomes erect (hence, the term *erection*). When the penis is erect, muscle contractions temporarily close off the urinary duct, allowing the ejaculation of semen.

The penis consists of three main parts: the root, the shaft, and the glans penis. The *root* connects the penis to the pelvis. The *shaft*, which is the spongy body of the penis, hangs free. At the end of the shaft is the *glans penis*, the rounded tip of the penis. The opening at the tip of the glans is called the *urethral meatus.* The glans penis is especially important in sexual arousal because it contains a high concentration of nerve endings, making it erotically sensitive. The *frenulum*, a small area of skin on the underside of the penis where the glans and shaft meet, is especially sensitive. The glans is covered by a thin sleeve of skin called the *foreskin.* Circumcision, the surgical removal of the foreskin, may damage the frenulum.

When the penis is flaccid, blood circulates freely through its veins and arteries, but as it becomes erect, the circulation of blood changes dramatically. The arteries expand and increase the flow of blood into the

Figure A.2
Cross Section of the Female Reproductive System

1. A follicle matures in the ovary and releases an ovum. 2. The fimbriae trap the ovum and move it into the fallopian tube. 3. The ovum travels through the fallopian tube to the uterus. 4. If the ovum is fertilized, the resulting blastocyst descends into the uterus. 5. If not fertilized, the ovum is discharged through the cervix into the vagina along with the shed uterine lining during the menstrual flow. 6. The vagina serves as a passageway to the body's exterior.

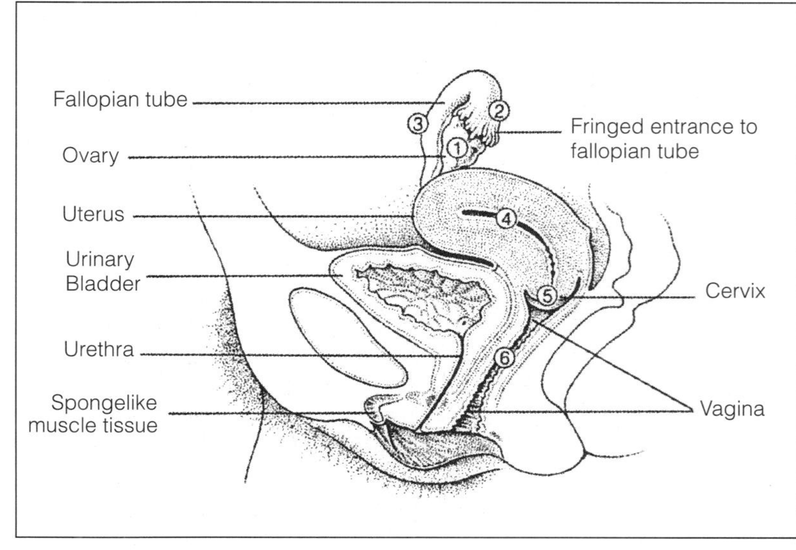

Figure A.3
External Male Genitalia

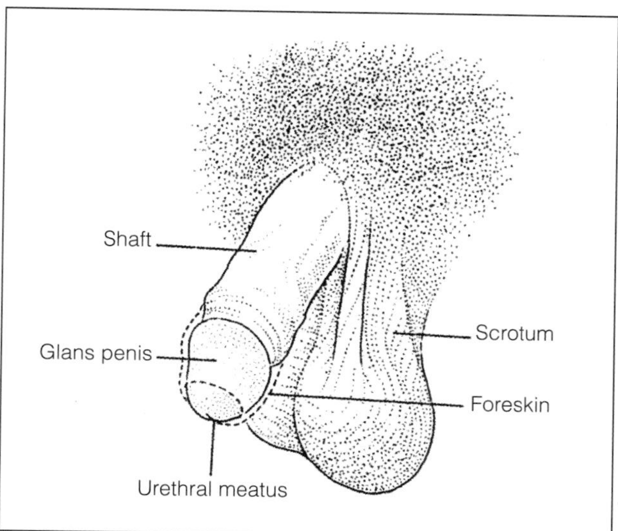

penis. The spongelike tissue of the shaft becomes engorged and expands, compressing the veins within the penis so that the additional blood cannot leave it easily. As a result, the penis becomes larger, harder, and more erect.

The Testes

Hanging behind the male's penis is his *scrotum*, a pouch of skin holding his two *testes* (singular *testis*; also called *testicles*). The testes are the male reproductive glands (also called *gonads*), which produce both sperm and the male hormone *testosterone*. The testes produce sperm through a process called *spermatogenesis*. Each testis produces between 100 million and 500 million sperm daily. Once the sperm are produced, they move into the *epididymis*, where they are stored prior to ejaculation.

The Path of the Sperm

The epididymis merges into the tubular *vas deferens* (plural *vasa deferentia*). The vasa deferentia can be felt easily within the scrotal sac. Extending into the pelvic cavity, each vas deferens widens into a flasklike area called the *ampulla* (plural *ampullae*). Within the ampullae, the sperm mix with an activating fluid from the *seminal vesicles*. The ampullae connect to the *prostate gland* through the *ejaculatory ducts*. Secretions from the prostate account for most of the milky, gelatinous liquid that makes up the *semen* in which the sperm are suspended. Inside the prostate, the ejaculatory ducts join to the urinary duct from the bladder to form the urethra, which extends to the tip of the penis. The two *Cowper's glands*, located below the prostate, secrete a clear, sticky fluid into the urethra that appears as small droplets on the meatus during sexual excitement.

If the erect penis is stimulated sufficiently through friction, an ejaculation usually occurs. *Ejaculation* is the forceful expulsion of semen. The process involves rhythmic contractions of the vasa deferentia, seminal vesicles, prostate, and penis. Altogether, the expulsion of semen may last from three to fifteen seconds. It is also possible to have an orgasm without the expulsion of semen.

Figure A.4
Cross Section of the Male Reproductive System

1. The testis produces sperm. 2. Sperm mature in the epididymis. 3. During ejaculation, sperm travel through the vas deferens. 4. The seminal vesicles and the prostate gland provide fluids. 5. Sperm mix with the fluids, making semen. 6. Semen leaves the penis by way of the urethra.

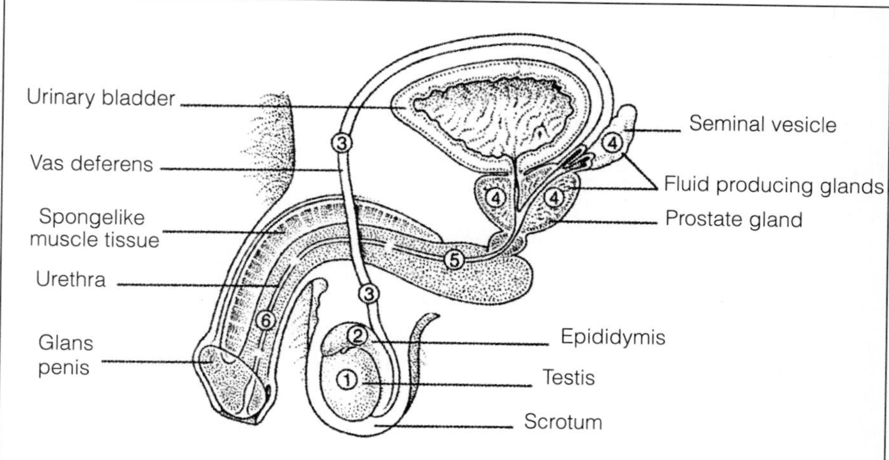

Appendix B

The Sexual Response Cycle

Psychological and Physiological Aspects

When we respond sexually, we begin what is known as the *sexual response cycle*. Helen Singer Kaplan (1979) developed a model to describe the sexual response cycle. According to this model, the cycle consists of three phases: the desire phase, the excitement phase, and the orgasmic phase. The desire phase represents the psychological element of the sexual response cycle; the excitement and orgasmic phases represent its physiological aspects.

Sexual Desire

Desire can exist separately from overtly physical sexual responses. It is the psychological component that motivates sexual behavior. We can feel desire but not be physically aroused. It can suffuse our bodies without producing explicit sexual stirrings. We experience sexual desire as erotic sensations or feelings that motivate us to seek sexual experiences. These sensations generally cease after orgasm.

Physiological Responses: Excitement and Orgasm

A person who is sexually excited experiences a number of bodily responses. Most of us are conscious of some of these responses: a rapidly beating heart, an erection or lubrication, and orgasm. Many other responses may take place below the threshold of awareness, such as curling of the toes, the ascent of the testes, the withdrawal of the clitoris beneath the hood, and a flush across the upper body.

The physiological changes that take place during sexual response cycle depend on two processes: vasocongestion and myotonia. *Vasocongestion* occurs when body tissues become engorged with blood. For example, blood fills the genital regions of both males and females, causing the penis and clitoris to enlarge. *Myotonia* refers to increased muscle tension as orgasm approaches. Upon orgasm, the body undergoes involuntary muscle contractions and then relaxes. (The word *orgasm* is derived from the ancient Sanskrit *urja*, meaning "vigor" or "sap.")

Excitement Phase

In women, the vagina becomes lubricated and the clitoris enlarges during the excitement phase. The vaginal barrel expands, and the cervix and uterus elevate, a process called "tenting." The labia majora flatten and rise; the labia minora begin to protrude. The breasts may increase in size, and the nipples may become erect. Vasocongestion causes the outer third of the vagina to swell, narrowing the vaginal opening. This swelling forms the *orgasmic platform*; during sexual intercourse, it increases the friction against the penis. The entire clitoris retracts but remains sensitive to touch.

In men, the penis becomes erect as a result of vasocongestion, and the testes begin to rise. The testes may enlarge to as much as 150 percent of their unaroused size.

Orgasmic Phase

Orgasm is the release of physical tensions after the buildup of sexual excitement; it is usually accompanied by ejaculation of semen in physically mature males. In women, the orgasmic phase is characterized by simultaneous rhythmic contractions of the uterus, orgasmic platform, and rectal sphincter. In men, muscle contractions occur in the vasa deferentia, seminal vesicles, prostate, and the urethral bulb, resulting in the ejaculation of semen; contractions of the rectal sphincter also occur. Ejaculation usually accompanies male orgasm, but ejaculation and orgasm are separate processes.

Following orgasm, one of the most striking differences between male and female sexual response occurs as males experience a *refractory period.* The refractory period denotes the time following orgasm during which male arousal levels return to prearousal or

Figure B.1
Stages of Female Sexual Response (internal left; external right)

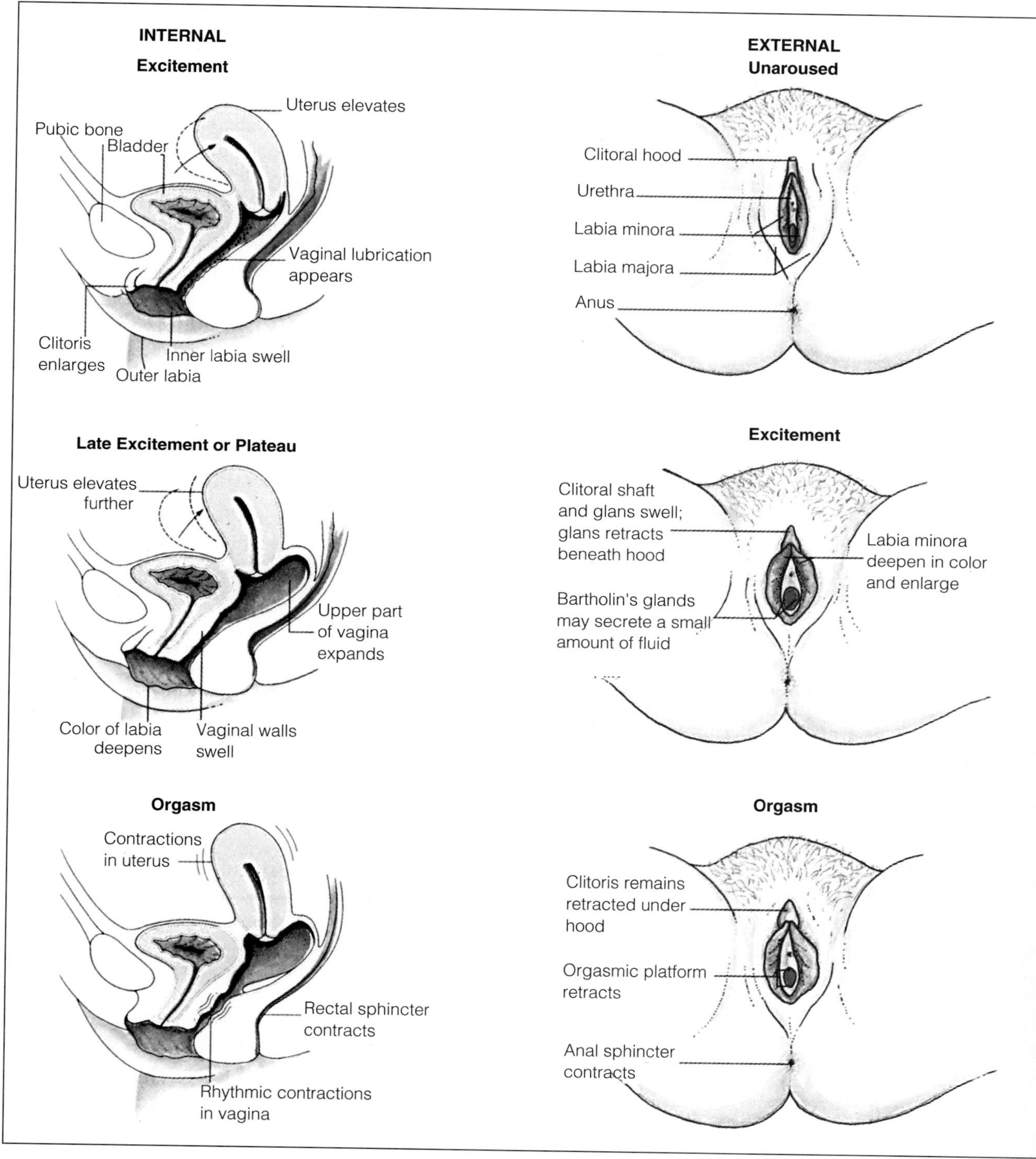
INTERNAL
Excitement
Uterus elevates
Pubic bone
Bladder
Vaginal lubrication appears
Clitoris enlarges
Inner labia swell
Outer labia
Late Excitement or Plateau
Uterus elevates further
Upper part of vagina expands
Color of labia deepens
Vaginal walls swell
Orgasm
Contractions in uterus
Rectal sphincter contracts
Rhythmic contractions in vagina
EXTERNAL
Unaroused
Clitoral hood
Urethra
Labia minora
Labia majora
Anus
Excitement
Clitoral shaft and glans swell; glans retracts beneath hood
Labia minora deepen in color and enlarge
Bartholin's glands may secrete a small amount of fluid
Orgasm
Clitoris remains retracted under hood
Orgasmic platform retracts
Anal sphincter contracts

Figure B.2
Stages of Male Sexual Response

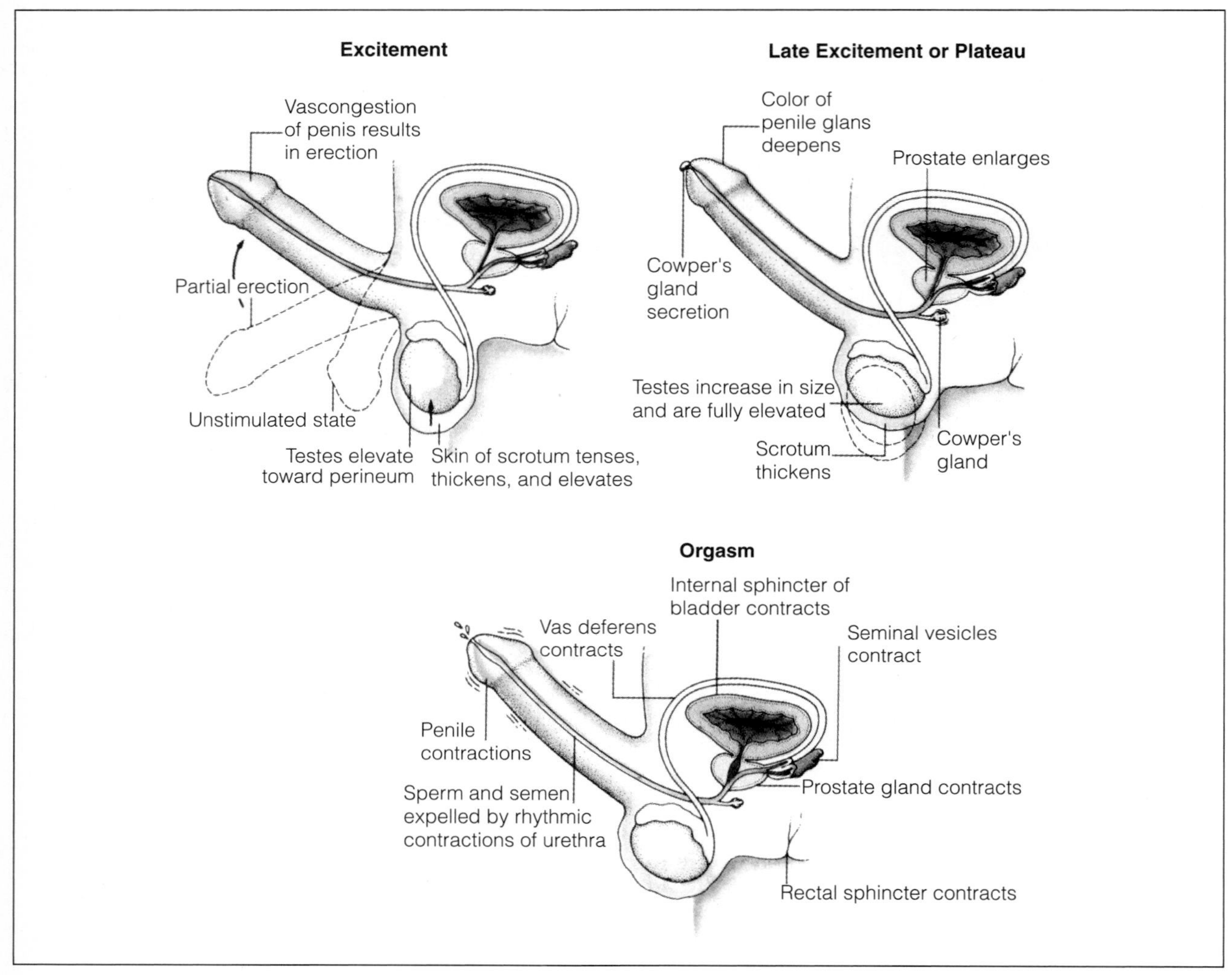

excitement levels. During the refractory period, additional orgasms are impossible. Females do not have any comparable period. As a result, they have greater potential for multiple orgasms—that is, for having a series of orgasms. Although most women have the potential for multiple orgasms, only about 13 to 16 percent regularly experience them. For multiple orgasms, women generally require continued stimulation of the clitoris. Most women (or their partners), however, do not seek additional orgasms after the first one because our culture uses the first orgasm (usually the male's) as a marker to end sexual activities.

In sexual intercourse, orgasm has many functions. For men, it serves a reproductive function by causing ejaculation of semen into a women's vagina. For both men and women, it is a source of erotic pleasure, whether it is an autoerotic or relational context; it is intimately connected with our sense of well-being. We may measure both our sexuality and ourselves in terms of orgasm. Did I have one? Did my partner have one? When we measure our sexuality by orgasm, however, we discount activities that do not necessarily lead to orgasm, such as touching, caressing, and kissing. We discount erotic pleasure as an end in itself.

Men tend to be more consistently orgasmic than women, especially in sexual intercourse. If all women are potentially orgasmic, why do smaller proportion of women have orgasms than men? An answer may be found in our dominant cultural model that calls for female orgasm to occur as a result of penile thrusting during heterosexual intercourse in the face-to-face, male-above position. This traditional American model calls for a "no-hands" approach. The women is supposed to be orgasmic without manual or oral stimulation by her partner or herself. If she is orgasmic during masturbation or cunnilingus, such orgasms are usually discounted because they aren't considered "real" sex—that is, heterosexual intercourse.

The problem for women in sexual intercourse is that the clitoris frequently does not receive sufficient stimulation from penile thrusting alone to permit orgasm. In an influential study on female sexuality, Shere Hite (1976) found that only 30 percent of her three thousand respondents experienced orgasm regularly through sexual intercourse "without more direct manual clitoral stimulation being provided at one time of orgasm." Hite concludes that many women need manual stimulation during intercourse to be orgasmic. They also need to be assertive. There is no reason why a women cannot be manually stimulated by herself or her partner to orgasm before or after intercourse. But to do so, a woman has to assert her own sexual needs and move away from the idea that sex is centered around male orgasm.

Appendix C

Fetal Development

Once fertilization of the ovum by a sperm occurs, the birth will take place in approximately 266 days, if the pregnancy is not interrupted. Traditionally, physicians count the first day of the pregnancy as the day on which the woman began her last menstrual period; thus, they calculate the gestation (pregnancy) period to be 280 days, which is also 10 lunar months.

Following fertilization, which normally occurs within the fallopian tube, the fertilized ovum, or *zygote*, undergoes a series of divisions during which the cells replicate themselves. After four or five days, the zygote contains about a hundred cells and is called *blastocyst*. On about the fifth day, the blastocyst arrives in the uterine cavity, where it floats for a day or two before implanting itself in the soft, blood-rich uterine wall (endometrium), which has spent the past three weeks preparing for its arrival. This process of *implantation* takes about a week. The hormone human chorionic gonadotropin (HCG), which is secreted by the blastocyst, maintains the uterine environment in an "embryo-friendly" condition and prevents the shedding of the endometrium that would normally occur during menstruation.

The blastocyst, or pre-embryo, rapidly grows into an *embryo* (which will, in turn, be referred to as a *fetus* around the eighth week of development). During the first two or three weeks of development, the embryonic membranes, including the *amnion*—a membranous sac that will contain the embryo and *amniotic fluid*—and the *yolk sac* are formed.

During the third week, extensive cell migration occurs and the stage is set for the development of the organs. The first body segments and the brain begin to be formed. The digestive and circulatory systems begin to develop in the fourth week; the heart begins to pump blood. By the end of the first month, the spinal cord and nervous system have also begun to develop.

The fifth week sees the formation of arms and legs. In the sixth week, the eyes and ears form. At seven weeks, the reproductive organs begin to differentiate in the males; female reproductive organs continue to develop. At eight weeks, the fetus is about the size of a thumb, although the head is nearly as large as the body. The brain begins to function to coordinate the development of the internal organs. Facial features begin to form, and bones begin to develop.

Arms, hands, fingers, legs, feet, toes, and eyes are almost fully developed at twelve weeks. At fifteen weeks, the fetus has a strong heartbeat, fair digestion, and active muscles. Most bones are developed by then, and the eyebrows appear. At this stage, the fetus is covered with a fine, downy hair called *lanugo*. (Figure C1 shows the actual size of the developing embryo and fetus through its first sixteen weeks.)

Throughout its development, the fetus is nourished through the *placenta*. The placenta begins to develop from part of the blastocyst following implantation. This organ grows larger as the fetus does, passing nutrients from the mother's bloodstream to the fetus, to which it is attached by the umbilical cord. The placenta blocks blood corpuscles and large molecules.

By five months, the fetus is 10 to 12 inches long and weighs between one-half and one pound. The internal organs are well developed, although the lungs cannot function well outside the uterus. At six months, the fetus is 11 to 14 inches long and weighs more than a pound. At seven months, it is 13 to 17 inches long, weighing about three pounds. At this point, most healthy fetuses are viable—capable of surviving outside the womb. (Although some fetuses are viable at five or six months, they require specialized care to survive.) The fetus spends the final two months of gestation growing rapidly. At term (nine months), it will be about 20 inches long and weigh about seven pounds. (Photographs in Chapter 8 show embryonic and fetal development.)

Figure C.1
Embryo and Fetus Growth

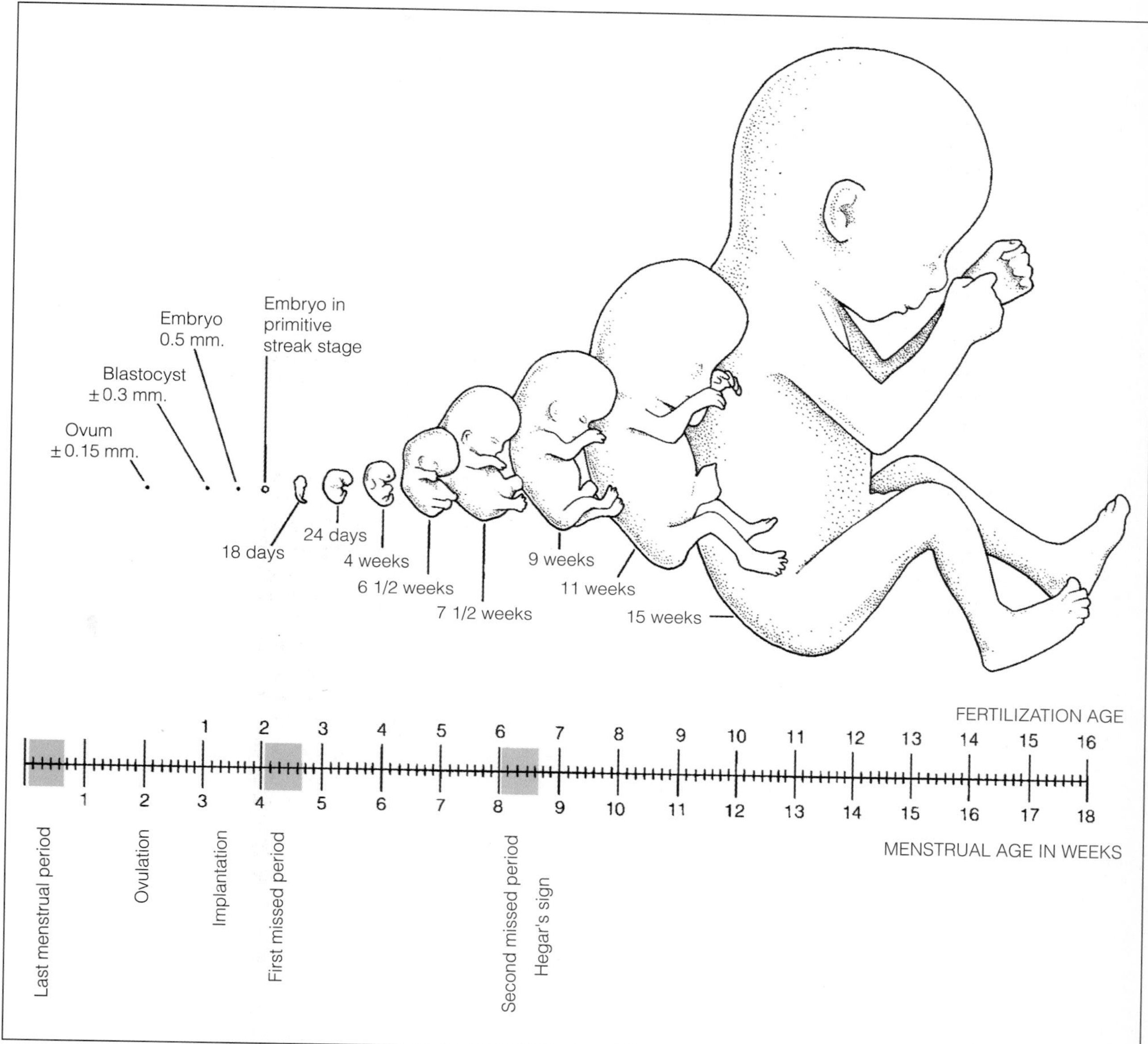

Ovum
±0.15 mm.
Blastocyst
±0.3 mm.
Embryo
0.5 mm.
Embryo in
primitive
streak stage
18 days
24 days
4 weeks
6 1/2 weeks
7 1/2 weeks
9 weeks
11 weeks
15 weeks
FERTILIZATION AGE
1 2 3 4 5 6 7 8 9 10 11 12 13 14 15 16
1 2 3 4 5 6 7 8 9 10 11 12 13 14 15 16 17 18
MENSTRUAL AGE IN WEEKS
Last menstrual period
Ovulation
Implantation
First missed period
Second missed period
Hegar's sign

Resource Center

Reference Guides for Studying Marriage and the Family

The following reference guides are designed to assist the student in pursuing information or research in various aspects of marriage and the family.

Marriage and Family Studies

Representative subject headings in library databases for marriage include the following:

- Betrothal.
- Common law marriage.
- Communication in marriage.
- Courtship.
- Deaf—Marriage.
- Divorce.
- Domestic relations.
- Endogamy and exogamy.
- Family.
- Family life education.
- Home.
- Honeymoon.
- Husbands.
- Love.
- Marital status.
- Marriage counseling.
- Married people.
- Mate selection.
- Matrimonial advertisements.
- Matrimony.
- Polyandry.
- Polygamy.
- Posthumous marriage.
- Remarriage.
- Sacraments.
- Sex.
- Sex in marriage.
- Sexual ethics.
- Teenage marriage.
- Weddings.
- Wives.

Representative subject headings for family include the following:

- Adoption.
- Birth order.
- Black families.
- Brothers and sisters.
- Children.
- Clans and clan system.
- Daughters-in-law.
- Divorce.
- Familial behavior in animals.
- Family size.
- Fathers.
- Grandparents.
- Homelessness.
- Households.
- Jewish families.
- Joint family.
- Kinship.
- Matriarchy.
- Mothers.
- Mothers-in-law.
- Only child.
- Parent and child.
- Parenthood.
- Parents.
- Parents-in-law.
- Polygamy.
- Poverty.
- Problem family.
- Rural families.
- Single-parent family.
- Tribes and tribal system.
- Twins.
- Unmarried couples.
- Widowers.
- Widows.

Journals

The following periodicals are of major interest to researchers, teachers, and students of marriage and family:

American Journal of Family Therapy focuses on issues of particular interest to counselors and family therapists.

American Journal of Orthopsychiatry is a multidisciplinary journal focusing on human development and mental health.

Archives of Sexual Behavior is a major journal in sex research.

The Black Scholar frequently contains articles regarding family, gender roles, feminism, and so on that are relevant to African Americans.

Canadian Journal of Sociology frequently contains articles about Canadian marriages and families.

Current Population Reports, Series P-20, is published several times annually by the U.S. Bureau of the Census; it provides the latest demographic statistics on various aspects of marriage and the family.

Family Law Quarterly reviews legal issues and the family.

Family Law Reporter is a weekly newsletter dealing with the legal aspects of family life, such as rights, violence, divorce, and abortion.

Family Planning Perspectives deals with birth control, sex education, and other aspects of family planning.

Family Process focuses on family psychotherapy, especially from a general systems approach.

Family Relations is a major journal emphasizing the implications of research on family counseling and services.

Hispanic Journal of Behavioral Science is a major journal addressing Latino issues, including the family.

Home Economics Research Journal is published bimonthly to encourage dialogue between home economists and scholars in related fields concerned with the well-being of the family and individuals.

Journal of Black Studies often includes articles relating to family issues.

Journal of Comparative Family Studies is dedicated to the cross-cultural study of marriage and the family.

Journal of Divorce focuses on divorce issues for counselors, therapists, lawyers, and family-life professionals.

Journal of Family History publishes historical articles; it focuses mostly on families in the United States, Canada, and Europe.

Journal of Family Issues is a multidisciplinary journal that alternates between general family issues and specific themes under a guest editor.

Journal of Family Violence is devoted to the emerging field of family violence.

Journal of the History of Sexuality is an interdisciplinary journal.

Journal of Home Economics focuses on general home economics themes and devotes a portion of each issue to a specific topic.

Journal of Marital and Family Therapy focuses on therapy issues, often from a family systems perspective.

Journal of Marriage and the Family is a multidisciplinary research journal devoted to the study of marriage and the family. It is the leading journal in the field.

Journal of Personal and Social Relationships focuses on personal relationships from an interdisciplinary perspective.

Journal of Sex and Marital Therapy focuses on therapy issues.

Journal of Sex Research is an interdisciplinary journal that contains articles, commentary, and book reviews.

Marriage and Family Review contains current abstracts and review articles of scholarly interest.

Mediation Quarterly: Journal of the Academy of Family Mediators focuses on specific mediation topics in each issue.

SAGE: A Scholarly Journal of Black Women devotes each issue to various aspects of women's lives, such as mother-daughter relations, health, and education.

Sex Roles: A Journal of Research is an interdisciplinary journal on research and theory regarding gender roles.

Signs: Journal of Women in Culture and Society is devoted to research, essays, reports, and commentaries about women.

Women and Society is a sociologically oriented journal dealing with women's issues.

Bibliographies and Indexes

The most useful general bibliographies that index scholarly articles from the United States and Canada include *Sociological Abstracts* (published six times annually), *Psychological Abstracts* (published quarterly), *Social Science Index* (published quarterly), and *Social Sciences Citation Index*. Specific bibliographies on marriage and the family include *Inventory of Marriage and Family Literature* (published annually) and *Sage Family Studies Abstracts* (published quarterly). Each issue of the journal *Family Relations* contains a bibliography (compiled from Family Resources Database) and a review essay on a specific topic.

Readers' Guide to Periodical Literature contains subject and author indexes of more than 180 general publications (such as *Time*, *Newsweek*, and *Harper's*) in the United States. *The New York Times Index* is a guide to the contents of the *New York Times;* it includes brief summaries of articles. *Newspaper Index* lists the contents of the *New York Times*, *Los Angeles Times*, *Wall Street Journal*, and *Washington Post*. *Black Newspaper Index* indexes articles appearing in African-American newspapers. *HAPI: Hispanic American Periodical Index* indexes articles in Spanish and supplements index articles in English.

Another helpful source is *Guide to Reference Books*, 11th ed., edited by Robert Balay (Chicago: American Library Association, 1996). It lists bibliographies, encyclopedias, handbooks, periodicals, and general works on hundreds of topics relevant to marriage and the family.

Databases

Many college, university, and public libraries have facilities for making computer searches on DIALOG and BRS information retrieval systems, which index U.S., Canadian, and many foreign language journals. These services will help you quickly compile bibliographies on virtually any topic in marriage and the family. You may request printouts of relevant bibliographic citations and journal abstracts. Check with your reference librarian about the available information retrieval services.

The most useful databases are PsycLit (based on *Psychological Abstracts*), Sociofile (based on *Sociological Abstracts*), Family Resources Database (based on *Inventory of Marriage and Family Literature*), Wilson Line (based on *Social Sciences Index*), and CenData (which retrieves published and unpublished census data). Chicano Database (based on the *Hispanic American Periodical Index* and other sources) retrieves data on Mexican Americans in both English and Spanish. *Index Medicus* indexes medical publications. *Inventory of Marriage and Family Literature: Family Studies Database, 1970–1995* is available on CD-ROM.

Dictionaries and Encyclopedias

The Encyclopedia of Marriage, Divorce, and the Family. By Margaret DiCanio. N.Y.: Facts on File, 1989.

A comprehensive treatment of present-day family life in North America, presented through more than 500 entries on such subjects as living wills, throwaway children, bisexual spouses, and homeless families. Detailed index, bibliography, and appendices.

The Dictionary of Family Psychology and Family Therapy: A Dictionary of Concepts and Terms. By Luciano L'Abate. Newbury Park, CA: Sage Publications, 1993.

An encyclopedic dictionary with definitions, examples, sources, and references to uses of the term.

Family Therapy Glossary. Ed. by Craig A. Everett [Rev. ed.]. Wash.: American Assoc. for Marriage and Family Therapy, 1992.

Alphabetical arrangement of terms, with three- to six-sentence definitions.

Handbooks

American Families: A Research Guide and Historical Handbook. Ed. by Joseph M. Hawes and Elizabeth I. Nybakken. N.Y.: Greenwood, 1991.

An introductory guide emphasizing recent work.

Book Reviews

A book review is an article that describes and evaluates a book soon after its publication. Popular books may be published and reviewed almost simultaneously. More scholarly reviews may appear two or three years after publication.

To find a book review, you'll need the book's author, title, and date of publication. If you're not sure of this information, check the library catalog or ask at the reference desk for assistance. Some indexes list reviews under an author's name, whereas others may group them under the headings "Books" or "Book Reviews."

General Book Review Indexes

- *Book Review Digest.* 1905–. Lists and quotes from reviews in ninety-five popular journals from all disciplines.
- *Book Review Index.* 1965–.
- *Combined Retrospective Index to Book Reviews in Scholarly Journals.* 1886–1974. Indexes over 1 million reviews in 459 journals in history, political science, and sociology.

General Periodical Indexes That Include Book Reviews

- *Arts and Humanities Citation Index.*
- *General Science Index.*
- *Humanities Index.*
- *Magazine Index.*
- *Readers' Guide to Periodical Literature.*
- *Social Sciences Citation Index.*
- *Social Sciences Index.*

Specialized Periodical Indexes That Include Book Reviews

- *Alternative Press Index.*
- *America: History and Life.*
- *Chicano Periodical Index.*
- *Education Index.*
- *Guide to Reviews of Books from and about Hispanic America.*
- *Hispanic American Periodical Index (HAPI).*
- *Historical Abstracts.*
- *Index to Black Periodicals.*
- *Index to Legal Periodicals.*
- *Women's Studies Abstracts.*

Women's Studies

Representative subject headings used in catalogs and periodical indexes include the following:

- Afro-American women.
- Education of women.
- Family Planning.
- Feminism.
- Hispanic American women.
- Jewish women.
- Lesbianism.
- Matriarchy.
- Mexican American women.
- Midwives.
- Minority women.
- Mothers.
- Rape.
- Sex role.
- Single women.
- Women—Crimes against.
- Women—Employment.
- Women-Legal status.
- Women-Services for.
- Women and religion.
- Women scientists.
- Women's rights.
- Women's studies.

Consult the *Library of Congress Subject Headings* to identify other headings.

Encyclopedias and Handbooks

- *American Black Women in the Arts and Social Sciences.* 1994.
- *American Women Writers.* Abridged edition. 1988.
- *Atlas of American Women.* 1987.
- *The Nature of Woman: An Encyclopedia and Guide to the Literature.* 1980. Short essays and quotations.
- *Resourceful Woman.* 1994.
- *Social Issues Resources Series—Women.* Vol. 4. 1994.
- *Women's Sourcebook: Resources and Information to Use Everyday.* 1994.
- *Women's Studies Encyclopedia.* 1989.
- *Women's World: A Timeline of Women in History.* 1995.

Guides to the Literature

- *American Indian Women: A Guide to Research.* 1991.
- *Introduction to Library Research in Women's Studies.* 1985.
- *Lesbianism: An Annotated Bibliography.* 1992.

- *Powerful Images: A Woman's Guide to Audio-visual Resources.* 1986.
- *Women of Color in the United States: A Guide to the Literature.* 1989.
- *Women's Collections; Libraries; Archives; and Consciousness.* 1986.
- *Women's Studies: A Guide to Information Resources.* 1990.

Directories, Indexes, and Abstracts

- *A Directory of Women's Media.* 1992.
- *Encyclopedia of Women's Associations Worldwide.* 1993.
- *National Women of Color Organizations.* 1991.
- *NWO: A Directory of National Women's Organizations.* 1992.
- *The Chicana/Latina Directory.* 1993.
- *Women's Health Perspectives: Abstracts and Bibliography.* 1988.
- *Women's Information Directory.* 1993.

Check *Humanities Index, Alternative Press Index, Left Index, Psychological Abstracts, Social Sciences Index, Sociological Abstracts, Chicano Periodicals Index,* and *Index to Articles by and about Blacks.* Many of these indexes can be searched on computer or CD-ROM.

Bibliographies

- *Athena Meets Prometheus: Gender, Science and Technology: A Selected Bibliography.* 1988.
- *Feminism and Women's Issues.* 1989.
- *Feminist Legal Literature: A Selective Annotated Bibliography.* 1991.
- *500 Great Books by Women: A Reader's Guide.* 1994.
- *Mothers and Mothering: An Annotated Feminist Bibliography.* N.Y.: Garland, 1991, compiled by Penelope Dixon. Eleven topical chapters treating mothering and daughters, sons, family, children, feminism, psychoanalysis, and reproductive issues; also covers single, working, lesbian, and black mothers. Includes 351 annotated entries for articles and books. 1970–90.
- *The Jewish Woman, 1900–1980: A Bibliography.* 1982.
- *Native American Women: A Contextual Bibliography.* 1993.
- *Older Women in 20th Century America: A Selected Annotated Bibliography.* 1982.
- *The State-by-State Guide to Women's Legal Rights.* 1987.
- *Violence against Women: An Annotated Bibliography.* 1992.
- *Women: A Bibliography of Bibliographies.* 1986.
- *Women and Aging: A Bibliography.* 1994.
- *Women of Color and Southern Women: A Bibliography of Social Science Research.* 1975–1992.
- *Women's Studies: A Bibliography.* 1980–1988.
- *Women's Studies: A Bibliography of Dissertations* 1870–1982.
- *Women's Studies: A Recommended Core Bibliography.* 1987.

Biographical Sources

- *Index to Women of the World from Ancient to Modern Times: Biographies and Portraits.* 1970. Supplement, 1988.
- *Notable American Women,* 1607–1950. 3 vols. 1971.
- *Notable American Women: The Modern Period.* 1980.
- *Through a Woman's I: An Annotated Bibliography of American Women's Autobiographical Writings, 1946–1976.* 1983.

Ethnic Studies Periodicals

This list contains the titles of periodicals primarily concerned with the following American ethnic groups:

- African Americans
- Asian Americans
- Latinos
- Native Americans
- Ethnic studies in general

Although these periodicals contain much material that is useful to the ethnic studies student, remember that there are hundreds of other general and scholarly periodicals containing thousands of other articles on these ethnic groups.

Many of these sources, as well as others, are available on-line and on CD-ROM.

Browsing through these materials directly will lead you to many valuable articles, but the following periodical indexes are useful:

- *Chicano Periodical Index.*
- *HAPI: Hispanic American Periodical Index.*
- *Index to Periodical Articles by and about Blacks.*
- *Sage Race Relations Abstracts.*

Other more general indexes include the following:

- *Alternative Press Index.*
- *America: History and Life.*
- *American Statistics Index.*
- *Education Index.*
- *Left Index.*
- *PAIS International in Print.*
- *Readers' Guide to Periodical Literature.*
- *Social Sciences Citation Index.*
- *Social Sciences Index.*
- *Sociological Abstracts.*
- *Women's Studies Abstracts.*

African Americans

- *Black American Literature Forum.*
- *Black Enterprise.*
- *Black Law Journal.*
- *The Black Nation.*
- *Black Perspective in Music.*

- *Black Scholar.*
- *Burning Spear.*
- *CAAS Newsletter.* University of California, Los Angeles. Center for Afro-American Studies.
- *Callaloo: Journal of Afro-American and African Arts and Letters.*
- *College Language Association (CLA) Journal: Afro-American, African and Caribbean Literature.*
- *The Crisis.*
- *Dollars & Sense.*
- *Ebony.*
- *Essence.*
- *Freedomways.*
- *Journal of Black Psychology.*
- *Journal of Black Studies.*
- *Journal of Negro Education.*
- *Journal of Negro History.*
- *Kokay I.*
- *Living Blues.*
- *Negro History Bulletin.*
- *Obsidian II: Black Literature in Review.*
- *Phylon, a Review of Race & Culture.*
- *Review of Black Political Economy.*
- SAGE: *A Scholarly Journal of Black Women.*
- *Slavery & Abolition.*
- *Southern Exposure.*
- *Truth: Newsletter of the Association of Black Women Historians.*
- *Western Journal of Black Studies.*

Asian Americans

- *Amerasia Journal.*
- *Asiaweek: Journal for the Asian American Community.*
- *East West.*
- *East Wind.*
- *Focus on Asian Studies.*
- *Katipunan.*
- *P/AAMHRC Research Review* (Pacific/Asian American Mental Health Research Center).
- *Pacific Citizen.*
- *Pacific Ties.*
- *Vietnam Forum.*

Latinos

- *Aztlan.*
- *Bilingual Review, La Revista Bilingue.*
- *El Chicano.*
- *Chicano Law Review.*
- *Hispanic Journal of Behavioral Sciences.*
- *Hispanic Link Weekly Report.*
- *Imagine: International Chicano Poetry Journal.*
- *Noticiero.*
- *Nuestro.*
- *Renato Rosaldo Lecture Series Monographs.*
- *Research Bulletin.* University of California, Los Angeles. Spanish Speaking Mental Health Research Center.
- *Revista Chicana-Requena.*
- *Revista Mujeres.*
- *El Tecolote.*

Native Americans

- *Akwesasne Notes.*
- *American Indian Art Magazine.*
- *American Indian Culture and Research Journal.*
- *American Indian Law Review.* (Also on-line).
- *American Indian Libraries Newsletter.*
- *American Indian Quarterly.*
- *Daybreak Star.*
- *Early American* (newsletter of the California Indian Education Association).
- *Indian Affairs* (newsletter of the Association of American Indian Affairs).
- *Native Americas (Akwe-Kon Journal)* (Akwe-Kon Press/Cornell University. N.Y.)
- *Native Press Research Journal* (University of Arkansas at Little Rock.)
- *Native Studies Review* (Native Studies Dept., Univ. of Saskatchewan, Saksatoon, Canada.)
- *NCAI News* (National Congress of American Indians).
- *News from Native California.*
- *Treaty Council News.*
- *Wassaja* (national newspaper of Indian America).

General Ethnic Studies

- *Ethnic Affairs.*
- *Ethnic Forum.*
- *Ethnic Groups.*
- *Ethnic and Racial Studies.*
- *Ethnic Reporter.*
- *Explorations in Ethnic Studies.*
- *Immigrants and Minorities.*
- *Interracial Books for Children Bulletin.*
- *Journal of American Ethnic History.*
- *Journal of Ethnic Studies.*
- *Race & Class.*

African-American Studies

Representative subject headings used in catalogs and periodical indexes include the following:

- African Americans.
- Afro-American Families.
- Black Muslims.
- Black power.
- Blacks.

- Civil rights.
- Discrimination.
- Ethnic attitudes.
- Minorities.
- Negroes.
- Race.
- Race discrimination.
- Race relations.
- Segregation.

Consult the *Library of Congress Subject Headings* to identify other subject headings.

Reference Books

- *The African-American Almanac* (formerly *The Negro Almanac*). 1997.
- *The African American Encyclopedia.* Michael W. Williams, ed. (N.Y.: Marshall Cavendish Corp., 1993).
- *Black Americans Information Directory: A Guide to Approximately 53,000 Organizations, Agencies, Institutions, Programs, and Publications* (Detroit: Gale). This is a biennial publication currently in its 3rd edition (1993).
- *Dictionary of American Negro Biography.* 1982.
- *In Black and White.* 1980, 1985.
- *We the People: An Atlas of America's Ethnic Diversity.* 1988. See especially Chapter 10.
- *Who's Who among Black Americans.* 1992. (Also available on-line).

Bibliographies

- *Afro-American History: A Bibliography.* By Dwight LaVern Smith. Vol. 1, 1974, and vol. 2, 1981.
- *Black American Families, 1965–1984: A Classified, Selectively Annotated Bibliography.* By W. Allen, 1965–1984 et al. Westport, CT: Greenwood Press, 1986.
- *The Black Family in the United States: A Revised, Updated, Selectively Annotated Bibliography.* By L. Davis. Westport, CT: Greenwood Press, 1986.
- *Black Lesbians: An Annotated Bibliography.* 1981.
- *Index to Afro-American Reference Resources.* New York: Greenwood, 1988.
- *The Progress of Afro-American Women: A Selected Bibliography and Resource Guide.* By Janet L. Sims-Wood. 1980.
- *Resource Guide on Black Families in America.* By C. S. Howard. Washington, DC: Institute for Urban Affairs and Research, Howard University, 1980.

Indexes and Abstracts

- *Alternative Press Index.* 1969–.
- *America: History and Life.* 1964–.
- *Biography Index.* 1946–. (Also on CD-ROM, 1984–.)
- *Education Index.* 1929–. (Also on CD-ROM, 1983–.)
- *Index to Black Periodicals* (formerly *Index to Periodical Articles by and about Blacks*). 1984–.
- *Index to Legal Periodicals.* 1926–. (Also on CD-ROM, 1981–, and on-line.)
- *Left Index.* 1982–.
- *PAIS International in Print,* 1915–. (Also on CD-ROM and on-line, 1972–.)
- *Readers' Guide to Periodical Literature.* 1900–. (Also on CD-ROM and on-line.)
- *Sage Race Relation Abstracts.* 1975–.
- *Social Sciences Index.* 1974–. See also the earlier title: *Social Sciences and Humanities Index.* 1965–1974; *International Index.* 1907–1965. (Also on CD-ROM and on-line, 1983–.)
- *Women's Studies Abstracts.* 1972–.

Computer Databases

- *America: History and Life.* 1964–.
- *ERIC.* 1966–.
- *Ethnic NewsWatch,* 1992–.
- *Magazine Index.* 1959–1970, 1973–.
- *National Newspaper Index.* 1979–.
- *P.A.I.S.* 1976–.
- *Social Scisearch.* 1972–.
- *Sociological Abstracts.* 1963–.

The following databases are available on CD-ROM.

- *The African American Experience: A History on CD-ROM.* (Minneapolis, MN: Quanta Press, 1993–).
- *Black Studies on Disc.* (N.Y.: G.K. Hall, annual).
- *DISCovering Multicultural America* (Detroit: Gale, 1995–).
- *Ethnic NewsWatch.* (Stamford, CT: Softline Information, 1992–).

Latino Studies

Representative subject headings for Latino studies include the following:

- Agricultural laborers—United States.
- American drama—Mexican-American authors.
- Bilingualism.
- Cuban Americans.
- Education—Bilingual.
- Ethnic attitudes.
- Mexican-American literature.
- Mexican-American women.
- Mexican Americans.
- Mexicans in California.
- Minorities.
- Puerto Ricans.
- Race discrimination.

Consult the *Library of Congress Subject Headings* for other headings.

A decade review of scholarship on Latino families may be found in William Vega, "Hispanic Families," in Alan Booth, ed., *Contemporary Families: Looking Forward, Looking Back* (Minneapolis: NCFR, 1991).

Dictionaries and Encyclopedias

- *El diccionario del español chicano (Dictionary of Chicano Spanish).* 1985.
- *Dictionary of Mexican American History.* 1981.
- *Harvard Encyclopedia of American Ethnic Groups.* 1980.

Handbooks, Guides, and Directories

- *The Hispanic Almanac.* 1990.
- *The Hispanic-American Almanac.* 1993.
- *Hispanic Americans Information Directory.* 1990.
- *Hispanic Children and Youth in the United States: A Resource Guide.* By Angela Carrasquillo. 1991.
- *The Hispanic Presence in North American from 1492 to Today.* By Carlos M. Fernández-Shaw. 1991. A general history with extensive appendices including associations and periodicals.
- *Hispanic Resource Directory.* 1988–.
- *Latinos of the Americas: A Source Book.* 1989. By Lynn K. Stoner.
- *Multiculturalism in the United States: A Comparative Guide to Acculturation and Ethnicity.* Edited by John D. Buenker and Lorman A. Ratner, 1992.
- *Sourcebook of Hispanic Culture in the United States.* 1982. Short essays and bibliographies.
- *Statistical Record of Hispanic Americans.* 1993.
- *The Mexican Americans: A Critical Guide to Research Aids.* 1980. An extensive and annotated listing of sources and bibliographies.
- *Who's Who: Chicano Office Holders.* 1977/78–.

Bibliographies

- *Arte Chicano: A Comprehensive Annotated Bibliography of Chicano Art. 1965–1981.* 1985.
- *Bibliography of Mexican American History.* 1984.
- *Literatura Chicana: Creative and Critical Writings through 1984.*
- *Mexican Americans: An Annotated Bibliography of Bibliographies.* 1984.
- *Mexican Americans in Urban Society: A Selected Bibliography.* 1986. By Albert Camarillo.

Indexes and Abstracts

- *Alternative Press Index.* 1969–.
- *America: History and Life.* 1964–.
- *American Statistics Index.* 1973–.
- *Chicano Periodical Index.* 1967–.
- *The Chicano Studies Index.* 1971–1991. (Chicano Studies Library Publications, Univ. of California, Berkeley, CA.)
- *Education Index.* 1929–.
- *HAPI: Hispanic American Periodicals Index.* 1970–.
- *Index to Legal Periodicals.* 1926–.
- *Left Index.* 1982–.
- *Psychological Abstracts.* 1927–.
- *PAIS International in Print.* 1915–. (Also on CD-ROM and on-line.)
- *Reader's Guide to Periodical Literature.* 1900–. (Also on CD-ROM and on-line).
- *Sage Race Relations Abstracts.* 1973–.
- *Social Sciences Citation Index.* 1966–.
- *Social Sciences Index.* 1974–.
- *Sociological Abstracts.* 1953–.
- *Women's Studies Abstracts.* 1972–.

Computer Databases

Chicano Database. 1990–. (CD-ROM. Available from Chicano Studies Library Publications, Univ. of California, Berkeley, CA.)

- *Chicano Database*, *ERIC* (education), *PsychInfo* and *PsycLIT* (psychology), *SocioFile* (sociology), and GPO (U.S. government documents) may be searched on-line.
- *ERIC*
- *Ethnic Newswatch.* 1991–. (CD-ROM. Available from Softline Information, Inc., Stamford, CT.)
- *GPO* (U.S. government documents)
- *Psych Info*
- *PsycLIT*
- *SocioFile*

Asian-American Studies

Subject headings for Asian-American groups are complex and inconsistent. They may appear in any of these forms: *Asian Americans; Asians—United States; Asians in the United States;* or, for example, *Chinese Americans; Chinese—United States; Chinese in the United States.* For the most complete results, check all of these forms for the groups that interest you. Other representative subject headings used in catalogs and periodicals indexes include the following:

- Chinese in California.
- Chinese in New York City.
- Chinese-American women.
- Discrimination.
- Ethnic attitudes.
- Filipino Americans.
- Filipinos in San Francisco.
- Japanese in Oregon.
- Japanese-American art.

- Manzanar, California.
- Vietnamese Americans.

Consult the *Library of Congress Subject Headings* to find other titles and subject headings.

Dictionaries, Encyclopedias, and Handbooks

- *The Asian-American Almanac: A Reference Work on Asians in the United States.* Edited by Susan Gall, 1995.
- *The Asian American Encyclopedia*, 6 vols. Edited by Franklin Ng, 1994.
- *Asian American Information Directory.* By Karen Backus and Julia C. Furtaw, 1991.
- *Asian Americans in the United States*, 2 vols. By Alexander Yamato, et al., 1993.
- *Dictionary of Asian American History.* 1986.
- *Handbook of Social Services for Asian and Pacific Islanders.* Edited by Noreen Mokuau, 1991.
- *Harvard Encyclopedia of American Ethnic Groups.* Edited by Stephan Thernstrom, 1980.
- *Multiculturalism in the United States: A Comparative Guide to Acculturation and Ethnicity.* Edited by John D. Buenker and Lorman A. Ratner, 1992.
- *Social Work Practice with Asian Americans.* Sourcebooks for the Human Services Series, 1992.
- *A Statistical Record of Asian Americans: Comprehensive Coverage of Americans and Canadians of Asian-Pacific Islander Descent.* Edited by Susan & Timothy L. Gall, 1993.
- *We the People: An Atlas of America's Ethnic Diversity.* 1988.

Bibliographies

- *Amerasia Journal.* Annual selected bibliography.
- *Asian American Studies: An Annotated Bibliography and Research Guide.* Edited by Hyung-Chan Kim, 1989.
- *Asians in America: A Selected Annotated Bibliography.* 1983.
- *Asians in California.* By Sucheng Chan. 1986.
- *Bibliography of Pacific/Asian American Materials in the Library of Congress.* By Elena S. H. Yu, et al., 1982.
- *Pacific/Asian Americans.* By Indu Vohra-Sahu. 1983.
- *Pacific/Asian Lesbian Book Collection.* By Alison Kim. 1987.
- *A Research Bibliography of California's Chinese Americans.* Compiled by Anna Chan, 1991.
- *The Social Sciences: A Cross-Disciplinary Guide to Selected Sources.* Edited by Nancy L. Herron, 1989.
- *Women, Race, and Ethnicity: A Bibliography.* By Susan Searing, et al., 1991.

Guides and Directories

- *An Asian American Internet Guide.* Compiled by Wataru Ebihara, 1995.
- *Author's Guide to Journals in Behavioral Sciences.* By Alum Yafu Wang, 1989.
- *Encyclopedic Directory of Ethnic Newspapers and Periodicals in the United States.* By Lubomyr R. Wynar, 1992.
- *Understanding Asian Americans: A Curriculum Resource Guide.* Edited by Marjorie & Peter Li, 1990.

Indexes and Abstracts

- *Alternative Press Index.* 1969–.
- *America: History and Life.* 1964–.
- *Asians in the United States: Abstracts of the Psychological and Behavioral Literature, 1967–1991.* Edited by Frederick T. Leong and James R. Whitfield, 1992.
- *Biography Index.* 1946–.
- *Education Index.* 1929–.
- *Index to Legal Periodicals.* 1926–.
- *Left Index.* 1982–.
- *PAIS International in Print.* 1915.
- *Readers' Guide to Periodical Literature.* 1900–.
- *Sage Race Relation Abstracts.* 1973–.
- *Social Sciences Index.* 1974–. See also the earlier titles: *Social Sciences and Humanities Index.* 1965–1974; *International Index.* 1907–1965.
- *Women's Studies Abstracts.* 1972–.

Computer Databases and CD-ROMs

- *America: History and Life.* 1964-.
- *ERIC.* 1966–.
- *Inventory of Marriage and Family Literature: Family Studies Database*, 1970–1995 (CD-ROM).
- *Magazine Index.* 1959–1970, 1973–.
- *Mental Health Abstracts*, 1969–.
- *National Newspaper Index.* 1979–.
- *P.A.I.S.* 1976–.
- *PsycLIT-Silver Platter*, 1974–.
- *Social Scisearch.* 1972–.
- *SocioFile-Silver Platter*, 1974–.
- *Sociological Abstracts.* 1963–.

Native American Studies

Representative subject headings used in catalogs and periodical indexes include the following:

- Civil rights.
- Discrimination.
- Ethnic attitudes.
- Folklore, Indian.
- Indians, treatment of.
- Indians of North America.
- Minorities.
- Navajo Indians.
- Race.
- Race relations.

Consult the *Library of Congress Subject Headings* to identify other subject headings. To find tribal names, look in the subject catalog.

Reference Books

- *American Indian and Alaska Native Newspapers and Periodicals, 1826-1924.* By Daniel F. Littlefield, Jr. and James W. Parins. Westport, CT: Greenwood, 1984.
- *Great North American Indians: Profiles in Life and Leadership.* By Frederick J. Dockstader. 1977.
- *Guide to Research on North American Indians.* 1983.
- *Handbook of North American Indians.* 1978.
- *Handbook of the American Frontier: Four Centuries of Indian-White Relationships.* Metuchen, NJ: Scarecrow, 1987–1991. v. 1–3 (1987-1993).
- *The Native American Almanac: A Reference Work on Native Americans in the United States and Canada.* Ed by Duane Champagne. Detroit: Gale, 1994.
- *Native American Directory: Alaska, Canada, U.S.* National Native American Co-Op. Tucson, AZ.
- *Native Americans Information Directory.* Gale Research. Detroit. MI.
- *Reference Encyclopedia of the American Indian.* Vol. 1, 1986.
- *Statistical Record of Native North Americans.* Ed by Marlita A. Reddy. Detroit: Gale, 1993.

Bibliographies

- *A Bibliographical Guide to the History of Indian-White Relations in the United States.* (1977 and suppl.). By Frances Paul Prucha. 1982.
- *Indians of North America: Methods and Sources for Library Research.* By Marilyn L. Haas. Hamden, CT: Library Professional Publ., 1983.
- *Native American Periodicals and Newspapers.* 1828–1982.
- *Native American Women.* 1983.
- *Native Americans: An Annotated Bibliography.* By Frederick E. Hoxie and Harvey Markowitz. Pasadena, CA: Salem Pr., 1991.
- *The Urbanization of American Indians.* 1982.

Indexes and Abstracts

- *Abstracts in Anthropology.* 1970–.
- *Alternative Press Index.* 1969–.
- *America: History and Life.* 1964–.
- *Education Index.* 1929–.
- *Index to Legal Periodicals.* 1926–.
- *National Newspaper Index.*
- *PAIS International in Print* 1915–.
- *Readers' Guide to Periodical Literature.* 1900–.
- *Social Sciences Index.* 1974–. See also the earlier titles: *Social Sciences and Humanities Index.* 1965–1974; *International Index.* 1907–1965.
- *Sociological Abstracts.* 1953–.
- *Women's Studies Abstracts.* 1972–.

Computer Databases and CD-ROMs

- *America: History and Life.* 1964.
- *ERIC.* 1966–.
- *Inventory of Marriage and Family Literature: Family Studies Database*, 1970–1995 (CD-ROM).
- *Magazine Index.* 1959–1970, 1973–.
- *Mental Health Abstracts*, 1969–.
- *National Newspaper Index.* 1979–.
- *P.A.I.S.* 1976–.
- *PsycLIT-Silver Platter*, 1974–.
- *Social Scisearch.* 1972–.
- *SocioFile-Silver Platter*, 1974–.
- *Sociological Abstracts.* 1963–.

Ethnic Studies Internet Sites

American Studies Web: Race and Ethnicity
http://pantheon.cis.yale.edu/~davidp/race.html

Armenian Research Center: Information and Resources
http://www.umd.umich.edu/dept/armenian/

Asian American Resources
http://www.mit.edu:8001/afs/athena.mit.edu/user/i/r/irie/www/aar.html

Cuban Research Institute
http://burrow.fiu.edu/~lacc/cri/

Harvard's Center for Middle Eastern Studies
http://fas-www.harvard.edu/~mideast/

The Journal of Latin American Perspectives
http://wizard.ucr.edu/~asampaio/lap.html

Latin American and Caribbean Center: Research and Programs
http://www.fiu.edu/~lacc/

Native American Issues
http://web.maxwell.syr.edu/nativeweb/

Race, Sex, and Religion in the 90's
http://members.aol.com/sshell8192/race/INDEX.HTML

UCLA Chicano Studies Research Center
http://www.sscnet.ucla.edu/csrc/

Self-Help Resource Directory

The information resources that follow include information centers and counseling centers, as well as support groups, prerecorded tape services, and crisis centers. Use them, share the questions that led you to call, and discover new information on numerous topics.

The directory is easy to use. Each resource is located under a general subject heading. You'll find a description of the service, its hours, and its location by city and state. To call a resource, follow the code found by its telephone number.

- LD—long distance. Many important resources must be dialed at your expense. Dial direct, if you can, and remember the difference in time.
- TF—toll free. Some organizations have toll-free numbers. The 800 prefix designates that convenience. Telephone these resources at no charge.

Need information on a topic that isn't included? Check your telephone directory for important local resources. If you can't find the information you need, check at the reference desk of your local public, college, or university library.

The following reference guides may also be helpful:

- *Directory of National Helplines: A Guide to Toll Free Public Service Numbers, 1995.* Pierian Press, 1995.
- *Psychology: A Guide to Reference and Information Resources.* By Pam M. Baxter, 1993.

Adoption

ALMA Society
P.O. Box 154
Radio City Station
New York, NY 10101-0727
LD 212-581-1568

The Adoptees' Liberty Movement Association is a nonprofit organization with chapters throughout the United States. ALMA believes in the right of adult adoptees and the biological parents of adoptees to locate one another. It supports and assists biologically related people searching for each other. Call or write for membership information, a newsletter, or a chapter list.

Adoptive Families of America (AFA), Inc.
3333 Hwy. 100 North
Minneapolis, MN 55422
LD 612-535-4829
TF 800-372-3300

AFA is a private, nonprofit membership organization that provides problem-solving assistance and information to adoptive and prospective adoptive individuals and families. U.S. membership is $24 yearly and includes a subscription to *Adoptive Families* magazine.

International Concerns Committee for Children
911 Cypress Dr.
Boulder, CO 80303
LD 303-494-8333
Monday–Friday 9 AM–4 PM

This organization provides information about ways to help homeless children, including foster parenting and adoption. It publishes the annual *Report on Foreign Adoption* (with frequent updates) containing current information on many aspects of adoption. It also maintains a listing service of children in urgent need of homes. In addition, it publishes a newsletter and provides information and counseling to adoptive families. There is a nominal charge for publications. Write or call for a brochure. Leave a message, and your call will be returned collect.

North American Council on Adoptable Children
970 Raymond Ave., Suite 106
St. Paul, MN 55114-1147
LD 612-644-3036
Monday–Friday 9 AM–5 PM

The council provides a wealth of information and support services to adoption agencies and parents. Call or write for publications.

You can also contact the appropriate local government agency for adoption information. Check your telephone directory for your city's or county's social services, family services, public welfare, child welfare, or social welfare department. Check the Yellow Pages under "Adoption."

AIDS and HIV

CDC National AIDS Hotline
TF 800-342-AIDS (800-342-2437) - English
TF 800-344-7432 - Spanish
24 hours

The hotline can give you general information on HIV and AIDS and refer you to doctors, clinics, testing services, legal services, or counselors in your area. Spanish-speaking operators are available. Free brochures in a number of languages are available on request. For brochures in Spanish, call 800-344-SIDA; for others, call the above number.

Good Samaritan Project Teen Hotline
Kansas City, MO
TF 800-234-TEEN (800-234-8336) - English
Monday–Saturday 4PM–8PM

Staffed by trained high school students, the hotline gives accurate information to teens and their parents.

San Francisco AIDS Foundation
25 Van Ness Ave.
San Francisco, CA 94102
TF 800-FOR-AIDS (800-367-2437)
Monday–Friday 9 AM–9 PM
Saturday and Sunday 11 AM–5 PM

The foundation promotes AIDS education and provides information and referrals to people with AIDS and their families. English, Spanish, and Filipino are spoken. Support services are provided in the San Francisco area. Publications are available.

Basic Satistics on HIV and AIDS
http://www.cdc.gov/nchstp/hiv_aids/stats/hasrlink.htm

The Centers for Disease Control posts tables containing statistics on the number of AIDS cases in the United States, by a broad range of categories.

Alcohol and Drug Abuse

Al-Anon Family Groups
1372 Broadway
New York, NY 10018
TF 800-356-9996
Monday–Friday 8 AM–5 PM

Al-Anon is a worldwide fellowship for relatives and friends of alcoholics who "share their experience, strength and hope in order to solve their common problems and to help others do the same." There are over 40,000 Al-Anon groups, including 3,500 Alateen groups for teenagers. They publish numerous pamphlets, books, newsletters, and a magazine. Material is available in thirty languages, as well as in Braille and on tape. Check your telephone directory for local listings, or call the above toll-free number for information.

Alcoholics Anonymous (AA)
P.O. Box 459, Grand Central Station
New York, NY 10163
LD 212-870-3400
Monday–Friday 9 AM–4:45 PM

AA is a voluntary worldwide fellowship of men and women who meet together to attain and maintain sobriety. There are no dues or fees for membership; the only requirement is a desire to stop drinking. There are over 89,000 groups and more than 2,490,000 members in 134 countries. Information on AA and chapter locations worldwide is available at the telephone number listed above. Check your telephone directory for local listings.

There is a growing number of alternative groups to AA. These groups emphasize personal responsibility and reject some of AA's views. Secular Organization for Sobriety (SOS), Rational Recovery (RR), and Women for Sobriety (WFS) are among the groups with increasing numbers of local chapters. Check your telephone directory or dial Information.

Mothers Against Drunk Driving (MADD)
511 E. John Carpenter Fwy., Suite 700
Irving, TX 75062
TF 800-438-6233

MADD's membership is made up of drunk-driving victims and concerned citizens. Its concerns are to speak on behalf of victims, to reform drunk-driving laws, and to educate the public. It assists victims with the legal process, conducts research, and sponsors workshops. It has local chapters throughout the United States. Publications are available. Write or call for program and membership information.

National Clearinghouse for Alcohol and Drug Information (NCADI)
P.O. Box 2345
Rockville, MD 20847
TF 800-729-6686
Monday–Friday 8 AM–7 PM

NCADI provides a referral service, answering inquiries on alcohol- and drug-related subjects by telephone or mail. The clearinghouse also gathers and disseminates current information (including books, curriculum guides, directories, and posters) free of charge.

National Institute on Drug Abuse
http://www.nida.nih.gov/

This site provides basic information on a wide range of drugs.

Phoenix House Foundation
164 W. 74th St.
New York, NY 10023
LD 212-595-5810
Monday–Friday 9 AM–5 PM

Phoenix House is a multiservice drug abuse agency offering residential treatment in New York City and southern California. It has numerous educational and treatment programs for youth and adults. Information and referrals are available on request.

Web of Addictions
http://www.well.com/user/woa/

The Web of Addictions provides accurate information about alcohol and drug addictions.

Breastfeeding (SEE Infant Care)

Child Abuse

Child Help USA Information Center
TF 800-422-4453
24 hours

The center provides general information and referrals.

Clearinghouse on Child Abuse and Neglect Information
P.O. Box 1182
Washington, DC 20013
LD 703-385-7565
FAX 703-385-3206
Monday–Friday 8:30 AM–5:30 PM

The clearinghouse provides information for professionals and concerned citizens interested in child maltreatment issues. A database of resources for professionals is available; the staff also performs searches for specific topics. Numerous publications, including bibliographies and research reviews, are available. Write or call for a free catalog.

National Committee to Prevent Child Abuse (NCPCA)
332 Michigan Ave., Suite 1600
Chicago, IL 60604-4357
LD 312-663-3520

NCPCA is a volunteer-based organization dedicated to reducing child abuse. It offers educational programs and materials and support and self-help groups for parents. There are numerous local chapters. Write or call for information and a chapter list.

Parents Anonymous (PA)
7120 Franklin Ave.
Los Angeles, CA 90046
LD 213-388-6685

PA has over 1,200 local chapters that provide support for parents who abuse, or fear they may abuse, their children. Parents Anonymous also publishes materials on child abuse; send an SASE. Look for a local chapter in your telephone directory, or check with your local department of social services.

Child Support

Office of Child Support Enforcement
Administration for Children & Families
US Dept. of Health & Human Service
370 L'Enfant Promenade, S.W., 4th Fl.
Washington, DC 20447

Carried out by state and local child-support enforcement, this program ensures that children are financially supported by both of their parents. The program helps locate an absent parent for child-support enforcement, establish paternity if necessary, determine child-support obligations, and enforce child-support orders. In most states, CSE offices are listed under human service agencies in the local government section of the telephone directory.

Conflict Resolution

Children's Creative Response to Conflict (CCRC)
Box 271
Nyack, NY 10960
LD 914-353-1796
FAX 914-358-4924
Monday–Friday 9 AM–5 PM

CCRC's trained facilitators conduct workshops for students, teachers, and parents to learn cooperation and conflict-resolution skills. A teacher's handbook (in English or Spanish), newsletter, and other publications are available. Write or call for a brochure and a list of local branches.

Consumer Services

Consumer Information Catalog
Pueblo, CO 81009
LD 719-948-4000

Numerous booklets on a variety of topics—careers, children, health, housing, money, and many others—are available. Booklets are free or in the 50¢-to-$3 range. Write or call for a catalog.

Consumer Product Safety Commission (CPSC)
4330 E. West Hwy.
Bethesda, MD 20814-4408
LD 301-504-0990
Monday–Friday 9 AM–5 PM
TF (Automated machine) 800-638-2772
TTY for the deaf 800-492-8140

The CPSC is involved in the evaluation of the safety of products sold to the public. Commission staff members will answer questions and provide free printed materials on different aspects of consumer product safety, such as the safety of children's toys or household appliances. The commission does not answer questions from consumers on automobiles, cosmetics, drugs, prescriptions, warranties, advertising, repairs, or maintenance.

FDA Consumer Affairs Office
5600 Fishers Ln.
Rockville, MD 20857
LD 301-443-5006
Monday–Friday 9 AM–5 PM

The Consumer Affairs Office provides information and referrals on food products, pharmaceutical drugs, and cosmetics.

Occupational Safety and Health Administration (OSHA)
200 Constitution Ave. NW, Rm. N3647
Washington, DC 20210
LD 202-219-8148
Monday–Friday 9 AM–5 PM

The OSHA Office of Information provides a referral service and basic information on job safety and the dangers posed by toxic substances in the workplace. There is no charge for pamphlets and booklets and a nominal charge for a catalog.

Disability

Mainstream, Inc.
3 Bethesda Metro Center, Suite 830
Bethesda, MD 20814
LD 301-654-2400 (voice/TDD)
Monday–Friday 9 AM–5 PM

Mainstream provides information on affirmative action for the disabled and will make referrals to agencies or state offices. In Washington, D.C., and Dallas, Project LINK places workers with disabilities, free of charge. Workshops, publications, and a newsletter are available. Write or call for information.

National Information Center for Children and Youth with Disabilities (NICHCY)
P.O. Box 1492
Washington, DC 20013
TF 800-695-0285
LD 202-884-8200
Monday–Friday 9 AM–5 PM

NICHCY is a national clearinghouse and referral source for parents of disabled children and professionals who work with them. Publications, including an informative newsletter and extensive resource list, are provided free of charge.

Parents Helping Parents
3041 Olcott St.
Santa Clara, CA 95054
LD 408-288-5010

Parents Helping Parents is a support group for parents of children who are physically and/or mentally disabled. The group provides telephone counseling, in-home visits if possible, information, and referrals to local agencies and support groups. The staff speaks Spanish and Vietnamese. A brochure is available.

The Disability Rag
gopher://gopher.etext.org/11/Politics/Disability.Rag

This is an on-line magazine that deals with issues disabled people face, such as access to public accomodations and discrimination in hiring practices.

Divorce

Joint Custody Association
10606 Wilkins Ave
Los Angeles, CA 90024
LD 310-475-5352

The association provides information on joint custody; assists parents, counselors, and others implementing joint custody practices; and surveys court decisions and their consequences.

Mothers Without Custody
P.O. Box 27418
Houston, Texas 77227-7418
LD 713-840-1622

This association is for women living apart from one or more of their children. Helps establish local self-help groups and offers support to women exploring their child custody options.

Parents Sharing Custody
420 South Beverly Dr., Suite 100
Beverly Hills, CA 90212-4410
LD 310 286-9171

An association of parents sharing custody of children after divorce. It educates parents on maintaining their parental roles and works to protect the rights of children.

Domestic Violence (SEE ALSO Rape)

National Coalition Against Domestic Violence
P.O. Box 18749
Denver, CO 80218-0749
LD 303-839-1852

This organization provides support to battered women and teens who are in abusive relationships.

Family Service America
11700 West Lake Park Drive
Milwaukee, WI 53224
TF 800-221-2681

Family Service America offers problem solving and referrals to victims of family violence.

Incest Survivors Resource Network International
P.O. Box 7375
Las Cruces, NM 88006-7375
LD 505-521-4260
2-4 PM and 11 PM-midnight (EST)

This is a survivor-run educational resource center helpline.

Drug Abuse (SEE Alcohol and Drug Abuse)

Environment

Earthworks Press
1400 Shattuck Ave., Box 25
Berkeley, CA 94709
LD 510-841-5866
FAX 510-841-7121

Earthworks publishes material for those concerned with protecting the environment. Check your local bookstore, or write for such titles as *50 Simple Things You Can Do to Save the Earth* ($4.95), *30 Simple Energy Things You Can Do to Save the Earth* ($3.95), and *The Recycler's Handbook* ($4.95). Include $1.00 postage per book. Bulk rates are available.

Energy Efficiency and Renewable Energy Clearinghouse (EREC)
P.O. Box 3048
Merrifield, VA 22116
TF 800-523-2929

EREC provides general and specific information on renewable energy technologies and energy conservation techniques for residential and commercial needs. It provides technical assistance for obtaining financing for energy conservation construction. EREC offers more than 150 publications, fact sheets, bibliographies, and 500 computerized letter units free of charge.

Family Planning

Planned Parenthood Federation of America
810 Seventh Ave.
New York, NY 10019
LD 212-541-7800

Most cities have a Planned Parenthood organization listed in the telephone directory. Planned Parenthood provides information, counseling, and medical services related to reproduction and sexual health to *anyone* who wants them. No one is denied because of age, social group, or inability to pay. Information can also be obtained through the national office, listed above.

Planned Parenthood's Web Page
http://www.igc.apc.org/ppfa/choices.html

Information to help you decide which method of birth control suits your needs.

A Woman's Guide to Sexuality
http://www.ppfa.org./ppfa/wmngdl.html

Provided by Planned Parenthood, this guide explores sexuality and relationships.

Public Health Departments

Counties throughout the United States, regardless of size, have county health clinics that will provide low-cost family planning services. Some cities also have public health clinics. To locate them, look up the city or county in the telephone directory, and then check under headings such as these:

Department of Health
Family Planning
Family Services
Health
Public Health Department
(City or County's Name) Health (or Medical) Clinic

Zero Population Growth (ZPG)
1400 16th St. NW, Suite 320
Washington, DC 20036
LD 202-332-2200
Monday–Friday 9 AM–5 PM
http://www.zpg.org.zpg/

ZPG is a national, nonprofit membership organization that works to achieve a sustainable balance between the earth's population and its environment and resources. Teaching materials, numerous publications, and a newsletter are available. There is no charge for multiple copies.

Fatherhood

The Fatherhood Project
Families and Work Institute
330 7th Ave., 14th Floor
New York, NY 10001
LD 212-465-2044

This is a national research, demonstration, and dissemination project designed to encourage wider options for male involvement in child rearing. Pamphlets are available on various subjects related to families and work.

Gay and Lesbian Resources

Gay & Lesbian Parents Coalition International
P.O. Box 50360
Washington, DC 20091
LD 202-583-8029

This is a coalition of lesbian and gay parenting groups across the country. It provides information, education, and support services. It also sponsors an annual conference and publishes a newsletter for parents and their children.

National Federation of Parents and Friends of Lesbians and Gays (PFLAG)
P.O. Box 96519
Washington, DC 20090-6519
LD 202-638-4200

PFLAG provides information and support for those who care about gay and lesbian individuals. Write or call for the number of a parent contact in your area, or check the telephone directory for regional offices. Publications are available.

National Gay and Lesbian Task Force
2320 17th St. NW
Washington, DC 20009-2702
LD 202-332-6483
FAX 202-332-0207
TTY 202-332-6219
Monday–Friday 9 AM–5 PM

The National Gay and Lesbian Task Force provides information and referrals on gay and lesbian issues and rights.

Indiana University's Gay, Lesbian, Bisexual, and Transexual Home Page
http://www.Indiana.edu/%7Eglbserv/
Access to global information on GBLT issues.

Health Information

Alzheimer's Disease and Related Disorders Association
919 N. Michigan Ave.
Chicago, IL 60611
LD 312-335-8700
TF 800-272-3900
Monday–Friday 8:30 AM–5 PM

This organization supports family members of those affected by Alzheimer's disease and related disorders. It also promotes research and education and represents patients' continuing care needs to health-care agencies, government, business, and communities. A newsletter is available.

American Institute for Preventative Medicine (AIPM)
30445 Northwestern Hwy., Suite 350
Farmington Hills, MI 48334
LD 810-539-1800

For $2.95, the AIPM will send you a directory of toll-free numbers to call for advice and referrals on numerous health-related issues, such as alcohol abuse and cancer.

Centers for Disease Control (CDC)
1600 Clifton Rd. NE
Atlanta, GA 30333
LD 404-639-3534
Monday–Friday 9 AM–4:30 PM

Inquiries from the public on topics such as preventative medicine, health education, immunization, and communicable diseases can be directed to the Public Inquiries Office at the CDC. The office will also answer questions on occupational safety and health issues, family planning, and public health problems such as lead-based paint and rodent control. Inquiries are answered directly or referred to an appropriate resource.

Department of Health and Human Services
http://www.dhhs.gov/

A wide range of information ranging from adolescent health to substance abuse.

Family Caregivers of the Aging (SEE National Council on the Aging under the "Seniors" heading)

Healthwise, Columbia University Health Service
http://www.cc.columbia.edu:80/cc/healthwise/

With a wide range of topics available, the "Go Ask Alice" option offers an interactive question and answer line.

Medicaid/Medicare
Office of Beneficiary Services
Health Care Financing Administration
Rm. 648, East High Rise Bldg.
6325 Security Blvd.
Baltimore, MD 21207
LD 410-966-3206
Monday–Friday 9 AM–5 PM

This office answers questions on child health, Social Security, and all aspects of Medicaid and Medicare. The office also provides a referral service that includes listings of local welfare offices.

National Cancer Institute
Cancer Information Service
Bldg. 31, Rm. 10-A07
31 Center Dr.
Bethesda, MD 20892-2580
TF 800-4-CANCER (800-422-6237)
Monday–Friday 9 AM–7 PM

Calls are automatically directed to regional service centers. Counselors answer questions and give referrals to local doctors and support groups. For free brochures and other information, call or write.

National Cancer Institute's CancerNet
http://www.icic.nci.nih.gov/clinpdg/risk.html

This site provides comprehensive information about cancer.

National Health Information Center (NHIC)
P.O. Box 1133
Washington, DC 20013-1133
LD 301-565-4167
TF 800-336-4797 (referral database)

The NHIC is a central clearinghouse designed to refer consumers to health information resources. The clearinghouse has identified many groups and organizations that provide health information to the public. All health-related questions are welcomed, although the NHIC is unable to respond to questions requiring medical advice or diagnosis. A variety of publications are available.

Statistical Abstracts of the United States
http://www.medaccess.com/census/c_03.htm

Need numerical data? This site offers current vital statistics on a broad number of health-related issues.

Stress Management: A Comprehensive Review of Principles
http://www.unl.edu/stress/mgmt/#toc

Basic information on stress and stress management techniques.

Tel-Med
952 S. Mount Vernon
Colton, CA 92324
LD 909-825-7000
Monday–Friday 8 AM–4:30 PM

Tel-Med is a tape library of recorded medical messages. There are over six hundred tapes, three to seven minutes long, on different medical subjects. Titles include "Accidents in the Home," "Heart Attack," "Sleep," "Where Did I Come From, Mama?" and many others. Check your telephone directory for similar services in your area.

University of Wisconsin-Stevens Point Home Page
http://welness.uwsp.edu/College_Health/

The top 100 wellness sites are offered through this web page.

Infant Care

La Leche League
http://www.Prairienet.org/111i/FAQMain.html

Everything you ever wanted to know about breastfeeding is provided in this site.

La Leche League International
1400 N. Meacham Rd.
Schaumberg, IL 60173-4840
TF 800-LA-LECHE (800-525-3243)
LD 708-519-7730 (counseling hotline)

La Leche League provides advice and support for nursing mothers. Write for a brochure or catalog of its numerous publications. For local groups, check your telephone directory or call 708-519-7730.

National Organization of Circumcision Information Resource Centers (NOCIRC)
P.O. Box 2512
San Anselmo, CA 94979
LD 415-488-9883
Monday–Friday 9 AM–5 PM

NOCIRC is a nonprofit health resource center that provides medical and legal information on the subject of circumcision and female genital mutilation. It is against routine hospital circumcision of newborns. Pamphlets, newsletters, and referrals to physicians and lawyers are available. Send an SASE with 2-ounce postage. A videotape is available for rent or purchase. NOCIRC sponsors a biannual international symposium.

Infertility

The American Fertility Society
2140 Eleventh Ave. South, Suite 200
Birmingham, AL 35205-2800
LD 205-933-8494
Monday–Friday 8 AM–5 PM

The American Fertility Society provides up-to-date information on all aspects of infertility, reproductive endocrinology, conception control, and reproductive biology. Booklets, pamphlets, a selected reading list, resource lists, a medical journal, and postgraduate courses are available. Also available are ethical guidelines for new reproductive technologies, a position paper on insurance for infertility services, and revised procedures for semen donation. Write or call for further information.

The Center for Surrogate Parenting, Inc.
8383 Wilshire Blvd., Suite 750
Beverly Hills, CA 90211
LD 213-655-1974 (collect accepted)
TF (California only) 800-696-4664
Monday–Friday 9 AM–5 PM

The Center for Surrogate Parenting is a private organization that matches prospective parents and surrogates for a fee. Extensive screening is involved. Write or call for further information.

Infertility Resources
http://www.ihr.com/infertility/index.html

Designed for people experiencing infertility difficulties, this site provides information for individuals and professionals.

Resolve
1310 Broadway
Somerville, MA 02144-1731
LD 617-643-2424
FAX 617-623-0252
Monday–Friday 9 AM–4 PM

Resolve, a national nonprofit organization focusing on the problem of infertility, refers callers to chapters across the nation, provides fact sheets on male and female infertility, and publishes a newsletter and a director of infertility resources. Phone counseling is also offered. There is a small charge for publications.

Marriage Enhancement

The Association for Couples in Marriage Enrichment (ACME)
P.O. Box 10596
Winston-Salem, NC 27108
TF 800-634-8325
Monday–Friday 8 AM–5 PM

ACME is a nonprofit, nonsectarian organization that promotes activities to strengthen marriage. It offers weekend retreats, local chapter meetings, workshops, and conferences throughout the United States and Canada. Members may purchase books, tapes, and other materials at a discount. Write for brochures and a resource list.

Worldwide Marriage Encounter
1908 E. Highland Ave., #A
San Bernardino, CA 92404
LD 909-881-3456
TF 800-795-LOVE (800-795-5683)
Monday–Friday 8:30 AM–5 PM

Although Marriage Encounter retreats are designed with Catholic couples in mind, spaces are also reserved for couples of other faiths. Encounter weekends are offered in many parts of the country to enhance communication in marriage. A donation is asked but not required. Call or write for a brochure and application.

Parenting Resources on the Internet

Childbirth Organization
http://www.childbirth.org/

Medical Information and discussions on a wide range of issues.

International Council on Infertility Information Dissemination
http://www.mnsinc.com/inciid.html

Current research and infromation on infertility treatment.

ParenTalk Newsletter
http://www.tnpc.com/parentalk/index/html

Numerous articles by physicians and psychologists.

ParenthoodWeb
http://parenthoodweb.com

Doctors answer your e-mail questions

Parenting Q & A
http://www.parenting-qa.com/

Essays on many parenting topics plus answers to your e-mail questions.

Parent Soup
http://parentsoup.com/

Discussion forums on many parenting issues.

Peace

Foundation for Global Community
222 High St.
Palo Alto, CA 94301-1097
LD 415-328-7756
FAX 415-328-7785

This nonprofit, nonpartisan educational foundation of volunteers in more than forty states and six countries is dedicated to "living in concert with the planet and with each other and contributing to the continuity of life."

The Friendship Force
57 Forsyth St. NW, Suite 900
Atlanta, GA 30303
LD 404-522-9490
FAX 404-688-6148
Monday–Friday 8:30 AM–5 PM

This nonprofit, nonpolitical organization is designed to bring people of many countries together through international exchanges of "citizen ambassadors." There are over one hundred chapters nationwide. Its quarterly magazine is $5.00 per year.

Peace Action
1819 H St. NW, Suite 640
Washington, DC 20006
LD 202-862-9740

Peace Action seeks a comprehensive worldwide nuclear test ban treaty, a multilateral halt to the nuclear arms race, deep reductions in nuclear weapons, and an end to U.S. military intervention, arms sales, and military aid abroad. Publications are available.

20/20 Vision
1828 Jefferson Place NW
Washington, DC 20036
LD 202-833-2020
TF 800-669-1782

For $20 a year, 20/20 Vision sends a monthly postcard with local and national military, environmental, and economic justice issues, including all information needed to write or call legislators. Local chapters are in more than thirty states.

Pregnancy (ALSO SEE Family Planning, Infant Care)

American College of Nurse-Midwives
818 Connecticut Ave. NW, Suite 900
Washington, DC 20006
LD 202-728-9860

Write or call the college for a directory of certified nurse-midwives in your area or to get information on accredited university-affiliated nurse-midwifery education programs.

California Teratogen Information Service
University of California at San Diego Medical Center
TF 800-532-3749

When you're living for two, that means being twice as careful. Pregnant women and health-care professionals seeking the latest information about the effects of drugs and medications on unborn babies may use this free and confidential service. Pregnant women who have been exposed to suspected teratogens (substances harmful to the fetus) can also participate in a free follow-up program after birth.

National Abortion Rights Action League (NARAL)
1156 15th St. NW
Washington, DC 20005
LD 202-823-9300
Monday–Friday 9 AM–5:30 PM

This political organization is concerned with family planning issues and is dedicated to making abortion "safe, legal, and accessible" for all women. It is affiliated with many state and local organizations. A newsletter and brochures are available.

National Life Center
686 N. Broad St.
Woodbury, NJ 08096
TF 800-848-5683
Monday–Friday 9:30 AM–12:30 PM, 7 PM–9 PM

The National Life Center is an assistance and counseling service for pregnant women. Services include telephone counseling in any aspect of pregnancy and referral to chapters and clinics throughout the United States and Canada. The center counsels against abortion and does not give out information concerning birth control.

The Olen Interactive Pregnancy Calendar
http://www.olen.com/baby/

This site will provide a calendar which describes the development of a baby from conception to birth.

Online Birth Center
http://www.efn.org/%7Edjz/birth/birthindex.html

Information on a wide range or pregnancy- and birth-related topics, including high risk situations and alternative health resources.

Pregnancy Loss and Infant Death

National SHARE Office
St. Joseph's Health Center
300 First Capitol Dr.
St. Charles, MO 63301
LD 314-947-6164
Monday–Friday 9 AM–5 PM

SHARE and its affiliate groups offer support for parents who have experienced miscarriage, stillbirth, or the death of a baby. It offers a manual on starting your own SHARE group, a national listing of groups and parent contacts, a selected bibliography, a resource manual on farewell rituals, children's books, and a newsletter. Check your local directory for groups, which may also be listed under "Sharing Parents," "Hoping and Sharing," "HAND," or a similar heading.

Sudden Infant Death Syndrome Alliance
10500 Little Patuxent Pkwy., Suite 420
Columbia, MD 21044
LD 301-964-8000 (Maryland)
TF 800-221-SIDS (800-221-7437)

The SIDS Alliance provides emotional support to families who have lost a child to sudden infant death syndrome. It has chapters throughout the United States and provides free crisis counseling, referrals, and information. It also supports research into SIDS. Free literature is available on request.

Rape

National Clearinghouse on Marital and Date Rape
2325 Oak St.
Berkeley, CA 94708
LD 510-524-1582
24 hours (message tape)

The clearinghouse provides information, referrals, seminars, and speakers covering numerous aspects of marital and date rape. There is a nominal charge for publications and consultation services. Calls are returned collect.

Check your telephone directory for "Rape Crisis Center" or a similar listing to find the local crisis center nearest you; the service is available in almost all cities.

Runaways

National Runaway Switchboard
TF 800-621-4000

The National Runaway Switchboard provides toll-free telephone services for young people who have run away from home, those considering leaving home, and parents. The service helps young people define their problems, determines if an emergency exists, and refers callers to programs that provide free or low-cost help. Complete confidentiality is guaranteed. If the caller wants to reestablish communication with his or her family, a message can be taken for delivery within 24 hours.

Self-Help

National Self-Help Clearinghouse
25 W. 43rd St., Rm. 620
New York, NY 10036
LD 212-354-8525

The clearinghouse maintains a databank and referral service to provide information on thousands of self-help groups throughout the United States. It publishes manuals and a newsletter on self-help. Write or call for information; send an SASE if you want a list of local clearinghouse throughout the country.

The Consumer Information Center, part of the U.S. General Services Administration, offers numerous booklets on a variety of useful subjects. *See* the "Consumer Services" section of this "Self-Help Resource Directory."

Seniors

Gray Panthers
2025 Pennsylvania Ave. NW, Suite 821
Washington, DC 20006
LD 202-466-3132
FAX 202-466-3133
Monday–Friday 9 AM–5 PM

Gray Panthers is a multigenerational, nonpartisan membership organization with a global reach and a vision of peace and justice for all. Issues of concern include peace, health, housing, education, justice, the environment, and human potential. Annual dues of $15.00 include a bimonthly newsletter.

National Council on the Aging (NCA)
409 3rd St. SW, Suite 200
Washington, DC 20024
LD 202-479-1200
Monday–Friday 9 AM–5 PM

The NCA promotes the concerns of older Americans in many areas and develops methods and resources for meeting their needs. Consultations, speakers, meetings, programs, and many publications are available. Family Caregivers of the Aging, part of the NCA, provides information and support for those caring for aging parents.

The Administration on Aging
http://www.aoa.dhhs.gov/

Information available on older persons, families, and health care.

National Institute on Aging
http://www.nih.gov/nia

Helpful information on physical, behavioral, and social aspects of aging.

Seniors Site
http://Seniors_Site.com

For individuals 50 and over and those who care for older people, this entertaining site provides informative and practical information.

Sex Education and Sex Therapy

American Association of Sex Educators, Counselors, and Therapists (AASECT)
435 N. Michigan Ave., Suite 1717
Chicago, IL 60611-4067
LD 312-644-0828
FAX 312-644-8557
Monday–Friday 9 AM–5 PM

AASECT is an interdisciplinary professional organization devoted to the promotion of sexual health through the development and advancement of the fields of sex therapy, sex counseling, and sex education. It provides professional education and certification for sex educators, counselors, and therapists. It publishes a newsletter and the *Journal of Sex Education and Therapy.*

Sex Information and Education Council of the United States (SIECUS)
130 W. 42nd St., Suite 2500
New York, NY 10036
LD 212-819-9770

SIECUS is a nonprofit educational organization that promotes "healthy sexuality as an integral part of human life." It provides information or referrals to *anyone* who requests it. It maintains an extensive library and computer database. Publications include sex education guides for parents (in English and Spanish) and a comprehensive bimonthly journal, *SIECUS Report.* Memberships are available.

Sexual Abuse

Parents United
232 E. Gish Rd.
San Jose, CA 95112
LD 408-453-7616
LD 408-279-8228 (crisis counseling)
Monday–Friday 9 AM–5 PM

Parents United and its related groups, Daughters and Sons United and Adults Molested as Children United, are self-help groups for all family members affected by incest and child sexual abuse. They provide referrals to treatment programs and self-help groups throughout the country. A number of informative publications are available, including a bimonthly newsletter. Write or call for a literature list. The Giarretto Institute—at the above address and phone number—is a separate nonprofit organization that provides treatment and therapist training.

The Sexual Assault Information Page
http://www.cs.utk.edu/-bartley/sainfoPage.html

This site provides information concerning date rape, child sexual abuse and assault, incest, sexual assault, and harassment.

Sexually Transmissible Diseases (ALSO SEE AIDS and HIV AND Health Information)

Herpes Resource Center
P.O. Box 13827
Research Triangle Park, NC 27709-9940
LD 919-361-8488
Monday–Friday 9 AM–7 PM

The center provides referrals for people who are infected with the herpes virus or who think they may be. Services are confidential.

STD Hotline (American Social Health Association)
TF 800-227-8922
Monday–Friday 8 AM–11 PM (EST)

The STD Hotline provides information on all aspects of sexually transmitted diseases. It will describe symptoms but cannot provide diagnosis. It provides referrals for testing and further information. Confidentiality is maintained.

Single Parents

Parents Without Partners (PWP)
401 N. Michigan Ave.
Chicago, IL 60611
LD 312-644-6610
TF 800-637-7974
Monday–Friday 9 AM–5 PM

Parents Without Partners is a mutual support group for single parents and their children. It has numerous local groups with more than 85,000 members. It offers educational programs and literature, including *The Single Parent* magazine. It also offers scholarships for PWP children.

Single Mothers by Choice (SMC)
P.O. Box 1642
Gracie Station
New York, NY 10028
LD 212-988-0993

SMC is a national group that supports single women over thirty who are considering or have chosen motherhood. It offers workshops, support groups, a newsletter, a bibliography, and information about donor insemination and adoption. Based in New York City, it has chapters throughout the United States and Canada.

Single Parents Association
TF 800-704-2102
Monday–Friday 9 AM–6 PM CST

This line helps parents find support groups and resources and answers parenting questions.

Stepfamilies

The Stepfamily Association of America
215 Centennial Mall South, Suite 212
Lincoln, Nebraska 68508
LD 402-477-7837

Offers a support network and serves as a national advocate for stepparents and their children. Publisher *Stepfamilies*, a quarterly. Produces books, research reports, articles covering issues of importance to stepfamilies.

The Stepfamily Foundation
333 West End Ave.
New York, NY 10023
LD 212 877-3244 or 1 800 SKY-STEP

Gathers and disseminates information on stepfamilies and stepfamily relations. Offers over-the-phone counseling and makes referrals. Publishes *Step News*, a quarterly newsletter. Offers books, audiocassettes, and videos on common stepfamily concerns.

Suicide Prevention

The Samaritans
500 Commonwealth Ave.
Boston, MA 02215
LD 617-247-0220 (24-hour hotline)
LD 617-536-2460 (office)

The Samaritans is a worldwide nonreligious organization of trained volunteers who talk with, and listen to, anyone who is suicidal, lonely, or depressed. The confidential service is dedicated to the prevention of suicide and the alleviation of loneliness and depression. Although the service is based in Massachusetts, staffers will help anyone in any area of the country. Brochures and pamphlets are available. Check your telephone directory white pages (or Yellow Pages) under "Suicide Prevention" for local listings.

Samariteen
LD 617-247-8050
3 PM–9 PM (Eastern time)

Samariteen offers the same services as The Samaritans (above) but is staffed by trained sixteen- to nineteen-year olds.

Terminal Illness

National Hospice Organization (NHO)
1901 N. Moore St., Suite 901
Arlington, VA 22209
LD 703-243-5900
TF 800-658-8898 (referrals only)
Monday–Friday 8:30 AM–5:30 PM

Hospices provide support and care for people in the final phase of terminal disease. There are over 2,000 hospice programs in the United States, over 1,800 of which are NHO members. The hospice "team" provides personalized care to minister to the physical, spiritual, and emotional needs of the patient and family. Call or write for literature and referral information.

The Webster
http://www.cyberspy.com/-webster/death.html

A comprehensive list of resources on death and dying.

Women

National Organization for Women (NOW)
1000 16th St. NW, Suite 700
Washington, DC 20036
LD 202-331-0066

NOW is an organization of women and men who support "full equality for women in truly equal partnership with men." NOW promotes social change through research, litigation, and political pressure. A newspaper and other publications are available. Many cities have local chapters of NOW. Membership dues vary.

The National Organization for Women (NOW) Home Page
http://now.org/now/

WomensNet @igc (Women's Issues on-line)
http://www.igc.apc.org/womensnet/

Work and Family

Families and Work Institute
330 7th Ave., 14th Floor
New York, NY 10001
LD 212-465-2044
FAX 212-465-8637
Monday–Friday 9 AM–5 PM

This nonprofit research and planning organization is committed to developing new approaches for balancing the changing needs of U.S. families with the continuing need for workplace productivity.

Women's Bureau
http://www.dol.gov/dol/wh/

An agency within the Department of Labor, from this site you can access press releases related to women's issues in the workplace.

Additional Internet Resources

Adoption Options
http://www.intergate.net/uhtml/sam/Adoption/

The Faces of Adoption: Information on Adoption
http://www.inetcom.net/adopt/

CTW (Children's Television Workshop) Family Corner
http://www.ctw.org/

The Divorce Page
http://www1.primenet.com/~dean/

Collected Domestic Partner Information
http://www.cs.cmu.edu/afs/cs.cmu.edu/user/scotts/domestic-partners/mainpage.html

Domestic Violence Information Center
http://www.feminist.org/other/dv/dvhome.html

QUIRX: Gay/Lesbian/Bisexual Generation X Group
http://www.youth.org/loco/quirx/index.html

Men's Issues Page
http://info-sys.home.vix.com/pub/men/

At-Home Dad: Home Page
http://www.parentsplace.com/readroom/athomedad/index.html

Sexual Assault Information Page
http://www.cs.utk.edu/~bartley/sainfopage.html

Single Parent Resource Center
http://rampages.onramp.net/~beuhamil/singleparentresourcece_478.html

Welfare and Families
http://epn.org/idea/welfare.html

Sexual Life—Taking Care of Our Bodies

Choosing a Method of Birth Control

To be fully responsible in using birth control, a person must know the options available, how reliable these methods are, and the advantages and disadvantages (including possible side effects) of each method. Choosing the best form of birth control for yourself and your partner, especially if you have not been practicing contraception, is not easy. But knowing the facts about the methods gives you a solid basis from which to make decisions and more security once a decision is reached. If you need to choose a birth control method for yourself, remember that *the best method is the one you will use consistently.* When you are having intercourse, a condom left in a purse or wallet, a diaphragm in the bedside drawer, or a forgotten pill in its packet on the other side of town is not an effective means of birth control.

The following material is written in sufficient detail to serve as a reference guide for those who are seeking a birth control method or who have questions about their current method. Of course, a person with painful or unusual symptoms or with questions not dealt with here should seek professional advice from a physician or family planning clinic.

In the following text, where we discuss method effectiveness, *theoretical effectiveness* implies perfectly consistent and correct use; *user effectiveness* refers to *actual* use (and misuse) based on studies by health-care organizations, medical practitioners, academic researchers, and pharmaceutical companies. User effectiveness is sometimes significantly lower than theoretical effectiveness because of factors that keep people from using a method properly or consistently. These factors may be inherent in the method or may be the result of a variety of influences on the user.

Hormonal Methods: The Pill, Implants, and Depo-Provera

Birth Control Pills

"The Pill" is actually a series of pills (twenty, twenty-one, or twenty-eight to a package) containing the synthetic hormones estrogen and progestin, which regulate egg production and the menstrual cycle. When taken for birth control, they accomplish some or all of the following: inhibit ovulation, thicken cervical mucus (preventing sperm entry), inhibit implantation of the blastocyst, or promote early spontaneous abortion of the blastocyst. The pill produces basically the same chemical conditions that would exist in a woman's body if she were pregnant.

Oral contraceptives must be prescribed by a physician or family planning clinic. There are a number of brands available that contain varying amounts of hormones. Most commonly prescribed are the combination pills, which contain a fairly standard amount of estrogen and differing doses of progestin according to the pill type. In the triphasic pill, the amount of progestin is altered during the cycle, purportedly to approximate the "normal" hormonal pattern. There is also a minipill containing progestin only, but it is generally prescribed only in cases where the woman should not take estrogen. It is considered slightly less effective than the combination pill, and it must be taken with precise and unfailing regularity to be effective.

With the twenty- and twenty-one-day pills, one pill is taken each day until they are all used. Two to five days later, the woman should begin her menstrual flow. Commonly, it is quite light. (If the flow does not begin, the woman should start the next series of pills seven days after the end of the last series. If she repeatedly has no flow, she should talk to her health-care practitioner.) On the fifth day of her menstrual flow, the woman should start the next series of pills.

The twenty-eight-day pills are taken continuously. Seven of the pills have no hormones. They are there simply to avoid breaks in the routine; some women prefer them because they find them easier to remember.

Along with contraceptive implants, the pill is considered the most effective birth control method available (except for sterilization) when used correctly. It is not effective when used carelessly. The pill must be taken every day, as close as possible to the same time each day. If one is missed, it should be taken as soon as the woman remembers, and the next one should be taken on schedule. If two are missed, the method cannot be relied on, and an additional form of contraception should be used for the rest of the cycle.

It should be remembered that the pill in no way protects against sexually transmitted diseases. Women on the pill should consider the additional use of a condom to reduce the risk of sexually transmitted diseases (STDs).

Effectiveness. The pill is more than 99.5 percent effective theoretically. User effectiveness (the rate shown by actual studies) is between 95 and 98 percent.

Advantages. Pills are easy to take. They are dependable. No applications or interruptions are necessary before or during intercourse. Some women experience side effects that please them, such as more regular or reduced menstrual flow or enlarged breasts.

Possible Problems. There are many possible side effects, which may or may not bother the user, that can occur from taking the pill. Those most often reported are the following:

- Change (usually a decrease) in menstrual flow
- Breast tenderness
- Nausea or vomiting
- Weight gain or loss

Some others include the following:

- Spotty darkening of the skin
- Nervousness or dizziness

- Loss of scalp hair
- Change in appetite
- Change (most commonly, a decrease) in sex drive
- Increase in body hair
- Increase in vaginal discharge and yeast infections

These side effects can sometimes be eliminated by changing the prescription, but not always. Certain women react unfavorably to the pill because of existing health factors or extra sensitivity to female hormones. Women with heart or kidney diseases, asthma, high blood pressure, diabetes, epilepsy, gallbladder diseases, or sickle-cell disease or those prone to migraine headaches or mental depression are usually considered poor candidates for the pill.

The pill also creates health risks, but to what extent is a matter of controversy. Women taking the pill stand a greater chance of problems with circulatory diseases, blood clotting, heart attack, and certain kinds of liver tumors. There is also an increased risk of contracting chlamydia. The health risks are low for the young (about half the number of risks encountered in childbirth), but they increase with age. The risk for smokers, women over thirty-five, and those with certain other health disorders is about four times as great as childbirth. For women over forty, the risks are considered high. Current literature on the pill especially emphasizes the risks for women who smoke. Definite risks of cardiovascular complications and various forms of cancer exist due to the synergistic action of the ingredients of cigarettes and oral contraceptives.

A number of studies have linked pill use with certain types of cancer, but they have not been conclusive. The risk of some types of cancer, such as ovarian and endometrial, appears to be significantly *reduced* by pill use. However, a link between cervical cancer and long-term pill use has been suggested by several studies. Regular Pap smears are recommended as an excellent defense against cervical cancer, for pill users and nonusers alike.

Certain other factors may need to be taken into account in determining if oral contraception is appropriate. Young girls who have not matured physically may have their development slowed by early pill use. Nursing mothers cannot use pills containing estrogen because the hormone inhibits milk production. Some lactating women use the minipill successfully.

Millions of women (approximately 10.7 million in the United States) use the pill with moderate to high degrees of satisfaction. For many women, if personal health or family history does not contraindicate it, oral contraception is both effective and safe.

Implants

In December 1990, the Food and Drug Administration (FDA) approved a new contraceptive—thin, matchstick-sized capsules containing progestin that are implanted under a woman's skin. Over a period of up to five years, the hormone is slowly released. When the implants are removed, fertility is restored. A set of soft tubes is surgically implanted under the skin of the upper arm in a simple doctor's-office procedure with local anesthesia. Once implanted, the capsules are not visible but may be felt under the skin. The implants, under the trade name Norplant, are currently on the market in more than fourteen countries. The cost may run as high as several hundred dollars, but it is probably less expensive than a five-year supply of birth control pills.

Effectiveness. Although Norplant has not been observed or tested to the degree that other contraceptives have, initial reports indicate a failure rate one-tenth to one-twentieth that of the pill. It may be the most effective contraceptive ever marketed.

Advantages. Convenience is clearly a big advantage of the implant. Once the implant is in, there's nothing to remember, buy, do, or take care of.

Possible Problems. The chief side effect, experienced by about half of implant users, is a change in the pattern of menstrual bleeding, such as lengthened periods or spotting between periods. Norplant should not be used by women with acute liver disease, breast cancer, blood clots, or unexplained vaginal bleeding. Possible long-term negative effects are not known at this time. Removal of the device has caused problems for some women.

Depo-Provera (DMPA)

The injectable contraceptive medroxy-progesterone acetate, Depo-Provera, or DMPA, provides protection from pregnancy for three months. It is used in over eighty countries throughout the world and has recently been approved for use in the United States. Generally speaking, DMPA has been shown to be remarkably free of serious side effects and complications. It does cause cessation of the menstrual period while it is being administered, and it may cause delayed fertility (up to a year in some cases) until the effects wear off. Sometimes "breakthrough bleeding" (bleeding not associated with the menstrual period) is a problem. DMPA is usually given in injections at three-month intervals.

Barrier Methods: Condom, Diaphragm, Cervical Cap, and Women's Condom

Barrier methods are designed to keep sperm and egg from getting together. The barrier device used by men is the condom. Barrier methods available to women are the diaphragm, the cervical cap, and the women's condom. The effectiveness of all barrier methods is increased by use with spermicides (sperm-killing chemicals).

Condom

A condom is a thin sheath of rubber (or processed sheep's intestine) that fits over the erect penis and thus prevents semen from being transmitted. Condoms are available in a variety of sizes, shapes, and colors. Some are lubricated, and some are treated with spermicides. They are available from drugstores, and most kinds are relatively inexpensive. Condoms are the third most widely used form of birth control in the United States. It is estimated that 5 billion condoms are used annually worldwide.

In late 1994, a condom made of polyurethane (Avanti) was introduced. The advantages of polyurethane are that it is both thin and strong so that it may provide for more sensitive contact for some people. It may be a good alternative for those who are allergic to latex. Laboratory tests on polyurethane show it to be an effective barrier against particles as small as sperm and disease-causing organisms.

Condoms not only provide effective contraception when properly used, but they also guard against the transmission of a number of sexually transmissible diseases, such as chlamydia, gonorrhea, genital herpes, and AIDS *if* they are made of latex rather than sheep's intestine. Due to increased publicity regarding the transmissibility of the AIDS virus (HIV), both heterosexuals and gay men are increasingly using latex condoms as prophylactic (disease-preventing) devices. Condom use does not guarantee *total* safety from STDs, however, as sexual partners may also transmit certain diseases by hand, mouth, and external genitalia other than the penis; also, condoms may occasionally tear or leak.

Today an estimated 40 percent of condoms are purchased by women, and condom advertising and packaging increasingly reflect this trend. Even if a woman regularly uses another form of birth control, such as the pill or intrauterine device (IUD), she may want to have the added protection provided by a condom, especially if it has been treated with spermicide (which provides further protection against disease organisms). Contraceptive aerosol foam and contraceptive film (a thin, translucent square of tissue that dissolves into a gel) are the most convenient spermicidal preparations to use with condoms.

Effectiveness. Condoms are 98 to 99 percent effective theoretically. User effectiveness is 88 percent. Failures sometimes occur from mishandling the condom, but they are usually the result of not putting it on until after some semen has leaked into the vagina or simply not putting it on at all.

Advantages. Condoms are easy to obtain. They are easy for men and women to carry in a wallet or purse. They help protect against STDs, including herpes and HIV.

Possible Problems. The chief drawback to a condom is that it must be put on after the man has been aroused but before he enters his partner. The interruption is the major reason for users neglecting or "forgetting" to put a condom on. Some men complain that sensation is dulled, and (very rarely) cases of allergy to rubber are reported. Both problems can be remedied by the use of polyurethane or animal tissue condoms. These are thinner and conduct heat better. *Animal tissue condoms, however, should not be used for prophylaxis; viral STDs, such as genital herpes and HIV, may easily penetrate them.* The condom user must take care to hold the sheath at the base of his penis when he withdraws in order to avoid leakage.

Diaphragm

The diaphragm is a rubber cup with a flexible rim. It, along with a spermicide, is placed in the vagina, blocking the cervix, to prevent sperm from entering the uterus and fallopian tubes. Different women require different sizes, and a woman may change sizes, especially after a pregnancy; the size must be determined by an experienced practitioner. Diaphragms are available by prescription from doctors and family planning clinics. Somewhat effective by itself, the diaphragm is highly effective when used with a spermicidal cream or jelly. (Creams and jellies are considered more effective than foam for use with a diaphragm.) Diaphragm users should be sure to use an adequate amount of spermicide and to follow their practitioner's instructions with care.

The diaphragm with spermicide can be put in place up to two hours before intercourse. It should be left in place six to eight hours afterward. A woman should not dislodge it or douche before it is time to remove it. If intercourse is repeated within six hours, the diaphragm should be left in place, but more spermicide should be inserted with an applicator. However, a diaphragm should not be left in place longer than twenty-four hours.

Diaphragms need to be replaced about once a year, because the rubber deteriorates, losing elasticity and increasing the chance of splitting. Any change in the way the diaphragm feels, as well as any dramatic gain or loss of weight, calls for a visit to a doctor or clinic to check the fit.

Effectiveness. When properly used, *with spermicide*, diaphragms are 98 percent effective. Numerous studies of diaphragm effectiveness have yielded varying results. Typical user effectiveness (actual statistical effectiveness) is in the 81 to 83 percent range. The lowest effectiveness ratings appear among young diaphragm users and inconsistent users.

Advantages. The diaphragm can be placed well before the time of intercourse. For most women, there are few health problems associated with its use. It helps protect against diseases of the cervix and pelvic inflammatory disease (PID).

Possible Problems. Some women dislike handling or placing diaphragms, or the mess or smell of the spermicide used with them. Some men complain of rubbing or other discomfort caused by the diaphragm. Occasionally a woman is allergic to rubber. Some women become more prone to urinary tract infections. As there is a small risk of toxic shock syndrome (TSS) associated with diaphragm use, a woman should not use a diaphragm under the following conditions.

- During menstruation or other bleeding.
- Following childbirth (for several months).
- During abnormal vaginal discharge.
- If she has had TSS or if *Staphylococcus aureus* bacteria are present.

She should also

- Never wear the diaphragm for more than twenty-four hours.
- Learn to watch for the warning signs of TSS (fever [temperature of 101 degrees or more], diarrhea, vomiting, muscle aches, and sunburnlike rash).

Cervical Cap

The cervical cap is a small rubber barrier device that fits snugly over the cervix; it can be filled with spermicidal cream or jelly. Cervical caps come in different sizes and shapes; proper fit is extremely important, and not everyone can be fitted. Fitting must be done by a physician or at a health clinic.

Effectiveness. There is a limited amount of research data in this country regarding the cervical cap's effectiveness. Reported user effectiveness ranges from 73 to 92 percent.

Advantages. The cervical cap may be more comfortable and convenient than the diaphragm for some women. It does not interfere with the body physically or hormonally.

Possible Problems. Some users are bothered by an odor that develops from the interaction of the cap's rubber and either vaginal secretions or the spermicide. There is some concern that the cap may contribute to erosion of the cervix. If the penis of a woman's partner touches the rim of the cap, the cap can become displaced during intercourse. Theoretically, the same risk of TSS exists for the cervical cap as for the diaphragm. The precautions discussed in the section on diaphragms also apply.

Women's Condom

There are two types of condoms designed for women's use. The most common is a disposable, soft, loose-fitting polyurethane sheath with a diaphragmlike ring at each end. One ring is inside the sheath and is used to insert and anchor the condom next to the cervix. The larger, outer ring remains outside the vagina and acts as a barrier, protecting the vulva and the base of the penis. The other type, made of latex, is secured by a G-string. A condom pouch in the crotch of the G-string unfolds as the penis pushes into the vagina.

Because of the newness of this product, its effectiveness is not known at this time, although clinical tests have indicated that it is less likely to leak than men's latex condoms. In laboratory trials, the women's condom was not permeated by HIV or cytomegalovirus, indicating that it may be a promising alternative for both contraception and the control of STDs.

Spermicides

A spermicide is a substance that is toxic to sperm. The most commonly used spermicide in products sold in the United States is the chemical nonoxynol-9.

Spermicidal preparations are available in a variety of forms: foam, film, jelly, cream, tablets, and suppositories. Some condoms are also treated with spermicides. Spermicidal preparations are considered most effective when used in combination with a barrier method.

A further benefit of spermicides is that they significantly reduce STD risk. Nonoxynol-9 has been demonstrated to have some protection against a number of disease agents, including HIV. Of course, the use of spermicides does not entirely eliminate the risk of contracting an STD, although it may lower the risk.

Contraceptive Foam

Contraceptive foam is a chemical spermicide sold in aerosol containers. Methods of application vary with each brand, but usually foam is released deep in the vagina either directly from the container or with an applicator. It forms a barrier to the uterus and inactivates sperm in the vagina. It is most effective if inserted no more than half an hour before intercourse. Shaking the container before applying the foam increases its foaminess so that it spreads farther. The foam begins to go flat after about thirty minutes. It must be reapplied when intercourse is repeated.

Effectiveness. Foam has a theoretical effectiveness rate of 98.5 percent. User failure brings its effectiveness down to as low as 71 percent. User failures tend to come from not applying the foam every single time the couple has intercourse, from relying on foam inserted hours before intercourse, or from relying on foam placed hurriedly or not placed deep enough. If used properly and consistently, however, it is quite reliable. Used with a condom, it is highly effective.

Advantages. There are almost no medical problems associated with the use of foam. Foam can help provide protection against certain sexually transmitted diseases.

Possible Problems. Some women dislike applying foam. Some complain of messiness, leakage, odors, or stinging sensations. Occasionally, a woman or man may have an allergic reaction to it.

Contraceptive Film

Contraceptive film is a relatively new spermicidal preparation. It is sold in packets of small (2-inch-square), translucent tissues. This thin tissue contains the spermicide nonoxynol-9; it dissolves into a sticky gel when inserted into the vagina. It should be inserted directly over the cervix, not less than five minutes or more than one and a half hours before intercourse. It remains effective for two hours after insertion. Contraceptive film works effectively in conjunction with the condom.

Effectiveness. Extensive research has not been done on the effectiveness of film due to its relative newness on the market. The highest effectiveness rates reported are from 82 to 90 percent. Of course, proper and consistent use with a condom highly increases the effectiveness.

Advantages. Film is easy to use for many women. It is available from a drugstore and is easy to carry in a purse, wallet, or pocket.

Possible Problems. Some women may not like inserting the film into the vagina. Some women may be allergic to it. Increased vaginal discharge and temporary pain while urinating after using contraceptive film have been reported.

Creams, Jellies, Suppositories, and Tablets

Creams and jellies are chemical spermicides that come in tubes and are inserted with applicators or placed inside diaphragms or cervical caps. They can be bought without a prescription at most drugstores. They work in a manner similar to foams but are considered less effective when used alone. Like foam, jellies and creams seem to provide some protection against STDs. This factor makes their use with a diaphragm even more attractive.

Other chemical spermicides are inserted into the vagina as suppositories or tablets before intercourse. Body heat and fluids dissolve the ingredients, which will inactivate sperm in the vagina after ejaculation. They must be inserted early enough to dissolve completely before intercourse.

Effectiveness. Reports on these methods vary widely. It is suspected that the variations are connected with each user's technique of application. Directions for use that come with these contraceptives are not always clear. Jellies, creams, vaginal tablets, and suppositories should be used in conjunction with a barrier method for maximum effectiveness.

Advantages. The methods are simple, easily obtainable, and associated with virtually no medical problems. They *help* protect against STDs, including chlamydia, trichomoniasis, gonorrhea, genital herpes, and AIDS.

Possible Problems. Some people have allergic reactions to the spermicides. Some women dislike the messiness, smells, or necessity of touching their own genitalia. Others experience irritation or inflammation, especially if they use the method frequently. A few women lack the vaginal lubrication to dissolve the tablets in a reasonable amount of time. And a few women complain of having anxiety about the method's effectiveness during intercourse.

The Intrauterine Device (IUD)

The intrauterine device (IUD) is a tiny plastic and/or metal device that is inserted into the uterus through the cervical opening. The particular type of device determines how long it may be left in place—one to four years.

Although most IUDs have been withdrawn from the U.S. market because of the proliferation of lawsuits against their manufacturer, they are still considered a major birth control method. There clearly are some risks involved with the IUD, especially for certain women. Nevertheless, it remains the birth control method of about 70 million women throughout the world, including 40 million to 45 million women in China.

Strictly speaking, the IUD is not a contraceptive. It is a birth control device that works principally by preventing the blastocyst from implanting in the uterine wall or by disrupting a blastocyst that has already implanted. The IUD may also serve to immobilize sperm and (in the case of progestin-bearing IUDs) induce thickened cervical mucus. IUDs must be inserted and removed by a trained practitioner.

Effectiveness. IUDs are 97 to 99 percent effective theoretically. User effectiveness is 90 to 96 percent. The copper T-380A IUD has the lowest failure rate of any IUD developed to date.

Advantages. Once inserted, IUDs require little care. They don't interfere with spontaneity during intercourse.

Possible Problems. Insertion of an IUD is often painful. Heavy cramping usually follows and sometimes persists. Menstrual flow usually increases. Up to one-third of users, especially women who have never been pregnant, expel the device within the first year. This usually happens during menstruation. The IUD can be reinserted, however, and many women retain it the second time.

The IUD is associated with increased risk of pelvic inflammatory disease (PID). Because of the risk of sterility induced by PID, many physicians recommend that women planning to have children use alternative methods. Women who have had PID or who have multiple sex partners should be aware that an IUD will place them at significantly greater risk of PID, STDs, and other infections.

The IUD cannot be inserted in some women due to unusual uterine shape or position. Until recently, the IUD was considered difficult to insert and less effective for teenage girls. But newer, smaller IUDs have been found to provide good protection even for young teenagers.

Occasionally the device perforates the cervix. This usually happens at the time of insertion, if it happens at all. Removal sometimes requires surgery.

Sometimes pregnancy occurs and is complicated by the presence of the IUD. If the IUD is not removed, there is a 50 percent change of spontaneous abortion; if it is removed, the chance is 25 percent. Furthermore, spontaneous abortions that occur when an IUD is left in place are likely to be septic (infection bearing) and possibly life threatening.

Fertility Awareness Methods

Fertility awareness methods of contraception require substantial education, training, and planning. They are based on a woman's knowledge of her body's reproductive cycle. These methods require high motivation and self-control; they are not for everyone.

Fertility awareness is also referred to as "natural family planning." Some people make the following distinction between the two. With fertility awareness, the couple may use an alternate method (such as a diaphragm with jelly or condom with foam) during the fertile part of the woman's cycle. Natural family planning does not include the use of any contraceptive device and is thus considered more "natural;" it is approved by the Roman Catholic church.

Fertility awareness methods include the rhythm (calendar) method, the basal body temperature (BBT) method, the mucus (also called the Billings or ovulation) method, and the symptothermal method, which combines the latter two. These methods are not recommended for women who have irregular menstrual cycles, including postpartum and lactating mothers.

All women can benefit from learning to recognize their fertility signs. It is useful to know when the greatest likelihood of pregnancy occurs, both for women who wish to avoid pregnancy and for those who want to become pregnant.

Basal Body Temperature (BBT) Method

A woman's temperature tends to be slightly lower during menstruation and for about a week afterward. Just before ovulation it dips, then rises sharply (one-half to one whole degree) following ovulation. It stays high until just before the next menstrual period.

A woman practicing the basal body temperature (BBT) method must record her temperature every morning upon waking for six to twelve months to have an accurate idea of her temperature pattern. When she is quite sure she recognizes the rise in temperature and can predict about when in her cycle it will happen, she can begin using the method. She will abstain from intercourse or use an alternative contraceptive method for three to four days before the expected rise and for four days after it has taken place. If she limits intercourse to only the "safe" time after her temperature has risen, the method is more effective. This method requires high motivation and control. For greater accuracy, it may be combined with the mucus method described in the next section.

Mucus Method (Billings or Ovulation Method)

In many women, there is a noticeable change in the appearance and character of cervical mucus prior to ovulation. After menstruation, most women experience a moderate discharge of cloudy, yellowish or white mucus. Then, for a day or two, a clear, slippery mucus is secreted. Ovulation occurs immediately after the clear, slippery mucus secretions. The preovulatory mucus is elastic in consistency, rather like raw egg white, and a drop can be stretched between two fingers into a thin strand (at least 2¼ inches, or 6 centimeters). This elasticity is called *spinnbarkeit.* Following ovulation, the amount of discharge decreases markedly. The four days before and the four days after these secretions are considered the unsafe days. An alternate contraceptive method may be used during this time. The method requires training and high motivation to be successful. Clinics are offered in some cities. (For more information, check with your local family planning or women's health clinic.) This method may be combined with the BBT method for greater effectiveness.

Sympto-Thermal Method

When the BBT and mucus methods are used together, it is called the sympto-thermal method. Additional signs that may be useful in determining ovulation are mid-cycle pain in the lower abdomen on either side *(mittelschmerz)* and a very slight discharge of blood from the cervix ("spotting"). Women who wish to rely on fertility awareness methods of contraception should enroll in a class at a clinic. Learning to read one's own unique fertility signs is a complex process requiring one-to-one counseling and close monitoring.

Calendar Method

The calendar (or "rhythm") method relies on calculating safe days based on the range of a woman's longest and shortest menstrual cycles. It is not practical or safe for women with irregular cycles. For women with regular cycles, the calendar method is reasonably effective because the period of time when an ovum is receptive to fertilization is only about twenty-four hours. As sperm generally live two to four days, the maximum period of time in which fertilization could be expected to occur may be calculated with the assistance of a calendar.

Ovulation generally occurs fourteen (plus or minus two) days before a woman's menstrual period. Taking this into account, and charting her menstrual cycles for a minimum of eight months to determine the longest and shortest cycles, a woman can determine her expected fertile period.

During the fertile period, a woman must abstain from sexual intercourse or use an alternative method of contraception. A woman using this method must be meticulous in her calculations, keep her calendar up to date, and be able to maintain an awareness of what day it is. Statistically only about one-third of all women have cycles regular enough to employ this method satisfactorily.

Effectiveness of Fertility Awareness Methods

It is hard to calculate the effectiveness of fertility awareness methods. With this type of contraception, in a sense, the user is the method; the method's success or failure rests largely on the woman's diligence. For those who have used fertility awareness with unfailing dedication, it has been demonstrated to be as much as 99 percent effective. Many studies, however, show fairly high failure rates. Some researchers believe that this is due to risk taking during the fertile phase.

Additionally, there is always some difficulty in predicting ovulation with pinpoint accuracy; thus, there is a better chance of pregnancy as a result of intercourse prior to ovulation than as a result of intercourse following it. Furthermore, there is evidence that sperm may survive as long as five days. This suggests that unprotected intercourse any time prior to ovulation may be risky. All fertility awareness methods increase in effectiveness if intercourse is unprotected only during the safe period *following* ovulation.

Advantages. Fertility awareness (or natural family planning) methods are acceptable to most religious groups. They are free and pose no health risks. If a woman wishes to become pregnant, awareness of her own fertility cycles is very useful.

Possible Problems. Fertility awareness methods are not suitable for women with irregular menstrual cycles or couples who are not highly motivated to use the methods. Some couples who practice abstinence during fertile periods may begin to take risks out of frustration. These couples may benefit from exploring other forms of sexual expression; counseling can help.

Sterilization

Among married couples in the United States, sterilization (of one or both partners) is the most popular form of birth control. Approximately 27 percent of couples have chosen to be sterilized.

Sterilization for Women

Approximately 9.6 million American women have chosen sterilization as their preferred form of birth control. Most female sterilizations are tubal ligations, or "tying the tubes." The two most common operations are done through laparoscopy and minilaparotomy. Less commonly performed types of sterilization for women are culpotomy, culdoscopy, and hysterectomy. Generally, sterilization surgery is not reversible; only women who are completely certain that they want no (or no more) children should choose this method.

Sterilization for women is quite expensive. Surgeon, anesthesiologist, and hospital fees are substantial. The newer procedures in which the woman returns home on the same day have the advantage of being one-half to one-third as expensive as abdominal surgery but still may cost several thousand dollars. Many health insurance policies will cover all or part of the cost of sterilization for both men and women. In some states, Medicaid pays for sterilization for certain patients.

Laparoscopy. Sterilization by laparoscopy usually requires a day or less in the hospital or clinic. General anesthesia is usually recommended. The woman's abdomen is inflated with gas to make the organs more visible. The surgeon inserts a rodlike instrument with a viewing lens (the laparoscope) through a small incision at the edge of the navel and locates the fallopian tubes (the ducts between the ovaries and the uterus). Through this incision or a second one, the surgeon inserts another instrument that closes the tubes. The tubes are usually closed by electrocauterization. Special small forceps that carry an electric current clamp the tubes and cauterize (burn) them. The tubes may also be closed off or blocked with tiny rings, clips, or plugs. There is a recovery period of several days to a week. During this time, the women will experience some tenderness and some bleeding from the vagina. Rest is important.

Minilaparotomy. Local or general anesthesia is used with minilaparotomy. A small incision is made in the lower abdomen, through which the fallopian tubes are brought into view. They are then tied off or sealed with electric current, clips, or rings. Recovery is the same as with laparoscopy.

Culpotomy and Culdoscopy. In culpotomy and culdoscopy, an incision is made at the back of the vagina. In culpotomy, the tubes are viewed through the incision, then tied or otherwise blocked, and then cut. Culdoscopy is the same procedure, but it uses a viewing instrument called a culdoscope. The advantage of these procedures is that they leave no visible scars. They require more expertise on the part of the surgeon, however, and have higher complication rates than laparoscopy and minilaparotomy.

Hysterectomy. Hysterectomy is not performed for sterilization except under special circumstances. Because it involves removal of the entire uterus, it is both riskier and more costly than other methods. It involves greater recovery time and, for some women, is potentially more difficult psychologically. It may be appropriate for women who have a uterine disease or other problem that is likely to require a future hysterectomy anyway.

Effectiveness of Sterilization for Women. Surgical contraception is essentially 100 percent effective. In *extremely* rare instances (less than ¼ of a percent), probably due to improperly done surgery, a tube may reopen or grow back together, allowing an egg to pass through.

Once sterilization is done, no other method of birth control will ever be necessary. (A woman who risks exposure to STDs, however, may wish to protect herself with a spermicide, condom, or women's condom.)

Sterilization does not reduce or change a woman's feminine characteristics. It is not the same as menopause and does not hasten the approach of menopause, as some people believe. A woman still has her menstrual periods until whatever age menopause naturally occurs for her. Her ovaries, uterus (except in the case of hysterectomy), and hormonal system have not been changed. The only difference is that sperm cannot now reach her eggs. (The eggs, which are released every month as before, are reabsorbed by the body.) Sexual enjoyment is not diminished. In fact, a high percentage of women report that they feel more relaxed during intercourse because anxiety about pregnancy has been eliminated. There do not seem to be any harmful side effects associated with female sterilization.

Sterilization should be considered irreversible. The most recently developed methods of ligation using clips, rings, or plugs may be more reversible than those employing electrocauterization, but the overall success rate for reversals is quite low; it is also very costly and not covered by most insurance plans. Only between 15 and 25 percent of women who seek to have their tubal ligations reversed succeed in conceiving.

The tubal ligation itself is a relatively safe procedure. With electrocauterization, there is a slight chance that other tissues in the abdomen may be damaged, especially if the surgeon is not highly skilled. Anesthesia complications are the most serious risk. The risk of death is quite low—four deaths per 100,000 operations. Infection is also a possibility; it may be treated with antibiotics.

Sterilization for Men

About 4.1 American men have chosen sterilization. A vasectomy is a minor surgical procedure that can be done in a doctor's office under local anesthesia. It takes approximately half an hour. In this procedure, the physician makes a small incision (or two incisions) in the skin of the scrotum. Through the incision, each vas deferens (sperm-carrying tube) is lifted, cut, tied, and often cauterized with electricity. After a brief rest, the man is able to walk out of the office; complete recuperation takes only a few days.

A man may retain some viable sperm in his system for days or weeks following a vasectomy. He should use other birth control methods until his semen has been checked, about eight weeks following the operation.

Effectiveness. Vasectomies are 99.85 percent effective. In very rare cases, the ends of a vas deferens may rejoin; this is virtually impossible if the operation is correctly performed.

Advantages. After vasectomy, no birth control method will ever be needed again, unless the man wishes to use a condom to prevent getting or spreading an STD. Sexual enjoyment will not be diminished; he will still have erections and orgasms and ejaculate semen. Vasectomy is relatively inexpensive, as surgical procedures go.

Possible Problems. Compared with other birth control methods, the complication rates for vasectomy are very low. Most problems occur when proper antiseptic measures are not taken during the operation or when the man exercises too strenuously in the few days after it. Hematomas (bleeding under the skin) and granulomas (clumps of sperm) can be treated with ice packs and rest. Epididymitis (inflammation of the tiny tubes that connect the testis and vas deferens) can be treated with heat and scrotal support.

One-half to two-thirds of men develop sperm antibodies following vasectomy. The body produces these antibodies in response to the presence of sperm that have been absorbed by body tissues. There is no physiological evidence that this poses any threat to health.

A few men—those who equate fertility with virility and potency—may experience psychological problems following vasectomy. However, most men have no adverse psychological changes following a vastectomy if they understand what to expect and have the opportunity to express their concerns and ask questions.

Among men who seek to have their vasectomies reversed, about 50 percent experience success. The cost is high and not covered by insurance in many cases. Vasectomy should be considered permanent.

Postcoital Birth Control

A controversial issue, postcoital birth control (also known as morning-after birth control) involves the expulsion of an ovum that may have been fertilized.

The Morning After Pill (ECP)

The most common type of postcoital birth control, the emergency contraceptive pill (ECP) is generally a combined estrogen-progestin pill which is given in a larger than normal dose within 72 hours of unprotected intercourse. Compared to the alternatives of abortion or childbirth, ECPs are both cost-effective and relatively safe.

Menstrual Extraction and Postcoital IUD Insertion

Menstrual extraction involves the removal of the endometrial contents by suction through a small tube attached to a vacuum pump. This procedure may be performed until an expected menstrual period is two weeks late and can be used with or without a positive diagnosis of pregnancy.

Another form of postcoital birth control is to have an IUD inserted 5-7 days after unprotected intercourse.

RU-486

In 1986, a study of a new drug, mifepristone, known as *RU-486*, concluded that it is an effective and safe method for termination of very early pregnancy but that it should be used under close medical supervision. RU-486 prevents the cells of the uterine lining from getting the progesterone they need to support a blastocyst (fertilized ovum). This "tricks" the body into thinking it's at the end of a menstrual cycle; it sheds the uterine lining in what appears to be a heavy menstrual period. If RU-486 is taken within a few days of unprotected intercourse, the oocyte, if present, will be expelled. It can also be taken within the first 5-6 weeks of pregnancy to induce a very early abortion.

Unreliable Methods of Contraception

Coitus Interruptus (Withdrawal)

The oldest form of contraception known is coitus interruptus, which involves the withdrawal of the penis from the vagina prior to ejaculation. This method is widely used throughout the world today and can be considered somewhat successful for *some* people. Success may depend on technique, on combination with rhythm methods, or on physical characteristics of the partners (such as the tendency toward infertility in one or both partners).

A problem with this method is its riskiness. Secretions from the man's Cowper's glands, urethra, or prostate, which sometimes seep into the vagina before ejaculation, can carry thousands of healthy sperm. Also, the first few drops of ejaculate carry most of the sperm. If the man is slow to withdraw or allows any ejaculate to spill into (or near the opening of) the vagina, the woman may get pregnant. The highest observed effective rate using coitus interruptus is 84 percent. The actual user effectiveness rate is 77 percent. Although it is generally considered an unreliable method of birth control, coitus interruptus is certainly better than nothing.

Douching

To douche, a woman flushes the vagina with liquid. As a contraceptive method it is faulty because after ejaculation, douching is already too late. By the time a woman can douche, the sperm may already be swimming through the cervix into the uterus. The douche liquid may even push the sperm into the cervix. Douching with any liquid, especially if done often, tends to upset the normal chemical balance in the vagina and may cause irritation or infection. There is also evidence that frequent douching can result in ectopic pregnancy.

Lactation

When a woman, after giving birth, breastfeeds her child, she may not begin to ovulate as long as she continues to nourish her child exclusively by breastfeeding. However, although some women do not ovulate while lactating, others do. Cycles may begin immediately after delivery or in a few months. The woman never knows when she will begin to be fertile. Breastfeeding is considered a method in some countries, but the success rates are extremely low.

Mythical Methods of Contraception

There are many myths among young and old about contraception. The young hear many rumors. Some of the old, who should know better, still believe misconceptions hatched in "the old days" when contraceptive devices were not as easily obtained or as dependable as they are today. (The dependability of condoms, always the most easily purchased device, was greatly improved when the product came under the supervision of the Federal Trade Commission.) Today, when satisfactory methods are so easy to obtain, it is senseless for anyone to use risky ones.

Widely known methods that are *totally* useless include the following:

1. Standing up during or after intercourse (sperm have no problem swimming "upstream").
2. Taking a friend's pill the day of, or the day after, intercourse (this doesn't work—and may even be dangerous).
3. Only having intercourse occasionally (it is when, not how often, that makes a difference; once is enough if the woman is fertile at that time).
4. Using plastic wrap or plastic bags for condoms (they are too loose, undependable, and unsanitary).

Abortion

Most people, when they hear the word *abortion*, think of a medical procedure or operation to end an unwanted pregnancy. But the word carries a wider range of meaning than most people realize. Any time a growing embryo or fetus is expelled from the uterus, an abortion has taken place. Such expulsion can happen naturally, or it can be made to happen in one of several ways. Many abortions happen spontaneously because the woman wears an IUD, because she suffers physical shock, because the fetus is not properly developed, or more commonly because physical conditions within the uterus change and end the development of the embryo or fetus. Approximately one-third of all abortions reported in a year are spontaneous ("miscarriages"). Of the approximately 6 million pregnancies that occur in this country each year, more than half are unintentional. Of these, almost half are terminated by induced abortions.

Abortions can be induced in several ways. Surgical methods are the most common, but the use of medications to induce abortion is also possible. Methods for early abortions (those performed in the first three months) differ from those for late abortions (those performed after the third month).

Surgical Methods of Abortion

Surgical methods of abortion include vacuum aspiration, dilation and curettage (D&C), dilation and evacuation (D&E), and hysterotomy, as well as several other methods.

To facilitate the dilation of the cervix, many physicians insert a laminaria, a small stick of seaweed, into the cervical opening. The laminaria expands gradually, dilating the cervix gently in the process. It must be placed at least six hours prior to the abortion.

Vacuum Aspiration (First-Trimester Method)

Vacuum aspiration is performed under local anesthesia. The cervix is dilated with a series of graduated rods (laminaria may have been used to begin dilation). Then a small tube attached to a vacuum is inserted through the cervix. The uterus is gently vacuumed, removing the fetus, placenta, and endometrial tissue. The patient returns home the same day. She will experience cramping, bleeding, and possibly, emotional reactions over the following days. Serious complications are unusual for a legal, properly performed abortion. Those problems that may arise are discussed in the section entitled "Adjustment Following Abortion."

Dilation and Curettage (D&C; First-Trimester Method)

In a D&C, the cervix is dilated and the uterine wall scraped with a small spoon-shaped instrument (curette). Local or general anesthesia is given. This method is generally considered less desirable than vacuum aspiration, as it causes more bleeding, is more painful, and is sometimes less effective. Possible problems are discussed in the section "Adjustment Following Abortion."

Dilation and Evacuation (D&E Second-Trimester Method)

A D&E is usually performed between the thirteenth and twentieth weeks of pregnancy. Local or general anesthesia is used. The cervix is slowly dilated and the fetus removed by alternating curettage with the injection of a solution toxic to the fetus. Patients are usually given an intravenous solution of the hormone oxytocin to encourage contractions and limit blood loss. Because it is a second-trimester procedure, a D&E is somewhat riskier and often more traumatic than a first-trimester abortion.

Hysterotomy (Second-Trimester Method)

In hysterotomy, the fetus is removed through an incision made in the woman's abdomen. This is essentially a cesarean section—major surgery, requiring several days in the hospital. Its use is limited.

Other Methods of Abortion

Abortion can also be induced medically with injections or suppositories containing certain substances. These abortions are performed during the second trimester and generally require hospitalization. Prostaglandins, saline solutions, and urea are common abortifacients. They are used singly or in varying combinations. Many cells of the body contain prostaglandins. A prostaglandin is a type of fatty acid that is active in many kinds of body processes, including reproduction. Prostaglandins are injected into the amniotic sac or administered as vaginal suppositories to induce abortion. Some of the side effects associated with their use are gastrointestinal symptoms, cervical lacerations, and temperature elevation. Saline solutions and solutions of urea are toxic to the fetus and can be injected amniotically. These solutions are relatively inexpensive; they are generally considered more effective when used in combination with prostaglandins.

Adjustment Following Abortion

Following an abortion, most women have some bleeding and cramps for about two weeks. The only cause for worry about bleeding is if it is heavier two days in a row than the heaviest day of normal menstrual flow. To protect herself against infection, a woman should not have sexual intercourse until she has had her follow-up visit. She should not use tampons for a week (sanitary napkins may be used instead) and should not douche for a week. Her usual period will begin about four to six weeks following the abortion. It is important that a woman use contraception even if her period has not yet begun again since it is still possible to get pregnant. If her period does not begin within eight weeks of her abortion, she should call her clinic or physician.

Although many women feel a great sense of relief once an abortion is completed, many also experience a profound sense of loss. Some women are surprised at the intensity of their feelings. They need to understand that this grieving process is normal. After the abortion, many men also feel residual guilt, sadness, and remorse. Many abortion clinics now provide counseling for men, as well as women, involved in abortion.

Complication rates are low for legal abortions performed by qualified practitioners, especially for first-trimester abortions. About 91 percent of abortions are performed before the thirteenth week of pregnancy; 99 percent are performed by twenty weeks, before the fetus is viable (able to survive outside the womb).

Properly done, first-trimester abortions are statistically six times safer than childbirth. Later abortions are significantly more dangerous. The possible complications include infection and hemorrhage, often due to retained material in the uterus.

Sexually Transmissible Diseases (STDs)

Chlamydia

Chlamydia is probably the most common sexually transmitted disease in the United States, affecting as many as 4 million people each year. It is caused by the organism *Chlamydia trachomatis*, which has properties of both a bacterium and a virus and affects the urinary tract and reproductive organs of both women and men. Greater awareness and large-scale screening have contributed to recognition and treatment of the disease.

The symptoms of chlamydia are similar to those of gonorrhea in both men and women, appearing seven to twenty-one days after infection. However, chlamydia is asymptomatic (showing no symptoms) in 30 to 50 percent of men, and a full 80 percent of women show no symptoms until serious complications have arisen. Symptoms in men are a whitish discharge from the penis, burning urination, pain and swelling in the testes, or a persistent low-grade fever. Women may have vaginal discharge or pain with urination. Pain in the lower abdomen or low-grade fever may indicate pelvic inflammatory disease (PID), infection of the fallopian tubes and pelvic cavity. The result may be severe damage to the tubes or ovaries, with risk of sterility or future ectopic pregnancy. Infected women may pass the infection to babies during delivery, causing dangerous eye, ear, and lung infections.

Because of the high incidence of chlamydia in the United States and the fact that most cases are asymptomatic, routine screening is strongly advised for sexually active individuals. (Testing every three to six months is recommended for those with multiple partners.) Chlamydia responds well to antibiotic therapy, usually doxycycline or tetracycline. Erythromycin is also sometimes used, particularly for pregnant women, who should not take tetracycline.

Genital Warts

An estimated 500,000 to 1 million new cases of genital warts are detected each year. Genital warts are caused by the human papilloma virus (HPV), which is in the same class of viruses as the common skin wart. An estimated two-thirds of the sexual partners of people with HPV contract the infection. The warts appear one to six months from exposure, and usually within three months. Genital warts are usually small, ranging in size from a pencil point to a quarter inch in diameter; they may even be invisible to the naked eye. They may be white, gray, pink, or brown, and round, smooth, flat, or bumpy. They may form clusters like miniature cauliflowers or tiny fingers. In men, the warts are most commonly found on the shaft or glans of the penis or around the anus. In women, they are found on the cervix, vaginal wall, vulva, or anus.

Although genital warts are generally considered more of a nuisance than a danger, HPV has been shown to be linked with cervical cancers in a number of studies. The nature of the relationship is disputed, however; HPV may be a contributing factor but may not actually cause cancer.

Diagnosis of genital warts is done by visual inspection and reaction to a vinegar wash. For women, HPV can also be detected on a Pap smear. Because warts on the vaginal wall or cervix can remain unnoticed in women, routine exams are recommended.

Warts are removed by freezing (cryosurgery), cutting, laser therapy, or chemical treatment by direct application of a solution of podophyllin. Warts may be removed for cosmetic reasons or for comfort; removal, however, does not eliminate HPV from the person's system. The extent to which a person can still transmit HPV after the visible warts have been removed is unknown. In 80 percent of cases, the warts eventually reappear, making repeated treatments necessary.

Gonorrhea

Affecting an estimated 800,000 to 1 million Americans anually, gonorrhea is caused by the bacterium *Neisseria gonorrhoeae*. A man will usually notice symptoms two to twenty-one days after infection: a discharge of fluid from his penis, a frequent desire to urinate, and an itching or burning sensation when he does so. If not treated at this stage, complications may develop: inflammation of the testes, arthritis, skin infections, urinary disorders, and sterility.

Symptoms in women include vaginal discharge or pain with urination. However, 50 to 80 percent of women are asymptomatic. For this reason, if a man discovers that he has gonorrhea, it is imperative that he tell his partner or partners so that they can be treated. Many cases in women go undetected until secondary symptoms develop. Like chlamydia, untreated gonorrhea may progress to pelvic inflammatory disease and possible infertility in women.

The infection can usually be halted in men and women by penicillin, tetracycline, or other antibiotics. However, some strains of the gonococcus bacterium are resistant to the usual cures and are thus difficult to treat.

Genital Herpes

Genital herpes, caused by the herpes simplex virus (HSV), is carried by an estimated 31 million Americans. The number of cases is increasing at the rate of 200,000 to 500,000 per year. There are two strains of HSV: HSV type 1, which is usually responsible for cold sores and fever blisters around the mouth; and HSV type 2, which is associated with genital lesions. Both types of HSV, however, can and do develop equally well on the mouth or genitals, and both can be transmitted during oral sex.

For most people with HSV, the initial infection is the most severe. Sometimes it is the only outbreak a person experiences. Within three to twenty days after exposure, small bumps or blisters appear on the genitals (the penis, anus, perineum, vulva, cervix, or vagina). The sores may be quite painful or itchy. They may be accompanied by swollen lymph nodes in the groin and flulike symptoms, particularly during the first outbreak. The first outbreak lasts an average of twelve days.

Although there are some treatments for HSV, no cure has been found. HSV remains in the body by hiding dormant in the nerve cells, and outbreaks of contagious sores may recur. Recurrences typically last four to ten days. Just prior to the outbreak is a period of a few days known as the *prodrome*, often with itchiness, tingling, or nerve pain at the site where the lesions will appear. The virus may be "shedding" on the surface of the skin from the prodrome period until the lesion is completely healed. To avoid passing the infection, contact with the affected area should be avoided during this entire period. Some people with HSV may shed the virus without experiencing symptoms and unwittingly infect others (asymptomatic transmission).

Serious complications from HSV are rare in adults, but in individuals with depressed immune systems (such as in AIDS), HSV may cause severe outbreaks, urethral blockages, or meningitis. It can also cause severe infection if spread to the eye. In women there appears to be some association between HSV and cervical cancer, although the nature of the connection is debated. Newborns may contract HSV if they come into contact with active lesions during birth. HSV is a very serious disease in newborns, causing infections of the eyes, skin, mucous membranes, or central nervous system, or even death. Fortunately, childbirth can be managed to prevent newborn infection in most cases.

The antiviral drug acyclovir (brand name Zovirax) is often helpful in reducing or suppressing HSV symptoms. It can be administered either orally (as a pill) or topically (as an ointment). Acyclovir is especially effective during initial infections, but some people with recurrent episodes find the oral medication helpful, either in low "maintenance" doses to ward off outbreaks or in higher doses when they feel an outbreak coming on. Informed health-care practitioners can provide information regarding other strategies to help prevent recurrences.

Syphilis

The incidence of syphilis is increasing among young heterosexuals, particularly minorities. An estimated 101,000 new cases were reported in the United States in 1995. Syphilis is caused by a bacterium (*Treponema pallidum*) that is easy to identify with a blood test. Not only is it spread through sexual contact, but an infected mother can pass it to the fetus through the placenta. Because syphilis can lead to brain damage and death, it is imperative for pregnant women to be screened for it within the first trimester. If the mother is treated during this period, the newborn will not be affected.

Syphilis is easily cured with antibiotics, usually penicillin. Untreated syphilis can lead to brain damage, insanity, heart disease, blindness, or death. The disease has four distinct stages that are alike for both women and men. The early stages of the disease are not painful, and it is not easily noticed at these times.

Primary Syphilis Within one to twelve weeks after exposure, a small, red, painless sore called a *chancre* (pronounced "shanker") appears at the site where the bacteria entered the body, usually in the mucous membranes of the genitals or mouth. The chancre may look like a pimple, a blister, an ulcer, or a cold sore. Without treatment, it will heal itself in one to five weeks, but the bacteria remain in the body, and the person is still highly contagious.

Secondary Syphilis Six weeks to six months after the chancre heals, the disease spreads to other parts of the body. The principal symptom at this stage is a skin rash that neither itches nor hurts; it is found on the palms of the hands, the soles of the feet, or other parts of the body. The individual may also experience flulike symptoms. The rash or other symptoms may be very mild or may pass unnoticed. The person is still ontagious. If secondary syphilis is not treated, the symptoms disappear within two to six weeks.

Latency Beginning six months to two years after infection, the disease becomes latent. During this time, the person still tests positive but does not infect others, although a pregnant woman can still transmit the disease to the fetus.

Tertiary Syphilis The symptoms of tertiary syphilis may appear years after the initial infection. Tissue damage and several disabling diseases can develop, including ulcers of internal organs and eyes, heart disease, neurological disorders, insanity, and death.

Hepatitis

Hepatitis is a viral disease affecting the liver. There are two types of the virus that may be sexually transmitted. Infectious hepatitis (hepatitis A) is carried in fecal matter, and serum hepatitis (hepatitis B) is carried in the blood.

Symptoms of hepatitis usually develop within one to four months of exposure. They include jaundice (a yellowing of the skin caused by the accumulation of blood pigments normally destroyed by the liver), fatigue, diarrhea, nausea, abdominal pain, and darkened urine. There is no medical treatment for hepatitis. Rest and fluids are recommended until the disease runs its course, generally in a few weeks. Occasionally, serious liver damage or death results.

Hepatitis A is most often contracted through unsanitary conditions, through contaminated food or water, or from surfaces in a bathroom used by an infected person to hands to mouth. It is believed to be transmitted sexually mainly via infected fecal matter—for example, during oral-anal sex ("rimming"). Although the symptoms of hepatitis A are similar to those of hepatitis B, the disease is not considered as dangerous. Immune serum globulin injections provide some immunity to hepatitis A; the injection is recommended for anyone known to have been exposed to an infected individual (through physical contact or sharing food, for example).

Hepatitis B is commonly spread through sexual contact, in blood, semen, vaginal secretions, saliva, and urine. The virus can enter the body through a break in the skin or through mucous membranes in the genitals or mouth; it is also spread by infected needles and transfusions. Hepatitis B affects an estimated 53,000 Americans annually and leads to about 5,000 deaths. The incidence is currently declining among gay men (probably due to safer sex practices) and increasing among heterosexuals. Hepatitis B can be prevented by a simple, widely available vaccination. Many health authorities recommend routine vaccination for those at high risk for exposure, such as those with multiple sex partners, sexually active teenagers, gay men, IV drug users, and some health-care workers. Screening for pregnant women is recommended.

HIV and AIDS

Acquired immune deficiency syndrome (AIDS) is caused by the human immunodeficiency virus (HIV), which attacks the body's immune system. The course of the disease varies from person to person. HIV may remain latent (causing no symptoms) for over ten years in some people, or it may cause symptoms within several months or several years of exposure. An AIDS diagnosis is made when a person has a positive blood test indicating the presence of HIV antibodies and a T-cell count below 200. If the T-cell count is higher, the person must have one or more other diseases or conditions associated with AIDS, as defined by the Centers for Disease Control. (T-cells are part of the immune system; a low count is one sign of how much HIV has suppressed the immune system.) Some people are sick off and on long before they receive an AIDS diagnosis. Length of life after AIDS diagnosis also varies, from a few days or weeks to many years; many people with AIDS have alternating periods of health and illness.

Possible symptoms of HIV/AIDS are the following:

- Unexplained persistent fatigue
- Unexplained fever, chills, or night sweats lasting several weeks or more
- Unexplained weight loss greater than 10 pounds or 10 percent of body weight in less than two months
- Unexplained swollen lymph nodes in the neck, armpits, or groin, lasting more than two months.
- Pink, purple, or brown blotches on or under the skin or inside the mouth, nose, eyelids, or rectum, which may initially resemble bruises but do not disappear
- Persistent fuzzy, white spots or other sores in the mouth
- Persistent dry cough and shortness of breath
- Persistent diarrhea

In addition, women may experience the following:

- Abnormal Pap smears
- Persistent vaginal candidiasis
- Abdominal cramping (due to PID)

Remember, however, that these are also symptoms of other common and less serious infections. *AIDS cannot be self-diagnosed.* A person concerned with symptoms he or she thinks might be AIDS should arrange to be tested.

Testing and Treatment

There are no widely available tests for the presence of HIV itself. The HIV antibody tests, however, are simple tests that show whether the body has developed antibodies to HIV. Used together, the ELISA and Western Blot antibody tests give an accurate result. Home test kits may now be purchased at many drug stores.

If HIV antibodies are absent, the test will be negative (HIV-). A negative test means either that the person has not been infected with HIV or that the person has been infected within the last six months, but the body has not yet produced the antibodies. A person who is infected but not yet showing antibodies can nevertheless infect others. Persons with a negative test result should continue to practice safer sex (which will be discussed shortly) and be tested again in six months to confirm the result.

If HIV antibodies are present, the test will be positive (HIV+). This indicates that the person is infected with HIV and is capable of transmitting it to others. A person with a positive HIV diagnosis should consult right away with a physician experienced in treating HIV and AIDS. Early intervention is very important in maximizing the quality of health and life. Although no cure for AIDS has been found, several treatments are used to improve the course of the disease. To delay the onset of symptoms, AZT and other drugs used to treat AIDS are often given even before an HIV-positive individual shows symptoms. New classes of drugs, including protease inhibitors and antiretroviral drugs, have restored health and hope to many people living with HIV. These drugs reduce symptoms and prolong life in many (but not all) of the patients who can afford them.

Transmission

HIV is not transmitted by casual contact—such as hugging, playing, shaking hands, sharing food or glasses, sitting on toilet seats, or even biting and spitting. Large numbers of people who have had casual contact with people with HIV or AIDS, such as friends, family members, co-workers, or fellow students, have not tested positive for HIV. Antibody tests of family members show no transmission through casual contact; only sexual partners have become infected.

Nonsexually, HIV is transmitted through blood, primarily from the use of unsterilized needles by injection drug users. Previously, it was transmitted through infected donor blood; since 1985, however, the U.S. donor supply has been tested for HIV, and the risk of infection is considered extremely low. HIV can also be passed from an infected mother to a fetus in utero (a 20 to 50 percent chance), via blood during delivery, or to an infant via breastfeeding.

Sexually, HIV is transmitted through semen, vaginal secretions, and blood. Any sexual practice that involves the contact of one of these fluids with the mucous membranes or broken skin of the other partner can transmit the virus—unprotected vaginal or anal intercourse, oral sex, and so on. Anal intercourse is considered especially risky because of the ease with which the delicate membranes of the rectum can be ruptured. The virus then enters the bloodstream via infected semen. Although gay men are the largest group that has been struck by the disease, it is not a "gay disease." Its association with gay men is a result of their engaging in anal intercourse significantly more frequently than heterosexuals. With the increased practice of safer sex among many gay men, the rate of new AIDS cases in this group has declined, whereas the rate among heterosexuals, especially women and youth, is rising rapidly.

Safer and Unsafe Sexual Practices

Safer sex practices are an integral part of good health practices to prevent the spread of HIV and other sexually transmitted diseases. (Many people prefer the term *safer sex* to *safe sex* because all sexual contact carries at least a slight risk—a condom's slipping off, perhaps—no matter how careful we try to be.) For safer sex, do not allow semen, vaginal secretions, or blood to come into contact with the penis, vagina, anus, mouth, any other mucous membrane, or broken skin.

Safer practices include the following:

- Hugging, touching, caressing, and massaging
- Kissing (but possibly not deep French kissing)
- Masturbation (solo or mutual, unless there are sores or abrasions on the genitals or hands)
- Erotic videos, books, and so on

The following are possibly safe practices:

- Deep French kissing, unless there are sores in the mouth
- Vaginal intercourse with a latex condom and spermicide
- Fellatio with a latex condom
- Cunnilingus, if the woman does not have her period or a vaginal infection (a latex dental dam or a square of plastic wrap provides extra protection)
- Anal intercourse with a latex condom and spermicide (authorities disagree as to whether this should be considered "safer" even with a condom because it is the riskiest sexual behavior without one)

These are unsafe practices:

- Vaginal or anal intercourse without a latex condom
- Fellatio without a latex condom
- Cunnilingus, if the woman has her period or a vaginal infection and a dental dam is not used
- Oral-anal contact
- Contact with blood, including menstrual blood
- Sharing vibrators or such things as dildos without washing them between uses

Hints for Effective Condom Use

- Use condoms every time you have sexual intercourse; this is the key to successful contraception and disease prevention.
- Use a spermicide with the condom. Foam and film are both easy to apply. Spermicide helps protect against pregnancy and STDs, including chlamydia, gonorrhea, genital herpes, and HIV infection.
- Always put the condom on before the penis touches the vagina. Even if the man has great "control," there is always the possibility of leakage prior to ejaculation.
- Leave about a half inch of space at the condom tip, and roll the condom all the way down to the base of the penis.
- Soon after ejaculation, the penis should be withdrawn. Make sure someone holds the base of the condom firmly against the penis as it is withdrawn.
- After use, check the condom for possible torn spots. If you are not using a spermicide and you find a tear or hole, immediately insert foam or jelly containing nonoxynol-9 into the vagina. This may reduce the chance of pregnancy or disease. If torn condoms are a persistent problem, use a water-based lubricant, such as K-Y jelly, or a spermicide to reduce friction.
- Do not reuse a condom.
- Keep condoms in a cool, dry, and convenient place.
- To help protect against HIV and other organisms, always use a latex rubber condom, *not* one made of animal tissue.

Here are a few further suggestions to assist you in using condoms safely and pleasurably:

- Some brands are more highly rated for safety than others. For up-to-date studies, check your library, family planning clinic, or AIDS resource center.
- To maximize your pleasure, try different kinds of condoms. Some brands and styles may fit more comfortably than others. Because some brands of thinner condoms may be more prone to tearing than others, try a different one.
- If breakage is a problem no matter what kind of condom you use (some people are just friskier than others), consider "double bagging"—using two condoms. Some folks swear by this.

There is no product or practice (except abstinence) that can guarantee 100 percent safety from HIV infection and other STDs, but with a little latex and common sense, we can provide a lot of protection for ourselves and those we care about.

If you or your partner are uncomfortable with condom use, consider the following:

- *Stand your ground.* (This is mainly for women, as it is generally men who object to condoms.) Unless you want to be pregnant and are sure your partner is free of STDs, you need protection during sex. If he says no to condoms, you can say no to him. If he cares about you, he will work with you to find birth control and safer sex methods that suit you both.
- *Communication is crucial.* It may seem "unromantic," but planning your contraception strategy before you are sexually entangled is essential. Giving or getting a disease or worrying about pregnancy is about as unromantic as you can get. Consider visiting a family planning clinic for counseling—together. Neither partner should be forced to use a form of birth control he or she is truly unhappy with. But the issue of protection must be dealt with—by both of you.
- *Don't forget your sense of humor and playfulness.* Condoms can actually provide lots of laughs, and laughter and sex go well together.

FIGURE R.1
Using a Condom

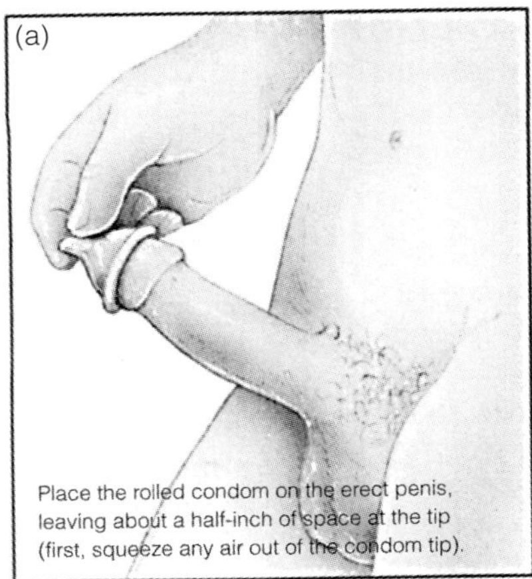
(a) Place the rolled condom on the erect penis, leaving about a half-inch of space at the tip (first, squeeze any air out of the condom tip).

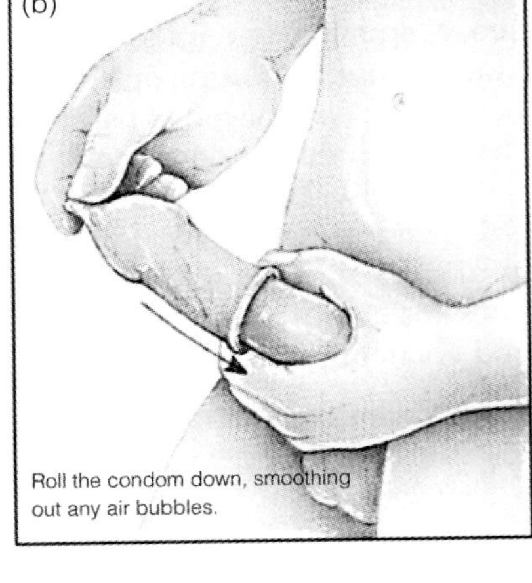
(b) Roll the condom down, smoothing out any air bubbles.

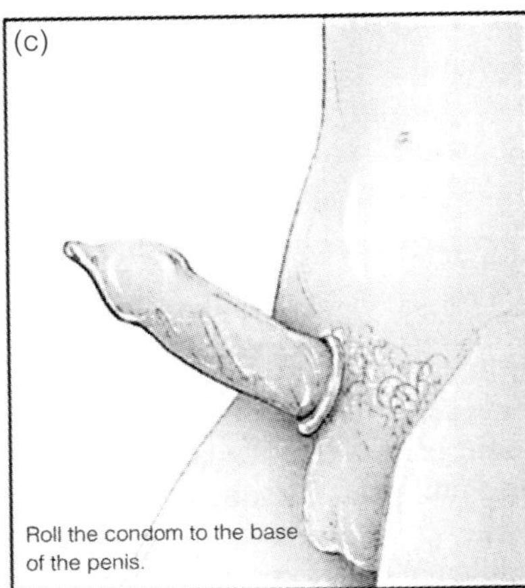
(c) Roll the condom to the base of the penis.

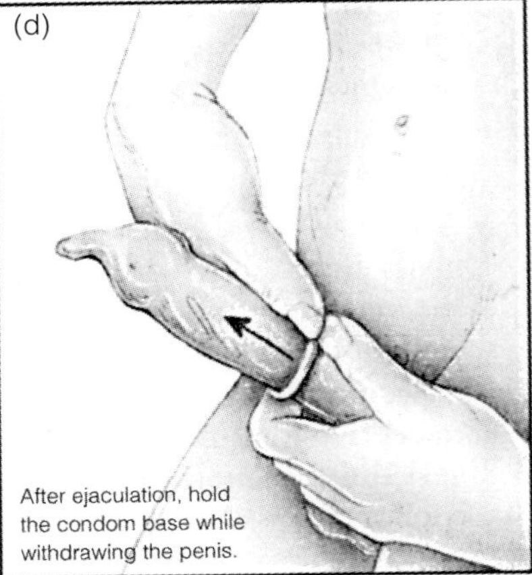
(d) After ejaculation, hold the condom base while withdrawing the penis.

SOURCE: © Bryan Strong and Christine DeVault. From Bryan Strong and Christine DeVault, *Human Sexuality.* Mayfield Publishing Co., 1997.

For Women

Breast Self-Examination

Follow the steps in Figure R.2 to examine your breast regularly, once a month. This examination is best done after your menstrual period. If you find something you consider abnormal, contact your doctor for an examination. Most breast lumps are not serious, but all should come to the doctor's attention for an expert opinion after appropriate examination. You may have a condition that will require treatment or further study. If necessary, your doctor may recommend laboratory tests or X-ray examination. Follow your doctor's advice—your early recognition of a change in your breast and the doctor's thoughtful investigation will determine the safest course. Keep up this important health habit even during pregnancy and after menopause.

Maintaining Sexual Health

Vaginitis

A variety of urethral and vaginal infections are considered to be STDs because they are aggravated or perpetuated by sexual contacts. Sometimes only one partner suffers from the discomfort, and the other seems to be resistant. Partner 2 acts as a carrier and continually reinfects Partner 1. These diseases fall under the general category of vaginitis and include trichomoniasis, candidiasis (yeast infection), and bacterial infections such as *Gardnerella* infection (a man who is carrying this type of organism may have symptoms of a urinary tract infection, such as painful urination).

Here are some hints that may help a woman avoid vaginitis:

- Do not use vaginal deodorants, especially deodorant suppositories or tampons, because they upset the natural chemical balance within the vagina. Use douches rarely, and then only for good reason, such as a diagnosed yeast infection. Despite what pharmaceutical companies advertise, the healthy vagina does not have an unpleasant odor. If the vagina does have an unpleasant odor, then something is wrong, and you should check with a physician or health practitioner. (See Table R.1.)
- Regularly wash and rinse the vulva and anus. Be careful about using bubble baths or strongly perfumed soaps because they may cause irritation.
- After a bowel movement, wipe the anus from the front to the back to prevent the spreading of germs from the anus to the vagina.
- Wear cotton underpants that have a cotton crotch, since nylon underpants and pantyhose keep heat and moisture in, encouraging bacterial growth.
- When lubricating the vagina, use only water-soluble lubricants. Vaseline is not water soluble and tends to form a film on the walls of the vagina, which keeps heat and moisture in, encouraging infection.
- Have your male partner treated also. Vaginitis is often sexually transmitted. Be on the safe side and have your partner treated to prevent reinfection.

Yeast Infection (Candidiasis)

If you suspect you have a *Candida* (yeast) infection, you may want to try treating it yourself.

- Plain yogurt applied to the vulva and vagina twice a day may eliminate symptoms. (The yogurt must contain live lactobacillus culture; this means that pasteurized supermarket yogurt won't work. A trip to a health-food store may be in order. Health-food stores may also stock homeopathic treatments for yeast.)
- Douching with a weak vinegar and water solution (2 tablespoons of vinegar to 1 quart of water) once every three days may also eliminate the symptoms.
- Nutrition can play a part. If you have recurring candidiasis, you might try increasing your intake of plain (health-food type) yogurt; you can also cut down on foods with high sugar or carbohydrate content.
- Stress can upset the vagina's chemical state, as can a change in the body's hormone levels. Pregnant women and women who use oral contraception are more susceptible to yeast infections, as are women who are taking antibiotics.
- Antifungal medicines such as clotrimazole (Gyne-Lotrimin) and miconazole nitrate (Monistat) are now available over the counter in cream or vaginal suppository form. If you have had a yeast infection before and recognize the symptoms, you may want to purchase one of these products. If you are not sure that you have a yeast infection, see your doctor or clinic.
- Because women with candidiasis may have trichomoniasis as well, they should be tested for both.

Symptoms often disappear in two or three days, but curing candidiasis usually requires one to four weeks of treatment.

Urinary Tract Infections (Cystitis)

Frequent, painful urination often indicates the presence of a urinary tract infection.

Preventing Cystitis

The following measures will lessen the likelihood of a cystitis attack:

- Urinate frequently. Holding back urine can weaken the muscles involved in urination; it can also allow the buildup of concentrated, highly acidic urine.
- Urinate before intercourse and soon afterwards to flush bacteria from the urethra.
- Drink plenty of fluids (water is best), especially just prior to and just after intercourse.

FIGURE R.2
Breast Self Examination

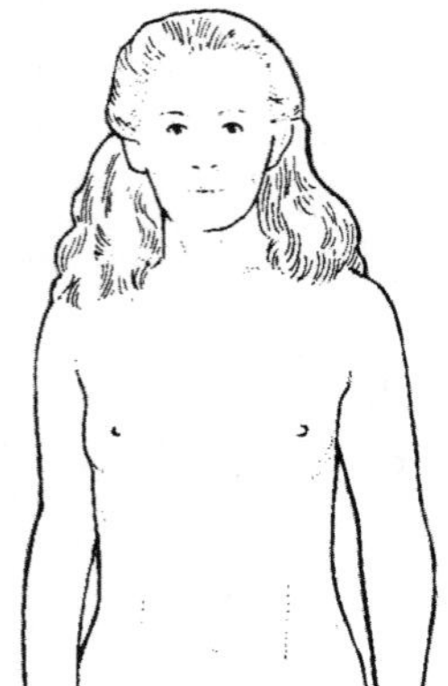

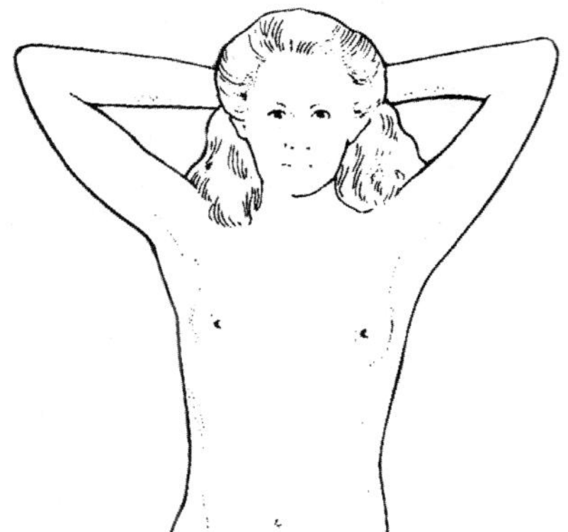

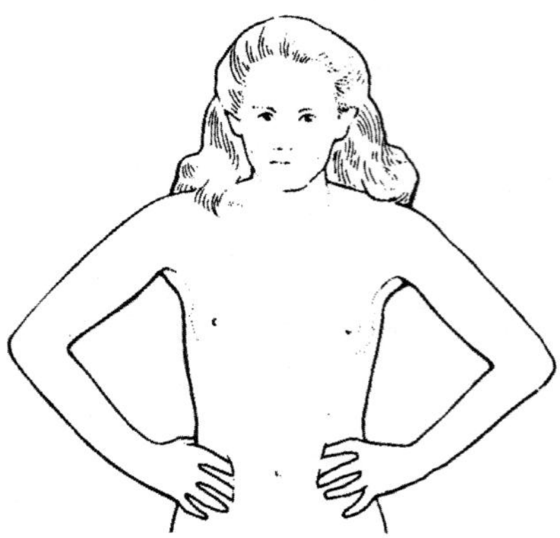

1. Stand before a mirror. Inspect both breasts for anything unusual, such as any discharge from the nipples or puckering, dimpling, or scaling of the skin.

The next two steps are designed to emphasize any changes in the shape or contour of your breasts. As you do them, you should be able to feel your chest muscles tighten.

2. Watch closely in the mirror. Clasp hands behind your head, and press hands forward.

3. Next, press hands firmly on hips and bow slightly toward the mirror as you pull your shoulders and elbows forward.

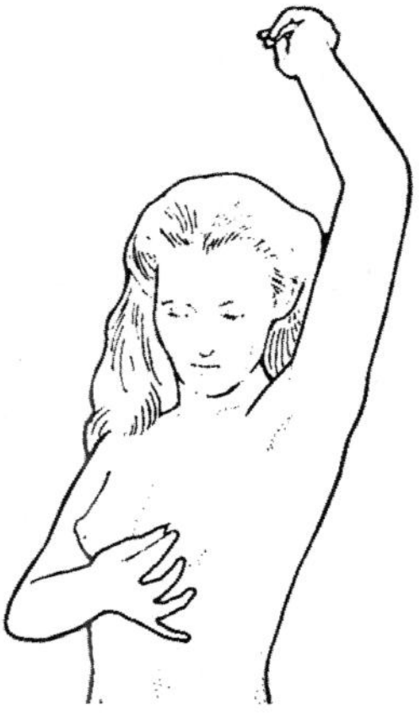

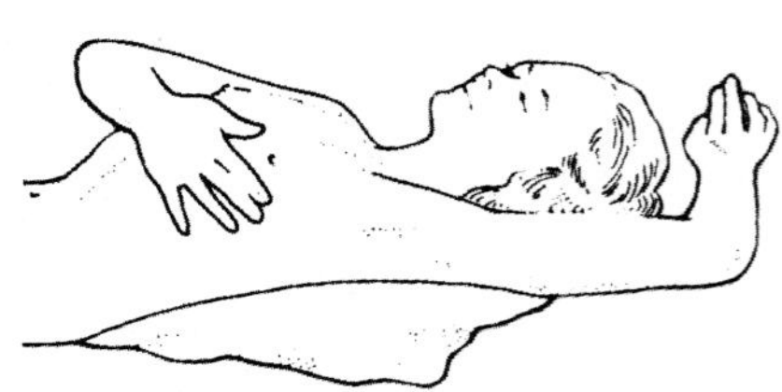

Some women do the next part of the exam in the shower. Fingers glide over soapy skin, making it easy to concentrate on the texture underneath.

4. Raise your left arm. Use three or four fingers of your right hand to explore your left breast firmly, carefully, and thoroughly. Beginning at the outer edge, press the flat part of your fingers in small circles, moving the circles slowly around the breast. Gradually work toward the nipple. Be sure to cover the entire breast. Pay special attention to the area between the breast and the armpit, including the armpit itself. Feel for any unusual lump or mass under the skin.

5. Gently squeeze the nipple, and look for a discharge. Repeat the exam on your right breast.

6. Repeat Steps 4 and 5 lying down. Lie flat on your back with your left arm over your head and a pillow or folded towel under your left shoulder. This position flattens the breast and makes it easier to examine. Use the same circular motion described earlier. Repeat on your right breast.

SOURCE: National Institutes of Health, *Breast Cancer: We're Making Progress Every Day,* NIH publication no. 83-2409, 1983 (a leaflet available from the National Cancer Institute).

- Eat a well-balanced diet and get plenty of rest, especially if your resistance is low.
- If you use a diaphragm and are troubled by cystitis, consider trying another form of contraception.

Treating Cystitis

The following are ways you can treat cystitis:

- If you feel an attack coming on, drink copious amounts of water (at least sixteen glasses a day). It won't hurt you as long as you urinate frequently. Drinking cranberry juice may also help; it has been found to have antibacterial properties. A pinch of baking soda in a glass of water can help neutralize the urine's acidity and ease the burning sensation. Avoid coffee, tea, colas, and alcohol because they may irritate the urinary tract.
- Itchiness may be relieved by spraying or sponging water on the vulva and urethral opening. If urination is especially painful, try urinating while sitting in a few inches of warm water (in the bathtub or a large pan).
- If symptoms persist or in case of fever, consult your doctor or clinic. Sulfa drugs or other specific antibiotics will usually clear up the symptoms in several days. (A caution about using sulfa drugs: about 10 to 14 percent of African Americans have an inherited deficiency of a blood enzyme called glucose-6-phosphate-dehydrogenase [G6PD]; sulfa drugs may cause a serious anemic disease in these people.)

Prenatal Nutrition

A nutritious diet throughout pregnancy is essential for both the fetus and the mother. Not only does the baby get all its nutrients from the mother, but it also competes with her for nutrients not sufficiently available to meet both their needs. When a woman's diet is low in iron or calcium, the fetus receives most of it and the mother may become deficient in it. To meet the increased nutritional demands of her body, a pregnant woman shouldn't just eat more; she should make sure that her diet is adequate in all the basic nutritional categories. In the early weeks of pregnancy, high intake of the B vitamin folic acid has been shown to decrease the risk of neural tube defects, including spina bifida. For this reason, it is recommended that all women of childbearing age consume 200–400 micrograms of folic acid per day from foods and/or supplements. (Foods such as bread, flour and pasta will be fortified with folic acid beginning in 1998.) In the second and third trimesters, requirements increase for calories and most nutrients, including protein, calcium, the B vitamins, vitamins A, C, D, and E, iodine, iron, magnesium, and zinc. With the possible exception of iron, for which many authorities recommend a supplement, all these nutrients can be obtained from a sensible, varied diet designed for a healthy pregnancy. Table R.2 will help you select foods that provide the nutrients needed for good nutrition during pregnancy.

Table R.1 Vaginal Mucus and Secretions Chart

Color	Consistency	Odor	Other Symptoms	Possible Cause	What To Do
Clear	Slightly rubbery, stretchy	Normal	—	Ovulation; sexual stimulation	Nothing
Milky	Creamy	Normal	—	Preovulation	Nothing
White	Sticky, curdlike	Normal	—	Postovulation; the pill	Nothing
Brownish	Watery and sticky	Normal or slightly different	—	Last days of period; spotting	Nothing
White	Thin, watery, creamy	Normal to foul or fishy	Itching	*Gardnerella* bacteria or nonspecific bacterial infection	See health practitioner
White	Curdlike or flecks, slight amount of discharge	Yeasty or foul	Itching or intense itching	Overgrowth of yeast cells, yeast infection	Apply yogurt or vinegar solution; see health practitioner
Yellow, Yellow-green	Smooth or frothy	Usually foul	Itching; may have red dots on cervix	Possible *Trichomonas* infection	See health practitioner
Yellow, Yellow-green	Thick, mucous	None to foul	Pelvic cramping or pelvic pain	Possible infection of fallopian tubes	See health practitioner *right away*

Practical Hints on Breastfeeding

Some women start breastfeeding with perfect ease and hardly any discomfort. For others, it can be frustrating and sometimes painful, but it need not be. Midwife Raven Lang tells us that the following method will lead to successful breastfeeding:

When you first put the baby to your breast, limit her to one minute per breast. Try not to nurse again for a half hour to an hour. If the baby fusses, you can give her the end of your little finger (or a pacifier) to suck.

The second hour, let her nurse two minutes at each breast; the third hour, three minutes; the fourth hour, four minutes; and so on. Of course, your baby will not want to nurse every hour of the day and night. Thus, the basic rule to follow is to increase your nursing time by only one minute per breast with each subsequent feedings until you are nursing comfortably for as long a session as you and the baby both enjoy. Remember that for the first three days, the baby is getting colostrum only. By the time your true milk comes in, on the third day, things should be going smoothly. Also, even a slow-nursing infant gets about four-fifths of her nourishment during the first five minutes.

Lang says that although mothers are generally most effective when they care for their babies "by feel" rather than "by the book" (or, in this case, "by the clock"), the process of establishing breastfeeding is an exception to this "rule." Try it; you'll agree.

Table R.2 Basic Foods You Need Each Day during Pregnancy

	Under 18 Years		Over 18 Years		
	Before and during First Three Months of Pregnancy	Last Six Months of Pregnancy	Before and during First Three Months of Pregnancy	Last Six Months of Pregnancy	Counts as One Serving
Meat, fish, poultry, eggs, or alternates	2–3 servings	3 servings	2–3 servings	3 servings	2-3 ounces cooked lean meat, fish, or poultry without bone; 2 medium eggs; 4 tablespoons peanut butter; 1 cup cooked dried beans or peas; 1 1/2 cups, split pea or bean soup; 2 or 3 ounces cheddar-type cheese; 1/2 to 3/4 cup cottage cheese; 1/4 to 1/2 cup nuts or seeds.
Milk and milk products	3–4 cups	4–5 cups	3 cups	4 cups	1 cup (8 fluid ounces) of skim or whole milk, buttermilk, or diluted evaporated milk. The following foods provide as much calcium as a cup of whole milk: 1 1/2 ounces cheddar-type cheese; 1 cup plain yogurt; 3 tablespoons regular nonfat dry milk; 6 tablespoons instant nonfat dry milk solids; 1 1/2 cups cottage cheese; 1 cup custard or pudding made with milk; 1 1/2 to 2 cups soup made or diluted with milk.
Fruits and vegetables (Vitamin C rich)	1 serving	1–2 servings	1 serving	1–2 servings	1/2 cup citrus juice or 1 medium orange; 1/2 grapefruit; 1/2 cantaloupe; 1/2 cup strawberries; 1/2 cup broccoli; 1/2 green pepper. You will need to eat 2 servings of foods that are fair sources of vitamin C. These foods include tomatoes, tomato juice, tangerines, tangerine juice, asparagus tips, raw cabbage, brussels sprouts, watermelon, and dark leafy greens.

Table R.2 *(continued)*

	Under 18 Years		Over 18 Years		
	Before and during First Three Months of Pregnancy	**Last Six Months of Pregnancy**	**Before and during First Three Months of Pregnancy**	**Last Six Months of Pregnancy**	**Counts as One Serving**
Fruits and vegetables (Vitamin A rich)	1 serving	2 servings	1 serving	2 servings	1/2 cup deep yellow fruits and vegetables such as apricots, cantaloupe, carrots, pumpkins, sweet potatoes, and winter squash; 1/2 to 3/4 cup dark green leafy vegetables such as collard greens, mustard greens, chard, kale, turnip tops, spinach, broccoli, and watercress. In addition to vitamin A, dark green leafy vegetables supply folacin, magnesium, and iron.
Other fruits and vegetables	2 servings	1 serving	2 servings	1 serving	1/2 cup of other fruits and vegetables such as green beans, wax beans, celery, corn, mushrooms, cauliflower, green peas, cucumbers, potatoes, lettuce, beets, pears, apples, bananas, pineapple, prunes, cherries, etc.
Breads and cereals (whole grain, enriched, or restored)	4 servings	5–6 servings	4 servings	4–5 servings	1 slice bread; 1 muffin; 1 hamburger or hot dog roll; 4 to 5 saltine crackers; 1/2 to 3/4 cup cooked cereals, rice, macaroni, noodles, spaghetti, and other pastas; 3/4 cup (1 ounce) ready-to-eat cereal. Read labels and select whole grain or fortified breads and cereals. Avoid presweetened cereals.
Fats and sweets	These energy foods supply mostly calories. Eat them only in amounts to meet your energy needs after your nutritional requirements have been met.				
Supplements	Your health practitioner will probably recommend dietary supplements, especially folic acid.				

Most women find that a good nursing bra—one that provides good uplift and that opens easily for nursing—makes breastfeeding easier and more comfortable. Many wear such a bra day and night during the months they are nursing.

Rest and relax as much as possible during the months that you are breastfeeding, especially at the beginning. Your body is doing a tremendous amount of work and needs extra care.

While nursing, find a position that is comfortable for you and your baby. A footstool, a pillow, and a chair with arms are often helpful.

Touch the baby's cheek with the nipple to start. She will turn her head to grasp the nipple. (If you try to push her to the nipple with a finger touching her other cheek or chin, she will turn away from the nipple toward the finger.)

Allow her to grasp the entire darker-colored part of the breast in her mouth. She gets the milk by squeezing it from the nipple, not by actually sucking. Her grasp on your nipple may hurt for the first few seconds, but the pain should disappear once she is nursing in a good rhythm. When you want to remove her mouth from your breast, first break the suction by inserting your finger into the corner of her mouth. This will save you from sore nipples.

A small amount of milk may come out of your nipples between feedings. A small nursing pad or piece of sanitary napkin inserted in your bra over the nipple will absorb this milk, keeping the bra clean and preventing irritation of the nipple.

If your entire breast becomes sore, you may be able to relieve the painfulness simply by lifting and supporting the breast with one hand during nursing. Hot compresses between nursing sessions may further relieve soreness.

If you notice a spot of tenderness or redness or a hard lump on your breast or nipple that persists for more than two feedings, be sure to seek advice promptly from your breastfeeding support group or physician.

It's important to note that even with a plugged milk duct or mastitis (an infected milk duct), a woman can usually continue to breastfeed. In fact, having the ducts emptied regularly is important in maintaining good breast health during nursing.

If you have difficulty beginning to breastfeed, don't give up! Ask friends, women's centers, clinics, or the local La Leche League chapter for help. Don't worry about not having enough milk; the more your baby nurses, the more you'll produce.

For Men: Testicular Self-Examination

Just as women practice breast self-examination each month, men should practice preventive medicine by doing testicular self-examination regularly. (See Figure R.3.)

The best time to discover any small lumps is right after a hot shower or bath, when the skin of the scrotum is most relaxed. Each testicle should be gently examined with the fingers of both hands, slowly and carefully. Learn what the collecting structure at the back of the testicle (the epididymis) feels like so that you won't mistake it for an abnormality. If you find any lump or growth, it most often will be on the front side of the testicle. Any lumps or suspicious areas should be reported to your urologist promptly.

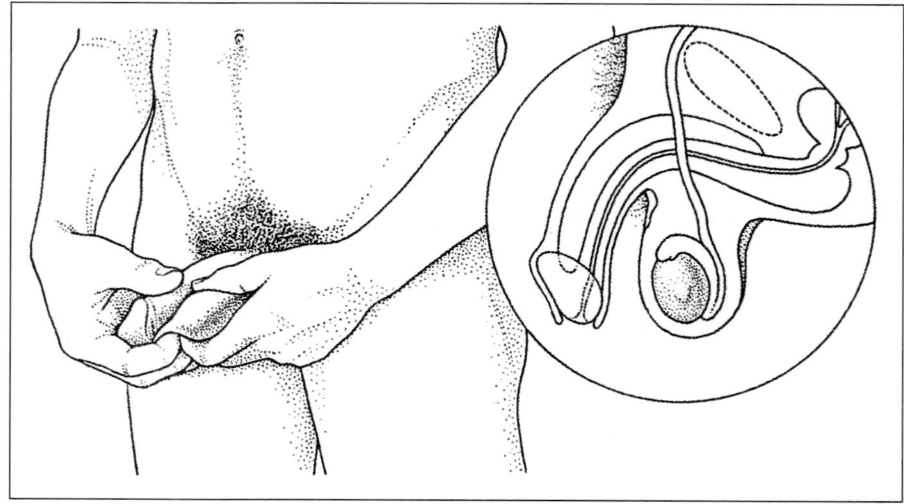

FIGURE R.3
Testicular Self-Examination

Roll each testicle between the thumb and fingers; the testicles should feel smooth, except for the epididymis at the back of each. A hard lump, an enlargement, or contour changes should be reported to your health-care provider.

Family Matters

Table R.3 Marriage Laws by State

	Physical exam and blood test for male and female					Waiting period		
	Age with parental consent		Age without consent		Max. period between exam and license	Scope of medical exam	Before license	After license issuance (expiration)
State	Male	Female	Male	Female				
Alabama*	14a,t	14a,t	18	18	—	b	—	30 days
Alaska	16z	16z	18	18	—	—	3 days, w	—
Arizona	16z	16z	18	18	—	—	—	—
Arkansas	17c, z	16c, z	18	18	—	—	v	—
California	aa	aa	18	18	30 days, w, h	jj	—	90 days
Colorado*y	16z	16z	18	18	—	—	—	30 days
Connecticut	16z	16z	18	18	—	bb	4 days, w	65 days
Delaware	18c	16c	18	18	—	—	24 hr, kk	30 days, e
Florida	16a, c	16a, c	18	18	—	—	—	—
Georgia*	aa, j	aa, j	16	16	—	bb	3 days, g	30 days
Hawaii	15j	15j	16	16	—	p	—	—
Idaho*	16z	16z	18	18	—	s, zzz	—	—
Illinois	16pp	16pp	18	18	30 days	n	1 day	60 days
Indiana	17c	17c	18	18	—	rr	72 hr, w	60 days
Iowa*	aa, j	aa, j	18	18	—	—	3 days	20 days
Kansas*y	aa, j	aa, j	18	18	—	—	3 days, w	—
Kentucky	aa, j	aa, j	18	18	—	—	—	—
Louisiana	18z	18z	18	18	10 days	—	72 hr, w	—
Maine	16z	16z	18	18	—	—	3 days, v, w	90 days
Maryland	16c, f	16c, f	18	18	—	—	48 hr, w	6 mo
Massachusetts	14j	12j	18	18	3–60 days, u	—	3 days, v	—
Michigan	16	16	18	18	—	—	3 days, w	—
Minnesota	16j	16j	18	18	—	—	5 days, w	—
Mississippi	aa, j	aa, j	17	15	30 days	b	3 days, w	—
Missouri	15d	15d	18	18	—	—	—	—
Montana*yy	16j	16j	18	18	—	b	—	180 days

*Indicates common-law marriage recognized. (a)Parental consent not required if minor was previously married. (aa)No age limits. (b)Venereal diseases. In WV and OK, Circuit Court judge may waive requirement. (bb)Venereal diseases and rubella (for female). (bbb)No exam required, but parties must file affidavit of nonaffliction with contagious venereal disease. (c)Younger parties may obtain license in case of pregnancy or birth of child. (cc)Unless parties are over 18 yrs. of age. (d)Younger parties may obtain license in special circumstances. (e)Residents before expiration of 24-hr waiting period; nonresidents formerly residents, before expiration of 96-hr waiting period; others 96 hr. (ee)License effective 1 day after issuance, unless court orders otherwise; valid for 60 days only. (f)If parties are at least 16 yrs. of age, proof of age and the consent of parents in person are required. If a parent is ill, an affidavit by the incapacitated parent and a physician's affidavit to that effect required. (ff)If one or both parties are below the age for marriage without parental consent, 3-day waiting period. (g)Unless parties are 18 yrs. of age or more, or female is pregnant, or applicants are the parents of a living child born out of wedlock. (h)When unmarried man and unmarried woman, not minors, have been living together as man and wife, they may, without health certificate, be married upon issuance of appropriate authorization. (hh)Parties must sign affidavit affirming that they have received and discussed brochure prepared by Division of Public Health Services, Dept. of Health and Human Services. (j)Parental consent and/or permission of judge required. (jj)Medical examination for syphilis (and for female, rubella), with required offer of HIV test. (k)Below age of consent parties need parental consent and permission of judge. (kk)Medical examination not required but certificate evidencing HIV counseling required. (l)If both parties are residents, 96 hr. (m)Mental incompetence, infectious tuberculosis, venereal diseases. (n)Venereal diseases; test for sickle cell anemia given at request of examining physician. (nn)Tests for sickle cell anemia may be required for certain applicants. (p)Rubella for female, except under limited circumstances

Source: World Almanac - 1997, from Gary N. Skoloff, Skoloff & Wolfe, Livingston, NJ: 1996.

Table R.3 *(continued)*

	Physical exam and blood test for male and female				Waiting period			
	Age with parental consent		Age without consent		Max. period between exam and license	Scope of medical exam	Before license	After license issuance (expiration)
State	Male	Female	Male	Female				
Nebraska[yy]	17	17	19	19	—	bb	—	1 yr
Nevada	16z	16z	18	18	—	—	—	1 yr
New Hampshire	14k	13k	18	18	—	hh	3 days, v, w	90 days
New Jersey	16z, c	16z, c	18	18	30 days	b	72 hr, w	30 days
New Mexico	16d, c	16d, c	18	18	30 days	b	—	—
New York	16k	16k	18	18	—	nn	24 hr, ee	60 days
North Carolina	16c	16c	18	18	—	m	—	—
North Dakota	16	16	18	18	—	—	—	60 days
Ohio	aa, j	16c, z	18	18	30 days	b	5 days,w,r	30 days
Oklahoma*	16c, z	16c, z	18	18	30 days, w	b	ff	30 days
Oregon	17tt	17tt	18	18	—	—	3 days, w	—
Pennsylvania*	16d	16d	18	18	30 days	b	3 days, w	60 days
Rhode Island*	d	16d	18	18	—	rrr	—	—
South Carolina*	16c	14c	18	18	—	—	1 day	—
South Dakota	16c	16c	18	18	—	—	—	20 days
Tennessee	16d	16d	18	18	—	—	3 days, cc, w	30 days
Texas*y	14j, k	14j, k	18	18	—	—	zzzz	30 days
Utah*	14a	14a	18x	18x	—	—	—	30 days
Vermont	14j	14j	18	18	30 days, w	b	1 day, w	—
Virginia	16a, c	16a, c	18	18	—	zz	—	60 days
Washington	17d	17d	18	18	—	bbb	3 days	60 days
West Virginia	18c	18c	18	18	—	b	3 days, w	—
Wisconsin	16	16	18	18	—	zzz	5 days, w	30 days
Wyoming	16d	16d	18	18	—	bb	—	—
Dist. of Columbia*	16a	16a	18	18	30 days	b	3 days, w	—
Puerto Rico	18c, d, z	16c, d, z	21	21c	—	b	—	—

(pp)Judicial consent may be given when parents refuse to consent. (r)Applicants under age 18 must state that they have had marriage counseling. (rr)Any unsterilized female under 50 must submit with application for license a medical report stating whether she has immunological response to rubella, or a written record that the rubella vaccine was administered on or after her 1st birthday. Judge may by order dispense with these requirements. (rrr)Physical examination and blood test required; offer of HIV counseling required. (s)Rubella for female; there are certain exceptions, and district judge may waive medical examination on proof that emergency exists. (t)Other statutory requirements apply. (tt)If a party has no parent residing within state, and one party has residence within state for 6 mo, no permission required. (u)Doctor's certificate must be filed 30 days prior to notice of intention. (v)Parties must file notice of intention to marry with local clerk. (w)Waiting period may be avoided. (x)Authorizes counties to provide for premarital counseling as a requisite to issuance of license to persons under 19 and persons previously divorced. (y)Marriages by proxy are valid. (yy)Proxy marriages are valid under certain conditions. (z)Younger parties may marry with parental consent and/or permission of judge. In CT, judicial approval. (zz)Required offer of HIV test, and/or must be provided with information on AIDS and tests available. (zzz)Applicants must receive information on AIDS and certify having read it. (zzzz)72-hr waiting period following issuance of license.

Source: World Almanac - 1997, from Gary N. Skoloff, Skoloff & Wolfe, Livingston, NJ: 1996.

Table R.4 Divorce Laws by State

Important: Almost all states also have other laws as well as qualifications of the laws shown below and have proposed divorce-reform laws pending. It would be wise to consult a lawyer in conjunction with the use of this chart.

Some Grounds for Absolute Divorce[1]

Residence		Adultery	Mental or physical cruelty	Desertion	Alcoholism	Impotence	Nonsupport	Insanity	Bigamy	Felony Conviction or Imprisonment	Drug addiction	Fraud, force, duress
AL	6 mo*	Yes	Yes	1 yr	Yes	Yes	2 yr	5 yr	A	2 yr*	Yes	A
AK*		Yes	Yes	1 yr	1 yr	Yes	No	18 mo	A	Yes	Yes	A
AZ	90 days	No	No	No	No	No	No	No	No	No	No	No
AR	60 days*	Yes	Yes	No	1 yr	Yes	Yes	3 yr	No	Yes	No	A
CA	6 mo*	No	No	No	No	A	No	Yes*	A	No	No	A
CO	90 days	No	No	No	A	A	No	No	A	No	A	A
CT	1 yr*	Yes	Yes	1 yr	No	No	Yes	5 yr	A	life*	No	Yes
DE	6 mo	Yes	Yes	Yes	Yes	A	No	Yes	Yes	Yes	Yes	A
FL	6 mo	No	No	No	No	No	No	3 yr	No	No	No	A
GA	6 mo	Yes	Yes	1 yr	Yes	Yes	No	2 yr	A	Yes*	Yes	Yes
HI	6 mo	No	No	No	No	No	No	No	A	No	No	A
ID	6 wk	Yes	Yes	Yes	No	A	No	3 yr	A	Yes	No	A
IL	90 days	Yes	Yes	1 yr	2 yr	Yes	No	No	Yes	Yes	2 yr	No
IN	6 mo*	No	No	No	No	Yes	No	2 yr	A	Yes	No	A
IA	1 yr*	No	No	No	No	A	No	A	A	No	No	No
KS	60 days	No	No	No	No	No	Yes	2 yr	A	No	No	A
KY	180 days	No	No	No	No	A	No	No	No	No	No	A
LA	6 mo*	Yes	No	No	No	No	No	No	A	Yes*	No	A
ME	6 mo*	Yes	Yes	3 yr	Yes	Yes	Yes	A	A	No	Yes	No
MD	*	Yes	†	1 yr†	No	No	No	3 yr	A	1 yr*	No	No
MA	1 yr*	Yes	Yes	1 yr	Yes	Yes	No†	A	A	5 yr*	Yes	No
MI	180 days*	No	No	No	No	No	No	No	No	No	No	A
MN	180 days	No	No	No	No	No	No	No	No	No	A	A
MS	6 mo	Yes	Yes	1 yr	Yes	Yes, A	No	3 yr, A	Yes	Yes	Yes	A
MO	90 days	No	No	No	No	No	No	No	A	No	No	A
MT	90 days	No	No	No	A	A	No	No	A	No	A	A
NE	1 yr*	No	No	No	No	A	No	A	A	No	No	A
NV	6 wk	No	No	No	No	No	No	2 yr	A	No	No	A
NH	1 yr*	Yes	Yes	2 yr	2 yr	Yes	2 yr	No	A	1 yr*	No	No
NJ	1 yr*	Yes	Yes	1 yr	1 yr	A	No	2 yr	A	18 mo	1 yr	A
NM	6 mo	Yes	Yes	Yes	No	No	No	No	No	No	No	No
NY	1 yr*	Yes†	Yes	1 yr†	No	No	†	A	A	3 yr†	No	A
NC	6 mo	No†	No†	No†	No†	A	No	3 yr	A	No	No†	No
ND	6 mo	Yes	Yes	1 yr	No	A	1 yr	5 yr	A	Yes	No	A
OH	6 mo	Yes†	Yes†	1 yr†	Yes†	No	Yes†	No	Yes†	Yes†	No	Yes†
OK	6 mo	Yes	Yes	1 yr	Yes	Yes	Yes	5 yr	Yes	Yes	No	Yes
OR	6 mo*	No	No	No	No	No	No	No	No	No	Yes	No
PA	6 mo	Yes	Yes	1 yr	No	No	No	18 mo*	Yes	Yes	No	No
RI	1 yr	Yes	Yes	5 yr*	Yes	Yes	1 yr	No	No	Yes	Yes	No
SC	1 yr*	Yes	Yes	1 yr	Yes	No	No	No	A	No	Yes	No
SD	*	Yes†	Yes†	1 yr†	1 yr†	A	1 yr†	5 yr†	A	Yes†	No	A
TN	6 mo*	Yes	Yes	1 yr	Yes	Yes	†	No	Yes	Yes	Yes	A
TX	6 mo*	Yes	Yes	1 yr	No	A	No	3 yr	No	1 yr	No	A
UT	3 mo*	Yes	Yes	1 yr	Yes	Yes	Yes	Yes*	A	Yes	No	No
VT	6 mo*	Yes	Yes	7 yr	No	No	Yes	5 yr†	A	3 yr	No	A
VA	6 mo*	Yes	Yes†	1 yr†	No	A	†	A	A	1 yr	No	A

Table R.4 *(continued)*

Important: Almost all states also have other laws as well as qualifications of the laws shown below and have proposed divorce-reform laws pending. It would be wise to consult a lawyer in conjunction with the use of this chart.

Some Grounds for Absolute Divorce[1]

Residence		Adultery	Mental or physical cruelty	Desertion	Alcoholism	Impotence	Non-support	Insanity	Bigamy	Felony Conviction or Imprisonment	Drug addiction	Fraud, force, duress
WA	bona fide resident	No	No	No	No	No	No	No	A*	No	No	A
WV	1 yr*	Yes	Yes	6 mo	Yes	A	No	3 yr	A	Yes	Yes	No
WI	6 mo	No	No	No	A	A	No	No	A	No	A	A
WY	2 mo*	No	No	No	No	No	No	2 yr	A	No	No	A
DC	6 mo	No	No	No	No	A	No	A	A	No	No	A
PR	1 yr	Yes	Yes	1 yr	Yes	Yes	No	Yes	A	Yes*	Yes	No

(1)Almost all states have "no-fault" divorce laws. Conduct that constitutes "no-fault" divorce may vary from state to state. (*)Indicates qualification; check local statutes. (A)Indicates grounds for annulment. (†)Indicates grounds for divorce or legal separation.

Source: World Almanac - 1997, from Gary N. Skoloff, Skoloff & Wolfe, Livingston, NJ: 1996.

Infertility

Infertility is generally defined as the inability to conceive a child after trying for a year or more. Until recently, the problem of infertility attracted little public attention. In the last few years, however, numerous couples, many of whom have deferred pregnancy because of career plans or later marriages, have discovered that they are unable to conceive or that the woman is unable to carry the pregnancy of live birth. Still, recent studies show that, overall, the fertility rate for American women is not declining. Of women aged fifteen to forty-years, 8.4 percent had an "impaired ability" to have children. About one out of twelve or thirteen American couples is involuntarily childless. The greatest increase in infertility is among young couples in the twenty- or twenty-four-year-old bracket. Infertility among young African-American couples is almost twice that of white couples. Every year, about 2.3 million American couples seek help for infertility.

Female Infertility

The leading cause of female infertility is blocked fallopian tubes, generally the result of *pelvic inflammatory disease (PID)*—an infection of the fallopian tubes or uterus that is usually the result of a sexually transmitted disease. It can be caused by gonococcus, chlamydia, or several other organisms. About 2 million cases of PID are treated annually; doctors estimate that about half of the cases go untreated because PID is often symptomless, especially in the early stages. Generally, only the woman with PID seeks treatment, and the man from whom she contracted the sexually transmitted disease that caused it may continue to pass it on. Surgery, including laser surgery, may restore fertility if the damage has not progressed too far. Septic abortions, abdominal surgery, and certain types of older intrauterine devices can also cause infections that lead to PID.

The second leading cause of infertility in women is *endometriosis*—a disease in which uterine tissue grows outside the uterus, often appearing on the ovaries, in the fallopian tubes (where it may also block the tubes), and in the abdominal cavity. In its most severe from it may cause painful menstruation and intercourse, but most women with endometriosis are unaware that they have it. Hormone therapy and sometimes surgery are used to treat endometriosis, which is sometimes called the career woman's disease because it is most prevalent in women aged thirty and over, many of whom have postponed childbirth.

Benign growths such as fibroids and polyps on the uterus, ovaries, or fallopian tubes may affect a woman's fertility; surgery can restore fertility in many of these cases. There may also be hormonal reasons for infertility. The pituitary gland may fail to produce sufficient hormones to stimulate ovulation, or it may release them at the wrong time. Stress, which may be increased by the anxiety of trying to achieve a pregnancy, may also contribute to lowered fertility. Occasionally, immunological causes may be present, the most important of which is the production of sperm antibodies by the woman. For an unknown reason, a woman may be allergic to her partner's sperm and her immune system will produce antibodies to destroy them.

Environmental factors can also affect fertility. Toxic chemicals and exposure to radiation threaten a woman's reproductive capability. Smoking appears to reduce fertility in women. Increasing evidence indicates that the daughters of mothers who were prescribed diethylstilbestrol (DES)—a drug that was once thought to reduce the risk of miscarriage—to increase their fertility have a significantly higher infertility rate, although studies remain somewhat contradictory. Nature also plays a part. Beginning around age thirty, women's fertility naturally begins to decline. By age thirty-five, about one-fourth of women are infertile.

Male Infertility

The primary causes of male infertility are low sperm count, lack of sperm motility, or blocked passageways. Some studies show that men's sperm counts have dropped by as much as 50 percent over the last thirty years. As with women, environmental factors may contribute to men's infertility. Increasing evidence suggests that toxic substances—such as lead and chemicals found in some solvents and herbicides—are responsible for decreased sperm counts. Smoking may produce reduced sperm counts or abnormal sperm. Some prescription drugs have also been shown to affect the number of sperm a man produces; among them are cimetidine (Tagamet, for ulcers), prednisone (a corticosteroid that reduces tissue inflammation), and some medications for urinary tract infections. Large doses of marijuana cause decreased sperm counts and suppression of certain reproductive hormones. These effects are apparently reversed when marijuana smoking stops. Men are more at risk than women from environmental factors because they are constantly producing new sperm cells; for the same reason, men may also recover faster once the factor has been removed.

Sons of mothers who took DES may have increased sperm abnormalities and fertility problems. Too much heat may temporarily reduce a man's sperm count (the male half of a couple trying to conceive may want to stay out of the hot tub for a while). A fairly common problem is the presence of a *varicocele*—a varicose vein above the testicle. Because it impairs circulation to the testicle, the varicocele causes an elevated scrotal temperature and thus interferes with sperm development. The varicocele may be surgically removed, but unless the man had a fairly good sperm count to begin with, his fertility may not improve.

Infertility Treatment

Almost without exception, fertility problems are physical, not emotional, despite myths to the contrary that often prevent infertile couples from seeking medical treatment. The two most popular myths are that anxiety over becoming pregnant leads to infertility and that if an infertile couple adopts a child, the couple will then be able to conceive on their own. Neither has any basis in medical fact, although it is true that some presumably infertile couples have conceived following an adoption. (This does not mean, however, that one should adopt a child to remedy infertility.) Approximately 10 percent of cases of infertility are unexplained. In some of these cases, fertility is restored for no discernible reason; in others, the infertility remains a mystery. About 60 percent of couples with serious infertility problems will eventually achieve a pregnancy.

Couples undergoing treatment for infertility often experience periods of depression and anxiety; they may have feelings of inadequacy, inferiority, or loss of control over their lives. Fertility specialists and other health-care providers can help alleviate such stress by increasing their patients' sense of control by such means as listening; offering encouragement; explaining risks, costs, and success rates in detail; and allowing the patients to make informed decisions as to tests and procedures. It has been suggested that those who are having difficulty conceiving for reasons as yet unexplained be referred to as "subfertile" instead of "infertile" and that their condition not be thought of as permanent.

Medical Intervention

In cases where infertility is a result of impaired ovulatory function in women, it may be remedied with medication. Treatment may include hormones that stimulate the ovarian follicles or regulate the menstrual cycle. Hormone therapy may be used alone or in combination with the techniques discussed in the following pages. Of the pregnancies achieved with the help of "fertility drugs," 10 to 20 percent result in multiple births because more than one egg is released by the ovary. Multiple births pose higher risks to both mothers and infants. Twins are about ten times more likely than babies born alone to have very low birth weights (3.3 pounds or less); triplets are over thirty times more likely to have very low birth weights and three times more likely to have severe disabilities than are infants born alone.

Current medical research is focusing increasingly on "male factor infertility." Approaches for treating men include new methods of sperm evaluation and processing and the use of medication to improve sperm velocity.

Intrauterine Insemination

When childlessness is the result of male infertility or low fertility or of a genetically transmitted disorder carried by the male, couples may try *intrauterine insemination (IUI)*, also known as *artificial insemination (AI)*. Single women who want children but who have not found an appropriate partner or who wish to avoid emotional entanglements have also made use of this technique, as have lesbian couples. The American Fertility Society estimates that there are about thirty thousand births a year from intrauterine insemination.

During ovulation, semen is deposited by syringe near the cervical opening. The semen may come from the partner; if he has a low sperm count, several collections of semen may be taken and frozen, then collectively deposited in the woman's vagina, improving the odds of conception. If the partner had a vasectomy earlier, he may have had the semen frozen and stored in a sperm bank. If the man is sterile or has a genetically transferable disorder, *therapeutic donor insemination (TDI)*, also known as artificial insemination by donor (AID), may be used. Anonymous donors—often medical students—are paid nominal amounts for their deposits of semen. Intrauterine insemination tends to produce males rather than females; in the general population, about 51 percent of the children born are male, but this figure reaches almost 60 percent when conception is achieved artificially. Intrauterine insemination has a success rate of about 60 percent for infertile couples.

Most doctors inseminate women at least twice during the preovulatory phase of their menstrual cycle; on the average, women who become pregnant have received inseminations over a period of two to four months.

Some lesbians, especially those in committed relationships, are choosing to create families through artificial insemination. To date, there are no reliable data on the number of such births, but anecdotal information indicates that it is in the thousands.

The practices of fertility clinics and sperm banks vary widely with respect to medical and genetic screening and limitations on the number of times a particular donor may be used. In a study by the Congressional Office of Technology Assessment in 1988, more than half the physicians surveyed said they did not screen donors for HIV. The commercial sperm banks surveyed did screen for HIV, however, as well as for most other sexually transmissible diseases. There has been one reported case in the United States of a woman contracting HIV through donated semen. The American Fertility Society issued guidelines for screening semen donors for HIV in 1992. The society has concluded that "under present circumstances the use of fresh semen for donor insemination is no longer warranted and that all frozen specimens should be quarantined for 180 days and the donor retested and found to be seronegative for HIV before the specimen is released." Although the viability of sperm is reduced by freezing, techniques of cryopreservation (superfreezing) minimize the deleterious effects. Studies show that TDI with frozen semen can have a success rate comparable to that with fresh semen.

In Vitro Fertilization and Embryo Transfer

In vitro fertilization (IVF) entails combining sperm and oocyte in a laboratory dish and subsequently implanting the blastocyst, or pre-embryo, into the uterus of the mother or a surrogate. To help increase the chances of pregnancy, several oocytes are generally collected, fertilized, and implanted. The mother takes hormones to regulate her menstrual cycle so the uterus will be prepared for the fertilized ova. Sometimes the blastocysts are frozen and stored, to be implanted at a later date.

The egg can come from the mother or from a donor. If it is from a donor, the donor may be artificially inseminated with the father's sperm and the embryo removed from the donor and transplanted into the mother-to-be's uterus. More commonly, donor oocytes are "harvested" from the donor's ovary, which has been hormonally stimulated to produce twelve to eighteen eggs instead of the usual one. In this procedure, a sixteen-inch needle is inserted into the donor's ovary by way of the vagina and the egg cells are drawn out and then mixed with the father's sperm. In two days, three or four blastocysts are implanted in the mother's uterus.

The procedure is quite costly and must usually be repeated a number of times before a viable pregnancy results. Varying success rates—from 12 to about 20 percent—have been reported. The number of IVF births is increasing yearly.

There is a high rate of multiple births (twins or triplets) with IVF. Some authorities recommend that the number of embryos transferred be limited to a maximum of three in order to reduce the health risks to mother and fetuses.

Recently, women in their late fifties and sixties have given birth to healthy babies—mostly as the result of IVF treatments in Europe. Some American IVF clinics also will treat women over age fifty if they are in good health. Although relatively few women past menopause seek IVF, such cases have caused public controversy.

GIFT and ZIFT

Another fertilization technique, gamete intrafallopian transfer (GIFT), may be recommended for couples who have no known reason for their infertility. In this process, sperm and eggs are collected from the parents and deposited together in the fallopian tube.

In zygote intrafallopian transfer (ZIFT), eggs and sperm are united in a petri dish and then transferred immediately to the fallopian tube to begin cell division. As with IVF, both GIFT and ZIFT carry an increased risk of multiple births.

Direct Sperm Injection

A revolutionary new method of fertilization developed by a Belgian physician entails the direct injection of a single sperm into an ovum in a laboratory dish. As in IVF, the blastocyst is then implanted in the uterus. As of this writing, about 1000 procedures are performed in the U.S. annually with a success rate of 24 percent. Direct sperm injection appears to hold promise for men who have low sperm counts or large numbers of abnormal sperm.

Surrogate Motherhood

The idea of one woman bearing a child for another is not new. In the Old Testament (Genesis 16:1–15), Abraham's wife, Sarah, finding herself unable to conceive, arranged for her husband to impregnate their servant Hagar. These days the procedures are considerably more complex and the issues are definitely cloudier. Some people question the motives of surrogate mothers. There have been cases of women having babies for their friends or even for their relatives. In 1991, a South Dakota woman became the first American "surrogate granny" when she was implanted with the fertilized ova of her daughter, who had been born without a uterus. Some women simply extend this kind of altruism to women they don't know. A study of 125 surrogate candidates found their major motivations to be money (they are usually paid $10,000 to $25,000 or more), liking to be pregnant, and unreconciled birth traumas, such as abortion or relinquishing a child for adoption.

Well-publicized court battles involving surrogacy point to the need for some kind of regulation. In the celebrated "Baby M" case, after numerous legal twists and turns, the New Jersey Supreme Court ruled in 1988 that contracts involving surrogate motherhood in exchange for money were illegal. It also ruled that a woman could volunteer to become a surrogate as long as no money was paid and she was allowed to revoke her decision to give up the child. The baby in question was allowed to remain in the custody of her biological father and his wife, a situation previously determined to be in the child's best interests. A lower court subsequently awarded the surrogate mother six hours a week of unsupervised visitation with the child.

Laws regarding surrogacy vary from state to state. Some states have laws prohibiting women from taking payment in exchange for giving a child up for adoption, and many have not as yet enacted legislation specifically relating to surrogacy. Exceptions are Nevada, which has legitimized surrogacy contracts involving payment, and Arkansas, which specifies that the couple who contract with the surrogate are the legal parents. Several other states have declared such contracts unenforceable, whereas at least one state is considering making surrogacy for pay a felony. Proposed legislation in other states would require all parties to undergo counseling and the surrogate and the biological father to be screened for genetic defects and sexually transmitted diseases.

There are thought to be several hundred births to surrogate mothers in the United States each year. A number of new, privately run agencies have been created to match surrogates and couples, highlighting the commercial potential of surrogate motherhood.

Sex Preselection

Sex preselection techniques—choosing the baby's sex—are a subject of much interest, although consistently reliable methods for assuring the gender of one's choice do not yet exist. Research indicates that the odds may be swayed slightly by such techniques as using a mild vinegar douche, shallow penetration by the penis, and no orgasm to increase the chances for a girl and a baking soda douche, deep penetration, and orgasm for a boy.

Genetic Counseling

About 2 to 4 percent of American children are born with birth defects. Each year, 1 million to 2 million infants, children, and adults are hospitalized for treatment of birth defects. These birth defects involve abnormalities of body structure or function, which may be genetically caused, the result of environmental influence on the fetus (such as smoking, drinking, diet, drugs, or exposure to toxic chemicals), or both. About 20 percent are inherited; 20 percent are caused by environmental influences on the fetus. The remainder may result from heredity and the environment interacting with each other or from causes which remain unknown.

Hereditary defects result from the interaction of the mother's and father's genes. Not all genes have an equal effect. Some genes are dominant over other genes, which are called recessive. The odds that a child will inherit a particular trait and the degree to which that trait will appear depend on many interrelated factors. These include (1) whether the trait is dominant or recessive, (2) the degree to which either parent has the trait, (3) the child's sex, and (4) the overall genetic makeup of the parents.

Individuals with hereditary defects are at significant risk of passing their disorders on to their children. Others are healthy themselves but carry a recessive abnormal gene; if they mate with another carrier of the same abnormal gene, there is a 25 percent risk in each pregnancy of having a child with that particular birth defect. (Each of us probably carries two to eight abnormal recessive genes.) Finally, there are some female carriers of sex-linked recessive traits who themselves are healthy but will pass the abnormal gene to half their children. Theoretically, half their sons will inherit the defect (such as hemophilia), whereas half their daughters will be carriers.

Although any couple may have a child with a birth defect or hereditary disease, some individuals or couples are at higher risk. Some can be identified during routine medical examination. Women over thirty-five years old, persons with congenital defects or hereditary diseases, and women who have had multiple miscarriages are all at higher risk than the rest of the population. Others can be identified by a careful review of their family medical history. Each person should obtain a family medical history from his or her parents and keep it as part of his or her permanent records.

Factors and risks that indicate the need for genetic counseling include those shown in Table R.5.

In genetic counseling, persons or couples who are at high risk are interviewed to determine whether they are potential

Table R.5 Indications of Need for Genetic Counseling

Factor	Risk to Child
Maternal age 35 years or older	Chromosomal anomaly
Previous child with chromosomal abnormality	Chromosomal anomaly
Adult with congenital abnormality	Occurrence in child
Previous child with congenital abnormality	Occurrence in child
Previous child with autosomal recessive gene (such as cystic fibrosis) or sex-linked conditions such as hemophilia	Occurrence in child
Family history of autosomal recessive gene or sex-linked condition	Occurrence in child
Adult with known hereditary syndrome	Occurrence in child
Ashkenazi Jew	Tay-Sachs disease
African heritage	Sickle-cell disease
Mediterranean ethnic group	Thalassemia
Infertility or multiple miscarriages	Chromosomal abnormality
Parent exposed to teratogen	Child with multiple congenital abnormalities
Family history of diabetes mellitus	Child with congenital malformation, diabetes mellitus
Deafness	Deafness
Psychosis	Psychosis

carriers of birth defects to their unborn children. Blood tests or other tests may be performed. After the diagnosis is made, the meaning of the disorder, its prognosis, and its treatment are explained. Finally, the genetic cause of the disorder is determined, along with an estimate of the risks of passing on a birth defect. If the disorder can be detected prenatally, the risks and benefits of amniocentesis and other techniques are explained. The couple's feelings about abortion are discussed so that they can make an informed and appropriate (for them) decision in the event that the fetus is affected by the condition. Genetic counseling may be time consuming and expensive, but for those at risk, it can improve the chances of giving birth to healthy children.

Making a Birth Plan

Prospective parents must make many important decisions. The more informed they are, the better able they will be to decide what is right for them. If you are planning a birth, how would you answer the following questions?

1. Who will be the birth attendant—a physician, a nurse-midwife? Do you already have someone in mind? If not, what criteria are important to you in choosing a birth attendant? Have you considered hiring a labor assistant, or doula—a professional childbirth companion employed to guide the mother during labor?
2. Who will be present at the birth—the husband or partner? Other relatives or friends? Children? How will these people participate? Will they provide emotional support and encouragement? Will they provide practical help, such as "coaching" the mother, giving massages, fetching supplies, taking photographs or videos? Can these people be sensitive to the needs of the mother?
3. Where will the birth take place—in a hospital, a birth center, at home? If in a hospital, is there a choice of rooms?
4. What kind of environment will you create in terms of lighting, room furnishings, and sounds? Is there special music you would like to hear?
5. What kinds of medication, if any, do you feel comfortable with? Do you know what the options are for pain-reducing medications? What about hormones to speed up or slow down labor? How do you feel about having an IV tube inserted as a precaution, even if medication is not planned? If you should change your mind about medication part of the way through labor, how will you communicate this to your attendants?
6. What about fetal monitoring? Will there be machines attached to you or the baby? What types and degree of monitoring do you feel comfortable with?
7. What is your attendant's policy regarding food and drink during labor? What kinds of foods or drinks, such as ice cream, fruit, juices, or ice chips do you think you (or your partner) might want to have?
8. What about freedom of movement during labor? Will you (or your partner) want the option of walking around during labor? Will there be a shower or bath available? Will the baby be delivered with the mother lying on her back with her feet in stirrups, or will she be free to choose her position, such as squatting or lying on her side?
9. Do you want a routine episiotomy? Under what conditions would it be acceptable?
10. What do you wish the role of instruments or other interventions, such as forceps or vacuum extraction, to be? Who will determine if and when they are necessary?
11. Under what conditions is a cesarean section acceptable? Who will decide?
12. Who will "catch" the baby as she or he is born? Who will cut the umbilical cord, and at what point will it be cut?
13. What will be done with the baby immediately after birth? Will he or she be with the mother or the father? Who will bathe and dress the baby? What kinds of tests will be done on the baby, and when? What other kinds of procedures, such as shots and medicated eyedrops, will be administered, and when?
14. Will the baby stay in the nursery, or is there rooming-in? Is there a visiting schedule?
15. How will the baby be fed—by breast or bottle? Will feeding be on a schedule or "on demand"? Is there someone with breastfeeding experience available to answer questions, if necessary? Will the baby have a pacifier between feedings?
16. If the baby is a boy, will he be circumcised? If so, when?

Choosing Day Care

Day care refers to any formal arrangement in which someone who is not a child's parent takes care of that child during the day. Traditionally, a child spent time with a parent—usually the mother—during the day. Now, this is not always possible.

Some families depend on day care to make life livable. Single-parent families, as well as families with both parents working, usually need to find some sort of day care for their children. If the children are school-aged, especially in the upper grades, the parent may ask a friend, neighbor, or relative to watch the child until after work. This arrangement might be quite informal. Sometimes such convenience is impossible. Then a parent must make formal arrangements. The varieties of child-care possibilities are unending, but day-care arrangements usually fall into one of the general categories listed here.

Neighborhood Child-Care Cooperatives

Cooperatives work under the principle that each person involved will watch other people's children for a certain amount of time in exchange for someone else watching theirs. Each cooperative has its own rules. Generally, neighborhood cooperatives are only suitable for people who work part-time or for people who have definite periods of time available and a suitable home to offer the group. An advantage is that cooperatives may take very young children, even infants. It depends on the group. There usually is no exchange of money.

Freelance or Licensed Baby-Sitters

People who like to baby-sit in their homes can often be found. They may sit for one or many children. They usually charge an hourly rate. Some states require sitters, especially those who take in more than one child, to be licensed. Sitters also may take very young children and infants; it depends on the sitter. Costs depends on the sitter's rates.

Day-Care Centers

Day-care centers are organizations that provide day care, usually from 7:30 or 8 AM to 6 PM, for a fee, at a particular site. They are often located in churches or community centers. Most require children to be toilet trained before joining. The age of the oldest children allowed varies. Generally, day-care centers tend to serve either the two- to six-year-old age group or the five-to-nine age group. Costs vary and always go up with inflation and more respectable salaries for the staff. Currently, monthly costs average between $250 and $500 per child for full-time care. Some centers are partial cooperatives, requiring parental participation in some way. Some have scholarship funds. Most are acceptable institutions to the government, which may provide day-care assistance to some low-income families.

The demand for day care today is greater than the spaces available. The best sources usually fill their lists for fall by June or midsummer. It is wise to check the possibilities well before your personal needs arise. Nothing is as informative as a visit to the day-care facility or home.

Here is a checklist of questions to help you judge each place:

- What is the ratio of adults to children?
- How large are the grounds?
- What kinds of toys, games, or playground equipment are provided?
- What activities, if any, are led (music, dance, art)?
- Do the children have nap time? Is it required?
- Does the caregiver have pets?
- Does the facility meet your standards of cleanliness?
- Is it run cooperatively?
- Will parents be asked to provide snacks or participate in work weekends? If so, how often?
- What food is served?
- Do the sitter, staff, or members of the cooperative seem generally to hold your values?
- Do you trust the judgment of the adult or adults running the show?
- When do half-days end and start?
- Can the caregiver accommodate irregular hours (if you need to work an extra or different shift, can your child stay, on short notice)?
- What does the caregiver do when faced with medical emergencies?
- What funds might be available to children attending (city, county, state, federal, or other)?

Rights of the Disabled

Is Your Disability Covered?

In its Section 504 regulation, the U.S. Department of Health and Human Services (HHS) identified a "handicapped" person as anyone with a physical or mental disability that substantially impairs or restricts one or more of such major life activities as walking, seeing, hearing, speaking, working, or learning. A history of such disability, or the belief on the part of others that a person has such a disability, whether it is so or not, also is recognized as a "handicap by the regulation." Handicapping conditions include, but are not limited to the following:

- Alcoholism.†
- AIDS (presumptive disability).
- Cancer.
- Cerebral palsy.
- Deafness or hearing impairment.
- Diabetes.
- Drug addiction.†
- Epilepsy.
- Heart disease.
- Mental or emotional illness.
- Mental retardation.
- Multiple sclerosis.
- Muscular dystrophy.
- Orthopedic, speech, or visual impairment.
- Perceptual handicaps such as dyslexia, minimal brain dysfunction, and developmental aphasia.

† The U.S. attorney general has ruled that alcoholism and drug addiction are physical or mental impairments that are "handicapping conditions" if they limit one or more of life's major activities.

What You Can Do

If you believe that your rights have been violated because of your disability or your child's disability by a business, hospital, physician, school, college, or any other institution receiving HHS assistance, write, giving details, to: Office for Civil Rights, Dept. of Health and Human Services, in your region.

Nutrition Guidelines

Of all habits and behaviors, our choices related to nutrition may have the greatest impact on our health of any decisions we make. Though many factors influence when we eat, what we eat, and how much we eat, finding the right balance between eating to maintain body functions and eating to satisfy our appetites is a problem for many.

Our traditional American "diet of affluence" coupled with lack of exercise and chronic stress, have not contributed to our positive well-being. Though an awareness of the need to modify our habits has influenced some of our decisions relating to nutrition, major health risks such as heart disease, certain types of cancer, hypertension, cirrhosis of the liver, osteoporosis, tooth decay, and chronic obesity still plague us. Excessive calorie consumption and a high concentration of fats, particularly saturated fats, and sugars appear to be the major contributors to obesity and early death in this country. Education about nutrition, resources to support good nutrition, and a commitment to our health are necessary elements for positive change.

The human body requires about 45 essential nutrients (substances which come from food because your body is unable to manufacture them). These essential nutrients can be found in carbohydrates, proteins, fats, vitamins, minerals, and water. A diet containing adequate amounts of each of these is vital because nutrients provide energy, help regulate body functions, and help build and maintain tissues. The energy in these foods is expressed in kilocalories; and three classes of nutrients that supply calories are protein (4 calories per gram), carbohydrates (4 calories per gram), and fats (9 calories per gram). Alcohol, though not an essential nutrient, provides 7 calories per gram.

Just meeting energy requirements is not enough. Our bodies require additional amounts of a wide variety of nutrients to grow and function properly. To illustrate graphically the importance of variety in our diet, including recommended servings, the United States Department of Agriculture (USDA) developed the Food Guide Pyramid. Note that it is not a rigid prescription but rather a general guide that lets us choose a healthful diet that is right for us and our family.

Additional Dietary Guidelines for Americans, published by the U.S. Department of Agriculture and the Department of Health and Human Services include:

Eat a variety of foods. Choose an appropriate number of servings from each group in the Food Guide Pyramid. Everyone, especially adolescent girls and women, should take special care to consume adequate amounts of calcium and iron.

Maintain healthy weight. Emphasize balancing food intake with regular physical activity. Weight loss can be accomplished by increasing physical activity and eating low-calorie, nutrient-rich foods. Diets with fewer than 800-1,000 calories per day can be hazardous and should be followed only under medical supervision.

Choose a diet low in fat, saturated fat, and cholesterol. Limit your fat intake to 30 percent or less of total calories. Limit your intake of saturated fat to one-third of your total fat intake (10 percent of total calories) and dietary cholesterol to 300 mg per day.

Choose a diet with plenty of vegetables, fruits, and grain products. Emphasize complex carbohydrates (starches and most types of dietary fiber) rather than simple ones (those that provide sweetness in our foods).

Use sugars only in moderation. Moderation here means less than 15 percent of total calories. Reducing sugar consumption means cutting back foods with added sugar.

Use salt and sodium only in moderation. You need only about 500 milligrams of sodium (about 1/4 tsp.) per day. It is recommended that you limit your sodium intake to no more than 3 grams per day.

If you drink alcoholic beverages, do so in moderation. Men should consume no more than two drinks per day, women, no more than one drink daily. Alcohol should not be consumed during pregnancy.

One additional word of advice: Use moderation when consuming salt-cured, smoked, and nitrate-cured foods because these may increase the risk of certain types of cancers.

Current patterns of eating may be influencing your health more than you realize. Being aware of the foods you consume and taking steps to improve the quality of nutrition in your life may be the first (and most important) steps toward improving your health and well-being.

Figure R.4
Food Guide Pyramid

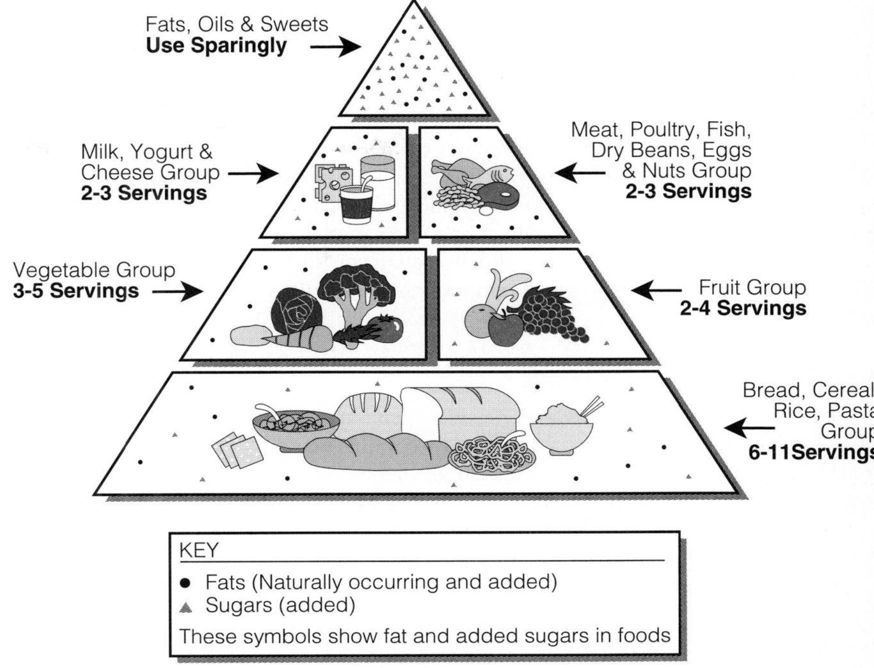

Money Matters

Managing Your Money*

The Management Process

A plan for money management is important for everyone. However, no ready-made plan fits every family, couple, or individual. Every household is different—not only in the number and characteristics of its members but also in its values, needs, wants, and resources. Only you can decide how your money should be spent, taking into consideration your income, the number and ages of your household members, where you live and work, your preferences, your responsibilities, and your goals for the future. By following a money management plan, you can be confident that expenses will be met and savings will be available. A plan can let you know where you stand financially and prevent emergencies from causing a financial strain.

Remember three basic concepts when developing a personal money management plan:

1. *Set realistic objectives.* Objectives set too high may lead to frustrations that could cause you to abandon your plan.
2. *Be flexible.* Your plan will require adjustments to keep up with your changing life cycle and financial situation. Do not make a plan so tight that each new development requires an entirely new plan.
3. *Be specific.* State your objectives concisely. If goals are vague, objectives may never be met, and you and other household members may have different ideas of what the end product will be.

* SOURCE: Joyce M. Pitts, *The Principles of Managing Your Finance*, U.S. Department of Agriculture, Agricultural Research Service (1986).

Evaluating Your Current Situation with a Net Worth Statement

The first step in developing a money management plan is to evaluate your current situation. An excellent way to do this is to prepare a *net worth statement.* Worksheet 1 can be used to add together your assets (what you own) and subtract from that the sum of your liabilities (what you owe). Space is provided for you to calculate your net worth statement now and again one year from now.

To determine the value of your *assets,* start with cash available. Include your checking and savings accounts, as well as your home bank. Now list all your investments, including bonds, mutual funds, life insurance cash values, and others. List additional assets using the current market value for your house, real estate, automobiles, jewelry, antiques, and other personal items.

To determine your *liabilities,* list the amount that you owe to all your creditors and lenders. Remember to include current bills, charge accounts, mortgage balance, and loans against your life insurance.

By subtracting liabilities from assets, you can determine your *net worth*. Look closely at your final net worth figure. Do you own more than you owe? Consider what you would like this figure to be a year from now. What do you need to do to achieve that goal?

Prepare a new net worth statement at the same time each year to reflect the changes in your finances. The market value of your assets (house, stock, or cars) may have declined or risen. You may have paid off one loan or gained another. A new net worth statement next year will help you decide if the money management plan you are developing now has helped put you ahead.

Planning to Reach Your Goals

An important step in developing a money management plan is to set household and individual goals. Goals are wants, needs, and future objectives for your household and its members. Goals may be long term, intermediate, or immediate.

Long-term goals are those you hope to reach in ten to twenty years or perhaps even longer. Long-term goals are often considered first so that they can be incorporated into the plan from the start. They are guided by expected changes in your household's life cycle and must sometimes be adjusted for future expected income and price changes. Long-term goals include such things as paying off a mortgage, putting children through college, or providing for a comfortable retirement.

Intermediate goals are those to be reached within the next five years or so. These goals may reflect the changes that will occur due to your increased income or larger family. Intermediate goals might include such things as a down payment on a house, a new car, or increased life insurance.

Immediate goals are required now—this week, this month, or this year. These are basic needs that must be met even at the expense of some future goals. Immediate goals may include paying current bills, maintaining health insurance, and buying food and clothing.

Addressing goals in this order ensures savings for long-term and intermediate goals and prevents immediate goals from pushing future ones aside. If your income is low, you may be able to meet only immediate goals. But you still should make intermediate and long-term plans that could help you to get ahead in the future.

Decide Which Goals Are Most Important to You

Think carefully about your financial goals. Many of us would like to be financially secure, own a large home, drive a fancy car, educate ourselves or our children, take long vacations, and so on. Realistically, however, most of us cannot have it all. We must select and work toward those goals that are most important to us and the ones we will be able to obtain. The following process may help you work through your goal-selection decisions:

- *Set goals.* Keep a listing of the goals that you and your family hope to achieve. Use Worksheet 2.
- *Rank goals.* List your goals in their order of importance to you and your household.
- *Assign dollar values to goals.* You will not be able to assign a dollar value to all goals now, but to most you will. For example, if you plan to buy a new car next year and you know the amount of the down payment, put this on your worksheet.
- *Reevaluate your goals.* After developing your budget, take another look at Worksheet 2. You may need to drop, postpone, or revise some of your goals. Decide how much change you are willing and able to make. For example, are you willing to change jobs or give up other goals in order to achieve a goal that is more important to you? Think about the trade-offs of saving for long-term and intermediate goals versus using income for current expenses. How much choice do you have? Can you, and do you want to, cut down on some immediate goals (current expenses) to improve your chances of meeting your long-term goals? Refer to your goals often as you plan.

Common Goals throughout the Life Cycle

Some goals are universal to all households, such as providing for sufficient food, comfortable shelter, and financial security. But most goals change as household members progress through the life cycle. When you are young and single, goals generally relate to your own personal development. When you are married and have children, your first priorities may switch from yourself to your children—establishing an education fund, for example. Different types of households use different ways to meet the same goal. For example, a retired couple may provide for their continued financial security during times of high inflation by cutting expenses. A young couple may seek higher-paying employment instead.

Look over the following description of household types, and determine which group your household resembles. Do you have goals similar to the ones stated for your group? Look at the goals for the other groups. They may help you anticipate future needs. This listing is not meant to be all-inclusive. However, it can be a starting point in determining your current and long-term goals and how those goals may change in years to come.

Household Type 1: Singles

This household type consists of adults who have never married or who are widowed or divorced. It includes persons from age eighteen to fifty-four who are considered to be self-supporting, even though they may be living with relatives or friends and sharing some household expenses. Income for this group may not be high, particularly for the younger members. Important goals involve the individual's personal, educational, and financial development.

Household Type 2: Single Parents

Members of this household group may also have never married or be widowed or divorced. Unlike Type 1 households, they are parents living with dependent children. The critical financial goals for single parents often relate to the care of their children and themselves.

Worksheet 1 Net Worth Statement

Assets (What You Own)	First Year	Second Year	Liabilities (What You Owe)	First Year	Second Year
Liquid Assets			Current bills		
Cash			Charge accounts		
Bank			Credit cards		
Checking			Taxes		
Savings			Installments		
Certificates			Mortgages		
Other			Other		
Savings bonds					
Other bonds					
Corporate					
Municipal					
Utility					
Life Insurance					
Mutual funds					
Other					
Other Assets					
Retirement plans					
Private pension plan					
Profit sharing plan					
Home					
Other real estate					
Car(s)					
Furniture					
Large appliances					
Antiques and art					
Jewelry					
Silverware					
Stamp or coin collection					
Debts others owe you					
Other					
Total Assets			**Total Liabilities**		

Net Worth (total assets minus total liabilities)

First year: ________ – ________ = ________

Second year: ________ – ________ = ________

Worksheet 2 Projecting Goals

Long-Term and Intermediate Goals

Goal	Cost of Goal	Number of Years Before Needed	Amount Already Saved	Amount to Save Each Year	Amount to Save Each Month
1					
2					
3					
4					
5					
6					
7					
8					
9					
10					
Total					

Immediate Goals

Goal	Cost of Goal	Number of Years Before Needed	Amount Already Saved	Amount to Save Each Year	Amount to Save Each Month
1					
2					
3					
4					
5					
6					
7					
8					
9					
10					
Total					

Household Type 3: Young Couples

This household type is often called the beginning family or the beginning marriage stage. It is a period of personal and financial adjustment for two persons. Ages of couples in this group typically range from about eighteen to thirty-four. There are no children, and there are often two incomes. Important goals involve setting up a household and adjusting to each other's needs.

Household Type 4: Young Families

In this growing-family stage, parents are typically young—age eighteen to thirty-four—and have dependent children in the household. There may be two incomes. Critical goals include protecting the family income and rearing the children.

Household Type 5: Middle Families

This household type is sometimes referred to as the contracting family. Parents are typically thirty-five to fifty-four years old. Children are "leaving the nest" for college, careers, and marriage. Unique goals for this household include providing for children's college or vocational education, weddings, and the parents' eventual retirement.

Household Type 6: Middle Couples

This group consists of persons thirty-five to fifty-four years old without children. This type of household often contains two earners. Income is often quite high, making investment maximization and tax minimization important financial goals.

Household Type 7: Older Singles

This group contains persons fifty-five years old and older who may be retired. There is no spouse present in the household. The majority of older singles are females. The major financial goal is to provide adequate income and reserves that will last for the balance of the older single's lifetime.

Household Type 8: Older Couples

This group consists of married couples of ages fifty-five and over who also may be retired. Their major financial goal is maintenance of an adequate standard of living for both persons for life.

The Budgeting Process

A budget is a plan for spending and saving. It requires you to estimate your available income for a particular period of time and decide how to allocate this income toward your expenses. A working budget can help you implement your money management plan. A well-planned budget does several things for you and your household. It can help you do the following:

- Prevent impulse spending.
- Decide what you can or cannot afford.
- Know where your money goes.
- Increase savings.
- Decide how to protect against the financial consequences of unemployment, accidents, sickness, aging, and death.

A working budget need not be complicated or rigid. However, preparing one takes planning, and following one takes determination. You must do several things to budget successfully.

First, communicate with other members of your household, including older children. Consider each person's needs and wants so that all family members feel they are a part of the plan. Everyone may work harder to make the budget a success and be less inclined to overspend if they realize the consequences. When families fail to communicate about money matters, it is unlikely that a budget will reflect a workable plan.

Second, be prepared to compromise. This is often difficult. Newlyweds, especially, may have problems. Each may have been living on an individual income and not be accustomed to sharing or may have been in school and dependent on parents. If, for example, one wants to save for things and the other prefers buying on credit, the two will need to discuss the pros and cons of both methods and decide on a middle ground that each can accept. A plan cannot succeed unless there is a financial partnership.

Third, exercise willpower. Try not to indulge in unnecessary spending. Once your budget plan is made, opportunities to overspend will occur daily. Each household member needs to encourage the others to stick to the plan.

Fourth, develop a good record-keeping system. At first, all members of the household may need to keep records of what they spend. This will show how well they are following the plan and will allow intermediate adjustments in the level of spending. Record keeping is especially important during the first year of a spending plan, when you are trying to find a budget that works best for you. Remember: a good budget is flexible, requires little clerical time, and, most important, works for you.

Choosing a Budget Period

A budget may cover any convenient period of time—a month, three months, or a year, for example. Make sure the period you use is long enough to cover the bulk of household expenses and income. Remember: not all bills come due monthly, and every household experiences some seasonal expenses. Most personal budgets are for twelve months. You can begin the twelve-month period at any time during the year. If this is your first budget, you may want to set up a trial plan for a shorter time to see how it works.

After setting up your plan, subdivide it into more manageable operating periods. For a yearly budget, divide income and expenses by 12, 24, 26, or 52, depending on your pay schedule or when your bills come due. Most paychecks are received weekly or every two weeks. Although most bills come due once a month, not all are due at the same time in the month. Try using each paycheck to pay your daily expenses and expenses that will be due within the next week or two. This way you will be able to pay your bills on time. You may also want to allocate something from each paycheck toward large expenses that will be coming due soon.

Developing a Successful Budget

Step 1: Estimate Your Income

Total the money you expect to receive during the budget period. Use Worksheet 3 as a guide in estimating your household income. Begin with regular income that you and your family receive—wages, salaries, income earned from a farm or other business, Social Security benefits, pension payments, alimony, child support, veterans' benefits, public assistance payments, unemployment compensation, allowances, and any other income. Include variable income, such as interest from bank accounts and investments, dividends from stock and insurance, rents from property you own, gifts, and money from any other sources.

If your earnings are irregular, it may be difficult to estimate your income. It is better to underestimate than to overestimate income when setting up a budget. Some households have sufficient income, but its receipt does not coincide with the arrival of bills. For these households, planning is very important.

Step 2: Estimate Your Expenses

After you have determined how much your income will be for the planning period, estimate your expenses. You may want to group expenses into one of three categories: fixed, flexible, or set-asides. Fixed expenses are payments that are basically the same amounts each month. Fixed regular expenses include such items as rent or mortgage payments, taxes, and credit installment payments. Fixed irregular expenses are large payments due once or twice a year, such as insurance premiums. Flexible expenses vary from one month to the next, such as amounts

Worksheet 3 Estimating Your Income

Source	January	February	March	April	May	June
Net salary:[a]						
Household member 1						
Household member 2						
Household member 3						
Household member 4						
Social Security payments						
Pension payments						
Annuity payments						
Veterans' benefits						
Assistance payments						
Unemployment compensation						
Allowances						
Alimony						
Child support						
Gifts						
Interest						
Dividends						
Rents from real estate						
Other						
Monthly Totals						

[a]Net salary is the amount that comes into the household for spending and saving after taxes, Social Security, and other deductions.

spent on food, clothing, utilities, and transportation. Set-asides are variable amounts of money accumulated for special purposes, such as for seasonal expenses, savings and emergency funds, and intermediate and long-term goals.

Use old records, receipts, bills, and canceled checks to estimate future expenses, if you are satisfied with what your dollars have done for you and your family in the past. If you are not satisfied, now is the time for change. Consider which expenses can be cut back and which expenses need to be increased. If you spent a large amount on entertainment, for example, your new budget may reallocate some of this money to a savings account to contribute to some of your future goals.

If you do not have past records of spending, or if this is your first budget, the most accurate way to find out how much you will need to allow for each expense is to keep a record of your household spending. Carry a pocket notebook in which you jot down expenditures during a week or pay period, and total the amounts at the end of each week. You may prefer to keep an account book in a convenient place at home and make entries in it. Kept faithfully for a month or two, the record can help you find out what you spend for categories such as food, housing, utilities, household operation, clothing, transportation, entertainment, and personal items. Use this record to estimate expenses in your plan for future spending. You also need to plan for new situations and changing conditions that increase or decrease expenses. For example, the cost of your utilities may go up.

Total your expenses for a year and divide to determine the amounts that you will have to allocate toward each expense during the budgeting period. Record your estimate for each budgetary expense in the space provided on Worksheet 4. Begin with the regular fixed expenses that you expect to have. Next, enter those fixed expenses that come due once or twice

Worksheet 3 *(continued)*

July	August	September	October	November	December	Yearly Totals

a year. Many households allocate a definite amount each budget period toward these expenses to spread out the cost.

One way to meet major expenses is to set aside money regularly before you start to spend. Keep your set-aside funds separate from other funds so you will not be tempted to spend them impulsively. If possible, put them in an account where they will earn interest. You may also plan at this point to set aside a certain amount toward the long-term and intermediate goals you listed on Worksheet 2. Saving could be almost as enjoyable as spending, once you accept the idea that saving money is not punishment but instead a systematic way of reaching your goals. You do without some things now in anticipation of buying what will give you greater satisfaction later.

You may want to clear up debts now by doubling up on your installment payments or putting aside an extra amount in your savings fund to be used for this purpose. Also, when you start to budget, consider designating a small amount of money for emergencies. Extras always come up at the most inopportune times. Every household experiences occasional minor crises too small to be covered by insurance but too large to be absorbed into the day-to-day budget. Examples may be a blown-out tire or an appliance that needs replacing. Decide how large a cushion you want for meeting emergencies. As your fund reaches the figure you have allowed for emergencies, you can start saving for something else. Now, record money allocated for occasional major expenses, future goals, savings, emergencies, and any other set-asides in the space provided for them on Worksheet 4.

After you have entered your fixed expenses and your set-asides, you are ready to consider your flexible expenses. Consider including here a personal allowance, or "mad money," for each member of the household. A little spending money that does not have to be accounted for gives everyone a sense of freedom and takes some of the tedium out of budgeting.

Step 3: Balance

Now you are ready for the balancing act. Compare your total expected income with the total of your planned expenses for the budget period. If your planned budget equals your estimated future income, are you satisfied with this outcome? Have you left enough leeway for emergencies and errors? If your expenses add up to more than your income, look again at all parts of the plan. Where can you cut down? Where are you overspending? You may have to decide which things are most important to you and which ones can wait. You may be able to do some trimming on your flexible expenses.

Once you have cut back your flexible expenses, scan your fixed expenses. Maybe you can make some sizable reductions here, too. Rent is a big item in a budget. Some households may want to consider moving to a lower-priced apartment or making different living arrangements. Others may turn in a too-expensive car and seek less expensive transportation. Look back at Worksheet 2. You may need to reallocate some of this income to meet current expenses. Perhaps you may have to consider saving for some of your goals at a later date.

If you have cut back as much as you think you can or are willing to do and your plan still calls for more than you make, consider ways to increase your income. You may want to look for a better-paying job, or a part-time second job may be the answer. If only one spouse is employed, consider becoming a dual-earner family. The children may be able to earn their school lunch and extra spending money by doing odd jobs in your neighborhood, such as cutting grass or baby-sitting. Older children can work part-time on weekends to help out. Another possibility, especially for short-term problems, is to draw on savings. These are decisions each individual household has to make.

If your income exceeds your estimate of expenses—good! You may decide to satisfy more of your immediate wants or to increase the amount your family is setting aside for future goals.

Carrying Out Your Budget

After your plan is completed, put it to work. This is when your determination must really come into play. Can you and your family resist impulse spending?

Become a Good Consumer

A vital part of carrying out a budget is being a good consumer. Learn to get the most for your money, to recognize quality, to avoid waste, and to realize time costs as well as money costs in making consumer decisions.

Keep Accurate Records

Accurate financial records are necessary to keep track of your household's actual money inflow and outgo. A successful system requires cooperation from everyone in the household. Receipts can be kept and entered at the end of each budget period in a "Monthly Expense Record" (Worksheet 5 on pages

Worksheet  **Expense Estimate and Budget Balancing Sheet, Fixed Expenses (Prepare for Each Month)**

Month:

	Amount Estimated	Amount Spent	Difference
Rent			
Mortgage			
Installments:			
Credit card 1			
Credit card 2			
Credit card 3			
Automobile loan			
Personal loan			
Student loan			
Insurance:			
Life			
Health			
Property			
Automobile			
Disability			
Set-asides:			
Emergency fund			
Major expenses			
Goals			
Savings and investments			
Allowances			
Education:			
Tuition			
Books			
Transportation:			
Repairs			
Gas and oil			
Parking and tolls			
Bus and taxi			
Recreation			
Gifts			
Other			
Total Fixed Expenses for Month			

R-62–R-63). It is sometimes a good idea to write on the back of each receipt what the purchase was for, who made it, and the date. Decide which family member will be responsible for paying bills or making purchases, and decide who will keep the record system up to date.

The household business record-keeping system does not need to be complex. The simpler it is, the more likely it will be kept current. Store your records in one spot—a set of folders in a file drawer or other fire-resistant box is a good place. You can assemble a folder for each of several categories, including budget, food, clothing, housing, insurance, investments, taxes, health, transportation, and credit. Use these folders for filing insurance policies, receipts, warranties, cancelled checks, bank statements, purchase contracts, and other important papers. Many households also rent a safe deposit box at the bank for storing deeds, stock certificates, and other valuable items.

Evaluating Your Budget

The information on Worksheet 4 can help you determine whether your actual spending follows your plan. If your first plan did not work in all respects, do not be discouraged. A budget is not something you make once and never touch again. Keep revising until the results satisfy you.

Dealing with Unemployment

Step 1: Take Time to Talk

Come right out and let your family know what's going on. Lay-off? Plant closing? Depressed economy? Business down? Explain what happened. Break down the big words so that everyone understands, especially the kids.

Fill in everyone at a family meeting or on a one-to-one basis. The important thing is not to leave anyone in the dark. If a family meeting seems out of the question, take time to talk when cleaning up after meals, cutting or raking the lawn, or taking trips to the store. Don't sugar coat the facts or tell "fairy tales." Living with less money will force your family to make hard changes. Yet let your kids know that even though there's less money, they can still count on a loving family—maybe more loving than ever.

Step 2: Take Time to Listen

Let everyone have a say about what these changes mean to him or her. Especially now, kids should be seen *and* heard.

Listen to words *and* actions. Is someone suddenly having a lot of crying spells, sleeping in late all the time, acting mean, drinking heavily, withdrawing, abusing drugs, complaining of stomach pains?

Step 3: Find Out Who's Hurting

Let everyone say what he or she is *really* feeling from time to time.

Just repeat whatever you hear, right when it's said. Then look for a nod to see if you heard it right. Is someone feeling helpless, sad, unloved, confused, worried, frightened, angry, like a burden to the family?

Try not to say "You shouldn't feel that way" because someone may be in real pain. The best you can do is let your loved ones have their say and get it off their chests.

Step 4: Let Your Feelings Out, Together or Alone

Give everyone in your family a space and time to let deep feelings out. Don't bottle them up or hide them from yourselves. If you're not comfortable showing others how you feel or fear you may strike someone who's dear to you, consider getting out of the house for a run or a brisk walk; having a good cry, alone; hitting a cushion or pillow; going to your room, shutting the door, and screaming; or all of the above.

Step 5: Solve Problems Together

Every week, look at the changes taking place in your household, and work out ways to deal with them. Working together as a team, your family can do more than survive. It can grow together and come through stronger.

Decide together things like these: what we can't afford now; what things we can do for family fun that don't cost a lot of money; who will do what chores around the house; how we'll all get by with less. If your discussions break down, go back to Step 1.

If you have a lot of trouble going through these steps, professional help may be what you need. Call and make an appointment with the family service agency nearest you. Whether or not you have money to pay for the services, the agency will do its best to help your family. Remember: you're not alone.

Sources of Help in Financial Crises

TANF/AFDC

You may be entitled to Temporary Assistance to Needy Families (TANF) (formerly Aid to Families with Dependent Children [AFDC]) if you have dependent children up to age twenty living with you who are your own or who are related to you. The program provides money and services to needy families with children until the families become self-supporting. The money for this program comes from federal, state, and county governments. The program is governed by federal and state laws but is administered through the county welfare department. Each state has different eligibility regulations, and they constantly change. Contact your local county welfare office to find out if your family qualifies.

Food Stamps

You may be eligible for food stamps, which will help stretch your food dollar. Eligibility is based on the household's net monthly income and your assets. If you live in your own home, its value is not included in your assets. If you qualify for food stamps, your allotment will be based on the number of people in your family, current or expected income, costs of shelter and dependent care, and other factors. These regulations are constantly changing, so check with your county welfare office.

Worksheet 5 Monthly Expense Record

Month:
Year:

Date	Expense Item	Mortgage/ Rent	Household	Utilities	Food	Clothing	Trans- portation	Child Care	Medical
Total									

Worksheet 5 *(continued)*

Savings	Debts	Insurance	Education	Personal Care	Recreation	Gifts	Business Related	Other: Taxes, Allowances, Legal Fees

Emergency Needs Program

Any person may apply for emergency help under the Emergency Needs programs administered by the Department of Social Services in some states. Emergency needs for food, clothing, rent, house payments, shelter, utility payments, taxes, security deposits, home repairs, appliances, furniture, transportation, and certain other necessities are considered under this program.

Medical Assistance

Some states or counties have medical assistance programs to help needy persons pay for a variety of medical services.

Your eligibility for medical assistance is determined according to your particular situation and income. You may still be eligible if you own certain types of property. You may be allowed to have (1) a homestead, (2) household items, (3) any tangible personal item you use in earning money, and (4) one passenger car per family. However, you usually cannot claim exemptions for intangible property such as securities, bonds, or cash that's invested or deposited in a savings account if the value exceeds a specific (generally low) limit.

Unemployment Insurance

You will probably qualify for unemployment insurance, which is paid through a payroll tax by your employer, if you have been laid off or lost your job. Also, you may qualify if you are a veteran or a retiree. Unemployment benefits, however, usually do not last longer than six months to a year, depending on state and federal policies.

Mortgage Arrangements

If you are unable to make your mortgage payment because of unemployment, an extended strike, illness, or other circumstances beyond your control, you should contact your mortgage lender *immediately* to discuss your situation. It is best to talk in person with your lender, if you can. You may be able to make special arrangements during the period that your income is reduced. If you cannot make such arrangements, you may be able to get other help as suggested below.

If you have a Federal Housing Administration (FHA)-insured mortgage, ask your mortgage lender to refer you to a Department of Housing and Urban Development (HUD)-approved home ownership counseling agency. The agency will discuss your problem with you and try to find solutions. If you have a Veterans Administration (VA) mortgage or land contract, contact the VA Loan Service and Claims Section.

You may find that payment of other bills is falling behind in addition to your mortgage payments. If so, credit counseling can help you.

Credit Counseling

If you are worried about past-due bills, wage garnishment, repossessions, or mortgage foreclosure, help is available. Nonprofit family financial counseling services can help you work out your financial problems and help you get back on your feet with dignity and a minimum of confusion. They will assist you in working out a budget and a debt repayment schedule. Also, they provide professional counseling on money management, family budgeting, and wise use of credit. If needed, they will provide debt management services in which they negotiate with your creditors and forward your payments to them.

Counseling services are often free. For debt management services, fees are based on your ability to pay. No one is refused service because of inability to pay a fee. For referral to the office nearest you, call your local family services or consumer affairs offices.

Utility Assistance

You may qualify for help with your utility bills if you are having trouble paying them or are threatened with a shutoff. There are programs in many states that help those on a limited income to meet their utility payments. Also, assistance in weatherizing homes is available to keep utility costs down. To see if you are eligible for such assistance, contact your local department of social services.

The federal government offers a residential energy tax credit. If you have added insulation or certain other energy-conserving measures to your home since April 1977, you should apply for this credit on your federal income tax return.

Utility Shutoffs (Gas, Electric, Telephone)

In most states, utility companies must allow you fourteen to twenty-one days to pay your bills. If your bill is not paid within this period, you will receive a shutoff notice and ten additional days to pay or to register a complaint. If there is a valid medical emergency in your home, the companies may be prevented temporarily from shutting off your service. They may be able to shut off service only between the hours of 8 AM and 4 PM and not before a day when reconnection cannot be made. You have a right to challenge your bills, if you think they are too high or incorrect. Ask for a hearing before the utility company's hearing examiner. If you are not satisfied with the examiner's decision, you can usually appeal it.

On any question or complaint you have about a utility bill, or if you cannot make full payment on any of your utilities, be sure to call the company's local office before your bill is due and ask for credit arrangements for partial payments. Approval of such a request usually depends on your payment history.

Legal Services

Persons with limited incomes can get legal counseling by contacting legal aid offices. Check in the Yellow Pages under "Attorneys."

Consumer Information

Shopping for Goods and Services

There are many criteria that you as a consumer can use to choose any product. The first step is to decide what your needs are. Is the purchase necessary at all? If the answer is "yes," you can form a clearer idea of your choice by listing your criteria before looking at the market.

For example, suppose you are choosing a blanket for your child's bed. You might make a list like this one:

Necessary?	Yes, child complains of cold
Size?	Twin (single bed)
Long lasting?	Yes, she's young (10 to 15 years)
Color?	Anything but white
Dry-clean?	No
Warmth?	Very warm
Resale possibilities?	Not important

Equipped with these decisions, you can buy the quality you need and avoid features you do not need. Sometimes comparative shopping is not reasonable. If your car is sputtering, for example, immediate necessity has the strongest claim, and you are likely to pull into the nearest gas station, not the cheapest. But for many purchases, especially the larger ones (rugs, appliances, automobiles, house paint, and so on), comparative shopping is worth the time and effort it requires. Telephoning two or three stores for price checks will usually give you an idea of high and low prices, and it only takes ten minutes.

Many people are unsure of the quality or special features of merchandise they are considering. Brand-name advertising and store clerks' sales pitches are not necessarily reliable sources of information. In our example, how would you know what the warmest, longest-lasting, machine-washable blanket on the market might be? The publications of the Consumers Union and Consumers Research, Inc., are a good source of this type of information. These two nonprofit organizations for consumer protection and enlightenment test products at random and publish their findings. They are mostly testing for safety and efficiency. If you are basically interested in economy or style, your opinions may differ.

The *Consumer Reports Buying Guide* and *Consumer Research Annual Guide*, although they sometimes list brand names that have gone off the market during the year, have a wealth of information on a wide range of goods. You can receive a free copy of "Consumer's Resource Handbook" by writing to the U.S. Consumer Information Center, Pueblo, CO 81009. (Also see the Resource Center's "Self-Help Resource Directory" under "Consumer Services.")

Secondhand Goods

Many durable goods (as opposed to food or services) can be purchased secondhand and serve your need just as well and more economically. A rake, a rug, a bicycle, or a car would be some examples of this sort of purchase. Many people discard goods before they have worn out. If newness is not a major criterion, you can save many dollars by buying used items.

It is also wise to be cautious in making these purchases. Some people get carried away by the prospect of picking up something for almost nothing. Are you getting your money's worth? Most people can judge for themselves on many items—a rake or a bed frame, say. But if the item requires repair or involves mechanical or electrical equipment, you can judge well only according to how much you know about that item. Do you want to spend six hours and several dollars repainting that chest of drawers? If you know nothing about cars, it is wise to consult or bring someone you trust with you to judge the car before you buy it. Buying secondhand goods you are unable to judge can be a way of throwing away money.

Newspaper want ads, garage or tag sales, and secondhand stores are the most common sources for finding secondhand goods.

Buying Services

In buying services—insurance, gardening, hauling, or plumbing—there is nothing like shopping around. Services are almost always competitive. There may be a going rate in the area for a particular service, but there is usually some company that hopes to get business by being cheaper. And for many services, the price range varies widely. This may or may not reflect the quality of the service you will get. It is wise to ask lots of questions: Will repair-people make house calls? Will the garden clippings be hauled away? What does the insurance cover, exactly? Companies are used to getting these calls. You may run into some grouches, but most businesspeople will be happy to answer your questions. They are offering a service. You are paying. You have the right to know how your money will be spent.

Consumer Complaints

What do you do when you are tricked, cheated, or robbed as a consumer? What do you do when a service is not provided or an item is no good?

The first step is to call or write the place of business. State the problem, and state what you want (I want the item replaced. I want my money back. I expect you to do the work this week.) If you write, you should make sure that the letter has your address and is dated, and you should make a copy of it to keep. Often the problem is unintentional. The storekeeper is happy to replace the item; the businessperson is happy to provide the service. If you cannot get prompt, courteous service from an employee, contact the manager. If the manager does not provide satisfaction, write directly to the president of the company or corporation, describing the problem. Letters to the president often produce quick action. Send a copy to the person with whom you were dealing.

You usually want to avoid calling in a third party. It always means time and trouble on your part. However, if you feel that you have been treated unfairly or that the provider does not respond to your complaint, you should tell the company that you intend to call in a consumer agency, and then do it.

Every state and most counties across the United States have a consumer affairs agency listed in the phone book under the name of the county in which you live. This is the best place to start. Tell the person at the agency what sort of complaint you have. You will probably be referred to another number. There are many branches of consumer protection. These agencies usually carry weight with businesspeople. They are your "big stick." Often their involvement will produce the results you want if you can prove your case. However, it may be months before this happens. Sometimes, though, these agencies can do nothing (if the company has gone bankrupt, for instance). This possibility must be accepted.

Buying on Credit

Credit buying is how most people purchase houses, automobiles, and other large consumer goods. This term also refers to some credit card buying. Buying on credit is always more expensive than purchasing with cash. Besides paying for the item, you are paying to use someone's money. This raises the price of the purchase considerably. Why does anyone do it, then? Why not wait until you have the cash to buy?

Buying on credit gives a person the advantage of using something he or she does not have the money to buy at the moment. If you buy a car on credit and your payments are $200 a month, you have the use of a car for about $6 a day for several years. And at the end of those years, you own it.

How to Obtain Credit

Remember that credit is something like rented money. You want to rent it from the person who will charge you the least for borrowing. The lender wants to have some assurance that you will pay back the loan plus a little something for letting you use the money in the first place. Some lenders can be quite mercenary about the "little something." This is why you want to know your options.

The Lender's Questions

The lender will want to know how able or likely you are to pay back the debt:

- What your yearly income is.
- How long you have worked at the same job.
- How long you have lived where you live now.
- What your normal expenses are per month.
- How much will be available for paying a loan.
- What your past record for repaying loans shows (called a *credit rating**).
- What assets (called *collateral*) you have that can cover the debt if you cannot repay. Assets include real estate (land or houses), savings accounts, stocks, cars, and other material goods.

*Paying your bills on time does not give you a good credit rating—you must prove you are capable of repaying *loans.*

The Borrower's Rights

The 1968 Truth-in-Lending Act, which applies to loans of $25,000 or less from most regular institutions, requires the lender to tell the borrower exactly how much interest is being charged on a loan. Ask.

The 1975 Equal Credit Opportunity Act prohibits discrimination in lending on the basis of sex, age, or marital status. It is now illegal for lenders to demand that a person provide his or her spouse's name, salary, or job description when the person wants credit in his or her own name. A parent's or spouse's co-signature cannot be required if the loan does not involve him or her. Alimony and child support payments must be considered regular income. Young couples do not need to divulge their methods of birth control or their intentions to bear children. And a lender cannot change the terms of credit because of a borrower's change in marital status, age, or job status.

The Equal Credit Opportunity Act was chiefly designed to end unfair practices against women. Banks and other lenders advise women to establish their own credit histories to avoid problems when borrowing. This can be done by obtaining credit cards in one's own name (Janet Doe, not Mrs. John Doe) or by taking out and repaying a small loan.

Where to Get Money

The cheapest way to borrow money is from yourself (if you can).

Savings

If you have a savings account, you can use it as collateral for a loan. You might be able to borrow up to 95 percent of the value of the account. If the loan is at 16 percent and your account makes 6 percent, the net interest on the loan is 10 percent.

Life Insurance

Certain life insurance policies have a pool of money—the cash value—from which you can borrow. It is your money, so you cannot be turned down for the loan. If you do not or cannot repay, the debt is repaid from your insurance policy.

Credit Unions

Credit unions are cooperatives, so you must belong to use one. They are very attractive, however. The average interest on loans runs 2 to 3 percent less than commercial banks. If the cooperative has a good year, you will also get some money back at the end.

You may need to turn to other sources of credit from time to time.

Commercial Banks

Banks are the usual source for many business and personal loans, especially for cars and homes.

Sales Finance Companies

Sales finance companies buy installment contracts, and their risks, from retail merchants. Most car loans that are not paid to commercial banks are paid to finance companies. About one-third of all personal loans also come from these finance companies.

Consumer Finance Companies

Consumer finance companies make small loans to consumers, usually at a very high interest rate. The loans are usually made for items other than the "biggies" (cars, homes, stereos)—for furniture, perhaps. The companies usually advertise their loan consolidation services—that is, paying several small loans with one bigger one—on television and radio.

Credit Cards

Credit cards issued by retail stores or national companies can be used in two ways. If you use your card only for purchases within your budget, you can always pay your credit card bill in full on time. Depending on the particular card, there may be no finance charge and no extra charges. You have used it like a check or like cash. It has the advantage of delaying cash payment, since it may take a month or more for a purchase to show on the bill. But you are not paying to borrow money, only for the item. Credit cards can also be used to buy items on credit—hence, the name. You can buy $200 speakers today, pay $20 plus finance charges for the next ten months, and use the speakers. Many credit card holders receive various "enhancements" along with their cards, such as travel and accident insurance, emergency cash, airline tickets, travel discounts, and bonus merchandise programs.

The pitfall of credit card buying is that these cards are easy to use but the finance charges are high—usually equivalent to 18 or 19 percent interest. Also, late charges are added if payments are not made within a certain number of days. Many people buy items on credit that are not worth the high price of the money they are using. Many people also overbuy without considering their income and so find themselves continually in debt. Only one-third of credit card users pay their bills before incurring finance charges (that is, interest).

Credit card charge accounts are also called "revolving charge plans." There is a top limit, but purchases are added as they are made without a new agreement's being written. A credit card's top limit, however, is usually no more than several thousand dollars. Most large purchases must be financed differently.

If you need to make a purchase on credit, check with banks, savings and loan associations, finance companies, your credit union, your life insurance company, and possibly the company from which you are making the purchase to determine which one offers the lowest annual percentage rate (APR). Also compare the annual fees of credit cards. Some cards are available at no yearly charge.

Fair Credit Reporting Act*

If you have a charge account, a mortgage on your home, or a life insurance policy, or if you have applied for a personal loan or a job, it is almost certain that somewhere there is a "file" that shows how promptly you pay your bills, whether you have been sued or arrested, if you have filed for bankruptcy, and so on. Such a file may include your neighbors' and friends' views of your character, general reputation, or manner of living.

The companies that gather and sell such information to creditors, insurers, employers, and other businesses are called *consumer reporting agencies*, and the legal term for the report is a *consumer report.* If, in addition to credit information, the report includes interviews with a third person about your character, reputation, or manner of living, it is referred to as an *investigative consumer report.*

The Fair Credit Reporting Act was passed by Congress to protect consumers against the circulation of inaccurate or obsolete information and to ensure that consumer reporting agencies adopt fair and equitable procedures for obtaining, maintaining, and giving out information about consumers. Under this law, you can take steps to protect yourself if you have been denied credit, insurance, or employment or if you believe you have had difficulties because of an inaccurate or unfair consumer report.

Your Rights

You have the following rights under the Fair Credit Reporting Act:

1. To be told the name and address of the consumer reporting agency responsible for preparing a consumer report that was used to deny you credit, insurance, or employment or to increase the cost of credit or insurance.
2. To be told by a consumer reporting agency the nature, substance, and sources (except investigative-type sources) of the information (except medical) collected about you.
3. To take anyone of your choice with you when you visit the consumer reporting agency to check on your file.
4. To obtain free of charge all information to which you are entitled if the request is made within thirty days after receipt of a notification that you have been denied credit, insurance, or employment because of information contained in a consumer report. Otherwise, the consumer reporting agency is permitted to charge a reasonable fee for giving you the information.
5. To be told who has received a consumer report on you within the preceding six months, or within the preceding two years if the report was furnished for employment purposes.
6. To have incomplete or incorrect information reinvestigated unless the consumer reporting agency has reasonable grounds to believe that the dispute is frivolous or irrelevant. If the information is investigated and found to be inaccurate, or if the information cannot be verified, you have the right to have such information removed from your file.
7. To have the consumer reporting agency notify those you name (at no cost to you), who have previously received the incorrect or incomplete

* SOURCE: Office of Consumer Affairs, Federal Deposit Insurance Corporation (1987).

information, that this information has been deleted from your file.

8 When a dispute between you and the reporting agency about information in your file cannot be resolved, you have the right to have your version of such dispute placed in the file and included in future consumer reports.

9 To request the reporting agency to send your version of the dispute to certain businesses without charge, if requested within thirty days of the adverse action.

10 To have a consumer report withheld from anyone who under the law does not have a legitimate business need for the information.

11 To sue a report agency for damages if the agency willfully or negligently violates the law, and, if you are successful, to collect attorneys' fees and court costs.

12 Not to have adverse information reported after seven years. One major exception is bankruptcy, which may be reported for ten years.

13 To be notified by a business that it is seeking information about you that would constitute an investigative consumer report.

14 To request from the business that ordered an investigative consumer report information about the nature and scope of the investigation.

15 To discover the nature and substance (but not the sources) of the information that was collected for an investigative consumer report.

What the Fair Credit Reporting Act Does Not Do

The Fair Credit Reporting Act does not do the following:

1. Require the consumer reporting agency to provide you with a copy of your file, although some agencies will voluntarily give you a copy.
2. Compel anyone to do business with an individual consumer.
3. Apply when you request commercial (as distinguished from consumer) credit or business insurance.
4. Authorize any federal agency to intervene on behalf of an individual consumer.
5. Require a consumer reporting agency to add new accounts to your file; however, some may do so for a fee.

How to Deal with Consumer Reporting Agencies

If you want to know what information a consumer reporting agency has collected about you, either arrange for a personal interview at the agency's office during normal business hours or call in advance for an interview by telephone. Some agencies will voluntarily make disclosure by mail. The consumer reporting agencies in your community can be located by consulting the Yellow Pages of your telephone book under such headings as "Credit" or "Credit Rating or Reporting Agencies."

If you decide to visit a consumer reporting agency to check on your file, the following checklist may be of help. For instance, in checking your credit file, did you

1. Learn the nature and substance of all the information in your file?
2. Find out the name of each of the businesses (or other sources) that supplied information on you to the reporting agency?
3. Learn the name of everyone who received reports on you within the past six months (or the last two years, if the reports were for employment purposes)?
4. Request the agency to reinvestigate and correct or delete information that was found to be inaccurate, incomplete, or obsolete?
5. Follow up to determine the results of the reinvestigation?
6. Ask the agency, at no cost to you, to notify those you name who received reports within the past six months (two years, if for employment purposes) that certain information was deleted?
7. Follow up to make sure that those named by you did, in fact, receive notices from the consumer reporting agency?
8. Demand that your version of the facts be placed in your file if the reinvestigation did not settle the dispute?
9. As the agency to send your statement of the dispute to those you name who received reports containing the disputed information within the past six months (two years, if received for employment purposes)? A reasonable fee may be charged for this service if you have not incurred adverse action from a creditor within the last thirty days.

The federal agency that supervises consumer reporting agencies is the Federal Trade Commission (FTC). Questions or complaints concerning consumer reporting agencies should be directed to the Federal Trade Commission, Division of Credit Practices, Washington, DC 20580.

Buying A Home*

What Can You Afford?

You've probably heard various ways to estimate what you can afford to spend on a home. These methods can be useful in arriving at approximate figures, but they overlook the variables that can affect your financial capability.

* SOURCE: Reprinted by permission from the CIRcular™ Consumer Information Report, "Steps to Buying a Home," copyright Bank of America NT&SA 1982, 1985.

Generally, the ideal monthly payments should equal about 25 percent of your gross monthly income, minus any outstanding debts. But you may be able to manage a monthly payment of up to 40 percent of your gross monthly income, depending on other factors. For example, you may be willing to cut back on other nonessential costs, or you may be at the start of a promising career.

To figure what you can spend on a home, you need to make two basic calculations. How much can you pay each month for the long-term expenses of owning a home? How much cash can you spend for the initial costs of buying a home?

Monthly Housing Costs

You can calculate how much you have to spend by preparing a personal financial statement that details total income and expenses. You'll also need this information when you apply for a loan. Begin preparing your statement by listing monthly income after taxes and other deductions. You should include your income and the income of anyone else participating in the purchase. Use an average figure if the income varies from month to month or year to year, and exclude any irregular income.

Next, estimate your average monthly expenses for all non-housing items—food, clothing, savings, debts, and so on—and subtract them from your monthly net income. What's left is the maximum amount you can pay each month for all long-term home ownership costs.

Remember: in addition to loan payments, your monthly costs also include payments for taxes and assessments, insurance, maintenance, and utilities. Unless you're willing to stick to a very strict budget, you'll probably be more comfortable with a home loan payment that's less than the maximum amount you can afford. When you find a home in which you're interested, get estimates of monthly costs for the following:

Home Loan Payments. You'll probably take out a loan to pay a major part of the purchase price, so it's a good idea to shop for a loan before you look for a home. Talk to several lenders about your eligibility for a loan, the maximum amount you can reasonably expect, their current loan terms, and the monthly payments for different loan amounts, repayment periods, and interest rates.

Property Taxes. [Laws vary by state.] Improving the home can affect its tax valuation.

Property Insurance. The cost of insuring a home varies with the home's age, type of construction, and location. As a general estimate, the annual insurance premium is one-third of 1 percent of the home's price. For a more accurate figure, call several local insurance agents, describe the home, and ask what you must pay to insure it. Lenders usually require you to carry enough insurance to cover the amount of your loan, but you may consider getting more, based on the cost of replacing your home.

Repair and Utilities. These costs vary with the home's age, size, design, and condition.

Tax Considerations. At the present time, you can deduct your property taxes, the interest payments on your home, and the loan origination fee you pay your lender on your federal income tax returns. [State laws may vary.]

Cash Needed

To calculate how much you have available to spend on a home, add up savings (other than an emergency reserve) and investments you might cash in. You'll need money for the following costs:

Professional's Fees. You might hire professionals such as a housing inspector and an attorney during the home-buying process. Ask them for fee estimates first.

Closing Costs. These are fees for services, including those performed by the lender, escrow agent, and title company. Closing costs can range from several hundred to several thousand dollars. Federal law requires the lender to send you an estimate of the closing costs within three days after you've applied for the loan. Although local custom usually determines who—you or the seller—pays for what costs, you may be able to negotiate some of the fees. Include the results of any negotiations in your written purchase contract.

For a full explanation of various closing fees, read the booklet on settlement costs prepared by the U.S. Department of Housing and Urban Development (HUD). It's available free from lenders and HUD offices.

Down Payment. The usual down payment required by many lenders is 20 percent of the home's total cost. The actual amount depends on the type of loan, your lender's policy, and current economic conditions. Typically, for down payments of less than 20 percent, the lender will require that the buyer purchase private mortgage insurance (PMI). PMI protects the lender against loss if you don't pay as agreed.

It is possible to reduce or eliminate your need for down payment cash. For instance, you can:

- Apply for a Federal Housing Administration [or a] Veterans Administration loan, which require relatively low down payments. . . .
- Lease a home with an option to buy it at a later date for an agreed-upon price. Usually, some or all of the rent you pay is credited against the purchase price. The buyer may have to pay an added charge for this option.

Select an Area

The area you choose can greatly affect your pocketbook as well as your personal happiness. For instance, you should consider how far the home is from your job and what distance you're willing to commute each day. Drive around and note the neighborhoods that appeal to you. Ask city officials, real estate agents, local businesspeople, and your prospective neighbors about the following points:

Public Services

How close is the fire station? Where is the nearest hospital? Is reliable public transportation available? Are good schools nearby? And can your children safely and easily walk or take transportation to get to them?

Public Safety

Get crime statistics from the local police. Ask for a report or map indicating the crime rates for various areas.

Zoning and Taxes

Contact the city or county planning department about plans for your area. Are there plans to widen the streets or add new buildings nearby? Ask the local tax assessor about assessments—charges for local public improvement such as paving, street lighting, and public transit. Have they been rising sharply, and are they likely to continue doing so? Find out about any local homeowner's tax exemptions or other tax credits you may be entitled to receive.

Environmental Conditions

City or county planning officials can tell you about such problems as flooding, erosion, smog, fire hazards, and earthquake fault lines that are present in your area.

Look for Homes

Begin looking for houses that best meet your needs. Consider the following:

Type of Ownership

Do you want to live in a single- or multiple-family residence? Or are you interested in a condominium or a planned unit development (PUD)? With a condominium or PUD, you and the other owners share rights to some parts of the property, called common areas. Usually, you'll also have to pay homeowner's dues.

Length of Use

Many people stay in their homes longer than they originally planned. Look ahead at least five years and try to anticipate changes—such as family size—that might affect your housing needs.

Space

Measure your present home's rooms, storage areas, and work surfaces, noting which spaces are large enough and which aren't. Then look for houses that are designed to meet your needs.

Where to Look

Find out about homes for sale by reading newspaper ads and by consulting real estate agents recommended by your friends, other agents, or the local real estate board. Pick up buyers' guides from realty and builders' associations, lenders, and stores. And ask friends living in the area to watch for home sales.

Inspect the Home

Inspect thoroughly any home you're interested in buying. Read books on homes and consult knowledgeable friends to learn how to inspect a home and judge the quality of the workmanship, materials, and design.

Professional Inspectors

It's generally a wise investment to hire a housing inspector to confirm your own judgment about the home. A housing inspector—unlike an appraiser, who judges the dollar value of a home—provides a detailed, written evaluation of the home's condition. Fees typically range from $100 to $200. Before hiring an inspector, make sure he or she is licensed and bonded. Find out whether the inspector's work is guaranteed and, if so, for how long.

You also should have a licensed pest control inspector check the home whether or not the lender requires such an inspection. The seller usually pays the cost.

Warranties

The seller may provide a home protection contract (home warranty). Or you can purchase one from a home protection company. A typical new-home warranty, whose term may range from one to ten years, covers the home's structure, its major systems (plumbing, heating, and electrical), and any appliances sold with the home. On an existing home, the warranty typically covers major systems and appliances for one year.

Make an Offer

Consider making your first offer for less than the asking price if you think the home is overpriced for the market or the circumstances are favorable—for example, if the seller seems eager to close the sale.

The Purchase Contract

When you decide what price to offer, you draw up a contract stating the sale terms. You submit your offer to the seller, who either accepts it as is or makes changes and sends it back to you. The contract goes back and forth as many times as necessary to reach an agreement. You should sign the contract only when both of you are satisfied.

According to state law, no agreement for the sale of real estate can be enforced unless it's in writing. Look over the contract carefully—with your legal adviser if possible—to make sure it covers all the sale conditions you want included. Following are some points you may wish to cover:

- The conditions under which the contract may be canceled without penalty—for instance, if you can't get the financing you want or if the home doesn't pass professional inspection.
- The closing costs you'll pay and those the seller will pay.
- An itemized list describing furnishings, appliances, and other personal property the sale includes and excludes.
- The date on which you'll check the home's condition before the sale is final.
- The date you get possession of the home.

The Deposit

At the time you sign the contract, you'll be asked for a deposit, sometimes referred to as earnest money. The amount can range from hundreds to thousands of dollars, depending on what you're willing to give and what the seller is willing to accept. The deposit usually is applied to the down payment or to your share of the closing costs. If the sale falls through, the deposit either will be kept by the seller or returned to you, according to the terms of your purchase contract.

Escrow

Once you and the seller have signed a purchase agreement, you're ready to begin escrow—a procedure in which your deposit and any other pertinent documents are placed in the keeping of a neutral third party called the escrow agent. You and the seller must agree on the agent, who may be from a title insurance company, an escrow company, or the lender's own escrow department.

Escrow can begin before or after you've arranged financing. You and the seller negotiate and sign a set of escrow instructions listing the conditions (including financing) that must be met before the sale is finalized. The escrow agent distributes the money and documents according to the escrow instructions.

Financing: Getting a Loan

Before you choose a loan, it's critical that you compare the following loan terms for similar types of loans.

Down Payment and Loan Fees. Those vary with the lender and type of loan.

Interest Rate. This is the cost of borrowing the money, usually a percentage of the loan amounts. A small variation in the interest rate can add up to thousands of dollars in the total loan payment amount.

The lender is required to tell you the annual percentage rate (APR). This is the cost of the loan per year including interest and additional finance charges, such as loan origination and certain closing fees. The APR expresses these charges as a percentage.

Repayment Period. With a fixed-rate loan, the longer the repayment period, the higher the total cost of the loan, but a shorter repayment period generally means a larger monthly payment. With an adjustable rate loan, the total cost and the monthly payment are affected by interest rate changes as well as by the repayment period. If your loan rate isn't fixed, you'll want to know whether you can extend the repayment period to reduce any increase in your monthly payment.

Prepayment. A lender may reserve the right to charge a fee—called a prepayment premium—if you pay back all or part of your loan early. Your promissory note (loan contract) usually will contain a clause describing under what conditions you must pay this premium. If the promissory note isn't specific, ask what these conditions are.

Other Financing

Instead of—or in addition to—getting a new loan from a lender, you may be able to obtain financing in one of the following ways:

Assumptions. Federal law permits lenders to make most loans nonassumable. To find out whether you can assume (take over) the seller's loan, check with the seller's lender. If the loan is assumable, you may be able to pay the seller the difference between the amount still owed on the loan and the purchase price and take over payments where the seller left off.

You make payment either to the seller or directly to the seller's lender. In the latter case, you may have to pay any loan fees and provide whatever credit information the lender requires.

An assumption can be a good arrangement if you can take over the loan at a lower-than-current interest rate. Some lenders, however, may require you to assume the loan at the current rate.

You may need more financing to make up the difference between the purchase price and the amount assumed. The seller often may carry the loan—grant you credit—for a short time (usually three to five years). This way, depending on the amount and the credit terms, you could have a large balloon payment. If you're thinking about having the seller carry the loan, consider whether you'll be able to meet the credit terms and make any balloon payment when it comes due. You'll also need to determine whether refinancing will be available, and, if not, whether the seller will extend the financing agreement.

As an alternative to assuming the seller's loan, you might negotiate with the seller's lender to give you a loan for the difference between the purchase price of the home and the down payment. In many cases, you can obtain an interest rate that's between the rate on the seller's original loan and the current rate.

Buy Downs. With a buy-down arrangement, the seller pays the lending institution an amount to lower the interest rate on your loan. Usually, the term is for a specified period of time—typically one to five years. After that, you pay the rate the lender was charging at the time you took out the loan.

Equity Sharing. Consider arranging for other investors to pay part of the loan, the down payment, or closing costs in exchange for part of the equity in your home. Many real estate agents and some states and local government agencies offer this kind of financing arrangement—sometimes called a shared-appreciation program.

In addition to all these financing alternatives, the seller may offer a variety of other arrangements. When considering any type of loan, be sure to get professional legal, tax, and real estate advice.

Close the Deal

Closing—also called settlement or closing escrow—is the final step. Before the sale is finalized, you must deposit in escrow all of the down payment and your closing costs. At the close of escrow, the agent will give your deposit and loan funds to the seller and have the deed recorded. After the recording, you'll receive the deed by mail in about 30 days.

Glossary

A

ABC-X model of family stress See *double ABC-X model* of family stress.

aberration A departure from what is culturally defined as "normal" behavior.

abortifacient A substance that can induce abortion.

abortion The termination of a pregnancy either through miscarriage (spontaneous abortion) or through human intervention (induced abortion).

abstinence Refraining from sexual intercourse, often on religious or moral grounds.

abuse Mistreatment; wrong, bad, injurious, or excessive use.

accommodation According to Jean Piaget, the process by which a child makes adjustments in his or her cognitive framework in order to incorporate new experiences.

acquaintance rape Rape in which the assailant is personally known to the victim, usually in the context of a dating relationship. Also known as *date rape*.

acquired immune deficiency syndrome (AIDS) An infection caused by the human immunodeficiency virus (HIV), which suppresses and weakens the immune system, leaving it unable to fight opportunistic infections.

adaptability Ability to adjust relationships, roles, and rules to changing circumstances.

adolescence The social and psychological state occurring during puberty.

adoption The process by which an individual or couple legally become the parents of a child not biologically their own.

advice/information genre Media such as self-help books, newspaper advice columns, radio and TV talk shows, and women's magazines that purport to offer factual and accurate information but are actually motivated by considerations such as the need to entertain and to make a profit.

AFDC See *Aid to Families with Dependent Children.*

affiliated kin Unrelated individuals who are treated as if they were related.

affinity 1. Relationship by marriage. 2. A close relationship.

affirmative action Programs that attempt to place qualified members of minorities in government, corporate, and educational institutions from which they have been historically excluded because of their minority status.

afterbirth The placenta and fetal membranes expelled from the uterus during the third stage of labor.

agape [AH ga pay] According to sociologist John Lee's styles of love, altruistic love.

agglutination test A urine analysis test used to determine the presence of human chorionic gonadotropin (HCG) secreted by the placenta, which is an indication of pregnancy.

AIDS See *acquired immune deficiency syndrome.*

Aid to Families with Dependent Children (AFDC) A government program designed to assist families with children financially during times of poverty. Cf. *TANF*

alcohol abuse Continued use of alcohol despite consequences.

alcohol dependence Extensive problems associated with the use of alcohol, including tolerance and withdrawal. Used interchangeably with alcoholism.

alcoholism A disorder characterized by preoccupation with alcohol and loss of control over its consumption so as to lead usually to intoxication, repetition, progression, and a tendency for relapse.

alimony Court-ordered monetary support to a spouse or former spouse following separation or divorce.

alpha-feto protein (AFP) screening A fetal diagnostic method involving testing a pregnant woman's blood to reveal neural tube defects in the fetus.

amniocentesis A method of fetal diagnosis in which amniotic fluid is withdrawn by syringe from a pregnant woman's uterus to determine possible birth defects.

anal eroticism Sexual activities involving the anus.

anal intercourse Penetration of the anus by the penis.

androgyny The state of having flexible gender roles combining instrumental and expressive traits in accordance with unique individual differences.

annulment The legal invalidation of a marriage as if the marriage never occurred.

anti-abortion movement Social movement that advocates against abortion. Also known as *pro-life movement.*

anti-gay prejudice Strong dislike, fear, or hatred of gay men and lesbians because of their homosexuality. See also *homophobia.*

Apgar score The rating given a newborn immediately after birth, indicating its overall condition based on heart rate, respiration, coloring, reflexes, and muscle tone.

artificial insemination (AI) See *intrauterine insemination.*

artificial insemination by donor (AID) See *therapeutic donor insemination.*

Asian American Collective term relating to Americans of Asian descent, such as Chinese American, Japanese American, Korean American, Vietnamese American, or Cambodian American.

asymptomatic Not showing symptoms.

attachment Close, enduring emotional bonds, especially those forged between an infant and his or her primary caregiver(s).

attachment theory of love A theory maintaining that the degree and quality of an infant's attachment to his or her primary caregiver is reflected in his or her love relationships as an adult.

authoritarian child rearing A parenting style characterized by the demand for absolute obedience.

authoritative child rearing A parenting style that recognizes the parent's legitimate power and also stresses the child's feelings, individuality, and need to develop autonomy.

autoeroticism Erotic behavior involving only the self; usually refers to masturbation, but also includes erotic dreams and fantasies.

B

barrier method Any of a number of contraceptive methods that place a physical barrier between sperm and egg, such as the condom, diaphragm, and cervical cap.

basal body temperature (BBT) method A contraceptive method based on variations of the woman's resting body temperature, which rises 24 to 72 hours before ovulation.

basic conflict Pronounced disagreement about fundamental roles, tasks, and functions. Cf. *nonbasic conflict.*

battering A violent act directed against another, such as hitting, slapping, beating, stabbing, shooting, or threatening with weapons.

BBT method See *basal body temperature method.*

behaviorism A model of human development that explains behavior solely on the basis of that which can be observed.

bereavement The response to a loved one's death, including customs, rituals, and the grieving process.

bias A personal leaning or inclination.

Billings method See *mucus method.*

binuclear family A postdivorce family with children, consisting of the original nuclear family divided into two families, one headed by the mother, the other by the father; the two "new" families may be either single-parent or stepfamilies.

bipolar gender role model The traditional view of masculinity and femininity in which male and female gender roles are seen as polar opposites, with males possessing exclusively instrumental traits and females possessing exclusively expressive traits.

birth control Devices, drugs, techniques, or surgical procedures used to prevent conception or implantation or to terminate pregnancy.

birth plan A written plan made by an expectant mother or couple and shared with the birth attendant, detailing expectations regarding birth setting, medications, father's participation, visitors, circumcision, breastfeeding, and so on.

birth rate The number of births per year per thousand of population in a given community or group. Cf. *fertility rate.*

bisexuality Sexual involvement with both sexes, usually sequentially rather than during the same time period.

blastocyst An early stage of the fertilized ovum, containing about 100 cells, that implants itself in the uterine wall; a pre-embryo.

blended family A family in which one or both partners have a child or children from an earlier marriage or relationship; a stepfamily. See also *binuclear family.*

bondedness The degree of emotional bonding or closeness within a family.

boomerang generation Individuals who, as adults, return to their family home and live with their parents.

boundary In systems theory, the emotional, psychological, or physical separation between subsystems or roles (such as between family members) required for adequate functioning.

Braxton Hicks contractions Uterine contractions that occur periodically throughout pregnancy and also initiate effacement and dilation of the cervix at the beginning of labor.

breech presentation A fetal position in which the baby enters the birth canal buttocks or feet first.

bride price The goods, services, or money a family receives in exchange for giving their daughter in marriage.

brit milah In Judaism, the circumcision ceremony for an infant boy. Also known as *bris.*

bundling A colonial Puritan courtship custom in which a couple slept together with a board separating them.

C

calendar method A fertility awareness method based on calculating "safe" days according to the range of a women's longest and shortest menstrual cycles; also known as the *rhythm method.*

candidiasis A yeast infection caused by the *Candida albicans* organism; also called *moniliasis* and *yeast infection.*

caregiver role In family caregiving, the role of the person who provides the most ongoing physical work and decision making relating to the one who is being cared for.

case-study method In clinical research, the in-depth examination of an individual or small group in some form of psychological treatment in order to gather data and formulate hypotheses.

celibacy Abstinence.

cervical cap A thimble-shaped cap that fits snugly over the cervix (uterine opening) to prevent conception.

cervix The opening of the uterus within the vagina.

Cesarean section (C-section) Surgical delivery of the child through an incision in the mother's abdominal and uterine walls.

chancre A painless sore or ulcer that may be the first symptom of syphilis.

chastity The state of being morally or sexually pure.

Chicano [fem. **-a**] A Mexican American born in the United States.

child-free marriage A marriage in which the partners have chosen not to have children.

child neglect Failure to provide adequate or proper physical or emotional care for a child.

child sexual abuse Any sexual interaction, including fondling, erotic kissing, oral sex, or genital penetration, that occurs between an adult (or older adolescent) and a prepubertal child.

child snatching The kidnapping of one's own children, usually by the noncustodial parent.

child support Court-ordered financial support by the noncustodial parent to pay or assist in paying child-rearing expenses incurred by the custodial parent.

chlamydia A common sexually transmitted disease caused by *Chlamydia trachomatis* affecting the urinary tract or other organs.

chorionic villus sampling (CVS) A method of fetal diagnosis involving the surgical removal through the cervix of a tiny piece of embryonic membrane to be analyzed for genetic defects.

Christmas A Christian holiday commemorating the birth of Jesus, celebrated on December 25, around the time of the winter solstice.

circumcision The surgical removal of the foreskin of the penis.

Circumplex Model See *Family Circumplex Model.*

clan A group of families related along matrilineal or patrilineal descent lines, regarded as the basic family unit in some cultures.

clinical research The in-depth examination of an individual or small group in clinical treatment in order to gather data and formulate hypotheses. See also *case-study method.*

closed field A setting in which potential partners may meet, characterized by a small number of people who are likely to interact, such as a class, dormitory, or party. Cf. *open field.*

coalition Within a family, an alliance between two or more family members.

cognition The mental processes, such as thought and reflection, that occur between the moment we receive a stimulus and the moment we respond to it.

cognitive developmental theory A model of human development in which growth is viewed as the mastery of specific ways of perceiving, thinking, and doing that occur at discrete stages. See also *accommodation.*

cohabitation The sharing of living quarters by two heterosexual, gay, or lesbian individuals who are involved in an ongoing emotional and sexual relationship. The couple may or may not be married.

cohesion Emotional closeness or connectedness among family members.

cohort A group of persons experiencing a specific event at the same time, such as a birth cohort consisting of persons born in the same year. See also *generation.*

coitus The insertion of the penis into the vagina and subsequent stimulation; sexual intercourse.

coitus interruptus The withdrawal of the penis from the vagina immediately before ejaculation; considered ineffective as a contraceptive measure.

colostrum A nutritionally rich fluid secreted by the breasts the first few days following childbirth.

coming out For gay, lesbian, and bisexual individuals, the process of publicly acknowledging one's sexual orientation.

compadrazgo [Spanish] The Latino institution of godparentage (literally, co-parentage).

compadre [Spanish] The godfather of one's child (literally, co-father).

commuter marriage A marriage in which couples who prefer living together live apart in pursuit of separate goals.

companionate love A form of love emphasizing intimacy and commitment.

companionate marriage A marriage characterized by shared decision making and emotional and sexual expressiveness.

comparable worth An economic model arguing that occupations traditionally employing women are compensated at a lower rate than those traditionally employing men as a result of gender discrimination and that to overcome income differences between men and women, pay should be based on experience, knowledge and skills, mental demands, accountability, and working conditions, not on specific occupations per se.

complementary marriage model A model in which male employment outside the home and female work within the home are viewed as separate but interdependent.

conception The union of sperm and ovum; impregnation.

concubine In polygamous societies, a secondary wife.

condom A sheath made from latex rubber or animal intestine that fits over the erect penis to prevent the deposit of sperm in the vagina; used as a contraceptive device and also as protection against sexually transmitted diseases (latex condoms only).

conflict theory A social theory that views individuals and groups as being basically in competition with each other. Power is seen as the decisive factor in interactions.

congenital From birth.

conjugal extended family Extended family formed through marriages.

conjugal family Family consisting of husband, wife, and children. See also *nuclear family.*

conjugal relationship A relationship formed by marriage.

consanguineous extended family An extended family formed through blood ties.

consanguineous relationship A relationship formed by common bloodlines.

contraception Devices, techniques, or drugs used to prevent conception (fertilization of the ovum by a sperm).

contraction In childbirth, the action of the uterine muscles that open the cervix and expel the fetus. See also *labor.*

contraindication A symptom, sign, or condition that indicates that a particular drug or device should not be used.

conventionality In marriage and family research, the tendency of subjects to give conventional or conformist responses.

coping The process of utilizing resources in response to stress.

correlational study A study (clinical, survey, or observational) that measures, but does not manipulate, two or more naturally occurring variables.

courtship The process by which a commitment to marriage is developed.

couvade The psychological or ritualistic assumption of symptoms of pregnancy and childbirth by the male.

crisis A turning point; a crucial time, stage, or event. See also *predictable crisis* and *unpredictable crisis.*

crisis model A theoretical construct for studying family strengths, based on the family's ability to function during times of high stress.

C-section See *Cesarean section.*

cultivation theory In media research, a theory asserting that there are consistent images, themes, and stereotypes across all media genres that form a more or less consistent world view.

cultural relativity The view that the practices of a particular culture should be evaluated in terms of how they fit within the culture as a whole and not judged in terms of another culture's standards or values.

culture of poverty The view that the poor form a qualitatively different culture from the larger society and that their culture accounts for their poverty.

cunnilingus Oral stimulation of the female genitals.

custodial Having physical or legal custody of a child. Cf. *noncustodial.*

custody Legal responsibility for certain aspects of a child's well-being. See also *joint custody, joint legal custody, joint physical custody, sole custody,* and *split custody.*

cycle of violence According to Lenore Walker's research, the recurring three-phase battering cycle of (1) tension building, (2) explosion, and (3) reconciliation.

cystitis A urinary tract infection usually affecting women.

D

date rape Rape in which the assailant is personally known to the victim, usually in the context of a dating relationship. Also known as *acquaintance rape.*

dating A process in which two individuals meet to engage in activities together; dating may be either exclusive or nonexclusive.

Day of the Dead See *Dia de los Muertos.*

D&C Dilation and curettage.

D&E Dilation and evacuation.

death rate The number of deaths per year per thousand of population in a given community or group.

deferred parenthood The intentional postponement of child-bearing until after certain goals have been fulfilled.

demographics The demographic characteristics of a population, such as family size, marriage and divorce rates, and ethnic and racial composition.

demography The study of population and population characteristics, such as family size, marriage and divorce rates, and ethnic and racial composition.

denial The conscious or unconscious refusal to recognize painful acts, situations, or ways of being.

dependent variable A variable that is observed or measured in an experiment and may be affected by an independent variable.

desire Erotic sensations or feelings which motivate a person to seek out or receive sexual experiences.

developmental systems approach An approach to human development that recognizes the importance of the individual's interactions within a complex and changing family system and within the numerous systems of the larger society.

developmental task Appropriate activities and responsibilities individuals learn at different stages in the life cycle.

deviant Departing from social or cultural norms.

Dia de los Muertos, el [Spanish] The Day of the Dead, a Latino and Latin American holiday with Indian and Catholic roots honoring the dead, celebrated on November 1.

Diagnostic and Statistical Manual of Mental Disorders (DSM-IV) A manual published by the American Psychiatric Association establishing categories of psychiatric disorders and listing criteria for diagnosing such disorders.

diaphragm A flexible rubber cup placed in the vagina to block the passage of sperm, preventing conception.

dilation The opening up of the cervix.

dilation and curettage (D&C) First trimester abortion technique in which the embryo is removed from the uterus with a sharp instrument (curette).

dilation and evacuation (D&E) Second trimester abortion technique in which suction and forceps are used to remove the fetus.

direct sperm injection Fertilization technique involving the injection of a single sperm into an ovum in a laboratory dish and subsequent implantation of the blastocyst in the mother's uterus.

disability A physical or developmental limitation. Preferred usage over "handicap."

discrimination The process of acting differently toward a person or group because the individual or group belongs to a minority.

disengagement An extremely low level of family cohesion occurring when families do not feel close to each other and are unable to communicate or to carry out minimum family tasks.

displaced homemaker A full-time homemaker who has lost economic support from her husband as a result of divorce or widowhood.

distress Negative stress that may result in intense upset, mood changes, headaches, and muscular tension. Often used interchangeably with stress.

division of labor The interdependence of persons with specialized tasks and abilities. Within the family, labor is traditionally divided along gender lines. See also *complementary marriage model.*

divorce The legal dissolution of marriage. Cf. *separation.*

divorce mediation The process in which a mediator (counselor) assists a divorcing couple in resolving personal, legal, and parenting concerns in a cooperative manner.

domestic partners act Law granting certain legal rights similar to those of married couples to committed cohabitants, whether heterosexual, gay, or lesbian.

domestic partnership Cohabiting couples—lesbian, gay, or heterosexual—in committed relationships. Domestic partners are legally recognized in some cities and countries and have some of the protections enjoyed by married partners, such as shared insurance benefits.

double ABC-X model A model describing stress, in which Aa represents stressor pileup, B represents family coping resources, C represents perception of stressor pileup, and X represents outcome.

double standard of aging The devaluation of women in contrast to men in terms of attractiveness as they age.

douching Introducing water or liquid into the vagina for medical, hygienic, or contraceptive reasons.

Down syndrome A chromosomal error characterized by mental retardation.

dowry The property a woman brings to her husband upon marriage. [Archaic] A man's gift to his bride. Cf. *bride price.*

DSM-IV See *Diagnostic and Statistical Manual of Mental Disorders.*

dual-career family A type of dual-earner family in which both husband and wife are committed to careers.

dual-earner family A family in which both husband and wife are employed. Also known as *dual-worker family.*

dual-worker family A dual-earner family.

duration-of-marriage effect The accumulation over time of various factors, such as poor communication, unresolved conflicts, role overload, heavy work schedules, and child-rearing responsibilities, that negatively affect marital satisfaction.

dyad A two-member group; a couple.

Dyadic Adjustment Scale A survey instrument developed by Graham Spanier which measures relationship satisfaction.

dysfunction Impaired or inadequate functioning, as in sexual dysfunction.
dyspareunia Painful sexual intercourse.

E

eclampsia A potentially life-threatening condition of late pregnancy brought on by untreated high blood pressure.
economic adequacy The psychological perception that one has sufficient income and economic resources.
economic distress Stressful aspects of the economic life of individuals or families, including unemployment, poverty, and worrying about money.
economies of scale Situations in which an economic unit, such as a family, is able to economize because the cost per individual goes down as the number of individuals increases.
ectopic pregnancy A pregnancy in which the fertilized egg implants in a fallopian tube instead of in the uterus; tubal pregnancy.
edema Fluid retention and swelling.
effacement Thinning of the cervix during labor.
egalitarian gender roles Gender roles in which men and women are treated equally.
ego In psychoanalytic theory, the part of the personality that is rational and mediates between the demands of the id and the constraints imposed by society. See also *id* and *superego.*
egocentric fallacy The mistaken belief that one's own personal experience and values are those of others in general.
ejaculate [noun] Semen. [verb] To expel semen.
ejaculation The expulsion of semen during orgasm.
embryo In human beings, the early development of life between about one week and two months after conception.
emission The first stage of male orgasm during which the semen moves into the urethra.
embryo transplant The implantation of the blastocyst into the mother's (or surrogate's) uterus following a fertilization procedure such as in vitro fertilization or direct sperm injection.
empty nest The experience of parents when the last grown child has left home. The "empty nest syndrome," in which the mother becomes depressed after the children have gone, is believed to be more of a myth than a reality.
endogamy Marriage within a particular group. Cf. *exogamy.*
engagement A pledge to marry.
enmeshment An extremely high level of family cohesion resulting from overidentification with the family and resulting in a lack of individual independence.
episiotomy An incision from the vagina toward the anus made during childbirth.
equity theory A theory emphasizing that social exchanges must be fair or equally beneficial over the long run.
erectile dysfunction Inability or difficulty in achieving erection.
erection An erect penis.
eros 1. From the Greek *eros* [love], the fusion of love and sexuality. 2. According to sociologist John Lee's styles of love, the passionate love of beauty.
erotic Pertaining to sexuality, sensuality, or sexual sensations.
ethnic group A large group of people distinct from others because of cultural characteristics, such as language, religion, and customs, transmitted from one generation to another. See also *minority group* and *racial group.*
ethnic stratification The hierarchical ranking of groups in superior and inferior positions according to ethnicity.
ethnicity Ethnic affiliation or identity.
ethnocentric fallacy The belief that one's own ethnic group, nation, or culture is inherently superior to others. See also *ethnocentrism* and *racism.*
ethnocentrism The emotionally charged belief that one's ethnic group, nation, or culture is superior to all others. See also *ethnocentric fallacy* and *racism.*
eustress Good or positive stress that results in a person feeling motivated or inspired.
exchange theory See *social exchange theory.*
excitement phase In the sexual response cycle, the second stage, denoting sexual arousal.
exogamy Marriage outside a particular group. Cf. *endogamy.*
experimental research Research method involving the isolation of specific factors (variables) under controlled circumstances to determine the effects of each factor.
expressive trait A supportive or emotional personality trait or characteristic.
extended family The family unit of parent(s), child(ren), and other kin, such as grandparents, uncles, aunts, and cousins. See also *conjugal extended family* and *consanguineous extended family.*
extended household A household composed of several different families.
extrafamilial sexual abuse Child sexual abuse that is perpetrated by nonrelated individuals. Cf. *intrafamilial sexual abuse.*
extramarital sex Sexual activities, especially sexual intercourse, occurring outside the marital relationship.

F

FAE See *fetal alcohol effect.*
fallacy A fundamental error in reasoning that affects our understanding of a subject.
familialism A pattern of social organization in which family loyalty and strong feelings for the family are important.
family A unit of two or more persons, of which one or more may be children who are related by blood, marriage, or affiliation and who cooperate economically and may share a common dwelling place.
Family Circumplex Model Model of family functioning in which cohesion, adaptability, and communication are the most important dimensions.
family ecosystem Family interactions and adaptations with the larger social environment, such as schools, neighborhoods, or the economy.
family hardship Difficulties specifically associated with a stressor, such as loss of income in the case of unemployment. See also *stressor.*
family life cycle A developmental approach to studying families, emphasizing the family's changing roles and relationships at various stages, beginning with marriage and ending when both spouses have died.

family of cohabitation The family formed by two people living together whether married or unmarried; may include children or stepchildren.

family of marriage The family formed through marriage. See also *family of procreation.* Cf. *family of orientation.*

family of orientation The family in which a person is reared as a child. Cf. *family of procreation.*

family of origin See *family of orientation.*

family of procreation The family formed by a couple and their child or children. See also *family of cohabitation.*

family policy A set of objectives concerning family well-being and specific measures initiated by government to achieve them.

family power Power exercised by individuals in their family roles as mother, father, child, or sibling.

family role A social role within the family, such as husband or wife, father or mother. See also *kinship system.*

family rule A family's patterned or characteristic response to events, situations, or persons. See also *family systems theory, meta-rule,* and *hierarchy of rules.*

family strengths Those relationship patterns, interpersonal skills, and social and psychological characteristics that create fulfillment and satisfaction for a family as individuals and as a whole.

family stress An upset in the steady functioning of the family.

family systems theory A theory viewing family structure as created by the pattern of interactions between its various subsystems, and individual actions as being strongly influenced by the family context.

family tree Diagrammatic representation of family and ancestors.

family work The unpaid work that is undertaken by family members to sustain the family, such as housework, laundry, shopping, yard maintenance, budgeting and bill-paying, and care of children, the sick, and the elderly.

FAS See *fetal alcohol syndrome.*

fecundity A person's maximum biological capacity to reproduce.

feedback In communication, an ongoing process in which participants and their messages produce a result and are subsequently modified by the result.

fellatio Oral stimulation of the male genitals.

feminism 1. The principle that women should have equal political, social, and economic rights with men. 2. The social movement to obtain for women political, social, and economic equality with men.

feminization of poverty The shift of poverty to females, primarily as a result of high divorce rates and births to unmarried women.

feral child A child purportedly nursed and reared in the wild by animals such as wolves, bears, or lions.

fertility The ability to conceive; a person's actual reproductive performance.

fertility awareness methods Contraceptive method based on predicting a woman's fertile period and either avoiding intercourse or using an additional method of contraception during that interval.

fertility rate In a given year, the number of live births per 1,000 women aged 15–44 years. See also *birth rate.*

fertilization Union of the egg and sperm. Also known as *conception.*

fetal alcohol effect (FAE) Growth retardation caused by the mother's chronic ingestion of alcohol during pregnancy.

fetal alcohol syndrome (FAS) A syndrome that may be characterized by unusual facial characteristics, small head and body size, poor mental capacities, and abnormal behavior patterns and is caused by the mother's chronic ingestion of alcohol during pregnancy.

fetus The unborn young of a vertebrate; the human embryo becomes a fetus at about the eighth week following fertilization.

field of eligibles A group of individuals of the same general background and age who are culturally approved potential marital partners.

filial crisis Psychological conflict and stress experienced by adult children when aged parents become dependent on them.

flextime Flexible work schedules determined by employee/employer agreement.

foreplay Erotic activity prior to coitus, such as kissing, caressing, sex talk, and oral/genital contact; petting.

foreskin The sleeve of skin covering the tip of the penis; prepuce.

friendship An attachment between people; the foundation for a strong love relationship.

G

gamete A reproductive cell (sperm or ovum) that can unite with another gamete to form a zygote.

gamete intrafallopian transfer (GIFT) Fertilization technique in which sperm and ova are collected from the parents and deposited together in the mother's fallopian tube for fertilization.

gay Pertaining to same-sex relationships, especially among males.

gay male A male sexually oriented toward other males. Cf. *lesbian.*

gender The division into male and female, often in a social sense; sex.

gender differences The orienting focus in most feminist writing, research, and advocacy.

gender identity The psychological sense of whether one is male or female.

gender role The culturally assigned role that a person is expected to perform based on male or female gender.

gender-role attitude A personal belief regarding appropriate male and female personality traits and behaviors.

gender-role behavior An actual activity or behavior in which males or females engage according to their gender role.

gender-role stereotype A rigidly held and oversimplified belief that all males and females possess distinctive psychological and behavioral traits as a result of their gender.

gender schema The cognitive organization of individuals, behaviors, traits, objects, and such by gender.

gender theory A theory in which gender is viewed as the basis of hierarchal social relations that justify greater power to males.

generation 1. A group of people born and living during the same general time period. 2. The approximately 30-year period between the birth of one generation and the next. See also *cohort.*

genital herpes A sexually transmitted disease caused by the herpes simplex virus type II, similar to cold sores or fever blisters but appearing on the genitals.

genitalia The external sex or reproductive organs; genitals.

genital warts Warts on the genitals caused by human papilloma virus (HPV), a sexually transmitted virus.

genogram A diagram of the emotional relationships of a family through several generations.

gestation The period of carrying young in the uterus from conception to birth.

gestational edema-proteinuria-hypertension complex A condition of pregnancy characterized by high blood pressure, toxemia, or preeclampsia.

getting together A courtship process in which men and women congregate ("get together") in groups to socialize or engage in common activities or projects. Cf. *dating.*

ghetto poor Inner-city residents, primarily African American and Latino, who live in poverty.

GIFT See *gamete intrafallopian transfer.*

gonorrhea A sexually transmitted disease caused by the *Neisseria gonorrhoeae* bacterium that initially infects the urethra in males and the cervix in females or the throat or anus in either sex, depending on the mode of sexual interaction.

grandparenting Performing the functions of a grandparent.

grieving process Emotional responses and expressions of feeling over the death of a loved one.

H

halo effect The tendency to infer positive characteristics or traits based on a person's physical attractiveness.

Hana Matsuri The Japanese celebration of the Buddha's birthday; literally, Flower Festival.

handicap See *disability.*

health The state of physical and mental well-being.

Heger's sign A softening of the uterus just above the cervix which may be an early indication of pregnancy.

hepatitis A liver disease causing blood pigments to accumulate. Hepatitis B (serum hepatitis) may be sexually transmitted.

herpes simplex type II See *genital herpes.*

heterogamy Marriage between those with different social or personal characteristics. Cf. *homogamy.*

heterosexuality Sexual orientation toward members of the other sex.

heterosociality Close association with members of the other sex.

Hispanic Of Spanish or Latin American origin or background; may be of any race. See also *Latino.*

HIV See *human immunodeficiency virus.*

HIV-positive Infected with human immunodeficiency virus.

Holī A spring festival originating in Northern India.

homemaker role A family role usually allocated to women, in which they are primarily responsible for home management, child rearing, and the maintenance of kin relationships. Traditionally the role is associated with economic dependency and has primacy over other female roles.

homeostasis A social group's tendency to maintain internal stability or balance and to resist change.

homoeroticism Erotic attraction to members of the same sex.

homogamy Marriage between those with similar social or personal characteristics. Cf. *heterogamy.*

homophobia Irrational or phobic fear of gay men and lesbians.

homosexuality Sexual orientation toward members of the same sex. See also *gay male* and *lesbian.*

homosociality The tendency to associate mostly with members of the same sex.

honeymoon effect The tendency of newly married couples to overlook problems, including communication problems.

hormone A chemical substance, secreted by the endocrine glands into the bloodstream, which organizes and regulates physical development.

hospice A place or program caring for the terminally ill, emphasizing both patient care and family support.

hostile environment An environment created through sexual harassment in which the harassed person's ability to learn or work is negatively influenced by the harrasser's actions.

housewife 1. A wife who manages the home. 2. A homemaker.

human chorionic gonadotropin (HCG) A hormone secreted by the placenta that helps sustain pregnancy.

human immunodeficiency virus (HIV) The virus causing AIDS.

human papilloma virus (HPV) The virus causing warts, including genital warts.

hypoactive sexual desire Inactive or limited sexual desire. Also known as *inhibited sexual desire.*

hypothesis An unproven theory or proposition tentatively accepted to explain a collection of facts.

hysterectomy The surgical removal of the uterus or part of the uterus.

hysterotomy A surgical method of abortion in which the fetus is removed through an abdominal incision.

I

id In psychoanalytic theory, the part of the personality that seeks to gratify pleasurable needs, especially sexual ones. See also *ego* and *superego.*

identity An individual's core sense of self.

identity bargaining The process of role adjustment in a relationship, involving identifying with a role, having the role validated by others, and negotiating with the partner to make changes in the role.

ie [EE eh] The basic family unit in traditional Japanese society consisting of past, present, and future members of the extended family and their households.

illegitimate 1. Not based on law, right, or custom. 2. Born outside of marriage (sometimes used in a derogatory manner).

incest Sexual intercourse between individuals too closely related to marry, usually interpreted to mean father/daughter, mother/son, or brother/sister. See also *intrafamilial sexual abuse.*

independent variable A variable that may be changed or manipulated in an experiment.

induced abortion The termination of a pregnancy through human intervention.

induction The formation of arguments whose premises are intended to provide some (but not conclusive) support for their conclusions.

institution An enduring social structure built around a significant and distinct cluster of social values. Institutions include the family, religion, education, and government.

instrument In social science, a research tool or device, such as a questionnaire, used to gather data about behaviors, attitudes, beliefs, or other such dimensions of an individual, group, or society.

instrumental trait A practical or task-oriented personality trait or characteristic.

interaction In communication, a reciprocal act that takes place between at least two people.

intermarriage Marriage between people of different ethnic or racial groups.

intermittent extended family The family that is formed when a family takes in other relatives in times of need.

interpersonal competence The ability to develop and share an intimate, growing relationship.

interrole conflict Conflict experienced when the role expectations of two or more roles are contradictory or incompatible. Also known as *role interference.* See also *role strain.*

intrafamilial sexual abuse Child sexual abuse that is perpetrated by related individuals, including steprelatives. See also *incest.* Cf. *extrafamiliar sexual abuse.*

intrauterine device (IUD) A device inserted into the uterus to prevent conception or implantation of the fertilized egg.

involution Following childbirth, the contracting process—over a period of about six weeks—by which the uterus returns to its prebirth state.

Issei First-generation Japanese (born in Japan).

I-statement In communication, a statement beginning with "I" that describes the speaker's feelings, such as "I feel upset when I see last week's dishes in the sink."

IUD See *intrauterine device.*

IUI See *intrauterine insemination.*

J

jealousy An aversive response occurring because of a partner's or other significant person's real, imagined, or likely involvement with or interest in another person.

joint custody Custody arrangement in which both parents are responsible for the care of the child. Joint custody takes two forms: *joint legal custody* and *joint physical custody.* See also *sole custody* and *split custody.*

joint legal custody Joint custody in which the child lives primarily with one parent but both parents jointly share in important decisions regarding the child's education, religious training, and general upbringing.

joint physical custody Joint custody in which the child lives with both parents in separate households and spends more or less equal time with each parent.

Juneteenth An African-American holiday commemorating freedom and celebrating black history and accomplishments.

K

kaddish In Judaism, a form of prayer.

kin Relatives.

kinship Family relationship.

kinship system The social organization of the family conferring rights and obligations based on an individual's status.

Kwanzaa An African-American harvest festival, observed between Christmas and New Year's Day and celebrating the culture and heritage of American blacks.

L

labor The physical efforts of childbirth.

lactation The production of milk.

Lamaze method A childbirth method in which the mother uses exercises and breathing techniques to assist her labor.

laparoscopy Tubal ligation technique using a laparoscope (viewing instrument) to locate the fallopian tubes, which are then closed or blocked.

latchkey children See *self-care.*

Latino [fem. **-a**] A person of Latin American origin or ancestry; may be of any race. See also *Hispanic* and *Spanish-speaking.*

legitimacy The state or quality of being sanctioned by custom, rights, or law.

lesbian A female sexually oriented toward other females.

lesbian separatist A lesbian interested in creating a separate "womyn's" culture distinct from both heterosexual and gay culture. The lesbian separatist movement was strongest in the late 1960s through the early 1980s.

life course A developmental perspective of individual change focusing on (1) individual time, an individual's own life span, (2) social time, social transition points, such as marriage, and (3) historical time, the times in which a person lives.

life cycle The developmental stages, transitions, and tasks individuals undergo from birth to death.

lochia A bloody vaginal discharge that appears for several weeks following childbirth.

low birth weight (LBW) Generally, a weight of less than 5.5 pounds at birth, often as a result of prematurity.

ludus [LOO dus] According to sociologist John Lee's styles of love, playful love.

M

macho [Spanish] In traditional Latin American usage, masculine, strong, or daring. In popular U.S. usage, excessively or stereotypically masculine.

madrina [Spanish] Godmother.

majority group A social category composed of people holding superordinate status and power and having the ability to impose their will on less powerful minority groups. Cf. *minority group.*

mania According to sociologist John Lee's styles of love, obsessive love.

marital disruption Marital instability that includes marital separation as well as divorce.

marianismo [Spanish] In Latin American culture, the idealized mother role as represented by the Virgin Mary.

marital exchange The process by which individuals trade resources with each other to secure the best marital partner. Traditionally, men exchanged their higher status and greater economic resources for women's physical attractiveness, expressive qualities, and childbearing and housekeeping abilities.

marital power The power exercised by individuals as husband or wife. Cf. *family power.*

marital rape Forced sexual contact by a husband with his wife; legal definitions of marital rape differ among states.

marriage The legally recognized union between a man and woman in which economic cooperation, legitimate sexual interactions, and the rearing of children may take place.

marriage contract 1. The legal and moral rights and responsibilities entailed by marriage. 2. An explicit contract delineating specific terms of marriage which, depending on the terms, may be legally binding. 3. A nonlegally binding agreement between partners, covering such areas as conflict resolution, division of household labor, employment, and child-rearing responsibilities.

marriage gradient The tendency for men to marry younger women of lower socioeconomic status and for women to marry older men of higher socioeconomic status. See also *marriage squeeze.*

marriage market An exchange process in which individuals bargain with each other using their resources in order to find the best available partner for marriage. See also *marital exchange.*

marriage squeeze The phenomenon in which there are greater numbers of marriageable women than marriageable men, particularly among older women and African-American women. See also *marriage gradient.*

masturbation Manual or mechanical stimulation of the genitals by self or partner; a form of autoeroticism.

matriarchal Pertaining to the mother as the head and ruler of a family. Cf. *patriarchal.*

matriarchy A form of social organization in which the mother or eldest female is recognized as the head of the family, kinship group, or tribe, and descent is traced through her. Cf. *patriarchy.*

matrilineal Descent or kinship traced through the mother. Cf. *patrilineal.*

mean world syndrome The belief, resulting from television viewing, that the world is more dangerous and violent than it is in actuality.

menarche [MEN ar kee] The first menstrual period, beginning in puberty.

menopause Cessation of menses for at least one year as a result of aging.

menses The monthly menstrual flow.

menstruation The discharge of blood and built-up uterine lining through the vagina that occurs approximately every four weeks among nonpregnant women between puberty and menopause.

meta-analysis The reanalysis of combined statistical data from previous studies.

meta-rule An abstract, general, unarticulated rule at the apex of the hierarchy of rules upon which other rules are based.

Mexican American A U.S. citizen of Mexican ancestry.

midwife A person who attends and facilitates birth.

mifepristone See *RU-486.*

minority group A social category composed of people whose status places them at economic, social, and political disadvantage. Cf. *majority group.* See also *ethnic group.*

minority status Social rank having unequal access to economic and political power.

miscarriage A spontaneous abortion.

model 1. A person who demonstrates a behavior observed and imitated by others. 2. Prototype.

modeling The process of teaching or learning using imitation.

monogamy 1. The practice of having only one husband or wife at a time. 2. [colloq.] Sexual exclusiveness.

morality A set of social, cultural, or religious norms defining right and wrong.

morning-after pill See *postcoital birth control.*

morning sickness Nausea experienced by many women during the first trimester of pregnancy.

mucus method A contraceptive method that relies on predicting a woman's fertile period by observing changes in the appearance and character of her cervical mucus; also called *Billings* or *ovulation method.*

N

natural childbirth See *prepared childbirth.*

natural family planning A fertility awareness method of birth control that relies solely on predicting a woman's fertile period and avoiding intercourse on those days.

neonate The newborn infant.

neurosis A psychological disorder characterized by anxiety, phobias, and so on.

neustress Stimuli that have no consequential effect.

Nisei Second-generation Japanese, whose parents were born in Japan.

nocturnal orgasm Involuntary orgasm occurring in both females and males during sleep, usually accompanied by erotic dreaming. In males, it is usually accompanied by ejaculation ("wet dream").

no-fault divorce The dissolution of marriage because of irreconcilable differences for which neither party is held responsible.

nonbasic conflict Pronounced disagreement about nonfundamental or situational issues. Cf. *basic conflict.*

noncustodial Not having physical or legal custody of a child. Cf. *custodial.*

nonmarital sex Sexual activities, especially sexual intercourse, that take place among older single individuals. Cf. *premarital sex* and *extramarital sex.*

nonverbal communication Communication of emotion by means other than words, such as touch, body movement, and facial expression.

norm A cultural rule or standard.

normal Conforming to group or cultural norms.

normative Establishing or representative of a norm or standard.

nuclear family The basic family building block, consisting of a mother, father, and at least one child; in popular usage, used interchangeably with *traditional family.* Some anthropologists argue that the basic nuclear family is the mother and child dyad.

nursing Breastfeeding.

O

objectivity Suspending the beliefs, biases, or prejudices we have about a subject until we have really understood what is being said.

observational research Research method using unobtrusive, direct observation.

obstetrician A physician specializing in pregnancy and childbirth.

occupational stratification The hierarchal ranking of jobs in superior and inferior positions based on pay and status.

open adoption A form of adoption in which the birth mother has an active part in choosing the adoptive parents; there is a certain amount of information exchanged between the birth mother and the adoptive parents, and there may be some form of continuing contact between the birth mother and the child or adoptive family following adoption.

open field A setting in which potential partners may not be likely to meet, characterized by large numbers of people who do not ordinarily interact, such as a beach, shopping mall, or large university campus. Cf. *closed field.*

open marriage Marriage in which the partners agree to allow one another to have openly acknowledged and independent sexual relationships outside the marriage.

operant conditioning A behavioral technique that uses a reinforcing stimulus to increase the frequency of a desired behavior.

opinion An unsubstantiated belief or conclusion based on personal values or biases.

opportunistic disease An infection that is normally resisted by the healthy immune system, such as Kaposi's sarcoma and *Pneumocystis carinii* pneumonia associated with AIDS.

oral contraception Contraceptive taken orally; the pill.

oral-genital sex The erotic stimulation of the genitals by the tongue or mouth; fellatio, cunnilingus, or mutual oral stimulation.

orgasm The release of physical tensions after the build up of sexual excitement; usually accompanied by ejaculation in physically mature males.

orgasmic dysfunction Inability to have orgasm.

orgasmic phase The phase of the sexual response cycle characterized by orgasm.

ovulation method See *mucus method.*

ovum The egg produced by the ovary. [plural] Ova.

P

Pacific Islander Collective term referring to those of native Hawaiian, Fijian, Guamanian, Samoan, or other Melanesian, Micronesian, or Polynesian descent.

padrino [Spanish] Godfather.

Pap smear See *Pap test.*

Pap test The sampling of cervical cells to diagnose cancer or a precancerous condition.

parenting The rearing of children.

parturition The process of childbirth.

passing [colloq.] Pretending to be heterosexual when actually gay or lesbian.

passionate love Intense, impassioned love. Cf. *companionate love.*

patriarchal Pertaining to the father as the head and ruler of a family. Cf. *matriarchal.*

patriarchy A form of social organization in which the father or eldest male is recognized as the head of the family, kinship group, or tribe, and descent is traced through him. Cf. *matriarchy.*

patrilineal Descent or kinship traced through the father. Cf. *matrilineal.*

pedophilia Adult sexual attraction to prepubescent children that is intense and recurring; an adult's use of children for sexual purposes.

peer A person of equal status, as in age, class, position, or rank.

permissive child rearing A parenting style stressing the child's autonomy and freedom of expression, often over the needs of the parents.

permissiveness with affection Sexual norm permitting nonmarital sexual activity for both men and women in an affectionate relationship.

permissiveness without affection Sexual standard permitting nonmarital sexual activity without regard as to the nature of the relationship.

personality conflict Conflict based on personality characteristics; such conflicts are unlikely to be resolved. Cf. *situational conflict.*

Pesach A Jewish festival celebrated after the first full moon of spring, commemorating the flight of Moses and the Jews from Egypt; the Feast of Passover.

petting Foreplay; sexual contact usually referring to the manual or oral stimulation of the genitals or breasts.

phallus Penis.

phenotype A set of genetically determined anatomical and physical characteristics, such as skin and hair color and facial structure.

physical abuse Intentional violent mistreatment. See *violence.*

PID See *pelvic inflammatory disease.*

placenta The organ of exchange between the fetus and the pregnant female through which nutrients and waste pass. It also serves as an endocrine gland producing large amounts of the hormones progesterone and estrogen to maintain pregnancy; it is attached to the mother's uterine wall and connected to the fetus by the umbilical cord.

pleasuring The giving and receiving of sensual pleasure through nongenital touching.

plural marriage The practice of having more than one husband or wife at the same time; polygamy.

polyandry The practice of having more than one husband at the same time. See also *polygamy;* cf. *polygyny.*

polygamy The practice of having more than one husband or wife at the same time; plural marriage. See also *polyandry, polygyny,* and *consanguineous extended family.*

polygyny The practice of having more than one wife at the same time. See also *polygamy;* cf. *polyandry.*

POSSLQ Person of opposite sex sharing living quarters in U.S. Census Bureau terminology.

postcoital birth control Birth control that is administered after intercourse has taken place but before a diagnosis of pregnancy is possible; usually involves the administering of high-estrogen oral contraceptives. Also called "morning-after birth control."

postmarital sex Sexual intercourse among previously married individuals. See also *nonmarital sex.*

postpartum period A period of about three months following childbirth during which critical family adjustments are made.

power The ability to exert one's will, influence, or control over another person or group.

power conflict Pronounced disagreements concerning dominance.

powwow A Native America intertribal social gathering centering around drumming and traditional dances.

pragma According to sociologist John Lee's styles of love, practical love.

predictable crisis Within the individual or family life cycle, normal but critical events, such as birth or death (of the elderly). Cf. *unpredictable crisis.*

preeclampsia Increasingly high blood pressure during late pregnancy; if untreated, it may lead to eclampsia. See also *gestational edema-proteinuria-hypertension complex.*

pre-embryo See *blastocyst.*

premarital sex Sexual activities, especially sexual intercourse, prior to marriage, especially among young, never-married individuals.

prematurity Birth before the normal gestation period has elapsed, often complicated by low birth weight.

prenatal Before birth.

prepared childbirth Birth philosophy stressing education and minimal use of anesthetics or other drugs; natural childbirth.

principle of least interest A theory of power in which the person less interested in sustaining a relationship has the greater power.

pro-choice movement Social movement that advocates women's right to choose abortion.

profamily movement A social movement emphasizing conservative family values, such as traditional gender roles, authoritarian child rearing, premarital virginity, and opposition to abortion.

pro-life movement Social movement that advocates against abortion. Also known as *anti-abortion movement.*

prophylactic [adjective] Protecting against disease. [noun] Condom.

prostate gland A gland at the base of a man's bladder which produces most of the seminal fluid in the ejaculate.

prototype In psychology, concepts organized into a mental model.

proximity Nearness to another in terms of both physical space and time.

pseudokin See *affiliated kin.*

psychoanalytic theory The Freudian model of personality development, in which maturity is seen as the ability to gain control over one's unconscious impulses.

psychosexual Pertaining to the psychological aspects of sexuality.

psychosexual development The growth of the psychological aspects (such as attitudes and emotions) of sexuality that accompany physical growth.

psychosocial theory A theory of human psychological development that emphasizes the role of family and society in such development.

puberty The period in which the individual develops secondary sex characteristics and becomes capable of reproduction.

PWA Person with AIDS.

Q

qualitative research Small groups or individuals are studied in an in-depth fashion.

quantitative research Samples taken from a great number of persons.

quickening The time of first movement of the fetus which may be felt by the pregnant woman.

quinceañera, la [Spanish] The traditional Latin American celebration of a girl's fifteenth birthday, formally introducing her into society.

R

racial group A large group of people defined as distinct because of their phenotype (genetically transmitted anatomical and physical characteristics, especially facial structure and skin color). Cf. *ethnic group.*

racism The practice of discrimination and subordination based on the belief that race determines character and abilities. See also *ethnocentric fallacy* and *ethnocentrism.*

Ramadan A month-long Islamic holiday commemorating the prophet Mohammed's illumination; observed by extensive fasting and prayer, followed by feasting and celebration at the month's end.

rape Sexual act against a person's will or consent as defined by law, usually including sexual penetration by the penis or other object; it may not, however, necessarily include penile penetration of the vagina. Also known as *sexual assault.* See also *acquaintance rape* and *marital rape.*

rape trauma syndrome A group of symptoms experienced by a rape survivor, including fear, self-blame, anxiety, crying, sleeplessness, anger, or rage.

rating and dating game The process described by sociologist Willard Waller in which men and women rate potential dates on a scale of one to ten and then try to date the highest-rated individuals.

reactive jealousy Jealousy that occurs when a partner's past, present, or anticipated involvement with another is revealed. Cf. *suspicious jealousy.*

reconstituted family See *stepfamily.*

refractory period Following orgasm, the period during which the penis cannot respond to additional stimulation.

reinforcement The process of influencing (increasing or decreasing) a behavior by adding or withholding a stimulus.

relative love and need theory A theory of power in which the person gaining the most from a relationship is the most dependent.

relocation camps During World War II, camps in which Japanese Americans of all ages were imprisoned without cause by the U.S. government.

remarriage A marriage in which one or both partners have been previously married.

reproductive organs External and internal structures involved in reproduction.

resource Anything that can be called into use or used to advantage, such as love, money, or approval, to exert influence or power.

rhythm method See *calendar method.*

Roe v. Wade U.S. Supreme Court decision (1973) affirming a woman's constitutional right to abortion based on the right to privacy.

role The pattern of behavior expected of a person in a group or culture as a result of his or her social position, such as husband or wife in a family.

role conflict See *interrole conflict.*

role interference See *interrole conflict.*

role modeling A significant means by which children are taught role attitudes and behavior by learning to imitate adults whom they admire.

role overload The experience of having more prescribed activities in one or more roles than can be comfortably or adequately performed. See also *role strain.*

role strain Difficulties, tensions, or contradictions experienced in performing a role, often because of multiple role demands. See also *interrole conflict* and *role overload.*

roleless role A role for which there are no clear guidelines for behavior, such as stepparent, widow, and ex-in-law roles.

romantic love Intense, passionate love. Cf. *companionate love.*

RU-486 An effective oral, postcoital birth control method containing mifepristone.

rule of thumb Prior to the nineteenth century, the legally sanctioned practice of disciplining one's wife with a rod, provided it was not wider than the husband's thumb.

S

safer sex Sexual practices, including the use of latex condoms, intended to prevent the transmission of bodily fluids, especially semen, that may contain HIV.

salpingitis Infection of a fallopian tube. See *pelvic inflammatory disease.*

sample A group randomly and systematically selected from a larger group.

sandwich generation Individuals and families who care for both their own children and their aging parents at the same time.

Sansei Third-generation Japanese, whose grandparents were born in Japan.

scapegoating The conscious or unconscious singling out and blaming of an individual or group.

schema The cognitive organization of knowledge according to particular criteria.

scientific method Method of investigation in which a hypothesis is formed on the basis of impartially gathered data and is then tested empirically.

script A mental map, plan, or pattern of behavior. See also *role.*

secondary data analysis Use of research gathered by public sources of information.

secondary sex characteristic A physical characteristic other than external genitals that distinguishes the sexes from each other, e.g., breasts and body hair.

self-care Children under age fourteen caring for themselves at home without supervision by an adult or older adolescent.

self-disclosure The revelation of deeply personal information about oneself to another.

self-esteem Feelings about the value of the self; high self-esteem includes feeling unique, having a sense of power, and feeling connected to others.

semen The fluid containing sperm which is ejaculated, produced mostly by the prostate gland. Also known as *ejaculate.*

seminal fluid The fluid containing sperm; semen.

separation The state or condition of a married couple who have chosen to live together no longer. Cf. *divorce.*

separation distress A psychological state following separation that may be characterized by depression, anxiety, intense loneliness, or feelings of loss.

sequential work/family role staging A pattern of combining employment and family work in dual-earner families in which women leave employment during pregnancy and while their children are young and return to it at a later time.

sex 1. Biologically, the division into male and female. 2. Sexual activities.

sex hormones Hormones such as testosterone and estrogen which are responsible for the development of secondary sex characteristics and for activating sexual behavior.

sexism 1. The belief that biological differences between males and females provide legitimate bases for female subordination. 2. The economic and social domination of women by men.

sex organs Internal and external reproductive organs; commonly refers only to the penis, vulva, and vagina.

sex ratio The ratio of men to women in a group or society.

sex role See *gender role.*

sexual assault A legal term referring to rape. See also *acquaintance rape*, *marital rape*, and *rape.*

sexual behavior Behavior that is characterized by conscious psychological/erotic arousal (such as desire) and that may also be accompanied by physiological arousal (such as erection or lubrication) or activity (such as masturbation or coitus).

sexual desire The psychological component that motivates sexual behavior.

sexual dysfunction Recurring problems in sexual functioning that cause distress to the individual or partner; may have a physiological or psychological basis.

sexual enhancement Any means of improving a sexual relationship, including developing communication skills, fostering a positive attitude, giving a partner accurate and adequate information, and increasing self-awareness.

sexual harassment Deliberate or repeated unsolicited verbal comments, gestures, or physical contact that is sexual in nature and unwelcomed by the recipient. Two types of sexual harassment involve (1) the abuse of power and (2) the creation of a hostile environment. See also *hostile environment.*

sexual identity The individual's sense of his or her sexual self.

sexual intercourse Coitus; heterosexual penile/vaginal penetration and stimulation.

sexual orientation Sexual identity as heterosexual, gay, lesbian, or bisexual.

sexual script A culturally approved set of expectations as to how one should behave sexually as male or female and as heterosexual, gay, or lesbian.

sexuality The state of being sexual, which encompasses the biological, social, and cultural aspects of sex.

sexually transmissible disease (STD) An infection that can be transmitted through sexual activities, such as sexual intercourse, oral/genital sex, or anal sex.

sexual preference See *sexual orientation.*

sexual stratification The hierarchical ranking in superior and inferior positions according to gender.

sexual variation A departure from sexual norms; atypical sexual behavior.

shiva In Judaism, a seven-day period of mourning for the dead during which certain austerities are practiced.

sibling A brother or sister.

SIDS See *sudden infant death syndrome.*

single-parent family A family with children, created by divorce or unmarried motherhood, in which only one parent is present. A family consisting of one parent and one or more children.

situational conflict Conflict arising as the result of specific acts, events, behaviors, or situations, which are amenable to resolution. Cf. *personality conflict.*

social exchange theory A theory which emphasizes the process of mutual giving and receiving of rewards, such as love or sexual intimacy, in social relationships, calculated by the equation Reward - Cost = Outcome.

social integration The degree of interaction between individuals and the larger community.

social learning theory A theory of human development that emphasizes the role of cognition (thought processes) in learning.

social role A socially established pattern of behavior that exists independently of any particular person, such as the husband or wife role or the stepparent role.

social support Instrumental and emotional assistance, such as physical care and love.

socialization The shaping of individual behavior to conform to social or cultural norms.

socioeconomic status Social status ranking determined by a combination of occupational, educational, and income levels.

sole custody Child custody arrangement in which only one parent has both legal and physical custody of the child. See also *joint custody* and *split custody.*

sonogram The image produced by ultrasonography.

Spanish-speaking Pertaining to Hispanic origin or ancestry.

spells of poverty The periodic movement in and out of poverty.

sperm The male gamete produced by the testis.

spermicide A substance toxic to sperm and used for contraception.

spinnbarkeit The elastic condition of cervical mucus just prior to ovulation.

spirit marriage In Canton, China, a marriage of two deceased persons, arranged by their families to provide family continuity.

split custody Custody arrangement when there are two or more children in which custody is divided between the parents, the mother generally receiving the girls and the father receiving the boys.

spontaneous abortion The natural but fatal expulsion of the embryo or fetus from the uterus; miscarriage.

status The position an individual occupies within a social hierarchy.

STD Sexually transmissible disease.

stepfamily A family in which one or both partners have a child or children from an earlier marriage or relationship. Also known as a *blended family.* See also *binuclear family.*

stepparent role The role a stepparent forges for her- or himself within the stepfamily as there is no such role clearly defined by society.

stereotype A rigidly held, simplistic, and overgeneralized view of individuals, groups, or ideas that fails to allow for individual differences and is based on personal opinion and bias rather than critical judgment.

sterilization Intervention (usually surgical) making a person incapable of reproducing.

stigmatization The process of labeling and internalizing perceptions of self, other individuals, groups, behaviors, feelings, or ideas as deviant.

stimulus-value-role theory A three-stage theory of romantic development proposed by Bernard Murstein: (1) stimulus brings people together; (2) value refers to the compatibility of basic values; (3) role has to do with each person's expectations of how the other should fulfill his or her roles.

strain Tension lingering from previous stressors or tensions inherent in family roles. See also *stressor.*

storge [STOR gay] According to sociologist John Lee's styles of love, companionate love.

stress Psychological or emotional distress or disruption. Cf. *eustress.*

stressor A stress-causing event.

stressor pileup The occurrence of a number of stresses, hardships, and strains within a short period of time which together can severely test family coping abilities.

structural functionalism A sociological theory that examines how society is organized and maintained by examining the functions performed by its different structures. In marriage and family studies, structural functionalism examines the functions the family performs for society, the functions the individual performs for the family, and the functions the family performs for its members.

subsystem A system that is part of a larger system, such as family, and religious and economic systems being subsystems of society and the parent/child system being a subsystem of the family.

sudden infant death syndrome (SIDS) The death of an apparently healthy infant during its sleep from unknown causes.

superego In psychoanalytic theory, the part of the personality that has internalized society's demands and acts as a sort of conscience to control the id. See also *ego* and *id.*

surrogate mother A woman who bears a child for another woman (often for money) and relinquishes custody upon birth; the pregnancy usually results from artificial insemination, in vitro fertilization, or embryo transplant.

survey research Research method using questionnaires or interviews to gather information from small, representative groups and to infer conclusions that are valid for larger populations.

suspicious jealousy Jealousy that occurs when there is either no reason for suspicion or only ambiguous evidence that a partner is involved with another. Cf. *reactive jealousy.*

symbiotic personality A personality characterized by excessive dependency and need for closeness.

symbol In communication, a word or gesture that represents something more than itself.

symbolic interaction A theory which focuses on the subjective meanings of acts and how these meanings are communicated through interactions and roles to give shared meaning.

symmetrical work/family role allocation The interface between family and employment in which family work is divided more equitably and females have greater commitment to work roles than in the traditional division of labor. Cf. *sequential work/family role staging* and *traditional-simultaneous work/family life cycle.*

syphilis A sexually transmitted disease whose first symptom is a painless chancre on the genitals, anus, or mouth; caused by the *Treponema pallidum* bacterium. Life-threatening if untreated.

T

TANF See *Temporary Assistance to Needy Families.*

TDI See *therapeutic donor insemination.*

Temporary Assistance to Needy Families (TANF) The current government program designd to financially assist families with children during times of poverty.

teratogen A substance capable of producing fetal defects if ingested by the mother.

Tet The Vietnamese New Year, which coincides with the beginning of the lunar year.

thanatology The study of death and dying.

theoretical effectiveness The maximum effectiveness of a drug, device, or method if used consistently, correctly, and according to instructions.

theory A set of general principles or concepts used to explain a phenomenon and to make predictions that may be tested and verified experimentally.

toxemia A condition of pregnancy characterized by high blood pressure and edema. See also *gestational edema-proteinuria-hypertension complex.*

traditional family In popular usage, an intact, married two-parent family with at least one child, which adheres to conservative family values; an idealized family. Popularly used interchangeably with *nuclear family.*

traditional-simultaneous work/family life cycle The interface between family and employment in which the husband is regarded as the primary economic provider with little responsibility for family work, and the wife, regardless of her employment status, is primarily responsible for family work. Cf. *sequential work/family role staging* and *symmetrical work/family role allocation.*

trait A distinguishing personality characteristic or quality.

transactional pattern In communication, a habitual pattern of interaction.

transition 1. Passing from one stage or phase to another. 2. During childbirth, the process during which the fetus's head enters the birth canal, marking the end of the first stage of labor.

traumatic sexualization The process of developing inappropriate or dysfunctional sexual attitudes, behaviors, and feelings by a sexually abused child.

traveling time The time immediately following the Civil War when former slaves traveled throughout the South in search of relatives separated by sale.

trial marriage Cohabitation with the purpose of determining compatibility prior to marriage.

triangular theory of love A theory developed by Robert Sternberg emphasizing the dynamic quality of love as expressed by the interrelationship of three elements: intimacy, passion, and decision/commitment.

trust Belief in the reliability and integrity of another.

tubal ligation A surgical method of female sterilization in which the fallopian tubes are tied off or closed, usually by laparoscopy.

tubal pregnancy See *ectopic pregnancy.*

typology A systematic categorization according to types, such as common traits or qualities.

U

ultrasonography Technological method used to view the fetus in utero by reflecting high frequency sound waves off it; ultrasound.

ultrasound See *ultrasonography.*

umbilical cord A hollow cord which connects the circulation system of the embryo or fetus to the placenta.

underclass The socioeconomic class marked by persistent poverty and poor employability.

unpredictable crisis An unforeseen crisis, such as terminal illness in a child. Cf. *predictable crisis.*

unrequited love Love that is not returned.

urethritis An infection of the urethra.

user effectiveness The actual effectiveness of a drug, device, or method based on statistical information.

uterus A hollow, muscular organ within the pelvic cavity of a female in which the fertilized egg develops into the fetus; the womb.

V

vacuum aspiration First trimester abortion method in which the contents of the uterus are removed by suction.

vagina The passage leading from the vulva to the uterus that expands during intercourse to receive the erect penis or during childbirth to permit passage of the child; the birth canal.

vaginismus The involuntary constriction of the vaginal muscles which prohibits penetration.

vaginitis Vaginal infection, most commonly caused by *Trichomonas vaginalis, Candida albicans,* or *Gardnerella vaginalis,* which may be sexually transmitted. Men may also acquire these infections but often remain asymptomatic.

value judgment An evaluation based on ethics or morality rather than on objective observation.

values The social principles, goals, or standards held as acceptable by an individual, family, or group.

variable In experimental research, a factor, such as a situation or behavior, that may be manipulated. See also *independent variable* and *dependent variable.*

variation Departure from social or cultural norms.

vas deferens One of two ducts that carry sperm from the testes to the seminal vesicles. [plural] Vasa deferentia.

vasectomy A surgical form of male sterilization in which the vas deferens is severed.

VD (Venereal disease) See *sexually transmitted disease.*

venereal warts Warts in the genital or anal area which may be sexually transmitted; caused by the human papilloma virus (HPV).

vernix A milky substance often covering infants at birth.

viability Ability to live and continue to grow outside the uterus.

violence An act carried out with the intention of causing physical pain or injury to another.

virginity The state of not having engaged in sexual intercourse.

vulva The external female genitalia, including the mons veneris, labia majora and labia minora, clitoris, and the vaginal and urethral openings.

W

wealth Net worth, including income, savings, investments, property, and inheritances.

wedding 1. The act of marrying. 2. A marriage ceremony or celebration.

wellness Optimal health and well-being. Components of wellness include physical, emotional, intellectual, spiritual, social, and environmental health.

wheel theory of love A theory developed by Ira Reiss holding that love consists of four interdependent processes: rapport, self-revelation, mutual dependency, and intimacy fulfillment.

withdrawal See *coitus interruptus.*

womb Uterus.

work spillover The effect that employment has on time, energy, activities, and psychological functioning of workers and their families.

workplace/family linkages Ways in which employment affects families.

Y

yeast infection See *candidiasis.*

Yonsei Fourth generation Japanese, whose great-grandparents were born in Japan.

Yuan Tan The Chinese New Year, beginning on the first day of the lunar year.

Z

ZIFT See *zygote intrafallopian transfer.*

zygote The fertilized ovum (egg).

zygote intrafallopian transfer (ZIFT) Fertilization technique in which ova and sperm are united in a laboratory dish and then transferred to the fallopian tube to begin cell division.

Bibliography

Abbey, A., and R. J. Harnish. "Perception of Sexual Intent: The Role of Gender, Alcohol Consumption, and Rape Supportive Attitudes." *Sex Roles* 32, 5–6 (March 1995): 297–313.

Abbey, A., L. J. Halman, and F. M. Andrews. "Psychosocial, Treatment, and Demographic Predictors of the Stress Associated with Infertility." *Fertility and Sterility* 57, 1 (January 1992): 122–128.

Abbey, Antonia. "Acquaintance Rape and Alcohol Consumption on College Campuses: How Are They Linked?" *Journal of American College Health* 39, 4 (January 1991): 165–169.

———. "Maternal Risk Factors and Fetal Alcohol Syndrome: Provacative and Permissive Influences." *Neurotoxicology and Teratology* 17, 4 (July 1995): 445.

Abel, E. P. "An Update on Incidence of Fetal Alcohol Syndrome: Fetal Alcohol Syndrome is Not an Equal Opportunity Birth Defect." *Neurotoxicology and Teratology* 17, 4 (July 1995): 437.

Abelsohn, David. "A 'Good Enough' Separation: Some Characteristic Operations and Tasks." *Family Process* 31, 1 (1992): 61–83.

Absi-Semaan, Nada, Gail Crombie, and Corinne Freeman. "Masculinity and Femininity in Middle Childhood: Developmental and Factor Analyses." *Sex Roles: A Journal of Research* 28 (1993): 187–207.

Acitelli, Linda K. "Gender Differences in Relationship Awareness and Marital Satisfaction among Young Married Couples." *Personality and Social Psychology Bulletin* 18, 1 (1992): 102–110.

Acker, Joan. "From Sex Roles to Gendered Institutions." *Contemporary Sociology* 21, 5 (1992): 565–570.

Acker, Michele, and Mark H. Davis. "Intimacy, Passion, and Commitment in Adult Romantic Relationships: A Test of the Triangular Theory of Love." *Journal of Social and Personal Relationships* 9, 1 (1992): 21–50.

Ackerman, Diane. *A Natural History of the Senses.* New York: Random House, 1990.

———. *The Natural History of Love.* New York: Random House, 1994.

Acock, Alan C., and David H. Demo. *Family Diversity and Well-Being.* Thousand Oaks, CA: Sage Publications, 1994.

Adams, Bert. "The Family Problems and Solutions." *Journal of Marriage and the Family* 47, 3 (August 1985): 525–529.

Adams, Candace B., et al. "Young Adults' Expectations about Sex-Roles in Midlife." *Psychological Reports* 69, 3 (1991): 823–830.

Adams, David. "Identifying the Assaultive Husband in Court: You Be the Judge." *Response to the Victimization of Women and Children* 13, 1 (1990): 13–16.

Adelmann, Pamela K. "Psychological Well-Being and Homemaker vs. Retiree Identity among Older Women." *Sex Roles* 29, 3–4 (1993): 195–212.

Adler, Alfred. "Individual Psychology Therapy." In *Psychotherapy and Counseling*, edited by W. S. Sahakian. Chicago: Rand McNally, 1976.

Adler, Jerry. "Learning from the Loss." *Newsweek* (March 24, 1986): 66–67.

Adler, Jerry, et al. "The Joy of Gardening." *Newsweek* (July 26, 1982).

Adler, N. A., and J. Schultz. "Sibling Incest Offenders." *Child Abuse and Neglect.* 19, 7 (July 1995): 811–819.

Adler, Nancy, Susan Hendrick, and Clyde Hendrick. "Male Sexual Preference and Attitudes toward Love and Sexuality." *Journal of Sex Education and Therapy* 12, 2 (September 1989): 27–30.

Adler, Nancy E., H. P. David, B. N. Major, S. H. Roth, N. F. Russo, and G. E. Wyatt . "Psychological Responses after Abortion." *Science* 246 (April 1990): 41–44.

"Advance Report of Final Natality Statistics, 1991." *Monthly Vital Statistics Report* (Centers for Disease Control and Prevention) 42, 3 (Supplement) (September 9, 1993): 1–6.

Ahn, Helen Noh, and Neil Gilbert. "Cultural Diversity and Sexual Abuse Prevention." *Social Service Review* 66, 3 (September 1992): 410–428.

Ahrons, Constance, and Roy Rodgers. *Divorced Families: A Multidisciplinary View.* New York: Norton, 1987.

Ahrons, Constance, and Lynn Wallisch. "The Relationship between Former Spouses." In *Intimate Relationships: Development, Dynamics, and Deterioration*, edited by Daniel Perlman and Steven Duck. Newbury Park, CA: Sage Publications, 1987.

"AIDS and Children: A Family Disease." *World AIDS Magazine*, (November 1989): 12–14.

Ainsworth, Mary D., M. D. Blehar, E. Waters, and S. Wall. *Patterns of Attachment: A Psychological Study of the Strange Situation.* Hillsdale, NJ: Erlbaum, 1978.

Alapack Richard. "The Adolescent First Kiss." *Humanistic Psychologist* 19, 1 (March 1991): 48–67.

Aldous, Joan. "American Families in the 1980s: Individualism Run Amok?" *Journal of Family Issues* 8, 4 (December 1987): 422–425.

———. *Family Careers: Developmental Change in Families.* New York: Wiley, 1978.

———. "New Views on the Family Life of the Eldery and Near-Elderly." *Journal of Marriage and the Family* 49, 2 (May 1987): 227–234.

———. "Perspectives on Family Change." *Journal of Marriage and the Family* 52, 3 (August 1990): 571–583.

Aldous, Joan, ed. *Two Paychecks.* Beverly Hills, CA: Sage Publications, 1982.

Aldous, Joan, and Wilfried Dumon. "Family Policy in the 1980s: Controversy and Consensus." In *Contemporary Families: Looking Forward, Looking Back*, edited by Alan Booth. Minneapolis: National Council on Family Relations, 1991.

Aldous, Joan, and David Klein. "Sentiment and Services: Models of Intergenerational Relationships in Midlife." *Journal of Marriage and the Family* 53, 3 (August 1991): 595–608.

Allen, Katherine. *Single Women/Family Ties.* Newbury Park, CA: Sage Publications, 1989.

Allen, Katherine R., and Elizabeth B. Farnsworth. "Reflexivity in Teaching about Families." *Family Relations* 42, 3 (July 1993): 351–356.

Allen, Katherine, and Robert Pickett. "Forgotten Streams in the Family Life Course." *Journal of Marriage and the Family* 49, 3 (August 1987): 517–528.

Allgeier, Elizabeth, and Naomi McCormick, eds. *Gender Roles and Sexual Behavior.* Palo Alto, CA: Mayfield, 1982.

Allport, Gordon. *The Nature of Prejudice.* Garden City, NY: Doubleday, 1958.

Altman, Dennis. *AIDS in the Mind of America.* Garden City, NY: Anchor/Doubleday, 1985.

———. *The Homosexualization of America, the Americanization of the Homosexual.* New York: St. Martin's Press, 1982.

Altschul, Martin S. "Cultural Bias and the Urinary Tract Infection (UTI) Controversy." *Truth Seeker* (July 1989): 43–45.

Amato, Paul. "Parental Divorce and Attitudes toward Marriage and Family Life." *Journal of Marriage and the Family* 50 (May 1988): 453–461.

Amato, Paul R. "Father-Child Relations, Mother-Child Relations, and Offspring Psychological Well-Being in Early Adulthood." *Journal of Marriage and the Family* 56, 4 (November 1994): 1031–1042.

Amato, Paul R., and Alan Booth. "A Prospective Study of Divorce and Parent-Child Relationships." *Journal of Marriage and the Family* 58, 2 (May 1996): 356–365.

Amato, Paul R., and Bruce Keith. "Parental Divorce and the Well-Being of Children: A Meta-analysis." *Psychological Bulletin* 110 (1991): 26–46.

Ambry, Margaret K. "Childless Chances." *American Demographics* 14, 4 (April 1992): 55.

American Academy of Child and Adolescent Psychiatry. "Making Day Care A Good Experience." No. 20 (October 1992).

_____. "The Influence of Music and Rock Videos." No. 40 (October 1992).

American College of Obstetricians and Gynecologists. "ACOG technical bulletin number 205: Preconception care." *International Journal of Gynaecology and Obstetrics* 50 (1995): 201–207.

American Fertility Society. "New Guidelines for the Use of Semen Donor Insemination: 1990." *Fertility and Sterility* 53, 3 (Supplement 1) (March 1990): 1S–13S.

American Psychological Association. *Interim Report of the APA Working Group on Investigation of Memories of Childhood Abuse.* Washington, DC: American Psychological Association, 1994.

American Psychiatric Association. *Diagnostic and Statistical Manual of Mental Disorders.* 4th ed. Washington, DC: American Psychiatric Association, 1994.

Andersen, D. A., M. W. Lustig, and J. F. Andersen. "Regional Patterns of Communication in the United States: A Theoretical Perspective." *Communication Monographs,* 54 (1987): 128–144.

Anderson, Elaine A., and Margaret Feldman. "Family-Centered Health Policy." *In Vision 2010: Families and Health Care,* edited by Barbara A. Elliott. Minneapolis: National Council on Family Relations, 1993.

Anderson, Stephen. "Parental Stress and Coping during the Leaving Home Transition." *Family Relations* 37 (April 1988): 160–165.

Andreasen, Margaret S. "Patterns of Family Life and Television Consumption from 1945 to the 1990s." In *Media, Children and the Family: Social Scientific Psychodynamic and Clinical Perspectives,* edited by J. Bryant, D. Zillman, and A.C. Huston. Hillsdale, NJ: Erlbaum, 1994.

Andrews, F. M., A. Abbey, and L. J. Halman. "Stress from Infertility, Marriage Factors, and Subjective Well-Being of Wives and Husbands." *Journal of Health and Social Behavior* 32, 3 (September 1991): 238–253.

Aneshensel, Carol, Eva Fielder, and Rosina Becerra. "Fertility and Fertility-Related Behavior among Mexican-American and Non-Hispanic White Females." *Journal of Health and Social Behavior* 30, 1 (March 1989): 56–78.

Aneshensel, C., and L. I. Pearlin. "Structural Contexts of Sex Differences in Stress." In *Gender and Stress,* edited by R. C. Barnett et al. New York: MacMillan, 1987.

Annon, Jack. *The Behavioral Treatment of Sexual Problems.* Honolulu, HI: Enabling Systems, 1974.

———. *Behavioral Treatment of Sexual Problems: Brief Therapy.* New York: Harper and Row, 1976.

Anson, Ofra. "Marital Status and Women's Health Revisited: The Importance of a Proximate Adult." *Journal of Marriage and the Family* 51, 1 (February 1989): 185–194.

"A.P.A. Says Television Has Potential to be Beneficial." *The Brown University Child and Adolescent Behavior Letter* (March 1992).

Applebome, Peter. "Although Urban Blight Worsens, Most People Don't Feel Its Impact." *New York Times* (January 26, 1991): 1, 12.

Arditti, Joyce A. "Factors Related to Custody, Visitation, and Child Support for Divorced Fathers: An Exploratory Analysis." *Journal of Divorce and Remarriage* 17, 3–4 (1992): 23–42.

———. "Noncustodial Fathers: An Overview of Policy and Resources." *Family Relations* 39, 4 (October 1990): 460–465.

Arditti, Joyce A., and Katherine R. Allen. "Understanding Distressed Fathers' Perceptions of Legal and Relational Inequities Postdivorce." *Family & Conciliation Courts Review* 31, 4 (1993): 461–476.

Arendell, Terry. *Mothers and Divorce: Legal, Economic, and Social Dilemmas.* Berkeley, CA: University of California, 1987.

Arms, Karen G., et al., eds. *Cultural Diversity and Families.* Dubuque, IA: William C. Brown, 1992.

Arms, Suzanne. *Adoption: A Handful of Hope.* Berkeley, CA: Celestial Arts, 1990.

Armstrong, Penny, and Sheryl Feldman. *A Wise Birth.* New York: Morrow, 1990.

Armsworth, Mary W. "Psychological Response to Abortion." *Journal of Counseling and Development,* 65, 4 (March 1991): 377–379.

Aron, Arthur. "Unrequited Love as Self-Expansion." Paper presented at Second Iowa Conference on Personal Relationships, Iowa City, Iowa, May 12, 1989.

Aron, Arthur, and Elaine Aron. "Love and Sexuality." In *Sexuality in Close Relationships,* edited by Kathleen McKinney and Susan Sprecher. Hillsdale, NJ: Erlbaum, 1991.

Aron, Arthur, et al. "Experiences of Falling in Love." *Journal of Social and Personal Relationships* 6 (1989): 243–257.

Arond, M., and S. I. Parker. *The First Year of Marriage.* New York: Warner, 1987.

Aswad, Barbara. "Arab American Families." In *Families in Cultural Context,* edited by Mary Kay DeGenova. Mountain View, CA: Mayfield, 1997.

"At a Glance." *OURS: The Magazine of Adoptive Families* 23, 6 (November 1990): 63.

Athey, Jean L. "HIV Infection and Homeless Adolescents." *Child Welfare* 70, 5 (September 1991): 517–528.

Atkinson, Alice M. "Rural and Urban Families' Use of Child Care." *Family Relations* 43, 1 (January 1994): 16–22.

Atkinson, Alice, and Diedre James. "The Transition between Active and Adult Parenting: An End and a Beginning." *Family Perspective* 25, 1 (1991): 57–66.

Atkinson, Maxine P., and Stephen P. Blackwelder. "Fathering in the 20th Century." *Journal of Marriage and the Family* 55, 4 (November 1993): 975–986.

Atwood, J. D., and J. Gagnon. "Masturbatory Behavior in College Youth." *Journal of Sex Education and Therapy* 13 (1987): 35–42.

Avruch, S., and A. P. Cackley. "Savings Achieved by Giving WIC Benefits to Women Prenatally," *Public Health Reports* 110 (1995): 27–34.

Axelson, Leland, and Paula Dail. "The Changing Character of Homelessness in the United States." *Family Relations* 37, 4 (October 1988): 463–469.

Axelson, Marta, and Jennifer Glass. "Household Structure and Labor Force Participation of Black, Hispanic, and White Mothers." *Demography* 22 (1985): 381–394.

Axinn, William G., and Arland Thornton. "Mothers, Children, and Cohabitation: The Intergenerational Effects of Attitudes and Behavior." *American Sociological Review* 58, 2 (1993): 233–246.

Baber, Kristen M., and Katherine R. Allen. *Women and Families: Feminist Reconstructions.* New York: Guilford Press, 1992.

Bachu, Amara. *Current Population Reports*, (Series P20–482), Washington, DC: U.S. Government Printing Office, 1995.

Baggett, Courtney R. "Sexual Orientation: Should It Affect Child Custody Rulings?" *Law and Psychology Review* 16 (1992): 189–200.

Bahr, Kathleen. "Student Responses to Genogram and Family Chronology." *Family Relations* 39, 3 (July 1990): 243–249.

Baird, Donna D., and Allen J. Wilcox. "Cigarette Smoking Associated with Delayed Conception." *Journal of the American Medical Association* 253, 20 (May 1985): 2979–2983.

Baird, M. A., and W. J. Doherty. "Risks and Benefits of a Family Systems Approach to Health Care." *Family Medicine* 18 (1990): 5–17.

Baker, Sharon, Stanton Thalberg, and Diane Morrison. "Parents' Behavioral Norms as Predictors of Adolescent Sexual Activity and Contraceptive Use." *Adolescence* 23 (June 1988): 265–282.

Balay, Robert, ed. *Guide to Reference Books.* 11th ed. Chicago: American Library Association, 1996.

Baldwin, J. D., S. Whitely, and J. I. Baldwin. "The Effect of Ethnic Group on Sexual Activities Related to Contraception and STDs." *Journal of Sex Research* 29, 2 (May 1992): 189–206.

Bandura, Albert. *Social Learning Theory.* Englewood Cliffs, NJ: Prentice Hall, 1977.

Bane, Mary Jo, and Paul A. Jargowsky. "The Links Between Public Policy and Family Structure: What Matters and What Doesn't." In *The Changing American Family and Public Policy*, edited by Andrew J. Cherlin. Washington, DC: The Urban Institute Press, 1988.

Barbach, Lonnie. *For Each Other: Sharing Sexual Intimacy.* Garden City, NY: Doubleday, 1982.

Barber, B.L., and J.M. Lyons. "Family Processes and Adolescent Adjustment in Intact and Remarried Families. *Journal of Youth and Adolescence* 23, 4 (August 1994): 421–436.

Barden, J. C. "Many Parents in Divorces Abduct Their Own Children." *New York Times* (May 6, 1990): 10.

Barkan, Susan, and Michael Bracken. "Delayed Childbearing: No Evidence for Increased Low Risk of Low Birth Weight and Preterm Delivery." *American Journal of Epidemiology* 125, 1 (1987): 101–109.

Barkas, J. L. *Single in America.* New York: Atheneum, 1980.

Barnett, R.C., and G. K. Baruch. "Determinants of Father's Participation in Family Work." *Journal of Marriage and Family* 49 (1987): 29–40.

Barret, R. L., and B. E. Robinson. *Gay Fathers.* Lexington, MA: Lexington Books, 1990.

Barrow, Georgia. *Aging, Ageism, and Society.* St. Paul, MN: West, 1989.

Bartholomew, Kim. "Avoidance of Intimacy: An Attachment Perspective." *Journal of Social and Personal Relationships* 7, 2 (1990): 147–178.

Baruch, Elaine Hoffman et al., eds. *Embryos, Ethics, and Women's Rights: Exploring the New Reproductive Technologies.* New York: Harrington Park Press, 1988.

Bashshur, R. L., R. K. Homan, and D. G. Smith. "Beyond the Uninsured: Problems in Access to Care." *Medical Care* 32, 5 (1994): 409–419.

Basow, S. A. *Gender: Stereotyping and Roles.* Pacific Grove, CA: Brooks/Cole, 1992.

Bass, Ellen, and Louise Thornton, eds. *I Never Told Anyone: Stories and Poems by Survivors of Child Sexual Abuse.* New York: Harper & Row, 1983.

Bassuk, Ellen, and Lenore Rubin. "Homeless Children: A Neglected Population." *American Journal of Orthopsychiatry* 57, 2 (April 1987): 279–286.

Bassuk, Ellen, et al. "Characteristics of Sheltered Homeless Families." *American Journal of Public Health* 76 (1986): 1097–1101.

Baumann, P., et al. "Risk of Miscarriage after Transcervical and Transabdominal CVS in Relation to Bacterial Colonization of the Cervix." *Prenatal Diagnosis* 11, 8 (August 1991): 551–557.

Baumeister, Roy F., Sara R. Wotman, and Arlene M. Stillwell. "Unrequited Love: On Heartbreak, Anger, Guilt, Scriptlessness, and Humiliation." *Journal of Personality and Social Psychology* 64, 3 (March 1993): 377–394.

Baumrind, Diana. "Current Patterns of Parental Authority." *Developmental Psychology Monographs* 4, 1 (1971): 1–102.

———. "Parental Disciplinary Patterns and Social Competence in Children." *Youth and Society* 9, 3 (March 1978): 239–276.

———. "Rejoinder to Lewis's Reinterpretation of Parental Firm Control Effects: Are Authoritative Families Really Harmonious?" *Psychological Bulletin* 94, 1 (July 1983): 132–142.

Baxter, L. A. "Cognition and Communication in Relationship Process." In *Accounting for Relationships: Explanation, Representation and Knowledge*, edited by R. Burnett, P. McGhee, and D. Clarke. London: Meuthen, 1987.

Baxter, Richard L., Cynthia De Riemer, Ann Landini, and Larry Leslie. "A Content Analysis of Music Videos." *Journal of Broadcasting and Electronic Media* 29, 3 (June 1985): 333–340.

Bean, Frank, and Marta Tienda. *The Hispanic Population of the United States.* New York: Russell Sage Foundation, 1987.

Beavers, Robert W. "Healthy, Midrange, and Severely Dysfunctional Families." In *Normal Family Processes*, edited by Froma Walsh. New York: Guilford Press, 1982.

Beavers, Robert W., and Robert Hampson. *Successful Families.* New York: Norton, 1990.

Becerra, Rosina. "The Mexican American Family." In *Ethnic Families in America: Patterns and Variations*, 3d ed., edited by Charles Mindel et al. New York: Elsevier North Holland, 1988.

Bechhofer, L., and L. Parrot. "What Is Acquaintance Rape?" In *Acquaintance Rape: The Hidden Crime*, edited by A. Parrott and L. Bechhofer. New York: Wiley, 1991.

Beck, Joyce W., and Barbara M. Heinzerling. "Gay Clients Involved in Child Custody Cases: Legal and Counseling Issues." *Psychotherapy in Private Practice* 12, 1 (1993): 29–41.

Beck, Melinda. "Miscarriages." *Newsweek* (August 15, 1988): 46–49.

Beck, Melinda, and Geoffrey Cowley. "Mother Nature?" *Newsweek* (January 17, 1994): 54–58.

Beck, Rubye, and Scott Beck. "The Incidence of Extended Households among Middle-Aged Black and White Women." *Journal of Family Issues* 10, 2 (June 1989): 147–168.

Becker, Howard. *Outsiders.* New York: Free Press, 1963.

Beggs, Joyce M., and Dorothy C. Doolittle. "Perceptions Now and Then of Occupational Sex Typing: A Replication of Shinar's 1975 Study." *Journal of Applied Social Psychology* 23, 17 (1993): 1435–1453.

Beitchman, Joseph H., et al. "A Review of the Long-Term Effects of Child Sexual Abuse." *Child Abuse and Neglect* 16, 1 (January 1992): 101.

———. "A Review of the Short-Term Effects of Child Sexual Abuse." *Child Abuse and Neglect* 15, 4 (1991): 537–556.

Belcastro, Philip. "Sexual Behavior Differences between Black and White Students." *Journal of Sex Research* 21, 1 (February 1985): 56–67.

Belkin, Lisa. "Bars to Equality of Women Seen as Eroding Slowly." *New York Times* (August 20, 1989).

Bell, Alan, and Martin Weinberg. *Homosexualities: A Study of Diversities among Men.* New York: Simon & Schuster, 1978.

Bell, Alan, et al. *Sexual Preference: Its Development in Men and Women.* Bloomington, IN: Indiana University Press, 1981.

Bellah, Robert, et al. *Habits of the Heart.* Berkeley, CA: University of California Press, 1985.

Belsky, Jay. "Infant Day Care, Child Development, and Family Policy." *Society* 27, 5 (July 1990): 10–12.

———. "Patterns of Marital Change and Parent-Child Interaction." *Journal of Marriage and the Family* 53, 2 (May 1991): 487–498.

———. "Stability and Change in Marriage across the Transition to Parenthood: A Second Study." *Journal of Marriage and the Family* 47, 4 (November 1985b): 855–865.

Bem, Sandra. "Androgyny versus the Tight Little Lives of Fluffy Women and Chesty Men." *Psychology Today* 9, 4 (September 1975): 58–59 ff.

———. *Bem Sex-Role Inventory: Professional Manual.* Palo Alto, CA: Consulting Psychologists Press, 1981.

———. "Gender Schema Theory: A Cognitive Account of Sex Typing." *Psychological Review* 88 (1981): 354–364.

———. "Gender Schema Theory and Its Implications for Child Development: Raising Gender-Aschematic Children in a Gender Schematic Society." *Signs* 8, 4 (June 1983): 598–616.

———. "The Measurement of Psychological Androgyny." *Journal of Consulting and Clinical Psychology* 42 (1974): 155–162.

———. "Sex Role Adaptability: One Consequence of Psychological Androgyny." *Journal of Personality and Social Psychology* 31, 4 (1975): 634–643.

Bengston, Vern, and Joan Robertson, eds. *Grandparenthood.* Beverly Hills, CA: Sage Publications, 1985.

Benin, Mary, and Joan Agostinelli. "Husbands' and Wives' Satisfaction with the Division of Labor." *Journal of Marriage and the Family* 50, 2 (May 1988): 349–361.

Benini, E., C. C. Johnston, D. Faucher, and J. V. Aranda. "Topical Anesthesia During Circumcision in Newborn Infants." *Journal of the American Medical Association* 270, 7 (August 18, 1993): 850–853.

Benson, D., C. Charlton, and F. Goodhart. "Acquaintance Rape on Campus: A Literature Review." *Journal of American College Health* 40 (1992): 157–165.

Bera, W., et al. *Male Adolescent Sexual Abuse.* Newbury Park, CA: Sage Publications, 1991.

Berardo, Felix. "Trends and Directions in Family Research in the 1980s." In *Contemporary Families: Looking Forward, Looking Back*, edited by Alan Booth. Minneapolis: National Council on Family Relations, 1991.

Bergen, David J., and John E. Williams. "Sex Stereotypes in the United States Revisited: 1972–1988." *Sex Roles* 24, 7/8 (1991): 413–423.

Berger, C. R. "Planning and Scheming: Strategies for Initiating Relationships." In *Accounting for Relationships: Explanations, Representation and Knowledge*, edited by R. Burnett, P. McChee, and D. Clarke. New York: Methuen, 1987.

Berger, Mark J., and Donald P. Goldstein. "Infertility Related to Exposure to DES in Utero: Reproductive Problems in the Female." In *Infertility: Medical, Emotional, and Social Considerations*, edited by Miriam Mazor and Harriet Simons. New York: Human Sciences Press, 1984.

Berk, Richard A. "What the Scientific Evidence Shows: On the Average, We Can Do No Better Than Arrest." In *Current Controversies in Family Violence*, edited by Richard Gelles and Donileen Loseke. Newbury Park, CA: Sage Publications, 1993.

Berk, Richard A., Sarah F. Berk, Phyllis J. Newton, and Donileen R. Loseke. "Cops on Call: Summoning the Police to the Scene of Spousal Violence." *Law & Society Review* 18, 3 (1984): 479–498.

Berk, Richard A., Phyllis J. Newton, and Sarah F. Berk. "What a Difference a Day Makes: An Empirical Study of the Impact of Shelters for Battered Women." *Journal of Marriage and the Family* 48 (August 1986): 481–490.

Berkowitz, Alan. "College Men as Perpetrators of Acquaintance Rape and Sexual Assault: A Review of Recent Research." *Journal of American College Health* 40, 4 (January 1992): 175–181.

Berman, William. "Continued Attachment After Legal Divorce." *Journal of Family Issues* 6, 3 (September 1985): 375–392.

Bernard, Jessie. *The Future of Marriage*, 2d ed. New York: Columbia University Press, 1982.

Berrick, J. D., and R. P. Barth. "Child Sexual Abuse Prevention—Research Review and Recommendations." *Social Work Research and Abstracts* 28 (1992): 6–15.

Berscheid, Ellen. "Emotion." In *Close Relationships*, edited by H.H. Kelley et al. New York: Freeman, 1983.

———. "Interpersonal Attraction." In *Handbook of Social Psychology*, edited by G. Lindzey and Elliot Aronson. New York: Random House, 1985.

———. "Interpersonal Relationships." *Annual Review of Psychology* 45 (1994): 79–129.

———. "Some Comments on Love's Anatomy: Or Whatever Happened to Old-fashioned Lust?" In *The Psychology of Love*, edited by Robert Sternberg and Michael Barnes. New Haven, CT: Yale University Press, 1988.

Berscheid, Ellen, and J. Frei. "Romantic Love and Sexual Jealousy." In *Jealousy*, edited by G. Clanton and L. Smith. Englewood Cliffs, NJ: Prentice Hall, 1977.

Berscheid, Ellen, and Elaine H. Walster. *Interpersonal Attraction*. Reading, MA: Addison-Wesley, 1978.

———. "A Little Bit About Love." In *Foundations of Interpersonal Attraction*, edited by T. L. Huston. New York: Academic Press, 1974.

Betchen, Stephen. "Male Masturbation as a Vehicle for the Pursuer/Distancer Relationship in Marriage." *Journal of Sex and Marital Therapy* 17, 4 (December 1991): 269–278.

Bianchi, S. "America's Children." *Population Bulletin* 45, 1 (June 1990): 3–41.

"Bibulous America: Over Half of All Adults are Drinkers." *Dialogue* 5, 1 (February 1995).

Bichoff, Gary P., Sandra M. Stith, and Stephan M. Wilson. "A Comparison of the Family Systems of Adolescent Sexual Offenders and Non-Sexual Offending Delinquents." *Family Relations* 41, 3 (July 1992): 318–323.

Bielby, William, and James Baron. "Woman's Place Is with Other Women: Sex Segregation in the Workplace." National Research Council, Workshop on Job Segregation by Sex, 1982. Unpublished paper.

Bigler, Rebecca S., and Lynn S. Liben. "Cognitive Mechanisms in Children's Gender Stereotyping: Theoretical and Educational Implications of a Cognitive-Based Intervention." *Child Development* 63, 6 (1992): 1351–1364.

Billingsley, Andrew. "The Impact of Technology on Afro-American Families." *Family Relations* 37, 4 (October 1988): 420–425.

Billy, John, Nancy Landale, William Grady, and Denise Zimmerle. "Effects of Sexual Activity on Adolescent Social and Psychological Development." *Social Psychology Quarterly* 51, 3 (September 1988): 190–212.

Billy, John, Koray Tanfer, William R. Grady, and Daniel H. Klepinger. "The Sexual Behavior of Men in the United States." *Family Planning Perspectives* 25, 2 (March 1993): 52–60.

Binion, Victoria. "Psychological Androgyny: A Black Female Perspective." *Sex Roles* 22, 7–8 (April 1990): 487–507.

Binsacca, B. D., et al. "Factors Associated with Low Birthweight in an Inner-City Population." *American Journal of Public Health* 77, 4 (April 1987): 505–506.

Bird, Chloe E., and Catherine E. Ross. "Houseworkers and Paid Workers: Qualities of the Work and Effects on Personal Control." *Journal of Marriage and the Family* 55, 4 (November 1993): 913–925.

Bird, Gerald, and Gloria Bird. "The Determinants of Mobility in Two-Earner Families: Does the Wife's Income Count." *Journal of Marriage and the Family* 47, 3 (August 1985): 753–758.

Bivas, N.K. "Letter to our Son's Grandparents: Why We Decided Against Circumcision." *Humanistic Judaism* 26, 3 (1988): 11–13.

Black, Rita D. "Women's Voices after Pregnancy Loss: Couples' Patterns." *Social Work in Health Care* 16, 2 (1991): 19–36.

Blair, Sampson Lee. "Employment, Family, and Perceptions of Marital Quality among Husbands and Wives." *Journal of Family Issues* 14, 2 (1993): 189–212.

———. "The Sex-Typing of Children's Household Labor: Parental Influence on Daughters' and Sons' Housework." *Youth and Society* 24, 2 (1992): 178–203.

Blair, Sampson, and Daniel Lichter. "Measuring the Division of Household Labor: Gender Segregation and Housework among American Couples." *Journal of Family Issues* 12, 1 (March 1991): 91–113.

Blakely, Mary Kay. "Surrogate Mothers: For Whom Are They Working?" *Ms.* (March 1987): 18, 20.

Blakeslee, Sandra. "Scientists Find Key Biological Causes of Alcoholism." *New York Times* (August 14, 1984): 19 ff.

Blau, Elizabeth. "Study Finds Barrage of Sex on TV." *New York Times* (January 27, 1988).

Blieszner, Rosemary, and Janet Alley. "Family Caregiving for the Elderly: An Overview of Resources." *Family Relations* 39, 1 (January 1990): 97–102.

Block, Jeanne. "Differential Premises Arising from Differential Socialization of the Sexes: Some Conjectures." *Child Development* 54 (December 1983): 1335–1354.

Bloom, Bernard, et al. "Sources of Marital Dissatisfaction Among Newly Separated Persons." *Journal of Family Issues* 6, 3 (September 1985): 359–373.

Blum, R., et al. "American Indian—Alaska Native Youth Health." JAMA: *Journal of American Medical Association* 267, 12 (March 25, 1992): 1637-1644.

Blumberg, Rae Lesser, ed. *Gender, Family, and the Economy*. Newbury Park, CA: Sage Publications, 1990.

Blumenfeld, Warren, and Diane Raymond. *Looking at Gay and Lesbian Life*. Boston: Beacon Press, 1989.

Blumstein, Philip. "Identity Bargaining and Self-Conception." *Social Forces* 53, 3 (1975): 476–485.

Blumstein, Philip, and Pepper Schwartz. *American Couples*. New York: McGraw-Hill, 1983.

Bogert, Carroll. "Bringing Back Baby." *Newsweek* (November 21, 1994): 78–79.

Bohannan, Paul. "The Six Stations of Divorce." In *Divorce and After*, edited by P. Bohannan. New York: Doubleday, 1970a.

Bohannan, Paul, ed. *Divorce and After*. New York: Doubleday, 1970b.

Boken, Halcyone. "Gender Equality in Work and Family." *Journal of Family Issues* 5, 2 (June 1984): 254–272.

Boland, Joseph, and Diane Follingstad. "The Relationship between Communication and Marital Satisfaction: A Review." *Journal of Sex and Marital Therapy* 13, 4 (December 1987): 286–313.

Boles, Abner J., and Harriet Curtis-Boles. "Black Couples and the Transition to Parenthood." *American Journal of Social Psychiatry* 6, 1 (December 1991): 314–318.

Bollen, N., M. Camus, C. Staessen, H. Tournaye, P. Devroey, and A. C. Van Steirteghem. "The Incidence of Multiple Pregnancy after In Vitro Fertilization and Embryo Transfer, Gamete, or Zygote Intrafallopian Transfer." *Fertility and Sterility* 55, 2 (February 1991): 314–318.

Bonney, Lewis A. "Planning for Postdivorce Relationships: Factors to Consider in Drafting a Transition Plan." *Family and Conciliation Courts Review* 31, 3 (1993): 367–372.

Book, Cassandra L., et al. *Human Communication: Principles, Contexts, and Skills.* New York: St. Martin's Press, 1980.

Boone, Margaret S. *Capital Crime: Black Infant Mortality in America.* Newbury Park, CA: Sage Publications, 1989.

Booth, A., and J. N. Edwards. "Starting Over: Why Remarriages Are More Unstable. *Journal of Family Issues* 13, 2 (June 1992): 179–194.

Booth, Alan. "Who Divorces and Why: A Review." *Journal of Family Issues* 6, 3 (September 1985): 255–293.

Booth, Alan, ed. *Contemporary Families: Looking Forward, Looking Back.* Minneapolis: National Council on Family Relations, 1991.

———. *Child Care in the 1990s: Trends and Consequences.* Hillsdale, NJ: Erlbaum, 1992.

Booth, Alan, and John Edwards. "Age at Marriage and Marital Instability." *Journal of Marriage and the Family* 47, 2 (February 1985): 67–74.

Booth, Alan, and David R. Johnson. "Declining Health and Marital Quality." *Journal of Marriage and the Family* 56, 1 (February 1994): 218–223.

Booth, Alan, et al. "Divorce and Marital Instability over the Life Course." *Journal of Family Issues* 7 (1986): 421–442.

———. "Predicting Divorce and Permanent Separation." *Journal of Family Issues* 6, 3 (September 1985): 331–346.

Borhek, Mary. "Helping Gay and Lesbian Adolescents and Their Families: A Mother's Perspective." *Journal of Adolescent Health Care* 9, 2 (March 1988): 123–128.

Borland, Dolores. "A Cohort Analysis Approach to the Empty-Nest Syndrome among Three Ethnic Groups of Women: A Theoretical Position." *Journal of Marriage and the Family* 44 (February 1982): 117–129.

Bosse, Raymond, et al. "Change in Social Support after Retirement: Longitudinal Findings from the Normative Aging Study." *Journal of Gerontology* 48, 4 (1993): 210–217.

Boston Women's Health Book Collective. *The New Our Bodies, Ourselves.* New York: Simon & Schuster, 1992.

Bostwick, Homer. *A Treatise on the Nature and Treatment of Seminal Disease, Impotency, and Other Kindred Affections.* 12th ed. New York: Burgess, Stringer, 1860.

Boswell, John. *Christianity, Social Tolerance, and Homosexuality.* Chicago: University of Chicago Press, 1980.

Bourne, Richard, and Eli Newberger, eds. *Critical Perspectives on Child Abuse.* Lexington, MA: Lexington Books, 1979.

Bowker, Lee. *Beating Wife Beating.* Lexington, MA: Lexington Books, 1983.

Bowlby, John. *Attachment and Loss.* 3 vols. New York: Basic Books, 1969, 1973, 1980.

Bowman, Madonna, and Constance Ahrons. "Impact of Legal Custody Status on Father's Parenting Post-Divorce." *Journal of Marriage and the Family* 47, 2 (May 1985): 481–485.

Bozett, F. W. and M. B. Sussman, editors. *Homosexuality and Family Relations.* New York: Harrington Park Press, 1990.

Bozett, Frederick W. "Children of Gay Fathers." In *Gay and Lesbian Parents,* edited by Frederick W. Bozett. New York: Praeger, 1987c.

Bozett, Frederick W. "Gay Fathers." In *Gay and Lesbian Parents,* edited by Frederick W. Bozett. New York: Praeger, 1987b.

Bozett, Frederick W., ed. *Gay and Lesbian Parents.* New York: Praeger, 1987a.

Brand, H. J. "The Influence of Sex Differences on the Acceptance of Infertility." *Journal of Reproductive and Infant Psychology* 7, 2 (April 1989): 129–131.

Brandenburg, H., et al. "Fetal Loss Rate after Chorionic Villus Sampling and Subsequent Amniocentesis." *American Journal of Medical Genetics* 35, 2 (February 1990): 178–180.

Braun, J., and H. Chao. "Attitudes Toward Women: A Comparison of Asian-Born Chinese and American Caucasians." *Psychology of Women Quarterly* 2 (1978): 195–201.

Braver, Sanford L., et al. "A Longitudinal Study of Noncustodial Parents: Parents without Children." *Journal of Family Psychology* 7, 1 (June 1993b): 9–23.

———. "A Social Exchange Model of Nonresidential Parent Involvement." In *Nonresidential Parenting: New Vistas in Family Living,* edited by C. E. Depner, and J. H. Bray. Newbury Park, CA: Sage Publications, 1993a.

Bray, James H., and Charlene Depner. "Nonresidential Parents: Who Are They?" In *Nonresidential Parenting: New Vistas in Family Living,* edited by C. E. Depner and J. H. Bray. Newbury Park, CA: Sage Publications, 1993.

Bray, James H., and E. Mavis Hetherington. "Families in Transition: Introduction and Overview" [Special Section: "Families in Transition"]. *Journal of Family Psychology* 7, 1 (1993): 3–8.

Bray, James H., and Sandra H. Berger. "Noncustodial Father and Paternal Grandparent Relationship in Stepfamilies." *Family Relations* 39, 4 (October 1990): 414–419.

———. "Nonresidential Parent-Child Relationships Following Divorce and Remarriage: A Longitudinal Perspective." In *Nonresidential Parenting: New Vistas in Family Living,* edited by C. E. Depner and J. H. Bray. Newbury Park, CA: Sage Publications, 1993.

Brayfield, April A. "Employment Resources and Housework in Canada." *Journal of Marriage and the Family* 54, 1 (February 1992): 19–30.

Brayfield, April A. "Juggling Jobs and Kids: The Impact of Employment Schedules on Fathers' Caring for Children." *Journal of Marriage and the Family* 57, 2 (May 1995): 321–332.

Breault, K. D., and Augustine Kposowa. "Explaining Divorce in the United States: A Study of 3,111 Counties, 1980." *Journal of Marriage and the Family* 49, 3 (August 1987): 549–558.

Brenner, Harvey. "Influence of the Social Environment on Psychopathology: The Historic Perspective." In *Stress and Mental Disorder,* edited by

James Barrett et al. New York: Raven, 1979.

———. *Mental Illness and the Economy.* Cambridge, MA: Harvard University Press, 1973.

Bretschneider, Judy, and Norma McCoy. "Sexual Interest and Behavior in Healthy 80- to 102-Year-Olds." *Archives of Sexual Behavior* 17, 2 (April 1988): 109–128.

Bridges, Judith S. "Pink or Blue: Gender Congratulations Cards." *Psychology of Women* 17, 2 (1993): 193–205.

Bringle, Robert G., and Glenda J. Bagby. "Self-Esteem and Perceived Quality of Romantic and Family Relationships in Young Adults." *Journal of Research in Personality* 26, 4 (1992): 340–356.

Bringle, Robert, and Bram Buunk. "Extradyadic Relationships and Sexual Jealousy." In *Sexuality in Close Relationships,* edited by Kathleen McKinney and Susan Sprecher. Hillsdale, NJ: Erlbaum, 1991.

———. "Jealousy and Social Behavior: A Review of Person, Relationship, and Situational Determinants." In *Review of Personality and Social Psychology Vol 6: Self, Situation, and Social Behavior,* edited by P. Shaver. Newbury Park, CA: Sage Publications, 1985.

Britton, D. M. "Homophobia and Homosociality: An Analysis of Boundary Maintenance." *Sociological Quarterly* 31, 3 (September 1990): 423–439.

Broderick, Carlfred, and James Smith. "The General Systems Approach to the Family." In *Contemporary Theories About the Family,* edited by Wesley Burr et al. New York: Free Press, 1979.

Brody, Elaine. "Parent Care as a Normative Family Stress." *Gerontologist* 25 (1985): 19–29.

Brody, Jane E. "Assessing the Question of Male Circumcision." *New York Times* (August 14, 1985): 18.

———. "Despite All the Benefits, Many Mothers Decide Against Breastfeeding." *New York Times* (April 6, 1994): B9.

———. "Estrogen is Found to Improve Mood, Not Just Menopause Symptoms." *New York Times* (January 1, 1992a): 141.

———. "Personal Health: Maintaining Friendships for the Sake of Good Health." *New York Times* (February 5, 1992b): B8.

Brodzinsky, David M., and Marshall D. Schecter, eds. *The Psychology of Adoption.* New York: Oxford University Press, 1990.

Broman, Clifford. "Satisfaction among Blacks: The Significance of Marriage and Parenthood." *Journal of Marriage and the Family* 50, 1 (February 1988): 45–51.

Bronfenbrenner, Urie. *The Ecology of Human Development.* Cambridge, MA: Harvard University Press, 1979.

Bronstein, Phyllis, JoAnn Clauson, Miriam F. Stoll, and Craig L. Abrams. "Parenting Behavior and Children's Social, Psychological, and Academic Adjustment in Diverse Family Structures." *Family Relations* 42, 3 (July 1993): 268–276.

Brooks, Jane B. *Parenting in the 90s.* Mountain View, CA: Mayfield, 1994.

Brown, Elizabeth, and William R. Hendee. "Adolescents and Their Music." *Journal of the American Medical Association* 262, 12 (September 22, 1989): 1659–1663.

Brown, J. D., and L. Schulze. "The Effects of Race, Gender, and Fandom on Audience Interpretations of Madonna's Music Videos." *Journal of Communication* 40 (1990): 88–102.

Brown, Jane D., and Kenneth Campbell. "Race and Gender in Music Videos: The Same Beat but a Different Drummer." *Journal of Communications* 36, 1 (December 1986): 94–106.

Brown, Jane D., and Susan F. Newcomer. "Television Viewing and Adolescents' Sexual Behavior." *Journal of Homosexuality* 21, 1/2 (1991): 77–91.

Brown, Lyn Mikel, and Carol Gilligan. *Meeting at the Crossroads: Women's Psychology and Girl's Development.* Cambridge, MA: Harvard University Press, 1992.

Brown, Patricia Leigh. "Where to Put the TV Set?" *New York Times* (October 4, 1990): B4.

Browne, Angela, and David Finkelhor. "Initial and Long-Term Effects: A Review of the Research." In *Sourcebook on Child Sexual Abuse,* edited by David Finkelhor. Beverly Hills, CA: Sage Publications, 1986.

Browne, Jane. "Terrific Tips from the Marriage Doctor." *New Woman.* (July 1997): 76-79ff.

Brownmiller, Susan. *Femininity.* New York: Faucett Columbine, 1983.

Bryant, Lois, et al. "Race and Family Structure Stereotyping: Perceptions of Black and White Nuclear Families and Stepfamilies." *Journal of Black Psychology* 15 (1988): 1–16.

Bryant, Z. Lois, and Marilyn Coleman. "The Black Family as Portrayed in Introductory Marriage and Family Textbooks." *Family Relations* 37, 3 (July 1988): 255–259.

Bryon, K. "Family Composition Changing." Washington, DC: U.S. Census Bureau, 1996.

Buchholz, Ester, and Barbara Gol. "More than Playing House: A Developmental Perspective on the Strengths in Teenage Motherhood." *American Journal of Orthopsychology* 56, 3 (July 1986).

Buchta, Richard. "Attitudes of Adolescents and Parents of Adolescents Concerning Condom Advertisements on Television." *Journal of Adolescent Health Care* 10, 3 (May 1989): 220–223.

Budiansky, Stephen. "The New Rules of Reproduction." *U.S. News and World Report* (April 18, 1988): 66–69.

Buehler, Cheryl, and Bobbie H. Legg. "Selected Aspects of Parenting and Children's Social Competence Postseparation: The Moderating Effects of Child's Sex, Age, and Family Economic Hardship" [Special Issue: "Divorce and the Next . . ."]. *Journal of Divorce & Remarriage* 18, 3–4 (1992): 177–195.

Bullough, Vern. *Sexual Variance in Society and History.* New York: Wiley, 1976.

Bumpass, Larry L., Teresa C. Martin, and James A. Sweet. "The Impact of Family Background and Early Marital Factors on Marital Disruption." *Journal of Family Issues* 12, 1 (1991): 22–42.

Bumpass, Larry, and James A. Sweet. "Children's Experience in Single Parent Families: Implications of Cohabitation and Marital Transition." *Family Planning Perspectives* 21, 6 (November 1989): 256–260.

Bumpass, Larry, James Sweet, and Teresa Castro Martin. "Changing Patterns of Remarriage." *Journal of Marriage and the Family* 52, 3 (August 1990): 747–756.

Burckly, William, Nicholas Reuterman, and Sondra Kopsky. "Dating Violence Among High School Students." *School Counselor,* 35, 5 (May 1988): 353–358.

Burgess, Ann W., ed. *Rape and Sexual Assault II.* New York: Garland Press, 1988.

Burgess, Ernest. "The Family as a Unity of Interacting Personalities." In *Family Roles and Interaction*, edited by Jerold Heiss. Chicago: Rand McNally, 1968.

———. "The Family as a Unity of Interacting Personalities." In *The Family* 7, 1 (March 1926): 3–9.

Burns, Ailsa. "Perceived Causes of Marriage Breakdown and Conditions of Life." *Journal of Marriage and the Family* 46, 3 (August 1984): 551–562.

Burns, Ailsa, and Cath Scott. *Mother-Headed Families and Why They Have Increased.* Hillsdale, NJ: Erlbaum, 1994.

Burns, Scott. *The Household Economy.* New York: Harper & Row, 1972.

Burr, W. R., and S. R. Klein. *Reexamining Family Stress.* Thousand Oaks, CA: Sage Publications, 1994.

Burr, Wesley, et al. *Contemporary Theories About the Family.* 2 vols. New York: Free Press, 1979.

Bush, Catherine R., Joseph P. Bush, and Joyce Jennings. "Effects of Jealousy Threats on Relationship Perceptions and Emotions." *Journal of Social and Personal Relationships* 5, 3 (August 1988): 285–303.

Buss, David M., Randy J. Larsen, Drew Westen, and Jennifer Semmelroth. "Sex Differences in Jealousy: Evolution, Physiology, and Psychology." *Psychological Science* 3, 4 (1992): 251–255.

Butter, I. H., and B. J. Kay. "Self Certification in Lay Midwives Organizations—A Vehicle for Professional Autonomy." *Social Science and Medicine* 30, 12 (1990): 1329–1339.

Butts, June Dobbs. "Adolescent Sexuality and Teenage Pregnancy from a Black Perspective." In *Teenage Pregnancy in a Family Context*, edited by T. Ooms. Philadelphia: Temple University Press, 1981.

Buunk, Bram, and Ralph Hupka. "Cross-Cultural Differences in the Elicitation of Sexual Jealousy." *Journal of Sex Research* 23, 1 (February 1987): 12–22.

Buunk, Bram, and Barry van Driel. *Variant Lifestyles and Relationships.* Newbury Park, CA: Sage Publications, 1989.

Buzawa, Eve S., and Karl G. Buzawa. "The Scientific Evidence Is Not Conclusive: Arrest Is No Panacea." In *Current Controversies in Family Violence*, edited by Richard Gelles and Donileen Loseke. Newbury Park, CA: Sage Publications, 1993.

Byers, E. S., and L. Heinlein. "Predicting Initiations and Refusals of Sexual Activities in Married and Cohabiting Heterosexual Couples." *Journal of Sex Research* 26 (1989): 210–231.

Byrd, W., et al. "A Prospective Randomized Study of Pregnancy Rates Following Intrauterine and Intracervical Insemination Using Frozen Donor Sperm." *Fertility and Sterility* 53, 3 (March 1990): 521–527.

Byrne, Donn, and Karen Murnen. "Maintaining Love Relationships." In *The Psychology of Love*, edited by Robert Sternberg and Michael Barnes. New Haven, CT: Yale University Press, 1988.

Cabai, Robert. "Gay and Lesbian Couples: Lessons on Human Intimacy." *Psychiatric Annals* 18, 1 (January 1988): 21–25.

Cado, Suzana, and Harold Leitenberg. "Guilt Reactions to Sexual Fantasies during Intercourse." *Archives of Sexual Behavior* 19, 1 (1990): 49–63.

Cahoon, D. E. M., Edmonds, R. M. Spaulding, and J. C. Dickens. "A Comparison of the Opinions of Black and White Males and Females Concerning the Occurrence of Rape." *Journal of Social Behavior and Personality* 10, 1 (March 1995): 91–100.

Calfin, Matthew S., James L. Carroll, and Jerry Schmidt. "Viewing Music-Videotapes Before Taking a Test of Premarital Sexual Attitudes." *Psychological Reports* 72, 2 (April 1993): 475–481.

Callan, Victor. "The Personal and Marital Adjustment of Mothers and of Voluntarily and Involuntarily Childless Wives." *Journal of Marriage and the Family* 47, 4 (November 1985): 1045–1050.

Camarillo, Albert. *Chicanos in a Changing Society.* Cambridge, MA: Harvard University Press, 1979.

———. *Latinos in the United States.* Santa Barbara, CA: ABC-Clio, 1986.

———. *Mexican-Americans in Urban Society: A Selected Bibliography.* Berkeley, CA: Floricanto Press, 1986.

Campbell, Jacquelyn. "Violence Toward Women: Homicide and Battering." In *Vision 2010: Families and Violence, Abuse and Neglect*, edited by Richard J. Gelles. Minneapolis: National Council on Family Relations, 1995.

Campbell, Thomas L. "Health Promotion/Disease Prevention and the Family." In *Vision 2010: Families and Health Care*, edited by Barbara A. Eliott. Minneapolis: National Council on Family Relations, 1993.

Canadian Paediatric Society, Fetus and Newborn Committee. "Neonatal Circumcision Revisited," *Canadian Medical Association Journal* 154, 6 (March 15, 1996): 769–780.

Cancian, F. M. "Gender Politics: Love and Power in the Private and Public Spheres." In *Family in Transition*, edited by Arelene S. Skolnick and J. H. Skolnick. Glenview, IL: Scott, Foresman, 1989.

Canter, Lee, and Marlene Canter. *Assertive Discipline for Parents.* Santa Monica, CA: Canter and Associates, 1985.

Cantor, Muriel. "The American Family on Television: From Molly Goldberg to Bill Cosby." *Journal of Comparative Family Studies* 22, 2 (June 1991): 205–216.

———. "Popular Culture and the Portrayal of Women: Content and Control." In *Analyzing Gender*, edited by Beth Hess and Myra Marx Ferree. Newbury Park, CA: Sage Publications, 1987.

Cappell, Charles, and Robert B. Heiner. "The Intergenerational Transmission of Family Aggression." *Journal of Family Violence* 5, 2 (June 1990): 121–134.

Cargan, Leonard, and Matthew Melko. *Singles: Myths and Realities.* Beverly Hills, CA: Sage Publications, 1982.

Carl, Douglas. "Acquired Immune Deficiency Syndrome: A Preliminary Examination of the Effects on Gay Couples and Coupling." *Journal of Marital and Family Therapy* 12, 3 (July 1986): 241–247.

Carrasquillo, Hector. "Puerto Rican Families in America." In *Families in Cultural Context*, edited by Mary Kay DeGenova. Mountain View, CA: Mayfield, 1997.

Carroll, Jerry. "Tracing the Causes of Infertility." *San Francisco Chronicle* (March 5, 1990): B3 ff.

Carroll, Jerry, K. D. Volk, and J. J. Hyde. "Differences in Males and Females in Motives for Engaging in Sexual Intercourse." *Archives of Sexual Behavior* 14 (1985): 131–139.

Carson, David K., et al. "Family of Origin Characteristics and Current Family Relationships of Female

Adult Incest Victims." *Journal of Family Violence* 5, 2 (June 1990): 153–172.

Carter, Betty, and Monica McGoldrick, eds. *The Changing Family Life Cycle,* 2d ed. Boston: Allyn and Bacon, 1989.

Carter, D. Bruce. *Current Conceptions of Sex Roles and Sex Typing.* New York: Praeger, 1987b.

———. "Sex Role Research and the Future New Directions for Research." In *Current Conceptions of Sex Roles and Sex Typing,* edited by D. Bruce Carter. New York: Praeger, 1987a.

———. "Societal Implications of AIDS and HIV Infections, HIV Antibody Testing, Health Care, and AIDS Education." *Marriage and Family Review* 13, 1 (1989): 129–188.

Cary, Alice. "Big Fans on Campus." *TV Guide* (April 18, 1992): 26–31.

Cassell, Carole. *Swept Away.* New York: Simon & Schuster, 1984.

Cate, Rodney M., and Sally A. Lloyd. *Courtship.* Newbury Park, CA: Sage Publications, 1992.

Cates, Jim A., Linda Graham, Donna Boeglin, and Steven Tielker. "The Effect of AIDS on the Family System." *Families in Society* 71, 4 (April 1990): 195–201.

CDC. *See* Centers for Disease Control and Prevention.

Celis, William III. "More College Women Drinking to Get Drunk." *New York Times* (June 8, 1994): B8.

Centers for Disease Control and Prevention. "HIV Survey in Childbearing Women." *National AIDS Hotline Training Bulletin* 103 (June 15, 1994): 2.

———. "Statewide Prevalence of Illicit Drug Use by Pregnant Women—Rhode Island." *Morbidity and Mortality Weekly Report* 39, 14 (April 3, 1990): 225–227.

———. "Surveillance Report: U.S. AIDS Cases Reported through June 1994." *HIV/AIDS Surveillance Report,* 1994.

———. "Surveillance Report: U.S. AIDS Cases Reported through December 1996." *HIV/AIDS Surveillance Report,* 1996.

Chasnoff, Ira J. "Drug Use in Pregnancy: Parameters of Risk." *Pediatric Clinics of North America* 35, 6 (December 1988): 1403–1412.

Chatters, Linda M., Robert J. Taylor, and Harold W. Neighbors. "Size of Informal Helper Network Mobilized during a Serious Personal Problem among Black Americans." *Journal of Marriage and the Family* 51, 3 (August 1989): 667–676.

Cherlin, Andrew. *The Changing American Family and Public Policy.* Washington, DC: Urban Institute Press, 1988.

———. *Marriage, Divorce, and Remarriage.* Rev. ed. Cambridge, MA: Harvard University Press, 1992.

Cherlin, Andrew, and Frank Furstenberg, Jr. *The New American Grandparent.* New York: Basic Books, 1986.

Cheseboro, J. "Communication, Values, and Popular Television Series—A Four Year Assessment." In *Television: The Critical View,* 4th ed., edited by H. Newcomb. New York: Oxford University Press, 1987.

Chesser, Barbara Jo. "Analysis of Wedding Rituals: An Attempt to Make Weddings More Meaningful." *Family Relations* 29, 2 (April 1980).

Chiasson, M. A., R. L. Stoneburner, and S. C. Joseph. "Human Immunodeficiency Virus Transmission through Artificial Insemination." *Journal of Acquired Immune Deficiency Syndromes* 3, 1 (1990): 69–72.

"Child Abuse and Neglect Still a Widespread Problem in America." *Nation's Health* (May/June 1997): 9.

Children's Defense Fund. "Moments in America for Children." Washington, DC, 1996.

"Children's Health." Washington, DC: *Nation's Health* (May/June 1997): 1.

Chilman, Catherine. "Working Poor Families: Trends, Causes, Effects, and Suggested Policies." *Family Relations* 40, 2 (April 1991): 191–198.

Chilman, Catherine Street. "Hispanic Families in the United States: Research Perspectives." In *Family Ethnicity: Strength in Diversity,* edited by Harriette Pipes McAdoo. Newbury Park, CA: Sage Publications, 1993.

Chilman, Catherine, et al., eds. *Variant Family Forms.* Beverly Hills, CA: Sage Publications, 1988.

Chira, Susan. "Years after Adoption, Adults Find Past, and New Hurdles." *New York Times* (August 30, 1993): B1, B6.

Chojnacki, Joseph T., and W. Bruce Walsh. "Reliability and Concurrent Validity of the Sternberg Triangular Love Scale." *Psychological Reports* 67, 1 (August 1990): 219–224.

Christensen, Andrew. "Dysfunctional Interaction Patterns in Couples." In *Perspectives on Marital Interaction,* edited by Patricia Noller and Mary Anne Fitzpatrick. Philadelphia: Multilingual Matters, 1988.

Christensen, Donna, et al. "Noncustodial Mothers and Child Support: Examining the Larger Context." *Family Relations* 39, 4 (October 1990): 388–394.

Christian-Smith, Linda K. *Becoming a Woman through Romance.* New York: Routledge, 1990.

Christopher, F., and R. Cate. "Anticipated Influences on Sexual Decision-Making for First Intercourse." *Family Relations* 34 (1985): 265–270.

———. "Factors Involved in Premarital Decision-Making." *Journal of Sex Research* 20 (1984): 363–376.

Christopher, F. S., and M. M. Frandsen. "Strategies of Influence in Sex and Dating." *Journal of Social and Personal Relationships* 7 (1990): 89–105.

Christopher, F. Scott, Richard A. Fabes, and Patricia M. Wilson. "Family Television Viewing: Implications for Family Life Education." *Family Relations* 38, 2 (April 1989): 210–214.

Ciancannelli, Penelope, and Bettina Berch. "Gender and the GNP." In *Analyzing Gender,* edited by Beth Hess and Myra Marx Ferree. Newbury Park, CA: Sage Publications, 1987.

Cimons, Marlene. "American Infertility Rate Not Growing, Study Finds." *New York Times* (December 7, 1990): A3.

Claes, Jacalyn A., and David M. Rosenthal. "Men Who Batter Women: A Study in Power." *Journal of Family Violence* 5, 3 (September 1990): 215–224.

Clanton, Gordon and Lynn Smith. *Jealousy.* Englewood Cliffs, NJ: Prentice Hall, 1977.

Clark, Danae. "Cagney and Lacey: Feminist Strategies of Detection." In *Television and Women's Culture: The Politics of the Popular,* edited by Mary Ellen Brown. Newbury Park, CA.: Sage Publications, 1992.

Clark-Nicolas, Patricia, and Bernadette Gray-Little. "Effect of Economic Resources on Marital Quality in Black Married Couples." *Journal of Marriage and the Family* 53, 3 (August 1991): 645–656.

Cleek, Margaret, and T. Allan Pearson. "Perceived Causes of Divorce: An Analysis of Interrelationships." *Journal of Marriage and the Family* 47, 2 (February 1985): 179–191.

Clemes, Harris, and Reynold Bean. *How to Raise Children's Self-Esteem.* San Jose, CA: Enrich, 1983.

Cleveland, Peggy, et al. "If Your Child Has AIDS . . .: Response of Parents with Homosexual Children." *Family Relations* 37, 2 (April 1988): 150–153.

Clingempeel, W. Glenn, and Eulalee Brand. "Quasi-kin Relationships, Structural Complexity, and Marital Quality in Stepfamilies: A Replication, Extension, and Clinical Implications." *Family Relations* 34, 3 (July 1985): 401–409.

Clingempeel, W. Glenn, et al. "Stepparent-Stepchild Relationships in Stepmother and Stepfather Families: A Multimethod Study." *Family Relations* 33 (1984): 465–473.

Cochran, Susan D., Vickie M. Mays, and Laurie Leung. "Sexual Practices of Heterosexual Asian-American Young Adults: Implications for Risk of HIV Infection." *Archives of Sexual Behavior* 20, 4 (August 1991): 381–394.

Cochran, Susan, et al. "Sexually Transmitted Diseases and Acquired Immunodeficiency Syndrome (AIDS): Changes in Risk Reduction Behaviors among Young Adults." *Sexually Transmitted Diseases* 16, 1 (January 1989): 80–86.

Coggle, Frances, and Grace Tasker. "Children and Housework." *Family Relations* 31 (July 1982): 395–399.

Cohen, T.F. "What Do Fathers Provide? Reconsidering the Economic and Nurturant Dimensions of Men as Parents." In *Men, Work and Family*, edited by J.C. Hood Newbury Park, CA: Sage Publications, 1993.

Cohler, Bertram, and Scott Geyer. "Psychological Autonomy and Interdependence within the Family." In *Normal Family Processes*, edited by Froma Walsh, New York: Guilford Press, 1982.

Cole, Louanne. "Sex Matters: Is Circumcision Correct for Newborn Boys?" *San Francisco Examiner* (August 11, 1993): Living, B7.

Cole, R., and D. Reiss. *How Do Families Cope with Chronic Illness?* Hillsdale, NJ: Erlbaum, 1993.

Cole, Robert. "Mental Illness and the Family." In *Vision 2010: Families & Health Care*, edited by Barbara A. Elliott. Minneapolis: National Council on Family Relations (1993): 18–19.

Cole, W. "Incest Perpetrators: Their Assessment and Treatment." *Psychiatric Clinics of North America* 15, 3 (September 1992): 689–701.

Coleman, Marilyn, and Lawrence Ganong. "The Cultural Stereotyping of Stepfamilies." In *Remarriage and Stepparenting: Current Research and Theory*, edited by Kay Pasley and Marilyn Ihinger-Tallman. New York: Guilford Press, 1987.

———. "Remarriage and Stepfamily Research in the 1980s: Increased Interest in an Old Form." In *Contemporary Families: Looking Forward, Looking Back*, edited by Alan Booth. Minneapolis: National Council on Family Relations, 1991.

Coles, Robert. *The Spiritual Life of Children.* Boston: Houghton Mifflin, 1990.

Collier, J., M. Z. Rosaldo, and S. Yanagisako. "Is There A Family? New Anthropological Views." In *Rethinking the Family: Some Feminist Questions*, edited by B. Thorne and M. Yalom. New York: Longman, 1982.

Collins, Glenn. "U.S. Day-Care Guidelines Rekindle Controversy." *New York Times* (February 4, 1985): 20.

Collins, Patricia Hill. "The Meaning of Motherhood in Black Culture." In *The Black Family*, edited by Robert Staples. 4th ed. Belmont, CA: Wadsworth Publishing, 1991.

Coltrane, Scott, and Neal Hickman. "The Rhetoric of Rights and Needs: Moral Discourse in the Reform of Child Custody and Child Support Laws." *Social Problems* 39, 4 (1992): 400–420.

"A Comparative Survey of Minority Health." The Commonwealth Fund. New York, 1996.

Comstock, Jamie, and Krystyna Strzyzewski. "Interpersonal Interaction on Television: Family Conflict and Jealousy on Primetime." *Journal of Broadcasting and Electronic Media* 34, 3 (1990): 263–282.

Condry, J., and S. Condry. "The Development of Sex Differences: A Study of the Eye of the Beholder." *Child Development* 47, 4 (1976): 812–819.

Condron, John, and Jerry Bode. "Rashomon, Working Wives, and Family Division of Labor: Middletown." *Journal of Marriage and the Family* 44, 2 (May 1982): 421–426.

"Contract with America." *New York Times* (November 11, 1994): A10.

Conway, Colleen. "Psychophysical Preparations for Childbirth." In *Contemporary Obstetric and Gynecological Nursing*, edited by Leota McNall. St. Louis: Mosby, 1980.

Cook, Alicia, et al. "Changes in Attitudes toward Parenting among College Women: 1972 and 1979 Samples." *Family Relations* 31, 1 (January 1982): 109–113.

Cook, Mark, ed. *The Bases of Human Sexual Attraction.* New York: Academic Press, 1981.

Cooney, Teresa M. "Young Adults' Relations with Parents: The Influence of Recent Parental Divorce." *Journal of Marriage and the Family* 56, 1 (February 1994): 45–56.

Coontz, Stephanie. *The Way We Never Were: American Families and the Nostalgia Trap.* New York: Basic Books, 1992.

Cooper, A., and C. D. Stoltenberg. "Comparison of a Sexual Enhancement and a Communication Training Program on Sexual and Marital Satisfaction." *Journal of Counseling Psychology* 34 (July 1987): 309–314.

Cooper, P. F., and A. C. Monheit. "Does Employment-Related Health Insurance Inhibit Job Mobility?" *Inquiry* 30, 4 (1993): 400–416.

Corby, Nan, and Judy Zarit. "Old and Alone: The Unmarried in Later Life." In *Sexuality in the Later Years: Roles and Behavior*, edited by Ruth Weg. New York: Academic Press, 1983.

Corea, Gena. *The Mother Machine: Reproductive Technology from Artificial Insemination to Artificial Wombs.* New York: Harper & Row, 1985.

Cortese, Anthony. "Subcultural Differences in Human Sexuality: Race, Ethnicity, and Social Class." In *Human Sexuality: The Societal and Interpersonal Context*, edited by Kathleen McKinney and Susan Sprecher. Norwood, NJ: Ablex, 1989.

Cosby, Bill. "Someone at the Top Has to Say: 'Enough of This'." *Newsweek* (December 6, 1993): 60.

———. "The Regular Way." *Playboy* (December 1968): 288–289.

———. "We Are Losing." *Newsweek* (March 17, 1997): 58.

Costantini, Maria F., Laurie Wermuth, James L. Sorensen, and John S. Lyons. "Family Functioning as a Predictor of Progress in Substance Abuse Treatment." *Journal of Substance Abuse Treatment* 9, 4 (1992): 331–335.

Coulanges, Fustel de. *The Ancient City.* 1867. Reprint, New York: Anchor Books, 1960.

Courtright, Joseph, and Stanley Baran. "The Acquisition of Sexual Information by Young People." *Journalism Quarterly* 57, 1 (March 1980): 107–114.

Cowan, Alison Leigh. "Can a Baby-Making Venture Deliver?" *New York Times* (June 1, 1992): C1, C4.

Cowan, Philip, and Carolyn Cowan. "Becoming a Family: Research and Intervention." In *Methods of Family Research: Biographies of Research Projects,* edited by Irving Sigel and Gene Brody. Hillsdale, NJ.: Erlbaum, 1990.

Cowley, Geoffrey. "AIDS: The Next Ten Years." *Newsweek* (June 25, 1990): 20–8.

Craig, R. Stephen. "The Effect of Television Day Part on Gender Portrayals in Television Commercials: A Content Analysis." *Sex Roles: A Journal of Research* 26, 5–6 (1992): 197–211.

Cramer, D. "Gay Parents and Their Children: A Review of Research and Practical Implications." *Journal of Counseling and Development* 64 (1986): 504–507.

Cramer, David, and Arthur Roach. "Coming Out to Mom and Dad: A Study of Gay Males and Their Relationships with Their Parents." *Journal of Homosexuality* 14, 1–2 (1987): 77–88.

Cramer, Robert Ervin, M. Dragna, R. G. Cupp, and P. Stewart. "Contrast Effects in the Evaluation of the Male Sex Role." *Sex Roles* 24, 3/4 (1991): 181–193.

Creti, L., and E. Libman. "Cognition and Sexual Expression in the Aging." *Journal of Sex and Marital Therapy* 15, 2 (June 1989): 83–101.

Crohan, Susan E. "Marital Quality and Conflict Across the Transition to Parenthood in African American and White Couples." *Journal of Marriage and Family* 58, 4 (November 1996): 933–944.

Crosbie-Burnett, Margaret, and Edith Lewis. "Use of African-American Family Structures and Functioning to Address the Challenges of European-American Postdivorce Families." *Family Relations* 42, 3 (July 1993): 245–248.

Crosby, John, ed. *Reply to Myth: Perspectives on Intimacy.* New York: Wiley, 1985.

Crouter, Ann C., and Beth Manke. "The Changing American Workplace: Implications for Individuals and Families." *Family Relations* 43, 2 (April 1994): 117–124.

Crouter, Ann C., and Susan M. McHale. "The Long Arm of the Job: Influences of Parental Work on Child Rearing." In *Parenting: An Ecological Perspective,* edited by Tom Luster and Lynn Okagaki. Hillsdale, NJ: Erlbaum, 1993.

Culp, Rex E., Alicia S. Cook, and Pat C. Housley. "A Comparison of Observed and Reported Adult-Infant Interactions: Effects of Perceived Sex." *Sex Roles* 9 (April 1983): 475–479.

Cupach, William, and J. Comstock. "Satisfaction with Sexual Communication in Marriage." *Journal of Social and Personal Relationships* 7 (1990): 179–186.

Cupach, William, and Sandra Metts, "Sexuality and Communication in Close Relationships." In *Sexuality in Close Relationships,* edited by Kathleen McKinney and Susan Sprecher. Hillsdale, NJ: Erlbaum, 1991.

Curran, Dolores. *Traits of a Healthy Family.* New York: Ballantine, 1983.

Cushman, John H., Jr. "E.P.A.'s New Focus on Threat to the Poor." *New York Times* (January 21, 1992): B7.

Dail, Paula W. "Prime-time Television Portrayals of Older Adults in the Context of Family Life." *Gerontologist* 28, 5 (1988): 700–706.

"The Daily Record." *National Law Journal.* (March 12, 1997).

Dainton, M. "The Myths and Misconceptions of the Stepmother Identity: Descriptions and Prescriptions for Identity Management. *Family Relations* 42, 1 (January 1993): 93–98.

Daley, Sandra P. "Health Experience of Minority Families." In *Vision 2010: Families and Health Care,* edited by Barbara A. Elliott. Minneapolis: National Council on Family Relations, 1993.

Daley, Suzanne. "Girl's Self-Esteem Is Lost on Way to Adolescence, New Study Finds." *New York Times* (January 9, 1991): B1.

Daly, Kerry. "Reshaping Fatherhood: Finding the Models." *Journal of Family Issues* 14, 4 (December 1993): 510–530.

Danziger, S. "Antipoverty Policies and Child Poverty." *Social Work Research & Abstracts* 26, 4 (December 1990): 17–24.

Danziger, S. K., and S. Danziger. "Child Poverty and Public Policy—Toward a Comprehensive Antipoverty Agenda." *Daedalus* 122, 1 (December 1993): 57–84.

Darling, Carol A., and J. Kenneth Davidson. "Enhancing Relationships: Understanding the Feminine Mystique of Pretending Orgasm." *Journal of Sex and Marital Therapy* 12 (1986): 182–196.

Darling, Carol A., J. Kenneth Davidson, and Ruth P. Cox. "Female Sexual Response and the Timing of Partner Orgasm." *Journal of Sex & Marital Therapy,* 17, 1 (March 1991): 3–21.

Darling, Carol A., J. Kenneth Davidson, and D. A. Jennings. "The Female Sexual Response Revisited: Understanding the Multiorgasmic Experience in Women." *Archives of Sexual Behavior* 20, 6 (December 21, 1991): 527–540.

Darling-Fisher, Cynthia, and Linda Tiedje. "The Impact of Maternal Employment Characteristics on Fathers' Participation in Child Care." *Family Relations* 39, 1 (January 1990): 20–26.

Davidson, J. Kenneth, Carol Darling, and Colleen Conway-Welch. "The Role of the Grafenberg Spot and Female Ejaculation in the Female Orgasmic Response: An Empirical Analysis." *Journal of Sex and Marital Therapy* 15 (June 1989): 102–120.

Davidson, Kenneth, and Carol Darling. "Changing Autoerotic Attitudes and Practices among College Females: A Two-Year Follow-up Study." *Adolescence* 23 (December 1988a): 773–792.

———. "The Stereotype of Single Women Revisited." *Health Care for Women International* 9, 4 (October 1988b): 317–336.

Davidson, Kenneth, and Linda Hoffman. "Sexual Fantasies and Sexual Satisfaction: An Empirical Investigation of Erotic Thought." *Journal of Sex Research* 22 (May 1986): 184–205.

Davis, Karen. "The Nation's Health Care Safety Net: The Leadership Challenge for Health Philanthropy." Speech delivered at the Grantmakers in Health 1996 National Meeting on Health & Human Services Philanthropy, Los Angeles, February 1996.

Davis, L. *The Black Family in the United States.* Westport, CT: Greenwood Press, 1986.

Davis, S. "Men as Success Objects and Women as Sex Objects: A Study of Personal Advertisements." *Sex Roles* 23 (July 1990): 43–50.

Davis, Sally M., and Mary B. Harris. "Sexual Knowledge, Sexual Interest, and Sources of Sexual Information of Rural and Urban Adolescents from Three Cultures." *Adolescence* 17, 66 (June 1982): 471–492.

Dawson, Deborah. "The Effects of Sex Education on Adolescent Behavior." *Family Planning Perspectives* 18, 4 (July 1986): 162 ff.

"A Day in the Life of China." *Time* (October 2, 1989).

Deal, James E., Karen S. Wampler, and Charles F. Halverson. "The Importance of Similarity in the Marital Relationship." *Family Process* 31, 4 (1992): 369–382.

De Armand, Charlotte. "Let's Listen to What the Kids Are Saying." *SIECUS Reports* (March 1983): 3–4.

"Death Rates for Minority Infants Were Underestimated, Study Says." *New York Times* (January 8, 1992): A10.

Deaux, K. "From Individual Differences to Social Categories: Analysis of a Decade's Research on Gender." *American Psychologist* 39, 2 (1984): 105–116.

DeBuono, Barbara, et al. "Sexual Behavior in College Women in 1975, 1986, and 1989." *New England Journal of Medicine* 322, 12 (March 22, 1990): 821–825.

DeCecco, John, ed. *Gay Relationships.* New York: Haworth Press, 1988.

DeCecco, John, and Michael Shively. "From Sexual Identity to Sexual Relationships: A Conceptual Shift." *Journal of Homosexuality* 9, 2–3 (December 1983): 1–26.

DeCecco, John P., and J. P. Elia. "A Critique and Synthesis of Biological Essentialism and Social Constructionist Views of Sexuality and Gender. Introduction." *Journal of Homosexuality* 24, 3–4 (1993): 1–26.

Declerq, Eugene R. "The Transformation of American Midwifery: 1975–1988." *American Journal of Public Health* 82, 5 (May 1992): 680 ff.

DeFleur, M., and S. Ball-Rokeach. *Theories of Mass Communication.* New York: Longman, 1989.

Degler, Carl. *At Odds.* New York: Oxford University Press, 1980.

DeGroot, J. M., et al. "Correlates of Sexual Abuse in Women with Anorexia Nervosa and Bulimia Nervosa." *Canadian Journal of Psychiatry* 37, 7 (September 1992): 516–518.

Delamater, J. D., and P. MacCorquodale. *Premarital Sexuality: Attitudes, Relationships, Behavior.* Madison: University of Wisconsin Press, 1979.

DelCarmen, Rebecca. "Assessment of Asian-Americans for Family Therapy." In *Mental Health of Ethnic Minorities,* edited by Felicisima Serafica et al. New York: Praeger, 1990.

De Leon, Brunilda, "Sex Role Identity among College Students: A Cross-Cultural Analysis." *Hispanic Journal of the Behavioral Sciences* 15, 4 (1993): 476–489.

Demarest, Jack, and Jeanette Garner. "The Representation of Women's Roles in Women's Magazines over the Past 30 Years." *Journal of Psychology* 126, 4 (July 1992): 357–369.

DeMaris, Alfred, and K. Vaninadha Rao. "Premarital Cohabitation and Subsequent Marital Stability in the United States: A Reassessment." *Journal of Marriage and the Family* 54, 1 (February 1992): 178–190.

D'Emilio, John, and Estelle Freedman. *Intimate Matters: A History of Sexuality in America.* New York: Harper and Row, 1988.

Deming, Robert. "The Return of the Unrepressed: Male Desire, Gender, and Genre (New Directions in Television Studies: Essays in Honor of Beverle Ann Houston)." *Quarterly Review of Film and Video* 14, 1–2 (1992): 125–148.

Demo, David, and Alan Acock. "The Impact of Divorce on Children." In *Contemporary Families: Looking Forward, Looking Back,* edited by Alan Booth. Minneapolis: National Council on Family Relations, 1991.

Demo, David H., and Alan C. Acock. "Family Diversity and the Division of Domestic Labor: How Much Have Things Really Changed? *Family Relations* 42, 3 (July 1993): 323–331.

Demos, Vasilikie. "Black Family Studies in the *Journal of Marriage and the Family* and the Issue of Distortion: A Trend Analysis." *Journal of Marriage and the Family* 52, 3 (August 1990): 603–612.

Denniston, George C. "Unnecessary Circumcision." *Female Patient* 17 (July 1992): 13–14.

Dentzer, Susan. "Do the Elderly Want to Work?" *U.S. News and World Report* (May 14, 1990): 48–50.

Denzin, Norman. "Toward a Phenomenology of Domestic Family Violence." *American Journal of Sociology* 90, 30 (1984): 483–513.

DeParle, Jason. "The New Majority's Agenda: Welfare." *New York Times* (November 11, 1994): A14.

———. "Suffering in the Cities Persists as U.S. Fights Other Battles." *New York Times* (January 27, 1991): 15.

"Depiction of Women in Mainstream TV Shows Still Stereotypical, Sexual." *Media Report to Women* 21, 1 (1993): 4.

Depner, Charlene, and James Bray, eds. "Modes of Participation for Noncustodial Parents: The Challenge for Research, Policy, Practice, and Education." *Family Relations* 39, 4 (October 1990): 378–381.

Depner, Charlene, and James Bray, eds. *Nonresidential Parenting: New Vistas in Family Living.* Newbury Park, CA: Sage Publications, 1993.

Derdeyn, A., and E. Scott. "Joint Custody: A Critical Analysis and Appraisal." *American Journal of Orthopsychiatry* 54 (April 1984): 199–209.

Derlega, Valerian J., Sandra Metts, Sandra Petronio, and S. Margulis. *Self-Disclosure.* Newbury Park, CA: Sage Publications, 1993.

DeSpelder, Lynne Ann, and Albert Strickland. *The Last Dance: Encountering Death and Dying.* 5th ed. Mountain View, CA: Mayfield, 1996.

DeWitt, P.M. "Breaking Up Is Hard To Do." *American Demographics,* reprint package (1994): 14–16.

DiBlasio, Frederick A., and Brent B. Benda. "Gender Differences in Theories of Adolescent Sexual Activity." *Sex Roles* 27, 5/6 (1992): 221–236.

Dick-Read, Grantly. *Childbirth without Fear.* 4th ed. New York: Harper & Row, 1972.

Dilworth-Anderson, Peggye, Linda M. Burton, and William L. Turner. "The Importance of Values in the Study of Culturally Diverse Families." *Family Relations* 42, 3 (July, 1993): 238–242.

Dilworth-Anderson, Peggye, and Harriette Pipes McAdoo. "The Study of Ethnic Minority Families: Implications for Practitioners and Policymakers." *Family Relations* 37, 3 (July 1988): 265–267.

Dinnerstein, Leonard, and David Reimers, eds. *Ethnic Americans: A History of Immigration.* 3d ed. New York: Harper and Row, 1988.

Dion, Karen. "Physical Attractiveness, Sex Roles, and Heterosexual Attraction." In *The Bases of Human Sexual Attraction*, edited by Mark Cook. New York: Academic Press, 1981.

Dion, Karen, et al. "What Is Beautiful Is Good." *Journal of Personality and Social Psychology* 24 (1972): 285–290.

Dodson, Fitzhugh. "How to Discipline Effectively." In *Experts Advise Parents*, edited by Eileen Shiff. New York: Dell, 1987.

Dodson, Jualynne. "Conceptualizations of Black Families." In *Black Families*, edited by Harriette Pipes McAdoo. 2d ed. Newbury Park, CA: Sage Publications, 1988.

Doherty, William J. "Research Update for Practitioners: Families and Health." Paper presented at the annual conference of the National Council on Family Relations. Baltimore: November 12, 1993a.

———. "Training Health Professionals about Families." In *Vision 2010: Families and Health Care*, edited by Barbara A. Elliott. Minneapolis: National Council on Family Relations, 1993b.

Doherty, William J., and Thomas Campbell. *Families and Health.* Newbury Park, CA: Sage Publications, 1988.

Donnelly, Kathleen. "Breaking the Barriers." *San Jose Mercury News* (September 27, 1993): 1C, 8C.

Donovan, P. "New Reproductive Technologies: Some Legal Dilemmas." *Family Planning Perspectives* 18 (1986): 57 ff.

Dooley, Karen, and Ralph Catalano. "Stress Transmission: The Effects of Husbands' Job Stressors on the Emotional Health of Their Wives." *Journal of Marriage and the Family* 53, 1 (February 1991): 165–177.

Dorr, Aimee, Peter Kovaric, and Catherine Doubleday. "Age and Content Influences on Children's Perceptions of the Realism of Television Families." *Journal of Broadcasting and Electronic Media* 34, 4 (1990): 377–397.

Dorr, Aimee, et al. "Parent-Child Coviewing of Television." *Journal of Broadcasting and Electronic Media* 33, 1 (December 1989): 35–51.

Douglass, Frederick. *The Life and Times of Frederick Douglass.* New York: Collier Books, 1962. Originally published 1845.

Dreikurs, Rudolph, and V. Soltz. *Children: The Challenge.* New York: Hawthorne Books, 1964.

Drugger, Karen. "Social Location and Gender-Role Attitudes: A Comparison of Black and White Women." *Gender and Society* 2, 4 (December 1988): 425–448.

Duck, Steve, ed. *Dynamics of Relationships.* Thousand Oaks, CA: Sage Publications, 1994.

Duncan, Greg, and Willard Rodgers. "Longitudinal Aspects of Childhood Poverty." *Journal of Marriage and the Family* 50, 4 (November 1988): 1007–1022.

Duncombe, Jean, and Dennis Marsden. "Love and Intimacy: The Gender Division of Emotion and 'Emotion Work': A Neglected Aspect of Sociological Discussion of Heterosexual Relationships." *Sociology* 27, 2 (May 1993): 221–242.

Durant, Robert, R. Pendergast, and C. Seymore. "Contraceptive Behavior among Sexually Active Hispanic Adolescents." *Journal of Adolescent Health* 11, 6 (November 1990): 490–496.

Duvall, Evelyn. "Family Development's First Forty Years." *Family Relations* 37, 2 (April 1988): 127–134.

Duvall, Evelyn, and Brent Miller. *Marriage and Family Development.* 6th ed. New York: Harper & Row, 1985.

Dworetsky, John P. *Introduction to Child Development.* 4th ed. St. Paul, MN: West, 1990.

Eagly, Alice H. *Sex Differences in Social Behavior: A Social-Role Interpretation.* Hillsdale, NJ: Erlbaum, 1987.

Eakins, P. S. "Freestanding Birth Centers in California." *Journal of Reproductive Medicine* 34, 12 (December 1989): 960–970.

Edelman, Marian Wright. "Children at Risk." In *Caring for America's Children*, edited by Frank J. Macchiarola and Alan Gartner. New York: Academy of Political Science, 1989.

———. "Children at Risk." *Proceedings of the Academy of Political Science* 37, 2 (1989a): 20–30.

———. "Forward." In *Vanishing Dreams: The Growing Economic Plight of America's Young Families*, edited by Clifford Johnson et al. Washington, DC: Children's Defense Fund, 1988.

Edelman, Marian, and Lisa Mihaly. "Homeless Families and the Housing Crisis in the United States." *Children and Youth Services Review* 11, 1 (1989): 91–108.

Edleson, Jeffrey et al. "Men Who Batter Women." *Journal of Family Issues* 6, 2 (June 1985): 229–247.

Edmondson, Brad, Judith Waldrop, Diane Crispell, and Linda Jacobsen. "The Big Picture." *American Demographics* 15, 12 (December 1993): 28–30.

Edwards, Daniel. "Native American Elders: Current Issues and Social Policy Implications." In *Aging in Minority Groups*, edited by R. L. McNeely and John N. Colen. Beverly Hills, CA: Sage Publications, 1983.

Edwards, Gwenyth H. "The Structure and Content of the Male Gender Role Stereotype: An Exploration of Subtypes." *Sex Roles: A Journal of Research* 27, 9–10 (November 1992): 533–553.

Egan, Timothy. "Old, Ailing and Finally a Burden Abandoned." *New York Times* (March 26, 1992): A1, A9.

Egeland, Byron. "A History of Abuse Is a Major Risk Factor for Abusing the Next Generation." In *Current Controversies in Family Violence*, edited by Richard Gelles and Donileen Loseke. Newbury Park, CA: Sage Publications, 1993.

Ehrenkranz, Joel R., and Wylie C. Hembree. "Effects of Marijuana on Male Reproductive Function." *Psychiatric Annals* 16, 4 (April 1986): 243–248.

Ehrenreich, Barbara. *The Hearts of Men.* Garden City, NY: Anchor/Doubleday, 1984.

———. "Housework Is Obsolescent." *Time* (October 25, 1993): 92.

Eichler, Margrit. "Reflections on Motherhood, Apple Pie, the New Reproductive Technologies and the Role of Sociologists in Society." *Society-Société* 13, 1 (February 1989): 1–5.

Ekerdt, David J., and Stanley DeViney. "Evidence for a Preretirement Process among Older Male Workers." *Journals of Gerontology* 48, 2 (March 1993): S35–S43.

Elders, M. Joycelyn, and Jennifer M. Tuteur. "Public Health." In *Vision 2010: Families and Health Care*, edited by Barbara A. Elliott. Minneapolis: National Council on Family Relations, 1993.

Ellard, G. A., et al. "Smoking During Pregnancy: The Dose Dependence of Birthweight Deficits." *The British Journal of Obstetrics and Gynaecology* 103, 8 (1996): 806–813.

Elliott, Barbara, ed. *Vision 2010: Families & Health Care.* Minneapolis: National Council on Family Relations, 1993.

Elliott, Diana M., and John Briere. "Sexual Abuse Trauma Among Professional Women: Validating the Trauma Symptom Checklist (TSC-40)." *Child Abuse and Neglect* 16, 3 (May 1992): 391 ff.

Ellis, Albert. "The Justification of Sex Without Love." In *Reply to Myth: Perpectives on Intimacy,* edited by John Crosby. New York: Wiley, 1985.

Ellison, Carol. "Intimacy-Based Sex Therapy." In *Sexology,* edited by W. Eicher and G. Kockott. New York: Springer-Verlag, 1985.

Ellison, Christopher. "Family Ties, Friendships, and Subjective Well-Being among Black Americans." *Journal of Marriage and the Family* 52, 2 (May 1990): 298–310.

Ellwood, David. *Poor Support: Poverty in the American Family.* New York: Basic Books, 1988.

Emanuele, M. A., J. Tentler, N. V. Emanuele, and M. R. Kelley. "In-vivo Effects of Acute Etoh on Rat Alpha-Luteinizing and Beta-Luteinizing Hormone Gene Expression." *Alcohol* 8, 5 (September 1991): 345–348.

Emery, Robert E., Sheila G. Matthews, and Katherine M. Kitzmann. "Child Custody Mediation and Litigation: Parents' Satisfaction and Functioning One Year after Settlement." *Journal of Consulting and Clinical Psychology* 62, 1 (1994): 124–129.

Emery, Robert, and Melissa Wyer. "Child Custody Mediation and Litigation: An Experimental Evaluation of the Experience of Parents." *Journal of Consulting and Clinical Psychology* 55, 2 (1987): 179–186.

Erikson, Erik. *Childhood and Society.* New York: Norton, 1963.

———. *Vital Involvements in Old Age: The Experience of Old Age in Our Time.* Boston: W. W. Norton, 1986.

Erikson, Erik H. *Identity and the Life Cycle.* New York: International Universities Press, 1959.

———. *Identity, Youth, and Crisis.* New York: Norton, 1968.

Espín, Olivia M. "Cultural and Historical Influences on Sexuality in Hispanic/Latin Women: Implications for Psychotherapy." In *Pleasure and Danger: Exploring Female Sexuality,* edited by Carole Vance. Boston: Routledge & Kegan Paul, 1984.

Evans, H. L., et al. "Sperm Abnormalities and Cigarette Smoking." *Lancet* 1, 8221 (March 21, 1981): 627–629.

Evans-Pritchard, E. E. *Kinship and Marriage among the Nuer.* Oxford: Oxford University Press, 1951.

Fabes, Richard, and Jeremiah Strouse. "Perceptions of Responsible and Irresponsible Models of Sexuality." *Journal of Sex Research* 23, 1 (February 1987): 70–84.

Fagot, Beverly, and Mary Leinbach. "Socialization of Sex Roles within the Family." In *Current Conceptions of Sex Roles and Sex Typing,* edited by D. Bruce Carter. New York: Praeger, 1987.

Fagot, Beverly I., Mary D. Leinbach, and Cherie O'Boyle. "Gender Labeling, Gender Stereotyping and Parenting Behaviors." *Developmental Psychology* 28, 2 (1992): 225–231.

Falik, Marilyn and Karen Scott Collins, eds. *Women's Health.* Baltimore, MD: Johns Hopkins University Press, 1996.

Faludi, Susan. *Backlash: The Undeclared War Against American Women.* New York: Crown, 1991.

Farber, Bernard, Charles H. Mindel, and Bernard Lazerwitz. "The Jewish American Family." In *Ethnic Families in America: Patterns and Variations,* edited by Charles H. Mindel et al. 3d ed. New York: Elsevier North Holland, 1988.

Fay, Robert, Charles Turner, Albert Klassen, and John Gagnon. "Prevalence and Patterns of Same-Gender Sexual Contact among Men." *Science* 243, 4889 (January 20, 1989): 338–348.

Feeney, Judith A., and Patricia Noller. "Attachment Style as a Predictor of Adult Romantic Relationships." *Journal of Personality and Social Psychology* 58, 2 (February 1990): 281–291.

———. "Attachment Style and Verbal Descriptions of Romantic Partners." *Journal of Social and Personal Relationships* 8, 2 (1991): 187–215.

Feeney, Judith A., and Beverley Raphael. "Adult Attachments and Sexuality: Implications for Understanding Risk Behaviors for HIV Infection." *Australian and New Zealand Journal of Psychiatry* 26, 3 (1992): 399–407.

Fehr, Beverly. "Prototype Analysis of the Concepts of Love and Commitment." *Journal of Personality and Social Psychology* 55, 4 (1988): 557–579.

Fein, Robert. "Research on Fathering." In *The Family in Transition,* edited by Arlene Skolnick and Jerome Skolnick. Boston: Little, Brown, 1980.

Feirstein, Bruce. *Real Men Don't Eat Quiche.* New York: Pocket Books, 1982.

Feldman, Shirley, and Sharon Churnin. "The Transition from Expectancy to Parenthood." *Sex Roles* 11, 1/2 (1984): 61–78.

"Female Role Models Busy with Romance, Hair." *San Francisco Chronicle* (May 2, 1997).

Ferman, Lawrence. "After the Shutdown: The Social and Psychological Costs of Job Displacement." *Industrial and Labor Relations Report* 18, 2 (1981): 22–26.

Ferree, Myra Marx. "Beyond Separate Spheres: Feminism and Family Research." In *Contemporary Families: Looking Forward, Looking Back,* edited by Alan Booth. Minneapolis: National Council on Family Relations, 1991.

Field, Tiffany. "Quality Infant Day Care and Grade School Behavior and Performance." *Child Development* 62 (1991): 863–870.

Figley, Charles, ed. *Treating Families Under Stress.* New York: Brunner/Mazel, 1989.

Fine, Mark A. "Current Approaches to Understanding Family Diversity: An Overview of the Special Issue." *Family Relations* 42, 3 (July 1993): 235–237.

Fineman, Martha Albertson, and Roxanne Mykitiuk. *The Public Nature of Private Violence: The Discovery of Domestic Abuse.* New York: Routledge, 1994.

Finkel, J., and P. Roberts. *The Incomes of Noncustodial Fathers.* Washington, DC: Center for Law and Social Policy, 1994.

Finkel, Judith A., and Finy J. Hansen. "Correlates of Retrospective Marital Satisfaction in Long-Lived Marriages: A Social Constructivist Perspective." *Family Therapy* 19, 1 (1992): 1–16.

Finkelhor, David. *Child Sexual Abuse: New Theory and Research.* New York: Free Press, 1984.

———. "Common Features of Family Abuse." In *The Dark Side of Families,* edited by David Finkelhor et al. Beverly Hills, CA: Sage Publications, 1983.

———. "Prevention: A Review of Programs and Research." In *Sourcebook on Child Sexual Abuse,* edited by David Finkelhor. Beverly Hills, CA: Sage Publications, 1986a.

———. "Prevention Approaches to Child Sexual Abuse." In *Violence in the Home: Interdisciplinary Perspectives,* edited by Mary Lystad. New York: Brunner/Mazel, 1986b.

———. "Sexual Abuse of Children." In *Vision 2010: Families and Violence, Abuse and Neglect,* edited by Richard J. Gelles. Minneapolis: National Council on Family Relations, 1995.

———. "Sexual Abuse in a National Survey of Adult Men and Women." *Child Abuse and Neglect* 14, 1 (1990): 19–28.

———. *Sexually Victimized Children.* New York: Free Press, 1979.

Finkelhor, David, and Larry Baron. "High Risk Children." In *Sourcebook on Child Sexual Abuse,* edited by David Finkelhor. Beverly Hills, CA: Sage Publications, 1986.

Finkelhor, David, and Angela Browne. "Initial and Long-Term Effects: A Conceptual Framework." In *Sourcebook on Child Sexual Abuse,* edited by David Finkelhor. Beverly Hills, CA: Sage Publications, 1986.

Finkelhor, David, G. Hotaling, I. A. Lewis, and C. Smith. *Missing, Abducted, Runaway, and Throwaway Children in America.* Washington, DC: U.S. Department of Justice, 1990.

Finkelhor, David, Gerald Hotaling, and Andrea Sedlak. "Abduction of Children by Family Members." *Journal of Marriage and the Family* 53, 3 (August 1991): 805–817.

Finkelhor, David, and Karl Pillemer. "Elder Abuse: Its Relationship to Other Forms of Domestic Violence." In *Family Abuse and Its Consequences,* edited by Gerald T. Hotaling et al. Newbury Park, CA: Sage Publications, 1988.

Finkelhor, David, and Kersti Ÿllo. *Forced Sex in Marriage: A Preliminary Research Report.* National Institute of Mental Health. Washington, DC: U.S. Government Printing Office, 1980.

———. *License to Rape: The Sexual Abuse of Wives.* New York: Free Press, 1987.

———. "Rape in Marriage." In *Abuse and Victimization across the Life Span,* edited by Martha Straus. Baltimore, MD: Johns Hopkins University Press, 1988.

Finkelhor, David, et al. *The Dark Side of Families.* Beverly Hills, CA: Sage Publications, 1983.

———. "Sexual Abuse in Day Care: A National Study." *National Center on Child Abuse and Neglect.* University of New Hampshire: Family Research Laboratory, 1988.

Finlay, B., and Scheltema, K.E. "The Relation of Gender and Sexual Orientation to Measures of Masculinity, Femininity, and Androgyny: A Further Analysis." *Journal of Homosexuality* 21, 3 (1991): 71–85.

Fischer, Lucy R. "Mothers and Mothers-in-Law." *Journal of Marriage and the Family* 45, 1 (February 1983): 187–192.

Fischl, Margaret, et al. "Evaluation of Heterosexual Partners, Children and Household Contacts of Adults with AIDS." *Journal of the American Medical Association* 257, 5 (February, 1987): 640–647.

Fisher, W. A., et al. "Erotophobia-Erotophilia as a Dimension of Personality." *Journal of Sex Research* 25, 1 (1988): 123–151.

Fisher, William, and Donn Byrne. "Social Background, Attitudes, and Sexual Attraction." In *The Bases of Human Sexual Attraction,* edited by Mark Cook. New York: Academic Press, 1981.

Fishman, Barbara. "The Economic Behavior of Stepfamilies." *Family Relations* 32 (July 1983): 356–366.

Fitting, Melinda, Peter Rabins, M. Jane Lucas, and James Eastham. "Caregivers for Demented Patients: A Comparison of Husbands and Wives." *Gerontologist* 26 (1986): 248–252.

Fitzpatrick, Joseph. "The Puerto Rican Family." In *Ethnic Families in America,* edited by Charles Mindel and Robert Habenstein. 2nd ed. New York: Elsevier-North Holland, 1981.

Fitzpatrick, Joseph, and Lourdes Parker. "Hispanic-Americans in the Eastern United States." *Annals of the American Academy of Political and Social Science* 454 (March 1981): 98–110.

Flaherty, Mary Jean, Lorna Facteau, and Patricia Garver. "Grandmother Functions in Multigenerational Families." In *The Black Family: Essays and Studies,* edited by Robert Staples. 4th ed. Belmont, CA: Wadsworth Publishing, 1990.

Flaks, David K., Ilda Ficher, Frank Masterpasqua, and G. Joseph. "Lesbians Choosing Motherhood: A Comparative Study of Lesbians and Heterosexual Parents and Their Children." *Developmental Psychology* 31, 1 (January 1995): 105–114.

Floge, Liliane. "The Dynamics of Child-Care Use and Some Implications for Women's Employment." *Journal of Marriage and the Family* 47, 1 (February 1985): 143–154.

Floyd, Frank J., Stephen N. Haynes, Elizabeth R. Doll, David Winemiller et al. "Assessing Retirement Satisfaction and Perceptions of Retirement Experiences." *Psychology & Aging* 7, 4 (1992): 609–621.

Flynn, Clifton P. "Sex Roles and Women's Response to Courtship Violence." *Journal of Family Violence* 5 1 (March 1990): 83–94.

Foa, Uriel G., Barbara Anderson, J. Converse, and W. A. Urbansky, "Gender-Related Sexual Attitudes: Some Cross-Cultural Similarities and Differences." *Sex Roles* 16, 19–20 (May 1987): 511–519.

Fogel, Daniel. *Junipero Serra, the Vatican, and Enslavement Theology.* San Francisco: ISM Press, 1988.

Foley, L., et al. "Date Rape: Effects of Race of Assailant and Victim and Gender of Subjects." *Journal of Black Psychology* 21, 1 (February 1995): 6–18.

Folk, Karen F., and Andrea H. Beller. "Part-Time Work and Child Care Choices for Mothers of Preschool Children." *Journal of Marriage and the Family* 55, 1 (February 1993): 147–157.

Folk, Karen Fox and Yunae Yi. "Piecing Together Child Care with Multiple Arrangements: Crazy Quilt or Preferred Pattern for Employed Parents of Preschool Children." *Journal of Marriage and the Family* 56, 3 (August 1994): 669–680.

Follingstad, Diane R., L. L. Rutledge, B. J. Berg, and E. S. Haure. "The Role of Emotional Abuse in Physically Abusive Relationships." *Journal of Family Violence* 5, 2 (June 1990): 107–120.

Foote, Nelson, and Leonard Cottrell. *Identity and Interpersonal Competence.* Chicago: University of Chicago Press, 1955.

Ford, Donna. "An Exploration of Alternative Family Structures among University Students." *Journal of Marriage and the Family* 43, 1 (January 1994): 68–73.

Forehand, Rex L., Page B. Walley, and William M. Furey. "Prevention in the Home: Parent and Family." In *Prevention of Problems in Childhood: Psychological Research and Application*, edited by Michael C. Roberts and Lizette Peterson. New York: Wiley, 1984.

Forste, Renata, and Tim Heaton. "Initiation of Sexual Activity among Female Adolescents." *Youth and Society* 19, 3 (March 1988): 250–268.

Foucault, Michel. *The History of Sexuality: An Introduction.* Vol. I., New York: Pantheon Books, 1980.

Fowler, Orison S. *Amativeness: or, Evils and the Remedies of Excessive Perverted Sexuality.* New York: Fowler and Wells, 1878.

Fox, Freer L., and Robert F. Kelly. "Determinants of Child Custody Arrangements at Divorce." *Journal of Marriage and the Family* 57, 3 (August 1995): 693–708.

"Fractional Families (Dear Dr. Demo)." *American Demographics* 14, 12 (December 1992): 6.

Francke, Linda Bird. "Childless by Choice." *Newsweek* (January 14, 1980): 96.

Franklin, D. L. "The Impact of Early Childbearing on Development Outcomes: The Case of Black Adolescent Parenting." *Family Relations* 37 (1988): 268–274.

Franks, Lucinda J., James P. Hughes, Linda H. Phelps, and D. G. Williams. "Intergenerational Influences on Midwest College Students by Their Grandparents and Significant Elders." *Educational Gerontology* 19, 3 (1993): 265–271.

Franks, P., C. M. Clancy, M. R. Gold, and P. A. Nutting. "Health Insurance and Subjective Health Status." *American Journal of Public Health* 83, 9 (1993): 1295–1299.

Franz, Wanda, and David Readon. "Differential Impact of Abortion on Adolescents and Adults." *Adolescence* 27, 105 (March 1992): 161–172.

Freed, Doris, and Timothy Walker. "Family Law in the Fifty States." *Family Law Quarterly* 21 (1988): 417–573.

Freeman, Edith M., ed. "Substance Abuse Treatment: A Family Systems Perspective." Sage Sourcebooks for the Human Services Series, No. 25. Newbury Park, CA.: Sage Publications, 1993.

Freeman, H. E., and C. R. Corey. "Insurance Status and Access to Health Services among Poor Persons." *Health Services Research* 28, 5 (1993): 531–541.

French, J. P., and Bertram Raven. "The Bases of Social Power." In *Studies in Social Power*, edited by L. Cartwright. Ann Arbor: University of Michigan Press, 1959.

Friday, Nancy. *Men in Love.* New York: Delacorte, 1980.

Friedman, Rochelle, and Bonnie Gradstein. *Surviving Pregnancy Loss.* Boston: Little, Brown, 1982.

Frisbie, W. Parker. "Variation in Patterns of Marital Instability among Hispanics." *Journal of Marriage and the Family* 48, 1 (February 1986): 99–106.

Fromm, Erich. *The Art of Loving.* New York: Perennial Library, 1974.

Fu, Haishan, and Noreen Goldman. "Incorporating Health into Models of Marriage Choice: Demographic and Sociological Perspectives." *Journal of Marriage and the Family* 58, 3 (August 1996): 740–758.

Fuller, Mary Lou. "Help Your Family Understand 'It's a Small, Small World.'" *PTA Today* (December 1989): 9–10.

Furstenberg, F. F., and A. J. Cherlin. *Divided Families: What Happens to Children When Parents Part.* Cambridge, MA: Harvard University Press, 1991.

Furstenberg, F. F., and J. O. Teitler. "Reconsidering the Effects of Marital Disruption: What Happens to Children of Divorce in Early Adulthood?" *Journal of Family Issues*, 15, 2 (June 1994): 173–190.

Furstenberg, Frank K., Jr. "The New Extended Family: The Experience of Parents and Children after Remarriage." In *Remarriage and Stepparenting: Current Research and Theory*, edited by Kay Pasley and Marilyn Ihinger-Tallman. New York: Guilford Press, 1987.

———. "Reflections on Remarriage." *Journal of Family Issues* 1, 4 (1980): 443–453.

Furstenberg, Frank K., Jr., and Andrew Cherlin. *Divided Families.* Cambridge, MA: Harvard University Press, 1991.

Furstenberg, Frank K. Jr., and Christine Nord. "Parenting Apart: Patterns in Childrearing after Marital Disruption." *Journal of Marriage and the Family* 47, 4 (November 1985): 893–904.

Furstenberg, Frank K., Jr., and Graham Spanier, eds. *Recycling the Family—Remarriage after Divorce.* Rev ed. Newbury Park, CA: Sage Publications, 1987.

Furstenberg, Frank K. Jr., R. Lincoln, and J. Menken, eds. *Teenage Sexuality, Pregnancy, and Childbearing.* Philadelphia: University of Pennsylvania Press, 1981.

Furukawa, S. "The Diverse Living Arrangements of Children." Series P70-38. Washington, DC: Current Population Reports, 1993.

Gagnon, John. "Attitudes and Responses of Parents to Pre-Adolescent Masturbation." *Archives of Sexual Behavior* 14, 5 (1985): 451–466.

———. "Sexual Scripts: Permanence and Change." *Archives of Sexual Behavior* 15, 2 (April 1986): 97–120.

Gagnon, John, and William Simon. "The Sexual Scripting of Oral Genital Contacts." *Archives of Sexual Behavior* 16, 1 (February 1987): 1–25.

———. *Sexual Conduct: The Social Sources of Human Sexuality.* Chicago: Aldine Publishing Co., 1973.

Gaines, Judith. "A Scandal of Artificial Insemination." *Good Health Magazine/New York Times Magazine* (October 7, 1990): 23 ff.

Gallup, George H., Jr. *The Gallup Poll: Public Opinion 1986.* Wilmington, DE: Scholarly Resources, 1987.

Gallup, George H., Jr., and Frank Newport. "Parenthood—A Nearly Universal Desire." *San Francisco Chronicle* (June 4, 1990), B3.

Ganong, Lawrence, et al. "A Meta-Analytic Review of Family Structure Stereotypes." *Journal of Marriage and the Family* 52, 2 (May 1990): 287–289.

———. "Stepparent: A Pejorative Term?" *Psychological Reports* 53, 3 (June 1983): 919–922.

Ganong, Lawrence, and Marilyn Coleman. "A Comparison of Clinical and Empirical Literature on Children in Stepfamilies." *Journal of Marriage and the Family* 48 (May 1986): 309–318.

———. "The Effects of Remarriage on Children: A Review of the Empirical Literature." *Family Relations* 33 (1984): 389–405.

———. "Gender Differences in Expectations of Self and Future Partner." *Journal of Family Issues* 13, 1 (March 1992): 55–64.

———. *Remarried Family Relationships.* Newbury Park, CA: Sage Publications, 1994.

———. "Sex, Sex Roles, and Family Love." *Journal of Genetic Psychology* 148 (March 1987): 45–52.

Gao, Ge. "Stability of Romantic Relationships in China and the United States." In *Cross-Cultural Interpersonal Communication*, edited by Stella Ting-Toomey and Felipe Korzenny. Newbury Park, CA: Sage Publications, 1991.

Garbarino, James. *Children and Families in the Social Environment.* Hawthorne, NY: Aldine De Gruyter, 1982.

Garfinkel, I., S. S. McLanahan, and P. R. Robins, eds. *Child Support and Child Well-Being.* Washington, DC: Urban Institute Press, 1994.

Garfinkel, Irwin, and Sara McLanahan. *Single Mothers and Their Children: A New American Dilemma.* Washington, DC: Urban Institute Press, 1986.

Garfinkel, Irwin, Donald Oellerich, and Philip K. Robins. "Child Support Guidelines: Will They Make a Difference?" *Journal of Family Issues* 12, 4 (December 1991), 404–429.

Garnets, L., et al. "Violence and Victimization of Lesbians and Gay Men: Mental Health Consequences." *Journal of Interpersonal Violence* 5 (1990): 366–383.

Garrison, James E. "Sexual Dysfunction in the Elderly: Causes and Effects." *Journal of Psychotherapy and the Family* 5, 1–2 (1989): 149–162.

Gary, Lawrence. "Strong Black Families: Models of Program Development for Black Families." In *Family Strengths: Positive Models for Family Life*, edited by Sally Van Sandt et al. Lincoln, NE: University of Nebraska Press, 1980.

Gay, Peter. *The Bourgeois Experience: The Tender Passion.* New York: Oxford University Press, 1986.

Gecas, Viktor, and Monica Seff. "Families and Adolescents." In *Contemporary Families: Looking Forward, Looking Back*, edited by Alan Booth. Minneapolis: National Council on Family Relations, 1991.

———. "Social Class, Occupational Conditions, and Self-Esteem." *Sociological Perspectives* 32 (1989): 353–364.

Geis, Frances, and Joseph Geis. *Marriage and the Family in the Middle Ages.* New York: Harper and Row, 1987.

Geist, Christopher. "Violence, Passion, and Sexual Racism: The Plantation Novel." *Southern Quarterly* 18, 2 (December 1980): 60–72.

Gelfand, Donald and Charles Barresi, eds. *Ethnic Dimensions of Aging.* New York: Springer Publishing, 1987.

Gelles, Richard J. "Alcohol and Other Drugs Are Associated with Violence—They Are Not Its Cause." In *Current Controversies in Family Violence*, edited by Richard Gelles and Donileen Loseke. Newbury Park, CA: 1993.

———. "Child Abuse and Violence in Single-Parent Families: Parent Absence and Economic Deprivation." *American Journal of Orthopsychiatry* 59, 4 (October 1989): 492–501.

———. *Intimate Violence in Families.* Thousand Oaks, CA: Sage Publications, 1997.

———. "Through a Sociological Lens: Social Structure and Family Violence." In *Current Controversies in Family Violence*, edited by Richard Gelles and Donileen Loseke. Newbury Park, CA: Sage Publications, 1993.

Gelles, Richard J., and Jon R. Conte. "Domestic Violence and Sexual Abuse of Children: A Review of Research in the Eighties." In *Contemporary Families: Looking Forward, Looking Back*, edited by Alan Booth. Minneapolis: National Council on Family Relations, 1991.

Gelles, Richard J., and Claire Pedrick Cornell. *Intimate Violence in Families.* 2nd ed. Newbury Park, CA: Sage Publications, 1987.

Gelles, Richard, and Donileen Loseke, eds. *Current Controversies in Family Violence.* Newbury Park, CA: Sage Publications, 1993.

Gelles, Richard J., and Murray A. Straus. "Determinants of Violence in the Family: Toward a Theoretical Integration." In *Contemporary Theories About the Family*, edited by Wesley Burr. New York: Free Press, 1979.

Genevie, Lou, and Eva Margolies. *The Motherhood Report: How Women Feel about Being Mothers.* New York: Macmillan, 1987.

Genovese, Eugene. *Roll, Jordan, Roll.* New York: Harper & Row, 1976.

George, Kenneth, and Andrew Behrendt. "Therapy for Male Couples Experiencing Relationship Problems and Sexual Problems." *Journal of Homosexuality* 14, 1–2 (1987): 77–88.

Geraghty, Christine. *Women and Soap Opera: A Study of Prime Time Soaps.* London: Polity Press, 1990.

Gernber, G., L. Gross, M. Morgan, and N. Signorielli. "Living with Television: The Dynamics of the Cultivation Process." In *Perspectives in Media Effects*, edited by J. Bryant and D. Zillman. Hillsdale, NJ: Erlbaum, 1986.

Gerson, Kathleen. *Hard Choices: How Women Decide About Work, Career, and Motherhood.* Berkeley: University of California Press, 1985.

Gerstel, Naomi. "Divorce and Stigma." *Social Forces* 34 (April 1987): 172–186.

———. *Families and Work.* Philadelphia: Temple University Press, 1987.

Gerstel, Naomi, and Harriet Gross. *Commuter Marriage: A Study of Work and Family.* New York: Guilford Press, 1984.

Getlin, Josh. "Legacy of a Mother's Drinking." *Los Angeles Times* (July 24, 1989): V1 ff.

Giarretto, Henry. "Humanistic Treatment of Father-Daughter Incest." In *Child Abuse and Neglect: The Family and the Community*, edited by Ray Helfer and Henry C. Kempe. Cambridge, MA: Ballinger, 1976.

Gibson, Rose. "Blacks at Middle and Late Life: Resources and Coping." *Annals of the American Academy* 464 (November 1982): 79–90.

Gilbert, Lucia, et al. "Perceptions of Parental Role Responsibilities: Differences between Mothers and Fathers." *Family Relations* 31 (April 1982): 261–269.

Gillen, K., and S. J. Muncher. "Sex Differences in the Perceived Casual Structure of Date Rape: A Preliminary Report." *Aggressive Behavior* 21, 2 (1995): 101–112.

Gilligan, Carol. *In a Different Voice: Psychological Theory and Women's Development.* Cambridge, MA: Harvard University Press, 1982.

Gilliss, Catherine L. "Health Professionals Serving Families." In *Vision 2010: Families and Health Care*, edited by Barbara A. Elliott. Minneapolis: National Council on Family Relations, 1993.

Gilman, Lois. *The Adoption Resource Book.* Rev. ed. New York: Harper & Row, 1987.

Givens, Ron. "A New Prohibition." *Newsweek on Campus* (April 1985), 7–13.

Glaser, Ronald, and Janice Kiecolt-Glaser. *Handbook of Human Stress and Immunity.* San Diego: Academic Press, 1994.

Glass, Robert H., and Ronald J. Ericsson. *Getting Pregnant in the 1980s.* Berkeley: University of California Press, 1982.

Glazer-Malbin, Nona, ed. *Old Family/New Family.* New York: Van Nostrand, 1975.

Glenn, Evelyn N., and Stacey G. H. Yap. "Chinese American Families." In *Minority Families in the United States: A Multicultural Perspective*, edited by R. L. Taylor. Englewood Cliffs, NJ: Prentice Hall, 1994.

Glenn, Norval. "Duration of Marriage, Family Composition, and Marital Happiness." *National Journal of Sociology* 3 (1989): 3–24.

Glenn, Norval, and Kathryn Kramer. "The Marriages and Divorces of the Children of Divorce." *Journal of Marriage and the Family* 49, 4 (November 1987): 811–825.

Glenn, Norval, and Beth Ann Shelton. "Regional Differences in Divorce in the United States." *Journal of Marriage and the Family* 47, 3 (August 1985): 641–652.

Glenn, Norval, and Michael Supancic. "The Social and Demographic Correlates of Divorce and Separation in the United States: An Update and Reconsideration." *Journal of Marriage and the Family* 46, 3 (August 1984): 563–575.

Glick, Jennifer E., Frank D. Bean, and Jennifer V. W. Van Hook. "Immigration and Changing Patterns of Extended Family Household Structure in the United States: 1970–1990." *Journal of Marriage and the Family* 59, 1 (February 1997): 177–191.

Glick, Paul. "American Household Structure in Transition." *Family Planning Perspectives* 16, 5 (September/October 1984a): 205–211.

———. "Marriage, Divorce, and Living Arrangements: Prospective Changes." *Journal of Family Issues* 4, 1 (March 1984b): 7–26.

———. "The Family Life Cycle and Social Change." *Family Relations* 38, 2 (April 1989): 123–129.

———. "Fifty Years of Family Demography." *Journal of Marriage and the Family* 50, 4 (November 1988): 861–873.

———. "Remarried Families, Stepfamilies and Stepchildren: A Brief Demographic Analysis." *Family Relations* 38 (1989): 24–27.

Glick, Paul, and Sung Ling Lin. "Recent Changes in Divorce and Remarriage." *Journal of Marriage and the Family* 49, 4 (November 1986): 737–747.

Glick, Paul, and Graham Spanier. "Married and Unmarried Cohabitation in the United States." *Journal of Marriage and the Family* 42, 1 (February 1980): 19–30.

Gnezda, Therese. *The Effects of Unemployment on Family Functioning.* Prepared Statement to the Select Committee on Children, Youth and Families, House of Representatives, at Hearings on the New Unemployed, Detroit, March 4, 1984. Washington, DC: Government Printing Office, 1984.

Go, K. J. "Recent Advances in the Treatment of Male Infertility." *Naacogs Clinical Issues in Perinatal and Women's Health Nursing* 3, 2 (1992): 320–327.

Gochoros, J. S. *When Husbands Come Out of the Closet.* New York: Harrington Park Press, 1989.

Goelman, Hillel, et al. "Family Environment and Family Day Care." *Family Relations* 39, 1 (January 1990): 14–19.

Goetting, Ann. "The Developmental Tasks of Siblingship over the Life Cycle." *Journal of Marriage and the Family* 48, 4 (November 1986): 703–714.

———. "Divorce Outcome Research." *Journal of Family Issues* 2, 3 (September 1981): 20–25.

———. "Patterns of Support among In-Laws in the United States: A Review of Research." *Journal of Family Issues* 11, 1 (1990): 67–90.

———. "The Six Stages of Remarriage: Developmental Tasks of Remarriage after Divorce." *Family Relations* 31 (April 1982): 213–222.

Goetz, Kathryn W. and Cynthia J. Schmiege. "From Marginalized to Mainstreamed: The HEART Project Empowers the Homeless." *Family Relations* 45, 4 (October 1996).

Goldenberg, H., and I. Goldenberg. *Counseling Today's Families.* 2nd ed. Pacific Grove, CA: Brooks/Cole, 1994.

Goldsmith, Jean. "The Postdivorce Family." In *Normal Family Process*, edited by Froma Walsh. New York: Guilford Press, 1982.

Goldstein, David, and Alan Rosenbaum. "An Evaluation of the Self-Esteem of Maritally Violent Men." *Family Relations* 34, 3 (July 1985): 425–428.

Goleman, Daniel. "Gay Parents Called No Disadvantage." *New York Times* (March 11, 1992).

———. "How Viewers Grow Addicted to Television." *New York Times* (October 16, 1990): C1, 8.

———. "A Modern Tradeoff: Longevity for Health." *New York Times* (May 16, 1991): B10.

———. "Marriage: Research Reveals Ingredients of Happiness." *New York Times* (April 16, 1985a): 19–20.

———. "Spacing of Siblings Strongly Linked to Success in Life." *New York Times* (May 28, 1985b): 17–18.

Golombok, S. "Psychological Functioning of Infertility Patients." *Human Reproduction* 7, 2 (February 1992): 208–212.

Gondolf, Edward. "Male Batterers." In *Family Violence: Prevention and Treatment*, edited by R. L. Hampton et al. Newbury Park, CA: Sage Publications, 1993.

Gondolf, Edward W. "The Effect of Batterer Counseling on Shelter Outcome." *Journal of Interpersonal Violence* 3, 3 (September 1988): 275–289.

———. "Evaluating Progress for Men Who Batter." *Journal of Family Violence* 2 (1987): 95–108.

Gongla, Patricia, and Edward Thompson, Jr., "Single Parent Families." In *Handbook of Marriage and the Family*, edited by Marvin Sussman and Suzanne Steinmetz. New York: Plenum Press, 1987.

Goode, William. *World Revolution and Family Patterns.* New York: Free Press, 1963.

———. "Force and Violence in the Family." *Journal of Marriage and the Family* 33 (November 1971): 624–636.

Goode, William, ed. *The Family.* 2d ed. Englewood Cliffs, NJ: Prentice Hall, 1982.

Goodenow, Carol, and Oliva M. Espín. "Identity Choices in Immigrant Adolescent Females." *Adolescence* 28, 109 (1993): 173–185.

Goodman, Ellen. "Welfare Reform Plank Won't Work." *San Jose Mercury News* (December 9, 1994): A1.

———. "Where Family Values Begin—And End." *Washington Post* (September 24, 1994): A27.

Goodman, Walter. "TV's Sexual Circus Has a Purpose." *New York Times* (August 4, 1992): B2.

Gordon, Jacob U. "A Culturally Specific Approach to Ethnic Minority Young Adults." In *Substance Abuse Treatment: A Family Systems Perspective*, edited by Edith M. Freeman. Newbury Park, CA: Sage Publications, 1993.

Gordon, Sol. "Parents as Sexuality Educators." *SIECUS Report* (March 1984): 10–11.

Gordon, Thomas. *P.E.T. in Action*. New York: Bantam Books, 1978.

Gottman, John M. *Why Marriages Succeed or Fail and How You Can Make Yours Work*. New York: Simon and Schuster, 1995.

Gottman, John M., and Lynn F. Katz. "Effects of Marital Discord on Young Children's Peer Interaction and Health." *Developmental Psychology* 25, 3 (May 1989): 373–381.

Gough, Kathleen. "Is the Family Universal: The Nayer Case." In *A Modern Introduction to the Family*, edited by Norman Bell and Ezra Vogel. New York: Free Press, 1968.

Gould, Jeffrey B., Becky Davey, and Randall S. Stafford. "Socioeconomic Differences in Rates of Cesarean Section." *New England Journal of Medicine* 321, 4 (July 27, 1989): 233–239.

Gove, Walter, et al. "Does Marriage Have Positive Effects on the Psychological Well-Being of the Individual?" *Journal of Health and Social Behavior* 24 (1983): 122–131.

Greeley, Andrew. "Sit-Coms as Modern Morality Plays." *New York Times* (May 17, 1987).

Greenberg, B. S. "Content Trends in Media Sex." In *Media, Children, and the Family: Social Scientific, Psychodynamic, and Clinical Perspectives*, edited by D. Zillman, J. Bryant and A.C. Huston. Hillsdale, NJ: Erlbaum, 1994.

Greenberg, Bradley S. *Life on Television: Content Analysis of U.S. TV Drama*. Norwood, NJ: Ablex Publishing Co., 1980.

Greenberg, Bradley S., and D. D. D'Alessio. "Quantity and Quality of Sex in the Soaps." *Broadcasting and Electronic Media* 29 (1985): 309–321.

Greenberg, Brigitte. "Barbie, GI Joe Get New Lines from a Doll-Liberation Group." *San Jose Mercury News* (December 29, 1993): 8A.

Greenberg, Dan, and Marsha Jacobs. *How to Make Yourself Miserable*. New York: Random House, 1966.

Greenberger, Ellen. "Explaining Role Strain: Intrapersona." *Journal of Marriage and the Family* 52, 1 (February 1994): 115–118.

Greenberger, Ellen, and Robin O'Neil. "Parents' Concern about Their Child's Development: Implications for Father's and Mother's Well-Being and Attitudes toward Work." *Journal of Marriage and the Family* 52, 3 (August 1990): 621–635.

Greenfield, Patricia M., Lisa Buzzone, Kristi Koyamatsu, and Wendy Satuloff. "What Is Rock Music Doing to the Minds of Our Youth? A First Experimental Look at the Effects of Rock Music Lyrics and Music Videos." *Journal of Early Adolescence* 7, 3 (1987): 315–329.

Greenstein, Theodore. "Marital Disruption and the Employment of Married Women." *Journal of Marriage and the Family* 52 (1990): 657–676.

Greenstein, Theodore N. "Husbands' Participation in Domestic Labor: Interactive Effects of Wives' and Husbands" Gender Ideologies." *Journal of Marriage and the Family* 58, 3 (August 1996): 585–595.

Greeson, Larry E. "Recognition and Ratings of Television Music Videos: Age, Gender, and Sociocultural Effects." *Journal of Applied Social Psychology* 21, 23 (1991): 1908–1920.

Greeson, Larry, and Rose Ann Williams. "Social Implications of Music Videos for Youth: An Analysis of the Content and Effects of MTV." *Youth and Society* 18, 2 (December 1986): 177–189.

Gregor, Thomas. *Anxious Pleasures: The Sexual Lives of an Amazonian People*. Chicago: University of Chicago Press, 1985.

Greif, Geoffrey. "Children and Housework in the Single Father Family." *Family Relations* 34, 3 (July 1985a): 353–357.

———. "Single Fathers Rearing Children." *Journal of Marriage and the Family* 47, 1 (February 1985b): 185–191.

Greif, Geoffrey L., and Rebecca L. Hegar. "Impact on Children of Abduction by a Parent: A Review of the Literature." *American Journal of Orthopsychiatry* 62, 4 (1992): 599–604.

Greif, Geoffrey L., and Joan Kristall. "Common Themes in a Group for Noncustodial Parents." *Families in Society* 74, 4 (1993): 240–245.

Gringlas, Marcy, and Marsha Weinraub. "The More Things Change . . . Single Parenting Revisited." *Journal of Family Issues* 16, 1 (January 1995): 29–52.

Grossman, Michele, and Wendy Wood. "Sex Differences in Intensity of Emotional Experience: A Social Role Interpretation." *Journal of Personality and Social Psychology* 65, 5 (1993): 1010–1023.

Grotevant, H.D., R.G. McCoy, C. Elde, and D.L. Frave. "Adoptive Family System Dynamics: Variations by Level of Openness in the Adoption." *Family Process* 33 (1994): 125–146.

Groth, Nicholas. *Men Who Rape: The Psychology of the Offender*. New York: Plenum Press, 1980.

Gruson, Lindsey. "Groups Play Matchmaker to Preserve Judaism." *New York Times* (April 1, 1985).

Gubman, Gayle, and Richard Tessler. "The Impact of Mental Illness on Families." *Journal of Family Issues* 8, 2 (June 1987): 226–245.

Guelzow, Maureen, et al. "Analysis of the Stress Process for Dual-Career Men and Women." *Journal of Marriage and the Family* 53, 1 (February 1991): 151–164.

Guerrero Pavich, Emma. "A Chicana Perspective on Mexican Culture and Sexuality." In *Human Sexuality, Ethnoculture, and Social Work*, edited by Larry Lister. New York: Haworth, 1986.

Guidubaldi, John. "The Status Report Extended: Further Elaborations on the American Family." *School Psychology Review* 9, 4 (September 1980): 374–379.

Gump, J. "Reality and Myth: Employment and Sex Role Ideology in Black Women." In *The Psychology of Women*, edited by F. Denmark and J. Sherman. New York: Psychological Dimensions, 1980.

Gunn, C. Douglas. "Family Identity Creation." In *Family Strengths: Positive Models for Family Life*, edited by Nick Stinnett. Lincoln, NE: University of Nebraska Press, 1980.

Guralnik, J. M., K. C. Land, D. Blazer, C. G. Fillenbaum, and L. G Branch. "Educational Status and Active Life Expectancy among Older Blacks and Whites." *New England Journal of Medicine* 329, 2 (1993): 110–116.

Gutek, Barbara. *Sex and the Workplace.* San Francisco: Jossey-Bass, 1985.

Gutek, Barbara, et al. "Sexuality and the Workplace." *Basic and Applied Social Psychology* 1, 3 (1980): 255–265.

Guttentag, M., and P. Secord. *Too Many Women.* Newbury Park, CA: Sage Publications, 1983.

Guttman, Herbert. *The Black Family: From Slavery to Freedom.* New York: Pantheon, 1976.

Guttman, Joseph. *Divorce in Psychosocial Perspective: Theory and Research.* Hillsdale, NJ: Erlbaum, 1993.

Gwartney-Gibbs, Patricia. "The Institutionalization of Premarital Cohabitation: Estimates from Marriage License Applications." *Journal of Marriage and the Family* 48, 2 (May 1986): 423–434.

Haaga, D. A. "Homophobia?" *Journal of Behavior and Personality* 6, (1991): 171–174.

Haferd, Laura. "Paddling Returns to Child Rearing." *San Jose Mercury News* (December 20, 1986): D12.

Hafstrom, Jeanne, and Vicki Schram. "Chronic Illness in Couples: Selected Characteristics, Including Wife's Satisfaction with and Perception of Marital Relationships." *Family Relations* 33 (1984): 195–203.

Hajal, Fady, and Elinor B. Rosenberg. "The Family Life Cycle in Adoptive Families." *American Journal of Orthopsychiatry* 61, 1 (January 1991): 78–85.

Hales, Dianne. *An Invitation to Health.* 7th ed. Pacific Grove, CA: Brooks/Cole Publishing Company, 1997.

Haley, Alex. "Counselors Can Improve Society through Families." Paper presented at the annual meeting of the American Association for Counseling and Development, Los Angeles, 1986.

———. *Roots.* New York: Dell, 1976.

Hall, D. R., and J. Z. Zhao. "Cohabitation and Divorce in Canada: Testing the Selectivity Hypothesis." *Journal of Marriage and the Family* 57 (May 1995): 421-427.

Hallstrom, Tore, and Sverker Samuelsson. "Changes in Women's Sexual Desire in Middle Life: The Longitudinal Study of Women in Gothenburg [Sweden]." *Archives of Sexual Behavior* 19, 3 (1990): 259–267.

Hamburg, David A. *Today's Children.* New York: Times Books, 1992.

Hamilton D. L. "A Cognitive-Attributional Analysis of Stereotyping." *Advances in Experimental Psychology* 12 (1979): 53–81.

Hampson, Robert, Robert Beavers, and Yosaf Hulgus. "Cross-Ethnic Family Differences: Interaction Assessments of White, Black, and Mexican-American Families." In *Cultural Diversity and Families,* edited by Karen G. Arms et al. Dubuque, IA: William C. Brown, 1992.

Hampton, Richard. "Violence in Families of Color." In *Vision 2010: Families and Violence, Abuse and Neglect,* edited by Richard J. Gelles. Minneapolis: National Council on Family Relations, 1995.

Hampton, Robert L., ed. *Black Family Violence: Current Research and Theory.* Lexington, MA: Lexington, 1991.

Hampton, Robert L., Richard J. Gelles, and John W. Harrop. "Is Violence in Black Families Increasing? A Comparison of 1975 and 1985 National Survey Rates." *Journal of Marriage and the Family* 51, 4 (November 1989): 969–980.

Hampton, Robert L., Thomas P. Gullotta, Gerald R. Adams, Earl H. Potter III, and Roger P. Weissberg, eds. *Family Violence: Prevention and Treatment.* Newbury Park, CA: Sage Publications, 1993.

Handy, Bruce. "Roll Over, Ward Cleaver." *Time* (April 14, 1997): 78–85.

Hanley, Robert. "Surrogate Deals for Mothers Held Illegal in New Jersey." *New York Times* (February 4, 1988): 1 ff.

Hanline, Mary F. "Transitions and Critical Events in the Family Life Cycle: Implications for Providing Support to Families of Children with Disabilities." *Psychology in the Schools* 28, 1 (1991): 53–59.

Hansen, Christine H., and Ranald D. Hansen. "The Influence of Sex and Violence on the Appeal of Rock Music Videos." *Communication Research* 17, 2 (1990): 212–234.

Hansen, Gary. "Dating Jealousy among College Students." *Sex Roles* 12, 7–8 (April 1985): 713–721.

———. "Balancing Work and Family: A Literature and Resource Review." *Family Relations* 40, 3 (July 1991): 348–353.

———. "Extradyadic Relations during Courtship." *Journal of Sex Research* 23, 3 (August 1987): 383–390.

Hanson, B., and C. Knopes. "Prime Time Tuning Out Varied Cultures." *USA Today* (July 6, 1993).

Hanson, F. W., et al. "Ultrasonography—Guided Early Amniocentesis in Singleton Pregnancies." *American Journal of Obstetrics and Gynecology* 162, 6 (June 1990): 1381–1383.

Hanson, Sandra L., David Myers, Alan L. Ginsburg. "The Role of Responsibility and Knowledge in Reducing Teenage Out-of-Wedlock Childbearing." *Journal of Marriage and the Family* 49, 2 (May 1987): 241–256.

Hanson, Sandra, and Theodora Oooms. "The Economic Costs and Rewards of Two-Earner, Two-Parent Families." *Journal of Marriage and the Family* 53, 3 (August 1991): 622–644.

Hanson, Thomas L., Sara S. McLanahan, and Elizabeth Thomson. "Double Jeopardy: Parental Conflict and Stepfamily Outcomes for Children." *Journal of Marriage and the Family,* 58, 1 (February 1996): 141–154.

Hare-Mustin, Rachel T., and Jeanne Marecek. "Beyond Difference." In *Making a Difference: Psychology and the Construction of Gender,* edited by Rachel T. Hare-Mustin and Jeanne Marecek. New Haven, CT: Yale University Press, 1990.

———. "On Making a Difference." In *Making a Difference: Psychology and the Construction of Gender*, edited by Rachel T. Hare-Mustin and Jeanne Maracek. New Haven, CT: Yale University Press, 1990b.

Hare-Mustin, R. T., and J. Marecek, eds. *Making a Difference: Psychology and the Construction of Gender,* New Haven, CT: Yale University Press, 1990a.

Haring-Hidore, Marilyn, et al. "Marital Status and Subjective Well-Being: A Research Synthesis." *Journal of Marriage and the Family* 47, 4 (November 1985): 947–953.

Harriman, Lynda. "Marital Adjustment as Related to Personal and Marital Changes Accompanying Parenthood." *Family Relations* 35, 2 (April 1986): 233–239.

———. "Personal and Marital Changes Accompanying Parenthood." *Family Relations* 32, 3 (July 1983): 387–394.

Harris, Kathleen, and S. Philip Morgan. "Fathers, Sons, and Daughters: Differential Paternal Involvement in Parenting." *Journal of Marriage and the Family* 53, 3 (August 1991): 531-544.

Harris, Sandra. "The Family and the Autistic Child." *Family Relations* 33, 1 (January 1984): 67–77.

Harris, Shanette M. "The Influence of Personal and Family Factors on Achievement Needs and Concerns of African-American and Euro-American College Women." *Sex Roles* 29:9–10 (November 1993): 671–689.

Harrison, Margaret. "The Reformed Australian Child Support Scheme." *Journal of Family Issues* 12, 4 (December 1991): 430–449.

Harrison, Patricia, et al. "Differential Drug Use Patterns among Sexually Abused Adolescent Girls in Treatment for Chemical Dependency." *International Journal of the Addictions* 24, 6 (June 1989): 499–514.

Harry, Joseph. "Decision Making and Age Differences among Gay Male Couples." In *Gay Relationships*, edited by John DeCecco. New York: Haworth, 1988.

Hart, J., E. Cohen, A. Gingold, and R. Homburg. "Sexual Behavior in Pregnancy: A Study of 219 Women." *Journal of Sex Education and Therapy* 17, 2 (June 1991): 88–90.

Haskell, Molly. "2000-Year-Old Misunderstanding: Rape Fantasy." *Ms.* (November 1976): 84–86 ff.

———. *From Reverence to Rape*. 2d ed. Chicago: University of Chicago Press, 1987.

Hatcher, Robert, et al. *Contraceptive Technology*. New York: Irvington, 1986.

Hatchett, Shirley J. "Women and Men." In *Life in Black America*, edited by James S. Jackson. Newbury Park, CA: Sage Publications, 1991.

Hatchett, Shirley, and James S. Jackson. "African-American Extended Kin System: An Assessment." In *Family Ethnicity: Strength in Diversity*, edited by Harriette Pipes McAdoo. Newbury Park, CA: Sage Publications, 1993.

Hatfield, Elaine. "Passionate and Companionate Love." *The Psychology of Love*, edited by Robert Sternberg and Michael Barnes. New Haven, CT: Yale University Press, 1988.

Hatfield, Elaine, and Richard Rapson. "Passionate Love/Sexual Desire: Can the Same Paradigm Explain Both?" *Archives of Sexual Behavior* 16, 3 (June 1987): 259–278.

Hatfield, Elaine, and Susan Sprecher. *Mirror, Mirror: The Importance of Looks in Everyday Life*. New York: State University of New York, 1986.

Hatfield, Elaine, and G. William Walster. *A New Look at Love*. Reading, MA: Addison-Wesley, 1981.

Havemann, Joel. "Diagnosis: Healthier in Europe." *Los Angeles Times* (December 30, 1992): A1, A9.

Hawkins, Alan J., and Tomi-Ann Roberts. "Designing a Primary Intervention to Help Dual-Earner Couples Share Housework and Childcare." *Family Relations* 41, 2 (April 1992): 169–177.

Hawkins, Alan J., Tomi-Ann Roberts, Shawn Christiansen, and C. M. Marshall. "An Evaluation of a Program to Help Dual-Earner Couples Share the Second Shift." *Family Relations* 43, 2 (April 1994): 213–220.

Hayes, Robert M. "Homeless Children." In *Caring for America's Children*, edited by Frank J. Macchiarola and Alan Gartner. New York: The Academy of Political Science, 1989.

Hays, Dorothea, and Aurele Samuels. "Heterosexual Women's Perceptions of their Marriages to Bisexual or Homosexual Men." *Journal of Homosexuality* 18, 2 (1989): 81–100.

Hazan, Cindy, and Philip Shaver. "Romantic Love Conceptualized as an Attachment Process." *Journal of Personality and Social Psychology* 52 (March 1987): 511–524.

Healy, Jane M. "Preventing Birth Defects of the Mind." *Parents' Magazine*, 63 (November 1988): 176 ff.

Healy, Joseph M., Abigail, J. Stewart, and Anne P. Copeland. "The Role of Self-Blame in Children's Adjustment to Parental Separation." *Personality and Social Psychology Bulletin* 19, 3 (1993): 279–289.

Heaton, Tim, and Stan Albrecht. "Stable Unhappy Marriages." *Journal of Marriage and the Family* 53, 3 (August 1991): 747–758.

Heaton, Tim B., and E. L. Pratt. "The Effects of Religious Homogamy on Marital Satisfaction and Stability." *Journal of Family Issues* 7, 2 (June 1990): 191–207.

Heaton, Tim, et al. "The Timing of Divorce." *Journal of Marriage and the Family* 47, 3 (August 1985): 631–639.

Hecht, Michael L., Peter J. Marston, and Linda Kathryn Larkey. "Love Ways and Relationship Quality in Heterosexual Relationships." *Journal of Social & Personal Relationships* 11, 1 (February 1994): 25–43.

Heer, D. M. "The Prevalence of Black-White Marriage in the United States, 1960 and 1970." *Journal of Marriage and the Family* 36 (1974): 246–259.

Hefner, R., et al. "Development of Sex-Role Transcendence." *Human Development* 18 (1975): 143–158.

Heilbrun, Alfred, et al. "Parent Identification and Gender Schema Development." *Journal of Genetic Psychology* 150, 3 (September 1989): 293–300.

Heilbrun, Carolyn. *Toward a Recognition of Androgyny*. New York: Norton, 1982.

Heiman, Julia, et al. *Becoming Orgasmic: A Sexual Growth Program for Women*. Englewood Cliffs, NJ: Prentice Hall, 1976.

Hemstrom, Orjan. "Is Marriage Dissolution Linked to Differences in Morbidity Risks for Men and Women?" *Journal of Marriage and the Family* 58, 2 (May 1996): 366–378.

Hendrick, Clyde, and Susan Hendrick. "Attachment Theory and Close Adult Relationships." *Psychological Inquiry* 5, 1 (1994): 38–41.

———. "Research on Love: Does It Measure Up?" *Journal of Personality and Social Psychology* 56, 5 (May 1989): 784–794.

———. "A Theory and Method of Love." *Journal of Personality and Social Psychology* 50 (February 1986): 392–402.

Hendrick, Clyde, et al. "Do Men and Women Love Differently?" *Journal of Personality and Social Psychology* 48 (1984): 177–195.

Hendrick, Susan. "Self-Disclosure and Marital Satisfaction." *Journal of Personality and Social Psychology* 40 (1981): 1150–1159.

Hendrick, Susan, and Clyde Hendrick. "Multidimensionality of Sexual Attitudes." *Journal of Sex Research* 23, 4 (November 1987): 502–526.

Hendricks, Glenn, et al., eds. *The Hmong in Transition*. Staten Island, NY: Center for Migration Studies of New York, 1986.

Henley, Nancy. *Body Politics: Power, Sex, and Nonverbal Communication.* Englewood Cliffs, NJ: Prentice Hall, 1977.

Henton, June, et al. "Romance and Violence in Dating Relationships." *Journal of Family Issues* 4, 3 (September 1983): 467–482.

Hepworth, J., S. McDaniel, and W. Doherty. "Medical Family Therapy with Families Coping with AIDS." In *Counseling Families with Chronic Illness.* American Counseling Association, 1993.

Herbert, Bob. "Scapegoat Time." *New York Times* (December 16, 1994): A19.

Herek, Gregory. "Beyond Homophobia: A Social Psychological Perspective on Attitudes toward Lesbians and Gay Men." *Journal of Homosexuality* 10, 1–2 (September 1984): 1–21.

———. "Hate Crimes against Lesbians and Gay Men: Issues for Research and Policy." *American Psychologist* 44 (June 1989): 948–955.

———. "On Heterosexual Masculinity: Some Phsychical Consequences of the Social Construction of Gender." *American Behavioral Scientist* 29, 5 (May 1986): 563–567.

———. "The Social Psychology of Homophobia: Toward a Practical Theory." *Review of Law and Social Change* 14, 4 (1986): 923–934.

Herold, Edward, and Leslie Way. "Oral-Genital Sexual Behavior in a Sample of University Females." *Journal of Sex Research* 19, 4 (November 1983): 327–338.

Herring, Cedric and Karen R. Wilson-Sadberry. "Preference or Necessity? Changing Work Roles of Black and White Women, 1973–1990." *Journal of Marriage and the Family* 55, 2 (May 1993): 314–325.

Hetherington, E. Mavis. "Divorce: A Child's Perspective." *American Psychologist* 34, 10 (October 1979): 851–858.

Hetherington, E. Mavis, et al. "Family Interactions and the Social, Emotional, and Cognitive Development of Children Following Divorce." In *The Family Setting Priorities,* edited by V. C. Vaugh and T. B. Brazelton. New York: Science and Medicine Publishers, 1979.

Hetherington, S. E. "A Controlled Study of the Effect of Prepared Childbirth Classes on Obstetric Outcomes." *Birth* 17, 2 (June 1990): 86–90.

Hevesi, Dennis. "Homeless in New York City: A Day on the Streets." *New York Times* (November 17, 1986): 13.

———. "TV News: Children's Scary Window on New York." *New York Times* (September 11, 1990): A 21.

Hewitt, J. "Preconceptual Sex Selection." *British Journal of Hospital Medicine* 37, 2 (February 1987): 149 ff.

Heyl, Barbara. "Homosexuality: A Social Phenomenon." In *Human Sexuality: The Societal and Interpersonal Context,* edited by Kathleen McKinney and Susan Sprecher. Norwood, NJ: Ablex, 1989.

Hicks, Mary. "Dual Career/Dual Worker Families: A Systems Approach." In *Contemporary Families and Alternative Lifestyles,* edited by Eleanor Macklin and Roger Rubin. Beverly Hills, CA: Sage Publications, 1983.

Higginbottom, Susan F., Julian Barling, and E. Kevin Kelloway. "Linking Retirement Experiences and Marital Satisfaction: A Mediational Model." *Psychology and Aging* 8, 4 (1993): 508–516.

Hill, Charles, et al. "Breakups before Marriage: The End of 103 Affairs." *Journal of Social Issues* 32 (1976): 147–168.

Hill, Ivan. *The Bisexual Spouse.* McLean, VA: Barlina Books, 1987.

Hill, Martha. "The Changing Nature of Poverty." *Annals of the American Academy of Political Science* 479 (May 1985): 31–37.

Hill, Reuben. "Generic Features of Families Under Stress." *Social Casework* 49 (February 1958): 139–150.

Hilton, Jeanne, and Virginia Haldeman. "Gender Differences in the Performance of Household Tasks by Adults and Children in Single-Parent and Two-Parent, Two-Earner Families." *Journal of Family Issues* 12, 1 (March 1991): 114–130.

Hilton, N. Zoe. "When Is an Assault Not an Assault? The Canadian Public's Attitudes Toward Wife and Stranger Assault." *Journal of Family Violence* 4, 4 (December 1989): 323–337.

Hilts, Philip. Growing Concern over Pelvic Infection in Women. *New York Times* (October 11, 1991): B7.

———. "Life Expectancy for Blacks in U.S. Shows Sharp Drop." New York Times (November 20, 1990): A1, B7.

Hiltz, Roxanne. "Widowhood: A Roleless Role." *Marriage and Family Review* 1, 6 (November 1978).

Hite, Shere. *The Hite Report.* New York: Macmillan, 1976.

Hobart, Charles. "Interest in Marriage among Canadian Students at the End of the Eighties." *Journal of Comparative Family Studies* 24, 1 (1993): 45–61.

Hobart, Charles, and Frank Griegel. "Cohabitation among Canadian Students at the End of the Eighties." *Journal of Comparative Family Studies* 23, 3 (1992): 311–338.

Hochschild, Arlie. *The Second Shift: Working Parents and the Revolution at Home.* New York: Viking Press, 1989.

Hodgson, Lynne G. "Adult Grandchildren and Their Grandparents: Their Enduring Bond." *International Journal of Aging and Human Development* 34, 3 (1992): 209–225.

Hoegerman, G., et al. "Drug Exposed Neonates." *Western Journal of Medicine* 152, 1 (May 1990): 559 ff.

Hoelter, Jon W. "Factoral Invariance and Self-Esteem-Reassessing Race and Sex Differences." *Social Forces* 61, 3 (March 1983): 834–846.

Hoelter, Jon, and Lynn Harper. "Structural and Interpersonal Family Influences on Adolescent Self-Conception." *Journal of Marriage and the Family* 49, 1 (February 1987): 129–139.

Hofferth, Sandra. "Updating Children's Life Course." *Journal of Marriage and the Family* 47, 1 (February 1985): 93–115.

Hoffman, Lois Wladis. "Maternal Employment: 1979." *American Psychologist* 34 (1979): 859–865.

Holder, Kathleen. "Drugs on the Rise among Students, State Study on Substance Abuse Finds." *Mercury News* (June 22, 1994): C6.

Holmes, Steven. "On the Edge of Despair When Jobless Benefits End." *New York Times* (January 28, 1991): A11.

Holmes, T., and R. Rahe. "The Social Readjustment Rating Scale." *Journal of Psychosomatic Medicine* 11 (1967): 213–218.

Holtzen, D. W., and A. A. Agresti. "Parental Responses to Gay and Lesbian Children." *Journal of Social and Clinical Psychology* 9, 3 (September 1990): 390–399.

Honeycutt, James. *Memory, Gender, and Relationships.* New York: Guilford Press, 1994a.

Honeycutt, James, Lynn B. Wellman, and Mary S. Larson. "Social Learning Theory and the Calculation of Televised Family Communication Influence: A Time-Series Analysis of Turn-at-Talk on a Popular Family Program." New Orleans, LA: Annual Speech Communication Association Conference, November 19, 1992.

Hong, George K., and MaryAnna D. Ham. "Impact of Immigration of the Family Life Cycle: Clinical Implications for Chinese Americans." *Journal of Family Psychotherapy* 3, 3 (1992): 27–40.

Hopkins, Ellen. "Childhood's End." *Rolling Stone* (October 18, 1990): 66–72 ff.

Horowitz, R. *Honor and the American Dream.* New Brunswick, NJ: Rutgers University Press, 1983.

Hort, Barbara, et al. "Are People's Notions of Maleness More Stereotypically Framed Than Their Notions of Femaleness?" *Sex Roles* 23, 3 (February 1990): 197–212.

Hort, Barbara E., M. D. Leinbach, and B. I. Fagot. "Is There Coherence among the Cognitive Components of Gender Acquisition?" *Sex Roles* 24, 3–4 (1991): 195–207.

Hotaling, Gerald T., and David Finkelhor. "Estimating the Number of Stranger-Abduction Homicides of Children: A Review of Available Evidence." *Journal of Criminal Justice* 18, 5 (1990): 385–399.

Hotaling, Gerald T., and David B. Sugarman. "An Analysis of Risk Markers in Husband to Wife Violence: The Current State of Knowledge." *Violence & Victims* 1, 2 (June 1986): 101–124.

———. "A Risk Marker Analysis of Assaulted Wives." *Journal of Family Violence* 5, 1 (March 1990): 1–14.

Hotaling, Gerald T., et al., eds. *Coping with Family Violence: Research and Perspectives.* Newbury Park, CA: Sage Publications, 1988.

———. *Family Abuse and Its Consequences.* Newbury Park, CA: Sage Publications, 1988.

Houseknecht, Sharon K. "Voluntary Childlessness." In *Handbook of Marriage and the Family,* edited by Marvin B. Sussman and Suzanne K. Steinmetz. New York: Plenum Press, 1987.

Houseknecht, Sharon K., Suzanne Vaughan, and Ann Statham. "The Impact of Singlehood on the Career Patterns of Professional Women." *Journal of Marriage and the Family* 49, 2 (May 1987): 353–366.

Howard, Judith. "A Structural Approach to Interracial Patterns in Adolescent Judgments about Sexual Intimacy." *Sociological Perspectives* 31, 1 (January 1988): 88–121.

Howes, Carollee. "Can the Age of Entry into Child Care and the Quality of Child Care Predict Adjustment in Kindergarten?" *Developmental Psychology* 26 (1990): 292–303.

Huang, Lucy. "The Chinese American Family." In *Ethnic Families in America: Patterns and Variations,* edited by Charles Mindel et al. 3rd ed. New York: Elsevier-North Holland, 1988.

Hudak, Mary A. "Gender Schema Theory Revisited: Men's Stereotypes of American Women." *Sex Roles: A Journal of Research* 28, 5–6 (1993): 279–293.

Hughes, Jean O'Gorman, and Bernice R. Sandler. "Friends Raping Friends—Could It Happen to You?" Project on the Status and Education of Women, Association of American Colleges, 1987.

Hunter, Andrea G., and James Earl Davis. "Constructing Gender: An Exploration of Afro-American Men's Conceptualization of Manhood." *Gender and Society* 6, 3 (September 1992): 464–479.

Hupka, Ralph B. "Cultural Determinants of Jealousy." *Alternative Lifestyles* 4, 3 (August 1981): 310–356.

Hurst, Marsha, and Pamela S. Summey. "Childbirth and Social Class: The Case of Cesarean Delivery." *Social Science and Medicine* 18, 8 (1984): 621–631.

Huston, Ted, S. M. McHale, and A. C. Crouter. "When the Honeymoon's Over: Changes in the Marriage Relationship over the First Year." In *The Emerging Field of Personal Relationships,* edited by Steve Duck and Robin Gilmour. Hillsdale, NJ: Erlbaum, 1986.

Huston, Ted, et al. "From Courtship to Marriage: Mate Selection as an Interpersonal Process." In *Personal Relationships 2: Developing Personal Relationships,* edited by Steve Duck and Robin Gilmour. London: Academic Press, 1981.

Huston, Ted L., and Gilbert Geis. "In What Ways Do Gender-Related Attributes and Beliefs Affect Marriage?" *Journal of Social Issues* 49, 3 (1993): 87–106.

Hutchins, Loraine, and Lani Kaahumanu, eds. *Bi Any Other Name: Bisexual People Speak Out.* Boston: Alyson Publications, 1991a.

Hutchins, Loraine, and Lani Kaahuman. "Overview." In *Bi Any Other Name: Bisexual People Speak Out,* edited by Loraine Hutchins and Lani Kaahumanu. Boston: Alyson Publications, 1991b.

Ickes, William. "Traditional Gender Roles: Do They Make, and Then Break, Our Relationships?" *Journal of Social Issues* 49, 3 (1993): 71–77.

Ihinger-Tallman, Marilyn. "Sibling and Stepsibling Bonding in Stepfamilies." In *Remarriage and Stepparenting Current Research and Theory,* edited by Kay Pasley and Marilyn Ihinger-Tallman. New York: Guilford Press, 1987.

Ihinger-Tallman, Marilyn, and Kay Pasley. "Divorce and Remarriage in the American Family: A Historical Review." In *Remarriage and Stepparenting: Current Research and Theory,* edited by Kay Pasley and Marilyn Ihinger-Tallmann. New York: Guilford Press, 1987a.

———. *Remarriage.* Newbury Park, CA: Sage Publications, 1987b.

"Immunization Information". Washington, DC: Centers For Disease Control and Prevention (March 9, 1995).

Indivik, Julie, and Mary Fitzpatrick. "'If You Could Read My Mind, Love . . .,' Understanding and Misunderstanding in the Marital Dyad." *Family Relations* 44, 4 (November 1982): 43–51.

Irving, Howard, et al. "Shared Parenting: An Empirical Analysis Utilizing a Large Data Base." *Family Process* 23 (1984): 561–569.

Isensee, Rik. *Love Between Men: Enhancing Intimacy and Keeping Your Relationship Alive.* New York: Prentice Hall Press, 1990.

Ishii-Kuntz, Masako. "Japanese American Families." In *Families in Cultural Context,* edited by Mary Kay DeGenova. Mountain View, CA: Mayfield, 1997.

Itard, Jean. "The Wild Boy of Aveyron." In *Wolf Children and the Problem of Human Nature,* edited by Lucien Malson. New York: Monthly Review Press, 1972. Originally published 1801.

Ja, Davis Y., and Bart Aoki. "Substance Abuse Treatment: Cultural Barriers in the Asian-American Community." *Journal of Psychoactive Drugs* 25, 1 (1993): 61–71.

Jackson, Robert L. "Panel Calls for U.S. to Curb Infant Deaths." *Los Angeles Times* (December 16, 1993): A37.

Jacoby, Arthur, and John Williams. "Effects of Premarital Sexual Standards and Behavior on Dating and Marriage Desirability." *Journal of Marriage and the Family* 47, 4 (November 1985): 1059–1065.

Janus, Samuel and Cynthia Janus. *The Janus Report.* New York: Wiley, 1993.

Jeffrey, T. B., and L. K. Jeffrey. "Psychologic Aspects of Sexual Abuse in Adolescence." *Current Opinion in Obstetrics and Gynecology* 3, 6 (December 1991): 825–831.

Jencks, Christopher. "The Homeless." *The New York Review of Books* 41, 8 (April 1994): 20–27.

———. *The Homeless.* Cambridge, MA: Harvard University Press, 1994.

———. "Housing the Homeless." *The New York Review of Books* 41, 9 (May 1994): 39–46.

Jenks, Richard. "Swinging: A Replication and Test of a Theory." *Journal of Sex Research* 21, 2 (May 1985): 199–210.

Jensen, Larry C., and Janet Jensen. "Family Values, Religiosity, and Gender." *Psychological Reports* 73, 2 (October 1993): 429–430.

Jensen, M. A. *Love's Sweet Return: The Harlequin Story.* Toronto: Women's Press, 1984.

Jensen-Scott, Rhonda L. "Counseling to Promote Retirement Adjustment." *Career Development Quarterly* 1, 3 (1993): 257–267.

Joanning, Harvey, William Quinn, Frank Thomas, and Robert Mullen. "Treating Adolescent Drug Abuse: A Comparison of Family Systems Therapy, Group Therapy, and Family Drug Education." *Journal of Marital and Family Therapy* 18, 4 (1992): 345–356.

Joe, Tom, and Douglas W. Nelson. "New Future for America's Children." In *Caring for America's Children,* edited by Frank J. Macchiarola and Alan Gartner. New York: Academy of Political Science, 1989.

Joesch, Jutta. "The Effects of the Price of Child Care on AFDC Mothers' Paid Work Behavior." *Family Relations* 40, 2 (April 1991): 161–166.

John, Robert. "The Native American Family." In *Ethnic Families in America: Patterns and Variations,* edited by Charles Mindel et al. 3d ed. New York: Elsevier North Holland, Inc., 1988.

Johnson, Beverly. "Single Parent Families." *Family Economics Review* (June 1980): 22–27.

Johnson, Catherine B., Margaret S. Stockdale, and Frank E. Saal. "Persistence of Men's Misperceptions of Friendly Cues across a Variety of Interpersonal Encounters." *Psychology of Women Quarterly* 15, 3 (September 1991): 463–475.

Johnson, Clifford, et al. *Vanishing Dreams: The Growing Economic Plight of America's Young Families.* Washington, DC: Children's Defense Fund, 1988.

Johnson, David, Lynn White, John Edwards, and Alan Booth. "Dimensions of Marital Quality: Toward Methodological and Conceptual Refinement." *Journal of Family Issues* 7 (1986): 31–49.

Johnson, Dirk. "At Colleges, AIDS Alarms Muffle Older Dangers." *New York Times* (March 8, 1990): B8.

Johnson, Leanor Boulin. "Perspectives on Black Family Empirical Research: 1965–1978. In *Black Families,* edited by Hariette Pipes McAdoo. Newbury Park, CA: Sage Publications, 1988.

Johnson, Michael P., Ted L. Huston, Stanley O. Gaines, and George Levinger. "Patterns of Married Life among Young Couples." *Journal of Social and Personal Relationships* 9, 3 (1992): 343–364.

Johnston, Thomas F. "Alaskan Native Social Adjustment and the Role of Eskimo and Indian Music." *Journal of Ethnic Studies* 3/4 (December 1976): 21–36.

Jones, Carl. *Mind Over Labor.* New York: Penguin, 1988.

Jones, Jennifer, and David Barlow. "Self-Reported Frequencies of Sexual Urges, Fantasies, and Masturbatory Fantasies in Heterosexual Males and Females." *Archives of Sexual Behavior* 19, 3 (1990): 269–279.

Jones, Maggie. *A Child by Any Means.* London: Piatkus, 1989.

Jones, T. S., and M. S. Remland. "Sources of Variability in Perceptions of and Responses to Sexual Harassment." *Sex Roles* 27, 3–4 (August 1992): 121–142.

Jorgensen, Stephen, and Russell Adams. "Predicting Mexican-American Family Planning Intentions: An Application and Test of a Social Psychological Model." *Journal of Marriage and the Family* 50, 1 (February 1988): 107–119.

Jorgensen, Stephen R., and A. C. Johnson. "Correlates of Divorce Liberality." *Journal of Marriage and the Family* 42 (1980): 617–622.

Juni, Samuel, and Donald W. Grimm. "Marital Satisfaction and Sex-Roles in a New York Metropolitan Sample." *Psychological Reports* 73, 1 (1993): 307–314.

Jurich, Anthony, and Cheryl Polson. "Nonverbal Assessment of Anxiety as a Function of Intimacy of Sexual Attitude Questions." *Psychological Reports* 57 (3, Pt. 2), (December 1985): 1247–1243.

Justice, Blair, and Rita Justice. *The Abusing Family.* Rev. ed. New York: Insight Books, 1990.

———. *The Broken Taboo: Incest.* New York: Human Sciences Press, 1979.

Kach, Julie, and Paul McGee. "Adjustment to Early Parenthood." *Journal of Family Issues* 3, 3 (September 1982): 375–388.

Kagan, Jerome, and N. Snidman. "Temperamental Factors in Human Development." *American Psychologist* 46, 8 (1991): 856–862.

Kain, Edward. *The Myth of Family Decline: Understanding Families in a World of Rapid Change.* Lexington, MA: Lexington Books, 1990.

Kalin, Tom. "Gays in Film: No Way Out." *US (Special Issue: The Sexual Revolution in Movie, Music & TV),* 175 (August 1992): 68–70.

Kalis, Pamela, and Kimberly Neuendorf. "Aggressive Cue-Prominence and Gender Participation in MTV." *Journalism Quarterly* 66, 1 (March 1989): 148–154, 229.

Kalmuss, Debra. "The Intergenerational Transmission of Marital Aggression." *Journal of Marriage and the Family* 46, 1 (February 1984): 11–20.

Kalof, Linda. "Dilemmas of Femininity: Gender and the Social Construction of Sexual Imagery." *Sociological Quarterly* 34, 4 (November 1994): 639–651.

Kamerman, Sheila, and C. D. Hayes. *Families That Work: Children in a Changing World.* Washington, DC: National Academy Press, 1982.

Kane, N., ed. *The Hispanic American Almanac: A Reference Work on Hispanics in the United States.* Detroit: MI: Gale Research, 1993.

Kantor, David, and William Lehr. *Inside Families.* San Francisco, CA: Jossey-Bass, 1975.

Kantrowitz, Barbara. "Gay Families Come Out." *Newsweek* (November 4, 1996): 51–57.

———. "Who Keeps 'Baby M'?" *Newsweek* (January 19, 1987): 44–49.

Kantrowitz, Barbara, and David A. Kaplan. "Not the Right Family." *Newsweek* (March 19, 1990): 50–51.

Kaplan, A., and J. P. Bean. "From Sex Stereotypes to Androgyny: Considerations of Societal and Individual Change." In *Beyond Sex-Role Stereotypes,* edited by A. Kaplan and J. P. Bean. Boston: Little, Brown, 1976.

Kaplan, Helen Singer. *Disorders of Desire.* New York: Simon & Schuster, 1979.

Karsenty, N. "A Mother Questions Brit Milah." *Humanistic Judaism* 16, 3 (1988): 14–20.

Kassop, Mark. "Salvador Minuchin: A Sociological Analysis of His Family Therapy Theory." *Clinical Sociological Review* 5 (1987): 158–167.

Katchadourian, Herant. *Midlife in Perspective.* San Francisco: Freeman, 1987.

Katz, Michael B. *The Undeserving Poor: From War on Poverty to War on Welfare.* New York: Pantheon, 1990.

Katzen, A.R., R.L. Warner, and A.C. Acock. "Girls or Boys? Relationship of Child Gender to Marital Instability." *Journal of Marriage and the Family* 56 (February 1994): 89–100.

Kaufman, Joan, and Edward Zigler. "The Intergenerational Transmission of Abuse Is Overstated." In *Current Controversies in Family Violence,* edited by Richard Gelles and Donileen Loseke. Newbury Park, CA: Sage Publications, 1993.

Kavanaugh, Robert. *Facing Death.* Baltimore: Penguin, 1972.

Kawamoto, Walter T., and Tamara C. Chesire. "American Indian Families." In *Families in Cultural Context,* edited by Mary Kay DeGenova. Mountain View, CA: Mayfield, 1997.

Kayal, P. M. "Healing Homophobia: Volunteerism and Sacredness in AIDS." *Journal of Religion and Health* 31, 2 (June 1982): 113–128.

Kaye, K., et al. "Birth Outcomes for Infants of Drug Abusing Mothers." *New York State Journal of Medicine* 144, 7 (May 1989): 256–261.

Kearl, Michael C. *Endings: A Sociology of Death and Dying.* New York: Oxford University Press, 1989.

Kehoe, Monika "Lesbians Over 60 Speak for Themselves." *The Gerontologist* 32 (April 1992): 280-282.

Keilor, Garrison. "It's Good Old Monogamy That's Really Sexy." *Time* (October 17, 1994): 71.

Keith, Pat, et al. "Older Men in Employed and Retired Families." *Alternative Lifestyles* 4, 2 (May 1981): 228–241.

Keller, David, and Hugh Rosen. "Treating the Gay Couple within the Context of Their Families of Origin." *Family Therapy Collections* 25 (1988): 105–119.

Kellett, J. M. "Sexuality of the Elderly." *Sexual and Marital Therapy* 6, 2 (1991): 147–155.

Kelley, Douglas L., and Judee K. Burgoon. "Understanding Marital Satisfaction and Couple Type as Functions of Relational Expectations." *Human Communication Research* 18, 1 (1991): 40–69.

Kelley, Harold. "Love and Commitment." In *Close Relationships,* edited by Harold Kelley et al. San Francisco: Freeman, 1983.

Kelley, Robert, and Patricia Voydanoff. "Work/Family Role Strain Among Employed Parents." *Family Relations* 34, 3 (July 1985): 367–374.

Kelly, Joan B. "Current Research on Children's Postdivorce Adjustment: No Simple Answers." *Family & Conciliation Courts Review* 31, 1 (1993): 29–49.

Kelly, Mary P., D. S. Strassberg, and J.R. Kircher. "Attitudinal and Experiential Correlates of Anorgasmia." *Archives of Sexual Behavior,* 19, 2 (April 1990): 165–167.

Kempe, C. Henry, and Ray Helfer, eds. *The Battered Child.* Rev. ed. Chicago: University of Chicago Press, 1980.

Kendall-Tackett, Kathleen, L. M. Williams, and D. Finkelhor. "Impact of Sexual Abuse on Children: A Review and Synthesis of Recent Empirical Studies." *Psychological Bulletin,* 113, 1 (January 1993): 164 ff.

Kennedy, Gregory E. "College Students' Expectations of Grandparent and Grandchild Role Behaviors." *Gerontologist* 30, 1 (1990): 43–48.

———. "Grandchildren's Reasons for Closeness with Grandparents." *Journal of Social Behavior and Personality* 6, 4 (1991): 697–712.

———. "Quality in Grandparent/Grandchild Relationships." *International Journal of Aging and Human Development* 35, 2 (1992b): 83–98.

———. "Shared Activities of Grandparents and Grandchildren." *Psychological Reports* 70, 1 (1992a): 211–227.

Kennedy, Gregory E., and C. E. Kennedy. "Grandparents: A Special Resource for Children in Stepfamilies." *Journal of Divorce and Remarriage* 19, 3–4 (1993): 45–68.

Keshet, Jamie. "From Separation to Stepfamily." *Journal of Family Issues* 1, 4 (December 1980): 517–532.

Kessler, R. C., et al. "Lifetime and 12-Month Prevalence of DSM-III-R Psychiatric Disorders in the United States. Results From the National Comorbidity Survey." *Archives of General Psychiatry* 51, 1 (January 1994): 8–19.

Kessler-Harris, Alice. *A History of Wage-Earning Women in America.* New York: Oxford University Press, 1982.

———. *Women Have Always Worked: A Historical Overview.* New York: McGraw-Hill, 1981.

Kett, Joseph. *Rites of Passage: Adolescence in America, 1970 to the Present.* New York: Basic Books, 1977.

Kibria, Nazli. "Vietnamese Families in the United States." In *Minority Families in the United States: A Multicultural Perspective,* edited by R. L. Taylor. Englewood Cliffs, NJ: Prentice Hall, 1994.

Kikumura, Akemi, and Harry Kitano. "The Japanese American Family." In *Ethnic Families in America: Patterns and Variations,* edited by Charles Mindel et al. 3rd ed. New York: Elsevier North Holland, 1988.

Kilborn, Peter T. "Day Care: Key to Welfare Reform." *San Francisco Chronicle* (June 1, 1997): A-4.

Kilborn, Peter T. "Job News Grim for High School Grads." *New York Times* (May 20, 1994): 1A.

Kilzer, Louis. "Kid Fingerprinting a Sham, Foes Claim." *Denver Post* (May 13, 1985a): A1, 14.

———. "Public Often Not Told Facts in Missing Children Cases." *The Denver Post* (September 22, 1985b): A1, 14.

Kilzer, Louis, and Diana Griego. "Missing-Child Reports Bring Out Best, Worst." *The Denver Post* (May 13, 1985): A1, 14.

King, Laura A. "Emotional Expression, Ambivalence over Expression, and Marital Satisfaction." *Journal of Social and Personal Relationships* 10, 4 (1993): 601–607.

King, Patricia. "Not So Different After All: Domestic Violence within the Gay Community." *Newsweek* (October 4, 1993): 75.

Kingston, Paul, and Stephen Nock. "Consequences of the Family Work Day." *Journal of Marriage and the Family* 47, 3 (August 1985): 619–629.

Kinnunen, Ulla, Jan Gerris, and Ad Vermulst. "Work Experiences and Family Functioning Among Employed Fathers with Children of School Age." *Family Relations* 45, 4 (October 1996) 449–455.

Kinsey, Alfred, Wardell Pomeroy, and Clyde Martin. *Sexual Behavior in the Human Male.* Philadelphia: Saunders, 1948.

Kinsey, Alfred, Wardell Pomeroy, Clyde Martin, and P. Gebhard. *Sexual Behavior in the Human Female.* Philadelphia: Saunders, 1953.

Kirkpatrick, Lee A., and Phillip R. Shaver. "An Attachment—Theoretical Approach to Romantic Love and Religious Belief." *Personality and Social Psychology Bulletin* 18, 3 (1992): 266–275.

Kirkpatrick, Martha, et al. "Lesbian Mothers and Their Children: A Comparative Study." *American Journal of Orthopsychiatry* 51, 3 (July 1981): 545–551.

Kissman, Kris and JoAnn Allen. *Single Parent Families.* Newbury Park, CA: Sage Publications, 1993.

Kitson, G. C., and L.A. Morgan. "The Multiple Consequences of Divorce: A Decade Review. In *Contemporary Families: Looking Forward, Looking Back*, edited by A. Booth. Minneapolis: National Council on Family Relations, 1991.

Kitson, Gay. "Marital Discord and Marital Separation: A County Survey." *Journal of Marriage and the Family* 47 (August 1985): 693–700.

Kitson, Gay, and Leslie Morgan. "Consequences of Divorce." In *Contemporary Families: Looking Forward, Looking Back,* edited by Alan Booth. Minneapolis: National Council on Family Relations, 1991.

Kitson, Gay, and Marvin Sussman. "Marital Complaints, Demographic Characteristics, and Symptoms of Mental Distress in Divorce." *Journal of Marriage and the Family* 44, 1 (February 1982): 87–101.

Kitzinger, Sheila. *The Complete Book of Pregnancy and Childbirth.* New York: Knopf, 1989.

———. *Woman's Experience of Sex.* New York: Penguin, 1985.

Kivett, Vira R. "Racial Comparisons of the Grandmother Role: Implications for Strengthening the Family Support System of Older Black Women." *Family Relations* 42, 2 (April 1993): 165–172.

———. "The Grandparent-Grandchild Connection." *Marriage and Family Review* 16, 3–4 (1991): 267–290.

Klein, Alan M. "Of Muscles and Men: Anthropological Study of the Culture of Bodybuilding." *Science* 33, 6 (1993): 32–38.

Klein, David M. and James M. White. *Family Theories: An Introduction.* Thousand Oaks, CA: Sage Publications, 1996.

Kline, A., E. Kline, and E. Oken. "Minority Women and Sexual Choice in the Age of AIDS." *Social Science and Medicine* 34, 4 (February 1992): 447–57.

Klinman, Deborah, et al. *Fatherhood, USA: The First National Guide to Programs, Services and Resources for and about Fathers.* New York: Garland Publishing Co., 1984.

Klor de Alva, J. "Telling Hispanics Apart: Latino Sociocultural Diversity." In *The Hispanic Experience in the United States: Contemporary Issues and Perspectives,* edited by E. Acosta-Belen and B. Sjostrom. New York: Praeger, 1988.

Knafo, D., and Y. Jaffe. "Sexual Fantasizing in Males and Females." *Journal of Research in Personality* 18 (1984): 451–462.

Knapp, J., and R. Whitehurst. "Sexually Open Marriage and Relationships: Issues and Prospects." In *Marriage and Alternatives: Exploring Intimate Relationships,* edited by R. Libby and R. Whitehurst. Glenview, IL: Scott, Foresman, 1977.

Knapp, Mark, et al. "Compliments: A Descriptive Taxonomy." *Journal of Communication* 34, 4 (1984): 12–31.

Knaub, P. K., et al. "Strengths of Remarriage." *Journal of Divorce* 7 (1984): 41–55.

Knox, David, and Caroline Schacht. "Sexual Behaviors of University Students Enrolled in a Human Sexuality Course." *College Student Journal* 26, 1 (March 1992): 38–40.

Knox, David, and K. Wilson. "Dating Behaviors of University Students." *Family Relations* 30 (1981): 83–86.

Koblinsky, Sally, and Christine Todd. "Teaching Self-Care Skills to Latchkey Children: A Review of the Research." *Family Relations* 38, 4 (October 1989): 431–435.

Koch, Liz. "Mothering: An Honorable Profession." *The Doula* 2, 2 (September 1987): 4–6.

Kohlberg, Lawrence. "The Cognitive-Development Approach to Socialization." In *Handbook of Socialization Theory and Research,* edited by A. Goslin. Chicago: Rand McNally, 1969.

Kohn, M. L. "Social Class and Parental Values." *American Journal of Sociology* 64 (1959): 337–351.

Kolata, Gina. *The Baby Doctors.* New York: Delacorte Press, 1989.

———. "Anti-Acne Drug Faulted in Birth Defects." *New York Times* (April 22, 1988): 1.

———. "Early Warnings and Latent Cures for Infertility." *Ms.* (May 1979): 86–89.

———. "Experts Declare Temperance Is Back in Style in America." *New York Times* (January 1, 1991): B3.

———. "A Major Operation on a Fetus Works for the First Time." *New York Times* (May 31, 1990a): B1.

———. "New Pregnancy Hope: A Single Sperm Injected." *New York Times* (August 11, 1993): B7.

———. "Racial Bias Seen in Prosecuting Pregnant Addicts." *New York Times* (July 20, 1990b): A10.

Kolker, A. "Advances in Prenatal Diagnosis: Social-Psychological and Policy Issues." *International Journal of Technology Assessment in Health Care* 5, 4 (1989): 601–617.

Komarovsky, Mirra. *Blue-Collar Marriage.* 2d ed. New Haven, CT: Yale University Press, 1987.

———. *Women in College.* New York: Basic Books, 1985.

Konker, C. "Rethinking Child Sexual Abuse: An Anthropological Perspective." *American Journal of Orthopsychiatry* 62, 1 (January 1992): 147–153.

Konner, Melvin. *Childhood.* Boston: Little, Brown, 1991.

Kortenhaus, Carole M., and Jack Demarest. "Gender Role Stereotyping in Children's Literature: An Update." *Sex Roles: A Journal of Research* 28, 3–4 (1993): 219–323.

Koss, Mary. "Hidden Rape: Sexual Aggression and Victimization in a National Sample of Students in Higher Education." In *Rape and Sexual Assault*, Vol. 2, edited by A. W. Burgess. New York: Garland, 1988.

Koss, Mary, Thomas Dinero, Cynthia Seibel, and Susan Cox. "Stranger and Acquaintance Rape: Are There Differences in the Victim's Experience?" *Psychology of Women* 12, 1 (March 1988): 1–24.

Koss, Mary P., and Sarah L. Cook. "Facing the Facts: Date and Acquaintance Rape Are Significant Problems for Women." In *Current Controversies in Family Violence*, edited by Richard Gelles and Donileen Loseke. Newbury Park, CA: Sage Publications, 1993.

Koss, Mary P., and T. E. Dinero. "Discriminant Analysis of Risk Factors for Sexual Victimization among a National Sample of College Women." *Journal of Consulting and Clinical Psychology* 57 (April 1989): 242–250.

Kozol, Jonathan. *Rachel and Her Children: Homeless Families in America.* New York: Crown, 1988.

Kranichfeld, Marion. "Rethinking Family Power." *Journal of Family Issues* 8, 1 (March 1987): 42–56.

Krause, Neal. "Race Differences in Life Satisfaction among Aged Men and Women." *Journals of Gerontology* 48, 5 (1993): S235–S244.

Kruk, Edward. "Promoting Co-operative Parenting after Separation: A Therapeutic/Interventionist Model of Family Mediation." *Journal of Family Therapy* 15, 3 (1993): 235–261.

Kubey, Robert. "Media Implications for the Quality of Family Life" in *Media, Children and the Family: Social Scientific, Psychodynamic and Clinical Perspectives*, edited by Dolf Zillman, Jennings Bryant, and Aletha C. Huston. Hillsdale, NJ: Erlbaum, 1994.

Kubler-Ross, Elisabeth. *AIDS: The Ultimate Challenge.* New York: Macmillan, 1987.

———. *On Death and Dying.* New York: Macmillan, 1969.

———. *Working It Through.* New York: Macmillan, 1982.

Kulhanjian, J. A., et al. "Identification of Women at Unsuspected Risk of Primary Infection with Herpes Simplex Virus Type 2 during Pregnancy." *New England Journal of Medicine* 320, 4 (April 2, 1992): 916–920.

Kumpfer, Karol L., and Rodney Hopkins. "Prevention: Current Research and Trends." *Psychiatric Clinics of North America* 16, 1 (1993): 11–20.

Kurdek, L. A. "Predicting Marital Dissolution: A 5-Year Prospective Longitudinal Study of Newlywed Couples." *Journal of Personality and Social Psychology* 64, 2 (1993): 221–242.

Kurdek, Lawrence, and Albert Siesky. "Children's Perceptions of Their Parents' Divorce." *Journal of Divorce* 3, 4 (June 1980): 339–378.

Kurdek, Lawrence, et al. "Correlates of Children's Long-Term Adjustment to Their Parents' Divorce." *Developmental Psychology* 17, 5 (September 1981): 565–579.

Kurdek, L. A., and M. A. Fine. "The Relation Between Family Structure and Young Adolescents' Appraisals of Family Climate and Parenting Behavior." *Journal of Family Issues* 14 (June 1993): 279–290.

Kutscher, Austin H., Arthur C. Carr, and Lillian G. Kutscher, eds. *Principles of Thanatology.* New York: Columbia University Press, 1987.

Ladas, Alice, et al. *The G Spot.* New York: Holt, Rinehart & Winston, 1982.

Laing, R. D. *The Politics of the Family and Other Essays.* New York: Random House, 1972.

———. *The Politics of Experience.* New York: Random House, 1967.

Lainson, Suzanne. "Breast-Feeding: The Erotic Factor." *Ms.* 11, 8 (February 1983): 66 ff.

Lamaze, Fernand. *Painless Childbirth.* 1st ed. 1956. Chicago: Regnery, 1970.

Lamanna, Mary Ann, and Agnes Riedmann. *Marriages and Families; Making Choices in a Diverse Society.* Belmont, CA: Wadsworth Publishing Company, 1997.

Lamb, Michael. "Book Review." *Journal of Marriage and the Family* 55, 4 (November 1993): 1047–1049.

Landau, Rivka. "Affect and Attachment: Kissing, Hugging, and Patting as Attachment Behaviors." *Infant Mental Health Journal* 10, 1 (March 1989): 59–69.

Laner, Mary R. "Violence or Its Precipitators: Which Is More Likely to Be Identified as a Dating Problem." *Deviant Behavior* 11, 4 (October 1990): 319–329.

Lantz, Herman. "Family and Kin as Revealed in the Narratives of Ex-Slaves." *Social Science Quarterly* 60, 4 (March 1980): 667–674.

Larsen, Andrea S., and David H. Olson. "Predicting Marital Satisfaction Using PREPARE: A Replication Study." *Journal of Marital and Family Therapy* 15, 3 (1989): 311–322.

Larson, Jeffry H., Stephan M. Wilson, and Rochelle Beley. "The Impact of Job Insecurity on Marital and Family Relationships." *Family Relations* 43, 2 (April 1994): 138–143.

Larson, Mary S. "Interaction between Siblings in Primetime Television Families." *Journal of Broadcasting and Electronic Media* 33, 3 (1989): 305–315.

Larson, Reed, Robert W. Kubey, and Joseph Colletti. "Changing Channels: Early Adolescent Media Choices and Shifting Investments in Family and Friends" [Special issue: "The Changing Life Space of Early Adolescence"]. *Journal of Youth and Adolescence* 18, 6 (1989): 583–599.

Lasch, Christopher. *Haven in a Heartless World.* New York: Basic Books, 1977.

Lauer, Robert, and Jeanette Lauer. "The Long-Term Relational Consequences of Problematic Family Backgrounds." *Family Relations* 40, 3 (July 1991): 286–291.

Laurent, S. L., S. J. Thompson, C. Addy, C. Z. Garrison, and E. E. Moore. "An Epidemiologic Study of Smoking and Primary Infertility in Women." *Fertility and Sterility* 57, 3 (March 1992): 565–572.

Lavee, Yoav, and David Olson. "Family Types and Response to Stress." *Journal of Marriage and the Family* 53, 3 (August 1991): 786–798.

Laviola, Marisa. "Effects of Older-Brother Younger-Sister Incest: A Review of Four Cases." *Journal of Family Violence* 4, 3 (September 1989): 259–274.

Lawler, F. H., R. S. Bisonni and D. R. Holtgrave. "Circumcision: a Decision Analysis of its Medical Value." *Family Medicine* 23, 8 (November 1990): 587–93.

Lawson, Erma Jean, and Aaron Thompson. "Black Men's Perceptions of Divorce-Related Stressors and Strategies for Coping with Divorce." *Journal of Family Relations* 17, 2 (March 1996): 249–273.

Leboyer, Frederick. *Birth Without Violence.* New York: Knopf, 1975.

Lederer, William, and Don Jackson. *Mirages of Marriage.* New York: Norton, 1968.

Lee, John A. *The Color of Love.* Toronto: New Press, 1973.

———. "Love Styles." In *The Psychology of Love*, edited by Robert Sternberg and Michael Barnes. New Haven, CT: Yale University Press, 1988.

Lee, Thomas, Jay Mancini, and Joseph Maxwell. "Contact Patterns and Motivations for Sibling Relations in Adulthood." *Journal of Marriage and the Family* 52, 2 (May 1990): 431–440.

Lehrer, E., and C. Chiswick. "The Religious Composition of Unions." *Demography* 30 (1993): 385–404.

Leiblum, Sandra R. "Sexuality and the Midlife Woman." [Special Issue: "Women at Midlife and Beyond."]. *Psychology of Women Quarterly* 14, 4 (December 1990): 495–508.

Leifer, Myra. *Psychological Effects of Motherhood: A Study of First Pregnancy.* New York: Praeger, 1990.

Leigh, Barbara C. "Alcohol Expectancies and Reasons for Drinking: Comments from a Study of Sexuality." *Psychology of Addictive Behaviors* 4, 2 (1990): 91–96.

Leitenberg, H., M. J. Detzer, and D. Srebnik. "Gender Differences in Masturbation and the Relation of Masturbation Experience in Preadolescence and or Early Adolescence to Sexual Behavior and Sexual Adjustment in Young Adulthood." *Archives of Sexual Behavior* 22, 2 (April 1993): 87–98.

Lemkau, Jeanne Parr "Emotional Sequelae of Abortion: Implications for Clinical Practice. Special Issue: Women's Health: Our Minds, Our Bodies." *Psychology of Women Quarterly* 12 (December 1988): 461–472.

Leonard, Kenneth E., and Theodore Jacob. "Alcohol, Alcoholism, and Family Violence." In *Handbook of Family Violence*, edited by Vincent B. Van Hasselt, et al. New York: Plenum Press, 1988.

Leslie, Leigh, and Katherine Grady. "Changes in Mothers' Social Networks and Social Support Following Divorce." *Journal of Marriage and the Family* 47, 3 (August 1985): 663–673.

Levenson, Robert W., Laura L. Carstensen, and John M. Gottman. "Long-Term Marriage: Age, Gender, and Satisfaction." *Psychology and Aging* 8, 2 (1993): 301–313.

Levin, Irene. "The Model Monopoly of the Nuclear Family." Paper presented at the National Conference on Family Relations, Baltimore, MD, November 1993

Levin, Nora J. *How to Care for Your Parents.* Friday Harbor, WA: Storm King Press, 1987.

Levine, James. *Working Fathers: Strategies for Balancing Work and Family.* Reading, MA: Addison Wesley Longman, 1997.

Levine, Linda, and Lonnie Barbach. *The Intimate Male.* New York: Signet Books, 1983.

Levinger, George. "Marital Cohesiveness and Dissolution: An Integrative Review." *Journal of Marriage and the Family* 27, 1 (February 1965): 19–28.

Levinger, George. "A Social Psychological Perspective on Marital Dissolution." In *Divorce and Separation*, edited by George Levinger and O. C. Moles. New York: Basic Books, 1979.

Levinger, George, and O. C. Moles, eds. *Divorce and Separation: Context, Causes, and Consequences.* New York: Basic Books, 1979.

Levinson, Daniel J. *The Seasons of a Man's Life.* New York: Ballantine, 1977.

Levy, Barrie, ed. *Dating Violence: Young Women in Danger.* Seattle: Seal Press, 1991.

Levy, Gary D. "High and Low Gender Schematic Children's Release from Proactive Inference." *Sex Roles* 30, 1–2 (January 1994): 93–108.

Levy-Shiff, R. "Individual and Contextual Correlates of Marital Change Across the Transition to Parenthood." *Developmental Psychology*, 30, 4 (1994): 591–601.

Levy-Shiff, Rachel, Ilana Goldshmidt, and Dov Har-Even. "Transition to Parenthood in Adoptive Families." *Developmental Psychology* 27, 1 (1991): 131–140.

Lewes, K. "Homophobia and the Heterosexual Fear of AIDS." *American Image* 49, 3 (September 1992): 343–356.

Lewin, Bo. "Unmarried Cohabitation: A Marriage Form in a Changing Society." *Journal of Marriage and the Family* 44, 3 (August 1982): 763–773.

Lewin, M. "Unwanted Intercourse: The Difficulty of Saying No." *Psychology of Women Quarterly* 9 (1985): 184–192.

———. "Drug Use During Pregnancy: New Issue before the Courts." *New York Times* (February 5, 1990): A1, 12.

———. "Syphilis Among Babies Climbs with Crack Use." *New York Times* (November 23, 1992): A9.

Lewis, Jerry M., ed. *The Birth of the Family: An Empirical Inquiry.* New York: Brunner/Mazel, 1989.

Lewis, Karen. "Children of Lesbians: Their Point of View." *Social Work* 25, 3 (May 1980): 198–203.

Lewis, Lisa. "Consumer Girl Culture: How Music Video Appeals to Girls." In *Television and Women's Culture: The Politics of the Popular*, edited by Mary Ellen Brown. Newbury Park, CA: Sage Publications, 1992.

Lewis, Robert. "The Family and Addictions: An Introduction." *Family Relations* 38, 3 (July 1989): 254–257.

Lewis, Ronald, and Wallace Gingerich. "Leadership Characteristics: Views of Indian and Non-Indian Students." *Social Casework* 61, 10 (1980).

Libman, E. "Sociocultural and Cognitive Factors in Aging and Sexual Expression: Conceptual and Research Issues." *Canadian Psychology* 30, 3 (July 1989): 560–567.

Lieberman, B. "Extrapremarital Intercourse: Attitudes toward a Neglected Sexual Behavior." *Journal of Sex Research* 24 (1988): 291–299.

Lieberson, Stanley, and Mary Waters. *From Many Strands: Ethnic and Racial Groups in Contemporary America.* New York: Russell Sage Foundation, 1988.

Liem, Ramsay, and J. Liem, "Social Support and Stress: Some General Issues and Their Application to the Problem of Unemployment." In *Mental Health and the Economy*, edited by L. Ferman and J. Gordus. Kalamazoo, MI: Upjohn Institute, 1979.

Lieu, Tracy A., Paul W. Newacheck, and Margaret A. McManus. "Race, Ethnicity, and Access to Ambulatory Care among U.S. Adolescents." *American Journal of Public Health* 83, 7 (1993): 960–965.

Liu, W. T. "Family Interactions Among Local and Refugee Chinese Families in Hong King." *Journal of Marriage and the Family*, 28 (1966): 314-323.

Lillie-Blanton, Marsha, Janice Bow, and Margurite Ro. "Social Factors and the Use of Preventive Services." In *Women's Health*, edited by Marilyn Falik and Karen Scott Collins. Baltimore, MD: Johns Hopkins University Press, 1996.

Lin, Chien, and William T. Liu. "Intergenerational Relationships among Chinese Immigrant Families from Taiwan." In *Family Ethnicity: Strength in Diversity*, edited by Harriette Pipes McAdoo. Newbury Park, CA: Sage Publications, 1993, 109–119.

Lindbohm, M. L., M. Hietanan, P. Kyronen, and M. Sallmen. "Magnetic Fields of Video Display Terminals and Spontaneous Abortion." *American Journal of Epidemiology* 136 (1992): 1041–1051.

Lindsey, Karen. *Friends as Family.* Boston: Beacon Press, 1982.

Lingren, Herbert, et al. "Enhancing Marriage and Family Competencies Through Adult Life Development." *Family Strengths 4: Positive Support Systems*, edited by Nick Stinnet et al. Lincoln: University of Nebraska Press, 1982.

Lino, Mark. "Expenditures on a Child by Husband-Wife Families." *Family Economics Review* 3, 3 (1990): 2–12.

Lips, Hilary. "Expenditures on a Child by Two-Parent Families." *Family Economics Review* 4, 1 (1991): 2–38.

———. *Sex and Gender.* 3rd ed. Mountain View, CA: Mayfield, 1997.

Little, Margaret A. "The Impact of the Custody Plan on the Family: A Five-Year Follow-up: Executive Summary." *Family and Conciliation Courts Review* 30, 2 (1992): 243–251.

Litwack, Leon. *Been in the Storm So Long.* New York: Alfred A. Knopf, 1979.

Livernois, Joe. "County Gears Up for Huge Welfare Shift." *Monterey County Herald* (June 1, 1997): A-1, A-8.

Lloyd, Sally A. "Conflict Types and Strategies in Violent Marriages." *Journal of Family Violence* 5, 4 (December 1990): 269–284.

———. "The Darkside of Courtship: Violence and Sexual Exploitation." *Family Relations* 40, 1 (January 1991): 14–20.

———. "Physical and Sexual Violence During Dating and Courtship." In *Vision 2010: Families and Violence, Abuse and Neglect*, edited by Richard J. Gelles. Minneapolis: National Council on Family Relations, 1995.

Lloyd, Sally A., and Beth C. Emery. "The Dynamics of Courtship Violence." Paper presented at the annual meeting of the National Council on Family Relations, Seattle, WA, November 1990.

Locke, D. *Increasing Multicultural Understanding.* Newbury Park, CA: Sage Publications, 1992.

Lockhart, Lettie. "A Reexamination of the Effects of Race and Social Class on the Incidence of Marital Violence: A Search for Reliable Differences." *Journal of Marriage and the Family* 49, 3 (August 1987): 603–610.

Lombardo, John P., and T. R. Kemper. "Sex Role and Parental Behaviors." *Journal of Genetic Psychology* 153, 1 (1992): 103–114.

Lombardo, W. K., et al. "Fer Cryin Out Loud—There Is a Sex Difference." *Sex Roles* 9 (1983): 987–995.

London, Richard, James Wakefield, and Richard Lewak. "Similarity of Personality Variables as Predictors of Marital Satisfaction." *Personality and Individual Differences* 11, 1 (1990): 39–43.

Long, Vonda O. and Estella A. Martinez. "Masculinity, Femininity, and Hispanic Professional Women's Self-Esteem and Self-Acceptance." *Journal of Counseling & Development* 73, 3 (November/December 1994): 183–186.

LoPresto, C., M. Sherman, and N. Sherman. "The Effects of a Masturbation Seminar on High School Males' Attitudes, False Beliefs, Guilt, and Behavior." *Journal of Sex Research* 21 (1985): 142–156.

Lorch, Barbara D., and Richard L. Dukes. "Factors Related to Youths' Perception of Needing Help for an Alcohol or Drug Problem." *Journal of Alcohol & Drug Education* 34, 3 (March 1989): 38–47.

Lord, Lewis. "Desperately Seeking Baby." *U.S. News and World Report* 5 (October 1987): 58–65.

Lord, Lewis, et al. "Coming to Grips with Alcoholism." *U.S. News and World Report* (November 30, 1987): 56–62.

Losh-Hesselbart, S. "Development of Gender Roles." In *Handbook of Marriage and the Family*, edited by M.B. Sussman and S. Steinmetz. New York: Plenum, 1987, 535-563.

Lott, B. *Women's Lives: Themes and Variations in Gender Learning.* 3rd ed. Pacific Grove, CA: Brooks/Cole, 1994.

Loulan, JoAnn. *Lesbian Sex.* San Francisco: Spinsters Books, 1984.

Lowry, Dennis, and David Towles. "Prime Time TV Portrayals of Sex, Contraception, and Venereal Disease." *Journalism Quarterly* 66, 2 (June 1989): 347–352.

———. "Soap Opera Portrayals of Sex, Contraception, and Venereal Disease." *Journal of Communication* 39, 2 (March 1989): 76–83.

Luker, Kristin. *Abortion and the Politics of Motherhood.* Berkeley: University of California Press, 1984.

———. *Taking Chances.* Berkeley: University of California Press, 1975.

Lunneborg, Patricia. *Abortion: The Positive Decision.* New York: Bergin & Garvy, 1992.

Lutz, Patricia. "The Stepfamily: An Adolescent Perspective." *Family Relations* 32, 3 (July 1983): 367–375.

Lyon, Jeff. *Playing God in the Nursery.* New York: Norton, 1985.

Macciarola, Frank J., and Alan Gartner, eds. *Caring for Americas Children.* New York: The Academy of Political Science, 1989.

Maccoby, Eleanor, and Carol Jacklin. *The Psychology of Sex Differences.* Stanford, CA: Stanford University Press, 1974.

Maccoby, Eleanor E., Christy M. Buchanan, Robert H. Mnookin, and Sanford M. Dornbusch. "Postdivorce Roles of Mothers and Fathers in the Lives of Their Children." *Journal of Family Psychology* 7, 1 (1993): 24–38.

MacCorquodale, Patricia. "Gender and Sexuality." In *Human Sexuality: The Societal and Interpersonal Context*, edited by Kathleen McKinney and Susan Sprecher. Norwood, NJ: Ablex, 1989.

Macdonald, Patrick T., Dan Waldorf, Craig Reinarman, and Sheigla Murphy. "Heavy Cocaine Use and Sexual Behavior." *Journal of Drug Issues* 18, 3 (June 1988): 437–455.

MacDonald, W. L., and A. DeMaris. "Remarriage, Stepchildren, and Marital Conflict. Challenges to the Incomplete Institutionalization Hypothesis. *Journal of Marriage and the Family* 57, 2 (May 1995): 387–398.

Mace, David, and Vera Mace. "Enriching Marriage." In *Family Strengths*, edited by Nick Stinnet et al. Lincoln: University of Nebraska Press, 1979.

———. "Enriching Marriages: The Foundation Stone of Family Strength." In *Family Strengths: Positive Models for Family Life*, edited by Nick Stinnet et al. Lincoln: University of Nebraska Press, 1980.

MacFarlane, Robin. "Summary of Adolescent Pregnancy Research: Implications for Prevention." In *The Prevention Researcher.* Eugene, OR: Integrated Research Services, 1997.

Macklin, Eleanor. "AIDS: Implications for Families." *Family Relations* 37, 2 (April 1988): 141–149.

———. "Nonmarital Heterosexual Cohabitation." *Marriage and Family Review* 1 (March 1978): 1–12.

———. "Nontraditional Family Forms." In *Handbook of Marriage and the Family*, edited by Marvin Sussman and Suzanne Steinmetz. New York: Plenum Press, 1987.

Macklin, Eleanor, ed. *AIDS and Families.* New York: Harrington Park, 1989.

Madsen, William, ed. *Mexican-American Youth of South Texas.* 2d ed. New York: Holt, Rinehart, & Winston, 1973.

Magno, Josefina. "The Hospice Concept of Care: Facing the 1990s." *Death Studies* 14, 2 (1990): 109–119.

Major, Brenda, and Catherine Cozzarelli. "Psychosocial Predictors of Adjustment to Abortion." *Journal of Social Issues* 48, 3 (1992): 121–142.

Makepeace, James. "Courtship Violence Among College Students." *Family Relations* 30, 1 (January 1981): 97–102.

———. "Dating, Living Together, and Courtship Violence." In *Violence in Dating Relationships: Emerging Social Issues*, edited by Maureen Pirog-Good and Jan Stets. New York: Praeger, 1989.

———. "Gender Differences in Courtship Violence Victimization." *Family Relations* 35, 3 (July 1986): 383–388.

Malatesta, Victor, Dianne Chambless, Martha Pollack, and Alan Cantor. "Widowhood, Sexuality, and Aging: A Life Span Analysis." *Journal of Sex and Marital Therapy* 14, 1 (March 1989): 49–62.

Maloney, Lawrence. "Behind Rise in Mixed Marriages." *U.S. News and World Report* (February 10, 1986): 68–69.

Malson, Lucien. *Wolf Children and the Problem of Human Nature.* New York: Monthly Review Press, 1972.

Malveaux, Julianne. "The Economic Status of Black Families." In *Black Families*, edited by Harriette Pipes McAdoo. 2nd ed. Newbury Park, CA: Sage Publications, 1988.

Mancini, Jay, and Rosemary Blieszner. "Aging Parents and Adult Children: Research Themes in Intergenerational Relations." In *Contemporary Families: Looking Forward, Looking Back*, edited by Alan Booth. Minneapolis: National Council on Family Relations, 1991.

———. "Research on Aging Parents and Adult Children." *Journal of Marriage and the Family* 51, 2 (May 1989): 275–290.

———. "Social Provisions in Adulthood: Concept and Measurement in Close Relationships." *Journal of Gerontology* 47, 1 (1992): 14–20.

Marecek, Jeanne, et al. "Gender Roles in the Relationships of Lesbians and Gay Men." In *Gay Relationships*, edited by J. DeCecco. New York: Haworth Press, 1988.

Marek, Lynne. "U.S. Vows War on Domestic Violence." *Chicago Tribune* (March 12, 1994).

Margolin, Gayla, Linda Gorin Sibner, and Lisa Gleberman. "Wife Battering." In *Handbook of Family Violence*, edited by Vincent B. Van Hasselt et al. New York: Plenum Press, 1988.

Margolin, Leslie. "Sexual Abuse by Grandparents." *Child Abuse and Neglect* 16, 5 (September 1992): 735.

Margolin, Leslie, and Lynn White. "The Continuing Role of Physical Attractiveness in Marriage." *Journal of Marriage and the Family* 49, 1 (February 1987): 21–27.

Margolin, Malcolm. *The Ohlone Way.* Berkeley: Heydey Books, 1978.

Marin, Peter. "Helping and Hating the Homeless." *Harper's* (January 1987): 39–49.

Marker, Nadine F. "Flying Solo at Midlife: Gender, Marital Status, and Psychological Well-Being." *Journal of Marriage and the Family* 58 (November 1996): 917–932.

Markman, Howard. "Application of a Behavioral Model of Marriage in Predicting Relationship Satisfaction of Couples Planning Marriage." *Journal of Consulting and Clinical Psychology* 47 (1979): 743–749.

———. "The Longitudinal Study of Couples Interactions: Implications for Understanding and Predicting the Development of Marital Distress." In *Marital Interaction*, edited by K. Hahlweg and N. S. Jacobsen. New York: Guilford Press, 1984.

———. "Prediction of Marital Distress: A Five-Year Follow-up." *Journal of Consulting and Clinical Psychology* 49 (1981): 760–761.

Markman, Howard, et al. "The Prediction and Prevention of Marital Distress: A Longitudinal Investigation." In *Understanding Major Mental Disorders: The Contribution of Family Interaction Research*, edited by Kurt Hahlweg and Michael Goldstein. New York: Family Process Press, 1987.

Marks, Stephen. "What Is a Pattern of Commitment?" *Journal of Marriage and the Family* 52, 1 (February 1994): 112–115.

Markstrom-Adams, C. "Coming of Age Among Contemporary American Indians as Portrayed in Adolescent Fiction." *Adolescence* 25 (1990): 225–237.

Marmor, Judd. "Homosexuality and the Issue of Mental Illness." In *Homosexual Behavior*, edited by Judd Marmor. New York: Basic Books, 1980a.

———. "The Multiple Roots of Homosexual Behavior." In *Homosexual Behavior*, edited by J. Marmor. New York: Basic Books, 1980b.

Marmor, Judd, ed. *Homosexual Behavior.* New York: Basic Books, 1980.

Marrero, M. A., and S. J. Ory. "Unexplained Infertility." *Current Opinion in Obstetrics and Gynecology* 3, 2 (April 1991): 211–218.

"Married without Children." *American Demographics* 14, 7 (July 1992): A10.

Marsiglio, William. "Male Procreative Consciousness an Responsibility: A Conceptual Analysis and Research Agenda." *Journal of Family Issues* 12, 3 (September 1992): 268–290.

Marsiglio, William, and Denise Donnelly. "Sexual Relations in Later Life: A National Study of Married Persons." *Journal of Gerontology* 46, 6 (November 1991): S338–S344.

Martelli, Leonard, et al. *When Someone You Know Has AIDS: A Book of Hope for Family and Friends.* New York: Crown, 1987.

Martin, Del. *Battered Wives.* San Francisco: New Glide, 1981.

Martin, Douglas. "Many Dads Struggle to Fit New Roles." *New York Times* (June 20, 1993): 11.

Martin, Peter, et al. "Family Stories: Events (temporarily) Remembered." *Journal of Marriage and the Family* 50, 2 (May 1988): 533–541.

Martin, T. C., and L. L. Bumpass. "Recent Trends in Marital Disruption." *Demography* 26 (February 1989): 37–51.

Maruta, Toshihiko, and Mary Jane McHardy. "Sexual Problems in Patients with Chronic Pain." *Medical Aspects of Human Sexuality* 17, 2 (February 1983): 68J–68U ff.

Marwell, G., et al. "Legitimizing Factors in the Initiation of Heterosexual Relationships." First International Conference on Personal Relationships, Madison, WI, July 1982.

Mason, Karen, and Yu Hsia Lu. "Attitudes toward Women's Familial Roles: Changes in the United States,

1977–1985." *Gender & Society* 2, 1 (March 1988): 39–57.

Masse, Michelle, and Karen Rosenblum. "Male and Female Created They Them: The Depiction of Gender in the Advertising of Traditional Women's and Men's Magazines." *Women's Studies International Forum* 11, 2 (1988): 127–144.

Masters, John, et al. "The Role of the Family in Coping with Childhood Chronic Illness." In *Coping with Chronic Disease*, edited by Thomas Burish and Laurence Bradley. New York: Academic Press, 1983.

Masters, William, et al. *Masters and Johnson on Sex and Human Loving*. Boston: Little, Brown, 1986.

Mathis, Richard D., and Zoe Tanner. "Cohesion, Adaptability, and Satisfaction of Family Systems in Later Life." *Family Therapy* 18, 1 (1991): 47–60.

Maticka-Tyndale, Eleanor. "Sexual Scripts and AIDS Prevention: Variations in Adherence to Safer-Sex Guidelines." *Journal of Sex Research* 28, 1 (February 1991): 145–166.

Matiella, Ana Consuelo. *Positively Different: Creating a Bias-Free Environment for Children*. Santa Cruz, CA: Network Publications, 1991.

Mattessich, Paul, and Reuben Hill. "Life Cycle and Family Development." In *Handbook of Marriage and the Family*, edited by Marvin Sussman and Suzanne Steinmetz. New York: Plenum Press, 1987.

Matthews, Sarah, and Tana Rosner. "Shared Filial Responsibility: The Family as the Primary Caregiver." *Journal of Marriage and the Family* 50, 1 (February 1988): 185–195.

Mayers, Raymond Sanchez, Barbara Lynn Kail, and Thomas D. Watts, eds. *Hispanic Substance Abuse*. Springfield, IL: Charles C Thomas, 1993.

Mays, Vickie M., S. D. Cochran, G. Bellinger, and R. G. Smith. "The Language of Black Gay Men's Sexual Behavior: Implications for AIDS Risk Reduction." *Journal of Sex Research* 29, 3 (August 1992): 425–434.

Mazor, Miriam. "Barren Couples." *Psychology Today* (May 1979): 101–108, 112.

Mazor, Miriam, and Harriet Simons, eds. *Infertility: Medical, Emotional and Social Considerations*. New York: Human Sciences Press, 1984.

McAdoo, Harriette Pipes. "Changes in the Formation and Structure of Black Families: The Impact on Black Women." Working paper no. 182, Center for Research on Women, Wellesley College, Wellesley, MA, 1988.

———. "Ethnic Family Strengths that Are Found in Diversity." In *Family Ethnicity: Strength in Diversity*, edited by Harriette Pipes McAdoo. Newbury Park, CA: Sage Publications, 1993.

McAdoo, Harriette Pipes, ed. *Black Families*. 2d ed. Beverly Hills, CA: Sage Publications, 1988.

———. *Black Families*. 3rd ed. Thousand Oaks, CA: Sage Publications, 1996.

———. *Family Ethnicity: Strength in Diversity*. Newbury Park, CA: Sage Publications, 1993.

McAdoo, Harriette Pipes, and John McAdoo, eds. *Black Children: Social, Educational, and Parental Environments*. Beverly Hills, CA: Sage Publications, 1985.

McAdoo, John Lewis. "Decision Making and Marital Satisfaction in African American Families." In *Family Ethnicity: Strength in Diversity*, edited by Harriette Pipes McAdoo. Newbury Park, CA: Sage Publications, 1993.

McClary, Susan. "Living to Tell: Madonna's Resurrection of the Flesh." *Genders* 7 (March 1990): 1–21.

McCormack, M. J., et al. "Patient's Attitudes Following Chorionic Villus Sampling." *Prenatal Diagnosis* 10, 4 (April 1990): 253–255.

McCormick, John. "Why Parents Kill." *Newsweek* (November 14, 1994): 31–35.

McCubbin, Hamilton, and Joan Patterson. "Family Adaptation to Crises." In *Family Stress, Coping and Social Support*, edited by H. I. McCubbin, A. Cauble, and J. Patterson. Springfield, IL: Charles C. Thomas, 1982.

McCubbin, Hamilton I., Constance Joy, and Elizabeth Cauble. "Family Stress and Coping: A Decade Review." *Journal of Marriage and the Family* 42, 4 (November 1980): 855–871.

McCubbin, Hamilton I., and Marilyn A. McCubbin. "Typologies of Resilient Families: Emerging Roles of Social Class and Ethnicity." *Family Relations* 37, 3 (July 1988): 247–254.

McCubbin, Marilyn A. "Family Stress, Resources, and Family Types: Chronic Illness in Children." *Family Relations* 37, 2 (April 1988): 203–210.

McCubbin, Marilyn A., and Hamilton I. McCubbin. "Family Stress Theory and Assessment." In *Family Assessment Inventories for Research and Practice*, edited by Hamilton I. McCubbin and Anne I. Thompson. Madison: University of Wisconsin Press, 1987.

———. "Theoretical Orientations to Family Stress and Coping." In *Treating Families under Stress*, edited by Charles Figley. New York: Brunner/Mazel, 1989.

McEwan, K. L., C. G. Costello, and P. J. Taylor. "Adjustment to Infertility." *Journal of Abnormal Psychology* 96, 2 (May 1987): 108–116.

McGoldrick, Monica. "Ethnicity and the Family Life Cycle." In *The Changing Family Life Cycle*, edited by Betty Carter and Monica McGoldrick. 2nd ed. Boston: Allyn and Bacon, 1989.

———. "Normal Families: An Ethnic Perspective." In *Normal Family Processes*, edited by Froma Walsh. New York: Guilford Press, 1982.

McGoldrick, Monica, and Randy Gerson. *Genograms in Family Assessment*. New York: Norton, 1985.

McGoldrick, Monica, J. K. Pearce, and J. Giordano, eds. *Ethnicity and Family Therapy*. New York: Guilford Press, 1982.

McGraw, J. Melbourne, and Holly A. Smith. "Child Sexual Abuse Allegations amidst Divorce and Custody Proceedings: Refining the Validation Process." *Journal of Child Sexual Abuse* 1, 1 (1992): 49–62.

McHale, Susan, and Ted Huston. "The Effect of the Transition to Parenthood on the Marriage and Family Relationship: A Longitudinal Study." *Journal of Family Issues* 6, 4 (December 1985): 409–433.

McIntosh, Everton. "An Investigation of Romantic Jealousy among Black Undergraduates." *Social Behavior and Personality* 17, 2 (1989): 135–141.

McIntosh, Everton, and Douglas T. Tate. "Correlates of Jealous Behaviors." *Psychological Reports* 66, 2 (April 1990): 601–602.

McIntosh, Everton G., and Calvin O. Matthews. "Use of Direct Coping Resources in Dealing with Jealousy." *Psychological Reports* 70, 3 pt. 2 (1992): 1037–1038.

McKim, Margaret K. "Transition to What? New Parents' Problems in the First Year." *Family Relations* 36, 1 (January 1987): 22–25.

McKinney, Kathleen, and Susan Sprecher, eds. *Human Sexuality: The Societal and Interpersonal Context.* Norwood, NJ: Ablex, 1989.

McKinney, Kathleen, and Susan Sprecher. *Sexuality in Close Relationships.* Hillsdale, NJ: Erlbaum, 1991.

McLanahan. S. S., and G. Sandefur. *Growing Up with a Single Parent: What Hurts, What Helps.* Cambridge, MA: Harvard University Press, 1994.

McLanahan, Sara, and Karen Booth. "Mother-Only Families: Problems, Prospects, and Politics." In *Contemporary Families: Looking Forward, Looking Back,* edited by Alan Booth. Minneapolis: National Council on Family Relations, 1991.

McLanahan, Sara, et al. "Network Structure, Social Support, and Psychological Well-Being." *Journal of Marriage and the Family* 43, 3 (August 1981): 601–612.

McLeer, S. V., et al. "Sexually Abused Children at High Risk for Post-Traumatic Stress Disorder." *Journal of the American Academy of Child and Adolescent Psychiatry* 31, 5 (September 1992): 875–879.

McLeod, B. "The Oriental Express." *Psychology Today* 20 (July 1986): 48–52.

McMahon, Kathryn. "The Cosmopolitan Ideology and the Management of Desire." *Journal of Sex Research* 27, 3 (August 1990): 381–396.

McNeely, R. L., and John N. Colen, eds. *Aging in Minority Groups.* Beverly Hills, CA: Sage Publications, 1983.

McNeely, R. L., and Barbe Fogarty. "Balancing Parenthood and Employment Factors Affecting Company Receptiveness to Family-Related Innovations in the Workplace." *Family Relations* 37, 2 (April 1988): 189–195.

McRoy, R. G., H. D. Grotevant, S. Ayers-Lopez. *Changing Patterns in Adoption.* Austin, TX: Hogg Foundation for Mental Health, 1994.

Meacham, R. E., and L. I. Lipshultz. "Assisted Reproductive Technologies for Male Factor Infertility." *Current Opinion in Obstetrics and Gynecology* 3, 5 (October 1991): 656–661.

Mead, Barbara J., and Arlene A. Ignicio. "Children's Gender-Typed Perceptions of Physical Activity: Consequences and Implications." *Perceptual and Motor Skills* 75, 3 (1992): 1035–1042.

Mead, Margaret. *Male and Female.* New York: Morrow, 1975.

Mederer, Helen J. "Division of Labor in Two-Earner homes: Task Accomplishment versus Household Management as Critical Variables in Perceptions about Family Work." *Journal of Marriage and the Family* 55, 1 (February 1993): 133–145.

Melichor, Joseph, and David Chiriboga. "Significance of Time in Adjustment to Marital Separation." *American Journal of Orthopsychiatry* 58, 2 (April 1988): 221–227.

Melito, Richard. "Adaptation in Family Systems: A Developmental Perspective." *Family Processes* 24 (1985): 89–100.

Melson, Gail F., Susan Peets, and Cheryl Sparks. "Children's Attachment to Their Pets: Links to Socio-Emotional Development." *Children's Environments Quarterly* 8, 2 (1991): 55–65.

Menaghan, Elizabeth, and Toby Parcel. "Parental Employment and Family Life Research in the 1980s." In *Contemporary Families: Looking Forward, Looking Back,* edited by Alan Booth. Minneapolis: National Council on Family Relations, 1991.

Menning, Barbara Eck. *Infertility: A Guide for Childless Couples.* 2d ed. New York: Prentice Hall, 1988.

Merritt, Bishetta, and Carolyn A. Stroman. "Black Family Imagery and Interactions on Television." *Journal of Black Studies* 23, 4 (June 1993): 492–499.

Metts, Sandra, and William Cupach. "The Role of Communication in Human Sexuality." In *Human Sexuality: The Social and Interpersonal Context,* edited by Kathleen McKinney and Susan Sprecher. Norwood, NJ: Ablex, 1989.

Meyer, Daniel R., and Judi Bartfeld. "Compliance with Child Support Orders in Divorce Cases." *Journal of Marriage and the Family* 58, 1 (February 1996): 201–212.

Meyer, Daniel R., Elizabeth Phillips, and Nancy L. Maritato. "The Effects of Replacing Income Tax Deductions with Children's Allowance." *Journal of Family Issues* 12, 4 (December 1991): 467–491.

Meyers, Marcia K. "Child Care in JOBS Employment and Training Program: What Difference Does Quality Make?" *Journal of Marriage and the Family* 55, 3 (August 1993): 767–783.

Meyrowitz, J. *No Sense of Place: The Impact of Electronic Media on Social Behavior.* New York: Oxford University Press, 1985.

Miall, Cherlene E. "The Stigma of Adoptive Parent Status: Perceptions of Community Attitudes Toward Adoption and the Experience of Informal Self-Sanctioning." *Family Relations* 36, 1 (January 1987): 34–39.

Michael, Robert, John Gagnon, Edward Laumann, and Gina Kolata. *Sex in America: The Definitive Survey.* Boston: Little, Brown, 1994.

Michaels, D., and C. Levine. "Estimates of the Number of Motherless Youth Orphaned by AIDS in the United States." *JAMA* 268, 40 (December 23-30, 1992): 3456–3461.

Milan, Richard Jr., and Peter Kilmann. "Interpersonal Factors in Premarital Contraception." *Journal of Sex Research* 23, 3 (August 1987): 321–389.

Milardo, Robert. "Changes in Social Networks of Women and Men Following Divorce: A Review." *Journal of Family Issues* 8 (March 1987): 78–96.

Miller, Brent. *Family Research Methods.* Beverly Hills, CA: Sage Publications, 1986.

Miller, Brent C., and G. L. Fox. "Theories of Adolescent Heterosexual Behavior." *Adolescent Research,* 2 (1987): 269–282.

Miller, Glenn. "The Psychological Best Interests of the Child." *Journal of Divorce and Remarriage* 19, 1–2 (1993): 21–36.

Miller, Jean B. "Psychological Recovery in Low-Income Single Parents." *American Journal of Orthopsychiatry* 52, 2 (April 1982): 346–352.

Miller, Randi, and Michael Gordon. "The Decline of Formal Dating: A Study in Six Connecticut High Schools." *Marriage and Family Relationships* 10, 1 (April 1986): 139–154.

Miller, Ron. "Black & White Television." *San Jose Mercury News* (June 15, 1992): D1, D5.

Miller, Susan. "Viewer Discretion." *Newsweek* (December 23, 1996): 60.

Mills, David. "A Model for Stepparent Development." *Family Relations* 33 (1984): 365–372.

Milos, Marilyn Fayre. "Infant Circumcision: What I Wish I Had Known." *Truth Seeker* (July 1989): 3.

Milos, Marilyn Fayre, and Donna Macris. "Circumcision: A Medical or Human Rights Issue?" *Journal of Nurse-Midwifery* 37, 2 (Supplement) (March 1992): 87S–96S.

Min, Pyong Gap. "The Korean American Family." In *Ethnic Families in America: Patterns and Variations*, edited by Charles Mindel, Robert Habenstein, and Roosevelt Wright. 3rd ed. New York: Elsevier-North Holland, 1988.

Mindel, Charles H., Robert W. Habenstein, and Roosevelt Wright Jr. eds. *Ethnic Families in America: Patterns and Variations.* 3rd. ed. New York: Elsevier North Holland, Inc., 1976, 1981, 1988.

Minkler, Meredith, and Kathleen M. Roe. *Grandmothers as Caregivers: Raising Children of the Crack Cocaine Epidemic.* Family Caregiver Applications Series 2. Newbury Park, CA: Sage Publications, 1993.

Minkler, Meredith, et al. "Profile of Grandparents: Raising Grandchildren in the United States." *The Gerontologist* 37, 3 (June 1997): 400–411.

Mintz, Steven, and Susan Kellogg. *Domestic Revolutions: A Social History of American Family Life.* New York: Free Press, 1988.

Minuchin, Salvador. *Family Therapy Techniques.* Cambridge, MA: Harvard University Press, 1981.

Mirandé, Alfredo. *The Chicano Experience: An Alternative Perspective.* Notre Dame, IN: University of Notre Dame Press, 1985.

Moffatt, Betty Clare, et al. *AIDS: A Self-Care Manual.* Los Angeles: IBS Press, 1987.

Moffatt, Michael. *Coming of Age in New Jersey: College and American Culture.* New Brunswick, NJ: Rutgers University Press, 1989.

Moller, Lora C., S Hymel, and K. H. Rubin. "Sex Typing in Play and Popularity in Middle Childhood." *Sex Roles* 26, 7/8 (1992): 331–335.

Monahan, Thomas. "Are Interracial Marriages Really Less Stable?" *Social Forces* 48 (1970): 461–473.

Money, John. *Love and Lovesickness.* Baltimore: Johns Hopkins University Press, 1980.

Montagu, Ashley. *Touching*, 3d ed. New York: Columbia University Press, 1986.

Montgomery, M. J., E. R. Anderson, E. M. Hetherington, and W. G. Clingempeel. "Patterns of Courtship for Remarriage: Implications for Child Adjustment and Parent-Child Relationships." *Journal of Marriage and the Family*, 54 (August 1992): 686–698.

Montgomery, Marilyn J., and Gwendoly T. Sorell. "Differences in Love Attitudes Across Family Life Stages." *Family Relations* 46, 1 (January 1997): 55–61.

Moore, Dianne, and Pamela Erickson. "Age, Gender, and Ethnic Differences in Sexual and Contraceptive Knowledge, Attitudes, and Behaviors." *Family and Community Health* 8, 3 (November 1985): 38–51.

Moore, Lisa J. "Protecting Babies from Hepatitis-B." *U.S. News and World Report* (May 9, 1988): 85.

Moore, M. M. "Nonverbal Courtship Patterns in Women: Context and Consequences." *Ethology and Sociobiology* 6, 2 (1985): 237–247.

Morgan, Carolyn, and Alexis Walker. "Predicting Sex Role Attitudes." *Social Psychology Quarterly* 46 (1983): 148–153.

Morgan, Edmund. *The Puritan Family.* New York: Harper & Row, 1966. (1944).

Morgan, Leslie. "Outcome of Marital Separation: A Longitudinal Test of Predictors." *Journal of Marriage and the Family* 50, 2 (May 1988): 493–498.

Morgan, S. Philip, Diane Lye, and Gretchen Condran. "Sons, Daughters, and the Risk of Marital Disruption." *American Journal of Sociology* 94 (July 1988): 110–129.

Morrison, Donna R. and Andres J. Cherlin. "The Divorce Process and Young Children's Well-Being: A Prospective Analysis." *Journal of Marriage and the Family* 57, 3 (August 1995): 800–812.

Morrison, Donna, and Daniel Lichter. "Family Migration and Female Employment: The Problem of Underemployment among Migrant Women." *Journal of Marriage and the Family* 50, 1 (February 1988): 161–172.

Mosher, D. L., and S. S. Tomkins. "Scripting the Macho Man: Hypermasculine Socialization and Enculturation." *Journal of Sex Research* 25 (February 1988): 60–84.

Mosher, Donald. "Sex Guilt and Sex Myths in College Men and Women." *Journal of Sex Research* 15, 3 (August 1979): 224–234.

Moss, Peter, et al. "Marital Relations During the Transition to Parenthood." *Journal of Reproductive and Infant Psychology* 4, 1–2 (September 1986): 57–67.

Moynihan, Daniel Patrick. *The Negro Family: The Case for National Action.* Washington, DC: U.S. Government Printing Office, 1965.

Muehlenhard, Charlene. "Misinterpreted Dating Behaviors and the Risk of Date Rape." *Journal of Social and Clinical Psychology* 9, 1 (1988): 20–37.

Muehlenhard, Charlene L., and S. W. Cook. "Men's Self-Reports of Unwanted Sexual Activity." *Journal of Sex Research* 24 (1988): 58–72.

Muehlenhard, Charlene L., and L. C. Hollabaugh. "Do Women Sometimes Say No When They Mean Yes? The Prevalence and Correlates of Women's Token Resistance to Sex." *Journal of Personality and Social Psychology* 54 (May 1988) 872–879.

Muehlenhard, Charlene L., and M. Linton. "Date Rape and Sexual Aggression in Dating Situations." *Journal of Consulting Psychology* 34 (April 1987): 186–196.

Muehlenhard, Charlene L., and M. L. McCoy. "Double Standard/Double Bind." *Psychology of Women Quarterly* 15 (1991): 447–461.

Muehlenhard, Charlene L., I. G. Ponch, J. L. Phelps, and L. M. Giusti. "Definitions of Rape: Scientific and Political Implications." *Journal of Social Issues* 48, 1 (Spring 1992): 23–44.

Muehlenhard, Charlene L., and J. Schrag. "Nonviolent Sexual Coercion." In *Acquaintance Rape: The Hidden Crime*, edited by A. Parrot and L. Bechhofer. New York: Wiley, 1991.

Mueller, B. A., et al. "Risk Factors for Tubal Infertility: Influence of History of Prior Pelvic Inflammatory Disease." *Sexually Transmitted Diseases* 19, 1 (January 1992): 28–34.

Mullan, Bob. *The Mating Trade.* Boston: Routledge & Kegan Paul, 1984.

Mullen, Paul E. "The Crime of Passion and the Changing Cultural Construction of Jealousy." *Criminal Behavior and Mental Health* 3, 1 (1993): 1–11.

Mulligan, Thomas, and C. Renee Moss. "Sexuality and Aging in Male Veterans: A Cross-Sectional Study of Interest, Ability, and Activity." *Archives of Sexual Behavior* 20, 1 (February 1991): 17–25.

Mulligan, Thomas, and Robert Palguta. "Sexual Interest, Activity, and Satisfaction among Male Nursing Home Residents." *Archives of Sexual Behavior* 20, 2 (April 1991): 199–204.

"Muppet Gender Gap." *Media Report to Women* 21, 1 (1993): 8.

Muram, David, K., Miller, and A. Cutler. "Sexual Assault of the Elderly Victim." *Journal of Interpersonal Violence* 7, 1 (March 1992): 70–76.

Murdock, George. *Social Structure.* New York: Free Press, 1967.

———. "World Ethnographic Sample." *American Anthropologist* 59 (1957): 664–687.

Murnen, S. K., A. Perot, and D. Byrne. "Coping with Unwanted Sexual Activity: Normative Responses, Situational Determinants, and Individual Differences." *Journal of Sex Research*, 26 (1989): 85–106.

Murray, Charles. *Losing Ground: American Social Policy, 1950–1980.* New York: Basic Books, 1984.

Murray, Maresa, and Kathleen Gilbert. "Images of African American Families on Prime-Time Situation Comedies: Are Our 'Roots' There?" Poster session, annual meeting of the National Council on Family Relations, Baltimore, MD, November 11, 1993.

Murry, Velma McBride. "Socio-Historical Study of Black Female Sexuality: Transition to First Coitus." In *The Black Family*, edited by Robert Staples. 4th ed. Belmont, CA: Wadsworth Publishing, 1991.

Murstein, Bernard. "A Clarification and Extension of the SVR Theory of Dyadic Pairing." *Journal of Marriage and the Family* 49 (1987): 929–933.

———. *Who Will Marry Whom: Theories and Research in Marital Choice.* New York: Springer Publishing, 1976.

Mydans, Seth. "Surrogate Losses Custody Bid in Case Defining Motherhood." *New York Times* (October 23, 1990).

Myers-Walls, Judith, and Fred Piercy. "Mass Media and Prevention: Guidelines for Family Life Professionals." *Journal of Primary Prevention* 5, 2 (December 1984): 124–136.

Nadelson, Carol, and Maria Sauzier. "Intervention Programs for Individual Victims and Their Families." In *Violence in the Home: Interdisciplinary Perspectives*, edited by Mary Lystad. New York: Brunner/Mazel, 1986.

Nadler, Arie, and Iris Dotan. "Commitment and Rival Attractiveness: Their Effects on Male and Female Reactions to Jealousy-Arousing Situations." *Sex Roles* 26, 7–8 (1992): 293–310.

Napier, Augustus, and Carl Whitaker. *The Family Crucible.* New York: Harper & Row, 1978.

"A Nation Out of Balance." *Health* (October 1994).

National Center for Health Statistics. "Advance Report of Final Natality Statistics, 1991." *Monthly Vital Statistics Report* 42, 3 (Supplement) (September 9, 1993).

———. *Healthy People 2000 Review.* Hyattsville, Maryland: Public Health Service, 1994.

Needle, Richard, et al. "Drug Abuse: Adolescent Addictions and the Family." In *Stress and the Family: Coping with Catastrophe*, Vol. 2, edited by Charles R. Figley and Hamilton McCubbin. New York: Brunner/Mazel, 1983.

Nelsen, Jane. *Positive Discipline.* New York: Ballantine, 1987.

Nelson, Margaret, and Gordon Nelson. "Problems of Equity in the Reconstituted Family: a Social Exchange Analysis." *Family Relations* 31, 2 (April 1982): 223–231.

"Neonatal Herpes Is Preventable." *U.S.A. Today* (February 1984): 8–9.

Nevid, Jeffrey. "Sex Differences in Factors of Romantic Attraction." *Sex Roles* 11, 5/6 (1984): 401–411.

"New APA Position Statement Urges Actions to Reduce High Rates of Nicotine Dependence." *Psychiatric Services* 46, 2 (February 1995).

Newcomb, H. ed. *Television: The Critical View.* 4th ed. New York: Oxford University Press, 1987.

Newcomb, Michael. "Cohabitation in America: An Assessment of Consequences." *Journal of Marriage and the Family* 41, 3 (August 1979): 597–603.

Newcomb, Nora. *Child Development: Change Over Time.* 8th ed. New York: HarperCollins, 1996.

Newcomer, Susan, and Richard Udry. "Oral Sex in an Adolescent Population." *Archives of Sexual Behavior* 14, 1 (February 1985): 41–46.

Newton, Niles. *Maternal Emotions.* New York: Basic Books, 1955.

Ney, Philip G. "Transgenerational Abuse." In *Intimate Violence: Interdisciplinary Perspectives*, edited by Emilio C. Viano. Washington, DC: Hemisphere, 1992.

NiCarthy, Ginny. *Getting Free: A Handbook for Women in Abusive Relationships.* Seattle, WA: Seal Press, 1986.

Nichols, William C., and Mary A. Pace-Nichols. "Developmental Perspectives and Family Therapy: The Marital Life Cycle." *Contemporary Family Therapy: An International Journal* 15, 4 (1993): 299–315.

Nickens, H. W. "The Health Status of Minority Populations in the United States." *Western Journal of Medicine* 155, 1 (1991): 27–32.

Nielsen, Joyce McCarl, Russell K. Endo, and Barbara L. Ellington. "Social Isolation and Wife Abuse: A Research Report." In *Intimate Violence*, edited by Emilio Viano. Washington, DC: Hemisphere Publishing, 1992.

Noble, Elizabeth. *Having Your Baby by Donor Insemination.* Boston: Houghton Mifflin, 1987.

Nobles, Wade W. "African-American Family Life: An Instrument of Culture." In *Black Families*, edited by Harriette Pipes McAdoo. 2nd ed. Newbury Park, CA: Sage Publications, 1988.

Nock, Steven. "The Symbolic Meaning of Childbearing." *Journal of Family Issues* 8, 4 (December 1987): 373–393.

Noller, Patricia, and Mary Anne Fitzpatrick. "Marital Communication." In *Contemporary Families: Looking Forward, Looking Back*, edited by Alan Booth. Minneapolis: National Council on Family Relations, 1991.

Noller, Patricia, and Mary Ann Fitzpatrick, eds. *Perspectives on Marital Interaction.* Philadelphia: Multilingual Matters, 1988.

Norton, Arthur J. "Family Life Cycle: 1980." *Journal of Marriage and the Family* 45 (1983): 267–275.

Norton, Arthur, and Jeanne Moorman. "Current Trends in American Marriage and Divorce Among American Women." *Journal of Marriage and the Family* 49 (February 1987): 3–14.

Notarius, Clifford, and Jennifer Johnson. "Emotional Expression in Husbands and Wives." *Journal of Marriage and the Family* 44, 2 (May 1982): 483–489.

Notman, Malkah T., and Eva P. Lester. "Pregnancy: Theoretical Considerations." *Psychoanalytic Inquiry* 8, 2 (1988): 139–159.

Nugent, R., and J. Gramick. "Homosexuality: Protestant, Catholic, and Jewish Issues: A Fishbone Tale." *Journal of Homosexuality* 18 (1989): 7–46.

Nuland, Sherwin B. *How We Die: Reflections on Life's Final Chapter.* New York: Knopf, 1994.

Nurmi, Jari Erik. "Age Differences in Adult Life Goals, Concerns, and Their Temporal Extension: A Life Course Approach to Future-Oriented Motivation." *International Journal of Behavioral Development* 15, 4 (1992): 487–508.

Nye, F. Ivan. "Fifty Years of Family Research." *Journal of Marriage and the Family* 50, 2 (May 1988): 305–316.

Nye, F. Ivan, and Felix Berardo, eds. *Emerging Conceptual Frameworks in Family Analysis.* 2 vols. New York: Praeger, 1981.

Nyquist, Linda, et al. "Household Responsibilities in Middle-Class Couples: The Contribution of Demographic and Personality Variables." *Sex Roles* 12, 1/2 (1985): 15–34.

Oakley, Ann, ed. *Sex, Gender, and Society.* Rev. ed. New York: Harper & Row, 1985.

Odent, Michel. *Birth Reborn.* New York: Pantheon, 1984.

O'Farrell, Timothy J., Keith A. Choquette, and Gary R. Birchler. "Sexual Satisfaction and Dissatisfaction in the Marital Relationships of Male Alcoholics Seeking Marital Therapy." *Journal of Studies on Alcohol* 52, 5 (September 1991): 441–447.

O'Flaherty, Kathleen, and Laura Eells. "Courtship Behavior of the Remarried." *Journal of Marriage and the Family* 50, 2 (May 1988): 499–506.

Oggins, Jean, Joseph Veroff, and Douglas Leber. "Perceptions of Marital Interaction among Black and White Newlyweds." *Journal of Personality and Social Psychology* 65, 3 (1993): 494–511.

Ohninger, S., and N. J. Alexander. "Male Infertility: The Focus Shifts to Sperm Manipulation." *Current Opinion in Obstetrics and Gynecology* 3, 2 (1991): 182–190.

Oldenberg, Don. "Watch TV with Kids, Experts Say." *San Jose Mercury* (April 12, 1992).

Olds, S. W. *The Eternal Garden: Seasons of Our Sexuality.* New York: Times Books, 1985.

O'Leary, K. Daniel. "Through a Psychological Lens: Personality Traits, Personality Disorders, and Levels of Violence." In *Current Controversies in Family Violence*, edited by Richard Gelles and Donileen Loseke. Newbury Park, CA: Sage Publications, 1993.

Oliver, Mary Beth, and Janet Shibley Hyde. "Gender Differences in Sexuality: A Meta-analysis." *Psychological Bulletin* 114, 1 (1993): 29–51.

Olshansky, E. F. "Redefining the Concepts of Success and Failure in Infertility Treatment." *Naacogs Clinical Issues in Perinatal and Women's Health Nursing* 3, 2 (1992): 343–346.

Olshansky, S. Jay, Bruce A. Carnes, and Christine Cassel. "In Search of Methuselah: Estimating the Upper Limits to Human Longetivity." *Science* 250, 4981 (November 2, 1990): 634–641.

Olson, David. "Insiders' and Outsiders' Views of Relationships: Research Studies." In *Close Relations*, edited by George Levinger and H. Rausch. Amherst, MA: University of Massachusetts Press, 1977.

Olson, David H., and John DeFrain. *Marriage and the Family; Diversity and Strengths.* 2nd ed. Mountain View, CA: Mayfield, 1997.

Olson, David H., Hamilton I. McCubbin, Howard Barnes, Andrea Larson, Maria Muxen, and Marc Wilson. *Families: What Makes Them Work.* Beverly Hills, CA: Sage Publications, 1983.

Olson, E. J. "No Room at the Inn: A Snapshot of an American Emergency Room." *Stanford Law Review* 46, 2 (1994): 449–501.

Olson, Myrna R., and Judith A. Haynes. "Successful Single Parents." *Families in Society* 74, 5 (1993): 259–267.

O'Neil, Robin, and Ellen Greenberger. "Patterns of Commitment to Work and Parenting: Implications for Role Strain." *Journal of Marriage and the Family* 52, 1 (February 1994): 101–115.

Opie, Anne. "Ideologies of Joint Custody." *Family and Conciliation Courts Review* 31, 3 (1993): 313–326.

O'Reilly, Jane. "Wife Beating: The Silent Crime." *Time* (September 5, 1983).

Orthner, D. K., and J. F. Pittman. "Family Contributions to Work Commitment." *Journal of Marriage and the Family* 48 (1986): 573–581.

Ortiz, Silvia, and Jesus Manuel Casas. "Birth Control and Low-Income Mexican-American Women: The Impact of Three Values." *Hispanic Journal of the Behavioral Sciences* 12, 1 (February 1990): 83–92.

Ortiz, Vilma, and Rosemary Santana Cooney. "Sex Role Attitudes and Labor Force Participation among Young Hispanic Females and Non-Hispanic White Females." *Social Science Quarterly* 65, 2 (June 1984): 392–400.

Oster, Sharon. "A Note on the Determinants of Alimony." *Journal of Marriage and the Family* 49, 1 (February 1987): 81–86.

O'Sullivan, Denis A., and Eleanor O'Leary. "Love: A Dimension of Life." *Counseling and Values* 37, 1 (1992): 32–38.

O'Sullivan, Lucia, and E. Sandra Byers. "College Students' Incorporation of Initiator and Restrictor Roles in Sexual Dating Interactions." *Journal of Sex Research* 29, 3 (August 1992): 435–446.

"Outlook." *U.S. News and World Report* (June 6, 1994): 12.

Padilla, E. R., and K. E. O'Grady. "Sexuality among Mexican Americans: A Case of Sexual Stereotyping." *Journal of Personality and Social Psychology* 52 (1987): 5–10.

Pais, Shoba. "Asian Indian Families in America." In *Families in Cultural Context*, edited by Mary Kay DeGenova. Mountain View, CA: Mayfield, 1997.

Pallow-Fleury, Angie. "Your Hospital Birth: Questions to Ask." *Mothering* (December 1983): 83–85.

Paludi, M. A. "Sociopsychological and Structural Factors Related to Women's Vocational Development." *Annals of New York Academy of Sciences*, 602 (1990): 157–168.

"Panel Says Nation Is Lagging in Children's Health." *New York Times* (March 29, 1992): A16.

Panuthos, Claudia, and Catherine Romeo. *Ended Beginnings: Healing Childbearing Losses.* New York: Warner, 1984.

Papernow, Patricia L. *Becoming a Stepfamily.* San Francisco: Jossey-Bass, 1993.

Pareles, Jon. "Indians' Heritage Survives in Songs." *New York Times* (December 4, 1990): B1.

Parker, Philip. "Motivation of Surrogate Mothers—Initial Findings." *American Journal of Psychiatry* 140, 1 (1983): 117–118.

Parsons, Jacqueline, ed. *The Psychobiology of Sex Differences and Sex Roles.* Washington, DC: Hemisphere, 1980.

Parsons, Talcott. "Family Structure and the Socialization of the Child." In *Family Socialization and Interaction Process*, edited by Talcott Parsons and R. F. Bales. Glencoe, IL: Free Press, 1955.

Parsons, Talcott, and R. F. Bales. *Family Socialization and Interaction Processes.* Glencoe, IL: Free Press, 1955.

Pasley, Kay. "Family Boundary Ambiguity: Perceptions of Adult Stepfamily Family Members." In *Remarriage and Stepparenting: Current Research and Theory,* edited by Kay Pasley and Marilyn Ihinger-Tallman. New York: Guilford Press, 1987.

Pasley, Kay, and Marilyn Ihinger-Tallman, eds. *Remarriage and Stepparenting: Current Research and Theory.* New York: Guilford Press, 1987.

Patterson, C. J. "Children of Lesbian and Gay Parents." *Child Development* 63 (October 1992): 1025–1042.

Patterson, G. *Families Applications of Social Learning to Family Life.* Champaign, IL: Research Press, 1971.

Patterson, Joan, and Hamilton McCubbin. "Chronic Illness: Family Stress and Coping." In *Stress and the Family: Coping with Catastrophe,* edited by Charles R. Figley and Hamilton McCubbin. New York: Brunner/Mazel, 1983.

Patton, Michael. "Twentieth-Century Attitudes toward Masturbation." *Journal of Religion and Health* 25, 4 (December 1986): 291–302.

Paulson, David. "Hot Tubs and Reduced Sperm Counts." *Medical Aspects of Human Sexuality* 14 (September 1980): 121.

Paulson, Morris J., Robert H. Coombs, and John Landsverk. "Youth Who Physically Assault Their Parents." *Journal of Family Violence* 5, 2 (June 1990): 121–134.

Payton, Isabelle. "Single-Parent Households: An Alternative Approach." *Family Economics Review* (December 1982): 11–16.

Pear, Robert. "Rich Got Richer in 80's: Others Held Even." *New York Times* (January 11, 1991): A1, A11.

———. "U.S. Reports Poverty Is Down but Inequality Is Up." *New York Times* (September 27, 1990): A12.

———. "U.S. Reports Rise in Low-Weight Births." *New York Times* (April 22, 1992): A18.

Pearl, D., and L. Bouthilet, and J. Lazar, eds. *Television and Behavior: Ten Years of Scientific Progress and Implications for the Eighties.* Vol I (DHHS Publication No. ADM 82-1196). Washington, DC: U.S. Government Printing Office, 1982.

Peck, M. Scott. *The Road Less Traveled: A New Psychology of Love, Traditional Values, and Spiritual Growth.* New York: Simon & Schuster, 1978.

Peek, Charles, et al. "Teenage Violence Toward Parents: A Neglected Dimension in Family Violence." *Journal of Marriage and the Family* 47, 4 (November 1985): 1051–1058.

Peirce, Kate. "Sex Role Stereotyping of Children: A Content Analysis of the Roles and Attributes of Child Characters." *Sociological Spectrum* 9, 3 (September 1989): 321–328.

Pennington, Saralie Bisnovich. "Children of Lesbian Mothers." In *Gay and Lesbian Parents,* edited by Frederick W. Bozett. New York: Praeger, 1987.

Peplau, Letitia. "Research on Homosexual Couples." In *Gay Relationships,* edited by John DeCecco. New York: Haworth Press, 1988.

———. "What Homosexuals Want." *Psychology Today,* 15, 3 (March 1977): 28–38.

Peplau, Letitia, and Susan Cochran. "Value Orientations in the Intimate Relationships of Gay Men." *Gay Relationships,* edited by John DeCecco. New York: Haworth Press, 1988.

Peplau, Letitia Anne, and Steven Gordon. "The Intimate Relationships of Lesbians and Gay Men." In *Gender Roles and Sexual Behavior,* edited by Elizabeth Allgeier and Naomi McCormick. Palo Alto, CA: Mayfield, 1982.

Peplau, Letitia Anne, Charles T. Hill, and Zick Rubin. "Sex Role Attitudes in Dating and Marriage: A 15-Year Follow-Up of the Boston Couples Study." *Journal of Social Issues* 49, 3 (1993): 31–53.

Peplau, Letitia Anne, et al. "Sexual Intimacy in Dating Relationships." *Journal of Social Issues* 33, 2 (March 1977): 86–109.

Perkins, Kathleen. "Psychosocial Implications of Women and Retirement." *Social Work* 37, 6 (1992): 526–532.

Perlman, Daniel, and Steve Duck, eds. *Intimate Relationships: Development, Dynamics, and Deterioration.* Beverly Hills, CA: Sage Publications, 1987.

Perlman, David. "Brave New Babies." *San Francisco Chronicle* (March 5, 1990): B3.

Perry-Jenkins, Maureen, and Karen Folk. "Class, Couples, and Conflict: Effects of the Division of Labor on Assessments of Marriage in Dual-Earner Marriages." *Journal of Marriage and the Family* 56, 1 (February 1994): 165–180.

Petchesky, R. P. *Abortion and Woman's Choice: The State, Sexuality, and Reproductive Freedom,* rev. ed. Boston: Northeastern University Press, 1990.

Peters, Stefanie, et al. "Prevalance." In *Sourcebook on Child Sexual Abuse,* edited by David Finkelhor. Beverly Hills, CA: Sage Publications, 1986.

Petersen, James R., et al. "Playboy's Readers' Sex Survey." *Playboy* (March 1983): 178–184.

Petersen, Larry. "Interfaith Marriage and Religious Commitment among Catholics." *Journal of Marriage and the Family* 48, 4 (November 1986): 725–735.

Petersen, Virginia, and Susan B. Steinman. "Helping Children Succeed after Divorce: A Court-Mandated Educational Program for Divorcing Parents." *Family and Conciliation Courts Review* 32, 1 (1994): 27–39.

Peterson, Gary W., and Boyd C. Rollins. "Parent-Child Socialization." In *Handbook of Marriage and the Family,* edited by Marvin B. Sussman and Suzanne K. Steinmetz. New York: Plenum Press, 1987.

Peterson, Marie Ferguson. "Racial Socialization of Young Black Children." In *Black Children: Social, Educational, and Parental Environments,* edited by Harriette Pipes McAdoo and John McAdoo. Beverly Hills, CA: Sage Publications, 1985.

Peterson, Richard R. "A Reevaluation of the Economic Consequences of Divorce." *American Sociological Review* 61, 3 (June 1996).

Petit, Charles. "New Study to Ask Why So Many Infants Die." *San Francisco Chronicle* (May 9, 1990): A3.

———. "Why Both Parents Should Be Tested for Herpes." *San Francisco Chronicle* (April 2, 1992): D7.

Petit, Ellen, and Bernard Bloom. "Whose Decision Was It? The Effect of Initiator Status on Adjustment to Marital Disruption." *Journal of Marriage and the Family* 46, 3 (August 1984): 587–595.

Peyrot, Mark, et al. "Marital Adjustment to Adult Diabetes: Interpersonal Congruence and Spouse Satisfaction." *Journal of Marriage and the Family* 50, 2 (May 1988): 363–376.

Pfaus, James, Myronuk, and Jacobs. "Soundtrack Contents and Depicted Sexual Violence." *Archives of Sexual Behavior* 15, 3 (June 1986): 231–237.

Pies, Cheri. "Considering Parenthood: Psychological Issues for Gay Men and Lesbians Choosing Alternative Fertilization." In *Gay and Lesbian Parents*, edited by Frederick W. Bozett. New York: Praeger, 1987.

Pilisuk, Marc. "The Delivery of Social Support: The Social Innoculation." *American Journal of Orthopsychiatry* 52, 1 (January 1982): 20–31.

Pill, Cynthia. "Stepfamilies: Redefining the Family." *Family Relations* 39, 2 (April 1990): 186–193.

Pillemer, Karl. "The Abused Offspring Are Dependent: Abuse Is Caused by the Deviance and Dependence of abusive Caregivers." In *Current Controversies in Family Violence*, edited by Richard Gelles and Donileen Loseke. Newbury Park, CA: Sage Publications, 1993.

Pillemer, Karl, and J. Jill Suitor. "Elder Abuse." In *Handbook of Family Violence*, edited by Vincent B. Van Hasselt et al. New York: Plenum Press, 1988.

Pink, Jo Ellen, and Karen Smith Wampler. "Problem Areas in Stepfamilies: Cohesion, Adaptability, and the Stepfather-Adolescent Relationship." *Family Relations* 34, 3 (July 1985): 327–335.

Pirog-Good, Maureen A., and Jane. Stets, eds. *Violence in Dating Relationships: Emerging Social Issues.* New York: Praeger, 1989.

Pistole, M. Carole. "Attachment in Adult Romantic Relationships: Style of Conflict Resolution and Relationship Satisfaction." *Journal of Social Personal Relationships* 6, 4 (1989): 505–512.

Piven, Frances Fox, and Richard Cloward. *Regulating the Poor.* New York: Vintage Books, 1972.

Pleck, Joseph. *Working Wives/Working Husbands.* Beverly Hills, CA: Sage Publications, 1985.

Plummer, William, and Margaret Nelson. "A Mother's Priceless Gift." *People Weekly* (August 26, 1991).

Pogrebin, Letty Cottin. *Growing Up Free: Raising Your Child in the 1980s.* New York: McGraw-Hill, 1980.

———. *Family Politics.* New York: McGraw-Hill, 1983.

Poland, Ronald L. "The Question of Routine Neonatal Circumcision." *New England Journal of Medicine* 322, 18 (May 3, 1990): 1312–1315.

"Poll on Families: Small Is Best." *New York Times* (May 25, 1986): L22.

Popovich, Paula M., et al. "Assessing the Incidence and Perceptions of Sexual Harassment Behaviors among American Undergraduates." *Journal of Psychology* 120 (1986): 387–396.

———. "Perceptions of Sexual Harassment as a Function of Sex of Rater and Incident Form and Consequence." *Sex Roles* 27 (December 1992): 609–625.

Porter, Nancy L., and F. Scott Christopher. "Infertility: Towards An Awareness of a Need Among Family Life Practitioners." *Family Relations* 33, 2 (April 1984): 309–315.

Portes, P. R., S. C. Howell, J. H. Brown, S. Eichenberger, and C. A. Mas. "Family Functions and Children's Postdivorce Adjustment. *Journal of Orthopsychiatry* 62 (October 1992): 613–617.

"Poverty Helps Break Up Families, Report Says." *New York Times* (January 15, 1993): A7.

Powell, Rachel. "It's One Party Even the Recession Can't Spoil." *New York Times* (June 23, 1991): 10.

Pratt, Clara, and Vicki Schmall. "College Students' Attitudes toward Elderly Sexual Behavior: Implications for Family Life Education." *Family Relations* 38, 2 (April 1989): 137–141.

Press, Robert. "Hunger in America." *Christian Science Monitor* (February 15, 1985): 18–19.

Presser, Harriet. "Shift Work Among American Women and Child Care." *Journal of Marriage and the Family* 48, 3 (August 1986): 551–663.

———. "Some Economic Complexities of Child Care Provided by Grandmothers." *Journal of Marriage and the Family* 51, 3 (August 1989): 581–591.

Price, J. H., and P. A. Miller. "Sexual Fantasies of Black and White College Students." *Psychological Reports* 54 (1984): 1007–1014.

Price, Jane. "Who Wants to Have Children? and Why?" In *Relationships: The Marriage and Family Reader*, edited by Jeffrey Rosenfeld. Glenview, IL: Scott, Foresman, 1982.

Price, John. "North American Indian Families." In *Ethnic Families in America: Patterns and Variations*, edited by Charles Mindel and Robert Haberstein. 2nd ed. New York: Elsevier-North Holland, 1981.

Price-Bonham, S., and J. O. Balswick. "The Noninstitutions: Divorce, Desertion, and Remarriage." *Journal of Marriage and the Family* 42 (1980): 959–972.

Priest, Ronnie. "Child Sexual Abuse Histories among African-American College Students: A Preliminary Study." *American Journal of Orthopsychiatry* 62, 3 (July 1992): 475.

Prober, Charles, et al. "Low Risk of Herpes Simplex Virus Infections in Neonates Exposed to the Virus at the Time of Vaginal Delivery to Mothers with Recurrent Genital Herpes Simplex Virus Infections." *New England Journal of Medicine* 316 (January 29, 1987): 129–138.

Pruett, Kyle. *The Nurturing Father: Journey toward Complete Man.* New York: Warner, 1987.

Puglisi, J. T., and D.W. Jackson. "Sex Role Identity and Self-Esteem in Adulthood." *Journal of Aging and Human Development* 12 (1981): 129–138.

Radway, J. A. *Reading the Romance.* Chapel Hill, NC: University of North Carolina, 1984.

Ramirez de Arrellano, Annette. "Latino Women—Health Status and Access to Care." In *Women's Health*, edited by Marilyn Falik and Karen Scott Collins. Baltimore, MD: Johns Hopkins University Press, 1996.

Ramsey, Patricia G. *Teaching and Learning in a Diverse World: Multicultural Education for Young Children.* New York: Teachers College Press, 1987.

Rando, Therese A. "Death and Dying Are Not and Should Not Be Taboo Topics." In *Principles of Thanatology*, edited by Austin H. Kutscher et al. New York: Columbia University Press, 1987.

Rank, Mark R. and Li-Chen Cheng. "Welfare Use Across Generations: How Important Are the Ties That Bind?" *Journal of Marriage and the Family* 57, 3 (August 1995): 673–684.

"Rates of Cesarean Delivery—United States, 1993." Centers for Disease Control and Prevention. *Morbidity and Mortality Weekly Report*, 44 (1995): 303–307.

Rao, Kavitha, et al. "Child Sexual Abuse of Asians Compared with Other Populations." *Journal of the American Academy of Child and Adolescent Psychiatry* 31, 5 (September 1992): 880 ff.

Rapp, Carol, and Sally Lloyd. "The Role of 'Home as Haven' Ideology in Child Care Use." *Family Relations* 38, 4 (October 1989): 427–430.

Raschke, Helen. "Divorce." In *Handbook of Marriage and the Family*, edited by Marvin Sussman and Suzanne Steinmetz. New York: Plenum Press, 1987.

Raval, H., et al. "The Impact of Infertility on Emotions and the Marital and Sexual Relationship." *Journal of Reproductive and Infant Psychology* 5, 4 (October 1987): 221–234.

Raven, Bertram, et al. "The Bases of Conjugal Power." In *Power in Families,* edited by Ronald Cromwell and David Olson. New York: Halstead Press, 1975.

Ravinder, Shashi. "Androgyny: Is It Really the Product of Educated Middle-Class Western Societies?" *Journal of Cross-Cultural Psychology* 18, 2 (June 1987): 208–220.

Rawlings, Steve "Households and Families." *Current Population Reports* (Series P-20, No. 483). Washington, DC: U.S. Government Printing Office, 1994.

———. "Household and Family Characteristics, March 1993." *Current Population Reports.* Population Characteristics. (Series P20, No. 477). Washington, DC: U.S. Government Printing Office, 1994.

———. "Studies in Marriage and the Family: Single Parents and Their Children." *Current Population Reports.* (Series, P-23.) Washington, DC: U.S. Government Printing Office, 1989.

Real, Michael R. *Exploring Media Culture.* Thousand Oaks, CA: Sage Publications, 1996.

Reed, David, and Martin Weinberg. "Premarital Coitus: Developing and Establishing Sexual Scripts." *Social Psychology Quarterly* 47, 2 (June 1984): 129–138.

Reedy, M., et al. "Age and Sex Differences in Satisfying Love Relationships across the Adult Life Span." *Human Development* 24 (1981): 52–86.

Regan, Mary, and Helen Roland. "Rearranging Family and Career Priorities: Professional Women and Men of the Eighties." *Journal of Marriage and the Family* 47, 4 (November 1985): 985–992.

Reid, Pamela, and L. Comas-Diaz. "Gender and Ethnicity: Perspectives on Dual Status." *Sex Roles* 22, 7 (April 1990): 397–408.

Reilly, Mary E., et al. "Tolerance for Sexual Harassment Related to Self-Reported Sexual Victimization." *Gender and Society* 6, 1 (March 1992): 122–138.

Reisman, C. *Divorce Talk: Women and Men Make Sense of Personal Relationships.* New Brunswick, NJ: Rutgers University Press, 1990.

Reisman, C. K., and N. Gerstel. "Marital Dissolution and Health: Do Males or Females Have Greater Risk?" *Social Science and Medicine* 20 (1985): 627–635.

Reiss, Ira. *Family Systems in America.* 3d ed. New York: Holt, Rinehart, & Winston, 1980a.

———. "A Multivariate Model of the Determinants of Extramarital Sexual Permissiveness." *Journal of Marriage and the Family* 42, 2 (May 1980b): 395–411.

Renzetti, Claire. "Violence in Gay and Lesbian Relationships." In *Vision 2010: Families and Violence, Abuse and Neglect,* edited by Richard J. Gelles. Minneapolis: National Council on Family Relations, 1995.

Resnick, Sandven. "Informal Adoption among Black Adolescent Mothers." *American Journal of Orthopsychiatry* 60 (April 1990): 210–224.

Retik, Alan B., and Stuart B. Bauer. "Infertility Related to DES Exposure in Utero: Reproductive Problems in the Male." In *Infertility: Medical, Emotional, and Social Considerations,* edited by Miriam Mazor and Harriet Simons. New York: Human Sciences Press, 1984.

Rettig, K. D., and M. M. Bubolz. "Interpersonal Resource Exchanges as Indicators of Quality of Marriage." *Journal of Marriage and the Family* 45 (1983): 497–510.

Rexroat, Cynthia. "Race and Marital Status Differences in the Labor Force Behavior of Female Family Heads: The Effect of Household Structure." *Journal of Marriage and the Family* 52, 3 (August 1990): 591–601.

Rexroat, Cynthia, and Constance Shehan. "The Family Life Cycle and Spouses' Time in Housework." *Journal of Marriage and the Family* 49, 4 (November 1987): 737–750.

Rice, Susan. "Sexuality and Intimacy for Aging Women: A Changing Perspective." *Journal of Women and Aging* 1, 1–3 (1989): 245–264.

Rich, Adrienne. *Of Woman Born.* New York: Norton, 1976.

Richards, Leslie N., and Cynthia J. Schmiege. "Problems and Strengths of Single-Parent Families: Implications for Practice and Policy." *Family Relations* 42, 3 (July 1993): 277–285.

Richardson, Diana. "The Dilemma of Essentiality in Homosexual Theory." *Journal of Homosexuality* 9, 2/3 (December 1983): 79–90.

Richmond-Abbott, M. *Masculine and Feminine: Gender Roles Over the Life Cycle.* 2nd ed. New York: McGraw-Hill, 1992.

Ridley, Jane, and Michael Crowe. "The Behavioural-Systems Approach to the Treatment of Couples." *Sexual and Marital Therapy* 7, 2 (1992): 125–140.

Riegle, Donald. "The Psychological and Social Effects of Unemployment." *American Psychologist* 37, 10 (October 1982): 1113–1115.

Riffer, Roger, and Jeffrey Chin. "Dating Satisfaction among College Students." *International Journal of Sociology and Social Policy* 8, 5 (1988): 29–36.

Riggs, David S. "Relationship Problems and Dating Aggression: A Potential Treatment Target." *Journal of Interpersonal Violence* 8, 1 (1993): 18–35.

Riley, Alan J. "Sexuality and the Menopause." *Sexual and Marital Therapy* 6, 2 (1991): 135–146.

Riportella-Muller, Roberta. "Sexuality in the Elderly: A Review." In *Human Sexuality: The Societal and Interpersonal Context,* edited by Kathleen McKinney and Susan Sprecher. Norwood, NJ: Ablex, 1989.

Riseden, Andrea D., and Barbara E. Hort. "A Preliminary Investigation of the Sexual Component of the Male Stereotype." Unpublished paper, 1992.

Rivlin, Leanne. "The Significance of Home and Homelessness." *Marriage and Family Review* 15, 1/2 (1990): 39–57.

Roberts, Linda, and Lowell Krokoff. "A Time Series Analysis of Withdrawal, Hostility, and Displeasure in Satisfied and Dissatisfied Marriages." *Journal of Marriage and the Family* 52, 1 (February 1990): 95–105.

Roberts, Michael C., Kristi Lekander, and Debra Fanurik. "Evaluation of Commercially Available Materials to Prevent Child Sexual Abuse and Abduction." *American Psychologist* 45, 6 (June 1990): 782–783.

Roberts, Sam. "The Hunger beneath the Statistics." *New York Times* (May 23, 1988): A13.

Robertson, Elizabeth, et al. "The Costs and Benefits of Social Support in Families." *Journal of Marriage and the Family* 53, 2 (May 1991): 403–416.

Robinson, Bryan. *Teenage Fathers.* Lexington, MA: Lexington Books, 1987.

Robinson, I., K. Ziss, B. Ganza, and S. Katz. "20 Years of the Sexual Revolution, 1965–1985—An Update." *Journal of Marriage and the Family* 53, 1 (February 1991): 216–220.

Robinson, Linda C., and Pricilla White Blanton. "Marital Strengths in Enduring Marriages." *Family Relations* 42 (January 1993): 38–45.

Robinson, Pauline. "The Sociological Perspective." In *Sexuality in the Later Years: Roles and Behavior*, edited by Ruth Weg. New York: Academic Press, 1983.

Rodgers, Roy, and L. Conrad. "Courtship for Remarriage: Influences on Family Reorganization after Divorce." *Journal of Marriage and the Family* 48 (1986): 767–775.

Rodman, Hyman, and Cynthia Cole. "When School-Age Children Care for Themselves: Issues for Family Life Educators and Parents." *Family Relations* 36, 1 (January 1987): 92–96.

Roen, Philip. *Male Sexual Health*. New York: 1974.

Roenrich, L., and B. N. Kinder. "Alcohol Expectancies and Male Sexuality: Review and Implications for Sex Therapy." *Journal of Sex and Marital Therapy* 17 (1991): 45–54.

Rogers, Stacy J. "Mothers' Work Hours and Marital Quality: Variations by Family Structure and Family Size." *Journal of Marriage and the Family* 58, 3 (August 1996): 606–617.

Rogers, Susan M., and Charles F. Turner. "Male-Male Sexual Contact in the U.S.A.: Findings from Five Sample Surveys, 1970–1990." *Journal of Sex Research* 28, 4 (November 1991): 491–519.

Rolland, John. *Families, Illness, and Disability*. New York: Basic Books, 1994.

Rooks, J., et al. "Outcomes of Care in Birth Centers." *New England Journal of Medicine* 321, 26 (December 28, 1989): 1804–1811.

Roopnarine, J. L., and N. S. Mounts. "Current Theoretical Issues in Sex Roles and Sex Typing." In *Current Conceptions of Sex Roles and Sex Typing: Theory and Research*, edited by D. B. Carter. New York: Praeger, 1987.

Roopnarine, Jaipaul, et al. "Mothers' Perceptions of Their Children's Supplemental Care Experience Correlation with Spousal Relationship." *American Journal of Orthopsychiatry* 56, 4 (October 1986): 581–588.

Roos, Patricia, and Lawrence Cohen. "Sex Roles and Social Support as Moderates of Life Stress Adjustment." *Journal of Personality and Social Psychology* 52, 3 (March 1987): 576–585.

Root, Maria P., ed. *Filipino Americans: Transforming Identity*. Thousand Oaks, CA: Sage, 1997.

Roscoe, Bruce, and Nancy Benaske. "Courtship Violence Experienced by Abused Wives: Similarities in Patterns of Abuse." *Family Relations* 34, 3 (July 1985): 419–424.

Rose, Suzanna, and Irene Hanson Frieze. "Young Singles' Contemporary Dating Scripts." *Sex Roles* 28, 9–10 (May 1993): 499–510.

Rosen, M. P., et al. "Cigarette Smoking: An Independent Risk Factor for Atherosclerosis in the Hypograstric-Cavernous Arterial Bed of Men with Arteriogenic Impotence." *Journal of Urology* 145, 4 (April 1991): 759–776.

Rosenberg, Joshua D. "In Defense of Mediation." *Family and Conciliation Courts Review* 30, 4 (1992): 422–467.

Rosengren, A., H. Wedel, and L. Wilhelmsen. "Marital Status and Mortality in Middle-Aged Swedish Men." *American Journal of Epidemiology* 129, 1 (January 1989): 54–64.

Rosenthal, Carolyn. "Kinkeeping in the Familial Division of Labor." *Journal of Marriage and the Family* 47, 4 (November 1985): 965–947.

Rosenthal, Elisabeth. "Cost of High-Tech Fertility: Too Many Tiny Babies." *New York Times* (May 26, 1992): B5, B7.

———. "When a Pregnant Woman Drinks." *New York Times Magazine* (February 4, 1990): 30.

Rosenthal, Kristine, and Harry F. Keshet. "The Not Quite Stepmother." *Psychology Today* 12 (1979): 82–86.

Rosenweig, P. M. *Married and Alone: The Way Back*. New York: Plenum, 1992.

Ross, Catherine, John Mirowsky, and Karen Goldsteen. "The Impact of the Family on Health." In *Contemporary Families: Looking Forward, Looking Back*, edited by Alan Booth. Minneapolis: National Council on Family Relations, 1991.

Ross, L., et al. "Television Viewing and Adult Sex-Role Attitudes." *Sex Roles* 8 (1982): 589–592.

Rothenberg, M. "Ending Circumcision in the Jewish Community?" Syllabus of Abstracts, Second International Symposium on Circumcision, San Francisco (April 30, 1991).

Rothenberg, R. B. "Those Other STDs." *American Journal of Public Health* 81 (October 1991): 1250–1251.

Rothschild, B. S., P. J. Fagan, and C. Woodall. "Sexual Functioning of Female Eating-Disordered Patients." *International Journal of Eating Disorders* 10 (1991): 389–394.

Rotter, Julian, B. Liverant, and D. P. Crowne. "The Growth and Extinction of Expectancies in Chance Controlled and Skilled Tests." *Journal of Psychology* 52 (1961): 161–177.

"Routine AZT Use Cuts Babies' HIV Risk, Study Finds." *San Mateo County Times* (July 10, 1996): A-6

Rowan, Edward. "Editorial: Masturbation According to the Boy Scout Handbook." *Journal of Sex Education and Therapy* 15, 2 (June 1989): 77–81.

Rowland, Robyn. "Technology and Motherhood: Reproductive Choice Reconsidered." *Signs: Journal of Women in Culture and Society* 12, 3 (1987): 512–528.

Rubenstein, Carin, and Carol Tavris. "Special Survey Results: 26,000 Women Reveal the Secrets of Intimacy." *Redbook* (September 1987): 147–149 ff.

Rubin, A. M., and J. R. Adams. "Outcomes of Sexually Open Marriages." *Journal of Sex Research* 22 (1986): 311–319.

Rubin, Lillian. *Erotic Wars*. New York: Farrar, Straus & Giroux, 1990.

———. *Families on the Faultline: America's Working Class Speaks about the Family, the Economy, Race, and Ethnicity*. New York: HarperCollins, 1994.

Rubin, Sylvia. "Women's Health Goes Mainstream." *San Francisco Chronicle* (March 21, 1994): E9.

Rubin, Zick. *Liking and Loving: An Invitation to Social Psychology*. New York: Holt, Rinehart, & Winston, 1973.

———. "Loving and Leaving: Sex Differences in Romantic Attachments." *Sex Roles* 7 (1981): 821–835.

———. "Self Disclosure in Dating Relationships: Sex Roles and the Ethic of Openness." *Journal of Marriage and the Family* 42 (1980): 305–317.

Rubin, Zick, F. Provenzano, and Z. Luria. "The Eye of the Beholder: Parents' View of Sex of Newborn." *American Journal of Orthopsychiatry* 44 (September 1974): 512–519.

Runyan, William. "In Defense of the Case Study Method." *American Journal of Orthopsychiatry* 52, 3 (July 1982): 440–446.

Rush, D., et al. "The National WIC Evaluation." *American Journal of Clinical Nutrition* 48, 2 (Supplement) (August 1988): 439–483.

Russel, C. "Transition to Parenthood." *Journal of Marriage and the Family* 36, 2 (May 1974): 294–302.

Russel, Diana. *Sexual Exploitation: Rape, Child Sexual Abuse, and Workplace Harassment.* Beverly Hills, CA: Sage Publications, 1984.

Russell, Diana E. H. *Rape in Marriage.* Revised ed. Bloomington, IN: Indiana University Press, 1990.

———. *The Secret Trauma: Incest in the Lives of Girls and Women.* New York: Basic Books, 1986.

Russo, Nancy Felipe, J. D. Horn, and R. Schwartz. "U.S. Abortions in Context: Selected Characteristics." *Journal of Social Issues* 48, 3 (1992): 183–202.

Rutter, V. "Lessons From Stepfamilies." *Psychology Today* (May 1994): 30–33, 60.

Ryan, Michael. "We Need to Teach Doctors to Care." *Parade* (July 3, 1994): 8, 10.

Sabatelli, Ronald, and Erin Cecil-Pigo. "Relational Interdependence and Commitment in Marriage." *Journal of Marriage and the Family* 47, 4 (November 1985): 931–938.

Safilios-Rothschild, Constantina. "Family Sociology or Wives' Sociology? A Cross-Cultural Examination of Decisionmaking." *Journal of Marriage and the Family* 38 (1976): 355–362.

———. "The Study of the Family Power Structure." *Journal of Marriage and the Family* 32, 4 (November 1970): 539–543.

Saitoti, Tepelit Ole. *The Worlds of a Masai Warrior: An Autobiography.* Berkeley: University of California Press, 1986.

Salgado de Snyder, V. Nelly, Richard Cervantes, and Amado Padilla. "Gender and Ethnic Differences in Psychosocial Stress and Generalized Distress among Hispanics." *Sex Roles* 22, 7 (April 1990): 441–453.

Salholz, E. "The Future of Gay America." *Newsweek* (March 12, 1990): 20–25.

Salholz, Eloise. "The Marriage Crunch: If You're a Single Woman, Here Are Your Chances of Getting Married." *Newsweek* (June 2, 1986): 54–58.

Salovey, P., and J. Rodin. "Provoking Jealousy and Envy: Domain Relevance and Self-Esteem Threat." *Journal of Social and Clinical Psychology* 10, 4 (December 1991): 395–413.

Saluter, Arlene. "Marital Status and Living Arrangements: March 1989." *Bureau of the Census Current Population Reports.* (Population Characteristics Series P-20, No. 468.) Washington, DC: U.S. Government Printing Office, 1992.

———. "Marital Status and Living Arrangements: March 1992." *Current Population Reports.* Bureau of the Census. Population Characteristics (Series P-20). Washington, DC: U.S. Government Printing Office, 1993.

———. "Marital Status and Living Arrangements: March 1993." *Bureau of the Census Current Population Reports.* (Population Characteristics Series P-20 No. 478). Washington, DC: U.S. Government Printing Office, 1994.

———. "Marital Status and Living Arrangements." Bureau of the Census. *Current Population Reports* (Series P20-483). U.S. Government Printing Office, Washington, DC: 1994.

Sampson, Ronald. *The Problem of Power.* New York: Pantheon, 1966.

Samuels, M., and N. Samuels. *The New Well Pregnancy Book.* New York: Simon and Schuster, 1996.

Samuels, Shirley. *Ideal Adoption: A Comprehensive Guide to Forming an Adoptive Family.* New York: Insight Books, 1990.

Sánchez-Ayéndez, Melba. "The Puerto Rican American Family." In *Ethnic Families in America: Patterns and Variations,* edited by Charles Mindel et al. 3rd ed. New York: Elsevier-North Holland, Inc., 1988.

———. "Puerto Rican Elderly Women: Shared Meanings and Informal Supportive Networks." In *All American Women: Lives That Divide, Ties That Bind,* edited by J. B. Cole. New York: Free Press, 1986.

Sander, William. "Catholicism and Intermarriage in the United States." *Journal of Marriage and the Family* 55, 4 (November 1993): 1037–1041.

Sanders, Gregory F., and Debra W. Trygstad. "Strengths in the Grandparent-Grandchild Relationship." *Activities, Adaptation and Aging* 17, 4 (1993): 43–53.

Sanders, Stephanie A., June M. Reinisch, and D. P. McWhirter. "Homosexuality/Heterosexuality: An Overview." In *Homosexuality/Heterosexuality: Concepts of Sexual Orientation,* edited by D. P. McWhirter, S. A. Sanders, and J. M. Reinisch. New York: Oxford University Press, 1990.

Sanderson, Bettie, and Lawrence A. Kurdek. "Race and Gender as Moderator Variables in Predicting Relationship Satisfaction and Relationship Commitment in a Sample of Dating Heterosexual Couples." *Family Relations* 42, 3 (July 1993): 263–267.

Santrock, John, and Karen Sitterle. "Parent-Child Relationships in Stepmother Families." In *Remarriage and Stepparenting: Current Research and Theory,* edited by Kay Pasley and Marilyn Ihinger-Tallman. New York: Guilford Press, 1987.

Satir, Virginia. *The New Peoplemaking.* Rev. ed. Mountain View, CA: Science and Behavior Books, 1988.

Peoplemaking. Palo Alto, CA: Science and Behavior Books, 1972.

Saunders, Daniel. "When Battered Women Use Violence: Husband-Abuse or Self-Defense." *Victim and Violence* 1, 1 (1986): 47–60.

Savin-Williams, Ritch, and Richard G. Rodriguez. "A Developmental, Clinical Perspective on Lesbian, Gay Male, and Bisexual Youths." In *Adolescent Sexuality,* edited by Thomas P. Gullotta et al. Newbury Park, CA: Sage Publications, 1993.

Scanzoni, John. "Reconsidering Family Policy: Status Quo or Force for Change?" *Journal of Family Issues* 3, 3 (September 1982): 277–300.

———. *Sexual Bargaining.* 2d ed. Englewood Cliffs, NJ: Prentice Hall, 1980.

Schaap, Cas, Bram Buunk, and Ada Kerkstra. "Marital Conflict Resolutions." In *Perspectives on Marital Interaction,* edited by Patricia Noller and Mary Anne Fitzpatrick. Philadelphia: Multilingual Matters, 1988.

Schaap, Cas, et al. "Marital Conflict Resolution." In *Perspectives on Marital Interaction,* edited by Patricia Noller and Mary Anne Fitzpatrick. Philadelphia: Multilingual Matters, 1988.

Schechter, S., and L. Gary. "A Framework for Understanding and Empowering Battered Women." In *Abuse and Victimization Across the Life Span,* edited by Martha B. Straus. Baltimore: Johns Hopkins University Press, 1988.

Schenden, Laurie. "Gay Couples Making Adoption Gains." *Los Angeles Times* (September 29, 1993).

Schiavi, Raul C., P. Schreiner-Engle, J. Mandeli, J. Schanzer, and E. Cohen. "Chronic Alcoholism and Male Sexual Dysfunction." *Journal of Sex & Marital Therapy* 16, 1 (March 1990): 23–33.

Schickedanz, Judith A., David I. Schickedanz, Karen Hansen, and Peggy O Forstyh. *Understanding Children.* Mountain View, CA: Mayfield, 1993.

Schmalz, J. "Homosexuals Wake to See a Referendum: It's on Them." *New York Times* (January 31, 1993).

Schmitt, Bernard, and Robert Millard. "Construct Validity of the Bem Sex Role Inventory: Does the BSRI Distinguish between Gender-Schematic and Gender Aschematic Individuals?" *Sex Roles* 19, 9–10 (November 1988): 581–588.

Schneck, M. E., et al. "Low-Income Adolescents and Their Infants: Dietary Findings and Health Outcomes." *Journal of the American Dietetic Association* 90, 4 (April 1990): 555–558.

Schneider, John. *Stress, Loss, and Grief: Understanding Their Origins and Growth Potential.* Baltimore: University Park Press, 1984.

Schneider, Keith. "Minorities Join to Fight Toxic Waste." *New York Times* (October 25, 1991): A12.

Schneider, Peter. "Lost Innocents: The Myth of Missing Children." *Harper's Magazine* (February 1987): 47–53.

Schoen, Edgar J. "The Status of Circumcision of Newborns." *New England Journal of Medicine* 322, 18 (May 3, 1990): 1308–1311.

Schooler, Carmi. "Psychological Effects of Complex Environments during the Life Span: A Review and Theory." In *Cognitive Functioning and Social Structure over the Life Course,* edited by Carmi Schooler and K. Warner Schaie. Norwood, NJ: Ablex, 1987.

Schooler, Carmi, and K. Warner Schaie, eds. *Cognitive Functioning and Social Structure over the Life Course.* Norwood, NJ: Ablex, 1987.

Schooler, Carmi, et al. "Work for the Household: Its Nature and Consequences for Husbands and Wives." *American Journal of Sociology* 90, 1 (July 1984): 97–124.

Schor, Juliet B. *The Overworked American: The Unexpected Decline of Leisure.* New York: Basic Books, 1991.

Schrager, Cynthia D. "Questioning the Promise of Self-Help: A Reading of 'Women Who Love Too Much.'" *Feminist Studies* 19, 1 (1993): 176–192.

Schudson, Charles B. "Antagonistic Parents in Family Courts: False Allegations or False Assumptions about True Allegations of Child Sexual Abuse?" *Journal of Child Sexual Abuse* 1, 2 (1992): 113–116.

Schumm, Walter, et al. "His and Her Marriage Revisited." *Journal of Family Issues* 6, 2 (June 1985): 211–227.

Schvaneveldt, Jay. "The Interactional Framework in the Study of the Family." In *Emerging Conceptual Frameworks in Family Analysis,* edited by F. Ivan Nye and Felix Berardo. 2nd ed. New York: Praeger, 1981.

Schvaneveldt, Jay, Shelley Lindauer, and Margaret Young. "Children's Understanding of AIDS: A Developmental Viewpoint." *Family Relations* 39, 3 (July 1990): 330–335.

Schvaneveldt, Jay, and Margaret H. Young. "Strengthening Families: New Horizons in Family Life Education." *Family Relations* 41, 4 (October 1992): 385–389.

Schwartz, Lita L. "Enabling Children of Divorce to Win." *Family and Conciliation Courts Review* 32, 1 (1994): 72–83.

Schwartz, Pepper. Interview with Bryan Strong. December 20, 1992.

Schwebel, Andrew, Mark Fine, and Maureena Renner. "A Study of Perceptions of the Stepparent Role." *Journal of Family Issues* 23, 1 (March 1991): 43–57.

Scott, Clarissa, Lydia Shifman, Lavenda Orr et al. "Hispanic and Black American Adolescents' Beliefs Relating to Sexuality and Contraception." *Adolescence* 23, 91 (September 1998): 667–688.

Scott, Jacqueline. "Conflicting Beliefs about Abortion: Legal Approval and Moral Doubts." *Social Psychology Quarterly* 52, 4 (December 1989): 319–328.

Scott, Janny. "Low Birth Weight's High Cost." *Los Angeles Times,* (December 24, 1990a): 1.

———. "Trying to Save the Babies." *Los Angeles Times* (December 21, 1990b): 1, 18ff.

Scott, Joan. "Gender: A Useful Category of Historical Analysis." *American Historical Review* 91 (1986): 1053–1075.

Scott, S. G., et al. "Therapeutic Donor Insemination with Frozen Semen." *Canadian Medical Association Journal* 143, 4 (August 15, 1990): 273–278.

Scott-Jones, Diane, and Sherry Turner. "Sex Education, Contraceptive and Reproductive Knowledge, and Contraceptive Use among Black Adolescent Females." *Journal of Adolescent Research* 3, 2 (June 1988): 171–187.

Sears, David. "College Sophomores in the Laboratory: Influences of Narrow Data Base on Social Psychology's View of Human Nature." *Journal of Personality and Social Psychology* 51 (Augsut 1986): 515–530.

Seaward, Brian Luke. *Managing Stress: Principles and Strategies for Health and Well-Being.* 2nd ed. Boston: Jones & Barlett Publishers, 1997.

Seidman, Steven. "Constructing Sex as a Domain of Pleasure and Self-Expression: Sexual Ideology in the Sixties." *Theory, Culture, and Society* 6, 2 (May 1989): 293–315.

Select Committee on Children, Youth and Families. *Families and Child Care: Improving the Options.* Rept. 98–1180, 98th Cong., 2d session, 1985.

Semler, Tracy Chutorian. *All About Eve: The Complete Guide to Women's Health and Mental Well-Being.* New York: HarperCollins, 1995.

Sennett, Richard. *Authority.* New York: Knopf, 1980.

Serdahely, William, and Georgia Ziemba. "Changing Homophobic Attitudes through College Sexuality Education." *Journal of Homosexuality* 10, 1 (September 1984): 148 ff.

Serovich, Julianne M., Sharon J. Price, and Steven F. Chapman. "Former In-Laws as a Source of Support." *Journal of Divorce and Remarriage* 17, 1–2 (1991): 17–25.

Settles, Barbara H. "A Perspective on Tomorrow's Families." In *Handbook of Marriage and the Family,* edited by Marvin Sussman and Suzanne Steinmetz. New York: Plenum Press, 1987.

Sexton, Christine, and Daniel Perlman. "Couples' Career Orientation, Gender Role Orientation, and Perceived Equity as Determinants of Marital Power." *Journal of Marriage and the Family* 51, 4 (November 1989): 933–941.

Sgoutas-Emch, Sandra, et al. "The Effects of an Acute Psychological Stressor on Cardiovascular, Endocrine and Cellular Immune Response." *Psychophysiology* 31, 3 (August 1994).

Shanis, B. S., et al. "Transmission of Sexually Transmitted Diseases by Donor Semen." *Archives of Andrology* 23, 3 (1989): 249–257.

Shannon, Thomas. *Surrogate Motherhood.* New York: Crossroad Publishing, 1988.

Shapiro, Adam D. "Explaining Psychological Distress in a Sample of Remarried and Divorced Persons. *Journal of Family Issues* 17, 2 (March 1996): 186–203.

Shapiro, David. "No Other Hope for Having a Child." *Newsweek* (January 19, 1987): 50–51.

Shapiro, Jerrold Lee. *The Measure of a Man: Becoming the Father You Wish Your Father Had Been.* New York: Delacorte, 1993.

Shapiro, Joanna. "Family Reactions and Coping Strategies in Response to the Physically Ill or Handicapped Child: A Review." *Social Science and Medicine* 17 (1983): 913–931.

Shapiro, Johanna, and Ken Tittle. "Maternal Adaptation to Child Disability in a Hispanic Population." *Family Relations* 39, 2 (April 1990): 179–185.

Sharpsteen, Don J. "Romantic Jealousy as an Emotion Concept: A Prototype Analysis." *Journal of Social and Personal Relationships* 10, 1 (1993): 69–82.

Shaver, Phillip R., and Kelly A. Brennan. "Attachment Styles and the 'Big Five' Personality Traits: Their Connections with Each Other and with Romantic Relationship Outcomes." *Personality and Social Psychology Bulletin* 18, 5 (1992): 536–545.

Shaver, Phillip, and Cindy Hazan. "Being Lonely, Falling in Love: Perspectives from Attachment Theory." *Journal of Social Behavior and Personality* 2, 2 Pt. 2 (1987): 105–124.

———. "A Biased Overview of the Study of Love." *Journal of Social and Personal Relationships* 5, 4 (1988): 473–501.

Shaver, Phillip, Cindy Hazan, and D. Bradshaw. "Love as Attachment: The Integration of Three Behavioral Systems." In *The Psychology of Love,* edited by Robert Sternberg and Michael Barnes. New Haven, CT: Yale University Press, 1988.

Shaw, David. "Despite Advances Stereotypes Still Used by Media." *Los Angeles Times* (December 12, 1990): A31.

Shaw, G. M., et al. "Preconceptional Vitamin Use, Dietary Folate, and the Occurrence of Neural Tube Defects." *Epidemiology* 6, 3 (1995): 219–226.

Shelp, Earl. *Born to Die?* New York: The Free Press, 1986.

Shelton, Beth A., and Daphne John. "Does Marital Status Make a Difference? Housework among Married and Cohabiting Men and Women." *Journal of Family Issues* 14, 3 (September 1993): 401–420.

Sherman, Barry, and Joseph Dominick. "Violence and Sex in Music Videos: TV and Rock 'n' Roll." *Journal of Communication* 36, 1 (December 1986): 79–93.

Sherman, Beth. "Spanking Experts Say No." *San Jose Mercury News* (June 24, 1985).

Shifren, Kim, Robert Bauserman, and D. Bruce Carter. "Gender Role Orientation and Physical Health: A Study among Young Adults." *Sex Roles: A Journal of Research* 29, 5–6 (1993): 421–432.

Shilts, Randy. *And the Band Played On: Politics, People, and the AIDS Epidemic.* New York: St. Martins Press, 1987.

Shon, Steven, and Davis Ja. "Asian Families." In *Ethnicity and Family Therapy,* edited by Monica McGoldrick et al. New York: Guilford Press, 1982.

Shostak, Arthur B. "Singlehood." In *Handbook of Marriage and the Family,* edited by Marvin Sussman and Suzanne Steinmetz. New York: Plenum Press, 1987.

———. "Tommorow's Family Reforms: Marriage Course, Marriage Test, Incorporated Families, and Sex Selection Mandate." *Journal of Marital and Family Therapy* 7, 4 (October 1981): 521–528.

Sidorowicz, Laura, and G. Sparks Lunney. "Baby X Revisited." *Sex Roles* 6, 1 (February 1980): 67–73.

SIDS Resource Center. Personal communication with author, Christine DeVault, 1996.

Sigelman, C. K., et al. "Courtesy Stigma: The Social Implications of Associating with a Gay Person." *Journal of Social Psychology* 131 (1991): 45–56.

Signorielli, Nancy. "Children, Television, and Gender Roles—Messages and Impact." *Journal of Broadcasting and Electronic Media* 33, 3 (June 1989a): 325–331.

———. "Television and Conceptions about Sex Roles—Maintaining Conventionality and the Status Quo." *Sex Roles* 21, 5 (September 1989b): 341–350.

Signorielli, Nancy, and Margaret Lears. "Children, Television, and Conceptions about Chores: Attitudes and Behaviors." *Sex Roles: A Journal of Research* 27, 3–4 (1992): 157–170.

Signorielli, Nancy, and N. Morgan. eds. *Cultivation Analysis: New Directions in Media Research.* Newbury Park, CA: Sage Publications, 1990.

Silver, Donald, and B. Kay Campbell. "Failure of Psychological Gestation." *Psychoanalytic Inquiry* 8, 2 (1988): 222–223.

Silverstein, Judith L. "The Problem with In-Laws." *Journal of Family Therapy* 14, 4 (1992): 399–412.

Silverstein, Meril, Xuan Chen, and Kenneth Heller. "Too Much of a Good Thing? Intergenerational Social Support and the Psychological Well-Being of Older Parents." *Journal of Marriage and the Family* 58, 4 (November 1996): 970-982.

Simon, William. "Letters to the Editor: Reply to Muir and Eichel." *Archives of Sexual Behavior* 21, 6 (December 1992): 595–597.

Simon, William, and John Gagnon. "Sexual Scripts." *Society* 221, 1 (November 1984): 53–60.

Simpson, George Eaton, and J. Milton Yinger. *Racial and Cultural Minorities: An Analysis of Prejudice and Discrimination.* New York: Plenum Press, 1985.

Simpson, Jeffrey. "The Dissolution of Romantic Relationships: Factors Involved in Relationship Stability and Emotional Distress." *Journal of Personality and Social Psychology* 53, 4 (1987): 683–694.

Simpson, Jeffrey, B. Campbell, and E. Berscheid. "The Association between Romantic Love and Marriage: Kephart (1967) Twice Revisited." *Personality and Social Psychology Bulletin* 12 (1986): 363–372.

Simpson, William S., and Joanne A. Ramberg. "Sexual Dysfunction in Married Female Patients with Anorexia and Bulimia Nervosa." *Journal of Sex & Marital Therapy* 18, 1 (March 1992): 44–54.

Singer, Dorothy, et al. *Use TV to Your Child's Advantage*. Washington, DC: Acropolis Books, 1990.

Singer, Peter, and Deane Wells. *Making Babies: The New Science and Ethics of Conception*. New York: Charles Scribner's Sons, 1985.

Sinnott, Jan. *Sex Roles and Aging: Theory and Research from a Systems Perspective*. New York: Karger, 1986.

Skitka, L. J., and C. Maslach. "Gender Roles and the Categorization of Gender-Relevant Information." *Sex Roles* 22 (1990): 3–4.

Slater, Alan, Jan A. Shaw, and Joseph Duquesnel. "Client Satisfaction Survey: A Consumer Evaluation of Mediation and Investigative Services: Executive Summary." *Family and Conciliation Courts Review* 30, 2 (1992): 252–259.

Slater, Suzanne, and Julie Mencher. "The Lesbian Family Life Cycle: A Contextual Approach." *American Journal of Orthopsychiatry* 61, 3 (1991): 372–382.

"Sleeping on Back Saves 1,500 Babies." *San Mateo County Times* (June 25, 1996): A-4.

Sluzki, Carlos. "The Latin Lover Revisited." In *Ethnicity and Family Therapy*, edited by Monica McGoldrick, et al. New York: Guilford Press, 1982.

Small, Stephen, and Dave Riley. "Toward a Multidimensional Assessment of Work Spillover into Family Life." *Journal of Marriage and the Family* 52, 1 (February 1990): 51–61.

Smeeding, T. M. O'Higgins, and L. Rainwater, eds. *Poverty, Inequality and Income Distribution in Comparative Perspective; The Luxembourg Income Study*. Washington, DC: The Urban Institute Press, 1990.

Smelser, Neil, and Erik Erikson, eds. *Themes of Work and Love in Adulthood*. Cambridge, MA: Harvard University Press, 1980.

Smith, Corless. "Sex and Genre on Prime Time." *Journal of Homosexuality* 21, 1–2 (1991): 119–138.

Smith, E. J. "The Black Female Adolescent: A Review of the Educational, Career, and Psychological Literature." *Psychology of Women Quarterly* 6 (1982): 261–288.

Smith, Herbert L., and S. Philip Morgan. "Children's Closeness to Father as Reported by Mothers, Sons and Daughters." *Journal of Family Issues* 15, 1 (March 1994): 3–29.

Smith, Ken, and Cathleen Zick. "The Incidence of Poverty among the Recently Widowed: Mediating Factors in the Life Course." *Journal of Marriage and the Family* 48 (1986): 619–630.

Smith, T. W. "Changing Racial Labels: From Negro to Black to African American." *Public Opinion Quarterly* 56, 4 (1992): 496–514.

SmithBattle, Lee. "Intergenerational Ethics of Caring for Adolescent Mothers and Their Children." *Family Relations* 45, 1 (January 1996): 56–64.

Smits, Jeroen, Wout Ultee, and Jan Lammers. "Effects of Occupational Status Differences Between Spouses on the Wife's Labor Force Participation and Occupational Achievement: Findings from 12 European Countries." *Journal of Marriage and the Family* 58, 1 (February 1996): 101–115.

Smock, P. J. "The Economic Costs of Marital Disruption for Young Women Over the Past Two Decades." *Demography 30, 3* (August 1993): 353–371.

Snarey, John, et al. "The Role of Parenting in Men's Psychosocial Development." *Developmental Psychology* 23, 4 (July 1987): 593–603.

Snitow, Ann. "Mass Market Romance: Pornography for Women is Different." In *Powers of Desire: The Politics of Sexuality*, edited by Ann Snitow et al. New York: Monthly Review Press, 1983.

Sobol, Thomas. "Understanding Diversity." *Educational Leadership* 48, 3 (November 1990): 27–31.

Solon, G. "Intergenerational Income Mobility in the United States." *American Economic Review* 82, 3 (June 1992): 393–408.

Sommers-Flanagan, Rita, John Sommers-Flanagan, and Britta Davis. "What's Happening on Music and Television? A Gender Role Content Analysis." *Sex Roles* 28, 11–12 (June 1993): 745–753.

Sonenstein, Freya L. "Rising Paternity: Sex and Contraception Among Adolescent Males." In *Adolescent Fatherhood*, edited by A. B. Elster and M. E. Lamb. Hillsdale, NJ: Erlbaum, 1986.

Sonenstein, Freya L., Joseph H. Pleck, and L. C. Ku. "Sexual Activity, Condom Use, and AIDS Awareness among Adolescent Males." *Family Planning Perspectives* 21, 4 (July 1989): 152–158.

Sontag, Susan. "The Double Standard of Aging." *Saturday Review* 55 (September 1972): 29–38.

Sophie, Joan. "Internalized Homophobia and Lesbian Identity." *Journal of Homosexuality* 14, 1–2 (September 1987): 53–65.

South, Scott, and Richard Felson. "The Racial Patterning of Rape." *Social Forces* 69, 1 (September 1990): 71–93.

South, Scott, and Glenna Spitze. "Determinants of Divorce over the Marital Life Course." *American Sociological Review* 47 (1986): 583–590.

Spallone, P., and D. L. Steinberg. *Made to Order: The Myth of Reproductive and Genetic Progress*. New York: Pergamon Press, 1987.

Spanier, Graham B. "Improve, Refine, Recast, Expand, Clarify—But Don't Abandon." *Journal of Marriage and the Family* 47, 4 (November 1985): 1073–1074.

———. "Measuring Dyadic Adjustment." *Journal of Marriage and the Family* 38, 1 (February 1976): 15–28.

Spanier, Graham B., and Frank Furstenberg Jr. "Remarriage After Divorce: A Longitudinal Analysis of Well-being." *Journal of Marriage and the Family* 44, 3 (August 1982): 709–720.

———. "Remarriage and Reconstituted Families." In *Handbook of Marriage and the Family*, edited by Marvin Sussman and Suzanne Steinmetz. New York: Plenum Press, 1987.

Spanier, Graham B., and R. L. Margolis. "Marital Separation and Extramarital Sexual Behavior." *Journal of Sex Research* 19 (1983): 23–48.

Spanier, Graham B., and Linda Thompson. "A Confirmatory Analysis of the Dyadic Adjustment Scale." *Journal of Marriage and the Family* 44, 3 (August 1982): 731–738.

———. "Relief and Distress after Marital Separation." *Journal of Divorce* 7 (1983): 31–49.

Spanier, Graham B., and Linda Thompson, eds. *Parting: The Aftermath of Separation and Divorce*. Beverly Hills, CA: Sage Publications, 1984.

Spanier, Graham B., et al. "Marital Trajectories of American Women: Variations in the Life Course." *Journal of Marriage and the Family* 47, 4 (November 1985): 993–1003.

Spector, I. P., and M. P. Carey. "Incidence and Prevalence of the Sexual Dysfunctions—A Critical Review of the Empirical Literature." *Archives of Sexual Behavior* 19, 4 (August 1990): 389–408.

Spence, Janet T. "Gender-Related Traits and Gender Ideology: Evidence for a Multifactorial Theory (Personality Processes and Individual Differences)." *Journal of Personality and Social Psychology* 64, 4 (1993): 624–636.

Spence, Janet, and L. L. Sawin. "Images of Masculinity and Femininity." In *Sex, Gender, and Social Psychology,* edited by V. O'Leary et al. Hillsdale, NJ: Erlbaum, 1985.

Spence, Janet, et al. "Ratings of Self and Peers on Sex Role Attributes and the Relation to Self-Esteem and Conceptions of Masculinity and Femininity." *Journal of Personality and Social Psychology* 32 (1975): 29–39.

———. "Sex Roles in Contemporary Society." In *Handbook of Social Psychology,* edited by G. Lindzey and Elliot Aronson. New York: Random House, 1985.

Spencer, Paul. *The Masai of Matapato.* Bloomington: Indiana University Press, 1988.

Spencer, S. L., and A. M. Zeiss. "Sex Roles and Sexual Dysfunction in College Students." *Journal of Sex Research* 23, (1987): 338–347.

Spiegel, Lynn. *Installing the Television Set: Television and the Family Ideal in Postwar America.* Chicago: University of Chicago Press, 1991.

Spitze, Glenna. "The Division of Task Responsibility in U.S. Households: Longitudinal Adjustments to Change." *Social Forces* 64 (1986): 689–701.

Spitze, Glenna, and Scott South. "Women's Employment, Time Expenditure, and Divorce." *Journal of Family Issues* 6, 3 (September 1985): 307–329.

Spock, Benjamin, and Michael Rothenberg. *Dr. Spock's Baby and Child Care.* New York: Pocket Books, 1985.

Sponaugle, G. C. "Attitudes Toward Extramarital Relations." In *Human Sexuality: The Societal and Interpersonal Context,* edited by Kathleen McKinney and Susan Sprecher. Norwood, NJ: Ablex, 1989.

Sprecher, Susan. "Influences on Choice of a Partner and on Sexual Decision Making in the Relationship." In *Human Sexuality: The Social and Interpersonal Context,* edited by Kathleen McKinney and Susan Sprecher. Norwood, NJ: Ablex, 1989.

———. "A Revision of the Reiss Premarital Sexual Permissiveness Scale." *Journal of Marriage and the Family* 50, 3 (August 1988): 821–828.

———. "Sex Difference in Bases of Power in Dating Relationships." *Sex Roles* 12, 34 (1985): 449–462.

Sprecher, Susan, and Kathleen McKinney. *Sexuality.* Newbury Park, CA: Sage Publications, 1993.

Sprecher, Susan, et al. "Sexual Relationships." In *Human Sexuality: The Societal and Interpersonal Context,* edited by Kathleen McKinney and Susan Sprecher. Norwood, NJ: Ablex, 1989.

Sprey, Jetse. "Current Theorizing on the Family: An Appraisal." *Journal of Marriage and the Family* 50, 4 (November 1988): 875–890.

Springer, D., and D. Brubaker. *Family Caregivers and Dependent Elderly: Minimizing Stress and Maximizing Independence.* Beverly Hills, CA: Sage Publications, 1984.

Stack, C. B., and L. M. Burton. "Kinscripts; Reflections on Family, Generation, and Culture." In *Mothering: Ideology, Experience, and Agency,* edited by E. Glenn, G. Chang, and L. R. Forcey. New York: Routledge, 1992.

Stack, Carol B. *All Our Kin: Strategies for Survival in a Black Community.* New York: Harper & Row, 1974.

Stallones, Lorann, Martin B. Marx, Thomas F. Garrity, and Timothy P. Johnson. "Pet Ownership and Attachment in Relation to the Health of U.S. Adults, 21 to 64 Years of Age." *Anthrozoos* 4, 2 (1990): 100–112.

Stanton, M. Colleen. "The Fetus: A Growing Member of the Family." *Family Relations* 34, 3 (July 1985): 321–326.

Staples, Robert. "The Black American Family." In *Ethnic Families in America,* edited by Charles Mindel and Robert Habenstein. New York: Elsevier-North Holland, 1976.

———. "The Black American Family." In *Ethnic Families in America: Patterns and Variations,* edited by Charles Mindel et al. 3rd ed. New York: Elsevier-North Holland, 1988a.

———. "The Emerging Majority: Resources for Nonwhite Families in the United States." *Family Relations* 37, 3 (July 1988c): 348–354.

———. "The Sexual Revolution and the Black Middle Class." In *The Black Family,* edited by Robert Staples. 4th ed. Belmont, CA: Wadsworth Publishing, 1991.

———. "Substance Abuse and the Black Family Crisis: An Overview." *Western Journal of Black Studies* 14, 4 (1990): 196–204.

Staples, Robert, ed. *The Black Family: Essays and Studies.* 3d ed. Belmont, CA: Wadsworth Publishing, 1988b.

Staples, Robert, and Leanor Boulin Johnson. *Black Families at the Crossroads: Challenges and Prospects.* San Francisco: Jossey-Bass, 1993.

Staples, Robert, and Alfredo Mirandé. "Racial and Cultural Variations among American Families: A Decennial Review of the Literature." *Journal of Marriage and the Family* 42 (1980): 887–922.

Starrels, Marjorie E. "Gender Differences in Parent-Child Relations." *Journal of Family Issues* 15, 1 (March 1994): 148–165.

Starrels, Marjorie E., Sally Bould, and Leon J. Nicholas. "The Feminization of Poverty in the United States: Gender, Race, Ethnicity, and Family Factors." *Journal of Family Issues* 15, 4 (December 1994): 590–607.

Steck, L., D. Levitan, D. McLane, and H. H. Kelley. "Care, Need, and Conceptions of Love." *Journal of Personality and Social Psychology* 43 (1982): 481–491.

Steele, Brandt F. "Psychodynamic Factors in Child Abuse." In *The Battered Child,* edited by C. Henry Kempe and Ray Helfer. Chicago: University of Chicago Press, 1980.

Steelman, Lala Carr, and Brian Powell. "The Social and Academic Consequences of Birth Order: Real, Artificial, or Both?" *Journal of Marriage and the Family* 47 (1985): 117–124.

Stein, Peter, ed. *Single.* Englewood Cliffs, NJ: Prentice Hall, 1976.

Steinberg, Laurence, and Susan Silverberg. "Marital Satisfaction in Middle Stages of Family Life Cycle." *Journal of Marriage and the Family* 49, 4 (November 1987): 751–760.

Steinglass, Peter. "Families and Substance Abuse." In *Vision 2010: Families and Health Care,* edited by Sharon Price and Barbara Elliott. Minneapolis: National Council of Family Relations, 1993.

———. "A Life History Model of the Alcoholic Family." In *Family Studies Yearbook,* Vol. 1, edited by David Olson and Brent Miller. Beverly Hills, CA: Sage Publications, 1983.

Steinglass, Peter, L. A. Bennett, S. J. Wolin, and D. Reiss. *The Alcoholic Family.* New York: Basic Books, 1987.

Steinman, Susan. "The Experience of Children in a Joint-Custody Arrangement: A Report of a Study." *American Journal of Orthopsychiatry* 51, 3 (July 1981): 403–414.

Steinmetz, Suzanne. "Family Violence." In *Handbook of Marriage and the Family,* edited by Marvin Sussman and Suzanne Steinmetz. New York: Plenum Press, 1987.

Steinmetz, Suzanne K. "The Abused Elderly Are Dependent: Abuse Is Caused by the Perception of Stress Associated With Providing Care." In *Current Controversies in Family Violence,* edited by Richard Gelles and Donileen Loseke. Newbury Park, CA: Sage Publications, 1993.

Stephen, Timothy. "Fixed-Sequence and Circular-Causal Models of Relationship Development: Divergent Views on the Role of Communication in Intimacy." *Journal of Marriage and the Family* 47, 4 (November 1985): 955–963.

Sternberg, Robert, and Michael Barnes, eds. "A Triangular Theory of Love." *Psychological Review* 93 (1986): 119–135.

———. "Real and Ideal Others in Romantic Relationships: Is Four a Crowd?" *Journal of Personality and Social Psychology* 49 (1985): 1589–1596.

———. *The Psychology of Love.* New Haven, CT: Yale University Press, 1988.

Sternberg, Robert, and S. Grajek. "The Nature of Love." *Journal of Personality and Social Psychology* 47 (1984): 312–327.

Stets, Jan E. "Verbal and Physical Aggression in Marriage." *Journal of Marriage and the Family* 52, 2 (May 1990): 501–514.

Stets, Jan E., and Maureen A. Pirog-Good. "Patterns of Physical and Sexual Abuse for Men and Women in Dating Relationships: A Descriptive Analysis." *Journal of Family Violence* 4, 1 (March 1989): 63–76.

———. "Violence in Dating Relationships." *Social Psychology Quarterly* 50, 3 (September 1987): 237–246.

Stets, Jan E., and Murray A. Straus. "The Marriage License as a Hitting License: A Comparison of Assaults in Dating, Cohabitation and Married Couples." *Journal of Family Violence* 4, 2 (June 1989): 161–180.

Stevens, Gillian, and Robert Schoen. "Linguistic Intermarriage in the United States." *Journal of Marriage and the Family* 50, 1 (February 1988): 267–280.

Stevenson, M. "Tolerance for Homosexuality and Interest in Sexuality Education." *Journal of Sex Education and Therapy* 16 (1990): 194–197.

Stinnett, Nick, and John DeFrain. *Secrets of Strong Families.* Boston: Little, Brown, 1985.

Stockard, Janice. *Daughters of the Canton Delta: Marriage Patterns and Economic Strategies in South China, 1860–1930.* Stanford, CA: Stanford University Press, 1989.

Stockdale, M. S. "The Role of Sexual Misperceptions of Women's Friendliness in an Emerging Theory of Sexual Harassment." *Journal of Vocational Behavior* 42, 1 (February 1993): 84–101.

Stoddard, J. J., R. F. St. Peter, and P. W. Newacheck. "Health Insurance Status and Ambulatory Care for Children." *New England Journal of Medicine* 330, 20 (1994): 1421–1425.

Stone, Katherine M., et al. "National Surveillance for Neonatal Herpes Simplex Virus Infections." *Sexually Transmitted Diseases* 16, 3 (July 1989): 152–160.

Strachan, Catherine E. "The Role of Power and Gender in Anger Responses to Sexual Jealousy." *Journal of Applied Social Psychology* 22, 22 (1992): 1721–1740.

Strasser, Susan. *Never Done: A History of American Housework.* New York: Pantheon Books, 1982.

Straus, Murray. "Preventing Violence and Strengthening the Family." In *Family Strengths, 4,* edited by Nick Stinnett et al. Lincoln: University of Nebraska Press, 1980.

Straus, Murray A. "Physical Assaults by Wives: A Major Social Problem." In *Current Controversies in Family Violence,* edited by Richard Gelles and Donileen Loseke. Newbury Park, CA: Sage Publications, 1993.

Straus, Murray, Richard Gelles, and Suzanne Steinmetz. *Behind Closed Doors.* Garden City, NY: Anchor Books, 1980.

Strom, Robert, Pat Collinsworth, Shirley Strom, and D. Griswold. "Strengths and Needs of Black Grandparents." *International Journal of Aging and Human Development* 36, 4 (1992–1993): 255–268.

Stroman, C. "The Socialization Influence of Television on Black Children." *Journal of Black Studies* 15 (September 1989): 79–100.

Strong, Bryan, and Christine DeVault. *Human Sexuality: Diversity in Contemporary America.* 2nd edition. Mountain View, CA: Mayfield, 1997.

Strube, Michael J., and Linda Barbour. "Factors Related to the Decision to Leave an Abusive Relationship." *Journal of Marriage and the Family* 46, 4 (November 1984): 837–844.

"Study Blames MTV for Video Violence." *San Francisco Chronicle* (May 15, 1997): E-1.

"Study Reveals Deep Scars of Divorce." *San Francisco Chronicle* (June 3, 1997): Al ff.

"Study Says 20% of Kids Get Poor Preventive Care." *San Jose Mercury News* (May 30, 1992): 8A.

Suarez, Zulema. "Cuban American Families." In *Families in Cultural Context,* edited by Mary Kay DeGenova. Mountain View, CA: Mayfield, 1997.

Sugarman, David B., and Gerald T. Hotaling. "Dating Violence: Prevalence, Context, and Risk Markers." In *Violence in Dating Relationships: Emerging Social Issues,* edited by Maureen Pirog-Good and Jan Stets. New York: Praeger, 1989.

Suitor, J. Jill. "Marital Quality and Satisfaction with Division of Household Labor." *Journal of Marriage and the Family* 53, 1 (February 1991): 221–230.

Sulloway, Frank J. *Born to Rebel; Birth Order, Family Dynamics, and Creative Lives.* New York: David McKay Company, 1996.

Sung, B. L. *Mountains of Gold.* New York: Macmillan, 1967.

Sunoff, Alvin. "A Conversation with Jerome Kagan." *U.S. News and World Report* (March 25, 1985): 63–64.

Suro, Roberto. "Pollution-Weary Minorities Try Civil Rights Tack." *New York Times* (January 11, 1993): A1, A12.

Surra, Catherine. "Research and Theory on Mate Selection and Premarital Relationships in the 1980s." In *Contemporary Families: Looking Forward, Looking Back,* edited by Alan Booth. Minneapolis: National Council on Family Relations, 1991.

Surra, Catherine, P. Arizzi, and L. L. Asmussen. "The Association between Reasons for Commitment and the Development and Outcome of Marital Relationships." *Journal of Social and Personal Relationships* 5 (1988): 47–63.

"Survey of 80,000 Cases Calls Birth-defect Test Safe." *San Jose Mercury News* (August 4, 1992): 3F.

Sussman, Marvin B. and Susan K. Steinmetz, eds. *Handbook of Marriage and the Family.* New York: Plenum Press, 1987.

Sweet, J.A., L.L. Bumpass, and V.R.A. Call. *The Design and Content of the National Survey of Families and Households.* (Working Paper NSFH-1). Madison: University of Wisconsin, Center for Demography and Ecology, 1988.

Swensen, C. H., Jr. "The Behavior of Love." In *Love Today: A New Exploration,* edited by H. A. Otto. New York: Association Press, 1972.

Swenson, Clifford, and Geir Trahaug. "Commitment and the Long-Term Marriage Relationship." *Journal of Marriage and the Family* 47, 4 (November 1985): 939–945.

Swift, Carolyn. "Preventing Family Violence: Family-Focused Programs." In *Violence in the Home: Interdisciplinary Perspectives,* edited by Mary Lystad. New York: Brunner/Mazel, 1986.

Symons, Donald. *The Evolution of Human Sexuality.* New York: Oxford University Press, 1979.

Szapocznik, Jose, and Roberto Hernandez. "The Cuban American Family." In *Ethnic Families in America: Patterns and Variations,* edited by Charles Mindel et al. 3rd ed. New York: Elsevier North Holland, Inc., 1988.

Szinovacz, Maximiliane. "Family Power." In *Handbook of Marriage and the Family,* edited by Marvin Sussman and Suzanne Steinmetz. New York: Plenum Press, 1987.

Takagi, Diana Y. "Japanese American Families." In *Minority Families in the United States: A Multicultural Perspective,* edited by R. L. Taylor. Englewood Cliffs, NJ: Prentice Hall, 1994.

"Talking to Children about Prejudice." *PTA Today* (December 1989–1990): 7–8.

Tallmer, Margot. "Grief as a Normal Response to the Death of a Loved One." In *Principles of Thanatology,* edited by Austin H. Kutscher et al. New York: Columbia University Press, 1987.

Tanfer, Koray. "Patterns of Premarital Cohabitation among Never-Married Women in the United States." *Journal of Marriage and the Family* 49, 3 (August 1987): 683–697.

Tanke, E. D. "Dimensions of the Physical Attractiveness Stereotype: A Factor/Analytic Study." *Journal of Psychology* 110 (1982): 63–74.

Tanouye, Elyse. "Safety of Tests for Birth Defect Backed in Study." *Wall Street Journal* (August 27, 1992): B1.

Tavris, Carol. *The Mismeasure of Woman.* New York: Norton, 1992.

Tavris, Carol, and Carol Wade, eds. *The Longest War: Sex Differences in Perspective.* 2d ed. New York: Harcourt Brace Jovanovich, 1984.

Taylor, Ella. "From the Nelsons to the Huxtables: Genre and Family Imagery in American Network Television." *Qualitative Sociology* 12, 1 (Spring 1989a): 13–28.

———. *Prime-Time Families: Television Culture in Postwar America.* Berkeley, CA: University of California Press, 1989b.

Taylor, J. R., A. P. Lockwood, and A. J. Taylor. "The Prepuce: Specialized Mucosa of the Penis and Its Loss to Circumcision," *British Journal of Urology* 77 (1996): 291–295.

Taylor, Patricia. "It's Time to Put Warnings on Alcohol." *New York Times* (March 20, 1988): B2.

Taylor, Robert J. "Black American Families." In *Minority Families in the United States: A Multicultural Perspective,* edited by R. L. Taylor. Englewood Cliffs, NJ: Prentice Hall, 1994.

———. "Minority Families in America." In *Minority Families in the United States: A Multicultural Perspective,* edited by R. L. Taylor. Englewood Cliffs, NJ: Prentice Hall, 1994b.

———. "Need for Support and Family Involvement among Black Americans." *Journal of Marriage and the Family* 52, 3 (August 1990): 584–590.

———. "Receipt of Support from Family among Black Americans." *Journal of Marriage and the Family* 48, 1 (February 1986): 67.

Taylor, Robert J., Linda M. Chatters, and James S. Jackson. "A Profile of Familial Relations Among Three-Generation Black Families." *Family Relations* 42, 3 (1993): 332–341.

Taylor, Robert J., Linda M. Chatters, Belinda Tucker, and Edith Lewis. "Developments in Research on Black Families." In *Contemporary Families: Looking Forward, Looking Back,* edited by Alan Booth. Minneapolis: National Council on Family Relations, 1991.

Taylor, Ronald L., ed. *Minority Families in the United States: A Multicultural Perspective.* Englewood Cliffs, NJ: Prentice Hall, 1994a.

Teachman, Jay. "Receipt of Child Support in the United States." *Journal of Marriage and the Family* 53, 3 (August 1991): 759–772.

Teachman, Jay, and Alex Heckert. "The Impact of Age and Children on Remarriage." *Journal of Family Issues* 6, 2 (June 1985): 185–203.

Teachman, Jay D., and Karen A. Polonko. "Cohabitation and Marital Stability in the United States." *Social Forces* 69, 1 (September 1990): 207–220.

———. "Timing of the Transition to Parenthood: A Multidimensional Birth-interval Approach." *Journal of Marriage and the Family* 47, 4 (November 1985): 867–879.

Teachman, Jay D., R. Vaughn, A. Call, and Karen P. Carver. "Marital Status and Duration of Joblessness Among White Men." *Journal of Marriage and the Family* 56, 2 (May 1994): 415–428.

Teeser, Abraham, and Richard Reardon. "Perceptual and Cognitive Mechanisms in Human Sexual Attraction." In *The Bases of Human Sexual Attraction,* edited by Mark Cook. New York: Academic Press, 1981.

Terry, K. "How Insurers Try to Curtail Doctor Visits." *Medical Economics* 71, 2 (1994): 36–38.

Tessina, Tina. *Gay Relationships: For Men and Women. How to Find Them, How to Improve Them, How to Make Them Last.* Los Angeles: Jeremy P. Tarcher, 1989.

Testa, Ronald J., Bill N. Kinder, and G. Ironson. "Heterosexual Bias in the Perception of Loving Relationships of Gay Males and Lesbians." *Journal of Sex Research* 23, 2 (May 1987): 163–172.

Textor, M. R. "Family Therapy with Drug Addicts: An Integrated Approach." *American Journal of Orthopsychiatry* 57 (1987): 495–507.

Thayer, Leo. *On Communication.* Norwood, NJ: Ablex, 1986.

Thoits, Peggy A. "Identity Structures and Psychological Well-Being: Gender and Marital Status Comparisons." *Social Psychology Quarterly* 55, 3 (1992): 236–256.

Thomma, Steven, and Angie Cannon. "Mood of Gloom and Doom Persists, But America Is Better Than It Was." *San Jose Mercury* (October 30, 1994): C 1, 6.

Thompson, Anthony. "Emotional and Sexual Components of Extramarital Relations." *Journal of Marriage and the Family* 46, 1 (February 1984): 35–42.

———. "Extramarital Sex: A Review of the Research Literature." *Journal of Sex Research* 19, 1 (February 1983): 1–22.

Thompson, Edward, Christopher Grisanti, and Joseph Pleck. "Attitudes toward the Male Role and Their Correlates." *Sex Roles* 13, 7 (October 1985): 413–427.

Thompson, Linda. "Conceptualizing Gender in Marriage: The Case of Marital Care." *Journal of Marriage and the Family* 55, 3 (August 1993): 557–569.

Thompson, Linda, and Alexis Walker. "Gender in Families: Women and Men in Marriage, Work, and Parenthood." In *Contemporary Families: Looking Forward, Looking Back*, edited by Alan Booth. Minneapolis: National Council on Family Relations, 1991.

Thompson, Linda, et al. "Developmental Stage and Perceptions of Intergenerational Continuity." *Journal of Marriage and the Family* 47, 4 (November 1985): 913–920.

Thomson, Keith. "Research on Human Embryos: Where to Draw the Line." *American Scientist* (March 1985): 187–189.

Thorne, B. and M. Yalom, eds. *Rethinking the Family: Some Feminist Questions*. New York: Longman, 1982.

Thornton, Arland. "Changing Attitudes toward Family Issues in the United States." *Journal of Marriage and the Family* 51, 4 (November 1989): 873–893.

———. "The Courtship Process and Adolescent Sexuality." *Journal of Family Issues* 11, 3 (September 1990): 239–273.

Thornton, Arland, William G. Axinn, and Daniel H. Hill. "Reciprocal Effects of Religiosity, Cohabitation, and Marriage." *American Journal of Sociology* 98, 3 (1992): 628–651.

Thornton, Arland, et al. "Causes and Consequences of Sex-Role Attitudes and Attitude Change." *American Sociological Review* 48 (1983): 211–227.

Thorton-Dill, B. "Fictive Kin, Paper Sons, and *Compadrazgo:* Women of Color and the Struggle for Family Survival." In *Women of Color in U.S. Society*, edited by M. Baca-Zinn and B. Thorton-Dill. Philadelphia: Temple University Press, 1994.

Tienda, Marta, and Ronald Angel. "Headship and Household Composition among Blacks, Hispanics, and Other Whites." *Social Forces* 61 (1982): 508–531.

Tienda, Marta, and Jennifer Glass. "Household Structure and Labor Force Participation of Black, Hispanic, and White Mothers." *Demography* 22 (1985): 281–394.

Tietjen, Anne, and Christine F. Bradley. "Social Support and Maternal Psychosocial Adjustment during the Transition to Parenthood." *Canadian Journal of Behavioral Science* 17, 2 (April 1985): 109–121.

Ting-Toomey, Stella. "An Analysis of Verbal Communication Patterns in High and Low Marital Adjustment Groups." *Human Communications Research* 9, 4 (June 1983): 306–319.

Toback, B. M. "Recent Advances in Female Infertility Care." *Naacogs Clinical Issues in Perinatal and Women's Health Nursing* 3, 2 (1992): 313–319.

Toback, James. "James Toback on 'The Hunger.'" *US* (Special Issue: "The Sexual Revolution in Movie, Music & TV") (August 1992): 56–58.

Tolman, Richard M. "Treatment Program for Men Who Batter." In *Vision 2010: Families and Violence, Abuse and Neglect*, edited by Richard J. Gelles. Minneapolis: National Council on Family Relations, 1995.

Toney, G., and J. Weaver. "Effects of Gender and Gender Role Self-Perceptions on Affective Reactions to Rock Music Videos." *Sex Roles* 30, 7–8 (April 1994): 567–583.

Tooth, Geoffrey. "Why Children's TV Turns Off So Many Parents." *U.S. News and World Report* (February 18, 1985): 65.

Torres, Aida, and Jacqueline D. Forrest. "Why Do Women Have Abortions?" *Family Planning Perspectives* 20, 4 (July-August 1988): 169-176.

Torres, José. "A Letter to a Child Like Me." *Parade* (February 26, 1991): 8–9.

Torrey, Barbara. "Aspects of the Aged: Clues and Issues." *Population and Development Review* 14, 3 (September 1988): 489–497.

Toufexis, Anastasia. "Older—But Coming on Strong." *Time* (February 22, 1988): 76–79.

Trafford, Abigail. "Medical Science Discovers the Baby." *U.S. News and World Report* (November 10, 1980): 59–62.

Tran, Than Van. "The Vietnamese American Family." In *Ethnic Families in America: Patterns and Variations*, edited by Charles H. Mindel et al. 3rd ed. New York: Elsevier, 1988.

Treas, Judith, and Vern L. Bengtson. "The Demography of Mid- and Late-Life Transitions." *Annals of the American Academy* 464 (November 1982): 11–21.

———. "The Family in Later Years." In *Handbook of Marriage and the Family*, edited by Marvin Sussman and Suzanne Steinmetz. New York: Plenum Press, 1987.

Treaster, Joseph B. "Drug Use Making Comeback, Study Says." *New York Times* (July 16, 1993): C18.

Tribe, Laurence, *Abortion: Clash of Absolutes*. New York: Norton, 1990.

Troiden, Richard. *Gay and Lesbian Identity: A Sociological Analysis*. New York: General Hall, 1988.

Troiden, Richard, and Erich Goode. "Variables Related to the Acquisition of a Gay Identity." *Journal of Homosexuality* 5 (June 1980): 383–392.

Troll, Lillian. "The Contingencies of Grandparenting." In *Grandparenthood*, edited by Vern Bengston and Joan Robertson. Beverly Hills, CA: Sage Publications, 1985a.

Trost, Jan. "Abandon Adjustment." *Journal of Marriage and the Family* 47,4 (November 1985): 1072–1073.

True, Reiko Homma. "Psychotherapeutic Issues with Asian American Women." *Sex Roles* 22, 7 (April 1990): 477–485.

Trzcinski, Eileen, and Matia Finn-Stevenson. "A Response to Arguments against Mandated Parental Leave: Findings from the Connecticut Survey of Parental Leave Policies." *Journal of Marriage and the Family* 53, 2 (May 1991): 445–460.

Tsuda, T., H. Aoyama, and J. Froom. "Primary Health Care in Japan and the United States." *Social Science and Medicine* 38, 4 (1994): 489–495.

Tucker, M. B., and R. J. Taylor. "Demographic Correlates of Relationship Status among Black Americans." *Journal of Marriage and the Family* 51 (August 1989): 655–665.

Tucker, Raymond K., M. G. Marvin, and B. Vivian. "What Constitutes a Romantic Act." *Psychological Reports* 89, 2 (October 1991): 651–654.

Tucker, Sandra. "Adolescent Patterns of Communication about Sexually Related Topics." *Adolescence* 24, 94 (June 1989): 269–278.

Tuleja, Tad. *Curious Customs.* New York: Harmony Books, 1987.

Turner, P. H., et al. "Parenting in Gay and Lesbian Families." Paper presented at the first meeting of the Future of Parenting Symposium, Chicago, IL, March 1985.

Turner, R. "Rising Prevalence of Cohabitation in the United States May Have Partially Offset Decline in Marriage Rates." *Family Planning Perspectives* 22, 2 (March 1990): 90–91.

Turner, R. Jay, and William R. Avison. "Assessing Risk Factors for Problem Parenting: The Significance of Social Support." *Journal of Marriage and the Family* 47, 4 (November 1985): 881–892.

Turner, Robert L., and M. E. Fakouri. "Androgyny and Differences in Fantasy Patterns." *Psychological Reports* 73, 3 (1993): 1164–1166.

Udry, J. Richard. "Marriage Alternatives and Marital Disruption." *Journal of Marriage and the Family* 43 (November 1981): 889–897.

———. *The Social Context of Marriage.* Philadelphia: Lippincott, 1974.

Umberson, Debra. "Parenting and Well-Being: The Importance of Context." *Journal of Family Issues* 10, 4 (December 1989): 427–439.

Umberson, Debra, and Walter R. Gove. "Parenthood and Psychological Well-Being: Theory, Measurement, and Stage in the Family Life Course." *Journal of Family Issues* 10, 4 (December 1989): 440–462.

Unger, Donald G., and Marvin B. Sussman. "Introduction: A Community Perspective on Families." *Marriage and Family Review* 15, (1990): 1–2.

Unger, R. K. "Toward a Redefinition of Sex and Gender." *American Psychologist* 34 (1979): 1085–1094.

"Update: Trends in Fetal Alcohol Syndrome—United States, 1979–1993." Centers for Disease Control and Prevention. *Morbidity and Mortality Weekly Report* 44, 13 (1995): 249–251.

Urwin, Charlene "AIDS in Children: A Family Concern." *Family Relations* 37, 2 (April 1988): 154–159.

U.S. Bureau of the Census. "The Hispanic Population in the United States: March 1990." In *Current Population Reports,* Series P-20. Washington, DC: U.S. Government Printing Office, 1991.

———. *Infant Mortality*. International Data Base, 1996.

———. *Statistical Abstract of the United States.* 114th ed. Washington, DC: U.S. Government Printing Office, 1994.

———. *Statistical Abstract of the United States.* 115th ed. Washington, DC: U.S. Government Printing Office, 1995.

———. *Statistical Abstract of the United States.* 116th ed. Washington, DC: U.S. Government Printing Office, 1996.

U.S. Commission on Civil Rights. *Child Care and Equal Opportunity for Women.* Washington, DC: U.S. Government Printing Office, 1981.

U.S. Department of Justice. "Sexual Offenses and Offenders." Washington, DC: Bureau of Justice Statistics, 1997.

———. "Women Usually Victimized by Offenders They Know." Washington, DC: Bureau of Justice Statistics, 1995.

Vaillant, Caroline O., and George E. Vaillant. "Is the U-Curve of Marital Satisfaction an Illusion? A 40-Year Study of Marriage." *Journal of Marriage and the Family* 55, 1 (February 1993): 230–240.

Valentine, Deborah. "The Experience of Pregnancy: A Developmental Process." *Family Relations* 31, 2 (April 1982): 243–248.

Van Buskirk. "Soap Opera Sex: Tuning In, Tuning Out." *US* (August 1992): 64–67.

Vande Berg, Leah R., and Diane Streckfuss. "Prime-Time Television's Portrayal of Women and the World of Work: A Demographic Profile." *Journal of Broadcasting and Electronic Media* (March 1992): 195–207.

Vandell, Deborah. "After School Care: Choices and Outcome for Third Graders." Paper presented to the Association for the Advancement of Science, May 27, 1985.

Van Hasselt, Vincent B., et al. *Handbook of Family Violence.* New York: Plenum Press, 1988.

Vaselle-Augenstein, Renata, and Annette Ehrlich. "Male Batterers: Evidence for Psychopathology." In *Intimate Violence: Interdisciplinary Perspectives,* edited by Emilio C. Viano. Washington, DC: Hemisphere, 1992.

Vasquez-Nuthall, E., et al. "Sex Roles and Perceptions of Femininity and Masculinity of Hispanic Women: A Review of the Literature." *Psychology of Women Quarterly* 11 (1987): 409–426.

Vasta, Ross. "Physical Child Abuse: A Dual-Component Analysis." *Developmental Review* 2, 2 (June 1992): 125–149.

Vaughan, Diane. *Uncoupling: Turning Points in Intimate Relationships.* New York: University Press, 1986.

Veevers, Jean. *Childless by Choice.* Toronto: Butterworth, 1980.

Vega, W. A. "The Study of Latino Families." In *Understanding Latino Families,* edited by R. Zambrana. Thousand Oaks, CA: Sage Publications, 1995.

Vega, William. "Hispanic Families." In *Contemporary Families: Looking Forward, Looking Back,* edited by Alan Booth. Minneapolis: National Council on Family Relations, 1991.

Velsor, Ellen, and Angela O'Rand. "Family Life Cycle, Work Career Patterns, and Women's Wages at Midlife." *Journal of Marriage and the Family* 46, 2 (May 1984): 365–373.

Vemer, Elizabeth, et al. "Marital Satisfaction in Remarriage: A Meta-analysis." *Journal of Marriage and the Family* 53, 3 (August 1989): 713–726.

Ventura, Jacqueline N. "The Stresses of Parenthood Reexamined." *Family Relations* 36, 1 (January 1987): 26–29.

Vera, Hernan, et al. "Age Heterogamy in Marriage." *Journal of Marriage and the Family* 47, 3 (August 1985): 553–566.

Verbrugge, Lois. "From Sneezes to Adieu: Stages of Health for American Men and Women." *Social Science and Medicine* 22 (1986): 1195–1212.

———. "Marital Status and Health." *Journal of Marriage and the Family* 41, 2 (May 1979): 267–285.

Viano, Emilio C., ed. *Intimate Violence: Interdisciplinary Perspectives.* Washington, DC: Hemisphere, 1992.

Vincent, Richard, et al. "Sexism on MTV: The Portrayal of Women in Rock Video." *Journalism Quarterly* 64, 4 (December 1987): 750–755, 941.

"Violence Kills More U.S. Kids." *San Francisco Chronicle* (February 7, 1997): A-1.

Visher, Emily B., and John S. Visher. *How to Win as a Stepfamily.* New York: Brunner/Mazel, 1991.

———. *Stepfamilies: A Guide to Working with Stepparents and Stepchildren.* New York: Brunner/Mazel, 1979.

Vobejda, Barbara. "Census Bureau Says Rapid Changes in Family Size, Style Are Slowing." *Washington Post* (June 24, 1993): A21.

Voeller, Bruce. "Society and the Gay Movement." In *Homosexual Behavior,* edited by Judd Marmor. New York: Basic Books, 1980.

Vogel, D. A., M. A. Lake, and S. Evans. "Children's and Adults' Sex-Stereotyped Perceptions of Infants." *Sex Roles* 24 (1991): 605–616.

Voydanoff, Patricia. "Economic Distress and Family Relations: A Review of the Eighties." In *Contemporary Families: Looking Forward, Looking Back,* edited by Alan Booth. Minneapolis: National Council on Family Relations, 1991.

———. *Work and Family Life.* Newbury Park, CA: Sage Publications, 1987.

———. "Work Role Characteristics, Family Structure Demands, and Work/Family Conflict." *Journal of Marriage and the Family* 50, 3 (August 1988): 749–761.

Voydanoff, Patricia, and Brenda Donnelly. *Adolescent Sexuality and Pregnancy.* Newbury Park, CA: Sage Publications, 1990.

———. "Work and Family Roles and Psychological Distress." *Journal of Marriage and the Family* 51, 4 (November 1989): 933–941.

Voydanoff, Patricia, and Linda Majka, eds. *Families and Economic Distress: Coping Strategies and Social Policy.* Beverly Hills, CA: Sage Publications, 1988.

Vredevelt, P. *Empty Arms: Emotional Support for Those Who Have Suffered Miscarriage or Stillbirth.* Sisters, OR: Questar, 1994.

Wade, R. C. *For Men about Abortion.* Boulder, CO: R. C. Wade, 1978.

Wagner, Roland M. "Psychosocial Adjustments during the First Year of Single Parenthood: A Comparison of Mexican-American and Anglo Women." *Journal of Divorce and Remarriage* 19, 1–2 (1993): 121–142.

Waite, Linda. "Does Marriage Matter?" *Demography* 32, 4 (November 1995): 483–507.

Waldman, Steven, and Lucy Shackelford. "Welfare Booby Traps." *Newsweek* (December 12, 1994): 34–35.

Waldman, Steven, and Lincoln Caplan. "The Politics of Adoption." *Newsweek* (March 21, 1994): 64–65.

Walker, Alexis J. "Reconceptualizing Family Stress." *Journal of Marriage and the Family* 47, 4 (November 1985): 827–837.

———. "Teaching About Race, Gender, and Class Diversity in United States Families." *Family Relations* 42, 3 (1993): 342–350.

Walker, Alexis J., et al. "Feminist Programs for Families." *Family Relations* 37, 1 (January 1988): 17–22.

———. "Perceptions of Relationship Change and Caregiver Satisfaction." *Family Relations* 39, 2 (April 1990): 147–152.

Walker, E. A., et al. "Medical and Psychiatric Symptoms in Women with Childhood Sexual Abuse." *Psychosomatic Medicine* 54, 6 (November 1992): 658–664.

Walker, Janet. "Co-operative Parenting Post-Divorce: Possibility or Pipedream?" *Journal of Family Therapy* 15, 3 (1993): 273–292.

Walker, Lenore. *The Battered Woman.* New York: Springer Publishing, 1984.

———. *The Battered Woman Syndrome.* New York: Harper Colophon, 1979.

———. "The Battered Woman Syndrome Is a Psychological Consequence of Abuse." In *Current Controversies in Family Violence,* edited by Richard Gelles and Donileen Loseke. Newbury Park, CA: Sage Publications, 1993.

———. "Psychological Causes of Family Violence." In *Violence in the Home: Interdisciplinary Perspectives,* edited by Mary Lystad. New York: Brunner/Mazel, 1986.

Wall, Jack C. "Maintaining the Connection: Parenting as a Noncustodial Father." *Child and Adolescent Social Work Journal* 9, 5 (1992): 441–456.

Waller, Willard, and Reuben Hill. *The Family: A Dynamic Interpretation.* New York: Dryden Press, 1951.

Wallerstein, Edward. *The Circumcision Decision.* Pamphlet. Seattle: Pennypress, 1990.

Wallerstein, Judith. "Children of Divorce: Report of a Ten-Year Follow-up of Early Latency-Age Children." *American Journal of Orthopsychiatry* 57, 2 (April 1987): 199–211.

Wallerstein, Judith, and Sandra Blakeslee. *Second Chances: Men, Women, and Children a Decade after Divorce.* New York: Ticknor & Fields, 1989.

Wallerstein, Judith, and Joan Kelly. "Effects of Divorce on the Visiting Father-Child Relationship." *American Journal of Psychiatry* 137, 12 (December 1980a): 1534–1539.

———. *Surviving the Breakup: How Children and Parents Cope with Divorce.* New York: Basic Books, 1980b.

Walling, Mary, et al. "Hormonal Replacement Therapy for Postmenopausal Women: A Review of Sexual Outcomes and Related Gynecologic Effects." *Archives of Sexual Behavior* 19, 2 (1990): 119–127.

Wallis, Claudia. "Children Having Children." *Time* (December 9, 1985): 78–79.

Walsh, Froma, ed. *Normal Family Processes.* New York: Guilford Press, 1982.

Walster, Elaine, and G. William Walster. *A New Look at Love.* Reading, MA: Addison-Wesley, 1978.

Wardle, Francis. "Helping Children Respect Differences." *PTA Today* (December 1989): 5–6.

Warren, Jennifer A. and Phyllis J. Johnson. "The Impact of Workplace Support on Work-Family Role Strain." *Family Relations* 44, 2 (April 1995): 163–169.

Waterson, E. J., and I. M. Murray-Lyon. "Preventing Alcohol Related Birth Damage: A Review." *Social Science and Medicine* 30, 3 (1990): 349–364.

Watkins, Susan Cotts, and Jane Menken. "Demographic Foundations of Family Change." *American Sociological Review* 52 (1987): 346–358.

Watkins, William G., and Arnon Bentovim. "The Sexual Abuse of Male Children and Adolescents: A Review of Current Research." *Journal of Child Psychology & Psychiatry & Allied Disciplines* 33, 1 (January 1992): 197–248.

Weatherstone, K. B., L. B. Rasmussen, A. Erenberg, E. M. Jackson, K. S. Claflin, and R. D. Leff. "Safety and Efficacy of a Topical Anesthetic for Neonatal Circumcision." *Pediatrics* 92, 5 (November 1993): 710–4.

Wechler, H., et al. "Health and Behavioral Consequences of Binge Drinking in College: A National Survey of Students at 140 Campuses." *Journal of the American Medical Association* 272, 21 (December 7, 1994): 1672–1677.

Wedemeyer, Nancy, and Harold Grotevant. "Mapping the Family System: A Technique for Teaching Family Systems Theory Concepts." *Family Relations* 31, 2 (April 1982): 185–193.

Weeks, Jeffrey. *Sexuality and Its Discontents.* London: Routledge, 1985.

Weeks, M. O'Neal, and Bruce Gage. "A Comparison of the Marriage-Role Expectations of College Women Enrolled in a Functional Marriage Course in 1961, 1972, and 1978." *Sex Roles* 11, 5/6 (1984): 377–388.

Weg, Ruth. "The Physiological Perspective." In *Sexuality in the Later Years: Roles and Behavior,* edited by Ruth Weg. New York: Academic Press, 1983.

———. *Sexuality in the Later Years: Roles and Behavior.* New York: Academic Press, 1983a.

Wegscheider, Sharon. *Another Chance: Hope and Health for the Alcoholic Family.* 2d ed. Palo Alto, CA: Science and Behavior Books, 1989.

Weinberg, Martin, and Colin Wilson. "Black Sexuality: A Test of Two Theories." *Journal of Sex Research* 25, 2 (May 1988): 197–218.

Weinberg, Martin S., C. J. Williams, and Douglas W. Pryor. *Dual Attraction: Understanding Bisexuality.* New York: Oxford University Press, 1994.

Weiner, L., B. A. More, and P. Garrido. "FAS/FAE: Focusing Prevention on Women at Risk." *International Journal of the Addictions* 24, 5 (May 1989): 385–395.

Weingarten, Helen. "Marital Status and Well-Being: A National Study Comparing First-Married, Currently Divorced and Remarried Adults." *Journal of Marriage and the Family* 47, 3 (August 1985): 653–662.

Weingarten, Helen. "Remarriage and Well-Being." *Journal of Family Issues* 1, 4 (December 1980): 533–559.

Weir, John. "Gay-Bashing, Villainy and the Oscars." *New York Times* (March 29, 1992): H17.

Weishaus, Sylvia, and Dorothy Field. "A Half Century of Marriage: Continuity or Change?" *Journal of Marriage and the Family* 50, 3 (August 1988): 763–774.

Weiss, David. "Open Marriage and Multilateral Relationships: The Emergence of Nonexclusive Models of the Marital Relationship." In *Contemporary Families and Alternative Lifestyles,* edited by Eleanor Macklin and Roger Rubin. Beverly Hills, CA: Sage Publications, 1983.

Weiss, David, and Joan Jurich. "Size of Community as a Predictor of Attitudes toward Extramarital Sexual Relations." *Journal of Marriage and the Family* 47, 1 (February 1985): 173–178.

Weiss, Peter. "The Bond of Mother's Milk." *San Jose Mercury News* (August 18, 1992): 1–2E.

Weiss, Robert. "The Fund of Sociability." *Transactions* (July 1969): 36–43.

———. *Marital Separation.* New York: Basic Books, 1975.

———. "Men and the Family." *Family Processes* 24 (1985): 49–58.

Weitzman, Lenore. *The Divorce Revolution: The Unexpected Social and Economic Consequences for Women and Children in America.* New York: Free Press, 1985.

———. *The Marriage Contract : Spouses, Lovers, and the Law.* New York: Macmillan, 1981.

Weitzman, Lenore, and Ruth Dixon. "The Transformation of Legal Marriage through No-Fault Divorce." In *Family in Transition,* edited by Arlene Skolnick and Jerome Skolnick. Boston: Little, Brown, 1980.

Weizman, R., and J. Hart. "Sexual Behavior in Healthy Married Elderly Men." *Archives of Sexual Behavior* 16, 1 (February 1987): 39–44.

Werner, Carol M., Barbara B. Brown, Irwin Altman, and Brenda Staples. "Close Relationships in Their Physical and Social Contexts: A Transactional Perspective." *Journal of Social and Personal Relationships* 9, 3 (1992): 411–431.

Wertz, Richard, and Dorothy Wertz *Lying-In: A History of Childbirth in America.* New York: Harper & Row, 1978.

Whelan, Elizabeth. *Boy or Girl?* New York: Pocket Books, 1986.

"When Parenthood Extends to Raising Grandchildren." *San Francsico Chronicle* (June 3, 1997): A-1.

Whitbourne, Susan, and Joyce Ebmeyer. *Identity and Intimacy in Marriage: A Study of Couples.* New York: Springer-Verlag, 1990.

White, Charles. "Sexual Interest, Attitudes, Knowledge, and Sexual History in Relation to Sexual Behavior of the Institutionalized Aged." *Archives of Sexual Behavior* 11 (February 1982): 11–21.

White, Gregory. "Inducing Jealousy: A Power Perspective." *Personality and Social Psychology Bulletin* 6, 2 (June 1980a): 222–227.

———. "Jealousy and Partner's Perceived Motives for Attraction to a Rival." *Social Psychology Quarterly* 44, 1 (March 1981): 24–30.

———. "Physical Attractiveness and Courtship Progress." *Journal of Personality and Social Psychology* 39, 4 (October 1980b): 660–668.

White, Jacquelyn W. "Feminist Contributions to Social Psychology." *Contemporary Social Psychology* 17, 3 (September 1993): 74–78.

White, Jacquelyn W., and R. Farmer. "Research Methods: How They Shape Views of Sexual Violence." *Journal of Social Issues* 48 (Spring 1992): 45–59.

White, James. "Premarital Cohabitation and Marital Stability in Canada." *Journal of Marriage and the Family* 49 (August 1987): 641–647.

White, Joseph, and Thomas Parham. *The Psychology of Blacks: An African-American Perspective.* 2d ed. Englewood Cliffs, NJ: Prentice Hall, 1990.

White, Lynn. "Determinants of Divorce." In *Contemporary Families: Looking Forward, Looking Back,* edited by Alan Booth. Minneapolis: National Council on Family Relations, 1991.

White, Lynn, and Alan Booth. "Divorce over the Life Course: The Role of Marital Happiness." *Journal of Family Issues* 12, 1 (March 1991): 5–22.

———. "The Transition to Parenthood and Marital Quality." *Journal of Family Issues* 6 (1985): 435–449.

White, Lynn, and Bruce Keith. "The Effect of Shift Work on the Quality and Stability of Marital Relations." *Journal of Marriage and the Family* 52, 2 (May 1990): 453–462.

Whitehead, Barbara Dafoe. "How to Rebuild a 'Family Friendly' Society." *Des Moines Sunday Register* (October 7, 1990): C3.

Whiteside, Mary F. "Family Rituals as a Key to Kinship Connections in Remarried Families." *Family Relations* 38, 1 (January 1989): 34–39.

Whitman, David. "The Coming of the 'Couch People.'" *U.S. News and World Report* (August 3, 1987): 19–21.

Wiehe, Vernon. *Sibling Abuse: The Hidden Physical, Emotional, and Sexual Trauma.* Lexington, MA: Lexington, Books, 1990.

Wiehe, Vernon R. "Religious Influence on Parental Attitudes Toward the Use of Corporal Punishment." *Journal of Family Violence* 5, 2 (June 1990): 173 ff.

Wilcox, Allen, et al. "Incidents of Early Loss of Pregnancy." *New England Journal of Medicine* 319, 4 (July 28, 1988): 189–194.

Wilkerson, Isabel. "Infant Mortality: Frightful Odds in Inner City." *New York Times* (June 26, 1987): 1.

Wilkie, Colleen F., and Elinor W. Ames. "The Relationship of Infant Crying to Parental Stress in the Transition to Parenthood." *Journal of Marriage and the Family* 48, 3 (August 1986): 545–550.

Wilkinson, Doris. "Family Ethnicity in America." In *Family Ethnicity: Strength in Diversity*, edited by Harriette Pipes McAdoo. Newbury Park, CA: Sage Publications, 1993.

Wilkinson, Doris Y. "American Families of African Descent." In *Families in Cultural Context*, edited by Mary Kay DeGenova. Mountain View, CA: Mayfield, 1997.

Wilkinson, Doris, Maxine Baca Zinn, and E. N. L. Chow. "Race, Class, and Gender: Introduction." *Gender and Society* 6, 3 (September 1992): 341–345.

Wilkinson, Doris, et al., eds. "Transforming Social Knowledge: The Interlocking of Race, Class, and Gender." *Gender and Society* [Special issue] (September 1992).

Willemsen, Tineke M. "On the Bipolarity of Androgyny: A Critical Comment on Kottke (1988)." *Psychological Reports* 72, 1 (February 1993): 327–332.

Williams, John, and Arthur Jacoby. "The Effects of Premarital Heterosexual and Homosexual Experience on Dating and Marriage Desirability." *Journal of Marriage and the Family* 51 (May 1989): 489–497.

Williams, Juanita. "Middle Age and Aging." In *Women: Behavior in a Biosocial Context*, edited by Juanita Williams. New York: Norton, 1977.

Wilson, Glenn. *The Secrets of Sexual Fantasy*. London: J. M. Dent, 1978.

Wilson, Melvin N., et al. "Flexibility and Sharing of Childcare Duties in Black Families." *Sex Roles* 22, 7–8 (April 1990): 409–425.

Wilson, Pamela. "Black Culture and Sexuality." *Journal of Social Work and Human Sexuality* 4, 3 (March 1986): 29–46.

Wilson, S. M., and N. P. Medora. "Gender Comparisons of College Students' Attitudes toward Sexual Behavior." *Adolescence* 25, 99 (September 1990): 615–627.

Wilson, Yumi. "This Time Domestic Violence Stayed in the Spotlight." *San Francisco Chronicle* (February 6, 1997): A-10.

Winch, Robert. *Mate Selection: A Study of Complementary Needs*. New York: Harper & Row, 1958.

Wineberg, H. "Marital Reconciliation in the United States: Which Couples Are Successful?" *Journal of Marriage and the Family* 56, 1 (February 1994): 80–88.

Winn, Rhoda, and Niles Newton. "Sexuality and Aging: A Study of 106 Cultures." *Archives of Sexual Behavior* 11, 4 (August 1982): 283–298.

Wise, P. H., and A. Meyer. "Poverty and Child Health." *Pediatric Clinics of North America* 35, 6 (December 1988): 1169–1186.

Wisensale, Steven. "Approaches to Family Policy in State Government: A Report on Five States." *Family Relations* 39, 2 (April 1990b): 136–140.

———. "The Family in the Think Tank." *Family Relations* 40, 2 (April 1990a): 199–207.

Wisensale, Steven, and Michael Allison. "Family Leave Legislation State and Federal Initiatives." *Family Relations* 38, 2 (April 1989): 182–189.

Wishy, Bernard. *The Child and the Republic*. Philadelphia: Lippincott, 1968.

Wiswell, T. E. "Routine Neonatal Circumcision: a Reappraisal." *American Family Physician* 41, 3 (March 1990): 859–63.

Woititz, Janet. *Adult Children of Alcoholics*. Deerfield, FL: Health Communications, 1983.

Wolf, Michelle, and Alfred Kielwasser. "Introduction: The Body Electric: Human Sexuality and the Mass Media." *Journal of Homosexuality* 21, 1/2 (1991): 7–18.

Wolf, Rosalie S. "Abuse and Neglect of the Elderly." In *Vision 2010: Families and Violence, Abuse and Neglect*, edited by Richard J. Gelles. Minneapolis: National Council on Family Relations, 1995.

Wolf, Rosalie S., and Karl A. Pillemer. "Intervention, Outcome, and Elder Abuse." In *Coping with Family Violence*, edited by Gerald T. Hotaling et al. Newbury Park, CA: Sage Publications, 1988.

Wolkind, Stephen and Eva Zajicek, eds. *Pregnancy: A Psychological and Social Study*. New York: Grune and Stratton, 1981.

Wong, Morrison G. "The Chinese American Family." In *Ethnic Families in America: Patterns and Variations*, edited by Charles H. Mindel et al. 3rd ed. New York: Elsevier, 1988.

Wood, J.T. *Gendered Lives: Communication, Gender, and Culture*. Belmont, CA: Wadsworth, 1994.

Woods, Stephen C., and James Guy Mansfield. "Ethanol and Disinhibition: Physiological and Behavioral Links." Proceedings of Alcoholism and Drug Abuse Conference. Berkeley/Oakland (February 11, 1981). Edited by Robin Room. Washington, DC: U.S. Department of Health and Human Services.

Woodward, K. L., V. Quade, and B. Kantrowitz. "Q: When Is a Marriage Not a Marriage?" *Newsweek* (March 13, 1995): 58–59.

Woodward, Kenneth. "New Rules for Making Love and Babies." *Newsweek* (March 23, 1987): 42–43.

Wright, J., C. Duchesne, S. Sabourin, F. Bissonette, J. Benoit, and Y. Girard. "Psychosocial Distress and Infertility: Men and Women Respond Differently." *Fertility and Sterility* 55, 1 (January 1991): 100–108.

Wright, Julia. "Getting Engaged: A Case Study and a Model of the Engagement Period as a Process of Conflict-Resolution." *Counselling Psychology Quarterly* 3, 4 (1990): 399–408.

Wu, Zheng, and T. R. Balakrishnan. "Attitudes towards Cohabitation and Marriage in Canada." *Journal of Comparative Family Studies* 23, 1 (1992): 1–12.

Wyatt, Gail E. "The Sociocultural Context of African American and White American Women's Rape." *Journal of Social Issues*, 48, 1 (March 1992): 77–91.

Wyatt, Gail Elizabeth. "The Aftermath of Child Sexual Abuse of African American and White Women: The Victim's Experience." *Journal of Family Violence* 5, 1 (March 1990): 61–81.

Wyatt, Gail, and Sandra Lyons-Rowe. "African American Women's Sexual Satisfaction as a Dimension of Their Sex Roles," *Sex Roles* 22, 7–8 (April 1990): 509–524.

Wyatt, Gail, Stephanie Peters, and Donald Guthrie. "Kinsey Revisited I: Comparisons of the Sexual Socialization and Sexual Socialization and Sexual Behavior of Black Women over 33 Years." *Archives of Sexual Behavior* 17, 3 (June 1988a): 201–239.

———. "Kinsey Revisited II: Comparison of the Sexual Socialization and Sexual Socialization and Sexual Behavior of Black Women over 33 Years." *Archives of Sexual Behavior* 17, 4 (August 1988b): 289–332.

Wyatt, Gail Elizabeth, et al. "Differential Effects of Women's Child Sexual Abuse and Subsequent Sexual Revictimization." *Journal of Consulting and Clinical Psychology* 60, 2 (April 1992): 167.

Wyche, Karen F. "Psychology and African-American Women: Findings from Applied Research." *Applied & Preventive Psychology* 2, 3 (June 1993): 115–121.

Wyn, Roberta, Richard Brown, and Hongjian Yu. "Women's Use of Preventive Health Services." In *Women's Health*, edited by Marilyn Falik and Karen Scott Collins. Baltimore, MD: Johns Hopkins University Press, 1996.

Yawn, Barbara P., and Roy A. Yawn. "Adolescent Pregnancy: A Preventable Consequence?" *The Prevention Researcher.* Eugene, OR: Integrated Research Services, 1997.

Ybarra, Lea. "When Wives Work: The Impact on the Chicano Family." *Journal of Marriage and the Family* 44, 1 (February 1982): 169–178.

Yellowbird, Michael, and C. Matthew Snipp. "American Indian Families." In *Minority Families in the United States: A Multicultural Perspective*, edited by R. L. Taylor. Englewood Cliffs, NJ: Prentice Hall, 1994.

Yerby, J., N. Buerkel-Rothfuss, and A. P. Bochner. *Understanding Family Communication.* Scottsdale, AZ: Gorsuch Scarisbrick, 1990.

Ÿllo, Kersti. "Marital Rape." In *Vision 2010: Families and Violence, Abuse and Neglect*, edited by Richard J. Gelles. Minneapolis: National Council on Family Relations, 1995.

———. "Through a Feminist Lens: Gender, Power, and Violence." In *Current Controversies in Family Violence*, edited by Richard Gelles and Donileen Loseke. Newbury Park, CA: Sage Publications, 1993.

Yogev, Sara, and Jane Brett. "Perceptions of the Division of Housework and Child Care and Marital Satisfaction." *Journal of Marriage and the Family* 47, 3 (August 1985): 609–618.

Young, L. "Sexual Abuse and the Problem of Embodiment." *Child Abuse and Neglect* 16, 1 (1992): 89–100.

Yulsman, Tom. "A Little Help for Creation." *Good Health Magazine (New York Times)* (October 7, 1990): 22ff.

Yura, Michael. "Family Subsystem Functions and Disabled Children: Some Conceptual Issues." *Marriage and Family Review* 11, 1 (1987): 135–151.

Zaslow, Martha, et al. "Depressed Mood in New Fathers." Unpublished paper, Society for Research in Child Development, April 1981.

Zavella, Patricia. *Women's Work and Chicano Families.* Ithaca, NY: Cornell University Press, 1987.

Zayas, Luis H., and Josephine Palleja. "Puerto Rican Familism: Considerations for Family Therapy." *Family Relations* 37, 3 (1988): 260–264.

———. "Puerto Rican Familism: Considerations for Family Therapy." In *Minority Families in the United States: A Multicultural Perspective*, edited by Ronald L. Taylor. Englewood Cliffs, NJ: Prentice Hall, 1994.

Zelnik, Melvin. *Sex and Pregnancy in Adolescence.* Beverly Hills, CA: Sage Publications, 1981.

Zilbergeld, Bernie. *The New Male Sexuality.* New York: Bantam Books, 1992.

Zill, Nicholas, Donna R. Morrison, and Mary J. Coiro. "Long-Term Effects of Parental Divorce on Parent-Child Relationships, Adjustment, and Achievement in Young Adulthood." *Journal of Family Psychology* 7, 1 (1993): 91–103.

Zimmerman, Shirley L. "State Level Public Policy Choices as Predictors of State Teen Birth Rates." *Family Relations* 37 (July 1988a): 315–321.

———. *Understanding Family Policy: Theoretical Approaches.* Beverly Hills, CA: Sage Publications, 1988b.

———. "The Welfare State and Family Breakup: The Mythical Connection." *Family Relations* 40, 2 (April 1991): 139–147.

Zimmerman, Shirley L., and Phyllis Owens. "Comparing the Family Policies of Three States: A Content Analysis." *Family Relations* 38, 2 (April 1989): 190–195.

Zinn, Maxine Baca. "Adaptation and Continuity in Mexican-Origin Families." In *Minority Families in the United States: A Multicultural Perspective*, edited by R. L. Taylor. Englewood Cliffs, NJ: Prentice Hall, 1994.

———. "Family, Feminism, and Race." *Gender and Society* 4 (1990): 68–82.

Zvonkovic, Anisa N., Kathleen M. Greaves, Cynthia J. Schmiege, and Leslie D. Hall. "The Marital Construction of Gender through Work and Family Decisions: A Qualitative Analysis." *Journal of Marriage and the Family* 58, 1 (February 1996): 91–100.

Zvonkovic, Anisa, et al. "Making the Most of Job Loss: Individual and Marital Features of Job Loss." *Family Relations* 37, 1 (January 1988): 56–61.

Index

Boldface pages indicate definitions of terms.

D

N

O

P

Q

R

S

T

U

V

W

Y

Z

Photo Credits

xxvii ©Norma Rannie/Christine's World Photography
xxvii ©Barbara Sayad
2 Steinbaum Kraus Gallery
5a Shooting Star
5b Shooting Star
5c Shooting Star
5d ©Archive Photos ™/Twentieth Century Fox Film Corp. 1991
7 Hamilton Projects
15 John Hannaford
17 ©Christine DeVault/Christine's World Photography
22 Bert Meyer
24 Carmen Lomas Garza
26 Addison-Wesley, *America: In Space and Time*, By Bryan Strong
27 The Image Works
30 American Stock/©Archive Photos ™
32 National Park Service: Statue of Liberty National Monument
38 ©Norma Rannie/Christine's World Photography
41 Laurie DeVault Photography
50 Laurie DeVault Photography
64 *left* Richard Hutchings/Photo Researchers, Inc.
64 *right* ©Christine DeVault/Christine's World Photography
66 Bob Torrez/Tony Stone Images
76 Laurie DeVault Photography
79 Ellis Herwig/Stock Boston
86 *left* David Young-Wolff/PhotoEdit
86 *right* ©Barbara Sayad
90 ©Fotex/Kuhmstedt/Shooting Star
92 Penny Gentien/Tony Stone Images
104 ©Comstock
107 Jeff Greenberg/PhotoEdit
112 Laurie DeVault Photography
124 ©Comstock
130 ©Jean Higgins/Envision
134 Donald W. Elliot
137 Barrow
148 Laurie DeVault Photography
159 Laurie DeVault Photography
164 King Features Syndicate
168 Vic Bader/PhotoEdit
171 Barrow
176 Claudia Kunin/©Tony Stone Images
191 ©Archive Photos ™
197 David Young-Wolff/PhotoEdit
202 ©Ron Chapple/FPG International Corp.
205 ©Christine DeVault 1995/Christine's World Photography
212 ©Miriam Grosman
216 Laurie DeVault Photography
219 Keri Pickett
223 Laurie DeVault Photography
227 Photo Researchers, Inc.
242 *left* ©Laima Druskis '89/Photo Researchers, Inc.
242 *right* ©Robert Brenner/PhotoEdit
254 Laurie DeVault Photography
257 Barrow
259 **a–f** Albert Bonniers Forlag AB
261 Laurie DeVault Photography
270 Fine Arts Museum of San Francisco
275 ©Gary Buss/FPG International Corp.
278 Richard T. Nonitz/Photo Researchers, Inc.
281 Laurie DeVault Photography
283 Laurie DeVault Photography
286 ©Raphael Macia 1993/Photo Researchers, Inc.
289 David Hanover/©Tony Stone Images
294 Laurie DeVault Photography
298 Mike Davis
300 ©Esbin Anderson/The Image Works
301 PhotoEdit
316 Laurie DeVault Photography
319 Ken Fisher/©Tony Stone Images
324 ©Janet Boynton/Christine's World Photography
327 Laurie DeVault Photography
342 Laurie DeVault Photography
343 ©Christine DeVault/Christine's World Photography
347 ©Christine DeVault/Christine's World Photography
348 Paul Conklin/PhotoEdit
357 Karen T. Borchers
364 Michael Newman/PhotoEdit
367 David Young-Wolff/PhotoEdit
371 Tom McCarthy/PhotoEdit
375 Laurie DeVault Photography
389 Laurie DeVault Photography
392 ©Tony Freeman/PhotoEdit
408 Carole Simons
411 Ken Fisher/©Tony Stone Images
417 ©Mark Richards/PhotoEdit
431 ©Michael Newman/PhotoEdit
439 *top* ©Emily Strong/The Picture Cube, Inc.
439 *bottom* Laurie DeVault Photography
450 Christine's World Photography
453 Laurie DeVault Photography
462 ©Mark Peterson/Saba Press Photos
466 Michael Newman/PhotoEdit
468 ©Barbara Sayad
478 Reprinted by permission: Sexual Diagnostic Center, Children's Institute
486 Barrow
489 ©Lee F. Snyder/Photo Researchers, Inc.
508 Kindra Clineff/The Picture Cube, Inc.
516 Uniphoto, Inc.
522 Laurie DeVault Photography
525 Laurie DeVault Photography
537 ©Christine DeVault/Christine's World Photography
553 Mark Richards/PhotoEdit
558 Comstock
561 ©Michael Newman/PhotoEdit
567 Laurie DeVault Photography
568 Laurie DeVault Photography
575 Andy Sacks/©Tony Stone Images
579 David Young-Wolff/PhotoEdit
584 Paul Conklin/PhotoEdit
586 Courtesy of Keegan Family/Christine's World Photography